D0302722

Deformation and Fracture Mechanics of Engineering Materials

Fifth Edition

Deformation and Fracture Mechanics of Engineering Materials

Fifth Edition

Richard W. Hertzberg
Richard P. Vinci
Jason L. Hertzberg

WILEY

JOHN WILEY & SONS, INC.

Vice President & Executive Publisher:	Don Fowley
Senior Editor and Product Designer:	Jennifer Welter
Assistant Editors:	Samantha Mandel, Alexandra Spicehandler
Marketing Manager:	Christopher Ruel
Media Specialist:	Andre Legaspi
Senior Production Manager:	Janis Soo
Associate Production Manager:	Joyce Poh
Assistant Production Editor:	Annabelle Ang-Bok
Cover Photo:	© Scott Leigh/iStockphoto, Kim Bisek Williams
Designer:	Seng Ping Ngieng

This book was set in 10.5/11.5 NimbusRomanNo.9L by Thomson Digital and printed and bound by Courier Westford. The cover was printed by Courier Westford.

This book is printed on acid free paper.

Founded in 1807, John Wiley & Sons, Inc. has been a valued source of knowledge and understanding for more than 200 years, helping people around the world meet their needs and fulfill their aspirations. Our company is built on a foundation of principles that include responsibility to the communities we serve and where we live and work. In 2008, we launched a Corporate Citizenship Initiative, a global effort to address the environmental, social, economic, and ethical challenges we face in our business. Among the issues we are addressing are carbon impact, paper specifications and procurement, ethical conduct within our business and among our vendors, and community and charitable support. For more information, please visit our website: www.wiley.com/go/citizenship.

Evaluation copies are provided to qualified academics and professionals for review purposes only, for use in their courses during the next academic year. These copies are licensed and may not be sold or transferred to a third party. Upon completion of the review period, please return the evaluation copy to Wiley. Return instructions and a free of charge return mailing label are available at www.wiley.com/go/returnlabel. If you have chosen to adopt this textbook for use in your course, please accept this book as your complimentary desk copy. Outside of the United States, please contact your local sales representative.

Library of Congress Cataloging-in-Publication Data

Hertzberg, Richard W., 1937-
  Deformation and fracture mechanics of engineering materials / Richard W. Hertzberg, Richard P. Vinci, Jason L. Hertzberg.— Fifth edition.
      pages cm
    Includes bibliographical references and indexes.
    ISBN 978-0-470-52780-1
1. Deformations (Mechanics) 2. Fracture mechanics. I. Vinci, Richard Paul. II. Hertzberg, Jason L., 1969- III. Title.
  TA417.6.H46 2012
  620.1^{1}123—dc23

2011051145

Printed in the United States of America
10 9 8 7 6 5 4 3 2

*To our wives Linda, Michelle, and Amy,
our children Michelle, Jason, Nicholas, Sofia,
Julia, Ryan, and Molli, and grandchildren
Henry, Abigail, Ryan, and Molli.*

Table of Contents

ABOUT THE COVER   xvii

FOREWORD   xix

PREFACE TO THE FIFTH EDITION   xxi

      Objectives   xxi
      Organization   xxi
      New to This Edition   xxii
      End-of-Chapter Problems   xxiv
      Acknowledgments   xxiv

ABOUT THE AUTHORS   xxv

SECTION ONE   **RECOVERABLE AND
NONRECOVERABLE DEFORMATION**   1

CHAPTER 1   **ELASTIC RESPONSE OF SOLIDS**   3

    1.1   Mechanical Testing   3
    1.2   Definitions of Stress and Strain   4
    1.3   Stress–Strain Curves for Uniaxial Loading   8
        1.3.1   Survey of Tensile Test Curves   8
        1.3.2   Uniaxial Linear Elastic Response   9
        1.3.3   Young's Modulus and Polymer Structure   13
            1.3.3.1   Thermoplastic Behavior   13
            1.3.3.2   Rigid Thermosets   14
            1.3.3.3   Rubber Elasticity   15
        1.3.4   Compression Testing   17
        1.3.5   Failure by Elastic Buckling   17
        1.3.6   Resilience and Strain Energy Density   19
        1.3.7   Definitions of Strength   19
        1.3.8   Toughness   22
    1.4   Nonaxial Testing   23
        1.4.1   Bend Testing   23
        1.4.2   Shear and Torsion Testing   26
    1.5   Multiaxial Linear Elastic Response   27
        1.5.1   Additional Isotropic Elastic Constants   27
        1.5.2   Multiaxial Loading   28
            1.5.2.1   Thin-Walled Pressure Vessels   30
            1.5.2.2   Special Cases of Multiaxial Loading   32
        1.5.3   Instrumented Indentation   33
    1.6   Elastic Anisotropy   34

1.6.1   Stiffness and Compliance Matrices   34
            1.6.1.1   Symmetry Classes   36
            1.6.1.2   Loading Along an Arbitrary Axis   37
1.6.2   Composite Materials   40
1.6.3   Isostrain Analysis   41
1.6.4   Isostress Analysis   43
1.6.5   Aligned Short Fibers   44
1.6.6   Strength of Composites   47
            1.6.6.1   Effects of Matrix Behavior   47
            1.6.6.2   Effects of Fiber Orientation   48
1.7   Thermal Stresses and Thermal Shock-Induced Failure   50
1.7.1   Upper Bound Thermal Stress   50
1.7.2   Cooling Rate and Thermal Stress   54
References   55
Further Readings   56
Problems   56
      Review   56
      Practice   57
      Design   59
      Extend   60

CHAPTER 2   YIELDING AND PLASTIC FLOW   63

2.1   Dislocations in Metals and Ceramics   63
2.1.1   Strength of a Perfect Crystal   63
2.1.2   The Need for Lattice Imperfections: Dislocations   65
2.1.3   Observation of Dislocations   67
2.1.4   Lattice Resistance to Dislocation Movement: The Peierls Stress   69
            2.1.4.1   Peierls Stress Temperature Sensitivity   70
            2.1.4.2   Effect of Dislocation Orientation on Peierls Stress   71
2.1.5   Characteristics of Dislocations   72
2.1.6   Elastic Properties of Dislocations   75
2.1.7   Partial Dislocations   78
            2.1.7.1   Movement of Partial Dislocations   80
2.2   Slip   81
2.2.1   Crystallography of Slip   81
2.2.2   Geometry of Slip   84
2.2.3   Slip in Polycrystals   87
2.3   Yield Criteria for Metals and Ceramics   88
2.4   Post-Yield Plastic Deformation   90
2.4.1   Strain Hardening   90
2.4.2   Plastic Instability and Necking   93
            2.4.2.1   Strain Distribution in a Tensile Specimen   94
            2.4.2.2   Extent of Uniform Strain   95
            2.4.2.3   True Stress Correction   95
            2.4.2.4   Failure of the Necked Region   96
2.4.3   Upper Yield Point Behavior   99
2.4.4   Temperature and Strain-Rate Effects in Tension   100
2.5   Slip in Single Crystals and Textured Materials   102
2.5.1   Geometric Hardening and Softening   103
2.5.2   Crystallographic Textures (Preferred Orientations)   105
2.5.3   Plastic Anisotropy   108

2.6   Deformation Twinning   111
    2.6.1   Comparison of Slip and Twinning Deformations   111
    2.6.2   Heterogeneous Plastic Tensile Behavior   113
    2.6.3   Stress Requirements for Twinning   113
    2.6.4   Geometry of Twin Formation   114
    2.6.5   Elongation Potential of Twin Deformation   116
    2.6.6   Twin Shape   116
    2.6.7   Twinning in HCP Crystals   117
    2.6.8   Twinning in BCC and FCC Crystals   120
2.7   Plasticity in Polymers   120
    2.7.1   Polymer Structure: General Remarks   120
        2.7.1.1   Side Groups and Chain Mobility   121
        2.7.1.2   Side Groups and Crystallinity   123
        2.7.1.3   Morphology of Amorphous and Crystalline Polymers   124
        2.7.1.4   Polymer Additions   127
    2.7.2   Plasticity Mechanisms   128
        2.7.2.1   Amorphous Polymers   128
        2.7.2.2   Semi-crystalline Polymers   130
    2.7.3   Macroscopic Response of Ductile Polymers   131
    2.7.4   Yield Criteria   133
References   136
Problems   139
    Review   139
    Practice   140
    Design   141
    Extend   141

CHAPTER 3   **CONTROLLING STRENGTH**   **143**
3.1   Strengthening: A Definition   143
3.2   Strengthening of Metals   143
    3.2.1   Dislocation Multiplication   143
    3.2.2   Dislocation–Dislocation Interactions   146
3.3   Strain (Work) Hardening   151
3.4   Boundary Strengthening   155
    3.4.1   Strength of Nanocrystalline and Multilayer Metals   156
3.5   Solid Solution Strengthening   158
    3.5.1   Yield-Point Phenomenon and Strain Aging   161
3.6   Precipitation Hardening   164
    3.6.1   Microstructural Characteristics   164
    3.6.2   Dislocation–Particle Interactions   167
3.7   Dispersion Strengthening   170
3.8   Strengthening of Steel Alloys by Multiple Mechanisms   172
3.9   Metal-Matrix Composite Strengthening   175
    3.9.1   Whisker-Reinforced Composites   175
    3.9.2   Laminated Composites   176
3.10   Strengthening of Polymers   177
3.11   Polymer-Matrix Composites   182
References   184
Further Reading   185
Problems   186
    Review   186
    Practice   186

Design   187
Extend   188

CHAPTER 4   **TIME-DEPENDENT DEFORMATION**   **189**

4.1   Time-Dependent Mechanical Behavior of Solids   189
4.2   Creep of Crystalline Solids: An Overview   191
4.3   Temperature–Stress–Strain-Rate Relations   195
4.4   Deformation Mechanisms   202
4.5   Superplasticity   205
4.6   Deformation-Mechanism Maps   208
4.7   Parametric Relations: Extrapolation Procedures for Creep Rupture Data   215
4.8   Materials for Elevated Temperature Use   220
4.9   Viscoelastic Response of Polymers and the Role of Structure   227
   4.9.1   Polymer Creep and Stress Relaxation   229
   4.9.2   Mechanical Analogs   235
   4.9.3   Dynamic Mechanical Testing and Energy-Damping Spectra   239
References   243
Problems   245
   Review   245
   Practice   246
   Design   247
   Extend   248

SECTION TWO   **FRACTURE MECHANICS
OF ENGINEERING MATERIALS**   **249**

CHAPTER 5   **FRACTURE: AN OVERVIEW**   **251**

5.1   Introduction   251
5.2   Theoretical Cohesive Strength   253
5.3   Defect Population in Solids   254
   5.3.1   Statistical Nature of Fracture: Weibull Analysis   255
      5.3.1.1   Effect of Size on the Statistical Nature of Fracture   258
5.4   The Stress-Concentration Factor   260
5.5   Notch Strengthening   264
5.6   External Variables Affecting Fracture   265
5.7   Characterizing the Fracture Process   266
5.8   Macroscopic Fracture Characteristics   269
   5.8.1   Fractures of Metals   269
   5.8.2   Fractures of Polymers   271
   5.8.3   Fractures of Glasses and Ceramics   273
   5.8.4   Fractures of Engineering Composites   277
5.9   Microscopic Fracture Mechanisms   278
   5.9.1   Microscopic Fracture Mechanisms: Metals   279
   5.9.2   Microscopic Fracture Mechanisms: Polymers   282
   5.9.3   Microscopic Fracture Mechanisms: Glasses and Ceramics   287
   5.9.4   Microscopic Fracture Mechanisms: Engineering Composites   289
   5.9.5   Microscopic Fracture Mechanisms: Metal Creep Fracture   291
References   294
Problems   295
   Review   295
   Practice   296

Design   297
Extend   297

CHAPTER 6   **ELEMENTS OF FRACTURE MECHANICS**   **299**

6.1   Griffith Crack Theory   299
        6.1.1   Verification of the Griffith Relation   301
        6.1.2   Griffith Theory and Propagation-Controlled Thermal Fracture   301
        6.1.3   Adapting the Griffith Theory to Ductile Materials   304
        6.1.4   Energy Release Rate Analysis   305
6.2   Charpy Impact Fracture Testing   307
6.3   Related Polymer Fracture Test Methods   311
6.4   Limitations of the Transition Temperature Philosophy   312
6.5   Stress Analysis of Cracks   315
        6.5.1   Multiplicity of Y Calibration Factors   323
        6.5.2   The Role of K 326
        FAILURE ANALYSIS CASE STUDY 6.1: Fracture Toughness of Manatee
        Bones in Impact   327
6.6   Design Philosophy   328
6.7   Relation Between Energy Rate and Stress Field Approaches   330
6.8   Crack-Tip Plastic-Zone Size Estimation   332
        6.8.1   Dugdale Plastic Strip Model   335
6.9   Fracture-Mode Transition: Plane Stress Versus Plane Strain   336
        FAILURE ANALYSIS CASE STUDY 6.2: Analysis of Crack Development
        during Structural Fatigue Test   339
6.10   Plane-Strain Fracture-Toughness Testing of Metals and Ceramics   341
6.11   Fracture Toughness of Engineering Alloys   344
        6.11.1   Impact Energy—Fracture-Toughness Correlations   347
        FAILURE ANALYSIS CASE STUDY 6.3: Failure of Arizona Generator
        Rotor Forging   354
6.12   Plane-Stress Fracture-Toughness Testing   355
6.13   Toughness Determination from Crack-Opening Displacement Measurement   358
6.14   Fracture-Toughness Determination and Elastic-Plastic Analysis with the J
        Integral   360
        6.14.1   Determination of J_{IC} 362
6.15   Other Fracture Models   368
6.16   Fracture Mechanics and Adhesion Measurements   371
References   375
Further Readings   378
Problems   378
        Review   378
        Practice   379
        Design   380
        Extend   381

CHAPTER 7   **FRACTURE TOUGHNESS**   **383**

7.1   Some Useful Generalities   383
        7.1.1   Toughness and Strength   383
        7.1.2. Intrinsic Toughness   385
        7.1.3   Extrinsic Toughening   387
7.2   intrinsic Toughness of Metals and Alloys   389
        7.2.1   Improved Alloy Cleanliness   389

          7.2.1.1   Cleaning up Ferrous Alloys   390
          7.2.1.2   Cleaning up Aluminum Alloys   394
      7.2.2   Microstructural Refinement   398
  7.3   Toughening of Metals and Alloys Through Microstructural Anisotropy   402
      7.3.1   Mechanical Fibering   402
      MICROSTRUCTURAL TOUGHENING CASE STUDY 7.1: The *Titanic*   404
      7.3.2   Internal Interfaces and Crack Growth   406
      7.3.3   Fracture Toughness Anisotropy   410
  7.4   Optimizing Toughness of Specific Alloy Systems   411
      7.4.1   Ferrous Alloys   411
      7.4.2   Nonferrous Alloys   414
  7.5   Toughness of Ceramics, Glasses, and Their Composites   416
      7.5.1   Ceramics and Ceramic-Matrix Composites   416
      7.5.2   Glass   422
  7.6   Toughness of Polymers and Polymer-Matrix Composites   426
      7.6.1   Intrinsic Polymer Toughness   426
      7.6.2   Particle-Toughened Polymers   427
      7.6.3   Fiber Reinforced Polymer Composites   432
  7.7   Natural and Biomimetic Materials   434
      7.7.1   Mollusk Shells   434
      7.7.2   Bone   437
      7.7.3   Tough Biomimetic Materials   438
  7.8   Metallurgical Embrittlement of Ferrous Alloys   440
      7.8.1   300 to 350°C or Tempered Martensite Embrittlement   441
      7.8.2   Temper Embrittlement   442
      7.8.3   Neutron-Irradiation Embrittlement   444
  7.9   Additional Data   449
  References   453
  Problems   459
      Review   459
      Practice   460
      Design   461
      Extend   461

CHAPTER 8   ENVIRONMENT-ASSISTED CRACKING   463

  8.1   Embrittlement Models   465
      8.1.1   Hydrogen Embrittlement Models   465
      8.1.2   Stress Corrosion Cracking Models   468
          8.1.2.1   SCC of Specific Material–Environment Systems   470
      8.1.3   Liquid-Metal Embrittlement   471
      8.1.4   Dynamic Embrittlement   472
  8.2   Fracture Mechanics Test Methods   472
      8.2.1   Major Variables Affecting Environment-Assisted Cracking   480
          8.2.1.1   Alloy Chemistry and Thermomechanical Treatment   480
          8.2.1.2   Environment   483
          8.2.1.3   Temperature and Pressure   485
      8.2.2   Environment-Assisted Cracking in Plastics   487
      8.2.3   Environment-Assisted Cracking in Ceramics and Glasses   489
  8.3   Life and Crack-Length Calculations   492
  References   493
  Problems   496

Review   496
Practice   497
Design   497
Extend   497

CHAPTER 9   **CYCLIC STRESS AND STRAIN FATIGUE**   **499**

9.1   Macrofractography of Fatigue Failures   499
9.2   Cyclic Stress-Controlled Fatigue   503
       9.2.1   Effect of Mean Stress on Fatigue Life   506
       9.2.2   Stress Fluctuation, Cumulative Damage, and Safe-Life Design   508
       9.2.3   Notch Effects and Fatigue Initiation   511
       9.2.4   Material Behavior: Metal Alloys   516
               9.2.4.1   Surface Treatment   520
       9.2.5   Material Behavior: Polymers   523
       9.2.6   Material Behavior: Composites   526
               9.2.6.1   Particulate Composites   526
               9.2.6.2   Fiber Composites   527
9.3   Cyclic Strain-Controlled Fatigue   529
       9.3.1   Cycle-Dependent Material Response   531
       9.3.2   Strain Life Curves   538
9.4   Fatigue Life Estimations for Notched Components   541
9.5   Fatigue Crack Initiation Mechanisms   545
9.6   Avoidance of Fatigue Damage   547
       9.6.1   Favorable Residual Compressive Stresses   547
       9.6.2   Pretensioning of Load-Bearing Members   550
References   554
Problems   556
       Review   556
       Practice   556
       Design   557
       Extend   557

CHAPTER 10   **FATIGUE CRACK PROPAGATION**   **559**

10.1   Stress and Crack Length Correlations with FCP   559
       10.1.1   Fatigue Life Calculations   563
       10.1.2   Fail-Safe Design and Retirement for Cause   567
10.2   Macroscopic Fracture Modes in Fatigue   568
FATIGUE FAILURE ANALYSIS CASE STUDY 10.1: Stress Intensity Factor
Estimate Based on Fatigue Growth Bands   571
10.3   Microscopic Fracture Mechanisms   572
       10.3.1   Correlations with the Stress Intensity Factor   575
10.4   Crack Growth Behavior at ΔK Extremes   578
       10.4.1   High ΔK Levels   578
       10.4.2   Low ΔK Levels   583
               10.4.2.1   Estimation of Short-Crack Growth Behavior   590
10.5   Influence of Load Interactions   592
       10.5.1   Load Interaction Macroscopic Appearance   596
10.6   Environmentally Enhanced FCP (Corrosion Fatigue)   600
       10.6.1   Corrosion Fatigue Superposition Model   605
10.7   Microstructural Aspects of FCP in Metal Alloys   606
       10.7.1   Normalization and Calculation of FCP Data   615

10.8   Fatigue Crack Propagation in Engineering Plastics   618
    10.8.1   Polymer FCP Frequency Sensitivity   620
    10.8.2   Fracture Surface Micromorphology   625
10.9   Fatigue Crack Propagation in Ceramics   628
10.10 Fatigue Crack Propagation in Composites   632
References   635
Further Reading   641
Problems   641
    Review   641
    Practice   642
    Design   643
    Extend   644

CHAPTER 11   ANALYSES OF ENGINEERING FAILURES   645

11.1   Typical Defects   647
11.2   Macroscopic Fracture Surface Examination   647
11.3   Metallographic and Fractographic Examination   651
11.4   Component Failure Analysis Data   652
11.5   Case Histories   652
CASE 1:   Shotgun Barrel Failures   653
    Overview of Failure Events and Background Information   653
    Proposed Causation Theories   654
    Fractographic Evidence of Failed Gun Barrels   655
    Estimation of the Material's Fatigue Endurance Limit   655
    Microfractography of Fatigue Fracture in Gun Barrel Material   656
    The Verdicts   658
CASE 2:   Analysis of Aileron Power Control Cylinder Service Failure   658
CASE 3:   Failure of Pittsburgh Station Generator Rotor Forging   660
CASE 4:   Stress Corrosion Cracking Failure of the Point Pleasant Bridge   661
CASE 5:   Weld Cold Crack-Induced Failure of Kings Bridge, Melbourne, Australia   664
CASE 6:   Failure Analysis of 175-mm Gun Tube   665
CASE 7:   Hydrotest Failure of a 660-cm-Diameter Rocket Motor Casing   670
CASE 8:   Premature Fracture of Powder-Pressing Die   673
CASE 9:   A Laboratory Analysis of a Lavatory Failure   674
11.6   Additional Comments Regarding Welded Bridges   676
References   680
Further Reading   681

CHAPTER 12   CONSEQUENCES OF PRODUCT FAILURE   683

12.1   Introduction to Product Liability   683
12.2   History of Product Liability   684
    12.2.1   Caveat Emptor and Express Liability   685
    12.2.2   Implied Warranty   685
    12.2.3   Privity of Contract   686
    12.2.4   Assault on Privity Protection   687
    12.2.5   Negligence   691
    12.2.6   Strict Liability   694
    12.2.7   Attempts to Codify Product Liability Case Law   696
12.3   Product Recall   697
    12.3.1   Regulatory Requirements and Considerations   698
        12.3.1.1   Consumer Product Safety Commission   698

12.3.1.1.1   Defect   699
12.3.1.1.2   Substantial Product Hazard   700
12.3.1.1.3   Unreasonable Risk   700
12.3.1.2   International Governmental Landscape   701
12.3.2   Technical Considerations Regarding Potential Recalls   701
12.3.2.1   Determination of the Failure Process   702
12.3.2.2   Identification of the Affected Product Population   704
12.3.2.3   Assessment of Risk Association with Product Failure   705
12.3.2.4   Generation of an Appropriate Corrective Action Plan   707
12.3.3   Proactive Considerations   707
12.3.3.1   Think Like a Consumer   707
12.3.3.1   Test Products Thoroughly   707
12.3.3.3   Ensure Adequate Traceability   708
12.3.3.4   Manage Change Carefully   708
RECALL CASE STUDY: The "Unstable" Ladder   708
References   710
Problems   712
Review   712
Extend   712

APPENDIX A   **FRACTURE SURFACE PRESERVATION, CLEANING AND
REPLICATION TECHNIQUES, AND IMAGE INTERPRETATION**   **713**

A.1   Fracture Surface Preservation   713
A.2   Fracture Surface Cleaning   713
A.3   Replica Preparation and Image Interpretation   715
References   717

APPENDIX B   ***K* CALIBRATIONS FOR TYPICAL FRACTURE TOUGHNESS
AND FATIGUE CRACK PROPAGATION TEST SPECIMENS**   **719**

APPENDIX C   ***Y* CALIBRATION FACTORS FOR ELLIPTICAL AND
SEMI-CIRCULAR SURFACE FLAWS**   **723**

APPENDIX D   **SUGGESTED CHECKLIST OF DATA DESIRABLE
FOR COMPLETE FAILURE ANALYSIS**   **725**

AUTHOR INDEX   **729**
MATERIALS INDEX   **741**
SUBJECT INDEX   **747**

About the Cover

The Liberty Bell appears on the cover of this textbook because the crack is internationally known and perhaps recognized more than any other fracture. However, there is considerable confusion as to the history of the bell and how it gained such worldwide recognition. The following is intended to highlight major events in the bell's existence from the casting foundry to the present.

To commemorate the 50th anniversary of the granting of William Penn's Charter of Liberties, the Pennsylvania Assembly purchased a bell for the Statehouse. Since there were no qualified bell foundries in the region, the bell was cast at the Whitechapel Foundry in London, England. The inscription on the bell was to read "Proclaim liberty through all the land unto all the inhabitants thereof," (Leviticus 25:10). On its completion, the bell was shipped to Philadelphia and placed in the Statehouse belfry. To the dismay of all, the bell cracked the first time it was struck. John Pass and Charles Stow, two area residents, agreed to recast the bell in time for the Charter of Liberty's jubilee celebration. After adjusting the alloy chemistry and recasting the bell twice, these amateur bell founders produced a bell with an acceptable tone. For their services, Pass and Stow were paid $295.25 and given a free advertisement: note their names on the shoulder of the bell.

Not leaving anything to chance, the Pennsylvania Assembly commissioned a second bell from the Whitechapel Foundry, which arrived from England when Pass and Stow had completed the third casting of the original bell. What were they to do with two bells? It was ultimately decided that the original bell (also known as the Liberty Bell) be used for grand occasions such as convening townsfolk for the first public reading of the Declaration of Independence and the second Whitechapel bell be used as the town's clockbell.

During the Revolutionary War, the Liberty Bell was taken to Allentown, Pennsylvania, to safeguard it from the advancing British armies. The city fathers were less concerned with protecting an American historical treasure (the bell had no historical value at that time) than with preventing the British from melting such bells to produce new artillery pieces. Cannon metal (also known as Admiralty bronze) contains 88% copper and 12% tin whereas bell metal contains roughly twice as much tin. After the bell was returned to Philadelphia in 1778, it continued to ring until 1835 when it cracked while tolling the funeral of Chief Justice Marshall. (The second Whitechapel bell was given to a church in 1828, before being destroyed in a fire.) After grinding the mating surfaces of the crack to prevent them from rubbing together, the Liberty Bell was struck once again in 1846 to celebrate Washington's birthday. After ringing for several hours, the original crack extended into the shoulder region. Since that time, the bell has effectively remained silent.

After being sent on a series of national tours, beginning with a trip to New Orleans in 1885, the Liberty Bell has become a symbol of American independence. It is now on permanent exhibition in the historical section of Philadelphia.

BACK COVER IMAGES: Liberty Bell; helical fracture of human finger bone (Figure 5.23c); instrumented tensile test (Figure 1.1b); macroscopic fatigue fracture markings (Figure 9.3b).

Foreword

It has been said that no hypothesis can ever be proven with absolute certainty. Such a theoretical construct may stand the test of time for years with the benefit of supporting evidence. And yet, that same theory can be disproved by a single conflicting, valid observation. The annals of scientific writings bear witness to many theories that were struck down by some fortuitous and/or unanticipated finding.

As such, scientific concepts and associated theories undergo constant scrutiny and necessary revision as our knowledge base expands and new insights are formed. Surely, we authors have recognized the need to revise and/or augment this text as it has evolved during these past three-dozen years. Our fifth edition reflects such contemporary revision. The authors have sought to update the subject matter of this text to the best of our abilities. It may well be the responsibility of the reader to expand our collective knowledge base, thereby leading to new insights. The quest for knowledge never ends.

Preface to the Fifth Edition

OBJECTIVES

This book examines the macroscopic and microscopic aspects of the mechanical behavior of metals, ceramics, polymers, and their composites. Particular emphasis is given to the application of fracture mechanics and materials science principles toward the understanding of material stiffness, strength, toughness, and time-dependent mechanical response. This text is suitable for advanced undergraduate and first-level graduate courses in metallurgy and materials, mechanical engineering, and civil engineering curricula, and provides a combined fracture mechanics-materials approach to the fracture of engineering solids. The book also will be useful to working engineers who want to learn more about the mechanical properties of solids and, in particular, the fracture-mechanics approach to the fracture of solids. To that end, the book contains more than 1500 references that are cited throughout the text. Furthermore, all principal and secondary authors are identified in an author index along with separate material and subject indices.

ORGANIZATION

The book is divided into two sections. In Section One, the principles of elastic and plastic deformation are presented. Chapter 1 begins with a discussion of elastic deformation in solids. Concepts of stress, strain, and stiffness are introduced for both isotropic and anisotropic materials. Chapter 2 addresses the plastic deformation response of solids. Here, emphasis is placed on continuum aspects of irreversible plastic deformation and the role of micro- and nanostructures, crystallography, and crystal defects (e.g., dislocations) in explaining the material deformation process. Subsequently, these parameters are used in Chapter 3 to understand various strengthening mechanisms in different material systems. The time–temperature dependent nature of material deformation in metallic, ceramic, and polymeric materials is addressed in Chapter 4. While familiarity with the topics discussed in Section One will be useful to the reader in Section Two, readers with some prior exposure to mechanical behavior concepts may be able to proceed directly from Chapter 1 to Chapter 5.

Section Two deals with the application of fracture mechanics principles to the subject of fracture in solids. Chapter 5 begins with an overview of failed components, and discusses stress concentrations and theoretical fracture strength, notch strengthening, statistical aspects of fracture and fracture surface micromorphology. The importance of the stress intensity factor and the fracture mechanics approach in analyzing the fracture of solids is developed in Chapter 6 and is compared with the older transition temperature approach to engineering design. From this macroscopic viewpoint, the emphasis shifts in Chapter 7 to a consideration of the role of micro- and nanostructural variables in determining material fracture toughness and embrittlement susceptibility, such as temper, irradiation, and 300°C embrittlement. Environmental degradation (i.e., stress corrosion cracking and both hydrogen and liquid-metal embrittlement) is described in Chapter 8 in terms of stress–environment–material systems. Fatigue and associated crack propagation in solids is discussed at length in Chapters 9 and 10, emphasizing the cyclic stress life, cyclic strain life, and fatigue crack propagation philosophies pertaining to cyclically induced material damage. In Chapter 11, actual service failures are examined to demonstrate the importance of applying fracture mechanics principles in failure analysis. Several bridge, aircraft, firearm, and generator rotor shaft failures are analyzed. Finally, the consequences of component failure are introduced in Chapter 12 with a discussion of product liability and product recall.

NEW TO THIS EDITION

With the timely addition of many new topics, including a new chapter on product liability and product recall, this text continues to serve as an advanced undergraduate/early graduate level textbook and as a reference volume for practicing engineers/scientists. The additions reflect new developments pertaining to the mechanical behavior of engineering materials and address the associated societal consequences of product failure. Furthermore, a considerable reorganization of subject matter enhances the pedagogical effectiveness of the book. Of major import, this edition benefits by the addition of two co-authors, whose talents and varied experiences broaden the text's perspective. The authors believe that this latest edition maintains a good balance between discussions of the continuum mechanics understanding of the failure of solids and the roles of the material's nano- and microstructure as they influence the mechanical properties of materials. This fifth edition contains over 300 additional references, thereby raising the total to more than 1500. Several new examples have been added to the text along with over 80 new figures, raising the latter total to more than 550. Over 300 new problems have been added along with numerous problems being made available online. New additions/ modifications to this fifth edition include the following:

Elastic Behavior of Engineering Solids

- Chapter 1 has been reorganized to focus on the elastic behavior of engineering solids.
- Discussions about buckling failure of slender columns, compression testing, and the elastic properties of bone have been added.
- Sections on elastic anisotropy and fiber-reinforced composites have been expanded.
- Some topics, such as short-term polymer elasticity and thermal stress development, have been moved forward in the text to this initial chapter.
- Nonlinear irreversible deformation processes have been moved to Chapter 2.

Plastic Behavior of Engineering Solids

- Chapter 2 now deals with yielding and plastic deformation processes in solids.
- Coverage of widely used yield criteria in metals and in polymers has been expanded.
- Chapter 3 treats deformation micromechanisms in solids as they influence strengthening mechanisms, and combines elements of Chapters 2 and 4 from earlier editions.
- The former chapter on creep and portions of the former chapter on polymeric solids have been combined into a new Chapter 4 that addresses time-dependent deformation in solids in a comparative fashion.

Failure and Fracture Mechanics of Solids

- As in earlier editions, the second part of the book begins (Chapter 5) with an overview of fracture and includes such topics as a discussion of actual failure case histories, electron fractography, and the concept of stress concentration factors.
- New discussion pertaining to the fracture behavior and fractography of ceramics, glasses and composite materials has been added.
- Several actual case history failure analyses and discussion of several failure processes have been moved forward from the penultimate chapter to Chapters 6 and 7 to provide

useful examples of the concepts developed in these chapters and to more fully develop specific topics when they are introduced.

- The fourth edition chapter dealing with Charpy testing has been eliminated, but relevant topics have been relocated to Chapter 6.
- A new section on adhesion has been added to the discussion of fracture mechanics analysis.
- A section on natural materials, including bone and sea shells, has been added to Chapter 7 to introduce a number of natural toughening mechanisms that hold promise for the improvement of toughness in engineering solids.

Subcritical Flaw Growth in Solids

- Chapter 8, which is focused on environment-assisted cracking, has been revised to incorporate expanded discussion of environmental degradation mechanisms and current research findings, including dynamic embrittlement and testing procedures for both metal alloys and polymeric solids.
- The two fatigue chapters, 9 and 10, have been updated with regard to both S-N test results, based on high-frequency test methods, a reexamination of the concept of a meaningful endurance limit in ferrous alloys, and improved test methods designed to better define the effective stress intensity factor range at the advancing tip of a fatigue crack.
- The discussion of environmentally enhanced fatigue crack propagation processes has been updated.
- A detailed fracture mechanics–based failure analysis case history, concerning the failure of shotguns, has been introduced that describes the use of fractographic information and leak-before-break failure criteria in the analysis of these failures.

Product Liability and Product Recall—a new chapter

With widespread attention being paid to component failures and associated product liability litigation, the authors have concluded that students and engineers must be aware of their role in identifying critical details of such failures. Similarly, the steadily increasing drumbeat of product recalls, ranging from household items to automobiles, have thrust engineers into the middle of such regulatory and safety issues. Accordingly, a new Chapter 12 has been added that focuses on:

- A historical perspective of the law as it pertains to products liability litigation.
- An overview of regulatory guidelines pertaining to product recalls.
- Furthermore, the reader is exposed to useful methodologies with regard to potential product recall investigations.

Current circumstances dictate that engineers should become more familiar with these areas of the law/regulatory requirements, as they relate to product safety.

Revised and supplemented appendices

- The text still concludes with an Appendix that contains information pertaining to fracture surface preservation and image interpretation, K calibrations for typical fracture toughness and fatigue crack propagation test specimens.
- Several test specimen configurations that are new to this edition.

- Analytical formulas for calculating the Y calibration factors for elliptical and semi-circular surface flaws are new to the 5th edition.
- The Checklist for failure analysis has been relocated to an Appendix for more convenient access.

END-OF-CHAPTER PROBLEMS

The problems concluding each chapter have been updated, and a new organizational approach has been adopted that characterizes the problems in such a fashion that the student will

- *Review* the chapter concepts
- *Practice* methods, both qualitative and quantitative, introduced in the chapter
- *Design*, analyze, and modify structures based on chapter material
- *Extend* knowledge beyond that presented in the text by means of resources such as those available on the Internet.

The solutions manual has been updated and is available to qualified instructors.

ACKNOWLEDGMENTS

In addition to the many individuals who assisted with the completion of the first four editions of this text, a number of people provided assistance with the completion of this latest edition, including constructive criticism of new sections. To these individuals we express our sincere gratitude. Special thanks are extended to Attorneys Nicholas P. Scavone, Jr., Seymour Traub, Malcolm Wheeler, and Alan Schoem, former Director of the Office of Compliance at the U.S. Consumer Product Safety Commission, for useful discussions and critical reviews of the new chapter on product liability and product recall. In addition, Attorney Charles Joern kindly reviewed the newly added shotgun failure analysis case history; Dr. Richard Gangloff provided invaluable assistance and editorial skills regarding revisions to Chapter 8 and Section 10.6; Drs. Claude Bathias, Klaus Friedrich, John Mecholsky, Jr., and J. Keith Donald supplied several new figures and/or contributed new insights regarding the fatigue fracture of metals. Drs. Hongbin Bei, Sang Hoon Shim, George Pharr, and Easo George provided data and a figure pertaining to the behavior of microspecimens. Drs. Raymond Pearson, Clare Rimnac, Charles McMahon, and Reinhold Dauskardt reviewed new topics in several chapters. We are grateful to the students in our recent classes who reviewed several new sections of the book. Finally, the authors appreciate the editorial and production staffs at John Wiley and Sharon Siegler from Lehigh University for their help with completion of the manuscript.

Our wives and children have been exceedingly patient and understanding during the revision of this book. To these special individuals, we extend our love and appreciation.

Richard W. Hertzberg
Richard P. Vinci
Jason L. Hertzberg

About the Authors

Richard W. Hertzberg received his B.S. cum laude in Mechanical Engineering from the City College New York, his M.S. in Metallurgy from M.I.T. and his Ph.D. in Metallurgical Engineering from Lehigh University. A recipient of two Alcoa Foundation Awards of Outstanding Research Achievement, co-recipient of Lehigh University's Award of Outstanding Research, recipient of Lehigh University's College of Engineering Teaching Excellence Award, and co-recipient of Lehigh University's award in Recognition of Outstanding Contributions to the University, Dr. Hertzberg has served as Research Scientist for the United Aircraft Corporation Research Labs, and Visiting Professor at the Federal Institute of Technology, Lausanne, Switzerland. As an active member of several engineering societies, he has been elected as a Fellow of the American Society for Metals and was recipient of the TMS 2000 Educator Award as the most outstanding educator in the nation. He has authored approximately 230 scholarly articles, co-authored *Fatigue of Engineering Plastics* (Academic Press, 1980), and co-authored the fifth edition of *Deformation and Fracture Mechanics of Engineering Materials*. Dr. Hertzberg has also been an invited lecturer in the United States, Asia, Israel, and Europe, and has served as a consultant to government and industry. He was previously Chair, Materials Science and Engineering Dept., and Director of the Mechanical Behavior Laboratory of the Materials Research Center at Lehigh University. Currently, he is New Jersey Zinc Professor Emeritus of Materials Science and Engineering.

Richard P. Vinci received his B.S. degree in 1988 from the Massachusetts Institute of Technology, and his M.S. and Ph.D. degrees in 1990 and 1994, respectively, from Stanford University, all in Materials Science and Engineering. After holding postdoctoral and Acting Assistant Professor appointments at Stanford University, in 1998 he joined Lehigh University where he is currently a Professor of Materials Science and Engineering and the Director of the Mechanical Behavior Laboratory. His research focuses on the mechanical properties of thin metallic films and small-scale structures with applications such as metallization for Micro-ElectroMechanical Systems, substrates for solid-state optical devices, and synthetic biomaterials. He has published more than 70 technical papers and is the holder of one U.S. patent, with others pending. From 2001 to 2003, he held a P. C. Rossin Assistant Professorship. From 2004 to 2006, he was the Class of 1961 Associate Professor of Materials Science and Engineering. Dr. Vinci has been a recipient of the NSF CAREER Award, the ASM International Bradley Stoughton Award for Young Teachers, the Lehigh University Junior Award for Distinguished Teaching, the P. C. Rossin College of Engineering Teaching Excellence Award, and the Donald B. and Dorothy L. Stabler Award for Excellence in Teaching.

Jason L. Hertzberg received his B.S. in Metallurgical Engineering from the University Scholars Program at Pennsylvania State University and both a M.S.E. and Ph.D. in Materials Science and Engineering from the University of Michigan, having received numerous academic awards at both institutions. He is also a California-registered Professional Metallurgical Engineer. He currently serves as a Corporate Vice President, Director of the Mechanical Engineering Practice, and a Principal Engineer at Exponent, Inc., a leading engineering and scientific consulting firm. He has extensive experience solving complex technical problems in a variety of industries and routinely leads multidisciplinary failure analysis investigations. Dr. Hertzberg addresses issues related to the mechanical behavior and environmental degradation of materials, and often works with companies addressing the technical aspects of product recalls as well as interacting with the Consumer Product Safety Commission. His expertise includes analysis of products before they are sold, management of change during production, use of risk methodologies, substantiation of product performance claims, product recall investigations of a wide range of products, and evaluation of proposed correction action plans. Dr. Hertzberg also has a background in mobile computing and substantiation of claims, having served as the Director of Competitive Analysis and Strategy for Palm, Inc. Dr. Hertzberg often serves as an invited lecturer, and is a co-author of several patent applications in the area of mobile computing.

Section One

Recoverable and Nonrecoverable Deformation

Chapter 1

Elastic Response of Solids

What material should be chosen for a nuclear reactor pressure vessel to ensure 40 years of safe operation? How can an aircraft wing skin be made lighter without sacrificing load-bearing capacity? Is it safe to use glass as a structural material? How far can an aircraft reliably fly between safety inspections? Why did a particular power plant generator shaft break in service, and what could be done to prevent a recurrence? What makes natural seashell and bone possible models for lighter, tougher future engineered materials? The information needed to address these questions—and many more—is the subject of this book. We will examine the ways in which engineering materials, and the components made from them, succeed or fail under load-bearing conditions. Throughout, we will emphasize that a well-rounded understanding of the interplay between material properties and design choices is the path to safe, efficient, and effective engineered structures.

1.1 MECHANICAL TESTING

Material properties are determined using a wide variety of mechanical tests. Despite the variety of specimen shapes and test conditions, however, all mechanical tests may be reduced to one of two general descriptions: either a controlled load (or combination of loads) is imposed and the resulting displacements are measured, or a controlled displacement (or combination of displacements) is imposed and the load(s) developed in response to the imposed displacement state is measured. Which type of test to use, and under what conditions, depends on the objective of the test. On the one hand, one may wish to establish the load-bearing capability of an engineering component under its expected loading conditions. In this case, a combination of loads (or displacements) may be applied to a real component or assembly, and the overall response will be measured. Often the question to be answered is something like, "Can the component survive the design load plus some additional load increment (as a margin of safety) without failing?" This experimental process often falls under the category of *product testing* (Fig. 1.1*a*). On the other hand, one may wish to establish fundamental material properties that can subsequently be used in a more universal way for both evaluation of material properties and their use for design purposes. The question here may be something like, "How does this material compare to other materials evaluated in the same fashion?" In this case, it is often desirable to use well-defined, simple, standardized specimen shapes and simple loading conditions (e.g., along a single axis of the test specimen). This is the mode generally used for *material testing* (Fig. 1.1*b*). In the end, it can be argued that the ultimate objective of both test types (product or material) is to avoid *failure* of an engineering component in service. Before discussing engineering component or fundamental material mechanical behavior, however, it is necessary to establish some definitions of possible responses to loading, and some definitions of failure.

There are three basic categories of mechanical response to an applied load: *elasticity*, *plasticity*, and *fracture*. Elasticity is defined by a fully-recoverable response; that is, a component is loaded and unloaded without any permanent change to its shape or integrity. This is usually the desirable response of structural components in service. Plasticity and fracture both involve permanent shape changes under load, but are clearly distinct from one another. Plasticity is shape change without cracking, as one might require during forging of a metal component, whereas fracture involves the creation or propagation of a crack that separates a portion of the component from the remainder.

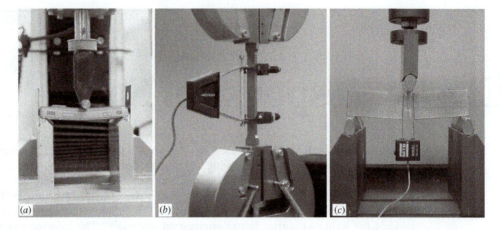

Figure 1.1   (*a*) Cellular phone product testing by bending (photo copyright Nokia 2011). (*b*) Tensile testing for fundamental material properties using a standardized tensile specimen. (Courtesy of Richard Vinci.). (*c*) Bend testing using a standardized fracture specimen (photo courtesy of Brett Leister, Lehigh University).

It is obvious to most people that fracture of an engineering component is undesirable, and is considered a failure. What may be less obvious is that failure can also occur under plastic or even elastic conditions. Consider, for example, a series of parallel columns topped by a crossbeam that is under load. If sufficient load is introduced so that the columns are plastically deformed without inducing total collapse, the shape of the overall structure will be altered even after the load is removed. This can constitute failure if retention of the original structure shape is important for aesthetic, functional, or safety reasons. If the columns are slender, it is also possible under certain conditions for them to buckle elastically, which means they suddenly bend outward to the side under loads that are too small to induce plasticity or fracture. In doing so, they lose essentially all of their load-bearing capability. If the crossbeam has other supports that can bear the load to avoid collapse, it may be possible to return the structure to its original shape by unloading (demonstrating that the phenomenon is elastic in nature). If there is no additional source of support, elastic buckling will rapidly progress to plastic deformation and/or fracture, accompanied by collapse of the structure. In either case, the columns will have failed to support the crossbeam in the intended manner, so we may say that they have failed.

Throughout the remainder of Chapter 1 we will primarily address the fundamentals of elastic material behavior. We will begin with behavior measured under tension because the tensile test is the most widely employed experimental test method. Following this, other modes of loading will be introduced, along with the elastic properties that can be measured under each mode. Given that a complete test record may contain important information concerning not only the material's elastic properties but also its strength, the character and extent of plastic deformation, and resistance to fracture, idealized stress–strain plots reflecting different deformation and failure characteristics of a wide variety of materials will also be introduced in this chapter. Finally, certain conditions that lead to failure (defined for the moment as a departure from purely elastic behavior) will be introduced as a precursor of things to come in later chapters.

1.2   DEFINITIONS OF STRESS AND STRAIN

Raw load and displacement information may be sufficient for certain product tests, but evaluations of material properties must use size-independent parameters: stress and strain. These essential terms may be defined in two generally accepted forms. The first definitions,

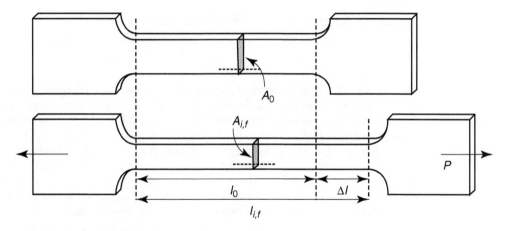

Figure 1.2   Schematic illustration of a rectangular tensile specimen before (subscript 0) and during (subscript i) or after (subscript f) loading. A cylindrical specimen with a circular cross section is also common. The same definitions apply to both shapes.

used extensively in engineering practice, are

$$\sigma_{\text{eng}} = \text{engineering stress} = \frac{\text{load}}{\text{initial cross-sectional area}} = \frac{P}{A_0} \qquad (1\text{-}1a)$$

$$\varepsilon_{\text{eng}} = \text{engineering strain} = \frac{\text{change in length}}{\text{initial length}} = \frac{l_f - l_0}{l_0} \qquad (1\text{-}1b)$$

where

l_f = final gage length
l_0 = initial gage length

as depicted in Fig. 1.2.

Alternatively, stress and strain may be defined as

$$\sigma_{\text{true}} = \text{true stress} = \frac{\text{load}}{\text{instantaneous cross-sectional area}} = \frac{P}{A_i} \qquad (1\text{-}2a)$$

$$\varepsilon_{\text{true}} = \text{true strain} = \ln\frac{\text{final length}}{\text{initial length}} = \ln\frac{l_f}{l_0} \qquad (1\text{-}2b)$$

The formula for true strain is derived by integrating the expression for engineering strain from l_0 to l_f so that accumulated strain is taken into account in each infinitesimal increment.

The fundamental distinction concerning the definitions for true stress and true strain is recognition of the interrelation between gage length (l) and cross-sectional area changes (ΔA) associated with plastic deformation. When a test specimen is deformed in tension, there is ordinarily a corresponding reduction in cross-sectional area, as depicted in Fig. 1.2. As discussed further in Section 1.5.1, when the deformation is purely elastic the volume of the specimen is unlikely to be conserved, the axial strain will probably be quite small, and the change in cross-sectional area is likely to be negligible (all good assumptions in most cases, but not for rubber-like materials). However, when the limit of pure elastic deformation is reached and plastic deformation begins, it is safe to assume that most of the subsequent deformation is a constant-volume process such that

$$A_1 l_1 = A_2 l_2 = \text{constant} \tag{1-3}$$

and plastic extension of the original gage length would produce a potentially significant contraction of the gage cross-sectional area. For example, if a 25-mm (1-in.)-long[i] sample were to be plastically deformed uniformly in length by 2.5 mm owing to a tensile load P, the real or *true* stress would have to be higher than that computed by the *engineering* stress formulation. Since $l_2/l_1 = 1.1$, from Eq. 1-3 $A_1/A_2 = 1.1$, so that $A_2 = A_1/1.1$. The *true* stress is then shown to be $\sigma_{\text{true}} = 1.1\, P/A_1$ and is larger than the *engineering* value. In fact, for common engineering materials the true stress in tension is always larger than the engineering stress. Although the engineering stress is more convenient in many ways, the material is actually responding to the true stress level, so it is often important to use the true stress when describing fundamental material behavior.

By combining Eqs. 1-1a, 1-2a, and 1-3, the relationship between true and engineering stresses is shown to be

$$\sigma_{\text{true}} = \frac{P}{A_0}(l_i/l_0) = \sigma_{\text{eng}}(l_i/l_0) = \sigma_{\text{eng}}(1 + \varepsilon_{\text{eng}}) \tag{1-4}$$

This expression is accurate once sufficient plastic deformation takes place so that the constant volume assumption is valid (approximately at a strain equal to twice the strain at yielding).

Equation 1-4 is based on the assumption that the same ΔA is occurring everywhere along the gage section during deformation. This is not always the case. For example, after the maximum engineering stress (the *tensile strength*) for a typical metal has been reached, one location along the gage section will reduce in cross-sectional area more quickly than the rest of the specimen. This phenomenon is known as *necking*; after it occurs the necked region rapidly reduces in size until failure occurs. After the unstable necking process has begun, Eqs. 1-3 and 1-4 are no longer useful, and actual measurements of the cross-sectional area must be made to determine the true stress.

True and engineering strains may be related by combining Eqs. 1-1b and 1-2b to yield

$$\varepsilon_{\text{true}} = \ln(\varepsilon_{\text{eng}} + 1) \tag{1-5}$$

The need to define true strain as in Eq. 1-2b stems from the fact that the actual strain at any given time depends on the instantaneous gage length l_i. Consequently, a fixed Δl displacement will result in a decreasing amount of incremental strain, since the gage length at any given time, l_i, will increase with each additional Δl increment. Furthermore, it should be possible to define the strain imposed on a rod (for instance) by considering the total change in length of the rod as having taken place in either one step or any number of discrete steps. Stated mathematically, $\sum_n \varepsilon_n = \varepsilon_T$. As a simple example, take the case of a wire drawn in two steps with an intermediate annealing treatment. On the basis of *engineering* strain, the two deformation strains would be $(l_1 - l_0)/l_0$ and $(l_2 - l_1)/l_1$. Adding these two increments does *not* yield a final strain of $(l_2 - l_0)/l_0$. On the other hand, a summation of *true* strains does lead to the correct result (as implied by the integral used to derive Eq. 1-2b). Therefore

$$\ln\frac{l_1}{l_0} + \ln\frac{l_2}{l_1} = \ln\frac{l_2}{l_0} = \varepsilon_{\text{true total}}$$

Note that Eq. 1-5, which links engineering and true strains, does not depend on constant volume deformation, so it is valid even during pure elastic loading. It does, however,

---

[i] See the inside cover for conversion factors.

EXAMPLE 1.1

A 25-cm (10-in.)-long rod with a diameter of 0.25 cm is loaded with a 4500-newton (1012-lb) weight. If the diameter decreases to 0.22 cm, compute the following, assuming that the elastic portion of the deformation may be neglected:

(a) The final length of the rod:

Since $A_1 l_1 = A_2 l_2$ (from Eq. 1-3),

$$l_2 = \frac{A_1}{A_2} l_1 = \frac{\frac{\pi}{4}(0.25)^2}{\frac{\pi}{4}(0.22)^2}(25)$$

$$l_2 = 32.3\,\text{cm}$$

(b) The true stress and true strain at this load:

$$\sigma_{\text{true}} = \frac{P}{A_i}$$

$$= \frac{4500}{(\pi/4)(2.2 \times 10^{-3})^2}$$

$$\sigma_{\text{true}} = 1185\,\text{MPa}(172{,}000\,\text{psi})$$

$$\varepsilon_{\text{true}} = \ln\frac{l_f}{l_0}$$

$$= \ln\frac{32.3}{25}$$

$$\varepsilon_{\text{true}} = 0.256 \text{ or } 25.6\%$$

(c) The engineering stress and strain at this load:

$$\sigma_{\text{eng}} = \frac{P}{A_0}$$

$$= \frac{4500}{\frac{\pi}{4}(2.5 \times 10^{-3})^2}$$

$$\sigma_{\text{eng}} = 917\,\text{MPa}$$

$$\varepsilon_{\text{eng}} = \frac{l_f - l_0}{l_0}$$

$$= \frac{32.3 - 25}{25}$$

$$\varepsilon_{\text{eng}} = 0.292 \text{ or } 29.2\%$$

The use of true strains offers an additional convenience when considering the constant-volume plastic deformation process in that $\varepsilon_x + \varepsilon_y + \varepsilon_z = 0$. In contrast, we find a less convenient relationship, $(1 + \varepsilon_x)(1 + \varepsilon_y)(1 + \varepsilon_z) = 1$, for the case of engineering strains.

depend on a homogeneous change in length everywhere in the test specimen or component, so it is invalid after necking begins. Once inhomogeneous deformation sets in, constant volume deformation dominates in the necked region (Eq. 1-3) so a local measurement of the cross-sectional area allows the true strain to be calculated using the expression $\varepsilon_{\text{true}} = \ln(initial\ area/final\ area) = \ln(A_0/A_f)$.

1.3   STRESS–STRAIN CURVES FOR UNIAXIAL LOADING

1.3.1 Survey of Tensile Test Curves

Before further addressing elastic behavior, it is useful to understand the full spectrum of material response to loading under tension. For structural purposes, materials may be categorized as belonging to one of four groups: ceramics and glasses, metals, polymers, and composites. There are other methods of grouping materials (e.g., solids vs. foams) but separating by chemistry and atomic bonding has a number of advantages, so this is the method that will be used here. Tensile test curves for the different categories of materials have characteristic shapes, several of which will be reviewed immediately as a way of gaining familiarity with the essential features of the curves and the typical behavior for each class. Discussion of less universal aspects of tensile curves will be reserved for later.

Figure 1.3 (*a–d*) shows schematic engineering stress–strain curves for common ceramics, glasses, and metals. As a steadily increasing tensile strain is applied, the tensile stress rises. (For many testing machines, it is most convenient to apply a displacement and then measure the resulting load that develops as the test specimen resists the imposed extension.[ii] This convention is followed in the tensile test descriptions, but most aspects of the curves would be identical under load control.).[1] The initial response of each of the materials depicted here is *linear-elastic* in nature; that is, the stress and the strain are linearly proportional to one another in the early part of each curve. If the displacement (or the strain) is reversed at any time while in the elastic regime, the unloading curve should exactly trace over the original loading curve; the values of stress and strain reach zero at the same moment, indicating that the strain has been fully recovered. It is this aspect of the tensile curve that allows elastic properties to be measured reliably.

There are two ways in which the *elastic limit* can be exceeded: immediate fracture, or plastic deformation followed eventually by fracture. In the case of curve *a* in Fig. 1.3, which would be typical of a ceramic, a silicate glass, or certain metals at low temperature, fracture occurs without any noticeable plastic deformation or other warning. In contrast, curves *b–d* become nonlinear as plastic deformation is introduced. After the onset of nonlinear behavior at the material's *proportional limit*, the curves follow different paths to a peak condition (defined as the *tensile strength*) at which necking commences, the engineering stress begins to fall, and the material ultimately fractures. Curves *b–d* are typical of most metals, and are fundamentally all the same, so there are really only two general types of behavior depicted here: curve *a* with *no ductility* (i.e., no capacity for plastic deformation), and a set of curves *b–d* with *moderate to high ductility*.

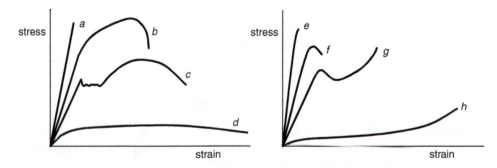

Figure 1.3   Schematic depictions of typical engineering stress–strain tensile curves for (*a*) ceramic and glass materials, (*b–d*) metals, (*e–h*) polymers. Not to scale.

[ii] Recommended specimen dimensions and conditions for testing are compiled in a series of standards by The American Society for Testing and Materials (ASTM International). ASTM is an organization comprised of volunteer engineers and scientists who develop standardized test procedures based on the consensus of experts in a particular field. ISO standards are similar procedures developed by an organization based outside the U.S.A. References to selected ASTM standards are presented throughout this text, but the reader is strongly advised to refer to the most recent book of standards to determine if a more relevant or updated standard is available for their situation before proceeding with a mechanical test.

Tensile curves typical of polymers are also depicted in Fig. 1.3. There are four distinct curves shown: (*e*) brittle, (*f*) plastic but with limited ductility, (*g*) plastic with significant ductility and strengthening, and (*h*) elastic (but nonlinear) to large strains. Whereas metal curves (*b–d*) would each depict a different metal or alloy, polymer curves (*e–g*) in Fig. 1.3 could possibly be either different polymers or the same polymer tested under different strain rate or temperature conditions. The polymer curves shown here have distinct linear (or nearly linear) elastic behavior at first, like the metals and ceramics, followed by a nonlinear response. Curves *e–g* owe their nonlinearity to viscoelastic or plastic behavior, but curve *h* depicts hyperelasticity (known as *rubber elasticity*) at large strains.

1.3.2 Uniaxial Linear Elastic Response

With some general understanding of the variety of stress–strain responses that are possible, we now turn to a closer study of linear elastic behavior. We will assume for the moment that the loading condition is *uniaxial*—that is, that the load is applied uniformly along a single axis. Furthermore, we will assume that the material under investigation is uniform in structure, and therefore will exhibit the same elastic behavior no matter what direction within the material is chosen as the loading axis. This uniform behavior is known as *isotropic*.

Over 300 years ago Robert Hooke reported in his classic paper "Of Spring" the following observations[2]:

> *Take a wire string of 20 or 30 or 40 feet long and fasten the upper part . . . to a nail, and to the other end fasten a scale to receive the weights. Then with a pair of compasses [measure] the distance [from] the bottom of the scale [to] the ground or floor beneath. Then put . . . weights into the . . . scale and measure the several stretchings of the said string and set them down. Then compare the several stretchings of the . . . string and you will find that they will always bear the same proportions one to the other that the weights do that made them.*

This observation may be described mathematically by the equation for an elastic spring:

$$F = kx \tag{1-6}$$

where

  $F =$ applied force
  $x =$ associated displacement
  $k =$ proportionality factor often referred to as the spring constant

When the force acts on a cross-sectional area A and the displacement x related to some reference gage length l, Eq. 1-6 may be rewritten as

$$\sigma = E\varepsilon \tag{1-7}$$

where

  $\sigma = F/A =$ stress
  $\varepsilon = x/l =$ strain
  $E =$ proportionality constant (often referred to as Young's modulus or the modulus of elasticity)

Equation 1-7—called Hooke's law—describes a material condition where stresses and strains are proportional to one another, leading to the initial stress–strain response shown for all of the curves in Fig. 1.3. In principle, Young's modulus can be measured during the initial loading behavior in a tensile test. In practice, it is often measured during unloading to ensure that no possibility of plasticity exists.

A wide range of values of the modulus of elasticity for many materials is shown in Table 1.1. Those with large elastic moduli are called *stiff* materials, and would provide significant resistance to elastic deformation. Those with low elastic moduli are called *compliant* materials, and their

Table 1.1a   Elastic Properties of Engineering Materials[a]

Material at 20°C	E (GPa)	G (GPa)	v
Metals			
Aluminum	70.3	26.1	0.345
Cadmium	49.9	19.2	0.300
Chromium	279.1	115.4	0.210
Copper	129.8	48.3	0.343
Gold	78.0	27.0	0.44
Iron	211.4	81.6	0.293
Magnesium	44.7	17.3	0.291
Nickel	199.5	76.0	0.312
Niobium	104.9	37.5	0.397
Silver	82.7	30.3	0.367
Tantalum	185.7	69.2	0.342
Titanium	115.7	43.8	0.321
Tungsten	411.0	160.6	0.280
Vanadium	127.6	46.7	0.365
Other Materials			
Aluminum oxide (fully dense)	~415	—	—
Diamond	~965	—	—
Glass (heavy flint)	80.1	31.5	0.27
Nylon 66	1.2–2.9	—	—
Polycarbonate	2.4	—	—
Polyethylene (high density)	0.4–1.3	—	—
Poly(methyl methacrylate)	2.4–3.4	—	—
Polypropylene	1.1–1.6	—	—
Polystyrene	2.7–4.2	—	—
Quartz (fused)	73.1	31.2	0.170
Silicon carbide	~470	—	—
Tungsten carbide	534.4	219.0	0.22

[a] G. W. C. Kaye and T. H. Laby, *Tables of Physical and Chemical Constants*, 14th ed., Longman, London, 1973, p. 31.

resistance to elastic deformation would be relatively low. The major reason for the large property variations seen in Table 1.1 is related to differences in the strength of the interatomic forces between adjacent atoms or ions. To illustrate this fact, let us consider how the potential energy $\mathscr{E}$ between two adjacent particles changes with their distance of separation x(Fig. 1.4a). The equilibrium distance of particle separation x_0, corresponding to a minimum in potential energy, is associated with a balance of the energies of repulsion and attraction between two adjacent atoms or ions. The form of this relationship is often given by $\mathscr{E} = -\alpha/x^m + \beta/x^n$, where $-\alpha/x^m$ and β/x^n correspond to the energies of attraction and repulsion, respectively, and $n > m$. At x_0, the force ($F = d\mathscr{E}/dx$) acting on the particles is equal to zero (Fig. 1.4b). The first derivative of the force with respect to distance of separation, dF/dx (i.e., $d^2\mathscr{E}/dx^2$), then describes the stiffness or relative resistance to separation of the two atoms or ions. As such, dF/dx is analogous to the Young's modulus quantity, E, given in Eq. 1-7. A simple analysis of bonding forces shows that the elastic stiffness is proportional to $1/x_0^n$. Examples of the strong dependence of elastic stiffness on x_0 for alkali metals are shown in Fig. 1.4c.

Since E depends on the strength of the interatomic forces that vary with the type of bonding found in a given material, it is relatively insensitive to alloying or changes in microstructure.

Table 1.1b   Elastic Properties of Engineering Materials[a]

Material at 68°F	E (10^6 psi)	G (10^6 psi)	v
Metals			
Aluminum	10.2	3.8	0.345
Cadmium	7.2	2.8	0.300
Chromium	40.5	16.7	0.210
Copper	18.8	7.0	0.343
Gold	11.3	3.9	0.44
Iron	30.6	11.8	0.293
Magnesium	6.5	2.5	0.291
Nickel	28.9	11.0	0.312
Niobium	15.2	5.4	0.397
Silver	12.0	4.4	0.367
Tantalum	26.9	10.0	0.342
Titanium	16.8	6.35	0.321
Tungsten	59.6	23.3	0.280
Vanadium	18.5	6.8	0.365
Other Materials			
Aluminum oxide (fully dense)	~60	—	—
Diamond	~140	—	—
Glass (heavy flint)	11.6	4.6	0.27
Nylon 66	0.17	—	—
Polycarbonate	0.35	—	—
Polyethylene (high density)	0.058–0.19	—	—
Poly(methyl methacrylate)	0.35–0.49	—	—
Polypropylene	0.16–0.39	—	—
Polystyrene	0.39–0.61	—	—
Quartz (fused)	10.6	4.5	0.170
Silicon carbide	~68	—	—
Tungsten carbide	77.5	31.8	0.22

[a] G. W. C. Kaye and T. H. Laby, *Tables of Physical and Chemical Constants*, 14th ed., Longman, London, 1973, p. 31.

Therefore, while heat treatment and minor alloying additions may cause the strength of a steel alloy to change from 210 to 2400 MPa, the modulus of elasticity of both materials remains relatively unchanged—about 200 to 210 GPa. The result is that all steel alloys have similar moduli, while all aluminum alloys have much lower, but also self-similar, moduli.

In many engineering materials, nonlinearity in the stress–strain plot is an indication that plasticity or fracture has occurred. However, many polymers, as well as soft copper and gray cast iron[3], display a certain degree of *nonlinear elasticity*. As such, the elastic modulus must be determined using either a *tangent modulus*[4]—the slope of a tangent line to the elastic portion of the curve at a chosen value of stress—or a *secant modulus*[5]—the slope of a line drawn from the origin to a chosen point on the stress–strain curve. The modulus is no longer a single value for a given material, but depends instead on the loading conditions. Nevertheless, a single value representing the initial slope is often reported and used for engineering purposes. It should be noted that elastomers like silicone rubber can exhibit highly nonlinear elastic behavior over a much wider range of strain than most other materials (e.g., recall Fig. 1.3*h*), so elastomer tangent or secant moduli may be reported at strains of 100, 200, or even 300%. (See Section 1.3.3 and Chapter 4 for more information about nonlinear behavior of polymers.)

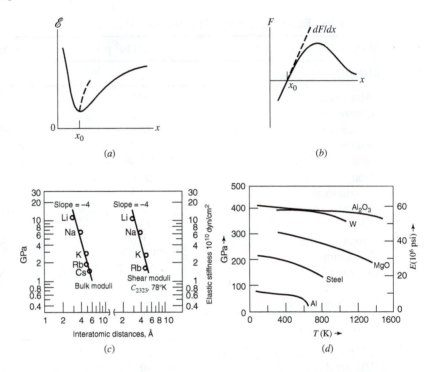

Figure 1.4   Dependence of elastic stiffness on interatomic spacing: (*a*) Potential energy versus interatomic spacing; (*b*) Force versus interatomic spacing; (*c*) Elastic stiffness of alkali metals versus interatomic spacing. (From J. J. Gilman, *Micromechanics of Flow in Solids,* McGraw-Hill, New York, 1969, with permission.); (*d*) Variation of Young's modulus with temperature in selected metals and ceramics. (From K. M. Ralls, T. H. Courtney, and J. Wulff, *Introduction to Materials Science and Engineering*, Wiley, 1976, with permission.)

There is another class of elastic behavior that is not linear, but in this case it is because the response to loading is not instantaneous. Upon loading and unloading it appears at first that plastic deformation has taken place because the strain is not zero at the time that the stress first reaches zero. Over time, however, the strain is fully recovered. This behavior is called *anelastic* deformation in metals, and *viscoelastic* deformation in polymers (although both indicate time-dependent-reversible strain), and is discussed in Chapter 4.

It is also possible for a metal that is ordinarily considered linear elastic to exhibit nonlinear elastic behavior under certain special conditions. For example, very high-strength metal fibers—often called whiskers—can exhibit tensile elastic strains in excess of 2%, as can tiny metal pillars tested in compression.[6,7] In this range of very large elastic strains, the modulus of elasticity reveals its weak dependence on strain—something that is completely obscured when strains are very small. This can be understood by noting that the slope dF/dx in Fig. 1.4*b* is not truly linear, but instead decreases with increasing distance of atom separation. If it is possible to separate the atoms of a material by a large distance without inducing plasticity or fracture, this fundamental nonlinearity at the atomic level becomes apparent at the macroscopic scale as a gradual reduction in the elastic modulus. As such, Hooke's law (Eq. 1-7) represents an empirical relationship, albeit a good one at the small strains (typically less than 0.2%) that mark the end of the elastic regime for many engineering materials.

From the preceding discussion regarding the atomic-level basis for elastic behavior, it follows that values of *E* for metals, ceramics, and glasses should decrease with increasing temperature (i.e, these materials become less stiff). This is related to the fact that the average distance of atom or ion separation increases with temperature, which manifests itself macroscopically as thermal expansion upon heating, and contraction upon cooling. Note the dashed line in Fig. 1.4*a*, which corresponds to the locus of values of the average separation at temperatures above absolute zero.

The loss of stiffness with increasing temperature is gradual, with only a small percent decrease occurring for a 100°C (180°F) temperature change (Fig. 1.4*d*).

1.3.3   Young's Modulus and Polymer Structure

The effect of temperature on the Young's modulus of polymers can be another story entirely. A polymer molecule is constructed with covalent bonds along its length, but links between polymer chains may be strong covalent bonds (cross-links), weak secondary bonds, or a mixture of the two. Polymers without cross-links are called *thermoplastics* because they have the ability to melt and remelt. Polymers with many cross-links cannot melt after they have been solidified, and so are called *thermosets*. In addition to this distinction, thermoplastic polymer structure can range from completely amorphous (a "pile of spaghetti") to mostly crystalline (folded lamellar regions separated by amorphous regions). The wide variability in bonding and structure gives the polymer class of materials the ability to take on many different properties, but also introduces *viscoelasticity* (addressed in detail in Chapter 4). This is time-dependent elasticity, in which the relationship between stress and strain changes over time. It is also highly sensitive to temperature. As a result, the concept of using a *simple* elastic modulus to describe the mechanical behavior of a particular polymer is dubious at best. But, despite their inadequacy, tangent or secant moduli are widely used to screen polymers during preliminary material selection procedures and for quality control.[5,8] The elastic behavior of all useful polymers tends to be nearly linear at small stresses and strains, so Hooke's law serves as a good first approximation of actual behavior for many applications as long as it is appreciated that a simple modulus is only relevant for a limited range of time, strain rate, and temperature.

1.3.3.1   Thermoplastic Behavior

When short-term tensile tests of amorphous or semi-crystalline thermoplastics are performed over a range of temperatures, it is found that the elastic modulus measured at a particular strain rate decreases with increasing temperature as shown schematically in Fig. 1.5. At low temperatures the modulus is relatively high, and is only mildly temperature-dependent. This is known as *glassy* behavior, although it is not unique to amorphous polymers. As the test temperature rises, there is a large transition in stiffness that occurs over a narrow temperature range. The midpoint in this range is called the *glass transition temperature*, T_g. (The actual temperature at which it occurs is a characteristic of a particular material.) The magnitude of the T_g modulus transition is strongly dependent on the structure. For amorphous materials the change is very large, and the stiffness rapidly declines to the point of melting (at which the elastic stiffness is zero). For this reason, load-bearing amorphous thermoplastics must have T_g values above the intended use temperature. This is true for common amorphous thermoplastics, including polystyrene (PS), unplasticized poly-vinylchloride (U-PVC), transparent polyethylene terephthalate (PET), polycarbonate (PC), and

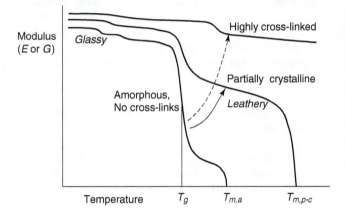

Figure 1.5   Schematic depiction of the temperature dependence of the short-term modulus for polymeric solids. The glass transition temperature, T_g, marks the largest modulus change, depicted here only for the amorphous case. The melting points of amorphous (*a*) and partially-crystalline (*p-c*) thermoplastics are also marked. The solid arrow indicates the trend with increasing degree of crystallinity, and the dashed arrow the trend with cross-linking.

acrylics like poly(methyl methacrylate) (PMMA), also known by such trademark names as Plexiglas and Lucite. As some degree of crystallinity is introduced, the drop in modulus at T_g becomes smaller, and a wide *leathery* (or *tough*) region appears before the stiffness final drop at the melt temperature. Semi-crystalline thermoplastics may therefore be used at temperatures above or below T_g. The short secondary bonds found in folded lamellar crystals better resist molecular reconfiguration than the longer secondary bonds between amorphous chains, which explains the observed trend of increasing stiffness with increasing crystallinity. Examples of common semicrystalline thermoplastics include low density polyethylene (LDPE), high-density polyethylene (HDPE), polytetrafluoroethylene (PTFE), and opaque versions of polyethylene terephthalate (PET). Plasticizers like dioctyl phthalate (DOP) may be added to thermoplastics to lower T_g and therefore alter room temperature behavior.[iii] They are short organic molecules that are soluble in the polymer and that make changes in molecular configuration easier at a given temperature. Whereas U-PVC is rigid at room temperature, plasticized PVC is flexible and tough.

Certain semicrystalline thermoplastics may be processed to form fibers in a manner that aligns the molecular chains along the fiber axis. Not surprisingly, this leads to direction-dependent elastic behavior (and strength). As the covalent bonds that make up the chains align with the tensile axis, these *oriented thermoplastics* have great longitudinal stiffness compared to their unoriented cousins. Ultra high molecular weight polyethylene (UHMWPE) can be processed in this way for use in lightweight armor, fishing line, surgical sutures, high-performance ropes, and fiber composite reinforcements.[iv] A class of thermoplastics called *liquid crystal polymers* has the unusual ability to retain their chain orientation even in the melt, which simplifies processing. Highly aligned liquid crystal polymer fibers (including aramids such as Kevlar) are also well known for extremely high stiffness (and strength) to weight ratio. Spider silk represents a biological example of a highly aligned thermoplastic that is renowned for its remarkable properties. However, it is difficult to produce artificial spider silk in bulk, so its use as an engineering material—although extremely attractive—is limited.

The temperature-dependent behavior of the shear modulus (proportional to E; see Sections 1.4.2 and 1.5.1) for many thermoplastic and thermoset materials is summarized in Fig. 1.6. The temperature axis is actually the reduced temperature, $T_{red} = 293\ \text{K}/T_g$. Depicted in this way, it is apparent that all of the polymers follow similar trends even though their actual T_g values differ. Any material with $T_{red} < 1.0$ has a $T_g > 293$ K (approximately room temperature), and is therefore typically used in its glassy mechanical state. Those with $T_{red} > 1.0$ have a $T_g < 293$ K, and are therefore typically used in the leathery state (or the rubbery state, in the case of the elastomers). Those with $T_{red} \approx 1.0$ would be expected to show strong temperature sensitivity near room temperature. For engineering design, it can be quite useful to know the T_g values of candidate polymers because they may limit the temperature range of safe operation.

1.3.3.2 Rigid Thermosets

Thermosets are highly cross-linked, so changes in molecular configuration are difficult. This gives the materials in this class fairly high moduli that are relatively insensitive to temperature. Furthermore, T_g tends to be high, and there is no melt transition possible without significant secondary bonding, so these materials retain much of their stiffness and strength up to the point where they begin to thermally degrade. This general behavior can be seen in Fig. 1.5. Common thermosets include epoxy resins, melamine resin, polyurethanes, and phenol-formaldehyde resins (Bakelite).

[iii] Health concerns about absorption of certain phthalates like DOP limit their use for medical devices and drinking vessels, though they are widely used for applications in which such exposure is unlikely.

[iv] UHMWPE is also used for hip prosthesis implants as acetabular socket replacements, but not in an oriented form. Unidirectional wear processes have been shown to plastically deform the implant surface, creating highly oriented fibrils that have great strength and stiffness along the longitudinal axis, but poor properties in the transverse axis. When the direction of joint motion is multi-directional, the fibrils can be ruptured, leading to the production of undesirable wear particles. Increased cross-linking can prevent the orientation process and thereby improve wear resistance.[37]

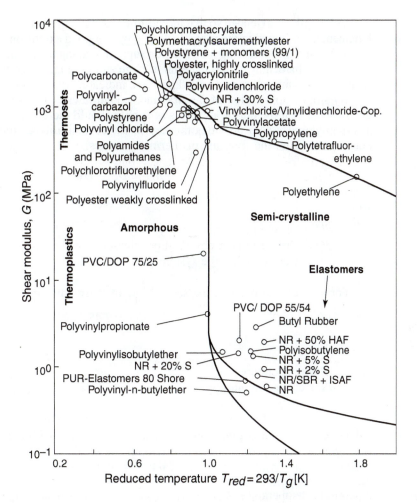

Figure 1.6   Shear modulus as a function of temperature for many polymer materials, plotted as a function of their reduced temperature. (Reprinted from Osswald and Menges,[9] with permission.)

1.3.3.3 Rubber Elasticity

Although elastomers are cross-linked and therefore have some of the characteristics of rigid thermosets, their tensile response is quite different from either the thermosets or thermoplastics.[10] Elastomer (or rubber) elasticity is distinguished by two basic characteristics: very large nonlinear elastic strains (often in excess of 100%) and elastic moduli that *increase* with increasing temperature. The latter response is opposite that found in other materials (including rigid polymers). Elastomers are polymers that contain moderate numbers of chemical or physical cross-links, and that are tested and used above their T_g values. Some degree of amorphous structure is essential to the ability to develop enormous elastic strains because it allows significant extension simply by chain straightening and recoiling. Cross-linking (e.g., by vulcanization) is also critical, as it prevents the possibility of plastic deformation associated with chains sliding past one another. In fact, the degree of cross-linking ultimately determines the extensibility of an elastomer, which is evident in the difference between the properties of moderately cross-linked rubber bands and heavily cross-linked bowling balls.

Rubber elasticity is related primarily to the straightening of amorphous polymer chains from their curled positions into partially extended conformations. As a result, the elastic moduli are very low because of the small contribution of actual polymer chain stretching. That is, a curled chain of length l is extended so that its end-to-end length approaches l with little additional chain lengthening attributed to the more difficult covalent bond extension mode. The straightening of the chains and the increasing load fraction supported along the covalent bonds is responsible for the apparent hardening of the material at large strains (Fig. 1.3h). When the applied loads are relaxed, the chains return to a curled position, indicating the latter conformation to be preferred.

By simple application of the first and second laws of thermodynamics, it is possible to demonstrate that the elastic modulus of rubber should increase with increasing temperature. Thermodynamics also provides the key to the reversibility of the large strain behavior of elastomers. From the first law of thermodynamics,

$$dU = \partial Q + \partial W \tag{1-8}$$

where

$dU =$ change in internal energy
$\partial Q =$ change in heat absorbed or released
$\partial W =$ work done on the system

For a reversible process, the second law of thermodynamics gives

$$dQ = TdS \tag{1-9}$$

where

$T =$ temperature
$dS =$ change in entropy

If an elastomeric rod of length l is extended by an amount dl owing to a tensile force F, the work ∂W done on the rod is Fdl. Combining Eqs. 1-8 and 1-9 with the expression for ∂W gives

$$dU = TdS + Fdl \tag{1-10}$$

At constant temperature

$$F = \left(\frac{\partial U}{\partial l}\right)_T - T\left(\frac{\partial S}{\partial l}\right)_T \tag{1-11}$$

where

$\left(\frac{\partial U}{\partial l}\right)_T =$ related to the strain energy associated with the application of a load

$\left(\frac{\partial S}{\partial l}\right)_T =$ related to the change in entropy or order of the rod as it is stretched

Since the chains prefer a random curled configuration, their initial degree of order is low and their entropy high. (Because of the very high degree of order of atoms in metals and ceramics, their entropy term by comparison is negligible.) However, when a tensile load is applied, the entropy decreases as the chains become straightened and aligned. As a consequence, $(\partial S/\partial l)_T$ is negative. The force required to extend the elastomer rod, therefore, increases with increasing temperature. By the same argument, it is entropy (not stored energy) that drives the chain recoiling process and the recovery of large elastic strains.

As expected, rubber stiffness increases with increasing cross-link density and corresponding decrease in the molecular weight of chain segments between cross-links (Mc). Regarding the latter, the modulus of rubber is found to vary inversely with Mc. Interestingly, rubbers are distinguished from most other materials in that their elastic moduli can be predicted from molecular structural details.[11]

The elastic response of elastomers is approximately linear up to about 1%, but decidedly nonlinear thereafter. Metals and ceramics typically undergo only small elastic strains, so the definition of strain given in Eq. 1-1b is applicable. Elastomers deform to large extensions for which the assumptions of small strain theory are invalid, hence these materials are often described as being *hyperelastic*, and the *stretch ratio* or *extension ratio*, λ, is used instead of strain:

$$\lambda = \frac{l_i}{l_0} = \varepsilon + 1 \tag{1-12}$$

Based on this definition, a nonlinear expression relating stress and extension ratio can be developed from kinetic theory of rubber elasticity, with

$$\sigma = \frac{E_0}{3}\left(\lambda - \frac{1}{\lambda^2}\right) \tag{1-13}$$

where E_0 is the elastic modulus as the extension ratio approaches 1 (i.e., zero extension). This model works well for the typical extension ranges expected of elastomers in service (on the order of 25–30%, or $\lambda = 1.25$ to 1.30).[12] For extension ratios much larger than 1.25 or so, the Mooney-Rivlin model based on strain energy considerations tends to fit better to experimental data. For simple uniaxial tension, this model takes the form

$$\sigma = 2\left(C_1 + \frac{C_2}{\lambda}\right)\left(\lambda - \frac{1}{\lambda^2}\right) \tag{1-14}$$

where the Mooney-Rivlin constants C_1 and C_2 are fitting coefficients associated with a particular material.

1.3.4 Compression Testing

Uniaxial compression tests provide much of the same information about material properties as tension tests. However, the compression test specimen is comparatively simple in shape: usually a cylinder with a ratio of length to diameter $L/D < 2$ to avoid non-axial motion. *Elastic behavior* in compression should ideally be the same as in tension, although in practice it is not always the case.[13] Caution must be taken during compression testing to minimize friction between the loading platen and the specimen because friction will provide an artificial resistance to ΔA, and will therefore make the material appear stiffer and stronger than it actually is. Even after plastic deformation has commenced, the true stress–true strain curve from a well-run compression test of a metal should closely match that of a tensile test, although the engineering curve will not because of tensile necking. The true stress–strain curve for a given polymer in tension is always lower than in compression since the chains are more mobile under tensile conditions. One potential advantage of compression testing is the avoidance of necking instability, so larger strains can often be imposed than are possible under tension. This can also be seen as a drawback if aspects of the necking behavior and ensuing tensile fracture are of interest. Compression testing also avoids early failure due to brittle cracking in ceramic materials.

1.3.5 Failure by Elastic Buckling

When a slender component (e.g., a column or pole) is under a compressive load along its long axis, an elastic instability can occur that leads to buckling under relatively low loads. As introduced in Section 1.1, buckling is manifested as excessive lateral deflection. It becomes a serious concern when the length/diameter ratio of the column is $L/D \geq 10$. If a column is aligned perfectly with the loading axis, the column will shorten with applied compressive load according to Hooke's law, regardless of aspect ratio. However, if the column is even slightly eccentric, the applied compressive load will generate a bending stress that can trigger the buckling response known as *Euler buckling*. For example, consider the case of a column of length L that has ends that are hinged and are therefore free to rotate but not to translate (i.e., ends that are *pinned*). The critical load P_{cr} (the *Euler buckling load*) for the onset of elastic buckling is given by

$$P_{cr} = L^{-2}\pi^2 EI = \frac{\pi^2 EI}{L^2} \qquad (1\text{-}15)$$

where E is the Young's modulus and I is the appropriate moment of inertia for the column cross-section shape. Note that the critical load for buckling decreases with the inverse square of L, the column length. For the case of a column with a circular cross section of diameter d, $I = \pi d^4/64$. If we rewrite the expression as a function of the cross-sectional area A, it is easy to show that the critical average stress is given by

$$\sigma_{cr} = \frac{P_{cr}}{A} = \frac{\pi^2 E}{(L/g)^2} \qquad (1\text{-}16)$$

where g is the radius of gyration of the column (the distance from the column's reference axis to the location where the column area is concentrated). For the circular column case, $g = d/4$.

It can be seen from Eq. 1-16 that the buckling stress decreases markedly with increasing slenderness ratio L/g. This is a strong motivation for the small L/D ratio recommended for compression testing. Also, the buckling stress varies with the elastic modulus of the column material, but is not dependent on the material strength (e.g., the yield strength). Therefore, two dimensionally similar columns, one of high-strength steel and the other a low-strength steel alloy, will buckle under the same critical load (recall that elastic modulus is not strongly affected by alloying or heat treatment). However, if the applied load can exceed the proportional limit prior to the onset of buckling, as may be the case for a column that is of intermediate length, then failure will occur by crushing rather than buckling and alloy strength comes back into play.

The critical buckling load also is affected strongly by the boundary conditions at the ends of the column. For the three cases of (a) both ends fixed, (b) one end fixed, one end pinned, and (c) one end fixed, one end free (e.g., like a flagpole), the critical load equations are nearly identical, but the effective unbuckled length between the points of zero moment, L_e, is used in place of the actual unbuckled length, L, as indicated in Fig. 1.7. After substitution into Eq. 1-15 we find that the critical loads for the three cases are given by

$$P_{cr, fixed-fixed} = \left(\frac{L}{2}\right)^{-2}\pi^2 EI = \frac{4\pi^2 EI}{L^2} \qquad (1\text{-}17a)$$

$$P_{cr, fixed-pinned} = \left(\frac{L}{\sqrt{2}}\right)^{-2}\pi^2 EI = \frac{2\pi^2 EI}{L^2} \qquad (1\text{-}17b)$$

$$P_{cr, fixed-free} = (2L)^{-2}\pi^2 EI = \frac{\pi^2 EI}{4L^2} \qquad (1\text{-}17c)$$

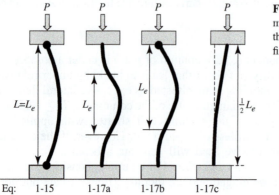

Figure 1.7 Schematic depictions of buckled slender members with four different boundary conditions and their corresponding equations: pinned-pinned, fixed-fixed, fixed-pinned, and fixed-free.

It is important to note that these equations provide upper-bound solutions for the critical buckling load. Actual critical loads are lower due to the small eccentricities that are inevitable in column construction. Hence, an appropriate safety factor should be used.

1.3.6 Resilience and Strain Energy Density

The *resilience* of a material is a measure of the amount of energy per unit volume (in units of $Pa = J/m^3$) that can be absorbed under elastic loading conditions and that is released completely when the loads are removed. From this definition, resilience may be measured from the area under the initial elastic portion of any curve in Fig. 1.3. If it is linear, then

$$\text{resilience} = \frac{1}{2}\sigma_{max}\varepsilon_{max} \tag{1-18}$$

where

$\sigma_{max} =$ maximum stress for elastic conditions
$\varepsilon_{max} =$ elastic strain limit

And, from Eq. 1-7,

$$\text{resilience} = \frac{\sigma_{max}^2}{2E} \tag{1-19}$$

Should an engineering design require a material that allows only for elastic response with large energy storage (such as in the case of a mechanical spring), the appropriate material to choose would be one possessing a high yield strength but low modulus of elasticity.

A similar analysis of stored elastic energy may be applied even after plastic deformation has commenced. For any point along the stress–strain curve, the *strain energy density* (SED) may be computed using Eq. 1-18, but with the substitution of the elastic portion of the total strain, $\varepsilon_{el} = \sigma/E$, and the stress at the point of interest, such that $SED = \frac{1}{2}\sigma\varepsilon_{el} = \sigma^2/2E$. The strain energy density describes the elastic energy that is stored in the material at any point in the load history, all of which is available for release either upon intentional unloading or upon sudden fracture.

1.3.7 Definitions of Strength

Strength is a measure of resistance to plastic deformation or fracture. Since there are several ways to define strength that are dependent on the nature of the testing mode, it is appropriate to introduce here those definitions that are relevant for uniaxial testing. The stress level indicated by the proportional limit has already been discussed, but it should be appreciated that this is not a very useful engineering measure. It is usually difficult to discern exactly where the transition from linear to nonlinear behavior occurs, and in some materials there is no such clean transition at all.

Instead, for ductile metals (Fig. 1.3 (*b–d*)) it is common to define a *yield strength* (or an *offset yield strength*) that is a measure of the stress needed to induce a very small amount of plastic strain (often 0.002, i.e., 0.2%). Either a specimen is loaded and unloaded to progressively larger strains until the desired permanent strain is evident after unloading, or a continuous test is performed to large strains and a line parallel to the elastic portion of the curve is drawn emanating from the desired point on the strain axis until it intercepts the loading curve.[v] This is the same path that would be followed if the material had been unloaded from that intercept point. Ideally, either method should lead to the same yield strength value. This value is very useful to an engineer because it reliably indicates the stress level at which plastic deformation can be said to begin in earnest. Values of tensile yield strength for selected materials are listed in Table 1.2.

---

[v] The recommended procedure for determining the "offset yield strength" is described in ASTM Standard E8.

Table 1.2a Tensile Properties for Selected Engineering Materials[a]

Material	Treatment	Yield Strength (MPa)	Tensile Strength (MPa)	Elongation in 5-cm Gage (%)	Reduction in Area (1.28-cm diameter) (%)
Steel Alloys					
1015	As-rolled	315	420	39	61
1050	"	415	725	20	40
1080	"	585	965	12	17
1340	Q + T (205°C)	1590	1810	11	35
1340	" (425°C)	1150	1260	14	51
1340	" (650°C)	620	800	22	66
4340	" (205°C)	1675	1875	10	38
4340	" (425°C)	1365	1470	10	44
4340	" (650°C)	855	965	19	60
301	Annealed plate	275	725	55	—
304	" "	240	565	60	—
310	" "	310	655	50	—
316	" "	250	565	55	—
403	Annealed bar	275	515	35	—
410	" "	275	515	35	—
431	" "	655	860	20	—
AFC-77	Variable	560–1605	835–2140	10–26	32–74
PH 15-7Mo	"	380–1450	895–1515	2–35	—
Titanium Alloys					
Ti-5Al-2.5Sn	Annealed	805	860	16	40
Ti-8Al-lMo-lV	Duplex annealed	950	1000	15	28
Ti-6A1-4V	Annealed	925	995	14	30
Ti-13V-llCr-3Al	Solution + age	1205	1275	8	—
Magnesium Alloys					
AZ31B	Annealed	103–125	220	9–12	—
AZ80A	Extruded bar	185–195	290–295	4–9	—
ZK60A	Artificially aged	215–260	295–315	4–6	—
Aluminum Alloys					
2219	-T31, -T351	250	360	17	—
2024	-T3	345	485	18	—
2024	-T6, -T651	395	475	10	—
2014	-T6, -T651	415	485	13	—
6061	-T4, -T451	145	240	23	—
7049	-T73	475	530	11	—
7075	-T6	505	570	11	—
7075	-T73	415	505	11	—
7178	-T6	540	605	11	—
Plastics					
ABS	Medium impact	—	46	6-14	—
Acetal	Homopolymer	—	69	25-75	—
Poly(tetra-fluorethylene)	—	—	14–48	100–450	—
Poly(vinylidene fluoride)	—	—	35–48	100–300	—
Nylon 66	—	—	59–83	60–300	—
Polycarbonate	—	—	55–69	130	—
Polyethylene	Low density	—	7–21	50–800	—
Polystyrene	—	—	41–54	1.5–24	—
Polysulfone	—	69	—	50–1000	—

[a] *Datebook 1974, Metal Progress* (mid-June 1974).

Table 1.2b   Tensile Properties for Selected Engineering Materials[a]

Material	Treatment	Yield Strength (ksi)	Tensile Strength (ksi)	Elongation in 2-in. Gage (%)	Reduction in Area (0.505-in. diameter) (%)
Steel Alloys					
1015	As rolled	46	61	39	61
1050	"	60	105	20	40
1080	"	85	140	12	17
1340	Q + T (400°F)	230	260	11	35
1340	" (800°F)	167	183	14	51
1340	" (1200°F)	90	116	22	66
4340	" (400°F)	243	272	10	38
4340	" (800°F)	198	213	10	44
4340	" (1200°F)	124	140	19	60
301	Annealed plate	40	105	55	—
304	" "	35	82	60	—
310	" "	45	95	50	—
316	" "	36	82	55	—
403	Annealed bar	40	75	35	—
410	" "	40	75	35	—
431	" "	95	125	20	—
AFC-77	Variable	81–233	121–310	10–26	32–74
PH 15-7Mo	"	55–210	130–220	2–35	—
Titanium Alloys					
Ti-5A1-2.5Sn	Annealed	117	125	16	40
Ti-8Al-lMo-lV	Duplex annealed	138	145	15	28
Ti-6A1-4V	Annealed	134	144	14	30
Ti-13V-11Cr-3Al	Solution + age	175	185	8	—
Magnesium Alloys					
AZ31B	Annealed	15–18	32	9–12	—
AZ80A	Extruded bar	27–28	42–43	4–9	—
ZK60A	Artificially aged	31–38	43–16	4–6	—
Aluminum Alloys					
2219	-T31, -T351	36	52	17	—
2024	-T3	50	70	18	—
2024	-T6, -T651	57	69	10	—
2014	-T6, -T651	60	70	13	—
6061	-T4, -T451	21	35	23	—
7049	-T73	69	77	11	—
7075	-T6	73	83	11	—
7075	-T73	60	73	11	—
7178	-T6	78	88	11	—
Plastics					
ABS	Medium impact	—	6.8	6–14	—
Acetal	Homopolymer	—	10	25–75	—
Poly(tetra-fluorethylene)	—	—	2–7	100–450	—
Poly(vinylidene fluoride)	—	—	5.1–7	100–300	—
Nylon 66	—	—	8.6–12	60–300	—
Polycarbonate	—	—	8–10	130	—
Polyethylene	Low density	—	1–3	50–800	—
Polystyrene	—	—	6–9	1.5–2.4	—
Polysulfone	—	10	—	50–100	—

[a] *Databook 1974, Metal Progress* (mid-June 1974).

Table 1.3 Elastic Modulus and Strength Properties of Selected Ceramics[a]

Material	Modulus of Elasticity [GPa (10^6 psi)]	Tensile Strength [MPa (ksi)]	Flexural Strength [MPa (ksi)]	Compressive Strength [MPa (ksi)]
Alumina (85% dense)	220 (32)	125 (18)	295 (42.5)	1620 (235)
Alumina (99.8% dense)	385 (56)	205 (30)	345 (60)	2760 (400)
Alumina silicate	55 (8)	17 (2.5)	62 (9)	275 (40)
Transformation toughened zirconia	200 (29)	350 (51)	635 (92)	1760 (255)
Partially stabilized zirconia + 9% MgO	205 (30)	—	690 (100)	1860 (270)
Cast Si_3N_4	115 (17)	24 (3.5)	69 (10)	138 (20)
Hot-pressed Si_3N_4	—	—	860 (125)	3450 (500)

[a] *Guide to Engineering Materials,* Vol. 1(1), ASM, Metals Park, OH, 1986, pp. 16, 64, 65.

It has also been mentioned that for ductile metals the stress corresponding to the maximum of the curve measured under tensile loading defines the *tensile strength* (or the *ultimate tensile strength*). This is a useful value because it corresponds to the maximum load sustainable by a particular material under tension, and also indicates the end of the plastic region in which uniform elongation and thinning of the gage section occurs. Nonuniform thinning is very dangerous when the component in question is bearing a fixed load (such as a cable from which a mass is suspended) because fracture is virtually inevitable shortly thereafter. Even when the component is not yet in service—for example, during fabrication of a thin sheet that will serve as the hood of an automobile—tensile instability is undesirable because it leads to locally thin areas that will act as points of weakness in the finished part (or, at the very least, will be unsightly). As a reminder, this condition is not observable from a compression test.

The yield strength of ductile polymers may be defined differently from that of metals. As seen in Fig. 1.3 (*f–g*), the yield strength can be defined at the first maximum in the curve (the point at which the true stress falls). If there is no local maximum then the yield strength can be defined by a 0.2% offset yield point, just as for metals.

Brittle materials, whether they are metals, ceramics, glasses, or polymers, have the stress at which sudden fracture occurs as their only measure of the strength. This is called the *fracture strength*. Fracture in brittle materials is strongly influenced by the size and character of surface flaws, so there is often a relatively large uncertainty associated with the fracture strength value. The fracture strength of ceramic materials may be much greater under compressive loading than under tensile loading, so *compressive strength* values are often reported for ceramic materials (as shown in Table 1.3). Of course, ductile materials will also fracture at sufficiently high strains, so the *true fracture stress* can be defined as the load at fracture divided by the final cross-sectional area of the test specimen.

1.3.8 Toughness

Toughness is another measure of resistance to fracture, but is measured in units of energy. We may define a brittle material as one absorbing little energy, while a tough material would require a large expenditure of energy in the fracture process. For a smooth tensile bar, the energy to break may be estimated from the area under the stress–strain curve.

$$\text{energy/volume} = \int_0^{\varepsilon_f} \sigma \, d\varepsilon \tag{1-20}$$

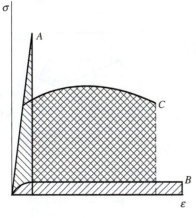

Figure 1.8   Stress–strain curves for strong material with little plastic flow capacity, *A*; low-strength but high-ductility material, *B*; and a metal with optimum combination of strength and ductility for maximum toughness, *C*.

Maximum toughness, therefore, is achieved with an optimum combination of strength and ductility; neither high strength (e.g., glass) nor exceptional ductility (e.g., taffy) alone provides for large fracture energy absorption (Fig. 1.8). Material toughness will be considered in much greater detail in Chapters 6–11.

1.4   NONAXIAL TESTING

In addition to axial tension and compression, engineering components may be subjected to bending, shearing, and torsion. Not surprisingly, standardized tests for evaluating material response under these loading conditions are well developed. While bend testing can be used to measure Young's modulus *E* much like tensile or compression testing, shear and torsion tests are used to measure *G*, a quantity known as the *shear modulus*.

1.4.1   Bend Testing

Although bend testing is an option for metals and polymers, flexural test methods are most frequently used to determine the elastic behavior and strength characteristics of ceramic and glass compounds. This arises from the fact that ceramics and glasses usually display essentially no plastic deformation and, as such, the mechanical response of these materials is very sensitive to the presence of complex sample shapes that introduce stress concentrations. (Such is the case with threaded grips that are sometimes machined into tensile bars.) Stress concentrations can cause premature failure, thereby limiting the usefulness of the standard tensile bar in this case. By contrast, bend bars have a smooth configuration, are easy to machine and test, and require simple load fixtures. The three-point and four-point methods represent two common loading configurations (see Fig. 1.9).

Under tensile or compressive loading parallel to the axis of some uniform load-bearing member, the stresses are typically constant over the entire component. In bending, however, where stresses are applied normal to the component main axis (as shown in Fig. 1.9), the axial stress (i.e., in the *L* direction) varies from one location to another within the beam. The surface on one side of the beam will be in compression, while the other side is under tension. The stress through the thickness of the beam varies linearly between the surface compression and tension stress values, with zero stress at the neutral axis. The stress along the beam surfaces will also vary, with maxima at the two surfaces either under the central load point for three-point bending, or everywhere between the inner load points for four-point bending (Fig. 1.9). There is no axial stress outside the outer load points in either case, even if the beam extends beyond these points.

The elastic modulus of a bend specimen can be measured using load (*P*) and midspan-deflection (*δ*) data collected at strains typically 20–50% of those needed to induce plasticity (if plasticity is possible for the material in question). As in uniaxial tension or compression, elastic loading in bending is linear for linear elastic materials. The elastic modulus is extracted from

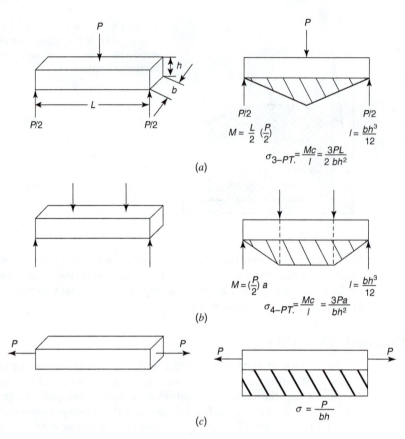

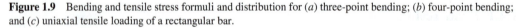

Figure 1.9   Bending and tensile stress formuli and distribution for (a) three-point bending; (b) four-point bending; and (c) uniaxial tensile loading of a rectangular bar.

the slope of the loading curve in an analogous fashion:

$$E_{b,3\text{-}pt.} = \frac{L^3}{4bh^3}\left(\frac{\Delta P}{\Delta\delta}\right) \tag{1-21a}$$

$$E_{b,4\text{-}pt.} = \frac{a(3L^2 - 4a^2)}{4bh^3}\left(\frac{\Delta P}{\Delta\delta}\right) \tag{1-21b}$$

As the reader may recall from his or her Strength of Materials or Mechanics courses, the flexural stress in a bend bar is given by

$$\sigma_{\max} = \frac{Mc}{I} \tag{1-22}$$

where M is the bending moment, c is the distance from the neutral axis to the outermost "fiber" surface, and I is the moment of inertia of the bar's cross section (just as for the columns discussed in Section 1.3.5). For a rectangular configuration,

$$I = \frac{bh^3}{12} \tag{1-23}$$

Table 1.4 Tensile and Bend Strengths of Ceramic Compounds[a]

Material	Tensile Strength MPa (ksi)	Modulus of Rupture[*] MPa (ksi)
Al$_2$O$_3$ (0–2% porosity)	200–310 (30–45)	350–580 (50–80)
Sintered BeO (3.5% porosity)	90–133 (13–20)	172–275 (25–40)
Sintered stabilized ZrO$_2$ (<5% porosity)	138 (20)	138–240 (20–35)
Hot-pressed Si$_3$N$_4$ (<1% porosity)	350–580 (50–80)	620–965 (90–140)
Fused SiO$_2$	69 (10)	110 (16)
Hot-pressed TiC (<2% porosity)	240–275 (35–40)	275–450 (40–65)

[a] D. W. Richerson, *Modern Ceramic Engineering*, Marcel Dekker, Inc., New York (1992).
[*] Values corresponding to three- and four-point bending samples.

where b and h are the beam width and height, respectively. It follows from Eq. 1.23 that for three-point loading, the bending moment increases linearly from the outer load points of the beam to the maximum value at the midspan location, given by

$$\sigma_{\text{3-pt.}} = \frac{3PL}{2bh^2} \tag{1-24}$$

For the four-point loading configuration, the bending moment increases linearly from either loading point at the ends of the beam to a constant maximum value within the region bounded by the interior loading points. Here the flexural stress is given by

$$\sigma_{\text{4-pt.}} = \frac{3Pa}{bh^2} \tag{1-25}$$

where a is the distance from the exterior to interior loading points.

If the material being tested is a ductile metal, it is possible to measure a bending *proof strength* that is analogous (but not identical) to the yield strength. It is defined as the minimum stress needed to induce a permanent strain of some chosen level (e.g., 0.01%).[14] If the material is a brittle ceramic, then the critical stress will correspond to the point of fracture. This value is called the *flexural strength* or the *modulus of rupture* (although it is not an elastic property of the material).[15] Several modulus of rupture values for common ceramic materials are listed in Tables 1.3 and 1.4.

EXAMPLE 1.2

A 50-mm-long rod of Si$_3$N$_4$ has a rectangular cross section with width and depth dimensions of 6 mm and 3 mm, respectively. When tested in 3-pt. bending, the rod fails with an applied load of 670 N. If the rod were tested in tension, the breaking load would be 10 kN. What are the modulus of rupture and tensile strength properties for this ceramic, and how well do these values agree with one another? Explain any property differences.

From Eq. 1-24, the modulus of rupture for the Si$_3$N$_4$ rod is

$$\sigma = \frac{3PL}{2bh^2}$$

$$\text{Modulus of Rupture} = \frac{3(670)(50 \times 10^{-3})}{2(6 \times 10^{-3})(3 \times 10^{-3})^2} = 930\,\text{MPa}$$

(Continued)

If the rod were pulled in tension and the load to break equal to $10\,\text{kN}$, then the tensile strength would be

$$\frac{P}{A} = \frac{10 \times 10^3}{(6 \times 10^{-3})(3 \times 10^{-3})} = 556\,\text{MPa}$$

We see that the modulus of rupture (bend strength) is considerably greater than the tensile strength for this material. This difference is supported by the numbers in Tables 1.3 and 1.4. Typically, this is the case for brittle solids, which occurs since the mechanical properties of such materials are extremely sensitive to the presence of defects in the sample. Correspondingly, the properties of a brittle solid depend strongly on the existence of a defect in the region of the highest stress level. Since the maximum stress is experienced across the entire cross section of a tensile bar, but is restricted to the surface layer beneath the center load point of a 3-pt. bend bar, it follows that there is a lower probability of finding a defect in the peak stress zone of the bend bar than in the tensile sample. Accordingly, the modulus of rupture in a brittle solid is higher than its corresponding tensile strength value. Likewise, the modulus of rupture measured in a three-point bend configuration is generally higher than in a four-point bend configuration.

1.4.2 Shear and Torsion Testing

Shearing is defined as the application of load in opposite directions along two parallel surfaces (rather than normal to them). As displacement occurs, the parallel surfaces remain parallel to one another, but are shifted such that a cube would become a parallelepiped, as shown in Fig. 1.10a. (Note that an "extra" pair of vertical shear forces in Fig. 1.10a must be present to avoid free-body rotation.) Hooke's law is still obeyed under shear by linear elastic materials, but for an isotropic material the *shear stress* τ and the *shear strain* γ are related by the *shear modulus*[vi] G such that

$$\tau = G\gamma \tag{1-26}$$

The shear stress is given by

$$\tau = F/A \tag{1-27}$$

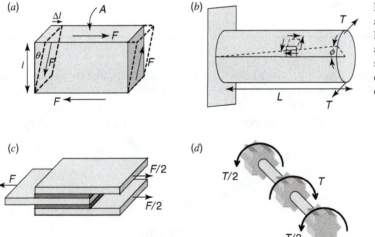

Figure 1.10   (a) Pure shear and (b) torsion loading. (c) Double lap joint with two adhesive pads in shear. (d) A shaft in torsion due to chain drives operating in different directions.

[vi] The symbol μ is also used to denote the shear modulus.

The shear strain calculation resembles that for axial loading with the form

$$\gamma = \Delta l / l \tag{1-28}$$

but the direction of l is perpendicular to that of Δl. It is equivalent to defining the shear strain by the shear angle θ, such that

$$\gamma = \tan \theta \tag{1-29}$$

There is no change in the area over which the force is imposed during pure shear loading, so there is no distinction made between engineering and true stresses.

Many cases of shear loading can be found in real components. As an example, shear loading is applied to the hinge pin in a pair of scissors or pliers that is closing on an object. Shear loading is also present in the rivets holding two overlapping plates together when the plates are pulled or pushed in opposite directions. In bonded lap joints, the shear load is borne by the adhesive layer holding two parts of the joint together (Fig. 1.10c).

Torsion loading is the application of a torque force such that a member is twisted about its axis. It is—not surprisingly—common in rotating shafts, from small screwdrivers to gigantic steam turbines. Torsion loading results in shear stresses and strains that are calculated in essentially the same manner as for pure shear, except that they are defined in terms of the torque T, the distance from the shaft axis r, and the rotational twist angle ϕ (in radians). The shear stress varies from zero along the axis to its maximum value at the outside surface of the shaft in the form

$$\tau = T r / I_p \tag{1-30}$$

where I_p is the polar moment of inertia. The polar moment of inertia for a circular solid shaft is

$$I_p = \pi D^4 / 32 \tag{1-31}$$

and for a circular hollow shaft

$$I_p = \pi (D^4 - d^4) / 32 \tag{1-32}$$

where D and d are the outer and inner shaft diameters, respectively. The twist angle ϕ varies with position along the length of the shaft, ranging from zero at the fixed end to a maximum at the end to which the twisting moment is applied, so the maximum shear strain in a circular solid shaft is given by

$$\gamma_{\max} = \left(\frac{D}{2} \right) \left(\frac{\phi}{L} \right) = \frac{r\phi}{L} \tag{1-33}$$

where L is the twisted length of the shaft and ϕ is in radians. It is somewhat difficult to create pure shear loading directly for ·measurement of elastic properties, so the shear modulus is often measured in torsion instead, combining Eqs. 1-26, 1-30, and 1-33 as $G = TL/I_p\phi$.[16] Values of G are listed in Table 1.1 for selected materials.

1.5 MULTIAXIAL LINEAR ELASTIC RESPONSE

1.5.1 Additional Isotropic Elastic Constants

Thus far we have only considered a stress coupled to a strain along the same axis, whether under uniaxial conditions, bending, or shear. However, we have also acknowledged that a strain applied in a uniaxial fashion along one axis will almost certainly result in orthogonal strains as the material attempts some degree of volume conservation. Hooke's law can be generalized to account for multiaxial effects with the addition of subscripts to indicate the direction so that

$$\varepsilon_{yy} = \frac{\sigma_{yy}}{E} \tag{1-34a}$$

Equivalent expressions exist for the X and Z directions. It is also necessary to introduce the Poisson's ratio v, the elastic constant that describes the proportionality between an imposed normal strain along one axis (e.g., Y in Eq. 1-34a) and a resulting normal strain (generally of opposite sign) along an orthogonal direction (e.g., X or Z) such that

$$\varepsilon_{xx} = \varepsilon_{zz} = -v\varepsilon_{yy} = -\frac{v}{E}\sigma_{yy} \tag{1-34b}$$

where

$$\sigma_{yy} = \text{stress acting normal to the } Y \text{ plane and in the } Y \text{ direction}$$
$$\varepsilon_{yy} = \text{normal strain in the direction of the applied stress}$$
$$\varepsilon_{xx}, \varepsilon_{zz} = \text{corresponding normal strains in orthogonal directions}$$
$$v = \text{Poisson's ratio } (= -\varepsilon_{xx}/\varepsilon_{yy} = -\varepsilon_{zz}/\varepsilon_{yy} \text{ for isotropic materials})$$
$$E = \text{modulus of elasticity}$$

This expression is only valid for small strains. At large strains, the behavior becomes nonlinear.

Poisson's ratio must be $v \leq 0.5$, and is generally greater than zero. At a value of 0.5, volume is conserved during elastic deformation (i.e., the material is *incompressible*). However, this is only the case for certain rubber-like materials. Most metals range from 0.25 to 0.45, with a typical value close to 0.33. Ceramics and glasses tend to be somewhat lower, in the range of 0.1–0.3, and polymers somewhat higher at 0.3–0.5. Natural cork has a Poisson's ratio close to zero. Values of v for specific materials are listed in Table 1.1. There is a small group of materials that are *auxetic* or *dilatational*. They actually expand laterally under longitudinal tension so that $-1 < v < 0$. Most of these unusual materials derive their odd behavior from the presence of internal pores with reentrant shapes.[17]

Although it appears that there are now three elastic constants (E, G, v), for isotropic materials only two of these constants are independent of one another. Once two are known, the third can be calculated by rearranging the expression

$$G = \frac{E}{2(1+v)} \tag{1-35}$$

There is another elastic constant called the *bulk modulus*[vii] K that describes the resistance to uniform compression along three orthogonal axes. This situation can be conceptualized as three applied stresses all causing Poisson responses simultaneously, so it may come as no surprise that it, too, can be determined by knowing any two of the other moduli:

$$K = \frac{E}{3(1-2v)} = \frac{EG}{3(3G-E)} \tag{1-36}$$

1.5.2 Multiaxial Loading

In practice, only certain structures are exposed to purely uniaxial or shear loading. Most are subjected to loads that occur simultaneously along multiple axes. Multiaxial loading can

---

[vii] The symbol B is sometimes used for the bulk modulus instead of K.

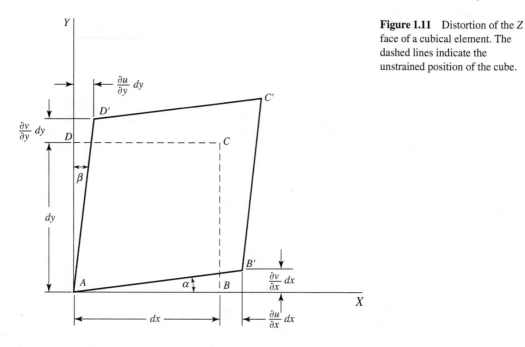

Figure 1.11 Distortion of the Z face of a cubical element. The dashed lines indicate the unstrained position of the cube.

be described by the superposition and interplay of the individual stresses and strain in a form sometimes called the *generalized Hooke's law*.

From Fig. 1.11, typical normal and shear strain components may be given by

$$\varepsilon_{xx} = \frac{\partial u}{\partial x} \tag{1-37a}$$

$$\varepsilon_{yy} = \frac{\partial v}{\partial y} \tag{1-37b}$$

$$\gamma_{xy} = \varepsilon_{xy} = \tan \alpha + \tan \beta = \frac{\partial v}{\partial x} + \frac{\partial u}{\partial y} \tag{1-37c}$$

with the other normal and shear strains defined in similar fashion. The double subscript for a normal component of the stress or strain has the same meaning as in Eq. 1-34b. For a shear component, the first subscript describes the face on which the load is imposed, and the second subscript indicates the direction (as shown in Fig. 1.12).

When multiaxial stresses are applied, the total strain in any given direction is the sum of all strains resulting from each normal and shear stress component. Thus for the case of *biaxial* loading with normal stresses σ_{xx} and σ_{yy} applied simultaneously along the X and Y axes, and no stress along Z, the resulting strain ε_{xx} along X is found to be the sum of the directly imposed strain and the strain from a Poisson contraction:

$$\varepsilon_{xx} = \frac{1}{E}\sigma_{xx} + \left(\frac{-\nu}{E}\sigma_{yy}\right) = \frac{1}{E}\left(\sigma_{xx} - \nu\sigma_{yy}\right) \tag{1-38}$$

It can be deduced from the final term in Eq. 1-38 that the strain along X will be smaller in this case than if it were loaded uniaxially. A biaxial stress state is also known as a state of *plane stress*. Stretched membranes, windows constrained by rigid frames, and pressure vessels are all typically loaded in plane stress.

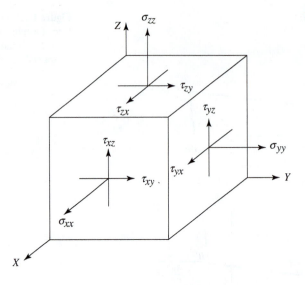

Figure 1.12   Stress components acting on a volume element.

For the most general case of *triaxial* loading, the full set of strains can be expressed for an isotropic material as

$$\varepsilon_{xx} = \frac{1}{E}\sigma_{xx} + \left(\frac{-\nu}{E}\sigma_{yy}\right) + \left(\frac{-\nu}{E}\sigma_{zz}\right) = \frac{\sigma_{xx} - \nu(\sigma_{yy} + \sigma_{zz})}{E}$$

$$\varepsilon_{yy} = \left(\frac{-\nu}{E}\sigma_{xx}\right) + \frac{1}{E}\sigma_{yy} + \left(\frac{-\nu}{E}\sigma_{zz}\right) = \frac{\sigma_{yy} - \nu(\sigma_{xx} + \sigma_{zz})}{E}$$

$$\varepsilon_{zz} = \left(\frac{-\nu}{E}\sigma_{xx}\right) + \left(\frac{-\nu}{E}\sigma_{yy}\right) + \frac{1}{E}\sigma_{zz} = \frac{\sigma_{zz} - \nu(\sigma_{xx} + \sigma_{yy})}{E} \tag{1-39}$$

$$\gamma_{xy} = \frac{\tau_{xy}}{G}$$

$$\gamma_{yz} = \frac{\tau_{yz}}{G}$$

$$\gamma_{xz} = \frac{\tau_{xz}}{G}$$

Note that there is no equivalent to a Poisson's ratio effect for the shear strains; a shear stress applied along one axis of an isotropic material does not cause shear strain along any other axis. Likewise, there is no coupling between the normal and shear terms.

1.5.2.1 Thin-Walled Pressure Vessels

The description of stresses in a *thin-walled pressure vessel* is introduced here to demonstrate an important application of biaxial stress, and to establish the basis for subsequent fatigue and fracture analyses that are discussed in later chapters. Consider a cylindrical vessel section of length L, internal diameter D, and wall thickness t that is subjected to a uniform gas or fluid pressure p (Fig. 1.13a). By examining a free-body diagram of the lower half of the cylinder (Fig. 1.13b), one sees that the summation of forces acting normal to the midplane is given by

$$\left[\sum Y = 0\right] F = pDL = 2P \tag{1-40}$$

or

$$P = \frac{pDL}{2} \tag{1-41}$$

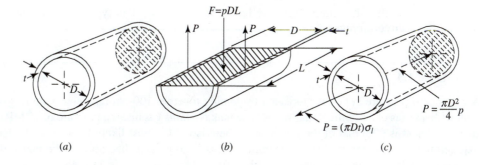

Figure 1.13   Thin-walled pressure vessel. (*a*) Overall shape; (*b*) free-body diagram of diametral section; and (*c*) free-body diagram of transverse section.

The tangential or "hoop" stress, σ_t, acting on the wall thickness is then found to be

$$\sigma_t = \frac{P}{A} = \frac{pDL}{2Lt} = \frac{pD}{2t} \tag{1-42}$$

or

$$\sigma_t = \frac{pr}{t} \tag{1-43}$$

where r is the vessel radius. For the case of thin-walled cylinders, where $r/t \geq 10$, Eq. 1-43 describes the hoop stress at all locations through the wall thickness. (The reader is referred to Strength of Materials texts for the more complex analysis of stresses in *thick*-walled cylinders where hoop and radial stresses are found to vary with location through the wall thickness.)

A second free-body diagram to account for cylindrical stresses in the longitudinal direction is shown in Fig. 1.13*c*. Here we see the "bursting force" across the end of the cylinder is resisted by the "tearing force" P acting over the vessel circumference. In this instance, the sum of forces acting along the axis of the cylinder is

$$\frac{\pi D^2 p}{4} = P \tag{1-44}$$

The cross-sectional area of the cylinder wall is characterized by the product of its wall thickness and the mean circumference [i.e., $\pi(D + t)t$]. For thin-walled pressure vessels where $D \gg t$, the cylindrical cross-sectional area may be approximated by πDt. Therefore, the longitudinal stress in the cylinder is given by

$$\sigma_l = \frac{\pi D^2 p}{4\pi Dt} = \frac{pD}{4t} = \frac{pr}{2t} \tag{1-45}$$

By comparing Eqs. 1-42 and 1-45, one finds that the tangential or hoop stress is twice that in the longitudinal direction. Therefore, vessel failure is likely to occur along a longitudinal plane oriented normal to the transverse or hoop stress direction. Furthermore, one can now calculate that the strains in the transverse and longitudinal directions under elastic loading are

$$\varepsilon_t = \frac{1}{E}\sigma_t + \left(\frac{-\nu}{E}\right)\sigma_l = \left(\frac{1 - \nu/2}{E}\right)\sigma_t \tag{1-46a}$$

and

$$\varepsilon_l = \frac{1}{E}\sigma_l + \left(\frac{-\nu}{E}\right)\sigma_t = \left(\frac{1/2 - \nu}{E}\right)\sigma_t \tag{1-46b}$$

so that the strain along the length of the cylinder is typically much smaller than around the circumference.

EXAMPLE 1.3

A cylindrical pressure vessel is fabricated by joining together 100-cm-diameter wide rings of 1015 as-rolled steel with a series of circumferential welds. The tank contains gas under a pressure of 15 MPa. If the strength of each weldment is 90% that of the base plate, where is failure most likely to occur and what is the minimum required thickness to ensure that the operating stress is no greater than 50% of the material's yield strength?

From Table 1.2, the yield strength of the 1015 alloy base plate is 315 MPa with the weldment strength estimated to be 283.5 MPa. For a design stress/yield strength ratio of 0.5, the hoop stress is computed to be

$$\sigma_t = \frac{pr}{t}$$

so that

$$0.5(315 \times 10^6) = \frac{15 \times 10^6 (50 \times 10^{-2})}{t}$$

$$\therefore t = 4.76\,\text{cm}$$

For the stresses in the longitudinal direction,

$$\sigma_l = \frac{pr}{2t}$$

so that

$$0.5(283.5 \times 10^6) = \frac{15 \times 10^6 (50 \times 10^{-2})}{2t}$$

$$\therefore t = 2.65\,\text{cm}$$

Therefore, the vessel must have a wall thickness of at least 4.76 cm, and any significant overpressurization will cause failure along a longitudinal plane normal to the hoop stress direction and not as a result of longitudinal stresses acting across the weaker circumferential weldments.

1.5.2.2 Special Cases of Multiaxial Loading

If biaxial stresses in an isotropic material happen to be equal in sign and magnitude so that $\sigma_{xx} = \sigma_{yy}$, then the in-plane strains will also be equal, and the out-of-plane strain will be generated by the sum of two equal Poisson contractions:

$$\varepsilon_{xx} = \frac{1}{E}\sigma_{xx} + \left(\frac{-\nu}{E}\sigma_{yy}\right) = \frac{1}{E}\left(\sigma_{xx} - \nu\sigma_{yy}\right) = \left(\frac{1-\nu}{E}\right)\sigma_{bi} = \varepsilon_{yy} \qquad (1\text{-}47\text{a})$$

$$\sigma_{bi} = \left(\frac{E}{1-\nu}\right)\varepsilon_{bi} \qquad (1\text{-}47\text{b})$$

$$\varepsilon_{zz} = \left(\frac{-\nu}{E}\sigma_{xx}\right) + \left(\frac{-\nu}{E}\sigma_{yy}\right) = \frac{-2\nu}{E}\sigma_{bi} \qquad (1\text{-}48)$$

This is a special case of *plane stress* loading, for which $\sigma_{zz} = 0$ always. The term $E/(1 - \nu)$ in Eq. 1-47b is known as the *biaxial modulus*, and is widely used for situations in which equal-biaxial loading applies. Examples include the behavior of spherical pressure vessels, pressure-loaded or centrally-loaded circular membranes, and thin films subjected to thermal stresses generated by thermal expansion mismatch between the film (the coating) and the substrate (the surface supporting the coating). For the case of *plane strain*, in which the strain along one axis is zero due to an applied constraint, a similar derivation gives a *plane strain modulus* of $E/(1 - \nu^2)$.

The stress state resulting from equal triaxial loading is known as *hydrostatic stress*. It is the situation found at great depths in the ocean, but also often at the tip of a growing crack. The latter situation can have a profound effect on plasticity and fracture behavior, as discussed in Chapters 2, 6, and 7.

1.5.3   Instrumented Indentation

Standard tests for measuring elastic properties work very well for macro-scale specimens, but are very difficult to perform on very small quantities of material. The ability to measure very small volumes is highly desirable for evaluating thin films and coatings, for establishing the properties of individual phases within a multiphase material, or for evaluating materials that have complex structures at the micrometer scale and below (e.g., bone). It is also the case that reliance on standard test specimens precludes testing *in situ*—that is, without removing a component (or a piece of a component) that is installed in its proper setting. *In situ* measurements are very useful for evaluating components in service from which small samples cannot be easily or safely removed (e.g., field testing of cooling pipes in a power plant). A technique known as *instrumented indentation* (or sometimes *nanoindentation*) is very useful in these cases.[18,19]

Instrumented indentation tests rely on measurement of load, P, and displacement, h, as a small indenter is driven into the surface of a specimen and then removed. The resulting load-displacement plot (see Fig. 1.14) may then be used to extract material properties including an elastic modulus and the hardness (yet another way to characterize strength). The initial slope of the unloading curve, dP/dh, is used to determine the elastic modulus, thereby ensuring that the material response is only due to elastic recovery. Because a triaxial stress state exists beneath the indenter (due to constraint of the material surrounding the indented region), an interplay among multiple stresses and strains exists, and the modulus extracted is the *plane strain modulus* $E/(1 - \nu^2)$. Values for E and ν cannot be independently determined using instrumented indentation, so it is common to either report the plane strain modulus directly, or to estimate ν (which does not vary much within a single class of materials) in order to gain a reasonable approximation for E.

The indenter tip is usually a three-sided shallow pyramid (a Berkovich indenter) or a sphere. Knowledge of the exact indenter shape allows calculation of the contact area, A, between the indenter and the material as a function of depth, so a mean stress under the indenter

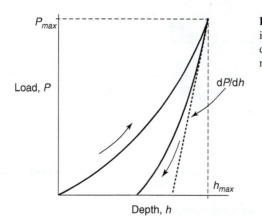

Figure 1.14   Schematic depiction of an instrumented indentation test performed with a pyramidal indenter in a ductile material. The loading and unloading directions are marked with curved arrows.

can be determined. This mean stress (or pressure, if you like) at the maximum depth of penetration, h_{max}, defines the indentation hardness, $H = P_{max}/A_{max}$.

1.6 ELASTIC ANISOTROPY

Amorphous materials that have no long-range structure and polycrystalline materials that have completely random grain orientations have elastic properties that do not vary as a function of direction. But what about single crystals that have distinct crystallographic directions, polymers with highly aligned chains, composites with aligned reinforcement fibers, or cortical bone with its highly oriented internal structure? All of these materials display *anisotropic* elastic properties. In order to address materials like these, we must modify our notation to take into account moduli that correspond to specific directions, and we must identify symmetry classes that will allow us to group together materials that show related degrees of anisotropy.

1.6.1 Stiffness and Compliance Matrices

Bone provides an excellent example of a material with strong elastic anisotropy that can be tied directly to an important functional purpose. Long bones such as the human femur (thigh) or humerus (upper arm) are designed to bear large loads, particularly in compression and bending. The shaft of such a bone is approximately cylindrical in shape, with a thin outer shell of hard, fairly dense *cortical bone* (also called *compact bone*) that surrounds highly porous *trabecular* (a.k.a. *cancellous*) bone, as shown in Fig. 1.15. Compact bone is the primary load-bearing component. It is actually made up of much smaller cylindrical units called osteons that are aligned with the long axis of the bone. The osteons themselves are made up of smaller units of longitudinally-aligned hydroxyapatite platelets (the hard ceramic constituent of bone) in a matrix of collagen (a soft polymeric constituent) arranged in concentric cylindrical lamellae.

If specimens are cut from compact bone so that behavior in three orthogonal directions can be tested, as shown schematically in Fig. 1.15, it is found that Young's modulus measured along the long axis (direction 3) differs significantly from the radial (1) and circumferential (2) directions, as listed in Table 1.5. (Note that the two-index letter notation introduced in Sections 1.5.1 and 1.5.2 has simply been replaced by direction numbers.) The values of Poisson's ratio and shear modulus also vary with direction. It can be seen from the table that the stiffness along the primary load-bearing axis, direction 3, is much greater than along the other two directions, which are similar to

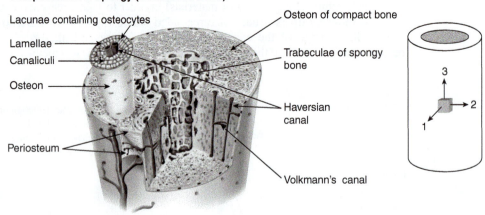

Figure 1.15 Graphical representation of long-bone construction.[viii] Also shown: three orthogonal directions relative to the bone axis.

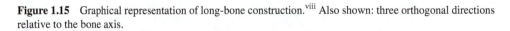

[viii] Image source: U.S. National Cancer Institute's Surveillance, Epidemiology and End Results (SEER) Program.

Table 1.5   Elastic properties of human femoral bone measured as a function of direction. Young's and shear moduli have GPa units.[20]

	E_{11}	E_{22}	E_{33}	ν_{12}	ν_{13}	ν_{23}	ν_{21}	ν_{31}	ν_{32}	G_{12}	G_{13}	G_{23}
Modulus	12.0	13.4	20.0	0.38	0.22	0.24	0.42	0.37	0.35	4.5	5.6	6.2

one another (but not identical). From a physiological point of view, it makes great sense to have the greatest stiffness align with the direction that will bear the greatest loads. The actual physical basis for this phenomenon is clearly connected to the cylindrical symmetry of the individual osteons and the alignment of the hard platelets within them.

At first glance it appears that there are now 12 distinct elastic moduli and their ratios that must be determined to fully describe the elastic properties of cortical bone using Eq. 1-39. Thankfully, symmetry considerations and a slightly modified depiction of the generalized Hooke's law reduce this somewhat. A closer look at Eq. 1-39 reveals that the three equations for the normal strains and stresses, and the moduli that link them, can be expressed in matrix notation as

$$
\begin{bmatrix} \varepsilon_1 \\ \varepsilon_2 \\ \varepsilon_3 \end{bmatrix} =
\begin{bmatrix}
\dfrac{1}{E_{11}} & \dfrac{-\nu_{12}}{E_{22}} & \dfrac{-\nu_{13}}{E_{33}} \\[2ex]
\dfrac{-\nu_{21}}{E_{11}} & \dfrac{1}{E_{22}} & \dfrac{-\nu_{23}}{E_{33}} \\[2ex]
\dfrac{-\nu_{31}}{E_{11}} & \dfrac{-\nu_{32}}{E_{22}} & \dfrac{1}{E_{33}}
\end{bmatrix}
\begin{bmatrix} \sigma_1 \\ \sigma_2 \\ \sigma_3 \end{bmatrix}
\tag{1-49}
$$

with $1 = x$, $2 = y$, $3 = z$. The pairs of Poisson's ratio indices indicate the strain number in the first position, and the stress component that contributes to that strain in the second position. For symmetry reasons, it makes physical sense that a strain along axis 2 (ε_2) induced by a stress along axis 1 (σ_1) must be equal to a strain along axis 1 (ε_1) induced by an identical stress along axis 2 ($\sigma_2 = \sigma_1$). Thus the terms $-\nu_{12}/E_{22}$ and $-\nu_{21}/E_{11}$ must be numerically equal. The same can be said of all terms reflected across the matrix diagonally. Thus instead of 9 distinct terms, there are only 6 that are unique.

Extending this matrix notation, the shear stresses and strains can be identified as $4 = yz = 23$, $5 = xz = 13$, and $6 = xy = 12$, with σ and ε in place of τ and γ for uniformity. Wherever there is no connection between a certain imposed stress, σ_j, and the strain of interest, ε_i, a zero is inserted. The full matrix expression for all six strain equations is therefore written as

$$
\begin{bmatrix} \varepsilon_1 \\ \varepsilon_2 \\ \varepsilon_3 \\ \varepsilon_4 \\ \varepsilon_5 \\ \varepsilon_6 \end{bmatrix} =
\begin{bmatrix}
\dfrac{1}{E_{11}} & \dfrac{-\nu_{12}}{E_{22}} & \dfrac{-\nu_{13}}{E_{33}} & 0 & 0 & 0 \\[2ex]
\dfrac{-\nu_{21}}{E_{11}} & \dfrac{1}{E_{22}} & \dfrac{-\nu_{23}}{E_{33}} & 0 & 0 & 0 \\[2ex]
\dfrac{-\nu_{31}}{E_{11}} & \dfrac{-\nu_{32}}{E_{22}} & \dfrac{1}{E_{33}} & 0 & 0 & 0 \\[2ex]
0 & 0 & 0 & \dfrac{1}{G_{23}} & 0 & 0 \\[2ex]
0 & 0 & 0 & 0 & \dfrac{1}{G_{13}} & 0 \\[2ex]
0 & 0 & 0 & 0 & 0 & \dfrac{1}{G_{12}}
\end{bmatrix}
\begin{bmatrix} \sigma_1 \\ \sigma_2 \\ \sigma_3 \\ \sigma_4 \\ \sigma_5 \\ \sigma_6 \end{bmatrix}
\tag{1-50}
$$

It can be seen from this layout that there is no additional symmetry reduction possible in the number of shear terms since they all lie on the diagonal. Thus symmetry considerations reduce the modulus matrix to 9 independent terms. Not an enormous reduction, but better than 12!

In this format, it is clear how the individual measured moduli would be combined to build the full matrix. This approach is unnecessarily bulky, however, because all 12 constants must be provided to perform useful calculations, even though there are only 9 distinct combinations. A less cluttered notation, called Einstein summation, is therefore widely used to describe the elastic behavior of anisotropic materials. The shorthand form of this notation is $\varepsilon_i = S_{ij}\sigma_j$, where a 6×6 S_{ij} matrix substitutes for the matrix of elastic constants in Eq. 1-50. In Einstein notation it is understood that summation over all $j = 1$–6 is implied for each i value. The most general expression for ε_1 would therefore be written

$$\varepsilon_1 = S_{11}\sigma_1 + S_{12}\sigma_2 + S_{13}\sigma_3 + S_{14}\sigma_4 + S_{15}\sigma_5 + S_{16}\sigma_6 \tag{1-51}$$

and for the specific case of bone, it would reduce to

$$\varepsilon_1 = S_{11}\sigma_1 + S_{12}\sigma_2 + S_{13}\sigma_3 \tag{1-52}$$

because $S_{14} = S_{15} = S_{16} = 0$. Similar equations would apply for the other components of strain. Matching matrix positions between Eqs. 1-50 and 1-51, the S_{11} term would have a numerical value equal to $1/E_{11}$, $S_{12} = -v_{12}/E_{22}$, and so forth.

Recalling that the corresponding isotropic version of Hooke's law is $\varepsilon = (1/E)\sigma$, and that $1/E$ is measure of a material's compliance, the matrix of moduli represented by S_{ij} is called the *compliance matrix*. Of course, it is also possible to write the anistropic version of Hooke's law to solve for the stresses, the equivalent of $\sigma = E\varepsilon$. In this case, it takes on the form $\sigma_i = C_{ij}\varepsilon_j$, with the counterpart to Eq. 1-51 written as

$$\sigma_1 = C_{11}\varepsilon_1 + C_{12}\varepsilon_2 + C_{13}\varepsilon_3 + C_{14}\varepsilon_4 + C_{15}\varepsilon_5 + C_{16}\varepsilon_6 \tag{1-53}$$

The C_{ij} matrix of moduli plays the role of E in Hooke's law, and is therefore called the *stiffness matrix* (thereby virtually assuring that generations of students will be highly confused about why $S = $ compliance and $C = $ stiffness). Applying symmetry considerations in the same way as before, it must be true that $S_{ij} = S_{ji}$ and $C_{ij} = C_{ji}$. Unfortunately, it is *not* true that $S_{ij} = 1/C_{ij}$.

1.6.1.1 Symmetry Classes

Bone is categorized as an *orthotropic* material[ix] because it has three elastically distinct a, b, and c axes that are each separated by interior angles of $\alpha = \beta = \gamma = 90°$. By definition, the components of the S_{ij} and C_{ij} matrices are aligned with the principal axes of the material. All orthotropic materials have 9 independent elastic constants (those for human femoral bone are listed in Table 1.6). Other examples include orthorhombic single crystals, wood, and many laminated aligned-fiber reinforced composites.

Table 1.6   Stiffness matrix components for human femoral bone. All elastic coefficients listed have GPa units.[20]

	C_{11}	C_{22}	C_{33}	C_{44}	C_{55}	C_{66}	C_{12}	C_{13}	C_{23}
Coefficient	18.0	20.2	27.6	6.23	5.61	4.52	9.98	10.1	10.7

[ix] Although cortical bone is orthotropic, it is sometimes approximated as *transversely isotropic* (i.e., isotropic in a certain plane) because the elastic constants for the radial and circumferential directions are nearly the same.

Cubic materials also have three axes separated by $\alpha = \beta = \gamma = 90°$, but the three axes are elastically equivalent. This means that $S_{11} = S_{22} = S_{33}, S_{44} = S_{55} = S_{66}$, and $S_{12} = S_{13} = S_{23}$, so a table describing cubic materials need only report the 3 independent S_{11}, S_{12}, and S_{44} values. The user of such tables is expected to understand the symmetry implications, and to fill out the matrix accordingly. Cubic single crystal metals and ceramics have cubic elastic symmetry, with the stiffness and compliance principal axes corresponding to the unit cell axes. It is possible to determine the conversion equations relating the S_{ij} and C_{ij} matrices by solving the linear equations simultaneously, then equating coefficients of like terms. For cubic symmetry, the conversion factors for compliance to stiffness are as follows:

$$C_{11} = \frac{S_{11} + S_{12}}{(S_{11} - S_{12})(S_{11} + 2S_{12})}$$

$$C_{12} = \frac{-S_{12}}{(S_{11} - S_{12})(S_{11} + 2S_{12})} \tag{1-54}$$

$$C_{44} = \frac{1}{S_{44}}$$

The conversions from stiffness to compliance have identical forms, but with S and C exchanged.

Hexagonal systems have 5 independent elastic constants, reflecting their somewhat lower symmetry than cubic crystals. Drawn fibers or wires tend to have *fiber symmetry* or *fiber texture* that has a distinct longitudinal axis, but no distinct radial directions. This *transversely isotropic* symmetry is essentially the same as hexagonal symmetry, and therefore also has 5 independent elastic constants. Trigonal and tetragonal systems have either 6 or 7 independent elastic constants, depending on the precise symmetry class within each system (e.g., Al_2O_3 is trigonal and has 6 independent elastic constants).

Finally, isotropic materials can also be described using stiffness and compliance matrix notation even though it usually isn't necessary to do so. For this special case, $S_{11} = S_{22} = S_{33} = 1/E$, $S_{44} = S_{55} = S_{66} = 1/G$ and $S_{12} = S_{13} = S_{23} = -v/E$, just as in Eq. 1-39. There are only 2 independent elastic constants, as previously described, so it must also be true that $S_{44} = 2(S_{11} - S_{12})$ and $C_{44} = 0.5(C_{11} - C_{12})$.

For materials with even lower symmetry than orthotropic, some of the locations filled with zeros in Eq. 1-50 will be nonzero. In these systems, imposed shear stresses can potentially induce normal strains, and vice versa. In the worst case, none of the elastic coupling constants are zero, but symmetry across the matrix diagonal still applies. The largest stiffness or compliance matrix therefore has 21 independent elastic constants.

The anisotropic elastic constants for several materials are given in Table 1.7, and certain values of relative elastic anisotropy as indicated by the ratio $2(S_{11} - S_{12})/S_{44}$ are tabulated in Table 1.8. Note the large anisotropy exhibited by many of these crystals as compared with the isotropic behavior of tungsten for which the elastic anisotropy ratio happens to be 1.

1.6.1.2 Loading Along an Arbitrary Axis

As previously indicated, the axes of the S_{ij} and C_{ij} matrices are aligned with the principal axes of the material in question. What if the loading direction of interest is along a different direction, such as the $\langle 111 \rangle$ direction in a cubic single crystal, or at a 45° angle to the fibers in an aligned fiber-reinforced composite? In these cases, it is possible to mathematically rotate the S_{ij} or C_{ij} matrix by Euler angles ψ, θ, and ϕ so that a new matrix S_{ij}' or C_{ij}' is created with directions $1'$, $2'$, and $3'$ that align with the loading coordinate system. Many nonzero matrix terms will be created in the rotation process, but the general form and usage remain unchanged.

In the special case of cubic crystals loaded uniaxially along an axis other than one of the unit cell axes, there is a simplified version of this matrix rotation process. It can be shown for

Table 1.7   Stiffness and Compliance Constants for Selected Crystals[a]

Material	$(10^{10}\,Pa)$			$(10^{-11}\,Pa^{-1})$		
Cubic	C_{11}	C_{12}	C_{44}	S_{11}	S_{12}	S_{44}
Aluminum, Al	10.82	6.13	2.85	1.57	−0.57	3.51
Copper, Cu	16.84	12.14	7.54	1.50	−0.63	1.33
Gallium arsenide, GaAs	11.8	5.35	5.94	1.18	−0.37	1.68
Gold, Au	18.60	15.70	4.20	2.33	−1.07	2.38
Iron, Fe	23.70	14.10	11.60	0.80	−0.28	0.86
Fe-17.5Cr-12Ni	21.6	14.4	12.9	1.00	−0.40	0.78
Lithium fluoride, LiF	11.2	4.56	6.32	1.16	−0.34	1.58
Magnesium oxide, MgO	29.3	9.2	15.5	0.401	−0.096	0.648
Molybdenum, Mo[b]	46.0	17.6	11.0	0.28	−0.08	0.91
Nickel, Ni	24.65	14.73	12.47	0.73	−0.27	0.80
Silicon, Si	16.4	6.35	7.96	0.77	−0.22	1.26
Silicon carbide, SiC (3C)[c]	35.2	14.0	23.3	0.37	−0.11	0.43
Sodium chloride, NaCl[b]	4.87	1.26	1.27	2.29	−0.47	7.85
Spinel, MgAl$_2$O$_4$	27.9	15.3	15.3	0.585	−0.208	0.654
Titanium carbide, TiC[b]	51.3	10.6	17.8	0.21	−0.036	0.561
Tungsten, W	50.1	19.8	15.14	0.26	−0.07	0.66
Zinc sulfide, ZnS	10.79	7.22	4.12	2.0	−0.802	2.43

Hexagonal	C_{11}	C_{12}	C_{13}	C_{33}	C_{44}	S_{11}	S_{12}	S_{13}	S_{33}	S_{44}
Cadmium, Cd	12.10	4.81	4.42	5.13	1.85	1.23	−0.15	−0.93	3.55	5.40
Cobalt, Co	30.70	16.50	10.30	35.81	7.53	0.47	−0.23	−0.07	0.32	1.32
Magnesium, Mg	5.97	2.62	2.17	6.17	1.64	2.20	−0.79	−0.50	1.97	6.10
Silicon carbide, SiC (4H, 6H)[c]	50.2	9.5	5.6	56.5	16.9	0.21	−0.04	−0.02	0.18	0.59
Titanium, Ti	16.0	9.0	6.6	18.1	4.65	0.97	−0.47	−0.18	0.69	2.15
Zinc, Zn	16.10	3.42	5.01	6.10	3.83	0.84	0.05	−0.73	2.84	2.61

[a] Data adapted from H. B. Huntington, *Solid State Physics*, Vol. 7, Academic, New York, 1958, p. 213, and K. H. Hellwege, *Elastic, Piezoelectric and Related Constants of Crystals*, Springer-Verlag, Berlin, 1969.
[b] Note that $E_{100} > E_{111}$.
[c] SiC has many polymorphs, including 3C (β-SiC, cubic), 4H (hexagonal), and 6H (α-SiC, hexagonal).

Table 1.8a   Elastic Anisotropy of Selected Materials

Metal	Relative Degree of Anisotropy $\left[\dfrac{2(s_{11}-s_{12})}{s_{44}}\right]$	E_{111} (GPa)	E_{100} (GPa)	$\left[\dfrac{E_{111}}{E_{100}}\right]$
Aluminum	1.219	76.1	63.7	1.19
Copper	3.203	191.1	66.7	2.87
Gold	2.857	116.7	42.9	2.72
Iron	2.512	272.7	125.0	2.18
Magnesium oxide	1.534	350.1	249.4	1.404
Spinel (MgAl$_2$O$_4$)	2.425	364.5	170.0	2.133
Titanium carbide	0.877	429.2	476.2	0.901
Tungsten	1.000	384.6	384.6	1.00

Table 1.8b   Elastic Anisotropy of Selected Materials

Metal	Relative Degree of Anisotropy $\left[\dfrac{2(s_{11}-s_{12})}{s_{44}}\right]$	E_{111} (10^6 psi)	E_{100} (10^6 psi)	$\left[\dfrac{E_{111}}{E_{100}}\right]$
Aluminum	1.219	11.0	9.2	1.19
Copper	3.203	27.7	9.7	2.87
Gold	2.857	16.9	6.2	2.72
Iron	2.512	39.6	18.1	2.18
Magnesium oxide	1.534	50.8	36.2	1.404
Spinel ($MgAl_2O_4$)	2.425	52.9	24.8	2.133
Titanium carbide	0.877	62.2	69.1	0.901
Tungsten	1.000	55.8	55.8	1.00

the cubic case that the modulus of elasticity in any given direction may be given by Eq. 1-55 in terms of the three independent elastic constants and the direction cosines of the crystallographic direction under study:

$$\frac{1}{E_{hkl}} = S_{11} - 2\left[(S_{11} - S_{12}) - {}^1/_2 S_{44}\right](l_1^2 l_2^2 + l_2^2 l_3^2 + l_1^2 l_3^2) \tag{1-55}$$

where l_1, l_2, l_3 are direction cosines. Direction cosine values for the principal crystallographic directions in the cubic lattice are given in Table 1.9. For example, the modulus in the $\langle 100 \rangle$ direction is given by $E_{100} = 1/S_{11}$, since $\sum l_i^2 l_j^2 = 0$. By comparison, $\sum l_i^2 l_j^2 = {}^1/_3$ (the maximum value) in the $\langle 111 \rangle$ direction so that $1/E_{111} = S_{11} - {}^2/_3\left[(S_{11} - S_{12}) - {}^1/_2 S_{44}\right]$. Depending on whether $(S_{11} - S_{12})$ is larger or smaller than ${}^1/_2 S_{44}$, the modulus of elasticity may be greatest in either the $\langle 111 \rangle$ or $\langle 100 \rangle$ direction. By comparison, the modulus in the $\langle 110 \rangle$ direction is in good agreement with the average value of $E_{isotropic}$ for a polycrystalline sample of the same material (see Example 1.4).

For the case of hexagonal crystals, the rotation of the compliance matrix reduces to

$$\frac{1}{E_{hkl}} = S_{11}(1 - l_3^2)^2 + S_{33}l_3^4 + (2S_{13} + S_{44})l_3^2(1 - l_3^2) \tag{1-56}$$

where l_1, l_2, l_3 are direction cosines for directions in the hexagonal unit cell. From Eq. 1-56 note that in hexagonal crystals E depends only on the direction cosine l_3, which lies normal to the basal plane. Consequently, the modulus of elasticity in hexagonal crystals is isotropic everywhere in the basal plane, as previously discussed.

Table 1.9   Direction Cosines for Principal Directions in Cubic Lattice

Direction	l_1	l_2	l_3
$\langle 100 \rangle$	1	0	0
$\langle 110 \rangle$	$1/\sqrt{2}$	$1/\sqrt{2}$	0
$\langle 111 \rangle$	$1/\sqrt{3}$	$1/\sqrt{3}$	$1/\sqrt{3}$

EXAMPLE 1.4

Compute the modulus of elasticity for tungsten and iron in the $\langle 110 \rangle$ direction.

From Tables 1.7 and 1.9 we obtain the necessary information regarding elastic compliance values and direction cosines. The modulus of elasticity in the $\langle 110 \rangle$ direction is then determined from Eq. 1-55. For tungsten,

$$\frac{1}{E_{110}} = 0.26 - 2\left\{[0.26 - (-0.07)] - \frac{1}{2}(0.66)\right\}\left(\frac{1}{4}\right) = 0.26 - (0)\left(\frac{1}{4}\right)$$

Therefore,

$$E_{110} = 384.6\,\text{GPa}$$

which is the same value given in Table 1.8a for E_{111} and E_{100}. For iron,

$$\frac{1}{E_{110}} = 0.80 - 2\left\{[0.80 - (-0.28)] - \frac{1}{2}(0.86)\right\}\left(\frac{1}{4}\right)$$

$$\therefore E_{110} = 210.5\,\text{GPa}$$

Note that $E_{111} > E_{110} > E_{100}$ and that E_{110} is in good agreement with the average value of E for a polycrystalline sample (Table 1.1).

1.6.2 Composite Materials

Composite materials are combinations of two or more materials that together provide properties not available when either of the individual materials are used alone. Referring to Fig. 1.3, we see that strong but brittle materials (curve *a*) can withstand large stresses prior to failure but possess limited ductility; soft, ductile materials (curve *d*) exhibit considerable plastic flow but little load-bearing capacity. Though only a relatively few materials exhibit both exceptional strength and ductility, a number of hybrid or composite materials have been developed to utilize the respective superior properties of the constituents of the composite material. For example, certain engineering plastics that possess considerable ductility are used as a *matrix* that is *reinforced* with high-strength glass, carbon, or aramid fibers to produce composite materials that possess both high strength and adequate ductility (see Table 1.10); such materials challenge metal alloys for use in numerous components, particularly since many offer excellent mechanical behavior at very low weight. It is now common for manufacturers to make significant use of polymer matrix composites in such items as automotive body frames and hood and door panels, aircraft wings and fuselage[x], boat hulls, and sporting equipment. In parallel fashion, metal–matrix composites reinforced with silicon carbide, silicon nitride, and/or alumina fibers are being used in a limited number of automobile engine components, fighter jet landing gear, and even bicycle frames where the performance gain (or at least the perceived performance gain) can justify the additional cost. Ceramic matrix composites are attractive for use in high-performance engines and gas turbines due to their desirable combination of temperature stability, hardness, and toughness. Tensile testing is widely used to characterize the behavior of composites, although different testing standards apply to each class of materials.[21-24]

Before analyzing the elastic response of composite materials in detail, it is appropriate to consider the respective functions of both the matrix and reinforcing phases in the composite. The many discrete fibers, filaments, or platelets are intended to carry much of the load applied to a composite structure. The fact that there are many discrete fibers in a given composite provides redundancy to the structure and precludes catastrophic fracture if one fiber were to contain a defect

---

[x] Notable examples include the Beechcraft Starship, the Airbus A320, and the Boeing 787.

Table 1.10   Tensile Properties of Selected Fibrous Composites and Reinforcements[a−d]

Materials	Modulus of Elasticity [GPa (10^6 psi)]	Tensile Strength [GPa (ksi)]	Elongation (%)
Composite			
Nylon 66 + 25 v/o carbon fibers	14 (2)	≈0.2 (29)	2.2
Epoxy resin + 60 v/o carbon fibers	220 (32)	1.4 (200)	0.8
Polyester resin + 50 v/o aligned glass fibers	38 (5.5)	0.75 (110)	1.8
Polyester resin + 20 v/o random glass fibers	8.5 (1.2)	0.11 (16)	2
Epoxy + 50 v/o boron fibers[b]	200 (29.2)	1.4 (200)	—
Epoxy + 72 v/o S-glass[b]	60.7 (8.8)	1.3 (187)	—
2024A1 + 25 v/o SiC[d]	124–172 (18–25)	0.53–0.64 (77–93)	≤1
Reinforcement			
Al_2O_3 whiskers[a]	415–485 (60–70)	7.0–21.0 (1000–3000)	—
Aramid (Kevlar 49)[c]	125 (18)	2.8–3.6 (400–520)	2–3
Boron	380 (55)	3.4 (500)	—
Carbon fiber,[c] Type I	390 (57)	2.2 (320)	0.5
Carbon fiber,[c] Type II	250 (36)	2.7 (390)	1.0
E glass[c]	76 (11)	1.4–2.5 (200–360)	2–3
S glass[a]	85 (12)	4.5 (650)	—
SiC whiskers[a]	485 (70)	20.7 (3000)	—
Si_3N_4 whiskers[a]	380 (55)	1.4 (200)	—

[a] Z. D. Jastrzebski, *The Nature and Properties of Engineering Materials,* Wiley, New York, 1977, p. 546.
[b] *Guide to Engineering Materials*, Vol. 1 (1) ASM, Metals Park, OH, 1986, p. 10.
[c] D. Hull, *An Introduction to Composite Materials,* Cambridge University Press, Cambridge, England, 1981.
[d] A. P. Divecha, C. R. Crowe, and S. G. Fishman, *Failure Modes in Composites IV*, J. A. Cornei and F. W. Crossman, Eds., AIME, 1979, p. 406.

and, therefore, fracture prematurely. The matrix phase serves to isolate the fibers from one another and to protect the fiber surface from damage. Of considerable importance, the matrix transmits the applied loads to the fiber through localized shear stresses acting along the fiber–matrix interface.

1.6.3 Isostrain Analysis

The stress–strain response of a composite material depends on the respective properties of the matrix and reinforcing phases, their relative volume fraction, the absolute length of the fibers, and the orientation of the fibers relative to the applied stress direction. We begin our analysis of the behavior of a reinforced composite by first assuming that the reinforcement phase consists of fibers that are continuous (i.e., they extend the entire length of the sample), possess uniform strength, and are oriented parallel to the applied stress direction as shown in Fig. 1.16*a*. If the fibers are properly bonded to the matrix and both phases behave elastically, the load applied to the composite in the direction of the fiber axes will be distributed such that

$$P_{c\parallel} = P_f + P_m \tag{1-57}$$

where

$P_{c,f,m} =$ load carried by the composite, fiber, and matrix, respectively

To avoid detachment of the fibers from the matrix, the strains experienced by the two phases must be identical so that $\varepsilon_c = \varepsilon_m = \varepsilon_f$. From the definition of stress and Eq. 1-57, this

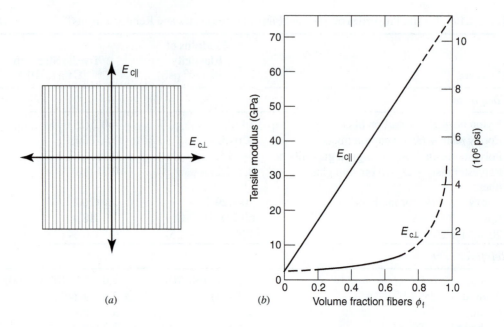

Figure 1.16   (*a*) Fiber orientation for isostrain ($\parallel$) and isostress ($\perp$) testing. (*b*) Tensile modulus in aligned glass-fiber reinforced epoxy resin as predicted from Eqs. 1-36 and 1-54.[25] (From N. G. McCrum, C. P. Buckley, and C. B. Bucknall, *Principles of Polymer Engineering,* Oxford Science Pub., Oxford, U.K. (1988). Reprinted by permission of Oxford University Press).

isostrain condition leads to

$$\sigma_{c\parallel}A_c = \sigma_f A_f + \sigma_m A_m \tag{1-58}$$

where

$\sigma_{c,f,m}=$ stress in composite, fibers, and matrix, respectively
$A_{c,f,m}=$ cross-sectional area of composite, fibers, and matrix, respectively

Since the area fraction of a continuous phase is equivalent to the volume fraction of the phase ($V_{f,m}$),

$$\frac{A_f}{A_c} = V_f \quad \text{and} \quad \frac{A_m}{A_c} = V_m \tag{1-59}$$

Therefore,

$$\sigma_{c\parallel} = \sigma_f V_f + \sigma_m V_m = \sigma_f V_f + \sigma_m(1 - V_f) \tag{1-60}$$

From Hooke's law (Eq. 1-7),

$$E_{c\parallel}\varepsilon_c = E_f \varepsilon_f V_f + E_m \varepsilon_m V_m \tag{1-61}$$

and so, finally, the Young's modulus for the isostrain case is

$$E_{c\parallel} = E_f V_f + E_m V_m \tag{1-62}$$

Also, the relative loads supported by the fibers and matrix are given by

$$\frac{P_f}{P_m} = \frac{E_f}{E_m}\frac{\varepsilon_f}{\varepsilon_m}\frac{V_f}{V_m} = \frac{E_f}{E_m}\frac{V_f}{V_m} \tag{1-63}$$

From Eq. 1-63, the load distributed to the fibers and matrix of the composite depends on the respective moduli and volume fractions of the two phases. Therefore, the fibers will assume much of the applied load when $E_f \gg E_m$ and $V_f \gg V_m$. The moduli of several reinforcement materials are listed in Table 1.10; for comparison, the modulus of a typical epoxy matrix may be 2–3 GPa, one to two orders of magnitude lower. If one assumes that the fibers are cylindrical in cross section, are close-packed in a hexagonal configuration, and that the gaps between them are completely filled with matrix material, it can be shown that it is impossible to exceed a fiber volume fraction of 0.9. In reality, $V_f \approx 0.6$ is more likely in a unidirectional aligned fiber composite, and even lower fractions are likely for woven fabrics. This underscores the importance of choosing a large modulus difference to ensure a favorable load distribution.

Assuming that the fibers are of uniform strength and are linear elastic when tested alone, the composite will exhibit linear elastic behavior to the point at which the fibers fail, the matrix fails, or the matrix deforms plastically. More will be said about failure in Section 1.6.6.

1.6.4 Isostress Analysis

When the applied stress direction is perpendicular to the fiber axes or reinforcing plates, the matrix and reinforcing phases are described as being in series with one another. Here the loads in the two phases are equal such that

$$P_c = P_f = P_m \tag{1-64}$$

where

$P_{c,f,m} =$ load carried by the composite, fiber, and matrix, respectively.

Unlike the isostrain analysis condition, displacements in the two phases under the *isostress* condition are additive, with the total composite strain being the weighted sum of strains in the matrix and reinforcing phases. Accordingly,

$$\varepsilon_{c\perp} = V_f \varepsilon_f + V_m \varepsilon_m = V_f \varepsilon_f + (1 - V_f)\varepsilon_m \tag{1-65}$$

By combining Eq. 1-65 with Hooke's law for the fiber and matrix components, one finds that the composite modulus in the direction normal to the aligned reinforcing phase is

$$E_{c\perp} = \frac{E_f E_m}{V_f E_m + (1 - V_f)E_f} \tag{1-66}$$

As shown by Eq. 1-62, it is obvious that the composite modulus $E_{c\parallel}$ increases linearly with fiber-volume fraction under isostrain conditions, whereas in Eq. 1-66 $E_{c\perp}$ increases nonlinearly for isostress loading conditions (Fig. 1.16b).[25] By examining moduli values (see Tables 1.1 and 1.10) for typical polymer matrices ($E \approx 2.5$ GPa) and glass fibers ($E \approx 76$ GPa), Eq. 1-66 can be approximated by

$$E_{c\perp} \approx \frac{E_m}{(1 - V_f)E_f} \tag{1-67}$$

We conclude that the elastic modulus of a composite loaded under isostress conditions is strongly dependent on the stiffness of the matrix, unlike the isostrain case where fiber stiffness dominates E_c (recall Eq. 1-62). Accordingly, the isostrain composite modulus is consistently larger than that associated with loading under isostress conditions (Fig. 1.16b).

In practice, unidirectional composites are only suitable for certain applications because of the enormous disparity in stiffness (and in strength) between the two orientations. To address this potential shortcoming, laminates can be formed from stacks of unidirectional plies intentionally rotated with respect to one another. This creates a sheet with fibers oriented in more than one direction in the plane of the sheet. The properties of the sheet will be determined by the fraction of plies oriented in any particular direction, and so may be tuned for a particular application. One common lay-up is called *quasi-isotropic*, in which 0°, 90°, and ±45° plies are included in equal proportions. Note that this gives the laminate nearly isotropic properties only in the plane of the plate, so it has *transversely isotropic* symmetry. Laminates with only 0° and 90° orientations are orthotropic. In both cases, elastic behavior must be described using the formalism already developed for anisotropic materials (i.e., with S_{ij} and C_{ij} matrices). In many cases it is important that a laminate is *balanced*—that is, laid up symmetrically with respect to the sheet thickness center plane (e.g., [0/90/90/0]) to avoid bending or twisting associated with low-symmetry elastic coupling.

EXAMPLE 1.5

For the case of an epoxy + 70 v/o S-glass long-fiber composite, what is the elastic modulus of the composite both parallel and perpendicular to the axis of the fibers? (Assume that $E_{epoxy} = 3$ GPa.)

Using Eq. 1-62 and mechanical property data from Tables 1.1 and 1.10,

$$E_{c\parallel} = E_f V_f + E_m V_m$$

so that $E_{c\parallel} = 85 \times 10^9(0.7) + 3 \times 10^9(0.3)$

Therefore, $E_{c\parallel} = 60.4$ GPa

Notice good agreement between this value and the elastic modulus for a 72 v/o S-glass + epoxy matrix composite (see Table 1.10).

For the case of loading perpendicular to the S-glass fibers, we see from Eq. 1-66 that

$$E_{c\perp} = \frac{85 \times 10^9(3 \times 10^9)}{0.7(3 \times 10^9) + 0.3(85 \times 10^9)}$$

$$E_{c\perp} = 9.24 \text{ GPa}$$

By comparison, Eq. 1-67 reveals the approximate elastic modulus to be

$$E_{c\perp} \approx \frac{3 \times 10^9}{0.3} \approx 10 \text{ GPa}$$

which is in relatively good agreement with the initial computation for $E_{c\perp}$.

1.6.5 Aligned Short Fibers

Continuous fiber composites may offer the greatest stiffness and strength possible for a given combination of matrix and reinforcement materials, but are not necessarily affordable for all applications. Processing techniques such as injection molding, for instance, are incompatible with continuous fibers, so short (discontinuous, chopped) fibers are used instead. It is often the case that a manufacturing process based on fluid flow can create preferential fiber alignment. The good news is that the behavior of aligned discontinuous fiber composites can be almost as good as for continuous fiber composites as long as the fibers exceed a certain critical length, l_c. Fibers that exceed this length carry the maximum load possible, whereas those that are shorter do not support loads as effectively. In order to determine the elastic behavior (and eventually the strength) of discontinuous fiber composites, we must first determine l_c.

An analysis of stress distributions within the composite reveals that the applied load is actually transmitted from the matrix to the short fibers by shear stresses acting along the fiber–matrix interface. This mode of fiber loading is often treated using *shear lag analysis*. For the case of a circular fiber with radius r, the interfacial shear stresses produce an axial stress along the fiber according to a force balance given by

$$\sigma_{zz}\pi r^2 = \tau_{rz}2\pi rz \qquad (1\text{-}68)$$

where

$\sigma_{zz} =$ axial stress along the fiber length
$\tau_{rz} =$ shear stress acting along the fiber–matrix interface at the ends of the fiber
$r =$ fiber radius
$z =$ shear stress transfer length (distance from each fiber end)

Upon rearranging Eq. 1-68,

$$\sigma_{zz} = \frac{2\tau_{rz}z}{r} \qquad (1\text{-}69)$$

Note that the axial stress σ_{zz} is zero at the end of the fiber where $z = 0$ and increases with increasing transfer length, as shown in Fig. 1.17a.[xi] However, there are limits to the maximum stress the fiber can support as the fiber length increases, as implied by Fig. 1.17b. One possibility is that the stress level σ_{zz} at the midpoint of the fiber will reach σ_{fc} (the fracture strength of the fiber) and fiber fracture will occur. This effectively reduces the fiber length, the maximum value of σ_{zz}, and the load supported by the fiber. The critical length for fiber fracture is one way to define l_c. Alternatively, strain compatibility between the fiber and matrix may prevent the fiber stress from rising beyond the level expected for a continuous fiber composite loaded in the isostrain configuration (the configuration in which the fiber bears the greatest possible fraction of the load). If this is the case, the stress profile will match Fig. 1.17b. The critical length in this case is the minimum length required to reach the isostrain condition. Note that if the limit imposed by strain compatibility is reached before the critical stress for fracture, much of the fiber strength is wasted. The best short-fiber performance is therefore achieved when the critical length as determined by the isostrain compatibility condition is the same as the

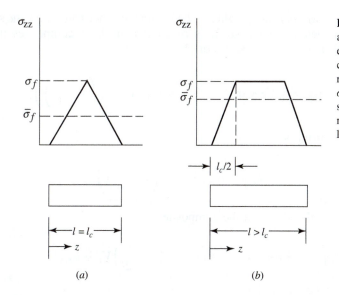

(a)   (b)

Figure 1.17   Axial stress distribution along fiber length when fiber strength is (a) equal to and (b) greater than the strain compatibility-limited critical length for reinforcement. In case (a), the critical stress σ_f is determined by the fiber fracture strength. In case (b), σ_f is determined by the maximum strain possible in the isostrain loading condition.

[xi] This theoretical analysis for the axial stress distribution along the length of the fiber (see H. L. Cox, *Brit. J. Appl. Phys.*, 3, 72 [1952]), was confirmed for the polydiacetylene fiber-epoxy matrix model composite system in experiments conducted by Robinson et al. (see I. M. Robinson, R. J. Young, C. Galiotis, and D. N. Batchelder, *J. Mater. Sci.*, 22, 3942 [1987]).

critical length for fiber fracture, thereby ensuring the greatest possible load-bearing benefit (i.e., fiber effectiveness is not preferentially limited by either fracture or strain compatibility). We can now define $z_c = l_c/2$, where $l_c/2$ is the critical transfer length for the particular stress-limiting mechanism operating (Fig. 1.17). Therefore, upon substitution in Eq. 1-69,

$$l_c = \frac{2r\sigma_{fc}}{2\tau_{rz}} \approx \frac{2r\sigma_{fc}}{\sigma_m} \tag{1-70}$$

where σ_m is the normal stress in the matrix.

To achieve a fiber stress of σ_{fc} at the center of the fiber (the ideal condition), the critical aspect length l_c must increase with increasing fiber strength and decreasing shear stress (or matrix normal stress)[xii] .For example, when the temperature of a polymer matrix composite is increased, τ_{rz} tends to decrease faster than σ_f with the result that l_c increases. If the actual fiber length $l < l_c$, then maximum load-bearing potential cannot be reached.

From Fig. 1.17a, the average axial stress on a fiber of length l_c is given by

$$\bar{\sigma}_f = \frac{\sigma_f(l_c/2)}{l_c} = \sigma_f/2 \tag{1-71}$$

When $l > l_c$, the maximum axial stress remains constant but now extends along the midregion of the fiber (Fig. 1.17b). For this condition (assuming that the axial stress increases linearly from zero at each end)

$$\bar{\sigma}_f = \frac{\sigma_f l - \sigma_f(l_c/2)}{l} = \sigma_f\left(1 - \frac{l_c}{2l}\right) \tag{1-72a}$$

And, if $l < l_c$, the maximum axial stress is never reached, so

$$\bar{\sigma}_f = \frac{l\sigma_m}{4r} \tag{1-72b}$$

Assuming that the short fibers are parallel to the loading axis, we can follow the same approach used previously to determine the stiffness of continuous fiber composites under isostrain loading (Eq. 1-60). For $l < l_c$, the stress on the composite is

$$\sigma_{c\parallel} = \bar{\sigma}_f V_f + \sigma_m(1 - V_f) = \sigma_m\left(1 - V_f\left(1 - \frac{l}{4r}\right)\right) \tag{1-73}$$

and the composite stiffness is

$$E_{c\parallel} = E_m\left(1 - V_f\left(1 - \frac{l}{4r}\right)\right) \tag{1-74}$$

whereas if $l > l_c$, then the stress on the composite is

$$\sigma_{c\parallel} = \bar{\sigma}_f V_f + \sigma_m(1 - V_f) = \sigma_f\left(1 - \frac{l_c}{2l}\right)V_f + \sigma_m(1 - V_f) \tag{1-75}$$

[xii] For short-fiber composites, strength and stiffness properties depend critically on the integrity of the fiber–matrix interface (e.g., see M. R. Piggott, *Polym. Compos.*, 3(4), 179 [1982]).

and the composite stiffness is

$$E_{c\parallel} = E_f\left(1 - \frac{l_c}{2l}\right)V_f + E_m(1 - V_f) \tag{1-76}$$

If the fibers are not all aligned with the loading axis, the stiffness is clearly reduced from that of the fully aligned case, as for continuous fibers.

1.6.6 Strength of Composites

Composite strength can be defined as the point at which the initial linear region of the stress–strain diagram ends, regardless of which phase is responsible for the change in behavior. As before, we will examine cases associated with continuous fibers before turning to discontinuous fibers.

1.6.6.1 Effects of Matrix Behavior

For some common polymer matrix composites, the matrix is fairly brittle (e.g., epoxy) and has low strength compared to that of the fiber phase. When loaded in the isostrain condition, the failure strain of the brittle matrix may be exceeded before the critical fiber strain is reached. In this case, the critical failure strain is that of the matrix, and the elastic stress in the composite at the point of fracture is described by

$$\sigma_{c\parallel} = E_{c\parallel}\varepsilon_m = \sigma_f V_f + \sigma_m(1 - V_f) = E_f\varepsilon_m V_f + E_m\varepsilon_m(1 - V_f) \tag{1-77}$$

where ε_m is the fracture strain of the brittle matrix. Once the matrix fractures, the fraction of the load formerly borne by the matrix transfers to the fibers. If the additional stress is sufficient to cause fiber fracture then the entire composite material will fail. This will almost certainly occur if the volume fraction of fibers is small because the load transfer will be very large. If the volume fraction of fibers is large then the load transferred upon matrix fracture will be relatively small and the composite may remain intact, although severely compromised. Either way, the composite stress at the point of matrix failure is much higher than the matrix material could ever withstand on its own, thanks to the load-sharing of the fibers.

If the matrix is ductile (e.g., thermoplastic or metal) then the critical elastic strain at failure is determined by the fiber properties, and the composite elastic stress at failure is

$$\sigma_{c\parallel} = E_{c\parallel}\varepsilon_f = \sigma_f V_f + \sigma_m(1 - V_f) = E_f\varepsilon_f V_f + E_m\varepsilon_f(1 - V_f) \tag{1-78}$$

This leads to a somewhat surprising situation: if V_f is less than some critical value $V_{f\text{crit}}$, the strength of the composite is actually less than that of the matrix alone (Fig. 1.18). This occurs because the matrix is capable of carrying a greater load than the fibers. In effect, the presence of a subcritical amount of a reinforcing phase reduces the overall load-bearing capacity of

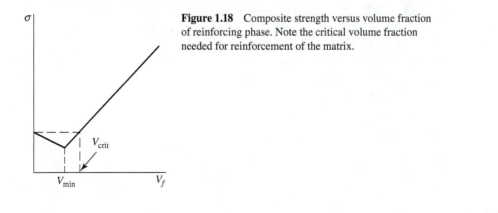

Figure 1.18 Composite strength versus volume fraction of reinforcing phase. Note the critical volume fraction needed for reinforcement of the matrix.

the matrix. Above this value the overall strength increases as the fibers carry a greater fraction of the total load so that yield in the matrix is less likely, and the composite strength will eventually exceed that of the matrix alone. For most practical composites the critical fiber volume fraction is fairly low, in the range of $\sim 10\%$.

If a continuous fiber composite is loaded in the isostress orientation then, by definition, the same stress exists in the fibers and the matrix. As a first approximation, one might expect that the composite strength would be identical to that of the weaker matrix phase. In reality, this is an upper bound because failure can occur at a fiber/matrix interface instead of within the matrix phase. If this is the case, the actual composite strength may be significantly lower than that of either phase.

Advantageously aligned discontinuous fiber composites with $l = l_c$ can be modeled in the same fashion as isostrain continuous fiber composites, but with the composite stress given by Eq. 1-75 such that

$$\sigma_{c\parallel} = \bar{\sigma}_f V_f + \sigma_m\left(1 - V_f\right) = \frac{\sigma_f}{2} V_f + \sigma_m\left(1 - V_f\right) \tag{1-79}$$

where σ_f is the fiber fracture stress that occurs at the same load as the matrix fracture stress σ_m. Appropriate substitutions for the average fiber stress at the point of failure from Eq. 1-72a or 1-72b allow calculation of the composite strength for fibers that are longer or shorter than the critical length.

It should be recognized that in all of the preceding discussion, the fiber strength parameter is treated as if all fibers are identical. In truth, the strength of any given fiber falls within a statistical distribution of fiber strengths. As a result, the weaker fibers fracture prematurely, and the associated load is transferred to the shorter, broken fiber segments and to the remaining unbroken fibers. After many local fracture events the average fiber length decreases. This results in a corresponding decrease in composite stiffness and strength even before catastrophic failure of the majority of fibers takes place.

1.6.6.2 Effects of Fiber Orientation

In addition to the factors just described, the strength of a composite varies with fiber orientation. Due to processing variations or to unintended loading, fibers may not always be oriented parallel to the primary stress axis. From Fig. 1.19, the cross-sectional area of the plane normal to the fiber axes is given by

$$A_\phi = \frac{A_0}{\cos\phi} \tag{1-80}$$

With the resolved load in the direction of the fiber axis (P_f) computed to be

$$P_f = P_0 \cos\phi \tag{1-81}$$

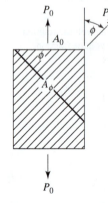

Figure 1.19   Applied load resolved in the direction of fiber orientation.

it follows that the stress acting parallel to the fibers is

$$\sigma_c = \frac{P_0 \cos\phi}{A_0/\cos\phi} = \sigma_0 \cos^2\phi \tag{1-82}$$

Note that—as expected—the axial load-bearing capacity of the composite with off-axis fibers is less than that associated with a composite that contains fibers aligned parallel to the axis of the bar. For the case of off-axis loading, macroscopic shear stresses are developed in the matrix parallel to the fibers. From Fig. 1.19, the resolved load parallel to the fibers is again given by Eq. 1-81, with the shear surface area A_s, described by

$$A_s = \frac{A_0}{\sin\phi} \tag{1-83}$$

The shear stress τ_m acting in the matrix parallel to the axes of the fibers is then found to be

$$\tau_m = \sigma_0 \sin\phi \cos\phi \tag{1-84}$$

Finally, fracture can occur by separation transverse to the fiber length with

$$\sigma_N = \frac{P_0 \sin\phi}{A_0/\sin\phi} = \sigma_0 \sin^2\phi \tag{1-85}$$

Whether shear fracture in the matrix takes place rather than fiber fracture depends on the angle of misorientation ϕ and the relative strengths of the matrix, interface, and fibers. For example, equating Eqs. 1-82 and 1-84 with respect to σ_0, shear failure will occur when

$$\tan\phi > \frac{\tau_m}{\sigma_c} \tag{1-86}$$

From Eq. 1-86, shear failure will occur at small misorientations when the matrix shear strength is small relative to that of the fiber breaking strength (e.g., see Fig. 1.20); it follows that shear failure

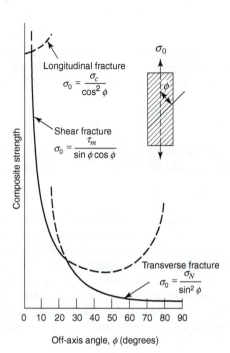

Figure 1.20   Competitive fracture processes depend on fiber misorientation.

parallel to the fiber axis is more likely to occur in off-axis composites at elevated temperatures since τ_m decreases more rapidly than σ_c with increasing temperature. Finally, transverse tensile fracture perpendicular to the fiber axes will occur at large angles of fiber misorientation.

1.7   THERMAL STRESSES AND THERMAL SHOCK-INDUCED FAILURE

We conclude this chapter with a discussion of elastic loading and brittle failure induced by temperature changes that generate residual stresses. Scientists, engineers, and homemakers have long known that when hot ceramic components are cooled quickly, some will crack, whereas others will not. The development of thermally induced stresses and the potential for thermal shock-induced fracture have been the subject of study for at least 2200 years! Indeed, Roman historians reported that Hannibal's military engineers used thermal shock to fracture rocks that blocked the path of the Carthaginian army during its advance across the Alps in 218 BC. Two hundred years later, Livy[26] reported that ". . . It was necessary to remove a rock. Trees were felled and the wood piled onto it in a huge pyre which was lighted and burned fiercely with the help of a fortunate breeze. The hot stone was then drenched with vinegar [presumably sour wine] to disintegrate it and attacked with pickaxes." In all likelihood, the hot rocks were doused with water to cause them to crack. For example, Pliny[27] reported that ". . . if fire has not disintegrated a rock, the addition of water makes it split." Other somewhat more recent reports of thermal shock-induced fracture of rocks were cited by Agricola[28] and de Beer.[29]

The modern technical literature provides more rigorous analyses of thermal stresses and thermal shock resistance than that found in early Roman documents. Of importance, Hasselman and co-workers[30,31] developed a unified theory of thermal stress fracture that accounts for both the initiation and propagation of cracks in a brittle solid. For the case of brittle materials, failure may occur when the thermal stress exceeds the material's fracture strength and a critical crack initiates. In this instance, fracture may be characterized by a "maximum stress" failure theory, and is suitable for analysis using the elasticity toolbox we have developed thus far. For the case of failure by propagation of an existing crack, the Griffith theory of failure by crack growth accounts for the fracture event. The crack propagation case requires an understanding of fracture mechanics, and so it is deferred until Chapter 6. In ductile materials, thermal stresses can also cause yielding, which can certainly be defined as failure if dimensional changes must be avoided.

1.7.1 Upper Bound Thermal Stress

Here, let us first consider the case where initiation-controlled fracture by thermal cracking cannot be tolerated, as in glasses, porcelain, electronic ceramics, and whiteware ceramics. If an unconstrained rod of length L is cooled from one temperature (T_1) to the other (T_2), it will contract by an amount

$$\Delta L = \alpha(T_1 - T_2)L \tag{1-87}$$

where $\alpha =$ linear coefficient of thermal expansion (CTE). If the rod were constrained between two rigid walls, a uniaxial thermal stress would be generated with the magnitude

$$\sigma_{th} = E\varepsilon_{th} = E\alpha(T_1 - T_2) \tag{1-88}$$

where $E =$ Young's modulus of elasticity and ε_{th} is the thermal strain. Although rigid walls are not truly found in practice, the approximation is reasonable if one component cools very quickly to match the ambient temperature (the rod) while the constraining component cools more slowly (the walls). Equation 1-88 is therefore an upper bound for the possible value of thermal stress. If the walls are also changing dimension with temperature, it is the difference between the thermal expansion of the wall material and that of the rod that would determine the difference in thermal strain, and hence the stress in the rod. The minimum thermal stress is generated when the two materials cool at the same rate. In this case, the difference in CTE, $\Delta\alpha$, is used in place of α in Eq. 1-88.

Note that if there is a mechanical stress also imposed on the rod, it is simply superimposed on the thermal stress so that $\sigma_{total} = \sigma_{mech} + \sigma_{th}$. If there are additional constraints, for example in the radial direction, then the generalized Hooke's law must be used to evaluate the total stress state, just as if the stresses were all generated mechanically. In the end, if the critical fracture stress for the brittle material is exceeded, regardless of the sources of that stress, failure will suddenly occur.

Thermal stresses pose particular problems for thin films and coatings adhered to thick substrates. If the film thickness is much smaller than the substrate thickness (by a factor of $100 \times$ or so) then most of the thermal strain mismatch is borne by the film. This can cause fracture if the film is brittle, or undesirable plasticity if the film is ductile. Even if failure by one of these modes is avoided, thermal stress in a film on only one side of a substrate causes a bending moment that can induce undesirable curvature in the pair.

An isotropic thin film that is constrained in two dimensions by a thick substrate is in a state of equal biaxial thermal stress, as described previously by Eq. 1-47. Rearranging this equation to solve for σ_{bi} (i.e., σ_{film}) and realizing that the thermal strain in this case is a biaxial strain, the film stress is given by

$$\sigma_{film} = \frac{E_{film}}{1 - \nu_{film}} \Delta\alpha\Delta T \tag{1-89}$$

where E_{film} and ν_{film} are the film's Young's modulus and Poisson's ratio, respectively. Recall from Section 1.5.2.2 that the elastic constant $E/(1 - \nu)$ is often called the biaxial modulus. Its usefulness is clear in this case. Equation 1-89 also assumes that the film and the substrate have the same temperature. This is a reasonable assumption for the small masses and slow heating/cooling rates typically involved in microelectronics materials.[xiii]

The CTE for a given material depends on the bond strength between atoms. As discussed in Section 1.3.2, increasing bond strength results in a decreasing interatomic distance of separation and a higher elastic stiffness. Likewise, increased interatomic bond strength (associated with higher melting points) leads to lower values of CTE[32] (Fig. 1.21). (For the case of ceramics, CTE decreases with both increasing bond strength and percentage of covalent bonding.) Additional CTE values are given in Table 1.11[33] for various metals, ceramics, and organic solids. These data

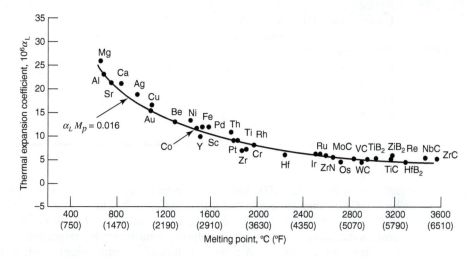

Figure 1.21   Relation between coefficient of thermal expansion (CTE) and melting point for metals, carbides, and borides with close-packed structures. (Reprinted from *Materials Research Bulletin*, vol. 12, L. G. Van Uitert, H. M. O'Bryan, M. E. Lines, H. J. Guggenheim, and G. Zydzik, Thermal expansion — An empirical correlation, p. 261, 1977, with permission from Elsevier.[32])

[xiii] It is also the case for the bi-metallic strips that are used as thermal indicator and control devices, and for micromachined thermal actuators used for micro-scale tweezers.

Table 1.11 Selected Coefficients of Thermal Expansion (CTE) for Metals, Ceramics, and Organic Solids ($10^{-6}/°C$).[33]

Metals	
Invar	2
Molybdenum	5.2
Alloy 42 (FeNi)	6
Titanium	10
Iron	12
Gold	14.2
Nickel	13–15
Gold-tin eutectic	16
Copper and its alloys	16–18
Silver	19
Lead-tin eutectic	21
Aluminum and its alloys	22–25
5-95 tin-lead	28
Lead	29
Ceramics, semiconductors, etc.	
Silica glasses	0.5–1.0
Silicon carbide	2.6
Silicon (single crystal)	2.8
Alumina	6.7
Beryllia	8
Gold-silicon eutectic	13
Gold-germanium eutectic	13
*Organics**	
Kevlar®	–2
Epoxy-glass (FR4)—horizontal	11–15
vertical	60–80
Polyimides	40–50
Polycarbonates	50–70
Epoxies	60–80
Polyurethanes	180–250
Sylgard®	300
RTV	800

*Below glass transition temperature.

are approximate since CTE values tend to increase moderately with temperature. Note that the highest CTE values are associated with organic solids, whereas ceramics display the lowest values of CTE. As previously discussed, these trends reflect the fact that ceramics and organic solids possess the highest and lowest interatomic bond strengths, respectively.

These trends imply that dissimilar material combinations, such as polymer coatings on ceramic substrates, or metal components bonded to ceramic components, are most likely to suffer from thermal stresses, particularly when elevated temperatures are involved in fabrication or bonding. Whichever phase has the larger value of α will be the phase in tension upon cooling from the fabrication temperature. For example, thermal barrier coatings in gas turbine engine components develop compressive residual thermal stresses in the outer ceramic layer due to differences in contraction between the intermediate metallic bond (higher CTE) coating layer (typically (Ni/Co)CrAlY) and the outer ceramic (lower CTE) thermal barrier coating (typically ZrO_2-based). Failure of the coating then occurs by delamination and subsequent spallation of the outer thermal barrier coating.[34] Such difficulties can clearly be reduced by selection of materials to minimize the CTE mismatch.

For the case of composite materials, internal thermal stresses are generated upon cooling from an elevated fabrication temperature or during in-service temperature fluctuations; such stresses will vary directly with differences in coefficients of thermal expansion between the matrix (α_m) and reinforcing phases (α_r). Thus metal matrix composites reinforced by ceramic fibers or particles and processed at elevated temperatures may be particularly susceptible to internal residual stress generation.

EXAMPLE 1.6

The temperature of a 10-cm-long rod of polycrystalline alumina (99.8% dense) decreases from 65 to 0°C.

(a) What would be the change in length if the rod were unconstrained?

(b) Would the rod survive this drop in temperature if it were fully constrained along its length?

(c) What is the critical temperature change to cause fracture?

(d) Having determined the critical ΔT, could the rod survive this same drop in temperature if it were constrained along all axes? Assume that $v = 0.22$.

To solve this problem, we must first collect relevant material property data. From Tables 1.3 and 1.11, we find thermal expansion, isotropic elastic modulus, and tensile strength[xiv] values:

$$\alpha = 6.7 \times 10^{-6}/°C$$

$$E = 385\,GPa$$

$$\sigma_{ts} = 205\,MPa$$

(a) When the rod is unconstrained, we see from Eq. 1-87 that the 200°C temperature change generates a length change of

$$\Delta L = 6.7 \times 10^{-6}\,°C^{-1}(65\,°C)(10\,cm)$$
$$\Delta L = 0.0044\,cm$$

which corresponds to a thermal strain of 0.00044, or 0.044%.

(b) If the rod were fully constrained along its length, the axial thermal stress is computed from Eq. 1-88, where

$$\sigma = 385 \times 10^9\,N/m^2(6.7 \times 10^{-6}\,°C^{-1})(65\,°C)$$
$$\sigma = 168\,MPa$$

Since the tensile strength of alumina is reported as 205 MPa, the rod would likely *not* have fractured under these conditions.

(c) Indeed, the maximum allowable temperature rise for an axially constrained rod of alumina is calculated to be

$$\Delta T = \frac{205 \times 10^6}{385 \times 10^9(6.7 \times 10^{-6})} = 79.5\,°C$$

(Continued)

[xiv] Recall from Section 1.4.1 that the fracture strength of a brittle material is not a constant, but instead depends on specimen size and testing method. Thus for this problem it is critical that the strength value used was measured in a configuration similar to that of the alumina rod in question, and not, for instance, in bending or compression. Furthermore, we must assume a similar specimen size and state of surface preparation for a reasonably accurate calculation.

(d) The thermal strains along all three axes are the same as that calculated in part (a) even though the dimensions of the specimen are not the same in the axial and radial directions. Because the material is elastically isotropic, the three stresses must also be equal. Applying the generalized Hooke's law,

$$\varepsilon_{th} = \alpha \Delta T = \frac{1}{E}(\sigma_{th} - v\sigma_{th} - v\sigma_{th}) = \frac{1 - 2v}{E}\sigma_{th}$$

and replacing the thermal stress with the tensile strength, we find

$$\Delta T = \frac{1 - 2(0.22)}{(6.7 \times 10^{-6} \,^{\circ}C^{-1})(385 \times 10^3 \, \text{MPa})}(205 \, \text{MPa}) = 44.5 \,^{\circ}C$$

because of the additional constraint in the radial direction that imposes a tensile Poisson strain in the axial direction. Thus the rod probably would fracture during a 65 °C temperature change if fully constrained.

1.7.2 Cooling Rate and Thermal Stress

As discussed, the stress level given by Eq. 1-88 corresponds to a maximum value for the case when the average temperature of a body after quenching is unchanged[xv] but its surface temperature matches that of the quenching medium. The latter condition is approached only under extreme quenching conditions and in materials possessing a very low coefficient of thermal conductivity.

The magnitude of thermal stress more typically depends on the heat transfer coefficient (h) between the cooling fluid and the quenched solid, the material's coefficient of thermal conductivity (k), and the geometry of the component. It is customary to compute the magnitude of the thermal stress by including a nondimensional parameter [defined as the Biot modulus (β)] that incorporates these factors where

$$\beta = xh/k \tag{1-90}$$

where $x =$ specimen dimension such as slab thickness or rod diameter. For large component dimensions, high heat transfer rates to the environment (corresponding to conductive rather than convective or radiative heat transfer), and/or low levels of thermal conductivity, β values are large ($\beta > 20$) and thermal stress levels approach those given by Eq. 1-88.

For the case of a coating on a substrate for which a temperature difference between the two materials exists, Eq. 1-89 is rewritten in the form

$$\sigma = \frac{c\beta E\alpha(\Delta T)}{1 - v} \tag{1-91}$$

where $c =$ function of specimen geometry. Jaeger[35] determined that values of $c\beta$ fall in the range $0 < c\beta \leq 1$. By rearranging Eq. 1-91 and combining with Eq. 1-90, the material's initiation-controlled thermal shock resistance is defined by the temperature change that can be tolerated without fracture. Hence,

$$\Delta T = \frac{\sigma_{fail}(1 - v)k}{E\alpha ch} \tag{1-92}$$

[xv] If the body of a component experiences a thermal gradient, there will exist a stress gradient from the surface to the core of the object.

Table 1.12 Mechanical and Thermal Properties of Selected Ceramics and Associated Thermal Shock Resistance Parameter[36]

Material	Bend Strength σ (MPa)	Young's Modulus E (GPa)	Poisson's Ratio ν	Thermal Expansion Coefficient α, 0–1000 °C (10^{-6}K^{-1})	Thermal Conductivity k at 500 °C $(\text{Wm}^{-1}\text{K}^{-1})$	$R' = \dfrac{\sigma k(1-\nu)}{E\alpha}$ (kWm^{-1})
Hot-pressed Si_3N_4	850	310	0.27	3.2	17	11
Reaction-bonded Si_3N_4	240	220	0.27	3.2	15	3.7
Reaction-bonded SiC	500	410	0.24	4.3	84	18
Hot-pressed Al_2O_3	500	400	0.27	9.0	8	0.8
Hot-pressed BeO	200	400	0.34	8.5	63	2.4
Sintered WC (6% Co)	1400	600	0.26	4.9	86	30

where σ_{fail} = fracture strength. Note that a material's resistance to thermal stress-induced failure is enhanced by a high fracture strength and coefficient of thermal conductivity, and low values of the modulus of elasticity, coefficient of thermal expansion, and heat transfer rate. It is ironic that those materials that possess superior high temperature stiffness and resistance to environmental degradation are most susceptible to catastrophic thermal-shock failure.

Table 1.12[36] lists relevant thermal and mechanical property data for selected ceramics, where R' represents the measure of thermal fracture resistance, comparable to that expressed by Eq. 1-92. Higher values of R' reflect greater thermal shock resistance. The relative ranking of these materials must be viewed with caution, however, since property values such as α and k increase and decrease, respectively, with temperature. As a result, R' values for a given material will vary as a function of temperature. Consequently, it is not possible to provide a simple ranking of materials in terms of their resistance to thermal shock.

REFERENCES

1. ASTM E8, ASTM International, West Conshohocken, PA.

2. R. Hooke, "Of Spring," 1678, as discussed in S. P. Timoshenko, *History of the Strength of Materials,* McGraw-Hill, New York, 1953, p. 18.

3. H. Morrogh, *J. Iron Steel Inst.*, (1968) 1.

4. ASTM E111, ASTM International, West Conshohocken, PA.

5. ASTM D638, ASTM International, West Conshohocken, PA.

6. S. S. Brenner, *J. Appl. Phys.* **27**, 1484 (1956).

7. H. Bei, S. Shim, G. M. Pharr, and E. P. George, *Acta Materialia* **56**, 4762 (2008).

8. ASTM D882, ASTM International, West Conshohocken, PA.

9. T. A. Osswald and G. Menges, *Materials Science of Polymers for Engineers*, Hanser, Munich, 1996.

10. ASTM D412, ASTM International, West Conshohocken, PA.

11. I. M. Ward and D. W. Hadley, *An Introduction to the Mechanical Properties of Solid Polymers*, Wiley, New York, 1993.

12. L. R. G. Treloar, *The Physics of Rubber Elasticity*, 3rd ed., Clarendon Press, Oxford, 1975.

13. ASTM E111, ASTM International, West Conshohocken, PA.

14. ASTM E855, ASTM International, West Conshohocken, PA.

15. ASTM C1161, ASTM International, West Conshohocken, PA.

16. ASTM E143, ASTM International, West Conshohocken, PA.

17. R. S. Lakes, *Science* **235**, 1038 (1987).

18. M. R. Van Landingham, *J. Res. Natl. Inst. Stand. Technol.* **108**, 249–265 (2003).

19. W. C. Oliver and G. M. Pharr, *J. Mater. Res.*, **19**(1), 3 (2004).

20. R. B. Ashman, S. C. Cowin, W. C.Van Buskirk and J. C. Rice, *J. Biomech.* **17** (5), 349 (1984), pp. 349–361.

21. ASTM D3039, ASTM International, West Conshohocken, PA.

22. ASTM D3552, ASTM International, West Conshohocken, PA.

23. ASTM C1275, ASTM International, West Conshohocken, PA.

24. ASTM C1359, ASTM International, West Conshohocken, PA.

25. N. G. McCrum, C. P. Buckley, and C. B. Bucknall, *Principles of Polymer Engineering*, Oxford Science Pub., Oxford, U.K. (1988).

26. Livy, xxi, 37.2.

27. Pliny, *Historia Naturalis,* xxiii. 27.57.

28. G. Agricola, *De Re Metallica*, translated from the first Latin edition of 1556 by H. C. Hooverand L. H. Hoover, Dover Publications, New York, 1950, p. 119.

29. G. de Beer, *Alps and Elephants—Hannibal's March*, Geoffrey Bles, London, 1955.

30. D. P. H. Hasselman, *J. Am. Ceram. Soc.* **52**, 600 (1969).

31. D. P. H. Hasselman, and J. P. Singh, *Thermal Stresses I*, Vol. **1**, R.B. Hetnarski, Ed., North-Holland, New York, Chap. 4, 264 (1986).

32. L. G. Van Uitert et al., *Mater. Res. Bull.* **12**, 261 (1977).

33. *Microelectronics Packaging Handbook*, R. R. Tummala and E. J. Rymaszewski, Eds., Van Nostrand Reinhold, New York, 278 (1989).

34. R. A. Miller and C. E. Lowell, *Thin Solid Films* **95**, 265 (1982).

35. J. C. Jaeger, *Phil. Mag.* **36**, 419 (1945).

36. R. W. Davidge, *Mechanical Behavior of Ceramics*, Cambridge University Press, Cambridge, 1979.

37. O. K. Muratoglu, in *UHMWPE Biomaterials Handbook, Second Edition: Ultra High Molecular Weight Polyethylene in Total Joint Replacement and Medical Devices*, S. M. Kurtz, Ed. Academic Press, 2009.

FURTHER READINGS

L. J. Broutman and R. H. Krock, *Modern Composite Materials*, Addison-Wesley, Reading, MA, 1967.

J. A. Cornie and F. W. Crossman, *Failure Modes in Composites IV*, Metallurgical Society, AIME, New York, 1979.

D. Hull, *An Introduction to Composite Materials*, Cambridge University Press, Cambridge, 1981.

R. M. Jones, *Mechanics of Composite Materials*, McGraw-Hill, New York, 1975.

PROBLEMS

Review

1.1   In your own words, what are two differences between product testing and material testing?

1.2   What are the distinguishing differences between *elasticity*, *plasticity*, and *fracture*?

1.3   Write the definitions for engineering stress, true stress, engineering strain, and true strain for loading along a single axis.

1.4   Under what conditions is Eq. 1-4 valid? What makes it no longer useful if those conditions are not met?

1.5   Sketch Figure 1.3, curve *b* (a ductile metal). Label it with the following terms, indicating from which location on the curve each quantity can be identified or extracted: elastic region, plastic region, proportional limit, tensile strength, onset of necking, fracture stress.

1.6   On a single set of axes, sketch schematic interatomic *force* vs. atom *separation distance* curves (like the one shown in Fig. 1.4*b*) for Li, Na, and Cs. Pay close attention to the point x_0 and the slope dF/dx for each of the curves you draw.

1.7   State the critical difference in the processing behavior of *thermoplastics* vs. *thermosets*.

1.8   What happens to the stiffness of a polymer as the temperature T_g is exceeded? For what group of polymers is this change the greatest? The smallest?

1.9   Write typical values of E for diamond, steel, aluminum, silicate glass, polystyrene, and silicone rubber subjected to small strains (note that the latter value is not included in this chapter, but is widely available). Clearly indicate the units for each value.

1.10   What is the purpose of a *plasticizer*, and what specific effect on room temperature behavior is likely when a plasticizer is added?

1.11   Identify a minimum of two structural characteristics and two mechanical characteristics that set *elastomers* apart from other classes of materials (including other polymers).

1.12   Define what is meant by *uniaxial, biaxial,* and *triaxial* loading.

1.13   State one advantage and disadvantage of compression testing.

1.14   Is *Euler buckling failure* initiated by an elastic, plastic, or cracking process? Explain.

1.15 What is the difference between the *resilience* and the *strain energy density* of a material under load? Illustrate your answer by reproducing Figure 1.3, curve *b* (a ductile metal), and annotating it appropriately.

1.16 Sketch Figure 1.3, curve *b* (a ductile metal) and show on the figure the difference between the *proportional limit* and the *offset yield strength*.

1.17 Describe when and why bend testing (flexural testing) is most advantageous.

1.18 Where can the maximum stress be found for a rectangular bar undergoing 3-point bending? 4-point bending?

1.19 Write the basic isotropic form of Hooke's law relating stress and strain for uniaxial tension/compression loading and shear loading. Define all quantities.

1.20 Why do we define engineering and true stresses for tension/compression loading but not for shear loading?

1.21 Sketch a pair of pliers squeezing an object and use the sketch to show why the hinge pin is under shear loading.

1.22 Write out the most general expression for *tension or compression strain* along a single axis resulting from all possible applied stresses, assuming that the material is elastically isotropic.

1.23 Write out the most general expression for *shear strain* along a single axis resulting from all possible applied stresses, assuming that the material is elastically isotropic.

1.24 Sketch and name the *stress state* present in the skin of a cylindrical thin-walled pressure vessel. Repeat for the *strain state*.

1.25 What is the name of the matrix, S_{ij}?

1.26 Why can the compliance and stiffness tensors for cubic and orthotropic materials be greatly simplified from the general case?

1.27 Describe the geometric criteria that differentiate *orthotropic* and *cubic* symmetry.

1.28 Define *hydrostatic stress state*.

1.29 What is the primary purpose of the fibers in a composite material? Of the matrix?

1.30 What does it mean for a fiber-reinforced composite to be quasi-isotropic, and how is this typically achieved?

1.31 Which is the stiffer orientation for a unidirectional fiber-reinforced composite, the isostress orientation or the isostrain orientation? Explain, and provide a sketch to support your answer.

1.32 Why are pairs of materials more likely to experience thermal stress problems when they represent two different material classes?

Practice

1.33 Sketch a tensile member with (a) a rectangular cross section, (b) a solid circular cross section, and (c) a circular tube cross section, and label the dimensions symbolically (e.g., label the radius for the solid circular case). For each

member, write out the definition of engineering stress in terms of the actual dimensions of the component. If the rectangular member has dimensions of width and thickness equal to 1 cm × 0.3 cm, what would be the radius of a solid circular member such that the stress is equal for an equal tensile load? If a tube has an outer radius equal to that of this same solid cylinder, what is the maximum inner radius such that the stress does not exceed 200% of the stress in the solid cylinder?

1.34 A commercially-pure copper wire originally 10.00 m long is pulled until its final length is 10.10 m. It is annealed, then pulled again to a final length of 10.20 m. What is the engineering strain associated with each of the two steps in the process? What is the true strain for each step? What are the total engineering and true strains for the combined steps? Finally, what agreement (if any) is there between the total strains calculated as the sum of two steps of 0.10 m vs. a single step of 0.20 m?

1.35 A 3-mm-long gold alloy wire intended to electrically bond a computer chip to its package has an initial diameter of 30 μm. During testing, it is pulled axially with a load of 15 grams-force. If the wire diameter decreases uniformly to 29 μm, compute the following:

    a. The final length of the wire.

    b. The true stress and true strain at this load.

    c. The engineering stress and strain at this load.

1.36 A cylindrical rod of Ni 200 alloy has the following properties: $E = 204$ GPa, $v = 0.31$. It is loaded *elastically* in compression at 12.5 kN. If the original rod length and diameter are 20 mm and 15 mm, respectively, determine the rod length and diameter under load.

1.37 A 0.5-m-long rod of annealed 410 stainless steel was loaded to failure in tension. The rod originally had a square cross section measuring 1.25 cm on a side. What was the load necessary to break the sample? If 85% of the total elongation occurred prior to the onset of localized deformation, compute the true stress at the point of incipient necking.

1.38 Natural rubber is tested in tension to a maximum extension ratio of $\lambda = 3$. The Mooney-Rivlin constants for this material are found to be $C_1 = 0.069$ MPa and $C_2 = 0.125$ MPa. Plot the corresponding uniaxial stress vs. extension ratio behavior over the tested range. Derive an expression for the slope of the function, then determine the secant and tangent moduli at 100% strain.

1.39 Compare the resilience of alloy Ti-6Al-4V (annealed) with that of stainless steel alloy 304 (annealed). Assume for the purpose of comparison that the elastic properties of these alloys are very similar to those of pure Ti and Fe, respectively. Then compare the elastic strain energy density just prior to the onset of necking for both alloys.

1.40 A cylindrical elastomeric rope is used to make a slingshot. The diameter is 15 mm and the original length is 1 m. It is stretched to twice its original length ($\lambda = 2$), then released. The behavior is fully elastic and not time-dependent over the time span of the slingshot's use. The stress-extension ratio behavior is shown in the plot below.

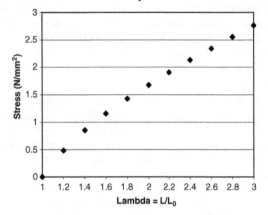

a. If the first two data points were used to calculate an initial linear elastic Young's modulus E_0, what would that value be? Answer in GPa or MPa units.

b. Based on the plot above, what is the diameter of the rope at $\lambda = 2$? Noting that this is a large strain, state and justify any assumption you must make to answer this question.

c. For a rubber material, one possible nonlinear relationship relating stress and extension ratio is given by Eq 1-13. Assume that this is a reasonable expression for the behavior depicted above, and calculate the expected stored energy density at $\lambda = 2$. Be sure to report units.

1.41   A rectangular plate 125 mm long, 10 mm wide, and 3 mm thick is formed from fused silica (aka fused quartz). It is tested in 3-point bending until it fails with a modulus of rupture of 110 MPa at a load of 66 N. Assume the central load point is on the top of the beam.

a. How far apart must have been the lower supports?

b. What was the maximum stress (magnitude and sign) on the top side of the beam halfway between the central load point and the left-hand lower load point?

c. What was the stress (magnitude and sign) directly beneath the central load point exactly 1.5 mm from the top surface?

d. If the same plate was tested in pure tension, would the stress at failure probably be higher or lower than measured by 3-point bending? Why?

1.42   A disk of SBR elastomer 3.0 cm in diameter × 0.5 cm thick is used as a cushioning surface between two steel rods of the same diameter, as shown below (not to scale).

	G	E	ν	SBR/steel interface shear strength	CTE
SBR	3.4 MPa	10 MPa	0.49	2 MPa	$220 \times 10^{-6}\,°C^{-1}$

a. If the rods are brought together with an axial force of 100 N such that the SBR is compressed elastically between them, what is the thickness of the SBR under load?

b. Under the same conditions as part (a), what is the greatest possible diameter of the SBR under load?

c. If the SBR is bonded to the rods, how far can one rod be rotated with respect to the other before the SBR/rod interface fractures? Assume that the rods are essentially rigid and the distance between them remains constant. Please answer in degrees of rotation.

1.43   A solid cylindrical rod 12 mm in diameter and 50 mm in length is attached to a rigid support at one end and twisted at its free end by 14°. If the Poisson's ratio for this isotropic material is 0.34 and Young's modulus is 70 GPa, what is the maximum shear stress induced?

1.44   Spinel ($MgAl_2O_4$) "optical ceramic" is a transparent *polycrystalline* ceramic with a combination of high hardness, light weight, and optical properties that make it very attractive for fracture-resistant windows (e.g., as armor or in a future manned space vehicle). It has the mechanical properties listed below.

	G	E	ν	Flexure strength	CTE
Spinel	192 GPa	277 GPa	0.26	200 GPa	$7 \times 10^{-6}\,°C^{-1}$

a. A rectangular plate 1 mm × 10 mm × 100 mm is mounted for use as a protective window over a sensor. In the course of mounting, a compressive load of 2.5 kN is exerted along the long (100-mm) axis. There is no constraint along the other two axes. What is the *stress* along the long axis?

b. What is the *strain* along the long axis?

c. What is the *strain* along the *width* (10-mm) axis?

d. What is the *strain* along the *thickness* (1-mm) axis?

e. Now the plate is rigidly constrained along its width (the 10-mm axis). This has the consequence that the plate cannot change length along that axis, although it is still free to change thickness dimension. The same 2.5 kN load is exerted along the long axis. Now what is the *strain* along the *long* axis?

1.45 Compute the moduli of elasticity for nickel and 3C silicon carbide single crystals in the <100>, <110>, and <111> directions. Compare these values with Young's modulus values reported for polycrystalline samples of Ni and β-SiC (204 GPa and 410 GPa, respectively). Then calculate the relative degree of anisotropy for both materials, and compare it to that of aluminum, spinel, and copper.

1.46 Assume the following *elastic* loading exists on a block of copper:

$$\sigma_X = 325\,\text{MPa}, \ \sigma_Y = 80\,\text{MPa}, \text{ and } \tau_{XY} = 40\,\text{MPa}$$

Calculate ε_X and ε_Z for this block, assuming

a. that it is a random polycrystalline material.

b. that it is a single crystal with the tensile and shear axes lining up along unit cell axes.

c. Explain why the relative strain values you calculated along the X axis make sense for the two cases, based on the elastic anisotropy of copper.

1.47 A weight lifter holds 300 pounds over his head, supporting the bar with both arms vertical.

a. What is the stress in each humerus (upper arm bone), assuming that it can be approximated as a solid cylindrical rod with cross-sectional area of 1.05 in^2? Use SI units.

b. What are the corresponding axial and radial strains?

c. If the humerus is 9 inches long, what are the length and diameter changes associated with this massive load? Please give this answer in inches.

1.48 A thin-walled pressure vessel is subjected to internal pressure such that a hoop stress of 100 MPa develops. Imagine that the vessel is made of an orthotropic continuous fiber composite with most of the fibers running around the circumference. The elastic constants for this material are given below, with direction 3 around the circumference, direction 2 along the length, and direction 1 through the thickness. What is the *strain* in the hoop direction?

	11	22	33	44	55	66	12	13	23
S (GPa^{-1})	0.083	0.075	0.05	0.161	0.178	0.221	0.031	0.019	0.018
C (GPa)	18.0	20.2	27.6	6.23	5.61	4.52	9.98	10.1	10.7

1.49 The mechanical properties of cobalt (Co) may be improved by incorporating fine particles of tungsten carbide (WC). Given that the moduli of elasticity of these materials are, respectively, 200 GPa and 700 GPa, plot modulus of elasticity vs. the volume percent of WC in Co from 0 to 100 vol% using both upper- and lower-bound expressions to form a performance envelope into which the material will fall. Please do this using plotting software, not by hand.

1.50 MgF$_2$ has the right refractive index to serve as an antireflective coating on fracture-resistant spinel windows (see the problem above). Assume that the MgF$_2$ can be deposited as a polycrystalline thin film on thick spinel. MgF$_2$ mechanical properties are listed below.

	G	E	ν	CTE
MgF$_2$	54.5 GPa	138.5 GPa	0.27	$10 \times 10^{-6}\,°\text{C}^{-1}$

a. If a thin coating of MgF$_2$ is deposited on a thick polycrystalline spinel substrate at a temperature of 200°C and then the film and substrate are cooled to 20°C, what is the *stress state* of the thin film? Use words like equal/unequal, uni/bi/triaxial, and tension/compression. Name the state and provide a supporting sketch.

b. What are the thermal *strains* induced in the MgF$_2$ film under the conditions from part (a)? Please give numerical answers for directions X, Y, and Z, where Z is the direction normal to the film surface.

c. What is the thermal *stress* induced? Please give numerical answers for directions X, Y, and Z.

d. If the MgF$_2$ were replaced by a fluoropolymer antireflective coating (like PTFE) deposited at the same temperature, would you expect the thermal *strain* in the film to be larger or smaller? Why?

Design

1.51 A solar panel is to be mounted at the top of a cylindrical post that is rigidly attached to the ground at its bottom, and that is protected from extreme bending by four guy wires strung from the top of the post to the ground. The post will be made of recycled polyethylene terephthalate (PET), which has an elastic modulus of approximately 3.5 GPa and a Poisson's ratio of 0.43. If the solar panel weighs 14.8 kg and the post must be 8 m tall to lift the panel above surrounding obstacles, what is the minimum post radius needed to avoid failure by buckling? Is this post diameter actually likely to be a safe design choice? Based only on the required post radius, what is your opinion about the choice of PET for this application?

1.52 You are in charge of designing a new fixture for a "universal testing machine" that will attach a tensile specimen to the machine using a clevis—a U-shaped piece with holes drilled through the two arms—and a cylindrical pin that passes through the clevis and the specimen. If the maximum load exerted by the machine is 30 kN and the pin is to be made of some sort of steel, what is the minimum pin diameter needed to ensure that the shear stress in the pin does not exceed 600 MPa? Assume that the steel has similar elastic properties to pure Fe.

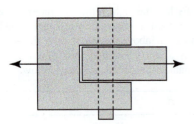

1.53 A 20-cm-diameter pipe is used to carry a pressure of 8.274 MPa without yielding. Assuming a safety factor of 2×, compute:

    a. the lighter

    b. and the less expensive pipe per unit length based on the following two possible material choices.

	Copper C71500	PVC
E (GPa)	150	3.25
σ_{ys} (MPa)	540	43
ρ (g/cm^3)	8.94	1.45
Cost (US$/kg)	27.00	1.75

1.54 A particular cylindrical rod will be subjected to axial cyclic compressive loads. It is designed to fit snugly through a hole in a separate plate, but it must not exert excessive pressure on the surrounding material while under load or a fatigue crack may develop in the plate. The diameter of the rod (and the hole) is 10 mm. The maximum compressive load the rod will experience is 24 kN. If the rod either yields plastically or increases in diameter by more than 0.008 mm, the design will not meet the specifications. Which of the four alloys listed below will satisfy these requirements at the lowest cost?

	E (GPa)	ν	σ_{ys} (MPa)	σ_{ts} (MPa)	US$/ kg	ρ (g/cm^3)
1020 alloy steel, normalized	207	0.30	340	440	1.35	7.85
304 stainless steel, cold worked	193	0.30	510	865	8.50	8.00
Al 6061-T6	69	0.33	275	310	7.75	2.70
Ti-6Al-4V, solution & aged	114	0.34	1100	1170	125.00	4.43

1.55 A 6061-T4 aluminum alloy is to be used to make a thin-walled cylindrical canister in which high-pressure chemical reactions will be performed. The design calls for a diameter of 50 cm, a length of 80 cm, and a maximum operating pressure of 5 MPa. Assume a safety factor of four is required (i.e., the maximum stress can never exceed one-quarter of the alloy's yield strength).

    a. What wall thickness is required to ensure safe operation?

    b. Is this wall thickness a maximum or a minimum? Explain.

    c. How do your answers change if the cylinder is made twice as long?

1.56 Imagine that you are designing a single crystal turbine blade for use in a jet engine. It will experience large tensile loads from the centripetal forces that exist during use. Minimizing the axial strain will allow for tighter gap tolerances between the turbine blade tips and the surrounding shroud; this leads to greater engine efficiency. You are restricted to using a Ni-based superalloy.

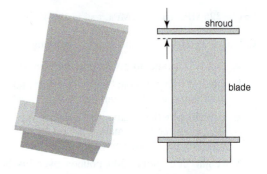

    a. Without performing any calculations, determine which orientation (<100>, <111> or <110>) you would choose along the tensile axis of the blade in order to minimize the strain during use? Why?

    b. Justify your choice by calculating the Young's modulus for each orientation and then calculating the corresponding strain at maximum load. For this problem, assume that the Ni-based superalloy in question has the same elastic behavior as pure Ni. Also assume that the blade experiences a maximum load of 10,000 lbs-force, and that the behavior is elastic. Consider only a simple uniaxial tensile load. Approximate the turbine blade airfoil cross section as an isosceles triangle 5 mm at its base by 50 mm tall. The blade length is 150 mm.

    c. Calculate the thermal strain imposed on a turbine blade after it has been heated from 25 °C to the engine operating temperature of 1100 °C. Assume a coefficient of thermal expansion of 13.5×10^{-6} C^{-1} for this particular superalloy.

    d. Calculate the best-case minimum gap size for a cold (25 °C) engine with no turbine rotation such that the blade will just barely touch the surrounding shroud material when the engine is operating at full rotation and maximum temperature.

Extend

1.57 Write a 1–2 page review of auxetic materials. Assume that you are writing a supplementary article for an introductory engineering text. Be sure to (1) define the term "auxetic material" and (2) explain what is unusual about the mechanical behavior of this class of materials. Include (3) a picture (sketch, diagram, or photograph) of an auxetic material. Also (4) describe at least two products that could (or do) benefit from the auxetic behavior. Provide full references for all of your information.

1.58 Select two thermoplastic materials from among those listed in Section 1.3.3.1. Using any resources available to you, determine a typical glass transition temperature, degree of crystallinity, and a common use for each of the polymer materials you selected. How does the use reflect the T_g value and the degree of crystallinity for each material?

1.59 Write a 1–2 page review of the *structure* and *elastic behavior* of natural, highly elastic materials. Assume that you are writing a supplementary article for an introductory engineering text. Choose two or more materials for comparison: dragline spider silk, non-dragline spider silk, collagen, elastin, mussel byssal threads, and resilin. In your review, be sure to (1) identify the natural use for each of the materials you selected, and (2) explain how the particular properties of the materials match their intended uses in nature. Mention (3) approximately how much of the behavior is purely elastic

(instantaneous recovery with no energy loss) and how much is viscoelastic (time-dependent recovery with some energy loss). Include (4) a picture (sketch, diagram, or photograph) or a plot that adds to the reader's understanding of the topic. Strength is also interesting and certainly worth mentioning, but is not the main focus of this paper. If you can find a case in which there has been an attempt to synthesize the material (s) for engineering purposes it would add much to this short article. Provide full references for all of your information.

1.60 Search published science and engineering literature to find an example of an engineered material used for bone replacement (partial or total). How well does the elastic behavior of the material match that of natural bone? Provide elastic property data from the source, a brief explanation of the potential advantages of this particular material, and a full reference for the source.

1.61 Search published science and engineering literature to find an example of a microelectromechanical device in which thermal mismatch strain is used to generate motion and/or force. Provide a figure from the source, a brief explanation of the device's purpose and design, and a full reference for the source.

Chapter 2

Yielding and Plastic Flow

In Chapter 1, the elastic limit of a given material was defined as the stress level above which strains are irreversible. Beyond this point, we may say that a ductile material has *yielded* and is undergoing *plastic deformation*. The objective of this chapter is to consider the manner by which permanent deformations are generated and to estimate the magnitude of stresses necessary for such movement.

2.1 DISLOCATIONS IN METALS AND CERAMICS

2.1.1 Strength of a Perfect Crystal

To begin our discussion of plasticity, we consider atom movements in a crystalline metal or ceramic. If we apply a *uniform* tensile or compressive force on a crystal in all directions, the atoms will move apart or move together, the bonds will lengthen or compress elastically (as described in Chapter 1), and a temporary change in the crystal dimensions will occur. When the hydrostatic force is removed, however, the crystal will return to its original dimensions; clearly no plastic deformation has been accomplished. If, however, the atoms move under an applied shearing force, something quite different can happen. Atom movements parallel to a particular crystallographic plane can lead to the displacement of the upper half of a cube relative to the bottom (Fig. 2.1). When the atoms in the upper half move along this *slip plane* by exactly one atomic spacing, the crystal will look perfect everywhere except at the ends. At this point the force can be removed and the displacement will be permanent. This will leave the crystal with a new external shape, and we can say that plastic deformation has occurred.

If atom A in Fig. 2.1 is to move to position B, atom B to position C, etc., a *simultaneous* translation of all atoms on the slip plane must occur. Since this would involve simultaneous rupture of all the interatomic bonds acting across the slip plane (e.g., bonds A–A′, B–B′, C–C′, etc.), the necessary stress would have to be very large. From Fig. 2.2a we see that the equilibrium atom positions within the crystalline lattice are located at P and R, with a separation of b units. Midway between P and R the energy is a maximum at Q, which represents a metastable equilibrium position. The exact shape of the energy curve shown in Fig. 2.2a depends on the nature of the interatomic bonds. Since this is not known precisely, a sinusoidal waveform is assumed for simplicity in this analysis. Locating the atoms on the slip plane anywhere other than at an equilibrium position, such as P and R, requires a force defined by the slope of the energy curve (Fig. 2.2b) at that position where $F \equiv -d\mathscr{E}/dx$. For example, to place

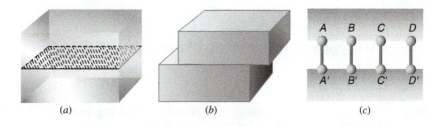

(a)   (b)   (c)

Figure 2.1 Movement of a solid cube along a particular slip plane. (*a*) Undeformed cube with anticipated slip plane; (*b*) slipped cube revealing relative translation of part of cube; (*c*) atom position showing bonds across slip plane (A–A′, B–B′, C–C′ D–D′).

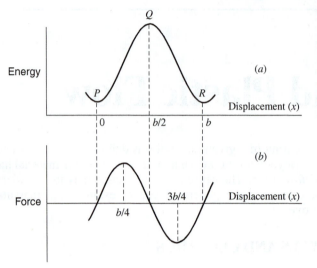

Figure 2.2 The periodic nature of a lattice. (*a*) Variation of energy with atom position in lattice. Preferred atom sites are at *P* and *R*, associated with minimum energy; (*b*) variation in force acting on atoms throughout lattice. Force is zero at equilibrium site positions and maximum at $b/4, 3b/4, 5b/4, \ldots, (2n-1)b/4$.

an atom between *P* and *Q* (that is, $0 < x < b/2$), a force acting to the right is required to counteract the tendency for the atom to move back to the equilibrium site at *P*. Note that between *P* and *Q*, the slope of the energy curve is everywhere positive so that the force is also positive from $0 \leq x \leq b/2$. When $b/2 < x < b$, the atom wants to slide into its new equilibrium position at *R* (that is, the energy decreases continually from *Q* to *R*). To prevent this, a force acting to the left is needed to keep the atom stationary at some location between *Q* and *R*. This force is in an opposite direction to that needed between *P* and *Q* and is, therefore, negative. Note that in the region between *Q* and *R*, the slope of the energy curve is negative. Therefore, the corresponding portion of the force curve in this same region must also be negative.

From the above discussion it is clear that the shear stress necessary to move the atoms on the slip plane varies periodically from zero at *P*, *Q*, and *R* to a maximum value at $b/4$ and $3b/4$. Therefore, the shear stress may be expressed in the following form (from an analysis due to Frenkel[1]), based on an assumed sinusoidal variation in energy throughout the lattice:

$$\tau = \tau_m \sin \frac{2\pi x}{b} \tag{2-1}$$

where

$$\tau = \text{applied shear stress}$$
$$\tau_m = \text{maximum theoretical strength of crystal}$$
$$x = \text{distance atoms are moved}$$
$$b = \text{distance between equilibrium positions}$$

Plastic flow (that is, irreversible deformation) will then occur when the upper part of the cube (Fig. 2.1) is translated a distance greater than $b/4$ because of an applied shear stress τ_m.

For elastic strains, the shear stress may also be defined by Hooke's law

$$\tau = G\gamma \tag{2-2}$$

and the shear strain may be approximated for small values by

$$\gamma \approx \frac{x}{a} \tag{2-3}$$

where $a = $ distance between slip planes. Combining Eqs. 2-1, 2-2, and 2-3 with $\sin(2\pi x/b)$ approximated by $2\pi x/b$ for small strains gives

$$G\frac{x}{a} \approx \tau_m \frac{2\pi x}{b} \tag{2-4}$$

Table 2.1   Theoretical and Experimental Yield Strengths in Various Materials[2]

Material	$G/2\pi$		Experimental Yield Strength		
	GPa	10^6 psi	MPa	psi	τ_m/τ_{exp}
Silver	4.6	0.67	0.37	55	$\sim 1 \times 10^4$
Aluminum	4.2	0.61	0.78	115	$\sim 5 \times 10^3$
Copper	7.2	1.05	0.49	70	$\sim 1 \times 10^4$
Nickel	12.2	1.78	3.2–7.35	465–1,065	$\sim 4 \times 10^3$
Iron	13.2	1.91	27.5	3,990	$\sim 5 \times 10^2$
Molybdenum	19	2.76	71.6	10,385	$\sim 3 \times 10^2$
Niobium	5.8	0.84	33.3	4,830	$\sim 2 \times 10^2$
Cadmium	3.8	0.56	0.57	85	$\sim 7 \times 10^3$
Magnesium (basal slip)	2.8	0.4	0.39	55	$\sim 7 \times 10^3$
Magnesium (prism slip)	2.8	0.4	39.2	5,685	$\sim 7 \times 10^1$
Titanium (prism slip)	6.3	0.92	13.7	1,985	$\sim 5 \times 10^2$
Beryllium (basal slip)	23.4	3.39	1.37	200	$\sim 2 \times 10^4$
Beryllium (prism slip)	23.4	3.39	52	7,540	$\sim 5 \times 10^2$

Upon rearranging,

$$\tau_m \approx \frac{Gb}{2\pi a} \tag{2-5}$$

For most crystals b is of the same order as a, so Eq. 2-5 may be rewritten in the form

$$\tau_m \approx \frac{G}{2\pi} \tag{2-6}$$

Because of the approximations made in this analysis, especially with regard to the form of the energy-displacement curve, the magnitude of the theoretical shear strength τ_m from Eq. 2-6 is of an approximate nature. More realistic estimates place τ_m in the range of $G/30$. Nevertheless, it is instructive to compare theoretical strength values calculated with Eq. 2-6 with experimentally determined shear strengths for single crystals of various materials. From Table 2.1, it is immediately obvious that very large discrepancies exist between theoretical and experimental values for all materials tabulated. Without question, the lack of precision regarding computations based on Eq. 2-6 is *not* responsible for these large errors. Rather, the discrepancies must be accounted for in a different manner.

2.1.2 The Need for Lattice Imperfections: Dislocations

In 1934 Taylor, Orowan, and Polanyi postulated independently the existence of a lattice defect that would allow the cube in Fig. 2.1 to slip at much lower stress levels[3,4]. By introducing an extra half plane of atoms into the lattice (Fig. 2.3), they were able to show that atom bond breakage on the slip plane could be restricted to the immediate vicinity of the bottom edge of the half plane. This linear defect where only one row of bonds is broken is called the *dislocation line*. As the dislocation line moves through the crystal, bond breakage across the slip plane occurs *consecutively* rather than *simultaneously* as was necessary in the perfect lattice (Fig. 2.1). The consecutive nature of bond breakage is shown in Fig. 2.4 where the dislocation line (and its associated extra half plane of atoms) is shown at different locations during its movement through the crystal. When the outer set of atoms becomes the extra half plane, the dislocation emerges from the crystal and no longer exists. The end result of the movement of this defect is the same as shown in Fig. 2.1—the upper half of the cube has been translated

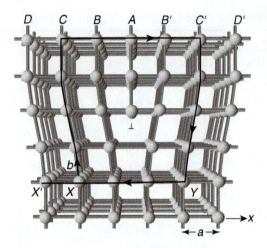

Figure 2.3 Lattice defect caused by introduction of an extra half plane of atoms, A. Note symmetrical displacement of planes B, B', C, C', etc. The dislocation line is defined as the edge of the half plane, A. The Burgers circuit XCC' YX' contains a closure failure $X'X$. (From Guy,[5] *Elements of Physical Metallurgy*, 2nd ed., Addison-Wesley, Reading, MA, 1959.)

relative to the bottom half by an amount equal to the distance between equilibrium atomic positions; this displacement is a characteristic of the dislocation designated as **b**, the *Burgers vector*. The major difference is the fact that it takes much less energy to break one bond at a time than all the bonds at once. This concept is analogous to moving a large floor rug across the room. If you have ever tried to grab the edge of the rug and pull it to a new position, you know

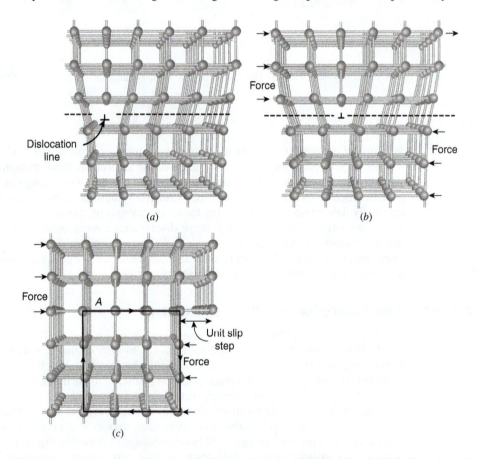

Figure 2.4 Successive positions of dislocation as it moves through crystal. Note that the final offset of crystal resulting from the passage of dislocation is the same as the simultaneous movement of the entire crystal. Also note the perfect Burgers circuit in (*c*) demonstrating that the crystal is perfect after the dislocation has passed through. (From Guy,[5] *Elements of Physical Metallurgy,* 2nd ed., Addison-Wesley, Reading, MA, 1959.)

Table 2.2   Theoretical and Experimental Strengths of Dislocation-Free Crystal (Whiskers)[7]

Material	Theoretical Strength ($G/2\pi$)		Experimental Strength		
	GPa	10^6 psi	GPa	10^6 psi	Error
Copper	19.1	2.77	3.0	0.44	~6
Nickel	33.4	4.84	3.9	0.57	~8.5
Iron	31.8	4.61	13	1.89	~2.5
B_4C	71.6	10.4	6.7	0.98	~10.5
SiC	132.1	19.2	11	1.60	~12
Al_2O_3	65.3	9.47	19	2.76	~3.5
C	156.0	22.6	21	3.05	~7

that it is nearly impossible to move a rug in this manner. In this case, the "theoretical shear stress" necessary to move the rug is strongly dependent on the frictional forces between the rug and the floor. If you persisted in your task you probably discovered that the rug could be moved quite easily in several stages by first creating a series of buckles at the edge of the rug and then propagating them, one at a time, across the rug by shuffling your feet behind each buckle. In this way you were able to move the rug by increments equal to the size of the buckle. Since the only part of the rug to move at any given time was the buckled segment, there was no need to overcome the frictional forces acting on the whole rug. Since the lattice dislocation is a similar work-saving "device," one may reconcile the large errors between theoretical and experimental yield strengths (Table 2.1) by assuming the presence of dislocations in the crystals that were examined.

Before we begin to deal with more detailed behavior of dislocations, it is natural to wonder whether the analysis leading to Eq. 2-6 is correct after all. What is needed, of course, are test data for crystals possessing *no* dislocations. Fortunately, such perfect crystals—produced in the form of fine wires called whiskers—have been prepared in the laboratory. The tensile strengths of these extraordinary crystals, shown in Table 2.2, are seen to be in close agreement with theoretical maximum values computed from Eq. 2-6. More recently, this result has been verified with compression tests conducted on ~500-nm-diameter molybdenum pillars that were prepared to be either dislocation-free or to contain pre-existing dislocations. The dislocation-free pillars yielded suddenly at 9.3 GPa, which corresponds to a critical shear stress of ~4.6 GPa (~1/25 of the shear modulus, within the expected range), and collapsed. In contrast, those pillars with preexisting dislocations yielded gradually at ~1.0 GPa.[6] Data and images are shown in Fig. 2.5. On this basis, the Frenkel analysis is verified.

2.1.3   Observation of Dislocations

Long after it was initially determined that dislocations must exist, laboratory techniques were developed to enable the direct examination of dislocations within a crystal. One successful technique involves chemical or electrolytic etching of polished free surfaces. By carefully controlling the strength of the etchant, the high-energy dislocation cores (i.e., the regions of greatest lattice distortion) exposed at the surface are attacked preferentially with respect to other regions on the polished surface. The result is the formation of numerous etch pits, each corresponding to one dislocation (Fig. 2.6). This technique has been used to study the effect of applied stress on dislocation velocity. Notable experiments by Johnston and Gilman[8] have identified the stress-induced, time-dependent change in dislocation position by repeated etching, with each new etch pit representing the new location of the moving dislocation. Note that this technique is extremely useful for establishing the relative arrangement and number (or density) of dislocations, but cannot elucidate the shape of the dislocations below the surface.

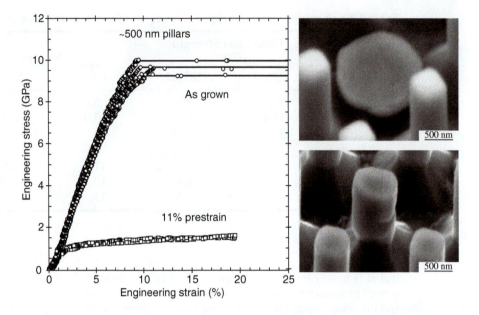

Figure 2.5   Engineering stress–strain data measured during compression of Mo micropillars either as-grown (initially dislocation-free) or prestrained (provided with preexisting dislocations). Accompanying images show the pillar shapes after yielding, demonstrating complete plastic collapse for an as-grown pillar (top), and moderate shape change for a prestrained pillar (bottom). (Data and images courtesy of H. Bei, S. Shim, G.M. Pharr and E.P. George, reprinted with permission.)

The most widely used technique for the study of dislocations involves their direct examination in the transmission electron microscope. Since electrons have little penetrating power, the specimens used for such studies are very thin films—only about 0.1 to 0.2 μm thick. Because dislocations are lattice defects, their presence perturbs the path of the diffracted electron beam relative to its path in a perfect crystal. As a result, various images are produced,

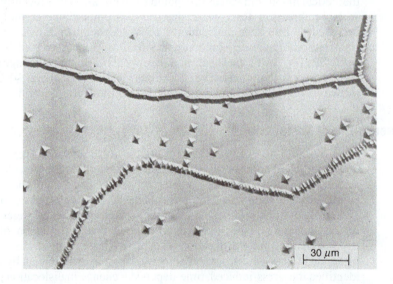

Figure 2.6   Etch pits on polished surface of lithium fluoride, each associated with an individual dislocation. The etch pit lineage indicates alignment of many dislocations in the form of low-angle boundaries (see Section 2.6). (From Gilman and Johnston[9]; reprinted by permission of General Electric Co.)

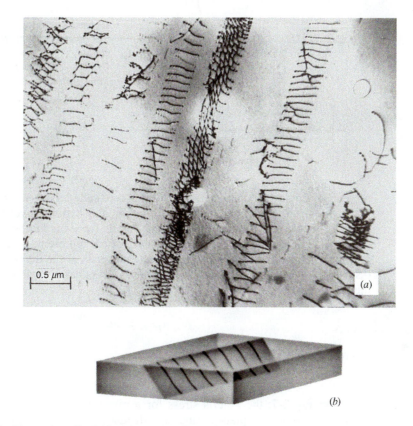

Figure 2.7   Observation of individual dislocations in thin foil. (*a*) Planar arrays of dislocations in 18Cr–8Ni stainless steels (from Michalak,[17] *Metals Handbook,* Vol. 8, copyright American Society for Metals, Metals Park, OH, 1973; used with permission); (*b*) diagram showing position of dislocations on the guide plane in the foil (after Hull[10]).

depending on the prevailing diffraction conditions. Often, dislocations appear as single dark lines like those shown in planar array in Fig. 2.7*a*. Each dislocation lies along a particular crystallographic plane and extends from the top to the bottom of the foil (Fig. 2.7*b*). As a result, the viewer sees only the projected length of the dislocation line, with the actual length being dependent on the foil thickness and angle of the plane containing the dislocations. Some important publications addressing interpretation of electron diffraction images and the identification of numerous dislocation configurations are cited at the end of the chapter.[10–16]

2.1.4   Lattice Resistance to Dislocation Movement: The Peierls Stress

From Fig. 2.3 it is clear that the "insertion" of the extra half plane of atoms has perturbed the lattice and caused atoms to be pushed aside laterally, particularly in the upper half of the crystal. For example, atoms along planes *B* and *C* are displaced to the left, while atoms in planes *B'* and *C'* are displaced to the right. Since the forces acting on these groups of atoms are equal and of opposite sign (that is, pairing atoms in plane *B* with those in *B'* and those in *C* with atoms in plane *C'*), movement of the extra half plane A either to the left or right would be met by self-balancing forces on the other atoms within the distorted region. On this basis, the force necessary to move a dislocation would be zero. However, Cottrell[4] pointed out that although the above situation should prevail when the dislocation occurs in a symmetrical position (such as the one shown for plane *A* in Fig. 2.3), it would not hold true when the dislocation passes through nonsymmetrical positions. Consequently, some force is necessary to move the dislocation through the lattice, even when no other impediments to dislocation motion are

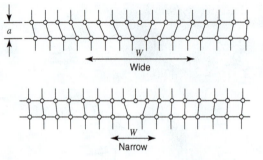

Figure 2.8 Characteristic width of an edge dislocation that affects the Peierls-Nabarro stress.[18] (Reproduced by courtesy of the Council of the Institution of Mechanical Engineers from *The Properties of Materials at High Rates of Strain* by A. H. Cottrell.)

present. An important characteristic of this force (called the *Peierls-Nabarro* or *Peierls force*) is that its magnitude varies periodically as the dislocation moves through the lattice.

It is known that the magnitude of the Peierls force (or *lattice friction*) depends to a large extent on (1) the width of the dislocation W, which represents a measure of the distance over which the lattice is distorted because of the presence of the dislocation (Fig. 2.8), and (2) the distance between similar planes a. The Peierls stress has been shown to depend on W and b in the form

$$\tau_{p-n} \propto Ge^{-2\pi W/b} \tag{2-7}$$

where

$$W = a/(1-v).$$

From Eq. 2-7, the Peierls stress for a given plane is seen to decrease with increasing distance between like planes. Since the distance between planes varies inversely with their atomic density, slip is preferred on closely packed planes. In addition, the Peierls stress depends on the dislocation width, which is dependent on atomic structure and the nature of the atomic bonding forces. For example, when the bonding forces are spherical in distribution and act along the line of centers between atoms, the dislocation width is large. Since this type of bonding is found in close-packed structures, it is seen that the Peierls stress in face-centered-cubic and close-packed hexagonal crystals is low. By contrast, when bonding forces are highly directional (as in the case of covalent, ionic, and body-centered-cubic crystals), the dislocation width is narrow and the Peierls stress correspondingly large. Although many attempts have been made to compute precisely the magnitude of the Peierls stress in a given lattice, considerable difficulties arise because the exact shape of the force–displacement curve is unknown. Foreman, Jaswon, and Wood[19] showed that when the amplitude of an assumed sinusoidal force–displacement law was reduced by half, the width of the dislocation increased fourfold. This, in turn, had the effect of reducing the computed Peierls stress value by more than *six orders of magnitude*. Until the force–displacement relation between atoms can be defined more precisely, the magnitude of the Peierls stress in crystals can be described only in qualitative terms. Nonetheless, even a qualitative description can be useful for understanding differences in plasticity behavior from one crystal structure to another.

2.1.4.1 Peierls Stress Temperature Sensitivity

One such qualitative characteristic of the Peierls stress relates to the temperature sensitivity of the yield strength. Since the Peierls stress depends on the short-range stress field of the dislocation core, it is sensitive to the thermal energy in the lattice and, hence, to the test temperature. At low temperatures, where thermal enhancement of dislocation motion is limited, the Peierls stress is larger than at higher temperatures. In crystals that have wide dislocations, however, the increase in Peierls stress with decreasing temperature is insignificant, since the Peierls stress is negligible to begin with.

Table 2.3   Relation Between Dislocation Width and Yield-Strength Temperature Sensitivity

Material	Crystal Type	Dislocation Width	Peierls Stress	Yield-Strength Temperature Sensitivity
Metal	FCC	Wide	Very small	Negligible
Metal	BCC	Narrow	Moderate	Strong
Ceramic	Ionic	Narrow	Large	Strong
Ceramic	Covalent	Very narrow	Very large	Strong

Accordingly, there is little yield strength-temperature dependence in FCC metals such as aluminum, copper, and austenitic stainless steel alloys. The situation is quite different in crystals that contain narrow dislocations. Although the Peierls stress in these materials may be small at elevated temperatures, it rises rapidly with decreasing temperature and represents a large component of the yield strength in the low-temperature regime. This can manifest itself as an increase in the elastic limit, an increase in hardness, and a decrease in ductility. The yield strength-temperature sensitivity of several engineering materials is shown in Table 2.3. The large Peierls stress in ceramic materials is partly responsible for their limited ductility at low and moderate temperatures. However, the Peierls stress decreases rapidly with increasing temperature, thereby enhancing plastic deformation processes in these materials at high temperatures.

2.1.4.2   Effect of Dislocation Orientation on Peierls Stress

The Peierls stress as described above represents an upper bound to the minimum stress necessary to move a dislocation through a crystal. In fact, dislocations will seldom lie completely along directions of lowest energy, or energy valleys, within the lattice. Rather, the dislocation line will contain bends or *kinks* that lie across energy peaks at some angle (Fig. 2.9a). The angle θ that the kink makes with the rest of the dislocation line, as well as its length l, is a direct consequence of the balance between two competing factors. On one hand, the dislocation will prefer to lie along the energy valleys such that the kink length is minimized and the kink angle maximized (90° (Fig. 2.9b)). (It should be noted that to create such a sharp kink angle will increase the energy of the dislocation, since the energy of any curved segment of a dislocation line increases with decreasing radius of curvature.) On the other hand, the dislocation line tries to be as short as

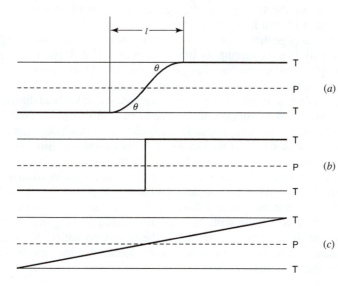

Figure 2.9   Position of dislocation line containing kinks with respect to energy troughs within lattice. (*a*) Typical configuration showing kink of length l with angle θ between kink segment and segment lying along energy trough; (*b*) sharp kink formed when magnitude of energy fluctuation in lattice is large. In this case $l \rightarrow 0$ and $\theta \rightarrow 90°$; (*c*) broad kink formed when energy fluctuation in lattice is small. Here $l \rightarrow \infty$ and θ becomes very small.

possible to minimize its self-energy and in the limit would prefer a straight-line configuration (Fig. 2.9c; see also Section 2.1.6). The degree to which the kinked dislocation line approaches either extreme will depend strongly on $\mathscr{E}$, the amplitude of the periodic energy change along the crystal. When this amplitude is large, the dislocation line will prefer to lie along energy troughs such that short sharp kinks will be formed. Alternatively, when $\Delta\mathscr{E}$ is small, long undulating kinks (more like gradual bends) will be observed.

The relative ease of movement of both the dislocation line segments lying along energy troughs and the kinked sections, respectively, is now considered. Since the kinked sections are located across higher energy portions of the crystal, they can move more easily than the line segments along the energy troughs, which must overcome the maximum energy barrier if they are to move. Upon application of a shear stress, the kinked segments shown in Fig. 2.9a move to the left or right (depending on the sense of the stress), which in effect allows the entire dislocation line to move in a perpendicular direction from one energy trough to the adjacent one. The lateral movement of such a kink may be likened to the motion of a whip that has been snapped. Consequently, the introduction of kinks into dislocations eases their movement through the lattice.

It may be concluded, then, that the lattice resistance to the movement of a dislocation depends on both the magnitude of the Peierls stress and the orientation of the dislocation line within the periodically varying energy field in the lattice. Since both factors will depend on $\Delta\mathscr{E}$, which depends on the force–displacement relation between atoms, the importance of the latter relation is emphasized. Unfortunately, current lack of specific knowledge concerning the force law severely hampers quantitative treatments of dislocation–lattice interactions, and we must be content for the moment with a qualitative understanding of structure-moderated trends in strength.

2.1.5 Characteristics of Dislocations

At this point, the reader should examine certain fundamental characteristics of dislocations that affect their motion in ways different from the Peierls stress. A dislocation is a lattice *line* defect that defines the boundary between slipped and unslipped portions of the crystal. Two basically different dislocations can be identified. The *edge* dislocation is defined by the edge of the extra half plane of atoms shown in Fig. 2.3. Note how this extra half plane is wedged into the top half of the crystal. As a result, the upper part of the crystal is compressed on either side of the half plane, while the region below the dislocation experiences considerable dilatation. By convention, the bottom edge of the half plane shown in Fig. 2.3 is defined as a *positive* edge dislocation. Had the extra half plane been introduced into the lower half of the crystal, the regions of localized compression and dilatation would be reversed and the dislocation line defined as a *negative* edge dislocation. Clearly, if a crystal contained both positive and negative edge dislocations lying on the same plane, their combination would result in mutual annihilation and the elimination of two high-energy regions of lattice distortion.

The movement of an edge dislocation and its role in the plastic deformation process may be understood more clearly by considering its Burgers circuit. (The Burgers circuit is a series of atom-to-atom steps along lattice vectors that generate a closed loop about any location in the lattice.) In a perfect lattice (Fig. 2.4c), the Burgers circuit beginning at A and progressing an equal and opposite number of lattice vectors in the horizontal and vertical directions will return to its starting position. When this occurs, the lattice contained within the circuit is considered perfect. When an edge dislocation is present in the lattice, the circuit does not close (Fig. 2.3). The vector needed to close the Burgers circuit ($X'X$) is called the Burgers vector **b** of the dislocation and represents both the magnitude and direction of slip of the dislocation.

Another important feature of **b** is its orientation relative to the dislocation line. For the edge dislocation, **b** is oriented normal to the line defect. Ordinarily, plastic flow via edge dislocation movement is restricted to that one plane defined by the dislocation line and its Burgers vector. Such *conservative* motion will occur with the edge dislocation moving in the same direction as **b** (i.e., the direction of slip). From Section 2.1.4, the planes on which dislocations move are usually those of greatest separation and atomic density. Recall that the Peierls-Nabarro stress is generally lowest on these planes.

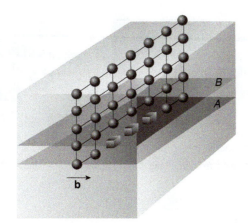

Figure 2.10   Dislocation climb involving vacancy (□) diffusion to an edge dislocation allowing its movement to climb from slip plane *A* to slip plane *B*. Note that the only atom (●) positions depicted are those in the extra half plane above the dislocation line.

It is possible for an edge dislocation to undertake *nonconservative* motion, that is, movement out of its normal glide plane. This can occur by removal of a row of atoms, such as by the diffusion of lattice vacancies to the bottom of the extra half plane (Fig. 2.10). In this manner, the dislocation *climbs* from one plane to another where conservative glide may occur once again. Since vacancy diffusion is a thermally activated process, dislocation climb becomes an important process only at elevated temperatures above about one-half the melting point of the material. This mechanism will be discussed again in Chapter 4.

The other line defect, called the *screw* dislocation, is defined by the line *AB* in Fig. 2.11, the latter being generated by displacement of one part of the crystal relative to the other. The Burgers circuit about the screw dislocation assumes the shape of a helix, very much like a spiral staircase, wherein a 360° rotation produces a translation equal to one lattice vector in a direction parallel to the dislocation line *AB*. A right-handed screw dislocation is defined when a clockwise 360° rotation causes the helix to advance one lattice vector. The same advance resulting from a 360° counterclockwise rotation is a left-handed screw dislocation. Thus, the screw dislocation and its Burgers vector are mutually parallel, unlike the orthogonal relationship found for the edge dislocation. Note that while the slip direction is again parallel to **b** as was found for the edge dislocation, the direction of movement of the screw dislocation is perpendicular to **b**. To better visualize this fact, take a piece of paper and tear it partway across its width. Note that the movement of your hands (the shear direction parallel to the Burgers vector) is perpendicular to the movement of the terminal point (the screw dislocation) of the tear.

Unlike the edge dislocation, a unique slip plane cannot be identified for the screw dislocation. Rather, an infinite number of potential slip planes may be defined, since the dislocation line and Burgers vector are parallel to one another. In fact, the movement of a screw dislocation is confined to those sets of planes that possess a low Peierls-Nabarro stress. Even so, the screw dislocation possesses greater mobility than the edge dislocation in moving through the lattice.

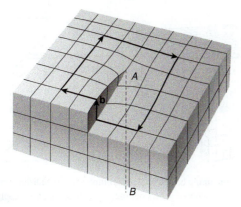

Figure 2.11   Screw dislocation *AB* resulting from displacement of one part of crystal relative to the other. Note that *AB* is parallel to **b**.

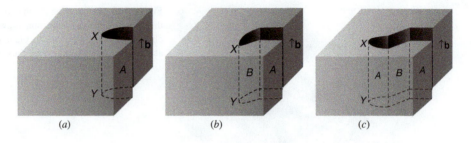

Figure 2.12 Cross-slip of a screw dislocation *XY* from (*a*) plane *A* to (*b*) plane *B* to (*c*) plane *A*. Slip always occurs in direction of Burgers vector **b**.

Table 2.4 Characteristics of Dislocations

	Type of Dislocation	
Dislocation Characteristic	**Edge**	**Screw**
Slip direction	Parallel to **b**	Parallel to **b**
Relation between dislocation line and **b**	Perpendicular	Parallel
Direction of dislocation line movement relative to **b**	Parallel	Perpendicular
Process by which dislocations may leave the glide plane	Nonconservative climb	Cross-slip

The movement of the screw dislocation from one slip plane to another takes place by a process known as *cross-slip* and may be understood by examining Fig. 2.12. At the onset of plastic deformation, the screw dislocation *XY* is seen to be moving on plane *A* (Fig. 2.12*a*). If continued movement on this plane is impeded by some obstacle, such as a precipitate particle, the screw dislocation can cross over to another equivalent plane, such as *B*, and continue its movement (Fig. 2.12*b*). Since the Burgers vector is unchanged, slip continues to occur in the same direction, though on a different plane. Movement of the screw dislocation may continue on plane *B* or return to plane *A* by a second cross-slip process (Fig. 2.12*c*). A summary of the basic differences between edge and screw dislocations is presented in Table 2.4.

Since many dislocations in a crystalline solid are curved like the one shown in Fig. 2.13*a*, they take on aspects of both edge and screw dislocations. With **b** the same along the entire length of the dislocation, the dislocation is seen to be pure screw at *A* and pure edge at *B*. The reader should verify this by constructing a Burgers circuit around the dislocation at *A* and *B*. It follows that at all points between *A* and *B* the dislocation possesses both edge and screw components. For this reason, *AB* is called a *mixed dislocation*. Another example of a mixed

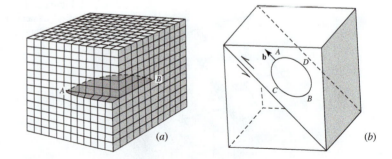

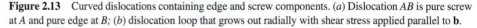

Figure 2.13 Curved dislocations containing edge and screw components. (*a*) Dislocation *AB* is pure screw at *A* and pure edge at *B*; (*b*) dislocation loop that grows out radially with shear stress applied parallel to **b**.

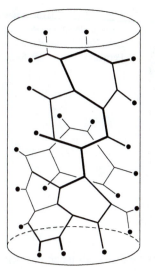

Figure 2.14   Network arrangement of dislocations in crystal. Dislocations can terminate only at a node, in a loop, or at a grain boundary or free surface.[18] (Reproduced by courtesy of the Council of the Institution of Mechanical Engineers from *The Properties of Materials at High Rates of Strain,* by A. H. Cottrell.)

dislocation is the *dislocation loop*. From Fig. 2.13*b* the loop is seen to be pure positive edge at *A* and pure negative edge at *B* while being pure right-handed screw at *D* and pure left-handed screw at *C*. Everywhere else the loop contains both edge and screw components. When a shear stress is applied parallel to **b**, we see from Table 2.4 that the loop will expand radially.

Dislocations can terminate at a free surface or at a grain boundary but never within the crystal. Consequently, dislocations either must form closed loops or networks with branches that terminate at the surface (Fig. 2.14). A basic characteristic of a network junction point or node involving at least three dislocation branches is that the sum of the Burgers vectors is zero:

$$\mathbf{b}_1 + \mathbf{b}_2 + \mathbf{b}_3 = 0 \tag{2-8}$$

Furthermore, when these dislocations are of the same sense,

$$\mathbf{b}_1 = \mathbf{b}_2 + \mathbf{b}_3 \tag{2-9}$$

and the Burgers vector of one dislocation is equal to the sum of the other two Burgers vectors. This holds when two dislocations combine to form a third or when one dislocation dissociates into two separate dislocations.

2.1.6 Elastic Properties of Dislocations

As might be expected, there is an elastic stress field associated with the distorted lattice surrounding a dislocation. It is easy to describe the stresses developed around a screw dislocation (Fig. 2.15*a*). By rolling the cylindrical element out flat (Fig. 2.15*b*), the shear strain $\gamma_{\theta z}$ is seen in polar coordinates to be

$$\gamma_{\theta z} = \frac{b}{2\pi r} \tag{2-10}$$

From Hooke's law, the corresponding stress is

$$\tau_{\theta z} = G\gamma_{\theta z} = \tau_{z\theta} = \frac{Gb}{2\pi r} \tag{2-11}$$

Since displacements are generated only in the *z* direction, the other stress components are zero. Equation 2-11 shows that the stress $\tau_{\theta z}$ becomes infinitely large as *r* approaches zero. Since this is

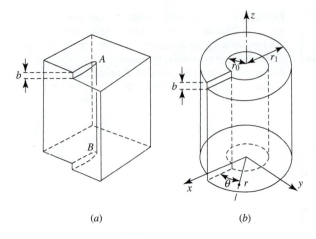

Figure 2.15   Elastic distortions surrounding screw dislocation. (Reprinted with permission from Hull,[10] *Introduction to Dislocations,* Pergamon Press, Elmsford, NY, 1965.)

(a)   (b)

unreasonable, there exists a limiting distance r_0 from the dislocation center (estimated to be 0.5 to 1 nm) within which Eq. 2-11 is no longer applicable. For the rectangular coordinates shown in Fig. 2.15*b*, the shear stresses surrounding the screw dislocation can also be given by

$$\tau_{xz} = \tau_{zx} = -\frac{Gb}{2\pi}\frac{y}{x^2 + y^2}$$
$$\tau_{yz} = \tau_{zy} = \frac{Gb}{2\pi}\frac{x}{x^2 + y^2}$$

$$(2\text{-}12)$$

Again all other stresses are zero.

The stress field surrounding an edge dislocation is more complicated, since both hydrostatic and shear stress components are present. In rectangular coordinates these stresses are given by[4]

$$\sigma_{xx} = -\frac{Gby}{2\pi(1-v)}\frac{(3x^2 + y^2)}{(x^2 + y^2)^2}$$
$$\sigma_{yy} = +\frac{Gby}{2\pi(1-v)}\frac{(x^2 - y^2)}{(x^2 + y^2)^2}$$
$$\tau_{xy} = \tau_{yx} = \frac{Gbx}{2\pi(1-v)}\frac{(x^2 - y^2)}{(x^2 + y^2)^2}$$
$$\sigma_{zz} = v(\sigma_{xx} + \sigma_{yy})$$
$$\tau_{xz} = \tau_{zx} = \tau_{yz} = \tau_{zy} = 0$$

$$(2\text{-}13)$$

where v = Poisson's ratio, Comparing Eq. 2-13 with Fig. 2.16, we find a region of pure compression directly above the edge dislocation ($X = 0$) and pure tension below the bottom edge of the extra half plane. Along the slip plane ($Y = 0$) the stress is pure shear. For all other positions surrounding the dislocation, the stress field is found to contain compressive and/or tensile components as well as a shear component.

The elastic strain energy is another elastic property of a dislocation. For the simple case of the screw dislocation, this quantity may be given by

$$E_{\text{screw}} = \frac{1}{2}\int_{r_0}^{r_1} \tau_{\theta z}\, b\, dr$$

$$(2\text{-}14)$$

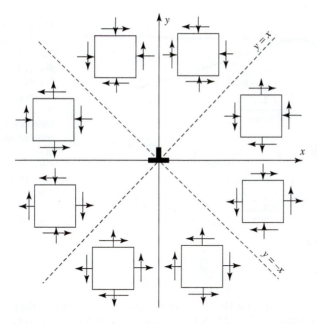

Figure 2.16   Elastic stress field surrounding an edge dislocation. (From Read,[3] *Dislocations in Crystals;* © McGraw-Hill Book Co., New York, 1953. Used with permission of McGraw-Hill Book Company.)

Note that the energy is defined for the region outside the core of the dislocation r_0 to the outer boundaries of the crystal r_1 (see Fig. 2.15). Combining Eqs. 2-11 and 2-14, we get

$$E_{\text{screw}} = \frac{1}{2} \int_{r_0}^{r_1} \frac{Gb^2}{2\pi} \frac{dr}{r} = \frac{Gb^2}{4\pi} \ln \frac{r_1}{r_0} \tag{2-15}$$

The elastic energy of the edge dislocation is slightly larger and given by

$$E_{\text{edge}} = \frac{Gb^2}{4\pi(1-\nu)} \ln \frac{r_1}{r_0} \tag{2-16}$$

Since a general dislocation contains both edge and screw components, its energy is intermediate to the limiting values given by Eqs. 2-15 and 2-16. For purposes of our discussion it is sufficient to note that

$$E = \alpha Gb^2 \tag{2-17}$$

where

E = energy of any dislocation
α = geometrical factor with α taken between 0.5 and 1.0

A particularly important consequence of Eqs. 2-15 to 2-17 is that slip will usually occur in close-packed directions so as to minimize the Burgers vectors of the dislocation. The preferred slip directions in major crystal types are given in Section 2.2. Equations 2-15 to 2-17 also allow one to determine whether or not a particular dislocation reaction will occur. From Eq. 2-9, such a reaction will be favored when

$$b_1^2 > b_2^2 + b_3^2$$

(neglecting possible anisotropy effects associated with G).

Two other elastic properties of a dislocation are its line tension and the force needed to move the dislocation through the lattice. The line tension T is described in terms of its energy per unit length and is given by

$$T \propto Gb^2 \tag{2-18}$$

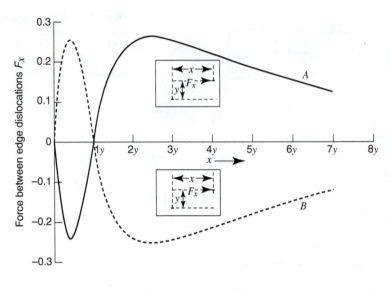

Figure 2.17 Force between parallel edge dislocations. Curve A corresponds to dislocations of the same sign. Force reversal when $X < Y$ causes like dislocations to become aligned as in a tilt boundary. Curve B corresponds to dislocations of opposite sign. Unit of force F_x is $Gb^2/2\pi(1-v)\cdot x$ $(x^2-y^2)/(x^2+y^2)^2$.[18] (From A. H. Cottrell, *The Properties of Materials at High Rates of Strain,* Institute of Mechanical Engineering, London, 1957.)

The line tension acts to straighten a dislocation line to minimize its length, thereby lowering the overall energy of the crystal (see Fig. 2.9c). Consequently, it is necessary to apply a stress τ so that the dislocation line remains curved. This stress is shown to increase with increasing line tension T and decreasing radius of curvature R where

$$\tau \propto \frac{T}{bR} \tag{2-19}$$

Combining Eqs. 2-18 and 2-19, we find that

$$\tau \propto \frac{Gb}{R} \tag{2-20}$$

This relationship will be referred to in Section 3.2.1.

Finally, the force acting on a dislocation is found to depend on the intrinsic resistance to dislocation movement through the lattice, the Peierls-Nabarro stress (Section 2.1.4), and interactions with other dislocations. As shown by Read[3] for the case of parallel dislocations, screw dislocations will always repel one another when the Burgers vectors of both dislocations are of the same sign; they will always attract one another when the signs of the Burgers vectors are opposite. In either case, the magnitude of the force is inversely proportional to the distance between the two dislocations. The force between two edge dislocations is complicated by a reversal in sign when the horizontal distance between two dislocations becomes less than the vertical distance between the two parallel slip planes (Fig. 2.17). Consequently, like edge dislocations are attracted to one another when $x < y$. As a result, like edge dislocations can form stable arrays of dislocations located vertically above one another in the form of simple tilt boundaries (Fig. 2.6).

2.1.7 Partial Dislocations

As noted in the previous section, the likelihood that dislocation $\mathbf{b_1}$ will dissociate into two dislocations $\mathbf{b_2}$ and $\mathbf{b_3}$, often referred to as Shockley partial dislocations, depends on whether the sum of the elastic energies of partial dislocations $\mathbf{b_2}$ and $\mathbf{b_3}$ is lower than the elastic energy associated with dislocation $\mathbf{b_1}$. From Eq. 2-17, the dissociation will occur when

$$G\mathbf{b}_1^2 > G\mathbf{b}_2^2 + G\mathbf{b}_3^2$$

Using the FCC lattice as a model, it can be shown that the whole dislocation $\mathbf{b}_1$, oriented in the $\langle 110 \rangle$ close-packed direction, can dissociate into two dislocations of type $\langle 112 \rangle$. We see

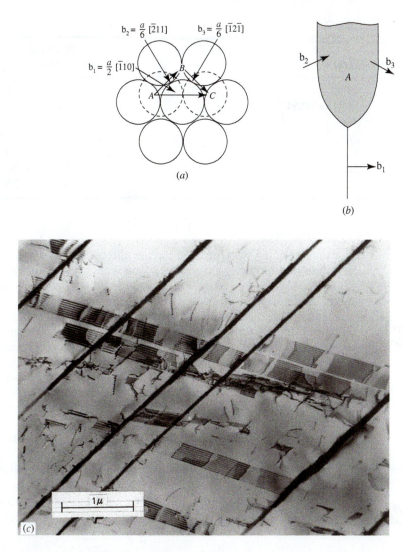

Figure 2.18   (a) Path of whole and partial (Shockley) dislocations; (b) Shockley $\mathbf{b}_2$ and $\mathbf{b}_3$ surrounding stacking fault region A; (c) Long stacking fault ribbons (bands of closely spaced lines) in low SFE 18Cr–8Ni stainless steel. Faults are bounded at ends by partial dislocations. Thin black bands are mechanical twins. (After Michelak[17]; reprinted with permission from *Metals Handbook,* Vol. 8, American Society for Metals, Metals Park, OH, © 1973.)

from Fig. 2.18a that the motion of the atoms on the slip plane is from A to B to C rather than directly in the close-packed direction AC. That is, the whole dislocation AC dissociates into two *partial dislocations*, AB and BC. For example,

$$\frac{a}{2}[\bar{1}01] \rightarrow \frac{a}{6}[\bar{2}11] + \frac{a}{6}[\overline{11}2]$$

From Eq. 2-17

$$\frac{a^2}{4}(1+0+1) > \frac{a^2}{36}[4+1+1] + \frac{a^2}{36}[1+1+4]$$

$$\frac{a^2}{2} > \frac{a^2}{3}$$

Table 2.5 Selected Stacking Fault Energies for FCC Metals

Metal	Stacking Fault Energy $(mJ/m^2 = ergs/cm^2)$
Brass	<10
Stainless steel	<10
Ag	~25
Au	~75
Cu	~90
Ni	~200
Al	~250

(For simplicity, any anisotropy in elastic shear modulus has been ignored.) Therefore, the dislocation reaction will proceed in the direction indicated. It is possible to sense this dislocation reaction in a tactile way. If you were to hold a sheet of close-packed Ping-Pong balls (glued together, of course) in one hand and then slide it across a second sheet of balls parallel to one of the close-packed directions, you will note that the sheets prefer to slide past one another in zigzag fashion along the troughs between the balls on the second sheet. In the FCC lattice, these troughs are parallel to $\langle 112 \rangle$ directions. Because of the reduction in strain energy and the fact that the partials have similar vector components, these partials will tend to repel one another and move apart. The extent of separation, denoted by area A in Fig. 2.18*b*, will depend on the nature of the change in stacking sequence that occurs between $\mathbf{b}_2$ and $\mathbf{b}_3$. Movement of these partial dislocations produces a change in the stacking sequence from the FCC type—*ABCABCABC*—to include a local perturbation involving the formation of a layer of HCP material—*ABCBCABC*. Examples of *stacking faults* are shown in Fig. 2.18*c*. For an FCC crystal, the layer of HCP material that is introduced will elevate the total energy of the system. Therefore, the equilibrium distance of separation of two partials reflects a balance of the net repulsive force between the two partial dislocations containing Burgers vector components of the same sign and the energy of the associated stacking fault. According to Cottrell,[4] this separation distance varies inversely with the *stacking fault energy* (SFE) and may be given by

$$d = \frac{G(\mathbf{b}_2\mathbf{b}_3)}{2\pi\gamma} \tag{2-21}$$

where

$d =$ partial dislocation separation
$\mathbf{b}_2, \mathbf{b}_3 =$ partial dislocation Burgers vectors
$G =$ shear modulus
$\gamma =$ stacking fault energy

The SFE of alloy crystals depends on their composition, and comparative values for pure metals also differ. Typical values for different elements and alloys are given in Table 2.5. For the case of copper-based alloys, Thornton el al.[20] showed SFE to be strongly affected by the material's electron/atom ratio. They found that when $e/a > 1.1$, the stacking fault energy usually decreased to below $20\,mJ/m^2$.

2.1.7.1 Movement of Partial Dislocations

The movement of the two Shockley partial dislocations is restricted to the plane of the fault, since movement of either partial on a different plane would involve energetically unfavorable atomic movements. Therefore, cross-slip of an extended screw dislocation around obstacles is not permitted without thermally activated processes and, as a consequence, the slip offsets seen

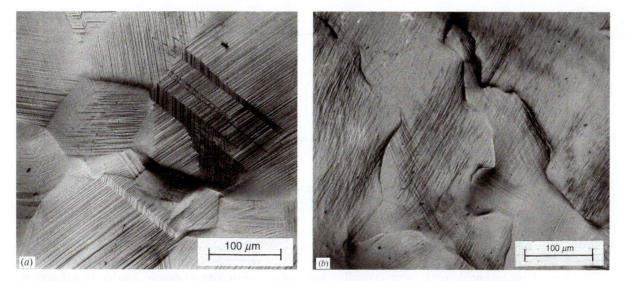

Figure 2.19   Photomicrographs revealing slip character. (*a*) Planar glide in low stacking fault energy material; (*b*) wavy glide in high stacking fault energy material.

on a polished surface will be straight (Fig. 2.19*a*). Such is the case for a material of low stacking fault energy and widely separated partial dislocations. This type of dislocation movement is called *planar glide*. By the application of a suitably large stress, however, it is possible to squeeze the partial dislocations together against a barrier to form a whole dislocation. If this recombined dislocation is of the screw type, it may cross-slip (recall Fig. 2.12). As you might imagine, the stress necessary to recombine the partial dislocations will depend on the equilibrium distance of separation of the partials, which in turn depends on the magnitude of the stacking fault energy (Eq. 2-21). For materials with low stacking fault energy, partial dislocation separation is large (on the order of 10 to 20**b**) and the force necessary for recombination is large. Conversely, little stress is necessary to recombine partial dislocations in a high stacking fault energy material where partial dislocation separation is small (on the order of 1**b** or less). When cross-slip is easy, slip offsets on a polished surface take on a wavy pattern (Fig. 2.19*b*), and this deformation is called *wavy glide*.

2.2   SLIP

Having introduced the concept of dislocation glide-induced slip in crystalline materials, it is appropriate to ask how this phenomenon can be used to predict the yield strength of a material at the macroscopic (continuum) level. Or, put another way, without attempting to determine the behavior of billions of individual dislocations, how can we determine when dislocation motion will commence, and the material will begin to yield? One key to understanding the onset of yield behavior is to recall that slip by dislocation glide can only occur when a shear stress is present at some critical fraction of the theoretical shear strength. All of the mechanical tests described in Chapter 1 can induce yielding in ductile materials, and yet most of them do not involve shear loading—or do they? By taking a closer look at the connection between slip and crystallography, it will become clear that uniaxial tensile or compressive loading can, indeed, cause shear stress and motion of dislocations.

2.2.1   Crystallography of Slip

We saw in the previous sections that plastic deformation occurs primarily by sliding along certain crystallographic planes, with one part of a crystal moving relative to another. This

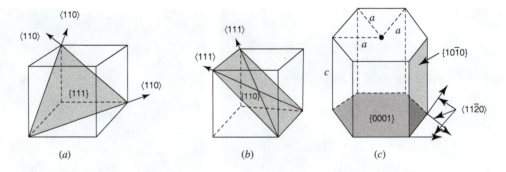

Figure 2.20 Diagram showing predominant slip systems in (*a*) FCC; (*b*) BCC; and (*c*) HCP crystals.

block-like nature of slip produces crystal offsets (called *slip steps*) in amounts given by multiples of the unit dislocation displacement vector **b**. To minimize the Peierls stress, slip occurs predominantly on crystallographic planes of maximum atomic density. In addition, slip will occur in the close-packed direction, which represents the shortest distance between two equilibrium atom positions and, hence, the lowest energy direction. As shown in Fig. 2.20, the dominant *slip systems* (combinations of slip planes and directions) vary with the material's crystal lattice, since the respective atomic density of planes and directions are different. Most metals used for load-bearing purposes have one of three crystal structures: face-centered-cubic (FCC), body-centered-cubic (BCC), and hexagonal close-packed (HCP). Ceramic materials generally have more complicated crystal structures, but the principles underlying the crystallography of slip are the same as for metals.

For the case of FCC crystals, slip occurs most often on {111} octahedral planes and in ⟨110⟩ directions that are parallel to cube face diagonals. In all, there are 12 such slip systems (four {111} planes and three ⟨110⟩ slip directions for each {111} plane). Other FCC slip systems have been found but will not be considered here since they are activated only by unusual test conditions.

In BCC crystals, slip occurs most easily in the ⟨111⟩ cube diagonal direction and on {110} dodecahedral planes. Slip may occur in the ⟨111⟩ directions on {112} and {123} planes as well, particularly at moderate to high temperatures. A total of 48 possible slip systems can be identified, based on combinations of these three slip planes and the common ⟨111⟩ slip direction. The fourfold greater number of slip systems in BCC as compared to FCC crystals does not mean that the former lattice provides more ductility; in fact, the reverse is true because FCC crystals have a much lower Peierls-Nabarro stress and thus contain more mobile dislocations.

Prediction of the preferred slip systems in HCP materials is not an easy task, despite the presence of clear close-packed planes in the ideal HCP structure. On the basis of the Peierls stress argument wherein the *most densely packed planes of greater separation* would be the preferred slip planes, one would expect the active slip planes in real hexagonal crystals to vary with the c/a ratio (see Fig. 2.20c). That is, if the c/a ratio is greater than that for ideal packing (1.633), the basal plane {0001} should be the preferred slip plane as seen for the case of zinc and cadmium, and so the slip system would be {0001} ⟨11$\bar{2}$0⟩. On the other hand, when $c/a < 1.633$, then the {1100} prism planes become atomically more dense relative to the basal plane. For this case, {10$\bar{1}$0} prism slip would be preferred in the ⟨11$\bar{2}$0⟩ close-packed direction (Fig 2.20). This has been found true for the case of zirconium and titanium (Table 2.6). Unfortunately, this neat trend is not true for cobalt, magnesium, or beryllium. Researchers have sought with little success other explanations to account for the observed slip behavior of these three metals.[21] On the other hand, when $c/a > 1.633$, the basal plane should be the preferred slip plane as shown for the case of zinc and cadmium.

Table 2.6   Observed Dominant Slip Planes in Hexagonal Crystals

Metal	*c/a* Ratio	Observed Slip Plane
Be	1.568	$\{0001\}$
Ti	1.587	$\{10\bar{1}0\}$
Zr	1.593	$\{10\bar{1}0\}$
Mg	1.623	$\{0001\}$
Co	1.623	$\{0001\}$
Zn	1.856	$\{0001\}$
Cd	1.886	$\{0001\}$

Thus far, our discussion has focused on the importance of the relative atomic density of crystallographic planes and directions in deciding whether a particular plane and direction combination could serve as a potential slip system. Certain slip-system combinations of ceramic crystals seem reasonable on the basis of density considerations, but are negated by the effects of strong directional bonding in covalent crystals or electrostatic interactions in ionic crystals. Since such atomic movements would be energetically unfavorable, the number of potential slip systems in these materials is restricted as is their overall ductility (except at relatively high test temperatures). A compilation of reported slip systems in selected ceramic crystals is given in Table 2.7.

The ductility of a material depends also on its ability to withstand a general homogeneous strain involving an "arbitrary shape change" of the crystal. An arbitrary shape change is defined as plastic deformation that can occur under the imposition of any combination of shear strains. Von Mises[23] showed that it is possible for a material to accommodate such a shape change when *five independent slip systems* are activated. If fewer than five independent slip systems are available, then some combinations of strains will cause immediate fracture instead of plasticity. If we allow one slip system to account for each of the six independent components of strain (Fig. 2.21), a total of six such systems would seem to be indicated; however, plastic deformation is a constant-volume process where $\varepsilon_{xx} + \varepsilon_{yy} + \varepsilon_{zz} = 0$, thereby reducing to five the number of independent slip systems.

An independent slip system is defined as one producing a crystal shape change that cannot be reproduced by any combination of other slip systems. On this basis, Taylor[24] showed that for the 12 possible $\{111\}\langle 110\rangle$ slip systems in FCC crystals, only five are independent. Furthermore, Taylor found there to be 384 different combinations of five slip systems that could produce a given strain, the activated combination being the one for which the sum of the glide shears is a minimum. Likewise, Groves and Kelly[25] found 384 combinations of five sets of $\{110\}\langle 111\rangle$ slip systems to account for slip in BCC metals. Since slip in BCC metals can occur

Table 2.7   Observed Slip Systems in Selected Ceramics[22]

Material	Structure Type	Preferred Slip System
C, Ge, Si	Diamond cubic	$\{111\}\langle\bar{1}10\rangle$
NaCl, LiF, MgO	Rock salt	$\{110\}\langle 1\bar{1}0\rangle$
CsCl	Cesium chloride	$\{110\}\langle 001\rangle$
CaF_2, UO_2, ThO_2	Fluorite	$\{001\}\langle 1\bar{1}0\rangle$
TiO_2	Rutile	$\{101\}\langle 10\bar{1}\rangle$
$MgAl_2O_4$	Spinel	$\{111\}\langle\bar{1}10\rangle$
Al_2O_3	Hexagonal	$\{0001\}\langle 11\bar{2}0\rangle$

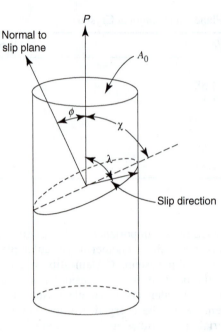

Figure 2.21 Diagram showing orientation of slip plane and slip direction in crystal relative to the loading axis.

also on $\{112\}$ $\langle 111 \rangle$ and $\{123\}$ $\langle 111 \rangle$ systems, the total number of combinations of five independent slip systems becomes incredibly large (as shown by Chin and coworkers,[26,27] who have relied on computer techniques to identify the preferred slip-system combinations for the case of BCC metals).

Difficulties arise when one seeks five independent slip systems in the hexagonal materials. Of the three possible $\{0001\}$ $\langle 11\bar{2}0 \rangle$ slip systems, only two are independent.[25] Similarly, only two independent $\{0010\}$ $\langle 11\bar{2}0 \rangle$ slip systems can be identified from the three possible prism slip systems. Although four independent pyramidal $\{00\bar{1}1\}\langle 11\bar{2}0 \rangle$ slip systems may be identified from a total of six such systems, the resulting deformations can be produced by simultaneous operation of the two independent basal and prism slip systems, respectively. Consequently, a fifth independent slip system is still needed. Besides some deformation twinning (see Section 2.6), additional non-basal slip with a c-axis Burgers vector component is necessary to explain the observed ductility in hexagonal engineering alloys.[28]

2.2.2 Geometry of Slip

It has been shown that the onset of plastic deformation in a single crystal takes place when the shear stress acting on the incipient slip plane and in the slip direction reaches a critical value. Recall from Chapter 1 that a shear stress is defined as a force applied parallel to, and distributed over, a certain plane area. From Fig. 2.21, we see that the cross-sectional area of the arbitrary slip plane depicted is given by

$$A_{\text{slip plane}} = \frac{A_0}{\cos\phi} \tag{2-22}$$

where

$A_0 = $ cross-sectional area of single crystal rod
$\phi = $ angle between the rod axis and the normal to the slip plane

Furthermore, we can treat the applied load, P, as a vector and therefore determine a component normal to the slip plane and one parallel to the slip plane; by definition, the latter component is a shear force. Thus, the shear component of the applied load that lies in this plane, resolved in the slip direction indicated, is given by

$$P_{\text{resolved}} = P\cos\lambda \tag{2-23}$$

where

$P =$ axial load
$\lambda =$ angle between load axis and slip direction

By combining Eqs. 2-22 and 2-23 the *resolved shear stress* acting on the slip system is

$$\tau_{RSS} = \frac{P}{A_0}\cos\phi\cos\lambda \tag{2-24}$$

where the term $\cos\phi\cos\lambda$ represents an orientation factor (often referred to as the *Schmid factor*). Note that P/A_0 is just the applied axial stress, σ, so the resolved shear stress, τ_{RSS}, will change as the applied stress changes. Plastic deformation will occur when τ_{RSS} reaches a critical value, τ_{CRSS}, which represents the shear yield strength of the single crystal. Unlike τ_{RSS}, the *critical resolved shear stress*, τ_{CRSS}, is a fixed value for a given material processed to a certain state of strength. The relationship between the uniaxial yield stress, σ_Y, and the critical resolved shear stress in a single crystal is therefore

$$\tau_{CRSS} = \sigma_Y\cos\phi\cos\lambda \tag{2-25a}$$

or

$$\sigma_Y = \tau_{CRSS}\left(\frac{1}{\cos\phi\cos\lambda}\right) \tag{2-25b}$$

From Eq. 2-24, we see that yielding will occur most readily on the slip system possessing the greatest Schmid factor. Consequently, if only a few systems are available, such as in the case of basal slip in hexagonal zinc and cadmium, the necessary load for yielding can vary dramatically with the relative orientation of the slip system with respect to the loading axis (i.e., the Schmid factor).[29,30] For example, the axial stress necessary for yielding anthracene[i] crystals (Fig. 2.22a, b) varies dramatically with crystal orientation, while the critical resolved shear stress is unchanged.[31] The invariance of τ_{CRSS} at the point of yielding for these anthracene crystals is best appreciated when the data from Fig. 2.22b are replotted as τ_{RSS} vs. the critical axial stress, P/A (Fig. 2.22c). (Note that the curve drawn in Fig. 2.22b was computed from Eq. 2-24 using a value of 137 kPa.) Clearly, yielding occurs when $(P/A_0)\cos\phi\cos\lambda$ values reach a critical level (i.e., τ_{CRSS}).

Furthermore, the stress normal to the slip plane

$$\sigma_n = \frac{P}{A_0}\cos^2\phi \tag{2-26}$$

can vary considerably without any correlation to the onset of yielding, as shown in Fig. 2.22d. As another example, Andrade and Roscoe[30] found for cadmium that measured values of τ_{CRSS} varied only by 2% for all crystal orientations examined, while the normal stress σ_n experienced a 20-fold change. This underscores the fact that it is the *shear stress* and *not the normal stress* on a particular slip system that is critical for yielding.

---

[i] A solid component of coal-tar constructed from benzene rings, and therefore possessing strong hexagonal symmetry.

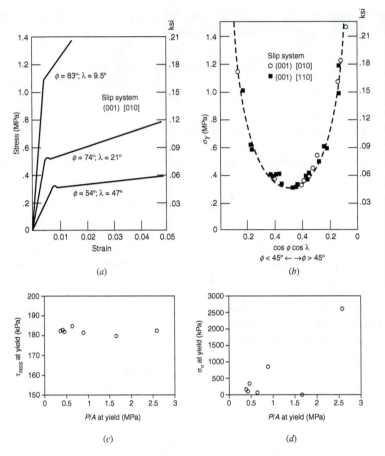

Figure 2.22   Yield behavior of anthracene single crystals, (*a*) Axial stress-strain curves for crystals possessing different orientations relative to the loading axis. (*b*) Axial stress for many crystals plotted versus respective Schmid factors. Dotted curve represents relation given by Eq. 2-24 where $\tau_{crss} = 137$ kPa. (After Robinson and Scott;[31] reprinted with permission from Robinson, *Acta Met* 15 (1967), Pergamon Press, Elmsford, NY.) (*c*) Invariance of the resolved shear stress at yield. (*d*) Lack of correlation between slip plane normal stress and axial yield stress.

EXAMPLE 2.1

Three cylindrically-shaped tensile samples, each 12 mm in diameter, were machined from three different spherically-shaped single crystals. Samples A and B yielded with applied loads of 77.1 and 56 N, respectively. Does the difference in load level indicate that the crystals possessed different strength levels? Also, what load level would be necessary to cause Sample C to deform and what is the controlling stress for yielding?

   The fact that different load levels were needed to cause yielding in Samples A and B *may* indicate that the materials in the two rods possessed different properties. Then, again, the materials may have been identical but, instead, machined at arbitrarily different orientations from the three spherically shaped single crystals. To resolve this issue, additional information is needed. Specifically, it is necessary to determine the crystallographic orientation of the three single crystals, relative to their respective loading directions. X-ray diffraction studies determined that the angles between the tensile axis and both slip plane normals and slip directions are

	ϕ	λ	P
Sample A	70.5	29	77.1
Sample B	64	23	56
Sample C	13	78	?

From Eq. 2-24, the resolved shear stresses for yielding in Samples A and B are

$$\tau_A = \frac{P}{A}\cos\phi\cos\lambda = \frac{77.1}{\pi(6\times10^{-3})^2}(\cos 70.5)(\cos 29) = 199{,}029\,\text{Pa}$$

$$\tau_B = \frac{P}{A}\cos\phi\cos\lambda = \frac{56}{\pi(6\times10^{-3})^2}(\cos 64)(\cos 23) = 199{,}803\,\text{Pa}$$

By contrast, the stress normal to the slip plane is given from Eq. 2-26 to be

$$\sigma_A = \frac{P}{A}\cos^2\phi = \frac{77.1}{\pi(6\times10^{-3})^2}\cos^2(70.5) = 75{,}961\,\text{Pa}$$

$$\sigma_B = \frac{P}{A}\cos^2\phi = \frac{56}{\pi(6\times10^{-3})^2}\cos^2(64) = 95{,}152\,\text{Pa}$$

We conclude that the strengths of the two samples are similar (approximately 199 kPa) and that yielding is controlled by the critical resolved shear stress acting on the slip system rather than the stress acting normal to the slip plane. It follows that the load needed to deform Sample C will generate a shear stress of 199 kPa on the active slip system.

Therefore,

$$\tau_C = \frac{P}{A}\cos\phi\cos\lambda = \frac{P}{\pi(6\times10^{-3})^2}(\cos 13)(\cos 78) \approx 199{,}000\,\text{Pa}$$

$$\therefore P = 111\,\text{N}$$

2.2.3 Slip in Polycrystals

Most load-bearing materials are not used in single crystal form, so what is the relevance of the preceding discussion to yield in a polycrystal that has an enormous number of grains, each with its own orientation with respect to the loading axis? Each grain has its own resolved shear stress on the most likely slip system, and so would be expected to yield at a different value of applied uniaxial stress. As such, some grains within the material will be more resistant to yield than others. Furthermore, if one grain yields and changes shape, that shape change must be accommodated in some way by the surrounding grains, or grain boundary cracks will develop. This *compatibility constraint* is an additional source of resistance to yield that is not found in single crystals.

Taylor addressed these concerns by calculating an average reciprocal Schmid factor (the *Taylor factor* $M = 1/\cos\phi\cos\lambda$) for a polycrystalline material, assuming that the grain orientations are random and that all grains experience the same strain. He also assumed that *multiple slip* could take place over the required five independent slip systems to allow for arbitrary shape changes.[24,32] In doing so, he showed that the value of the resulting average reciprocal Schmid factor for FCC materials deforming by $\{111\}\langle1\bar{1}0\rangle$ slip is $M \approx 3.1$. This calculation was confirmed by Bishop and Hill.[33] Kocks concluded that the Taylor factor for BCC materials deforming at low temperatures by $\{110\}\langle111\rangle$ slip was the same as for FCC.[34] Recall, however, that there are several slip systems in BCC metals that contain $\langle111\rangle$ directions and are of similar packing density. For BCC slip acting simultaneously on $\{110\}$, $\{112\}$, and $\{123\}$ planes in $\langle111\rangle$ directions, $M \approx 2.7$.[35] From Eq. 2-25b we can see that this decrease in the Taylor factor would correlate with a decrease in the uniaxial yield strength, even though the fundamental critical resolved shear strength of the individual grains is unchanged. The process of simultaneous slip on multiple BCC slip systems is called *pencil glide* because of the faceted slip surface that can result.

2.3   YIELD CRITERIA FOR METALS AND CERAMICS

The preceding discussion of yielding was framed in terms of a simple uniaxial load that causes the onset of plastic deformation when the resolved shear stress on a certain slip system exceeds the critical resolved shear stress for the material. This principle also can be used to predict yielding under complex loading conditions. In Chapter 1, the combined elastic behavior of multiple stresses was described using the generalized Hooke's law. It can now be appreciated that a different interplay of multiple normal loads must be assessed, along with the influence of constraints imposed by contiguous grains, to determine the resolved shear stress on potential slip systems in a given grain.

Figure 2.23 depicts three idealized loading scenarios, two of which are uniaxial and one of which is biaxial. In a simple *normal stress yield theory*, yielding would occur whenever any of the normal stress components (labeled σ in the figure) exceeds a certain critical tensile (or compressive) value. This does not match experimental observations of multiaxial loading, however.[ii] Instead, we must consider the shear stresses induced by the normal stresses. In each case depicted in Fig. 2.23, a plane and a slip direction within that plane 45° to the loading axis is shown along with the sign of the shear stress resolved onto that particular slip system. The two uniaxial cases create equal but opposite resolved shears on this particular plane. It can be seen for the case of the equal biaxial stress state that the resolved shear stresses on this slip system cancel. Assuming that no other slip systems exist, there is therefore no possibility of yielding regardless of how large the applied stresses become. If the biaxial stresses are not equal then only partial cancellation occurs, and yielding is possible but will occur at higher applied stresses than under uniaxial loading. (Note that the resolved shear stresses occur in a different coordinate system than the applied normal stresses, so there is no conflict with the lack of coupling between elastic normal and shear stresses described by the generalized Hooke's law for isotropic materials.)

The *Tresca yield criterion* uses this principle to predict the onset of yielding in a polycrystalline material when the *maximum resolved shear stress* in the material exceeds the shear strength associated with yielding measured by the uniaxial tension test:

$$\tau_{max} = \frac{\sigma_{max} - \sigma_{min}}{2} \geq \tau_{CRSS} \qquad (2\text{-}27)$$

where

$$\tau_{max} = \text{maximum shear stress possible for any slip system orientation}$$
$$\sigma_{max,min} = \text{largest and smallest principal tensile stresses, respectively}$$

For uniaxial loading conditions ($\sigma_2 = \sigma_3 = 0$), the shear yield strength is found to be equal to one-half the yield strength in tension. This is to be expected because the maximum possible value of the Schmid factor is (cos 45 cos 45) = 0.5.

The Tresca theory provides a reasonable description of yielding in ductile materials, but the *maximum distortion energy* or *von Mises yield criterion* is generally preferred because it is based on better correlation with actual test data. In this theory, yielding is assumed to occur when the

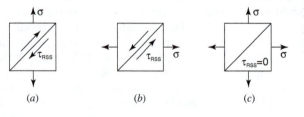

(a)   (b)   (c)

Figure 2.23   Resolved shear stress on a 45° plane under (*a*) vertical uniaxial loading, (*b*) horizontal unaxial loading, and (*c*) equal biaxial loading.

[ii] Although it is not sufficient for predicting multiaxial yielding, it may work as a criterion for fracture of brittle materials where the normal stress (and not the shear stress) may be the critical factor.

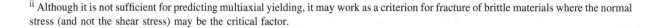

distortional energy in a component that experiences multiaxial loading is equal to the distortional energy required to induce a shape change during a tensile test. (Recall that plastic deformation conserves volume. The energy associated only with a volume change must be elastic in nature and is therefore not included in this analysis.) To characterize the distortional energy (proportional to σ^2/E) within a specimen or component that experiences multiaxial loading, it is convenient to consider an *equivalent tensile stress*, or a *von Mises stress*, σ_e, given by

$$\sigma_e = \frac{\sqrt{2}}{2}\sqrt{(\sigma_2 - \sigma_1)^2 + (\sigma_3 - \sigma_1)^2 + (\sigma_3 - \sigma_2)^2} \qquad (2\text{-}28)$$

where $\sigma_1, \sigma_2, \sigma_3 =$ principal stresses. The use of the normal stress differences ensures that only those components that lead to shear stress are considered. The equivalent stress can also be described in terms of the tensile and shear stresses acting on three arbitrary orthogonal planes wherein

$$\sigma_e = \frac{\sqrt{2}}{2}\left[(\sigma_{yy} - \sigma_{xx})^2 + (\sigma_{zz} - \sigma_{xx})^2 + (\sigma_{zz} - \sigma_{yy})^2 + 6\left(\tau_{yz}^2 + \tau_{xz}^2 + \tau_{xy}^2\right)\right]^{1/2} \qquad (2\text{-}29)$$

If σ_e equals the yield strength from a uniaxial tensile test, yielding in the multiaxially loaded sample is predicted.

Alternatively, the so-called *octahedral shear stress*, τ_{oct}, existing under multiaxial loading can be compared to the *critical octahedral shear stress*, τ_{oct0}, determined under uniaxial loading as another way to express the von Mises yield criterion. The octahedral shear stress is the resolved shear stress that exists in equal magnitude on each of the eight *octahedral planes* that make up a regular octahedron about the principal stress axes (each of which has a direction cosine with respect to the principal axes of $1/\sqrt{3}$). The octahedral shear stress is defined as

$$\tau_{oct} = \frac{1}{3}\sqrt{(\sigma_1 - \sigma_2)^2 + (\sigma_2 - \sigma_3)^2 + (\sigma_1 - \sigma_3)^2} \qquad (2\text{-}30)$$

and the critical octahedral shear stress is related to the uniaxial yield strength as

$$\tau_{oct0} = \frac{1}{3}\sqrt{(\sigma_Y)^2 + (\sigma_Y)^2} = \frac{\sqrt{2}}{3}\sigma_Y \qquad (2\text{-}31)$$

Although the form of the von Mises criterion is somewhat different from that of the Tresca criterion, the necessity of shear stress for inducing plasticity is retained. A two-dimensional depiction of yield loci for the normal stress, shear stress, and distortion-energy theories is shown in Fig. 2.24. The normal stress threshold is a square, the von Mises threshold an ellipse, and the

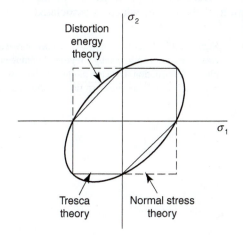

Figure 2.24   Failure envelopes for normal stress, Tresca, and distortion energy theories. Failure occurs when stress combinations fall outside the envelope for applicable theory.

Tresca threshold a six-sided polygon. Note that all three failure theories predict the same yielding conditions under uniaxial ($\sigma_1 = \sigma_1$; $\sigma_2 = \sigma_3 = 0$) and balanced biaxial ($\sigma_1 = \sigma_2$; $\sigma_3 = 0$) loading conditions; however, different failure conditions are predicted for conditions of pure shear ($\sigma_1 = -\sigma_2$; $\sigma_3 = 0$) with the distortion-energy theory predicting yielding when the applied stress is $\sigma_Y/\sqrt{3} = 0.577\sigma_Y$. A three-dimensional version of the yield loci would reveal that the von Mises yield surface is a cylinder with radius $\sigma_Y\sqrt{2/3}$ and an axis that lies along the hydrostatic axis ($\sigma_1 = \sigma_2 = \sigma_3$). The Tresca yield surface would be a hexagonal prism that fits just inside the von Mises cylinder, so it is slightly more conservative predictor of yield (i.e., it predicts slightly smaller yield stresses for most stress states). The two-dimensional depiction in Fig. 2.24 can therefore be understood as a projection of these shapes onto the σ_1 and σ_2 axes ($\sigma_3 = 0$). Most data conform best to predictions of the von Mises theory, but both options are generally available to users of finite element computing codes that calculate stresses and strains in components with complicated shapes. For a more detailed discussion of such classical failure theories, see the text by Juvinall.[36]

Regardless of which yield criterion is used, the prediction of yield under multiaxial loading is based on a comparison to an experimental measurement (usually uniaxial tension). What experimentally-derived value of strength should be used for this purpose? This quantity is difficult to define unequivocally, since the point where plastic flow appears to begin will depend on the sensitivity of the displacement transducer. The more sensitive the gage, the lower the stress level where some evidence of plastic flow is found. Special capacitance strain gages exist that have been used to measure strains in the range of 10^{-6}. In fact, a number of studies[37] dealing with the mechanical behavior of materials in this *microstrain* region have been undertaken as a result of this breakthrough in instrumentation. These investigations have shown, for example, that plastic deformation—the irreversible movement of dislocations—occurs at stress levels many times lower than the conventionally determined engineering yield strength. However, for engineering purposes it is usually the case that the 0.2% offset yield strength, as defined by ASTM Standard E8, is used.[38]

2.4 POST-YIELD PLASTIC DEFORMATION

Now that the micro- and macro-scale conditions associated with the onset of yielding have been established, it is time to examine what happens to a typical metal at the macro scale as plastic deformation progresses. Much more will be said about micro-scale plasticity processes in Chapter 3.

2.4.1 Strain Hardening

When a material has the capacity for significant plastic flow, the stress–strain curve often assumes the shape shown in Fig. 2.25. Here we see that the initial linear elastic region is followed by a smooth parabolic portion of the curve, which is associated with homogeneous

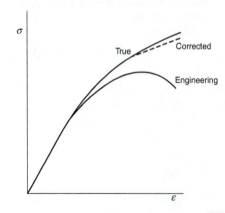

Figure 2.25 Ductile stress–strain behavior revealing elastic behavior followed by a region of homogeneous plastic deformation. Data are plotted on the basis of engineering and true stress–strain definitions, as labeled.

plastic deformation processes such as the irreversible movement of dislocations in metals and ceramics, and a number of other possible deformation mechanisms in polymers. That the curve continues to rise to a maximum engineering stress level (the tensile strength) reflects an increasing resistance on the part of the material to further plastic deformation—a process known as *strain hardening*. The homogeneous plastic portion of the true stress-strain curve (from the onset of yielding to the point of maximum load) may be described empirically by a relationship generally attributed to Hollomon[39], although Bülfinger[40] proposed a similar parabolic relationship between stress and strain almost 200 years earlier. Regardless of origin, the true stress is related to true plastic strain by:

$$\sigma_t = K\varepsilon_t{}^n \tag{2-32}$$

where

$\sigma_t =$ true stress
$\varepsilon_t =$ true plastic strain
$n =$ strain-hardening coefficient
$K =$ strength coefficient, defined as the true stress at a true strain of 1.0

To ensure that the strain hardening behavior is adequately represented, a minimum of five engineering stress and strain pairs are extracted at equal strain intervals from the portion of the stress–strain curve in which homogeneous plasticity occurs.[iii] Typically the data set begins shortly after the yield point (or the lower yield point if one exists—see Section 2.4.3) and ends at the onset of necking (the maximum engineering stress). The engineering values must be converted[iv] to true values using Eqs. 1-4 and 1-5. (It is also possible to evaluate strain hardening after necking commences if cross-section measurements are available.) Although Eq. 2-32 is written in terms of the true plastic strain, it is generally acceptable to use the true total strain as long as the elastic strain ($\varepsilon_{el} = \sigma_t/E$) makes up no more than 10% of the value. Noting that Eq. 2-32 may be rewritten as $\ln\sigma_t = \ln K + n\ln\varepsilon_t$, so parameters n and K may be determined by plotting[v] $\ln \sigma_t$ vs. $\ln \varepsilon_t$, and evaluating the slope and the intercept at $\varepsilon_t = 1$. If Eq. 2-32 describes the material behavior perfectly, the plot will be linear and a single n value will result. However, this is not always the case, which reflects the fact that this relationship is only an empirical approximation.[41] When a nonlinear log–log plot does result for a given material, the strain-hardening coefficient is often defined at a particular strain value or over a particular range of strain. In general, n increases with decreasing strength level and with decreasing mobility of certain dislocations in the crystalline lattice.

The magnitude of the strain-hardening coefficient reflects the ability of the material to resist further deformation. In the limit, n may be equal to unity, which represents ideally elastic behavior (such that K is essentially E), or equal to zero, which represents an ideally plastic material. Selected values of strain-hardening coefficients for some engineering metal alloys determined using Eq. 2-32 are given in Table 2.8. (Note that n values are sensitive to thermomechanical treatment; they are generally larger for materials in the annealed condition and smaller in the cold-worked state.)

Although the Holloman relation is widely used to describe the behavior of a strain-hardening material undergoing plastic deformation, it cannot be used to depict the entire stress–strain curve from beginning to end. Furthermore, it implies an abrupt transition from elastic to plastic behavior that is generally not seen with real materials. A convenient model that captures elastic and plastic behavior with a gradual transition between the two was proposed by Ramberg and Osgood in 1943 as an aid in the design of aircraft.[vi] The basic form of the Ramberg-Osgood

---

[iii] ASTM E646, ASTM International, West Conshohocken, PA.

[iv] Repeated here for convenience: $\sigma_{true} = \sigma_{eng}(1 + \varepsilon_{eng})$ and $\varepsilon_{true} = \ln(1 + \varepsilon_{eng})$

[v] Base 10 logs are also fine.

[vi] W. Ramberg and W. R. Osgood, NACE Technical Note No. 902, Washington, DC (1943).

Table 2.8   Selected Strain-Hardening Coefficients

Material	Strain-Hardening Coefficient, n
Stainless steel	0.45–0.55
Brass	0.35–0.4
Copper	0.3–0.35
Aluminum	0.15–0.25
Iron	0.05–0.15

relation reflects the fact that the total true strain is a simple sum of the true elastic and plastic strains, expressed in terms of the true stress as

$$\varepsilon_t = \varepsilon_{el} + \varepsilon_{pl} = \frac{\sigma_t}{E} + K_{RO}\left(\frac{\sigma_t}{E}\right)^{n_{RO}} \tag{2-33}$$

where

σ_t = true stress
K_{RO} = Ramberg-Osgood strength coefficient
n_{RO} = Ramberg-Osgood strain hardening exponent

When the true stress is very low, the elastic strain term dominates and the behavior is nearly linear elastic. As the stress increases, the plastic strain term plays an increasing role and the slope gradually changes, approaching a regime in which plasticity dominates.

It is also possible to introduce an explicit yield strength term, σ_y, to define a new parameter, $\alpha = K(\sigma_y/E)^{n-1}$, so that Eq. 2-33 can be expressed as

$$\varepsilon_t = \frac{\sigma_t}{E} + \alpha\left(\frac{\sigma_y}{E}\right)\left(\frac{\sigma_t}{\sigma_y}\right)^{n_{RO}} \tag{2-34}$$

There is no distinct yield point on the Ramberg-Osgood stress–strain curve, so any reasonable value for the yield stress may be chosen, but this choice also determines α. When $\sigma_t = \sigma_y$, the true plastic strain (i.e., the offset strain at yield) is given by $\varepsilon_{offset} = \alpha\sigma_y/E$. This makes it possible to select a yield strength value at a chosen strain offset (such as the oft-used 0.002).

The Ramberg-Osgood strain hardening exponent, n_{RO}, can be extracted from the slope of a $\ln\varepsilon_t$ vs. $\ln\sigma_t$ plot (note that the axes are reversed as compared to the Holloman expression). When evaluated for very large plastic strains, $n_{RO} \approx 1/n$, so values of 2–5 (or even larger) are common. In practice, it is often useful to rewrite the Ramberg-Osgood relation so that it more closely resembles the Holloman relation by subtracting the elastic strain from the total strain to leave only the true plastic strain

$$\varepsilon_{pl} = \varepsilon_t - \frac{\sigma_t}{E} = K_{RO}\left(\frac{\sigma_t}{E}\right)^{n_{RO}} \tag{2-35a}$$

$$\left(\frac{E^{n_{RO}}}{K_{RO}}\right)\varepsilon_{pl} = \sigma_t^{n_{RO}} \tag{2-35b}$$

$$\sigma_t = \left(\frac{E}{K_{RO}^{\frac{1}{n_{RO}}}}\right)\varepsilon_{pl}^{\frac{1}{n_{RO}}} = H\varepsilon_{pl}^n \tag{2-35c}$$

In this form (Eq. 2-35c), the strain-hardening exponent (n) and strength coefficient (H) values can be determined from a more conventional $\ln\sigma_t$ vs. $\ln\varepsilon_{pl}$ plot. Note that it is necessary to first subtract the elastic strains from the true strains to solve for the plastic strains before plotting, as in Eq. 2-35a—a step that is generally skipped when performing the Holloman analysis.

Table 2.9   Slip Character and Strain-Hardening Coefficients for Several Metals

Metal	Stacking Fault Energy (mJ/m^2)	Strain-Hardening Coefficient	Slip Character
Stainless steel	<10	~0.45	Planar
Cu	~90	~0.3	Planar/wavy
Al	~250	~0.15	Wavy

The direct connection between strain hardening and dislocation mobility makes it possible to relate strain-hardening coefficients (from Table 2.8) with stacking fault energy values (from Table 2.5) as shown in Table 2.9. One major implication of the dependence of cross-slip on stacking fault energy is the dominant role the latter plays in determining the strain-hardening characteristics of a material (discussed in more detail in Chapter 3). When the stacking fault energy is low, cross-slip is restricted so that barriers to dislocation movement remain effective to higher stress levels than in material of higher stacking fault energy. That is to say, the low stacking-fault-energy material strain hardens to a greater extent. Note that the strain-hardening coefficient increases with decreasing stacking fault energy while the slip character changes from a wavy to a planar mode.

2.4.2   Plastic Instability and Necking

The true and engineering stress–strain plots from a tensile test reveal basic differences, as shown in Fig. 2.25. While the engineering curve reaches a maximum at maximum load and decreases thereafter to fracture, the true curve rises continually to failure. The inflection in the engineering curve is due to the onset of *localized* plastic flow and the manner in which engineering stress is defined (recall Chapter 1). To understand this, consider for a moment the following sequence of events. When the stress reaches a critical level, plastic deformation will occur at the weakest part of the test sample, somewhere along the gage length. This local extension under tensile loading will cause a simultaneous area constriction so that the true local stress is higher at this location than anywhere else along the gage length. Consequently, all additional deformation would be expected to concentrate in this most highly stressed region. Such is the case in an ideally plastic material ($n = 0$). For all other materials, however, this localized plastic deformation strain hardens the material in the highly stressed region, thereby making it more resistant to further damage. At this point, the applied load must be increased to produce additional plastic deformation at the second weakest position along the gage length. Here again the material strain hardens and the process continues. On a macroscopic scale, it appears that the gage length extends uniformly and there is a uniform reduction in cross-sectional area. (Recall that plastic deformation is a constant-volume process.) Uniform plasticity proceeds in this fashion as long as the increase in load-bearing capacity associated with strain hardening is greater than the decreased load-bearing capacity due to the reduction in cross-section area anywhere along the gage length. However, as the load required for continued plasticity increases, the rate of strain hardening decreases. Eventually, the strain-hardening capacity of the material exactly balances the change in cross-section area at a certain spot in the material, and the maximum load-bearing capacity is achieved. After this point of maximum load, further plastic deformation is localized in the same spot, and the true stress at that spot increases continually with the subsequent areal contraction. The bar elongates preferentially in this necked region. As it does so, the applied load required to sustain the plastic deformation decreases as the material outside the necked area undergoes some degree of elastic unloading. Since engineering stress is based on A_0, the decreasing load on the sample after the neck has formed will result in the appearance of a decreasing stress. By comparison, the decreasing load value is more than offset by the decrease in instantaneous cross-sectional area such that the true stress continues to rise to failure even after the onset of necking, as shown in Fig. 2.25.

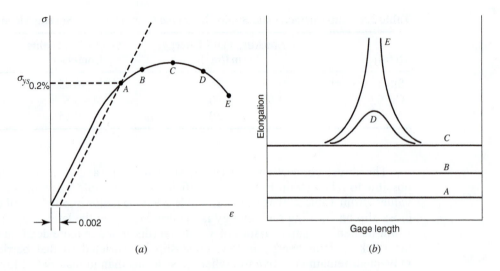

(a)   (b)

Figure 2.26 (a) Engineering stress–strain curve. Tensile yield strength is defined at intersection of stress–strain curve and 0.2% offset line. Points *A, B, C, D,* and *E* are the arbitrary stress levels shown in (b), a schematic representation of specimen elongation along the gage length. Uniform extension occurs up to the onset of necking (*C*). Additional displacements are localized in the necked region (*D* and *E*).

2.4.2.1 Strain Distribution in a Tensile Specimen

The total strain distribution along the specimen gage length is shown schematically in Fig. 2.26 for various stress levels, as indicated on the accompanying engineering stress–strain curve. Owing to the variation of elongation along the gage length of the tensile specimen, engineers occasionally report both the total strain, $(l_f - l_0)/l_0$ or $\ln(l_f/l_0)$, and the uniform strain, $(l_n - l_0)/l_0$ or $\ln(l_n/l_0)$, which is related to the elongation just prior to local necking (line *C* in Fig. 2.26b). It should be emphasized that the total strain reported for a given test result will depend on the gage length of the test bar. From Fig. 2.26, it is clear that as the gage length decreases, the elongation involved in the necking process becomes increasingly more dominant. Consequently, total strain values after necking commences will be larger for bars with shorter gage lengths (for which the length of the necked region is a larger fraction of the total length). For this reason, both specimen size and total strain data should be reported. ASTM has standardized specimen dimensions to minimize variability in test data resulting from such geometrical considerations. As noted in Table 2.10, the gage length to diameter ratio is standardized to a value of about 4.

Table 2.10 Round Tension Test Specimen Dimensions[38]

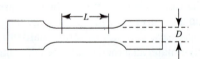

Diameter (D)		Gage Length (L)	
mm	(in.)	mm	(in.)
12.5	(0.5)	50	(2.0)
8.75	(0.345)	35	(1.375)
6.25	(0.25)	25	(1.0)
4.0	(0.16)	16	(0.63)
2.50	(0.1)	10	(0.394)

2.4.2.2 Extent of Uniform Strain

From the standpoint of material usage in an engineering component, it is desirable to maximize the extent of uniform elongation prior to the onset of localized necking. It may be shown that the amount of uniform strain is related to the magnitude of the strain-hardening exponent.

$$P = \sigma_t A_i$$
$$dP = \sigma_t dA_i + A_i d\sigma_t \qquad (2\text{-}36)$$

Recalling that necking occurs at maximum load

$$dP = 0$$

so that

$$\frac{d\sigma_t}{\sigma_t} = -\frac{dA_i}{A_i}$$

And, because constant volume requires that $A_i l_i$ is also constant (Eq. 1-3),

$$A_i dl_i + l_i dA_i = 0$$
$$-\frac{dA_i}{A_i} = \frac{dl_i}{l_i}$$

Since $dl/l \equiv d\varepsilon$, we find

$$\sigma_t = \frac{d\sigma_t}{d\varepsilon_t} \qquad (2\text{-}37)$$

The discovery of this relationship as the critical condition for the onset of necking is due to Considère in 1885. Connecting this relationship to the Hollomon relation (Eq. 2-32) we find that

$$K\varepsilon_t^n = Kn\varepsilon_t^{n-1}$$

Therefore,

$$n = \varepsilon_t \qquad (2\text{-}38)$$

Thus, the true plastic strain at the point of necking instability is numerically equal to the strain-hardening coefficient. The connection between n and ε_{neck} means that the capacity of a material to undergo large uniform elongations correlates directly with the material's strain-hardening capability. This is quite a useful relation because it can be difficult to unambiguously determine the strain at necking from a stress–strain plot that is relatively flat near its peak. Thus it is common to use a strain-hardening experiment to indirectly determine the limiting strain for homogeneous deformation.

2.4.2.3 True Stress Correction

In the preceding discussion, an implicit assumption was made regarding the stress state in an unstable neck: namely, that it is uniaxial just as it was prior to necking. In reality, the development of a neck creates a triaxial stress state in the local vicinity (Fig. 2.27). The new radial (σ_r) and transverse (σ_t) stresses that are induced are developed as a result of a Poisson effect. In order for the more highly stressed material within the neck to undergo a large local extension associated with an increase in load (as seen in Fig. 2.26), it should also undergo a large local decrease in cross-section area. However, the material immediately outside the necked area experiences a much lower stress level and smaller extension from the same increase in load, so the area change should be small. The expected radial and transverse contractions in these adjacent areas are therefore in conflict, so stresses in these directions must develop. Within the neck, tensile radial and transverse

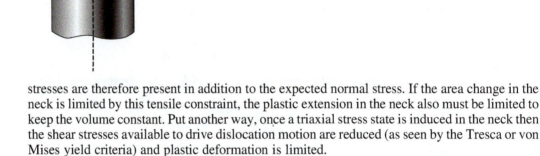

stresses are therefore present in addition to the expected normal stress. If the area change in the neck is limited by this tensile constraint, the plastic extension in the neck also must be limited to keep the volume constant. Put another way, once a triaxial stress state is induced in the neck then the shear stresses available to drive dislocation motion are reduced (as seen by the Tresca or von Mises yield criteria) and plastic deformation is limited.

To provide for additional plastic flow, the axial stress must be increased to overcome the effect of the radial and transverse stresses. The stress values recorded on the true stress–strain curve (Fig. 2.25) after the onset of necking reflect the higher axial stresses necessitated by the local triaxial stress condition. Bridgman was able to correct the applied axial stress (σ_{app}) to determine the true stress (σ_{true}) that would be necessary to deform the material were it not for the presence of the triaxial stresses in the neck. The corrected true stress–strain curve shown in Fig. 2.25 may be determined from the Bridgman[42] relation

$$\frac{\sigma_{true}}{\sigma_{app}} = \frac{1}{(1 + 2R/a)[\ln(1 + a/2R)]} \tag{2-39}$$

where R is the radius of curvature of the neck contour, and a is the radius of the minimum cross-sectional area, as shown in Fig. 2.27. It is seen from this formula that the axial stress necessary to produce a given level of plastic deformation will increase with increasing notch root acuity for a given notch depth.

2.4.2.4 Failure of the Necked Region

At some critical point, the triaxial tensile stress condition within the necked region causes small particles within the microstructure to either fracture or separate from the matrix. The resulting microvoids then undergo a period of growth and eventual coalescence, producing an internal, disk-shaped crack oriented normal to the applied stress axis. Final fracture then occurs by a shearing-off process along a conical surface oriented 45° to the stress axis. This entire process produces the classical cup–cone fracture surface appearance shown in Fig. 2.28. Sometimes the circular region in the middle of the sample (called the fibrous zone) is generated entirely by slow, stable crack growth, while the smooth shear walls are formed at final failure. Usually the fibrous zone contains a series of circumferential ridges reflecting slight undulations in the stable crack propagation direction. However, test conditions can be altered to suppress the extent of the slow, stable crack growth region; instead, the crack continues to grow on the same plane but in unstable fashion at a much faster rate. This new region, defined as the radial zone, contains radial markings (Fig. 2.29) often associated with the fracture of oriented inclusions in test bars prepared from rod stock. (More will be said of this fracture detail in Chapter 11.) The relative amount of fibrous, radial, and shear lip fracture zones has been found to depend on the strength of the material and the

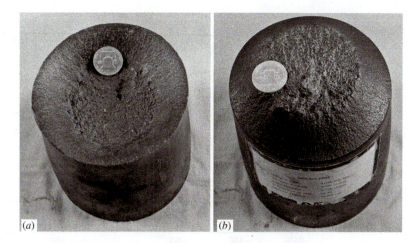

Figure 2.28   Typical cup–cone fracture appearance of unnotched tensile bar: (*a*) cup portion; (*b*) cone portion. (Courtesy of Richard Sopko, Lehigh University.)

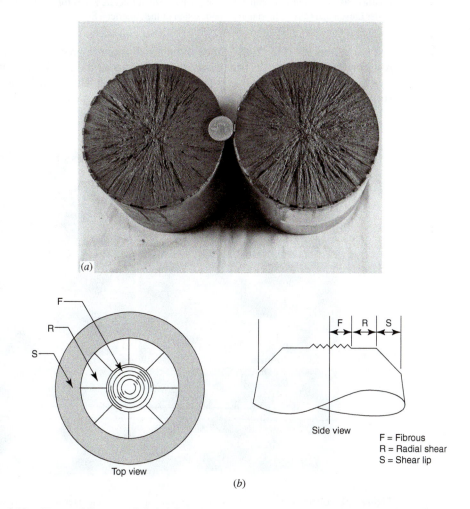

Figure 2.29a   Extent of fibrous, radial, and shear lip zones: (*a*) macrofractograph (Courtesy of Richard Sopko, Lehigh University); (b) schema showing zone location. (After Larson and Carr[43]; reprinted by permission of the American Society for Metals, Metals Park, OH.)

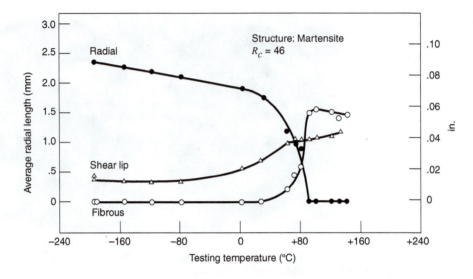

Figure 2.30 Effect of test temperature on relative size of fracture zones for AISI 4340 steel heat treated to R_c 46. (After Larson and Carr[43]; reprinted by permission of the American Society for Metals, Metals Park, OH.)

test temperature[43] (Fig. 2.30). Since the internal fracture process depends on plastic constraint resulting from the tensile triaxiality within the neck, the crack nucleation process could be suppressed by introducing hydrostatic pressure. Indeed, Bridgman[44] demonstrated that when a sufficiently large hydrostatic pressure is applied, necking can proceed uninterrupted almost to where the sample draws down to a point (Fig. 2.31).

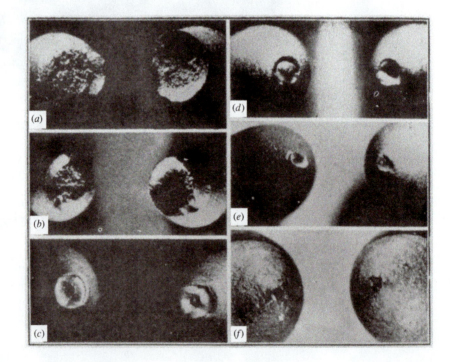

Figure 2.31 Effect of increasing hydrostatic pressure in suppressing internal void formation within necked region. (a) Atmospheric pressure, 10×; (b) 235-MPa hydrostatic pressure, 10×; (c) 1000 MPa, 12×; (d) 1290 MPa, 12×; (e) 1850 MPa, 12×; and (f) 2680 MPa, 18×. (After Bridgman[44]; reprinted by permission of the American Society for Metals, Metals Park, OH.)

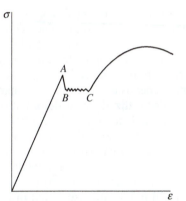

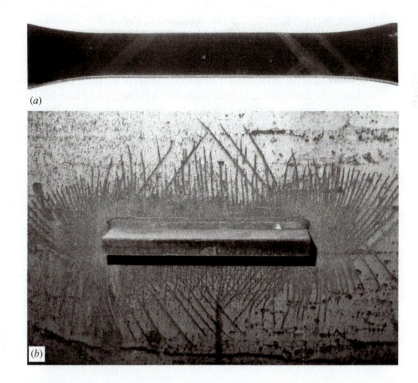

Figure 2.32 Stress–strain behavior exhibiting a narrow heterogeneous deformation region between initial elastic and final homogeneous flow regions. Onset of local yielding occurs at upper yield point A with corresponding load drop to B defined as the lower yield point. After passage of the Lüders bands throughout the gage section, homogeneous deformation commences at C.

2.4.3 Upper Yield Point Behavior

In many body-centered-cubic iron based alloys and some nonferrous alloys, a relatively narrow region of heterogeneous plastic deformation (with a range of approximately 1 to 3% strain) separates the elastic region from the homogeneous plastic flow portion of the stress–strain curve (Fig. 2.32). This segment of the curve is caused by interactions between dislocations and solute atoms. After being loaded elastically to A, defined as the *upper yield point*, the material is observed to develop a local deformation band (Fig. 2.33); the sudden onset of plastic deformation associated with this *Lüders band* is responsible for the initial load drop to B, defined as the *lower yield point*. Outside the Lüders band the material is still loaded elastically. Since the upper yield point is very sensitive to minor stress concentrations, alignment of the specimen in the test grips, and other related factors, measured values reflect

Figure 2.33 (*a*) Concentrated deformation (Lüders) bands formed in plain carbon steel test sample. The band will grow across the gage section before homogeneous deformation develops at point C from Fig. 1.15. (*b*) Lüders band development from weld-related residual tensile stresses. (Photo courtesy of P. Keating.)

considerable scatter. For this reason, the yield strength of materials exhibiting this type of behavior (Fig. 2.32) is usually reported as the lower yield-point value. The remainder of the heterogeneous segment of deformation (*B–C* in Fig. 2.32) is consumed in the passage of the Lüders band across the entire gage section. (Occasionally more than one band may propagate simultaneously during this period.) When deformation has spread to all parts of the gage length, the material then continues to deform in a homogeneous manner with work hardening, necking, and eventual failure, as described previously. This localized plastic deformation phenomenon is well known for low-C steels, in which interstitial carbon and nitrogen atoms can form "atmospheres" around dislocations. These atmospheres strongly pin the dislocations, making it difficult to move them initially (point *A*). However, once the dislocations break away from the pinning solute atoms and become mobile, it is relatively easy to continue their movement (*B* to *C*). Weaker yield point behavior associated with dislocation pinning by substitutional Mg atoms in Al-Mg alloys has also been observed.[45] Yield points are also found in ionic and covalent materials. These, and other related phenomena, are discussed more fully in Section 3.5.1.

The Lüders bands associated with the inhomogeneous plastic deformation phenomenon cause visible surface markings (or *stretcher strains*), often at ∼45° to the tensile axis, that are not desirable in parts for which surface finish is critical. The extent of the elongation associated with the discontinuous yielding process is called the *Yield Point Elongation* (YPE) or the *Lüders strain*, and is defined as the change in elongation between the upper and lower yield points. A good surface finish requires that the YPE be minimized or avoided by appropriate alloy design. Plastic deformation beyond the YPE stage also eliminates subsequent Lüders band formation, although with a mild heat treatment it is possible for the solute atoms to reform pinning atmospheres. This *strain aging* treatment restores the upper yield point behavior.

2.4.4 Temperature and Strain-Rate Effects in Tension

Brief mention was made in Section 2.4.2.4 of a temperature-induced transition in macroscopic fracture surface appearance (as indicated in Fig. 2.30). Since this transition most often parallels important changes in the strength and ductility of the material, some additional discussion is indicated. It is known that the general flow curve for a given material will decrease with increasing temperature T and decreasing strain rate $\dot{\varepsilon}$ (Figs. 2.34a, b). The magnitude of these changes varies with the material; body-centered-cubic metals (e.g., iron, chromium, molybdenum, and tungsten) and ceramic materials are much more sensitive to T and $\dot{\varepsilon}$ than are face-centered-cubic metals (e.g., aluminum, copper, gold, and nickel), with polymeric solids being especially sensitive. Over the years, a number of investigators have sought to define the overall response of a material in terms of some generalized equation of state reflecting the dependence of true stress on strain, strain rate, and temperature. The relationship between *true stress* and *strain rate* is of the same general parabolic

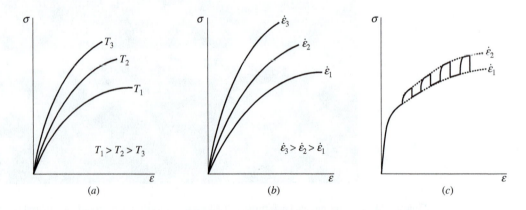

Figure 2.34 Yield strength change as a function of (*a*) temperature and (*b*) strain rate. When the strain rate is changed abruptly, (c) the strain rate sensitivity is evident in a corresponding change in stress.

form as noted by the Holloman relationship in Eq. 2-32, and is given by

$$\sigma_t = K\dot{\varepsilon}^m \tag{2-40}$$

where

$m =$ strain-rate sensitivity factor
$\dot{\varepsilon} =$ strain rate
$K =$ material constant
$\sigma_t =$ true stress

In practice, m is determined by a "step test" in which the specimen is deformed at a constant strain rate initially, then the strain rate is stepped up to a greater value for a short time before returning to the original strain rate, as shown in Fig. 2.34c. Because $\log K = \log\sigma_1 - m\log\dot{\varepsilon}_1$ and $\log K = \log\sigma_2 - m\log\dot{\varepsilon}_2$, the step test allows the determination of m from the expression $m = \log(\sigma_2/\sigma_1)/\log(\dot{\varepsilon}_2/\dot{\varepsilon}_1)$ when the stress values are measured at nominally the same strain (i.e., on the step down from $\dot{\varepsilon}_2$ to $\dot{\varepsilon}_1$). When a series of strain rate steps are performed, m is often found to be a function of the total strain.

For most metals m is low and varies between 0.02 and 0.2. Under certain conditions wherein $m > 0.3$, a given material may exhibit a significant degree of strain-rate sensitivity in association with superplastic deformation behavior (see Section 4.5). In the limit where $m = 1$, the stress–strain-rate material response is analogous to that associated with Newtonian viscous flow (see Section 4.9).

Depending on the nature of the test or service condition, strain rates may vary by more than a dozen orders of magnitude. At low strain rates, below about 10^{-3} s^{-1}, material behavior is characterized by its *creep* and *stress rupture* response (see Chapter 4). At strain rates between 10^3 and 10^5 s^{-1}, the material experiences *impact* conditions and may fail with reduced fracture energy (Chapter 6). *Ballistic conditions* occur with strain rates above 10^5 s^{-1} and involve the shockwave–material interactions associated with such circumstances as projectile impact, high-energy explosions, and meteorite impact with spacecraft (see Section 7.3.1).

Attempts have been made to characterize material properties in terms of parameters that include both test temperature and strain rate. For example, on the basis of simple rate theory, Bennett and Sinclair[46] proposed that the yield strength of iron and other body-centered-cubic transition metals be described in terms of a rate–temperature parameter $T\ln(A/\dot{\varepsilon})$, where A is a frequency factor with an approximate value of 10^8/sec for these materials. As seen in Fig. 2.35, the

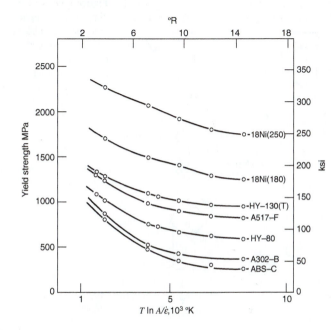

Figure 2.35 Yield strength for seven steels in terms of the Bennett-Sinclair parameter, $T\ln(A/\dot{\varepsilon})$. (Reprinted with permission from A. K. Shoemaker and S. I. Rolfe, *Engineering Fracture Mechanics*, 2, 319, © Pergamon Press, Elmsford, NY (1971.)

parameter provides good correlation for the case of seven steels. While the lower strength steels reveal a somewhat larger yield-strength sensitivity to T and $\dot{\varepsilon}$ than do the stronger alloys at low $T\ln(A/\dot{\varepsilon})$ levels, the seven curves are remarkably similar, reflecting comparable *absolute* changes in yield strength with temperature and strain-rate variations. It is important to recognize, however, that the *relative* change in yield strength with $T\ln(A/\dot{\varepsilon})$ is much greater in the lower strength alloys.

2.5   SLIP IN SINGLE CRYSTALS AND TEXTURED MATERIALS

Let us now consider in detail what happens to the shape and orientation of a single crystal once it begins to yield. From Fig. 2.36 we see that slip on planes oriented χ degrees away from the tensile axis can occur in two ways. First, the planes can simply slide over one another without changing their relative orientation to the load axis. This would be analogous to offsetting groups of playing cards on a table. Since such lateral movement of the crystal planes in a tensile bar would be forbidden by lateral constraints imposed by the specimen grips, the slip planes are forced to rotate with $\chi_i < \chi_0$. X-ray diffraction studies have shown that crystal planes undergo pure rotation in the middle of the gage length but experience simultaneous rotation and bending near the end grips. If we focus attention on the simpler midregion of the sample, it can be shown that the reorientation of the slip plane varies directly with the change in length of the specimen gage length according to the relationship[47]

$$\frac{L_i}{L_0} = \frac{\sin\chi_0}{\sin\chi_i} \tag{2-41}$$

where

$L_0, L_t =$ gage length before (0) and after (i) plastic flow

$\chi_0, \chi_i =$ angle between slip plane and stress axis before (0) and after (i) plastic flow

(Note that $\chi + \Phi = 90°$. However, $\lambda + \Phi = 90°$ *only* when the two vectors are coplanar.) By analogy, the deformation-induced rotation of slip planes is similar to the rotation of individual Venetian blind slats—the more you pull on the cord, the more the individual slats change orientation.

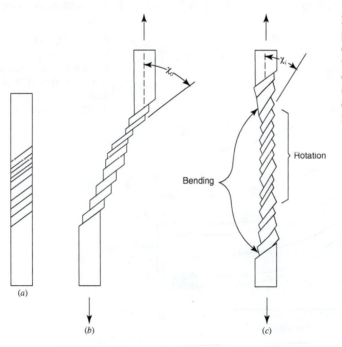

Figure 2.36   Orientation of crystal slip plane, (*a*) prior to deformation; (*b*) after deformation without grip constraint where crystal segments move relative to one another but with no slip rotation; (*c*) after deformation with grip constraint revealing slip plane rotation in gage section (note $\chi_i < \chi_0$).

Rotation

Bending

(*a*)

(*b*)

(*c*)

From the work of Schmid and Boas,[47] when $\chi_0 = \lambda_0$, the shear strain γ, after a given amount of extension, is found to be

$$\gamma = \frac{1}{\sin\chi_0} \left\{ \left[\left(\frac{L_i}{L_0}\right)^2 - \sin^2\lambda_0 \right]^{1/2} - \cos\lambda_0 \right\} \tag{2-42}$$

Note that γ is determined by the initial orientation of the glide elements and by the amount of extension. Furthermore, Eq. 2-42 is valid when only one slip system is active, since multiple slip involves an undefined amount of crystal rotation from each system. The resolved shear stress is given by

$$\tau = \frac{P}{A} \sin\chi_0 \left[1 - \frac{\sin^2\lambda_0}{(L_i/L_0)} \right]^{1/2} \tag{2-43}$$

For a detailed discussion of other relationships involving the shear stresses and strains in single crystals, see Schmid and Boas.[47]

2.5.1 Geometric Hardening and Softening

It is instructive to trace the path of rotation of the slip plane. This is accomplished most readily with the aid of a stereographic projection.[vii] For the purpose of this discussion, some basic understanding of this method is desirable. For the crystal block shown in Fig. 2.37a, imagine that a normal to each plane is extended to intersect an imaginary reference sphere that surrounds the block. Now place a sheet of paper (called the projection plane) tangent to the sphere. Next, take a position at the other end of the sphere diameter, which is oriented normal to

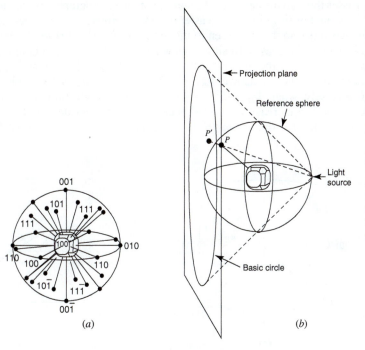

Figure 2.37   Geometric constructions to develop a stereographic projection. (*a*) Intersection of plane normals or reference sphere. (After C. W. Bunn, *Chemical Crystallography,* Clarendon Press, Oxford, England, 1946, p. 30.) (*b*) Projection of poles on the reference sphere to the projection plane. (After N. H. Polakowski and E. J. Ripling, *Strength and Structure of Engineering Materials,* p. 83. Reprinted by permission of Prentice-Hall, Inc., Englewood Cliffs, NJ, © 1966.)

(*a*)   (*b*)

[vii] See B. D. Cullity and S. R. Stock, *Elements of X-Ray Diffraction, 3e,* Addison-Wesley, Reading, MA, 2001, for a treatment of stereographic projections.

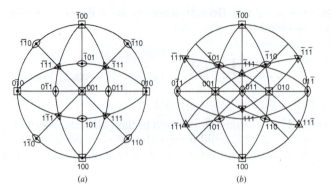

Figure 2.38 Standard stereographic projections for cubic crystals: (*a*) (001) and (*b*) (011) projections.

the projection plane. From this position (called the point of projection) draw lines through the points on the reference sphere and continue on to the projection plane (Fig. 2.37*b*). The points on the projection plane then reflect the relative position of various planes (or plane normals) with planar angle relationships faithfully reproduced. For convenience, standard stereographic projections are used to portray the relative positions of major planes, such as those shown in Fig. 2.38. Since a cubic crystal is highly symmetrical, the relative orientation of a crystal can be given with respect to any triangle within the stereographic projection. As a result, attention is usually focused on the central section of the projection. In Fig. 2.39*a*, for example, we see the axis of a rod in terms of its angular relationship with the (001), (011), and ($\bar{1}$11) planes, respectively. That is, *P* is the normal to the plane lying perpendicular to the rod axis. When this rod is stressed to τ_{CRSS}, the crystal will yield on that slip system possessing the greatest Schmid factor and begin to rotate. For all orientations within triangle I (sometimes referred to as the standard triangle), the (111) [$\bar{1}$01] slip system possesses the greatest Schmid factor and will be the first to operate.[48,49] The rotation occurs along a great circle (corresponding to the trace of a plane on the reference sphere that passes through the center of the sphere) of the stereographic projection and toward the [$\bar{1}$01] slip direction. For simplicity, it is easier to consider rotation of the stress axis relative to the crystal than vice versa, so that *P* is seen to move toward the [$\bar{1}$01] pole. As the crystal rotates, λ will decrease while Φ increases. In situations where $\lambda_0 > 45° > \Phi_0$, rotation of the crystal will bring about an increase in the Schmid factor, since both λ_i and Φ_i, would approach 45°. As a result, yielding can continue at a lower load and the crystal is said to have undergone *geometrical softening*. Conversely, when $\Phi_0 > 45° > \lambda_0$, crystal rotation will bring about a reduction in the Schmid factor, thereby increasing the load necessary for further deformation on the initial slip system. Bear in mind that this *geometrical hardening* is distinct from strain hardening, which

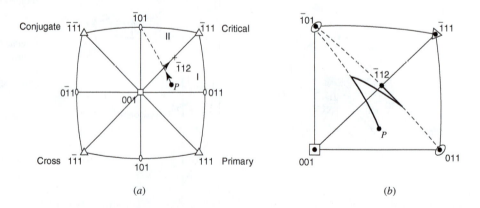

Figure 2.39 (*a*) A (001) stereographic projection showing lattice rotation for FCC crystals during tensile elongation. (*b*) Lattice rotation of FCC crystal involving "overshoot" of primary and conjugate slip systems.

involves dislocation–dislocation interactions (see Section 2.4.1). Geometrical hardening continues as the crystal axis moves toward the $[001]$–$[\bar{1}11]$ tie line. As soon as the relative crystal orientation crosses over into the adjacent triangle II, the Schmid factor on the primary slip system becomes less than that associated with the $(\bar{1}\bar{1}1)\,[011]$ system, the latter being the slip system that would have operated had the crystal been oriented initially within triangle II. This newly activated slip system (the conjugate system) now causes the crystal to rotate along a different great circle toward the $[011]$ direction of the conjugate slip system. Shortly, however, this movement returns the axis of the crystal to within the bounds of triangle I, where primary slip resumes. The ultimate effect of this jockeying back and forth between primary and conjugate slip systems is the movement of the crystal axis along the $[001]-[\bar{1}11]$ tie line to a location where further crystal rotations in either slip direction occur along the same great circle. This point is reached when the load axis is parallel to the $[\bar{1}12]$ direction.

The above analysis reflects the classic geometrical arguments proposed originally by Taylor and Elam.[48,49] In reality, some alloy crystals exhibit "overshooting," wherein the primary slip system continues to operate well into triangle II even though the Schmid factor of the conjugate system is greater. Similarly, the conjugate system, once activated, may continue to operate into triangle I (Fig. 2.39*b*). Koehler[50] proposed that overshooting was caused by weakening of the primary system by passage of dislocations that destroyed precipitates and other solute atom clusters. Consequently, he argued that slip would be easier if continued on the softened primary plane. Alternatively, Piercy et al.[51] argued that overshooting resulted from a "latent hardening" process involving increased resistance to conjugate slip movement resulting from the dislocation debris found on the already activated primary system. That is, for slip to occur on the conjugate system, dislocations on this plane would have to cut across many dislocations lying on the primary plane. By comparison, then, Koehler[50] argued that overshooting resulted from a relative weakening of the primary plane while Piercy et al.[51] argued that the conjugate plane was strengthened relative to the primary plane by a latent hardening mechanism. By careful experimentation, the latent hardening theory was proven correct.

From Fig. 2.39*a*, two other slip systems can be identified. These are denoted as the cross-slip system $(1\bar{1}1)\,[\bar{1}01]$ and the critical slip system $(\bar{1}11)\,[0\bar{1}1]$. The critical system is not encountered very often; the cross-slip system is the system involving the movement of screw dislocations that have cross-slipped out of the primary slip plane. Note that the slip direction is the same in this case (recall the discussion in Section 2.1.5 and Fig. 2.12).

2.5.2 Crystallographic Textures (Preferred Orientations)

From the previous section, it should not be surprising to find individual grains in a polycrystalline aggregate undergoing similar reorientation as a result of plastic deformation. If this is the case, a material with an initially-random orientation of grains can be transformed into one with a substantial fraction of grains aligned in a predictable way with the primary tensile axis (and perhaps other axes as well). This is often the case with drawn wires and rolled plates. Preferred grain orientations introduce anisotropic behavior, including direction-dependent elastic, plastic, and fracture properties.

As might be expected, lattice reorientation in a given grain is impeded by constraints introduced by contiguous grains, making the development of *crystallographic textures* (preferred grain orientations) in polycrystalline aggregates a complex process. In addition, the preferred orientation is found to depend on a number of additional variables, such as the composition and crystal structure of the metal, and the nature, extent, and temperature of the plastic deformation process.[52] As a result, the texture developed by a metal usually is not complete, but instead may be described by the strength of one orientation component relative to another.

Crystallographic textures are portrayed frequently by the pole figure, which is essentially a stereographic projection showing the distribution of *one* particular set of $\{hkl\}$ poles in orientation space. That is, X-ray diffractometer conditions are fixed for a particular diffraction angle and X-ray wavelength so that the distribution of one set of $\{hkl\}$ poles in the polycrystalline sample can be monitored. To illustrate, consider the single-crystal orientation

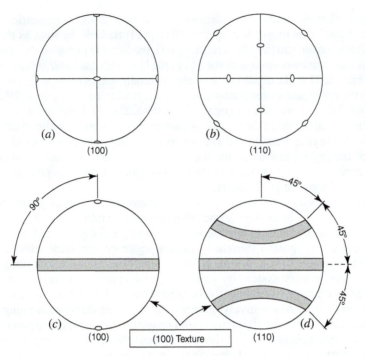

(a)   (100)

(b)   (110)

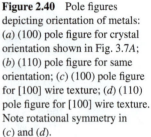

Figure 2.40   Pole figures depicting orientation of metals: (*a*) (100) pole figure for crystal orientation shown in Fig. 3.7*A*; (*b*) (110) pole figure for same orientation; (*c*) (100) pole figure for [100] wire texture; (*d*) (110) pole figure for [100] wire texture. Note rotational symmetry in (*c*) and (*d*).

(c)   (100)

(100) Texture

(d)   (110)

responsible for the (100) stereographic projection shown in Fig. 2.38*a*. The (100) pole figure for this crystal would reveal (100) diffraction spots at the north, south, east, and west poles and at the center of the projection (Fig. 2.40*a*). No information concerning the location of {110}, {111}, or {*hkl*} poles is collected, since diffraction conditions for these planes are not met. Their location would have to be surmised based on the position of the (100) poles and the known angular relation between the {100} and {*hkl*} poles. It is possible, of course, to change diffraction conditions to "see" the location of these other poles, but then the {100} poles would "disappear" from the {*hkl*} pole figure. Figure 2.40*b* shows the same crystal as in Fig. 2.40*a*, but with its orientation portrayed by a (110) pole figure. It is important to appreciate that although these two pole figures look different, they convey the same information—the orientation of the crystal. By analogy, different {*hkl*} pole figures represent different languages by which the same thought (the preferred orientation) is conveyed.

When wires or rods are produced, such as by drawing or swaging, a uniaxial preferred orientation may develop in the drawing direction, with other crystallographic poles distributed symmetrically about the wire axis. For a [100] crystallographic *wire texture*, such as heavily deformed silver wire, the texture is given by Fig. 2.40*c* as portrayed by a (100) pole figure. Note the presence of {100} poles at the north (and south) pole of the projection corresponding to the drawing direction and the smearing out of the other {100} poles across the equator, the latter reflecting the rotational symmetry found in wire textures. The same texture is shown in Fig. 2.40*d* via a (110) pole figure. Here the rotational symmetry of the wire texture is again evident while the [100] wire texture must be inferred from the relative position of the {110} poles. Naturally, a (111) pole figure would present yet another interpretation of the same [100] wire texture. (The reader is advised to sketch the (111) pole figure for the [100] wire texture for his or her edification.) Typical wire textures for a number of FCC metals and alloys are given in Fig. 2.41, where the variation of texture with a material's stacking fault energy (SFE) is shown clearly.[53] (Recall that low stacking fault energy tends to inhibit dislocation motion.) The explanation for the SFE dependence of texture transition and for the reversal in texture at very low stacking fault energies has been the subject of considerable debate. Cross-slip,[54–56] mechanical twinning,[57,58] overshooting,[59,60] and extensive movement of Shockley partial

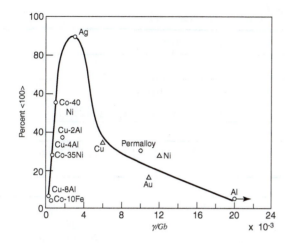

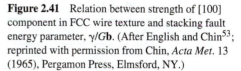

Figure 2.41   Relation between strength of [100] component in FCC wire texture and stacking fault energy parameter, γ/Gb. (After English and Chin[53]; reprinted with permission from Chin, *Acta Met.* 13 (1965), Pergamon Press, Elmsford, NY.)

dislocation[61] mechanisms have been proposed as possible contributing factors toward development of both wire and sheet textures.

Interestingly, wire texture (also called *fiber texture* in this context) is also typical of many polycrystalline thin films deposited by vapor deposition techniques on substrates such as silicon or glass.[viii] In this case, the texture development occurs during film growth rather than from plastic deformation, and the preferred orientation normal to the film plane is selected primarily by surface energy considerations. For FCC metals, a [111] fiber texture typically gives the lowest surface energy. As with drawn wires, there can be significant anisotropy when comparing the in-plane and normal directions of a fiber-textured thin film, but the average behavior everywhere in-plane is isotropic.

For the case of BCC metals, the drawn wire texture is uncomplicated and found to be [110]. In HCP metals, texture is found to vary with the *c/a* ratio. When *c/a* < 1.633, a $[10\bar{1}0]$ texture may be developed with the basal plane lying parallel to the rod axis.[52] By contrast, texture development is more complex when *c/a* > 1.633.

Although wire textures may be defined by one component—the direction parallel to the wire axis—sheet textures in rolled metal plates are given by both the crystallographic plane oriented parallel to the rolling plane and the crystallographic direction found parallel to the rolling direction. Hence, rolling textures are described by the notation (*hkl*) [*uvw*], where (*hkl*) corresponds to the plane parallel to the rolling plane and [*uvw*] to the direction parallel to the rolling direction. Rolling textures are very complex, with several different components often existing simultaneously (Table 2.11).

The (110) $[\bar{1}12]$ texture is often referred to as the *brass* or *silver* texture, typical of FCC materials that possess low stacking fault energy. In copper, nickel, and aluminum, which possess intermediate and high SFE, respectively, the major textural components are (123) $[\bar{4}1\bar{2}]$, (146) $[\bar{2}1\bar{1}]$, and (112) $[11\bar{1}]$. This more complicated preferred orientation is called the *copper* texture. Since the SFE for an alloy depends on solute content (i.e., changes in the electron to atom ratio), the texture of a metal can change from copper to brass type with increasing alloy additions. Some researchers have argued that the importance of SFE in controlling the type of deformation texture is related to the relative ease by which cross-slip of dislocations occurs, the brass texture being generated when cross-slip is more difficult. Others have suggested that the brass texture develops when mechanical twinning (see Section 2.6) or deformation faulting (the creation of stacking faults) is relatively easy. For this reason, the preferred orientation should also be sensitive to the temperature of deformation, since cross-slip, mechanical twinning, and

---

viii Metal films on silicon are used in the semiconductor industry for integrated circuit wiring; metal coatings on glass are often used for their optical properties on mirrors and on solar energy collectors.

Table 2.11   Typical Rolling Textures in Selected Engineering Alloys [52]

Alloy	Rolling Texture
FCC	
Brass, silver, stainless steel	$(110)[\bar{1}12] + (110)[001]$
Copper, nickel, aluminum	$(123)[\bar{4}12] + (146)[\bar{2}\bar{1}1] + (112)[11\bar{1}]$
BCC	
Iron, tungsten, molybdenum, tantalum, niobium	$(001)[\bar{1}10]\ to\ (111)[\bar{1}10] + (112)[\bar{1}10]\ to\ (111)[\bar{2}11]$
HCP	
Magnesium, cobalt (c/a $\approx$1.633)	$(0001)[2\bar{1}\bar{1}0]$
Zinc, cadmium (c/a > 1.633)	(0001) plane tilted ±20-$25°$ from rolling plane about a $[10\bar{1}0]$ transverse direction axis
Titanium, zirconium, beryllium (c/a < 1.633)	(0001) plane tilted ±30-$40°$ from rolling plane about a $[10\bar{1}0]$ rolling direction axis

faulting are thermally dependent processes. In studying the rolling texture in high-purity silver, Hu and Cline[62] found that cold rolling at 0°C produced a typical $\{110\}\langle211\rangle$ *brass* or *silver* texture. However, when the silver was rolled at 200°C, near $\{123\}\ \langle412\rangle$ and $\{146\}\ \langle211\rangle$ components were observed, reflecting a *copper*-type texture. A similar *brass- to copper-type* texture transition was found when 18-8 stainless steel was rolled at 200 and 800°C, respectively.[63] Conversely, a reverse *copper* to *brass* texture transition was realized for copper when the rolling temperature was reduced from ambient to-196°C.[64] Finally, by combining the effects of alloy content and deformation temperature on SFE, Smallman and Green[54] demonstrated for the silver–aluminum alloy that the *brass* to *copper* rolling texture transition temperature increased with decreasing initial stacking fault energy.

Vapor-deposited polycrystalline films can exhibit sheet textures when there is a particular crystallographic orientation between the film and an underlying crystalline substrate that minimizes interfacial energy. In the extreme, a single crystal film can be grown *epitaxially* on a single crystal substrate—the ultimate in film texture. This is the case for *epitaxial growth* of Si-Ge on Si single crystals, GaN on sapphire (single crystal Al_2O_3) substrates, and many other technologically important systems.

2.5.3 Plastic Anisotropy

It follows from the previous discussion that when a metal sheet or rod contains a preferred crystallographic orientation, the ability of the material to deform in an isotropic manner is altered. For example, assume that a sheet of α-titanium possesses an idealized texture with the (0001) basal planes oriented parallel plane of the sheet and $\langle1\bar{2}10\rangle$ directions aligned parallel to the rolling directions.[65–67] Slip can occur on (0001), $(10\bar{1}0)$, and $(10\bar{1}1)$ planes but only in the $\langle11\bar{2}0\rangle$ close-packed directions; therefore, no sheet thinning can occur in association with these slip systems. Recall that plastic deformation is a constant volume process (Eq. 1-3) where

$$\varepsilon_1 = -(\varepsilon_w + \varepsilon_t) \qquad (2\text{-}44a)$$

Since $\varepsilon_t = 0$,

$$\varepsilon_1 = -\varepsilon_w \qquad (2\text{-}44b)$$

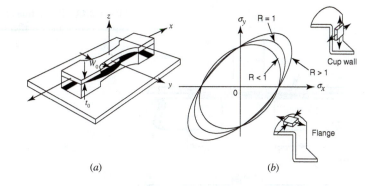

Figure 2.42   (*a*) Tensile coupon and dimensions as cut from sheet stock.[65] (Walter A. Backofen, Ed., et al., *Fundamentals of Deformation Processing* (Syracuse, NY: Syracuse University Press, 1964). By permission of the publisher.) (*b*) Yield loci for textured material.[66] When $R > 1$, material exhibits thinning resistance and high strength under biaxial tension; when $R < 1$, the material displays easy thinning and low biaxial tensile strength. (W. F. Hosford and R. M. Caddell, *Metal Forming: Mechanics, and Metallurgy*, 2d ed., Cambridge University Press, p. 273 (1993). Reprinted with the permission of Cambridge University Press.)

Therefore, tensile strains in a coupon prepared from a textured sheet (Fig. 2.42*a*) would be balanced only by a reduction in sample width. Accordingly,

$$\varepsilon_1 = \ln(1/l_0) = -\ln(w/w_0) \tag{2-44c}$$

whereas

$$\varepsilon_t = \ln(t/t_0) = 0 \tag{2-44d}$$

A useful parameter to quantify the amount of plastic strain anisotropy in a sheet is identified by R, where

$$R = \frac{\varepsilon_w}{\varepsilon_t} \tag{2-45a}$$

Since R and elastic modulus values typically vary within the plane of the textured sheet, it is common to describe an average R-value, $\bar{R}$, where

$$\bar{R} = \frac{R_0 + 2R_{45} + R_{90}}{4} \tag{2-45b}$$

with the subscripts corresponding to the orientation within the sheet.

For the ideal texture described previously, $\bar{R} = \infty$; alternatively, when no texture exists and the material behaves in an isotropic manner, $\bar{R} = 1$. For realistic crystallographic textures, such as in α-titanium alloys where basal planes are tilted at $\pm 30°$–$40°$ from the rolling plane (Table 2.11), $\bar{R}$ values of 3–7 are typically experienced; the higher the value of $\bar{R}$ the greater the sheet's resistance to thinning and the higher the material's yield strength under through-thickness compression or balanced biaxial tensile loading conditions (Fig. 2.42*b*). Note that the existence of texture and its influence on yield strength is obscured under uniaxial loading conditions and of limited importance in pure shear (i.e., where $\sigma_x = -\sigma_y$).

The influence of plastic anisotropy on metal forming is demonstrated by the deep drawing of flat sheets into cartridge cases, bathtubs, brass flashlight cases, and automobile panels.

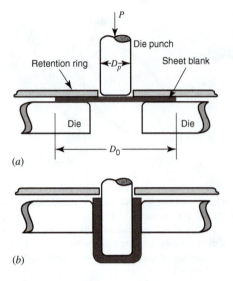

(a)

(b)

Figure 2.43 Illustration revealing deep drawing of a cylindrical cup (a) before and (b) after drawing.

In this process, a circular sheet of metal is clamped over a die opening and then pressed through the die with a punch (Fig. 2.43). The load from the punch is transmitted along the sidewall of the cup to the flange area, where most of the deformation takes place. Within the flange area, the stress state approaches that of pure shear, corresponding to tension in the radial direction and compression in the circumferential direction (Fig. 2.42b, Flange). By contrast, a plane strain biaxial tension condition exists in the cup wall (Fig. 2.42b, Cup wall). Failure occurs by localized necking within a narrow ring of material in the cup wall just above the radius of the punch. Analysis of this forming process reveals that the upper-bound theoretical limiting drawing ratio (LDR) is estimated to be[67]

$$\text{LDR} \approx \left(\frac{D_0}{D_p}\right)_{\text{max}} \approx e^{\eta} \qquad (2\text{-}46)$$

where D_0 and D_p are the original sheet and final cup diameters, respectively, and η is a parameter that accounts for frictional losses in the drawing process. For ideal efficiency, $\eta = 1$ and LDR ≈ 2.7. Typically, however, $\eta \approx 0.74$ to 0.79; hence, LDR ≈ 2.1 to 2.2.

The limiting drawing ratio can be increased—to permit the drawing of deeper cups—by restricting the material's ability to thin in the critical zone near the bottom of the cup wall. This can be achieved by strengthening the sheet in the thickness direction through the development of a crystallographic texture ($R > 1$) that limits deformation under the plane strain biaxial tension conditions experienced in the cup wall (recall Fig. 2.42b, Cup wall). Notice that LDR increases for several metal alloys with average plastic strain ratio, $\bar{R}$ (Fig. 2.44). For further discussion of the influence of texture on metal forming, the reader is referred to texts by Hosford and Caddell,[66] and Dieter.[68]

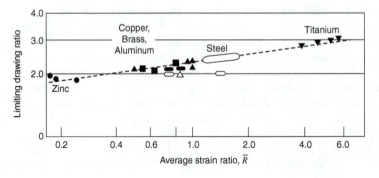

Figure 2.44 Influence of average strain ratio, $\bar{R}$ on limiting drawing ratio for several sheet metal alloys.[69] (M. Atkinson, *Sheet Metal Industries,* 44, 167 (1967) with permission.)

2.6 DEFORMATION TWINNING

As was noted in Section 2.2.1, the simultaneous operation of at least five independent slip systems is required to maintain continuity at grain boundaries in a polycrystalline solid. Failure to do so will lead to premature fracture. If a crystal possesses an insufficient number of independent slip systems, *twin modes* may be activated in some metals to provide the additional deformation mechanisms necessary to bring about an arbitrary shape change.

2.6.1 Comparison of Slip and Twinning Deformations

The most obvious difference between a slipped versus a twinned crystal is the external shape change resulting from these deformations. Whereas slip involves a simple translation across a slip plane such that one rigid portion of the solid moves relative to the other, the twinned body undergoes a significant shape change without any translation along the *twinning plane* (Fig. 2.45).

According to Bilby and Crocker,[70] "A deformation twin is a region of a crystalline body which had undergone a homogeneous shape deformation in such a way that the resulting product structure is identical with that of the parent, but oriented differently." As pointed out earlier in Section 2.1.2, dislocation movement associated with slip will take place in multiples of the unit displacement, *b*. By contrast, the shape change found in the twinned solid results from atom movements taking place on all planes in fractional amounts within the twin. In fact, we see from Fig. 2.45c that the displacement in any plane within the twin is directly proportional to its distance from the twin-matrix boundary. Upon closer examination of these twinning displacements in a simple cubic lattice, it is seen that the twinning process has effected a *rotation of the lattice* such that the atom positions in the twin represent a *mirror image* of those in the untwinned material (Fig. 2.46). By contrast, slip occurs by translations along widely spaced planes in whole multiples of the displacement vector, so that the relative orientation of different regions in the slipped cube remains unchanged.

The differences associated with these deformation mechanisms are revealed when one examines the deformed surface of a sample that was prepolished (Fig. 2.47). Offsets due to slip are revealed as straight or wavy lines (depending on the stacking fault energy of the material and the active slip systems) with no change in image contrast noted on either side of the slip offset. Twin bands do exhibit a change in contrast, since the associated lattice reorientation within the twin causes the incident light to be reflected away from the objective lens of the microscope. After repolishing and etching the sample, only twin band markings persist, since they were associated with a reorientation of the lattice, and not with surface features (Figs. 2.47b and 2.47d).

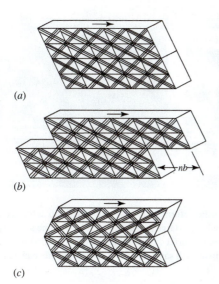

(a)

(b)

(c)

Figure 2.45 Shape change in a solid cube caused by plastic deformation. (*a*) Undistorted cube; (*b*) slipped cube with offsets *nb*; (c) twinned cube revealing reorientation within twin. Displacements are proportional to the distance from the twin plane.

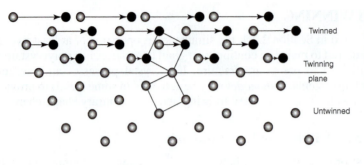

Figure 2.46 Twinning on the (120) plane in a simple cubic crystal. Gray circles represent original atom positions. Black circles are final atom positions.

Before proceeding further, it is appropriate to distinguish between *deformation* twins (Figs. 2.47*a* and 2.47*b*) and *annealing* twins (Figs. 2.47*c* and 2.47*d*). The deformation twins in the zinc specimen were generated as a result of plastic deformation, where the annealing twins in the brass sample *preexisted* plastic deformation. The annealing twins were formed instead during prior heat treatment of the brass in association with recrystallization and growth of new grains. During the formation of a new packing order in the new crystals, the emerging grains could have encountered packing sequence defects in the original grains (such as stacking faults); this interaction would result in the formation of annealing twins.

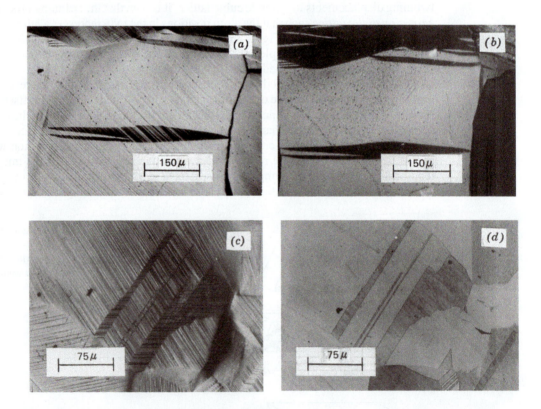

Figure 2.47 Surface markings resulting from plastic deformation. (*a*) Prepolished and deformed zinc revealing slip lines (upper left to lower right markings) and twin bands (large horizontal band); (*b*) same as (*a*) but repolished and etched to show only twin bands; (*c*) prepolished and deformed brass revealing straight slip lines (reflecting low stacking fault energy) and preexisting annealing twins; (*d*) same as (*c*) but repolished and etched to show only annealing twins.

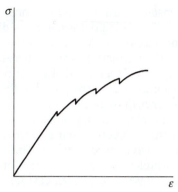

Figure 2.48   Type III stress–strain behavior reflecting elastic behavior followed by heterogeneous plastic flow. The latter can be caused by twin controlled deformation or solute atom-dislocation interactions.

For example, the error indicated by the vertical line in the following planar packing sequence—*ABCABCBACBA*—constitutes a twin boundary. Note also that the two planes on either side of the twin plane are similar: This arrangement constitutes a stacking fault (i.e., *BCB*). Without the preexistence of stacking faults in the old grains, annealing twins are unlikely to form. Hence, annealing twins are rarely seen in aluminum, which has a high stacking fault energy (low stacking fault probability). Conversely, annealing twins are observed readily in brass, which has a low stacking fault energy (high stacking fault probability). Since the stacking fault probability also depends on the extent of deformation, the number of annealing twins found in a given material should increase with increasing prior cold work. As such, the number of annealing twins found in a recrystallized material provides a clue as to the deformation history of the material.

2.6.2   Heterogeneous Plastic Tensile Behavior

Occasionally, a tensile test specimen will produce a stress–strain curve that exhibits a series of serrations that are superimposed on the parabolic portion of Fig. 2.25 after the normal range of elastic response. Such behavior, shown in Fig. 2.48, reflects nonuniform or heterogeneous deformation within the material. As we can now appreciate, when hexagonal close-packed metals are tested over a relatively wide temperature range, they tend to deform plastically by a combination of slip along glide planes and twinning in discrete zones within the specimen. When twinning occurs, extension of the gage length proceeds in discrete bursts that are associated with twin band nucleation and growth. Often, these bursts of deformation are associated with audible clicks emitted from within the sample. Whenever the instantaneous strain rate in the specimen exceeds the rate of motion of the test machine crosshead, a load drop will occur. A similar stress–strain response is found in body-centered-cubic metals tested at low temperatures and in face-centered-cubic metals tested under a combination of low temperatures and high strain rates.

However, it should be recognized that serrated stress–strain curves are also encountered in materials for which twinning is very unlikely, such as in room temperature body-centered-cubic iron alloys containing carbon in solid solution, and in dilute solid solutions of aluminum. In these cases the phenomenon is related to inhomogeneous dislocation motion (see Sections 2.4.3 and 3.5.1), not bursts of twinning activity, so serrated loading curves should not necessarily be interpreted as evidence of twinning without other substantiating evidence.

2.6.3   Stress Requirements for Twinning

The establishment of a critical resolved shear stress (CRSS) is the basis for predicting the conditions required for plastic deformation by dislocation motion. Given that mechanical twinning is a shear process as well, it is logical to assume that a similar CRSS criterion might apply to deformation by twinning. Unfortunately, the picture is quite cloudy even after decades of work, and there remains no consensus regarding the existence of an equivalent simple twinning criterion.[71] In a pure material undergoing deformation by dislocation glide, the shear stress needed to initiate dislocation motion is the same as the level needed to continue the motion. It is now known that the twin initiation stress is usually much greater than the stress needed to propagate a preexisting twin. In a related observation, the *nucleation* of a new twin is often associated with a sudden load drop (responsible for serrated stress-strain curves as shown in Fig. 2.48), while the *growth* of an existing twin exhibits smoother loading behavior. Furthermore, twinning is inherently "antisymmetric" in the sense that a certain shear stress may cause twinning in a certain direction, but a shear stress of identical magnitude applied in the opposite direction may have no such effect. This is quite different from the behavior of a gliding dislocation, for which forward and reverse deformation occurs under shear stresses of comparable magnitude. These observations cast serious doubt on the validity of a single "CRSS for twinning." Finally, even though there are studies that present experimental evidence of a critical twinning stress, it is often difficult to rule out other influences on the behavior, and the degree of scatter in the values is very large for a given material. Thus, although twinning stress values have been reported and are sometimes used for modeling deformation behavior, there is limited support for the concept of a universal rule for twinning based on a resolved shear stress.

2.6.4 Geometry of Twin Formation[72]

Consider the growth of a twin over the upper half of a crystalline unit sphere. Any point on the sphere will be translated from coordinates X, Y, Z to X', T', Z', where $X = X'$, $Z = Z'$ and $Y' = Y + SZ$ (Fig. 2.49). Since S represents the magnitude of the shear strain, we see that the shear displacement on any plane is directly proportional to the distance from the twinning plane (called the composition plane). Therefore, the equation for the distorted sphere is given by

$$X'^2 + Y'^2 + Z'^2 = 1 = X^2 + Y^2 + 2SZY + S^2Z^2 + Z^2 = 1 \tag{2-47}$$

or

$$X^2 + Y^2 + 2SZY + Z^2(S^2 + 1) = 1 \tag{2-48}$$

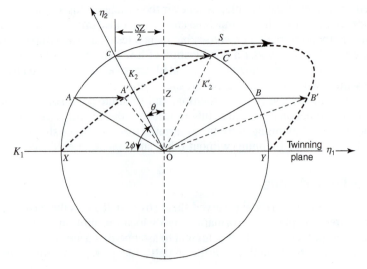

Figure 2.49 Crystal sphere distorted to that of an ellipsoid. Undistorted planes are K_1 and K_2, separated by angle 2ϕ. Note foreshortening of plane OA after twinning, while plane OB is extended.

Table 2.12 Observed Twin Elements *a* Metals[74,75]

Metal	Crystal Structure	c/a Ratio	K_1	K_2	η_1	η_2	S	$\frac{(l'-l)}{l_{max}}$
Al, Cu, Au, Ni, Ag, γ-Fe	FCC		$\{111\}$	$\{11\bar{1}\}$	$\langle 11\bar{2}\rangle$	$\langle 112\rangle$	0.707	41.4%
α-Fe	BCC		$\{112\}$	$\{\bar{1}\bar{1}2\}$	$\langle \bar{1}\bar{1}1\rangle$	$\langle 111\rangle$	0.707	41.4
Cd	HCP	1.886	$\{10\bar{1}2\}$	$\{\bar{1}012\}$	$\langle 10\bar{1}1\rangle$	$\langle 10\bar{1}1\rangle$	0.17	8.9
Zn	HCP	1.856	$\{10\bar{1}2\}$	$\{\bar{1}012\}$	$\langle 10\bar{1}1\rangle$	$\langle 10\bar{1}1\rangle$	0.139	7.2
Mg	HCP	1.624	$\{10\bar{1}2\}$	$\{\bar{1}012\}$	$\langle 10\bar{1}1\rangle$	$\langle 10\bar{1}1\rangle$	0.131	6.8
			$\{11\bar{2}1\}$	$\{0001\}$	$\langle 11\bar{2}6\rangle$	$\langle 11\bar{2}0\rangle$	0.64	37
Zr	HCP	1.589	$\{10\bar{1}2\}$	$\{\bar{1}012\}$	$\langle 10\bar{1}1\rangle$	$\langle 10\bar{1}1\rangle$	0.167	8.7
			$\{11\bar{2}1\}$	$\{0001\}$	$\langle 11\bar{2}6\rangle$	$\langle 11\bar{2}0\rangle$	0.63	36.3
			$\{11\bar{2}2\}$	$\{10\bar{2}4\}$	$\langle 11\bar{2}3\rangle$	$\langle 22\bar{4}3\rangle$	0.225	11.9
Ti	HCP	1.587	$\{10\bar{1}2\}$	$\{\bar{1}012\}$	$\langle 10\bar{1}1\rangle$	$\langle 10\bar{1}1\rangle$	0.167	8.7
			$\{11\bar{2}1\}$	$\{0001\}$	$\langle 11\bar{2}6\rangle$	$\langle 11\bar{2}0\rangle$	0.638	36.9
			$\{11\bar{2}2\}$	$\{10\bar{2}4\}$	$\langle 11\bar{2}3\rangle$	$\langle 22\bar{4}3\rangle$	0.255	11.9
Be	HCP	1.568	$\{10\bar{1}2\}$	$\{\bar{1}012\}$	$\langle 10\bar{1}1\rangle$	$\langle 10\bar{1}1\rangle$	0.199	10.4

which defines a quadric surface. Specifically, the distorted sphere forms an ellipsoid whose major axis is inclined to η_1 by an angle ϕ. It is clear from Fig. 2.49 that most planes contained within the sphere are either foreshortened or extended. For example, consider the movement of points A and B, which are translated by the twinning deformation to A' and B', respectively. If AO and BO represent the traces of two different planes, it is clear that AO has been foreshortened ($A'O$) while BO has been stretched ($B'O$). Only two planes remain undistorted after the twin shear has been completed. The first is the composition plane, designated as K_1; the direction of the shear is given by η_1. The second undistorted plane is the one shown in profile by the line OC. (Note that $OC = OC'$.) This plane is designated as the K_2 plane, where η_2 is defined by the line of intersection of the K_2 plane and the plane of shear (the plane of this page). The final position of this second undistorted plane is designated as the K_2 plane. Therefore, all planes located between X and C will be compressed, while all planes located between C and Y will be extended. Typical values for $K_1 K_2$, η_1, and η_2 are shown in Table 2.12 and discussed in the following sections. By definition,[73] when K_1 and η_2 are rational and K_2 and η_1 are not, we speak of this twin as being of the *first kind*. The orientation change resulting from this twin can be accounted for by reflection in the K_1 plane or by a 180° rotation about the normal to K_1. When K_2 and η_1 are rational but K_1 and η_2 are not, the twin is of the *second kind*. The twin orientation in this case may be achieved either by a 180° rotation about η_1 or by reflection in the plane normal to η_1. When all twin elements are rational, the twin is designated as *compound*. This occurs often in crystals possessing high symmetry (such as most metals), where the reflection and rotation operations are equivalent.

The magnitude of the shear strain S in the unit sphere is given by the angle 2ϕ between the two undistorted planes K_1 and K_2. From Fig. 2.49

$$\tan\theta = SZ/2/Z = \frac{S}{2} \tag{2-49}$$

Since

$$\theta + 2\phi = 90°$$

$$\cot 2\phi = \frac{S}{2} \tag{2-50}$$

2.6.5 Elongation Potential of Twin Deformation

Hall[72] has shown that the total deformation strain to be expected from a completely twinned crystal may be given by

$$\frac{l'}{l} = [1 + S\tan\chi]^{1/2} \tag{2-51}$$

where l, l' = initial and final lengths, respectively

$$\tan\chi = \frac{S \pm \sqrt{S^2 + 4}}{2}$$

From Eq. 2-51, the maximum potential elongation of the metals shown in Table 2.12 is quite small, particularly in HCP crystals, which undergo $\{10\bar{1}2\}$-type twinning. Although the twinning reaction contributes little to the total elongation of the sample, the rotation of the crystal within the twin serves mainly to reorient the slip planes so that they might experience a higher resolved shear stress and thereby contribute more deformation by slip processes.

2.6.6 Twin Shape

From the above geometrical analysis, one would assume twinned regions to be bounded by two parallel composition planes representing the two twin-matrix coherent interfaces. In practice, twins are often found to be lens-shaped, so that the interface must consist of both coherent and noncoherent segments. These noncoherent portions of the interface can be described in terms of particular dislocation arrays (Fig. 2.50). Mahajan and Williams[76] have reviewed the literature and found that twin formation has been rationalized both in terms of heterogeneous nucleation at some dislocation arrangement or by homogeneous nucleation in a region of high stress concentration. It is worth noting that dislocations are also needed to account for the requirement of a much lower stress to move a twin boundary than the theoretically expected value.

Cahn[73] postulated that the lens angle β should increase with decreasing shear strain. Since the magnitude of β controls the permissible thickness of the lens, Cahn's postulate correctly predicts the empirical fact that twin thickness increases with decreasing shear strain. More recently, Friedel[78] also concluded that the optimum lens thickness to length ratio should increase with decreasing shear strain. Since twin formation involves discontinuous deformations, some type of lattice accommodation is necessary along the perimeter of the twin lens. When the parent lattice possesses limited ductility, the lens angle β is kept small and the strain discontinuity accommodated by crack formation.[79] At the other extreme, lattice plane bending and/or slip may be introduced to "smear out" the strain discontinuity resulting from the twin. If the crystal is able to slip readily, the lens angle β can increase, thereby enabling the twin to thicken. Therefore, we find that in ductile crystals, the thickness of deformation twins increases with decreasing twin shear strain (Fig. 2.51). From this discussion, it follows that the twins seen in the brass sample in Figs. 2.47c and d were of the *annealing* type since the deformation twin strain ($S = 0.707$) in this material would have produced thin deformation twins similar to those shown in Fig. 2.51a.

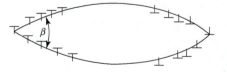

Figure 2.50   Diagram of a lens-shaped twin with dislocations to accommodate noncoherent twin-matrix interface regions. Lens angle β increases with decreasing twin shear and increasing ability of matrix to accommodate the twin strain concentration.

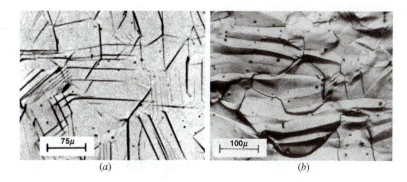

Figure 2.51 Prepolished and subsequently deformed surfaces revealing shape of twins. (*a*) Narrow deformation twins in α-Fe ($S = 0.707$); (*b*) broad deformation twins in Mg ($S = 0.131$). (After Eckelmeyer and Hertzberg[77]; American Society for Metals, Metals Park, OH, © 1970.)

2.6.7 Twinning in HCP Crystals

Among the three major unit cells found in metals and their alloys, twinning is most prevalent in HCP materials. Over a broad temperature range, twinning and slip are highly competitive deformation processes. We saw from Section 2.2.1 that regardless of the *c/a* ratio (that is, whether basal or prism slip was preferred), an insufficient number of independent slip systems can operate to satisfy the von Mises requirement.[80] Since alloys of magnesium, titanium, and zinc are known to possess reasonable ductility, some other deformation mechanisms must be operative. While combinations of basal, prism, and pyramidal slip do not provide the necessary five independent slip systems necessary for an arbitrary shape change in a polycrystalline material, deformation twinning often is necessary to satisfy von Mises' requirement.

Twinning in HCP metals and alloys has been observed on a number of different planes (Table 2.12). One twin mode common to many HCP metals is that involving $\{10\bar{1}2\}$ planes. One of three possible sets of these planes is shown in Fig. 2.52*a*. Activation of one particular set will depend on the respective Schmid factors. Naturally, twinning will occur on the $\{10\bar{1}2\}$ plane and $\langle\bar{1}011\rangle$ direction that experiences the highest resolved shear stress. For additional clarification, the angular relationships between the undistorted $\{10\bar{1}2\}$ planes and the prism and basal planes are shown in Fig. 2.52*b*, where

$$\tan\theta = \frac{c}{\sqrt{3}a} \tag{2-52}$$

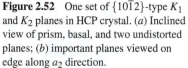

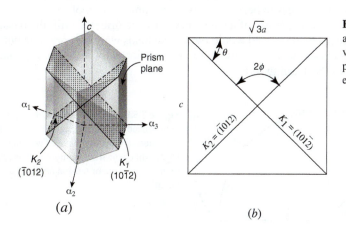

Figure 2.52 One set of $\{10\bar{1}2\}$-type K_1 and K_2 planes in HCP crystal. (*a*) Inclined view of prism, basal, and two undistorted planes; (*b*) important planes viewed on edge along a_2 direction.

Table 2.13   Interplanar Angles in Beryllium and Zinc

	$\{10\bar{1}2\} - \{0001\}$	$\{10\bar{1}2\} - \{\bar{1}012\}$	$\{10\bar{1}2\} - \{10\bar{1}0\}$
Beryllium	42°10′	84°20′	47°50′
Zinc	46°59′	86°02′	43°01′

Since $2\phi + 2\theta = 180°$ and $\tan 2\phi = 2/S$,

$$\tan 2\phi = 2/S = \tan(180 - 2\theta) \tag{2-53}$$

With trigonometric identities it may be shown that

$$S = \frac{\tan^2\theta - 1}{\tan\theta} \tag{2-54}$$

Combining Eqs. 2-52 and 2-54 and rearranging, we find

$$S = [(c/a)^2 - 3]\frac{\sqrt{3}a}{3c} \tag{2-55}$$

From Eq. 2-55, it is seen that the sense of the twin deformation is opposite for HCP metals exhibiting c/a ratios $\neq \sqrt{3}$. When $c/a = \sqrt{3}$, the analysis predicts that $S = 0$ and that twinning would not occur by the $\{10\bar{1}2\}$ mode. Stoloff and Gensamer[81] have verified this in a magnesium crystal alloyed with cadmium to produce a c/a ratio of $\sqrt{3}$. The reversal in sense of the twin deformation is seen when the responses of beryllium ($c/a = 1.568$) and zinc ($c/a = 1.856$) are compared using strain ellipsoid diagrams. The relevant interplanar angles in each metal are determined by

$$\cos\theta = \frac{h_1 h_2 + k_1 k_2 + \frac{1}{2}(h_1 k_2 + h_2 k_1) + \frac{3a^2}{4c^2}l_1 l_2}{\left[\left(h_1^2 + k_1^2 + h_1 k_1 + \frac{3a^2}{4c^2}l_1^2\right)\left(h_2^2 + k_2^2 + h_2 k_2 + \frac{3a^2}{4c^2}l_2^2\right)\right]^{1/2}} \tag{2-56}$$

and are given in Table 2.13.

For the case of beryllium, the basal plane bisects the acute angle separating the $\{10\bar{1}2\}$ planes, and the prism plane bisects its supplement. In addition, the prism plane may be positioned simply by the fact that it must lie 90° away from the basal plane. From Fig. 2.53, we see that the twinning process in beryllium involves compression of the basal plane and tension of the prism plane. Consequently, if a single crystal were oriented with the basal plane parallel to the loading direction, the crystal would twin if the loads were compressive but not if the loads were tensile. The crystal would be able to twin in tension only if the basal plane were oriented perpendicularly to the loading axis.

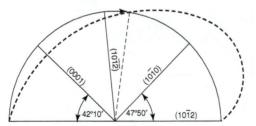

Figure 2.53   Strain ellipsoid for beryllium revealing twin-related foreshortening of the basal plane and extension of the prism plane. Twinning by $\{10\bar{1}2\}$ mode will occur when compression is applied parallel to the basal plane or tension applied parallel to the prism plane.

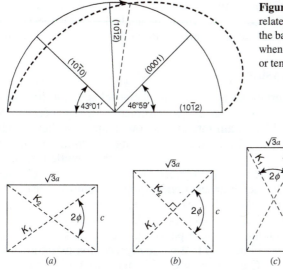

Figure 2.54 Strain ellipsoid for zinc revealing twin-related foreshortening of the prism plane and extension of the basal plane. Twinning by $\{10\bar{1}2\}$ mode will occur when compression is applied parallel to the prism plane or tension applied parallel to the basal plane.

Figure 2.55 Conditions for $\{10\bar{1}2\}$ twinning in hexagonal crystals. (a) When $c/a < \sqrt{3}$ twinning results when the basal plane is compressed or the $\{10\bar{1}0\}$ planes stretched; (b) when $c/a = \sqrt{3}$ no twinning by $\{10\bar{1}2\}$ mode occurs; (c) when $c/a > \sqrt{3}$ twinning occurs when the prism planes are compressed or the basal planes stretched.

The situation is completely opposite for zinc. Here, because the prism plane bisects the acute angle between K_1 and K_2, zinc will twin when the applied stress causes compression of the prism or extension of the basal plane (Fig. 2.54). The response of any HCP metal that twins by the $\{10\bar{1}2\}$ mode is summarized in Fig. 2.55. When $c/a < \sqrt{3}$ twinning will occur if compressive loads are applied parallel to the basal plane or tensile loads applied parallel to the prism planes. The opposite is true for the case of $c/a > \sqrt{3}$, where twinning occurs when the prism plane is compressed or the basal plane extended.

The other HCP twin modes shown in Table 3.4 may operate under certain conditions; however, they are generally not preferred since the strain energy of the twin increases with S.[23] Therefore, if the resolved shear stress for given K_1 and η_1 twin elements is sufficient, twinning will occur via the mode possessing the lowest shear strain. As might be expected, there is competition not only between twin modes but also between slip and twinning as the dominant deformation mechanism under specific test conditions. Reed-Hill[74] examined the likelihood of either prism slip or $\{10\bar{1}2\}$ twinning in zirconium and found these mechanisms to be complementary (Fig. 2.56). As such, slip will occur on a viable slip system if the resolved shear stress is high enough; twinning will occur if the resolved shear stress along the K_1 and η_1 twin elements is high enough *and* the direction of loading consistent with the twinning process. For example, Fig. 2.56 shows that the highest shear stress along the K_1 and η_1 elements, corresponding to the largest orientation factor, *may* generate

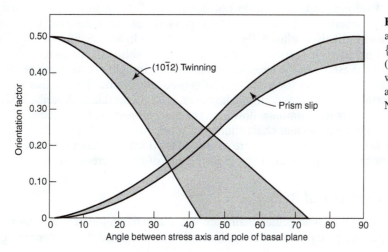

Figure 2.56 Competitive aspects of prism slip and $\{10\bar{1}2\}$ twinning in zirconium. (After Reed-Hill[74]; reprinted with permission from Gordon and Breach Science Publishers, New York.)

twinning when the angle between the applied stress axis and pole of the basal plane is zero degrees. In this situation, twinning will occur *only* when the applied stress is tensile; when this stress is compressive in nature, the resolved shear stress will be the same, but no twinning will occur.

2.6.8 Twinning in BCC and FCC Crystals

Twinning in BCC materials has been examined most closely for the case of ferritic steels, because of their engineering significance. Twin formation in steels (called *Neumann bands*) occurs most readily under high strain rate and/or low-temperature test conditions. The twin plane is found to be of type {112}, with the shearing direction parallel to [$\bar{1}\bar{1}1$] What is intriguing again is the fact that twinning will depend on the direction of shear; twinning will occur in the [$\bar{1}\bar{1}1$] direction but not in the opposite [$11\bar{1}$] direction.[82] Deformation twinning is found least frequently in FCC metals except under cryogenic temperature conditions, extremely high strain rates, and in certain alloys. Since the twin elements {111}, {$11\bar{1}$}, $\langle 11\bar{2} \rangle$, and $\langle 112 \rangle$ produce a large twin strain (0.707), it would appear that slip processes are more highly favored (i.e., partial dislocation motion along close-packed planes) than are twin-related movements. By comparison, it should be pointed out that while *deformation* twinning is found only under extreme conditions, some FCC alloys may exhibit many *annealing* twins. As discussed in Section 2.6.1, these twins result from accidents associated with the growth of recrystallized grains from previously deformed material possessing a high density of stacking faults.

2.7 PLASTICITY IN POLYMERS

In certain respects, the deformation of polymeric solids bears strong resemblance to that of metals and ceramics: Polymers become increasingly deformable with increasing temperature, as witnessed by the onset of additional flow mechanisms. Also, the extent of polymer deformation is found to vary with time, temperature, stress, and microstructure consistent with parallel observations for fully crystalline solids. In contrast, the macromolecular nature of polymeric solids leads to a set of plasticity mechanisms that are quite different in many respects from those typical of crystalline metals and ceramics. Unlike crystalline metals and ceramics, which can be pictured as regular arrays of individual atoms connected by bonds of a certain character, a polymer is made up of molecular chains that can be arranged either in a completely amorphous fashion (picture a jar full of spaghetti) or with regions of crystalline order (picture a rope folded neatly back and forth on itself). The bonds along the backbone of a chain are covalent in nature, and are therefore quite strong, while the bonds between separate chains in *thermoplastics* are of secondary character (e.g., Van der Waals) and are therefore relatively weak. In the class of polymers known as *thermosets*, there are strong covalent *cross-links* between the chains. As in ceramic materials, the covalent bonds in polymers are not easy to break and reform, and as such do not contribute directly to any significant plastic strain. Thermosets (whether stiff in the case of an epoxy or flexible in the case of a silicone elastomer) are therefore unlikely to plastically deform in tension in any significant way. The key to large plastic deformations in thermoplastic polymers is the ability of chains to slide past one another thanks to the presence of the weak secondary bonds. In light of this departure from dislocation-based plasticity, criteria for yielding and subsequent plastic flow of thermoplastic materials must be established. Before doing so, however, it is appropriate to describe basic features of the polymer structure that dominate flow (and fracture) properties. In particular, the conditions that enhance or hinder molecular chain sliding must be considered. In preparing this section, several excellent books about polymers were consulted to which the reader is referred.[83−88] Additional reading material on polymers is cited at the end of the reference section.

2.7.1 Polymer Structure: General Remarks

Individual macromolecular chains are formed by the union of two or more structural units of a simple compound (a *mer*). In polymeric materials used in engineering applications, the

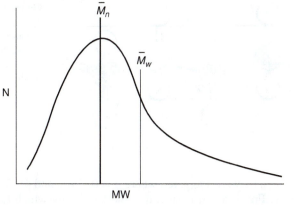

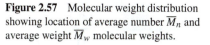

Figure 2.57 Molecular weight distribution showing location of average number $\overline{M}_n$ and average weight $\overline{M}_w$ molecular weights.

number of such unions—known as the degree of polymerization (DP)—often exceeds many thousands. The length of a given polymer chain is determined by the statistical probability of a specific activated mer attaching itself to a particular chain during a polymerization reaction. Some chains will be very short, while others might be very long. Consequently, the reader should appreciate a distinctive characteristic of a polymeric solid: There is no unique chain length for a given polymer and no specific molecular weight (MW). Instead, there is a distribution of these values. Contrast this with metal and ceramic solids that exhibit a well-defined lattice parameter and unit cell density. An example of the molecular weight distribution (MWD) for all the chains in a polymer is shown in Fig. 2.57; in this case, a larger number of small chains exist relative to the very long chains. The MWD will vary with the nature of the monomer and the conditions of polymerization so as to be skewed to higher or lower MW and/or made narrower or broader. For example, when the processing temperature is high and/or large amounts of initiator are added to the melt, MW will be low, and vice versa. Rather than referring to a molecular weight distribution curve to describe the character of a polymer, it is often more convenient to think in terms of an average molecular weight $\overline{M}$. Such a value can be described in a number of ways, but is usually described in terms of either a weight average or number average molecular weight where $\overline{M}_w$ emphasizes relatively high MW fractions and $\overline{M}_n$ emphasizes the importance of the smaller MW chains. The molecular weight distribution can be described by the ratio $\overline{M}_w/\overline{M}_n$. A narrow MWD prepared under carefully controlled conditions may have $\overline{M}_w/\overline{M}_n < 1.5$, while a broad MWD would reveal $\overline{M}_w/\overline{M}_n$ in excess of 25. As will be shown in later sections, MW exerts a very strong influence on a number of polymer physical and mechanical properties.

2.7.1.1 Side Groups and Chain Mobility

The ability of a polymer solid to plastically deform by chain sliding is determined by the mobility of its individual molecular chains. In order to understand the factors that determine chain mobility, we now look more closely at a segment of a polyethylene (PE) chain. In the fully extended conformation, the chain assumes a zigzag pattern, with the carbon–carbon bonds describing an angle of about 109° (Fig. 2.58). With the zigzag carbon main chain atoms lying in the plane of this page, the two hydrogen atoms are disposed above and below the paper. The chain is truly three dimensional, though it is often represented schematically in two-dimensional space only. Adjacent pairs of hydrogen atoms are positioned relative to one another so as to minimize their *steric hindrance* (i.e., interference due to their spatial arrangement). That is, as rotations occur about a C–C bond (permissible as long as the bond angle remains 109°), both favorable and unfavorable juxtapositions of the hydrogen atom pairs are experienced. This is perhaps more readily seen by examining the rotations about the C–C bond in ethane, C_2H_6 (Fig. 2.59), recalling that the hydrogen atoms do not lie in the plane of the page. We see that when the hydrogen atoms are located opposite one another, the potential energy of the system is

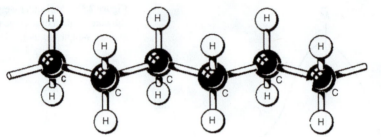

Figure 2.58 Extended chain of polyethylene showing coplanar zigzag arrangement of C–C bonds with hydrogen pairs located opposite one another.

maximized. Conversely, when they are staggered where $\varphi = 0$, $2\pi/3$, and $4\pi/3$, the configuration has lowest potential energy. From this, it is seen that the facility by which C–C bond rotation occurs will depend on the magnitude of the energy barrier in going from one low-energy configuration to another. For the two pairs of adjacent hydrogen atoms in the PE chain, the lowest potential energy trough occurs when the hydrogen atom pair associated with one carbon atom is 180° away from its neighboring hydrogen pairs (Fig. 2.58).

As might be expected, the extent of rotational freedom about the C–C bond depends on the nature of side groups often substituted for hydrogen in the PE chain. When one hydrogen atom is replaced, we have a vinyl polymer. As shown in Table 2.14, a number of different atoms or groups can be added to form a variety of vinyl polymers, which all act to restrict C–C rotation to a greater or lesser degree. Generally, the bigger and bulkier the side group and the greater its

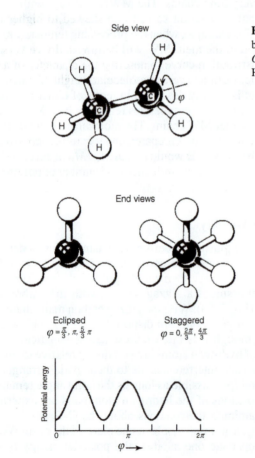

Figure 2.59 Potential energy variation associated with C–C bond rotation in ethane.[86] (Alfrey, T., and Gurnee, E. F., *Organic Polymers,* 1967. Reprinted by permission of Prentice-Hall Inc., Englewood Cliffs, NJ.)

Table 2.14   Selected Vinyl Polymers

Repeat Unit	Polymer
$\left[\begin{array}{c} \text{H} \quad \text{H} \\ -\overset{\mid}{\underset{\mid}{\text{C}}}-\overset{\mid}{\underset{\mid}{\text{C}}}- \\ \text{H} \quad \text{H} \end{array}\right]$	Polyethylene (PE)
$\left[\begin{array}{c} \text{H} \quad \text{Cl} \\ -\overset{\mid}{\underset{\mid}{\text{C}}}-\overset{\mid}{\underset{\mid}{\text{C}}}- \\ \text{H} \quad \text{H} \end{array}\right]$	Poly(vinyl chloride) (PVC)
$\left[\begin{array}{c} \text{H} \quad \text{F} \\ -\overset{\mid}{\underset{\mid}{\text{C}}}-\overset{\mid}{\underset{\mid}{\text{C}}}- \\ \text{H} \quad \text{H} \end{array}\right]$	Poly(vinyl fluoride) (PVF)
$\left[\begin{array}{c} \text{H} \quad \text{CH}_3 \\ -\overset{\mid}{\underset{\mid}{\text{C}}}-\overset{\mid}{\underset{\mid}{\text{C}}}- \\ \text{H} \quad \text{H} \end{array}\right]$	Polypropylene (PP)
$\left[\begin{array}{c} \text{H} \quad \bigcirc\!\!\!\!\!\!\text{O} \\ -\overset{\mid}{\underset{\mid}{\text{C}}}-\overset{\mid}{\underset{\mid}{\text{C}}}- \\ \text{H} \quad \text{H} \end{array}\right]$	Polystyrene (PS)

polarity, the greater the resistance to rotation, since the peak-to-valley energy differences (Fig. 2.59) would be greater. Restrictions to such movement may also be caused by double carbon bonds in the main chain, which rotate with much greater difficulty. Furthermore, the main chain in some polymers may contain flat cyclic groups (such as a benzene ring) which prefer to lie parallel to one another. Consequently, C–C bond rotation would be made more difficult by their presence.

C-C bond rotation is critical for the sliding mobility of polymer chains because steric hindrance occurs between the side groups of unconnected adjacent chains. At low temperatures, chain mobility is very limited and widespread interchain sliding is difficult. It is still possible, however, for some local sliding to occur, so a limited capacity for plastic deformation exists. As the temperature is elevated, thermal energy assists with C-C bond rotation, and side groups on adjacent chains are increasingly likely to be able to rotate away from one another to reduce the steric hindrance and enable chain sliding. Plastic deformation is therefore enhanced by increasing temperature. At a certain temperature, called the *glass transition temperature*, T_g, a significant change in chain mobility occurs, and the resistance to plastic deformation drops precipitously.

2.7.1.2   Side Groups and Crystallinity

Thus far, we have discussed the effect of side group size, shape, and polarity on main chain mobility. The *location* of these groups along the chain is also of critical importance, since it affects the relative packing efficiency of the polymer and, ultimately, the mechanical behavior. It is seen from Fig. 2.60 that the side groups can be arranged either randomly along the chain, only on one side, or on alternate sides of the chain. These three configurations are termed *atactic*, *isotactic*, and *syndiotactic*, respectively. Atactic polymers with large side groups (e.g., polystyrene) have low packing efficiency, with the chains arranged in a random array. Consequently, polystyrene, poly(methyl methacrylate), and, to a large extent, poly(vinyl chloride) are amorphous. In a regular and symmetric polymer (e.g., polyethylene), the chains can be packed close together, resulting in a high degree of crystallinity. In fact, the density of a

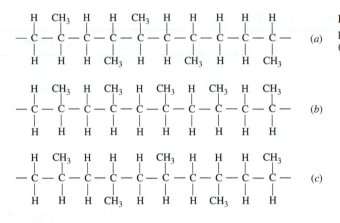

Figure 2.60 Location of side groups in polypropylene: (*a*) atactic; (*b*) isotactic; (*c*) syndiotactic.

given polymer serves as a useful measure of crystallinity; the higher the density, the greater the degree of crystallinity.

For the data shown in Table 2.15, densities were varied by the amount of main chain branching produced during polymerization (Fig. 2.61). Extensive branching reduces the opportunity for closer packing, and little branching promotes the polymerization of higher density polyethylene. Polypropylene represents an example of a stereoregular polymer that has a high packing efficiency and resultant crystallinity. Although it is not stereoregular, the propensity for crystallinity in nylon 66 is enhanced by the highly polar nature of the nylon chain. The H–N–C=O groups in adjacent chains have great affinity for one another, with the associated hydrogen bond providing additional cause for closer packing and chain alignment.

To summarize, the degree to which polymers will crystallize depends strongly on the polarity, symmetry, and stereoregularity of the chain and its tendency for branching. The extent of crystallinity of several polymers is given in Table 2.16, along with other material characteristics. In general, higher levels of crystallinity yield greater stiffness, strength, thermal stability, and chemical resistance. Conversely, elongation and toughness are enhanced by reduced levels of crystallinity.

2.7.1.3 Morphology of Amorphous and Crystalline Polymers

Certain polymers take on an amorphous structure when cooled from the melt. These include polymethylmethacrylate (PMMA), polystyrene (PS), and natural rubber. For many years, the structure of amorphous polymers was presumed to consist of a collection of randomly coiled molecules surrounding a certain unoccupied volume. (A coiled molecule can be created by a random combination of C–C bond rotations along the backbone of the molecule.) More recent studies have suggested that this simple view is not correct. Instead, Geil and Yeh[90–92] have proposed that seemingly amorphous polymers actually contain small domains in which the

Table 2.15 Relation between Density–Crystallinity and Ultimate Tensile Strength in Polyethylene[89]

Density (g/cm³)	Crystallinity (%)	Ultimate Tensile Strength	
		MPa	ksi
0.92	65	13.8	2
0.935	75	17.2	2.5
0.95	85	27.6	4
0.96	87	31.0	4.5
0.965	95	37.9	5.5

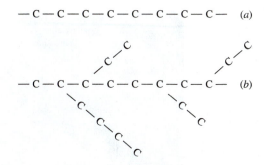

Figure 2.61   Degree of chain branching in polymeric solid. (*a*) Linear; (*b*) branched.

Table 2.16   Characteristics of Selected Polymers

Material	Repeat Unit	Major Characteristics	Applications
Low-density polyethylene (LDPE)	$\left[\begin{matrix} H & H \\ -C-C- \\ H & H \end{matrix}\right]$	Considerable branching; 55–70% crystallinity; excellent insulator; relatively cheap	Film; moldings; cold water plumbing; squeeze bottles
Polypropylene (PP)	$\left[\begin{matrix} H & CH_3 \\ -C-C- \\ H & H \end{matrix}\right]$	Extent of crystallinity depends on stereo-regularity; can be highly oriented to form integral hinge with extraordinary fatigue behavior	Hinges; toys; fibers; pipe; sheet; wire covering
Acetal copolymer	$\left[\begin{matrix} H & & H & H \\ -C-O- & C-C-O- \\ H & & H & H \end{matrix}\right]$	Highly crystalline; thermally stable; excellent fatigue resistance	Speedometer gears; instrument housing; plumbing valves; glands; shower heads
Nylon 66	$\left[-N-(CH_2)_6-N-\overset{\displaystyle O}{\overset{\|}{C}}-(CH_2)_4-\overset{\displaystyle O}{\overset{\|}{C}}- \right]$ with H below N	Excellent wear resistance; high strength and good toughness; used as plastic and fiber; highly crystalline; strong affinity for water	Gears and bearings; rollers; wheels; pulleys; power tool housings; light machinery components; fabric
Poly (tetrafluoroethylene) (PTFE, Teflon)	$\left[\begin{matrix} F & F \\ -C-C- \\ F & F \end{matrix}\right]$	Extremely high MW; high crystallinity; extraordinary resistance to chemical attack; nonsticking	Coatings for cooking utensils; bearings and gaskets; pipe linings; insulating tape; nonstick, loadbearing pads
Poly(vinyl chloride) (PVC)	$\left[\begin{matrix} H & Cl \\ -C-C- \\ H & H \end{matrix}\right]$	Primarily amorphous; variable properties through polymeric additions; fire self-extinguishing; fairly brittle when unplasticized; relatively cheap	Floor covering; film; handbags; water pipes; wiring insulation; decorative trim; toys; upholstery
Poly(methyl methacrylate) (PMMA, Plexiglas)	$\left[\begin{matrix} H & CH_3 \\ -C & C- \\ H & O=C-O-CH_3 \end{matrix}\right]$	Amorphous; brittle; general replacement for glass	Signs; canopies; windows; windshields; sanitary ware

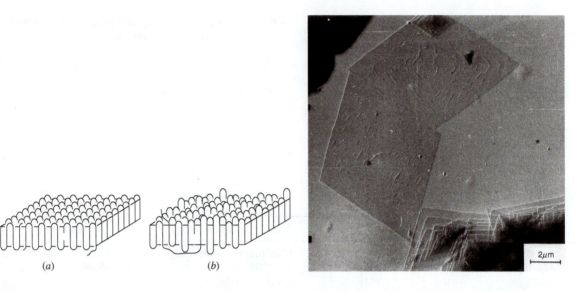

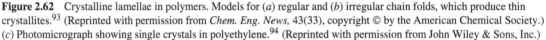

Figure 2.62 Crystalline lamellae in polymers. Models for (*a*) regular and (*b*) irregular chain folds, which produce thin crystallites.[93] (Reprinted with permission from *Chem. Eng. News,* 43(33), copyright © by the American Chemical Society.) (*c*) Photomicrograph showing single crystals in polyethylene.[94] (Reprinted with permission from John Wiley & Sons, Inc.)

molecules are aligned to some degree. However, it is undoubtedly the case that there is a lack of long-range order in amorphous polymers, and it is this factor that is used to explain their overall mechanical behavior.

Although it is possible to grow single crystal polymers (see Fig. 2.62*c*), in practice polymers that tend to develop crystalline order (such as linear polyethylene) consist of crystallites that are connected by amorphous regions. This mixed character has a profound effect on mechanical behavior. The degree of crystallinity can vary widely, and can sometimes exceed 90%. The crystalline structure of polymers can be described by two factors: chain conformation and chain packing. The conformation of a chain relates to its geometrical shape. In polyethylene, the chains assume a zigzag pattern, as noted above, and pack flat against one another. This is not observed in polypropylene, which has a single large methyl group, and in poly(tetrafluoroethylene), which contains four large fluorine atoms. Instead, these zigzag molecules twist about their main chain axis to form a helix. In this manner, steric hindrance is reduced between side groups along a chain. It is interesting to note that poly(tetrafluoroethylene)'s extraordinary resistance to chemical attack, mentioned in Table 2.16, is believed partly attributable to the sheathing action of the fluorine atoms that cover the helical molecule.

It is currently believed that chain packing in a crystalline polymer is achieved by repeated chain folding, such that highly ordered crystalline lamellae are formed. Two models involving extensive chain folding to account for the formation of crystalline lamellae are shown in Fig. 2.62*a,b*.[93,94] The thickness of these crystals is generally about 10 to 20 nm, while the planar dimensions can be measured in micrometers. Since chains are many times longer than the observed thickness of these lamellae, chain folding is required. Consequently, chains are seen to extend across the lamellae but reverse direction on reaching the crystallite boundary. In this manner, a chain is folded back on itself many times. The loose loop model depicting less perfect chain folding (some folds occurring beyond the nominal boundary of the lamellae; see Fig. 2.62*b*) is viewed by Clark[95] as being more realistic in describing the character of real crystalline polymers.

In addition to loose loops and chain ends (cilia) that lie on the surface of the lamellae, there exist *tie molecules* (*tie chains*) that extend from one crystal to another[96] (Fig. 2.63). The latter provide mechanical strengthening to the crystalline aggregate, as is discussed later. In the unoriented condition, crystalline polymers possess a *spherulitic* structure consisting of stacks of lamellae positioned along radial directions (Fig. 2.64). Since the extended chains within the

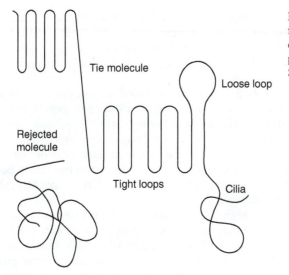

Figure 2.63   Schematic representation of chain-folded model containing tie molecules, loose loops, cilia (chain ends), and rejected molecules.[96] (By permission, from *Polymeric Materials,* © American Society for Metals, Metals Park, OH, 1975).

lamellae are normal to the lamellae surface, the extended chains are positioned tangentially about the center of the spherulite. Although this structure may bear some resemblance to that of a metal or ceramic grain structure, it should be noted that each spherulite contains multiple crystalline lamellae with many different orientations with respect to one another. A spherulite has a boundary with its neighbor that can act mechanically much like a grain boundary, but spherulites do not act like single crystals.

2.7.1.4 Polymer Additions

Often, commercial polymeric products contain a variety of additives that change the overall structure and associated properties. It is, therefore, appropriate to identify the major types of additives and their primary functions.

Pigments and Dyestuff. These materials are added to impart color to the polymer. They are not expected to cause significant changes in mechanical behavior at typical concentrations.

Figure 2.64   Sheaf-like stacks of crystal lamellae in polychlorotrifluoroethylene, which represent intersecting spherulites.[94] (Reprinted with permission from John Wiley & Sons, Inc.)

Stabilizers. Stabilizers suppress molecular breakdown in the presence of heat, light, ozone, and oxygen. One form stabilizes the chain ends so the chains will not "unzip," thereby reversing the polymerization process. Other stabilizers act as antioxidants and antiozonants that are attached preferentially by O_2 and O_3 relative to the polymer chain. They aid in long-term mechanical stability.

Fillers. Various ingredients are sometimes added to the polymer to enhance certain properties. For example, the addition of carbon black to automobile tires improves their strength and abrasion resistance. Fillers also serve to lower the volume cost of the polymer–filler aggregate, since the cost of the filler is almost always much lower than that of the polymer.

Blowing Agents. These substances are used to form expanded or foamed polymers by decomposing into gas bubbles within the polymer melt, producing stable holes. The porous nature of the resulting polymer solid creates substantially different mechanical properties than the solid version of the same material.

Plasticizers. Plasticizers are high-boiling-point, low-MW monomeric liquids that possess low volatility. They are added to a polymer to improve its processability and/or ductility. These changes arise for a number of reasons. Plasticizers add a low MW fraction to the melt, which broadens the MWD and shifts $\overline{M}$ to lower values. This enhances polymer processability. The liquid effectively shields chains from one another, thus decreasing their intermolecular attraction. Furthermore, by separating large chains, the liquids provide the chains with greater mobility for molecule segmental motion. The decrease in $\overline{M}$ and the lowering of intermolecular forces contribute toward improving polymer ductility and toughness. It should be recognized that these beneficial changes occur while stiffness and maximum service temperature decrease (as T_g decreases). Consequently, the extent of polymer plasticization is determined by an optimization of processability, ductility, strength, and stiffness, and service temperature requirements. Unplasticized PVC is rigid at room temperature (e.g., pipes for plumbing) whereas plasticitized PVC can be quite flexible (e.g., insulation for electrical wires). It is interesting to note that nylon 66 is inadvertently plasticized by the moisture it picks up from the atmosphere.

Cross-linking Agents. As previously discussed, the basic difference between thermoplastic and thermosetting polymers lies in the nature of the dominant intermolecular bonds linking adjacent chains. Sulfur is a classic example of a cross-linking agent as used in the vulcanization of rubber. Cross-linking reduces the effect of temperature on stiffness and strength, and severely restricts molecular mobility.

From this very brief description of polymer additives, it is clear that a distinction should be made between a pure polymer and a polymer plus assorted additives; the latter is often referred to as a *plastic,* though rubbers also may be compounded. Although the terms *plastic* and *polymer* are often used synonymously in the literature, they truly represent basically different entities.

2.7.2 Plasticity Mechanisms

We now consider the mechanisms by which amorphous and crystalline polymers deform. As discussed previously, significant plasticity is only possible in materials in which some degree of chain sliding can occur, and that chain mobility is a strong function of temperature. Furthermore, chain mobility and the details of the deformation mechanisms at work differ from amorphous to crystalline phases.

2.7.2.1 Amorphous Polymers

In amorphous polymers, shear yielding can occur in a homogeneous fashion, or in an inhomogenous fashion in the form of localized shear bands (somewhat reminiscent of Lüders bands in certain metals). At T_g and above, chain mobility is very high and homogeneous deformation with large plastic strains is favored. In this condition, amorphous polymers are *rubbery* (not to be confused with elastomeric behavior, which requires cross-links). At temperatures below T_g, amorphous thermoplastics are *glassy*. Chain mobility is reduced, and shear band formation is favored. Within a shear band, the actual process of chain sliding is similar to that of the homogeneous deformation mode, while outside a shear band little plastic deformation takes place. Shear bands form in directions parallel to maximum shear stress, so in a uniaxial test they

Figure 2.65   Schematic depiction of the cross section of an unplasticized PVC sheet deformed below T_g in 4-point bending. Under cross polarization, shear bands extending from the original surfaces are visible as bright streaks. The approximate shear band depth is marked on the compressive (d_c) and tensile (d_t) sides of the specimen. The shear band depth is different on the two sides due to pressure-dependent yield behavior.

appear at angles close to 45° to the tensile or compression axis. In bending, shear bands tend to nucleate at the outer, high-stress surfaces and extend inward,[97] as shown in Fig. 2.65.

In many polymers, there is a significant load drop almost immediately after plastic deformation begins. (Note the difference between this behavior and that of ductile metals for which unstable necking is delayed until substantial work hardening has occurred.) It is often easier to continue shear yielding than to initiate it, so there can be a local softening once yielding commences. In some cases the material quickly necks in an unstable fashion, and failure occurs. However, in many other cases the neck reduces to a certain cross section, stabilizes, and spreads, causing a large reduction in cross-sectional area along the gage length of the test specimen as shown in Fig. 2.66. Shear yielding mechanisms are associated with this phenomenon of *cold drawing*. The propensity for cold drawing depends on strong work hardening in the necked region that comes from an overall alignment of the chains along the principal stress axes. In some amorphous polymers, the degree of alignment may actually be significant enough to cause crystallization (e.g., in polyethylene terephthalate, PET) although this need not be the case to develop reasonable work hardening.

In the presence of tensile stresses (particularly triaxial tensile stresses) there is another form of inhomogeneous glassy polymer yielding called *crazing* that can compete with shear banding. Crazes are micrometer-scale crack-like features that form within the material along an axis perpendicular to the principal tensile stress direction. As shown in Fig. 2.67, a craze develops as microvoids nucleate in regions of high stress concentration, and surrounding chains rotate to align with the tensile axis. These *fibrils* bridge the craze, preventing the microvoids from

Figure 2.66   Cold drawing in polypropylene, which produces greater optical transparency in gage section as a result of enhanced molecular alignment.

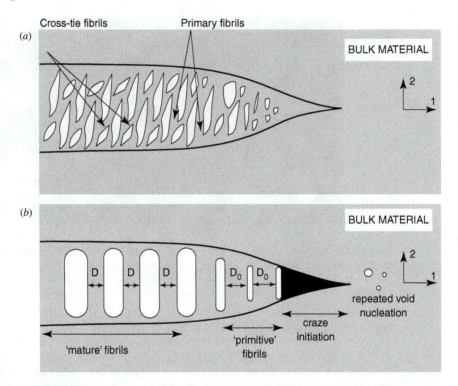

Figure 2.67 (*a*) Schematic depiction of a craze illustrating the fibrils oriented along direction 2, the principal stress axis, and (*b*) an idealization of a craze illustrating its evolution as it extends in direction 1.[109] (Reprinted from *Journal of the Mechanics and Physics of Solids*, 48, R. Estevez, M.G.A. Tijssens, and E. Van der Giessen, "Modeling of the competition between shear yielding and crazing in glassy polymers," p. 2585, 2000, with permission from Elsevier.)

immediately linking up to form a true crack. The growth of a craze occurs laterally by extension of the craze tip into uncrazed material. At the same time, the craze thickens by lengthening of the fibrils. If the applied load is increased, the fibrils will eventually break, which leads to macroscale fracture. Under conditions in which crazing dominates over shear yielding, failure by *craze embrittlement* precludes cold drawing and large plastic strain development in tension. In compression, craze formation cannot occur and so large plastic deformation by shear yielding is possible even in glassy polymers that tend to have poor ductility in tension. For several key reviews and current articles pertaining to craze formation and fracture, the reader is referred to several key references.[98–108]

2.7.2.2 Semi-crystalline Polymers

The amorphous regions of semi-crystalline polymers can undergo similar chain sliding and alignment to that seen in fully-amorphous material. However, the constraint imposed by the surrounding crystallites tends to make sliding more difficult than in the fully-amorphous case. Unoriented crystalline regions like those found within spherulites are found to deform by a complex process involving initial breakdown and subsequent reorganization.[110–112] After an initial stage of plastic deformation in the amorphous regions of the spherulites, the latter begin to break down. Lamellae packets oriented normal to the applied stress may separate along the amorphous boundary region between crystals, while others begin to rotate toward the stress axis (analogous to slip-plane rotation discussed in Section 2.5.1). The crystals themselves are now broken into smaller blocks, but the chains maintain their folded conformation. As this phase of the deformation process continues, these small bundles become aligned in tandem along the

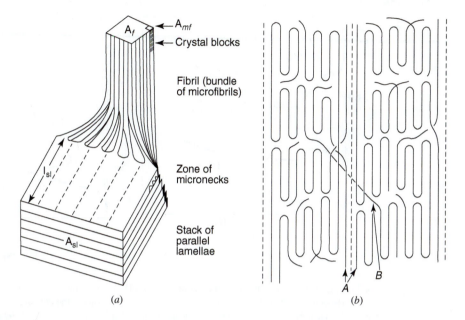

Figure 2.68 (*a*) Model depicting transformation of a stack of parallel lamellae into a bundle of densely packed and aligned microfibrils.[111] Crystal blocks oriented as shown in *b*. (Reprinted from *Macromol. Chem.* 8, 277 (1973), with permission.) (*b*) Alignment of crystal blocks in microfibrils. Intrafibrillar extended tie molecules shown at *A* with interfibrillar extended tie molecule at *B*.[112] (By permission from *Polymeric Materials*, copyright © American Society for Metals, Metals Park, OH, 1975.)

drawing direction, forming long microfibrils (Fig. 2.68). Note that the extended chains within each bundle are positioned parallel to the draw axis along with a large number of fully extended tie molecules. Since many tandem blocks are torn from the same lamellae, they remain connected through a number of tie molecules created by unfolding chains from the original lamellae. The combination of many more fully extended tie molecules and the orientation of the bundles within each fibril contributes toward a rapid increase in strength and stiffness. By contrast, few primary bonds join blocks in adjacent microfibrils, except those representing tie molecules from the original lamellae (Fig. 2.68*b*). It is this initial spherulite structure break- down, followed by microfibril formation, that gives rise to the substantial hardening associated with aligned semi-crystalline polymers. Continued deformation of the microfibrillar structure is extremely difficult because of the high strength of the individual microfibrils and the increasing extension of the interfibrillar tie molecules. These tie molecules become more extended as a result of microfibril shear relative to one another. This extreme work hardening is ideal for cold drawing, and tends to result in an aligned material that is even stronger along the drawing axis than aligned material produced from amorphous polymers.

2.7.3 Macroscopic Response of Ductile Polymers

Finally, we turn to the macroscopic deformation response of ductile polymers. For those polymers that cold draw, there is a significant drop in stress immediately after yielding begins, as seen in Fig. 2.69*a*. In this case, it is common to identify the peak stress as the yield stress. Not all materials deform in this fashion, however, so an offset yield criterion sometimes becomes necessary. When necking does occur, there is an apparent drop in the yield strength followed by a long plateau if the neck propagates. This is somewhat deceptive, however, because the large change in cross-sectional area associated with the neck makes engineering stress useless for evaluating the intrinsic strengthening behavior of the material. This is arguably an even greater challenge for polymers than for metals because the onset of necking in polymers can be

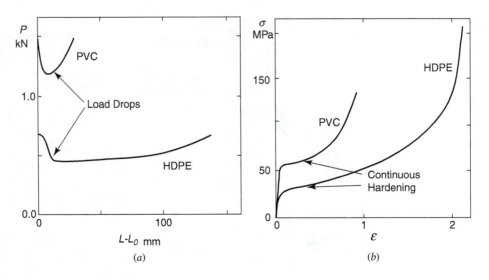

Figure 2.69 (*a*) Raw load-displacement data for PVC and HDPE specimens tested in tension. Both exhibit large load drops associated with necking. (*b*) True stress–true strain versions of the data demonstrating continuous hardening.[113] (Data from *Springer Science + Business Media: Journal of Materials Science*, 14, p. 583, 1979, C. G'Sell and J. J. Jones, figure numbers 1 and 4.)

coincident with yielding, in which case there is no extended period of homogeneous plastic deformation in which the equations relating engineering and true stress can be used (recall Eqs. 1-4 and 1-5). Figure 2.69*b* shows the true stress and true strain measured within the necked region using special instrumentation that tracked the local areal dimensions. It can be seen that for necked PVC and HDPE there is a change in slope associated with yielding, but no load drop. Furthermore, work hardening increases steadily to the point of failure. This is quite different than work hardening in metals, which begins at a high rate and then declines. It is this increasing rate of work hardening depicted in the PVC and HDPE curves that supports cold drawing.

Following the approach of G'Sell and Jonas,[113] a relative strain-hardening coefficient can be defined as $\gamma = (d\ln\sigma/d\varepsilon)_{\dot\varepsilon}$. For the curves in Fig. 2.69 it can be seen that γ increases with increasing strain. If one also defines a strain rate sensitivity coefficient $m = (d\ln\sigma/d\dot\varepsilon)_\varepsilon$, the flow curve can be described in the form

$$\sigma(\varepsilon, \dot\varepsilon) = K\dot\varepsilon^m\exp\left(\frac{\varepsilon^2\lambda_\varepsilon}{2}\right) \tag{2-57}$$

where K is a constant. G'Sell and Jonas found that m was equal to 0.06 for HDPE and 0.05 for PVC (although it diminished to nearly zero with increasing strain for PVC). These values are comparable to those of many metals (see Section 2.4.4).

Using hourglass-shaped tensile specimens and a video-based strain measurement instrument, G'Scll et al. have also evaluated the yield behavior of semi-crystalline and amorphous materials as a function of temperature. Applying a Bridgman correction for the triaxial stresses developed in this narrow region of their specimens (recall Section 2.4.2.3), they report corrected "effective" true stress values that are equivalent to pure uniaxial stresses in the absence of the artificial neck (Fig. 2.70). They found in this study that glassy polymers (PC, PS, PMMA, PVC) tested below T_g all showed a large true stress drop followed by increasing strain hardening. Semi-crystalline polymers tested below T_g (PA6, PEEK) yielded with small stress drops. Semi-crystalline polymers testing above T_g (PE, PP, PTFE, POM) showed no stress drop, and progressed smoothly from elastic to plastic deformation. They attributed these trends to the generation of transient defects that occur when an amorphous phase first yields in the glassy state. When an amorphous material is above T_g, viscoelastic effects (i.e., time-dependent elasticity) can smooth the transition (see Chapter 4 for discussion of viscoelasticity).

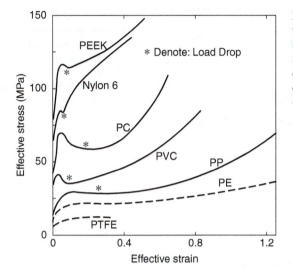

Figure 2.70 True stress and strain corrected for triaxial stresses in the necked region showing a true stress drop (∗) in some polymers (solid lines) and not in others (dashed lines). (Data from *Springer Science + Business Media: Journal of Materials Science*, 27, p. 5031, 1992, C. G'Sell, J. M. Hiver, A. Dahoun, and A. Souahi, figure number 10.)

2.7.4 Yield Criteria

We conclude this discussion by revisiting the question of appropriate yield criteria for polymer plasticity. Classical yield theories assume that (i) the material in question is isotropic and homogeneous, (ii) yield is insensitive to the presence of normal (or hydrostatic) stresses, (iii) yield is identical in tension and compression, and (iv) volume is conserved during plastic deformation. As shown in Section 2.5.3, yield behavior of textured metals must be altered to adjust for anisotropy. The same will be true for aligned polymers. The other assumptions hold reasonably well for metals,[ix] but are not necessarily true for polymers. In particular, it has been seen that polymers typically have very different yield behavior in tension and in compression, and that this can be attributed to hydrostatic pressure effects.

The effect of pressure on yielding has been studied by a number of research groups, and several trends have emerged. First, it is seen that yield in tension occurs at lower true stress than yield in compression. This is evident in the shear band size asymmetry depicted in Fig. 2.65 for a bend test of unplasticized PVC, and in the differences in tension and compression curves for PMMA in Fig. 2.71. As positive pressure (compressive stress) is exerted on these low-stiffness materials, the chains are pressed together and chain mobility is diminished. Conversely, under negative pressure (tension stress) the average spacing between the chains in the direction perpendicular to the principal normal stress is increased, and chain mobility is enhanced.

In light of the significant influence of pressure on polymer yielding, several pressure-dependent yield criteria have been proposed and compared to experimental results. The *Mohr-Coulomb model* and the *modified Tresca criterion* are based on the existence of a critical shear stress, just like the classical Tresca criterion. In the Mohr-Coulomb model, the pressure influence is described as a simple normal stress imposed on the plane of sliding, with a "frictional" term added to couple the normal stress to the critical shear stress. It takes the form

$$\tau_{MC} = \tau_c - \mu_{MC}\sigma_n \tag{2-58a}$$

where τ_{MC} is the critical shear stress in the presence of a normal stress, τ_c is the critical shear stress in the absence of normal stress, μ_{MC} is the coefficient of friction, and σ_n is the normal stress (negative for compression as written). This is a significant departure from the case of

---

[ix] Cold-working of metals can cause yield asymmetry in tension and compression due to dislocation back-stresses. This is known as the Bauschinger effect; it is a function of prior processing, not a fundamental property.

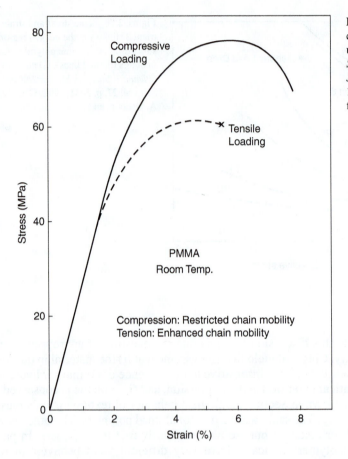

Figure 2.71 Asymmetry in tensile and compressive yield for PMMA tested under uniaxial conditions.[115] (Data from *Springer Science + Business Media: Journal of Materials Science*, 9, p. 81, 1974, S. Rabinowitz and P. Beardmore, figure number 14.)

metals and ceramics, for which the normal stress plays no role in determining yield (recall Section 2.2.2). The Mohr-Coulomb criterion is also sometimes written as

$$\tau_{MC} = \tau_c - (\tan\phi)\sigma_n \qquad (2\text{-}58b)$$

where ϕ is the angle that the yield surface makes with the classical failure surface, as shown in Fig. 2.72. Note that the Mohr-Coulomb criterion does not distinguish between yielding under

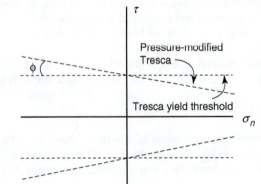

Figure 2.72 Failure envelopes for Tresca (short dash) and pressure-modified Tresca (long dash) yield criteria plotted on shear and normal stress axes. Note that the classical Tresca behavior is not dependent on the normal stress, and that an angle ϕ characterizes the deviation expressed by the introduction of normal stress dependence in the pressure-modified version.

uniaxial or multiaxial loading. The pressure-modified Tresca criterion, however, treats the pressure phenomenon in three dimensions, but has a similar form:

$$\tau_T = \tau_{T0} + \mu_T P$$
$$\frac{1}{2}(\sigma_{max} - \sigma_{min}) = \tau_{T0} + \mu_T P \tag{2-59}$$

where μ_T is the pressure coefficient and P is the mean (hydrostatic) pressure

$$P = -\frac{1}{3}(\sigma_1 + \sigma_2 + \sigma_3) \tag{2-60}$$

so that P is positive for hydrostatic *compression*.

The pressure-modified version of the von Mises yield criterion depends on a critical distortion energy, just the like the classical version, and takes essentially the same overall form as the modified Tresca:

$$\tau_{VM} = \tau_{oct0} + \mu_{VM} P \tag{2-61}$$

where τ_{VM} is the critical octahedral shear stress for yielding as a function of pressure, μ_{VM} is again a pressure coefficient, and P is the mean pressure. Here, τ_{oct0} is the critical octahedral shear stress determined in the *absence of hydrostatic pressure* (i.e., in pure shear). When the applied octahedral shear stress (Eq. 2-30) exceeds the pressure-modified critical value so that $\tau_{oct} \geq \tau_{VM}$, yielding occurs. This is identical in form to the linear *Drucker-Prager yield criterion* originally developed to describe the yield behavior of soil, which may be found as an option in finite element analysis software.

Note that for uniaxial loading,

$$\tau_{oct} = \frac{\sqrt{2}}{3}|\sigma|$$
$$P = -\frac{\sigma}{3} \tag{2-62}$$

All the aforementioned pressure-modified yield criteria assume a linear dependence on the pressure. It has been found that this works well at low pressures (e.g., 100 MPa) but may not be suitable for very high hydrostatic pressures. If this is the case, a nonlinear version can be applied such as the exponent Drucker-Prager criterion.[116]

Regardless of whether the maximum shear or distortion energy criterion is used, the effect on the yield surface is similar. In biaxial stress space, the symmetric yield surfaces depicted in Fig. 2.24 are shifted toward the biaxial compression quadrant, and are distorted as shown in Fig. 2.73. A tractable method for determining which yield criterion is best for a particular polymer is to test under several different stress states then to compare with the various model predictions. In Fig. 2.73 this approach is demonstrated for PMMA and PS at several different temperatures.[117] The tests consisted of uniaxial tension and compression (pure σ_1 and/or σ_2), pure shear ($\sigma_1 = -\sigma_2$; $\sigma_3 = 0$), and plane strain compression ($\sigma_1 = 0.5\sigma_2$; $\sigma_3 = 0$). Quinson et al. found that PMMA was best described by the modified von Mises criterion for all temperatures tested, while the PS was better described by the modified Tresca criterion at 90°C. Although not shown here, PC yield was matched best by the modified Tresca criterion at 20°C and 90°C. These results confirmed those of an earlier study by Bowden and Jukes.[97] The trends were explained by relating the assumptions underlying the models to the observed plasticity modes. The PMMA exhibited diffuse shear yielding over large deformation zones at all temperatures, whereas the PC showed thin, distinct shear bands (conceptually matching the expectations of the Tresca maximum shear criterion). The PS showed thin shear bands at 20°C but diffuse shear yielding at 90°C. Transmission optical microscopy images of the diffuse and sharp shear deformation modes (photographed between

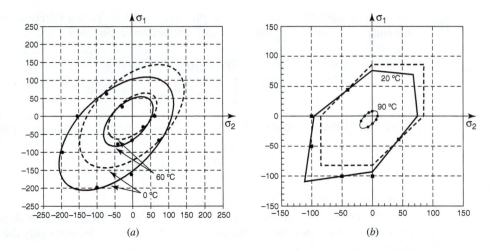

(a)   (b)

Figure 2.73 Solid lines depict pressure-modified von Mises and Tresca biaxial stress yield envelopes for (a) PMMA, and (b) PS as a function of temperature. It has been observed that the pressure modified von Mises criterion best matches PMMA behavior for the temperature range 0–90°C, while PS undergoes a transition from Tresca to von Mises as temperature increases.[117] For comparison, dashed lines indicate pressure-independent yield envelopes based on the same critical shear conditions. (Data from R. Quinson, J. Perez, M. Rink, and A. Pavan, *Journal of Materials Science*, 32, p. 1371, 1997.)

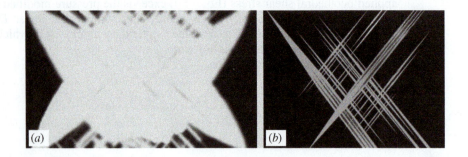

Figure 2.74 Schematic depictions of shear band appearance for two different polymers deformed at 20°C then imaged at the same magnification with transmission optical microscopy between crossed polarizers. (a) Diffuse shear yielding in PMMA, and (b) narrow shear bands in PS.

crossed polarizers) are shown in Fig. 2.74. In the earlier study by Bowden and Jukes, it was suggested that any polymer that deforms in a relatively homogeneous manner (e.g., PVC, epoxy resins, and HDPE) might be expected to match best with the modified von Mises criterion; those that deform inhomogeneously by distinct shear band formation (e.g., PET) might therefore be best described by the modified Tresca criterion. Confirmation of these trends awaits further experimental proof.

REFERENCES

1. J. Frenkel, *Z. Phys.* **37**, 572 (1926).

2. W. J. McG. Tegart, *Elements of Mechanical Metallurgy*, Macmillan, New York, 1966.

3. W. T. Read, Jr., *Dislocations in Crystals*, McGraw-Hill, New York, 1953.

4. A. H. Cottrell, *Dislocations and Plastic Flow in Crystals*, Clarendon Press, Oxford, 1953.

5. A. G. Guy, *Elements of Physical Metallurgy*, 2nd ed., Addison-Wesley, Reading, MA, 1959.

6. H. Bei et al., *Acta Materialia* **56**, 4762–4770 (2008).

7. W. H. Sutton, B. W. Rosen, and D. G. Flom, *SPE J.* **72**, 1203 (1964).

8. W. G. Johnston and J. J. Gilman, *J. Appl. Phys.* **30**, 129 (1959).

9. J. J. Gilman and W. G. Johnston, in *Dislocations and Mechanical Properties of Crystals*, J. C. Fisher, W. G. Johnston, R. Thomson, and T. Vreeland, Jr., Eds., Wiley, New York, 1957, p. 116.

10. D. Hull, *Introduction to Dislocations*, Pergamon, Oxford, 1965.

11. P. B. Hirsch, *J. Inst. Met.* **87**, 406 (1959).

12. P. B. Hirsch, *Metall. Rev.* **4**, 101 (1959).

13. A. Howie, *Metall. Rev.* **6**, 467 (1961).

14. P. Kelly and J. Nutting, *J. Inst. Met.* **87**, 385 (1959).

15. P. B. Hirsch, *Prog. Met. Phys.* **6**, 236 (1956).

16. J. B. Newkirk and J. H. Wernick, Eds., *Direct Observations of Imperfections in Crystals*, Interscience, New York, 1962.

17. J. T. Michalak, *Metals Handbook*, Vol. **8**, ASM, Metals Park, OH, 1973, p. 218.

18. A. H. Cottrell, *The Properties of Materials at High Rates of Strain*, Institute of Mechanical Engineering, London, 1957.

19. A. J. Forman, M. A. Jaswon, and J. K. Wood, *Proc. Phys. Soc. A* **64**, 156 (1951).

20. P. R. Thornton, T. E. Mitchell, and P. B. Hirsch, *Philos. Mag.* **7**, 1349 (1962).

21. P. G. Partridge, *Met. Rev.* **12**, 118 (1967).

22. A. Kelly and G. W. Groves, *Crystallography and Crystal Defects*, Addison-Wesley, Reading, MA, 1970, p. 175.

23. R. von Mises, *Z. Ang. Math. Mech.* **8**, 161 (1928).

24. G. I. Taylor, *J. Inst. Met.* **62**, 307 (1938).

25. G. W. Groves and A. Kelly, *Philos. Mag.* **8**, 877 (1963).

26. G. Y. Chin and W. L. Mammel, *Trans. Met Soc. AIME* **239**, 1400 (1967).

27. G. Y. Chin, W. L. Mammel, and M. T. Dolan, *Trans. Met Soc. AIME* **245**, 383 (1969).

28. W. J. McG. Tegart, *Philos. Mag.* **9**, 339 (1964).

29. D. C. Jillson, *Trans. Met Soc. AIME* **188**, 1129 (1950).

30. E. N. Andrade and R. Roscoe, *Proc. R. Soc.* **49**, 166 (1937).

31. P. M. Robinson and H. G. Scott, *Acta Met.* **15**, 1581 (1967).

32. G. I. Taylor, in *Deformation and Flow of Solids*, Springer, Berlin, 1956.

33. J. F. W. Bishop, and R. Hill, *Philos. Mag.* **42**, 1298 (1951).

34. U. F. Kocks, *Metall. Trans.* **1**, 1121 (1970).

35. J. M. Rosenberg and H. R. Piehler, *Metall. Trans. B* **2**(1), 257 (1971).

36. R. C. Juvinall, *Fundamentals of Machine Component Design*, John Wiley & Sons, New York, 1983.

37. C. J. McMahon, Jr., Ed., *Advances in Materials Research, Vol. 2, Microplasticity*, Interscience, New York, 1968.

38. ASTM E8, ASTM International, West Conshohocken, PA.

39. J. H. Hollomon, *Trans. AIME* **162**, 268 (1945).

40. G. B. Bülfinger, *Comm. Acad. Petrop.* **4**, 164 (1735).

41. A. W. Bowen and P. G. Partridge, *J. Phys. D: Appl. Phys.* **7**, 969 (1974).

42. P. W. Bridgman, *Trans. ASM* **32**, 553 (1944).

43. F. R. Larson and F. L. Carr, *Trans. ASM* **55**, 599 (1962).

44. P. W. Bridgman, *Fracturing of Metals*, ASM, Metals Park, OH, 1948, p. 246.

45. H. Inagaki and T. Komatsubara, *Mat. Sci. Forum* **331–337**, 1303 (2000).

46. P. E. Bennett and G. M. Sinclair, *Trans. ASME, J. Basic Eng.* **88**, 518 (1966).

47. E. Schmid and W. Boas, *Plasticity of Crystals*, Hughes, London, 1950, p. 55.

48. G. I. Taylor and C. F. Elam, *Proc. R. Soc. London Ser. A* **108**, 28 (1925).

49. C. F. Elam, *Distortion of Metal Crystals*, Oxford Universtiy Press, London, 1935.

50. J. S. Koehler, *Acta Met.* **1**, 508, (1953).

51. G. R. Piercy, R. W. Cahn, and A. H. Cottrell, *Acta Met.* **3**, 331 (1955).

52. H. Hu, *Texture* **1**(4), 233 (1974).

53. A. T. English and G. Y. Chin, *Acta Met.* **13**, 1013 (1965).

54. R. E. Smallman and D. Green, *Acta Met.* **12**, 145 (1964).

55. I. L. Dillamore and W. T. Roberts, *Acta Met.* **12**, 281 (1964).

56. N. Brown, *Trans. Met Soc. AIME* **221**, 236 (1961).

57. J. S. Kallend and G. J. Davies, *Texture* **1**, 51 (1972).

58. G. Wassermann, *Z. Met.* **54**, 61 (1963).

59. E. A. Calnan, *Acta Met.* **2**, 865 (1954).

60. J. F. W. Bishop, *J. Mech. Phys. Sol.* **3**, 130 (1954).

61. H. Hu, R. S. Cline, and S. R. Goodman, *Recrystallization Grain Growth and Textures*, ASM, Metals Park, OH, 1965, p. 295.

62. H. Hu and R. S. Cline, *J. Appl. Phys.* **32**, 760 (1961).

63. S. R. Goodman and H. Hu, *Trans. Met Soc. AIME* **230**, 1413 (1964).

64. H. Hu and S. R. Goodman, *Trans. Met. Soc. AIME* **227**, 627 (1963).

65. W. F. Hosford and W. A. Backofen, *Fundamentals of Deformation Processing*, Syracuse University Press, Syracuse, New York, 1964.

66. W. F. Hosford and R. M. Caddell, *Metal Forming, Mechanics and Metallurgy*, 2nd ed., PTR Prentice Hall, Englewood Cliffs, NJ, 1993.

67. W. A. Backofen, *J. Mech. Phys. Solids.* **14**, 233 (1966).

68. G. Dieter, *Mechanical Metallurgy*, 3rd ed., McGraw-Hill, New York, 1987.

69. M. Atkinson, *Sheet Met. Ind.* **44**, 167 (1967).

70. B. A. Bilby and A. G. Crocker, *Proc. R. Soc. London Ser. A* **288**, 240 (1965).

71. J. W. Christian and S. Mahajan, *Prog. Mat. Sci.* **39**, 1 (1995).

72. E. O. Hall, *Twinning and Diffusionless Transformations in Metals*, Butterworth, London, 1954.

73. R. W. Cahn, *Adv. Phys.* **3**, 363 (1954).

74. R. E. Reed-Hill, *Deformation Twinning*, Gordon & Breach, New York, 1964, p. 295.

75. P. G. Partridge, *Met Rev.* **12**, 169 (1967).

76. S. Mahajan and D. F. Williams, *Int. Metall. Rev.* **18**, 43 (1973).

77. K. E. Eckelmeyer and R. W. Hertzberg, *Met. Trans.* **1**, 3411 (1970).

78. J. Friedel, *Dislocations*, Pergamon, Oxford, England, 1964.

79. D. Hull, *Fracture of Solids*, Interscience, New York, 1963, p. 417.

80. R. von Mises, *Z Ang. Math. Mech.* **8**, 161 (1928).

81. N. S. Stoloff and M. Gensamer, *Trans. Met. Soc. AIME* **227**, 70 (1963).

82. R. Clark and G. B. Craig, *Prog. Met. Phys.* **3**, 115 (1952).

83. R. D. Deanin, *Polymer Structure, Properties and Applications*, Cahners, Boston, MA, 1972.

84. R. M. Ogorkiewicz, Ed., *Engineering Properties of Thermoplastics*, Wiley-Interscience, London, 1970.

85. M. Kaufman, *Giant Molecules*, Doubleday, Garden City, NY, 1968.

86. T. Alfrey and E. F. Gurnee, *Organic Polymers*, Prentice-Hall, Englewood Cliffs, NJ, 1967.

87. S. L. Rosen, *Fundamental Principles of Polymeric Materials for Practicing Engineers*, Barnes & Noble, New York, 1971.

88. L. E. Nielsen, *Mechanical Properties of Polymers*, Reinhold, New York, 1962.

89. H. V. Boening, *Polyolefins: Structure and Properties*, Elsevier, Lausanne, 1966, p. 57.

90. P. H. Geil, *Polymeric Materials*, ASM, Metals Park, OH, 1975, p. 119.

91. G. S. Y. Yeh, *Crit. Rev. Macromol Sci.* **1**, 173 (1972).

92. G. S. Y. Yeh and P. H. Geil, *J. Macromol Sci. B* **1**, 235 (1967).

93. P. H. Geil, *Chem. Eng. News* **43** (33), 72 (Aug. 16, 1965).

94. P. H. Geil, *Polymer Single Crystals*, Interscience, New York, 1963.

95. E. S. Clark, *Polymeric Materials*, ASM, Metals Park, OH, 1975, p. 1.

96. R. F. Boyer, *Polymeric Materials*, ASM, Metals Park, OH, 1975, p. 277.

97. P.B. Bowden and D. A. Jukes, *J. Mater. Sci.* **7**, 52 (1972).

98. S. Rabinowitz and P. Beardmore, *CRC Crit. Rev. Macromol. Sci.* **1**, 1 (1972).

99. E. H. Andrews, *The Physics of Glassy Polymers*, R. N. Haward, Ed., Wiley, New York, 1973, p. 394.

100. R. P. Kambour, *J. Polym. Sci. Macromol Rev.* **7**, 1 (1973).

101. E. J. Kramer, *Developments in Polymer Fracture*, E. H. Andrews, Ed., Applied Science Publ., London, 1979.

102. H. H. Kausch, *Polymer Fracture*, Springer-Verlag, Berlin, 1978.

103. E. J. Kramer and B. D. Lauterwasser, *Deformation Yield and Fracture of Polymers*, Plastics and Rubber Institute, London, 1979, p. 34–1.

104. N. Verheulpen-Heymans, *Polymer* **20**, 356 (1979).

105. A. S. Argon and J. G. Hannoosh, *Philos. Mag.* **36**(5), 1195 (1977).

106. J. A. Sauer and C. C. Chen, *Adv. Polym. Sci.* **52/53**, 169 (1983).

107. A. S. Argon, R. E. Cohen, O. S. Gebizlioglu, and C. E. Schwier, *Adv. Polym. Sci.* **52/53**, 276 (1983).

108. W. Döll, *Adv. Polym. Sci.* **52/53**, 105 (1983).

109. R. Estevez, M. G. A. Tijssens, and E. Van der Giessen. *J. Mech. Phys. Solids* **48** (2000) 2585–2617.

110. A. Peterlin, *Advances in Polymer Science and Engineering*, K. D. Pae, D. R. Morrow, and Y. Chen, Eds., Plenum, New York, 1972, p. 1.

111. A. Peterlin, *Macromol. Chem.* **8**, 277 (1973).

112. A. Peterlin, *Polymeric Materials*, ASM, Metals Park, OH, 1975, p. 175.

113. C. G'Sell and J. J. Jonas, *J. Mater. Sci.* **14**, 583 (1979).

114. C. G'Sell, J. M. Hiver, A. Dahoun, and A. Souahi, *J. Mat. Sci.* **27**, 5031 (1992).

115. S. Rabinowitz and P. Beardmore, *J. Mat. Sci.* **9**, 81 (1974).

116. G. Dean, L. Crocker, B. Read, and L. Wright, *Int. J. Adhesion Adhesives* **24**, 295 (2004).

117. R. Quinson, J. Perez, M. Rink, and A. Pavan, *J. Mater. Sci.* **32**, 1371 (1997).

PROBLEMS

Review

2.1 Is a *dislocation* a physical item or substance? If not, what is it?

2.2 Why are dislocations necessary for explaining the plasticity typically seen for crystalline materials?

2.3 Identify two techniques for observing dislocations, and describe at least one strength and one weakness of each technique.

2.4 Rank the relative Peierls force in different materials and material classes and briefly explain why you gave that rank, in each case.

2.5 Which can cross-slip—an edge dislocation, a screw dislocation, or a mixed dislocation? Why?

2.6 Sketch an edge dislocation and a screw dislocation as if you are looking directly along the dislocation line in each case. Clearly mark the line direction and the slip plane (if there is a unique slip plane). Indicate on your sketches where you will find regions of hydrostatic tension, hydrostatic compression, and pure shear stress surrounding the dislocation lines.

2.7 Sketch a representative portion of the TEM images in Figures 2.7 and 2.18, including only the dislocation lines. Indicate on your sketches which features are dislocation lines, which are stacking faults, and which are the top and bottom edges of the slip planes.

2.8 Identify the crystal structure in the faulted region of an FCC crystal. Why is this the case?

2.9 When an FCC material has high stacking fault energy, do you expect widely-spaced or closely-spaced leading and trailing partial dislocations? Do you expect wavy or planar glide? Briefly explain both trends.

2.10 Consider the following face-centered-cubic dislocation reaction:

$$\frac{a}{2}[110] \rightarrow \frac{a}{6}\left[21\overline{1}\right] + \frac{a}{6}[121]$$

    a. Prove that the reaction will occur.

    b. What kind of dislocations are the $(a/6)\langle121\rangle$?

    c. What kind of crystal imperfection results from this dislocation reaction?

    d. What determines the distance of separation of the $(a/6)[21\overline{1}]$ and the $(a/6)[121]$ dislocations?

2.11 List which main slip systems are active in FCC, BCC, and HCP metals, and explain why those particular planes/directions are favored.

2.12 Sketch a 3D FCC unit cell and indicate where all 12 FCC slip systems can be found.

2.13 What does it mean to be an *independent slip system*?

2.14 What is the effect of resolved normal stress on the yield behavior of crystalline metals and ceramics?

2.15 What is the role of the *Taylor factor*?

2.16 Reproduce Figure 2.23 twice, first adjusting it so that it accurately depicts the case in which the horizontal stress is half that of the vertical stress, and second so that the horizontal stress is twice that of the vertical stress. Use arrow length to indicate relative stress magnitude.

2.17 Which predicts the lower yield strength for most combinations of applied stress—the Tresca or the von Mises yield criterion? Under what stress conditions are the predictions equal?

2.18 Identify the trend between stacking fault energy and work-hardening coefficient, and then use it to predict which is likely to work harden more strongly: pure copper or pure nickel, the latter of which has a stacking fault energy of approximately $240\,mJ/m^2$.

2.19 What are the critical differences between work hardening and geometric hardening?

2.20 Describe how a wire texture is different from rolling texture, and sketch an example of each with arrows indicating the directions of preferred orientation.

2.21 After a dislocation has passed through a crystal, thereby causing plastic deformation, what does the inside of the crystal look like? Contrast this with the appearance of the interior of a crystal that has deformed by twinning.

2.22 Under what conditions is twinning favored in BCC and/or FCC crystals?

2.23 What is the basic molecular mechanism for polymer plasticity, and how does it differ from that of ductile crystalline metals?

2.24 What specific aspects of polymer molecule structure (e.g., side group size, shape, polarity, and location) favor chain sliding?

2.25 How does the structure of a crystalline polymer differ from that of a crystalline metal? What are the implications of this difference for plasticity in both classes of material?

2.26 What are the two micro-scale plasticity mechanisms active in amorphous polymers? Are they likely to occur simultaneously? Explain.

2.27 Explain the role of *crazing* in determining the extent of maximum plastic deformation for some polymers. Be sure to include both tensile and compressive loading in your answer.

2.28 What typically happens to the strength level of a polymer that has undergone *cold drawing*? Why?

2.29 What is the typical effect of resolved normal stress on the yield behavior of polymeric materials? Is this the same as for crystalline metals and ceramics?

Practice

2.30   The dislocations shown below (on three separate slip planes) represent different characters. Assume that no negative edge or left-hand screw dislocations are included.

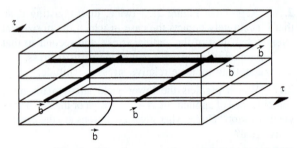

a. Sketch this diagram, then clearly identify the character of each dislocation by writing a label nearby. How do you know each type?

b. How does each dislocation behave under the applied shear shown on the diagram? Sketch the dislocation lines and indicate the direction of motion, if any, on three separate projections of the slip planes (i.e., as seen from above).

2.31   Two edge dislocations of opposite sign are found in a material separated by several planes of atoms as shown below.

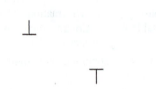

Please provide a helpful sketch and an explanation along with the answer for each of the following questions:

a. Without changing slip planes, will they spontaneously line up one under the other?

b. Under what circumstances could they move to the same slip plane?

c. If they did so, what would tend to happen once they were on the same plane?

2.32   For austenitic stainless steel, Cu, and Al (all FCC metals):

a. Calculate the actual magnitudes of the full and partial dislocations, assuming that the lattice parameters are 0.365 nm, 0.362 nm, and 0.405 nm, respectively.

b. Calculate the equilibrium partial dislocation separation distance d for all three materials.

c. Put the numbers from part (b) in context by comparing them to the atomic size (diameter) and lattice parameter for each material.

d. In which of the three material(s) is wavy glide very likely to be observed?

2.33   A cube of material is loaded triaxially, resulting in the following stresses at the point of plastic yielding: $\sigma_x = 140\,\text{MPa}$, $\sigma_y = 20\,\text{MPa}$, and $\sigma_z = 35\,\text{MPa}$.

a. What is the *shear strength* of the material according to the Tresca yield criterion?

b. If the stress in direction Z at failure were 70 MPa instead, how does this change your result? Explain.

2.34   A single-crystal rod of FCC nickel is oriented with the [001] direction parallel to the rod axis.

a. Identify the type of slip system involved in the plastic flow of nickel.

b. How many such slip systems are in a position to be activated at the same time when the load is applied parallel to this crystallographic direction?

c. What is the Schmid factor for this slip system? (The angles between the {100} and {110} and {100} and {111} planes are 45 and 54.7°, respectively.)

2.35   From the work of D. C. Jillson, *Trans. AIME* **188**, 1129 (1950), the following data were taken relating to the deformation of zinc single crystals.

ϕ	λ	F (newtons)
83.5	18	203.1
70.5	29	77.1
60	30.5	51.7
50	40	45.1
29	62.5	54.9
13	78	109.0
4	86	318.5

The crystals have a normal cross-sectional area of $122 \times 10^{-6}\,\text{m}^2$.

ϕ = angle between loading axis and normal to slip plane

λ = angle between loading axis and slip direction

F = force acting on crystal when yielding begins

a. Identify the slip system for this material.

b. Calculate the resolved shear τ_{RSS} and normal σ_n stresses acting on the slip plane when yielding begins.

c. From your calculations, does τ_{RSS} or σ_n control yielding?

d. Plot the Schmid factor versus the normal stress P/A_0 acting on the rod. At what Schmid factor value are these experimentally-measured yield loads at a minimum? Does this make sense?

2.36   Draw the (111) pole figure for the [100] wire texture in silver and for the [110] wire texture in iron wires.

2.37   A low-carbon steel alloy was loaded in tension until just after yielding took place. A few Lüders bands were visible on the surface. The bar can either be reloaded (a) immediately, (b) after a brief and moderate temperature aging treatment, or (c) after several weeks without any exposure to elevated temperature. In each of the three cases, how is the yield strength of the reloaded bar likely to compare to that of the original test?

2.38 The tensile strength for cold-rolled magnesium alloy AZ31B plate is approximately 160 MPa for specimens tested either parallel or perpendicular to the rolling direction. When similarly oriented specimens are compressed, the yield strength is only 90 MPa. Why? (*Hint:* Consider the possible deformation mechanisms available in the magnesium alloy and any crystallographic texture that might exist in the wrought plate.)

2.39 An HCP alloy, known as Hertzalloy 200, has a *c/a* ratio of 1.600.

a. Identify the most probable slip system for this material.

b. For each of the following diagrams, determine whether slip will occur and whether twinning will occur (consider only $\{10\bar{1}2\}$ twinning). Briefly justify your answers.

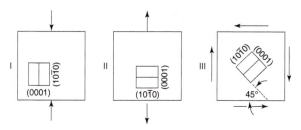

2.40 Assume that the yield behavior of PMMA is well described by the pressure-modified von Mises yield criterion, that yielding occurs under pure shear loading at $\sigma_1 = -\sigma_2$ 60 MPa, and that yielding occurs under pure tension loading at $\sigma_1 = 94.2$ MPa. Predict the stress needed to cause yielding (a) in uniaxial compression along direction 1 or 2, (b) under equal biaxial tension, and (c) under equal biaxial compression. Finally, plot these yield conditions in a fashion similar to that of Fig. 2.73.

Design

2.41 The design of a metallic component is undergoing a change such that the stress state will go from pure uniaxial compression to biaxial loading, with the secondary load applied at a 90° angle to the primary load, and with the secondary stress at 20% of the primary stress but of opposite sign. If the material in question is a plate of rolled 304 stainless steel in the annealed state, and the original uniaxial compressive design stress was 50% of the yield strength (to achieve a safety factor of 2×), what safety factor remains under the new, biaxial loading condition?

2.42 An unplasticized PVC component is intended to be loaded under uniaxial tension. After premature failure of a prototype component, a cross section observed under crossed polarizing lenses reveals that there are shear bands extending from the surfaces into the material, but none in the interior. What, if anything, does that tell you about the actual loading of the component in service that could be used to guide changes to the design?

2.43 You have been asked to use a finite element computer model to predict the yield condition for a PMMA component under a complex loading scenario. Among the many yield criteria that are likely to be available in the finite element software package, which would probably be the best choice for this case? Why?

Extend

2.44 Acquire a journal paper that uses Neumann bands as evidence in a failure analysis. Summarize the article, clearly identify the role that the discovery of Neumann bands played in the failure analysis, and provide a formal reference for the paper.

2.45 Find five examples of products made of plasticized PVC and five made of unplasticized PVC. How does the choice of plasticized vs. unplasticized PVC match the engineering requirements of the products in each category?

Chapter 3

Controlling Strength

3.1 STRENGTHENING: A DEFINITION

We are now in a position to examine the various ways by which a ductile material may be strengthened; stated another way, we now seek to control the material's resistance to plastic deformation, thereby increasing or decreasing the strain range that is purely elastic. This topic was introduced in Chapter 2, in which the post-yield and work hardening of metals were discussed. The effects of polymer chain orientation on yield strength were also introduced in Chapter 2. These effects are examples of *intrinsic* strengthening—the fundamental yield behavior of the materials is altered. It should be noted that intrinsic strengthening in metals, per se, primarily relates to processes by which dislocation motion is restricted within the lattice. Intrinsic polymer strengthening is largely achieved by reducing the tendency for chain sliding. Metals and polymers can also be strengthened by the addition of high-strength fibers, creating composite materials. In a sense, such strengthening can be viewed as being *extrinsic* in nature since the load on the matrix is transferred to the high-strength fibers while the *intrinsic* resistance to deformation in the matrix is not changed. Note that discussion of the strength of glass and ceramic materials will be deferred until Chapter 7 because the elastic limit in these materials is determined by the onset of fracture processes rather than by plasticity.

3.2 STRENGTHENING OF METALS

A number of metal-strengthening mechanisms have been identified, all of which are associated with control of dislocation density, and in particular the density of mobile dislocations. That dislocation density is a critical factor in determining the strength of a metal or metal alloys is evident in the trend shown in Fig. 3.1. It is interesting to note that the strength of a metal approaches extremely high levels when there are either no dislocations present (recall the behavior of dislocation-free metal whiskers and Mo micro-pillars in Chapter 2) or when the number of dislocations is extremely high ($\geq 10^{10}/cm^2$); low strength levels correspond to the presence of moderate numbers of dislocations ($\sim 10^3$–$10^5/cm^2$). The typical range of dislocation density for structural metals and alloys is shown in Fig. 3.1, within which an increase in the dislocation density monotonically increases the number of interaction events that inhibit subsequent plastic deformation. These events include dislocation interactions with other dislocations (strain hardening), grain boundaries, solute atoms (solid solution strengthening), precipitates (precipitation hardening), and dispersoids (dispersion strengthening). Some of these mechanisms work in pure metals, others only in alloys; some in single crystals, others only in polycrystals; some over a range of temperatures, others only at low temperatures relative to the metal's melting point. These are critical distinctions that affect the processing and use conditions of the material in question, so each mechanism will be discussed individually and these distinctions will be made clear.

3.2.1 Dislocation Multiplication

A fundamental question to address before proceeding on to a discussion of individual strengthening mechanisms is, "By what means does dislocation density change?" We can gain our first insight into this phenomenon by reconsidering the macroscopic evidence of slip presented in

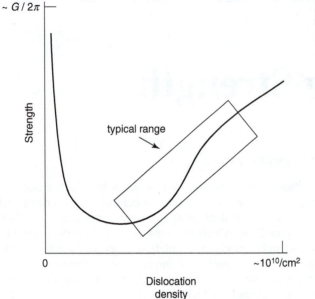

Chapter 2. Because slip offsets are clearly visible in a light microscope (e.g., see Fig. 2.19), they must be in the range of at least 1 μm in height. Since the typical Burgers vector for a dislocation is on the order of 0.2 to 0.3 nm, there is a requirement for approximately 10^4 dislocations on each slip plane to create the slip step. That so many dislocations of the same sign should lie on the same plane *before* the crystal is stressed is highly unlikely. A possible alternative explanation is that additional dislocations must have been generated during deformation. This view is supported by the observations of electron microscopists who have found the dislocation densities in thin metal films to increase from 10^4–10^5 to 10^{11}–10^{12} dislocations/cm^2 as one proceeds from the annealed to heavily cold-worked state. This corresponds to an evolution from a strength state near the minimum in Fig. 3.1 to a state at the far right.

A widely accepted mechanism for dislocation generation is based on the planar Frank-Read source. In this model, a segment of a dislocation line is considered to be pinned either by foreign atoms or particles, or by interactions with other dislocations (Fig. 3.2). Recalling Eq. 2-20, when a shear stress is applied to the crystal, the segment AB will bow out with a radius given by

$$R \propto \frac{Gb}{\tau} \tag{2-20}$$

Dislocation bowing will increase with increasing applied stress, while the radius of curvature decreases to the point where R equals half the pinned segment length l (Fig. 3.2b). At this point, the loop becomes unstable and begins to bend around itself (Fig. 3.2c). The stress necessary to produce this instability is given by

$$\tau \approx \frac{Gb}{l} \tag{3-1}$$

where l = distance between pinning points. Finally, the loop pinches off at C and C', since these two regions correspond to screw dislocations of opposite signs (Fig. 3.2d). After this has occurred, the loop and cusp ACB straighten, leaving the same segment AB as before but with an additional loop containing the same Burgers vector as the original segment (Fig. 3.2e). Upon further application of the stress, the segment AB can bow again to form a second loop while the initial loop moves out radially. With continued application of the shear stress, this source can generate an unlimited number of dislocations. In reality, however, the source is eventually shut down by "back stresses" produced by the pileup of dislocation loops against unyielding

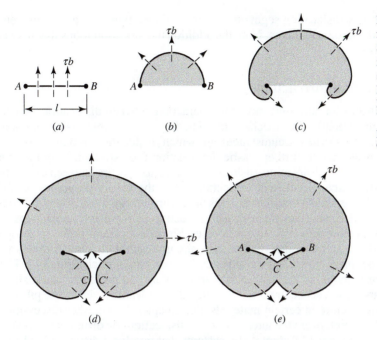

Figure 3.2 Frank-Read source for dislocation multiplication. Slipped area is shaded. Loop instability point is reached when shear stress $\tau \approx Gb/l$. (From Read;[1] *Dislocations in Crystals,* McGraw-Hill Book Co., New York, © 1953. Used with permission of McGraw-Hill Book Company.)

obstacles (e.g., grain boundaries as discussed in Section 3.4). The photograph shown in Fig. 3.3*a* represents a classic illustration of a Frank-Read source in a silicon crystal.

Another closely related dislocation generation mechanism has been suggested by Koehler[2] and modified by Low and Guard.[3] The basic feature of this model is that through the process of cross-slip, a screw dislocation can generate additional Frank-Read sources. From Fig. 3.3*b*, we

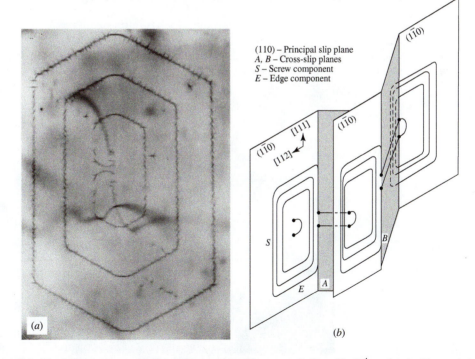

(110) – Principal slip plane
A, B – Cross-slip planes
S – Screw component
E – Edge component

Figure 3.3 Frank-Read sources. (*a*) Photomicrograph in silicon crystal. (From Dash;[4] reprinted with permission of General Electric Co.) (*b*) Dislocation multiplication by double cross-slip mechanism. (From Low and Guard;[3] reprinted with permission from Low, *Acta Met.* 7 (1959), Pergamon Press, Elmsford, NY.)

see that a screw dislocation segment has cross-slipped twice to resume movement on a plane parallel to the initial slip plane. Note the additional dislocation loops that may be generated by this process.

3.2.2 Dislocation–Dislocation Interactions

Like dislocation multiplication, the interaction between dislocations plays a fundamental role in many strengthening mechanisms. The key point is that dislocation–dislocation interactions tend to create circumstances in which dislocation mobility is reduced, thereby increasing resistance to further plastic deformation (i.e., strengthening the metal or alloy).

The first type of dislocation–dislocation interaction occurs between dislocations moving on the same slip plane. Recall that in Section 2.1.6, the elastic properties of dislocations were introduced. It was shown that elastic stress fields surround all dislocations due to the distortion of the nearby lattice, and that there can be attractive or repulsive interactions between certain dislocations in close proximity to one another. Since dislocations of the same sign will repel one another and not coalesce, they will tend to pile up (each with a unit Burgers vector) against a barrier on the slip plane such as a grain boundary or a hard particle (Fig. 3.4). The resulting *back stresses* can immobilize the dislocations so that they can no longer cause slip. (As one might expect, a large stress concentration is developed at the leading edge of the pileup, which can lead to premature fracture in certain materials. See Chapter 7 for further discussion of this point.)

Equally important are the intersections of dislocations on different slip planes. Given that in FCC metals there are 12 different slip systems that involve 4 distinct {111} planes oriented to form a tetrahedron, it is easy to imagine that most dislocations in a crystal do not lie in the same slip plane. In fact, one common model for hardening imagines that a single mobile dislocation traveling on a certain slip plane encounters a *forest* of dislocations passing through the active slip plane. As they are not on the most favorably oriented plane for slip, these forest dislocations are considered to be *sessile*—that is, they are not moving, and serve as barriers rather than as enablers of plastic deformation. There are four general categories of interaction between mobile

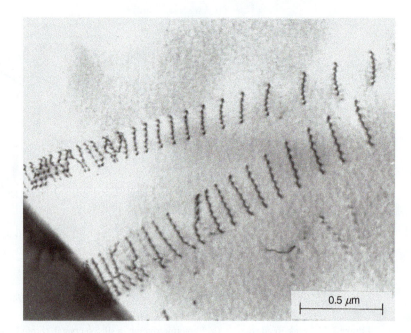

Figure 3.4   Dislocation pileups on two systems against a grain boundary in 309 stainless steel ($\gamma = 35\ \mathrm{mJ/m^2}$). (Courtesy of Anthony Thompson, Carnegie-Mellon University.)

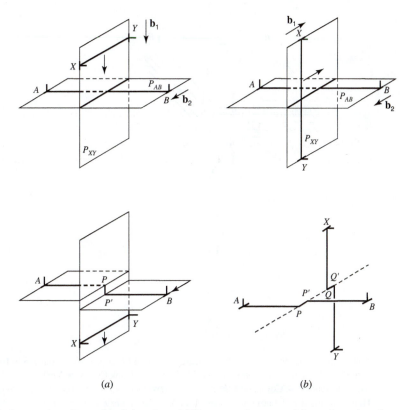

and forest dislocations: repulsion (similar to the case of like dislocations on a single plane), attraction and formation of *junctions* or *locks* between dislocations, and the formation of jogs.

Two different jog-forming edge dislocation interactions are shown in Fig. 3.5. In the first case, where the Burgers vectors of the two dislocations are at right angles, the intersection of dislocation *AB* leads to a simple lengthening of dislocation *XY*. On the other hand, dislocation *XY* with a Burgers vector b_1 cuts dislocation *AB*, producing a jog *PP'* that has a length equal to that of b_1. The Burgers vector of *PP'*, however, is b_2—the same as dislocation *AB* to which it belongs. Since b_2 and *PP'* are normal to one another, *PP'* is of the edge type. This jog moves under the same conditions as the original dislocation, and therefore its presence does not impede the movement of the dislocation *AB*.

When the Burgers vectors of the edge dislocations are parallel, the jogs produced are different in character. As shown in Fig. 3.5*b*, dislocation *XY* with its Burgers vector b_1 produces a jog *PP'* in dislocation *AB*. Since the Burgers vector b_2 in dislocation *AB* is parallel to the jog *PP'*, the jog is of the screw type. Similarly, the jog *QQ'* in dislocation *XY* is found also to be of the screw type. The screw jogs *PP'* and *QQ'* both have greater mobility than the edge dislocations to which they belong. Consequently, their presence does not impede the overall motion of the dislocation. In summary, *jogs generated in edge dislocations will not affect the movement of the dislocation.*

The same cannot be said for intersections involving screw dislocations. As illustrated in Fig. 3.6*a*, the intersection of an edge and screw dislocation will produce a jog *PP'* in the edge dislocation *AB* and another jog *QQ'* in the screw dislocation *XY*. Since each jog assumes the same Burgers vector as its dislocation, it may be seen that *PP'* and *QQ'* are both edge jogs. From

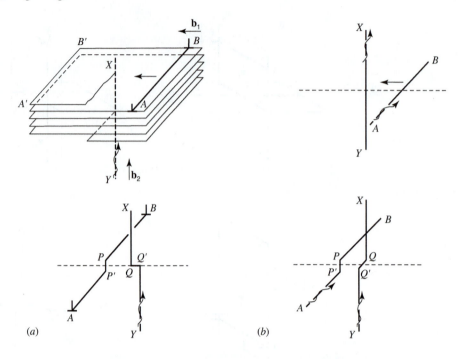

Figure 3.6 Intersection of screw dislocation XY with (a) edge dislocations AB to form two edge jogs PP' and QQ'. (From Read,[1] *Dislocations in Crystals;* McGraw-Hill Book Co., New York, © 1953. Used with permission of McGraw-Hill Book Company.) (b) Another screw dislocation AB which forms two edge jogs PP' and QQ'. (From Hull[5]; reprinted with permission from Hull, *Introduction to Dislocations,* Pergamon Press, Elmsford, NY, 1965.) Edge jogs in screw dislocations impede their motion.

the above discussion, PP' *will not* impede the motion of dislocation AB, whereas QQ' *will* restrict the movement of the screw dislocation XY. The same can be said for the edge type jogs PP' and QQ' found in the screw dislocations AB and XY, respectively, shown in Fig. 3.6b. The restriction placed on the mobility of the screw dislocations is caused by the inability of the edge jog to move on any plane other than that defined by the jog QQ' and $\mathbf{b}_2$ (i.e., the plane QQ' YZ; see Fig. 3.7). Consequently, when a shear stress is applied parallel to $\mathbf{b}_2$, the screw segments XQ and $Q'Y$ will produce displacements parallel to $\mathbf{b}_2$, while the screw dislocation lines move to DE and FG, respectively. The only way that the edge jog QQ' can follow along plane $EFQ'Q$ is by nonconservative motion involving vacancy-assisted dislocation climb. As shown schematically

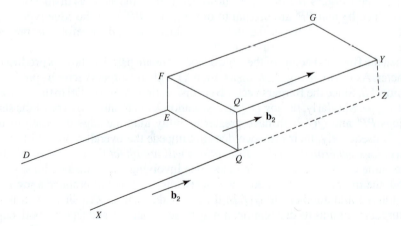

Figure 3.7 Screw dislocation XY containing an edge jog QQ', which can move conservatively on plane QQ' YZ but nonconservatively on plane EFQ' Q when screw components XQ and Q' Y move to DE and FG, respectively.

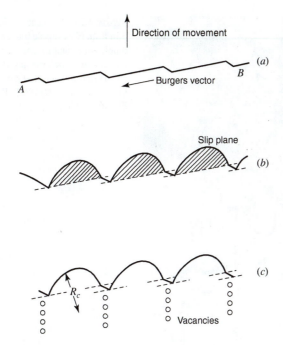

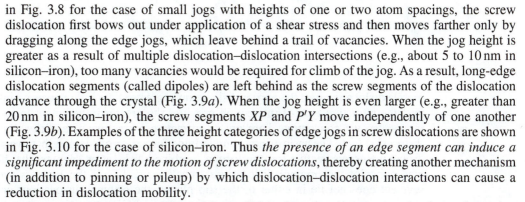

Figure 3.8   Detailed movement of jogged screw dislocation. (*a*) Jogged dislocation under zero stress; (*b*) applied shear stress causes screw component to bow out between edge jogs; (*c*) edge jogs follow screw segments by nonconservative climb, leaving behind a trail of vacancies. (From Hull;[5] reprinted with permission from Hull, *Introduction to Dislocations,* Pergamon Press, Elmsford, NY, 1965.)

in Fig. 3.8 for the case of small jogs with heights of one or two atom spacings, the screw dislocation first bows out under application of a shear stress and then moves farther only by dragging along the edge jogs, which leave behind a trail of vacancies. When the jog height is greater as a result of multiple dislocation–dislocation intersections (e.g., about 5 to 10 nm in silicon–iron), too many vacancies would be required for climb of the jog. As a result, long-edge dislocation segments (called dipoles) are left behind as the screw segments of the dislocation advance through the crystal (Fig. 3.9*a*). When the jog height is even larger (e.g., greater than 20 nm in silicon–iron), the screw segments *XP* and *P'Y* move independently of one another (Fig. 3.9*b*). Examples of the three height categories of edge jogs in screw dislocations are shown in Fig. 3.10 for the case of silicon–iron. Thus *the presence of an edge segment can induce a significant impediment to the motion of screw dislocations*, thereby creating another mechanism (in addition to pinning or pileup) by which dislocation–dislocation interactions can cause a reduction in dislocation mobility.

Interactions between mobile and forest dislocations that do not cause simple jog formation can be categorized as *repulsive* or *attractive*. In both cases, the stress required to move the

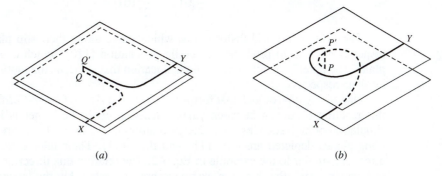

Figure 3.9   Effect of jog height on screw dislocation mobility. (*a*) Intermediate jog height *QQ'* causes long-edge segments (dipoles) to form as screw segments glide through crystal; (*b*) large jog height *PP'* allows screw segments *XP* and *YP'* to move independently of one another. (From Gilman and Johnston;[6] reprinted with permission of the authors and Academic Press, Inc., New York.)

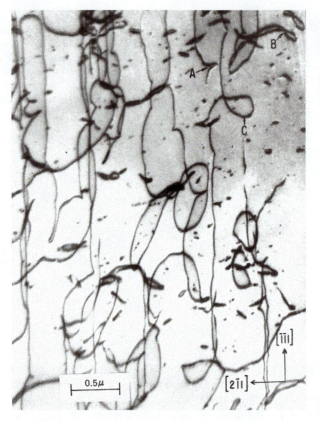

Figure 3.10   Dislocations in silicon–iron thin film. Note dipole trails at *A*, pinched-off dipoles at *B,* and independent dislocation movement at the large jog at *C.* (From Low and Turkalo;[7] reprinted with permission from Low, *Acta Met.* 10 (1962), Pergamon Press, Elmsford, NY.)

mobile dislocation past the barrier dislocation is greater than if there was no interaction. Of the two, the attractive type of interaction provides the greater impediment to further dislocation motion, so we concentrate here on the attractive formation of a *dislocation junction*, or a *dislocation lock*. The basic principle underlying formation of a lock is that two dislocations with different Burgers vectors react to form a combined dislocation segment that has lower energy than either of the original dislocations (recall Eq. 2-17). If the Burgers vector of the new segment does not lie in either of the slip planes, the original dislocations are pinned by the sessile segment. An example of such a situation in FCC materials is given by

$$\frac{a}{2}[01\bar{1}] + \frac{a}{2}[101] = \frac{a}{2}[110] \tag{3-2}$$

The $[01\bar{1}]$ and $[101]$ dislocations, which move along their slip planes, (111) and $(11\bar{1})$, respectively, join to produce the sessile dislocation [110], which cannot move along either plane. The latter is therefore a sessile dislocation that impedes the motion of other dislocations on their respective slip planes.

One strong type of lock that forms in FCC metals is the *Lomer-Cottrell lock*. An example of this reaction occurring between *partial dislocations* in FCC metals is shown via computer simulation of high stacking-fault energy aluminum in Fig. 3.11. In part *a* of the figure, the two glide planes depicted are the (111) and the $(\bar{1}11)$. Their intersection lies along the $[\bar{1}10]$ direction, similar to the example in Eq. 3-2. The relative line directions of the original partial dislocations and the junction segment are indicated by the arrows in the figure. The $\frac{a}{2}[01\bar{1}](111)$ dislocation has split into Shockley partial dislocation $A\delta$ of type $\frac{a}{6}[\bar{1}2\bar{1}]$ and δC of type $\frac{a}{6}[11\bar{2}]$. The splitting of the other dislocation is similar in nature. Where they react along the $[\bar{1}10]$ direction, two distinct junction segments are actually created. The first is a

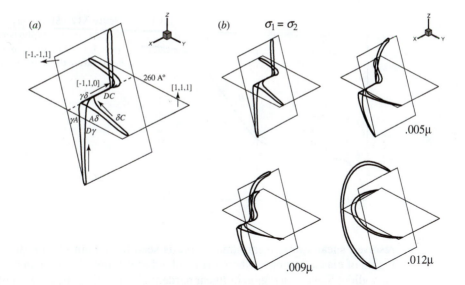

Figure 3.11 Computer simulation of (*a*) a junction formed between two dislocations in aluminum, and (*b*) the same dislocations bowing under increasing stress normalized by the shear modulus μ (usually *G* in this text). Note that the dislocations are pinned together until a shear stress of 0.009 μ (∼235 MPa for aluminum) causes the junction to break. (Adapted from Figure 1a and Figure 3 with permission from V. B. Shenoy, R. V. Kukta, and R. Phillips, *Physical Review Letters*, 84, 1491 (2000). © 2000 by the American Physical Society.)[8]

Lomer-Cottrell dislocation segment, denoted γδ with Burgers vector $\frac{a}{6}\langle 110 \rangle$, which has a stacking fault on either side.[i] The adjoining segment *DC* without the stacking faults is a dislocation identified as a sessile *Lomer lock*. Note that these two segments are quite short— 3.8 nm and 4.2 nm, respectively, according to the simulation results. Nevertheless, their influence on dislocation motion can be significant. The difficulty associated with moving these merged dislocations can be seen in Fig. 3.11*b*, in which the dislocations bow out under applied shear stress. The Lomer lock segment pins the two original dislocations along the $[\bar{1}10]$ direction. As the stress is increased, the Lomer lock segment slides to the left and reduces in length. It eventually breaks, allowing the two dislocations to move forward separately on their original slip planes. The stress required to break the lock is large, and represents a significant resistance to plastic deformation.

3.3 STRAIN (WORK) HARDENING

Strain hardening (also referred to as work hardening or cold working) results from a dramatic increase in the number of dislocation–dislocation interactions and the associated reduction in dislocation mobility that occurs during plastic deformation via the mechanisms just introduced in Section 3.2. As a result of these interactions, progressively larger stresses must be applied in order that additional deformation may take place. An awareness of strain hardening dates back to the Bronze Age, and is perhaps the first widely used strengthening mechanism for metals. Artisans hammered and bent metals to desired shapes and achieved superior strength in the process. Typical cold-worked commercial products that find use today include cold-drawn piano wire and cold-rolled sheet metal.

To characterize more clearly the general strain-hardening behavior of metals and their alloys, it is helpful to examine the stress–strain response of single crystals. From Fig. 3.12, the

---

[i] This arrangement of stacking faults on different planes meeting along a junction segment is a so-called *stair-rod* dislocation, in analogy to the (now) decorative horizontal metal rods used to hold a carpet runner securely against the risers of a set of stairs.

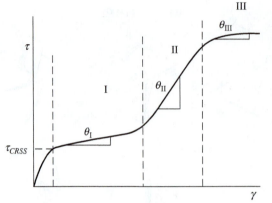

Figure 3.12 Shear stress–strain curve for single crystal revealing elastic behavior when $\tau < \tau_{crss}$ and Stage I, II, III plastic response when $\tau > \tau_{crss}$. θ_I, θ_{II}, θ_{III} measure the strain hardening rate in each region.

resolved shear stress–shear strain curve is seen to contain several distinct regions: an initial region of elastic response where the resolved shear stress is less than τ_{CRSS}; Stage I, a region of easy glide; Stage II, a region of linear hardening; and Stage III, a region of dynamic recovery or parabolic hardening. The latter three regions involve different aspects of the plastic deformation process for a given crystal. It is known that the extent of Stages I, II, and III depends on such factors as the test temperature, crystal purity, initial dislocation density, and initial crystal orientation.[9,10] It should be noted that Stage III closely resembles the stress–strain response of the polycrystalline form of the same material.

A number of theories have been proposed to explain the strain-hardening process in crystals, including the reason for the dramatic changes in strain-hardening rate associated with the three stages of plastic deformation. An extensive literature[11] has developed regarding these theories, all of which have focused on some of the dislocation interaction mechanisms described in the previous section. Unfortunately, a certain degree of confusion has arisen in this field because of the varying importance of certain dislocation interactions in different alloy crystals. One may wonder then why the three distinct stages of deformation are so reproducible from one material to another and why the work-hardening coefficient θ_{II} associated with Stage II deformation is almost universally constant at $G/300$. For these reasons, the "mesh length" theory of strain hardening proposed by Kuhlmann-Wilsdorf[12,13] is appealing pedagogically, since it does not depend on any specific dislocation model that might be appropriate for one material but not for another. Her theory may be summarized as follows: In Stage I a heterogeneous distribution of low-density dislocations exists in the crystal. Since these dislocations can move along their slip planes with little interference from other dislocations, the strain hardening rate θ_I is low. The easy glide region (Stage I) is considered to end when a fairly uniform dislocation distribution of moderate density is developed but not necessarily in lockstep with the onset of conjugate slip where a marked increase in dislocation–dislocation interactions would be expected. At this point Kuhlmann-Wilsdorf theorizes the existence of a quasi-uniform dislocation array with clusters of dislocations surrounding cells of relatively low dislocation density (Fig. 3.13a). It is believed that such cell structures represent a minimum energy and, hence, preferred dislocation configuration within the crystal.[14] Studies have shown that high stacking fault energy metals (e.g., aluminum) exhibit cell walls that are narrower and cell interiors that are more dislocation-free than in lower stacking fault energy metals (e.g., copper) (Fig. 3.13b). (In very low stacking fault energy metals (e.g., Cu–7%Al) the crystal substructure is characterized by dislocation planar arrays, consistent with the tendency for these materials to exhibit restricted cross-slip (Fig. 3.13c)). The stress necessary for further plastic deformation is then seen to depend on the mean free dislocation length $\bar{l}$ in a manner similar to that necessary for the activation of a Frank-Read source where

$$\tau \propto \frac{Gb}{\bar{l}} \tag{3-3}$$

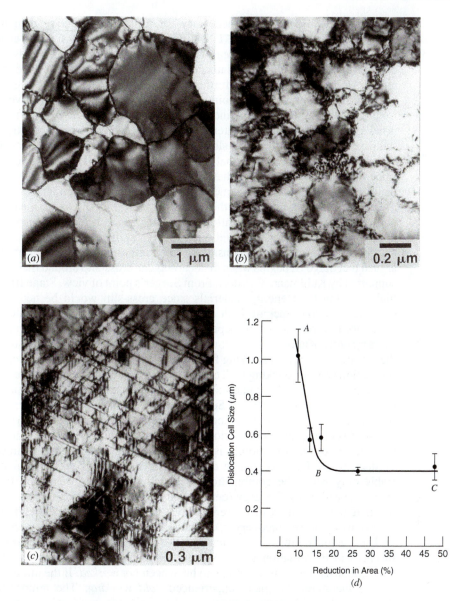

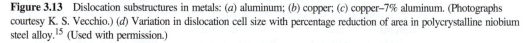

Figure 3.13 Dislocation substructures in metals: (*a*) aluminum; (*b*) copper; (*c*) copper–7% aluminum. (Photographs courtesy K. S. Vecchio.) (*d*) Variation in dislocation cell size with percentage reduction of area in polycrystalline niobium steel alloy.[15] (Used with permission.)

Since the dislocation density is proportional to $(\bar{l})^{-2}$ Eq. 3-3 may be written in the form

$$\Delta \tau \propto Gb\sqrt{\rho} \qquad (3\text{-}4)$$

where

ρ = dislocation density

$\Delta \tau$ = incremental shear stress necessary to overcome dislocation barriers

This relationship has been verified experimentally for an impressive number of materials[16] and represents a necessary requirement for any strain-hardening theory. With increasing plastic deformation, ρ increases resulting in a decrease in the mean free dislocation length $\bar{l}$. From

Eqs. 3-3 and 3-4, the stress necessary for further deformation then increases. Kuhlmann-Wilsdorf suggests[12] that there is a continued reduction in cell size and an associated increase in flow stress throughout the linear hardening region. In other words, the *character* of the dislocation distribution remains unchanged, and only the *scale* of the distribution changes (see region *AB* in Fig. 3.13*d*). With further deformation, the number of free dislocations within the cell interior decreases to the point where glide dislocations can move relatively unimpeded from one cell wall to another. Since the formation of new cell walls (and hence a reduction in $\bar{l}$) is believed to depend on such interactions, a point would be reached where the cell size $\bar{l}$ would stabilize or at best decrease slowly with further deformation. According to Kuhlmann-Wilsdorf,[13] this condition signals the onset of Stage III and a lower strain-hardening rate, since $\bar{l}$ would not decrease further. Bassin and Klassen[15] provided experimental confirmation that Stage III behavior corresponds to strain levels where l remains constant (see region *BC* in Fig. 3.13*d*). Of particular note, the data reported in Fig. 3.13*d* are measurements taken from a polycrystalline niobium steel alloy; as such, the mesh length theory of strain hardening is applicable for both single-crystal and polycrystalline commercial alloys.

Stacking fault energy is considered to be important to the onset of Stage III. Seeger[17] has argued that Stage III begins when dislocations can cross-slip around their barriers, a view initially supported by Kuhlmann-Wilsdorf. From Seeger's point of view, Stage III would occur sooner for high stacking fault energy materials since cross-slip would be activated at a lower stress. Conversely, a low stacking fault energy material, such as brass, would require a larger stress necessary to force the widely separated partial dislocations to recombine and hence cross-slip. More recently, Kuhlmann-Wilsdorf[13,14] suggested that the mesh length theory could also explain the sensitivity of τ_{III} to stacking fault energy by proposing that enhanced cross-slip associated with a high value of stacking fault energy would accelerate the dislocation rearrangement process. Consequently, $\bar{l}$ would become stabilized at a lower stress level. Setting aside for the moment the question of the correctness of the Seeger versus Kuhlmann-Wilsdorf interpretations, it is sufficient for us to note that both theories account for the inverse dependence of τ_{III} on stacking fault energy.

In discussing the deformation structure of metals, it is important to keep in mind the temperature of the operation. It is known that the highly oriented grain structure in a wrought product, which has a very high dislocation density (10^{11} to 10^{13} dislocations/cm^2), remains stable only when the combination of stored strain energy (related to the dislocation substructure) and thermal energy (determined by the deformation temperature) is below a certain level. If not, the microstructure becomes unstable and new strain-free equiaxed grains are formed by combined recovery, recrystallization, and grain growth processes. These new grains will have a much lower dislocation density (in the range of 10^4 to 10^6 dislocations/cm^2). When mechanical deformation at a given temperature causes the microstructure to recrystallize spontaneously, the material is said to have been *hot worked*. If the microstructure were stable at that temperature, the metal experienced *cold working*. The temperature at which metals undergo hot working varies widely from one alloy to another but is generally found to occur at about one-third the absolute melting temperature. Accordingly, lead is hot worked at room temperature, while tungsten may be cold worked at 1500 °C.

Before concluding the discussion of single-crystal stress–strain curves, it is appropriate to consider whether one can relate qualitative and quantitative aspects of the stress–strain response of single-crystal and polycrystalline specimens of the same material. For one thing, the early stages of single-crystal deformation would not be expected in a polycrystalline sample because of the large number of slip systems that would operate (especially near grain boundary regions) and interact with one another. Consequently, the tensile stress–strain response of the polycrystalline sample is found to be similar only to the Stage III single-crystal shear stress–strain plot. A number of attempts have been made to relate these two stress–strain curves. From Eq. 2-25 and Section 2.2.3, recall that

$$\sigma = \frac{P}{A} = \tau \frac{1}{\cos\phi\cos\lambda} = \tau M \tag{3-5}$$

where the Taylor factor $M = 1/(\cos\phi\,\cos\lambda)$. Assuming the individual grains in a polycrystalline aggregate to be randomly oriented, M would vary with each grain such that some average orientation factor $\overline{M}$ would have to be defined,[18] as discussed previously in Section 2.2.3. It may be shown[19] that the normal strain and the shear strain can be related by

$$\varepsilon = \gamma\overline{M} \tag{3-6}$$

By combining Eq. 3-5 and 3-6 it is seen that

$$\frac{d\sigma}{d\varepsilon} = \overline{M}^2\frac{d\tau}{d\gamma} \tag{3-7}$$

One can see from Eq. 3-7 that the strain-hardening rate of a polycrystalline material is many times greater than its single-crystal counterpart.

3.4 BOUNDARY STRENGTHENING

The presence of grain boundaries has an additional effect on the deformation behavior of a material by serving as an effective barrier to the movement of glide dislocations. From the work of Petch[20] and Hall,[21] the yield strength of a polycrystalline material could be given by

$$\sigma_{ys} = \sigma_i + k_y d^{-1/2} \tag{3-8}$$

where

$\sigma_{ys} =$ yield strength of the polycrystalline sample
$\sigma_i =$ overall resistance of the lattice to dislocation movement
$k_y =$ "locking parameter," a measure of the relative hardening contribution of grain boundaries
$d =$ grain size

Although Eq. 3-8 is simply a fit to the experimental results of Petch and Hall, it works very well for many materials. As a result, there have been many attempts to identify the underlying physics and therefore to justify the form of the equation. One such effort can be traced to the work of Eshelby et al.[22] In this model, the number of dislocations that can occupy the space between the dislocation source and the grain boundary is given by

$$n = \frac{\alpha\tau_s d}{Gb} \tag{3-9}$$

where

$n =$ number of dislocations in the pileup
$\alpha =$ constant
$\tau_s =$ average resolved shear stress in the slip plane
$d =$ grain diameter
$G =$ shear modulus
$b =$ Burgers vector

The stress acting on the lead dislocation is found to be n times greater than τ_s. When this local stress exceeds a critical value τ_c, the blocked dislocations are able to glide past the grain boundary. Hence

$$\tau_c = n\tau_s = \frac{\alpha\tau_s^2 d}{Gb} \tag{3-10}$$

Since the resolved shear stress τ_s is equal to the applied stress τ less the frictional stress τ_i associated with intrinsic lattice resistance to dislocation motion, Eq. 3-9 may be rewritten as

$$\tau_c = \frac{\alpha(\tau - \tau_i)^2 d}{Gb} \tag{3-11}$$

After rearranging,

$$\tau = \tau_i + k_y d^{-1/2} \tag{3-12}$$

which is the shear stress form of Eq. 3-8. The Hall-Petch relation also characterizes alloy yield strength in terms of other microstructural parameters such as the pearlite lamellae spacing and martensite packet size in steel (see Section 3.8). It is readily seen that grain refinement techniques (e.g., normalizing alloy steels) provide additional barriers to dislocation movement and enhance the yield strength. As will be shown in Chapter 7, improved toughness also results from grain refinement.

Conrad[23] has demonstrated clearly that σ_i may be separated into two components: σ_{ST}, which is not temperature sensitive but structure sensitive where dislocation–dislocation, dislocation–precipitate, and dislocation–solute atom interactions are important; and σ_T, which is strongly temperature sensitive and related to the Peierls stress. The yield strength of a material may then be given by

$$\sigma_{ys} = \sigma_T + \sigma_{ST} + k_y d^{-1/2} \tag{3-13}$$

where the σ_T term describes short-range order Peierls stress effects (< 1 nm), the σ_{ST} term describes long-range order dislocation stress field effects (10–100 nm), and the final term describes very long-range structural size effects (>1000 nm). Note that the overall yield strength of a material depends on both short- and long-range stress field interactions with moving dislocations.

The universal use of the Hall-Petch relation to characterize the behavior of metal alloys should be viewed with caution since other equations can sometimes better describe the observed strength–microstructural size relation.[24] There is some consensus that grain boundary-induced dislocation pileups may not always be responsible for the yield-strength-microstructural size relation described above. Instead, in this case thought focuses on the important role of the grain boundary as a source for dislocations, with the yield strength being given by

$$\tau = \tau_i + \alpha Gb\rho^{1/2} \tag{3-14}$$

Li[25] theorized that dislocations were generated at grain-boundary ledges and noted that the dislocation density ρ was inversely proportional to the grain size, d. Consequently, Eq. 3-14 has the same form as Eq. 3-12, with αGb standing in for k_y.

3.4.1 Strength of Nanocrystalline and Multilayer Metals

It has been noted that extrapolation of Eq. 3-8 to extremely small grain sizes leads to the prediction of yield strength levels that approach theoretical levels.[26] Much effort has been expended to explore the limits of the Hall-Petch relation as grain size shrinks to nanometer dimensions. This has been done experimentally with nanocrystalline (nc) metals, with nanotwinned (nt) metals, and with very thin metallic multilayers, and has also been examined using computer modeling and simulation techniques.[27] A key aspect of nc metals is that with grain sizes of less than 250 nm, and sometimes approaching 2–5 nm, the grain boundaries make up a large volume fraction of the overall material. Likewise, in nt metals and multilayers, the shortest distance between potential barriers to dislocation motion is only on the order of 10**b** to 1000**b**, where **b** is the Burgers vector.

In early work with pure Cu and Pd deposited by a technique called inert gas condensation[ii] it was found that porosity negatively affects certain mechanical characteristics of nc metals. However, even with some degree of porosity the materials exhibited compressive yield strength values ranging from 0.65 to over 1.1 GPa.[28] These are substantial increases over their larger-grained counterparts. The density of these nc materials was over 90%, and the grain size varied from approximately 20 nm to 65 nm. Unfortunately, the increase in strength in such nc metals tends to be accompanied by a decrease in tensile ductility. This is different from the trend of larger-grain metals (including those considered Ultra Fine Grained with sizes between 250 and 1000 nm) in which a grain size reduction often improves ductility. Some of the ductility penalty is probably due to defects introduced during processing, but some may be inherent to the grain structure of nc metals that discourages plasticity-enabling dislocation mobility and generates high local stress concentrations that can nucleate cracks in tension.[27]

While it appears that increased strengthening occurs with decreasing grain size well below 1 μm, the behavior may deviate from the linear $d^{-1/2}$ relationship, and may saturate or even reverse as grain size falls below approximately 100 nm when non-dislocation-based deformation mechanisms like grain boundary sliding may come into play. Many reasons for a breakdown of Hall-Petch behavior have been proposed, but one central argument is that grain sizes below a certain critical value cannot accommodate multiple dislocations and therefore cannot develop dislocation pileups in the classical sense. This can be seen from Eq. 3-9 when d is not much bigger than **b**. If this is the case, the strengthening associated with pileups should eventually level off as grain size decreases.[29]

As an alternative to the nc metal strengthening approach, nt metals have been proposed. It has been shown that coherent twin boundaries in Cu can block dislocation transmission while avoiding some of the instabilities that accompany nc grains. In one study,[30] it was shown that Cu with grains of ∼400–450 nm filled with twins of width 15–96 nm increased in tensile strength *and in ductility* as the nano-twin density increased (i.e., as the spacing between nano-twins decreased). This trend can be seen in the curves labeled *A–C* in Fig. 3.14. Coarse Grained (CG) and nc Cu fabricated by the inert gas condensation technique are shown for comparison. The nt Cu follows the Hall-Petch trend with the mean twin lamella spacing substituting for grain size.

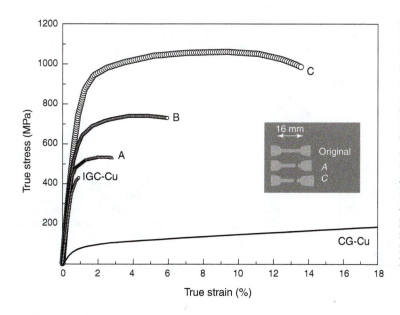

Figure 3.14 Curves A, B, and C: true stress–true strain tensile curves for nano-twinned (nt) Cu. Nano-twin density increases from curve A to curve C. Inert gas condensation (IGC) nc Cu and conventional Coarse Grained (CG) Cu specimens are shown for comparison.[30] (Reprinted from *Scripta Materialia*, vol. 52, Y. F. Shen, L. Lu, Q. H. Lu, Z. H. Jin, and K. Lu, "Tensile properties of copper with nano-scale twins," p. 989, 2005, with permission from Elsevier.)

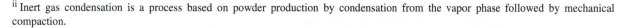

[ii] Inert gas condensation is a process based on powder production by condensation from the vapor phase followed by mechanical compaction.

Finally, multilayer composites of dissimilar metals (as well as metal/intermetallic and metal/ceramic combinations) have been used to explore the potential of a nanoscale multiphase approach to strengthening. Like nc and nt metals, nanoscale multilayered composites can demonstrate high strength levels because of the extremely high densities of interfaces that act as strong barriers to slip transmission. In Cu–Nb multilayers, for instance, it has been seen that the Hall-Petch model with its $h^{-1/2}$ dependence works for layer thickness values down to approximately $h = 75$–100 nm.[31] Below this layer thickness, classical dislocation pileups become increasingly unlikely, as noted above, and there is deviation from the $h^{-1/2}$ dependence. However, the strength for the Cu–Nb multilayer system continues to increase even as layer thickness is reduced to $h \approx 5$ nm, at which point the strength saturates. This trend, like those observed for nc and nt metals, offers intriguing possibilities for the employment of remarkably high-strength metal coatings and structural metals that derive their attractive properties completely from feature size, and not from complex chemistry.

3.5   SOLID SOLUTION STRENGTHENING

Up to this point, we have considered strain-hardening and grain-boundary strengthening mechanisms that would be operative both in pure metals and in alloys. When a metal is alloyed in such a way that the elements involved form a single-phase solid solution, another important strengthening mechanism comes into play that is not available in pure metals: solid solution strengthening. Common alloys that derive much of their strength from solid solution hardening include the Al 3xxx series (Mn and sometimes Mg in solution) and the Al 5xxx series (Mg in solution).

When two or more elements are combined such that a single-phase microstructure is retained, various elastic, electrical, and chemical interactions take place between the stress fields of the solute atoms and the dislocations present in the lattice.[32–35] Of these, elastic interactions are believed to be most important and will be the focus of our discussion.

With reference to the stress fields surrounding both edge and screw dislocations, we see from Figs. 2.15 and 2.16 that shear stresses are associated with a screw dislocation, whereas both shear and hydrostatic stress fields surround an edge dislocation. Regarding the latter, one finds that the edge dislocation is surrounded by combined shear/hydrostatic stress fields at all locations except along the Y and X axes. Along the Y axis, the stress field is one of hydrostatic

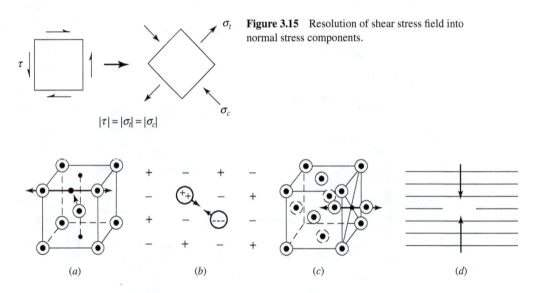

Figure 3.15   Resolution of shear stress field into normal stress components.

$|\tau| = |\sigma_t| = |\sigma_c|$

Figure 3.16   Nonsymmetrical stress fields in crystals, (*a*) Octahedral interstitial site in BCC crystal (<100> anisotropy); (*b*) divalent ion-vacancy pair (<110> anisotropy); (*c*) interstitial pair in FCC crystal (<100> anisotropy); (*d*) vacancy disk (<111> anisotropy).

compression above the dislocation line and of hydrostatic tension below the dislocation line. This should be intuitively obvious to the reader since the extra "half plane" associated with the edge dislocation is squeezed into the top half of the crystal and, as such, acts to dilate the bottom half of the crystal, much as an axe blade splits open a log of wood. Along the X axis, no hydrostatic stresses are present and the stresses are pure shear in nature.

When the shear stress fields associated with both edge and screw dislocations are resolved into their normal stress components (Fig. 3.15), note that the absolute magnitude of the shear stress is equal to the normal stress; of importance, however, is the fact that the sign of the normal stress is reversed along the $\pm 45°$ directions. It follows that the shear stress field surrounding a screw dislocation is distortional (i.e., stretched in one direction and compressed in the other), whereas the edge dislocation contains both distortional and dilatational components.

The potential interaction between an edge or screw dislocation with a solute atom depends on the stress field associated with the solute atom. For example, if an atom of chromium were to substitute for an atom of FCC nickel or BCC iron, the host lattices would experience a symmetrical (hydrostatic) misfit stress associated with differences in size between solute and solvent atoms.[36] Lattice distortion would be felt equally in all directions, with the strengthening contribution being proportional to the magnitude of the misfit ε_m such that

$$\varepsilon_m = \frac{1}{a}\frac{da}{dc} \qquad (3\text{-}15)$$

where

$a =$ lattice parameter
$c =$ solute concentration

The hydrostatic stress field of a substitutional solute atom interacts with the hydrostatic stress field associated with edge dislocations but not with the distortional stress field surrounding screw dislocations in the lattice. The level of hardening also depends on how much the local modulus G of the crystal is altered as a function of solute content. In the case of a symmetrical defect, a decrease in the local modulus associated with the addition of a relatively low modulus solute causes the local dislocation line tension (recall Eq. 2-18) to be reduced, which pins the dislocation at the site of the solute atom. In the case of a relatively high modulus solute atom the opposite situation occurs, and there is repulsion. If the fractional change in modulus is expressed as

$$\varepsilon_G = \frac{1}{G}\frac{dG}{dc} \qquad (3\text{-}16)$$

then the difference between the cases of low and high modulus solute atom additions is captured as a change in the sign of ε_G. Fleischer[37] showed that the concentration and modulus effects can be either synergistic or antagonistic depending on the solute–solvent pair in question, and that for many binary alloys the degree of strengthening correlates with

$$\frac{d\tau}{dc} \propto \left| \frac{\varepsilon_G}{\left(1 + \frac{1}{2}|\varepsilon_G|\right)} - \beta\varepsilon_m \right| \qquad (3\text{-}17)$$

where β is a fitting parameter. This relationship can therefore often be used to predict the relative strengthening effects of different solute atoms on a particular solvent. It can be seen that for a given misfit strain, ε_m (which is always positive), the choice of a solute with a lower modulus than the solvent causes ε_G to be negative, so the modulus and misfit terms in Eq. 3-17 are synergistic. In this case, the overall effect on the critical resolved shear stress is greater than if the solute modulus were larger than that of the solvent.

An even greater solute atom–dislocation interaction occurs when the misfit stress field associated with the solute atom interacts with both edge and screw dislocations. The stress fields associated with the four lattice defects shown in Fig. 3.16 satisfy this requirement in that they are nonsymmetrical and, as such, will interact with the nonsymmetrical stress components of both edge and screw dislocations. The defect type shown in Fig. 3.16*a* identifies one of the octahedral interstitial sites within the BCC iron lattice where carbon and/or nitrogen atoms are located. The size of this octahedral interstitial site along any edge in the BCC lattice (or its equivalent location in the middle of each cube face) is not symmetrical and provides insufficient room for carbon and nitrogen atoms in the (100) direction[38]; this arises from the fact that the site size is 0.038 and 0.156 nm in the (100) and (110) directions, respectively, whereas the diameter of the carbon atom is 0.154 nm. Theoretical considerations as well as experimental findings have shown that steel alloy strength increases rapidly at small carbon concentrations with a relationship of the form

$$\tau \propto c^{1/2} \tag{3-18}$$

as demonstrated in Fig. 3.18. (Note that the same proportionality exists between shear strength and concentration for symmetric defects even though the absolute magnitude is not as large.)

Such alloy strengthening is of great commercial interest to the steel industry. The insufficient amount of space available for the carbon atom in the BCC lattice also accounts for the very limited solid solubility of carbon in BCC iron (approximately 0.02%) and leads to the development of a body-centered-*tetragonal* lattice in high-carbon martensite rather than the body-centered-*cubic* crystal form for pure iron. It should be noted that the octahedral interstitial site in FCC iron is symmetrical and provides space for an atom whose diameter is as great as 0.102 nm. Since the extent of lattice distortion in the FCC lattice is much less than that found in the BCC form, the strengthening contribution of carbon in FCC iron (i.e., austenite) is low. (At the same time, the solubility limit of carbon in FCC iron is in excess of 2%—more than 100 times greater than that associated with carbon in the BCC ferrite phase.) To summarize, the strengthening potential for carbon in FCC iron is much less than that for carbon in BCC iron since the strain field surrounding the interstitial atom site is symmetrical in the FCC lattice; solute atom interaction with screw dislocations is then much weaker than for the placement of carbon atoms in the nonsymmetrical interstitial sites in the BCC lattice.

Other nonsymmetrical defects are shown in Fig. 3.16. The substitution of a divalent ion in a monovalent crystal requires that two monovalent ions be replaced by a single divalent ion; this is necessary to maintain charge balance. The divalent ion and the associated vacancy have an affinity for one another, which establishes a nonsymmetrical stress field in the ⟨110⟩ direction (Fig. 3.16*b*). Interstitial atom pairs such as those resulting from irradiation damage in an FCC crystal produce a stress field in the ⟨100⟩ direction (Fig. 3.16*c*). Finally, the collapsed vacancy disk in an FCC lattice produces a dislocation loop with asymmetry in the ⟨111⟩ direction (Fig. 3.16*d*).

From the above discussion, it is seen that the relative strengthening potential for a given solute atom is determined by the nature of the stress field associated with the solute atom. When the stress field is symmetrical, the solute atom interacts only with the edge dislocation and solid solution strengthening is limited. Examples of such symmetrical defects are shown in Table 3.1. In sharp contrast, when the stress field surrounding the solute atom is nonsymmetrical in character, the solute atom interacts strongly with both edge and screw dislocations; in this instance, the magnitude of solid solution strengthening is much greater (Table 3.1). *Note that the degree of solid solution strengthening depends on whether the solute atom possesses a symmetrical or nonsymmetrical stress field and not whether it is of the substitutional or interstitial type.* Examples of solid solution strengthening in both symmetrical (Pd or Pt in Cu) and asymmetrical distortional stress fields (C in Fe and N in Nb) are shown in Fig. 3.17. Finally, it is interesting to note that the addition of a given amount of solute atoms to the host metal may, in some instances, lead to solid solution hardening at one temperature and *softening* at another.[39,40] It has been suggested that this contrasting response is due to complex temperature-dependent interactions of screw dislocations with Peierls and solute misfit strain fields.

Table 3.1   Dislocation-Solute Interaction Potential[32]

Material	Defect	Hardening Effect $\frac{d\tau}{dc}$ as $f(G)$
Symmetrical Defects		
Al	Substitutional atom	$G/10$
Cu	Substitutional atom	$G/20$
Fe	Substitutional atom	$G/16$
Ni	Interstitial carbon	$G/10$
Nb	Substitutional atom	$G/10$
NaCl	Monovalent substitutional ion	$G/100$
Nonsymmetrical Defects		
Al	Vacancy disk (quenched)	$2G$
Cu	Interstitial Cu (irradiation)	$9G$
Fe	Interstitial carbon	$5G$
LiF	Interstitial fluorine (irradiation)	$5G$
NaCl	Divalent substitutional ion	$2G$

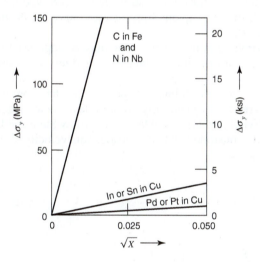

Figure 3.17   Alloy strength dependence on solute content. Greater strengthening associated with nonsymmetrical defect sites. (Reprinted with permission from K. M. Ralls, T. H. Courtney, and J. Wulff, *Introduction to Materials Science and Engineering,* Wiley, New York (1976).

3.5.1 Yield-Point Phenomenon and Strain Aging

We are now in a position to describe in detail the discrete load drops sometimes observed during tensile testing of certain metals as shown in Fig. 3.18. The curve marked *A* will be familiar as one associated with the formation of *Lüders bands*, as introduced in Section 2.4.3. As we noted in the previous section, carbon and nitrogen atoms possess a strong attraction for both edge and screw dislocations within the BCC iron lattice; accordingly, a solute *atmosphere* is formed around each dislocation core. A similar affinity appears to exist between interstitial oxygen atoms and dislocations in commercial-purity α-Ti.[41] Since these dislocations are pinned by such solute atmospheres, dislocation motion is severely restricted until a sufficiently high stress (the upper yield point on curve *A*) is applied to enable the dislocations to rip free and move through the lattice. According to theory,[36,42] these unpinned dislocations multiply rapidly by a multiple-cross-slip mechanism (Fig. 3.3). As a result, the number of mobile dislocations increases sharply, yielding becomes easier, and the load

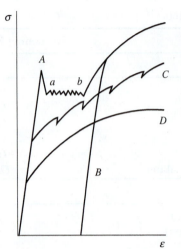

Figure 3.18 Stress-strain curves influenced by discontinuous yielding. Curve *A:* yield-point behavior; curve *B:* ordinary homogeneous yield and strain-hardening response after reloading; curve *C:* serrated yield behavior associated with dislocation-solute atom interactions leading to heterogeneous plastic deformation; curve *D:* ordinary strain-hardening behavior associated with homogeneous plastic deformation.

necessary for continued deformation decreases to the level associated with the lower yield point (marked as point *a* on curve *A*). As additional regions (i.e., the Lüders bands) deform in this manner, the stress level remains relatively constant until essentially all dislocations have broken free from their respective solute atom clusters. At this point continued deformation takes place by homogeneous plastic flow (curve *A* beginning at point *b*). Furthermore, if the test was interrupted after completion of the Lüders strain region (*ab*) and the load removed and then immediately reapplied, the subsequent stress–strain curve would not display any yield point (see Fig. 3.18, curve *B*).

Although this explanation for yield-point phenomenon may be appropriate for iron single crystals containing small solute additions of interstitial carbon and nitrogen as well as for α-Ti with oxygen in solution, it does not explain similar yield-point behavior in other material such as silicon, germanium, and lithium fluoride. Johnston[43] and Hahn[44] have proposed that yield-point behavior in these crystals is related to an initially low mobile dislocation density and a low dislocation-velocity stress sensitivity. Regarding the latter, studies by Stein and Low,[45] Gilman and Johnston,[46,47] and others demonstrated that the dislocation velocity v depends on the resolved shear stress as given by

$$v = \left(\frac{\tau}{D}\right)^m \tag{3-19}$$

where

$$v = \text{dislocation velocity}$$
$$\tau = \text{applied resolved shear stress}$$
$$D, m = \text{material properties}$$

Defining the plastic strain rate by

$$\dot{\varepsilon}_p \propto Nbv \tag{3-20}$$

where

$$\dot{\varepsilon}_p = \text{plastic strain rate}$$
$$N = \text{number of dislocations per unit area free to move about and multiply}$$
$$b = \text{Burgers vector}$$
$$v = \text{dislocation velocity}$$

Johnston[43] argued that when the initial mobile dislocation density in these materials is low, the plastic strain rate would be less than the rate of movement of the test machine crosshead and

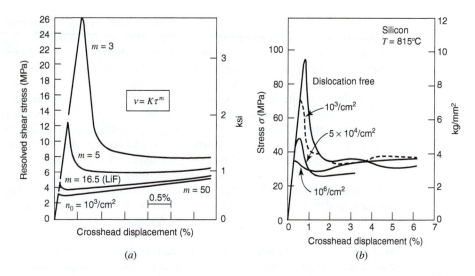

Figure 3.19   (a) Effect of stress sensitivity m in LiF. (From Johnston;[43] reprinted with permission of American Institute of Physics, New York.) (b) Effect of initial mobile dislocation density n_0 in Si on severity of yield drop. (From Patel and Chaudhuri;[48] reprinted with permission of American Institute of Physics, New York.)

little overall plastic deformation would be detected. At higher stress levels, the dislocations would be moving at a higher velocity and also begin to multiply rapidly such that the total plastic strain rate would then exceed the rate of crosshead movement. To balance the two rates, the dislocation velocity would have to decrease. From Eq. 3-19, this may be accomplished by a drop in stress, the magnitude of which would depend on the stress-sensitivity parameter m. If m were very small (less than 20 as in the case of covalent- and ionic-bonded materials as well as in some BCC metals), then a large drop in load would be required to reduce the dislocation velocity by the necessary amount. If m were large (greater than 100 to 200 as found for FCC metal crystals), only a small load drop would be required to effect a substantial change in dislocation velocity. The severity of the yield drop is depicted in Fig. 3.19a for a range of dislocation velocity stress sensitivity values. Note the magnitude of the yield drop increasing with decreasing m. If there are many free dislocations present at the outset of the test, they may multiply more gradually at lower stress levels, precluding the occurrence of a sudden avalanche of dislocation generation at higher stress levels. The corresponding decrease in magnitude of the yield drop with increasing initial mobile dislocation density is shown in Fig. 3.19b. From the above discussion, a yield point is pronounced in crystals that (1) contain few mobile dislocations at the beginning of the test, (2) have the potential for rapid dislocation multiplication with increasing plastic strain, and (3) exhibit relatively low dislocation-velocity stress sensitivity. Since many ionic- and covalent-bonded crystals possess these characteristics,[19] yield points are predicted and found experimentally in these materials.

For the case of carbon- and nitrogen-locked dislocations in iron, dislocation mobility is essentially zero prior to the upper yield point where dislocations are finally able to tear away from interstitial atmospheres. It is theorized that the unpinning of some dislocations, their rapid multiplication, and weak velocity stress sensitivity (i.e., low m value) all contribute to the development of a yield point in engineering iron alloys. By contrast, most FCC metals have an initially high mobile dislocation density and a very high dislocation-velocity stress sensitivity, thereby making a yield drop an unlikely event in most of these materials.

The serrated character of curve C in Fig. 3.18 sometimes observed in plain carbon steel alloys and certain aluminum alloys (e.g., Al-Mg alloys) can also be explained in terms of dislocation-solute atom interactions. Curve C is inhomogeneous in nature, but does not have the distinctive horizontal segment that characterizes the upper and lower yield point phenomenon seen in A. Bursts of mechanical twin formation and growth are one possible cause of the serrated behavior seen in curve *C*, as discussed in Section 2.6.2. However, in plain carbon steel and Al-Mg alloys dislocation motion is the dominant (or only) plasticity mechanism active under normal loading conditions. In these cases, the appearance of curve *C* is known as the *Portevin-Le Chatelier effect*, and is evidence of *dynamic strain aging*. It has been argued that the inhomogeneous behavior is due in these cases to solute atom or vacancy interactions with lattice dislocations just like those associated with yield point behavior.[71] When a sufficiently large stress is applied, dislocations can break free from solute clusters and cause a load drop and partial loss of strength. Recall that a metal component in this state can be unloaded and strain-aged (held at a slightly elevated temperature) to enable the solute atoms to diffuse to the unmoving dislocations, thereby restoring the upper yield point strength. In the case of dynamic strain-aging behavior, however, testing (or forming a component) in a particular range of strain rates and temperatures allows the solute atoms to diffuse quickly enough to retrap the moving dislocations dynamically as deformation occurs. When this happens, the stress must build up again to continue the deformation process. As long as the diffusion rate for the solute atoms is equal to or slightly greater than the rate of plastic deformation, dislocations will alternately break free from solute atmospheres and then be repinned, producing serrated curve *C*. If the strain rate and test temperature were outside the critical range, homogeneous dislocation flow would take place since solute atmosphere formation would no longer be favored; accordingly, the stress–strain curve would be smooth (Fig. 3.18, curve *D*). Dynamic strain aging can cause poor surface quality and reduced ductility associated with the formation of undesirable surface marks similar to Lüders bands, so it is preferable to avoid it through careful alloy selection and choice of processing conditions when possible.

3.6   PRECIPITATION HARDENING

Like solid solution hardening, precipitation hardening is active only in alloy systems. However, it differs significantly in that a precipitate is a particle comprised of multiple atoms in the form of a second phase within the parent matrix. As such, precipitates can vary widely in size, shape, volume fraction, composition, degree of atomic ordering, interphase boundary details, and location; with these parameters as "knobs for the engineer to turn," the effect of precipitates on the strength of an alloy can often be tailored through appropriate processing. Common precipitation hardened alloys include Al 2024 (with Cu and Mg as the key alloying elements that form Al_2Cu or Al_2CuMg precipitates), Al 6061 (alloyed primarily with Si and Mg), Al 7075 (alloyed with Zn, Mg, and Cu), maraging steel (alloyed with Ni, Mo, Ti, and Co), and Inconel 718 Ni-based superalloy (alloyed with Nb).

3.6.1 Microstructural Characteristics

When the solute concentration in an alloy exceeds the limits of solubility for the matrix phase, equilibrium conditions dictate the nucleation and growth of second-phase particles, provided that suitable thermal conditions are present. From Fig. 3.20, which shows a portion of an equilibrium phase diagram, we see that for an alloy of composition *X*, a single phase α is predicted at temperatures above T_s whereas two phases, α and β, are stable below the *solvus line* that separates the regions of phase stability. When such an alloy is heated into the single-phase field (called a *solution treatment*) and then rapidly quenched, the resulting microstructure contains only supersaturated solid solution α even though the phase diagram predicts a two-phase mixture; the absence of the β phase is attributed to insufficient atomic diffusion. If this alloy is heated to an intermediate temperature (called the *aging temperature*) below the solvus temperature, diffusional processes are enhanced and result in the precipitation of β particles either within α grains or at their respective grain boundaries.

The onset of precipitation depends strongly on the aging temperature itself (Fig. 3.21). At temperatures approaching the solvus temperature, there is little driving force for the precipitation process, even though diffusion kinetics are rapid. Alternatively, precipitation of the second phase proceeds slowly at temperatures well below T_s despite the large driving force for

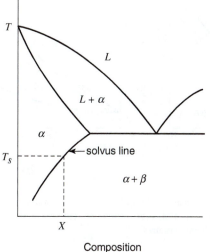

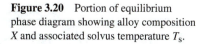

Figure 3.20   Portion of equilibrium phase diagram showing alloy composition X and associated solvus temperature T_s.

nucleation of the second phase; in this instance, diffusional processes are restricted. An optimal temperature for rapid precipitation is then identified at an intermediate temperature corresponding to an ideal combination of particle nucleation and growth rates.

The development of the two-phase mixture can most generally be described as taking place in three stages. After an *incubation period*, clusters of solute atoms form and second-phase particles nucleate and begin to grow either homogeneously within the host grains or heterogeneously along host grain-boundary sites. During the second stage of aging, particle nucleation continues along with the growth of existing precipitates; these processes continue until the equilibrium volume fraction of the second phase has been reached. In the third and final stage of aging, these second-phase particles *coarsen*, with larger particles growing at the expense of smaller ones. This process, referred to as *Ostwald ripening*, is diffusion-driven so as to reduce the total amount of interfacial area between the two phases.

For reasons to be addressed shortly, the precipitation of second-phase particles throughout the matrix increases the difficulty of dislocation motion through the lattice. (Conversely, little strengthening has been attributed to the presence of grain-boundary precipitates.) Typically, the hardness and strength of the alloy increases initially with time (and particle size) but may then decrease with further aging (Fig. 3.22). The strength and sense of the strength–time slope ($d\tau/dt$) depends on four major factors: the volume fraction, distribution, the nature of the precipitate, and

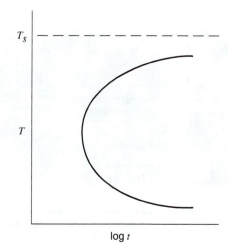

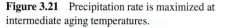

Figure 3.21   Precipitation rate is maximized at intermediate aging temperatures.

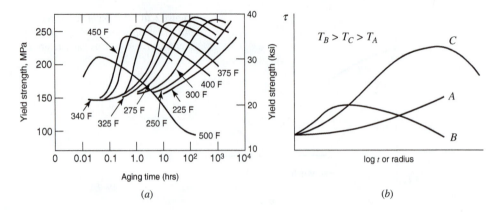

Figure 3.22 (a) Aging curves in 6061-T4 aluminum alloy. (From J. E. Hatch, Ed., *Aluminum Properties and Physical Metallurgy,* ASM, Metals Park, OH, 1984, p. 178; with permission.) (b) Schematic representation of aging process at low (A), high (B), and intermediate (C) temperatures.

the nature of the interphase boundary. Surely, were all things to remain constant, the resistance to dislocation motion through the lattice would be expected to increase with increasing volume fraction of the dislocation barrier (i.e., the precipitate). Accordingly, the first two stages of aging generally contribute to increased strengthening with time and/or particle dimension (i.e., positive $d\tau/dt$). On the other hand, Ostwald ripening, corresponding to long aging times and/or the growth of large second-phase particles, leads to negative $d\tau/dt$ conditions (see curves B and C in Fig. 3.22).

Whether the dislocation cuts through or avoids the precipitate depends on the structure of the second phase and the nature of the particle–matrix interface. The interface between the two phases may be *coherent*, which implies good registry between the two lattices. A dislocation moving through one phase would then be expected to pass readily from the matrix lattice into that of the precipitate. Such a coherent interface possesses a low surface energy. At the same time, however, lattice misfit (related to the difference in lattice parameters between the two phases) leads to the development of elastic strain fields surrounding the coherent phase boundary. Researchers have found that the shape of the precipitate particles depends on the degree of misfit. For example, when the misfit strain is small, spherical particles are formed such as in the case of the Al–Li binary alloy (Fig. 3.23a). When such particles grow in size and/ or when a large misfit is developed, cuboidal particles are formed as in nickel superalloys (Fig. 3.23b). With increasing particle size and/or misfit strain, the microstructure reveals aligned cubes or rodlike particles.[49] As these small coherent precipitates grow with time, their interfaces may become *semicoherent*, with the increased lattice misfit between the two phases

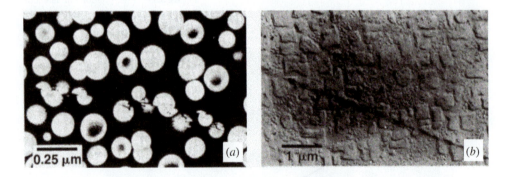

Figure 3.23 Precipitate morphology dependence on degree of lattice misfit. (a) Low misfit spherical particles in Al–Li alloy. (Courtesy of S. Baumann.) (b) Moderate misfit cuboidal particles in Ni–Al alloy. Both are sheared by dislocations.

being accommodated by the development of interface dislocations, which bring the two lattices back into registry. At this stage, misfit energy decreases markedly, whereas surface energy increases to a significant degree. Finally, in the latter stages of aging associated with the development of coarse particles, the interface between the two phases may break down completely and become *incoherent*; the surface energy associated with this interphase boundary is then increased whereas its strain field is essentially eliminated.

3.6.2 Dislocation–Particle Interactions

Why do the strength–time plots, as shown in Fig. 3.22, vary with aging temperature? Why does alloy strength increase with time (particle size) and then decrease after maximum strength has been achieved? The answers to these questions involve assessment of several dislocation–particle interactions that depend on whether dislocations are able to cut through precipitate particles or, instead, are forced to loop around them. When *particle cutting* occurs, hardening depends to some extent on the relative importance of elastic interactions between the dislocations and the precipitates. As previously noted, differences in lattice parameter between the host and precipitate phase will produce misfit strains that slow the movement of dislocations through the host lattice. The misfit strain is simply $\varepsilon = (a_{ppt} - a_{lattice})/a_{lattice}$. Researchers[49,50] have found the strengthening contribution of misfit hardening to be

$$\tau \propto G\varepsilon^{3/2}(rf)^{1/2} \tag{3-21}$$

where

$\varepsilon =$ misfit strain (proportional to difference in lattice parameter of the two phases)
$r =$ particle radius
$f =$ volume fraction of precipitated second phase
$G =$ shear modulus

For many nickel-based superalloys, however, metallurgists tinker with alloy composition to limit misfit strains so as to maintain coherency for larger precipitates. As a result, the strengthening contribution of lattice misfit in these alloys is relatively minor.[51] On the other hand, low misfit strains minimize Ostwald ripening, which leads to enhanced creep resistance. Other elastic interactions include those associated with differences in shear modulus and stacking fault energy between the two phases. Here, again, for a number of important precipitation-hardened commercial alloys, the strengthening contribution of these factors is relatively small.

A second group of dislocation–particle cutting interactions involves energy storing mechanisms associated with the generation of new interphase boundary and antiphase-domain boundary area. For example, when dislocations cut through a particle, additional precipitate–matrix interfacial area is created, which increases the overall energy of the lattice (recall Fig. 3.23); since the interfacial energy of coherent precipitates is small, this hardening mechanism contributes little to the strength of alloys that contain low misfit precipitates. On the other hand, if the precipitate has an *ordered lattice* (such as $CuAl_2$ particles in an aluminum alloy or Ni_3Al (γ') precipitates in a nickel-based superalloy) the character of the deformation process is altered. In the case of the intermetallic compound Ni_3Al, the aluminum atoms are located at the eight corner positions of the unit cell and the nickel atoms are located at the six cube faces (Fig. 3.24). Note that the ordered Ni_3Al phase is of the FCC type and contains four atoms (three nickel and one aluminum) per unit cell. The passage of a dislocation through half of a spherical particle containing this *superlattice* (consisting of Ni and Al atoms in specific lattice sites) generates an unfavorable rearrangement of the aluminum and nickel atoms, as shown in Fig. 3.25a. We see that the Ni (open circles) and Al (solid circles) atoms are opposite one another along that part of the slip plane that was traversed by the dislocation. This arrangement of non-preferred atom pairs on the slip plane creates an *antiphase domain boundary* (APB). Since there is an additional energy

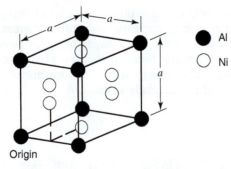

Origin

Figure 3.24 Nickel and aluminum atom locations in ordered Ni$_3$Al phase.

Al ●
Ni ○

associated with the APB, which depends on the degree of order in the lattice, dislocation motion is restricted. However, if a second identical dislocation were to sweep across the same plane, atomic disorder would be eliminated with atoms again assuming their preferred positions in the lattice (Fig. 3.25b). Note that the right side of the particle is still disordered since the second dislocation has passed through only half of the particle. The equilibrium distance separating these two dislocations (referred to as a *superlattice dislocation*) reflects a balance between an attractive force associated with minimization of APB energy and a repulsive force due to the stress fields of identical dislocations (recall Eq. 2-21). An example of superlattice dislocations (i.e., dislocation pairs) in Ni$_3$Al is shown in Fig. 3.26.

Since the APB energy is roughly ten times greater than the *interphase* boundary energy in nickel superalloys[53] and in Al–Li alloys, the strengthening contribution due to APB formation in these systems is significant. Gleiter and Hornbogen[49,54] and Ham[55] reported the strengthening contribution associated with this mechanism to be of the form

$$\tau \propto \gamma^{3/2} \left(\frac{rf}{G} \right)^{1/2} \tag{3-22}$$

where

$\gamma =$ APB energy.

It is interesting to note that the passage of superlattice dislocation pairs through a precipitate particle effectively reduces the length of the ordered path for subsequent dislocation pairs; for this reason, dislocation movement within microstructures containing ordered precipitates is of a heterogeneous nature and typically involves the activity of relatively few slip planes associated with large slip steps.

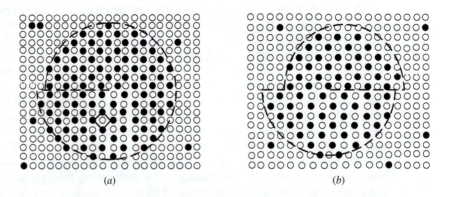

(a)   (b)

Figure 3.25 (100) planar view of spherical Ni$_3$Al particle in Ni lattice. (a) Initial superlattice dislocation disorders atom pairs along slipped portion of glide plane. Note orientation of cube face in Ni$_3$Al particle. (b) Passage of second superlattice dislocation reorders Ni$_3$Al lattice. Dashed horizontal line corresponds to APB. Nickel (○) and aluminum (●) atom locations noted. (From Gleiter and Hornbogen.[49])

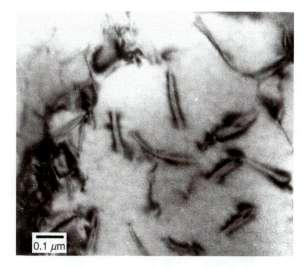

Figure 3.26   Superlattice dislocation pairs in fully ordered Ni$_3$Al. (Photo courtesy of M. Khobaib.[52])

If the misfit strain is large, the interface incoherent, or the average particle separation above a certain critical value, dislocations are unable to cut through the precipitate; instead they *loop* around individual particles as shown in Fig. 3.27. (In some alloy systems, both particle cutting and looping can occur simultaneously.) Note the strong similarity of such dislocation looping with the Frank-Read mechanism for dislocation multiplication (Fig. 3.2). The stress necessary for the dislocation to loop around the precipitate is the same as that given for activation of the Frank-Read source, where l is the distance between the particles:[56]

$$\tau = \frac{Gb}{l} \tag{3-1}$$

With the passage of subsequent dislocations, the *effective* distance between two adjacent precipitates l' decreases with the increasing number of dislocation loops surrounding the particles. As such, the dislocation looping mechanism provides a measure of strain hardening.[57] For a given volume fraction of second-phase particles, l increases as the precipitates grow larger with further aging. Consequently, the stress necessary for dislocations to loop around precipitates should decrease with increasing particle size. It is important to note that dislocation looping (often referred to as *Orowan looping*) is controlled by the spacing between particles and not by the nature

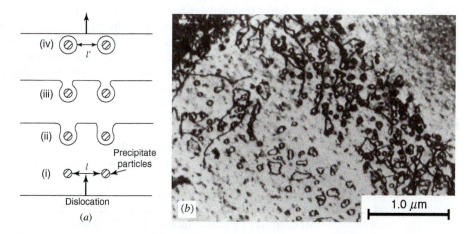

Figure 3.27   Dislocation looping around particles. (*a*) Schema revealing reduced "effective" particle spacing with looping; (*b*) looping in Al–Li alloy. (Photo courtesy of S. Baumann.)

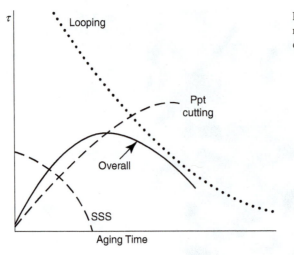

Figure 3.28   Schematic representation showing role of major hardening mechanisms in development of overall hardening response.

of the particle itself. Furthermore, such slip activity is of a more homogeneous nature than that described above for the case of deformation by particle cutting.

The complex interaction between this group of strengthening mechanisms is responsible for the strength–time relations shown in Fig. 3.22, which may be summarized in the following manner. After solution treatment and subsequent quenching, the alloy experiences the greatest *potential* for solid solution strengthening, since the greatest amount of solute is present in the host matrix. If the solute possesses a nonsymmetrical stress field, solid solution strengthening would be great, as in the case of as-quenched carbon martensite (see Section 3.8). In sharp contrast, were the solute to possess a symmetrical stress field (such as in aluminum- and nickel-based alloys), limited solid solution strengthening would be expected. With aging, and the associated precipitation of second-phase particles, the solute level in the host matrix would decrease along with the solid solution strengthening component. This relatively small reduction in absolute strength in aluminum- and nickel-based alloys is more than compensated for by several precipitation hardening mechanisms, such as dislocation interactions with precipitate misfit strain fields, particle cutting, and elastic modulus interaction effects. As noted above, the extent of such hardening *increases* with time (i.e., particle size). With further aging, leading to the loss of coherency, or a wide interparticle spacing, dislocation looping around the particles takes place. Alloy strength then *decreases* with further aging time and/or particle size. The overall aging response of the alloy may then be characterized by the attainment of maximum strength at intermediate aging times and particle dimensions. A schematic representation of the interplay between these hardening processes is shown in Fig. 3.28 and bears close resemblance to the experimental results given in Fig. 3.22. Regarding Fig. 3.22*b*, *underaged* conditions are associated with curve A and the left portions of curves *B* and C; *overaging* corresponds to aging times greater than those needed for peak strengthening. It should be noted that with the exception of the γ' phase in nickel superalloys, most homogeneous precipitates in other alloy systems are metastable.

3.7   DISPERSION STRENGTHENING

It was seen in the previous section that the growth of hard precipitates can provide significant strengthening, but it can also be appreciated that this approach is only possible for certain alloy systems, and is only useful at temperatures below which overaging (or even re-solutionizing) are possible. For use at elevated temperatures, or for material systems that are not thermodynamically conducive to controlled precipitate formation, alloys can also be strengthened by the addition of oxide particles that obstruct dislocation motion. By adding stable refractory Al_2O_3, TiO_2, ThO_2, or Y_2O_3 particles to a metal matrix such as aluminum, copper, or nickel, these metals achieve attractive strength properties at temperatures approaching their melting points.[58,59] The oxide particles block dislocation motion and also restrict high-temperature recrystallization that would lead to larger grain size and reduced strength. As expected, alloy strength increases with oxide volume fraction and decreasing particle spacing.[60,61] As such, it is critical that the particles are uniformly dispersed in the matrix. Since the strengthening potential for dislocation looping around noncoherent particles is less than that associated with particle cutting processes, such *oxide-dispersion-strengthened* (ODS) alloys are not among the strongest structural materials. On the other hand, the microstructures of dispersion-hardened alloys are more stable than those associated with precipitation-hardened alloys, thereby making them more suitable for load-bearing applications at elevated temperatures. As such, ODS alloys are discussed further in the context of high temperature time-dependent deformation in Chapter 4.

Processing of ODS alloys by ingot methods is generally not possible because desirable oxide particles are typically not soluble in the liquid metal. This necessitates the use of powder metallurgy techniques instead. Although straightforward mixing of metal and oxide powders is possible for very simple alloys, the high degree of microstructure control required for ODS superalloys has led to the use of a *mechanical alloying* process.[62] Mixtures of powder particles of different constituents are blended together in a dry, high-energy ball mill. The discrete particles are repeatedly welded together, fractured, and rewelded. Such intimate mechanical mixing leads to the formation of particles with a homogeneous phase distribution that is extremely fine grained and heavily cold worked. The powders are then hot compacted, hot extruded, and/or hot rolled to produce materials with attractive mechanical properties.[63,64] By choosing a matrix alloy of virtually any composition, it is possible to tailor a material to meet a wide range of property requirements. The compositions of three commercial nickel-based ODS alloys are given in Table 3.2. These alloys contain Cr in solid solution for elevated temperature corrosion resistance and Y_2O_3 for dispersion hardening. Alloy MA754 contains a mixture of yttrium oxides and yttria aluminates in a size range from 5 to 100 nm; these fine particles have a planar spacing of approximately 0.1 μm and constitute about 1 v/o of the alloy.[65] Inconel MA 6000 contains approximately 7 w/o Al + Ti, which introduces the precipitation-hardening γ' phase to the nickel matrix. These alloys are intended for gas turbine vanes, turbine blades, and sheets for use in oxidizing/corrosive atmospheres.

The elevated temperature strength and rupture behavior of ODS alloys are discussed in the next chapter. For the present, it is timely to examine the rupture strength of ODS alloys versus precipitation-hardened nickel-based superalloys (see Figs. 4.34 and 3.23). At 750 °C, the dispersion-strengthened ODS alloys are seen to exhibit lower strength levels than the precipitation-hardened superalloys. However, at temperatures in excess of 900 to 950 °C, the more stable dispersion-strengthened ODS alloys are found to possess superior strengths to those of the precipitation strengthened alloys; in the latter instance, the precipitate particles have begun to coarsen and/or redissolve in the matrix in this temperature range. The primary limits to the implementation of ODS alloys in a wide variety of high temperature applications include the inability to join components by traditional fusion welding techniques (during which the distribution of oxide particles is lost) and the high cost associated with the complicated mechanical alloying process.

There is one ODS alloy system that is widely used for structural purposes in which electrical and/or thermal conductivity is also critical — Al_2O_3 in copper. Because Al is soluble in Cu to a fraction greater than 18 at% (>7 wt%), it is possible to create a solid solution powder of the two metals. When exposed to oxygen, the Al inside the powder particles preferentially oxidizes to form nanometer-scale Al_2O_3 dispersoids in a nearly pure Cu matrix. The ODS

Table 3.2   Chemical Composition of Mechanically Alloyed ODS Superalloys

Alloy	Ni	Cr	Al	Ti	Y_2O_3	W	Mo	Ta
Inconel MA754	bal	20	0.3	0.5	0.6	—	—	—
Inconel MA 6000	bal	15	4.5	2.5	1.1	4	2	2
Alloy 51	bal	9.5	8.5	—	1.1	6.6	3.4	—

powder can then be consolidated without the need for mechanical alloying to form a uniform dispersion. Due to the high purity of the matrix, the properties of the alloy (with the notable exception of mechanical strength) closely resemble those of pure Cu. In contrast, conventional Cu alloys have much poorer electrical and thermal conductivities, and experience a significant drop in strength at temperatures greater than approximately 60% of the melting point (~400 °C). The superior properties of the ODS Cu alloys allow the use of smaller cross-section sizes without loss of strength, current-carrying capacity, or heat-transport capacity for applications such as resistance spot welding electrodes for the automotive industry.

3.8 STRENGTHENING OF STEEL ALLOYS BY MULTIPLE MECHANISMS

A brief discussion of the strengthening mechanisms associated with steel alloys is appropriate since this class of materials is of major commercial importance. In addition, different steel alloys derive their strength from various combinations of the strengthening mechanisms considered thus far; as such, an analysis of the strength of steel alloys provides pertinent examples of these strengthening mechanisms. Several review articles[66-68] concerning the strength of steel alloys point to the fact that some or all of the major strengthening mechanisms that we have studied (i.e., solid solution strengthening, strain hardening, grain-boundary hardening, and precipitation and dispersion strengthening) are operative in each alloy system, depending on the character of the transformation product(s).

The four major microstructural features found in steel alloys are ferrite, pearlite, bainite or lath martensite, and plate martensite (Fig. 3.29); these microstructural features are shown schematically in Fig. 3.30. Clearly, ferrite (single-phase solid solution of iron) exhibits the

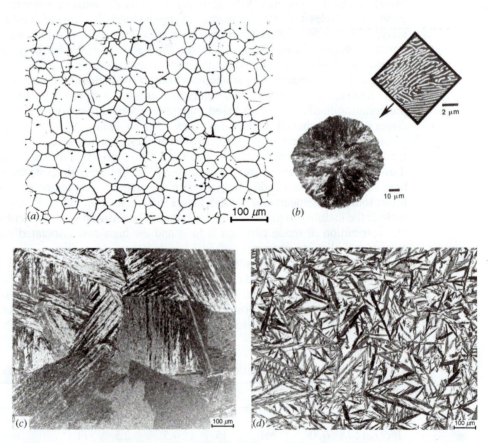

Figure 3.29 Typical microstructures in ferrous alloys: (*a*) ferrite; (*b*) pearlite; (*c*) lath martensite; (*d*) plate (acicular) martensite. (Courtesy of A. Benscoter and J. Ciulik.)

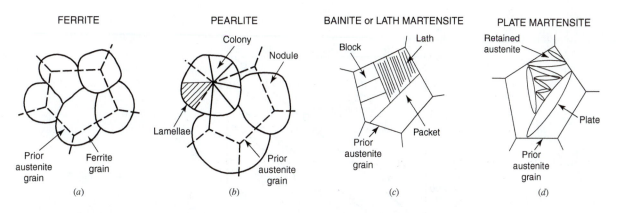

| FERRITE | PEARLITE | BAINITE or LATH MARTENSITE | PLATE MARTENSITE |

(a)   *(b)*   *(c)*   *(d)*

Figure 3.30   Schematic representation of microconstituents shown in Fig. 4.15.[68] (Reprinted with permission from the proceedings of an International Conference on Phase Transformations in Ferrous Alloys, A. R. Marder and J. I. Goldstein, Eds., 1984, The Metallurgical Society, 420 Commonwealth Drive, Warrendale, PA 15086.)

simplest microstructure, with the principle strengthening mechanism corresponding to grain refinement that follows the Hall-Petch trend of Eq. 3-8, with ferrite grain size d as the critical parameter.[69] Values of σ_i and k_y are found to vary with test temperature (Fig. 3.31a) and alloy content (Fig. 3.31b). Note the stronger dislocation locking tendency, and thus a greater k_y value, for nitrogen solute additions relative to that of carbon atoms. In pearlitic steels, the interlamellar spacing (S) between the ferrite and iron carbide lamellae is the critical dimension for boundary strengthening; little influence on alloy strength is noted with changes in austenite (the FCC form of iron) grain size and nodule diameter.[68] Langford[70] has shown that the strength of pearlitic steels may be described by a modified version of the Hall-Petch expression

$$\sigma_{ys} = \sigma_i + k_1 S^{-1/2} + k_2 S^{-1} \tag{3-23}$$

where

$$S = \text{interlamellar spacing}$$
$$k_1, k_2 = \text{constants}$$
$$\sigma_i = \text{resistance of lattice to dislocation movement}$$

The specific influence of S on alloy strength (i.e., $S^{-1/2}$ versus S^{-1} dependence) varies with the dominant hardening mechanism, which, in turn, reflects a change in the controlling free path for movement of dislocations; an $S^{-1/2}$ dependence of σ_{ys} corresponds to processes associated with the formation of *dislocation pileups* (the original basis for Eq. 3-8). When strength is controlled by the work required for the *generation* of dislocations, an S^{-1} lamellae size–strength dependence develops (similar to the behavior of a Frank-Read source described by Eq. 3-3).

When asked which strengthening mechanism controls the mechanical properties of martensitic steels, the reader would not err by replying, "All of the above." Indeed, the high strength of martensite draws upon several mechanisms, with solid solution strengthening exerting the greatest influence (recall Fig. 3.17). To illustrate, lath martensite contains up to 0.6 wt% carbon and possesses boundary obstacles (e.g., packet boundaries) along with a highly dislocated substructure ($>10^{10}$ dislocations/cm^2). Norstrom[71] has proposed a comprehensive relation to describe the yield strength of lath martensite:

$$\sigma_{ys} = \sigma_i + k\sqrt{c} + k_y d^{-1/2} + \alpha\, Gb\sqrt{\rho} \tag{3-24}$$

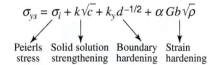

Peierls stress   Solid solution strengthening   Boundary hardening   Strain hardening

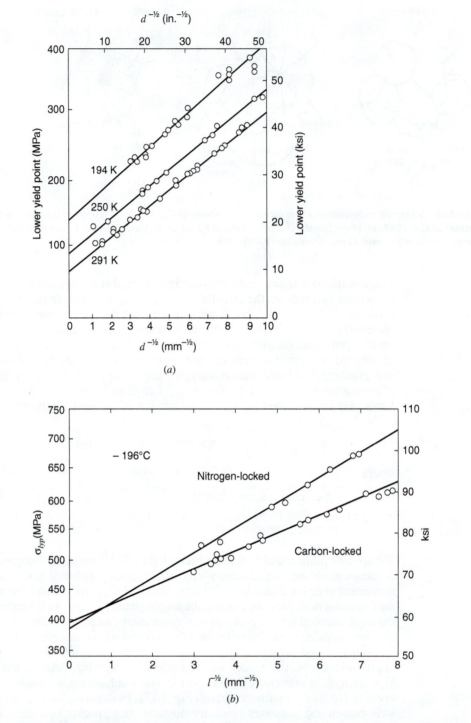

Figure 3.31 (a) Influence of grain size on yield strength in ferritic steel; (b) grain-size dependence of lower yield point in steel reflecting greater dislocation locking k_y with nitrogen interstitial.[69] (Reprinted with permission from MIT Press, Cambridge, MA.)

where

σ_i = resistance of lattice to dislocation movement
k, α = constants
c = solute atom concentration
k_y = locking parameter
d = packet size (recall Fig. 3.30)
G = shear modulus
$\mathbf{b}$ = Burgers vector
ρ = dislocation density

Furthermore, a dispersion-hardening component is introduced with the precipitation of iron carbide particles during tempering. It should be noted, however, that total alloy strength decreases with tempering since the solid solution strengthening contribution is correspondingly reduced. For this reason, carbon martensites are strongest after rapid quenching from the austenite region, but lose strength with tempering. Such behavior contrasts markedly with precipitation-hardening alloys (as discussed in Section 3.6), which are soft upon quenching from the solution treatment zone but then strengthen with aging. Maraging steels are alloys that conform to the latter strengthening type since precipitation of fine second-phase particles is responsible for strengthening in such alloys; correspondingly, extremely low carbon levels in these alloys preclude significant solid solution strengthening.

3.9 METAL-MATRIX COMPOSITE STRENGTHENING

Thus far, we have examined several *intrinsic* strengthening mechanisms in metal alloys. As discussed in Section 1.6.2, metal alloys may also be strengthened *extrinsically* through the addition of high-strength continuous fibers such as carbon, aramid (e.g., Kevlar), and boron, or reinforcement may be achieved with Al_2O_3 or SiC whiskers to form *discontinuously-reinforced composites* (DRC). Furthermore, *laminates* of metals and nonmetallic materials offer benefits for certain applications. The reader may recognize that at a certain level the ODS alloys introduced in Section 3.7 are composites too. However, the techniques used to fabricate ODS alloys—and the mechanisms behind their attractive properties—are quite different from those found for composites created with reinforcements of larger scale (nanometer vs. micrometer). As such, more conventional whisker-reinforced or laminated metal-matrix composites merit separate discussion.

3.9.1 Whisker-Reinforced Composites

Continuous boron fiber Al matrix MMCs have found use in space applications as tubular struts and in the landing gear of the Space Shuttle Orbiter. However, the high cost of such materials has precluded much wider acceptance. DRC materials are attractive because they can be fabricated at reasonable cost and can be processed by a variety of methods. Some metal–matrix DRCs have been fabricated by liquid infiltration of the matrix around the fibers; other composite systems have been prepared by extrusion of hot compacted matrix powders and high-strength whiskers. Such powder–metallurgy composites, consisting of aluminum alloys reinforced with SiC whiskers, have attracted considerable attention because they offer high strength (and good resistance to fatigue damage) in a lightweight material. When further processed by deformation techniques such as extrusion, a combination of plastic deformation of the matrix and whisker alignment produces particularly high longitudinal strength. Several investigators[72–75] have noted that the strength of such Al–SiC composites exceeds that predicted from conventional composite theory (recall Section 1.6.2). Transmission electron microscope studies have determined that these enhanced strength levels are attributed to the presence of relatively high dislocation densities in the aluminum alloy matrices examined (Fig. 3.32). Arsenault and coworkers[73,74] theorized and subsequently confirmed that such high dislocation-density levels resulted from the large difference (10:1) in coefficients of thermal expansion between the

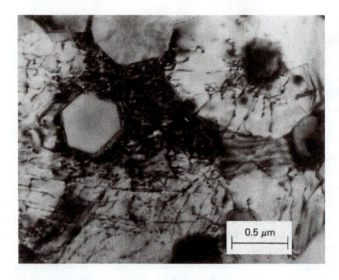

Figure 3.32 High dislocation density on 6061 aluminum alloy reinforced with 20 v/o SiC.[73] (Reprinted with permission from M. Vogelsang, R. J. Arsenault, and R. M. Fisher, *Metallurgical Transactions*, 17A, 379 (1986)).

aluminum alloy matrix and the SiC whiskers. Accordingly, when the composite is cooled from elevated temperatures, the misfit strains that develop are relieved by the generation of dislocations at the ends of the SiC whiskers. Studies have also shown that this increased dislocation density accelerates the aging process within the aluminum alloy matrix by shortening the aging time to achieve maximum strength relative to that associated with the unreinforced matrix alloy.[72,75] These findings are of major significance since computations of composite strength must account for alterations in matrix properties due to the presence of the reinforcing phase. As such, the high-strength phase serves to strengthen the matrix both *extrinsically* by load transfer to the fibers/whiskers and *intrinsically* by increasing the dislocation density.

3.9.2 Laminated Composites

A number of laminated metal composites have been used in various engineering components, with contiguous plies being joined together by such methods as diffusion bonding or with the use of adhesives. Two fiber-metal laminate (FML) aluminum composites have been designed specifically for aircraft structural components. The first material consists of thin aluminum alloy sheets (typically 2024-T3 or 7075-T6) and epoxy/aramid composite layers that are adhesively bonded together (Fig. 3.33a).[76-78] This material, called ARALL (ARamid ALuminum Laminate), can be machined and formed into useful shapes. Its specific weight (mass per unit volume) is lower than that of solid aluminum. Also, the aluminum alloy outer layers provide impact resistance and damage detectability. Furthermore, cracks that may initiate in the aluminum alloy surface layers are arrested when the crack front encounters the epoxy/aramid fiber layer. As discussed in Chapter 10, this greatly extends overall fatigue lifetime for components fabricated with this unusual material.

Panels of ARALL typically contain three layers of aluminum alloy (approximately 0.3 to 0.5 mm thick) that sandwich two epoxy/aramid fiber adhesive layers, each 0.25 mm thick. Since the volume fraction of the aligned aramid fibers in the epoxy resin is between 40 and 50 v/o, the overall volume fraction of aramid fibers in the hybrid composite is approximately 15%. Typical stress–strain curves for ARALL along with its constituent layers are shown in Fig. 3.33b as a function of loading direction relative to the alignment direction of the aramid fibers.[78] As with other aligned-fiber reinforced composites, the compressive strength of ARALL is inferior to that in tension, owing to buckling of the high-strength fibers. This tendency is reduced when the aramid fibers are replaced by glass fibers.[79] A successor to ARALL called GLARE (GLAss-REinforced fiber metal laminate) therefore uses unidirectional glass fiber/epoxy composite layers in place of aramid/epoxy. Each fiber/epoxy composite layer is a *prepreg* sheet (i.e., fibers

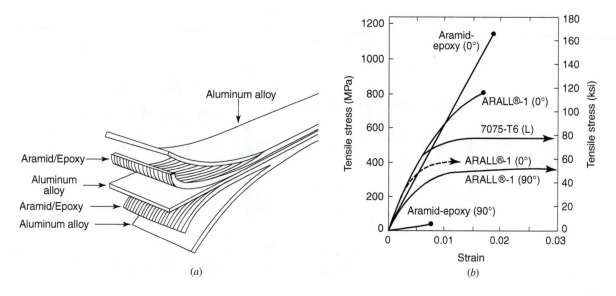

Figure 3.33 (a) Layup of ARALL hybrid composite, consisting of alternate layers of aluminum alloy and aramid/epoxy laminates. (b) Tensile stress–strain curves for ARALL and constituent layers as function of loading angle relative to fiber axis. Compression test results shown with dashed curve. (From R. J. Bucci, Alcoa Technical Center, with permission.)

pre-impregnated with resin) that can be made up of several unidirectional layers of different orientations, thereby tuning the prepreg anisotropy and the overall mechanical behavior of the laminate. It has been shown that a good approximation of static laminate mechanical properties can be calculated using a simple rule of mixtures, similar to that used in Chapter 1 when evaluating the elastic behavior of continuous-fiber composites.[80] In this case, however, the metal and composite layers are each treated as homogeneous, orthotropic sheets that contribute linearly to the total laminate in proportion to their volume fraction. In the case of yield or ultimate strength, this is simply

$$\sigma_{ult}^{Lam} = V_f^{Al}\sigma_{ult}^{Al} + (1 - V_f^{Al})\sigma_{ult}^{pre} \tag{3-25}$$

where σ_{ult}^{Lam} is the laminate ultimate strength, σ_{ult}^{Al} is the ultimate strength of the aluminum alloy, σ_{ult}^{pre} is the ultimate strength of the prepreg, and V_f^{Al} is the aluminum volume fraction (typically 50–70%). This expression has been shown to work well in compression as well as in tension over the range of volume fractions tested.[80] The first structural application of GLARE in a commercial aircraft was in the fuselage of the Airbus A380 airliner.[81]

3.10 STRENGTHENING OF POLYMERS

The mechanisms by which amorphous and semi-crystalline polymers yield were introduced in Chapter 2. As these mechanisms are quite distinct from those active in metals, different approaches to strengthening should also be expected.

Strengthening through changes in *chemistry* can also be brought about by the introduction of large side groups and intrachain groups that restrict C—C bond rotation (recall Section 2.7.1.1). Some polymers can also be cross-linked to lock their molecules together in rigid fashion, thereby precluding viscous flow (recall Section 2.7.1.4 and the effect of cross-linking agents). For example, a bowling ball contains a much higher cross-link density than a handball.

In another approach, the superstructure or *architecture* of a polymer can be modified to effect dramatic changes in mechanical strength greater than those made possible by chemistry adjustments. First, mechanical properties are found to increase with molecular weight (MW),

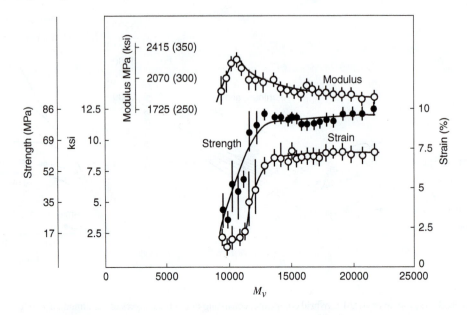

Figure 3.34   Mechanical properties in polycarbonate as a function of molecular weight.[83] (Reprinted with permission from John Wiley & Sons, Inc.)

and relations[82] often assume the forms

$$\text{mechanical property} = A - \frac{B}{\overline{M}} \tag{3-26a}$$

or

$$\text{mechanical property} = C + \frac{B}{\overline{M}} \tag{3-26b}$$

An example of such data[83] is shown in Fig. 3.34. It may be argued that as chain length increases beyond a critical length, the combined resistance to flow from chain entanglement and intermolecular attractions exceeds the strength of primary bonds, which can then be broken. Consequently, once the molecular weight exceeds a critical lower limit $\overline{M}_c$ entanglement and primary bond breakage occur. At this point the mechanical property becomes less sensitive to MW. Studies have shown that polymer viscosity depends on MW. When MW $< M_c$, the viscosity η is proportional to MW. When MW $> M_c$, η is proportional to MW[3.5] instead (Fig. 3.35).[84]

Mechanical properties are improved most dramatically by molecular and molecular segment alignment parallel to the stress direction. This stands to reason, since the loads would then be borne by primary covalent bonds along the molecule rather than by weak van der Waals forces between molecules. Figure 3.36 illustrates the rapid increase in polymer stiffness with increasing fraction of covalent bonds aligned in the loading direction.[85] A similar curve would describe the strength of the polymer. Such orientation hardening is distinct from the strain-hardening phenomenon found in metals. In the latter case, strength is lost when the sample is annealed due to the annihilation of dislocations. The strength of oriented polymers is high even after annealing (below T_g) since the polymer chain orientation is retained.

The alignment of molecules in an amorphous polymer is described with the aid of Fig. 3.37. Thermal energy causes the molecules in the polymer at $T > T_g$ to vibrate with relative ease in random fashion, but below T_g the randomness is "frozen in." If the material were drawn quickly at a temperature not too far above T_g (say, T_1), some chain alignment could be achieved and effectively "frozen in," provided the stretched polymer were to be quenched from that

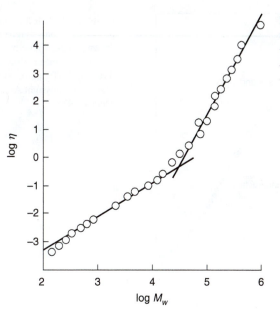

Figure 3.35   Dependence of melt viscosity on molecular weight in polydimethylsiloxane at 20 °C.[84] (With permission, N. J. Mills, *Plastics: Microstructure, Properties and Applications*, Edward Arnold, London, 1986.)

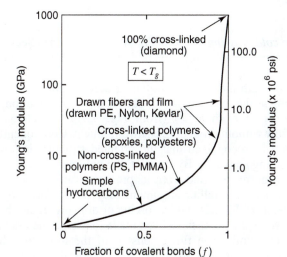

Figure 3.36   Polymer stiffness dependence on fraction of covalent bonds in the loading direction.[85] (Reprinted with permission from M. F. Ashby and D. R. H. Jones, *Engineering Materials 2*, Pergamon Press, Oxford, 1986.)

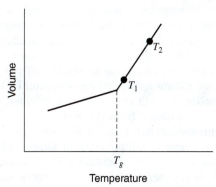

Figure 3.37   Rapid drawing at T_1 followed by quenching can produce molecular alignment and polymer strengthening. Viscous flow during drawing at T_2 precludes such alignment.

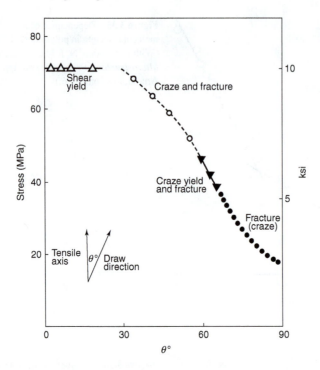

Figure 3.38 Effect of orientation on tensile properties of polystyrene tested at 20°C, hot drawn to a draw ratio of 2.6. Note deformation mechanism transition with test direction.[86] (By permission, from *Polymeric Materials,* copyright American Society for Metals, 1975.)

temperature. (Recall that the *cold drawing* phenomenon was introduced in Section 2.7.2.1.) The resulting material would be stronger in the direction of drawing and correspondingly weakened in the lateral direction. An example of this anisotropy is shown in Fig. 3.38 for drawn polystyrene. Note that the strength anisotropy parallels a deformation mechanism transition from crazing to shear yielding as the tensile axis approaches the draw direction. If the drawing were conducted slowly at T_1 or even at T_2 (from Fig. 3.37), the elongation could be accommodated by viscous flow without producing chain alignment. Consequently, no strengthening would result. It should be recognized that the oriented structure is unstable and will contract upon subsequent heating above T_g. By contrast, the polymer stretched at T_2 would be dimensionally stable, since it never departed from its preferred fully random state.

In semi-crystalline polymers, crystallite alignment may be produced by cold drawing spherulitic material (recall Section 2.7.2.2) and by forcing or drawing liquid through a narrow orifice. During the past few years, attempts have been made to extend the practice of polymer chain orientation to its logical limit—the full extension of the molecule chain—with the potential of producing a very strong and stiff fiber.[87−89] Indeed, this has been partially accomplished. Highly oriented and extended commercial fibers, such as DuPont Kevlar, possess a tensile modulus two-thirds that of steel but with a much lower density (recall Section 1.3.3.1). This is truly extraordinary, since unaligned commercial plastics generally exhibit elastic moduli fully two orders of magnitude smaller than steel. By converting the folded chain conformation to a fully extended one, the applied stresses are sustained by the very strong main chain covalent bonds, which are less compliant than the much weaker intermolecular van der Waals forces. An example of the effect of *draw ratio* (final length/initial length) on the tensile strength of a polycarbonate and liquid crystalline polymer (LCP) blend is shown in Fig. 3.39. It can be seen that higher draw ratios that cause greater LCP fiber formation and alignment correspond to greater tensile strength values. Highly oriented fibers have been produced both by cold forming and direct spinning from the melt.

While the formation of engineered high-strength crystalline fibers or filaments directly from the melt is an intriguing process, it is by no means a unique event in nature; rather, the materials scientist must yield to the long-recorded activities of arachnids and silkworms. For example, researchers have determined that spider silk is generated by the drawing of an amorphous protein

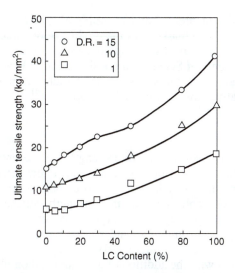

Figure 3.39   Dependence of tensile strength for a polycarbonate/liquid crystalline polymer blend. Increasing draw ratio corresponds to increasing LCP alignment.[90] (Reprinted with permission from S. H. Jung and S. C. Kim, *Polymer* 20(1), 73 (1988), The Society of Polymer Science, Japan.)

liquid from various glands, which then converts quickly to a highly oriented, very long crystalline filament with a diameter on the order of several tens of nanometers. The highly oriented and crystalline morphology of these filaments is believed responsible for their reported strengths that can sometimes exceed 1 GPa.[91] In one intriguing investigation, for instance, Lucas[92] showed that a certain variety of spider silk had twice the *tenacity*, defined as grams-force/denier, of a 2070-MPa steel wire and four times the extension at break.[iii] Experimentally-measured mechanical properties for two types of silk produced by the spider *Araneus diadematus* (anorb-web-weaving araneid spider) are shown in Table 3.3 along with typical order-of-magnitude properties for several other natural elastic materials as well as engineered fibers.

Table 3.3   Relative-magnitude tensile mechanical properties of spider silks and selected other fiber materials.[94] Reproduced with permission from J. M. Gosline, P. A. Guerette, C. S. Ortlepp, and K. N. Savage, *Journal of Experimental Biology* 202, 3295 (1999).

Material	Stiffness, E_{init} (GPa)	Strength, σ_{max} (GPa)	Extensibility, ε_{max}	Toughness (MJm^{-3})
Araneus MA silk*	10	1.1	0.27	160
Araneus viscid silk**	0.003	0.5	2.7	150
Bombyx mori cocoon silk	7	0.6	0.18	70
Tendon collagen	1.5	0.15	0.12	7.5
Bone	20	0.16	0.03	4
Wool, 100% RH	0.5	0.2	0.5	60
Elastin	0.001	0.002	1.5	2
Resilin	0.002	0.003	1.9	4
Synthetic rubber	0.001	0.05	8.5	100
Nylon fiber	5	0.95	0.18	80
Kevlar 49 fiber	130	3.6	0.027	50
Carbon fiber	300	4	0.013	25
High-tensile steel	200	1.5	0.008	6

*MA silk is produced by the spider's major ampullate (MA) gland, which forms the dragline used by the spider for suspension, and the web frame used to form the structure of the web.
**Viscid silk is produced by the spider's flagelliform (FL) gland, and is used as the glue-covered insect-catching spiral fibers of the web.

[iii] A denier is a unit that quantifies the fineness of fiber, filament or thread in terms of the mass in grams per 9000 meters. A lower denier number therefore indicates a finer yarn, and the tenacity is a dimensionless number analogous to specific strength (strength/weight ratio).

Such high strengths along with the apparent abundant supply of spider silk have prompted enterprising individuals to seek commercial markets for the product of our arachnid friends. In one such feasibility study in 1709, several pairs of stockings and gloves were woven from spider silk and presented to the French Academy of Science for consideration.[93] However, as spiders are not capable of producing sufficient silk to support the needs of industry, more recent efforts have sought artificial means of creating fibers with similar properties.[91]

3.11 POLYMER-MATRIX COMPOSITES

An extensive discussion of polymer-matrix composite strength was presented in Chapter 1, Section 1.6.6, in the context of elastic behavior. Here, we add a few refinements to our understanding. The addition of strong fibers such as glass and carbon to polymeric matrices (e.g., epoxy, polyester, and nylon 66) enhances polymer strength, stiffness, dimensional stability, and elevated temperature resistance at the expense of ductility. This strength enhancement may be realized to such a degree that under certain conditions fiber fracture determines the composite strength rather than polymer matrix yielding or cracking. The change in tensile strength of selected polymers with the addition of glass and carbon fibers is shown in Fig. 3.40, and tensile strength values for selected composites are listed in Table 1.11. Note the superior strengthening potential of graphite fibers relative to that of glass. Since the density of nonmetallic fibers is relatively low (see Table 3.4), the specific strength and stiffness of polymeric composites (σ_{TS}/ρ and E/ρ, where ρ = density) exceeds that of conventional metal alloys.[iv] For this reason, high-performance polymer composites are finding increasing use in transportation applications, and in aircraft in particular, as discussed in Section 1.6.2.

The properties of fiber-reinforced plastics depend in a complex manner on processing history. For example, during injection molding, the chopped fibers fracture, with the result that many glass and carbon fibers have lengths in the range of 100 to 500 μm. In addition, fiber orientation differs throughout the thickness of the injection-molded part. This follows from the

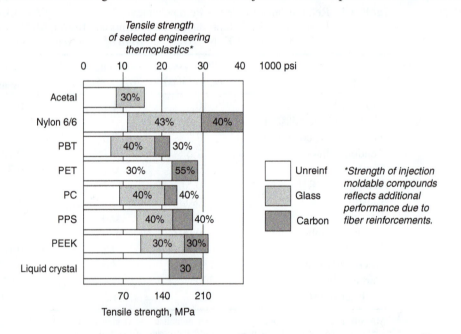

Figure 3.40 Tensile strength of selected engineering thermoplastics and their respective composites.[95] (Reprinted with permission from *Guide to Engineering Materials*, 2nd ed.; ASM International, Metals Park, OH, 1987.)

[iv] Manufacturers usually report fiber content in a composite by either weight or volume fraction. For the case of glass-reinforced plastics, the density of the glass is roughly twice that of the polymer matrix; consequently, a 30 w/o glass content corresponds to approximately 15 v/o.

Table 3.4   Density of Selected Fibers and Matrices[85]

Fibers	Density (g/cm³)	Matrices	Density (g/cm³)
Carbon Type 1	1.95	Epoxies	1.2–1.4
Carbon Type 2	1.75	Polyesters	1.1–1.4
E glass	2.56	Nylon	1.1–1.2
Kevlar	1.45	Concrete/cement	2.4–2.5
SiC	2.5–3.2	Aluminum alloys	2.6–2.9
Al_2O_3	3.9	Steel alloys	7.8–8.1

flow characteristics associated with injection molding.[97,98] Near the mold wall (S), the fibers tend to be aligned in the molding direction. In the interior core region (C), however, the fibers tend to be aligned parallel to the advancing liquid front (Fig. 3.41); as such, these fibers are nominally normal to the flow direction. Accordingly, the properties of an injection-molded component depend both on the relative thickness of the surface and core layers and the direction of loading. Likewise, extruded composites contain oriented fibers that were broken during the extrusion process into lengths smaller than their initial size. Note that fiber breakage during injection molding, extrusion, or any other process reduces the strength and stiffness of the composite below the theoretical potential for the composite (recall Section 1.6.6). Figure 3.42 shows the difference in strength of nylon 66 composites as a function of fiber length.

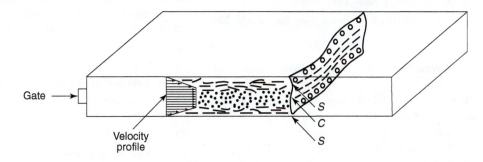

Figure 3.41   Through-thickness fiber orientation in injection-molded part.

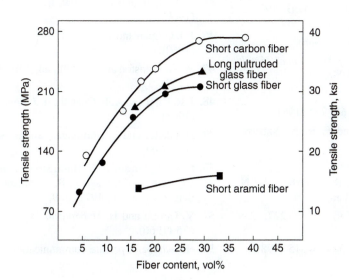

Figure 3.42   Strength in nylon 66 composites as a function of fiber content, type, and length.[96] (Reprinted with permission from *Advanced Materials and Processes*, 131(2), 57 (1987), ASM International, Metals Park, OH, 1987.)

REFERENCES

1. W. T. Read, Jr., *Dislocations in Crystals*, McGraw-Hill, New York, 1953.

2. J. S. Koehler, *Phys. Rev.* **86**, 52 (1952).

3. J. R. Low and R. W. Guard, *Acta Met.* **7**, 171 (1959).

4. W. C. Dash, *Dislocations and Mechanical Properties of Crystals*, J. C. Fisher, Ed., Wiley, New York, 1957.

5. D. Hull, *Introduction to Dislocations*, Pergamon, Oxford, 1965.

6. J. J. Gilman and W. G. Johnston, *Solid State Phys.* **13**, 147 (1962).

7. J. R. Low and A. M. Turkalo, *Acta Met.* **10**, 215 (1962).

8. V. B. Shenoy, R.V. Kukta, and R. Phillips, *Phys. Rev. Lett.* **84** (7), 1491 (2000).

9. J. Garstone and R. W. K. Honeycombe, *Dislocations and Mechanical Properties of Crystals*, J. Fisher, Ed., Wiley, New York, 1957, p. 391.

10. L. M. Clarebrough and M. E. Hargreaves, *Progress in Materials Science*, Vol. 8, Pergamon, London, 1959, p. 1.

11. J. P. Hirth and J. Weertman, Eds., *Work Hardening*, Gordon & Breach, New York, 1968.

12. D. Kuhlmann-Wilsdorf, *Trans. Met. Soc. AIME* **224**, 1047 (1962).

13. D. Kuhlmann-Wilsdorf, *Work Hardening*, J. P. Hirth and J. Weertman, Eds., Gordon & Breach, New York, 1968, p. 97.

14. D. Kuhlmann-Wilsdorf, *Work Hardening in Tension and Fatigue*, A. W. Thompson, Ed., AIME, New York, 1977, p. 1.

15. M. N. Bassin and R. J. Klassen, *Mater. Sci. Eng.* **81**, 163 (1986).

16. H. M. Otte and J. J. Hren, *Exp. Mech.* **6**, 177 (1966).

17. A. Seeger, *Dislocations and Mechanical Properties of Crystals*, J. C. Fisher, Ed., Wiley, New York, 1957, p. 243.

18. G. I. Taylor, *J. Inst. Met.* **62**, 307 (1938).

19. W. J. McG. Tegart, *Elements of Mechanical Metallurgy*, Macmillan, New York, 1966.

20. N. J. Petch, *JISI* **173**, 25 (1953).

21. E. O. Hall, *Proc. Phys. Soc. B* **64**, 747 (1951).

22. J. D. Eshelby, F. C. Frank, and F. R. N. Nabarro, *Philos. Mag.* **42**, 351 (1951).

23. H. Conrad, *JISI* **198**, 364 (1961).

24. R. W. Armstrong, Y. T. Chou, R. A. Fisher, and N. Lovat, *Philos. Mag.* **14**, 943 (1966).

25. J. C. M. Li, *Trans. Met. Soc. AIME* **227**, 239 (1963).

26. H. Gleiter, *Prog. Mater. Sci.* **33**, 223–315 (1989).

27. M. A. Meyers, A. Mishra, and D. J. Benson, *Prog. Mater. Sci.* **51**, 427–556 (2006).

28. J. Youngdahl, P. G. Sanders, J. A. Eastman, and J. R. Weertman, *Scripta Mat.* **37**, 809 (1997).

29. C. S. Pande and K. P. Cooper, *Prog. Mater. Sci.* **54**, 689–706 (2009).

30. Y. F. Shen, L. Lu, Q. H. Lu, Z. H. Jin, and K. Lu, *Scripta Mat.* **52**, 989 (2005).

31. A. Misra, J. P. Hirth, and R. G. Hoagland, *Acta Mat.* **53**, 4817–4824 (2005).

32. R. L. Fleischer, *The Strengthening of Metals*, D. Peckner, Ed., Reinhold, New York, 1964, p. 93.

33. P. Haasen, *Physical Metallurgy*, R. M. Cahn and P. Haasen, Eds., Vol. 2, Chap. 21, North-Holland, Amsterdam, 1983, p. 1341.

34. K. R. Evans, *Treatise on Materials Science and Technology*, H. Herman, Ed., Vol. 4, Academic, New York, 1974, p. 113.

35. R. L. Fleischer, *Acta Metall.* **11**, 203 (1963).

36. A. H. Cottrell, Report on the Conference on Strength of Solids, The Physical Society, London, 1948, p. 30.

37. R. L. Fleischer, *Acta Metall.* **11**, 203 (1963).

38. G. K. Williamson and R. E. Smallman, *Acta Crystallogr.* **6**, 361 (1953).

39. A. Sato and M. Meshii, *Acta Metall.* **21**, 753 (1973).

40. D. J. Quesnel, A. Sato, and M. Meshii, *Mater. Sci. Eng.* **18**, 199 (1975).

41. Donoso and Reed-Hill, *Met. Trans. A* **8A**, 945 (1977).

42. H. Conrad, *JISI* **198**, 364 (1961).

43. W. G. Johnston, *J. Appl. Phys.* **33**, 2716 (1962).

44. G. T. Hahn, *Acta Metall.* **10**, 727 (1962).

45. D. F. Stein and J. R. Low, Jr., *J. Appl. Phys.* **31**, 362 (1960).

46. J. J. Gilman and W. G. Johnston, *J. Appl. Phys.* **31**, 687 (1960).

47. W. G. Johnston and J. J. Gilman, *J. Appl. Phys.* **30**, 129 (1959).

48. J. R. Patel and A. R. Chaudhuri, *J. Appl. Phys.* **34**, 2788 (1963).

49. H. Gleiter and E. Hornbogen, *Mater. Sci. Eng.* **2**, 285 (1967/68).

50. L. M. Brown and R. K. Ham, *Strengthening Methods in Crystals*, A. Kelly and R. B. Nicholson, Eds., Applied Science, London, 1971, p. 9.

51. V. Gerald and H. Haberkorn, *Phys. Status Solidi* **16**, 675 (1966).

52. K. Khobaib, private communication.

53. E. Nembach and G. Neite, *Prog. Mater. Sci.* **29** (3), 177 (1985).

54. H. Gleiter and E. Hornbogen, *Phys. Status Solidi* **12**, 235 (1965).

55. R. K. Ham, *Trans. Japan Inst. Met.* **9** (supplement), 52 (1968).

56. E. Orowan, Discussions in Symposium on Internal Stresses in Metals and Alloys, Institute of Metals, London, 451 (1948).

57. J. C. Fisher, E. W. Hart, and R. H. Pry, *Acta Metall.* **1**, 336 (1953).

58. I. Irmann, *Metallurgia* **49**, 125 (1952).

59. G. B. Alexander, U. S. Patent No. 2,972,529, Feb. 21, 1961.

60. E. Gregory and N. J. Grant, *Trans. AIME* **200**, 247 (1954).

61. F. V. Lenel, A. B. Backensto, Jr., and M. V. Rose, *Trans. AIME* **209**, 124 (1957).

62. J. S. Benjamin, *Met. Trans.* **1**, 2943 (1970).

63. R. C. Benn, L. R. Curwick, and G. A. J. Hack, *Powder Metall.* **24**, 191 (1981).

64. R. Sunderesan and F. H. Froes, *J. Metals.* **39** (8), 22 (1987).

65. T. E. Howson, J. E. Stulga, and J. K. Tien, *Met. Trans.* **11A**, 1599 (1980).

66. E. Hornbogen, Strengthening Mechanisms in Steel, in *Steel-Strengthening Mechanisms*, Climax Molybdenum Co., Zurich, 1969, p. 1.

67. F. B. Pickering, *Physical Metallurgy and the Design of Steels*, Applied Science, London, 1978, Chap. 1.

68. A. R. Marder and J. I. Goldstein, Eds., *Phase Transformations in Ferrous Alloys*, AIME, Warrendale, PA, 1984.

69. N. J. Petch, Fracture, Proceedings, Swampscott Conference, Wiley, New York, 1959, p. 54.

70. G. Langford, *Met. Trans.* **8**, 861 (1977).

71. L. A. Norstrom, *Scand. J. Metall.* **5**, 159 (1976).

72. T. G. Nieh and R. F. Karlak, *Scripta Met.* **18**, (1984).

73. M. Vogelsang, R. J. Arsenault, and R. M. Fisher, *Met. Trans.* **17A**, 379 (1986).

74. R. J. Arsenault and R. M. Fisher, *Scripta Met.* **17**, 67 (1983).

75. T. Christman and S. Suresh, Brown University Report NSF-ENG-8451092/1/87, June 1987.

76. R. Marissen and L. B. Vogelesang, Int. SAMPE Meeting, January 1981, Cannes, France.

77. R. Marissen, DFVLR-FB 84-37, Institute Fur Werkstoff-Forschung, Koln, Germany, 1984.

78. R. J. Bucci, L. N. Mueller, R. W. Schultz, and J. L. Prohaska, 32nd Int. SAMPE Meeting, April 1987, Anaheim, CA.

79. J. L. Verolme, Report LR-666, Delft University of Technology, Delft, Netherlands, 1991.

80. H. F. Wu, L. L. Wu, W. J. Slagter, and J. L. Verolme, *J. Matl. Sci.* **29**, 4583–4591 (1994).

81. G. Wu and J.-M. Yang, *JOM* **57**, (1), 72–79 (2005).

82. P. J. Flory, *J. Am. Chem. Soc.* **67**, 2048 (1945).

83. J. H. Golden, B. L. Hammant, and E. A. Hazell, *J. Polym. Sci.* **2A**, 4787 (1974).

84. N. J. Mills, *Plastics: Microstructure, Properties and Applications*, Edward Arnold, London, 1986.

85. M. F. Ashby and D. R. H. Jones, *Engineering Materials 2*, Pergamon, Oxford, 1986.

86. D. Hull, *Polymeric Materials*, ASM, Metals Park, OH, 1975, p. 487.

87. F. C. Frank, *Proc. R. Soc. London, Ser. A* **319**, 127 (1970).

88. R. S. Porter, J. H. Southern, and N. Weeks, *Polym. Eng. Sci.* **15**, 213 (1975).

89. J. Preston, *Polym. Eng. Sci.* **15**, 199 (1975).

90. S. H. Jung, S. C. Kim, *Polymer J.* **20** (1), 73 (1988).

91. F. Vollrath and D. P. Knight, *Nature* **410**, 541–548 (2001).

92. F. Lucas, *Discovery* **25**, 20 (1964).

93. W. J. Gertsch, *American Spiders*, Van Nostrand, New York, 1949.

94. J. M. Gosline, P. A. Guerette, C. S. Ortlepp, and K. N. Savage, *J. Exp. Bio.* **202**, 3295 (1999).

95. *Guide to Engineering Materials*, Vol. 2 (1), ASM, Metals Park, OH, 1987.

96. *Advance Materials and Processing*, Vol. 131 (2), ASM, Metals Park, OH, 1987, p. 59.

97. Z. Tadmor, *J. Appl. Polym. Sci.* **18**, 1753 (1974).

98. S. S. Katti and J. M. Schultz, *Polym. Eng. Sci.* **22** (16), 1001 (1982).

FURTHER READING

99. M. F. Ashby and D. R. H. Jones, *Engineering Materials 2*, Pergamon, Oxford, 1986.

100. R. J. Crawford, *Plastics Engineering*, 3rd ed., Butterworth-Heinemann, 1998.

101. *Encyclopedia of Polymer Science and Engineering*, Wiley, New York, 1986.

102. *Engineering Design with Plastics; Principles and Practice*. Plastic and Rubber Institute, London, 1981, p. 1982.

103. A. J. Kinloch and R. J. Young, *Fracture Behavior of Polymers*, Applied Science, New York, 1983.

104. N. G. McCrum, C. P. Buckley, and C. B. Bucknall, *Principles of Polymer Engineering*, 2nd ed., Oxford University Press, Oxford, 1997.

105. N. J. Mills, *Plastic: Microstructure and Engineering Applications*, 3rd ed., Butterworth-Heinemann, 2005.

106. *Modern Plastics: Plastics Handbook*, McGraw-Hill, New York, 1994.

107. *Structural Plastics Design Manual*, FHWA-TS-79-203, U.S. Dept. of Transportation, 1979.

PROBLEMS

Review

3.1 List and briefly define the five main strengthening mechanisms in metals.

3.2 Calculate the approximate total dislocation line length (or range of lengths) expected in a cubic cm of a very highly cold-worked metal.

3.3 Does the strength of a metal always increase as dislocation density increases? Explain.

3.4 In Fig. 3.2, must it be true that the dislocation segments at points C and C' are screw dislocations? And must it be true that they are of opposite sign?

3.5 If a series of dislocations produced by a single Frank-Read source encounter an impassible barrier, a *back stress* is created. What is the reason for this back stress, and why is it very effective for dislocations produced by a single F-R source?

3.6 How do dislocation junctions contribute to strengthening?

3.7 Define *cell wall* and explain its role in strengthening.

3.8 Define *hot work* and *cold work* in the context of dislocations and microstructure stability, then give an estimate of the temperature (or fraction of the melting point) that usually marks the transition between the two.

3.9 What data would you collect, and what axes would you use, to make a linear plot for determining the grain-size dependence of yield strength according to the Hall-Petch relationship?

3.10 State two different possible functions of a grain boundary that are often invoked to explain Hall-Petch behavior.

3.11 Explain which has a larger effect on solid solution strengthening—symmetrical or asymmetrical point defects—and identify which specific defects lead to symmetrical or asymmetrical stress fields. List at least one example of an engineering material in which this factor comes into play.

3.12 What do the formation of *Lüders bands* and *dynamic strain aging* have in common, and how are they different?

3.13 Reproduce the binary phase diagram depicted in Fig. 3.20. For the composition shown at X, mark on the diagram approximate temperatures used for the three main thermal process steps used in precipitation hardening: solution treatment, quenching, and aging.

3.14 What is another name for *Ostwald ripening*, and what effect does the process have on mechanical strength?

3.15 Describe the conditions that favor cutting of a particle vs. looping of a dislocation around a particle.

3.16 Describe the trends between strength change and increasing particle spacing or particle size.

3.17 Why are Ni_3Al (γ') precipitates particularly effective at strengthening nickel-based superalloys?

3.18 Compare and contrast precipitation strengthening and dispersion strengthening.

3.19 Describe a method by which "ODS" alloys are produced and explain why it is not possible to cast these alloys using conventional techniques.

3.20 What is the difference between an *intrinsic* and an *extrinsic* strengthening mechanism, and on which do metal matrix composites depend?

3.21 What are three advantages of fiber-metal laminates over conventional metals for certain aircraft applications?

3.22 What aspect of thermoset polymers can be controlled to increase strength?

3.23 What are two fundamental differences between orientation strengthening of metals vs. polymers? The first should address the mechanism by which strengthening is achieved, and the second the thermal stability of the high strength characteristic.

Practice

3.24 A Frank-Read source created from a complete edge dislocation segment in Al is observed in a transmission electron microscope. Using a straining stage, it is possible to load the specimen to watch the source in action. If the pinning points are 55 nm apart, estimate the minimum resolved shear stress necessary to cause the dislocation segment to become unstable, thereby generating a new loop.

3.25 Experimentally, it has been observed for single crystals that the critical resolved shear stress τ_{CRSS} is a function of the dislocation density ρ_D as

$$\tau_{CRSS} = \tau_0 + A\sqrt{\rho_D}$$

where τ_0 and A are constants. For copper, the critical resolved shear stress is 0.69 MPa at a dislocation density of 10^4 mm^{-2}.

    a. If it is known that the value of τ_0 for copper is 0.069 MPa, please calculate the τ_{CRSS} at a dislocation density of 10^6 mm^{-2}.

b. Plot τ_{CRSS} for Cu over a dislocation density range of 10^3 to $10^{10}/cm^2$. Does your plot look like some or all of Fig. 3.1? What does this mean with regard to the validity of the equation?

3.26 When making hardness measurements, whether by nanoindentation or by conventional indentation testing, what will be the effect of making an indent very close to a preexisting indent? Why?

3.27 Summarize the general effect that Stacking Fault Energy has on the ability of an FCC metal to work harden, then briefly describe *three* mechanisms by which this influence occurs. The *first* of your answers should address the interaction between dislocations and second phase particles, the *second* should address dislocation junctions, and the *third* should address cell development.

3.28 The lower yield point for a certain plain carbon steel bar is found to be 135 MPa, while a second bar of the same composition yields at 260 MPa. Metallographic analysis shows that the average grain diameter is 50 μm in the first bar and 8 μm in the second bar.

    a. Predict the grain diameter needed to cause a lower yield point of 205 MPa.

    b. If the steel could be fabricated to form a stable grain structure of 500 nm grains, what strength would be predicted?

    c. Why might you expect the upper yield point to be more alike in the first two bars than the lower yield point?

3.29 A high-carbon steel with a fully pearlitic microstructure was used to form a high-strength bolt (H.-C. Lee et al., *J. Mater. Proc. Tech.* 211, 1044 (2011)). It was found that the bolt head had an average interlamellar spacing of 257 nm whereas the average spacing in the body of the bolt was 134 nm. Assuming that dislocation pileup is the primary mechanism responsible for the strength of this alloy, what ratio of strength (or hardness) might be expected in the head and body of the bolt?

3.30 A sketch of the Os-Pt binary phase diagram is provided below.

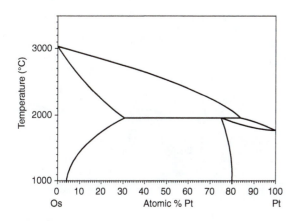

a. Reproduce the binary phase diagram and mark on it approximate candidate composition ranges for precipitation strengthened osmium-rich and platinum-rich Os-Pt alloys.

b. From a processing perspective alone, would precipitation strengthening be equally practical to achieve in osmium-rich and in platinum-rich Os-Pt alloys? Explain your reasoning.

c. Assume that an osmium-rich precipitation strengthened alloy is created. What other strengthening mechanisms are likely to be acting in the same alloy? List any critical assumptions behind the existence of each mechanism you believe is relevant.

3.31 The lattice parameters of Ni and Ni_3Al are 3.52×10^{-10} m and 3.567×10^{-10} m, respectively. The addition of 50 at% Cr to a Ni-Ni_3Al superalloy increases the lattice parameter of the Ni matrix to 3.525×10^{-10} m. Calculate the fractional change in alloy strength associated with the Cr addition, all other things being equal.

3.32 The addition of C to Fe greatly increases the room-temperature strength of the alloy, but an equal amount of C added to Ag has little effect. Why?

3.33 Some alloys use a combination of strain hardening and precipitation hardening to achieve particularly high strength levels. The usual order of strengthening is solution treatment, quenching, cold working, and finally precipitation heat treatment. Why not reverse the order of the cold working and precipitation heat treatment steps?

Design

3.34 Provide a reasonable explanation for the following observation: a welded component made of Al 6061-T6 alloy is routinely found to deform plastically first in the region adjacent to the weld joint despite the fact that the stress is nominally the same everywhere in the component. What solution would you propose to fix this problem, assuming that the weld joint cannot be eliminated from the design?

3.35 An aircraft fuselage design calls for a 2/1 layup of GLARE laminate (2 layers of Al and 1 layer of glass/epoxy prepreg). A study of this material (H. F. Wu, L. L. Wu, W. J. Slagter, and J. L. Verolme, *J. Matl. Sci.* 29, 4583–4591 (1994)) found that the 0.38-mm-thick glass prepreg layer had an ultimate strength of 1507 MPa in the longitudinal direction (along which 70% of the glass fibers were aligned), while the 0.3-mm-thick 2024-T3 aluminum layers each had an ultimate strength of 490 MPa. The density of the 2/1 layup was 2.45 g/cm^3. The density of 2024 alloy is 2.77 g/cm^3. The study also found that the laminate followed the rule of mixtures (Eq. 3-25) with regard to density and to ultimate tensile strength. It is proposed that a change from a 2/1 laminate to a 3/2 laminate for the aircraft fuselage offers the opportunity to reduce overall vehicle weight by reducing the composite density. Calculate the density of the 3/2 layup to check this assertion, then calculate the ultimate tensile strength of both laminates to ensure that there is no significant tradeoff with regard to ultimate strength. Finally, briefly discuss any other potential drawbacks that would have to be evaluated before selecting the 3/2 laminate over the 2/1 laminate.

Extend

3.36 A processing technique called Equal Channel Angular Pressing (ECAP) has been used for many metals and alloys to impose severe plastic deformation, and therefore create extreme dislocation densities. However, the primary purpose of ECAP is not to create high strength by severe work hardening. What is the main reason to perform ECAP processing, and why is the technique particularly attractive for this purpose?

3.37 Find a journal paper that describes either an experimental study or a simulation of the so-called "reverse/inverse Hall-Petch" phenomenon. To what mechanism does the paper attribute the phenomenon, and over what grain size range is it claimed to act? Provide a full reference for the paper in a standard reference format.

3.38 Look up the standard aluminum alloy heat treatment temper designations. Use the designations to justify the yield strength behavior of the following aluminum alloys listed in Table 1.2: 2024-T3 vs. 2024-T6 and 7075-T6 vs. 7075-T73.

3.39 Why are rivets of a 2017 aluminum alloy often refrigerated until the time they are used?

3.40 Find publications that describe the structure and properties of *Araneus* MA silk and *Araneus* viscid silk. Use what you learn about the structure of these materials to explain the differences in the stiffness, strength, and extensibility reported in Table 3.3. Provide full references in a standard reference format for any papers you used to develop your explanations.

Chapter 4

Time-Dependent Deformation

For the most part, our discussions of deformation in solids thus far have been limited to the *instantaneous (i.e., time-independent) deformation response*—elastic or plastic—to the application of a load. This enabled us to develop several relatively simple stress–strain relations that can describe material response under a number of different elastic and plastic loading conditions. However, prior discussion in Chapters 1 and 2 briefly introduced the effects of strain rate and temperature on the mechanical response of engineering materials. For example, in Sections 1.3.1–1.3.3 the possibility of time- and temperature-dependent elastic behavior was addressed, particularly for polymers. In Section 2.1.4.1, the temperature sensitivity of strength in crystalline solids was attributed to the role played by the Peierls-Nabarro stress in resisting dislocation movement through a given lattice. In addition, the potential importance of temperature in controlling crystalline deformation through thermally activated edge dislocation climb was mentioned in Section 2.1.5. Finally, in Section 2.4.4, it was also shown that the tensile strength of many materials increases with increasing strain rate and decreasing temperature.

As such, a *time-dependence of the observed stress or strain* in a structural component adds a new dimension to the problem, and requires a reformulation of some of the previously discussed phenomena in terms of stress–strain–time relations. It also opens up the possibility of a new mode of failure (*creep*) by time-dependent deformation. In the ensuing discussion, time and temperature effects on elastic, plastic, and fracture properties are explored more extensively. Characteristics common to crystalline materials (typical metals and ceramics) and to non-crystalline materials (certain polymers as well as glassy materials) are presented first, followed by separate discussions of the issues most relevant to these two categories of materials.

4.1 TIME-DEPENDENT MECHANICAL BEHAVIOR OF SOLIDS

Two consequences of time-dependent mechanical response to simple uniaxial tensile loading are shown in Fig. 4.1. In the first case, Fig. 4.1a, it is imagined that a fixed load is rapidly applied to the test bar, reaching a stress σ_0 in time t_0. In the second case, Fig. 4.1b, a fixed displacement is applied, inducing a strain ε_0 in the same time t_0. Unlike a purely elastic material that would exhibit a single, time-independent relationship between stress and strain as indicated by Hooke's law, $\sigma_0 = E\varepsilon_0$, a test bar undergoing time-dependent deformation processes would elongate with time under a fixed stress, causing the strain to increase over time (Fig. 4.1a). Under these conditions, the material is said to undergo *creep deformation*. Likewise, were the same bar to have been stretched to a certain length and then held firmly at constant strain, the necessary stress to maintain the stretch would gradually decrease (Fig. 4.1b). This is known as *stress relaxation*. In both cases depicted in Figs. 4a and b, slower loading might reach the target value of stress (or strain) at times t_1 or t_2. It can be seen from the dashed lines that this would effectively create a time-dependent modulus (the slope) that decreases in magnitude with decreasing strain rate (i.e., with increasing time to reach the fixed value of stress or strain). If some or all of the strain is recoverable after the load is removed, such response is said to be *viscoelastic*. If none of the strain is recovered, the length change is permanent and the material is *viscoplastic*.

Under the standard tensile test conditions in which the specimen displacement is applied at a constant rate, it is easy to imagine that dynamic stress relaxation occurring continuously throughout the test could lead to nonlinear elastic behavior as shown in Fig. 4.1c. As a result, in addition to having time-dependent strain or stress values, viscoelastic materials that are

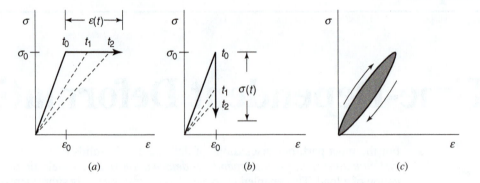

Figure 4.1 Time-dependent stress–strain response: (*a*) creep, (*b*) stress relaxation, and (*c*) hysteresis during loading/unloading.

elastically loaded and unloaded display *hysteresis* (i.e., a time-lag between the stress and strain responses that causes the creation of a loop instead of a single load/unloading path.) Because the area under the stress–strain curve is a measure of energy density, the shaded area between the loading and unloading curves depicted in Fig. 4.1*c* is a measure of the difference between the energy put into the material during loading and the energy returned during unloading. Therefore, unlike purely elastic materials that *store* energy in their stretched bonds, viscoelastic materials also *dissipate* energy. To illustrate, the next time you complete a relatively long auto trip, feel how warm your tires became as a result of the hysteretic heating of the tire rubber as you motored down the highway.

All solids have some capacity for time-dependent mechanical behavior. However, for many materials under normal use conditions it is acceptable to ignore the time-dependence, and instead to treat the material as we have done in Chapters 1–3. The onset of substantial viscoelastic or viscoplastic behavior can be estimated by comparing the operating temperature to the melting temperature, T_m, of a metal or a ceramic. For an amorphous polymer or a glass, the glass transition temperature, T_g, is also a critical factor. (Recall that T_g was introduced in Section 1.3.3 as the temperature at which a polymer or glass undergoes a transition between low-temperature glassy behavior and high-temperature rubbery behavior.) For crystalline metals, time-dependent deformation processes usually do not become significant until the operation temperature is approximately $0.3T_m$ (in Kelvin units) or higher.[i] For tin, then, creep may start to become important at temperatures as low as approximately $-120°C$ (153 K), and is certainly a factor at room temperature, whereas nickel and tungsten should not be expected to suffer in this way until at least $300°C$ (573 K) and $955°C$ (1228 K), respectively. Due to strong directional atomic bonding, ceramic materials may not experience creep or stress relaxation in any meaningful way until 0.4–$0.5T_m$, giving Al_2O_3 a creep threshold close to $1000°C$ ($\sim$1273 K). Of course, even exceedingly slow deformation processes may become important over very long time scales as observable in the creep of glaciers or the folding of sedimentary rock layers, but such long times are generally not the purview of materials engineers. However, since T_g and T_m of most polymeric materials are not much above ambient (and in fact may be lower as in the case of natural rubbers), these materials can often exhibit viscoelastic creep and relaxation phenomena at or below room temperature.

Other than the absolute temperature at which time-dependent deformation becomes important, there are substantial differences between material classes (and even within a single class of materials) with respect to the extent of strain recovery upon unloading. It is common to think of viscoelasticity as a mixture of elastic solid behavior and viscous liquid behavior. Whereas an elastic solid has strains that are instantaneously recoverable, and are predicted by an elastic modulus, a viscous liquid experiences strains that are nonrecoverable, and are

---

[i] A temperature expressed as a fraction of a material's melting point on the Kelvin scale is called the *homologous temperature*.

predicted by a viscosity. For the specific case of creep, this approach makes it reasonable to break up the total creep strain, ε_c, into perfectly elastic, time-dependent elastic (viscoelastic), and viscous (viscoplastic) components, here called ε_e, ε_{ve}, and ε_{vp}, respectively. The components simply add so that

$$\varepsilon_c = \varepsilon_e + \varepsilon_{ve} + \varepsilon_{vp} \tag{4-1}$$

It has been found that the behavior of metals and ceramics is dominated by the *elastic* term at low temperatures, and by the *elastic* and *viscoplastic* terms at high temperatures. There is no sharp onset of viscoplastic behavior in these materials; instead, its importance grows steadily with increasing temperature. A small *viscoelastic* component shows up as energy loss during hysteretic *anelastic damping* of vibrations in certain metals and alloys, but otherwise the fraction of recoverable time-dependent deformation is negligible in most cases.[ii] Polymers are a different story. Highly cross-linked and highly crystalline polymers have little or no *viscoplastic* contribution to their overall behavior at any temperature, so the time-dependent deformation that appears for $T \geq T_g$ is essentially all reversible. On the other hand, amorphous polymers (or lightly crystallized polymers with a large amorphous fraction) can show a range of behavior from a nearly elastic behavior for $T < T_g$ to a nearly viscous behavior for $T > T_g$ (but still below T_m), and they show a particularly strong viscoelastic contribution when $T \approx T_g$. With these thoughts in mind, it is now time to delve into the specific time-dependent behaviors that are important for users of structural metals, ceramics, and polymers.

4.2   CREEP OF CRYSTALLINE SOLIDS: AN OVERVIEW

The irreversible time-dependent deformation typical of metals and ceramics at high temperatures is most often characterized using the constant-load (or constant-stress) isothermal *creep test* in which strain is recorded as a function of time. As shown in Fig. 4.2a, after a load has been applied, the strain increases with time until failure finally occurs. For convenience, researchers have subdivided the creep curve into three regimes, based on the similar response of many materials (Fig. 4.2b). After the initial instantaneous strain ε_0, materials often undergo a period of transient response where the strain rate $d\varepsilon/dt$ decreases with time to a minimum

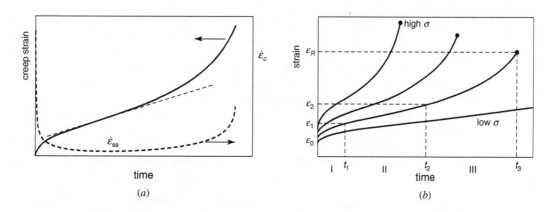

Figure 4.2   (a) Typical creep curve depicting creep strain as a function of time, and the corresponding creep strain rate curve. The steady state slope on the creep curve is fitted with a dashed line. The steady state creep rate is labeled as $\dot{\varepsilon}_{ss}$. (b) Four creep curves at different stress levels. One curve is marked to indicate the three characteristic stages of creep.

[ii] Certain metals of high purity have shown recoverable strains not found in the same metals in alloy form [T. S. Ke, *Phys. Rev.* **71**, 533–546 (1947); N. Nir, E. W. Hart, and C. Y. Li, *Scripta Met.* **10**, 189–194 (1976)]. Another exception to the relative unimportance of the viscoelastic term for crystalline materials may be found for certain metals in thin film form, for which it has been shown that surprising amounts of fully recoverable stress relaxation and creep can occur [S. Hyun, T. K. Hooghan, W. L. Brown, and R.P. Vinci, *Appl. Phys. Lett.* **87**, 061902 (2005)].

steady-state value that persists for a substantial portion of the material's life. Appropriately, these two regions are referred to in the literature as the *transient* or *primary* creep stage (Stage I) and the *steady-state* creep stage (Stage II), respectively. Final failure with a *rupture life* t_R then comes soon after the creep rate increases during the third, or *tertiary*, stage of creep (Stage III).

It is generally believed that the varying creep response of a material reflects a continually changing interaction between *strain hardening* and *softening (recovery)* processes, which strongly affect the overall strain rate of the material at a given temperature and stress. Strain hardening at elevated temperatures is believed to involve rearrangement of dislocations to form subgrains,[1] while thermally activated cross-slip and edge dislocation climb represent two dominant recovery processes (see Chapter 2). It is logical to conclude that the decrease in strain rate in Stage I (Fig. 4.2b) must be related to substructure changes that *increase overall resistance to dislocation motion*. Correspondingly, the low, approximately constant strain rate in Stage II would indicate a *stable substructure and a dynamic balance between hardening and softening processes*. Indeed, Barrett et al.[2] verified that the substructure in Fe-3Si was invariant during Stage II. (Note that the strain rate in Stage II is the *minimum creep rate* exhibited by the material during the test.) At high stress and/or temperature levels, the balance between hardening and softening processes is lost, and the accelerating creep strain rate in the tertiary stage is dominated by a number of *weakening metallurgical instabilities*. Among these microstructural changes are localized necking, corrosion, intercrystalline fracture, microvoid formation, precipitation of brittle second-phase particles, and dissolution of second phases that originally contributed toward strengthening of the alloy. In addition, the strain-hardened grains may recrystallize and thereby further destroy the balance between material hardening and softening processes.

The engineering creep strain curve shown schematically in Fig. 4.2a reflects the material response under constant tensile loading conditions and represents a convenient method by which most elevated temperature tests are conducted. However, from Eq. 1-2a, the true stress increases with increasing tensile strain. As a result, a comparable true creep strain–time curve should differ significantly if the test is conducted under constant *stress* rather than constant *load* conditions (Fig. 4.3). This is especially true for Stage II and III behavior. As a general rule, data being generated for engineering purposes are obtained from constant load tests, while more fundamental studies involving the formulation of mathematical creep theories should involve constant stress testing. In the latter instance, the load on the sample is lowered progressively with decreasing specimen cross-sectional area. This is done either manually or by the incorporation of computer-controlled load-shedding devices in the creep stand load train.

The creep response of materials depends on a large number of material and external variables. Certain material factors are considered in more detail later in this chapter. For the present, attention will be given to the two dominant external variables—stress (Fig. 4.2b) and test temperature (Fig. 4.4)—and how they affect the shape of the creep–time curve. Certainly, environment represents another external variable because of the importance of corrosion and oxidation in the fracture process. Unfortunately, consideration of this variable is not within the scope of this book.

The effect of temperature and stress on the minimum creep rate and rupture life are the two most commonly reported data for a creep or creep rupture test, although different material

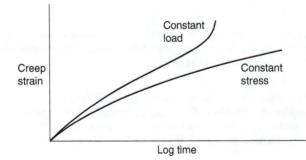

Figure 4.3 Creep curves produced under constant load and constant stress conditions.

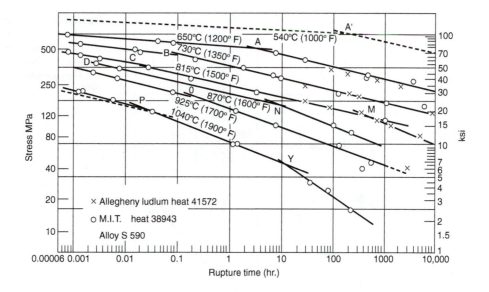

Figure 4.4   Stress–rupture life plot at several test temperatures for iron-based alloy S-590. (From N. J. Grant and A. G. Bucklin, copyright American Society for Metals, Metals Park, OH, © 1950.)

parameters are sometimes reported.[3,4] The rupture life at a given temperature and stress is obtained when it is necessary to evaluate the response of a material for use in a short-life situation, such as for a rocket engine nozzle ($t_R \approx 100$ s) or a turbine blade in a military aircraft engine ($t_R \approx 100$ hr). In such short-life situations, the dominant question is whether the component will or will not fail, rather than by how much it will deform. As a result, the details of the creep–time curve are not of central importance to the engineering problem. For this reason, *creep rupture* tests usually provide only one datum—the rupture life t_R. Rupture life information is sometimes used in the design of engineering components that will have a service life up to 10^5 hr. An example of such data is given in Fig. 4.4 for the high-temperature, iron-based alloy, S590. As expected, the rupture life t_R is seen to decrease with increasing test temperature and stress. When preparing this plot, Grant and Bucklin[5] chose to separate the data for a given temperature into several discrete regimes. This was done to emphasize the presence of several metallurgical instabilities that they identified metallographically and that they believed to be responsible for the change in slope of the $\log \sigma$–$\log t_R$ curve.

For long-life material applications, such as in a nuclear power plant designed to operate for several decades, component failure obviously is out of the question. However, it is equally important that the component not creep excessively. For long-life applications, the steady-state creep rate (or minimum creep rate) represents the key material response for a given stress and test temperature. To obtain this information, *creep* tests are performed into Stage II, where the steady-state creep rate $\dot{\varepsilon}_s$ can be determined with precision. Therefore, the *creep* test focuses on the early deformation stages of creep and is seldom carried to the point of fracture. As one might expect, the accuracy of $\dot{\varepsilon}_s$ increases with the length of time the specimen experiences Stage II deformation. Consequently, $\dot{\varepsilon}_s$ values obtained during instrumented creep rupture tests are not very accurate because of the inherently short time associated with the creep rupture test. The magnitude of $\dot{\varepsilon}_s$ often depends strongly on stress. As a result, steady-state creep rate data are usually plotted against applied stress, as shown in Fig. 4.5. The significance of the $\dot{\varepsilon}_s$ differences between α– and γ–iron[6] at the allotropic transformation temperature is discussed in Section 4.3.

Since the creep and creep rupture tests are similar (though defined over different stress and temperature regimes), it would seem reasonable to assume the existence of certain relations among various components of the creep curve (Fig. 4.2). In his text, Garofalo[3] summarized a number of log-log relations between t_R and other quantities, such as $t_2 - t_1, t_2$, and the steady-state

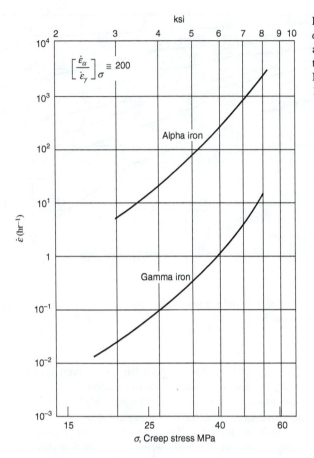

Figure 4.5 Stress–steady-state creep rate for α- and γ-iron at 910°C. (From O. D. Sherby and J. L. Lytton[6]; reprinted with permission of the American Institute of Mining, Metallurgical and Petroleum Engineers, 1956.)

creep rate $\dot{\varepsilon}_s$. Regarding the latter, Monkman and Grant[7] identified an empirical relation between t_R and $\dot{\varepsilon}_s$ with the form

$$log\ t_R + m\ log\ \dot{\varepsilon}_s = B \tag{4-2a}$$

where

$$t_R = \text{rupture life}$$
$$\dot{\varepsilon}_s = \text{steady-state creep rate}$$
$$m, B = \text{constants}$$

For a number of aluminum, copper, titanium, iron, and nickel base alloys, Monkman and Grant found $0.77 < m < 0.93$ and $0.48 < B < 1.3$. To a first approximation, then, the rupture life was found to be inversely proportional to $\dot{\varepsilon}_s$ such that

$$t_R = \frac{C_{MG}}{\dot{\varepsilon}_s^m} \approx \frac{C_{MG}}{\dot{\varepsilon}_s} \tag{4-2b}$$

where $C_{MG} = 10^B$ so that the typical range of C_{MG} is 3 to 20 when the unit of m is hours and C is %/hr. This relation allows t_R to be estimated as soon as $\dot{\varepsilon}_s$ is determined. Of course, the magnitude of t_R can be estimated from Eq. 4-2b only after the validity of the relation for the material in question is established and the constants m and B are identified.

A number of other empirical relations have been proposed to relate the primary creep strain to time at stress and temperature. Garofalo[3] summarized the work of others and showed that for low temperatures $(0.05 < T_h < 0.3)$[iii] and small strains, a number of materials exhibit *logarithmic creep*:

$$\varepsilon_t \propto \ln t \tag{4-3}$$

where

$\varepsilon_t =$ true strain
$t =$ creep time

In the range $0.2 < T_h < 0.7$, another relation has been employed with the form

$$\varepsilon_t = \varepsilon_{0_t} + \beta t^m \tag{4-4}$$

where

$\varepsilon_{0_t} =$ instantaneous true strain accompanying application of the load
$\beta, m =$ time-independent constants

Creep response in materials according to Eq. 4-4 is often referred to in the literature as *parabolic creep* or *β flow*. Since $0 < m < 1$ in transient creep, both Eqs. 4-3 and 4-4 reflect a decreasing strain rate with time. The strain rate $\dot{\varepsilon}$ can be derived from Eqs. 4-3 and 4-4 with the form

$$\dot{\varepsilon} \propto t^{-n} \tag{4-5}$$

as suggested by Cottrell,[8] where
$\dot{\varepsilon} =$ strain rate
$t =$ time
$n =$ constant

It is generally found that n decreases with increasing stress and temperature. At low temperatures when $n = 1$, Eq. 4-5 describes logarithmic creep (see Eq. 4-3). In the parabolic creep regime at higher temperatures, $m = 1 - n$. To provide a transition from Stage I to Stage II creep, another term $\dot{\varepsilon}_s t$, has to be added to Eq. 4-4 to account for the steady-state creep rate in Stage II. Hence

$$\varepsilon_t = \varepsilon_{0_t} + \beta t^m + \dot{\varepsilon}_s t \tag{4-6}$$

where

$\dot{\varepsilon}_s =$ steady-state creep rate in Stage II, reflecting a balance between strain hardening and recovery processes.

When $m = {}^1/_3$, Eq. 4-6 reduces to the relation originally proposed by Andrade[9] in 1910.

4.3 TEMPERATURE–STRESS–STRAIN-RATE RELATIONS

Since the creep life and total elongation of a material depends strongly on the magnitude of the steady-state creep rate $\dot{\varepsilon}_s$ (Eqs. 4-2 and 4-6), much effort has been given to the identification of those variables that strongly affect $\dot{\varepsilon}_s$. As mentioned in Section 4.2, the external variables, temperature and stress, exert a strong influence along with a number of material variables. Hence the steady-state creep rate may be given by

$$\dot{\varepsilon}_s = f(T, \sigma, \varepsilon, m_1, m_2) \tag{4-7}$$

[iii] T_h represents the homologous temperature.

where

$T =$ absolute temperature

$\sigma =$ applied tensile stress

$\varepsilon =$ creep strain

$m_1 =$ various intrinsic lattice properties, such as the elastic modulus G and the crystal structure

$m_2 =$ various metallurgical factors, such as grain and subgrain size, stacking fault energy, and thermomechanical history

It is important to recognize that m_2 also depends on T, σ, and ε. For example, subgrain diameter decreases markedly with increasing stress. Consequently, there exists a subtle but important problem of separating the effect of the major test variables on the structure from the deformation process itself that controls the creep rate. Dorn, Sherby, and coworkers[10-13] suggested that where $T_h > 0.5$ for the steady-state condition, the structure could be defined by relating the creep strain to a parameter θ

$$\varepsilon = f(\theta) \tag{4-8}$$

where

$\theta = te^{-\Delta H/RT}$ described as the temperature-compensated time parameter

$t =$ time

$\Delta H =$ activation energy for the rate-controlling process

$T =$ absolute temperature

$R =$ gas constant

The activation energy ΔH, shown schematically in Fig. 4.6, represents the energy barrier to be overcome so that an atom might move from A to the lower energy location at B. Upon differentiating Eq. 4-8 with respect to time, one finds

$$Z = f(\varepsilon) = \dot{\varepsilon}e^{\Delta H/RT} \tag{4-9a}$$

$$\dot{\varepsilon} = Ze^{-\Delta H/RT} \tag{4-9b}$$

$$\dot{\varepsilon} = K\sigma^n e^{-\Delta H/RT} \tag{4-9c}$$

which describes the strain-rate-temperature relation for a given stable structure and applied stress. The constant K and the *creep stress exponent n* are material constants. When the rate process is described by the logarithm of the minimum creep rate $\dot{\varepsilon}_s$ plotted against $1/T$, a series of parallel straight lines for different stress levels is predicted from Eq. 4-9c (Fig. 4.7). The slope[iv] of these lines, $\Delta H/R$ or $\Delta H/2.303R$, then defines the activation energy for the controlling

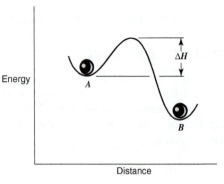

Figure 4.6   Diagram revealing significance of activation energy required in moving an atom from A to B.

Energy

ΔH

A

B

Distance

[iv] The slope is $\Delta H/R$ if $\ln(\dot{\varepsilon}_s)$ is plotted, or $\Delta H/2.303R$ if $\log(\dot{\varepsilon}_s)$ is plotted instead as in Fig. 4.7.

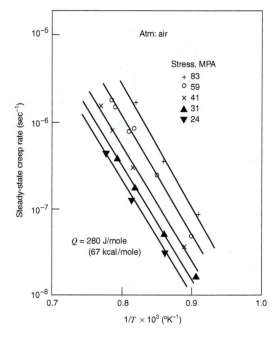

Figure 4.7   Log steady-state creep rate versus reciprocal of absolute temperature for rutile (TiO_2) at various stress levels. (From W. M. Hirthe and J. O. Brittain[14]; reprinted with permission from the American Ceramic Society, © 1963.)

creep process. The fact that the isostress lines were straight in Fig. 4.7 suggests that only one process had controlled creep in the TiO_2 single crystals throughout the stress and temperature range examined. Were different mechanisms to control the creep rate at different temperatures, the $log\,\dot{\varepsilon}_s$ vs. $1/T$ plots would be nonlinear. When multiple creep mechanisms are present and act in a concurrent and dependent manner, the slowest mechanism would control $\dot{\varepsilon}_s$. The overall strain rate would take the form

$$\frac{1}{\dot{\varepsilon}_T} = \frac{1}{\dot{\varepsilon}_1} + \frac{1}{\dot{\varepsilon}_2} + \frac{1}{\dot{\varepsilon}_3} + \cdots + \frac{1}{\dot{\varepsilon}_n} \tag{4-10}$$

where

$$\dot{\varepsilon}_T = \text{overall creep rate}$$
$$\dot{\varepsilon}_{1,2,3,\ldots,n} = \text{creep rates associated with } n \text{ mechanisms}$$

For the simple case where only two mechanisms act interdependently,

$$\dot{\varepsilon}_T = \frac{\dot{\varepsilon}_1 \dot{\varepsilon}_2}{\dot{\varepsilon}_1 + \dot{\varepsilon}_2} \tag{4-11}$$

Conversely, if the n mechanisms were to act independently of one another, the fastest one would control. For this case, $\dot{\varepsilon}_T$ would be given by

$$\dot{\varepsilon}_T = \dot{\varepsilon}_1 + \dot{\varepsilon}_2 + \dot{\varepsilon}_3 + \cdots + \dot{\varepsilon}_n \tag{4-12}$$

To determine the activation energy for creep over a small temperature interval, where the controlling mechanism would not be expected to vary, researchers often make use of the temperature differential creep test method. After a given amount of strain at temperature T_1, the temperature is changed abruptly to T_2, which may be slightly above or below T_1. The difference in the steady-state creep rate associated with T_1 and T_2 is then recorded (Fig. 4.8). If the stress is held constant and the assumption made that the small change in temperature does not change

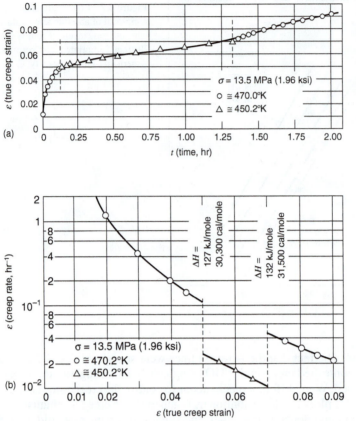

Figure 4.8 Incremental step test involving slight change in test temperature to produce change in steady-state creep rate in aluminum. (From J. E. Dorn, *Creep and Recovery*, reprinted with permission from American Society for Metals, Metals Park, OH, © 1957.)

the alloy structure, then Z is assumed constant. From Eq. 4-9 the activation energy for creep may then be calculated by

$$\Delta H_C = \frac{-R \ln \dot{\varepsilon}_2 / \dot{\varepsilon}_1}{1/T_2 - 1/T_1} \tag{4-13}$$

where

ΔH_C = activation energy for creep
$\dot{\varepsilon}_1, \dot{\varepsilon}_2$ = creep rates at T_1 and T_2, respectively

This value of ΔH_c should correspond to the activation energy determined by a data analysis like that shown in Fig. 4.7, as long as the same mechanism controls the creep process over the expanded temperature range in the latter instance. As shown in Fig. 4.9, this is not always the case. The activation energy for creep in aluminum is seen to increase with increasing temperature up to $T_h \approx 0.5$, whereupon ΔH_c remains constant up to the melting point. Similar results have been found in other metals.[15] It would appear that different processes were rate controlling over the test temperature range.[13] Furthermore, it should be recognized that ΔH_c may represent some average activation energy reflecting the integrated effect of several mechanisms operating simultaneously and interdependently (see Section 4.4).

Dorn,[12] Garofalo,[3] and Weertman[16] have compiled a considerable body of data to demonstrate that at $T_h \geq 0.5$, ΔH_C is most often equal in magnitude to ΔH_{SD}, the activation energy for self-diffusion (Fig. 4.10); this fact strongly suggests the latter to be the creep rate-controlling process in this temperature regime. While the approximate equality between ΔH_C and ΔH_{SD}

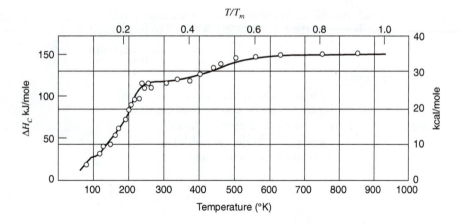

Figure 4.9   Variation of apparent activation energy for creep in aluminum as a function of temperature. (From O. D. Sherby, J. L. Lytton, and J. E. Dorn,[13] reprinted with permission from Sherby and Pergamon Press, Elmsford, NY, 1957.)

seems to hold for many metals and ceramics at temperatures equal to and greater than half the melting point, some exceptions do exist, particularly for the case of intermetallic and nonmetallic compounds. It is found that small departures from stoichiometry of these compounds have a pronounced effect on ΔH_C, which in turn affects the creep rate. For example, a reduction in oxygen content in rutile from TiO_2 to $TiO_{1.99}$ causes a reduction in ΔH_C from about 280 to

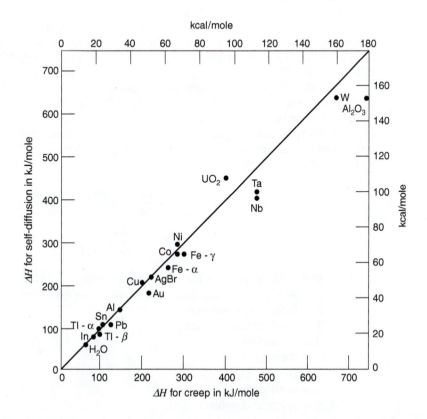

Figure 4.10   Correlation between activation energy for self-diffusion and creep in numerous metals and ceramics. (From J. Weertman,[16] reprinted with permission from American Society for Metals, Metals Park, OH, © 1968.)

120 kJ/mol (67–29 kcal/mol)[v] with an associated 100-fold increase in $\dot{\varepsilon}_s$.[14] For the more general case, however, the creep process is found to be controlled by the diffusivity of the material

$$D = D_0 e^{-\Delta H_{SD}/RT} \tag{4-14}$$

where

$$
\begin{aligned}
D &= \text{diffusivity, cm}^2/\text{s} \\
D_0 &= \text{diffusivity constant } \approx 1 \text{ cm}^2/\text{s} \\
\Delta H_{SD} &= \text{activation energy, J/mol} \\
R &= \text{gas constant, J/(mol K)} \\
T &= \text{absolute temperature, K}
\end{aligned}
$$

$$D = D_0 e^{-(K_0+V)T_m/T} \tag{4-15}$$

where

$$
\begin{aligned}
K_0 &= \text{dependent on the crystal structure and equal to 14 for BCC lattice, 17 for FCC and} \\
&\quad \text{HCP lattices, and 21 for diamond-cubic lattice} \\
V &= \text{valence of the material} \\
T_m &= \text{absolute melting temperature}
\end{aligned}
$$

The constants K_0 are estimates associated with an assumed diffusivity constant $\approx 1 \text{ cm}^2/\text{s}$.

By combining Eqs. 4-14 and 4-15

$$\Delta H_{SD} = RT_m(K_0 + V) \tag{4-16}$$

we see that the activation energy for self-diffusion increases (corresponding to a reduction in D) with increasing melting point, valence, packing density, and degree of covalency. Consequently, although refractory metals with high melting points, such as tungsten, molybdenum, and chromium, seem to hold promise as candidates for high-temperature service, their performance in high-temperature applications is adversely affected by their open BCC lattice, which enhances diffusion rates. From Eq. 4-16, ceramics are identified as the best high-temperature materials because of their high melting point and the covalent bonding that often exists.

It is important to recognize that creep rates for all materials cannot be normalized on the basis of D alone because other test variables affect the creep process in different materials. For example, Barrett and coworkers[19] noted the important influence of elastic modulus on the creep rate and on determination of the true activation energy for creep. A semi-empirical relationship with the form

$$\frac{\dot{\varepsilon}_s kT}{DG\mathbf{b}} = A\left(\frac{\sigma}{G}\right)^n \tag{4-17a}$$

$$\dot{\varepsilon}_s = \frac{ADG\mathbf{b}}{kT}\left(\frac{\sigma}{G}\right)^n \tag{4-17b}$$

has been proposed[1] to account for other factors where

$$
\begin{aligned}
\dot{\varepsilon} &= \text{steady-state creep rate} \\
k &= \text{Boltzman's constant} \\
T &= \text{absolute temperature} \\
D &= \text{diffusivity} \\
G &= \text{shear modulus} \\
\mathbf{b} &= \text{Burgers vector} \\
\sigma &= \text{applied stress} \\
A, n &= \text{material constants}
\end{aligned}
$$

---

[v] To convert from kcal to kJ, multiply by 4.184.

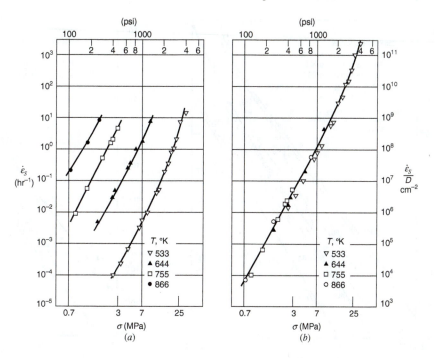

Figure 4.11   Creep data in aluminum. (*a*) Stress versus steady-state creep rate $\dot{\varepsilon}_s$ at various test temperatures; (*b*) data normalized by plotting stress versus $\dot{\varepsilon}_s$ divided by the diffusion coefficient. (From O. D. Sherby and P. M. Burke[17]; reprinted with permission from Sherby and Pergamon Press, Elmsford, NY, 1968.)

By combining Eqs. 4-9 and 4-14, the steady-state creep rate at different temperatures can be normalized with respect to D to produce a single curve, as shown in Fig. 4.11. This is an important finding since it allows one to conveniently portray a great deal of data for a given material. For example, we see from a reexamination of Fig. 4.5 that at the allotropic transformation temperature, the creep rate in γ-iron (FCC lattice) is found to be approximately 200 times slower than that experienced by α-iron (BCC lattice).[6] This substantial difference is traced directly to the 350-fold lower diffusivity in the close-packed FCC lattice in γ-iron. Similar findings were reviewed by Sherby and Burke[17] for the allotropic transformation from HCP to BCC in thallium. Therefore, it is appropriate to briefly consider those factors that strongly influence the magnitude of D. Sherby and Simnad[18] reported an empirical correlation showing D to be a function of the type of lattice, the valence, and the absolute melting point of the material.

Here again we see that creep is assumed to be diffusion controlled. Even after normalizing creep data with Eq. 4-17, a three-decade scatter band still exists for the various metals shown in Fig. 4.12. While some of this difference might be attributable to actual test scatter or relatively imprecise high-temperature measurements of D and G, other as yet unaccounted for variables most likely will account for the remaining inexactness. For example, there appears to be a trend toward higher creep rates in FCC metals and alloys possessing high stacking fault energy (SFE). Whether the SFE variable should be incorporated into either A or n is the subject of current discussion.[20–22] The role of substructure on A and n must also be identified more precisely.

One important factor in Eq. 4-17 is the stress dependency of the steady-state creep rate. It is now generally recognized that $\dot{\varepsilon}_s$ varies directly with σ at low stresses and temperatures near the melting point. At intermediate to high stresses and at temperatures above $0.5T_m$, where the thermally activated creep process is dominated by the activation energy for self-diffusion, $\dot{\varepsilon}_s \propto \sigma^{4-5}$ (so-called power law creep). It should be noted that this stress dependency holds for pure metals and their solid solutions. Much stronger stress dependencies of $\dot{\varepsilon}_s$ and t_R have been reported in oxide-dispersion-strengthened superalloys (see Section 4.8). At very high stress

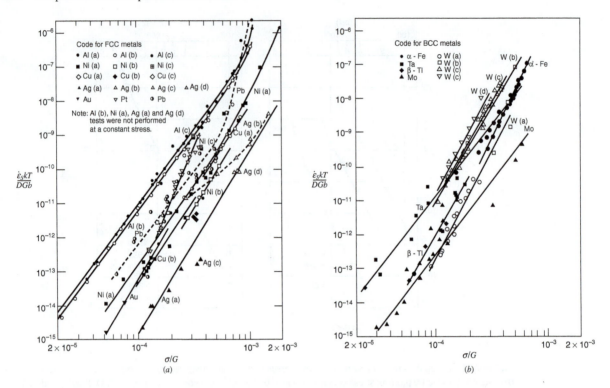

Figure 4.12 Creep data in metals. (*a*) Data for FCC metals; materials with high-stacking fault energy tend to have higher steady-state creep rates. (*b*) Data for BCC metals. (From A. K. Mukherjee, J. E. Bird, and J. E. Dorn[1]; copyright American Society for Metals, Metals Park, OH, © 1969.)

levels $\dot{\varepsilon}_s \propto e^{\alpha\sigma}$. Garofalo[23] showed that power law and exponential creep represented limiting cases for a general empirical relationship

$$\dot{\varepsilon}_s \propto (\sinh \alpha\sigma)^n \tag{4-18}$$

Equation 4-18 reduces to power law creep when $\alpha\sigma < 0.8$, but approximates exponential creep when $\alpha\sigma > 1.2$. An explanation for the changing stress dependence of $\dot{\varepsilon}_s$ in several operative deformation mechanisms is discussed in the next section.

4.4 DEFORMATION MECHANISMS

At low temperatures relative to the melting point of crystalline solids, the dominant deformation mechanisms are slip and twinning (Chapter 2). However, at intermediate and high temperatures, other mechanisms become increasingly important and dominate material response under certain conditions. It is with regard to these additional deformation modes that attention will now be focused.

Over the years a number of theories have been proposed to account for the creep data trends discussed in the previous sections. In fact, the empirical form of Eq. 4-17 takes account of mathematical formulations for several proposed creep mechanisms. At low stresses and high temperatures, where the creep rate varies with applied stress, Nabarro[24] and Herring[25] theorized that the creep process was controlled by stress-directed atomic diffusion. Such *diffusional creep* is believed to involve the migration of vacancies along a gradient from grain boundaries experiencing tensile stresses to boundaries undergoing compression (Fig. 4.13); simultaneously atoms would be moving in the opposite direction, leading to elongation of the grains and the test bar. This gradient is produced by a stress-induced decrease in energy to

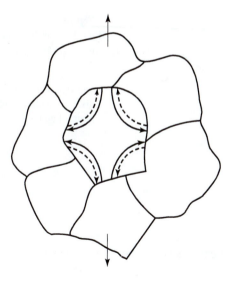

Figure 4.13   Stress-directed flow of vacancies (solid lines) from tensile to compressive grain boundaries and corresponding reverse flow of atoms or ions (dashed lines).

create vacancies when tensile stresses are present and a corresponding energy increase for vacancy formation along compressed grain boundaries. *Nabarro-Herring creep* can be described by Eq. 4-17 when $A \approx 7 \, (b/d^2)$ ($d =$ grain diameter) and $n = 1$, such that[21]

$$\dot{\varepsilon}_s \approx \frac{7\sigma D_v b^3}{kTd^2} \tag{4-19}$$

where $D_v =$ volume diffusivity through the grain interior. As expected, $\dot{\varepsilon}_s$ is seen to increase with increasing number of grain boundaries (i.e., smaller grain size).

A closely related *diffusional creep* process described by Coble[26] involves atomic or ionic diffusion along grain boundaries. Setting $A \approx 50(b/d)^3$ and $n = 1$, Eq. 4-17 reduces to the *Coble creep* relationship

$$\dot{\varepsilon}_s \approx \frac{50\sigma D_{gb} b^4}{kTd^3} \tag{4-20}$$

(Note that Coble creep is even more sensitive to grain size than is Nabarro-Herring creep.) In complex alloys and compounds there is a problem in deciding which particular atom or ion species controls the diffusional process and along what path such diffusion takes place. This is usually determined from similitude arguments. That is, if ΔH_C is numerically equal to ΔH_{SD} for element A along a particular diffusion path, then it is presumed that the self-diffusion of element A had controlled the creep process.

At intermediate to high stress levels and test temperatures above $0.5T_m$, creep deformation is believed to be controlled by diffusion-controlled movement of dislocations. Several of these theories have been evaluated by Mukherjee et al.,[1] with the Weertman[16,27] model being found to suffer from the least number of handicaps and found capable of predicting best the experimental creep results described in Section 4.3. Weertman proposed that creep in the above-mentioned stress and temperature regime was controlled by edge dislocation climb away from dislocation barriers. Again using Eq. 4-17 as the basis for comparison, Bird et al.[21] showed that when A is constant and $n \approx 5$, *dislocation creep* involving the climb of edge dislocations could be estimated by

$$\dot{\varepsilon}_s \approx \frac{ADGb}{kT}\left(\frac{\sigma}{G}\right)^5 \tag{4-21}$$

It should be noted that in many creep situations, the dislocation creep process dominates the elevated temperature ($T \geq 0.5T_m$) response of engineering alloys.

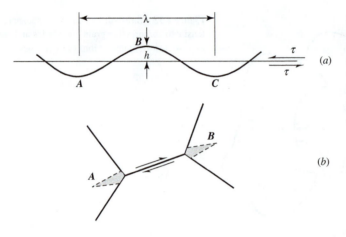

Figure 4.14 Accommodation mechanisms for grain-boundary sliding. (*a*) Shear along boundary accommodated by diffusional flow of vacancies of region *AB* to *BC*; (*b*) grain-boundary sliding accommodated by dislocation climb within contiguous grains *A* and *B*.

The actual Weertman relationship expresses the shear strain rate $\dot{\gamma}_s$, in terms of the shear stress τ by

$$\dot{\gamma}_s \propto \tau^2 \sinh \tau^{2.5} \tag{4-22}$$

As such, the transition from power law to exponential creep mentioned earlier is readily predicted from Eq. 4-22. Weertman[27] theorized that the onset of exponential creep ($\dot{\varepsilon}_s \propto e^{\alpha\sigma}$) at high stress levels was related to accelerated diffusion, because of an excess vacancy concentration brought about by dislocation-dislocation interactions.

Another high-temperature deformation mechanism involves grain-boundary sliding. The problem in dealing with grain-boundary sliding, however, is that it does not represent an independent deformation mechanism; it must be accommodated by other deformation modes. For example, consider the shear-induced displacement of the two grains in Fig. 4.14*a*. At sufficiently high temperatures, the local grain-boundary stress fields can cause diffusion of atoms from the compression region *BC* to the tensile region *AB* by either a Nabarro-Herring or Coble process. As might be expected, the rate of sliding should depend strongly on the shape of the boundary. Raj and Ashby[28] demonstrated that $\dot{\varepsilon}_s$ increased rapidly as the ratio of perturbation period λ to perturbation height h increased. Furthermore, when λ is small and the temperature relatively low, diffusion is found to be controlled by a grain-boundary path. On the other hand, when λ is large and the temperature relatively high, volume diffusion controls the grain-boundary sliding process.[28] Consequently, grain-boundary sliding may be accommodated by diffusional flow, which is found to depend on both the temperature and the grain-boundary morphology. For this case, the sliding rate would be directly proportional to stress (see Eqs. 4-19 and 4-20). By examining this problem from a different perspective, one finds that Nabarro-Herring and Coble creep models are themselves dependent on grain-boundary sliding! From Fig. 4.15, note that the

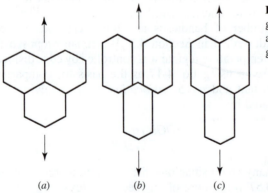

(*a*)   (*b*)   (*c*)

Figure 4.15 Stress-induced diffusional flow elongates grains and could lead to grain separation (*b*), but is accommodated by grain-boundary sliding, which brings grains together (*c*).

stress-directed diffusion of atoms from compression to tension grain boundaries causes the grain boundaries to separate from one another (Fig. 4.15b). Grain-boundary sliding is needed, therefore, to maintain grain contiguity during diffusional flow processes (Fig. 4.15c).[28–30] On the basis of this finding, Raj and Ashby concluded that Nabarro-Herring and Coble diffusional creep mechanisms were "identical with grain-boundary sliding with diffusional accommodation."[28]

For the internal boundary shown in Fig. 4.14b, grain-boundary sliding could be accommodated by dislocation creep within grains A and B. Matlock and Nix[31] examined this condition for several metals and found that the grain-boundary-sliding strain-rate contribution was proportional to σ^{n-1}, where n is the exponent associated with the dislocation creep mechanism ($n \approx 4 - 5$). Unfortunately this stress sensitivity does not agree with any presently known theoretical predictions.

It is apparent from the above discussion that these high-temperature deformation mechanisms all depend on atom or ion diffusion but differ in their sensitivity to other variables such as G, d, and σ. As such, a particular strengthening mechanism may strengthen a material *only* with regard to a particular deformation mechanism but not another. For example, an increase in alloy grain size will suppress Nabarro-Herring and Coble creep along with grain boundary sliding, but will not substantially change the dislocation climb process.[1] As a result, the rate-controlling creep deformation process would shift from one mechanism to another. Consequently, marked improvement in alloy performance requires simultaneous suppression of several deformation mechanisms. This point is considered further in Section 4.6.

4.5 SUPERPLASTICITY

As we have just seen, fine-grained structures are to be avoided in high-temperature, load-bearing components since this would bring about an increase in creep strains resulting from Nabarro-Herring, Coble, and grain-boundary-sliding creep mechanisms. In fact, experience in the turbine engine industry reveals improved creep response in alloys possessing either no grain boundaries (i.e., single-crystal alloys) or highly elongated boundaries (produced by uni-directional solidification) oriented parallel to the major stress axis.[33] However, where the opposite of creep resistance (i.e., easy flow) is required, such as in hot-forming processes, fine-grained structures are preferred. Some such materials are known to possess *superplastic* behavior[34] with total strains in excess of 1000% (Fig. 4.16). These large strains, generated at low stress levels, drastically improve the formability of certain alloys.

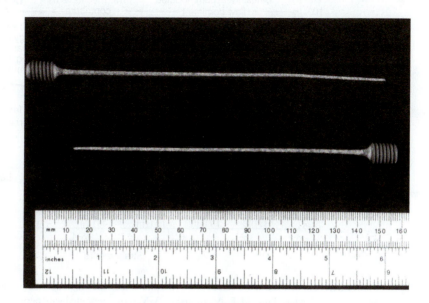

Figure 4.16   Tensile specimen having experienced superplastic flow.

By expressing the flow stress–strain-rate relation (Eq. 2-37) in the form

$$\sigma = \frac{F}{A} = K\dot{\varepsilon}^m \tag{4-23}$$

where

$F =$ applied force
$A =$ cross-sectional area
$K =$ constant
$\dot{\varepsilon} = \frac{1}{l}\frac{dl}{dt} = \frac{1}{A}\frac{dA}{dt}$
$m =$ strain-rate sensitivity factor

superplasticity is found when m is large[34–36] and approaches unity. Figures 4.17 and 4.18 show the normalized stress–strain-rate relation for loading in the superplastic region. After substituting for $\dot{\varepsilon}$ and rearranging, the change in cross-sectional area with time, dA/dt, is given by

$$\frac{-dA}{dt} = \frac{F^{(1/m)}}{K}\left[\frac{1}{A^{(1-m)/m}}\right] \tag{4-24}$$

In the limit, as the rate sensitivity factor m approaches unity, note that dA/dt depends only on the applied force and is independent of any irregularities in the specimen cross-sectional area, such as incipient necks and machine tool marks, which are maintained but not worsened. That is, the sample undergoes extensive deformation without pronounced necking.

Superplastic behavior has been reported in numerous metals, alloys, and ceramics[34] and associated in all cases with (1) a fine grain size (on the order of 1–10 μm), (2) deformation temperature $>0.5\ T_m$, and (3) a strain-rate sensitivity factor $m > 0.3$. The strain-rate range associated with superplastic behavior has been shown to increase with decreasing grain size and

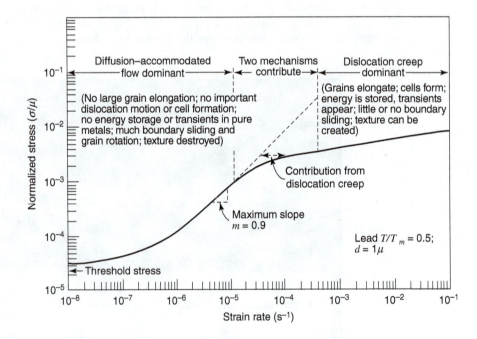

Figure 4.17 Normalized stress versus strain rate plot in lead showing intermediate region associated with superplastic behavior. (From M. F. Ashby and R. A. Verall;[37] reprinted with permission from Ashby and Pergamon Press, Elmsford, NY, 1973.)

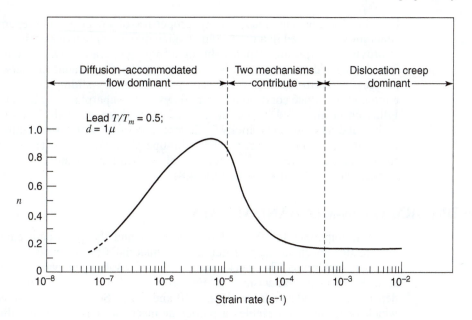

Figure 4.18   Strain-rate sensitivity factor versus strain rate in lead. (From M. F. Ashby and R. A. Verall;[37] reprinted with permission from Ashby and Pergamon Press, Elmsford, NY, 1973.)

increasing temperature, as shown schematically in Fig. 4.19. There has been considerable debate, however, regarding the mechanisms responsible for the superplastic process. Avery and Back-ofen[36] originally proposed that a combination of deformation mechanisms involving Nabarro-Herring diffusional flow at low stress levels and dislocation climb at higher stresses were rate controlling. The applicability of the Nabarro-Herring creep model in the low stress regime has been questioned, based on experimental findings and theoretical considerations. First, it is generally found that m is of the order 0.5 rather than unity, the latter being associated with Nabarro-Herring creep. Furthermore, Nabarro-Herring creep would lead to the formation of elongated grains proportional in length to the entire sample. To the contrary, equiaxed grain structures are preserved during superplastic flow. More recent theories have focused with greater success on grain-boundary-sliding arguments, with diffusion-controlled accommodation[37–39] as the operative deformation mechanism associated with superplasticity at low stress levels.

As mentioned above, the formability of a material is enhanced greatly when in the superplastic state, while forming stresses are reduced substantially. To this end, grain refinement is highly desirable. Grain sizes on the order of 1–3 μm are commonly needed to attain superplastic behavior. The alert reader will immediately recognize, however, that once an alloy is rendered superplastic through a grain-refinement treatment, it no longer possesses the optimum grain size for high-temperature load applications. To resolve this dichotomy, researchers are currently seeking to develop duplex heat treatments to optimize

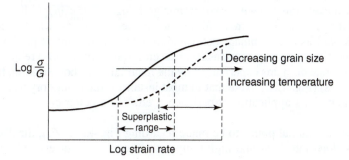

Figure 4.19   Temperature- and grain-size-induced shift in strain-rate range associated with superplastic behavior.

both hot-forming and load-bearing properties of an alloy. For example, a nickel-based superalloy to be used in a gas turbine engine may first receive a grain-refining heat treatment to provide superplastic response during a forging operation. Once the alloy has been formed into the desired component, it is given another heat treatment to coarsen the grains so as to suppress Nabarro-Herring, Coble, and grain-boundary-sliding creep processes during high-temperature service conditions. For reviews of the superplasticity literature, see the papers by Edington et al.[40] and Taplin et al.[41] along with an analysis of current problems in our understanding of superplasticity.[42] Several additional articles pertaining to the mechanical, microstructural, and fracture processes in superplastically formed materials are recommended for the reader's attention.[43] Commercial applications of superplasticity are described by Hubert and Kay[44] (also see Section 4.8).

4.6 DEFORMATION-MECHANISM MAPS

It is important for the materials scientist and the practicing engineer to identify the deformation mechanisms that dominate a material's performance under a particular set of boundary conditions. This can be accomplished by solving the various constitutive equations for each deformation mechanism (e.g., Eqs. 4-17 to 4-21) and recognizing their respective inter-dependence or independence (Eqs. 4-10 and 4-12). Solutions to these equations reveal over which range of test variables a particular mechanism is rate controlling. Ashby and cow-orkers[45-47] have displayed such results pictorially in the form of maps in stress–temperature space based on the original suggestion by Weertman.[16] Typical deformation-mechanism maps for pure silver and germanium are shown in Fig. 4.20, where most of the high-temperature deformation mechanisms discussed in Section 4.4 (as well as pure glide) are shown. Each mechanism is rate controlling within its stress–temperature boundaries. Consistent with the previous discussion, dislocation creep is seen to dominate the creep process in both materials at relatively high stresses and homologous temperatures above 0.5. For the FCC metal, diffusional creep by either Nabarro-Herring or Coble mechanisms dominates at high temperatures but lower stress levels. The virtual absence of these two diffusional flow mechanisms in covalently bonded diamond-cubic germanium is traced to its larger activation energy for self-diffusion and associated lower diffusivity. The boundaries separating each deformation field are defined by equating the appropriate constitutive equations (Eqs. 4-17 to 4-21) and solving for stress as a function of temperature. This amounts to the boundary lines representing combinations of stress and temperature, wherein the respective strain rates from the two deformation mechanisms are equal. Triple points in the deformation map occur when a particular stress and temperature produce equal strain rates from three mechanisms.

The maps shown in Fig. 4.20 do not portray a grain-boundary-sliding region, since uncertainties exist regarding the appropriate constitutive equation for this mechanism (see the discussion in Section 4.4). Studies[47] have shown, however, that the dislocation creep field can be subdivided with a grain-boundary-sliding contribution existing at the lower stress levels associated with lower creep strain rates. Regarding the latter point, it is desirable to portray on the deformation map the strain rate associated with a particular stress–temperature condition, regardless of the rate-controlling mechanism. This may be accomplished by plotting the diagram contours of isostrain rate lines calculated from the constitutive equations. Examples of such modified maps are given in Fig. 4.21 for pure nickel prepared with two different grain sizes. These maps allow one to pick any two of the three major variables—stress, strain rate, and temperature—which then identifies the third variable as well as the dominant deformation mechanism. This is particularly useful in identifying the location of testing domains (such as creep and tensile tests) relative to the stress–temperature–strain-rate domains experienced by the material (e.g., hot-working, hot torsion, and geological processes) (Fig. 4.22). Note that in most instances, the laboratory test domains do not conform to the material's application experience. Certainly a better correspondence would be more desirable.

There are two additional points to be made regarding Fig. 4.21. First, the dislocation climb field has been divided into low- and high-temperature segments, corresponding to dislocation

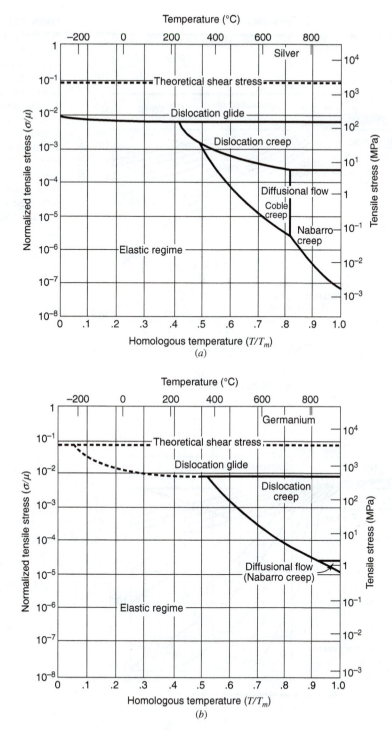

Figure 4.20 Deformation mechanism map for (*a*) pure silver and (*b*) germanium, showing stress–temperature space where different deformation mechanisms are rate controlling. Grain size in both materials is 32 μm. Elastic boundaries determined at a strain rate of 10^{-8}/s. (From M. F. Ashby[45]; reprinted with permission from Ashby and Pergamon Press, Elmsford, NY, 1972.)

climb controlled by dislocation core and lattice diffusion, respectively. Furthermore, since Coble creep involves grain boundary diffusion, three diffusion paths are represented on these maps. Second, a large change in grain size in pure nickel drastically shifts the isostrain rate contours and displaces the deformation field boundaries. For example, at $T_h = 0.5$ and a strain rate of 10^{-9}/s, a 100-fold decrease in grain size causes the creep rate-controlling process to shift from low-temperature dislocation creep to Coble creep. Furthermore, the stress necessary to produce this

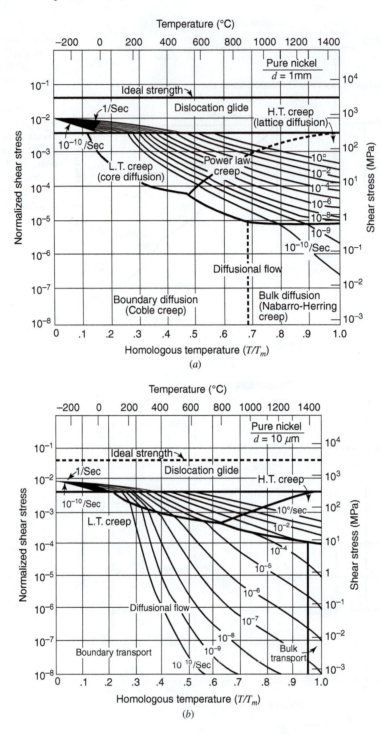

Figure 4.21 Deformation map for (*a*) 1-mm and (*b*) 10-μm grain-size nickel. Isostrain rate lines superimposed on map. Dislocation climb region divided into low-temperature (core diffusion) and high-temperature (volume diffusion) regions. Note lower strain rates in more coarsely grained material. (M. F. Ashby[46]; reprinted with permission of the Institute of Metals.)

strain rate decreases by almost three orders of magnitude! Both the expansion of the Coble creep regime and the much lower stress needed to produce a given strain rate reflect the strong inverse dependence of grain size on the rate of this mechanism (Eq. 4-20). The Nabarro-Herring creep domain also expands for the same reason (Eq. 4-19). Since grain size effects on deformation maps are large, some researchers[29,48] have further modified the maps to include grain size as one of the dominant variables along with stress and isostrain rate contour lines. The diagrams, such as the

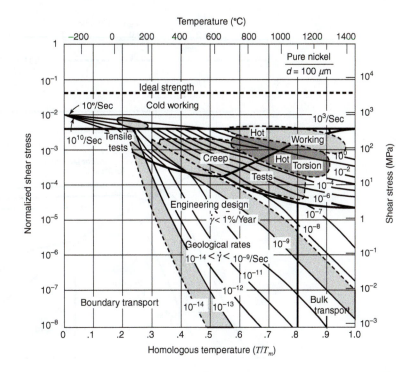

Figure 4.22   Deformation map for 100-μm nickel showing laboratory test regimes relative to deformation fields experienced by the material. (From M. F. Ashby;[46] reprinted with permission of the Institute for Metals.)

one shown in Fig. 4.23, portray the deformation field boundaries at a fixed temperature, where the grain-size dependence of each deformation mechanism is clearly indicated. (Note the lack of grain-size dependence in the dislocation creep region.)

Figure 4.24 provides one final map comparison by showing the effect of nickel-based superalloy (MAR-M200) multiple strengthening mechanisms in shrinking the dislocation climb domain relative to that associated with pure nickel. In addition, the creep strain rates in the stress–temperature region associated with gas turbine material applications are reduced substantially. By combining alloying additions *and* grain coarsening, the isostrain rate contours are further displaced, thereby providing additional creep resistance to the material.[33] In summary, it must be recognized that displacement of a particular boundary resulting from some specific strengthening mechanism does not in itself eliminate an engineering design problem. It may simply shift the rate-controlling deformation process to another mechanism. The materials designer then must suppress the strain rate of the new rate-controlling process with a different flow attenuation mechanism. As such, the multiple strengthening mechanisms built into high-temperature alloys are designed to counteract simultaneously a number of deformation mechanisms much in the same manner as an all-purpose antibiotic attacks a number of bacterial infections that may assault living organisms.

Other studies involving deformation maps have focused on new mechanism portrayal methods.[49–52] For example, deformation maps have been constructed as a function of the creep rate $\dot{\varepsilon}$ versus T_m/T as compared with normalized stress σ/G versus T/T_m diagrams (e.g., Figs. 4.20–4.22). Furthermore, three-dimensional maps have been developed using coordinates of $\dot{\varepsilon}$, T_m/T, and d/b or σ/G, T_m/T, and d/b where d is the grain size and b the atomic diameter[51,52]; as before, these maps identify those regions associated with a dominant deformation mechanism. For example, Fig. 4.25 reveals the individual regions corresponding to six different deformation mechanisms in a high stacking fault energy FCC alloy.[52] Oikawa suggested that $\dot{\varepsilon}$-based diagrams are useful in defining strain-rate conditions associated with

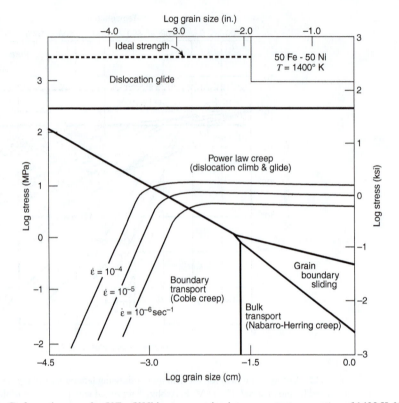

Figure 4.23 Deformation map for 50Fe–50Ni in stress–grain-size space at a temperature of 1400 K. Note inclusion of grain-boundary sliding field. (Courtesy of Michael R. Notis, Lehigh University.)

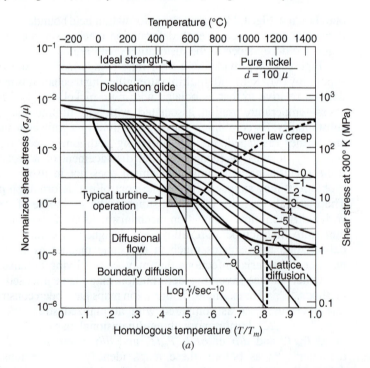

(a)

Figure 4.24 Deformation map for (a) nickel (100 μm), (b) MAR-M200 nickel-based alloy (100 μm), and (c) MAR-M200 (1 cm). Creep rate is suppressed by multiple strengthening mechanisms and grain coarsening. (From M. F. Ashby[46]; reprinted with permission from the Institute of Metals.)

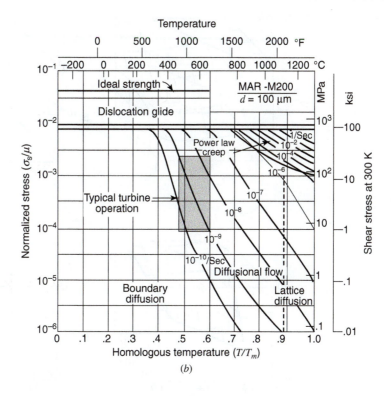

(b)

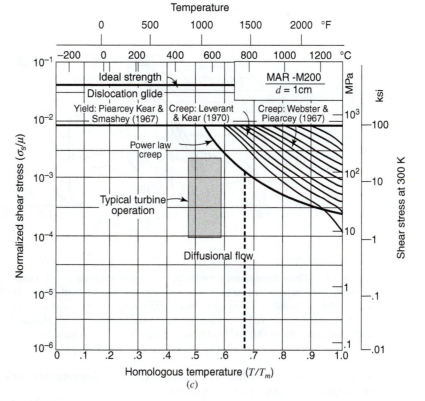

(c)

Figure 4.24   (*Continued*)

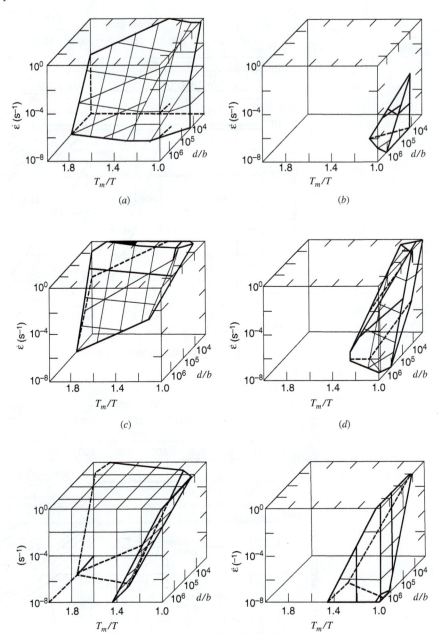

Figure 4.25 Three-dimensional deformation mechanism maps for a high stacking fault energy FCC metal. Each map reveals the conditions associated with a specific deformation mechanism: (*a*) Coble creep; (*b*) Nabarro-Herring creep; (*c*) grain-boundary sliding controlled by grain-boundary diffusion; (*d*) grain-boundary sliding controlled by lattice diffusion; (*e*) power-law creep controlled by dislocation-core diffusion; and (*f*) power-law creep controlled by lattice diffusion. (From Oikawa[52]; with permission from Pineridge Press Ltd.)

easier hot working. On the other hand, σ/G-based diagrams are useful in describing conditions associated with higher creep resistance.

Proceeding in another direction, Ashby and coworkers[53–55] have constructed fracture mechanism maps wherein the conditions for various failure mechanisms are defined. Thus far, fracture mechanism maps have been compiled for various FCC, BCC, and HCP metals and alloys and ceramics.[53-56] For a more detailed study of fracture micromechanisms in metals, ceramics, and engineering plastics, see Sections 5.7, 10.3, and 10.8.

4.7 PARAMETRIC RELATIONS: EXTRAPOLATION PROCEDURES FOR CREEP RUPTURE DATA

It goes without saying that an engineering alloy will not be used for a given elevated temperature application without first obtaining a profile of the material's response under these test conditions. Although this presents no difficulty in short-life situations, such as for the rocket engineer nozzle or military gas turbine blade, the problem becomes monumental when data are to be collected for prolonged elevated temperature exposures, such as those encountered in a nuclear power plant. If the component in question is to withstand 30 or 40 years of uninterrupted service, should there not be data available to properly design the part? If this were done, however, final design decisions concerning material selection would have to wait until all creep tests were concluded. Not only would the laboratory costs of such a test program be prohibitively expensive, but all plant construction would have to cease and the economies of the world would stagnate. In addition, while such tests were being conducted, superior alloys most probably would have been developed to replace those originally selected. Assuming that some of these new alloys were to replace the older alloys in the component manufacture, a new series of long-time tests would have to be initiated. Obviously, nothing would ever be built!

The practical alternative, therefore, is to perform certain creep and/or creep rupture tests covering a convenient range of stress and temperature and then to *extrapolate* the data to the time–temperature–stress regime of interest. A considerable body of literature has been developed that examines parametric relations (of which there are over 30) intended to allow one to extrapolate experimental data beyond the limits of convenient laboratory practice. A textbook[4] on the subject has even been written. Although it is beyond the scope of this book to consider many of these relations to any great length, it is appropriate to consider two of the more widely accepted parameters.

The Larson-Miller parameter is, perhaps, most widely used. Larson and Miller[57] correctly surmised creep to be thermally activated with the creep rate described by an Arrhenius-type expression of the form

$$r = Ae^{-\Delta H/RT} \tag{4-25}$$

where

$r =$ creep process rate
$\Delta H =$ activation energy for the creep process
$T =$ absolute temperature
$R =$ gas constant
$A =$ constant

Equation 4-25 also can be written as

$$\ln r = \ln A - \frac{\Delta H}{RT} \tag{4-26}$$

After rearranging and multiplying by T, Eq. 4-26 becomes

$$\Delta H/R = T(\ln A - \ln r) \tag{4-27}$$

Since $r \propto (1/t)$ (also suggested by Eq. 4-2), Eq. 4-25 can be written as

$$\frac{1}{t} = A'e^{-\Delta H/RT} \tag{4-28}$$

Therefore,

$$-\ln t = \ln A' - \frac{\Delta H}{RT} \tag{4-29}$$

and after rearranging Eq. 4-29, multiplying by T, and converting $\ln t$ to $\log t$,

$$\Delta H/R = T(C + \log t) \tag{4-30}$$

which represents the most widely used form of the Larson-Miller relation. Assuming ΔH to be independent of applied stress and temperature (not always true as demonstrated earlier) the material is thought to exhibit a particular Larson-Miller parameter $[T(C + \log t)]$ for a given applied stress. That is to say, the rupture life of a sample at a given stress level will vary with test temperature in such a way that the Larson-Miller parameter $T(C + \log t)$ remains unchanged. For example, if the test temperature for a particular material with $C = 20$ were increased from 800 to 1000°C, the rupture life would decrease from an arbitrary value of 100 hr at 800°C to 0.035 hr at 1000°C. The value of this parametric relation is shown by examining the creep rupture data in Fig. 4.26, which are the very same data used in Fig. 4.4. The normalization potential of the Larson-Miller parameter for this material is immediately obvious. Furthermore,

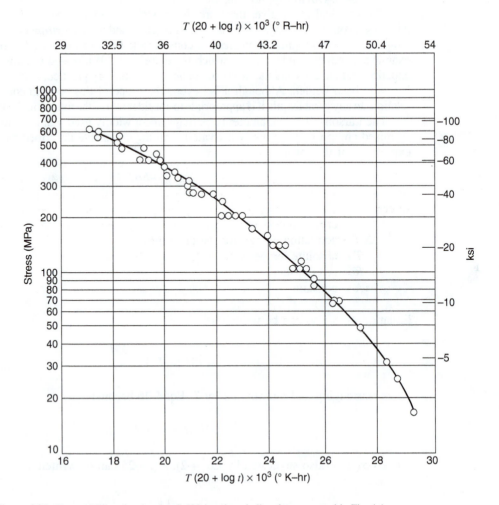

Figure 4.26   Larson-Miller plot showing S-590 iron-based alloy data presented in Fig. 4.4.

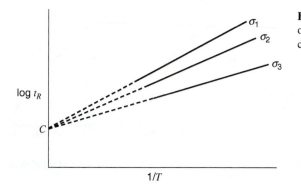

Figure 4.27   Convergence of isostress lines in plot of $\log t_R$ versus $1/T$ to determine magnitude of constant C in Larson-Miller parameter.

long-time rupture life for a given material can be estimated by extrapolating high-temperature, short rupture life response toward the more time-consuming low-temperature, long rupture life regime. It is generally found that such extrapolations to longer time conditions are reasonably accurate at higher stress levels because a smaller degree of uncertainty is associated with this portion of the Larson-Miller plot. Increased extrapolation error is found at lower stress levels where experimental scatter is greater. A comparison between predicted and experimentally determined rupture lives will be considered later in this section.

The magnitude of C for each material may be determined from a minimum of two sets of time and temperature data. Again, assuming $\Delta H/R$ to be invariant and rearranging Eq. 4-30,

$$C = \frac{T_2 \log t_2 - T_1 \log t_1}{T_1 - T_2} \tag{4-31}$$

It is also possible to determine C graphically based on a rearrangement of Eq. 4-30 where

$$\log t = -C + \frac{\text{constant}}{T} \tag{4-32}$$

When experimental creep rupture data are plotted as shown in Fig. 4.27, the intersection of the different stress curves at $1/T = 0$ defines the value of C. It is important to note that not all creep rupture data give the same trends found in Fig. 4.27. For example, isostress lines may be parallel, as shown in Fig. 4.7, for the case of rutile (TiO_2) and other ceramics and metals. Representative values of C for selected materials[57] are given in Table 4.1. For convenience, the constant is sometimes not determined experimentally but instead assumed equal to 20. Note that the magnitude of the material constant C depends on units of time. (Since practically all data reported in the literature give both the

Table 4.1   Material Constants for Selected Alloys[57]

Alloy	C	
	Time, hr	Time, s
Low carbon steel	18	21.5
Carbon moly steel	19	22.5
18–8 stainless steel	18	21.5
18–8 Mo stainless steel	17	20.5
$2\frac{1}{4}$ Cr–1 Mo steel	23	26.5
S-590 alloy	20	23.5
Haynes Stellite No. 34	20	23.5
Titanium D9	20	23.5
Cr–Mo–Ti–B steel	22	25.5

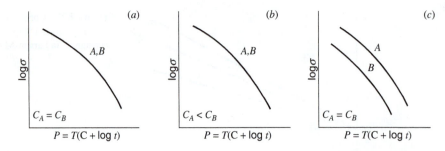

Figure 4.28   Parametric comparison of alloy behavior. (*a*) Alloy A = alloy B; (*b*) and (*c*) alloy A superior to alloy B.

material constant C and the rupture life in more convenient units of hours rather than in seconds—the recommended SI unit for time—test results in this section will be described in units of hours.)

In addition to being used for the extrapolation of data, the Larson-Miller parameter also serves as a figure of merit against which the elevated temperature response of different materials may be compared (e.g., in the case of alloy development studies). For example, when the curves for two materials with the same constant C are coincident, the materials obviously possess the same creep rupture behavior (Fig. 4.28*a*). The same conclusion does not follow, however, when the coincident curves result from materials with different values of C (Fig. 4.28*b*). When $C_A < C_B$, material A would be the stronger of the two. (For the same parameter P, and at the same test temperature, $\log t_{RA}$ for alloy A would have to be greater than $\log t_{RB}$ since $C_B > C_A$.) A direct comparison of material behavior is evident when C is the same but the parametric curves are distinct from one another (Fig. 4.28*c*). Here alloy A is clearly the superior material.

While such alloy comparisons for specified conditions of stress and temperature are possible using the Larson-Miller parameter (and other parameters as well), it should be understood that such parameters provide little insight into the mechanisms responsible for the creep response in a particular time-temperature regime. This is done more successfully by examining deformation maps (Section 4.5). The Sherby-Dorn (SD) parameter $\theta = t_R e^{-\Delta H/RT}$ (where $t = t_R$) described in Eq. 4-8 has been used to compare creep rupture data for different alloys much in the same manner as the Larson-Miller (LM) parameter. Reasonably good results have been obtained with this parameter in correlating high-temperature data of relatively pure metals[10] (Fig. 4.29). The reader should recognize that if the Sherby-Dorn parameter does apply for a given material, then when θ is constant, a plot of the logarithm of rupture life against $1/T$ should yield a series of straight lines corresponding to different stress levels. This is contrary to the response predicted by the Larson-Miller parameter, where the isostress lines converge when $1/T = 0$. The choice of the LM or SD parameters to evaluate a material's creep rupture response would obviously depend on whether the isostress lines converge to a common point or are parallel. In fact, the choice of a particular parameter (recall that over 30 exist) to correlate creep data for a specific alloy is a very tricky matter. Some parameters seem to provide better correlations than others for one material but not another. This may be readily seen by considering Goldhoff's tabulated results[58] for 19 different alloys (Table 4.2). Shown here are root-mean-square (RMS) values reflecting the accuracy of the LM, SD, and other parameters in predicting creep rupture life. The RMS value is defined as

$$\text{RMS} = \left[\frac{\sum (\log \text{ actual time to rupture} - \log \text{ predicted time to rupture})^2}{\text{number of long-time data points}} \right]^{1/2} \tag{4-33}$$

Note that for some metals, either the LM or SD parameter represented the best time–temperature parameter (TTP) of the four examined by Goldhoff and predicted actual test results

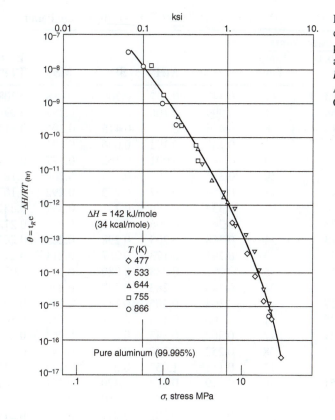

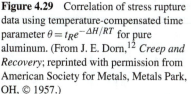

Figure 4.29   Correlation of stress rupture data using temperature-compensated time parameter $\theta = t_R e^{-\Delta H/RT}$ for pure aluminum. (From J. E. Dorn,[12] *Creep and Recovery*; reprinted with permission from American Society for Metals, Metals Park, OH, © 1957.)

most correctly. Alternatively, these two parameters provided poor correlations when compared to other parameters for different materials; the use of the LM or SD parameters in evaluating these alloys led to significant error in the prediction of actual rupture life.

The inconsistency with which a particular TTP predicts actual creep rupture life for different alloys represents a severe shortcoming of the parametric approach to creep design. These deficiencies may be traced in part to some of the assumptions underlying each parameter. For example, the LM and SD parameters are based on the assumption that the activation energy for the creep process is not a function of stress and temperature. Clearly, the test results shown in Fig. 4.9 and the extended discussion in Section 4.4 discredit this supposition. (Recall, however, that when $T \geq 0.5 T_m$, the activation energy for creep is essentially constant and equivalent to the activation energy for self-diffusion.) Furthermore, none of the TTP make provision for metallurgical instabilities.

Attempts are being made to standardize creep data parametric analysis procedures through the establishment of required guidelines by which an investigator arrives at the selection of a particular TTP. In this regard, the minimum commitment method (MCM)[59,60] holds considerable promise in that it presumes initially a very general time–temperature–stress relation. The precise form is obtained on the basis of actual test data. As such, the MCM can lead to the selection of a standard parametric relation, such as LM or SD, or it may define a new parameter that can reflect the possible existence of metallurgical instabilities. Note the reduced RMS values for the MCM method as compared to the LM, SD, or the other two TTP evaluated by Goldhoff (Table 4.2).

Another method, referred to as the graphical optimization procedure (GOP), also has been used to improve the accuracy of life predictions based on various extrapolation procedures.[61,62] To illustrate this point, Woodford employed the GOP to demonstrate that the material constant C used in the Larson-Miller parameter was a function of rupture life. For example, he found for

Table 4.2 Comparative RMS Values Reflecting Accuracy of Different Time–Temperature Parameters[58]

Data Set	Alloy	Data Points Short-Time	Long-Time	LM[a]	MH[b]	SD[c]	MS[d]	Best TTP[e]	MCM[f]
1	Al 1100-0	53	11	0.347	0.377	0.308	0.488	0.308	0.260
2	Al 5454-0	68	7	0.099	0.166	0.143	0.287	0.099	0.081
4	Carbon steel	18	8	0.456	0.313	0.415	0.396	0.313	0.084
5	Cr-Mo steel	23	10	0.152	0.102	0.056	0.191	0.056	0.122
6	Cr-Mo-V steel	17	9	0.389	0.091	0.162	0.477	0.091	0.102
7A	304 stainless steel	33	19	0.375	0.207	0.185	0.309	0.185	0.194
7B	304 stainless steel	41	11	0.454	0.167	0.272	0.292	0.167	0.179
8	304 stainless steel	26	13	0.334	0.349	0.237	0.457	0.237	0.228
9	316 stainless steel	28	10	0.244	0.296	0.212	0.323	0.212	0.073
11A	347 stainless steel	18	24	0.368	0.203	0.298	0.265	0.203	0.123
11B	347 stainless steel	31	13	0.291	0.173	0.267	0.211	0.173	0.107
12	A-286	19	5	0.097	0.338	0.089	0.111	0.089	0.220
13	Inco 625	78	21	0.343	0.283	0.337	0.329	0.283	0.317
14	Inco 718	17	9	0.104	0.565	0.110	0.100	0.100	0.084
15	René 41	26	11	0.106	0.144	0.139	0.113	0.106	0.131
16	Astroloy®	21	12	0.302	0.343	0.231	0.264	0.231	0.107
17A	Udimet 500	65	38	0.252	0.342	0.316	0.348	0.252	0.268
17B	Udimet 500	93	12	0.111	1.057	0.247	0.173	0.111	0.124
18A	L-605	51	49	0.319	0.652	0.420	0.261	0.261	0.247
18B	L-605	76	28	0.374	0.641	0.460	0.305	0.305	0.290
19	Al 6061-T651	74	25	0.361	0.382	0.217	0.473	0.217	0.311
Average of above 21 data sets				0.280	0.342	0.244	0.294	0.190	0.174
Average excluding B data sets				0.273	0.303	0.228	0.305	0.191	0.174

[a] Larson-Miller.
[b] Manson-Haferd parameter.
[c] Sherby-Dorn parameter.
[d] Manson-Succop parameter.
[e] Time-temperature parameter.
[f] Minimum commitment method.

the case of IN718 nickel-based alloy that C varied from 27.1 at short lives to 20 at a 10,000 rupture hour.[61,62] By utilizing the correct time-dependent value of C in the Larson-Miller formula, less scatter was observed in the data normalization procedure.

4.8 MATERIALS FOR ELEVATED TEMPERATURE USE

From the previous discussions, a material suitable for high-temperature service should possess a high melting point and modulus of elasticity, and low diffusivity. In addition, such materials must possess a combination of superior creep strength, thermal fatigue resistance, and oxidation and hot corrosion resistance. As a result, alloy development has focused primarily on nickel- and cobalt-based superalloys, with earlier iron-based alloys being replaced because of their relatively low melting point and high diffusivity.[63−68] These high-temperature alloys have been produced by several methods including casting, mechanical forming, powder metallurgy, directional solidification of columnar and single crystals, and mechanical alloying.

For the case of nickel-based superalloys, constituent elements are introduced to enhance solid solution properties, as precipitate and carbide formers, and as grain-boundary and free surface stabilizers.[69] Tungsten (W), molybdenum (Mo), and titanium (Ti) are very effective solid solution

Figure 4.30   Electron micrographs revealing Ni₃Al precipitates (γ') in a nickel solid solution (γ) matrix. (*a*) Cubic form in MAR M-200. (*b*) Rafted morphology in Ni-14.3Mo-6Ta-5.8Al (Alloy 143). Tensile stress axis is in vertical direction and parallel to [001] direction. Creep tested with 210 MPa at 1040C.[73] (Courtesy of E. Thompson.)

strengtheners; W and Mo also serve to lower the diffusion coefficient of the alloy. (There is a general inverse relation between the melting point and alloy diffusivity.) Though the incremental influence of chromium (Cr) on solid solution strengthening is small (i.e., $d\tau/dc$ is low), the overall solid solution strengthening potential of Cr in nickel (Ni) alloys is large since large amounts of Cr can be dissolved in the Ni matrix. Cobalt (Co) provides relatively little solid solution strengthening but serves to enhance the stability of the submicron-size Ni₃(Al,X) (γ') precipitates within the nickel solid solution (γ) matrix (Fig. 4.30*a*). Within the γ' phase, X corresponds to the presence of Ti, niobium (Nb), or tantalum (Ta). The difficulty of dislocation motion through the ordered γ' particles in these alloys is responsible for their high creep strength at elevated temperatures. Of particular note, the γ' phase exhibits unusual behavior in that strength *increases* by three-to sixfold with increasing temperature from ambient to approximately 700°C.[70-72]

Also noteworthy is the fact that γ' precipitates in single-crystal alloys tend to coarsen under stress at 1000°C and form thin parallel plate-like arrays that are oriented normal to the applied stress axis (Fig. 4.30*b*). Studies have confirmed that alloy creep resistance is enhanced by the development of this "rafted" microstructure[73,74], it is believed that the absence of dislocation climb around the γ' particles, due to their lenticular shape, forces dislocations to cut across the ordered γ' phase. As noted in Section 3.6.2, this dislocation path enhances the alloy's resistance to plastic flow.

The presence of carbides along grain boundaries in polycrystalline alloys serves to restrict grain-boundary sliding and migration. Carbide formers such as W, Mo, Nb, Ta, Ti, Cr, and vanadium (V) lead to the formation of M₇C₃, M₂₃C₆, M₆C, and MC, with MC carbides being most stable (e.g., TiC). When Cr levels are relatively high, Cr₂₃C₆ particles are formed.

Surface stabilizers include Cr, Al, boron (B), zirconium (Zr), and hafnium (Hf). The presence of Cr in solid solution allows for the formation of Cr₂O₃, which reduces the rate of oxidation and hot corrosion. Aluminum contributes to improved oxidation resistance and resistance to oxide spalling. Finally, B, Zr, and Hf are added to impart improved hot strength, hot ductility, and rupture life.[75] Cobalt-based alloys derive their strength from a combination of solid solution hardening and carbide dispersion strengthening. The mechanical properties of representative nickel-based and cobalt-based alloys are given in Table 4.3; references 63 to 68 provide additional information concerning these materials.

More recent efforts to improve the high-temperature performance of superalloys have tended more toward optimizing component design and making use of advanced processing techniques rather than tinkering with alloy chemistry.[76] One such technique involves the directional solidification of conventional superalloys to produce either highly elongated grain

Table 4.3 Mechanical Properties of Selected Superalloys

Alloy Designation	Yield Strength [MPa(ksi)]			100-hr Rupture Strength [MPa(ksi)]		1000-hr Rupture Strength [MPa(ksi)]	
	21°C (70°F)	760°C (1400°F)	982°C (1800°F)	760°C (1400°F)	982°C (1800°F)	760°C (1400°F)	982°C (1800°F)
Cast Alloys							
B1900	825 (120)	808 (117)	415 (60)	505 (73)[a]	170 (25)	380 (55)[a]	105 (15)
IN-100	850 (123)	860 (125)	370 (54)	625 (91)	170 (25)	515 (75)	105 (15)
MAR-M-200	840 (122)	840 (122)	470 (68)	635 (92)	179 (26)	580 (84)	130 (18.5)
MAR-M-200(DS)[b]	860 (125)	925 (134)	620 (90)	725 (105)	200 (29)	660 (96)	140 (20)
TRW-NASA VI A	940 (136)	945 (137)	520 (75)	725 (105)[c]	215 (31)	585 (85)	140 (20)
MAR-M 509	570 (83)	365 (53)	180 (26)	345 (50)	105 (15)	260 (38)	79 (11.5)
Wrought Alloys							
Astroloy	1050 (152)	910 (132)	275 (40)	540 (78)	105 (15)	430 (62)	55 (8.0)
Hastelloy X	360 (52)	260 (38)	110 (16)	145 (21)	26 (3.8)	100 (15)	14 (2.0)
Waspalloy	795 (115)	675 (98)	140 (20)	415 (60)	45 (6.5)	290 (42)	—
ODS Alloys							
MA 6000	1069 (155)	781 (113)	344 (50)	485 (70)	210 (30)	410 (59)	180 (26)
Alloy 51	903 (131)	972 (141)	517 (75)	600 (87)	221 (32)	469 (68)	186 (27)

[a] Data corresponds to 816°C (1500°F).
[b] Directionally solidified.
[c] Extrapolated values.
[d] Data courtesy of Inco Alloys Inc.

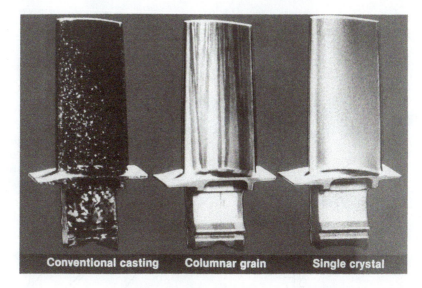

Figure 4.31   Conventional and directional solidification used to prepare gas turbine blades with equiaxed, columnar, and single-crystal morphologies. (F. L. VerSnyder and E. R. Thompson, *Alloys for the 80's,* R. Q. Barr, Ed., Climax Molybdenum Co., 1980, p. 69; with permission.)

boundaries or single-crystal components (Fig. 4.31). Helical molds are used to cast single-crystal turbine blades; multiple grains form initially and grow into the helical section of the mold. The faster growing ⟨100⟩-oriented grains then crowd out other grains until a single ⟨100⟩ grain is left to fill the mold cavity.[77–79] Current sophisticated mold designs now allow for the simultaneous growth of two turbine blades from the same single crystal.[79] The alignment of airfoils (turbine blades) along the ⟨100⟩ axis parallel to the centrifugal stress direction allows for a 40% reduction in the elastic modulus and associated lower plastic strain range during thermal fatigue cycling; a 6- to 10-fold improvement in thermal fatigue resistance is thus achieved. Since grain boundaries are eliminated, their influence on grain-boundary sliding, cavitation, and cracking is obviated.[77,78] Furthermore, it is no longer necessary to add such elements as hafnium, boron, carbon, and zirconium for the purpose of improving grain-boundary hot strength and ductility.[80] Without these elements, the incipient melting temperature of the alloy is increased by approximately 120°C and the alloy chemistry simplified. The development of cast superalloy turbine blades is shown in Fig. 4.32*a*; the relative ranking of the rupture lifetime for equiaxed and columnar polycrystalline alloys is compared with that of single-crystal alloys

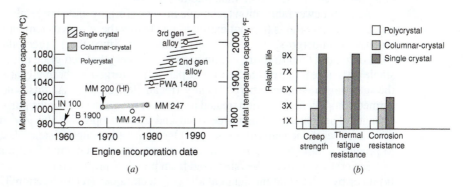

Figure 4.32   (*a*) Development of turbine blade temperature capability. (*b*) Comparative high temperature strength and corrosion resistance of equiaxed, columnar, and single-crystal superalloys.[79] (Reprinted with permission from *Journal of Metals,* 39(7), 11 (1987), a publication of the Metallurgical Society, Warrendale, PA. 15086.)

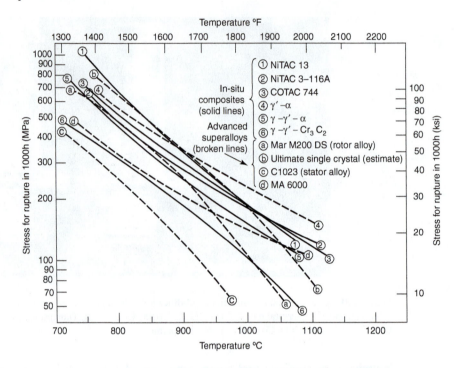

Figure 4.33 1000-hr strength as a function of temperature in eutectic superalloys and conventional directionally solidified single-crystal and oxide-dispersion-strengthened superalloys. In situ (eutectic) composites reveal generally superior stress rupture behavior. (From Lemkey[81]; reprinted by permission of the publisher from F. D. Lemkey, *Proceedings,* MRS Conference, CISC IV, Vol. 12, F. D. Lemkey, H. E. Cline, and M. McLean, Eds., copyright by Elsevier Science Publishing Co., Inc., Amsterdam, © 1982.)

in Fig. 4.32*b*. By applying unidirectional solidification to alloys of eutectic composition, it has been possible to produce eutectic composite alloys possessing properties superior to those found in conventional superalloys[81] (Fig. 4.33). A number of these alloys contain a γ/γ' matrix that is reinforced with high-strength whiskers of a third phase; these strong filamentary particles are oriented parallel to the maximum stress direction. Although the properties of these alloys are very good, the allowable solidification rates for their manufacture are much lower than those permissible in the manufacture of directionally solidified columnar or single-crystal micro-structures. One is then faced with a trade-off between the superior properties of eutectic composites and their higher manufacturing costs.

Another newer fabrication technique involves forging under superplastic conditions.[82] In this process, the material is first hot extruded just below the γ' solvus temperature, which causes the material to undergo spontaneous recrystallization. Since the γ' precipitates in the nickel solid solution matrix tend to restrict grain growth, the recrystallized grain diameter remains relatively stable in the size range of 1 to 5 μm. The part is then forged isothermally at a strain rate that enables the material to deform superplastically (recall Section 4.5). At this point, the superplastically formed component is solution treated to increase the grain size for the purpose of enhancing creep strength. The material is then quenched and aged to optimize the γ/γ' microstructure and the associated set of mechanical properties. One major advantage of superplastic forging is its ability to produce a part closer to its final dimensions, thereby reducing final machining costs.

Superalloys also can be fabricated from powders produced by vacuum spray atomization of liquid or by solid-state mechanical alloying techniques (recall Section 3.7). Powders then may be placed in a container that is a geometrically larger version of the final component shape. The can then is heated under vacuum and hydrostatically compressed to yield a fully dense component with dimensions close to the design values. The microstructure of hot isostatically

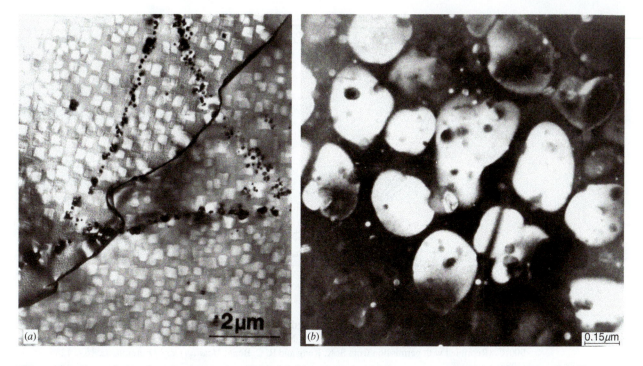

Figure 4.34 Transmission electron micrographs of P/M nickel-based alloys. (*a*) Microstructure of HIP'd Astroloy superalloy. Note persistent necklaces of prior particle boundary borides, carbides, and oxides.[83] (Reprinted with permission from J. S. Crompton and R. W. Hertzberg, *J. Mater Sci.,* 21, 3445 (1986), Chapman & Hall Pub.) (*b*) Microstructure of MA 6000 showing γ' precipitates (large light areas) and Y_2O_3 dispersoids (small dark regions). [(Photo courtesy of W. Hoffelner from W. Hoffelner and R. F. Singer, *Metallurgical Transactions* 16A, 393 (1985).)]

pressed (HIP) Astroloy superalloy is shown in Fig. 4.34*a*.[83] Note the persistence of the necklace of prior particle boundary borides, carbides, and oxides that surround the atomized powder particles. Hot isostatic pressing also is being used to heal defects in conventionally cast parts and to heal certain defects in parts that experience creep damage in service.

With significant additions of γ' formers, such as Al and Ti, mechanically alloyed oxide-dispersion-strengthened (MA/ODS) products possess attractive strength levels over a broad temperature range.[84,85] Two such alloys are MA6000 and Alloy 51, which contain approximately 55 v/o and 75 v/o γ', respectively (Fig. 4.34*b*).[84,85] The 1000-hr rupture strength (normalized with respect to density) of these alloys and others is shown in Fig. 4.35 as a function of temperature. As expected, directionally solidified (DS MAR-M200) and single-crystal (PWA 1480) cast alloys are superior to the two mechanically alloyed products at temperatures up to 900°C with the relative rankings being reversed above this temperature. At high temperatures near the γ' solvus temperature, the γ' particles that dominate the precipitation hardening process tend to coarsen and/or go back into solution. The superiority of MA materials relative to that of directionally solidified and single-crystal cast alloys at temperatures in excess of 900°C is due to the oxide-dispersion-strengthening influence of the Y_2O_3 particles that remain in the microstructure and do not coarsen to any significant degree.

Much attention has focused on the unusual creep rate and rupture-life stress dependence of ODS alloys (introduced in Section 3.7). Whereas most pure metals and associated solid solutions reveal a σ^{4-5} dependence of $\dot\varepsilon$ (recall Eq. 4-16 and 4-21), the steady-state creep rate in ODS alloys exhibits a stress dependency of 20 or more.[70,84,86] Furthermore, the apparent activation energy for the creep process is found to be two to three times greater than the activation energy for self-diffusion. Tien and coworkers[70,86] have suggested that these apparent differences in creep response can be rationalized by considering creep to be dominated by an *effective* stress rather than the applied stress; the effective stress is defined as the applied

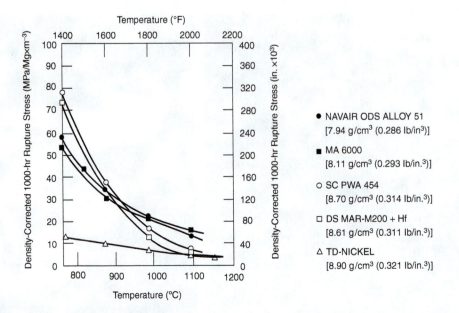

Figure 4.35 Comparison of 1000-hr rupture strength (density corrected) in directionally solidified and oxide-dispersion-strengthened nickel-based superalloys.[85] Note superior properties of ODS alloys at temperatures above 900°C. (Reprinted with permission from S. K. Kang and R. C. Benn, *Metallurgical Transactions,* 16A, 1285 (1985).)

stress minus a back stress that reflects dislocation interactions with Y_2O_3 dispersion strengthening particles. When the applied stress level is replaced by the effective stress value in Eq. 4-21, the stress dependency of $\dot{\varepsilon}_s$ and the apparent activation energy for creep are found to be similar to those values corresponding to pure metals (i.e., $n \sim 4$–5 and $\Delta H_c \sim \Delta H_{SD}$).

In corresponding fashion, the rupture life of ODS alloys can reveal a very strong applied stress dependency and an *upward* slope change with increasing rupture lifetime, opposite to that observed in many other alloys (e.g., recall Fig. 4.4). Figure 4.36 reveals that MA6000 and Alloy 51 exhibit two regions of behavior; Region I corresponds to high stress levels and intermediate temperatures and is dominated by the γ' precipitates. At higher temperatures, lower stress levels and longer times (Region II), stress rupture is dominated by the Y_2O_3 dispersoid phase. Note that ODS alloy MA754, which contains no γ' phase, does not exhibit Region I behavior; conversely, cast alloy IN939, which contains no dispersion strengthening phase, exhibits no Region II behavior. Recent studies have sought to clarify the nature of the dislocation–dispersoid particle interaction so as to better understand the unique phenomenological behavior of ODS alloys.[87]

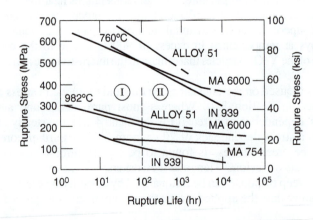

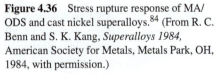

Figure 4.36 Stress rupture response of MA/ODS and cast nickel superalloys.[84] (From R. C. Benn and S. K. Kang, *Superalloys 1984,* American Society for Metals, Metals Park, OH, 1984, with permission.)

Finally, fiber-reinforced superalloys also make interesting candidate materials for structural use at elevated temperatures. Tungsten fibers hold promise as a suitable reinforcement for superalloys in that they possess superior high-temperature strength and creep resistance.[88] In addition, a good interface is developed between the superalloy matrix and the tungsten fibers without excessive surface reactions that degrade W-fiber mechanical properties. Preliminary studies have shown that operating temperatures of fiber-reinforced superalloys may be increased by 175°C over that of unreinforced superalloys.

While alloy development continues, there are other paths to achieving greater operating temperatures for metal components. For example, cooling channels integrated into a gas turbine engine blade can be used to extract heat. This allows the superalloy component to run at an effective temperature much lower than that of the surrounding environment, thereby improving its creep performance. Enormous gains in turbine engine operating temperature have also been achieved through the use of ceramic coatings that insulate the metal from the surrounding combustion gas. These *thermal barrier coatings* (TBCs) can reduce superalloy turbine blade surface temperatures by as much as 125–250°C. Because there can be a significant difference between the thermal expansion of the ceramic TBC layer and the underlying metal, the tendency for spallation due to thermally induced strains must be suppressed. As a result, porous yttria-stabilized zirconia (YSZ) is commonly used. However, this layer does not protect the underlying superalloy from oxidation damage. In the end, no single material can provide an optimum combination of both thermal insulation and oxidation resistance, so layered TBC systems have been developed that consist of a metallic bondcoat layer adhered directly to the superalloy, a very thin thermally grown aluminum oxide layer on top of the bondcoat, and a porous YSZ layer on the outside surface. The bondcoat is an *oxidation barrier coating* made up of MCrAlY (where M = Ni, Co, and Fe) or PtNiAl; surface coatings with such compositions promote the natural formation and retention of a thin layer of Al_2O_3, which serves as an effective barrier to the diffusion of oxygen into the component interior.[79] The bondcoat layer also serves to bond the YSZ layer securely to the superalloy.

Given that ceramic coatings can survive gas temperatures that would destroy uncoated metals, researchers looking beyond metal superalloys have focused attention on the development of a gas turbine engine using components made entirely of ceramic materials. Ceramics often possess higher melting points, higher moduli of elasticity, and lower diffusivities than metal systems, so they offer considerable potential in such applications. Unfortunately, monolithic ceramics such as SiC and Si_3N_4 suffer from low ductility and brittle behavior in tension (see Table 7.8). This serious problem must be resolved before the ceramic engine can become a reality. Significant progress toward this end has been made with the development of continuous ceramic fiber reinforced ceramic matrix composites. In particular, materials such as melt-infiltrated SiC/SiC composites exhibit many promising characteristics for gas turbine applications including relatively high creep rupture resistance and thermal conductivity, as well as enhanced thermal shock and oxidation resistance, compared to many other ceramic-based materials. Two significant challenges that affect creep performance in these materials are the tendency for matrix cracking and time-dependent crack growth that is exacerbated by oxidation, and time-dependent degradation of the fiber strength associated with creep-controlled flaw growth in the fiber material.[vi] The mechanisms behind the improved fracture properties of ceramic matrix composites are discussed in detail in Section 7.5.1.

4.9   VISCOELASTIC RESPONSE OF POLYMERS AND THE ROLE OF STRUCTURE

As discussed in Section 4.1, polymers are likely to exhibit significant viscoelastic behavior over a wide range of common operating temperatures. Thus, it is possible for a polymer to undergo *viscoelastic creep* or *viscoelastic stress relaxation*, unlike metals or ceramics for which such behavior is almost invariably viscoplastic. If the polymer has a significant amorphous volume fraction and is not heavily cross-linked, a viscoplastic component of the total strain is also likely, particularly at higher temperatures.

[vi] G. N. Morscher et al., AFRL Technical Report AFRL-RX-WP-TP-2009-4053, 2007.

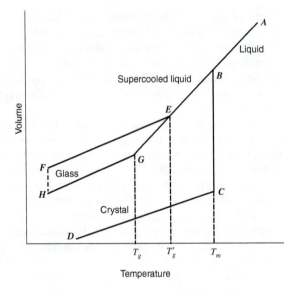

Figure 4.37   Change in volume as function of temperature. Crystalline melting point at $T_{m'}$, glass transition temperature at $T_{g'}$, and excess free volume *FH*.

The source of this strong tendency for time-dependent deformation on the part of polymer materials can be found at the molecular level. At all temperatures above absolute zero, the existing thermal energy causes the polymer chains to vibrate and wriggle about. First, small-scale vibrations are permitted. Then, with increasing temperature, molecule segments begin to move more freely. Finally, at sufficiently high temperatures associated with the molten state, entire chains are free to move about. It is seen from Fig. 4.37 that these large-amplitude molecular vibrations cause the polymer to become less dense. If crystallization is likely for the particular type of polymer in question, upon cooling the material undergoes a first-order transformation at *B* associated with the melting point T_m. Heat of fusion is liberated and the specific volume drops abruptly to *C*. Further cooling involves additional change in the specific volume (*D*) as molecular oscillations become increasingly restricted. When crystallization does not occur in the polymer, the liquid cools beyond T_m (location *B*) without event. However, a point is reached where molecular motions are highly restricted and the individual chains are no longer able to arrange themselves in equilibrium configurations within the supercooled liquid. Below this point (*G*) (the glass transition temperature T_g), the material is relatively frozen into a glassy state. The change from a supercooled liquid to glass represents a second-order transformation that does not involve a discrete change in specific volume or internal heat. From Fig. 4.37, it is seen that the polymer in the amorphous state *occupies more volume* than in the crystalline form. This is to be expected, since higher density forms of a particular polymer are associated with greater crystallinity as a result of greater chain-packing efficiency (recall Table 2.15). The relative difference in chain-packing density can be described in terms of the fractional unoccupied volume (the *free volume*) given by Litt and Tobolsky[89] as

$$\bar{f} = \frac{v_a - v_c}{v_a} = 1.0 - \left(\frac{d_a}{d_c}\right) \tag{4-34}$$

where
$$\bar{f} = \text{fractional unoccupied free volume}$$
$$v_a, d_a = \text{specific volume and density of amorphous phase}$$
$$v_c, d_c = \text{specific volume and density of crystalline phase}$$

For many polymers, $0.01 < \bar{f} < 0.1$. Greater free volume enables the molecules to slide more easily past one another.

Table 4.4   Comparison of Typical Creep Behavior in Metals, Ceramics, and Polymers

Creep Behavior	Metals and Ceramics	Polymers
Linear elastic	No	Sometimes
Recoverable	No	Partially
Temperature range	Temperatures above $\sim 0.3\, T_m$ (metals) or $\sim 0.5\, T_m$ (ceramics)	All temperatures above approximately $-200°C$

Since the glass transition occurs where molecular and segmental molecular motions are restricted, it is sensitive to cooling rate. Consequently, a polymer may not exist at its glassy equilibrium state. Instead, nonequilibrium cooling rates could preclude the attainment of the lowest possible free volume in the amorphous polymer. In Fig. 4.37, this would correspond to line *EF* with the glass transition temperature increasing to T_g'. Petrie[90] describes the difference between the equilibrium and actual glassy free volume as the *excess free volume* and postulates that this quantity is important in understanding the relation between polymer properties and their thermodynamic state. Note that the free volume will differ from one polymer to another; within the same polymer, the excess free volume is sensitive to thermal history.

In metals and ceramics, only the grain boundary regions can be considered to have significant free volume. As such, there is much less freedom for rearrangement at the atomic scale unless the material has extremely small grains (at the nanometer scale) and/or is operating at very high temperature. This goes a long way to explaining the greater tendency for time-dependent deformation in polymers. Also, as pointed out in Section 1.3.3.3, amorphous high-molecular-weight polymer chains are highly kinked in the unloaded state. When a chain is straightened under load, there is a strong entropic driving force to rekink it once the load is removed. This provides a driving force for viscoelastic strain recovery in amorphous polymers that is absent in metals and ceramics. In light of these differences, a comparison of creep behavior between metals and polymers is summarized in Table 4.4.

4.9.1 Polymer Creep and Stress Relaxation

In many circumstances, the viscoelastic response of a polymer exhibits a set of character-istics that together are called *linear viscoelasticity*. When the elastic strains and viscous flow rate are small (approximately 1 to 2% and $0.1\ \mathrm{s}^{-1}$, respectively), the viscoelastic strain may often be approximated by

$$\varepsilon = \sigma \cdot f(t) \tag{4-35}$$

That is, the stress–strain ratio is a function of time only, and no unique elastic modulus exists. This response can be described by the simple addition of *linear elastic* and *linear viscous* (Newtonian) flow components. When the stress–strain ratio of a material varies with time and *stress*

$$\varepsilon = g(\sigma, t) \tag{4-36}$$

the response is *nonlinear viscoelastic*.

On the basis of the simple creep test it is possible to define a linear viscoelastic *creep modulus* $E_c(t)$ or its inverse, a *creep compliance* $J_c(t)$, such that

$$E_c(t) = \frac{\sigma_0}{\varepsilon(t)} \tag{4-37a}$$

$$J_c(t) = \frac{\varepsilon(t)}{\sigma_0} \tag{4-37b}$$

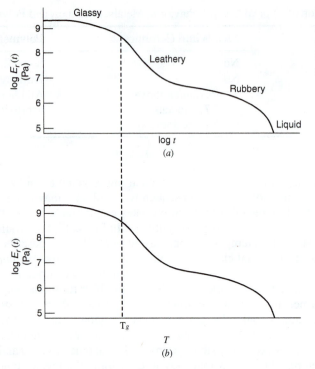

Figure 4.38 Time–temperature dependence of elastic modulus in thermoplastic polymeric solids: (*a*) change in relaxation modulus $E_r(t)$ as function of time; (*b*) change in tensile modulus as function of temperature.

where

$$\sigma_0 = \text{constant applied stress}$$
$$\varepsilon(t) = \text{time-dependent strain}$$

Likewise, in a stress relaxation test where the strain ε_0 is fixed and the associated stress is time dependent, a relaxation modulus $E_r(t)$ may be defined[vii]

$$E_r(t) = \frac{\sigma(t)}{\varepsilon_0} \tag{4-38}$$

The reader may recognize that these creep moduli are extracted from the time-dependent behavior shown in Fig. 4.1. They can be plotted against log time to reveal their strong time dependence, as shown schematically in Fig. 4.38*a* for $E_r(t)$. (For small strains and up to moderate temperatures, corresponding to linear viscoelastic behavior, $E_r \approx E_c$.) It is clear that material behavior changes radically from one region to another. For very short times, the relaxation modulus approaches a maximum limiting value where the material exhibits glassy behavior associated with negligible molecule segmental motions. At longer times, the material experiences a transition to leathery behavior associated with the onset of short-range molecule segmental motions. At still longer times, complete molecule movements are experienced in the rubbery region associated with a further drop in the relaxation modulus. Beyond this point, liquid flow occurs.

It is interesting to note that the same type of curve may be generated by plotting the modulus (from a simple tensile test) against test temperature (Fig. 4.38*b*; recall also Fig. 1.5). In this instance, the initial sharp decrease in E from its high value in the glassy state occurs at T_g. The shape of this curve can be modified by structural changes and polymer additions. For example, the entire curve is shifted downward and to the left as a result of plasticization (Fig. 4.39*a*). As $\bar{M}$ increases, the rubbery flow region is displaced to longer times (Fig. 4.39*b*), because molecular and

---

[vii] The notation $G(t)$ is often used for the stress relaxation modulus, but there is a danger of confusing it with the shear modulus G as defined in Chapter 1.

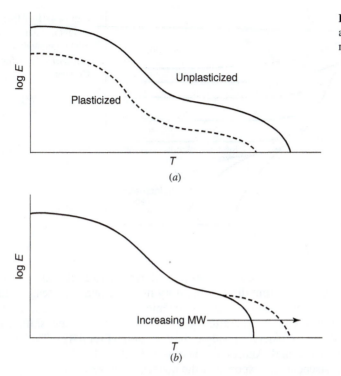

Figure 4.39   Effect of (*a*) plasticization and (*b*) molecular weight on elastic modulus as a function of temperature.

segmental molecular movements are suppressed when chain entanglement is increased. Molecular weight has relatively little effect on the onset of the leathery region, since T_g is relatively independent of $\bar{M}$ except at low $\bar{M}$ values (Fig. 4.40). The effect of $\bar{M}$ on T_g is believed to be related to the chain ends.[91] Since the ends are freer to move about, they generate a greater than average amount of free volume. Adjacent chains are then freer to move about and contribute to greater mobility of the polymer. Since the chain ends are more sensitive to $\bar{M}_n$ than $\bar{M}_w$, T_g is best correlated with the former measure of molecular weight. The leathery region is greatly retarded by cross-linking, while the flow region is completely eliminated, the latter being characteristic of thermosetting polymers (Fig. 4.41).

The temperature–time (i.e., strain rate^{-1}) equivalence seen in Fig. 4.38 closely parallels similar observations made earlier in this chapter. It is seen that the same modulus value can be obtained either at low temperatures and long times or at high test temperatures but short times. In fact, this equivalence is used to generate E_r versus $\log t$ curves as shown in Fig. 4.38*a*. The

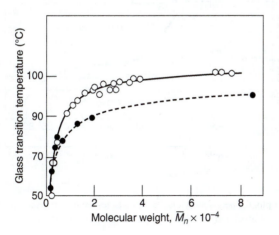

Figure 4.40   Glass transition temperature in PMMA (○) and polystyrene (●) as a function of $\bar{M}_n$. (M. Miller, *The Structure of Polymers,* © 1966 by Litton Educational Publishing by permission of Van Nostrand Reinhold.)

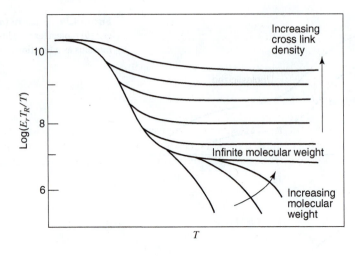

Figure 4.41 Effect of molecular weight and degree of cross-linking on relaxation modulus.[94] (Reprinted with permission from McGraw-Hill Book Company.)

reader should appreciate that since such plots extend over 10 to 15 decades of time, they cannot be determined conveniently from direct laboratory measurements. Instead, relaxation data are obtained at different temperatures over a convenient time scale. Then, after choosing one temperature as the reference temperature, the remaining curves are shifted horizontally to longer or shorter times to generate a single master curve (Fig. 4.42). This approach was first introduced by Tobolsky and Andrews[92] and was further developed by Williams et al.[93] Assuming that the viscoelastic response of the material is to be controlled by a single function of temperature (i.e., a single rate-controlling mechanism), Williams, Landel, and Ferry[93] developed a semiempirical relation for an amorphous material, giving the time shift factor a_T as

$$\log a_T = \log \frac{t_T}{t_{T_0}} = \frac{C_1(T - T_0)}{C_2 + T - T_0} \qquad (4\text{-}39)$$

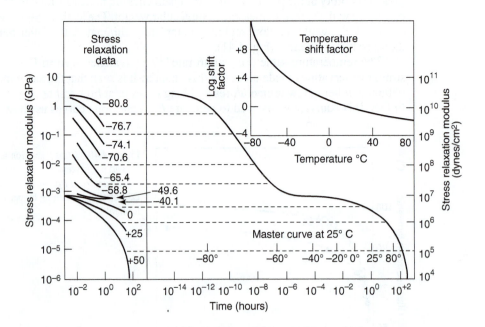

Figure 4.42 Modulus–time master plot for polyisobutylene based on time–temperature superposition of data to a reference temperature of 25°C. (From Catsiff and Tobolsky,[95] with permission from John Wiley & Sons, Inc.)

where

a_T = shift factor that is dependent on the difference between the reference and data temperatures $T - T_0$

t_T, t_{T_0} = time required to reach a specific E_r at temperatures T and T_0, respectively

C_1, C_2 = constants dependent on the choice of the reference temperature T_0

T = test temperatures where relaxation data were obtained, K

This relation is found to hold in the temperature range $T_g < T < T_g + 100\,\text{K}$, but is sometimes used beyond these limits on an individual basis as long as time–temperature superposition still occurs. This would indicate that the same rate-controlling processes were still operative. Two reference temperatures are often used to normalize experimental data—T_g and $T_g + 50\,\text{K}$—for which the constants C_1 and C_2 are given in Table 4.5.

Table 4.5   Constants for WLF Relationship

Reference Temperature	C_1	C_2
T_g	−17.44	51.6
$T_g + 50\,\text{K}$	−8.86	101.6

The shift function may be used to normalize creep data,[96] enabling this information to be examined on a single master curve as well. Furthermore, by normalizing the creep strain results relative to the applied stress σ_0, the normalization of both axes converts individual creep–time plots into a master curve of creep compliance versus adjusted time (Fig. 4.43). These curves can be used to demonstrate the effect of MW and degree of cross-linking on polymer mechanical response much in the manner as the modulus relaxation results described in Figs. 4.39 and 4.41. Note that viscous flow is eliminated and the magnitude of the creep compliance reduced with increasing cross-linking in thermosetting polymers. For the thermoplastic materials, compliance decreases with increasing viscosity, usually the result of increased MW.

As previously noted (e.g., see Eqs. 4-37 and 4-38), the elastic modulus of engineering plastics varies with time as a result of time-dependent deformation. For this reason, the designer of a plastic component must look beyond basic tensile data when computing the deformation response of a polymeric component. For example, if a designer were to limit component strain to less than some critical value ε_c, the maximum allowable stress would be given by $E\varepsilon_c$ so long as the material behaved as an ideally elastic solid. Since most engineering plastics experience creep, the level of strain in the component would increase with time as noted by the creep curves in Fig. 4.44a. To account for this additional deformation, designers often make use of *isochronous* stress–strain curves, which are derived from such creep data (e.g., see Fig. 4.44a line *XY*). Figure 4.44b shows three isochronous stress–strain curves corresponding to loading times of 10^2, 10^4, and 10^6 s, respectively. To illustrate the use of these curves, we see that to limit the strain in a component to no more than 0.02 after 10^4 s, the allowable stress must not exceed 32 MPa.

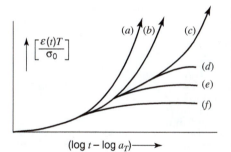

Figure 4.43   Master creep curve revealing effect of increasing MW ($a \rightarrow b \rightarrow c$) and degree of cross-linking ($d \rightarrow e \rightarrow f$) on creep strain. (T. Alfrey and E. F. Gurnee, *Organic Polymers*, © 1967. Reprinted by permission of Prentice-Hall Inc., Englewood Cliffs, NJ.)

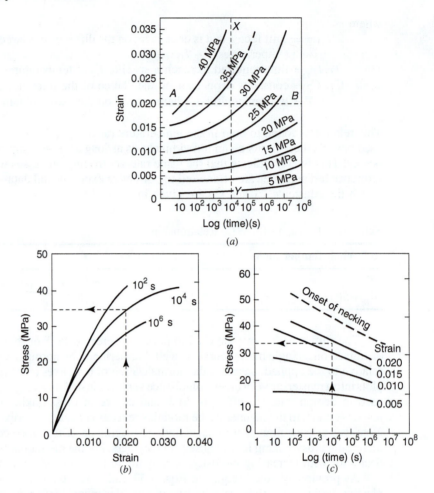

Figure 4.44 Creep response in PVC pipe resin at 20°C. (*a*) Creep curve; (*b*) isochronous stress–strain curves; (*c*) isometric stress–time curves.[97] (By permission of ICI Ltd.)

EXAMPLE 4.1

A PVC rod experiences a load of 500 N. An acceptable design calls for a maximum strain of 1% after one year of service. What is the minimum allowable rod diameter?

We will assume that the creep characteristics of the PVC pipe are identical to data shown in Fig. 4.44. Since one year is equal to 3.15×10^7 s, we see from Fig. 4.44*a* that an allowable strain of 1% would correspond to a stress of approximately 15 MPa. A similar result could have been identified with an isometric stress–time curve corresponding to 1% strain or with an isochronous stress–strain curve, corresponding to 3.15×10^7 s. The minimum rod diameter is then found to be

$$\sigma = \frac{P}{\frac{\pi}{4}d^2}$$

$$15 \times 10^6 = \frac{500}{\frac{\pi}{4}d^2}$$

$$\therefore d \sim 6.5\,\text{mm}$$

The creep data shown in Fig. 4.44a can be analyzed in alternative fashion by considering the stress–time relation associated with various strain levels (e.g., line AB, Fig. 4.44a). The resulting *isometric* curves provide stress–time plots corresponding to different strain levels (Fig. 4.44c). For example, if a component were designed for which strain must be less than 0.02 after 10^4 s, the maximum permissible stress level would again be 32 MPa.

4.9.2 Mechanical Analogs

The linear viscoelastic response of polymeric solids has for many years been described by a number of mechanical models (Fig. 4.45). Many, including these authors, have found that these models provide a useful physical picture of time-dependent deformation processes. The spring element (Fig. 4.45a) is intended to describe linear elastic behavior

$$\varepsilon = \frac{\sigma}{E} \quad \text{and} \quad \gamma = \frac{\tau}{G} \tag{1-7}$$

such that resulting strains are not a function of time. (The stress–strain–time diagram for the spring is shown in Fig. 4.46a.) Note the instantaneous strain upon application of stress σ_0, no further extension with time, and full strain recovery when the stress is removed. The dashpot (a piston moving in a cylinder of viscous fluid) represents viscous flow (Fig. 4.45b).

$$\dot{\varepsilon} = \frac{\sigma}{\eta} \quad \text{and} \quad \dot{\gamma} = \frac{\tau}{\eta} \tag{4-40}$$

where
$\dot{\varepsilon}, \dot{\gamma} =$ tensile and shear strain rates
$\sigma, \tau =$ applied tensile and shear stresses
$\eta =$ fluid viscosity in units of stress-time

The viscosity η varies with temperature according to an Arrhenius-type relation

$$\eta = Ae^{\Delta H/RT} \tag{4-41}$$

where
$\Delta H =$ viscous flow activation energy at a particular temperature
$T =$ absolute temperature

On the basis of time–temperature equivalence, η is seen, therefore, to depend strongly on time as well. For example, at $t = 0$ the viscosity will be extremely high, while at $t \to \infty$, η is

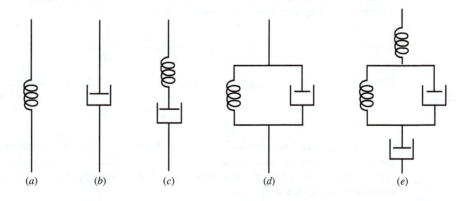

Figure 4.45   Mechanical analogs reflecting deformation processes in polymeric solids: (a) elastic; (b) pure viscous; (c) Maxwell model for viscoelastic flow; (d) Voigt model for viscoelastic flow; (e) four-element viscoelastic model.

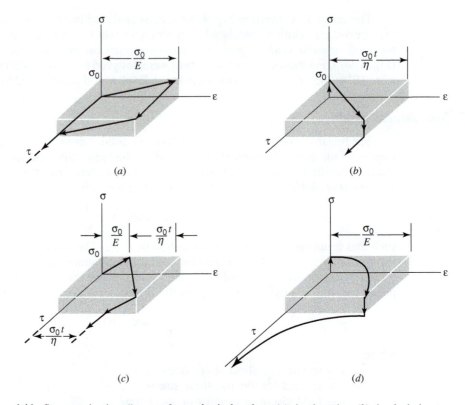

Figure 4.46 Stress-strain–time diagrams for mechanical analogs: (*a*) simple spring; (*b*) simple dashpot; (*c*) Maxwell model; (*d*) Voigt model.

small. The deformation response of a purely viscous element is shown in Fig. 4.46*b*. Upon loading ($t=0$), the dashpot is infinitely rigid. Consequently, there is no instantaneous strain associated with σ_0 (the same holds when the stress is removed). With time, the viscous character of the dashpot element becomes evident as strains develop that are directly proportional to time. When the stress σ_0 is removed these strains remain. When the spring and dashpot are in series, as in Fig. 4.45*c* (called the Maxwell model), we are able to describe the mechanical response of a material possessing both elastic and viscous components.

The stress–strain–time diagram for this model is shown in Fig. 4.46*c*. Note that all the elastic strains are recovered, but the viscous strains arising from creep of the dashpot remain. Since the elements are in series, the stress on each is the same, and the total strain or strain rate is determined from the sum of the two components. Hence

$$\frac{d\varepsilon}{dt} = \frac{\sigma}{\eta} + \frac{1}{E}\frac{d\sigma}{dt} \tag{4-42}$$

For stress relaxation conditions, $\varepsilon = \varepsilon_0$ and $d\varepsilon/dt = 0$. Upon integration, Eq. 4-42 becomes

$$\sigma(t) = \sigma_0 e^{-Et/\eta} = \sigma_0 e^{-t/\tau_r} \tag{4-43}$$

where $\tau_r \equiv$ relaxation time defined by η/E. From Eq. 4-43, the extent of stress relaxation for a given material will depend on the relationship between τ_r and t. When $t \gg \tau_r$ there is time for viscous reactions to take place so that $\sigma(t)$ will drop rapidly. When $t \ll \tau_r$ the material behaves elastically such that $\sigma(t) \approx \sigma_0$.

When the spring and dashpot elements are combined in parallel, as in Fig. 4.45*d* (the Voigt model), this unit predicts a different time-dependent deformation response. First, the strains in the two elements are equal, and the total stress on the pair is given by the sum of the two components

$$\varepsilon_T = \varepsilon_S = \varepsilon_D$$
$$\sigma_T = \sigma_S + \sigma_D \tag{4-44}$$

Therefore

$$\sigma_T(t) = E\varepsilon + \eta \frac{d\varepsilon}{dt} \tag{4-45}$$

For a creep test, $\sigma_T(t) = \sigma_0$ and after integration

$$\varepsilon(t) = \frac{\sigma_0}{E}\left(1 - e^{-t/\tau_r}\right) \tag{4-46}$$

The strain experienced by the Voigt element is shown schematically in Fig. 4.46*d*. The absence of any instantaneous strain is predicted from Eq. 4-46 and is related in a physical sense to the infinite stiffness of the dashpot at $t = 0$. The creep strain is seen to rise quickly thereafter, but reach a limiting value σ_0/E associated with full extension of the spring under that stress. Upon unloading, the spring remains extended, but now exerts a negative stress on the dashpot. In this manner, the viscous strains are reversed, and in the limit when both spring and dashpot are unstressed, all the strains have been reversed. Consequently, the Maxwell and Voigt models describe different types of viscoelastic response. A somewhat more realistic description of polymer behavior is obtained with a four-element model consisting of Maxwell and Voigt models in series (Fig. 4.45*e*). By combining Eqs. 1-7, 4-40, and 4-46, it can be readily shown that the total strain experienced by this model may be given by

$$\varepsilon(t) = \frac{\sigma}{E_1} + \frac{\sigma}{E_2}\left(1 - e^{-t/\tau_r}\right) + \frac{\sigma}{\eta_3}t \tag{4-47}$$

which takes account of elastic, viscoelastic, and viscous strain components, respectively (Fig. 4.47). Even this model is overly simplistic with many additional elements often required

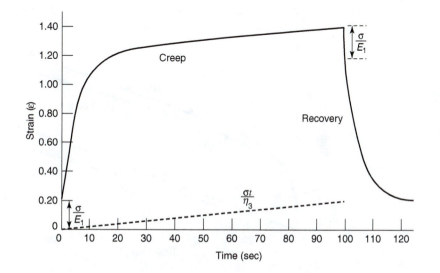

Figure 4.47   Creep response of four-element model with $E_1 = 5 \times 10^2$ MPa, $E_2 = 10^2$ MPa, $\eta_2 = 5 \times 10^2$ MPa-sec, $\eta_3 = 50$ GPa-sec, and $\sigma = 100$ MPa.[98] (L. Nielsen, *Mechanical Properties of Polymers,* © 1962 by Litton Educational Publishing Inc., reprinted by permission of Van Nostrand Reinhold.)

to adequately represent mechanical behavior of a polymer. For example, such a model might include a series of Voigt elements, each describing the relaxation response of a different structural unit in the molecule.

Even so, the four-element model is useful in characterizing the response of different types of polymers. For example, a stiff and rigid material, such as a polyester thermoset resin, can be simulated by choosing stiff springs and high-viscosity dashpots. These elements would predict high stiffness and little time-dependent deformation, characteristic of a thermoset material. On the other hand, a soft and flexible material such as low-density polyethylene could be simulated by choosing low stiffness springs and dashpots with low viscosity levels. Accordingly, considerable time-dependent deformation would be predicted. Finally, the temperature dependence of the mechanical response of a polymer can be modeled by appropriate adjustment in dashpot and spring values (i.e., lower spring stiffness and dashpot viscosity levels for higher temperatures and vice versa for lower temperature conditions).

EXAMPLE 4.2

Let us examine the viscoelasticity of a soft and flexible material—cheese. This edible commodity is composed primarily of protein substances that are polymeric in nature. Sperling and coworkers[viii] conducted experiments to examine the viscoelastic response of Velveeta® brand processed cheese. Such cheeses are plasticized or softened by the addition of water. A 15-cm-long block of this cheese, with cross-sectional dimensions of 4 cm × 6 cm, was supported in a slightly tilted holder and subjected to a compressive load of 4.9 N for approximately 2 h. The height of the cheese block was measured prior to loading and every 5 minutes thereafter. No additional displacement measurements were made after removal of the load. A duplicate experiment was conducted with a second cheese block under a compressive load of 6.85 N. The two creep curves from these experiments are illustrated below. With the exception of the unloading portion of the curve shown in Fig. 4.47, note the similarity in shape between the experimental Velveeta® creep curves and the computed curve for the stiffer polymer.

If we assume that the creep response of the Velveeta® cheese may be characterized by a four-element viscoelastic model (Fig. 4.45e), the strain–time plot is given by Eq. 4-47

$$\epsilon = \frac{\sigma}{E_1} + \frac{\sigma}{E_2}\left(1 - e^{-\left(\frac{E_2}{\eta_2}\right)t}\right) + \frac{\sigma}{\eta_3}(t)$$

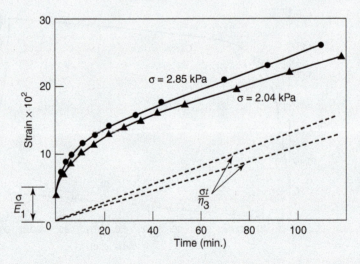

[viii] V. S. Chang, J. S. Guo, Y. P. Lee, and L. H. Sperling, *J. Chem. Ed.*, **63**, 1077 (1986).

For the two experiments, the applied stress, σ, is equal to

$$\sigma = \frac{P}{A} = \frac{4.9}{(4 \times 10^{-2})(6 \times 10^{-2})} = 2.04 \, \text{kPa}$$

Also

$$\sigma = \frac{P}{A} = \frac{6.85}{(4 \times 10^{-2})(6 \times 10^{-2})} = 2.85 \, \text{kPa}$$

As shown in Fig. 4.47, the elastic modulus, E_1, for the spring in series is determined by the strain at zero time (i. e., $E_1 = \sigma/\varepsilon$). The viscosity, η_3, of the dashpot in series is determined from the slope of the linear portion of the creep curve at long times. Finally, the strain associated with the viscoelastic Voigt elements is obtained from the total strain less that associated with the spring and dashpot series elements. By simple curve fitting, the Voigt elements, E_2 and η_2, can then be determined. The constants for the four-element model are listed in the accompanying table. We see relatively good agreement between the two sets of values. As expected, the elastic and viscous elements for the processed cheese are much lower than those associated with the engineering polymer, described in Fig. 4.47.

Experimentally Determined Constants for Four-Element Viscoelastic Model of Velveeta® Cheese*

	4.8 Newtons	6.85 Newtons
E_1 (kPa)	4.88×10^4	5.18×10^4
E_2 (kPa)	2.82×10^4	4.24×10^4
H_2 (MPa-s)	1.52×10^7	1.78×10^7
η_2 (MPa-s)	1.00×10^8	1.21×10^8

*V. S. Chang, J. S. Guo, Y. P. Lee, and L. H. Sperling, *J. Chem. Ed.*, **63**, 1077 (1986).

4.9.3 Dynamic Mechanical Testing and Energy-Damping Spectra

Another method by which time-dependent moduli and energy-dissipative mechanisms are examined is through the use of dynamic test methods. These studies have proven to be extremely useful in identifying the major molecular relaxation at T_g as well as secondary relaxations below T_g. It is believed that such relaxations are associated with motions of specific structural units within the polymer molecule. Two basically different types of dynamic test equipment have been utilized by researchers. One type involves the free vibration of a sample, such as that which takes place in the torsion pendulum apparatus shown in Fig. 4.48. A specimen is rotated through a predetermined angle and then released. This causes the sample to oscillate with decreasing amplitude resulting from various energy-dissipative mechanisms. The extent of mechanical damping is defined by the decrement in amplitude of successive oscillations as given by

$$\Delta = \ln \frac{A_1}{A_2} = \ln \frac{A_2}{A_3} = \ldots = \ln \frac{A_n}{A_{n+1}} \tag{4-48}$$

where

$\Delta = \log$ (base e) decrement which the amount of damping

$A_1, A_2 = $ amplitude of successive oscillations of the vibrating sample

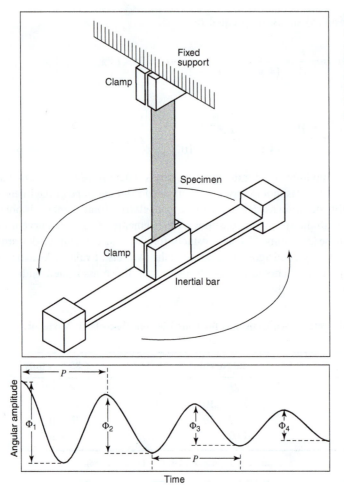

Figure 4.48   Simple torsion pendulum and amplitude-time curve for free decay of torsional oscillation.[98] (L. Nielsen, *Mechanical Properties of Polymers,* © 1962 by Litton Educational Publishing Inc., reprinted by permission of Van Nostrand Reinhold.)

From these same observations, stiffness of the sample is determined from the period of oscillation P, the shear modulus G increasing with the inverse square of P.

The other type of dynamic instruments introduces to the sample a forced vibration at different set frequencies. The amount of damping is found by noting the extent to which the cyclic strain lags behind the applied stress wave. The relation between the instantaneous stress and strain values is shown in Fig. 4.49. Note that the strain vector ε_0 lags the stress vector σ_0 by the phase angle δ. It is instructive to resolve the stress vector into components both in phase and 90° out of phase with ε_0. These are given by

$$\sigma' = \sigma_0 \cos \delta \quad \text{(in-phase component)}$$
$$\sigma'' = \sigma_0 \sin \delta \quad \text{(out-of-phase component)}$$

(4-49)

The corresponding in-phase and out-of-phase moduli are determined directly from Eq. 4-49 when the two stress components are divided by ε_0. Hence

$$E' = \frac{\sigma'}{\varepsilon_0} = \frac{\sigma_0}{\varepsilon_0} \cos \delta = E^* \cos \delta$$
$$E'' = \frac{\sigma''}{\varepsilon_0} = \frac{\sigma_0}{\varepsilon_0} \sin \delta = E^* \sin \delta$$

(4-50)

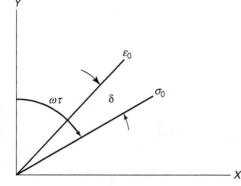

Figure 4.49 Forced vibration resulting in phase lag δ between applied stress σ_0 and corresponding strain ε_0.

where $E^* =$ absolute modulus $= (E'^2 + E''^2)^{1/2}$. E' reflects the elastic response of the material, since the stress and strain components are in phase. This part of the strain energy, introduced to the system by the application of stress σ_0, is stored but then completely released when σ_0 is removed. Consequently, E' is often referred to as the *storage* modulus. E'', on the other hand, describes the strain energy that is completely dissipated (mostly in the form of heat) and for this reason is called the *loss* modulus. The relative amount of damping or energy loss in the material is given by the loss tangent, $\tan \delta$:

$$\frac{E''}{E'} = \frac{E^* \sin \delta}{E^* \cos \delta} = \tan \delta \tag{4-51}$$

By comparison,[98]

$$\frac{G''}{G'} \approx \frac{\Delta}{\pi} \tag{4-52}$$

with the result that

$$\Delta \approx \pi \tan \delta \tag{4-53}$$

When dynamic tests are conducted, the values of the storage and loss moduli and damping capacity are found to vary dramatically with temperature (Fig. 4.50). Note the correlation between the rapid drop in G', the rise in G'', and the corresponding damping maximum. The relaxation time associated with these changes (occurring in Fig. 4.38b in the vicinity of T_g) is considered to have an Arrhenius-type temperature dependence associated with a specific activation energy. In turn, the activation energy is then used to identify the molecular motion responsible for the change in dynamic behavior. Dynamic tests can be conducted either over a range of test temperatures at a constant frequency or at different frequencies for a constant temperature. Since the fixed frequency tests are usually more convenient to perform, most studies employ this procedure. Experiments of this type are now conducted routinely in many laboratories to characterize polymers with regard to effects of thermal history, degree of crystallinity, molecular orientation, polymer additions, molecular weight, plasticization, and other important variables. Consequently, the extant literature for such studies is enormous. Fortunately, a number of books and review articles have been prepared[99,100,104] on the subject to which the interested reader is referred. Within the scope of this book, we can only highlight some of the major findings.

When dynamic tests are performed over a sufficiently large temperature range, multiple secondary relaxation peaks are found in addition to the T_g peak shown in Fig. 4.50. Boyer[99] has summarized some of these data in the schematic form shown in Fig. 4.51. He noted that

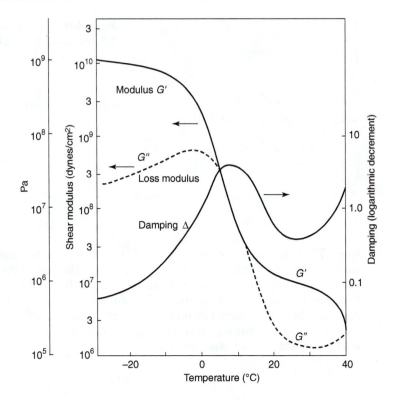

Figure 4.50   Dynamic mechanical response of un-cross-linked styrene and butadiene copolymer revealing temperature dependence of G', G'', and Δ.[98] (L. Nielsen, *Mechanical Properties of Polymers,* © 1962 by Litton Educational Publishing Inc., reprinted by permission of Van Nostrand Reinhold.)

relaxation response in amorphous and semicrystalline polymers could be separated conveniently into four regions, as summarized in Table 4.6. Furthermore, crude temperature relations between various damping peaks were identified (e.g., $T_m \approx 1.5 T_g$ and the $T < T_g$ transition (β) occurring at about $0.75 T_g$). The dynamic mechanical spectra for a given material characterizes localized molecular movement such as small-scale segmental motions and side-chain group

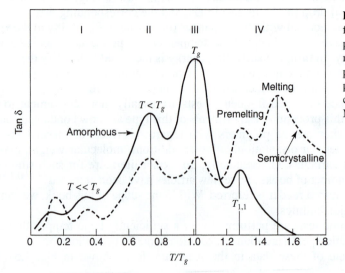

Figure 4.51   Energy damping spectra for semicrystalline and amorphous polymers at various temperatures normalized to T_g. Several damping peaks are found for each material.[99] (By permission, from *Polymeric Materials,* copyright American Society for Metals, Metals Park, OH, © 1975.)

Table 4.6 Transition Regions in Polymers[99]

Region	Temperature of Occurrence	Cause
I	$T \ll T_g$ (the γ peak)	Believed to be caused by movements of small groups involving only a few atoms
II	$T < T_g$ (the β peak)	Believed to be related to movement of 2–3 consecutive repeat units
III	T_g (the α peak)	Believed to be related to coordinated movements of 10–20 repeat units
IV	$T > T_g$	Large-scale molecular motions

rotations. These transitions have been described by different activation energy levels that increase with increasing temperature of the transition and size of the side group responsible for the transition. Boyer[104] and Heijboer[100] have considered possible correlations between the size of the β peak and impact resistance (toughness). The reader also is referred to Section 10.8 for a discussion on the influence of the β peak and test frequency on fatigue crack propagation rates in numerous polymeric solids.

REFERENCES

1. A. K. Mukherjee, J. E. Bird, and J. E. Dorn, *Trans. ASM* **62**, 155 (1969).

2. C. R. Barrett, W. D. Nix, and O. D. Sherby, *Trans. ASM* **59**, 3 (1966).

3. F. Garofalo, *Fundamentals of Creep and Creep-Rupture in Metals*, Macmillan, New York, 1965.

4. J. B. Conway, *Stress-Rupture Parameters: Origin, Calculation and Use*, Gordon & Breach, New York, 1969.

5. N. J. Grant and A. G. Bucklin, *Trans. ASM* **42**, 720 (1950).

6. O. D. Sherby and J. L. Lytton, *Trans. AIME* **206**, 928 (1956).

7. F. C. Monkman and N. J. Grant, *Proc. ASTM* **56**, 593 (1956).

8. A. H. Cottrell, *J. Mech. Phys. Sol.* **1**, 53 (1952).

9. E. N. DaC. Andrade, *Proc. R. Soc. London Ser. A* **84**, 1 (1910).

10. R. L. Orr, O. D. Sherby, and J. E. Dorn, *Trans. ASM* **46**, 113 (1954).

11. O. D. Sherby, T. A. Trozera, and J. E. Dorn, *Trans. ASTM* **56**, 789 (1956).

12. J. E. Dorn, *Creep and Recovery*, ASM, Metals Park, OH, 1957, p. 255.

13. O. D. Sherby, J. L. Lytton, and J. E. Dorn, *Acta Met.* **5**, 219 (1957).

14. W. M. Hirthe and J. O. Brittain, *J. Am. Ceram. Soc.* **46** (9), 411 (1963).

15. S. L. Robinson and O. D. Sherby, *Acta Met.* **17**, 109 (1969).

16. J. Weertman, *Trans. ASM* **61**, 681 (1968).

17. O. D. Sherby and P. M. Burke, *Prog. Mater. Sci.* **13**, 325 (1968).

18. O. D. Sherby and M. T. Simnad, *Trans. ASM* **54**, 227 (1961).

19. C. R. Barrett, A. J. Ardell, and O. D. Sherby, *Trans. AIME* **230**, 200 (1964).

20. C. R. Barrett and O. D. Sherby, *Trans. AIME* **230**, 1322 (1964).

21. J. E. Bird, A. K. Mukherjee, and J. E. Dorn, *Quantitative Relation Between Properties and Microstructure*, Israel Universities Press, Haifa, Israel, 1969, p. 255.

22. H. J. Frost and M. F. Ashby, NTIS Report AD-769821, August 1973.

23. F. Garofalo, *Trans AIME* **227**, 351 (1963).

24. F. R. N. Nabarro, *Report of a Conference on the Strength of Solids*, Physical Society, London, 1948, p. 75.

25. C. Herring, *J. Appl. Phys.* **21**, 437 (1950).

26. R. L. Coble, *J. Appl Phys.* **34**, 1679 (1963).

27. J. Weertman, *J. Appl. Phys.* **28**, 362 (1957).

28. R. Raj and M. F. Ashby, *Met. Trans.* **2**, 1113 (1971).

29. T. G. Langdon, *Deformation of Ceramic Materials*, R. C. Bradt and R. E. Tressler, Eds., Plenum, New York, 1975, p. 101.

30. L. M. Lifshitz, *Sov. Phys. JETP* **17**, 909 (1963).

31. D. K. Matlock and W. D. Nix, *Met Trans.* **5**, 961 (1974).

32. D. K. Matlock and W. D. Nix, *Met Trans.* **5**, 1401 (1974).

33. B. J. Piearcey and F. L. VerSnyder, *Met. Prog.*, 66 (Nov. 1966).

34. R. H. Johnson, *Met. Mater.* **4** (9), 389 (1970).

35. W. A. Backofen, I. R. Turner, and D. H. Avery, *Trans. ASM* **57**, 981 (1964).

36. D. H. Avery and W. A. Backofen, *Trans. ASM* **58**, 551 (1965).

37. M. F. Ashby and R. A. Verall, *Acta Met.* **21**, 149 (1973).

38. T. H. Alden, *Acta Met.* **15**, 469 (1967).

39. T. H. Alden, *Trans. ASM* **61**, 559 (1968).

40. J. W. Edington, K. N. Melton, and C. P. Cutler, *Prog. Mater. Sci.* **21**, 61 (1976).

41. D. M. R. Taplin, G. L. Dunlop, and T. G. Langdon, *Annu. Rev. Mater. Sci.* **9**, 151 (1979).

42. T. G. Langdon, *Creep and Fracture of Engineering Materials and Structures*, B. Wilshire and D. R. J. Owen, Eds., Pineridge Press, Swansea, U.K., 1981, p. 141.

43. T. G. Langdon, *Metallurgical Transactions* **13** (5), 689–701 (1982).

44. J. F. Hubert and R. C. Kay, *Met. Eng. Quart.* **13**, 1 (1973).

45. M. F. Ashby, *Acta Met.* **20**, 887 (1972).

46. M. F. Ashby, *The Microstructure and Design of Alloys*, Proceedings, Third International Conference on Strength of Metals and Alloys, Vol. 2, Cambridge, England, 1973, p. 8.

47. F. W. Crossman and M. F. Ashby, *Acta Met.* **23**, 425 (1975).

48. M. R. Notis, *Deformation of Ceramic Materials*, R. C. Bradt and R. E. Tressler, Eds., Plenum, New York, 1975, p. 1.

49. T. G. Langdon and F. A. Mohamed, *J. Mater. Sci.* **13**, 1282 (1978).

50. T. G. Langdon and F. A. Mohamed. *Mater. Sci. Eng.* **32**, 103 (1978).

51. H. Oikawa, *Scripta Met.* **13**, 701 (1979).

52. H. Oikawa, *Creep and Fracture of Engineering Materials and Structures*, B. Wilshire and D. R. J. Owen, Eds., Pineridge Press, Swansea, U.K. 1981, p. 113.

53. M. F. Ashby, *Fracture 1977*, Vol. 1, Waterloo, Canada, 1977, p. 1.

54. M. F. Ashby, C. Gandhi, and D. M. R. Taplin, *Acta Met.* **27**, 699 (1979).

55. C. Gandhi and M. F. Ashby, *Acta Met.* **27**, 1565 (1979).

56. Y. Krishna, M. Rao, V. Kutumba Rao, and P. Rama Rao, *Titanium 80,* H. Kimura and O. Izumi, Eds., AIME, Warrendale, PA, 1981, p. 1701.

57. F. R. Larson and J. Miller, *Trans. ASME*, **74**, 765 (1952).

58. R. M. Goldhoff, *J. Test. Eval.* **2** (5), 387 (1974).

59. S. S. Manson, *Time-Temperature Parameters for Creep-Rupture Analysis*, Publication No. D8-100, ASM, Metals Park, OH, 1968, p. 1.

60. S. S. Manson and C. R. Ensign, *NASA Tech. Memo TM X-52999,* NASA, Washington, DC, 1971.

61. D. A. Woodford, *Mater, Sci. Eng.* **15**, 69 (1974).

62. D. A. Woodford, *Creep and Fracture of Engineering Materials and Structures*, B. Wilshire and D. R. J. Owen, Eds., Pineridge Press, Swansea, U.K., 1981, p. 603.

63. C. T. Sims and W. C. Hagel, Eds., *Superalloys*, Wiley, New York, 1972.

64. B. H. Kear, D. R. Muzyka, J. K. Tien, and S. T. Wlodek, Eds., *Superalloys: Metallurgy and Manufacturer*, Claitors, Baton Rouge, LA, 1976.

65. J. K. Tien, S. T. Wlodek, H. Morrow III, M. Gell, and G. E. Mauer, Eds., *Superalloys 1980*, ASM, Metals Park, OH, 1980.

66. E. F. Bradley, Ed., *Source Book on Materials for Elevated Temperature Applications*, ASM, Metals Park, OH, 1979.

67. *High Temperature High Strength Nickel Base Alloys*, 3rd ed., International Nickel Co., New York, 1977.

68. M. Gell, C. S. Kortovich, R. H. Bricknell, W. B. Kent, and J. F. Radavich, Eds., *Superalloys 84*, AIME, Warrendale, PA, 1984, p. 357.

69. A. K. Jena and M. C. Chaturvedi, *J. Mater. Sci.* **19**, 3121 (1984).

70. R. R. Jensen and J. K. Tien, *Metallurgical Treatises*, J. K. Tien and J. F. Elliott, Eds., AIME, Warrendale, PA, 1981, p. 529.

71. N. S. Stoloff, *Strengthening Methods in Crystals*, A. Kelly and R. B. Nicholson, Eds., Wiley, New York, 1971, p. 193.

72. P. H. Thornton, R. G. Davies, and T. L. Johnston, *Met Trans.* **1**, 207 (1970).

73. E. R. Thompson, Private communication.

74. D. D. Pearson, F. D. Lemkey, and B. H. Kear, *Super-alloys 1980*, ASM, Metals Park, OH, 1980, p. 513.

75. R. F. Decker and J. W. Freeman, *Trans. AIME* **218**, 277 (1961).

76. B. H. Kear and E. R. Thompson, *Science* **208**, 847 (1980).

77. B. H. Kear and B. J. Piearcey, *Trans. AIME* **238**, 1209 (1967).

78. F. L. VerSnyder and M. E. Shank, *Mater. Set. Eng.* **6**, 213 (1970).

79. M. Gell, D. N. Duhl, D. K. Gupta, and K. D. Sheffler, *J. Met.* **39** (7), 11 (1987).

80. D. N. Duhl and C. P. Sullivan, *J. Met.* **23** (7), 38 (1971).

81. F. D. Lemkey, *Proceedings, MRS Conference 1982, CISC IV*, Vol. 12, F. D. Lemkey, H. E. Cline, and M. McLean, Eds., Elsevier Science, Amsterdam, 1982.

82. J. B. Moore and R. L. Athey, U.S. Patent 3,519,503, 1970.

83. J. S. Crompton and R. W. Hertzberg, *J. Mater. Sci.* **21**, 3445 (1986).

84. R. C. Benn and S. K. Kang, *Superalloys 1984*, ASM, Metals Park, OH, 1984, p. 319.

85. S. K. Kang and R. C. Benn, *Met. Trans.* **16A**, 1285 (1985).

86. T. E. Howson, J. E. Stulga, and J. K. Tien, *Met. Trans.* **11A**, 1599 (1980).

87. A. H. Cooper, V. C. Nardone, and J. K. Tien, *Super-alloys 1984*, ASM, Metals Park, OH, 1984, p. 319.

88. D. W. Petrasek, D. L. McDanels, L. J. Westfall, and J. R. Stephans, *Metal Prog.* **130** (2), 27 (1986).

89. M. H. Litt and A. V. Tobolsky, *J. Macromol Sci. Phys. B* **1** (3), 433 (1967).

90. S. E. B. Petrie, *Polymeric Materials*, ASM, Metals Park, OH, 1975, p. 55.

91. R. D. Deanin, *Polymer Structure, Properties and Applications*, Cahners, Boston, 1972.

92. A. V. Tobolsky and R. D. Andrews, *J. Chem. Phys.* **13**, 3 (1945).

93. M. L. Williams, R. F. Landel, and J. D. Ferry, *J. Appl. Phys.* **26**, 359 (1955).

94. F. Rodriguez, *Principles of Polymer Systems*, McGraw-Hill, New York, 1970.

95. E. Catsiff and A. V. Tobolsky, *J. Polym. Sci.* **19**, 111 (1956).

96. T. Alfrey and E. F. Gurnee, *Organic Polymers*, Prentice-Hall, Englewood Cliffs, NJ, 1967.

97. R. M. Ogorkiewicz, Ed., *Engineering Properties of Thermoplastics*, Wiley-Interscience, London, 1970.

98. L. E. Nielsen, *Mechanical Properties of Polymers*, Reinhold, New York, 1962.

99. R. F. Boyer, *Polymeric Materials*, ASM, Metals Park, OH, 1975, p. 277.

100. J. Heijboer, *J. Polym. Sci.* **16**, 3755 (1968).

101. R. P. Kambour and R. E. Robertson, *Polymer Science: A Materials Science Handbook*, Vol. 1, A. D. Jenkins, Ed., North Holland, 1972, p. 687.

102. N. G. McCrum, B. E. Read, and G. Williams, *Anelastic and Dielectric Effects in Polymeric Solids*, Wiley, London, 1967.

103. R. F. Boyer, *Rubber Chem. Tech.* **36**, 1303 (1963).

104. R. F. Boyer, *Poly. Eng. Set* **8** (3), 161 (1968).

PROBLEMS

Review

4.1 Reproduce Fig. 4.1*a*. Imagine that the material in question is *viscoelastic*, and that very rapid elastic loading is followed by creep deformation over a long period of time designated by t_2. Add to the sketch two additional curves showing the behavior (i) during rapid unloading to zero stress, and (ii) during a long period of time ($t \gg t_2$) at zero stress after unloading.

4.2 Follow the instructions for Review Problem 4.1, but for a material that is *viscoplastic*.

4.3 State at what homologous temperature creep may start to become an issue for metals and for ceramics. Then estimate the actual temperature for Al, Ti, ZrO_2, and SiC. Is the onset of creep behavior with respect to temperature sudden or gradual?

4.4 What factors determine whether polymer behavior is predominantly elastic, viscoelastic, or viscoplastic?

4.5 Describe the standard loading conditions for a creep test. Explain why these loading conditions are chosen.

4.6 What two general competing processes control the creep rate of a metal? What is the relative strength (or rate) of the two processes during Stage I and Stage II creep behavior?

4.7 What leads to a change from Stage II to Stage III behavior? Be specific about the mechanisms involved, and what can influence the time of this transition.

4.8 Describe under what circumstances the steady-state creep rate may be a more useful parameter than rupture life, and vice versa.

4.9 Describe the general trends that connect stress level and temperature with steady-state creep rate and rupture life.

4.10   Why do the curves in Fig. 4.4 have more than one linear segment, and what does the transition from one segment to the next indicate?

4.11   What data would have to be collected to determine the stress exponent for creep? What plot axes would give a linear relationship using this data set?

4.12   What data are typically collected to determine the activation energy for the controlling creep mechanism?

4.13   How would you know if there were different creep mechanisms acting at different temperatures?

4.14   The plot in Fig. 4.5 shows significant differences in the creep rates of α- and γ-iron tested at the same temperature, 910°C. How can this difference be explained?

4.15   List and briefly describe the four major creep deformation mechanisms active in crystalline materials.

4.16   Under what high temperature circumstances is a very small grain size detrimental, and under what circumstances is it advantageous?

4.17   What do the regions on a Deformation Mechanism Map (DMM) represent?

4.18   When designing an alloy for creep resistance, why is it generally advantageous to employ multiple composition and microstructure strategies?

4.19   Explain what the Larson-Miller parameter is used for, and what assumption underlies the form of the expression that makes use of this parameter.

4.20   What are the units of temperature and time for which the typical Larson-Miller parameter is equal to ~20 for many metallic alloys?

4.21   What is the advantage in casting turbine blades that have either highly aligned grain boundaries or no grain boundaries at all?

4.22   What is a TBC and what important roles does it play?

4.23   What is a fundamental difference between the degree of recoverable strain after creep of metals and ceramics vs. many polymers?

4.24   What is the link between polymer free volume and creep rate, and why is free volume a more important concept for polymers than for metals and ceramics?

4.25   What provides the driving force for rekinking of an amorphous polymer chain when loaded and unloaded?

4.26   What is an important consequence of time-temperature equivalence for testing of polymer time-dependent elastic moduli?

4.27   What data from a standard creep plot does an isochronous diagram extract? An isometric diagram?

4.28   What two mechanical components are often used to model the time-dependent behavior of polymers?

4.29   If a simple Maxwell model and a simple Voigt model for viscoelastic flow are loaded at time t_0, held for a certain time t_1, and then unloaded, what is different about their responses during the loading and unloading processes? Also, what is different about the final state of the two mechanical models? Sketch a plot of strain vs. time for both cases to illustrate your descriptions.

4.30   Of what phenomenon is tan δ a measure, and what does it mean when tan δ is large?

Practice

4.31   A study of creep in ODS-Al alloys[ix] found the following relationships between the minimum creep strain rate and the rupture life. Determine the Monkman-Grant constants m, B, and C_{MG} using units of hours and %/hr, then predict the rupture lifetime in hours for a minimum creep rate of 1.0×10^{-9} s^{-1}. How does this value of C_{MG} compare to typical values for other materials?

Strain Rate s^{-1}	t_r (ks)
3.00×10^{-8}	183.7
9.00×10^{-8}	55.3
1.00×10^{-7}	72.6
4.30×10^{-8}	133.4
5.90×10^{-9}	884
1.20×10^{-6}	13.8

4.32   Use the diagram below to answer the following questions.

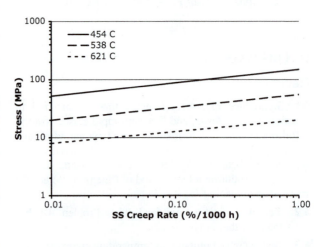

a. Determine the value of the creep exponent "n" at 454°C (850°F).

[ix] D. C. Dunand, B. Q. Han, and A. M. Jansen, *Metal. Mater. Trans. A* **30**, 829 (1999).

b. What mechanism or category of mechanism is implied by this "n" value—diffusion or dislocation creep?

c. Without calculating "n" for the other two temperatures, is "n" a strong function of temperature? How do you know?

d. Determine the activation energy for creep in units of kJ/mol, assuming that "n" is not a strong function of temperature (regardless of reality).

4.33 For a certain high-temperature alloy, failure was reported after 4100 hrs at 680°C when subjected to a stress level of 270 MPa. If the same stress were applied at 725°C, how long would the sample be expected to last? State any assumptions you must make to allow you to make this determination.

4.34 Construct a Larson-Miller plot using the following creep rupture data, assuming $C = 20$.

Temp. (°C)	Stress (MPa)	Rupture Time (hr)	Temp. (°C)	Stress (MPa)	Rupture Time (hr)
650	480	22	815	140	29
650	480	40	815	140	45
650	480	65	815	140	65
650	450	75	815	120	90
650	380	210	815	120	115
650	345	2700	815	105	260
650	310	3500	815	105	360
705	310	275	815	105	1000
705	310	190	815	105	700
705	240	960	815	85	2500
705	205	2050	870	83	37
760	205	180	870	83	55
760	205	450	870	69	140
760	170	730	870	42	3200
760	140	2150	980	21	440
			1095	10	155

a. Plot the data twice: (i) with axes of Stress vs. the LM parameter in thousands of hours, and (ii) with axes of Log(Stress) vs. the LM parameter in thousands of hours. Fit a second-order polynomial to the second plot.

b. Using your fitted line, determine the expected life for a sample tested with a stress of 240 MPa at 650°C, and with a stress of 35 MPa at 870°C. (This may be easiest if you add a curve to the first plot based on the polynomial fit to the log(stress) data.)

c. What is the maximum operational temperature such that failure should not occur in 5000 hr at stress levels of 140 and 420 MPa, respectively?

4.35 A 200-mm-long polypropylene rod, with a rectangular cross-section that is 20 mm by 4 mm, is subjected to a

tensile load of 300 N, directed along its length. If the rod extends by 0.5 mm after being under load for 100 s, determine the creep modulus.

4.36 Calculate the typical relaxation time for silicate glass and comment on its propensity for stress relaxation at room temperature. $E \approx 70$ GPa and $\eta \approx 1 \times 10^{12}$ GPa-s (10^{22} poise).

4.37 The deformation response of a certain polymer can be described by the Voigt model. If $E = 400$ MPa and $\eta = 2 \times 10^{12}$ MPa-s, compute the relaxation time. Compute $\varepsilon(t)$ for times to $5T$ when the steady stress is 10 MPa. How much creep strain takes place when $t = T$ and when $t = \infty$?

Design

4.38 A solder joint between a computer chip connector and a printed circuit board is typically subjected to shear due to thermal expansion differences. For optical communications devices (e.g., those that use lasers to transmit information), dimensional stability is critical or the components will lose alignment. A crude approximation of a circuit board, solder joint, and connector is shown here with the shear load indicated by arrows. For this problem, assume that the critical joint is 0.5 mm thick, 2 mm wide, and 4 mm long.

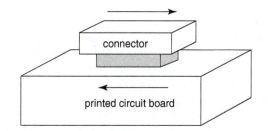

a. If the component is designed to last for five years under ordinary use, which is likely to be more important solder data: steady state creep rate or rupture life? Explain briefly.

b. It is found that accelerated creep tests of a particular solder give the following results. If it is known that the stress exponent for this particular solder is $n = 6$, calculate the steady-state creep rate at a stress level of 100 MPa and a temperature of 100°C.

Shear Stress (MPa)	Strain Rate (s^{-1})	T (°C)
70	1×10^{-5}	190
70	2.5×10^{-3}	225

c. What *assumption* did you make about the nature of the creep at 100, 190, and 225°C in order for the calculation in part (b) to be possible? How could you check the validity of this assumption experimentally?

d. If the maximum allowable shear displacement of the connector relative to the printed circuit board after 3 years of continuous use is 0.25 μm, will the joint meet the design criteria?

e. Possible solder materials for this application include Indium, Lead-Tin alloy, Bismuth-Tin alloy, and Tin-Silver-Copper alloy. Without doing any mechanical testing or looking up any mechanical data on these metals, how could you make a reasonable attempt at rank ordering them from slowest creep rate to fastest creep rate at 100°C? Don't actually make the list; just explain what information you would need and how you would use it.

4.39   A 200-mm-long polypropylene rod, with a rectangular cross-section that is 20 mm by 4 mm, is subjected to a tensile load of 300 N, directed along its length. If the rod extends by 0.5 mm after being under load for 100 s, determine the creep modulus. How does this value compare with the typical static modulus for polypropylene, and what does the comparison imply about the amount of creep that has probably taken place in 100 s?

4.40   A superalloy gas turbine component was originally designed to operate at temperatures up to 760°C and exhibited a stress rupture life of 900 h under this operating condition. An updated design calls for a thermal barrier coating to be added to the same component to allow engine operation under the same conditions, but with an increase in reliability. If an increase in rupture life to 1800 h is desired, what temperature differential between the inside and outside of the component must be achieved by the addition of the TBC?

4.41   A 10-cm-long cylindrical rod of polypropylene is subjected to a tensile load of 550 N. If the maximum allowable strain of 2% is experienced no earlier than after four months of service, what is the minimum required rod diameter? Also, what is the rod diameter after the four-month service period? You will find the plot in the next problem to be useful.

4.42   For safe and reliable operation, a certain polypropylene pipe must withstand an internal pressure of 0.5 MPa for a minimum of three years. If the pipe diameter is 100 mm, what is the minimum necessary wall thickness to ensure that the pipe will not experience a strain greater than 1.3%? To solve this problem, use the accompanying graph

that reveals the room temperature creep response for polypropylene. (R. J. Crawford, *Plastics Engineering*, SPE, Brookfield Center, CT (1981). Reprinted by permission.)

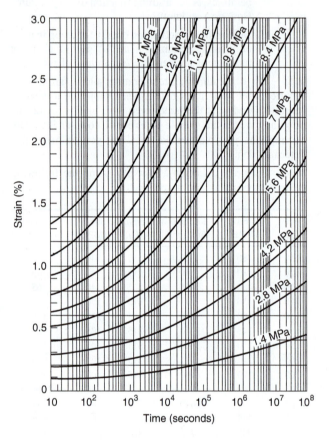

Extend

4.43   What are the dictionary definitions and the etymologies of the words *isochronous* and *isometric*?

4.44   Using information acquired from the National Transportation Safety Board (NTSB), explain the role that epoxy creep played in the 2006 collapse of a 3-ton concrete ceiling panel in Boston's Fort Point Channel Tunnel. What could have been done to prevent this disaster?

Chapter 5

Fracture: An Overview

5.1 INTRODUCTION

On January 15, 1919, something frightening happened on Commercial Street in Boston. A huge tank, 27 meters in diameter and about 15 meters high, fractured catastrophically, and over 7.5×10^6 liters (2×10^6 gallons) of molasses cascaded into the streets.

> *Without an instant's warning the top was blown into the air and the sides were burst apart. A city building nearby, where the employees were at lunch, collapsed burying a number of victims and a firehouse was crushed in by a section of the tank, killing and injuring a number of firemen.*[1]
>
> *On collapsing, a side of the tank was carried against one of the columns supporting the elevated structure [of the Boston Elevated Railway Co.]. This column was completely sheared off . . . and forced back under the structure . . . the track was pushed out of alignment and the superstructure dropped several feet. . . . Twelve persons lost their lives either by drowning in molasses, smothering, or by wreckage. Forty more were injured. Many horses belonging to the paving department were drowned, and others had to be shot.*[2]

The molasses tank failure dramatically highlights the necessity of understanding events that contribute to premature fracture of any engineering component. Other manufactured structures are susceptible to the same fate. For example, several bridges have fractured and collapsed in various countries throughout the world during the past 50 years, resulting in the loss of many lives. Two such case histories are described in Chapter 11. In addition, numerous cargo ship failures have occurred, dating from World War II to the present (Fig. 5.1*a*). Subsequent studies concluded that these failures, which broke the vessels in two, were primarily attributable to the presence of stress concentrations in the ship superstructure and the ability of cracks to traverse welds that joined adjacent steel plates; the existence of faulty weldments and inferior steel quality were also cited as contributing factors in the fracture process.[3] A number of more recent oil cargo ship failures have resulted in extensive pollution of rich fishing grounds and coastal resort beach areas.

It is intriguing to note the similar fracture path of the cargo tanker (Fig. 5.1*a*) with that of the passenger liner *Titanic*, which struck a large iceberg during its maiden voyage in 1912 and sank, causing the death of 1500 passengers and crew members (Fig. 5.1*b*). The remains of this vessel were first discovered in 1985 at a depth of 3.6 km beneath the surface of the Atlantic Ocean during a joint expedition by the French oceanographic agency IFREMER and the Woods Hole Oceanographic Institute, led by Dr. Robert Ballard and co-workers aboard the *Alvin*, *Angus*, and *Argo* minisubmarine research vessels.[4] Garzke et al.[5] speculated that the *Titanic*'s sinking was caused by a *brittle fracture of the steel superstructure* as a result of the ship having struck an iceberg in the North Atlantic Ocean. More recently, Foecke et al.[6,7] discounted the brittle steel idea, and instead proposed an alternative failure theory, based in part on new on-site observations and metallurgical findings taken from ship components. They argued that when the large iceberg (reported to be three to six times the mass of the ship) smashed along the starboard (right) side of the ship, hull plates were damaged, and *numerous rivet heads were popped off*. The latter condition enabled seawater to gush through six separate ruptured hull plate seams, thereby hastening the ship's sinking. Discussion as to why the rivet heads popped off will be deferred until the reader is familiar with certain metallurgical characteristics associated with the wrought iron rivets used to join the ship hull's steel plates (see Section 7.3).

(a)

(b)

Figure 5.1 (a) Fractured T-2 tanker, the S. S. *Schenectady,* which failed in 1941.[3] (Reprinted with permission of Earl R. Parker, *Brittle Behavior of Engineering Structures,* National Academy of Sciences, National Research Council, Wiley, New York, 1957.) (b) Bow portion of the *Titanic*. (Painting by Ken Marschall, based on photographs taken aboard *Alvin*, *Angus*, and *Argo* research vessels. (Courtesy of Dr. Robert D. Ballard, *The Discovery of the Titanic*.)

Various aircraft and rockets also are not immune to periodic failure. The debris from a ruptured 660-cm-diameter rocket motor casing is shown in Fig. 11.20, and the failure is analyzed in Section 11.5, Case 7. Additional fractures of domestic products are shown in Fig. 5.2.

It is quite apparent, then, that the subject of fracture in engineering components and structures is certainly a dynamic one, with new examples being provided continuously for

(a)　　　　　　(b)

Figure 5.2 Fractured components and devices. (a) Ruptured beer barrel; (b) fractured toilet seat.

evaluation. You might say that things are going wrong all the time. (In all seriousness, the reader should recognize that component failures are the exception and not the rule.)

5.2 THEORETICAL COHESIVE STRENGTH

Recall from Chapter 2 that the theoretical shear stress necessary to deform a perfect crystal was many orders of magnitude greater than values commonly found in engineering materials. It is appropriate now to consider how high the cohesive strength σ_c might be in an ideally perfect crystal. Again, using a simple sinusoidal force-displacement law with a half period of $\lambda/2$, we see from Fig. 5.3 that the shape of the curve may be approximated by

$$\sigma = \sigma_c \sin \frac{\pi x}{\lambda/2} \tag{5-1}$$

where σ reflects the tensile force necessary to pull atoms apart. For small atom displacements, Eq. 5-1 reduces to

$$\sigma = \sigma_c \frac{2\pi x}{\lambda} \tag{5-2}$$

and the slope of the curve in this region becomes

$$\frac{d\sigma}{dx} = \frac{2\sigma_c \pi}{\lambda} \tag{5-3}$$

Since Hooke's law (Eq. 1-7) applies to this region as well, the slope of the curve also may be given by

$$E = \frac{\text{stress}}{\text{strain}} = \frac{\sigma}{x/a_0} \tag{5-4}$$

where

$a_0 =$ equilibrium atomic separation
$E =$ modulus of elasticity

Upon differentiation,

$$\frac{d\sigma}{dx} = E/a_0 \tag{5-5}$$

By combining Eqs. 5-3 and 5-5 and solving for σ_c

$$\sigma_c = \frac{E\lambda}{2\pi a_0} \tag{5-6}$$

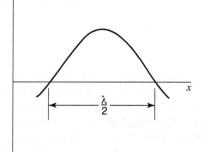

Figure 5.3 Simplified force vs. atom displacement relation.

Table 5.1   Maximum Strengths in Solids[8]

Material	σ_f GPa	σ_f (psi × 10^6)	E GPa	E (psi × 10^6)	E/σ_f
Silica fibers	24.1	(3.5)	97.1	(14.1)	4
Iron whisker	13.1	(1.91)	295.2	(42.9)	23
Silicon whisker	6.47	(0.94)	165.7	(24.1)	26
Alumina whisker	15.2	(2.21)	496.2	(72.2)	33
Ausformed steel	3.14	(0.46)	200.1	(29.1)	64
Piano wire	2.75	(0.40)	200.1	(29.1)	73

If we let $a_0 \approx \lambda/2$, a reasonably accurate assumption, then

$$\sigma_c \approx E/\pi \tag{5-7}$$

It is apparent from Table 5.1 that some materials have the potential for withstanding extremely high stresses before fracture.

If the energetics of the fracture process are considered, the fracture work done per unit area during fracture is given by

$$\text{fracture work} = \int_0^{\lambda/2} \sigma_c \sin\frac{\pi x}{\lambda/2} dx = \sigma_c \frac{\lambda}{\pi} \tag{5-8}$$

If all this work is set equal to the energy required to form two new fracture surfaces 2γ, we may substitute for λ and show from Eq. 5-6 that

$$\sigma_c = \sqrt{\frac{E\gamma}{a_0}} \tag{5-9}$$

EXAMPLE 5.1

What is the cohesive strength of fused silica? Using a value of $1750\,\text{mJ/m}^2$ for the estimated surface energy in fused silica, 0.16 nm for the equilibrium Si–O atomic separation, and 69 to 76 GPa for the elastic modulus, the cohesive strength is found from Eq. 5-9 to be approximately 28 GPa, in agreement with experimental results from carefully prepared specimens (see Table 5.1). Note that a comparable value could have been computed from Eq. 5-7 had the modulus been the only known quantity.

Regardless of the equation used to obtain σ_c, the problem discussed in Chapter 2 reappears—it is necessary to explain not the great strength of solids, but their weakness.

5.3   DEFECT POPULATION IN SOLIDS

Materials possess low fracture strengths relative to their theoretical capacity because most materials deform plastically at much lower stress levels and eventually fail by an accumulation of this irreversible damage. In addition, materials contain defects that are microstructural in origin or introduced during the manufacturing process. These include porosity, shrinkage cavities, quench cracks, grinding and stamping marks (such as gouges, burns, tears, scratches, and cracks), seams, and weld-related cracks. Other microconstituents, such as inclusions, brittle second-phase particles, and grain-boundary films, can lead to crack formation if the applied stress level

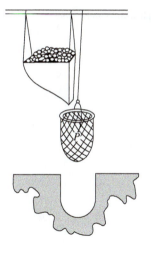

Figure 5.4   Sketch from the notebook of Leonardo da Vinci illustrating tensile test apparatus for iron wires.

exceeds some critical level. An illustration of the role of preexisting defects in the failure of engineering structures is the Duplessis Bridge failure in Quebec, Canada, in 1951, which was traced to a preexistent crack in the steel superstructure.[3] In fact, when a crack that had been sighted before the collapse was examined, *paint* was found on the fracture surface near the crack origin. Certainly, this crack had to be present some time before failure.

In light of such findings, is it not reasonable and even conservative to assume that an engineering component will fail as a consequence of preexistent defects and that this hypothesis provides the basis for fracture control design planning? To a first approximation, then, the problem reduces to one of statistics. How many defects are present in the component or structure, how big are they, and where are they located with respect to the highly stressed portions of the part? Certainly, component size should have some bearing on the propensity for premature failure if for no other reason than the fact that larger pieces of material should contain more defects than smaller ones. Indeed, Leonardo da Vinci used a simple wire-testing apparatus, shown in Fig. 5.4, over 500 years ago to demonstrate that short iron wires were stronger than long sections. In one of his manuscripts we find the following passage:[9]

> The object of this test is to find the load an iron wire can carry. Attach an iron wire 2 braccia [about 1.3 m] long to something that will firmly support it, then attach a basket or any similar container to the wire and feed into the basket some fine sand through a small hole placed at the end of a hopper. A spring is fixed so that it will close the hole as soon as the wire breaks. The basket is not upset while falling, since it falls through a very short distance. The weight of sand and the location of the fracture of the wire are to be recorded. The test is repeated several times to check the results. Then a wire of one-half the previous length is tested and the additional weight it carries is recorded; then a wire of one-fourth length is tested and so forth, noting each time the ultimate strength and the location of the fracture.

5.3.1   Statistical Nature of Fracture: Weibull Analysis

The fact that Leonardo repeated his experiments several times to verify his results reflected his concern with the statistical nature of the fracture event. In general, statistical variation in a given mechanical property value (e.g., fracture strength) depends on inherent measurement errors (including those resulting from variations in specimen alignment and test environment) and inherent property variations of the material. For purposes of this discussion, we will consider only the latter source of statistical variation in fracture properties. As such, the failure event depends on the probability that a flaw of a certain size and orientation is present when a specific stress is applied. For example, assume the existence of four different flaws as shown in

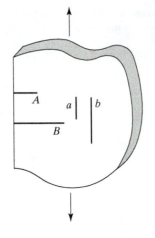

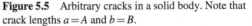

Figure 5.5   Arbitrary cracks in a solid body. Note that crack lengths $a = A$ and $b = B$.

Fig. 5.5. For argument's sake, let us further assume that lengths $a = A$ and $b = B$, and that each defect acts independently of the other three. It will be shown in Chapter 7 that if $\sigma\sqrt{B}$ reaches a critical value, failure will occur. Since $A < B$, failure will not occur in association with crack A. Furthermore, since cracks a and b are parallel to the direction of stress, failure will not occur either, even though length $b = B$. Consequently, the probability of failure will depend only on the existence of the B type crack.

Since a key element in the design and fabrication of a component pertains to its reliability (usually determined by safety and/or economic considerations), it is important that one knows the statistical probability of a given fracture event. For the case of the fracture of brittle wires, such as those studied by da Vinci, failure depends on the length of the wire: The longer the wire, the greater the likelihood that a critical defect is present to cause failure. This weakest-link theory pertains to fracture being dependent on the volume of material in the wire where the volume is proportional to the length of the wire of a given diameter. Proceeding further, if we assume that more defects in a solid are found on the surface of the body rather than at the interior (e.g., due to mechanical handling-induced damage and/or environmental degradation), it follows that the probability of failure will depend on the overall surface area of a component.

For design purposes, it is often more important to determine the probability that a property of a component will exceed some specified threshold value than to identify a particular value such as the average tensile strength. Experience has shown that a normal (Gaussian) distribution of property values for a *ductile* material yields a reasonably accurate characterization of material behavior; the same cannot be said for the case of a *brittle* solid. Instead, other statistical theories (e.g., the Weibull analysis[10-12]) are necessary to account for the variability of strength and the probability of survival of a particular component as a function of its volume and the applied stress.

For the case of a brittle rod of length, L, subjected to a stress σ, the survival probability is given as $S(L)$. For another piece of the same rod with length, L', the survival probability at the same stress is $S(L')$. If one were to attach x pieces of this rod, the overall probability of survival at the same stress would be $S(L_o)^x$ where L_o is a unit length of wire. [The multiplicity of survival probabilities of x segments of the rod is analogous to the probability of obtaining a specific result from a series of coin tosses. Since the probability of flipping a "head" is $\frac{1}{2}$, it follows that the probability of tossing two consecutive "heads" is $\frac{1}{4}$. For three "heads," the probability would be $\frac{1}{8}$, or $(\frac{1}{2})^3$.] If the flaw distribution in each volume is the same, then the probability of survival of the rod would be

$$S(V) = S(V_0)^x \tag{5-10}$$

where V_O is the unit volume of the solid. By taking logarithms and rearranging Eq. 5-10, we have

$$S(V) = \exp(x \ln[S(V_0)]) \tag{5-11}$$

Weibull defined the risk of failure F as

$$F = -x \ln[S(V_0)] \tag{5-12}$$

Therefore,

$$S(V) = \exp(-F) \tag{5-13}$$

For an infinitesimal volume dV, the risk of failure, F, is shown to depend only on stress where

$$F = \int f(\sigma) dV \tag{5-14}$$

Weibull postulated that

$$F = \left(\frac{\sigma - \sigma_u}{\sigma_0}\right)^m \tag{5-15}$$

where

$\sigma =$ applied stress
$\sigma_u =$ stress below which there is a zero probability of failure
$\sigma_0 =$ a characteristic strength (at which the material's failure probability is 63.2%) that is analogous to the mean value of a normal distribution
$m =$ Weibull modulus that characterizes the variability in the strength of the material, where $1/m$ is analogous to the standard deviation of the material's strength.

This implies an upper limit to the size of defects present in the material. [For the case of fatigue in steel alloys, $\sigma_u = \sigma_{fat}$, where σ_{fat} represents the fatigue endurance limit (see Section 9.2). For brittle ceramics, $\sigma_u = 0$ since any tensile stress might cause failure. Failure would not be expected when $\sigma < 0$ since compressive stresses would act to close the crack.]

Increasing m values reflect more homogeneous material behavior with strength levels for a given component being more predictable. In the limit where m approaches ∞, the probability of fracture is zero for all stress levels $< \sigma_0$. When $\sigma = \sigma_0$, the probability of fracture becomes unity. Conversely, when m approaches zero, F approaches unity and failure occurs with equal certainty at any stress level. For all other values ($0 < m < \infty$), the risk of failure also is a function of the Weibull modulus. For example, if $\sigma_u = 0$ and for the case of the average strength of a material (i.e., where the risk of failure is 0.5),

$$0.5 = (\sigma_{F=0.5}/\sigma_0)^m \tag{5-16}$$

It follows that the stress level corresponding to a risk of failure of 0.5 is given by

$$\sigma_{F=0.5} = (0.5)^{1/m}\sigma_0 \tag{5-17}$$

When $m = 2$ or 20, $\sigma_{F=0.5}$ equals 0.71 σ_0 and 0.97 σ_0, respectively. Note that the lower the value of m, the lower the allowable stress to ensure the same probability of failure. Furthermore, it is dangerous to assign the same factor of safety[i] to the average strengths in the two materials,

---

[i] A factor of safety essentially provides a margin of error in the design of a component. For example, by adding a factor of safety of two to the material's yield strength, the design stress is then set at 50% of the yield strength.

since the same probability of fracture of the two different brittle ceramics corresponds to different fractions of the average strength of the two materials. For example, when m approaches zero, the minimum strength approaches zero and the probability of fracture approaches unity for all stress levels. Conversely, when m approaches ∞, the scatter in strength values approach zero, $\sigma_{min} = \sigma_{avg}$, and $F = 0$ for $\sigma < \sigma_{avg}$.

5.3.1.1 Effect of Size on the Statistical Nature of Fracture

By combining Eqs. 5-13 to 5-16, the probability of survival for a component not of unit volume is

$$S(V) = \exp\left\{-V\left(\frac{\sigma - \sigma_u}{\sigma_0}\right)^m\right\} \qquad (5\text{-}18)$$

Alternatively, the risk of failure is given by $F = 1 - S(V)$ so that

$$F = 1 - \exp\left\{-V\left(\frac{\sigma - \sigma_u}{\sigma_0}\right)^m\right\} \qquad (5\text{-}19)$$

By taking double logarithms of Eq. 5-18 or 5-19 and rearranging, a linear relationship is found between the probability of survival (or failure) of a component and the applied stress with the slope of the data plot being characterized by the Weibull modulus, m (Eq. 5-20).

$$\ln\ln\left(\frac{1}{S(V)}\right) = \ln V + m\ln(\sigma - \sigma_u) - m\ln\sigma_0 \qquad (5\text{-}20a)$$

$$\ln\ln\left(\frac{1}{1 - F}\right) = \ln V + m\ln(\sigma - \sigma_u) - m\ln\sigma_0 \qquad (5\text{-}20b)$$

Tensile strength data[13] for 21 samples of SiC are shown in Table 5.2, where strength levels are ranked in ascending value. These results, assuming $\sigma_u = 0$, are plotted in Fig. 5.6, using Weibull-probability graph paper with $\ln\ln 1/S(V)$ as ordinate and $\ln \sigma$ as abscissa. The Weibull modulus is then given by the negative value of the slope of the plot and is 8.8 for this set of data, while the characteristic life is 327 MPa. (The negative of the slope is used since $\ln\ln 1/S(V)$ increases as $S(V)$ decreases.) Based on these findings, the probability of failure for a SiC specimen of the same type at a stress level of 200 MPa is

$$F = 1 - \exp\left\{-\left(\frac{\sigma - \sigma_u}{\sigma_0}\right)^m\right\} = 1 - \exp\left\{-\left(\frac{200\,\text{MPa} - 0}{327\,\text{MPa}}\right)^{8.8}\right\} = 0.013$$

Table 5.2 Tensile-Strength Data for Self-Bonded Silicon Carbide[13]

Rank (1–21)	1	2	3	4	5	6	7
Strength (MPa)	232	252	256	274	282	285	286
	8	9	10	11	12	13	14
	289	294	308	311	314	316	324
	15	16	17	18	19	20	21
	334	337	339	341	365	379	382

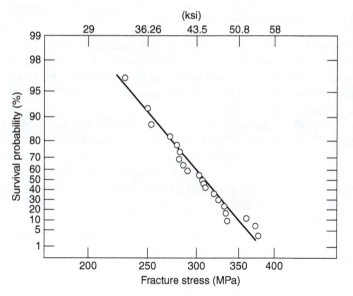

Figure 5.6 Survival probability in SiC as a function of fracture stress,[13] according to Eq. 5-20 with $\sigma_u = 0$. Data listed in Table 5.2. Weibull modulus, $m = 8.8$. (R. W. Davidge, *Mechanical Behaviour of Ceramics*, Cambridge University Press, Cambridge, p. 137 (1979). Reprinted with the permission of Cambridge University Press.)

or 1.3%, and the probability of survival is 98.7%. The same conclusion would be reached by extrapolating the line in Fig. 5.6 to 200 MPa.

Equation 5-18 reveals that for the same probability of survival (or failure), the fracture strength of a material varies with the volume of the component. For example, if $\sigma_u = 0$ and the volume of a real component (1) is greater than that of a test specimen (2), $V_1 > V_2$, we find

$$V_1(\sigma_1)^m = V_2(\sigma_2)^m \tag{5-21}$$

and

$$\sigma_1/\sigma_2 = (V_2/V_1)^{1/m} \tag{5-22}$$

Clearly, for an equal probability of survival (or failure), the larger the volume of the component, the lower the stress necessary for fracture (see Fig. 5.7).

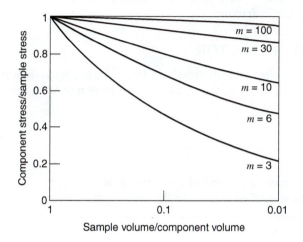

Figure 5.7 Influence of component volume on allowable stress in a component to maintain a given probability of survival or failure.

EXAMPLE 5.2

Two different brittle ceramic parts are being considered for use in a particular application. The Weibull moduli for the two materials are 2 and 10, respectively. If σ_A and σ_B represent the stresses necessary for the same probability of survival in laboratory specimens of materials A and B, respectively, by how much will the respective fracture stresses change if the volume of the full-scale components of each material are ten times larger than that of the laboratory specimens?

In each instance, the stress to produce the same probability of survival will vary with the specimen/component volume ratio (see Eq. 5-22). For the case of material B with the higher Weibull modulus, the stress to generate the same probability of survival will decrease by a factor of 1.26 when the component volume is ten times greater than that of the laboratory specimen. Of greater concern, the stress necessary for the same probability of survival in material A will decrease by a factor of 3.16 when the volume changes by a factor of ten. Clearly, the material with a larger degree of scatter in mechanical properties (i.e., lower Weibull modulus) exhibits a more pronounced size effect.

The probability of survival will also depend on the existence of stress gradients and the associated volume of material that experiences the maximum stress in the component or specimen. For example, it is to be expected that the probability of fracture would be greater under a tensile rather than a flexural loading condition (recall Section 1.4.1). Clearly, a tensile load generates the same stress across a component's entire cross-sectional area, whereas flexural loading generates a maximum stress only at the outermost fiber of the beam. By taking into consideration the respective volumes of the maximum tensile and flexural stresses, it can be shown that the same probability of survival would yield different fracture stress levels in the tensile versus the bend bar samples. For purposes of illustration, the stress ratio between bending and tensile loading for the same probability of survival is given by

$$\sigma_{\text{3-pt bend}}/\sigma_{\text{tensile}} = \{2(m+1)^2\}^{1/m} \tag{5-23}$$

If the Weibull modulus of a given material were 2 versus 10, as in the previous illustrative problem, the ratio of flexural/tensile strengths for an equal probability of survival would be 4.24 and 1.73, respectively. The bend bar is seen to possess the greater strength since its volume under maximum stress is smaller than that of the tensile sample. This is especially true for cases where m values are low. Conversely, as m approaches ∞, the properties of the tensile and bend samples converge to the same value.

As we will see in Chapter 7, however, the probability argument is not the complete explanation of the size effect in fracture.

5.4 THE STRESS-CONCENTRATION FACTOR

By analyzing a plate containing an elliptical hole, Inglis[14] was able to show that the applied stress σ_a was magnified at the ends of the major axis of the ellipse (Fig. 5.8) so that

$$\frac{\sigma_{\max}}{\sigma_a} = 1 + \frac{2a}{b} \tag{5-24}$$

where

$\sigma_{\max}$ = maximum stress at the end of the major axis
σ_a = applied stress applied normal to the major axis
a = half major axis
b = half minor axis

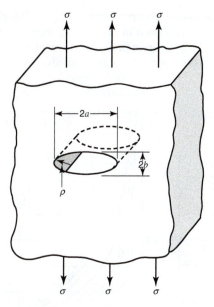

Figure 5.8   Elliptical hole in infinitely large panel produces stress concentration of $1 + 2a/b$.

Since the radius of curvature ρ at the end of the ellipse is given by

$$\rho = \frac{b^2}{a} \tag{5-25}$$

Eqs. 5-24 and 5-25 may be combined so that

$$\sigma_{max} = \sigma_a(1 + 2\sqrt{a/\rho}) = \sigma_a k_t \tag{5-26}$$

In most cases $a \gg \rho$, therefore

$$\sigma_{max} \approx 2\sigma_a\sqrt{a/\rho} \tag{5-27}$$

The term σ_{max}/σ_a is defined as the stress-concentration factor k_t and describes the effect of crack geometry on the local crack-tip stress level. Many textbooks and standard handbooks describe stress concentrations in components with a wide range of crack configurations; a few are listed in the references.[15,16] Although the exact formulations vary from one case to another, they all reflect the fact that k_t increases with increasing crack length and decreasing crack radius. Therefore, all cracks, if present, should be kept as small as possible. One way to accomplish this is by periodic inspection and replacement of components that possess cracks of dangerous length. Alternatively, once a crack has developed, the relative severity of the stress concentration can be reduced by drilling a hole through the crack tip. In this way, ρ is increased from a curvature associated with a natural, sharp crack tip to that of the hole radius. In fact, it was suggested by Wilfred Jordan that the crack in the Liberty Bell "might have been stopped by boring a very small hole through the metal a short distance beyond the former termination of the first crack. [However], it was never done, with the result that the original fracture, known to millions, has more than doubled in length and has extended up and around the shoulder of the Bell." (This can be seen in the book cover image.)[17]

   The latter procedure is used occasionally in engineering practice today but should be employed with caution. Obviously, a crack is still present after drilling and may continue to grow beyond the hole after possible reinitiation. Also, field failures have occurred earlier than expected, simply because the blunting hole was introduced *behind* the crack tip, and the sharp crack-tip radius was not eliminated.

Stress concentrations may also be defined for component configuration changes, such as those associated with section size changes. As shown in Fig. 5.9, k_t increases for a number of design configurations whenever there is a large change in cross-sectional area and/or where the associated fillet radius is small.[ii] For every book written on the analysis of stress concentrations,

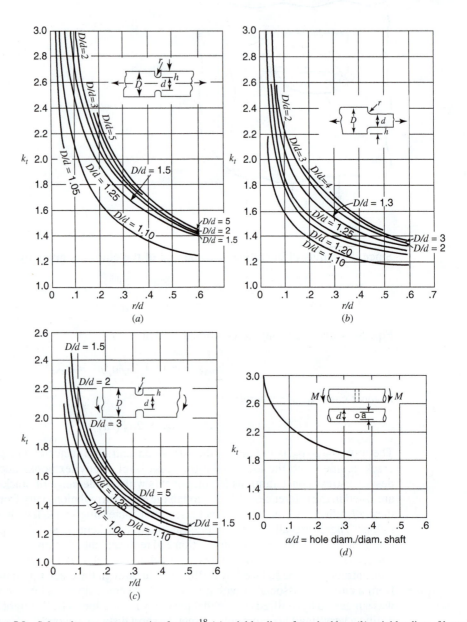

Figure 5.9 Selected stress-concentration factors:[18] (*a*) axial loading of notched bar; (*b*) axial loading of bar with fillet; (*c*) bending of notched bar; (*d*) bending of shaft with transverse hole; (*e*) axial loading of bar with transverse hole; (*f*) straight portion of shaft keyway in torsion. (From *Metals Engineering; Design* by American Society of Mechanical Engineers. Copyright © 1953 by the American Society of Mechanical Engineers. Used with permission of McGraw-Hill Book Company.)

[ii] Note that the solutions presented here for the stress concentration factor assume elastic loading. If plastic deformation occurs in the zone of high stress, the stress concentration will be lower than predicted by the elastic solution. For more information about estimating the local stress in this case, see T. Seeger and P. Heuler, *J. Test. Eval.* 8(4), 199–204 (1980).

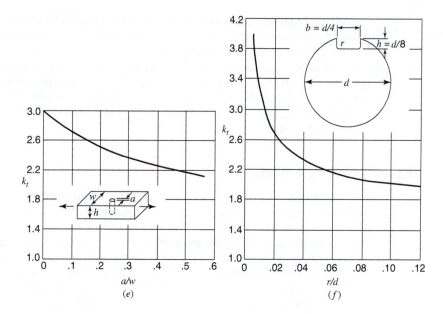

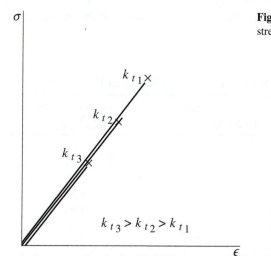

Figure 5.9   (*Continued*)

no doubt many others could be filled with case histories of component failures attributable to either ignorance or neglect of these same factors.

The completely elastic stress–strain material response is affected greatly by the presence of stress concentrations within the sample. As shown schematically in Fig. 5.10, the maximum stress and strain level that any component may support decreases with increasing k_t. This explains, in part, why many ceramic materials are much weaker in tension than in compression. The higher the stress concentration, the easier it is to break the sample. This fact is put to good use regularly by glaziers, who first score and then bend glass plate to induce fracture with minimal effort and along a desired path.

Fortunately, stress concentrations in most materials will not result in the escalation of the local crack-tip stress to dangerously high levels. Instead, this potentially damaging stress elevation is avoided by plastic deformation processes in the highly stressed crack-tip region.

Figure 5.10   Effect of stress concentration k_t on allowable stress and strain in completely elastic material.

As a result, the local stress does not greatly exceed the material's yield strength level as the crack tip blunts, thereby reducing the severity of the stress concentration. The ability of a component to plastically deform in the vicinity of a crack tip is the saving grace of countless engineering structures. It should also be recognized that a component that contains a stress concentration may still perform in a satisfactory manner. This follows from the fact that the local stress at the crack tip is represented by the product of the applied stress and k_t values (Eq. 5-27); as long as the applied stress level is sufficiently low, $\sigma_a k_t$ values will remain comfortably below the local stress level necessary for fracture.

5.5   NOTCH STRENGTHENING

An interesting turn of events may occur with regard to the failure behavior of notched components when an appreciable amount of plastic deformation is possible (assuming that plastic yielding is considered failure). We saw in Chapter 2 that plastic constraints are developed in the necked region of a tensile bar as a result of a triaxial stress state; the unnecked regions of the sample experience a lower true stress than the necked section and, therefore, restrict the lateral contraction of the material in the neck. Similar stress conditions exist in the vicinity of a notch in a round bar. When the net section stress reaches the yield-strength level, the material in the reduced section attempts to stretch plastically in the direction parallel to the loading axis. Since conservation of volume is central to the plastic deformation process, the notch root material seeks to contract also, but is constrained by the bulk of the sample still experiencing an elastic stress. The development of tensile stresses in the other two principal directions—the constraining stresses—makes it necessary to raise the axial stress to initiate plastic deformation. The deeper the notch, the greater is the plastic constraint and the higher the axial stress must be to deform the sample. Consequently, the yield stress (not necessarily the yield load) of a notched sample may be *greater* than the yield strength found in a smooth bar tensile test. The data shown in Table 5.3 demonstrate the "notch-strengthening" effect in 1018 steel bars, notched to reduce the cross-sectional area by up to 70%.

A laboratory demonstration is suggested to illustrate a seemingly contradictory test response in two different materials. First, austenitize and quench a high-strength steel, such as AISI 4340, to produce an untempered martensite structure, and then perform a series of notched tensile tests. You will note that the net section stress at failure *decreases* with increasing notch depth because of the increasing magnitude of the stress-concentration factor, as expected from the discussion in Section 5.4. Now conduct notch tests with a ductile material such as a low-carbon steel or aluminum alloy. In this case, note that the net section stress at failure will *increase* with increasing notch depth as a result of the increased plastic constraint. In this manner, you may prove to yourself that materials with limited deformation capacity will *notch weaken,* and highly ductile materials will *notch strengthen.*

Table 5.3   Notch Strengthening in 1018 Steel

Reduction of Area in Notched Sample	Yield Strength Ratio $\frac{\text{notched bar}}{\text{smooth bar}}$
0	1.00
20	1.22
30	1.36
40	1.45
50	1.64
60	1.85
70	2.00

EXAMPLE 5.3

Two 0.5-cm-diameter rods of 1020 steel ($\sigma_{ts} = 395$ MPa) are to be joined with a silver braze alloy 0.025 cm thick ($\sigma_{ys} = 145$ MPa) to produce one long rod. What will be the ultimate strength of this composite? The response of this bar may be equated to that of a notched rod of homogeneous material. In this instance, preferential yielding in the weaker braze material would be counterbalanced by a constraining triaxial stress field similar to that found in a notched bar of homogeneous material. As such, the strength of the joint will depend to a great extent on the geometry of the joint. Specifically, it would be expected that braze joint constraint would increase with increasing rod diameter and decreasing joint thickness. The experimental results by Moffatt and Wulff[19] (Fig. 5.11) reflect the importance of these two geometrical variables on the composite strength $\bar{\sigma}$. Accordingly, $\bar{\sigma}$ is found to be approximately 345 MPa.

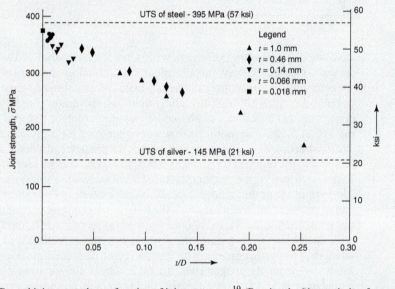

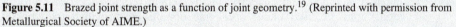

Figure 5.11 Brazed joint strength as a function of joint geometry.[19] (Reprinted with permission from Metallurgical Society of AIME.)

Two factors need to be emphasized when discussing the observed notch-strengthening effect. First, even though the notched component may have a *higher net section stress*, it still requires a *lower load* for failure than does the smooth sample when based on the *original* cross-sectional area. I trust that this should temper the enthusiasm of any overzealous student who might otherwise race about, hacksaw in hand, with the intent of "notch strengthening" all the bridges in town. Second, there is a limit to the amount of notch strengthening that a material may exhibit. From theory of plasticity considerations, it is shown that the net section stress in a deformable material may be elevated to $2\frac{1}{2}$ to 3 times the smooth bar yield-strength value. It would appear, then, that the brazed joint system described in the previous example represented an optimum matching of material properties. Using a higher strength steel with the same braze alloy would not have made the joint system stronger.

Why discuss the effect of a notch on yield strength here, when this is a chapter focused on fracture? It turns out that notch behavior has important implications for fracture strength as well, as discussed in the following section.

5.6 EXTERNAL VARIABLES AFFECTING FRACTURE

As mentioned in Section 5.4, the damaging effect of an existing stress concentration depends strongly on the material's ability to yield locally and thereby blunt the crack tip. Consequently,

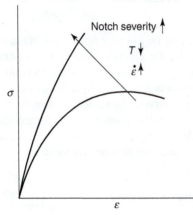

Figure 5.12   Effect of temperature, strain rate, and plastic constraint on flow curve.

anything that affects the deformation capacity of the material will affect its fracture characteristics as well. Obviously, any metallurgical strengthening mechanism designed to increase yield strength will simultaneously suppress plastic deformation capacity and the ability of the material to blunt the crack tip. For any given material, there are, in addition, a trilogy of external test conditions that contribute to premature fracture: notches, reduced temperatures, and high strain rates (Fig. 5.12). The more the flow curve is raised for a given material, the more likely brittle fracture becomes. As we saw in Section 5.5, the presence of a notch acts to plastically constrain the material in the reduced section and serves to elevate the net section stress necessary for yielding. Likewise, lowering the test temperature and/or increasing strain rate will elevate the yield strength (the magnitude of this change depends on the material). For example, yield-strength temperature and strain-rate sensitivity vary with the Peierls stress component of the overall yield strength of the material (Section 2.1.4). In BCC alloys, such as ferritic steels, the Peierls stress increases rapidly with decreasing temperature, thereby causing these materials to exhibit a sharp rise in yield strength at low temperatures. This can and often does precipitate premature fracture in structures fabricated from these materials. In other materials, such as FCC alloys, the Peierls stress component is small. Consequently, yield-strength temperature sensitivity is small, and these materials may be used under cryogenic conditions, provided they possess other required mechanical properties. Some thermoplastic polymers exhibit strong temperature and strain rate effects, particularly in the vicinity of the glass transition temperature. Ceramics and glassy solids, on the other hand, are relatively insensitive to temperature and strain rate over a wide operating range because they have an inherently limited capability for plastic deformation.

5.7   CHARACTERIZING THE FRACTURE PROCESS

Many words and phrases have been used to characterize failure processes in engineering materials. Since these terms are born of different disciplines, each having its own relatively unique jargon, confusion exists along with some incorrect usage. In this section, discussion of failure will be restricted to fracture (as distinct from elastic buckling, plastic yielding, or other processes that can cause a component to deviate from its intended performance). Complicating matters, even within this narrow definition several words are used in multiple contexts. In an attempt to simplify and clarify this situation, we have found it convenient to describe the fracture of a test sample or engineering component in terms of three general characteristics: energy of fracture, macroscopic fracture path and texture, and microscopic fracture mechanisms. Although details of fracture differ among the material classes, these general characteristics can be used to identify the degree of component toughness as well as the fracture origin and the loading mode. Here we primarily address mechanical overload failures—fractures that occur because on one specific occasion the applied stress exceeds a critical level. Unique characteristics of time-dependent failures that can occur as the result of much lower stresses are deferred to Chapters 8–10.

Often one of the first questions asked in a failure analysis is "what was the mode of the failure?" Already we may arouse confusion because there are several legitimate ways to interpret this question! One very useful approach is to designate the behavior of the material in the failure process as either tough or brittle. As discussed in Chapter 1, the toughness of a given material is a measure of the energy absorbed before and during the fracture process. The area under the uniaxial tensile stress–strain curve would provide one measure of toughness as described by Eq. 1-20, reproduced here for convenience:

$$\text{energy/volume} = \int_0^{\varepsilon_f} \sigma d\varepsilon \qquad (1\text{-}20)$$

If the energy is high (such as for curve C in Fig. 1.8), the material is said to be *tough* or to possess *high fracture toughness*. Conversely, if the energy is low (e.g., curves A and B in Fig. 1.8), the material is described as *brittle* or having *low fracture toughness*. It is the job of *fracture mechanics* to turn the general description of energy absorption into a quantitative measure of toughness that can be used in failure analysis or component design. That will become our focus of attention beginning in the next chapter.

The relative toughness or brittleness of a given material subjected to different conditions may be estimated by noting the extent of plasticity surrounding the crack tip. Since the stress concentration at a crack tip will often elevate the local applied stress above the level necessary for irreversible plastic deformation, a zone of plastically deformed material—the *process zone* in which the fracture process is actually taking place—will be found there. This region will be surrounded by elastically deformed media. Since much more energy is dissipated during plastic flow than during elastic deformation, the toughness of a sample should increase with the volume of the crack-tip plastic zone. As shown in Fig. 5.13, when the plastic zone is small just before failure, the overall toughness level of the sample is low and the material is classified as relatively brittle. On the other hand, were plasticity to extend far from the crack tip, the energy to break would be high and the material would be relatively tough. It is important to note that a normally tough material can fail with relatively low toughness if the geometry of the component reduces the capacity for plasticity (e.g., by introducing a triaxial stress state [recall Section 2.4.2.3] at a sharp notch) or the environment modifies the material behavior.

It is obvious that real failures do not exhibit conveniently shaded surface features that clearly delineate the plastic zones as sketched in Figure 5.13. So, how can we determine the extent of plastic deformation in practice? Examination of a failed component at a macroscopic level may be sufficient to answer this critical question. This is typically accomplished with the naked eye, or

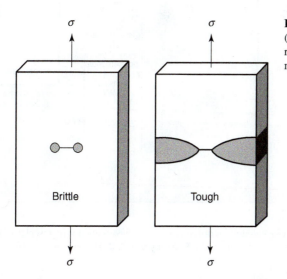

Figure 5.13   Extent of plastic zone development (shaded region) at the ends of a center crack for relatively brittle and tough fractures in the same material.

perhaps with the aid of a stereomicroscope for which magnifications are on the order of 1–100×. Visual evidence of plasticity (and therefore relative toughness for a given material) is available from the shape of the material in the vicinity of the fatal crack. Volume conservation during plastic deformation can cause a noticeable reduction in cross-sectional area, which will appear as a change in surface topology from flat (elastically loaded) to concave (plastically deformed) in the vicinity of the plastic zone.

Because there is a connection between plasticity and toughness, failures are often classified as macroscopically *ductile* or *brittle*; these terms are often assumed to be synonymous to the tough or brittle designations as previously defined. Given the preceding discussion, and the reuse of the word "brittle," this may appear to make good sense. However, a *ductile failure* is simply one that exhibits visible plasticity prior to failure, whereas in this particular nomenclature scheme a *brittle failure* does not. This may be a very useful distinction within a group of *similar* materials (e.g., steel alloys that undergo a temperature-induced transition from high toughness to low toughness) but it is dangerous to apply this terminology more broadly. For example, it is shown in Figure 5.14 that relatively ductile polymers such as polypropylene have

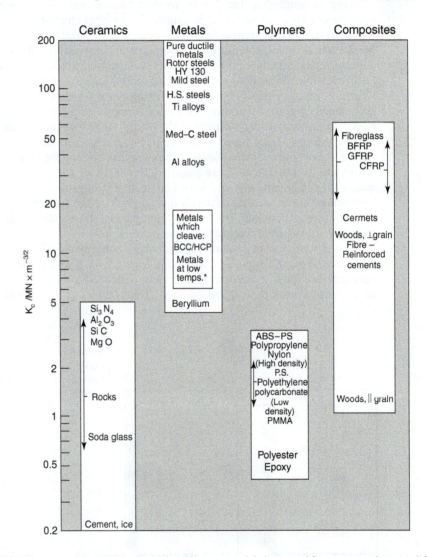

Figure 5.14 Fracture toughness (K_C) plotted for different material classes and for representative materials within each class. (Reprinted from *Engineering Materials*, M. F. Ashby and D. R. H. Jones, p. 126, 1980, with permission from Elsevier.)

lower absolute toughness values than metals that fail with very little plasticity (those that "cleave"). This is because toughness depends on a combination of strength and ductility. Evidence of ductility or lack thereof is therefore not a reliable indicator of absolute toughness; furthermore, the terms ductile (plastic) and brittle (non-plastic) in this context are not equivalent to the designations of tough (high energy absorption) and brittle (low energy absorption) in a fracture energy context. In this text we will attempt to avoid this confusing situation by using "brittle" and "tough" to indicate relatively low and high toughness, respectively, from an energy perspective.

Although we do not recommend the use of the ductile vs. brittle designations in describing fracture mode, they do have value in certain circumstances such as establishing a connection between the potential for component inspection and failure avoidance. Presumably a ductile failure would be preceded by visible plasticity, and would therefore allow for the possibility of visual detection and an appropriate engineering response before catastrophic fracture takes place. For example, the 2007 failure of the I-35 bridge in Minneapolis, Minnesota, has been attributed to fracture of certain steel gusset plates. (Gusset plates serve to connect two or more adjoining bridge members.) Following this tragic bridge collapse, examination of old inspection photographs revealed that some of the gusset plates exhibited visible bowing.[21] Increased loading over time, due to additions of concrete to the roadway and the presence of construction vehicles parked on the deck, caused already-bent gusset plates to fracture. Brittle fractures, on the other hand, offer few warning signs and thus appear to occur suddenly.

5.8   MACROSCOPIC FRACTURE CHARACTERISTICS

As is obvious from Figs. 5.1 and 5.2, engineering service failures can generate large areas of fracture surface. Since a key element in analyzing a given failure lies in identifying the mechanism(s) by which a critical crack developed, it is necessary for investigators to focus most of their attention on the small crack origin region rather than the vast areas associated with rapid, unstable fracture. A piecewise examination of the *entire* fracture surface of the T-2 tanker (Fig. 5.1a), a square millimeter at a time, would take years to complete, would be extremely costly, and would prove to be excessively redundant. Fortunately, the crack often leaves a series of *fracture markings* in its wake that may indicate the relative direction and character of crack motion. These markings differ from one material class to another and from one failure mode to another, but the stress level, the loading conditions, and the direction of crack growth are all often evident to the educated observer. These features will be described in the following sections, grouped by the material class in which they appear.

5.8.1   Fractures of Metals

Metals have the greatest potential for high fracture toughness of any material class, which is arguably the primary reason that metals are so widely used for structural applications. However, as seen in Figure 5.14, the fracture toughness of metals can vary by approximately two orders of magnitude. To a large degree this is due to the wide variation in ductility found within the class of metals. Metal toughness may also be sensitive to variations in processing or use conditions, and this sensitivity may be evident from the extent of plastic deformation present. The toughness of metal alloy sheet- or plate-type components often can be reflected by the relative amounts of macroscopically flat and slant fracture (Fig. 5.15). Here we find that toughness levels are higher in association with a slant-fractured appearance and correspondingly lower when the fracture surface is essentially flat. Since the explanation for this behavior requires a significant fracture mechanics toolbox, discussion of this behavior is deferred to Section 6.9.

Failures due primarily to shear or torsional loading do not create shear lips, but ductile tearing of the fracture surface tends to leave other evidence of plastic deformation: linear tearing marks in the case of shear and spiral marks in the case of torsion (Fig. 5.16).

Crack path offers another potential indicator of tough vs. brittle metal fracture. Tough metal failures tend to result in a single crack, simplifying determination of the origin. Brittle metal failures,

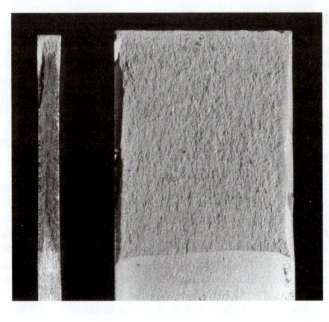

Figure 5.15 Fractured surfaces of aluminum test specimens revealing slant- and flat-type failure. Decreasing toughness level corresponds with increasing fraction of flat fracture, as on the right.

however, are essentially elastic to the point of fracture and may therefore release considerable energy with nowhere to go except into crack growth. In these cases there may be significant crack branching during failure, resulting in multiple fracture surfaces. Thankfully there are unique macroscopic fracture surface markings that are often visible in brittle metal failures that allow the failure process to be reconstructed. In some components, brittle failures leave radial markings that fan out from the crack origin, as shown in Fig. 5.17 for the case of a 6-cm-diameter steel reinforcing bar ($\sigma_{ys} = 550$ MPa) with a fracture origin near the center.

If the growth of radial lines is dimensionally constrained across the width of a plate or sheet, they develop a distinctive form (called *chevron* markings) with curved lines that diverge from the mid-thickness of the fracture surface as shown in Fig. 5.18. It is believed that small brittle cracks nucleate in the highly stressed zone ahead of the primary crack then grow back to meet the advancing crack, forming these curved tear lines. (Another example of chevron markings is shown in Fig. 6.33c.) The chevron markings in a highly branched failure may point in different directions

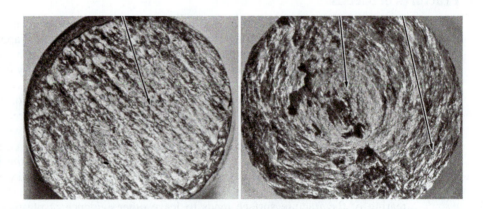

Figure 5.16 Visible evidence of plasticity in shear and torsion overload failures of annealed Ti-6Al-4V cylindrical rods. (A. Phillips, V. Kerlins, and B. V. Whiteson, *Electron Fractography Handbook*, Air Force Materials Laboratory Technical Report ML-TDR-64-416, 1965.)

Figure 5.17 Radial markings pointing back to an internal flaw in steel reinforcing bar. (Courtesy of Roger Slutter, Lehigh University.)

with regard to the component geometry, but it is important to recognize that the different sets of chevron markings all point in the same *relative* direction—back toward the origin (inset, Fig. 5.18).

5.8.2 Fractures of Polymers

Although the absolute toughness range for polymers is low compared to that of metals (Fig. 5.14), polymeric materials are ubiquitous in consumer products. They also play critical roles as medical device components and as structural adhesives for which failure can be catastrophic.[iii] Polymers, like metals, can vary considerably in their capacity for plastic deformation and therefore relative toughness, so visible evidence of plasticity can be important in evaluating a polymer fracture. However, the mechanisms responsible for plasticity (and energy dissipation in general) in polymers are quite different from those active in metals; as such, it is possible for a polymeric material to exhibit reasonable fracture toughness without macroscopic evidence of significant distortion by plastic flow (i.e., obvious cross-section size change). This is a significant departure from metal behavior.

In certain polymers, particularly some glassy thermoplastics, microscale plasticity mechanisms like crazing (Fig. 5.19, also discussed in Section 5.9.2) and shear band formation can occur without

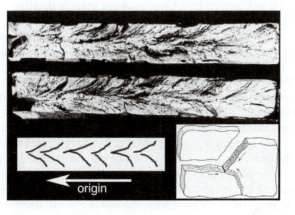

Figure 5.18 Chevron markings in metals curve in from the two surfaces and point back to the crack origin. (Courtesy of Roger Slutter, Lehigh University.)

origin

[iii] As in the case of the 2006 ceiling collapse in Boston's Ted Williams Tunnel.

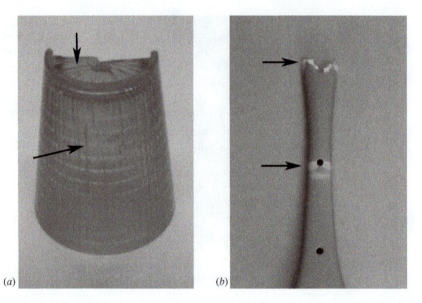

Figure 5.19   (*a*) Crazing in a plastic cup. (*b*) Stress whitening in a broken electric toothbrush handle.

any major distortion occurring at the macroscopic scale. The presence of crazes or shear bands is an indication of damage accumulation, highly localized yielding, and energy absorption prior to outright fracture. Stress whitening—induced scattering of light associated with microscopic damage—may also be visible macroscopically, but embrittling and toughening micromechanisms can both cause this visual effect so a simple conclusion regarding plasticity cannot be made from its presence. Thus the visual distinction between "ductile" and "brittle" fracture in polymers is problematic, and we should depend on the degree of energy absorption to determine if a failure involves relatively high or low fracture toughness. In light of the challenges involved in interpreting visual evidence, it is common to categorize brittle and tough polymer fractures as those with low strain at failure (e.g., less than 4% in tension) and those with greater strain at failure, respectively.[23]

Another critical difference between polymer fracture and metal fracture is the degree to which temperature, strain rate, and stress level can affect the toughness level and hence the fracture mode. Polymers may, for example, show enormous changes in toughness with fairly small changes in temperature. Likewise, the same polymer may fracture by brittle, plastic, or creep cracking modes when subjected to high, medium, or low constant stresses.[24] These phenomena are all closely tied to the substantial time-dependence of polymer behavior. The reader is encouraged to review Chapter 1 or Chapter 4 for greater details.

At room temperature, brittle fracture is typical for glassy thermoplastics such as polystyrene (PS), polymethylmethacrylate (PMMA, acrylic), and unplasticized polyvinylchloride (PVC), and for highly cross-linked thermosets such as epoxies. As discussed in Chapter 1, these materials are either operating below or very close to T_g at room temperature, or are so highly cross-linked that the temperature has little influence on behavior. Elastomers may accommodate very high strains, but there is generally very little plastic strain involved so fracture is also sudden. In brittle failures, fracture features identical to those in glasses and ceramics may be formed: *mirror*, *mist* and *hackle* zones, *Wallner lines*, and significant crack branching associated with high energy input. These features are described in detail in Section 5.8.3. Parabolic markings, somewhat reminiscent of chevrons in brittle metal failures, can also aid in determining the location of the crack origin (microscopic versions are pictured in Fig. 5.40). Crazes or microcracks that form in the process zone near the crack tip (and then link up to form additional crack length) are likely to underlie the brittle fracture process. It may be possible to distinguish between these two mechanisms macroscopically by identifying a rainbow sheen that is due to the presence of a polymer fibril "carpet" on the fracture surface. These fibrils are specifically associated with the crazing process.

Higher toughness at room temperature is typical of many semicrystalline thermoplastics such as polyethylene (PE), polypropylene (PP), para-aramid (PA, e.g., Kevlar), and nylon, and also of certain amorphous polymers including polycarbonate (PC), polyethylene terephthalate (PET), and plasticized PVC. There may be obvious tearing, shear, and plastic yielding at the macroscopic level, but clear indications of crack propagation direction are often absent. As with higher toughness metal fractures, the fragments are likely to be few in number so identification of the origin still may be tractable even without obvious directional indicators.

5.8.3   Fractures of Glasses and Ceramics

Glasses and ceramics must be used very cautiously in critical load-bearing applications[iv] due to their low absolute toughness under ordinary use conditions (Fig. 5.14) and their lack of capacity for plastic deformation. However, this lack of plasticity somewhat simplifies failure analysis by macroscopic observation because relative toughness within this material class will be dependent only on material strength, not ductility. One of the best indicators of the absolute energy level in a brittle ceramic or glass fracture, and therefore the strength of the material involved, is the macroscopic appearance of the crack path.

High-strength glass and ceramic materials can experience high stress levels before failure occurs. These relatively high-strength fractures release their stored elastic energy violently, resulting in significant *crack branching* (Fig. 5.20a) and numerous fragments that are themselves often rich in surface features. The distance between branches can sometimes be used to evaluate the failure stress level quantitatively, as will be discussed shortly. Tempered glass fracture is a special case of relatively high-energy failure in which most of the energy released comes from the residual stress present in the glass. Such a failure is virtually unmistakable, with an extensive network of cracks that can reduce an entire sheet of glass to small fragments (Fig. 5.21). In contrast, very low-energy fractures associated with weak materials result in very little crack branching (Fig. 5.20c) and few fracture markings on the fracture surfaces. A good example of this is a thermal shock stemming from a small temperature gradient and an edge flaw. In this case, a single crack typically extends straight out from the site of origin (Fig. 5.22). The crack may meander after a short distance, but branching is minimal. If the crack extends outside the region of the thermal gradient or if the edge flaw was severe, there may be insufficient driving force for continued growth and the crack can arrest within the material.

The crack path can also indicate the type of loading that existed at the time of fracture. Like the wide metal plate in Fig 5.15, a glass or ceramic fracture surface formed under pure tensile loading would appear as a flat plane aligned perpendicular to the tensile loading direction. Thermal loading can create this condition at an edge flaw, creating fractures that are initially perpendicular to the plate edge. Even after crack meandering begins, the fracture plane remains perpendicular to the plate surfaces. Aside from thermal fractures, however, flat cracks are not very common because pure tensile mechanical loading of glass or ceramic components rarely occurs.

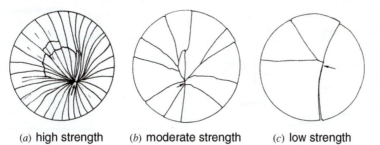

Figure 5.20 Energy level and crack branching patterns associated with three glass failures with central origins. (Adapted from G. D. Quinn, *Fractography of Ceramics and Glasses*, NIST Special Pub 960-16, 2007.)

(a) high strength   (b) moderate strength   (c) low strength

[iv] A spectacular example is the Grand Canyon Skywalk introduced in 2007, where visitors may walk on a glass pathway suspended 1100 meters (3600 feet) above the canyon floor.

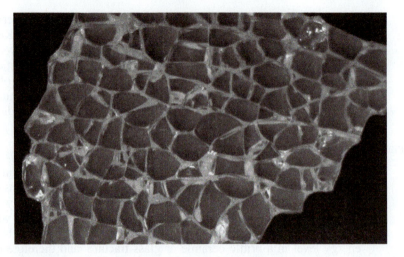

Figure 5.21   Typical fracture of tempered glass illustrating the network of cracks that forms when the stored energy in the glass is suddenly released. The many fragments, still loosely held together, are each about 1 cm in diameter. (Photo courtesy of Samuel Lawrence, Lehigh University.)

Much more common in service is bending failure, as shown in Fig. 5.23a, during which one side of the component is in tension and the other in compression. As might be expected, propagating cracks begin on the tension side then deviate from their perpendicular growth pattern as they approach the compressive side of the component. This results in a characteristic *cantilever curl*.[25] If the energy level is high enough, crack branching may occur and the curl may bifurcate. This creates a "Y" profile, with a chip formed between its arms. Somewhat less common than bending, torsional loading of a rod creates tensile stresses at an angle (45°, as discussed in Chapters 2 and 3) to the torsion axis. This often results in a distinctive helical fracture that can easily be modeled by twisting a piece of blackboard chalk (Fig. 5.23b); the similarity to other brittle materials can be seen by comparing Fig. 5.23b to the torsional fracture of a finger bone in Fig. 5.23c.

We now consider the case of a complicated fracture involving many fragments for which the primary crack path may not be immediately apparent. Undaunted, the intrepid fractographer looks for features on the fracture surfaces themselves that can offer important clues about the direction of crack propagation. Depending on the fracture conditions, these fracture markings

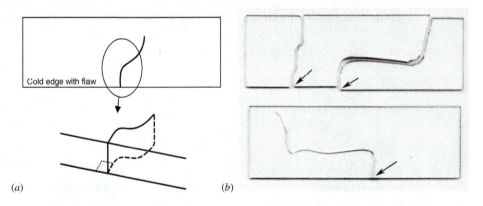

(a)   (b)

Figure 5.22   (a) Sketches of a typical thermal fracture crack in glass that emerges from an edge at 90°, meanders, and then terminates within the plate. (b) Glass microscope slides with complete (top) and incomplete (bottom) thermal fractures that were initiated at flaws on the bottom edges. The crack origins are marked with arrows.

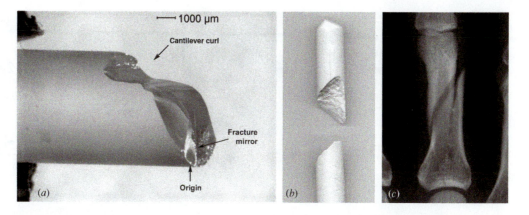

Figure 5.23   (*a*) Cantilever curl seen in glass caused by bending, (*b*) helical fracture of chalk caused by torsion, (*c*) helical fracture of a finger bone caused by torsion. (Photo (*a*) from G. D. Quinn, *Fractography of Ceramics and Glasses*, NIST Special Pub 960-16, 2007; x-ray image (*c*) courtesy of Michelle Geoffrion-Vinci.)

may be visible with the naked eye. Glass and ceramic fracture surfaces often exhibit *Wallner lines*, also known as *rib marks*, that are curved features spreading out in the crack growth direction like ripples from a stone dropped in a pond. Note that this is the opposite curvature with respect to the crack origin as displayed by metal chevron or polymer parabolic markings. In a bent plate failure like that in Fig. 5.24, for instance, asymmetric primary Wallner lines spread out from the tensile side (the bottom edge), clearly identifying the type of loading, the direction of bending, and the direction of crack propagation. Wallner lines are also frequently present in ceramics, but may be more difficult to observe than in glasses, particularly when the ceramic has a large grain size or lacks any glassy phase.

Once the macroscopic crack path has been identified, the fractographer's attention can turn to analyzing the nature of the crack initiation event. In many cases, features in the immediate vicinity of the origin can be used for this purpose. In particular, the fracture origin in glasses and ceramics often reveals several characteristic regions as shown in Fig. 5.25.[26–34] Surrounding the origin is a *mirror region* named after its highly reflective fracture surface. This relatively smooth area is bordered by a *misty region* that contains small visible radial ridges. The mist region in turn is surrounded by an area that is much rougher in appearance. Depending on the size of the sample, this *hackle region* may be bounded by macroscopic crack branching. The

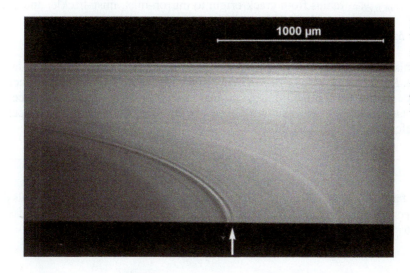

Figure 5.24   Primary Wallner lines. Tension side is on bottom edge and origin is to the left. (G. D. Quinn, *Fractography of Ceramics and Glasses*, NIST Special Pub 960-16, 2007.)

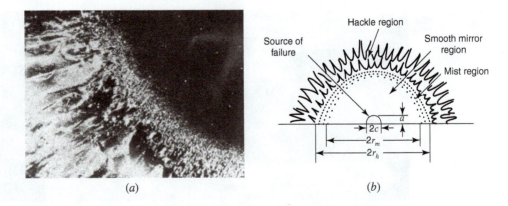

Figure 5.25 Fracture surface appearance in glassy ceramic revealing mirror, mist, and hackle regions. (*a*) Plate-glass fracture surface. Tensile fracture stress = 28.3 MPa, or 4.1 ksi. (From Orr.[33]) Arrow indicates approximate crack origin. (*b*) Schematic diagram showing different fracture regions and approximate textural detail. (From Mecholsky et al.[28] Copyright, American Society for Testing and Materials, Philadelphia, PA. Reprinted with permission.)

hackle marks are similar in shape, but not in genesis, to the wake hackle already described. The progression from mirror to mist to hackle occurs as the crack velocity increases from the initial rate to a material-dependent maximum, as it typically does in a catastrophic fracture. The size (and therefore the visible presence) of these different regions depends on such factors as the stress level, specimen size, and the material in question. For example, if the strength of the glass is low and the component size is small, it is not uncommon for the mirror zone to extend across the entire fracture surface.[32] In addition, the microstructural characteristics of a ceramic material (e.g., grain boundaries, pores, and second phases) can make observation of these features more difficult than in glassy materials.

Of great significance, the radii of the various fracture regions, as well as the distance between macroscopic crack branching points, have been shown to vary inversely with the square of the fracture stress with a relation of the form

$$r_{m,h,cb} = \left(M_{m,h,cb}/\sigma\right)^2 \tag{5-28}$$

where

σ = fracture stress, in MPa units

$r_{m,h,cb}$ = radius from crack origin to mirror–mist, mist–hackle, and crack-branching boundaries, respectively, in m units

$M_{m,h,cb}$ = "mirror constant" corresponding to r_m, r_h and r_{cb}

This, then, allows a reliable estimate to be made of the local stress state at the time of failure. Relations can also be drawn between the mirror constant and other fracture properties, as discussed in Section 6.15. For additional discussion of the fracture surface appearance of glasses and ceramics, see the extensive review by Rice and several other papers contained in that reference, as well as a more recent publication by Quinn.[25,35]

EXAMPLE 5.4

The fracture surface of a glass-fiber reinforced polyethylene terephthalate (PET) composite reveals the presence of ruptured glass fibers. Estimate the stress at fracture for the fiber identified by "A" in the accompanying photograph, Fig. 5.26.

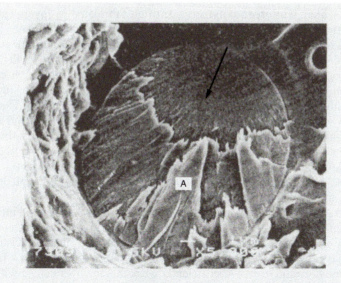

Figure 5.26 Fracture surface features appearing on a glass reinforcement fiber. (Courtesy of H. Azimi.)

This broken fiber shows an example of a mirror–mist zone (see the arrow) where the radius is measured to be 4.6 μm. Since the mirror constant for E-glass is approximately 2 MPa $\sqrt{m}$, it follows from Eq. 5-14 that the stress to cause fracture of this fiber is

$$\sigma = M_m / \sqrt{r_m}$$
$$\sigma = 2 \times 10^6 / \sqrt{4.6 \times 10^{-6}}$$
$$\therefore \sigma = 933 \, MPa$$

This value is smaller than estimates for the tensile strength of E-glass (see Table 1.11) (i.e., 1400–2500 MPa) and may reflect fiber damage during manufacture of the composite.

Interestingly, fractures in cortical bone (a natural ceramic/polymer composite structural material) show some of the same macroscopic fracture features as glasses and ceramics. Long bones broken in pure bending may show cantilever curl, in pure torsion helical fracture may be evident, and higher energy fractures (regardless of loading mode) cause crack branching and formation of multiple fragments (a.k.a. comminuted fractures).[36] A *spiral* or *torsion fracture* of a proximal phalanx (i.e., the finger bone closest to the hand) can be seen in Fig. 5.23c; the striking similarity to the chalk fracture in Fig. 5.23b is evident. Fracture markings on cortical bone may also include features such as hackle marks that allow reconstruction of crack growth direction.[37] Unlike engineered glasses and ceramics, however, bone is capable of limited plasticity (and is sometimes called *quasi-brittle*), so quantitative analysis of surface features must be applied with caution.

5.8.4 Fractures of Engineering Composites

The fracture morphologies of commercially-available composite materials vary markedly because they depend on the individual characteristics of the many reinforcements and matrices that are used. In addition, fracture surfaces and fracture paths are found to vary with the nature of the fiber–matrix interface and its dependence on the environment and temperature. Here, we choose to limit our discussion to those characteristics typical of fiber-reinforced polymer matrix composites. Examination of the macroscopic fracture features may provide information about the loading conditions that led to failure and the toughness level of the material, and may also help to identify the particular constituent that is the "weak link" in a certain composite structure.

Polymer matrix fiber reinforced polymers are often fabricated in layers (plies), each of which has fibers oriented in a single direction. The layers are bonded together by the matrix material. Fractures of laminated composites may be separated into translaminar (transverse, across the plies) and interlaminar (delamination between the plies) failures. Both may occur in a single failure, but one mode generally precedes the other. Determining which came first is a matter of deduction from observing the crack paths and fracture surfaces features.

Interlaminar failures appear as a separation of the plies from one another. The crack path is therefore primarily in the matrix, and the appearance of the matrix material is the key to understanding the failure conditions. The absolute toughness level of these failures can range from low to moderately high, depending on the nature of the matrix. Thermoset composites at room temperature are often very brittle, just as the neat epoxy phase would be (i.e., without any reinforcement phase). Thermoplastic matrices may be brittle or relatively tough depending on a number of factors, as described in Sections 5.8.2 and 5.9.2.

Translaminar fractures inherently involve fiber fracture, and often depend more heavily on fiber properties than on matrix properties. In tougher fractures, a significant amount of fiber microcracking may accumulate before loss of strength is so large that a catastrophic failure occurs. If the fibers break inside the matrix at some distance from the main crack face, and if the interface between the fiber and matrix is sufficiently weak that the fiber can be withdrawn like a sword from a sheath, the energy involved in the *fiber pullout* process can result in substantial toughening. In this case, the fracture surfaces will be covered with holes and protruding fibers. Tensile failures that are brittle—that is, have no capacity for fiber pullout and also no significant plasticity in the polymer matrix—will have a relatively flat surface on which the fibers are broken along the face of the crack. The fiber pullout process must be associated with tensile crack opening stresses, so the tensile side of a bending failure may show considerable evidence of fiber pullout (a highly fibrous appearance) whereas the compressive side of a bending failure would have much less pullout, and the surface would be flatter. Thus the fracture surface appearance may indicate the type and direction of loading as well as the relative toughness level.

5.9 MICROSCOPIC FRACTURE MECHANISMS

As is probably apparent from the discussions of macroscopic fracture appearance, there are many possible sources of plasticity that can lead to toughness, and of weakness that can lead to brittle failure. In the end, crack propagation is a microscopic process that occurs in the small process zone surrounding the crack tip. A full understanding of the causes behind a particular fracture must therefore depend on determining the specific micromechanism(s) involved in the crack growth process.

Not too long ago, the light microscope was the tool most often used in the microscopic examination of fracture processes. Because of the very shallow depth of focus, examination of the fracture surface was not possible except at very low magnifications. The understanding of fracture mechanisms in materials increased dramatically when the electron microscope was developed. Because its depth of field and resolution were superior to those of the light microscope, many topographical fracture surface features were observed for the first time. Many of these markings have since been applied to current theories of fracture. Much of the original high magnification fractographic work was conducted on transmission electron microscopes (TEM). Since the penetrating power of electrons is quite limited, fracture surface observations in a TEM require the preparation of a replica of the fracture surface that allows transmission of the high-energy electron beam. More recently, however, significant progress also has been made in the use of scanning electron microscopy in failure analysis. A major advantage of the scanning electron microscope (SEM) for some examinations is that the fractured sample may be viewed directly in the instrument, thereby obviating the need for replica preparation. When it is not possible to cut the fractured component to fit into the viewing chamber, replicas must be used instead.

Before one can proceed with an interpretation of high resolution fracture surface markings, it is necessary to review replication techniques and electron image contrast effects. To this end, the reader is referred to Appendix A. In addition, fractography handbooks[38−41] are available that contain both discussions of techniques and thousands of electron fractographs. For the purposes of the present discussion, it is important only to recognize that the most commonly employed replication technique for the TEM leads to a reversal in the "apparent" fracture surface morphology. That is, electron images may suggest that the fracture surface consists of mounds or hillocks, when in reality it is composed of troughs or depressions. *Everything on a replica that looks up is really down and vice versa.* On the other hand, SEM images do not possess this height deception. For completeness and comparative purposes, many of the microscopic fracture mechanisms discussed in this book are described with both TEM and SEM electron images.

5.9.1 Microscopic Fracture Mechanisms: Metals

Metals that are capable of plastic deformation tend to fail by a process called *microvoid coalescence* (MVC). It is present in high-toughness metal fractures that are obviously ductile at the macroscopic scale, and in geometrically-induced low toughness fractures for which visible plasticity is limited (e.g., thick plates with interior triaxial stresses that inhibit macroscopic plasticity as shown in Fig 5.15). This fracture mechanism, observed in most metallic alloys and many engineering plastics as well, takes place by the nucleation of microvoids, followed by their growth and eventual coalescence to form cracks. The microvoid initiation stage has largely been attributed to either particle cracking or interfacial failure between an inclusion or precipitate particle and the surrounding matrix. Accordingly, the spacing between adjacent microvoids is closely related to the distance between inclusions. Where a given material contains more than one type of inclusion, associated with a bimodal size distribution, microvoids with different sizes are often found on the fracture surface. These mechanically induced micropores should not be confused with preexisting microporosity sometimes present as a result of casting or powder sintering procedures.

The fracture surface appearance of microvoids depends on the state of stress.[42] Under simple uniaxial loading conditions, the microvoids will tend to form in association with fractured particles and/or interfaces and grow out in a plane generally normal to the stress axis. (This occurs in the fibrous zone of the cup–cone failure shown in Section 2.4.2.4.) The resulting micron-sized "equiaxed dimples" are generally spherical, as shown in Figs. 5.27a, b. Since the growth and coalescence of these voids involves a local plastic deformation process, it is to be expected that total fracture energy should be related in some fashion to the size of these dimples. In fact, it has been shown in laboratory experiments that fracture energy does increase with increasing depth and width of the observed dimples.[43,44]

When failure is influenced by shear stresses, the voids that nucleate in the manner cited above grow and subsequently coalesce along planes of maximum shear stress. (This behavior is

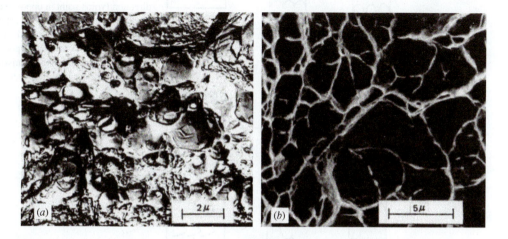

Figure 5.27   Microvoid coalescence under tensile loading, which leads to "equiaxed dimple" morphology: (*a*) TEM fractograph shows "dimples" as mounds; (*b*) SEM fractograph shows "dimples" as true depressions.

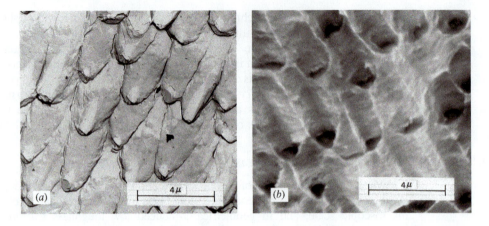

Figure 5.28 Microvoid coalescence under shear loading, which leads to "elongated dimple" morphology: (*a*) TEM fractograph shows "dimples" as raised parabolas; (*b*) SEM fractograph shows "dimples" as true elongated troughs.

found in the shear walls of the cup–cone failure.) Consequently, those voids tend to be elongated and result in the formation of parabolic depressions on the fracture surface, as shown in Figs. 5.28*a*, *b*. If one were to compare the orientation of these "elongated dimples" from matching fracture faces, one would find that the voids are elongated in the direction of the shear stresses and point in opposite directions on the two matching surfaces.

Finally, when the stress state is one of combined tension and bending, the resulting tearing process produces "elongated dimples," which can appear on gross planes normal to the direction of loading. The basic difference between these "elongated dimples" and those produced by shear is that the tear dimples point in the same direction on both halves of the fracture surface. It is important to note that these dimples point back toward the crack origin. Consequently, when viewing a replica that contains impressions of tear dimples, the dimples may be used to direct the viewer to the crack origin. A schematic diagram illustrating the effect of stress state on microvoid morphology is presented in Fig. 5.29.

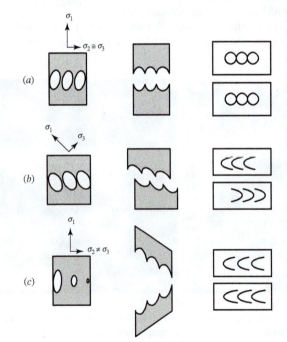

Figure 5.29 Diagrams illustrating the effect of three stress states on microvoid morphology: (*a*) tensile stresses produce equiaxed microvoids; (*b*) pure shear stresses generate microvoids elongated in the shearing direction (voids point in opposite directions on the two fracture surfaces); (*c*) tearing associated with nonuniform stress, which produces elongated dimples on both fracture surfaces that point back to crack origin.[42] (Reprinted with permission of the American Society of Mechanical Engineers.)

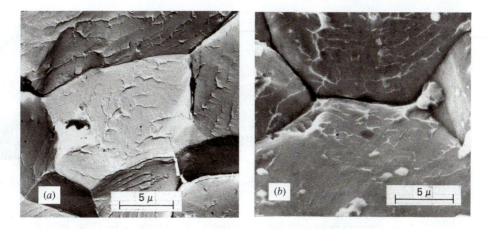

Figure 5.30   Intergranular fracture in maraging steel tested in the gaseous hydrogen environment: (*a*) TEM; (*b*) SEM.

In determining the cause of a particular failure, it may be desirable to establish the chemical composition of the particle responsible for the initiation of the voids. By selected area diffraction in the TEM of particles extracted from replicas or by X-ray detector instrumentation in the SEM, it often is possible to identify the composition of particles responsible for microvoid initiation. With this information, it may be possible to select a different heat-treating procedure and/or select an alloy of higher purity so as to suppress the void formation initiation process.

In addition to MVC, two other fast fracture micromechanisms can occur in metals: *intergranular fracture* and *cleavage fracture*. Both are usually associated with low-energy, brittle fracture caused by sources of weakness in the material. *Intergranular* failure is character- ized by crack growth primarily along grain boundaries. It leaves behind a fracture plane with exposed grain boundary surfaces and a distinctly faceted morphology as seen in Fig. 5.30. Intergranular fracture can result from a number of processes. These include microvoid nucleation and coalescence at inclusions or second-phase particles located along grain boundaries; grain- boundary crack and cavity formations associated with elevated temperature stress rupture conditions; decohesion between contiguous grains due to the presence of impurity elements at grain boundaries and in association with aggressive atmospheres such as gaseous hydrogen and liquid metals (Sections 7.8 and 8.1); and stress corrosion cracking processes associated with chemical dissolution along grain boundaries. Also, if the material has an insufficient number of independent slip systems (see Chapter 2) to accommodate plastic deformation between contigu- ous grains, grain-boundary separation may occur. Additional discussion of intergranular fracture under cyclic loading conditions is found in Section 10.7.

The process of *metal cleavage* involves *transgranular* fracture along specific crystallo- graphic planes (Fig. 5.31*a*). This mechanism is commonly observed in certain BCC and HCP metals (as indicated in Fig. 5.14), but can also occur in FCC metals when they are subjected to severe environmental conditions such as extremely high strain rates or very low temperatures. Cleavage is characterized by a relatively flat fracture surface with small converging ridges known as *river patterns* within many of the grains. The crack in Fig. 5.31*b* propagated from right to left across a grain boundary (probably high-angle), generating the river pattern as the advancing crack reoriented in search of weak cleavage planes in the new grain. It is also possible that the cleavage crack traversed a low-angle twist boundary, and the cleavage steps were produced by the intersection of the cleavage crack with screw dislocations.[45] Regardless of which specific mechanism was responsible, the progression from many small ridges along the grain boundary to fewer large ridges within the grain unambiguously indicates the local crack growth direction.

In some materials, such as ferritic steel alloys, the temperature and strain-rate regime necessary for cleavage formation is similar to that required to activate deformation twinning (see Chapter 2) so both may be present. Fine-scale height elevations (so-called tongues) seen in

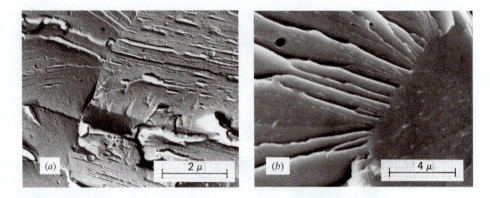

Figure 5.31   Cleavage fracture in a low-carbon steel. Note parallel plateau and ledge morphology and river patterns reflecting crack propagation along many parallel cleavage planes: (*a*) TEM; (*b*) SEM.

Figs. 5.32*a, b* provide proof of deformation twinning during or immediately preceding failure. In BCC iron, etch pit studies have verified that these fracture surfaces consist of {100} cleavage facets and {112} tongues, the latter representing failure along twin matrix interfaces.

Little information may be obtained from cleavage facets for use in quantitative failure analyses. However, one may learn something about the phase responsible for failure by noting the shape of the facet and comparing it to the morphology of different phases in the alloy. Furthermore, in materials that can undergo an environmentally-induced fracture mechanism transition, it is possible to relate the presence of the cleavage mechanism to a general set of external conditions. In most mild steel alloys (which can undergo a fracture mechanism transition from MVC to cleavage), the observation of cleavage indicates that the component was subjected to some combination of low temperature, high strain rate, and/or a high tensile triaxial stress condition. This point is discussed further in Chapter 7.

5.9.2 Microscopic Fracture Mechanisms: Polymers

As is the case for metal alloy fracture surfaces, the fracture surface micromorphology of engineering polymers reflects both the underlying microstructure and the deformation mechanisms active in these materials. Let us consider first the microscopic fracture surface appearance of an amorphous polymer.[46–50] Sternstein[51] found that under multiaxial stress conditions, glassy polymers could yield by two distinct mechanisms: normal yielding (crazing, as first introduced in Section 5.8.2) and shear yielding (Fig. 5.33). The term *crazing* was originally applied to the

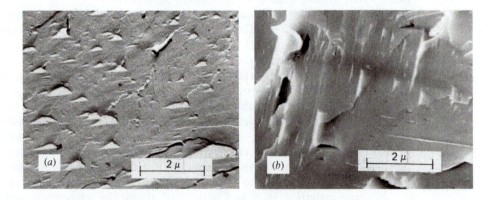

Figure 5.32   Cleavage facets revealing fine-scale height elevations caused by localized deflection of the cleavage crack along twin-matrix interfaces: (*a*) TEM; (*b*) SEM.

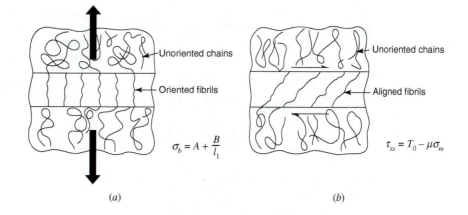

$$\sigma_b = A + \frac{B}{l_1}$$

$$\tau_{xx} = T_0 - \mu\sigma_m$$

(a)   (b)

Figure 5.33 Deformation mechanisms in amorphous polymers: (a) normal yielding (crazing) and (b) shear yielding.

intricate and often beautiful arrays of fine cracks that glassy pottery glazes tend to develop. When the study of polymer fracture began to be pursued intensely, it was recognized that similarly oriented crazes in glassy polymers were not, in fact, simple cracks as in ceramics; rather, they consisted of expanded material containing oriented fibrils interspersed with small (10 to 20 nm) interconnected voids[46,52−62] (Fig. 5.34a). As in the pottery glazes, polymer crazes tend to grow along planes normal to the principal tensile stress direction. The typical craze thickness in glassy polymers is on the order of $5\,\mu m$ or less, which corresponds in some cases to plastic strains in excess of 50%.[57] Little change in craze thickness occurs as the craze propagates (Fig. 5.34b). Since the refractive index of the craze is lower than that of the parent polymer, the presence of the craze is observed readily with the unaided eye, even though the details are impossible to resolve without the aid of a microscope. The combination of fibrils (extending across the craze thickness) and interconnected microvoids contributes toward an overall weakening of the material, though the craze is capable of supporting some level of stress unlike a true crack. The latter point is proven conclusively by the load-bearing capability of samples containing crazes that extend completely across the sample.[52,63] It is the crazing phenomenon that gives amorphous polymers somewhat greater toughness than typical silicate glasses in which no crack bridging is possible.

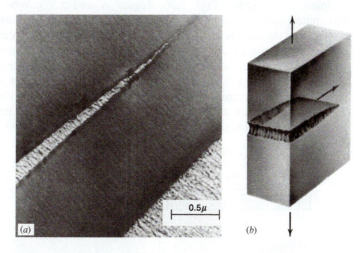

Figure 5.34 (a) Crazes in polyphenylene oxide revealing interconnected microvoids and aligned fibrils.[58] (Reprinted with permission from the Polymer Chemistry Division of ACS.) (b) Sketch showing craze development normal to applied principal stress. Note slight surface dimpling along the craze perimeter.

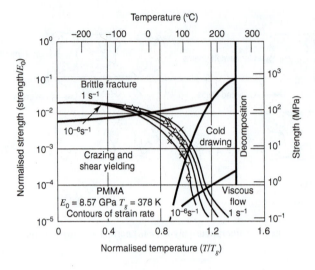

Figure 5.35 Deformation map for PMMA showing deformation regions as a function of normalized stress versus normalized temperature.[64] (Reprinted with permission from M. F. Ashby and D. R. H. Jones, *Engineering Materials 2*, Pergamon Press, 1986.)

Kausch[24] has pointed out that the initiation of a craze depends simultaneously on factors associated with the macroscopic state of stress and strain, the nature and distribution of heterogeneities within the solid, and the molecular behavior of the polymer for a given set of thermal and environmental conditions. Crazing should be favored by high temperatures and high stress concentrations, and restricted by applied hydrostatic pressure and the occurrence of creep or other types of flow that could decrease local stress concentrations. Because of its intrinsic weakness, the craze is an ideal path for crack propagation.

A deformation map for PMMA is shown in Fig. 5.35 that characterizes the regions where particular deformation mechanisms are dominant.[64] Note the analogous form of this map to the ones described in Section 4.6. While the presence of the craze is considered undesirable from the standpoint of introducing a likely crack path, the localized process of fibril and void formation does absorb considerable strain energy. This is particularly true in the case of rubber-toughened polymers (see Section 7.6.2).

Once crazes have formed, subsequent fracture then usually occurs in stages that generate mirror–mist–hackle markings. Crazed matter converts to a fully open crack in one of several ways. Initially, the crack grows by the nucleation, growth, and merging of conically shaped voids along the midplane of the craze (Figs. 5.36, Region A, and 5.37). When these pores are viewed with transmitted light, they appear as a series of concentric rings that correspond to changes in the crazed thickness on the fracture surface (Fig. 5.37*b*). As the crack velocity increases, the crack tends to grow alternately along one craze–matrix boundary interface and then the other (Fig. 5.36, Region B).[47,48] Consequently, the fracture surface contains islands or patches of craze matter attached to one-half of the fracture surface (Figs. 5.37*a* and 5.38*a*). If the craze matter is organized into bands that are parallel to the advancing crack front, they are referred to in the literature as *mackerel bands* (Fig. 5.38*b*), since they are reminiscent of scale markings along the flanks of the mackerel fish. Note evidence for fibril elongation along the walls of the craze patches (Fig. 5.37*a*). These detached craze

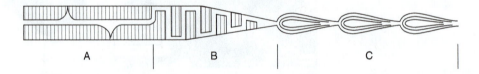

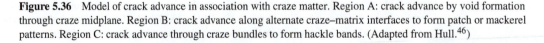

Figure 5.36 Model of crack advance in association with craze matter. Region A: crack advance by void formation through craze midplane. Region B: crack advance along alternate craze–matrix interfaces to form patch or mackerel patterns. Region C: crack advance through craze bundles to form hackle bands. (Adapted from Hull.[46])

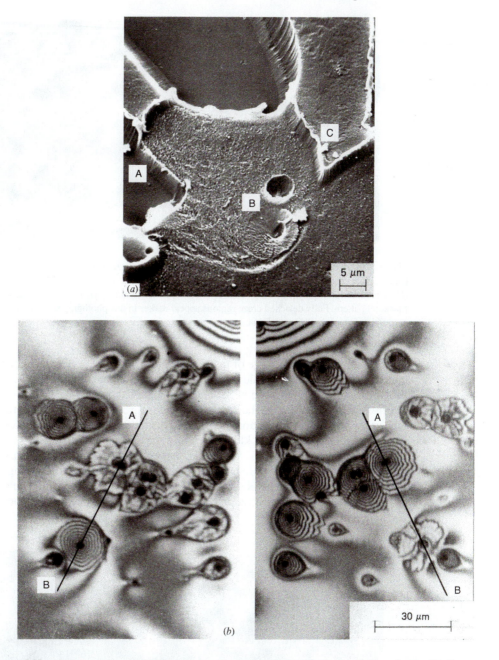

Figure 5.37 Void formation. (*a*) SEM image revealing two conical voids along midplane of craze (B). Craze–matrix detachment at A and C. Note fibril extension between AB and BC. (*b*) Transmitted light images revealing concentric rings on one surface that match with relatively featureless zones on the mating fracture surface. (From Hull,[46] "Nucleation and Propagation Processes in Fracture," *Polymeric Materials,* American Society for Metals, Metal Park, OH, 1975, p. 524.)

zones tend to decrease in size in the crack growth direction since crazes tend to get thinner toward the craze tip (Figs. 5.36, Region B, and 5.38*a*). At a macroscopic level, the mirror region often exhibits colorful patterns that reflect the presence of the layer of craze matter with a different refractive index on the fracture surface, as mentioned in Section 5.8.2.[65–67] One color indicates the existence of a single craze with a uniform thickness; packets of multicolor fringes reflect the presence of a few craze layers at the fracture surface or a single craze with variable thickness.

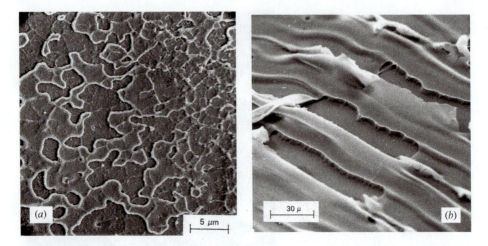

Figure 5.38 Crack advance from left to right along alternate craze–matrix interfaces in polystyrene. (*a*) Patch pattern. Note decreasing patch size in crack propagation direction. (Courtesy of Clare Rimnac, Lehigh University.) (*b*) Stereoscan micrograph showing bands of craze separated by areas where the craze has been stripped off. Mackerel pattern. Fully detached craze regions appear darker in this photograph. (Murray and Hull,[48] with permission from John Wiley & Sons, Inc.)

During the terminal stage of fast fracture, the crack front outpaces the main craze tip, with continued crack extension taking place by the interconnection of bundles of secondary crazes that form in the process zone at the crack tip. The banded nature of this surface (referred to as *hackle bands*) suggests that the crack propagates through one bundle of crazes, at which point a new craze bundle is formed and the process is repeated (Figs. 5.36, Region C, and 5.39*a*).[49] A close examination of the hackle zone at a higher magnification reveals the patch morphology described above (Fig. 5.39*b*). *Parabolic*-shaped voids are also found on some portions of the mist region. Similar to the elongated dimples associated with the tearing mode of fracture in metals (Fig. 5.40), these parabolic-shaped voids are produced by the nucleation and growth of cavities ahead of the advancing crack front, with the extent of cavity elongation depending on the relative rates of cavity growth and crack-front advance. Figure 5.40 reveals several

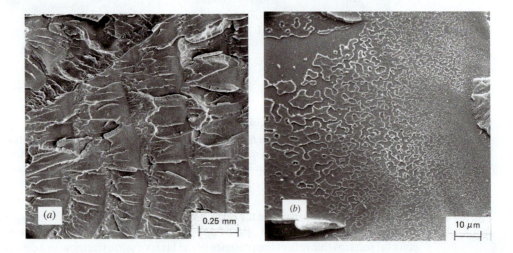

Figure 5.39 Banded hackle markings in fast fracture region. (*a*) Crack advances in jumps through craze bundles; (*b*) path appearance on hackle band surface. Crack propagation from left to right. (Courtesy of Clare Rimnac, Lehigh University.)

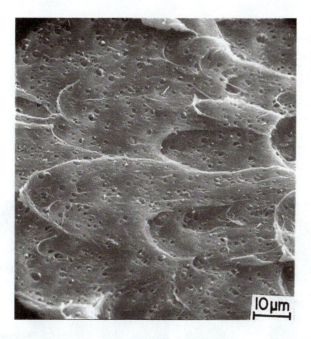

Figure 5.40   Tear dimples in Noryl polymer. Microvoid nucleation at butadiene–polystyrene duplex particles. (Courtesy of Clare Rimnac, Lehigh University.)

elongated tear dimples in ABS polymer that were nucleated at duplex butadiene-polystyrene particles (also see Fig. 7.47).

The fracture surface appearance of semicrystalline polymers depends on the crack path with respect to underlying microstructural features. A semicrystalline polymer may develop spherulites that are clusters of individual crystalline plates. These spherulites meet at boundaries that are analogous to grain boundaries in metals. In this case, a crack may choose an *interspherulitic* or *transspherulitic* crack path equivalent to the intergranular and transgranular paths in ceramics and brittle metals.[68] Four possible crack paths through or around a spherulite are shown in Fig. 5.41*a*, with respect to the orientation of crystal lamellae in the spherulite (Fig. 5.41*b*). A completely interspherulitic fracture path is observed when the crack velocity is low, shown for polypropylene in Fig. 5.41*d*. A fast-running crack tends to show a greater proportion of transspherulitic character, as seen in Fig. 5.41*c*. It should be noted that fractographic evidence for transspherulitic or interspherulitic failure may be obscured by extensive prior deformation of the polymer, which can distort beyond recognition the characteristic details of the underlying microstructure.

5.9.3 Microscopic Fracture Mechanisms: Glasses and Ceramics

Many of the glass and ceramic fracture markings discussed in Section 5.8.3 may be observed with electron microscopy techniques. Electron microscopy may actually be required when the component is small (as in the case of the glass fiber in Example 5.4). An exception is visualization of Wallner lines, which are generally very shallow and provide little contrast in the SEM.[25]

Examination of small-grained crystalline ceramics in the SEM may be particularly useful for differentiating between *intergranular* and *transgranular* fracture mechanisms. Perhaps the most immediately recognizable features are those of intergranular failure wherein the crack prefers to follow grain surfaces. The resulting fracture surface morphology immediately suggests the three-dimensional character of the grains that comprise the ceramic microstructure, as shown in Fig. 5.42a. The occurrence of intergranular fracture is indicative of a number of processes typically associated with slow crack growth. At elevated temperatures *creep damage* may dominate. In this mode, plastic flow by grain boundary sliding can occur and grain boundary cracks may eventually link up to cause failure. At room temperature, certain ceramics—particularly oxide ceramics or those with a glassy grain boundary phase—are susceptible to *stress corrosion cracking*. In this failure mode, tensile stress opens a flaw or preexisting crack allowing water or another corrosive

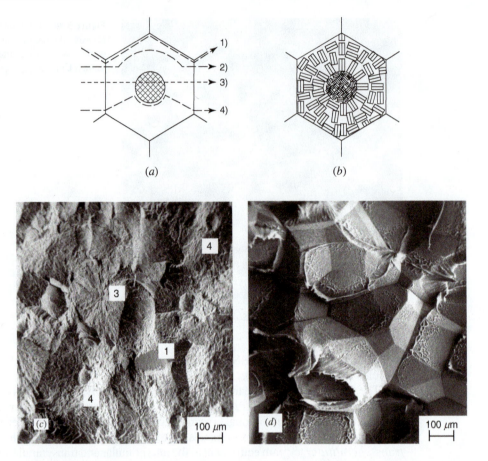

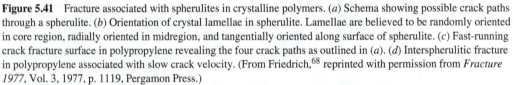

Figure 5.41 Fracture associated with spherulites in crystalline polymers. (*a*) Schema showing possible crack paths through a spherulite. (*b*) Orientation of crystal lamellae in spherulite. Lamellae are believed to be randomly oriented in core region, radially oriented in midregion, and tangentially oriented along surface of spherulite. (*c*) Fast-running crack fracture surface in polypropylene revealing the four crack paths as outlined in (*a*). (*d*) Interspherulitic fracture in polypropylene associated with slow crack velocity. (From Friedrich,[68] reprinted with permission from *Fracture 1977*, Vol. 3, 1977, p. 1119, Pergamon Press.)

Figure 5.42 (*a*) ZrO$_2$ intergranular fracture at high temperature and (*b*) transgranular fracture at room temperature. (Courtesy of J.D. French, Lehigh University.)

fluid to attack the material at the tip of the crack. The fracture often proceeds along grain boundaries where the material is weakest or most chemically susceptible to attack. When the crack grows to a critical size it will transition to a fast-fracture mode. As it does so, a considerable amount of transgranular fracture may develop. High-temperature intergranular creep fracture is discussed at greater length in Section 5.9.5. More information pertaining to creep and stress corrosion cracking mechanisms may also be found in Chapter 8.

Transgranular (or transcrystalline) fracture occurs along specific crystallographic planes just as it does in brittle metals. This *cleavage* process results in a morphology that is typically flat, although within individual grains it may reflect a parallel plateau and ledge morphology (Fig. 5.42b). Often these cleavage steps appear as *river patterns* wherein fine steps are seen to merge progressively into larger ones (see the example for metals in Fig. 5.31).

5.9.4 Microscopic Fracture Mechanisms: Engineering Composites

Too many material/test conditions combinations—too many fractographs—too little available space in the text! Some comments are appropriate, however, regarding the knowledge that can be gained by examining the microscopic fracture surface appearance of selected fibrous and particulate composites. For example, interlaminar fracture of composites, as defined in Section 5.8.4, can occur in a *cohesive* or *adhesive* manner. In the former, the matrix phase adheres well to the fibers and the crack passes through the matrix, so the properties of the matrix dominate the process. In the latter, the crack passes along the fiber/matrix interface. In this case, there is little or no matrix phase residue on the fibers, and it is the interface that dominates behavior. Examples of these two different fracture paths are shown in Fig. 5.43 for a nylon 66–glass fiber composite.[69] The strong bond between fiber and matrix in this composite is confirmed by the cohesive failure produced under monotonic tensile loading conditions (Fig. 5.43a). Note that the glass fibers are coated with the nylon 66 matrix material. When the sample was cyclically loaded to produce a fatigue crack, fracture took place along the fiber–matrix interface (Fig. 5.43b). Adhesive failures are expected to prevail when the fiber–matrix interface is weak. However, interfacial failure does also occur in fatigue, with monotonic loading at elevated temperatures, and in the presence of a moist environment, even though the fiber–matrix interface is deemed to be strong.

An additional example of interfacial fracture between reinforcement and matrix is shown in Fig. 5.44 for a particle-reinforced composite. Here we see the fatigue fracture surface of an epoxy resin containing hollow glass beads.[70] Figure 5.44a shows that some of the beads are

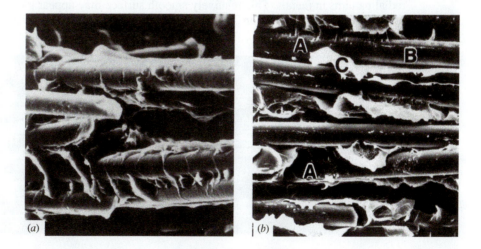

(a)   (b)

Figure 5.43 Fracture surface appearance in short glass-fiber reinforced nylon 66. (*a*) Fast-fracture appearance revealing matrix adhesion to glass fibers. (*b*) Fatigue fracture revealing fiber-matrix debonding (A and B) and matrix drawing (C). (Courtesy of R. Lang.)

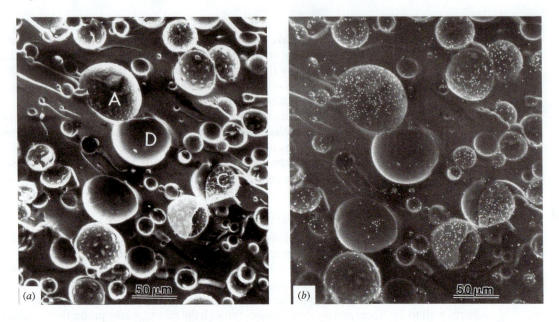

Figure 5.44 Fracture surface in hollow glass sphere-filled epoxy resin. (*a*) Glass particles located at A, detached at D, and cracked at C. (*b*) Silicon X-ray map showing location of glass spheres. (Courtesy of M. Breslaur.)

retained on this half of the fracture surface, whereas other particles were either pulled out or shattered. Note the *wake hackle* in the matrix on the trailing side of many glass beads, another indicator of crack growth direction. The silicon *x*-ray map shown in Fig. 5.44*b* clearly reveals the presence of both undamaged and fragmented hollow glass spheres on the fracture surface.

Figure 5.45 reveals that fracture surface features in graphite fiber–epoxy composites change with the mode of loading.[71] In the case of interlaminar fractures, the matrix is likely to exhibit most of the useful microfractographic evidence. Recall that epoxies are amorphous thermosets that typically have brittle fracture properties at room temperature. Under tensile loading conditions, interply failure reveals a cleavage-like appearance in the epoxy resin as evidenced by the presence of river patterns, which converge in the crack propagation direction (Fig. 5.45*a*). There is often also a *feathering* of the matrix that diverges in the direction of crack growth like radial patterns in metals. This relatively smooth surface may appear glossy at the macroscopic level, as described in Section 5.8.2. When shear stresses dominate the interlaminar fracture

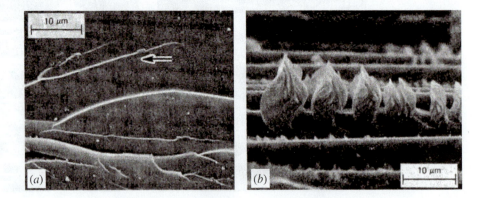

Figure 5.45 Fracture surface appearance in graphite fiber-reinforced epoxy. (*a*) Tensile failure reveals cleavage-like pattern. Arrow indicates crack direction. (*b*) Shear failure reveals presence of hackle markings in epoxy resin.[71] (Reprinted with permission. Copyright ASTM.)

process, the epoxy resin reveals a series of hackle markings on the fracture surface, which presumably reflect the coalescence of many tension-induced microcracks inclined at an angle to the overall fracture plane (Fig. 5.45*b*). This surface corresponds with a dull macroscopic appearance. Thermoplastic matrix composites are likely to be semicrystalline with relatively ductile matrices, so microscopic evidence like river patterns and hackle tends to be obscured by the matrix microstructure and gross deformation.

Transverse fractures inherently involve a great deal of fiber fracture, so the features on the fiber ends are often a good source of information about the nature of the activities that occurred in the process zone as the fracture progressed. In particular, glassy or ceramic fibers may show classic mirror, mist, and radial marks that indicate the local crack growth direction and the local stress state that existed at the point of fracture (as demonstrated in Example 5.4). The average crack growth direction across many fibers will indicate the overall crack propagation path. In compression (e.g., on the compressive side of a bending fracture), the fibers may have buckled under load prior to fracture. This local bending can cause features reminiscent of cantilever curl, with a progression from tensile to compressive behavior across each fiber fracture surface as the crack moves through.

5.9.5 Microscopic Fracture Mechanisms: Metal Creep Fracture

In closing this chapter, it is appropriate to consider mechanisms associated with the fracture of materials (particularly metals) at elevated temperatures. As noted in Chapter 2, slip and twinning deformation processes occur at high stresses and ambient temperatures, which can lead to the *transgranular* fracture of polycrystalline materials. With increasing temperatures and relatively low stress levels, however, *intergranular* fracture generally dominates material response. This change in fracture path occurs since grain boundaries become weaker with respect to the matrix as temperature is increased. The transition temperature for this crack path changeover is often referred to as the "equicohesive temperature." It is generally recognized that intergranular fracture takes place by a combination of grain boundary sliding and grain boundary cavitation associated with stress concentrations or structural irregularities such as grain boundary ledges, triple points, and hard particles.[72]

Grain boundary sliding (GBS), generally thought to occur by grain boundary dislocation motion, becomes operational at temperatures greater than approximately $0.4T_h$ and contributes to both creep strain and intergranular fracture in polycrystals (Fig. 5.46*a*). As

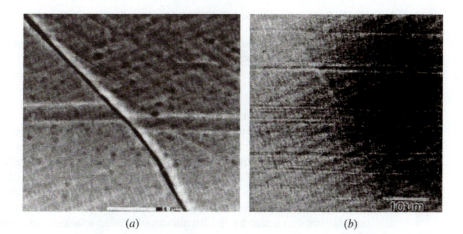

(*a*)   (*b*)

Figure 5.46   Influence of 300 wt-ppm carbon-doping of ultra-high purity Ni-16Cr-9Fe alloy after 20% elongation at 360°C in argon. (*a*) Clear evidence of grain boundary sliding in UHP alloy as noted by displacement of fiduciary line; (*b*) lack of grain boundary sliding when alloy contains 300 wt-ppm carbon in solid solution.

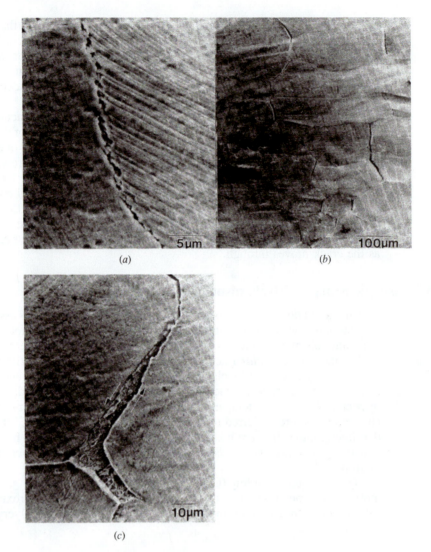

Figure 5.47   Cavitation and cracking in UHP Ni-16Cr-9Fe allow after 35% elongation at 360°C in argon. Initial strain rate was 3×10^{-7} sec^{-1}. (a) Slip-boundary induced cavitation; (b) intergranular cracking in UHP alloy; (c) triple point cracking. Note involvement of grain boundary sliding (i.e., displacement of fudiciary markings) and grain boundary microvoid coalescence on new fracture surface.

discussed in Section 4.4, the GBS process is accommodated by two major deformation processes; at high temperatures, diffusional creep (i.e., Nabarro-Herring and Coble creep) accommodates the sliding process, whereas at lower temperatures and elevated stresses, grain boundary sliding is controlled by dislocation creep, involving glide and climb of lattice dislocations. If neither diffusional mass transport or intragranular plastic flow is operable, grain boundary decohesion develops by the formation of a planar array of grain boundary cavities (Fig. 5.47a) that eventually coalesce to form grain boundary cracks (Fig. 5.47b); in addition, failure can occur by the formation of wedge cracks at grain boundary triple points (Fig. 5.47c).

The magnitude of the contribution of grain boundary sliding (GBS) to creep strain is a strong function of the stress level and, accordingly, is inversely proportional to the minimum creep rate (Fig. 5.48).[73] At low strain rates (corresponding to low stress levels) and high

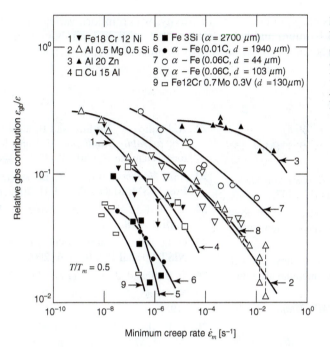

Figure 5.48 Relative contribution of grain boundary sliding as a function of minimum creep rate. Abscissa can also be viewed as describing influence of increasing stress level.[73] (J. Cadek, *Creep in Metallic Materials*, Elsevier, Amsterdam, 1988. Reprinted by permission.)

temperatures, the rate of grain boundary sliding represents a significant portion of the overall creep rate; at high strain rates (i.e., high stress levels), deformation within the grains occurs at a much faster rate than sliding, thereby leading to a negligible contribution of GBS on total creep strain. While it is generally agreed that grain boundary particles can reduce the amount of boundary sliding by their pinning of grain boundaries, the effects of other changes in boundary structure (e.g., solute segregation, grain misorientation) on sliding and associated cracking propensity have recently been investigated.[74,75] Figure 5.46a illustrates typical grain boundary sliding in an ultra-high purity (UHP) Ni-16Cr-9Fe alloy. However, when 300 ppm C was added, the carbon in solid solution was found to limit both grain boundary sliding and cavitation, thereby suppressing intergranular cracking during slow strain rate testing at 360°C ($\sim 0.38 T_h$)[75] (Fig. 5.46b).

Since cavity formation usually occurs by decohesion at grain boundary-particle interfaces, the cavity formation rate will depend on the extent of impurity segregation at such boundaries. In this regard, it has been shown that the critical radius, r_c, for stable cavity development is given by

$$r_c = 2\gamma_s/\sigma \tag{5-29}$$

where

$\gamma_s =$ grain boundary surface energy
$\sigma =$ local tensile stress normal to the grain boundary

Hence, the stability of a newly formed pore is not guaranteed but, instead, depends on the pore radius, relative to that of the critical pore size, r_c. For example, if the pore radius, r, is less than the critical value, r_c, the cavity surfaces will begin to sinter and close up. On the other hand, when the $r > r_c$, the pore is stable and will grow larger;[76,77] intergranular failure then results from grain boundary cavity growth and coalescence. The magnitude of the effect of impurity level on pore stability can depend upon impurity atom size, with interstitial impurities generally having a potentially greater stabilizing effect than substitutional atoms.[76,77] Furthermore, the cavity growth rate depends on grain boundary diffusivity, which in turn varies with solute levels. For example, beneficial hafnium and boron solute segregation to grain boundaries in nickel-based alloys reduces grain boundary diffusion rates by an order of magnitude, thereby slowing cavitation growth kinetics.[78]

REFERENCES

1. *Scientific American* **120**, 99 (Feb. 1, 1919).

2. *Engineering News-Record* **82** (20), 974 (May 15, 1919).

3. E. R. Parker, *Brittle Behavior of Engineering Structures*, John Wiley & Sons, New York, 1957.

4. W. J. Broad, "Wreckage of Titanic Reported Discovered 12,000 Feet Down," *New York Times*, September 3, 1985.

5. W. Garzke, D. Yoerger, and S. Harris, Society of Naval Architects and Marine Engineers, Centennial Meeting, New York, September 15, 1993.

6. T. Foecke, *Metallurgy of the RMS Titanic*, NIST, Gaithersburg, MD, 1998.

7. J. H. McCarty and T. Foecke, *What Really Sank the Titanic*, Citadel Press, Kensington Pub. Co., New York, 2008.

8. W. J. McGregor Tegart, *Elements of Mechanical Metallurgy*, Macmillan, New York, 1966.

9. W. B. Parsons, *Engineers and Engineering in the Renaissance*, Williams and Wilkins, Baltimore, 1939.

10. W. Weibull, *Proceedings 151*, Stockholm: Royal Swedish Academy of Engineering Sciences, 1939.

11. W. Weibull, *J. Appl. Mech.* **18** (3), 293 (1951).

12. W. Weibull, *J. Mech. Phys. Solids* **8**, 100 (1960).

13. R. W. Davidge, *Mechanical Behavior of Ceramics*, Cambridge University Press, Cambridge, 1979.

14. C. E. Inglis, *Proceedings, Institute of Naval Architects* **55**, 219 (1913).

15. R. E. Peterson, *Stress Concentration Design Factors*, John Wiley & Sons, New York, 1974.

16. H. Neuber, *Kerbspannungslehre*, Springer, Berlin; English translation available from Edwards Bros., Ann Arbor, MI, 1959.

17. W. Jordan, *Proceedings, American Numismatical and Antiquarian Society* **27**, 109 (1915).

18. *ASME Handbook—Metals Engineering—Design*, McGraw-Hill, New York, 1953.

19. W. Moffatt and J. Wulff, *J. Met.*, 440 (April 1957).

20. M. F. Ashby and D. R. H. Jones, *Engineering Materials*, Pergamon, Oxford, 1986.

21. National Transportation Safety Board, Highway Accident Report NTSB/HAR-08/03, Washington, DC, 2008.

22. A. Phillips, V. Kerlins, and B. V. Whiteson, *Electron Fractography Handbook*, Air Force Materials Laboratory Technical Report ML-TDR-64-416, 1965.

23. *ASM Handbook* 11, W. T. Becker and R. J. Shipley, Eds., ASM, Metals Park, OH, 2002, pp. 650–661.

24. H. H. Kausch, *Polymer Fracture*, Springer-Verlag, Berlin, 1978.

25. G. D. Quinn, *Fractography of Ceramics and Glasses*, NIST Special Pub 960-16, 2007.

26. E. B. Shand, *J. Am. Ceram. Soc.* **37** (12), 572 (1954).

27. N. Terao, *J. Phys., Proc. Phys. Soc., Japan* **8**, 545 (1953).

28. J. J. Mecholsky, S. W. Freiman, and R. W. Rice, ASTM, *STP 645*, 1978, p. 363.

29. J. J. Mecholsky and S. W. Freiman, ASTM, *STP 678*, 1979, p. 136.

30. A. I. A. Abdel-Latif, R. C. Bradt, and R. E. Tressler, *Int. J. Fract.* **13** (31), 349 (1977).

31. J. J. Mecholsky and S. W. Freiman, ASTM, *STP 733*, 1981, p. 246.

32. A. I. A. Abdel-Latif, R. C. Bradt, and R. E. Tressler, ASTM, *STP 733*, 1981, p. 259.

33. L. Orr, *Materials Research and Standards*, **12** (1), 21 (1971).

34. J. J. Mecholsky, S. W. Freiman, and R. W. Rice, *J. Mater. Sci.* **11**, 1310 (1976).

35. R. W. Rice, ASTM, *STP 827*, 1984, p. 5.

36. R. B. Martin, D. B. Burr, and N. A. Sharkey, *Skeletal Tissue Mechanics*, Springer, New York, 1998, p. 138.

37. K. B. Clifton, R. L. Reep, and J. J. Mecholsky Jr., *J. Mater. Sci.* **43**, 2026 (2008).

38. A. Phillips, V. Kerlins, and B. V. Whiteson, *Electron Fractography Handbook*, AFML TDR-64-416, WPAFB, Ohio, 1965.

39. *Metals Handbook,* Vol. 12, American Society of Metals, Metals Park, OH, 1987.

40. L. Englel, H. Klingele, G. W. Ehrenstein, and H. Schaper, *An Atlas of Polymer Damage*, Prentice-Hall, Englewood Cliffs, NJ, 1981.

41. *Fractography Handbook*, Chubu Keiei Kaihatsu Center, Nagoya, Japan, 1985.

42. C. D. Beachem, *Trans. ASME J. Basic Eng. Ser. D* **87**, 299 (1965).

43. A. J. Birkle, R. P. Wei, and G. E. Pellissier, *Trans. ASM* **59**, 981 (1966).

44. D. R. Passoja and D. C. Hill, *Met. Trans.* **5**, 1851 (1974).

45. J. J. Gilman, *Trans. Met. Soc. AIME* **212**, 310 (1958).

46. D. Hull, *Polymeric Materials*, ASM, Metals Park, OH, 1975, p. 487.

47. J. Murray and D. Hull, *Polymer* **10**, 451 (1969).

48. J. Murray and D. Hull, *J. Polym. Sci. A-2* **8**, 583 (1970).

49. D. Hull, *J. Mater. Sci.* **5**, 357 (1970).

50. R. P. Kusy and D. T. Turner, *Polymer* **18**, 391 (1977).

51. S. S. Sternstein, *Polymeric Materials*, ASM, Metals Park, OH, 1975, p. 369.

52. R. P. Kambour and R. E. Robertson, *Polymer Science: A Materials Science Handbook*, Vol. 1, A. D. Jenkins, Ed., North Holland, 1972, p. 687.

53. B. Maxwell and L. F. Rahm, *Ind. End. Chem.* **41**, 1988 (1949).

54. J. A. Sauer, J. Marin, and C. C. Hsiao, *J. Appl. Phys.* **20**, 507 (1949).

55. C. C. Hsiao and J. A. Sauer, *J. Appl. Phys.* **21**, 1071 (1950).

56. J. A. Sauer and C. C. Hsiao, *Trans. ASME* **75**, 895 (1953).

57. R. P. Kambour, *Polymer* **5**, 143 (1964).

58. R. P. Kambour and A. S. Holick, *Polym. Prepr.* **10**, 1182 (1969).

59. V. D. Frechette, ASTM *STP 827*, 1984, p. 104.

60. R. P. Kambour, *J. Polym. Sci.* Part A-2 4, 17 (1966).

61. J. P. Berry, *J. Polym. Sci.* **50**, 107 (1961).

62. J. P. Berry, *Fracture Processes in Polymeric Solids*, B. Rosen, Ed., Interscience, 1964, p. 157.

63. S. Rabinowitz and P. Beardmore, *CRC Crit. Rev. Macromol. Sci.* **1**, 1 (1972).

64. M. F. Ashby and D. R. H. Jones, *Engineering Materials 2*, Pergamon, Oxford, 1986.

65. J. P. Berry, *Nature (London)* **185**, 91 (1960).

66. R. P. Kambour, *J. Polym. Sci. Part A-3*, 1713 (1965).

67. R. P. Kambour, *J. Polym. Sci. Part A-2* **4**, 349 (1966).

68. K. Friedrich, *Fracture 1977*, Vol. 3, ICF4, Waterloo, Canada, 1977, p. 1119.

69. R. W. Lang, J. A. Manson, and R. W. Hertzberg, *Polym. Eng. Sci.* **22**, 982 (1982).

70. M. Breslaur, private communication.

71. B. W. Smith and R. A. Grove, *ASTM STP 948*, 1987, p. 154.

72. M. H. Yoo and H. Trinkhaus, *Metall. Trans.*, 14A, 547 (1983).

73. J. Cadek, *Creep in Metallic Materials*, Elsevier, Amsterdam, 1988.

74. T. Watanabe, *Mater. Sci. & Eng.*, A166, 11 (1993).

75. V. Thaveeprungsriporn, T. M. Angeliu, D. J. Paraventi, J. L. Hertzberg, and G. S. Was, *Proc. of 6th Int. Sym. on Env. Degrad. of Matl. in Nuclear Power Systems/ Water Reactors*, R. E. Gold and E. P. Simonen, Eds., TMS, Warrendale, PA, 721 (1993).

76. E. D. Hondros and M. P. Seah, *Metall Trans.* **8A**, 1363 (1977).

77. M. P. Seah, *Phil. Trans. Roy. Soc. London*, 295, 265 (1980).

78. J. H. Schneibel, C. L. White, and R. A. Padgett, *Proc. 6th Int. Conf. Strength of Metals and Alloys*, ICSMA6, 2, R. C. Gifkins, Ed., Pergamon Press, Oxford, 649 (1982).

PROBLEMS

Review

5.1 Why do most materials exhibit fracture strengths much lower than their theoretical capacities to support load?

5.2 If a set of small ceramic bars is tested to failure with the goal of predicting the service failure probability of a larger ceramic component, how does the probability of failure at a given stress level compare for the two components?

5.3 Is it preferable to have a larger or smaller Weibull modulus m for a structural product?

5.4 Name, define, and give an expression for k_t.

5.5 Write the equation that relates *applied stress* to *local maximum stress*.

5.6 Equation 5-26 can often be simplified by assuming that $a \gg \rho$. For what ratio of a/ρ can one make this assumption if the error introduced in σ_{max} must be less than 1%? If the crack tip radius is on the order of 10 nm, what is the corresponding minimum crack length?

5.7 Explain why boring a hole in a part may extend its lifetime.

5.8 Give three examples of design features that may lead to a reduction in component fracture strength.

5.9 Explain the connection between notch strengthening and material ductility; which is more likely to notch strengthen—pure Al or martensitic steel?

5.10 Identify the conditions under which a brazed joint is likely to be "stronger" than the brazing material by itself. Describe the trends in joint fracture strength as it becomes wider/narrower or thinner/thicker.

5.11 List the *microscopic* fracture surface markings for metals, state under what typical conditions each is produced, and identify a visible characteristic associated with each mechanism that would allow you to identify it from a fracture surface micrograph.

5.12   Sketch a typical fracture surface of a steel plate that displays chevron markings; identify the origin of the crack on the sketch.

5.13   In what class of material is stress whitening often seen, and what is the physical origin of the phenomenon?

5.14   What does the presence of significant crack branching in a glass fracture indicate about the fracture event?

5.15   Define the following terms, provide a sketch of each, and state for what material or materials they are relevant: mirror, mist, hackle, Wallner lines.

5.16   Sketch interlaminar and translaminar fractures of a laminated composite, clearly showing the relative orientation of the crack growth and the layers in both cases.

5.17   Is the presence of microvoid coalescence on a metal fracture surface inconsistent with macroscopic evidence of brittle fracture?

5.18   Identify a likely cause of intergranular fracture in metals or ceramics, and sketch a cross section of an intergranular fracture surface.

5.19   What role can craze formation play in toughening of an amorphous polymer? What role can crazing play in weakening an amorphous polymer?

5.20   Which microscopic fracture mechanism is associated with *river patterns*, and why do they form?

5.21   When a fiber-reinforced polymer matrix composite plate is fractured in bending, what evidence may exist that would indicate which side of the plate was in tension and which side was in compression?

5.22   What visual evidence supports the presence of significant grain boundary sliding in many creep failures?

Practice

5.23   Using data from Chapter 1, calculate the theoretical strengths of diamond and silicon carbide. If the experimentally-determined tensile strengths of certain diamond and silicon carbide fibers are 1000 MPa and 500 MPa, respectively, what is the ratio of Young's modulus to fracture strength for each material? How do they compare with the materials listed in Table 5.1? On the basis of this comparison, what conclusion might you come to regarding the quality of the diamond and silicon carbide fibers in question?

5.24   When a failure data set for a ceramic material processed in a certain facility is analyzed, it is found that the characteristic strength is 327 MPa and the Weibull modulus is 8.75. A nominally identical batch of material processed in a different facility is also tested and found to have essentially the same characteristic strength, but the Weibull modulus is 6.25. At what stress level is the

probability of failure equal to 50% for each set of material? What initial conclusion might you draw about the quality control procedures at the two facilities?

5.25   A thin plate of a ceramic material with E = 225 GPa is loaded in tension, developing a stress of 450 MPa. Is the specimen likely to fail if the most severe flaw present is an internal crack oriented perpendicular to the load axis that has a total length 0.25 mm and a crack tip radius of curvature equal to 1 μm?

5.26   A rectangular bar is notched on two sides as shown in Fig. 5.9a. The dimensions of the bar are thickness $t = 0.2$ cm, $D = 2$ cm, $d = 1.8$ cm, $h = 0.1$ cm, and $r = 0.15$ cm. If an elastic load of 15 kN is applied along the axis of the bar, what is the maximum stress in the vicinity of the notches? If the yield strength of the material is 950 MPa, will the material yield near the notches under this load?

5.27   Two 0.5-cm-diameter rods of 1020 steel ($\sigma_{ts} = 395$ MPa) are to be joined with a silver braze alloy 0.025 cm thick ($\sigma_{ys} = 145$ MPa) to produce one long rod. The ultimate strength of this brazed structure is found to be approximately 345 MPa. If it is necessary to reduce the diameter of the rods to 0.25 cm with no change to the braze joint itself, will the strength (in MPa units) increase, decrease, or remain the same? If it changes, by how much?

5.28   You are called as an expert witness to analyze the fracture of a sintered silicon carbide plate that was fractured in bending when a blunt load was applied to the plate center. Measurement of the distance between the fracture origin and the mirror/mist boundary on the fracture surface gives a radius of 0.796 mm. You are given three pieces of the same SiC to test, and you determine that the mirror radius is 0.603, 0.203, and 0.162 mm for bending failure stress levels of 225, 368, and 442 MPa, respectively. What is your estimate of the stress present at the time of fracture for the original plate?

5.29   Two rods of Ni are broken in tension. One rod is nearly pure Ni whereas the other has been previously doped with Bi—a silver-colored, low-melting-point metal known to segregate to Ni grain boundaries. If SEM images of the fracture surfaces are compared, what features should appear in the two cases?

5.30   Two panes of ordinary float glass originally mounted in metal frames were broken in service. The panes were removed from the metal frames and photographs were brought to you for analysis (shown below). Pane (a) was damaged but intact; it was photographed to show one edge. Pane (b) was broken into multiple pieces; it was photographed to show a closeup of one fracture surface. The owner claims that both panes cracked as a result of mechanical overloading in the pane center in a direction perpendicular to the surface. First, assess the truth of this claim based on the visual evidence. Assuming that one or both truly could have fractured as claimed, determine on which side of the plate the pressure was applied, and in what direction the crack grew. Explain your rationale.

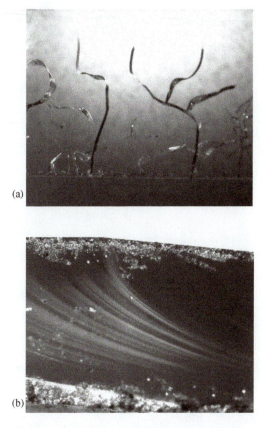

(a)

(b)

(Photographs courtesy of Anthony Spizzirri, Lehigh University)

5.31 A rod of PMMA is tested in tension at a strain rate of $1\,s^{-1}$ to determine the maximum elastic stress it can bear. At a temperature of 100°C, what would you expect the approximate maximum elastic stress to be, and what mechanism would you expect to take over at that stress level: brittle fracture, plastic deformation by cold drawing, or plastic deformation by crazing/shear yielding? How do you know?

Design

5.32 While preparing a SiC mirror for its role in an orbiting telescope, it becomes necessary for you to design a protective support system to ensure that the mirror does not fracture from the forces involved in boosting the satellite into orbit. For the purposes of this exercise, the mirror can be modeled as a flat disk 3.25 m in diameter and 3 mm thick. For evaluation, 20 small disks 100 mm in diameter × 3 mm thick prepared in the same fashion as the mirror are tested in bending. The results indicate that the characteristic strength is 474 MPa and the Weibull modulus is 16.9. What is the maximum bending stress that the full-scale mirror can be

allowed to experience so that the probability of failure is no greater than 1%? No greater than 0.1%? Based on your results, would you recommend designing for the lower of the two failure probabilities even if it would increase the cost?

5.33 In Problem 5.26, the existing design of the notches will lead to local plastic deformation even though the applied load is elastic elsewhere in the bar. If the load, the material, and the dimensions t, D, d, and h cannot be altered, what is the minimum notch radius r that would prevent yielding (with no safety factor)?

5.34 A long rod of solid 6061-T6 aluminum has a diameter of 2.54 cm. A hole 1.0 mm in diameter (intended for mounting a bracket to the rod) is drilled through the rod diameter as shown in Fig. 5.9d. It is calculated that when the wind blows, the rod will be subjected to a nearly pure bending moment. It is found during testing that the rod is cracking at 90% of the design stress right at the hole. If the rod is reconfigured to safely reach the design stress, should the hole diameter be increased or reduced? What hole diameter range would meet the design criteria?

Extend

5.35 Acquire the National Transportation Safety Board (NTSB) report describing the 2007 failure of the I-35 bridge in Minneapolis. What was the "generally accepted practice among Federal and State transportation officials" that may have contributed to the disaster? If this practice had not been common, what evidence might have been gathered that would have prevented the disaster?

5.36 In the National Transportation Safety Board (NTSB) report describing the 2007 failure of the I-35 bridge in Minneapolis it is reported that in October 1998, bridge inspectors found 12 fatigue cracks in 8 girders. The largest of these was over 50 inches long. Acquire the NTSB report and explain what was done to limit further growth of these cracks. Did this work indefinitely?

5.37 Acquire a journal paper of your choosing that describes a failure analysis involving fractography. Reproduce at least one image of the fracture surface from the paper, and describe what feature(s) were important to the conclusions of the failure analysis. Provide a full citation for the paper in a standard reference format.

5.38 Find a journal paper that describes the effect of notch sensitivity on the fracture behavior of Ultra High Molecular Weight Polyethylene (UHMWPE) components used in orthopedic implants. Summarize the findings of the paper, and provide a full citation for the paper in a standard reference format. Also explain why UHMWPE is chosen over other materials for the application described in the paper; be sure to include at least two design criteria and show how/why UHMWPE meets those criteria.

Chapter **6**

Elements of Fracture Mechanics

As outlined in the previous chapter, the fracture behavior of a given structure or material will depend on stress level, presence of a flaw, material properties, and the mechanism(s) by which the fracture proceeds to completion. The main purpose of this chapter is to develop quantitative relations between some of these factors. With knowledge of these relations, fracture phenomena may be better understood and design engineers more equipped to anticipate and thus prevent structural deficiencies. A secondary focus is the application of these fracture mechanics concepts to specific design codes. In addition to the sample problems given in this chapter, case histories are presented to illustrate the application of fracture mechanics to analysis of real structures.

6.1 GRIFFITH CRACK THEORY

The quantitative relations that engineers and scientists use today in determining the fracture of cracked solids were initially stated close to a century ago by A. A. Griffith.[1] Griffith noted that when a crack is introduced to a stressed plate of elastic material, a balance must be struck between the decrease in potential energy (related to the release of stored elastic energy and work done by movement of the external loads) and the increase in surface energy resulting from the presence of the crack. Likewise, an existing crack would grow by some increment if the necessary additional surface energy were supplied by the system. This "surface energy" arises from the fact that there is a nonequilibrium configuration of nearest neighbor atoms at any surface in a solid. For the configuration seen in Fig. 6.1, Griffith estimated the surface energy term to be the product of the total crack surface area $(2a \cdot 2 \cdot t)$, and the specific surface energy γ_s, which has units of energy/unit area. He then used the stress analysis of Inglis[2] for the case of an infinitely large plate containing an elliptical crack and computed the decrease in potential energy of the cracked plate to be $(\pi\sigma^2 a^2 t)/E$. Hence, the change in potential energy of the plate associated with the introduction of a crack may be given by

$$U - U_0 = -\frac{\pi\sigma^2 a^2 t}{E} + 4at\gamma_S \tag{6-1}$$

where

$U =$ potential energy of body with crack
$U_0 =$ potential energy of body without crack
$\sigma =$ applied stress
$a =$ one-half crack length
$t =$ thickness
$E =$ modulus of elasticity
$\gamma_S =$ specific surface energy

By rewriting Eq. 6-1 in the form

$$U = 4at\,\gamma_S - \frac{\pi\sigma^2 a^2 t}{E} + U_0 \tag{6-2}$$

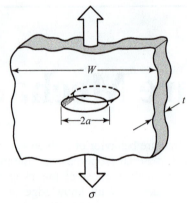

Figure 6.1 Through-thickness crack in a large plate.

and determining the condition of equilibrium by differentiating the potential energy U with respect to the crack length and setting equal to zero

$$\frac{\partial U}{\partial a} = 4t\gamma_S - \frac{2\pi\sigma^2 at}{E} = 0 \qquad (6\text{-}3)$$

($\partial U_0/\partial a = 0$, since U_0 accounts for the potential energy of the body without a crack and does not vary with crack length). Therefore

$$2\gamma_S = \frac{\pi\sigma^2 a}{E} \qquad (6\text{-}4)$$

which represents the equilibrium condition. The left side of the equation represents the energy required to create an additional unit area of crack surface while the right side is related to the elastic energy per unit volume (σ^2/E) available to drive crack extension.

The nature of the equilibrium condition described by Eqs. 6-3 and 6-4 is determined by the second derivative, $\partial^2 U/\partial a^2$. Since

$$\frac{\partial^2 U}{\partial a^2} = -\frac{2\pi\sigma^2 t}{E} \qquad (6\text{-}5)$$

and is negative, the equilibrium condition described by Eq. 6-3 is unstable, and the crack will always grow.

Griffith rewrote Eq. 6-4 in the form

$$\sigma = \sqrt{\frac{2E\gamma_S}{\pi a}} \qquad (6\text{-}6)$$

for the case of plane stress (biaxial stress conditions), and

$$\sigma = \sqrt{\frac{2E\gamma_S}{\pi a(1 - \nu^2)}} \qquad (6\text{-}7)$$

for the case of plane strain (triaxial stress conditions associated with the suppression of strains in one direction).

Since Poisson's ratio ν is approximately 0.25 to 0.33 for many materials, the difference in allowable stress level in a given material subjected to plane-strain or plane-stress conditions does not appear to be large. However, major differences do arise for other reasons, as will be discussed in Section 6.9.

It is important to recognize that the Griffith relation was derived for an elastic material containing a very sharp crack. Although Eqs. 6-6 and 6-7 do not explicitly involve the crack-tip radius ρ, as was the case for the stress concentration in Eq. 6-27, the radius is assumed to be very sharp. As such, the Griffith relation, as written, should be considered necessary but not sufficient for failure. The crack–tip radius also would have to be atomically sharp to raise the local stress above the cohesive strength.

6.1.1 Verification of the Griffith Relation

The other half of Griffith's classic paper was devoted to experimentally confirming the accuracy of Eqs. 6-6 and 6-7. Thin round tubes and spherical bulbs of soda lime silica glass were deliberately scratched or cracked with a sharp instrument, annealed to eliminate any residual stresses associated with the cracking process, and fractured by internal pressure. By recording the crack size and stress at fracture for these glass samples, Griffith was able to compute values of $\sigma\sqrt{a}$ in the range of 0.25 to 0.28 MPa $\sqrt{m}$ (0.23 to 0.25 ksi $\sqrt{in.}$),[i] which correspond to values of $\sqrt{2\gamma_S E/\pi}$ (Eq. 6-6). From experimentally determined surface tension values γ_S for glass fibers between 745 and 1110 °C, a room temperature value was obtained by extrapolation (risky business, but reasonable for a first approximation). By multiplying this value, 0.54 N/m, by the modulus of elasticity, 62 GPa, the value of $\sqrt{2\gamma_S E/\pi}$ determined from material properties was found to be 0.15 MPa $\sqrt{m}$. It should be recognized that the exceptional agreement between theoretical and experimental values may be somewhat fortuitous in light of some inaccuracies contained in the original development by Griffith.[1] For example, the second law of thermodynamics requires some inefficiency in transfer of energy from stored strain energy to surface energy. Also, recent estimates of the fracture resistance of soda lime glass are more than three times greater than that reported by Griffith (see Tables 7.8a and b). Nevertheless, the Griffith equation and its underlying premise are basically sound and represent a major contribution to the fracture literature.

6.1.2 Griffith Theory and Propagation-Controlled Thermal Fracture

With a clear connection between surface energy and crack propagation now established, it is possible to revisit the case of brittle thermal fracture originally introduced in Chapter 1. There, it was shown that thermal fractures could be avoided if crack initiation could be prevented. Maximizing the ratio of fracture strength to modulus, σ_f/E, was one possible strategy to achieve this (Eq. 1-92). Let us assume now that crack initiation cannot be avoided because particularly severe thermal conditions exist. If this is the case, the only way to avoid outright failure is to minimize crack extension. Following Griffith, Hasselman[3] determined that the extent of thermal crack propagation is proportional to the available amount of elastic strain energy at fracture (σ^2/E) and inversely proportional to the material's fracture energy necessary for the creation of new crack surface (γ_S). Correspondingly, he defined a *thermal stress damage resistance parameter*, R'''', that describes a material's resistance to crack instability as

$$R'''' = \frac{E\gamma_S}{\sigma_f^2}(1 - \nu) \tag{6-8}$$

Note that resistance to crack instability is improved by maximizing E/σ_f—the opposite approach to that needed to limit crack initiation.[ii] Proceeding further, Hasselman[3] computed the temperature change necessary for unstable crack extension as a function of crack size. For the case of a constrained flat plate (Fig. 6.2) with thermal expansion coefficient α, the critical temperature difference ΔT_c needed for simultaneous propagation of N embedded circular and

---

[i] To convert from ksi$\sqrt{in.}$ to MPa$\sqrt{m}$, multiply by 1.099.

[ii] Equation 6-8 loses relevancy as σ_f approaches zero.

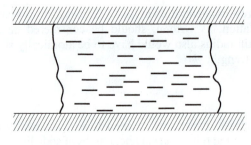

Figure 6.2 Crack distribution in rigidly constrained body subjected to a decrease in temperature.

noninteracting cracks of equal diameter(l)/unit volume, is given by

$$\Delta T_c = \left(\frac{2\gamma}{\pi l\alpha^2 E} \right)^{0.5} (1 + 2\pi N l^2) \tag{6-9}$$

Figure 6.3 shows the critical temperature difference (ΔT_c) necessary for crack instability as a function of initial crack length for two different crack densities. For the case of short cracks, ΔT_c decreases with increasing crack length (i.e., the maximum temperature change that can be withstood without causing crack growth decreases with increasing crack length) until a minimum is reached at $l = l_m$. At this point, l_m is found to be equal to $(6\pi N)^{-1/2}$. It would appear at first that this would cause crack growth to arrest at length l_m. However, when a critical temperature difference is reached and the crack ($l < l_m$) propagates in an unstable manner, the strain energy release rate becomes greater than that necessary to create additional new fracture surface as required by Griffith's energy criterion. The excess strain energy is then transformed into kinetic energy that drives the crack beyond l_m until the two energy terms are in balance and the crack arrests. Hasselman[3] determined that this energy balance corresponds to those arrested crack lengths defined by the dashed lines in Fig. 6.3, shifting the stable crack size to a larger value l_f that is dependent on the crack density (i.e., the total amount of fractured surface area).

The two solid lines in Fig. 6.3 where $l > l_m$ represent values of ΔT_c associated with conditions for quasi-stable crack growth in brittle bodies containing two different crack densities. In both cases a sufficiently large ΔT can cause crack growth, but as soon as l increases then ΔT_c also increases, halting further growth. Note that the two cases show that thermal shock resistance is expected to *increase* with *increasing* microcrack density. Although this may be counterintuitive at first, the trend springs directly from Griffith theory: the total surface energy cost associated with a certain increment of crack extension is greater when there

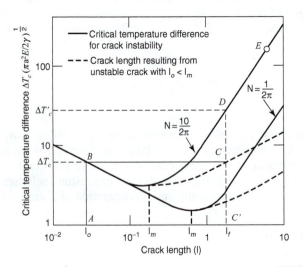

Figure 6.3 Thermal shock resistance described in terms of critical temperature difference ΔT_C versus crack length as a function of crack density N.[4] (D. R. Larson, J. A. Coppola, D. P. H. Hasselman, and R. C. Bradt, *Journal of the American Ceramic Society,* 57, 417 (1974). Reprinted by permission.)

are more cracks per unit volume, and thus more total crack area created for a given crack extension. Indeed, this has been demonstrated in BeO-matrix composites containing 15 w% SiC dispersed particles, and in MgO-3 v% W composites. Both of the composites showed significantly less thermal shock damage than the pure matrix materials.[5,6] In these instances, three-dimensional networks of microcracks between the matrix and reinforcement phases developed to relieve thermal stresses generated upon cooling from the fabrication temperature. These microcracks then made the composites less susceptible to further thermal shock damage.

The long crack behavior just described can now be used to glean one more critical insight from Fig. 6.3 regarding growth of short cracks. We see from the figure that for cracks of length l_0, instability occurs when the temperature difference increases to ΔT_c. As noted above, such cracks grow unstably until they reach a size l_f and then arrest. They then remain stable until the temperature difference is increased by an additional amount equal to $\Delta T'_c - \Delta T_c$. This creates a temperature difference regime in which fracture strength is constant. Once $\Delta T'_c$ is reached, the cracks grow in a quasi-static manner with increasing increments of ΔT as described previously. This is advantageous because thermal fracture damage will accumulate gradually rather than catastrophically.

Representative tensile strength data as a function of ΔT are shown in Fig. 6.4a for the case of industrial-grade polycrystalline aluminum oxide that had small initial cracks. Note the significant strength drop at $\Delta T \approx 300\,°C$ that corresponds to region BC in Fig. 6.3, and the strength plateau in the regime where $\Delta T_c \leq \Delta T \leq \Delta T'_c$ that matches region CD (or $C'D$) in the same figure. Alternatively, the strength-ΔT plot shown in Fig. 6.4b for a high-alumina refractory corresponds to material behavior shown in Fig. 6.3 along path CDE, with a plateau followed by stable crack growth. The absence of unstable crack extension in the refractory material is associated with its lower strength level (and hence greater initial microcrack size) as compared with that found in the industrial grade aluminum oxide (recall Eq. 6-8).

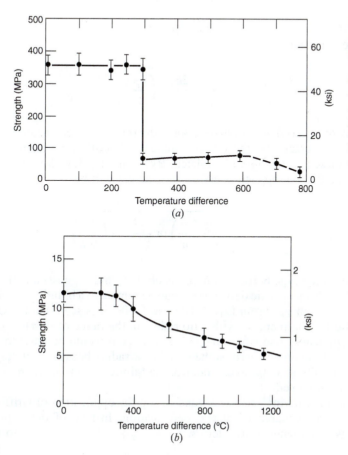

Figure 6.4   Tensile strength versus quench temperature difference for (a) industrial polycrystalline aluminum oxide[8]; and (b) high-alumina refractory[9]. (From D. P. H. Hasselman, *Journal of the American Ceramic Society*, 53, 490 (1979). Reprinted by permission. D. P. H. Hasselman and J. P. Singh, *Thermal Stresses*, R. B. Hetnaski et al., Eds., Vol 1, Chap 4, 264 (1986). Reprinted by permission.)

The choice of a suitable brittle solid to maximize thermal shock resistance may now be characterized in two fundamentally different ways:[7]

1. For relatively mild thermal environments, thermal shock fracture can be avoided by the selection of a material with low thermal expansion and elastic modulus, and high strength and thermal conductivity (to prevent crack nucleation; see Chapter 1).

2. For severe thermal fluctuations, optimal material properties should include low thermal expansion and fracture strength, and high elastic modulus. In addition, the material should be designed to contain a high density of microcracks such that $l_0 > l_m$. The latter will ensure stable crack propagation. Note that the choice of a low-strength ceramic in connection with its large population of microcracks must be restricted to low load-bearing applications.

6.1.3 Adapting the Griffith Theory to Ductile Materials

Since plastic deformation processes in ceramics and glasses are very limited, the difference in surface energy and fracture energy values is not expected to be great. This is not true for metals and polymers, where the fracture energy is found to be several orders of magnitude greater than the surface energy of a given material. Orowan[10] recognized this fact and suggested that Eq. 6-6 be modified to include the energy of plastic deformation in the fracture process so that

$$\sigma = \sqrt{\frac{2E(\gamma_S + \gamma_P)}{\pi a}} = \sqrt{\frac{2E\gamma_s}{\pi a}\left(1 + \frac{\gamma_P}{\gamma_S}\right)} \tag{6-10a}$$

where $\gamma_P =$ plastic deformation energy and $\gamma_P \gg \gamma_S$.
Under these conditions

$$\sigma \approx \sqrt{\frac{2E\gamma_S}{\pi a}\left(\frac{\gamma_P}{\gamma_S}\right)} \tag{6-10b}$$

The applicability of Eqs. 6-6 or 6-10 in describing the fracture of real materials will depend on the sharpness of the crack and the relative amount of plastic deformation. The following relation reveals these two factors to be related. By combining Eqs. 5-9 and 5-27 and letting $\sigma_{\max} = \sigma_c$, we see that the applied stress σ_a for fracture will be

$$\sigma_a = \frac{1}{2}\sqrt{\frac{E\gamma_S}{a}\left(\frac{\rho}{a_0}\right)} \text{ or } \sqrt{\frac{2E\gamma_s}{\pi a}\left(\frac{\pi\rho}{8a_0}\right)} \tag{6-10c}$$

The similarity between Eqs. 6-10b and 6-10c is obvious and suggests a correlation between γ_P/γ_S and $\pi\rho/8a_0$; that is, plastic deformation can be related to a blunting process at the crack tip; ρ will increase with γ_P. From Eqs. 6-10b and 6-10c, it is seen that the Griffith relation (Eq. 6-6) is valid for sharp cracks with a tip radius in the range of $(8/\pi)a_0$. Equation 6-6 is believed to be applicable also where $\rho < (8/\pi)a_0$, since it would be unreasonable to expect the fracture stress to approach zero as the crack root radius became infinitely small. When $\rho > (8/\pi)a_0$, Eq. 6-10b or 6-10c would control the failure condition where plastic deformation processes are involved.

At the same time, Irwin[11] also was considering the application of Griffith's relation to the case of materials capable of plastic deformation. Instead of developing an explicit relation in terms of the energy sink terms, γ_s or $(\gamma_s + \gamma_P)$, Irwin chose to use the energy source term (i.e., the elastic energy per unit crack-length increment $\partial U/\partial a$). Denoting $\partial U/\partial a$ as $\mathscr{G}$, Irwin showed that

$$\sigma = \sqrt{\frac{E\mathscr{G}}{\pi a}} \tag{6-11}$$

which is one of the most important relations in the literature of fracture mechanics.[iii] By comparison of Eqs. 6-10 and 6-11, it is seen that at equilibrium

$$\mathscr{G} = 2(\gamma_S + \gamma_P) \tag{6-12}$$

At the point of instability, the *elastic strain energy release rate* $\mathscr{G}$ (also referred to as the crack driving force) reaches a critical value $\mathscr{G}_c$, whereupon fracture occurs. This *critical strain energy release rate* may be interpreted as a material parameter, and can be measured in the laboratory (e.g., with sharply notched test specimens).

6.1.4 Energy Release Rate Analysis

In the previous section, the elastic energy release rate $\mathscr{G}$ was related to the release of strain energy and the work done by the boundary forces. The significance of these two terms will now be considered in greater detail. For an elastically loaded body containing a crack of length a (Fig. 6.5), the amount of stored elastic strain energy is given by

$$V = \frac{1}{2}P\delta \quad \text{or} \quad \frac{1}{2}\frac{P^2}{M_1} \tag{6-13}$$

where

$V =$ stored strain energy
$P =$ applied load
$\delta =$ load displacement
$M_1 =$ body stiffness for crack length a

If the crack extends by an amount da, the necessary additional surface energy is obtained from the work done by the external body forces $P\,d\delta$ and the release of strain energy dV.[12] As a result,

$$\mathscr{G} = \frac{dU}{da} = P\frac{d\delta}{da} - \frac{dV}{da} \tag{6-14}$$

with the stiffness of the body decreasing to M_2. Whether the body was rigidly gripped such that incremental crack growth would result in a load drop from P_1 to P_2 or whether the load was fixed such that crack extension would result in an increase in δ by an amount $d\delta$, the stiffness of the plate M would decrease. For the fixed grip case, both P and M would decrease but the ratio P/M would remain the same, since from Fig. 6.5

$$\delta_1 = \delta_2 = \frac{P_1}{M_1} = \frac{P_2}{M_2} \tag{6-15}$$

The elastic energy release rate would be

$$\left(\frac{\partial U}{\partial a}\right)_\delta = \frac{1}{2}\left[\frac{2P}{M}\frac{\partial P}{\partial a} + P^2\frac{\partial(1/M)}{\partial a}\right] \tag{6-16}$$

[iii] Note that the elastic strain energy release rate is written as a script $\mathscr{G}$ in this text. It is sometimes written as a standard G, but we have avoided this usage because of potential confusion with the shear modulus, also often called G.

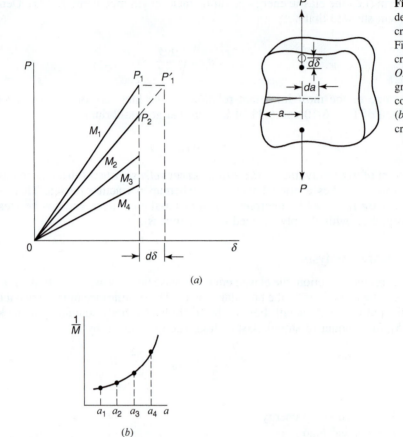

(a)

(b)

Figure 6.5 (a) Load-deflection response of a cracked plate such as shown in Fig. 6.1 for the case where the crack length increases by da. OP_2 corresponds to the fixed grip condition, while OP_1 corresponds to fixed load cases. (b) Compliance dependence on crack length.

By differentiating Eq. 6-15 to obtain

$$\frac{1}{M}\frac{\partial P}{\partial a} + P\frac{\partial(1/M)}{\partial a} = 0 \tag{6-17}$$

and substituting the result into Eq. 6-16

$$\left(\frac{\partial U}{\partial a}\right)_\delta = -\frac{1}{2}P^2\frac{\partial(1/M)}{\partial a} \tag{6-18a}$$

It may be shown[12] that under fixed load conditions

$$\left(\frac{\partial U}{\partial a}\right)_P = \frac{1}{2}P^2\frac{\partial(1/M)}{\partial a} \tag{6-18b}$$

Note that in both conditions, the elastic energy release rate is the same (only the sign is reversed), reflecting the fact that $\mathscr{G}$ is independent of the type of load application (e.g., fixed grip, constant load, combinations of load change and displacement, and machine stiffness). At instability, then, the critical strain energy release rate is

$$\mathscr{G}_c = \frac{P^2\,\text{max}}{2}\frac{\partial(1/M)}{\partial a} \tag{6-19}$$

where $1/M$ is the compliance of the cracked plate, which depends on the crack size. Once the compliance versus crack-length relation has been established for a given specimen configuration, $\mathscr{G}_c$ can be obtained by noting the load at fracture, provided the amount of plastic deformation at the crack tip is kept to a minimum. To illustrate, load–displacement plots corresponding to samples that contain cracks of different lengths are shown in Fig. 6.5a. Since compliance $(1/M)$ is given by δ/P, the crack-length dependence of $1/M$ takes the form given in Fig. 6.5b.

6.2   CHARPY IMPACT FRACTURE TESTING

The contributions of Griffith, Orowan, and Irwin make it possible to perform *and interpret* useful tests of fracture toughness for a wide variety of materials. This information, then, allows for design with fracture resistance in mind. Before advanced fracture mechanics concepts (such as those presented beginning in Section 6.5) were developed, engineers sought convenient laboratory-sized samples and suitable test conditions with which to simulate field failures without resorting to the forbidding expense of destructively testing full-scale engineering components. To anticipate the worst possible set of circumstances that might surround a potential failure, these laboratory tests employed experimental conditions that could *suppress the capacity of the material to plastically deform by elevating the yield strength*: low test temperatures, high strain rates, and a multiaxial stress state caused by the presence of a notch or defect in the sample. This approach is of considerable importance in pressure vessel, bridge, and ship structure design due to the fact that in body-centered-cubic metals, such as ferritic alloys, the yield strength is far more sensitive to temperature and strain-rate changes than it is in face-centered-cubic metals such as aluminum, nickel, copper, and austenitic steel alloys. As pointed out in Chapter 2, this increased sensitivity in BCC alloys can be related to the temperature-sensitive Peierls-Nabarro stress contribution to yield strength, which is much larger in BCC metals than in FCC metals.

To a first approximation, the relative *notch sensitivity* of a given material may be estimated from the yield- to tensile-strength ratio. When the ratio is low, the plastic constraint associated with a biaxial or triaxial stress state at the crack tip will elevate the entire stress–strain curve and allow for a net section stress greater than the smooth bar tensile strength value. As discussed in Sections 5.5 and 5.6, a 2.5 to 3-fold increase in net section strength is possible in ductile materials that "notch strengthen" (Fig. 6.6a). On the other hand, in

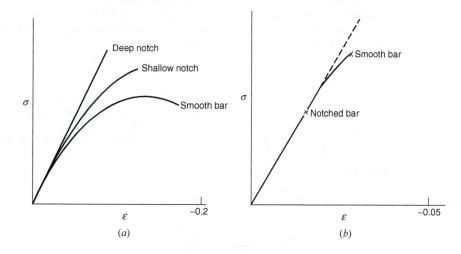

Figure 6.6   (*a*) Plastic constraint resulting from triaxial stresses at notch root produces elevation of flow curve in ductile material. (*b*) For material with little intrinsic plastic flow capacity, introduction of sharp crack induces premature brittle failure.

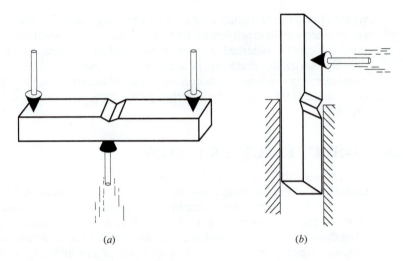

(a)   (b)

Figure 6.7 Flexed-beam impact samples. (*a*) Charpy type[13] (three-point loaded) used extensively with metal alloys. (*b*) Izod type[14] (cantilever-beam loaded) used extensively with polymers. Both samples contain 0.25-mm notch radius.

materials that have less inherent ability for plastic deformation, the stress concentration at a notch root is not offset by the necessary degree of crack-tip plasticity needed to blunt the crack tip. Consequently, the notch with its multiaxial stress state raises the local stress to a high level and suppresses what little plastic deformation capacity the material possesses, and brittle failure occurs (Fig. 6.6*b*).

The *Charpy specimen* (Fig. 6.7*a*) and associated test procedure provides a relatively severe test of material toughness. The notched sample is loaded at very high strain rates because the material must absorb the impact of a falling pendulum and is tested over a range of temperatures. Considerable data can be obtained from the impact machine reading and from examination of the broken sample.

First, the amount of energy absorbed by the notched Charpy bar can be measured by the maximum height to which the pendulum rises after breaking the sample (Fig. 6.8). If a typical 325-J (240-ft-lb)[iv] machine were used, the extreme final positions of the pendulum would be either at the same height of the pendulum before it was released (indicating no energy loss in breaking the sample), or at the bottom of its travel (indicating that the specimen absorbed the full 325 J of energy). The dial shown in Fig. 6.8 provides a direct readout of the energy absorbed

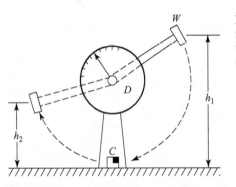

Figure 6.8 Diagram showing impact hammer *W* dropping from height h_1, impacting sample at *C* and rising to maximum final height h_2. Energy absorbed by sample, related to height differential $h_2 - h_2$, is recorded on dial *D*.

[iv] To convert from foot-pounds to joules, multiply by 1.356.

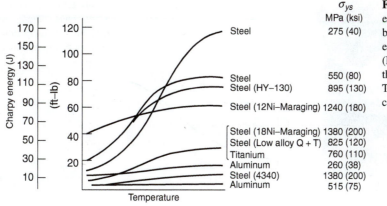

Figure 6.9 Charpy impact energy versus temperature behavior for several engineering alloys.[15] (Reprinted by permission of the American Society for Testing and Materials from copyright material.)

by the sample. Typical impact energy versus test temperature for several metals is plotted in Fig. 6.9. It is clearly evident from this plot that some materials show a marked change in energy absorption when a wide range of temperatures is examined. In fact, this sudden shift or transition in energy absorption with temperature has suggested to engineers the possibility of designing structural components with an operating temperature above which the component would not be expected to fail.

The effect of temperature on the energy to fracture has been related in low-strength ferritic steels to a change in the microscopic fracture mechanism: cleavage at low temperatures and void coalescence at high temperatures. The onset of microscopic cleavage and macroscopic brittle behavior in low-strength ferritic steels is so closely related that "cleavage" and "brittle" often are used synonymously in the fracture literature. This is unfortunate since in Chapter 5 *brittle* is defined as a low level of fracture *energy* or limited crack-tip plasticity, while *cleavage* describes a failure *micromechanism*. Confusion arises since brittle behavior can occur without cleavage, as in the fracture of high-strength aluminum alloys; alternatively, you can have 4% elongation (reflecting moderate energy absorption) in a tungsten-25 at% rhenium alloy specimen and still have a cleavage fracture.[v] Since a direct correlation does not always exist between a given fracture mechanism and the magnitude of fracture energy, it is best to treat the two terms separately.

Unless the fracture energy changes discontinuously at a given temperature, some criterion must be established to *define* the "transition temperature." Should it be defined at the 13.5-, 20-, or 27-J (10-, 15-, or 20-ft-lb)[vi] level as it is sometimes done, or at some fraction of the maximum (or "shelf") energy? The answer depends on how well the defined transition temperature agrees with the service experience of the structural component under study. For example, Charpy test results for steel plate obtained from failures of Liberty ships (recall Fig. 5.1a) revealed that plate failures never occurred at temperatures greater than the 20-J (15-ft-lb) transition temperature. Unfortunately, the transition temperature criterion based on such a specific energy level is not constant but varies with material. Specifically, Gross[16] has found for several steels with strengths in the range of 415 to 965 MPa that the appropriate energy level for the transition temperature criterion should increase with increasing strength.

The same problem arises when the transition temperature is estimated from other measurements. For example, if the amount of lateral expansion on the compression side of the bar is measured (Fig. 6.10), it is found that it, too, undergoes a transition from small values at low temperature to large values at high temperature. (This increase in observed plastic deformation is consistent with the absorbed energy–temperature trend.) Whether the correct

---

[v] P. L. Raffo, NASA TND-4567, May 1968, Lewis Research Center, Cleveland, OH.

[vi] Dual units are retained for reference since the specific foot-pound energy levels cited above represent long-standing design criteria.

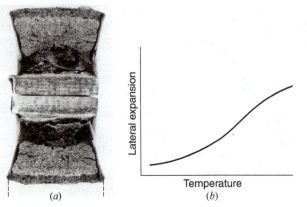

Figure 6.10 (*a*) Measurement of lateral expansion at compression side of Charpy bar; (*b*) schema of temperature dependence of lateral expansion revealing transition behavior.

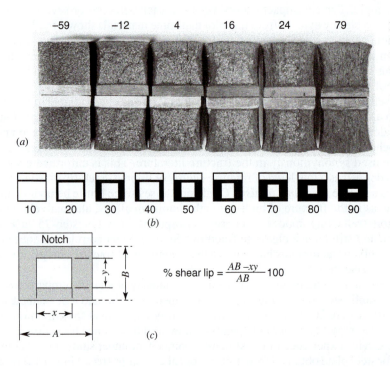

$$\% \text{ shear lip} = \frac{AB - xy}{AB} \cdot 100$$

Figure 6.11 Transition in fracture surface appearance as function of test temperature. (*a*) Actual fracture series for A36 steel tested in the transverse direction; (*b*) standard comparison chart showing percentage shear lip; (*c*) computation for percentage shear lip.

transition temperature conforms to an absolute or relative contraction depends on the material. Finally, transitional behavior is found when the amount of fibrous or cleavage fracture on the fracture surface is plotted against temperature. A typical series of fracture surfaces produced at different temperatures is shown in Fig. 6.11*a*. Here again, the appropriate percentage of cleavage or fibrous fracture (based on comparison with a standard chart such as in Fig. 6.11*b* or measured directly as in Fig. 6.11*c*) to use to define the transition temperature will depend on the material as well as other factors. To make matters worse, transition temperatures based on either energy absorption, ductility, or fracture appearance criteria do not agree even for the same material. As shown in Table 6.1, the transition temperatures defined by a 20-J energy criterion and by a 0.38-mm (15-mil) lateral expansion are in reasonably good agreement, but are consistently lower than the 50% fibrous fracture transition temperature. Which transition temperature to use "is a puzzlement!"

Table 6.1a Transition Temperature Data for Selected Steels[17]

Material	σ_{ys}, MPa $\overline{\sigma_{ts}, \text{MPa}}$	Transition Temperature, °C		
		20 J	**0.38 mm**	**50% fibrous**
Hot-rolled C–Mn steel	$\frac{210}{442}$	27	17	46
Hot-rolled, low-alloy steel	$\frac{385}{570}$	−24	−22	12
Quenched and tempered steel	$\frac{618}{688}$	−71	−67	−54

Table 6.1b Transition Temperature Data for Selected Steels[17]

Material	σ_{ys}, ksi $\overline{\sigma_{ts}, \text{ksi}}$	Transition Temperature, °F		
		15 ft-lb	**15 mil**	**50% fibrous**
Hot-rolled C–Mn steel	$\frac{30.5}{64.1}$	80	62	115
Hot-rolled, low-alloy steel	$\frac{55.9}{82.6}$	−12	−7	53
Quenched and tempered steel	$\frac{89.7}{99.8}$	−95	−88	−66

6.3 RELATED POLYMER FRACTURE TEST METHODS

Additional specimens and test methods have been developed to evaluate the toughness response of engineering plastics.[18] The Izod sample (Fig. 6.7*b*) is a notched bar that is fixed at one end and impacted on the unsupported section along the side of the bar that contains the notch.[14] Numerous studies have shown that the Izod impact energy of many plastics, usually defined by the energy absorbed per unit area of the net section, possesses a ductile–brittle transition response. Figure 6.12*a* shows the change in transition temperature for PVC with different notch radii. The large reduction in fracture energy at a given temperature with the presence of a notch is attributed to the high strain rate at the notch root and the virtual elimination of energy necessary to initiate a crack in the sample. A comparison of the room temperature impact resistance for several polymers is given in Fig. 6.12*b* and reveals the strong notch sensitivity of these materials.[19]

Two other impact test methods for engineering plastics are described by the drop-weight and tensile-impact test procedures. The drop-weight method (ASTM[vii] D3029[20]) measures the energy to initiate fracture in an unnotched sheet of material; a disk-shaped sample is supported horizontally by a steel ring and struck with different weights that are dropped from a given height. A mean failure energy is then defined on a statistical basis, which corresponds to a 50% failure rate of the disks. The high-speed tensile-impact method (ASTM D1822[21]) makes use of a small tensile specimen that is clamped to a pendulum hammer at one end and attached to a striker plate at the other. The pendulum is released and drops with the attached sample trailing behind. When the striker plate impacts a rigidly mounted stationary anvil, the sample experiences rapid straining in tension to failure. The fracture energy is determined from the differences in pendulum height before and after its release. Impact energies for many engineering plastics are given in the *Modern Plastics Encyclopedia*.[22]

[vii] References to selected ASTM standards are presented throughout this text, but the reader is strongly advised to refer to the most recent book of standards to determine if a more relevant or updated standard is available for their situation before proceeding with a mechanical test.

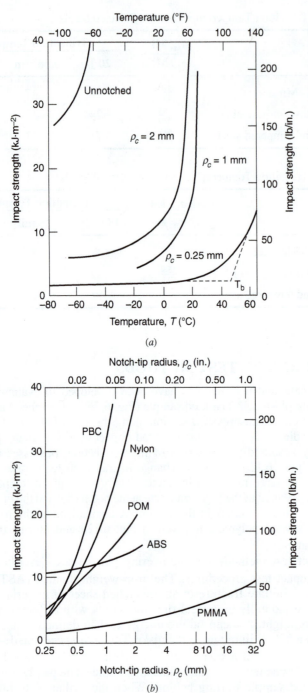

Figure 6.12 Impact strength in (*a*) PVC as a function of temperature for different notch-root radii and (*b*) selected engineering plastics as a function of notch-tip radius.[19] (Reprinted with permission from *Impact Tests and Service Performance of Thermoplastics*, courtesy of Plastics and Rubber Institute.)

6.4 LIMITATIONS OF THE TRANSITION TEMPERATURE PHILOSOPHY

It is important to recognize some limitations in the application of the transition temperature philosophy to component design. First, the absolute magnitude of the experimentally determined transition temperature, as defined by any of the previously described methods (energy absorbed, ductility, and fracture appearance), has been shown to depend on the *thickness* of the specimen used in the test program. (As explained later in Section 6.9, it is now understood that

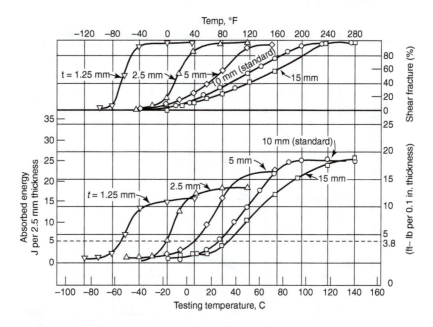

Figure 6.13 Adjusted energy-temperature curves and shear fracture–temperature curves for 38-mm-thick plate of A283 steel tested with Charpy V-notch specimens of various thicknesses. Absorbed energy defined at 5.2 J/2.5 mm (3.8 ft-lb/0.1 in.) of specimen thickness[23]. (Reprinted from *Welding Journal* by permission of the American Welding Society.)

this is due to the potential for a plane-strain/plane-stress, stress-state transition when sample thickness is varied.) In evaluating this effect, McNicol[23] found that the transition temperature in several steels, based on energy, ductility, and fracture appearance criteria, increased with increasing Charpy bar thickness t. Figure 6.13 shows temperature-related changes in energy absorbed per 2.5-mm sample thickness and percentage shear fracture as a function of sample thickness for A283, a hot-rolled carbon manganese steel. It is clear from this figure that the transition temperature increased with increasing thickness. Moreover, the transition temperature was different for the two criteria. With increasing sample thickness, it would be expected that the transition temperature would rise to some limiting value as full plane-strain conditions were met. This condition is inferred from Fig. 6.14, which shows the transition temperature reaching a maximum level with increasing thickness for three different steel alloys.

It is clear, then, that the defined transition temperature will depend not only on the measurement criteria but also on the thickness of the test bar. Therefore, laboratory results may bear no direct relation to the transition temperature characteristics of the engineering component if the component's thickness is different from that of the test bar. This difficulty in extending test results to service conditions is most disturbing when engineering design decisions must be made.

In addition to transition temperature–thickness effects, there are uncertainties relating to crack-length effects as well. This effect may be seen by considering Fig. 6.15. We see the general relation between flaw size and allowable stress level for a material with a given toughness level. The solid line represents the allowable stress level assuming ideal elastic conditions. Brittle fracture will therefore occur when the stress at a crack tip reaches the level of the solid curve. The dashed deviations from the toughness curve marked T_1 and T_2 indicate the stress required to initiate crack-tip plasticity at two different temperature levels. If the crack tip stress crosses the yield threshold before the expected transition between elastic loading and fracture is reached, plasticity will occur before fracture. It is seen that a notched bar with initial crack length a_1 would therefore be brittle at either test temperature, but the same material with a crack length a_3 would exhibit tough behavior at both temperatures. If the test temperature were to be reduced from T_1 to T_2, the response of a sample with crack length a_2 would change from

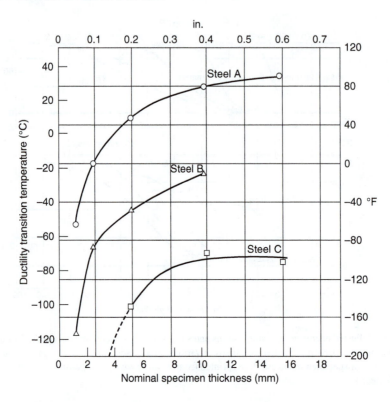

Figure 6.14 Effect of specimen thickness on Charpy V-notch ductility transition temperature of steels A, B, and C. The ductility transition temperature was selected with the same relative energy/unit thickness ratio given in Fig. 9.11.[23] (Reprinted from *Welding Journal* by permission of the American Welding Society.)

tough to brittle. An additional temperature reduction to well below T_2 would be necessary for the sample with crack length a_3 to exhibit brittle behavior.

From the above discussion, it becomes apparent that a wide range of "transition temperatures" can be obtained simply by changing the specimen thickness and/or the crack length of the test bar. For this reason, transition temperature values obtained in the laboratory bear little relation to the performance of the full-scale component, thereby necessitating a range of correction factors.

As mentioned previously, the onset of brittle fracture is not always accompanied by the occurrence of the cleavage microscopic fracture mechanism. Rather, it should be possible to choose a specimen size for a given material, and tailor both thickness and planar dimensions such that a temperature-induced transition in energy to fracture, amount of lateral contraction, and macroscopic fracture appearance would occur *without the need for a microscopic mechanism transition.* Figure 6.16, from the work of Begley,[24] is offered as proof of this statement. Substandard-

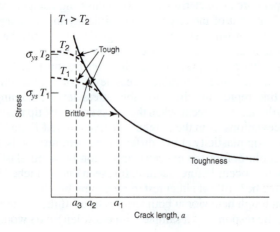

Figure 6.15 Schematic diagram showing relation between allowable stress level and flaw size. Solid line represents material fracture toughness; dashed lines show effect of plasticity.

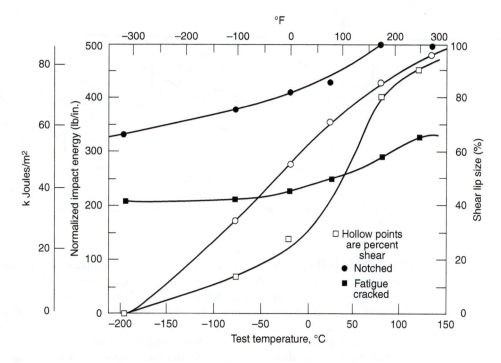

Figure 6.16 Charpy impact data for subsize specimen revealing transition temperature response in 7075-T651 aluminum alloy.[24] (Courtesy of James A. Begley.)

thickness Charpy bars of 7075-T651 aluminum alloy were tested and shown to exhibit a temperature-induced transition in impact energy and fracture appearance. From Fig. 6.9, no such transition was observed when standard Charpy specimens of an aluminum alloy were broken.

6.5 STRESS ANALYSIS OF CRACKS

Although the Griffith-Orowan-Irwin energy balance approach and the transition temperature testing methodology provide important conceptual and practical ways to inform design decisions, their limitations leave considerable uncertainty with regard to assessing the likelihood of component fracture. A more sophisticated and flexible approach to the fracture of flawed components is available through a stress analysis based on concepts of elastic theory. As an introduction to this body of work, it is useful to first consider three prototypical modes of loading that can drive crack growth in different ways. Using modifications of analytical methods described by Westergaard,[25] Irwin[26] published solutions for crack-tip stress distributions associated with the three major modes of loading shown in Fig. 6.17. Note that these modes involve different crack surface displacements.

Mode I. Opening or tensile mode, where the crack surfaces move directly apart.

Mode II. Sliding or in-plane shear mode, where the crack surfaces slide over one another in a direction perpendicular to the leading edge of the crack.

Mode III. Tearing or antiplane shear mode, where the crack surfaces move relative to one another and parallel to the leading edge of the crack.

Mode I loading is encountered in the overwhelming majority of actual engineering situations involving cracked components. Consequently, considerable attention has been given to both analytical and experimental methods designed to quantify Mode I stress–crack-length

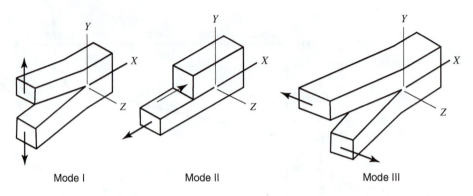

Mode I   Mode II   Mode III

Figure 6.17 Basic modes of loading involving different crack surface displacements.

relations. Mode II is found less frequently in structural components, but can play an important role in failures of adhesive joints (see Section 6.16). One example of mixed Mode I–II loading involves axial loading (in the Y direction) of a crack inclined as a result of rotation about the Z axis (Fig. 6.18). In this instance, analytical methods[27] show the Mode I contribution to dominate the crack-tip stress field when $\beta > 60°$. Mode III may be regarded as a pure shear problem such as that involving a notched round bar in torsion.

For the notation shown in Fig. 6.19, the crack-tip stresses are found to be

$$\sigma_y = \frac{K}{\sqrt{2\pi r}} \cos\frac{\theta}{2} \left(1 + \sin\frac{\theta}{2}\sin\frac{3\theta}{2}\right)$$

$$\sigma_x = \frac{K}{\sqrt{2\pi r}} \cos\frac{\theta}{2} \left(1 - \sin\frac{\theta}{2}\sin\frac{3\theta}{2}\right) \qquad (6\text{-}20)$$

$$\tau_{xy} = \frac{K}{\sqrt{2\pi r}} \left(\sin\frac{\theta}{2}\cos\frac{\theta}{2}\cos\frac{3\theta}{2}\right)$$

Figure 6.18 Crack inclined β degrees about the z axis. Mode I dominates when $\beta > 60°$.

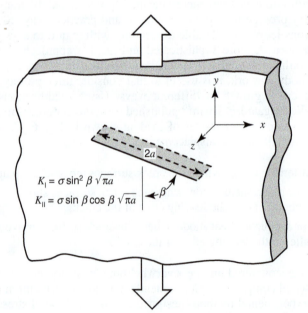

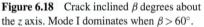

$$K_{\mathrm{I}} = \sigma \sin^2 \beta \sqrt{\pi a}$$
$$K_{\mathrm{II}} = \sigma \sin \beta \cos \beta \sqrt{\pi a}$$

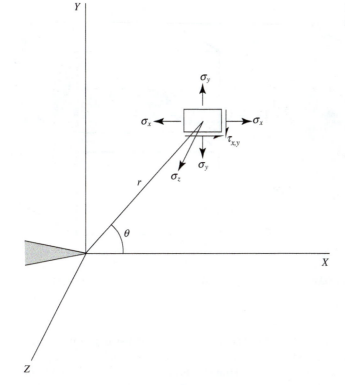

Figure 6.19 Distribution of stresses in the vicinity of the crack tip.

where K is a stress field parameter that essentially describes the severity of the local stress concentration. K_I in Fig. 6.18 describes the Mode I severity, and K_{II} the Mode II severity, for a given crack and loading geometry. It is apparent from Eq. 6-20 that the local stresses could rise to extremely high levels as r approaches zero. As pointed out earlier in the chapter, this circumstance is precluded by the onset of plastic deformation at the crack tip. Since this plastic enclave is embedded within a large elastic region of material and is acted upon by either biaxial $(\sigma_y + \sigma_x)$ or triaxial $(\sigma_y + \sigma_x + \sigma_z)$ stresses, the extent of plastic strain within this region is suppressed. For example, if a load were applied in the Y direction, the plastic zone would develop a positive strain ε_y and attempt to develop corresponding negative strains in the X and Z direction, thus achieving a constant volume condition required for a plastic deformation process $(\varepsilon_y + \varepsilon_z + \varepsilon_x = 0)$. However, σ_x acts to restrict the plastic zone contraction in the X direction, while the negative ε_z strain is counteracted by an induced tensile stress σ_z. Since there can be no stress normal to a free surface, the through-thickness stress σ_z must be zero at both surfaces but may attain a relatively large value at the midthickness plane. At one extreme, the case for a thin plate where σ_z cannot increase appreciably in the thickness direction, a condition of *plane stress* dominates, so

$$\sigma_z \approx 0 \tag{6-21}$$

In thick sections, however, a σ_z stress is developed, which creates a condition of triaxial tensile stresses acting at the crack tip and severely restricts straining in the z direction. This condition of *plane strain* can be shown to develop a through-thickness stress

$$\sigma_z \approx \nu(\sigma_x + \sigma_y) \tag{6-22}$$

The distribution of σ_z stress through the plate thickness is sketched in Fig. 6.20 for conditions of plane stress and plane strain.

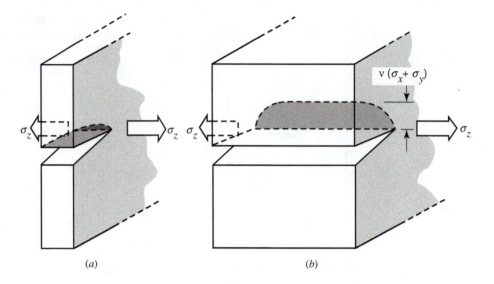

Figure 6.20   Through-thickness stress σ_z in (*a*) thin sheets under plane-stress state and (*b*) thick plates under plane-strain conditions.

An important feature of Eq. 6-20 is the fact that the stress distribution around any crack in a structure is similar and depends only on the parameters r and θ. The difference between one cracked component and another lies in the magnitude of the stress field parameter K, defined as the *stress-intensity factor*. In essence, K serves as a scale factor to define the magnitude of the crack-tip stress field. From Irwin's paper we see that

$$K = f(\sigma, a) \tag{6-23a}$$

where the functionality depends on the configuration of the cracked component and the manner in which the loads are applied. Note that the typical form of this expression as shown in Fig. 6.18 (and throughout this chapter) is

$$K = Y\sigma\sqrt{a} \quad \text{or} \quad Y'\sigma\sqrt{\pi a} \tag{6-23b}$$

where the geometric term $Y = Y'\sqrt{\pi}$. This term plays a role somewhat analogous to that of k_t in describing the local stress conditions at the crack tip. In this text, we use the first form in which the $\sqrt{\pi}$ term is included in Y, but the second form is equally legitimate. (Situations in which K is independent of crack length or varies inversely with a [e.g., Eq. 10-4] are reported elsewhere.[27–30])

In recent years, stress-intensity factor functions have been determined by mathematical procedures other than the Airy stress function approach used by Westergaard. Many such K functions have been determined for various specimen configurations and are available from handbooks and in the fracture mechanics literature.[27–30] Several solutions are shown in Fig. 6.21 for both commonly encountered cracked component configurations and standard laboratory test sample shapes. (Analytical expressions for these specimen configurations are given in Appendix B.) Consistent with Eq. 6-23, the stress-intensity factor is most often found to be a function of crack length and stress, and the behavior is defined by $Y(a/W)$ and $\sigma(P)$. When the crack size a is small compared to specimen width W, certain constant values of Y may be used. For a center-cracked panel like that examined by Griffith, $Y = 1.0\sqrt{\pi}$ as shown in Fig. 6.21a for the case where W is infinitely large. It can be seen from the solid line in Fig. 6.21c that this is also the limiting value of Y as $2a/W$ approaches zero for a center-cracked panel of finite width ($Y = 1.772$). When the crack is moved to the outer edges, as for the double edge-notched (DEN) finite width panel also shown in Fig. 6.21c, and the total crack length $2a$ is still much smaller than W, then the

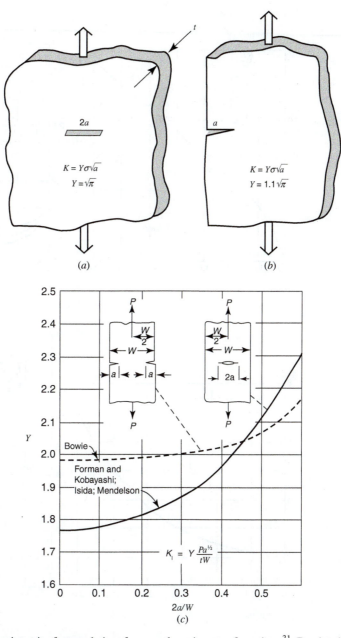

Figure 6.21 Stress-intensity-factor solutions for several specimen configurations.[31] (Reprinted by permission of the American Society for Testing and Materials from copyright material.)

geometric factor increases slightly to $Y = 1.12\sqrt{\pi}$. This corresponds to the small-crack limiting value of the dashed line shown in the same figure ($Y = 1.985$). This increase in Y is due to the reduced constraint on the opening of the crack associated with having free surfaces above and below the crack, and is present for all through-thickness small cracks ($a \ll W$) that open on a free surface of a plate that is uniformly loaded. The local constraints on crack opening and the characteristic crack length a are therefore identical for single and double edge-notched panels when the crack is very small. Thus this same limiting Y value of $1.12\sqrt{\pi}$ applies to the single edge-notched (SEN) infinitely-wide panel depicted in Fig. 6.21b and the SEN finite-width panel in

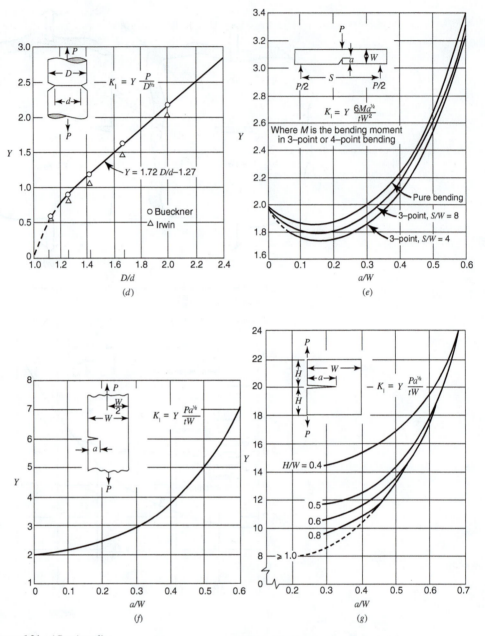

Figure 6.21 (*Continued*)

Fig. 6.21*f*. Note, however, that there is a greater increase in *Y* as a *f(a/W)* for the crack configuration shown in Fig. 6.21*f* (the SEN finite width panel) as compared with that shown in Fig. 6.21*c* (the DEN finite width panel) due to the presence of bending stresses present in the non-axisymmetrically loaded configuration of the SEN panel when *a/W* is not small.

At first glance the specimens depicted in Figs. 6.21f and 6.21g may appear to be identical, but a closer look reveals that the SEN panel in Fig. 6.21*f* is infinitely long and centrally-loaded whereas the compact tension [C(T)] specimen depicted in Fig. 6.21*g* has a finite length 2*H* and is loaded only on the cracked side. This combination of factors leads to significantly larger crack opening displacements for a given load in the C(T) configuration, so the *Y* factor is considerably larger than for the SEN panel at a given *a/W* ratio. The severity of the C(T)

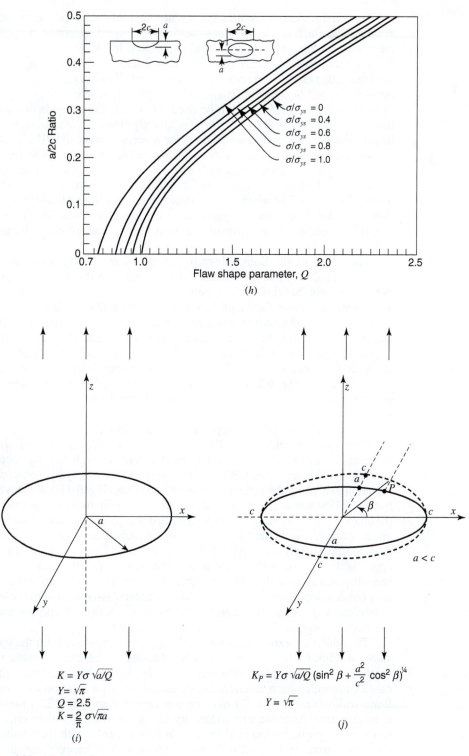

$$K = Y\sigma \sqrt{a/Q}$$
$$Y = \sqrt{\pi}$$
$$Q = 2.5$$
$$K = \frac{2}{\pi} \sigma \sqrt{\pi a}$$

(i)

$$K_P = Y\sigma \sqrt{a/Q} \left(\sin^2 \beta + \frac{a^2}{c^2} \cos^2 \beta\right)^{1/4}$$
$$Y = \sqrt{\pi}$$

(j)

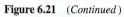

Figure 6.21 (*Continued*)

configuration makes crack propagation at low loads relatively easy, so this is a convenient arrangement for fracture toughness testing. (Standard specimen proportions are provided in Appendix B along with an analytical expression for K_I.)

Like the SEN and DEN configurations discussed thus far, Fig. 6.21e depicts a through-thickness edge crack in a rectangular component. However, the bar in Fig. 6.21e is exposed to bending instead of uniform tension. Once again, if a/W is sufficiently small, the edge crack approximation $Y = 1.12\sqrt{\pi}$ can be used. However, the stress is not constant throughout the thickness of a bent beam so Y deviates quickly from the small-crack value as a/W becomes larger (and actually decreases slightly for a certain range of a/W).

If an infinitesimally-thin SEN or DEN panel is rotated about its long axis, it will trace out the shape of a cylindrical shaft with a notch around its circumference. Perhaps, then, it is not surprising that the geometric factor of $Y = 1.12\sqrt{\pi}$ will also be reasonably accurate for the case depicted in Fig. 6.21d when D/d is small (analogous to small a/W for a plate). If the notch is deep (i.e., $2a/d \geq 0.2$) then a graphical or analytical solution for K_I such as the one given in Fig. 6.21d should be used instead (additional analytical solutions for tension or bending of a notched cylinder are provided in Appendix B).

The two configurations shown in Fig. 6.21h are analogous to those in Figs. 6.21a, b as seen looking directly at the fracture surface, but with the added complication of crack front curvature (note that this leads to neither having a through-crack). The curvature requires an additional correction factor of $Y = \sqrt{1/Q}$, where $Q = f(a/2c)$. As $a/2c$ increases, the flaw shape parameter Q also increases, and Y decreases (thus decreasing K for a given stress level and crack width $2c$). Note that as $a/2c$ approaches 0.5, $a/c \approx 1$, and the ellipse dimensions approach those of a circle. In this case, Q approaches 2.47, so that $\sqrt{1/Q}$ equals $2/\pi$, which is the limiting solution (Fig. 6.21i) for a small embedded circular flaw or a semicircular surface flaw. (Note that Fig. 6.21i depicts a circular crack, but at the angle of viewing it appears elliptical, whereas Fig. 6.21j depicts an elliptical crack as well as a circular crack.) For $a/2c < 0.5$, either curves like those in Fig. 6.21h or an analytical solution like the one given in Appendix C can provide an approximate value for Q.

Over time, an embedded elliptical flaw will actually grow such that $a/2c$ increases to a value of 0.5, assuming that the component is wide enough to fully accommodate this growth and that fast fracture does not occur before the shape change is complete. This results from a variation in K along the surface of an elliptical crack (Fig. 6.21j). K is maximized when $\beta = 90°$, but is smallest where $\beta = 0°$. As a result, the crack will always grow fastest where $\beta = 90°$ until the crack assumes a circular configuration. At this point, the K level is the same along the entire crack perimeter, with additional crack growth maintaining a circular shape. A corollary to this case is that embedded elliptical flaws will grow first into a circular configuration before appreciably increasing the crack size in the direction of the major axis of the ellipse. (For a semielliptical *surface* flaw, the equilibrium $a/2c$ is closer to 0.36 and a truly semicircular flaw may not develop, because an additional K influence associated with the free surface is present.) Analytical expressions for bending or tensile loading of components with elliptical and semicircular surface flaws are given in Appendix C.

The analytical expressions for $Y(a/W)$ given in Appendix B for the specimen configurations shown in Figs. 6.21c–g are not the only solutions for these configurations found in the fracture mechanics literature. Many others are compiled in handbooks, along with the accuracy and the range over which each solution can be used.[28–30] The solution for a center-cracked panel of finite width (i.e., large a/W) like the one depicted in Fig. 6.21c, however, deserves further comment here. Accurate stress-intensity-factor solutions in polynomial form for this configuration were originally reported by several investigators, with the results shown graphically as the solid curve in Fig. 6.21c.[31] More recently, Feddersen[32] noted that these polynomial expressions could be described with equal precision but much more conveniently by a secant expression with the form

$$K = \sqrt{\pi \sec(\pi a/W)} \cdot \sigma \sqrt{a} \qquad (6\text{-}24)$$

Equation 6-24 can be used also to describe the K conditions associated with a partial through-thickness flaw (i.e., one in which the through-thickness crack configuration shown in Fig. 6.21a is replaced by a semielliptical flaw on the front surface of the test panel that partially penetrates to the back face) where a corresponds to the depth of penetration of the crack through the component wall thickness; W is replaced by a quantity equal to two times the plate thickness dimension. It is intriguing to note that Eq. 6-24 is an *empirical* expression. Perhaps some future fracture mechanics analyst will provide the solution leading to this exact expression.

EXAMPLE 1.1

You are asked to confirm that the fracture toughness of a particular steel alloy is approximately $60\,\mathrm{MPa}\sqrt{\mathrm{m}}$. Furthermore, the test is to be conducted with the smallest possible load cell, using either a 1-cm-thick, center-cracked panel (CCT) or compact [$C(T)$] specimen. Which specimen would you choose for this property verification?

The stress intensity relations for these two specimen configurations are given in Appendix B. For the CCT panel, $a = 2\,\mathrm{cm}$ and $W = 10\,\mathrm{cm}$, whereas $a = 3\,\mathrm{cm}$ and $W = 6\,\mathrm{cm}$ for the $C(T)$ sample.

For the CCT panel,

$$K = \frac{P\sqrt{\pi a}}{BW} f(a/W)$$

where

$$f(a/W) = \sqrt{\sec \frac{\pi a}{W}}$$

Therefore,

$$K = 60\,\mathrm{MPa}\sqrt{\mathrm{m}} = \frac{P\sqrt{\pi 0.02}}{0.01(0.1)} \sqrt{\sec \frac{\pi 0.02}{0.1}}.$$

Hence, $P = 4.35 \times 10^4$ newtons.
For the $C(T)$ specimen,

$$K = \frac{P}{B\sqrt{W}} f(a/W)$$

where

$$f(a/W) = \frac{(2 + a/W)}{(1 - a/W)^{1.5}} \left[0.886 + 4.64a/W - 13.32(a/W)^2 + 14.72(a/W)^3 - 5.6(a/W)^4 \right]$$

By substitution, $P = 6.21 \times 10^3$ newtons.

We see that the maximum load needed for the $C(T)$ sample is seven times smaller than that required for the CCT configuration. It is also worth noting that much less material is needed to prepare the $C(T)$ sample than for the CCT panel. Indeed, the $C(T)$ sample was developed by Westinghouse engineers to minimize the volume of radioactive material needed to determine the fracture toughness of neutron irradiated material.

6.5.1 Multiplicity of Y Calibration Factors

It is important to recognize that the stress-intensity factor for a given flaw shape and loading configuration often may involve several Y calibration factors. The total Y is simply the product of the individual contribution terms. For example, an alternative form for the stress-intensity-factor solution for the elliptically shaped crack (Fig. 6.21j) can be written as

$$K = Y_1 \cdot Y_2 \cdot Y_3 \cdot \sigma\sqrt{a} \tag{6-25}$$

where
$$Y_1 = 1.0\sqrt{\pi}$$
$$Y_2 = \sqrt{1/Q}$$
$$Y_3 = \left(\sin^2\beta + \frac{a^2}{c^2}\cos^2\beta\right)^{1/4}$$

Furthermore, were this defect to grow to an appreciable size relative to that of the section thickness, then Eq. 6-25 also would have to include a finite width correction factor (e.g., $Y_4 = \sqrt{\sec(\pi a/2t)}$). Other stress-intensity-factor solutions involving multiple Y calibration factors are now estimated in the following three examples.

Case 1: *Crack Emanating from a Round Hole*[27,33]

This configuration (Fig. 6.22a) is commonly found in engineering practice (especially in aircraft components, which contain many rivet holes) since cracks often emanate from regions of high stress concentration. For the case of a round hole, stress concentration factor k_t (and hence the Y factor for a crack at a certain location on the hole surface) equals 3 at position A and is -1 at B. For a shallow crack ($L \ll R$) at location A, the crack tip is embedded within this local stress concentration and may be considered to be a shallow surface flaw like that in Fig. 6.21b. The stress-intensity factor is then estimated to be

$$K \approx 1.12(3)\sigma\sqrt{\pi L} \quad (L \ll R) \tag{6-26}$$

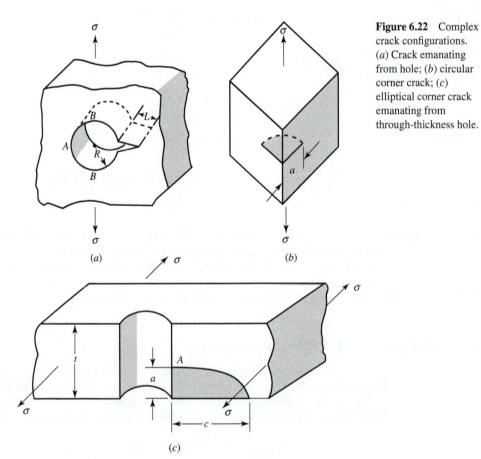

Figure 6.22 Complex crack configurations. (*a*) Crack emanating from hole; (*b*) circular corner crack; (*c*) elliptical corner crack emanating from through-thickness hole.

(a)

(b)

(c)

As one might expect, K drops quickly when L increases because the crack runs out of the high stress concentration region. Eq. 6-26, therefore, represents an upper bound solution for this crack configuration. A lower bound solution may be estimated from conditions where $L > R$. Here we may estimate the crack length to be $L + 2R$. Hence

$$K \approx \sigma \sqrt{\pi \left(\frac{L + 2R}{2} \right)} \quad (L > R) \tag{6-27}$$

Both upper and lower bound solutions are given in Table 6.2 for several L/R ratios. Also shown are correction factors $F(L/R)$ for this crack configuration based on the solution by Bowie[33] where

$$K = F(L/R)\sigma\sqrt{\pi L} \tag{6-28}$$

Note the excellent agreement between Eqs. 6-26 and 6-27 at both $L/R \approx 0$ and $L/R > 1$ extremes. Paris and Sih[27] also tabulated the correction factors for the case involving cracks emanating from both sides of the hole.

Table 6.2 Stress-Intensity Correction Factors for a Single Crack Emanating from a Hole[33]

L/R	Eq. 6-24	Eq. 6-25	$F(L/R)$
0.00	3.39	—	3.39
0.10	—	—	2.73
0.20	—	—	2.30
0.30	—	1.96	2.04
0.40	—	1.73	1.86
0.50	—	1.58	1.73
0.60	—	1.47	1.64
0.80	—	1.32	1.47
1.0	—	1.22	1.37
1.5	—	1.08	1.18
2.0	—	1.00	1.06
3.0	—	0.91	0.94
5.0	—	0.84	0.81
10.0	—	0.77	0.75
∞	—	0.707	0.707

The only difficulty in using the engineering approximations for K (i.e., Eqs. 6-26, 6-27) arises in the region $0 < L/R < 0.5$, where the stress concentration at the hole decays rapidly. Broek[34] considered this problem and concluded that the residual strength of cracked sheets and the crack propagation rate of cracks emanating from holes could be accounted for in reasonable fashion by considering the hole as being part of the crack (i.e., by using Eq. 6-27).

Case 2: Semicircular Corner Crack[28]

This configuration (Fig. 6.22b) involves geometries shown in Figs. 6.21b and 6.21i. Since the crack is circular and lies along two free surfaces, the prevailing stress-intensity level may be approximated by

$$K \approx (1.12)^2 \frac{2}{\pi} \sigma \sqrt{\pi a} \tag{6-29}$$

where $(1.12)^2$ represents two surface flaw corrections and $2/\pi$ represents the correction for a penny-shaped embedded crack.

Case 3: *An Elliptical Corner Crack Growing from One Corner of a Through-Thickness Hole*[28]

The solution to this crack configuration (Fig. 6.22c) incorporates many of the factors discussed in the previous two examples as well as some configurations shown in Fig. 6.21. The maximum stress-intensity condition in this instance is located at A, since this part of the crack experiences the maximum stress concentration caused by the hole and because A is located at $\beta = 90°$ (see Fig. 6.21j). An approximate solution may be given by

$$K_A \approx 1.12 \cdot \sqrt{1/Q} \cdot \sqrt{\sec \frac{\pi a}{2t}} \cdot 3\sigma\sqrt{\pi a} \tag{6-30}$$

where

$K_A =$ maximum stress-intensity condition along elliptical surface at A
$1.12 =$ surface flaw correction at A
$a =$ depth of elliptical flaw
$c =$ half width of elliptical flaw
$Q =$ elliptical flaw correction $= f(a/2c)$
$t =$ plate thickness
$\sqrt{\sec \frac{\pi a}{2t}} =$ finite panel width correction accounting for relatively large a/t ratio (recall Eq. 6-24)
$3\sigma =$ stress concentration effect at A

6.5.2 The Role of *K*

At this point, it is informative to compare the stress-intensity factor K and the stress-concentration factor k_t introduced in Chapter 5. Although k_t accounts for the geometrical variables, crack length and crack-tip radius, the stress-intensity factor K incorporates *both geometrical terms* (the crack length appears explicitly, while the crack-tip radius is assumed to be very sharp) *and the stress level*. As such, the stress-intensity factor provides more information than does the stress-concentration factor.

Once the stress-intensity factor K for a given test sample is known, it is then possible to determine the maximum stress-intensity factor that would cause failure. This critical value K_c is described in the literature as the *fracture toughness* of the material. When $K \geq K_c$, unstable crack growth commences and fast fracture occurs.

A useful analogy may be drawn between *stress and strength*, and the *stress-intensity factor and fracture toughness*. A component may experience many levels of stress, depending on the magnitude of load applied and the size of the component. However, there is a unique stress level that produces permanent plastic deformation and another stress level that causes failure. These stress levels are defined as the yield *strength* and fracture *strength*. Similarly, the stress-intensity level at the crack tip will vary with crack length and the level of load applied. That unique stress intensity level that causes failure is called the critical stress-intensity level or the *fracture toughness*. *Therefore, stress is to strength as the stress-intensity factor is to fracture toughness.*

Any specimen size and shape may be used to determine the fracture toughness of a given material, provided the stress-intensity-factor calibration is known. Obviously, some samples are more convenient and cheaper to use than others. For example, when the nuclear power plant manufacturers set out to test the steels to be used in nuclear reactors, they chose a small sample like the one shown in Fig. 6.21g so that fracture studies of neutron irradiated samples could be carried out in relatively small environmental chambers. You would use a similar sample in your laboratory if you had a limited amount of material available for the test program or if your testing machine had limited loading capacity. The notched bend bar (Fig. 6.21e) with a long span S also would be an appropriate sample to use when laboratory load capacity is limited. Of course, this sample would require much more material than the compact tension sample (Fig. 6.21g).

Failure Analysis Case Study 6.1: *Fracture Toughness of Manatee Bones in Impact*[35]

It is instructive to observe how the fracture mechanics concepts developed thus far in this chapter can be applied to the analysis of real fractures, even when the specimen conditions are not ideal. The following case study demonstrates the application of fractography and fracture mechanics to address the failure of a natural material: manatee rib bones. Manatees are large aquatic mammals that are often injured or killed as a result of collisions with watercraft. The goal of this study was to determine the energy level needed to cause rib fracture, thereby establishing a basis for controlling watercraft speed conditions on Florida waterways.[35]

The team performed fracture measurements of cleaned, wet, whole manatee rib bones bent under impact by a blunt projectile fired from a compressed air gun. They instrumented the bones with strain gages to measure the behavior during the impact event, allowing them to evaluate the amount of plasticity present, and the strain (ε_f) at the point of fracture. They also performed slow three-point bend tests on similar bones to determine a Young's modulus (E) value. It was observed that in both the slow and impact tests fracture occurred without any significant plasticity. This was not a foregone conclusion because bone, as a composite of mineral and soft tissue, is often only *quasi-brittle* (i.e., it can undergo energy-dissipating time-dependent plastic deformation). The authors of the study concluded that it was safe to assume purely elastic behavior for their analysis. From the measured strain, then, the stress at fracture was calculated using Hooke's law: $\sigma_f = E\varepsilon_f$.

In order to calculate a critical stress intensity factor value (K_C) in the manner of Eq. 6-23b, two other pieces of information were necessary: a Y factor and an initial crack length. Fortuitously, manatee ribs are made of solid cortical bone (i.e, there is no spongy tribecular core) and therefore develop fracture surface markings that can be read like those in engineered ceramics. The location and size of the preexisting critical flaw in each bone was determined, as shown in Fig. 6.23. The critical flaws were ellipsoidal surface cracks with depths much less than the rib thickness ($a \ll W$). On this basis, the geometric factor (Y) was estimated for all initial cracks as

$$Y = \frac{2}{\pi}(1.12)\sqrt{\pi} \approx 1.25$$

while an effective semicircular crack radius (a_C) was approximated from the depth and half width of each flaw as

$$a_C \approx \sqrt{a \cdot c}$$

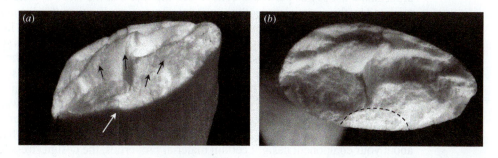

Figure 6.23   (*a*) The white arrow indicates a natural surface groove associated with the pre-crack in a typical specimen. The black arrows show the approximate crack growth directions during fracture. (*b*) The dashed line delineates the ellipsoidal precracked region on the same fracture surface. (Reproduced with permission from photographs provided by J. J. Mecholsky, Jr.)

The average fracture toughness of the 22 ribs tested was determined to be approximately 8 MPa√m, with nearly identical results from both the static and impact tests. In comparison to the fracture toughness values shown graphically in Fig. 6.19, it can be seen that the manatee bone is tougher than any of the ceramic and polymer materials depicted, but still fairly brittle on an absolute scale. The authors concluded that typical watercraft obeying existing Florida regulations (as of 2008) could easily cause rib fracture during a collision.

In concluding this case study, it is interesting to note an additional observation made by the authors: their fracture toughness value is greater than previously reported values for manatee rib bones (only 2 to 4.5 MPa√m).[36,37] This difference was attributed to the small size of the specimens used in the two prior studies, neither of which tested whole bones. This apparent dependence on sample size implies the likelihood that the bone displays something called "R-curve" behavior, where toughness is dependent on crack length (a topic to be discussed in Section 6.12).

6.6 DESIGN PHILOSOPHY

The interaction of material properties, such as the fracture toughness, with the design stress and crack size controls the conditions for fracture in a component. For example, it is seen from Fig. 6.21a that the fracture condition for an infinitely large center-cracked plate would be

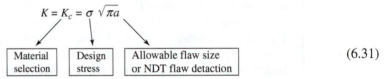

$$K = K_c = \sigma \sqrt{\pi a}$$

| Material selection | Design stress | Allowable flaw size or NDT flaw detection |

(6.31)

This relation may be used in one of several ways to design against a component failure. For example, if you are to build a system that must withstand the ravages of a liquid metal environment, such as in some nuclear reactors, one of your major concerns is the selection of a suitable corrosion-resistant material. Once done, you have essentially fixed K_c. In addition, if you allow for the presence of a relatively large stable crack—one that can be readily detected and repaired—the design stress is fixed and must be less than $K_c/\sqrt{\pi a}$.

A second example shows another facet of the fracture control design problem. A certain aluminum alloy was chosen for the wing skin of a military aircraft because of its high strength and light weight; hence, K_c was fixed. The design stress on the wing was then set at a high level to increase the aircraft's payload capacity. Having fixed K_c and σ, the allowable flaw size was defined by Eq. 6-31 and beyond the control of the aircraft designers. In one case history, a fatigue crack grew out from a rivet hole in one of the aluminum wing plates and progressed to the point where the conditions of Eq. 6-31 were met. The result—fracture. What was most unfortunate about this particular failure was the fact that the allowable flaw size that could be tolerated by the material under the applied stress was *smaller than the diameter of the rivet head covering the hole*. Consequently, it was impossible for maintenance and inspection people to know that a crack was growing from the rivet hole until it was too late. This situation could have been avoided in several ways. Had it been recognized beforehand that the wing plate should have tolerated a crack greater than the diameter of the rivet head, the stresses could have been reduced and/or a tougher material selected for the wing plates. It is worth noting that one of the difficulties leading to the early demise of the British Comet jet transport was the selection of a lower toughness 7000 series aluminum alloy for application in critical areas of the aircraft.

The significance of Eq. 6-31 lies in the fact that you must first decide what is most important about your component design: certain material properties, the design stress level as affected by many factors such as weight considerations, or the flaw size that must be tolerated for safe operation of the part. Once such a priority list is established, certain critical decisions can be made. However, once any combination of two of the three variables (fracture toughness, stress, and flaw size) is defined, the third factor is fixed.

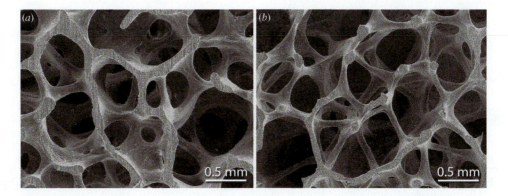

Figure 6.24 Aging-related degradation of human tribecular bone associated with low fracture toughness. (*a*) Healthy bone; (*b*) osteoporotic bone revealing substantial bone loss.[38] (Reprinted with permission from R. O. Ritchie, M. J. Buehler, and P. Hansma, *Physics Today*, 62 (6), 41 (2009). Copyright 2009, Dr. James Weaver.)

Looking ahead to later chapters in this text, we find that all of the variables in Eq. 6.31 may contain a time-dependent component. For example, stress levels will vary under cyclic loading conditions (see Chapters 9 and 10); the crack length may grow in the presence of an aggressive environment (see Chapter 8) or during cyclic loading conditions; and the material's fracture toughness may vary with time and/or temperature-induced changes in the material's micro-structure (see Chapter 7). On a biological level, the increased occurrence of hip fractures in aged senior citizens results in part from decreased toughness of human bone as a result of osteoporotic degradation (e.g., see Fig. 6.24).

If the component in question is a pressure vessel (e.g., a gas cylinder, an oil pipeline, or an aircraft fuselage), an additional design consideration would be to evaluate whether or not a leak-before-break condition[39] could develop in a manner shown in Fig. 6.25. If it were considered less dangerous to release some of the pressurized fluid than to suffer total failure of the vessel, it might be advantageous to intentionally design to meet this condition. Assume that a semielliptical surface flaw with dimensions $2c$ and a is located at the inner surface of a pressure vessel and is oriented normal to the hoop stress direction. Furthermore, we will allow that this crack can grow gradually in the combined presence of a sustained load and aggressive environment (see Chapter 8) or under cyclic loading conditions (see Chapter 10). Recall from Fig. 6.21*j* and the associated discussion that a semielliptical surface flaw tends to grow more rapidly in a direction parallel to the minor axis of the ellipse (where $\beta = 90°$) until the flaw approaches a semicircular configuration. The crack would then continue to grow as an ever-expanding semicircle until it breached the vessel's outer wall, thereby allowing fluid to escape. At this point, the crack would break through the remaining unbroken ligament (the fibrous zone between the crack front and vessel surface in Fig. 6.25) before assuming the configuration of a through-thickness flaw. Assuming that the crack remained semicircular to the point where breakthrough occurred (at which time a equals the wall thickness), the characteristic dimension of the resulting through-thickness flaw would be $2a$ or equivalent to twice the vessel wall thickness, $2t$. Hence, the stress-

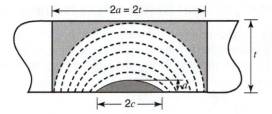

Figure 6.25 Diagram showing growth of semielliptical surface flaw to semicircular configuration. At leak condition ($a = t$), unbroken ligaments (fibrous areas) break open to form through-thickness crack.

intensity factor for this crack becomes

$$K = \sigma\sqrt{\pi a} = \sigma\sqrt{\pi t} \qquad (6\text{-}32)$$

Now, if $K < K_C$ (or K_{IC}), then fracture would not take place even though leaking had commenced. In general, then, the leak-before-break condition would exist when a crack of length equal to at least twice the vessel wall thickness could be tolerated (i.e., was stable) under the prevailing stresses. (See Case History 1 in Chapter 11 for an application of this concept.)

EXAMPLE 6.2

A 7049-T73 aluminum forging is the material of choice for an 8-cm-internal-diameter hydraulic actuator cylindrical housing that has a wall thickness of 1 cm. After manufacture, each cylinder is subjected to a safety check, involving a single fluid overpressurization that generates a hoop stress no higher than 50% σ_{ys}. The component design calls for an operating internal fluid pressure, corresponding to a hoop stress no higher than 25% σ_{ys}. Prior to overpressurization, a 2-mm-deep semicircular surface flaw that was oriented normal to the hoop stress direction was discovered in one cylinder. Given that $\sigma_{ys} = 460\,\text{MPa}$ and $K_{IC} = 23\,\text{MPa}\sqrt{\text{m}}$, would the cylinder have survived the overpressurization test and would the cylinder experience a leak-before-break condition? Also, what were the fluid pressure levels associated with the overpressurization cycle and design stress? Note that K_{IC} is simply K_C determined under a specific fracture condition as specified in Section 6.9.

The K level associated with the overpressurization test is given from Figs. 6.21b and 6.21i, where

$$K = 1.1\left(\frac{2}{\pi}\right)\sigma\sqrt{\pi a}$$

Since the maximum hoop stress level is 50% that of σ_{ys},

$$K = 1.1\left(\frac{2}{\pi}\right)230 \times 10^6 \sqrt{\pi(2 \times 10^{-3})} \therefore K = 12.77\,\text{MPa}\sqrt{\text{m}}$$

The prevailing K level is less than K_{IC} and so the cylinder would have survived the proof test. To determine whether the cylinder would experience leak-before-break conditions during normal service conditions (i.e., $\sigma_{\text{design}} = 115\,\text{MPa}$), we use Eq. 6-32 to find

$$K = \sigma\sqrt{\pi t}$$
$$\therefore K = 115 \times 10^6 \sqrt{\pi(1 \times 10^{-2})} = 20.4\,\text{MPa}\sqrt{\text{m}}$$

Since $K < K_{IC}$, leak-before-break conditions would exist, but with a relatively small margin of safety. Finally, we calculate from Eq. 1-43 that the overpressurization and design pressure levels are 57.5 and 28.75 MPa, respectively.

6.7 RELATION BETWEEN ENERGY RATE AND STRESS FIELD APPROACHES

Thus far, two approaches to the relation between stresses, flaw sizes, and material properties in the fracture of materials have been discussed. At this point, it is appropriate to demonstrate the similarity between the two. If Eq. 6-11 is rearranged so that

$$\sigma\sqrt{\pi a} = \sqrt{E\mathscr{G}} \qquad (6\text{-}33)$$

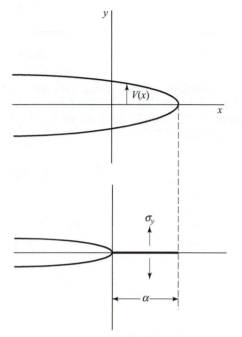

Figure 6.26   Diagram showing partial closure of crack over distance α.

it is seen from Eq. 6-31 that

$$K = \sqrt{E\mathcal{G}} \quad \text{(plane stress)} \tag{6-34}$$

for plane stress conditions, and

$$K = \sqrt{\frac{E\mathcal{G}}{(1 - \nu^2)}} \quad \text{(plane strain)} \tag{6-35}$$

for plane strain. This relation between K and $\mathcal{G}$ is not merely fortuitous but, rather, can be shown to be valid based on an analysis credited to Irwin.[40]

Consider the energy needed to reclose part of a crack that had formed in a solid. In reverse manner, then, once this energy was removed the crack should reopen. With the notation shown in Fig. 6.26, the work done per unit area (unit thickness) to close the crack by an amount α is given by

$$\mathcal{G} = \frac{2}{\alpha} \int_0^\alpha \frac{\sigma_y V(x)}{2} \, dx \tag{6-36}$$

The constant "2" accounts for the total closure distance, since $V(x)$ is only half of the total crack opening displacement; $1/\alpha$ relates to the average energy released over the total closure distance α; $\sigma_y V(x)/2$ defines the energy under the load deflection curve. From Eq. 6-20 where $\theta = 0$,

$$\sigma_y = \frac{K}{\sqrt{2\pi x}} \tag{6-37}$$

while it has been shown[27] that

$$V = \frac{2K}{E} \sqrt{\frac{2(\alpha - x)}{\pi}} \tag{6-38}$$

Combining Eqs. 6-37 and 6-38 with 6-36, note that

$$\mathcal{G} = \frac{2K^2}{\alpha E \pi} \int_0^\alpha \sqrt{\frac{\alpha - x}{x}}\, dx \tag{6-39}$$

Note that $\mathcal{G}$ represents an average value taken over the increment a, while K will vary with a because K is a function of crack length. These difficulties can be minimized by shrinking a to a very small value so as to arrive at a more exact solution for $\mathcal{G}$. Therefore, taking the limit of Eq. 6-39, where a approaches zero, and integrating, it is seen that

$$\mathcal{G} = \frac{K^2}{E} \quad \text{(plane stress)} \tag{6-40}$$

and

$$\mathcal{G} = \frac{K^2}{E}(1 - \nu^2) \quad \text{(plane strain)} \tag{6-41}$$

which is the same result given by Eqs. 6-34 and 6-35.

6.8 CRACK-TIP PLASTIC-ZONE SIZE ESTIMATION

As you know by now, a region of plasticity is developed near the crack tip whenever the stresses described by Eq. 6-20 exceed the yield strength of the material. An estimate of the size of this zone may be obtained in the following manner. First, consider the stresses existing directly ahead of the crack where $\theta = 0$. As seen in Fig. 6.27, the elastic stress $\sigma_y = K/\sqrt{2\pi r}$ will exceed the yield strength at some distance r from the crack tip, thereby truncating the elastic stress at that value. By letting $\sigma_y = \sigma_{ys}$ at the elastic-plastic boundary,

$$\sigma_{ys} = \frac{K}{\sqrt{2\pi r}} \tag{6-42}$$

and the plastic–zone size is computed to be $K^2/2\pi\sigma_{ys}^2$. Since the presence of the plastic region makes the material behave as though the crack were slightly longer than actually measured, the "apparent" crack length is assumed to be the actual crack length plus some fraction of the plastic-zone diameter. As a first approximation, Irwin[26] set this increment equal to the plastic-zone radius, so that the apparent crack length is increased by that amount. In effect, the plastic-zone diameter is a little larger than $K^2/2\pi\sigma_{ys}^2$ as a result of load redistributions around the zone

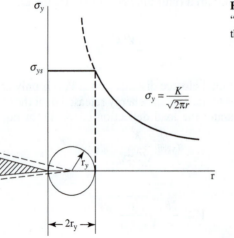

Figure 6.27 Onset of plastic deformation at the crack tip. "Effective" crack length taken to be initial crack length plus the plastic-zone radius.

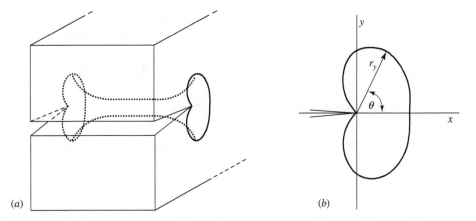

Figure 6.28 Schematic depiction of the crack-tip plastic-zone boundary (a) in the interior, (b) as it depends on θ.

and is estimated to be twice that value. Therefore

$$r_y \approx \frac{1}{2\pi} \frac{K^2}{\sigma_{ys}^2} \quad \text{(plane stress)} \tag{6-43}$$

For conditions of plane strain where the triaxial stress field suppresses the plastic-zone size, the plane-strain plastic-zone radius is smaller and has been estimated[41] to be

$$r_y \approx \frac{1}{6\pi} \frac{K^2}{\sigma_{ys}^2} \quad \text{(plane strain)} \tag{6-44}$$

By comparing Eqs. 6-43 and 6-44 it is seen that the size of the plastic zone varies along the crack front, being largest at the two free surfaces and smallest at the midplane (Fig. 6.28a). Note that there is an inverse relation between the size of the plastic zone and the magnitude of the through-thickness stress, σ_z (depicted in Figs. 6.20a, b).

The reader should recognize that the size of the plastic zone also varies with θ. If the plastic-zone size is determined for the more general case by the distortion energy theory (recall Section 2.3 where σ_x, σ_y, and σ_z are described in terms of r and θ, it can be shown that

$$r_y = \frac{K^2}{2\pi\sigma_{ys}^2} \cos^2\frac{\theta}{2}\left(1 + 3\sin^2\frac{\theta}{2}\right) \quad \text{(plane stress)} \tag{6-45}$$

where the zone assumes a shape as drawn in Fig. 6.28b. Hahn and Rosenfield[42] have confirmed this plastic-zone shape by way of etch pit studies in an iron–silicon alloy. Note that when $\theta = 0$, Eq. 6-45 reduces to Eq. 6-43. Equations 6-43 and 6-44 may now be used to determine the effective stress-intensity level K_{eff}, based on the effective or apparent crack length, so that

$$K_{\text{eff}} \approx Y\left(\frac{a+r_y}{W}\right)\sigma\sqrt{a+r_y} \tag{6-46}$$

Since the plastic-zone size is itself dependent on the stress-intensity factor, the value of K_{eff} must be determined by an iterative process that may be truncated at any given level to achieve the desired degree of exactness for the value of K_{eff}. For example, the iteration may be terminated when $(a+r_y)_2 - (a+r_y)_1 \leq X$, where X is arbitrarily chosen by the investigator. A special case is an infinite plate with a small central notch, where the stress-intensity factor is defined by Eq. 6-31. Iteration is not necessary in this case, and K_{eff} may be determined directly.

Substituting Eq. 6-43 into Eq. 6-31 yields

$$K_{\text{eff}} = \sigma \sqrt{\pi \left(a + \frac{1}{2\pi} \frac{K_{\text{eff}}^2}{\sigma_{ys}^2} \right)} \tag{6-47}$$

Upon rearranging Eq. 6-47, it is seen that

$$K_{\text{eff}} = \frac{\sigma \sqrt{\pi a}}{\left[1 - \frac{1}{2} \left(\frac{\sigma}{\sigma_{ys}} \right)^2 \right]^{1/2}} \tag{6-48}$$

so that K_{eff} will always be greater than K_{applied}, although the difference may be very small under low stress conditions.

EXAMPLE 6.3

A plate of steel with a central through-thickness flaw of length 16 mm is subjected to a stress of 350 MPa normal to the crack plane. If the yield strength of the material is 1400 MPa, what is the plastic-zone size and the effective stress-intensity level at the crack tip?

Assuming the plate to be infinitely large, r_y may be determined from Eqs. 6-31 and 6-43 so that

$$r_y \approx \frac{1}{2\pi} \left[\frac{350^2 \pi (0.008)}{1400^2} \right] \approx 0.25 \, \text{mm}$$

Since r_y/a is very small, it would not be expected that K_{eff} would greatly exceed K_{applied}. In fact, from Eq. 6-48

$$K_{\text{eff}} = \frac{350 \sqrt{\pi (0.008)}}{[1 - 1/2(350/1400)^2]^{1/2}} = 56.4 \, \text{MPa} \sqrt{\text{m}}$$

which is only about 2% greater than K_{applied}. When the plastic zone is relatively small in relation to the overall crack length, the plastic-zone correction to the stress-intensity factor is usually ignored in practice. This occurs often under fatigue crack propagation conditions, where the applied stresses are well below the yield strength of the material.

If, on the other hand, a second plate of steel with the same crack size and applied stress level were heat treated to provide a yield strength of 385 MPa, the plasticity correction would be substantially larger. The plastic-zone size would be

$$r_y = \frac{1}{2\pi} \left[\frac{350 \sqrt{\pi (0.008)}}{385} \right]^2 = 3.3 \, \text{mm}$$

or one-fifth the size of the total crack length. Correspondingly, the effective stress-intensity factor would be considerably greater than the applied level, wherein

$$K_{\text{eff}} = \frac{350 \sqrt{\pi (0.008)}}{[1 - 1/2(350/385)^2]^{1/2}} = 72.4 \, \text{MPa} \sqrt{\text{m}}$$

which represents a 30% correction. When the computed plastic zone becomes an appreciable fraction of the actual crack length, as found above, and generates a large correction for the stress-intensity level, the entire procedure of applying the plasticity correction becomes increasingly suspect. When such a large plasticity correction is made to the elastic solution, the assumptions of a dominating elastic stress field become tenuous.

6.8.1 Dugdale Plastic Strip Model

Another model of the crack-tip plastic zone has been proposed by Dugdale[43] for the case of plane stress. As shown in Fig. 6.29, Dugdale considered the plastic regions to take the form of narrow strips extending a distance R from each crack tip. For purposes of the mathematical analysis, the internal crack of length $2c$ is allowed to extend elastically to a length $2a$; however, an internal stress is applied in the region $|c| < |x| < |a|$ to reclose the crack. It may be shown that this internal stress must be equal to the yield strength of the material such that $|c| < |x| < |a|$ represents local regions of plasticity. By combining the internal stress field surrounding the plastic enclaves with the external stress field associated with a stress acting on the crack, Dugdale demonstrated that

$$c/a = \cos\left(\frac{\pi}{2}\frac{\sigma}{\sigma_{ys}}\right) \tag{6-49}$$

or, since $a = c + R$

$$R/c = \sec\left(\frac{\pi}{2}\frac{\sigma}{\sigma_{ys}}\right) - 1 \tag{6-50}$$

When the applied stress $\sigma \ll \sigma_{ys}$ Eq. 6-50 reduces to

$$R/c \approx \frac{\pi^2}{8}\left(\frac{\sigma}{\sigma_{ys}}\right)^2 \tag{6-51}$$

By rearranging Eq. 6-51 in the form of $D(K/\sigma_{ys})^2$, it is encouraging to note the reasonably good agreement between Eqs. 6-51 and 6-43 (i.e., $D = \pi/8 \approx 1/\pi$). (Note that $2r_y$ is used in Eq. 6-43 for

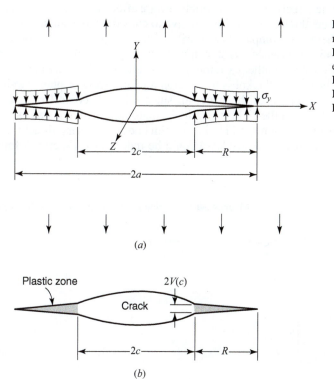

Figure 6.29 Dugdale plastic-zone strip model for non-strain-hardening solids. Plastic zones R extend as thin strips from each end of the crack. (After Hahn and Rosenfield;[42] reprinted with permission of Hahn, *Acta Metall.* 13 (1965), Pergamon Publishing Company.)

comparison with the Dugdale zone.) Dugdale plastic zone development in polymeric solids is discussed in Section 10.8.2.

6.9   FRACTURE-MODE TRANSITION: PLANE STRESS VERSUS PLANE STRAIN

As discussed in Section 6.8, the plastic-zone size depends on the state of stress acting at the crack tip. When the sample is thick in a direction parallel to the crack front, a large σ_z stress can be generated that will restrict plastic deformation in that direction. As shown by Eqs. 6-43 and 6-44, the plane-strain plastic-zone size is correspondingly smaller than the plane-stress counterpart (recall Fig. 6.28a). Since the fracture toughness of a material will depend on the volume of material capable of plastically deforming prior to fracture, and since this volume depends on specimen thickness, it follows that the fracture toughness K_c will vary with thickness as shown in Fig. 6.30. When the sample is thin (for example, at t_1) and the degrees of plastic constraint acting at the crack-tip minimal, plane-stress conditions prevail and the material exhibits maximum toughness. Alternatively, when the thickness is increased to bring about plastic constraint and plane-strain conditions at the crack tip, the toughness drops sharply to a level that may be one-third (or less) that of the plane-stress value. The fracture toughness under plane strain conditions is often measured using Mode I loading; if this is the case, then the critical value is called K_{IC}, which is pronounced "K-one-C" in recognition of the loading mode. One very important aspect of this lower level of toughness (i.e., the *plane-strain fracture toughness* K_{IC}) is that it does not decrease further with increasing thickness, thereby making this value a conservative lower limit of material toughness in any given engineering application.

Once K_{IC} is determined in the laboratory for a given material with a sample at least as thick as t_2 (Fig. 6.30), an engineering component much thicker than t_2 should exhibit the same toughness. To summarize, the plane-stress fracture toughness K_c is related to both metallurgical and specimen geometry, while the plane-strain fracture toughness K_{IC} depends only on metallurgical factors. Consequently, the best way to compare materials of different thickness on the basis of their respective intrinsic fracture-toughness levels should involve a comparison of K_{IC} values, since thickness effects may be avoided.

Since stress-state effects on fracture toughness are affected by the size of the plastic enclave in relation to the sheet thickness, it is informative to consider the change in stress state in terms of the ratio r_y/t, where r_y is computed arbitrarily with the plane-stress plastic-zone size relation as given by Eq. 6-43. Experience has shown that when $r_y/t \geq 1$, plane-stress conditions prevail and toughness is high. At the other extreme, plane-strain conditions will exist when $r_y/t < 1/10$. In either case, the necessary thickness to develop a plane-stress or plane-strain condition will depend on the yield strength of the material, since this will control r_y at any given stress-intensity level. Therefore, if the yield strength of a material were increased by a factor of two by some thermomechanical treatment (TMT), the thickness necessary to achieve a plane-strain condition for a given stress-intensity level could be reduced by a factor of four, assuming, of

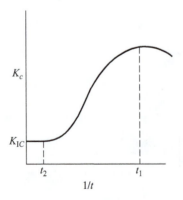

Figure 6.30   Variation in fracture toughness with plate thickness.

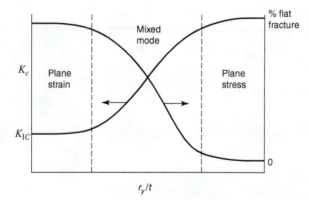

Figure 6.31   Effect of relative plastic zone size to plate thickness on fracture toughness and macroscopic fracture surface appearance. Plane-stress state associated with maximum toughness and slant fracture. Plane-strain state associated with minimum toughness and flat fracture.

course, that K_{IC} was not altered by the TMT. Clearly, very thin sections can still experience plane-strain conditions in high-yield-strength material, whereas very large sections of low-yield-strength material may never bring about a full plane-strain condition.

Another feature of the fracture-toughness–stress-state dependency is the commonly observed fracture mode transition mentioned in the previous chapter. As shown in Fig. 6.31, the relative degree of flat and slant fracture depends on the crack-tip stress state. When plane-stress conditions prevail and $r_y \geq t$, the fracture plane often assumes a $\pm 45°$ orientation with respect to the load axis and sheet thickness (Fig. 6.32a). This may be rationalized in terms of failure occurring on those planes containing the maximum resolved shear stress. (Since $\sigma_z = 0$ in plane stress, a Mohr circle construction will show that the planes of maximum shear will lie along $\pm 45°$ lines in the YZ plane.) In plane strain, where $\sigma_z \approx v(\sigma_y + \sigma_x)$ and $r_y \ll t$, the plane of maximum shear is found in the XY plane (Fig. 6.32b). ($\sigma_y, \sigma_x,$ and σ_z may be computed from Eqs. 6-20 and 6-22, where it may be shown, for example, that when $\theta = 60°$, $\sigma_y > \sigma_z > \sigma_x$.) Apparently, the fracture plane under plane-strain conditions lies midway between the two maximum shear planes. This compromise probably also reflects the tendency for the crack to remain in a plane containing the maximum net section stress.

The existence of a fracture-mode transition in many engineering materials such as aluminum, titanium, and steel alloys, and a number of polymers makes it possible to estimate the relative amount of energy absorbed by a component during a fracture process. When the fracture surface is completely flat (Fig. 6.33c), plane-strain test conditions probably prevail, and the observed

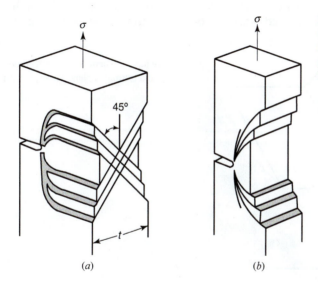

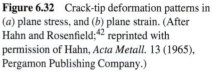

Figure 6.32   Crack-tip deformation patterns in (a) plane stress, and (b) plane strain. (After Hahn and Rosenfield;[42] reprinted with permission of Hahn, *Acta Metall.* 13 (1965), Pergamon Publishing Company.)

(a)   (b)

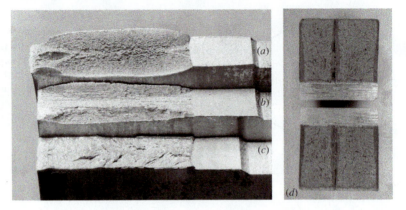

Figure 6.33 Fracture-mode transition in alloy steel induced by change in test temperature. (*a*) Slant fracture at high temperature; (*b*) mixed mode at intermediate temperature; (*c*) flat fracture at low temperature; (*d*) double set of shear lips in a steel Charpy sample. Internal shear lips are associated with the formation of center-line delamination. (Photograph courtesy of K. Vecchio.)

fracture toughness is low. If the fracture is completely of the slant or shear type (Fig. 6.33*a*), plane-stress conditions probably dominate to produce a tougher failure. Obviously, a mixed fracture appearance (Fig. 6.33*b*) would reflect an intermediate toughness condition.

By measuring the width of the shear lip and relating it to the size of the plastic-zone radius, it is often possible to estimate the stress intensity factor associated with a particular service failure.[12] Since $r_y \approx (1/2\pi)(K^2/\sigma_{ys}^2)$ at the surface of the plate and the shear lips form on $\pm 45°$ bands to the sheet thickness, it is seen from Fig. 6.34 that the depth D of the shear lips can be approximated by the plastic-zone radius. Hence

$$D \approx r_y \approx \frac{1}{2\pi}\left(\frac{K}{\sigma_{ys}}\right)^2 \tag{6-52}$$

Combining Eqs. 6-31, and 6-52, we find

$$\text{shear lip} \approx \frac{1}{2\pi}\frac{Y^2\sigma^2 a}{\sigma_{ys}^2} \tag{6-53}$$

The geometrical correction factor Y for the component and the crack length where the shear lip was measured can be used with Eq. 6-53 to estimate the prevailing stress level. This

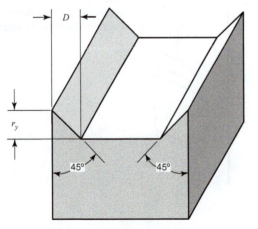

Figure 6.34 Schema showing relation between shear lip depth and estimated plane stress plastic-zone size.

approach to determining the stress level is highly empirical and appears to work satisfactorily only for certain materials such as high-strength aluminum and certain alloy steels, but not for lower strength steels. When the correlation does not hold, it usually results in too low an estimate of K from Eq. 6-52; that is, the shear lip is smaller than what would be expected from the actual plastic zone. Because Eq. 6-53 is a highly empirical determinant of stress level, computed values must be considered tentative until corroborated by additional findings. There are exceptions to this stress-state–fracture-mode correlation that should be recognized by the reader. For example, ferritic and pearlitic steels tend to exhibit a smaller shear lip zone than that expected based on estimates of the plane-stress plastic-zone size.[44] From this discussion, it is clear that shear lips form at the free surfaces of the specimen where plane-stress conditions prevail. A unique exception to this pattern is shown in Fig. 6.33d where *two* sets of shear lips are found on mating surfaces of a Charpy specimen (recall Fig. 6.7a). The external pair of shear lips correspond to the plane-stress condition that normally exists at the specimen's free surfaces. The internal pair of shear lips was produced as a result of the delamination created during the fracture process. In this instance, the through-thickness stress σ_z (recall Fig. 6.20) acted to split open the Charpy sample along a metallurgical plane of weakness. Once the delamination was created, two additional free surfaces were created, each corresponding to a plane-stress condition. Consequently, a second set of shear lips was produced in the middle of the specimen—a most unusual location for shear lip development.

Failure Analysis Case Study 6.2: *Analysis of Crack Development during a Structural Fatigue Test*[45]

Paris[45] reported an analysis of a laboratory fracture that presents excellent, well-documented illustrations of several different (and independent) fracture mechanics procedures in the solution of a fracture problem. A program load fatigue test was conducted on a 1.78-cm-thick plate of D6AC steel that had been tempered to a yield strength of 1500 MPa. Fracture of the plate occurred after fatigue cracks that had developed on both sides of a drilled hole grew into a semicircular configuration and reached a critical size, as shown in Fig. 6.35. The growth rings within the two corner cracks were produced by fatigue block loading conditions. (These may be compared with similar markings shown in Fig. 10.38a.) The stress at failure was reported to be 830 MPa. For the time being, discussion will be limited to estimates of the stress intensity factor at failure based on two analyses of the crack configuration and an analysis of shear lip depth. A stress intensity factor estimate based on the fatigue growth bands will be deferred to Chapter 10.

The stress intensity factor solution for the given crack configuration was estimated in two ways. The actual hole–crack combination was approximated first by a semicircular surface flaw with a

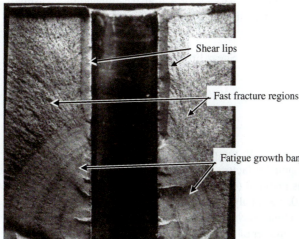

Shear lips

Fast fracture regions

Fatigue growth bands

Figure 6.35 Two corner cracks emanating from through-thickness hole, revealing fatigue growth bands and shear lips.[45] (After R. J. Gran, F. D. Orazio, Jr., P. C. Paris, G. I. Irwin, and R. W. Hertzberg, AFFDL-TR-70-149, March 1971.)

radius of 0.86 cm and then by a through-thickness flaw with a total length of 1.73 cm. These estimates reflect lower and upper bound solutions, respectively, since the former solution does not account for the hole passing through the entire plate thickness, and the latter solution indicates more fatigue crack growth than was actually observed. The lower bound of the stress intensity factor may be given by[27]

$$K_L = \left[1 + 0.12\left(1 - \frac{a}{c}\right)\right] \cdot \sqrt{\frac{1}{Q}} \cdot \sqrt{\sec\frac{\pi a}{2t}} \cdot \sigma\sqrt{\pi a}$$

where

K_L = lower bound stress intensity solution
a = crack depth, 0.86 cm or 0.0086 m
c = half-flaw width, 0.86 MPa
σ = applied stress, 830 MPa
Q = elliptical flaw correction, 2.5
t = plate thickness, 1.78 cm

$$K_L = \left[1 + 0.12\left(1 - \frac{0.86}{0.86}\right)\right] \cdot \sqrt{\frac{1}{2.5}} \cdot \sqrt{\sec\frac{\pi(0.86)}{2(1.78)}} \cdot [830]\sqrt{\pi(0.0086)}$$

$$K_L = 101.3 \, \text{MPa}\sqrt{m}$$

The upper bound solution is given by

$$K_U = \sigma\sqrt{\pi a}$$
$$= 830\sqrt{\pi(0.0086)}$$
$$= 136 \, \text{MPa}\sqrt{m}$$

From these results, the actual K level at fracture may be bracketed by

$$101 < K_c < 136 \, \text{MPa}\sqrt{m}$$

with the correct value being more closely given by the lower bound solution because of a smaller error in this estimation. Consequently, K_c (or K_{IC}) $\approx 110 \, \text{MPa}\sqrt{m}$.

For comparison, the stress-intensity factor was also estimated by measurement of the shear lip depth (about 0.8 mm) along the surface of the hole (Fig. 6.35). From Eq. 6-53

$$\text{Shear lip depth} \approx \frac{1}{2\pi}\frac{K^2}{\sigma_{ys}^2}$$

$$8 \times 10^{-4} \approx \frac{1}{2\pi}\left(\frac{K}{1500}\right)^2$$

$$K \approx 106 \, \text{MPa}\sqrt{m}$$

which agrees extremely well with the previous estimate of 110 MPa$\sqrt{m}$.

The critical stress intensity factors estimated by these methods are almost identical with the known K_{IC} level for this material (see Table 7.8). Coupled with the fatigue growth band analysis discussed in Chapter 10, the analysis of this laboratory failure demonstrates that *independent* approaches based on fracture mechanics concepts can be used in investigating a service failure. Ideally, one should use a number of these procedures to provide cross-checks for each computation.

6.10   PLANE-STRAIN FRACTURE-TOUGHNESS TESTING OF METALS AND CERAMICS

Since the plane-strain fracture toughness K_{IC} is such an important material property in fracture prevention, it is appropriate to consider the procedures by which this property is measured in the laboratory. Accepted test methods for most metallic materials have been set forth by the American Society for Testing and Materials under Standard E399 and for ceramic materials in Standard C1421.[46,47] Although the reader should examine these standards for precise details and for an understanding of the limits of each technique, the most important features of K_{IC} testing are summarized in this section.

Fracture toughness testing of metals under Standard E399 assumes that some degree of crack-tip plasticity is likely. As such, a recommended metallic test sample is initially fatigue-loaded to extend a machined notch a prescribed amount. [A three-point bend bar (Fig. 6.21*e*), a compact tension sample (Fig. 6.21*g*), and a C-shaped sample (Appendix B) are all considered acceptable for such a K_{IC} determination. The stress-intensity factor expressions for these three specimen configurations are tabulated in Appendix B.] A clip gage is placed at the mouth of the crack to monitor its displacement when the specimen load is applied. Typical load-displacement records for a K_{IC} test are shown in Fig. 6.36. From such curves, two important questions should be answered. First, what is the apparent plane-strain fracture-toughness value for the material? Second, is this value *valid* in the sense that a thicker or bigger sample might not produce a lower K_{IC} number for the same material? If a lower toughness level is achieved with a thicker sample, then the value obtained initially is not a valid number. Brown and Srawley[31] examined the fracture toughness of several high-strength alloys and found empirically that a valid plane-strain fracture-toughness test is performed when the specimen thickness and crack length are both greater than a certain minimum value. Specifically,

$$t \text{ and } a \geq 2.5 \left(\frac{K_{IC}}{\sigma_{ys}} \right)^2 \tag{6-54}$$

The ratio $(K_{IC}/\sigma_{ys})^2$ suggests that the required sheet thickness and crack length are related to some measure of the plastic-zone size, since Eqs. 6-43 and 6-44 are of the same form. Using the plastic-zone size determined from Eq. 6-44 and substituting into Eq. 6-54, it is seen that the criteria for plane-strain conditions reflect a condition where

$$a \text{ and } t \geq 50 r_y \tag{6-55}$$

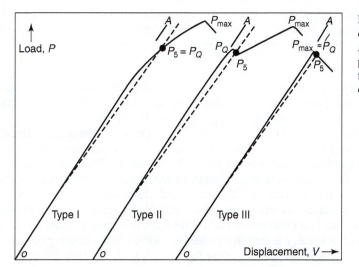

Figure 6.36   Major types of load-displacement records obtained during K_{IC} testing.[46] (Reprinted by permission of the American Society for Testing and Materials from copyright material.)

Under this condition K_{eff} as defined in Eq. 6-46 would not be significantly different from $K_{applied}$, such that the plastic-zone correction would be unnecessary.

EXAMPLE 6.4

Identical compact tension specimens have been prepared to determine the fracture toughness of the 7178 aluminum alloy, subjected to two different heat treatments. The crack length and thickness of the samples are 4 cm and 1 cm, respectively. From the data shown in Table 6.3 for 7178-T651 and 7178-T7651, would the specimen dimensions, described above, provide valid plane-strain fracture toughness conditions?

From Table 6.3, we see that for the two alloy conditions:

Condition	K_{IC}(MPa$\sqrt{m}$)	σ_{ys} (MPa)
7178-T651	23.1	570
7178-T7651	33	490

From Eq. 6-54, the plane-strain validity criteria require that

$$t \text{ and } a \geq 2.5\left(\frac{K_{IC}}{\sigma_{ys}}\right)^2$$

For alloy 7178-T651,

$$t \text{ and } a \geq 2.5\left(\frac{23.1}{570}\right)^2 = 0.004 \text{ m}$$

Since both crack length (0.04 m) and sheet thickness (0.01 m) are greater than 0.004 m, a valid K_{IC} value could be determined with this specimen.

For alloy 7178-T7651,

$$t \text{ and } a \geq 2.5\left(\frac{33}{490}\right)^2 = 0.011 \text{ m}$$

Since the specimen thickness (0.01 m) is less than that required according to Eq. 6-54 (i.e., 0.011 m), a valid fracture toughness value could not be determined with the specimen described above, even though the crack length dimension (0.04 m) exceeded the size requirement. Therefore, to provide for valid plane-strain conditions for the 7178-T7651 alloy, the specimen thickness would have to be increased to a value >0.011 m. Note that some trial is necessary sometimes before determining the proper specimen dimensions for a valid plane-strain experiment.

To arrive at a valid K_{IC} number, it is necessary to first calculate a tentative number K_Q, based on a graphical construction on the load-displacement test record. If K_Q satisfies the conditions of Eq. 6-54, then $K_Q = K_{IC}$. From ASTM Specification E-399, the graphical construction involves the following procedures.[46] On the load-deflection test record, draw a secant line OP_5 through the origin with a slope that is 5% less than the tangent OA to the initial part of the curve. For the three-point bend bar and the compact tension sample, a 5% reduction in slope is approximately equal to a 2% increase in the effective crack length of the sample, a level reflecting minimal crack extension and plasticity correction.[31] P_5 is defined as the load at the intersection of the secant line OP_5 with the original test record. (Note the similarity between this graphical method and the one described in Chapter 1 for the determination of the 0.2% offset yield strength.) The load P_Q, which will be used to calculate K_Q, is then determined as follows: If the load at every point on the record which precedes P_5 is lower than P_5, then P_Q is P_5 (Fig. 6.36, Type I); if, however, there is a

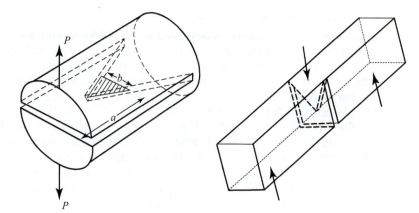

Figure 6.37 (*a*) Short rod specimen containing a deep chevron notch. Loading causes a stable crack to form (shaded area) before maximum load is reached. (*b*) Chevron notch beam for flexure testing.

maximum load preceding P_5 that exceeds it, then this maximum load is P_Q (Fig. 6.18, Types II and III).[46] If the ratio of P_{max}/P_Q is less than 1.1, it is then permissible to compute K_Q with the aid of the appropriate K calibration. If K_Q satisfies Eq. 6-54, then K_Q is equal to K_{IC}. If not, then a thicker and/or more deeply cracked sample must be prepared for additional testing so that a valid K_{IC} may be determined.

Another specimen configuration—the short rod or short bar sample—has been standardized to determine K_{IC} for metals under tensile loading, as described in ASTM E1304.[48] This specimen is either cylindrical or rectangular in shape and contains a deep, machined notch on one end with a chevron configuration at its root, as shown in Fig. 6.37*a*. In the fracture test, a wedge is pushed into the mouth of the chevron-shaped notch, thereby developing a Mode I tensile load at the tip of the chevron. Both the loads and crack-mouth displacements are monitored and analyzed according to methods described by Barker.[49,50] A crack initiates at the tip of the chevron notch with increasing load and grows to the same length at the time of the peak load, provided that conditions of linear elastic fracture mechanics apply. For the case of an ideally elastic material, K_{IC} is given by

$$K_{IC} = AP_c B^{-3/2} \tag{6-56}$$

where
$$A = \text{short rod calibration constant} \approx 22$$
$$P_c = \text{maximum load}$$
$$B = \text{short rod diameter}$$

Barker[50] also has modified Eq. 6-56 to account for localized crack-tip plasticity. Fracture toughness values determined by this method correspond to a slowly advancing steady-state crack that is initiated from a chevron-shaped notch. By contrast, K_{IC} values, obtained from E399 testing requirements, characterize the start of crack extension from a fatigue crack. These differences may cause reported fracture toughness values derived from Standard E1304 procedures to be larger in some materials than those corresponding to E399 test methods. Accordingly, K_{IC} values determined by these two geometries cannot be assumed to be equivalent. One obvious advantage of the short rod sample is its small and simple shape. In addition, the fatigue-precracking procedure, necessary in E399-type samples, is not required. This makes the chevron notch specimen particularly attractive for brittle alloys since they are difficult to fatigue precrack.

Ceramic and cermet (ceramic/metal composite) materials are also difficult to precrack using fatigue loading, so the configurations used for metallic materials may not always be easily adapted. Use of the chevron-notched rod test method for evaluation of ceramic materials has been reported,[51,52] but more convenient test configurations for ceramics are based on bent beams (3-point or 4-point), as described in ASTM Standard C1421.[47] These have a notch or sharp crack on the tensile side of a rectangular specimen. One such specimen, the chevron notch beam (CNB), is shown in Fig. 6.37b. A simpler specimen that uses a straight-edge notch rather than the chevron is called the single-edge vee-notch beam (SEVNB). In both cases, the notch is created either by an extremely sharp rotating diamond blade, or by reciprocating motion of a straight razor blade coated with diamond slurry.[53] The key to repeatable measurements is the development of a narrow notch with a tip that is sufficiently sharp to behave like a true sharp crack. On the basis of a round-robin test (i.e., identical specimens sent to multiple labs for measurement) it was concluded that the notch tip radius must be no larger than the grain size.[54] When the notch is insufficiently sharp, either from inadequate preparation or because the grain size is extremely small, a true crack must be initiated before it can propagate; this causes significant scatter in the results, and tends to overestimate the fracture toughness.

Recognizing the importance of creating a sharp notch, the surface crack in flexure (SCF) specimen has also been developed for K_{IC} testing of ceramic materials.[55,56] Here, a sharp precrack is introduced by pressing a Knoop (rhombohedral) indenter into the surface with the long axis of the indenter oriented across the beam width. When a small amount of surface material is ground away to eliminate local compressive residual stresses introduced by the indentation process, the remaining flaw will approximate a semielliptical surface crack with micrometer-scale radius. The fracture surface must be examined after the test to determine the initial flaw size, after which an appropriate Y factor can computed and our usual fracture mechanics approaches can be applied.[52]

Numerous attempts have also been made to directly measure toughness levels in brittle ceramics with the use of micro-hardness indentation. The simplicity of the indentation fracture (IF) technique and the small sample size required are both very attractive characteristics. The sample is loaded with a micro-hardness indenter such as an equiaxed Vickers type, an elongated Knoop indenter, or a sharp cube corner indenter, with the intention of inducing visible cracks at the corners of the indentation depression. The distance c from the center of the indentation to the end of each crack is measured, and a toughness value proportional to $c^{-3/2}$ is calculated as

$$K_C = \alpha \left(\frac{E}{H}\right)^{\frac{1}{2}} \left(\frac{P}{c^{3/2}}\right) \tag{6-57}$$

where α is a calibration constant (often given as 0.016), E is the Young's modulus, H is the indentation hardness, and P is the maximum indentation load. Initially, encouraging correlations were reported between fracture-toughness values obtained from IF measurements and those from conventional test methods.[57,58] Since then, the approach has been widely used to evaluate the toughness of ceramics, glasses, and hard biological materials, including bone. However, considerable doubt has more recently been cast on the underlying assumptions and the accuracy of the IF method.[52,59,60] The residual stress and fracture patterns beneath the indenter are often complex, and cannot always be reliably described using fracture mechanics analysis techniques. As such, the IF test method must be used with caution. It may be useful for comparative measurements in certain material systems, but is not recommended for quantitative K_C testing.

6.11 FRACTURE TOUGHNESS OF ENGINEERING ALLOYS

Typical K_{IC} values for various steel, aluminum, and titanium alloys are listed in Table 6.3 along with associated yield-strength levels. The table also provides a listing of critical flaw sizes for each material, based on a hypothetical service condition involving a through-thickness center notch of length $2a$ embedded in an infinitely large sheet sufficiently thick to develop plane-strain conditions at the crack tip. If it is assumed that the operating design stress is taken to be one-half the yield strength, then the critical crack length would equal $(1/\pi)(K_{IC}/(\sigma_{ys}/2))^2$. One basic data trend becomes immediately obvious: The fracture toughness and allowable flaw size of a given material decreases, often precipitously, when the yield strength is elevated. Consequently, there is a price to pay when one wishes to raise the strength of a material. More will be said about this in Chapter 7.

The fracture-toughness data for the aluminum alloys deserve additional comment at this time. It is seen that the allowable flaw size in 7075-T651 is only one-eighth that for the 2024-T3. Accordingly, a design engineer would be alerted to the greater propensity for brittle fracture in the 7075-T651 alloy. In actual engineering structures, such as the wing skins of commercial and military aircraft, the relative difference between the two materials is even greater. Since wing skins are about 1 to 2 cm thick, it may be shown by Eq. 6-54 that 7075-T651 would experience approximately plane-strain conditions, while 2024-T3 would be operating in a plane-stress environment, where K_c is about two or three times as large as K_{IC}. Consequently, there may be a 50-fold difference in allowable flaw size between the two materials, taking into account both the metallurgical and stress-state factors.

As might be expected, K_{IC} values for a particular material may decrease with increasing loading rate and decreasing test temperature. This is particularly true for the case of structural steels. Accordingly, the reader should recognize that the representative K_{IC} values given in Table 6.3 correspond to room temperature experiments conducted at slow loading rates (approximately $10^{-5}\,\text{s}^{-1}$). Additional insights regarding K_{IC} dependence on strain rate and temperature, as reflected by impact energy measurements, can be found in the following section (Section 6.11.1).

Table 6.3   Plane-Strain Fracture Toughness of Selected Engineering Alloys

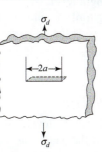

$$\sigma_d = \frac{\sigma_{ys}}{2};\, K_{IC} = \sigma_d \sqrt{\pi a}$$

$$a_c = \frac{1}{\pi}\left(\frac{K_{IC}}{\sigma_{ys}/2}\right)^2$$

Material	K_{IC}		σ_{ys}		a_c	
	MPa$\sqrt{\text{m}}$	ksi$\sqrt{\text{in.}}$	MPa	ksi	mm	in.
2014-T651	24.2	22	455	66	3.6	0.14
2024-T3	~44	~40	345	50	~21	~0.82
2024-T851	26.4	24	455	66	4.3	0.17
7075-T651	24.2	22	495	72	3.0	0.12
7178-T651	23.1	21	570	83	2.1	0.08
7178-T7651	33.	30	490	71	5.8	0.23
Ti-6A1-4V	115.4	105	910	132	20.5	0.81
Ti-6A1-4V	55.	50	1035	150	3.6	0.14
4340	98.9	90	860	125	16.8	0.66
4340	60.4	55	1515	220	2.	0.08
4335 + V	72.5	66	1340	194	3.7	0.15
17-7PH	76.9	70	1435	208	3.6	0.14
15-7Mo	49.5	45	1415	205	1.5	0.06
H-11	38.5	35	1790	260	<0.6	<0.02
H-11	27.5	25	2070	300	0.23	0.009
350 Maraging	55.	50	1550	225	1.6	0.06
350 Maraging	38.5	35	2240	325	<0.4	<0.02
52100	~14.3	~13	2070	300	~0.06	<0.002

EXAMPLE 6.5

Assume that a component in the shape of a large sheet is to be fabricated from 0.45–Ni–Cr–Mo steel. It is required that the critical flaw size be greater than 3 mm, the resolution limit of available flaw detection procedures. A design stress level of one-half the tensile strength is indicated. To save weight, an increase in the tensile strength from 1520 to 2070 MPa is suggested. Is such a strength increment allowable? (Assume plane-strain conditions in all computations.)

The answer to this question bears heavily upon the changes in fracture toughness of the material resulting from the increase in tensile strength. At the 1520 MPa strength level, it is found that the K_{IC} value is 66 MPa$\sqrt{m}$, while at 2070 MPa, K_{IC} drops sharply to 33 MPa$\sqrt{m}$ (Fig. 6.38). For a large sheet the stress-intensity factor is determined from Eq. 6-31, where the design stress is $\sigma_{ts}/2$. For the alloy heat treated to the 1520 MPa strength level,

$$66 \text{ MPa}\sqrt{m} = 760 \text{ MPa}\sqrt{\pi a}$$
$$2a = 4.8 \text{ mm}$$

which exceeds the minimum flaw size requirements. At the 2070-MPa strength level, however,

$$33 \text{ MPa}\sqrt{m} = 1035 \text{ MPa}\sqrt{\pi a}$$
$$2a = 0.65 \text{ mm}$$

which is five times smaller than the minimum flaw size requirement and approximately eight times smaller than the maximum flaw tolerated at the 1520-MPa strength level.

Therefore, it is not possible to raise the strength of the alloy to 2070 MPa to save weight and still meet the minimum flaw size requirement. Furthermore, using the flaw size found in the 1520-MPa material for the 2070-MPa alloy would necessitate a decrease in design stress from 1035 to 380 MPa.

$$\sigma = \frac{33 \text{ MPa}\sqrt{m}}{\sqrt{\pi(0.0024)}} \approx 380 \text{ MPa}$$

Thus, under similar flaw size conditions, the allowable stress level in the stronger alloy could be only half that in the weaker alloy, resulting in a twofold *increase* in the weight of the component!

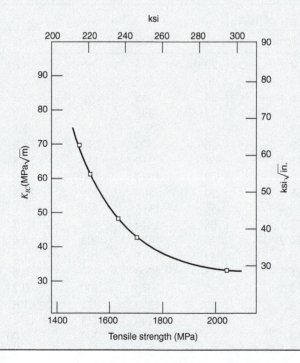

Figure 6.38 Fracture toughness in 0.45C–Cr–Mo steel containing 0.045% S. (After A. J. Birkle, R. P. Wei, and G. E. Pellissier, Trans. ASM 59, 981 (1966). Copyright American Society for Metals, Metals Park, OH.)

EXAMPLE 6.6

A plate of Incrediballoy 100, a steel alloy with a yield strength of 415 MPa, has been found to exhibit a K_{IC} value of 132 MPa$\sqrt{m}$. The material is available in various gages up to 250 mm but will be used in very wide 100-mm-thick plates for a given application. If the plate is subjected to a stress of 100 MPa, how large can a crack grow from a hole in the middle of the plate before catastrophic failure occurs?

    You would be correct in assuming a central crack configuration so that Eq. 6-31 could be used, but would be wrong if you substituted the K_{IC} value of 132 MPa$\sqrt{m}$ into the equation. In fact, not enough information has been provided to answer the question properly. To use K_{IC} in Eq. 6-31 presumes that plane-strain conditions exist in the component. It is seen from Eq. 6-54 that a plate 250 mm thick is required for plane-strain conditions to apply. Since the plate in question is only 100 mm thick, plane-*strain* conditions would not prevail. (In all likelihood, the valid K_{IC} value was obtained from a 250-mm plate). It would be possible to determine K_c for Incrediballoy 100 if data such as shown in Fig. 6.30 were available. In addition, an estimate of K_c may be obtained from the K_{IC} value in the region near plane strain by an empirical relation shown by Irwin,[61] where

$$K_c^2 = K_{IC}^2(1 + 1.4\beta_{IC}^2) \tag{6-58}$$

where

$$\beta_{IC} = \frac{1}{t}\left(\frac{K_{IC}}{\sigma_{ys}}\right)^2$$

$$t = \text{thickness}$$

6.11.1 Impact Energy—Fracture-Toughness Correlations

    Although handicapped by the inability to bridge the size gap between small laboratory sample and large engineering component, the Charpy test sample method (Section 6.2) does possess certain advantages over K_{IC} testing, such as ease of preparation, simplicity of test method, speed, low cost in test machinery, and low cost per test. Recognizing these factors, many researchers have attempted to modify the test procedure to extract more fracture information and seek possible correlations between Charpy data and fracture-toughness values obtained from fracture mechanics test samples. In one such approach, Orner and Hartbower[62] precracked the Charpy sample so that the impact energy for failure represented energy for crack propagation but not energy to initiate the crack.

$$E_T = E_i + E_p \tag{6-59}$$

where
        E_T = total fracture energy
        E_i = fracture initiation energy
        E_p = fracture propagation energy

    They found that a correlation could be made between the fracture toughness of the material $\mathcal{G}_c$ and the quantity W/A, where W is the energy absorbed by the precracked Charpy test piece and A the cross-sectional area broken in the test. Although good results have been observed for some materials (for example, see Fig. 6.39), the applicability of this test method should be restricted to those materials that exhibit little or no strain-rate sensitivity, since dynamic Charpy data are being compared with static fracture-toughness values. Also, the neglect of kinetic energy absorption by the broken samples as part of the energy-transfer process from the load pendulum to the specimen makes it impossible to develop good data in brittle materials where the kinetic energy component is no longer negligible.[31] Orner and Hartbower did point out, however, that the precracked Charpy sample could be used to measure the strain-rate sensitivity of a given material by conducting tests under

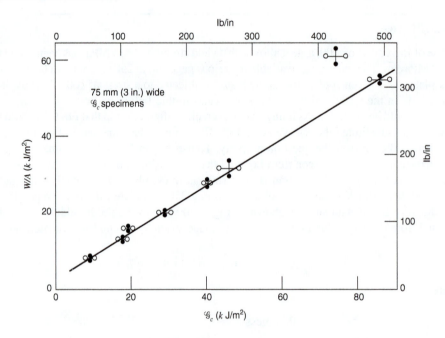

Figure 6.39 Relation between fatigue-cracked V-notch Charpy slow bend and $\mathscr{G}_c$ in a variety of 3.2-mm (0.125-in.)-thick aluminum alloys.[62] (Reprinted from *Welding Journal* by permission of the American Welding Society.)

both impact and slow bending conditions. Barsom and Rolfe[63] have verified this hypothesis with a direct comparison of static and dynamic test results from precracked Charpy V-notch (CVN) and plane-strain fracture-toughness samples, respectively. First, they established the strain-rate-induced shift in transition temperature for several steel alloys in the strength range of 275 to 1725 MPa (Figs. 6.40 and 6.41, and Table 6.4). They noted that the greatest transition temperature shift was found in the low-strength steels and no apparent strain-rate sensitivity was present in alloys with yield strengths in excess of 825 MPa. When these same materials were tested to determine their plane-strain fracture-toughness value, a corresponding shift was noted as a function of strain rate.

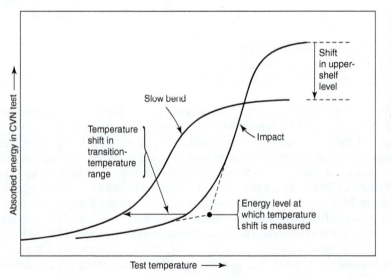

Figure 6.40 Diagram of impact energy versus test temperature revealing shift in transition temperature due to change in strain rate. (Note the higher shelf energy resulting from dynamic loading conditions, which may be related to a strain-rate-induced elevation in yield strength.)[63] (Reprinted by permission of the American Society for Testing and Materials from copyright material.)

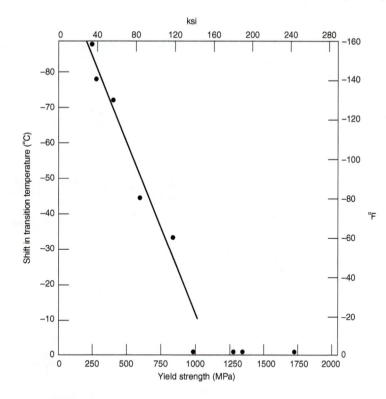

Figure 6.41 Effect of yield strength on shift in transition temperature between impact and slow bend CVN tests.[63] (Reprinted by permission of the American Society for Testing and Materials from copyright material.)

Figures 6.42*a* and *b* show static (K_{IC}) and dynamic (K_{ID}) plane-strain fracture-toughness values plotted as a function of test temperature. One additional point should be made with regard to these data. Although K_{IC} increased gradually with temperatures for the high-strength steels, a dramatic transition to higher values was observed for the low- and intermediate-strength alloys. It should be emphasized that this transition was not associated with the plane-strain to plane-stress transition, since all the data reported represented valid plane-strain conditions. A similar transition in plane-strain ductility (measured with a thin, wide sample) occurred in the same temperature region, but no such transition developed in axisymmetric ductility (measured with a conventional round tensile bar). This tentative correlation between

Table 6.4 Transition Temperature Shift Related to Change in Loading Rate[63]

Steel	σ_{ys}		Shift in Transition Temperature	
	MPa	**(ksi)**	**°C**	**(°F)**
A36	255	(37)	−89	(−160)
ABS-C	269	(39)	−78	(−140)
A302B	386	(56)	−72	(−130)
HY-80	579	(84)	−44	(−80)
A517-F	814	(118)	−33	(−60)
HY-130	945	(137)	0	(0)
10Ni–Cr–Mo–V	1317	(191)	0	(0)
18Ni (180)	1241	(180)	0	(0)
18Ni (250)	1696	(246)	0	(0)

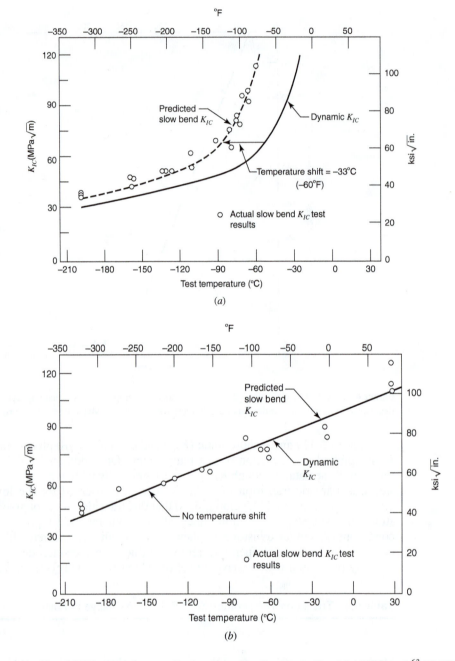

Figure 6.42   Use of CVN test results to predict the effect of loading rate on K_{IC}. (a) A517-F steel.[63] (b) 18Ni-(250) maraging steel. (Reprinted by permission of the American Society for Testing and Materials from copyright material.)

the K_{IC} and plane-strain ductility transitions was strengthened with the observation that both transitions were associated with a fracture mechanism transition from cleavage at low temperatures to microvoid coalescence at high test temperatures.[64,65]

It is seen that the toughness levels of both strain-rate sensitive and insensitive materials increased with increasing temperature (Figs. 6.42a and b). Of significance is the fact that the predicted static K_{IC} values (broken line), obtained by applying the appropriate temperature

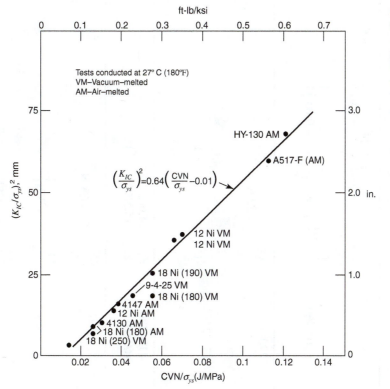

Figure 6.43 Relation between K_{IC} and CVN values in the upper shelf region.[63]
(Re-printed by permission of the American Society for Testing and Materials from copyright material.)

shift (Fig. 6.41) to the dynamic test results (solid line), were confirmed by experimentation. Since dynamic plane-strain fracture-toughness testing procedures are more complex and beyond the capability of many laboratories, estimation of K_{ID} from more easily determined K_{IC} values represents a potentially greater application of the strain-rate-induced temperature shift in the determination of fracture properties.

Additional efforts have focused on developing empirical relations between impact energy absorbed in Charpy[63] specimens and K_{IC} values. Such a relation is shown in Figure 6.43 with additional correlations given in Table 6.5. Note that these relations vary as a function of material, the test temperature range, notch acuity, and strain rate. For example, these correlations are different in the upper-shelf energy regime as compared with the transition zone; they depend also on whether the Charpy specimen is precracked and whether it is impacted or tested at slow strain rates. Roberts and Newton[66] examined the accuracy of 15 such relations and concluded that no single correlation could be used with any degree of confidence to encompass all possible test conditions and differences in materials. Furthermore, because of the intrinsic scatter associated with K_{IC} and CVN measurements, the correlations possessed a relatively wide scatter band. Roberts and Newton also pointed out that some of the K_{IC} values used to establish these correlations were invalid with respect to E399 test requirements, and that CVN values tended to vary according to the CVN specimen location in the plate.

In addition to these difficulties, certain additional basic problems must not be overlooked. For example, the K_{IC}–CVN correlation implies that you can directly compare data from blunt- and sharp-notched samples, and data from statically and dynamically loaded samples, respectively. The latter difficulty may not be too important for the materials shown in Fig. 6.43 since they all have yield strengths greater than 825 MPa (except A517-F) where strain-rate effects are minimized (see Fig. 6.41 and Table 6.4). When the material's fracture properties are sensitive to strain rate, however, a two-step correlation between impact CVN data and K_{IC} values is recommended. First, K_{ID} values are inferred from impact CVN data with the aid of an appropriate correlation (e.g., see Table 6.5). Then K_{IC} is estimated from K_{ID} data through the use of the

Table 6.5 Fracture-Toughness–Charpy Energy Correlations[66]

Material	Notch	Test	Temperature Range	Range of Charpy Results (J)	σ_{ys} (MPa)	Correlation[a]	References
A517D 4147 HY130 4130 12Ni–5Cr–3Mo 18Ni–8Co–3Mo	V-notch	Impact	Upper shelf	31–121 (23–89 ft-lb)	760–1700 (110–246 ksi)	$\left(\dfrac{K_{IC}}{\sigma_{ys}}\right)^2 = 0.64\left(\dfrac{CVN}{\sigma_{ys}} - 0.01\right)$ $\left[\left(\dfrac{K_{ID}}{\sigma_{ys}}\right)^2 = 5\left(\dfrac{CVN}{\sigma_{ys}} - 0.05\right)\right]$	b, c
A517F A3202B ABS-C HY-130 18Ni (250) Ni–Cr–Mo–V Cr–Mo–V Ni–Mo–V	V-notch	Impact	Transition	4–82 (3–60 ft-lb)	270–1700 (39–246 ksi)	$\dfrac{K_{IC}^2}{E} = 0.22\,(CVN)^{1.5}$ $\left[\dfrac{K_{IC}^2}{E} = 2\,(CVN)^{1.5}\right]$	c
A533B A517F A542	V-notch	Impact	Transition	7–68 (5–50 ft-lb)	410–480 (60–70 ksi)	$K_{IC} = 14.6(CVN)^{0.5}$ $[K_{IC} = 15.5(CVN)^{0.5}]$	d
ABS-C A305-B A517-F	V-notch	Impact Slow bend	Transition	2.7–61 (2–45 ft-lb)	250–345 (36–50 ksi)	$\dfrac{K_{ID}^2}{E} = 0.64\,CVN$ $\left[\dfrac{K_{ID}^2}{E} = 5\,CVN\right]$ $\dfrac{K_{IC}^2}{E} = 0.64\,CSB$ $\left[\dfrac{K_{IC}^2}{E} = 5\,CSB\right]$	e

Table 6.5 (*Continued*)

Material	Notch	Test	Temperature Range	Range of Charpy Results (J)	σ_{ys} (MPa)	Correlation[a]	References
ABS-C A302-B A517-F	Precrack	Impact	Transition	2.7–61	250–345	$\dfrac{K_{ID}^2}{E} = 0.52\,PCI$	e
				(2–45 ft-lb)	(36–50 ksi)	$\left[\dfrac{K_{ID}^2}{E} = 4\,PCI\right]$	
		Slow bend				$\dfrac{K_{IC}^2}{E} = 0.52\,PSB$ $\left[\dfrac{K_{IC}^2}{E} = 4\,PSB\right]$	

[a]Correlation in square brackets uses English units.
[b]J. M. Barsom and S. T. Rolfe, *ASTM STP 466*, 281 (1970).
[c]S.T. Rolfe and S. R. Novak, *ASTM STP 463*, 124 (1970).
[d]R. H. Sailors and H. T. Corten, *ASTM STP 514*, Part II, p. 164 (1972).
[e]J. M. Barsom, *Eng. Fract. Mech. 7*, 605 (1975).

temperature shift factor (Fig. 6.41). Finally, fracture mechanics-impact energy correlations for engineering plastics have been reported and are reviewed elsewhere.[18]

Failure Analysis Case Study 6.3: *Failure of Arizona Generator Rotor Forging*[67−69]

This case history does not describe a true service failure, since the power generator rotor shaft in question failed during a routine balancing test *before* it was placed in service and at an operating speed *less* than that for design operation. The forged rotor, manufactured more than 50 years ago, did not possess benefits derived from current vacuum degassing melting practices as described in Chapter 7; consequently, a large amount of hydrogen gas was trapped in the ingot as it solidified. With time, the hydrogen precipitated from the solid to form hydrogen flakes, evidenced by disk-shaped internal flaws such as the one shown in Fig. 6.44. Investigators[67,68] concluded that these 2.5- to 3.8-cm-diameter circular defects existed before the balancing test and were responsible for its failure, although no specific hydrogen flake could be identified as the critical nucleation site.

The forged material contained 0.3C, 2.5Ni, 0.5Mo, and 0.1V, exhibited room temperature tensile yield and ultimate strengths of 570 and 690 MPa, respectively, and a Charpy V-notch impact energy at the fracture temperature (27 °C) of 5.4 to 16.3 J. The rotor contained a central hole along its entire bore. This was done to remove the central section of the original ingot, which normally contains a relatively high percentage of inclusions and low melting point micro-constituents, and to permit a more thorough examination of the rotor for evidence of any defects.[67] Unfortunately, however, by introducing the borehole, the centrifugal tangential stresses at the innermost part of the rotor are approximately doubled even when the inner bore diameter is very small, as follows:

$$\sigma_{\max \text{(solid cylinder)}} = \frac{3+v}{8}\rho\omega^2 R_2^2$$
$$\sigma_{\max \text{(hollow cylinder)}} = \frac{3+v}{4}\rho\omega^2\left(R_2^2 + \frac{1-v}{3+v}R_1^2\right)$$

where

$v =$ Poisson's ration
$\rho =$ mass density
$\omega =$ rotational speed
$R_1 =$ inner radius
$R_2 =$ outer radius

Figure 6.44 Hydrogen flake (dark circle) that contributed to fracture of Arizona turbine rotor. [67] (Reprinted with permission from Academic Press.)

Although one would normally try to keep stresses as low as possible, the higher stress levels associated with introduction of the borehole are justified for the reasons cited above. Using these equations, Yukawa et al.[67] determined the maximum bore tangential stress to be 350 MPa at the fracture speed (3400 rpm).

From the above description of the Arizona rotor failure, the most reasonable stress intensity factor calibration would appear to be that associated with an internal circular flaw. Assuming the worst condition, where the flaw is oriented normal to the maximum bore tangential stress, we have from Fig. 6.21

$$K_{IC} = \frac{2}{\pi} \sigma \sqrt{\pi a}$$

K_{IC}–CVN relations[viii] proposed by Barsom and Rolfe[64] and Sailors and Corten[66] for the transition temperature regime are

$$\frac{K_{IC}^2}{E} = 2(CVN)^{3/2} \quad \text{(Barsom-Rolfe)} \tag{6-60}$$

$$\frac{K_{IC}^2}{E} = 8(CVN) \quad \text{(Sailors-Corten)} \tag{6-61}$$

So, for example, $K_{IC} = \sqrt{2E(CVN)^{3/2}} = \sqrt{2(30 \times 10^6 \text{ psi})(4 \text{ ft-lb})^{3/2}} = 21909 \text{ psi}\sqrt{\text{in}} \approx 22 \text{ ksi}\sqrt{\text{in}}$ for the lowest CVN value reported by Barsom and Rolfe. Other estimates of the K_{IC} value for the rotor material were obtained in a similar fashion, and are summarized in Table 6.6. These values must be considered as first-order approximations in view of normal test scatter in Charpy energy measurements and the empirical nature of both Eqs. 6-60 and 6-61, but they do provide a starting point from which critical flaw sizes may be computed and compared with experimentally observed hydrogen flake sizes. (Obviously it would have been more desirable to have actual fracture toughness values to use in these computations.) For example, using the K_{IC} values derived from the Sailors-Corten relation in Eq. 6-61, the critical flaw size range is calculated to be

$$34 \text{ to } 59 = \frac{2}{\pi}(350)\left(\sqrt{\pi a}\right)$$

$$\therefore a = 0.74 \text{ to } 2.2 \text{ cm}$$

Table 6.6 K_{IC}–CVN Correlations

| | Estimated K_{IC} | |
| | Barsom–Rolfe[64] | Sailors–Corten[66] |
CVN, J (ft-lb)	MPa$\sqrt{\text{m}}$(ksi$\sqrt{\text{in.}}$)	MPa$\sqrt{\text{m}}$(ksi$\sqrt{\text{in.}}$)
5.4–16.3 (4–12)	24–55 (22–50)	34–59 (31–54)

or a hydrogen flake *diameter* range of about 1.5 to 4.3 cm, in excellent agreement with the observed size of these preexistent flaws. The reader should take comfort in the knowledge that hydrogen flakes have been eliminated from modern large forgings by vacuum degassing techniques, and overall toughness levels of newer steels have been increased measurably.

6.12 PLANE-STRESS FRACTURE-TOUGHNESS TESTING

The graphical procedures set forth in the previous section for the determination of K_{IC} are sometimes confounded by the deformation and fracture response of the material. With rising load conditions, it is possible for the material to experience slow stable crack extension prior to failure,

---

[viii] Using English units

which makes it difficult to determine the maximum stress-intensity level at failure because the final stable crack length is uncertain. During the early days of fracture-toughness testing, a droplet of recorder ink was placed at the crack tip to follow the course of such slow crack extension and to stain the fracture surface, thereby providing an estimate of the final stable crack length. Since the fracture properties of some high-strength materials are adversely affected by aqueous solutions (see Chapter 8), this laboratory practice has long since been abandoned. In addition to uncertainties associated with the measurement of the stable crack growth increment Δa, a plastic zone may develop in tougher materials that also must be accounted for in the stress-intensity-factor computation. Consequently, different methods must be employed to determine the fracture-toughness value of a material when the final crack length is not clearly defined but is greater than the initial value.

Irwin[71] proposed that crack instability would occur in accordance with the requirements of the Griffith formulation. That is, failure should occur when the rate of change in the elastic energy-release rate $\partial \mathscr{G} / \partial a$ equals the rate of change in material resistance to such crack growth $\partial R / \partial a$. The material's resistance to fracture R is expected to increase with increasing plastic-zone development and strain hardening; for the case of ceramics, crack resistance increases as a result of strain-induced phase transformations and processes occurring in the wake of the advancing crack front (see Chapter 7). Consequently, both $\mathscr{G}$ and R increase with increasing stress level. The Griffith instability criterion is depicted graphically in Fig. 6.45 as the point of tangency between the $\mathscr{G}$ and R curves when plotted against crack length. Knowing the material's R curve and using the correct stress and crack-length dependence of $\mathscr{G}$ for a given specimen configuration, it would then be possible to determine $\mathscr{G}_c$ or K_c. This fracture-toughness value is generally designated as the *plane-stress* fracture-toughness level, since r_y is no longer much smaller than the sheet thickness and the criterion set forth in Eq. 6-54 is not met.

One should recognize that the fracture-toughness level will vary with the planar dimensions of the specimen. For example, it is seen from Fig. 6.46a that for a given material the fracture-toughness value will depend on the initial crack length, since the tangency point is displaced slightly when the starting crack length is changed. However, this effect is not too large and is minimized when large samples are used, since K_c reaches a limiting value with increasing initial crack length. A considerably larger potential change in K_c values is evident when samples of different planar dimensions are used. Since $\mathscr{G}$ and K depend on specimen configuration, the shape of the $\mathscr{G}$ curve is different for each case. Consequently, the point of tangency and, hence, $\mathscr{G}_c$ will change for a given material. As shown in Fig. 6.46b, the $\mathscr{G}_c$ value increases with increasing sample width and in the limit (the case of an infinitely large panel, where $Y(a/W) = \sqrt{\pi}$ and $\mathscr{G} = \sigma^2 \pi a / E$) reaches a maximum value. Again, it should be fully recognized that the plane-stress fracture toughness of a material is dependent on both metallurgical factors and specimen geometry, while the plane-strain fracture toughness relates only to metallurgical variables.

Although the $\mathscr{G}$ curve can be determined from known analytical relations, such as those provided in Fig. 6.21, the R curve is determined by graphical means. In one accepted procedure the initial step involves the construction of a series of secant lines on the load-displacement record from a test sample (Fig. 6.47a). The compliance values δ/P from these secant lines are

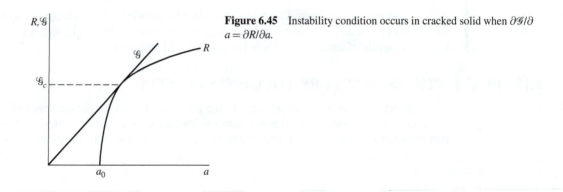

Figure 6.45   Instability condition occurs in cracked solid when $\partial \mathscr{G} / \partial a = \partial R / \partial a$.

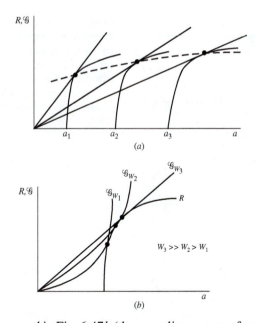

Figure 6.46   Effect of (*a*) initial crack size and (*b*) plate width on fracture toughness.

used in Fig. 6.47*b* (the compliance curve for the given specimen configuration) to determine the associated *a/W* values that reflect the *effective* crack length $a_{\text{eff}} = a_0 + \Delta a + r_y$. Each a_{eff}/W value is then used to determine a respective K_{eff} value from the calibration curve (Fig. 6.47*c*). Finally, the $\mathscr{G}$ or K^2/E curve is constructed from these K_{eff} and associated a_{eff} values (Fig. 6.47*d*). Once the *R* curve for a given material is determined from this graphical procedure, it becomes part of the information about the material. As such, the same *R* curve may be used with samples of different crack length and planar dimension, as discussed above and depicted in Figs. 6.46a and 6.46b. Once the *R* curve is known, the fracture toughness may be determined for an engineering component with any configuration, as long as an analytical expression for $\mathscr{G}$ is known. (Recall that $\mathscr{G}_c$ also may be determined from Eq. 6-19 even when the stress and crack-length dependence of $\mathscr{G}$ is not known.)

A recommended practice for the determination of *R* curves for a given material is described in ASTM Standard E561[72] for the middle-cracked tension (M(T)), compact tension (C(T)), and crack-line-wedge-loaded (C(W)) sample geometries. (Other related toughness tests have been developed for quality control screening purposes—e.g., ASTM Standard B646[73] for aluminum alloys.) From these K_R test values, the plane stress fracture toughness level (K_c), is defined by the K_R value corresponding to the maximum load level and the *effective* crack length (a_{eff}). The latter is identified from the compliance value corresponding to the secant line drawn from the origin to the P_{max} loading point. In addition, the *apparent* plane stress fracture toughness value (K_{app}) may be calculated using the *original* crack length and the maximum load observed during the test. The latter toughness metric provides an engineering estimate of toughness that may be used to calculate a component's residual strength. The reader should recognize two analogies between K_c and K_{app} with true stress and engineering stress values as defined in Chapter 1. Whereas K_c and K_{app} values are based on the *effective* and *original* crack lengths, respectively, σ_{true} and σ_{eng} values are based on the *actual* and *original* cross-sectional areas, respectively. Furthermore, K_c, K_{app}, and *R*-curve determinations all are derived from the load-displacement record for the precracked test sample; similarly, yield and ultimate tensile strengths and ductility are derived from the load-displacement record of the tensile test.

Recalling the strict specimen size requirement for a valid K_{IC} determination (i.e., Eq. 6.54), here, too, individual data points corresponding to the *R*-curve, must represent predominantly elastic testing conditions. For the M(T) sample, possessing a central-crack that is *W*/3 long, valid test results correspond to the net section stress, based on the physical crack size ($a_p = a_0 + \Delta a$) being less than the yield strength of the material. (The physical crack length

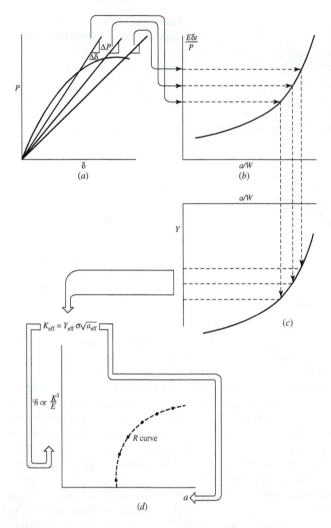

Figure 6.47 Graphical procedure to determine R curve. (a) Secant lines drawn on P–δ test record and plotted on compliance curve; (b) procedure to obtain effective a/W values; (c) calibration factors determined from a_{eff}/W and used to calculate K_{eff}^2/E values for (d) R-curve construction.

at any portion of the test may be determined from the slope of a partially unloaded point along the load-displacement curve.) For the case of the C(T) and C(W) samples, the remaining unbroken ligament also must remain predominantly elastic. In this instance, a predominantly elastic condition conforms to the remaining unbroken ligament, $W - a_p$, being equal to or greater than eight times the plastic-zone size or

$$ (W - a_p) \geq \frac{4}{\pi}\left(\frac{K_R}{\sigma_{ys}}\right)^2 \tag{6-62}$$

Even when validity size requirements are not met, such R-curve and associated K_R values do retain some engineering value. For example, these "nonvalid" test results may be used for quality assurance purposes and for comparative ranking and screening purposes between comparable materials.

6.13   TOUGHNESS DETERMINATION FROM CRACK-OPENING DISPLACEMENT MEASUREMENT

As demonstrated earlier in this chapter, there is a limit to the extent to which K_{eff} may be adjusted for crack-tip plasticity. Whether one uses Eqs. 6-43 or 6-51 in the computation of r_y, it becomes increasingly inaccurate to determine K_{eff} from Eq. 6-45 when r_y/a becomes large. The allowable loads or stresses in a component rapidly approach a limiting value upon general yielding and, while the associated strains or crack-opening displacements (**COD**) (i.e., how much the crack opens under load) increase continually to the point of failure, it is more reasonable to monitor the latter in a general yielding situation. Accordingly, the concept of a critical crack-opening displacement near the crack tip has been introduced to provide a fracture criterion and an alternative measure of fracture toughness.[74,75]

By using the Dugdale crack-tip plasticity model[31] for plane stress, it has been possible to conveniently compute the magnitude of the crack-opening displacement $2V(c)$ defined at the elastic-plastic boundary (Fig. 6.29b). Goodier and Field[76] demonstrated that

$$2V(c) = \frac{8\sigma_{ys}c}{\pi E} \ln\left(\sec\frac{\pi\sigma}{2\sigma_{ys}}\right) \tag{6-63}$$

When $\sigma \ll \sigma_{ys}$, $\ln x \approx x - 1$, so that Eqs. 6-50, 6-51, and 6-63 can be combined to yield

$$2V(c) = \frac{K^2}{E\sigma_{ys}} = \frac{\mathscr{G}}{\sigma_{ys}} \tag{6-64}$$

The fracture toughness of a material is then shown to be

$$\mathscr{G}_c \approx 2\sigma_{ys}V^*(c) \tag{6-65}$$

where $V^*(c) =$ critical crack-opening displacement. One important aspect of Eq. 6-64 is that $V(c)$ can be computed for conditions of both elasticity and plasticity, while $\mathscr{G}$ may be defined only in the former situation. The COD concept, therefore, bridges the elastic and plastic fracture conditions. It is important to recognize, however, that the strain fields and crack-opening displacements associated with the crack tip will vary with specimen configuration. Consequently, a single critical crack-opening displacement value for any given material cannot be defined, since it will be affected by the geometry of the specimen used in the test program.[77]

Standard test methods have been developed that describe the onset of fracture in terms of a critical crack-opening displacement.[78,79] These standards have been used extensively to characterize fracture in structural steel components. Experimental crack-opening values are measured at the specimen surface and then related mathematically to the crack-tip-opening displacement.

The physical significance of Eq. 6-65 warrants additional comment. Since the crack-opening displacement $2V(c)$ is related to the extent of plastic straining in the plastic zone, Eq. 6-65 is analogous to the measurement of toughness from the area under the stress–strain curve in a uniaxial tensile test. As discussed in Chapters 1 and 6, maximum toughness is achieved by an optimum combination of stress and strain. Since fracture toughness most often varies *inversely* with yield strength (see Table 6.3 and Chapter 7), Eq. 6-65 would appear to predict toughness–yield-strength trends incorrectly. This is not the case, since an increase in σ_{ys} of an alloy resulting from any thermomechanical treatment is offset normally by a proportionately greater *decrease* in $2V(c)$. The important point to keep in mind is that σ_{ys} and $2V(c)$ are interrelated. As a final note, some uncertainty exists concerning the correct value of yield strength to use in Eq. 6-65. When plane-strain conditions dominate at the crack tip, should not the tensile yield-strength value be elevated to account for crack-tip triaxiality? Certainly, this would be consistent with a similar previous justification to adjust the plastic-zone size estimate for plane-stress and plane-strain conditions (Eqs. 6-43 and 6-44). On this basis, Eq. 6-65 would be modified so that

$$\mathscr{G}_c \approx n\sigma_{ys}2V^*(c) \tag{6-66}$$

where $1 \leq n \leq 1.5$–2.0. Consequently, when plane-stress conditions are prevalent, $n = 1$, and n increases with increasing plane strain.

6.14 FRACTURE-TOUGHNESS DETERMINATION AND ELASTIC-PLASTIC ANALYSIS WITH THE *J* INTEGRAL

Another key parameter has been developed to define the fracture conditions in a component experiencing both elastic and plastic deformation. Using a line integral related to energy in the vicinity of a crack, Rice[80–82] was able to solve two-dimensional crack problems in the presence of plastic deformation. The form of this line integral, the *J* integral, is given in Eq. 6-67 with failure (crack initiation) occurring when *J* reaches some critical value.

$$J = \int_c \left(W dy - \mathbf{T} \cdot \frac{\partial \mathbf{u}}{\partial x} ds \right) \tag{6-67}$$

where

$x, y =$ rectangular coordinates normal to the crack front (see Fig. 6.48)
$ds =$ increment along contour C
$\mathbf{T} =$ stress vector acting on the contour
$\mathbf{u} =$ displacement vector
$W =$ strain energy density $= \int \sigma_{ij} d\varepsilon_{ij}$

For the case of a closed contour, Rice showed that the line integral is equal to zero. If we examine the contour $C + C' + Q + R$ in Fig. 6.48, we find that $dy = 0$ and $T_i = 0$ along the contour segments Q and R. Therefore, $J_Q = J_R = 0$, which leads to the fact that $J_c = J_{c'}$. Since the paths along contours C and C' are also opposite in sign, it is concluded that the J integral is path independent and that J can be determined from a stress analysis where σ and ε are defined on some arbitrary contour away from and surrounding a crack tip. For example, one could determine J by making use of a finite element analysis to determine σ and ε at locations other than at the crack tip.

In establishing the mathematical framework for the J-integral fracture analysis, Rice assumed that deformation theory of plasticity applied; that is, the stresses and strains in a plastic or elastic–plastic body are considered to be the same as for a nonlinear elastic body with the same stress–strain curve. This means that the stress–strain curve is nonlinear and reversible or that almost proportional straining has occurred throughout the body. (Recall the discussion pertaining to the nonlinear elastic stress–strain response of "whiskers" in Section 1.3.2.) As a result, determination of the crack-tip strain energy using the J integral is valid as long as no unloading occurs. It then follows that the J integral may not be defined rigorously under cyclic loading conditions. As such,

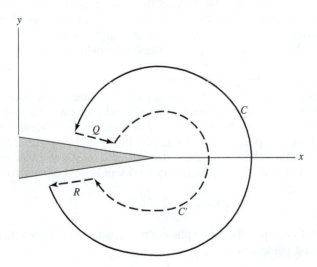

Figure 6.48 Line contour surrounding crack tip.

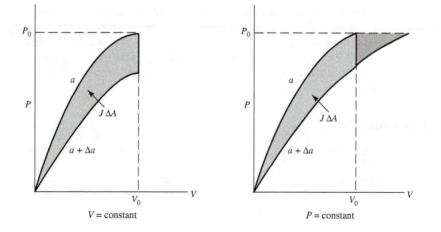

Figure 6.49   Determination of *J* integral based on fixed displacement and fixed load.

the *J* integral has significance in terms of defining the stress and strain conditions for crack *initiation* under monotonic loading conditions and also in the presence of a limited amount of stable crack extension. (Recall that K_{IC} is used to define the condition for unstable crack *propagation* or the value of *K* associated with a 2% apparent extension of the crack.) Subsequent studies have shown that this restriction is not as severe as first thought. For example, Hutchinson and Paris[83] presented a theoretical justification for use of the *J* integral in the analysis of stable crack growth, and Dowling[84] has used the change in *J* during cyclic loading (ΔJ) to correlate fatigue crack propagation rates.

Of particular significance, Rice[81,82] also presented an alternative and equivalent definition for *J* with the latter being defined as the pseudopotential energy difference between two identically loaded bodies possessing slightly different crack lengths. The change in strain energy ΔU associated with an incremental crack advance Δa is portrayed in Fig. 6.49 and equal to $J\Delta a$ or $J = (du/da)$, where

$$J = \int_0^P \left(\frac{\partial \delta}{\partial a}\right)_P dP$$

$$J = -\int_0^\delta \left(\frac{\partial P}{\partial a}\right)_\delta d\delta$$

(6-68)

From Eq. 6-68, it is seen that *J* can be determined from load versus load-point-displacement records for specimens containing slightly different crack lengths. Since, for elastic conditions, $\mathscr{G} = (du/da)$, the relation between *J* and $\mathscr{G}$ is apparent (see Section 6.1.4). For either linear or nonlinear elastic conditions, *J* is the energy made available at the crack tip per unit crack extension, *da*. That is, *J* is equivalent to the crack driving force. Therefore,

$$J = \mathscr{G} = \frac{K^2}{E'}$$

(6-69)

where

$$E' = E \,(\text{plane stress})$$
$$E' = E/(1 - v^2) \,(\text{plane strain})$$

For conditions of plasticity, however, energy is dissipated irreversibly during plastic flow within the solid (e.g., through dislocation motion), so that *J* can no longer represent energy made available at the crack tip for crack extension. However, *J* can be defined in the same manner for nonlinear elastic and elastic–plastic situations as long as the stress–strain curves are the same. J then becomes a measure of the intensity of the entire elastic–plastic stress–strain field that surrounds the crack tip. It is seen then that J is analogous to K for the case of elastic stress fields. On this basis, J can be used as a failure criterion in that fracture would be expected to occur in two different samples or components if they are subjected to an equal and critical J level.

6.14.1 Determination of J_{IC}

Analogous to K_{IC}, which represents the material's resistance to crack extension, one may define the value J_{IC}, which characterizes the toughness of a material near the outset of crack extension. Studies[85,86] have shown that J_{IC} is numerically equal to $\mathscr{G}_{IC}$ values determined from valid plane-strain fracture-toughness specimens for predominantly elastic conditions associated with sudden failure without prior crack extension (Fig. 6.50).[87] For this condition,

$$J_{IC}(\text{plastic test}) = \mathscr{G}_{IC} = \frac{K_{IC}^2}{E}(1 - \nu^2)(\text{elastic test}) \tag{6-70}$$

Rice et al.[88] developed a simple method for the determination of the plastic component of J_{IC}. For the case of a plate containing a deep notch and subjected to pure bending, Rice found that

$$J_{pl} = \frac{2}{b} \int_0^{\delta_c} P d\delta_c \tag{6-71}$$

where

  δ_c = displacement of sample containing a crack
  P = load/unit thickness
  b = remaining unbroken ligament

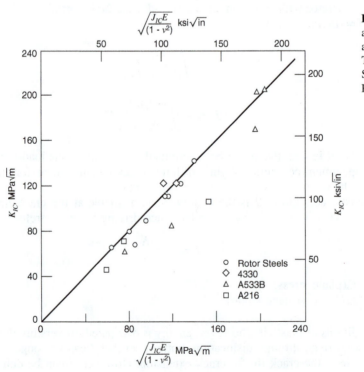

Figure 6.50   Comparison of K_{IC} and J_{IC} values for several steel alloys.[87] (© American Society for Testing and Materials, 1916 Race Street, Philadelphia, PA 19103. Reprinted with permission.)

Equation 6-71 then reduces to

$$J_{pl} = \frac{2A}{Bb} \tag{6-72}$$

where
$B =$ specimen thickness
$A =$ area under the load versus load-point displacement curve

Equation 6-72 also defines J for a three-point bend specimen with a span/width ratio of four. For the compact specimen, the above relation must be modified to account for the presence of a tensile loading component such that

$$J_{pl} = \frac{\eta A}{Bb} \tag{6-73}$$

where $\eta = 2 + 0.522(b/W)$. ASTM Standard E813[89] (now incorporated into ASTM 1820[90]) then defines

$$J_i = J_{el(i)} + J_{pl(i)} \tag{6-74}$$

which defines the J value at a point on the load-displacement line corresponding to a load P_i and displacement δ_l. For the conditions of pure bending and three-point bending with a span/width ratio of four, J is given as

$$J_i = \frac{K_i^2}{E}(1 - v^2) + \frac{2A_i}{Bb} \tag{6-75}$$

EXAMPLE 6.7

An SEN bend bar of a steel alloy was used to conduct a J Integral test. The specimen had dimensions of $B = 10\,\text{mm}$, $W = 20\,\text{mm}$, $S = 80\,\text{mm}$, and $a = 10\,\text{mm}$. The alloy possessed the following mechanical properties: $E = 205\,\text{GPa}$, and $v = 0.25$. The J test developed a load-displacement curve similar to that shown in Fig. 6.51c with the area within the curve $= 5\,\text{Nm}^2$ and the maximum load $= 15\,\text{kN}$. Calculate the value of J for this material.

From Eq. 6.75, the value of J is given by

$$J = \frac{K^2}{E}(1 - v^2) + \frac{2A}{B(W - a)}$$

Proceeding further, the K relation for this specimen configuration is (see Appendix B, Type 6):

$$K = \frac{PS}{BW^{1.5}} f(a/W)$$

where

$$f(a/W) = \frac{3(a/W)^{0.5}}{2(1 + 2a/W)(1 - a/W)^{1.5}} \times \left[1.99 - (a/W)(1 - a/W)(2.15 - 3.93a/W + 2.7a^2/W^2)\right]$$

From the dimensional values given above, $f(a/W) = 1.694$ and

$$K = \frac{15{,}000(80 \times 10^{-3})}{10 \times 10^{-3}(20 \times 10^{-3})^{1.5}}(1.694) = 71.8\,\text{MPa}\sqrt{\text{m}}$$

$$\therefore J = \frac{(71.8 \times 10^6)^2}{205 \times 10^9}(1 - v^2) + \frac{2(5)}{(0.01)(0.02 - 0.01)}$$

$$J = 25{,}148 + 100{,}000 = 125.1\,\text{kN/m}$$

Alternately,

$$J_i = \frac{K_i^2}{E}(1 - v^2) + \frac{A_i(2 + 0.522(b/W))}{Bb} \tag{6-76}$$

for the compact specimen design.

The determination of J_{IC} by this method then involves the measurement of J values from several different samples that have experienced varying amounts of crack extension, Δa.[87] These data are then plotted with J versus Δa and characterized by a best-fit power law relation; as such, this information represents a crack resistance curve R analogous to that described in Section 6.12. (See also ASTM Standard 1820 for the determination of J-R curves.[90]) J_{IC} is defined at some location on the $J - \Delta a$ curve corresponding to the onset of cracking. More specifically, ASTM Standard E813[89] defines J_{IC} for 0.2-mm crack extension as established from the following recommended test procedure:

1. Load several three-point bend bars or compact specimens to various displacements and then unload. These samples should reveal various amounts of crack extension from the fatigue-precracked starter notch (Fig. 6.51).

2. To record Δa for each sample, the sample is marked. For steel samples, heat tinting at 300 °C for 10 min per 25 mm of thickness will discolor the existing fracture surface. For

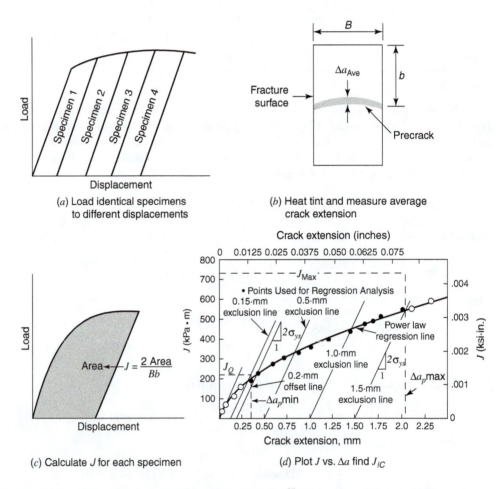

(a) Load identical specimens to different displacements

(b) Heat tint and measure average crack extension

(c) Calculate J for each specimen

(d) Plot J vs. Δa find J_{IC}

Figure 6.51 Procedures for multispecimen determination of J_{IC}.[87] (© American Society for Testing and Materials, 1916 Race Street, Philadelphia, PA 19103. Reprinted with permission.)

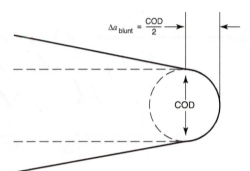

Figure 6.52 Effective crack extension approximately given by COD/2.

other materials that do not discolor readily after heat tinting (e.g., aluminum alloys), the crack-extension region can be marked by a short amount of fatigue crack growth.

3. The sample is then broken and Δa recorded from that region sandwiched between the fatigue–precrack zone and the marker zone (produced either by heat tinting or cyclic loading). Δa is given by the average of nine readings taken across the crack front from one surface to the other (Fig. 6.51*b*).

4. The area under the $P - \delta$ plot for each sample is measured, and J_{pl} is computed from either Eq. 6-72 or 6-73, depending on the specimen configuration (Fig. 6.51*c*).

5. To establish the R curve, $J - \Delta a$ data are chosen that conform to certain acceptance criteria based on the construction of a second $J - \Delta a$ line, called the blunting line. The blunting line is conceived from the following considerations: Consider that a sharp crack begins to blunt as the load on the component is increased. Blunting (i.e., *effective* crack extension) then increases until *actual* crack extension occurs. The amount of crack-opening displacement (COD) is given by Eq. 6-64 as $2V(c) = \mathcal{G}/\sigma_{ys}$ or $\mathcal{G} = 2V(c)\sigma_{ys}$. From Fig. 6.52, the *effective* crack extension is approximated by COD/2 where $\Delta a = \mathcal{G}/2\sigma_{ys}$ or

$$\mathcal{G} \approx J \approx 2\sigma_{ys}\Delta a \tag{6-77}$$

Since Δa values due to blunting are difficult to measure, especially for the case of brittle materials, it has been found more useful to describe the blunting line as $J = 2\sigma_{flow}\Delta a$, where $\sigma_{flow} = (\sigma_{ys} + \sigma_{ts})/2$, which takes into account strain hardening in the material. Additional lines parallel to the blunting line are then constructed with offsets of 0.15, 0.5, 1.0, and 1.0 mm, respectively. When selecting the $J - \Delta a$ data for the R curve, one data point must lie within the 0.15-and 0.5-mm offset lines as well as within the 1.0- and 1.5-mm lines. Several additional points are then taken from the acceptable region of valid data (between 0.15- and 1.5-mm offset lines). The value J_Q is then obtained by noting the intersection of the best-fit power law $J - \Delta a$ curve with a 0.2-mm offset line—the latter corresponding to a fracture-toughness determination for 0.2-mm crack extension. Finally, $J_Q = J_{IC}$ if B, $b > 25J_Q/\sigma_{ys}$. (Other validation criteria are noted in ASTM E813, and in the more recent ASTM E1820.)

From the above discussions, it is clear that the determination of J_{IC} involves the testing of numerous specimens, which makes the procedure both tedious and very expensive. Fortunately, another technique has been developed that enables J_{IC} to be determined from multiple loadings of a single sample.[91] After the sample is loaded to a certain load and displacement level, the load is reduced by approximately 10% (Fig. 6.53). By measuring the specimen compliance during this slight unloading period, the crack length corresponding to this compliance value can be defined. The crack length could also be inferred with electrical resistance measurements or with an ultrasonic detector. The load is then increased again until another slight unloading event is introduced. Here, again, the new crack length is inferred from any one of the above-mentioned techniques. From a number of such load interruptions, Δa values can be determined along with the associated values of J corresponding to the respective locations along the $P - \delta$ plot. This information is then used to obtain an R curve with a $J - \Delta a$ plot. As before, J_{IC} is defined at the

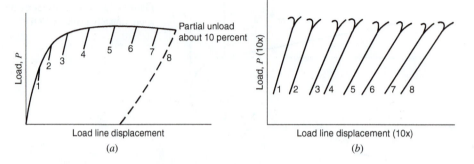

Figure 6.53   Single specimen compliance method for J_{IC} determination. (*a*) Load versus load-line displacement with several partial unloadings; (*b*) amplified segments of (*a*) revealing change in compliance associated with crack extension (ASTM Standard E813-89). (© American Society for Testing and Materials, 1916 Race Street, Philadelphia, PA 19103. Reprinted with permission.)

intersection point with the computed blunting line. From Fig. 6.54 it is seen that J_{IC}, determined from multiple specimens using heat tinting, and single specimens, using compliance and electrical potential techniques, are in excellent agreement. Apparently, the slight amount of unloading associated with the compliance technique does not compromise the value of J_{IC} even though the *J*-integral requirement for deformation plasticity conditions is violated. As noted above, Paris[83,92] has considered this apparent contradiction and concluded that some unloading may be permissible. Furthermore, fatigue crack growth processes may be characterized by a *J* parameter even though considerable unloading occurs during each loading cycle.[84]

It is instructive to compare specimen size requirements associated with valid J_{IC} and K_{IC} test procedures. Recall from Eq. 6-54 that the relevant specimen dimensions for the K_{IC} test must be

$$\geq 2.5 \left(\frac{K_{IC}}{\sigma_{ys}} \right)^2$$

whereas $J_Q = J_{IC}$ if $B, b > 25\, J_Q / \sigma_{ys}$. Since

$$J_{IC} = \frac{K_{IC}^2}{E / (1 - v^2)}$$

it follows that

$$\frac{K_{IC} \text{ specimen size}}{J_{IC} \text{ specimen size}} = \frac{2.5 \left(\frac{K_{IC}}{\sigma_{ys}} \right)^2}{25 \frac{K_{IC}^2}{E \sigma_{ys}}} \geq 20 \tag{6-78}$$

so that plane-strain fracture-toughness values can be determined by J_{IC} techniques with a much smaller specimen as compared with K_{IC} specimen size requirements.

It is important to recognize that the measurement point for J_{IC} corresponds to minimal (0.2-mm) crack extension while K_{IC} corresponds to 2% apparent crack extension. (The non-linearity in a $P - \delta$ plot is due to both crack growth and plastic-zone formation at the crack tip.) Depending on the shape of the *R* curve, the J_{IC} and K_{IC} values may or may not agree closely. Note from Fig. 6.55 that, for the case of a brittle material with a shallow *R* curve, K_{IC} and J_{IC} should bear good agreement. Conversely, when the *R* curve is steep, as in the case of a tough material, the agreement between J_{IC} and K_{IC} values would not be as good. In such a case, J_{IC} should be

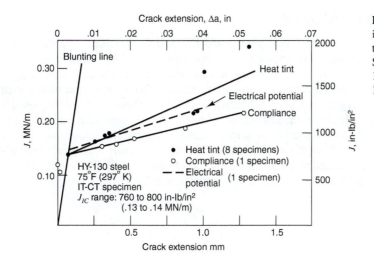

Figure 6.54   Determination of J_{IC} in HY-130 steel based on different test procedures.[87] (© American Society for Testing and Materials, 1916 Race Street, Philadelphia, PA 19103. Reprinted with permission.)

recognized as being a very conservative measurement of toughness, since tough material would be expected to exhibit a finite amount of crack extension at higher loads. In situations where J_{IC} values represent an ultraconservative measure of toughness, other techniques have been attempted to more correctly describe the critical set of conditions for fracture. Paris and co-workers have defined an additional material parameter, the tearing modulus T, which allows for stable tearing in addition to crack blunting.[93-95] They showed that the tearing modulus is given by

$$T = \frac{dJ}{da}\frac{E}{(\sigma_{ys})^2} \qquad (6\text{-}79)$$

and described a new instability criteria when $T_{applied} > T_{material}$. (Note that (dJ/da) corresponds to the slope of the $J - \Delta a$ curve.) Further studies are in progress to establish the general applicability of this new fracture criterion. (See also ASTM Standard E1820, which describes the determination of $J - \Delta a$ curves.[90])

To conclude our discussion of elastic–plastic fracture, we find for the case of ductile materials that fracture may be characterized by a critical crack-opening displacement (δ_c) value or when the J integral reaches the J_{IC} level. Both δ_c and J_{IC} measure the resistance to crack initiation whereas the tearing modulus $T(dJ/da)$ describes the material's resistance to crack extension. When conditions of full-scale plasticity are present, other methods are needed to assess the fracture process; in this regard, the reader is referred to a document entitled "An Engineering Approach to Elastic-Plastic Fracture Analysis."[96]

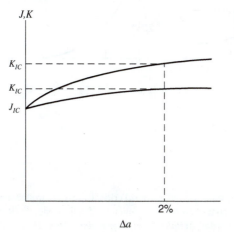

Figure 6.55   Difference between J_{IC} and K_{IC} measurement points based on the slope of R curve.

6.15   OTHER FRACTURE MODELS

Having now discussed widely-accepted K_{IC} and J_{IC} fracture toughness tests in detail, we turn briefly to other, more exploratory approaches. Since K_{IC} testing is fairly complex and expensive by most standards, it would be valuable to be able to determine K_{IC} on the basis of more readily obtained mechanical properties, such as those associated with a tensile test. Some progress in this regard has been achieved, with failure theories being proposed to account for cracking (e.g., cleavage or interfacial failure) and rupture (e.g., microvoid coalescence) processes.[97–107] For the case of slip-initiated transgranular cleavage, K_{IC} conditions are believed to conform to the attainment of a critical fracture stress σ_f^* over a characteristic distance l_0^* that is related to some microstructural feature.[97–101] The model proposed by Ritchie et al.[100] (Fig. 6.56a) characterizes cleavage fracture by

$$K_{IC} \propto \left[\frac{(\sigma_f^*)^{(1+n)/2}}{\sigma_{ys}^{(1-n)/2}} \right] (l_0^*)^{1/2} \tag{6-80}$$

where

$n=$ strain-hardening exponent
$\sigma_{ys}=$ yield strength

For the case of low-carbon steels that possess ferrite-pearlite microstructures, l_0^* is found to be equal to several grain diameters and conforms to the spacing between grain boundary carbide particles. Also note that K_{IC} increases with increasing strain-hardening coefficient. Hahn and Rosenfield[108] also noted a direct dependence of K_{IC} on n with

$$K_{IC} \approx n\sqrt{2E\sigma_{ys}\varepsilon_f/3} \tag{6-81}$$

where

$E=$ elastic modulus
$\varepsilon_f=$ true fracture strain in uniaxial tension

Knott has proposed a simpler form of Eq. 6-81 with $K_{IC} = \sqrt{2\pi l_0^*}$; this relation has been examined for different steel microstructures and chemistries.[99]

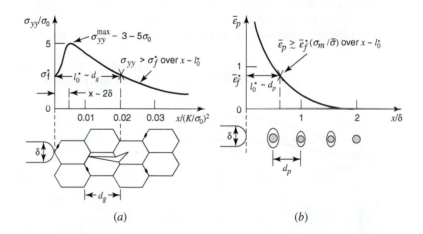

(a)   (b)

Figure 6.56   Schematic representations of theoretical models for (a) critical stress-controlled model for cleavage fracture and (b) stress-modified critical strain-controlled model for microvoid coalescence.[98] (Reprinted with permission from R. O. Ritchie and A. W. Thompson, *Metallurgical Transactions,* 16A, 233 (1985).)

For conditions involving ductile fracture, a stress-modified critical strain criterion has been formulated[101,103,105,106] with fracture defined when a critical strain ε_f^* exists over a critical distance l_0^* (Fig. 8.4b):

$$K_{IC} \propto \sqrt{E\sigma_{ys}\varepsilon_f^* l_0^*} \qquad (6\text{-}82)$$

where

$\varepsilon_f^* = $ critical fracture strain

$l_0^* = $ multiple of mean distance between microvoid-producing particles

Note that Eqs. 6-81 and 6-82 possess several similar components and that ε_f^* increases with the strain-hardening exponent. Chen and Knott[107] also determined for the case of several aluminum alloys that the critical COD value increased with ε_f^* and the square of the strain-hardening exponent. Other treatments of ductile fracture, involving microvoid coalescence, are discussed in the review by Van Stone et al.[104] The direct dependence of K_{IC} on $\sqrt{\sigma_{ys}}$ (Eqs. 6-81 and 6-82) deserves additional comment since empirical studies have generally noted an *inverse* dependency of K_{IC} on σ_{ys}. However, as discussed later in Section 6.13, increases in σ_{ys} are usually associated with proportionately greater decreases in critical crack opening displacement and critical fracture strain levels.

The models discussed above possess some merit, but it is obvious that further work is necessary before one is able to determine K_{IC} from more readily obtained mechanical properties.

Other attempts have been made to develop K_{IC} data based on certain fractographic measurements in conjunction with the fracture-toughness model that includes crack-opening displacement considerations. From Fig. 6.57, the sharp crack produced by cyclic loading (each line represents one load cycle [Region A]) is blunted by a stretching open process (Region B) prior to crack instability where microvoid coalescence occurs (Region C). Some investigators[61,109] have postulated that the width of the "stretch zone" reflects the amount of crack-opening displacement, but Broek[110] claims the depth of this zone to be the more relevant dimension. More recently, Kobayashi and coworkers[111] showed that the stretch zone width (SZW) at instability was related to J_{IC}. They also showed that at J values less than J_{IC} the SZW in several steel, aluminum, and titanium alloys varied with the quantity J/E (Fig. 6.58) in a relation of the form (see also Section 10.4.1)

$$SZW = 95\,J/E \qquad (6\text{-}83)$$

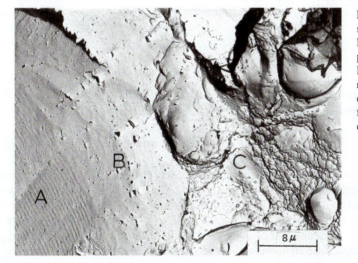

Figure 6.57 Precracked sample subsequently loaded to fracture. Region A represents fatigue precracking zone (see Section 10.3); Region B is the "stretch zone" representing crack blunting prior to crack instability; Region C is overload fracture region revealing microvoid coalescence.

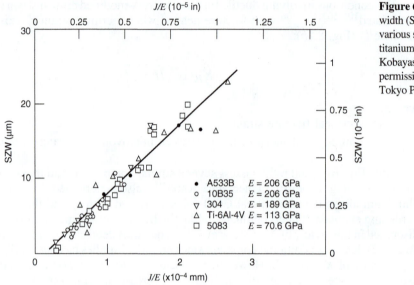

Figure 6.58 Stretch zone width (SZW) versus J/E for various steel, aluminum, and titanium alloys. (From Kobayashi et al.[111], with permission from University of Tokyo Press.)

In another study, Klassen et al.[112] characterized the effect of alloying elements on the fracture toughness in high-strength, low-alloy steels and determined J_{IC} values on the basis of stretch zone measurements. Since $J \propto \sigma_{\text{flow}} \cdot V(c)$, where SZW is proportional to $V(c)$, they determined that $J_{IC} \propto \sigma_{\text{flow}} \cdot$ SZW. As expected, they reported that J_{IC} increased with a decreasing grain size, and with volume fraction and size of inclusions. Since stretch zone measurements are difficult to make and vary widely along the crack front, their use in the determination of K_{IC} and J_{IC} should be approached with caution.

Before concluding this section, it is interesting to reexamine the significance of Eq. 5-28 in light of our subsequent development of the fracture-toughness parameter, K_{IC}. For the case of a semicircular surface flaw, the stress intensity factor at fracture may be given from Eq. 6-84.

$$K_{IC} = 1.1 \frac{2}{\pi} \sigma \sqrt{\pi a_c} \qquad (6\text{-}84)$$

where a_c is the depth of the semicircular crack at fracture. (Note that only one surface flaw correction is used in this crack configuration.) Combining Eqs. 6-84 and 5-28, we find

$$M_h = 0.81 \, K_{IC}(r_h/a_c)^{1/2} \qquad (6\text{-}85)$$

where

M_h = mirror constant corresponding to the mist–hackle boundary
K_{IC} = plane-strain fracture toughness
r_h = distance from crack origin to mist–hackle boundary
a_c = critical flaw depth of semicircular surface flaw

When available data corresponding to M_h and K_{IC} values for various ceramics are plotted together, the data cluster about a line corresponding to a mist–hackle boundary: critical flaw size ratio of 13 to 1 (Fig. 6.59). Note that this ratio appears to hold true for both glassy and crystalline ceramics. As such, the critical flaw size in a ceramic can be estimated from Eq. 6-85 if the mist–hackle radius can be measured on the fracture surface. Mecholsky et al.[113] have verified this procedure for a number of glasses and crystalline ceramics.

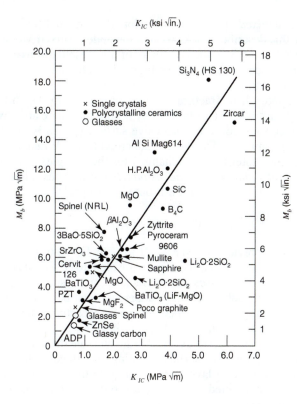

Figure 6.59 Mist–hackle constant M_h versus K_{IC} plotted for glassy and crystalline ceramics. Solid line corresponds to a mist–hackle-boundary–critical flaw size ratio of 13 to 1. (From Mecholsky et al.[113], reproduced with permission from *Journal of Materials Science* 11, 1310 (1976).)

6.16   FRACTURE MECHANICS AND ADHESION MEASUREMENTS

Throughout this chapter we have used fracture mechanics to analyze the conditions pertaining to fracture of monolithic materials. Fracture mechanics principles also can be applied to the failure of interfaces between dissimilar materials as found in adhesive bonds, surface coatings, and multilayer composite materials. We must introduce several new terms, however, before proceeding. First, an *adhesive joint* is made up of an *adhesive* layer and two or more *adherends*—that is, the solid pieces that are joined in the assembly. The primary purpose of the adhesive layer is, of course, to bond the adherends together by the establishment of two strong interfaces. In a similar fashion, a *coating* or *surface film* is adhered to an underlying *substrate*. The primary purpose of a coating may be to add certain chemical, thermal, optical, or electrical functionality to a component,[ix] while good adhesion is a secondary characteristic typically required for reliability. A coating may have a free upper surface, or it may be topped by another coating layer (a *superlayer*) thereby establishing a second interface between two solids. For simplicity, we will use the terms "adhesive" and "adherend" in most of the following discussion, keeping in mind that "coating" and "substrate" could generally be substituted as necessary.

If fracture occurs exactly along an interface between two materials such that two fracture surfaces are created with pure adhesive material on one side and pure adherend material on the other side, it is called an *adhesive failure*. Conversely, if the crack path travels through either the adhesive or adherend near the interface leaving a thin layer of the same material on both fracture surfaces, it is called a *cohesive failure* of either the adhesive or the adherend. Cohesive failure is generally considered the more favorable of the two because it means that sufficient interfacial strength was achieved during bonding to avoid creation of an easy fracture path. Not surprisingly, in some cases a failure can alternate between adhesive and cohesive depending on

---

[ix] For example, ceramic thermal barrier coatings that provide thermal insulation for turbine blades, polymer films that reduce friction during insertion of hypodermic needles, and metal films that act as electrical conductors between silicon integrated circuit devices.

local conditions, leading to a *mixed mode of failure*. Note the potential for confusion with the phrase *mixed mode failure* that describes crack propagation under a mixed Mode I-II-III loading condition, as explained in Section 6.5. In the remainder of the present discussion, the relevance of *mixed mode failure* to interface cracking will be highlighted, but no more will be said about mixed cohesive and adhesive failure.

The thermodynamic approach to fracture analysis originated by Griffith can be easily extended to address purely elastic adhesive failures by noting that there are now three surface energy terms that must be considered: a bare adhesive surface (γ_α), a bare adherend surface (γ_β), and a bonded interface between adhesive and adherend ($\gamma_{\alpha\beta}$). When an adhesive failure occurs, the energy of the intact interface is exchanged for the energies of the two individual fracture surfaces. The *thermodynamic work of adhesion* (also sometimes called the *true work of adhesion*) is therefore given by an expression attributed to Dupre[114]

$$W_{ad} = (\gamma_\alpha + \gamma_\beta) - \gamma_{\alpha\beta} \tag{6-86}$$

Thus the Griffith stress intensity previously given as $(2E\gamma_s)^{1/2}$ in Eqs. 6-6 and 6-7 can be replaced by $(EW_{ad})^{1/2}$ for bimaterial interfaces, and is therefore equivalent to $(E\mathcal{G})^{1/2}$ at equilibrium. As we well know by now, additional energy losses due to crack-tip processes such as microcracking, viscoelasticity, or plasticity in the adhesive and/or adherend must be added for many realistic material combinations, leading to a practical work of adhesion that actually determines the critical energy release rate measured in adhesion tests: $W_{ad,P} = W_{ad} + \gamma_p = \mathcal{G}_c$.

This energy analysis immediately gives us insight into an issue that can be important to the toughness of real adhesive joints or surface coating interfaces: the effect of layer thickness[115,116]. Adhesive and coating layers are often quite thin, ranging from millimeters to nanometers in scale. As was pointed out in Section 6.9 and shown in Fig. 6.30, as the thickness of a large homogeneous plate is reduced there is initially an increase in fracture toughness K_c. This is due to a decrease in crack-tip triaxiality and an accompanying increase in plasticity associated with approaching a plane-stress condition. However, there is a limit to this trend. If the plate thickness falls below the expected plastic-zone size then the energy dissipated in plastic deformation is also reduced, causing toughness to fall. The same phenomenon applies to adhesive joints if the adhesive material is capable of plastic deformation or other dissipative processes such as microcracking. Chai, for example, tested polymer adhesive joints with bondline thickness values ranging from approximately $1\,\mu m$ to several hundred micrometers.[115] He found that as bondline thickness increased, the practical work of adhesion under shear loading increased from less than 1 kJ/m^2 to more than 11 kJ/m^2 for a ductile BP-907 adhesive. A plateau in fracture energy was reached for joints with bondline thickness exceeding $220\,\mu m$. In a review of adhesion in thin film coating systems, Gerberich[116] compared $\mathcal{G}_c$ for Nb films and Ta$_2$N films on sapphire substrates. Nb is a refractory metal with reasonable ductility in bulk form. Ta$_2$N is an intermetallic compound with poor ductility. When the films were deposited with nanometer-scale thickness, the measured work of adhesion was $\sim$1 J/m^2 for both systems. When the film thickness was increased to the micrometer scale the Ta$_2$N/sapphire adhesion was unchanged, while the Nb/sapphire interface toughness rose to above 10 J/m^2. In multilayer coatings, such as Cu/TaN on SiO$_2$, it may be that plasticity in an adjacent superlayer (Cu in this example) can provide some energy dissipation even when the adhesive layer (TaN) is incapable of significant plastic deformation.[117] In this scenario, it is the thickness of the superlayer that rules the failure energy. In summary, we can draw two general conclusions about the adhesive thickness effect. First, caution must be used when designing with thin adhesives or coatings that depend on plasticity or other dissipative mechanisms for much of their failure resistance. Second, intentionally reducing adhesive thickness for testing purposes may provide a method for assessing the thermodynamic work of adhesion for a bimaterial interface by suppressing the activity of dissipative mechanisms.

A key characteristic of many adhesive joint fractures is that the crack tip may be subjected to mixed-mode loading. In bulk materials the crack path can often choose an orientation that maximizes the Mode I condition. However, the design of a structural component may require that

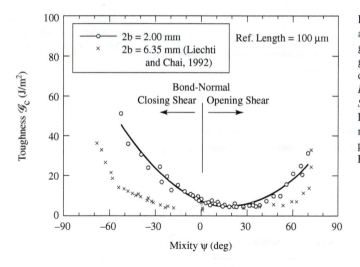

Figure 6.60 Interface toughness as a function of mixity (phase angle) for glass/epoxy joints. Toughness is greatest when the Mode II component dominates.[120] (Reprinted from *International Journal of Solids and Structures* 32, Y.-M. Liang and K. M. Liechti, "Toughening mechanisms in mixed-mode interfacial fracture," p. 957, 1995, with permission from Elsevier.)

the bimaterial interfaces are oriented at angles to the principle tensile load axis other than 90°. In this situation, a growing crack may be forced to follow a path with a significant non-Mode I contribution. *Mode mixity* (e.g., the ratio of Mode I/Mode II loading) can have a strong influence on $\mathcal{G}_c$.[118,119] Using the definitions of the Mode I and II stress intensity factors K_I and K_{II} as shown in Fig. 6.18, and defining the *phase angle of loading* as $\Psi = \tan^{-1}(K_{II}/K_I)$, we see that $\Psi = 0°$ indicates pure Mode I loading and $\Psi = \pm90°$ is pure Mode II. Any phase angle between these values indicates a mixed-mode condition. A high fraction of Mode II loading (large Ψ magnitude) generally encourages greater plastic deformation and energy dissipation, maximizing joint toughness. Conversely, increasing the relative Mode I component (small Ψ) tends to decrease $\mathcal{G}_c$. This trend is shown in Fig. 6.60 for adhesion tests of two different glass/epoxy joints.[120] This understanding of $\mathcal{G}(\Psi)$ leads directly to a general design criterion for bonded joints, which is that they should be constructed in such a manner that loading is as close to pure Mode II as possible.[121] This philosophy can be seen in some of the design recommendations of the Primary Adhesively Bonded Structure (PABST) program initiated by the U.S. Air Force in the 1970s that led to a significant increase in the use of adhesively-bonded joints in aerospace structures.[122,123] More recently, under the auspices of the Composite Affordability Initiative (CAI), a team representing the Air Force Research Laboratory, the Office of Naval Research, and several aircraft manufacturers used similar fracture mechanics principles to develop a robust bonded joint called the Pi joint that is expected to see wide use in composite aircraft structures.[124,125] One of the features of this joint is a large fraction of Mode II loading.

With these definitions under our belts, we can now turn to brief descriptions of the types of tests used to determine the quality of adhesive bonds. It is worth noting that many dozens of tests have been created to assess the myriad of adhesive joints and coatings found in engineering applications, but most of these are qualitative in nature. That is, they may be extremely useful for purposes of comparison, but cannot be used to measure the fundamental behavior of the materials involved in a joint. Here, we will restrict our discussion to just a few of the techniques that do provide the opportunity for fracture mechanics-based analysis: those that use laminated-beam specimens.

Laminated-beam, or *sandwich*, test specimens require two relatively thick, stiff, elastic adherends bonded on two sides of a planar adhesive layer. Four common forms of sandwich specimen are depicted in Fig. 6.61. The first is the Double Cantilever Beam (DCB) as described in ASTM D3433 for adhesives[126] and ASTM D5528 for laminated composites.[127] If the adherends are identical and the unbonded ends of the beams are pulled apart perpendicular to the adhesive interface, a pure Mode I condition applies. In this case, using simple beam theory it is found that

$$\mathcal{G}_{IC} = \frac{12P^2a^2}{b^2h^3E_s} \qquad (6\text{-}87)$$

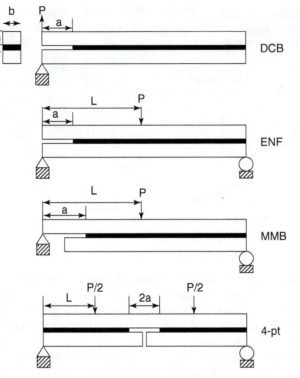

Figure 6.61   Laminated-beam sandwich specimens used for measuring fracture under different loading modes.

where P is the applied load at failure, a is the initial crack length (in this case the unbonded length plus any precrack length), E_S is the substrate (or adherend) modulus, and b and h are the width and height of one adherend, similar to the definitions used for 3-pt and 4-pt specimen bending in Chapter 1. When the critical load P is reached, the crack will grow. If this occurs in a stable fashion, growth will stop after a certain distance (i.e., after sufficient elastic energy has been released to reach equilibrium at the new crack length). The specimen can then be unloaded and reloaded multiple times to acquire additional failure load values for different initial crack lengths. If the joint is of uniform character, a single $\mathcal{G}_{IC}$ value should result. In a variation on the standard DCB test, a wedge can be pushed between the unbonded arms of the adherends and left there to apply a fixed displacement. This static Wedge Opening Load (WOL) or Boeing Wedge Test (BWT) configuration is convenient for measurement of interface time-dependent environment-assisted cracking (see Chapter 8 and Appendix B).[128]

The sandwich specimen used in the DCB test can be modified to determine $\mathcal{G}_{IIC}$ values as well. In this configuration it is known as an End-Notched Flexure (ENF) specimen.[129] The beam is loaded in a 3-point bend fixture with the crack tip between the outer and inner load points. We saw in Chapter 1 that the flexural stresses in a solid bend specimen are maximized at the outer surfaces, and that there is no flexural stress along the neutral axis. Shear stress in a 3-point bend configuration, however, is a maximum at the centerline and decreases nonlinearly to zero at the outer surfaces. This combination provides the pure Mode II loading condition along the ENF adhesive joint. If the adhesive of the ENF specimen is assumed to be rigid, the fracture energy as derived from beam theory can be given as:[130]

$$\mathcal{G}_{IIC} = \frac{9P^2a^2}{16b^2h^3E_s} \tag{6-88}$$

If the adhesive can deform to a large extent, as might be expected for many polymeric materials, Eq. 6-88 must be corrected to account for the additional deformation,[130] but the general form of the equation is unchanged.

A hybrid of the DCB and ENF tests, called the Mixed Mode Bending (MMB) test, can be used to apply a known mode mixity during fracture by simultaneously bending and opening a laminated-beam specimen.[131] This test as applied to laminated composites is summarized in ASTM D6671.[132]

An alternative to the laminated-beam specimens already discussed is a 4-point bend configuration.[133] It bears a resemblance to the MMB specimen, but has two crack fronts. Recall from Chapter 1 that the flexural stress along the surface between the inner load points is constant for 4-point bending. In a solid beam the center section would be in pure bending, and the shear would be zero everywhere. Introducing the slit in the lower beam alters this condition. When the fracture specimen is bent and the critical load is reached, the crack will advance and then arrest when the tip reaches an inner load point. This means that multiple tests can be carried out with a single specimen by progressively moving the load points apart, similar to the multiple tests than can be performed with a single DCB specimen. An advantage of the 4-point bend specimen is that the energy release rate is independent of crack length. The critical energy release rate for this structure must take into account the Poisson's ratio of the adherends, ν_s, as well as the Young's modulus, and is given by[133,134]

$$\mathscr{G}_C = \frac{21P^2L^2(1-\nu_s^2)}{16b^2h^3E_s} \tag{6-89}$$

While the sandwich structures may not appear at first glance to be useful for coatings, it is possible to fabricate a laminated-beam structure by bonding an appropriate adherend to the free surface of the coating for testing purposes. This approach has been very successful when applied to thin films for microelectronic applications,[134] but care must be taken to ensure that the fabrication process, the presence of an additional interface, and plasticity in an adjacent layer do not influence the behavior of the coating layer and its interface to the original substrate.

For more in-depth information on adhesion testing, adhesive joint design, and failure analysis, the reader is encouraged to consult the *Engineering Materials Handbook on Adhesives and Sealants*.[135]

REFERENCES

1. A. Griffith, *Philos. Trans. R. Soc. London, Ser. A* **221**, 163 (1920). (This article has been republished with additional commentary in *Trans. ASM* **61**, 871 (1968).)

2. E. Inglis, *Proceedings,* Institute of Naval Architects, Vol. 55, 1913, p. 219.

3. P. H. Hasselman, *J. Am. Ceram. Soc.* **52**, 600 (1969).

4. R. Larson, J. A. Coppola, D. P. H. Hasselman, and R. C. Bradt, *J. Amer. Ceram. Soc.* **57**, 417 (1974).

5. R. C. Rossi, *Ceram. Bull.* **48** (7), 736 (1969).

6. R. C. Rossi, *J. Amer. Ceram. Soc.* **52** (5), 290 (1969).

7. D. P. H. Hasselman, *Ceramics in Severe Environments*, W. W. Kriegel and H. Palmour, III, Plenum Press, New York, 1971, p. 89.

8. D. P. H. Hasselman, *J. Amer. Ceram. Soc.* **53**, 490 (1970).

9. D. P. H. Hasselman and J. P. Singh, *Thermal Stresses I*, Vol. **1**, R. B. Hetnarski, Ed., North-Holland, New York, 1986, Chap. 4, p. 264.

10. E. Orowan, *Fatigue and Fracture of Metals*, MIT Press, Cambridge, MA, 1950, p. 139.

11. G. R. Irwin, *Fracturing of Metals*, ASM, Cleveland, OH 1949, p. 147.

12. G. R. Irwin and J. A. Kies, *Weld. J. Res. Suppl.* **33**, 193s (1954).

13. ASTM E23-81, ASTM International, West Conshohocken, PA.

14. ASTM D256, ASTM International, West Conshohocken, PA.

15. W. T. Matthews, ASTM *STP 466*, 1970, p. 3.

16. J. Gross, ASTM *STP 466*, 1970, p. 21.

17. R. C. McNicol, *Weld. Res. Suppl.* 385s (Sept. 1965).

18. A. J. Kinloch and R. J. Young, *Fracture Behavior of Polymers*, Applied Science, London, 1983.

19. P. I. Vincent, *Impact Tests and Service Performance of Thermoplastics*, Plastics Institute, London, 1971.

20. ASTM D3029-84, ASTM International, West Conshohocken, PA.

21. ASTM D1822-84, ASTM International, West Conshohocken, PA.

22. *Modern Plastics Encyclopedia*, Vol. **62** (10A), McGraw-Hill, New York, 1985.

23. R. C. McNicol, *Weld. Res. Suppl.* 385s (Sept. 1965).

24. J. A. Begley, *Fracture Transition Phenomena*, Ph.D. Dissertation, Lehigh University, Bethlehem, PA, 1970.

25. H. M. Westergaard, *Trans. ASME, J. Appl. Mech.* **61**, 49 (1939).

26. G. R. Irwin, *Handbuch der Physik*, Vol. **6**, Springer, Berlin, 1958, p. 551.

27. P. C. Paris and G. C. M. Sih, ASTM *STP 381*, 1965, p. 30.

28. H. Tada, P. C. Paris, and G. R. Irwin, *The Stress Analysis of Cracks Handbook*, 3rd ed., ASME, 2000.

29. G. C. M. Sih, *Handbook of Stress Intensity Factors*, Lehigh University, Bethlehem, PA, 1973.

30. Y. Murakami, Ed., *Stress Intensity Factors Handbook*, Pergamon, Oxford, 1987.

31. W. F. Brown, Jr. and J. E. Srawley, ASTM *STP 410*, 1966.

32. C. Feddersen, ASTM *STP 410*, 1967, p. 77.

33. O. L. Bowie, *J. Math. Phys* **35**, 60 (1956).

34. D. Broek and H. Vliegar, *Int. J. Fract. Mech.* **8**, 353 (1972).

35. K. B. Clifton, R. L. Reep, and J. J. Mecholsky, *J. Mater. Sci.* **43**, 2026 (2008).

36. K. B. Clifton, J. Yan, J. J. Mecholsky, and R. L. Reep, *J. Zool.* **274** (2), 150 (2007).

37. J. Yan, K. B. Clifton, R. L. Reep, and J. J. Mecholsky, *J. Biomech.* **39**, 1066 (2006).

38. R. O. Ritchie, M. J. Buehler, and P. Hansma, *Phys. Today* **62**, 41 (2009).

39. G. R. Irwin, *Appl. Mater. Res.* **3**, 65 (1964).

40. G. R. Irwin, *Trans. ASME, J. Appl. Mech.* **24**, 361 (1957).

41. F. A. McClintock and G. R. Irwin, ASTM *STP 381*, 1965, p. 84.

42. G. T. Hahn and A. R. Rosenfield, *Acta Met.* **13**, 293 (1965).

43. D. S. Dugdale, *J. Mech. Phys. Solids* **8**, 100 (1960).

44. R. W. Hertzberg and R. H. Goodenow, Proceedings, Micro Alloying 75, Union Carbide, New York, 1977, p. 503.

45. R. J. Gran, F. D. Orazio, Jr., P. C. Paris, G. R. Irwin, and R. W. Hertzberg, AFFDL-TR-70-149, March 1971.

46. ASTM E399-09e1, ASTM International, West Conshohocken, PA.

47. ASTM C1421-09, ASTM International, West Conshohocken, PA.

48. ASTM E1304-97(2008)e1, ASTM International, West Conshohocken, PA.

49. L. M. Barker, *Eng. Fract. Mech.* **9**, 361 (1977).

50. L. M. Barker and F. I. Baratta, *J. Test. Eval.* **8** (3), 97 (1980).

51. J. J. Mecholsky and L. M. Barker, ASTM *STP 855*, 1984, p. 324.

52. R. Morrell, *Adv. Appl. Ceram.* **105** (2), 88 (2006).

53. T. Nishida, Y. Hanaki, and G. Pezzotti, *J. Am. Ceram. Soc.* **77**, 606 (1996).

54. J. J. Kübler, VAMAS Technical Report 37, EMPA, Dübendorf, Switzerland, 1999.

55. J. J. Petrovic and L. A. Jacobson, *J. Am. Ceram. Soc.* **59**, 34 (1976).

56. J. C. Newman and I. S. Raju, *Eng. Fract. Mech.* **12**, 20 (1988).

57. G. R. Anstis, P. Chantikul, B. R. Lawn, and D. B. Marshall, *J. Am. Ceram. Soc.* **64** (9), 533 (1981).

58. D. B. Marshall, *J. Am. Ceram. Soc.* **66** (2), 127 (1983).

59. G. D. Quinn and R. C. Bradt, *J. Am. Ceram. Soc.* **90** (3), 673 (2007).

60. J. J. Kruzic, D. K. Kim, K. J. Koester, and R. O. Ritchie, *J. Mech. Behav. Biomed. Mater.* **2**, 384–395 (2009).

61. G. R. Irwin, NRL Report 6598, Nov. 21, 1967.

62. G. M. Orner and C. E. Hartbower, *Weld. Res. Suppl.*, 405s (1961).

63. J. M. Barsom and S. T. Rolfe, ASTM *STP 466*, 1970, p. 281.

64. J. M. Barsom and S. T. Rolfe, *Eng. Fract. Mech.* **2** (4), 341 (1971).

65. J. M. Barsom and J. V. Pellegrino, *Eng. Fract. Mech.* **5** (2), 209 (1973).

66. R. H. Sailors and H. T. Corten, ASTM *STP 514*, Part II, 1972, p. 164.

67. R. Roberts and C. Newton, Bulletin 265, Welding Research Council, Feb. 1981.

68. S. Yukawa, D. P. Timio, and A. Rubio, *Fracture*, Vol. **5**, H. Liebowitz, Ed., Academic, New York, 1969, p. 65.

69. C. Schabtach, E. L. Fogelman, A. W. Rankin, and D. H. Winnie, *Trans. ASME* **78**, 1567 (1956).

70. R. J. Bucci and P. C. Paris, Del Research Corporation, Hellertown, PA, Oct. 23, 1973.

71. G. R. Irwin, *ASTM Bulletin*, Jan. 1960, p. 29.

72. ASTM E561-05, ASTM International, West Conshohocken, PA.

73. ASTM B646-06a, ASTM International, West Conshohocken, PA.

74. A. A. Wells, *Brit. Weld. J.* **13**, 2 (1965).

75. A. A. Wells, *Brit. Weld. J.* **15**, 221 (1968).

76. J. N. Goodier and F. A. Field, *Fracture of Solids*, Interscience, New York, 1963, p. 103.

77. D. C. Drucker and J. R. Rice, *Eng. Fract. Mech.* **1**, 577 (1970).

78. British Standard Institution BS5762, BSI, 1979, London.

79. ASTM E1290-93, ASTM International, West Conshohocken, PA.

80. J. R. Rice and E. P. Sorenson, *J. Mech. Phys. Solids* **26**, 163 (1978).

81. J. R. Rice, *J. Appl. Mech.* **35**, 379 (1968).

82. J. R. Rice, *Treatise on Fracture*, Vol. **2**, H. Liebowitz, Ed., Academic, New York, 1968, p. 191.

83. J. W. Hutchinson and P. C. Paris, ASTM *STP 668*, 1979, p. 37.

84. N. E. Dowling and J. A. Begley, ASTM *STP 590*, 1976, p. 82.

85. J. A. Begley and J. D. Landes, ASTM *STP 514*, 1972, p. 1.

86. J. A. Begley and J. D. Landes, ASTM *STP 514*, 1972, p. 24.

87. J. D. Landes and J. A. Begley, ASTM *STP 632*, 1977, p. 57.

88. J. R. Rice, P. C. Paris, and J. G. Merkle, ASTM *STP 532*, 1973, p. 231.

89. ASTM E813-89, ASTM International, West Conshohocken, PA.

90. ASTM El820, ASTM International, West Conshohocken, PA.

91. G. A. Clarke, W. R. Andrews, P. C. Paris, and D. W. Schmidt, ASTM *STP 590*, 1976, p. 27.

92. P. C. Paris, "Flaw growth and fracture," ASTM *STP 631*, 1977, p. 3.

93. P. C. Paris, H. Tada, A. Zahoor, and H. Ernst, ASTM *STP 668*, 1979, p. 5.

94. P. C. Paris, H. Tada, H. Ernst, and A. Zahoor, ASTM *STP 668*, 1979, p. 251.

95. H. Ernst, P. C. Paris, M. Rossow, and J. W. Hutchinson, ASTM *STP 667*, 1979, p. 581.

96. V. Kumar, M. D. German, and C. F. Shih, EPRI NP-1931, Res. Proj. 1237-1, Elec. Pow. Res. Inst., 1981.

97. R. O. Ritchie, *Metals Handbook, Vol. 8, Mechanical Testing*, Metals Park, OH, 1985, p. 465.

98. R. O. Ritchie and A. W. Thompson, *Met. Trans.* **16A**, 233 (1985).

99. J. F. Knott, *Advances in Fracture Research*, Vol. **1**, S. R. Valluri, D. M. R. Taplin, P. Rama Rao, J. F. Knott, and R. Dubey, Eds., Pergamon, Oxford, England, 1984, p. 83.

100. R. O. Ritchie, J. F. Knott, and J. R. Rice, *J. Mech. Phys. Solids* **21**, 395 (1973).

101. R. O. Ritchie, W. L. Server, and R. A. Wullaert, *Met. Trans.* **10A**, 1557 (1979).

102. F. A. McClintock, *J. Appl. Mech. Trans. ASME Ser. H* **25**, 363 (1958).

103. R. C. Bates, *Metallurgical Treaties*, J. K. Tien and J. F. Elliott, Eds., AIMI, Warrendale, PA, 1982, p. 551.

104. R. H.Van Stone, T. B. Cox, J. R. Low, and J. A. Psioda, *Int. Met. Rev.* **30** (4), 157 (1985).

105. S. Lee, L. Majno, and R. J. Asaro, *Met. Trans.* **16A**, 1633 (1985).

106. J. R. Rice and M. A. Johnson, *Inelastic Behavior of Solids*, M. F. Kanninen, W. F. Adler, A. R. Rosenfield, and R. I. Jaffee, Eds., McGraw-Hill, New York, 1970, p. 641.

107. C. Q. Chen and J. F. Knott, *Met. Sci.* **15**, 357 (1981).

108. G. T. Hahn and A. R. Rosenfield, ASTM *STP 432*, 1968, p. 5.

109. R. C. Bates and W. G. Clark, Jr., *Trans. ATM* **62**, 380 (1969).

110. D. Broek, *Eng. Fract. Mech.* **6**, 173 (1974).

111. H. Kobayashi, H. Nakamura, and H. Nakazawa, *Recent Research on Mechanical Behavior of Solids*, University of Tokyo Press, 1979, p. 341.

112. R. J. Klassen, M. N. Bassim, M. R. Bayoumi, and H. G. F. Wilsdorf, *Mater. Sci. Eng.* **80**, 25 (1986).

113. J. J. Mecholsky, S. W. Freiman, and R. W. Rice, *J. Mater. Sci.* **11**, 1310 (1976).

114. Dupre, Theorie Mecanique de la Chaleur, Ed. Gauthier-Villars, Paris 1869.

115. H. Chai, *Inter. J. of Fracture* **37**, 137 (1988).

116. W. W. Gerberich and M. J. Cordill, *Rep. Prog. Phys.* **69**, 2157 (2006).

117. M. Lane, R. H. Dauskardt, A. Vainchtein, and H. Gao, *J. Mater. Res.* **15** (12), (2000).

118. J. R. Rice, *J. Appl. Mech.* **55**, 98 (1988).

119. V. Tvergaard and J. W. Hutchinson, *J. Mech. Phys. Solids* **41**, 1119 (1993).

120. Y.-M. Liang and K. M. Liechti, *Int. J. Solids Structures* **32**, 957 (1995).

121. A. V. Pocius, *Adhesion and Adhesives Technology*, 2nd ed., Hanser Gardner, Cincinnati, 2002.

122. *Primary Adhesively Bonded Structure Technology (PABST) Design Handbook for Adhesive Bonding*, Technical Report AFFDL-TR-79-3129, 1979.

123. L. J. Hart-Smith, "Adhesive bonding of aircraft primary structures," SAE Technical Paper 801209, 1980.

124. J. D. Russell, "Composites Affordability Initiative," *AMMTIAC Quarterly* **1** (3), (2006), pp. 3–6.

125. B. D. Flansburg, S. P. Engelstad, and J. Lua, Proc. 50th AIAA/ASME/ASCE/AHS/ASC Structures, Structural Dynamics, and Materials Conference, May 2009, Palm Springs, CA.

126. ASTM D3433, ASTM International, West Conshohocken, PA.

127. ASTM D5528-01, ASTM International, West Conshohocken, PA.

128. R. D. Adams, J. W. Cowap, G. Farquharson, G. M. Margary, and D. Vaughn, *Inter. J. Adhesion & Adhesives* **29**, 609 (2009).

129. H. Chai and S. Mall, *I. J. Fract.* **36**, R3 (1988).

130. K. S. Alfredsson, *Inter. J. Solids Struct.* **41**, 4787 (2004).

131. J. R. Reeder and J. H. Crews, Jr., *AIAA Journal* **28** (7), 1270 (1990).

132. ASTM D6671-06, ASTM International, West Conshohocken, PA.

133. P. G. Charalambides, J. Lund, A. G. Evans, and R. M. McMeeking, *J. Appl. Mech.*, **56**, 1989, pp. 77–82.

134. R. H. Dauskardt, M. Lane, Q. Ma, and N. Krishna, *Engin. Fracture Mech.* **61**, 141 (1998).

135. *Adhesives and Sealants: Engineered Materials Handbook*, Vol. **3**, ASM International, 1990.

FURTHER READINGS

J. M. Barsom and S. T. Rolfe, *Fracture and Fatigue Control in Structures*, 3rd ed., ASTM International, West Conshohocken, PA, 1999.

D. Broek, *Elementary Engineering Fracture Mechanics*, 4th ed., Springer, 1982.

N. E. Dowling, *Mechanical Behavior of Materials: Engineering Methods for Deformation, Fracture, and Fatigue*, 3rd ed., Prentice Hall, Upper Saddle River, NJ, 2007.

H. L. Ewalds and R. J. H. Wanhill, *Fracture Mechanics*, Arnold, London, 1984.

J. F. Knott, *Fundamentals of Fracture Mechanics*, Butterworth & Co. Publishers Ltd, London, 1976.

A. Saxena, *Nonlinear Fracture Mechanics for Engineers*, CRC Press, Boca Raton, FL, 1998.

H. Tada, P. C. Paris, and G. R. Irwin, *The Stress Analysis of Cracks Handbook*, 3rd ed., ASME, New York, 2000.

PROBLEMS

Review

6.1 Summarize the fundamental concept behind Griffith's major contribution to understanding the fracture process.

6.2 Write Griffith's expression for the fracture stress of a material under plane stress conditions. Identify which parameters are *applied* and which are *material constants*.

6.3 State the limitation of Griffith's analysis with regard to crack geometry, and explain why it is a limitation.

6.4 What is the initially counterintuitive relationship between thermal shock resistance and microcrack density in a precracked brittle material? Why, after some thought, docs this relationship makes sense after all?

6.5 Summarize the two recommendations for choosing a thermal shock resistant material, making it clear under what condition(s) each recommendation applies.

6.6 State the limitation of Griffith's analysis that led to the analyses of Orowan and Irwin, and explain why the additional work was necessary.

6.7 Define the *strain (or elastic) energy release rate* first in words, then as an energy-based equation.

6.8 What are the units of the *elastic strain energy release rate*?

6.9 Sketch a load-displacement plot for a linear elastic material, then illustrate on the plot how V (the stored strain energy) is represented in the sketch.

6.10 As a crack advances, what happens to the stiffness of the cracked body? What happens to the compliance?

6.11 What important information can be learned from a Charpy impact test that can be used for design purposes, particularly with steels? What phenomenon limits the direct applicability of the transition temperature derived from standardized Charpy impact test results to the design of many structures?

6.12 Sketch the Mode I, Mode II, and Mode III crack opening geometries, and clearly state which is the most commonly found in engineering structures.

6.13 Define *plane stress* and *plane strain*, making clear which, if any, of the stresses are zero in each case.

6.14 Make a table that clearly shows the constant geometric Y values that apply to very short cracks that are (i) center through-cracks, (ii) edge through-cracks, (iii) circular embedded cracks, and (iv) semicircular edge cracks.

6.15 Name and define K, K_C, and K_{IC}. Explain the differences and the conditions under which each parameter applies. State the units for each parameter.

6.16 Starting with Eq. 6-42, derive the expression for the plastic zone radius. Why is the cross section of the plastic zone not really a circle?

6.17 What must be true about the dimensions of a plate in order for plane-strain conditions to apply?

6.18 What can be learned by comparing the relative size of the shear lips and the flat surface of a metal failure?

6.19 Sketch an adhesive joint, labeling the adherends and the adhesive. Then sketch the same joint as it would look if it had been damaged in (a) an adhesive failure, (b) a cohesive failure, and (c) a mixed mode of failure.

Practice

6.20 Use the Griffith analysis to calculate the critical stress required for the propagation of an edge crack of length 0.05 mm in a thin plate of soda-lime glass, assuming that the specific surface energy of this glass is 0.30 J/m^2.

6.21 Two sets of Charpy specimens from two different materials (*X* and *Y*) and of different sizes were tested over a wide temperature range. Unfortunately, the pictures were not labeled, so we do not know from which samples they were obtained.

    a. If photographs *a* and *b* (see below) were both taken from Material *X*, speculate on the type of material that was tested and what the relative temperatures were corresponding to photographs *a* and *b*.

    b. If photograph *a* is representative of the fracture surface in Material *Y* at all test temperatures, speculate on the identity of Material *Y*.

    c. Given your answers to the previous questions, speculate on the reason or reasons why Materials *X* and *Y* both showed a tough-to-brittle transition in the specimen sets tested.

    d. Name the fracture mechanisms shown in photographs *a* and *b*, and describe the process for their formation.

6.22 Consider a steel plate with a through-thickness edge crack like the one shown in Fig. 6.21f. The plate width (*W*) is 75 mm, and its thickness (*t*) is 12.0 mm. Furthermore, the plane-strain fracture toughness and yield strength values for this material are 80 MPa m$^{1/2}$ and 1200 MPa, respectively. If the application in which the plate is used is expected to cause a stress of 300 MPa along the axis perpendicular to the crack, would you expect failure to occur if the crack length *a* is 15 mm? Explain.

6.23 Two square steel rods were brazed end-to-end to produce a rod with dimensions 6.25 × 6.25 × 30 cm. The silver braze is 0.063 cm thick and was produced with material possessing an ultimate strength of 140 MPa in bulk form. The steel rod sections have yield and tensile strengths of 690 and 825 MPa, respectively, and a plane-strain fracture-toughness value of 83 MPa $\sqrt{m}$.

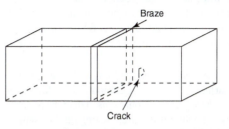

Braze

Crack

    a. If the rod/braze assembly is loaded in tension perpendicular to the joint plane, will failure occur in the braze joint or in the steel? Assume that the steel rod contains an elliptical surface flaw 1.25 cm deep and 3.75 cm wide that is oriented normal to the stress axis. Determine the stress necessary for failure.

    b. If the same rod had instead contained a through-thickness crack of depth 2.5 cm at the same crack location and orientation, determine where failure will occur and at what stress level.

6.24 Cortical bone is decidedly anisotropic in its fracture behavior. Crack propagation parallel to the long axis of a bone is much easier than perpendicular to the axis (i.e., "across" the bone). When fracture of a human tibia

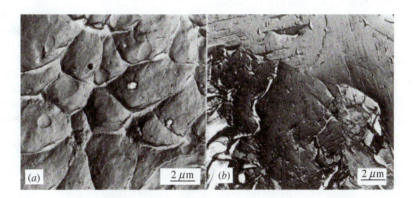

(*a*)   2 μm   (*b*)   2 μm

is measured in the parallel crack configuration, an average value of $K_{IC} = 3.95$ MPa-m$^{1/2}$ is found. Imagine that a hole is drilled through cortical bone in a direction perpendicular to the bone axis. The hole is intended to be used for insertion of a pin associated with repair of a nearby fracture. Imagine that in the hole drilling process, a *semicircular* crack is created emanating from the drilled hole aligned with the osteon cement lines (see Figure 6.22c for clarification). Assuming that the cortical bone is 3 mm thick, the modulus can be estimated as $E = 20$ GPa, and that during the surgery the worst-case tension strain that the bone will experience perpendicular to the crack is 2×10^{-3}, what is the maximum tolerable crack radius to avoid fracture at the drilled hole?

6.25 A thin-walled pressure vessel 1.25 cm thick originally contained a small semicircular flaw (radius 0.25 cm) located at the inner surface and oriented normal to the hoop stress direction. Repeated pressure cycling enabled the crack to grow larger. If the fracture toughness of the material is 88 MPa $\sqrt{m}$, the yield strength equal to 825 MPa, and the hoop stress equal to 275 MPa, would the vessel leak before it ruptured?

6.26 A 3-cm-diameter penny-shaped slag inclusion is found on the fracture surface of a very large component made of steel alloyed with Ni, Mo, and V. Could this defect have been responsible for the fracture if the stress acting on the component was 350 MPa? The only material data available are Charpy results in the transition temperature regime where impact energy values of 7–10 ft-lb were reported. Be careful of units when you consult the K_{IC}–CVN conversions.

6.27 A compact tension test specimen ($H/W = 0.6$), is designed and tested according to the ASTM E399-90 procedure. Accordingly, a Type I load versus displacement (P vs. δ) test record was obtained and a measure of the maximum load P_{max} and a critical load measurement point P_Q were determined. The specimen dimensions were determined as $W = 10$ cm, $t = 5$ cm, $a = 5$ cm, the critical load-point measurement point $P_Q = 100$ kN, and $P_{max} = 105$ kN. Assuming that all other E399 requirements regarding the establishment and sharpness of the fatigue starter crack are met, determine the critical value of stress intensity. Does it meet conditions for a valid K_{IC} test if the material yield stress is 700 MPa? If it is 350 MPa?

6.28 An infinitely large sheet is subjected to a gross stress of 350 MPa. There is a central crack $5/\pi$ cm long and the material has a yield strength of 500 MPa.

   a. Calculate the stress-intensity factor at the tip of the crack.

   b. Calculate the plastic-zone size at the crack tip.

   c. Comment upon the validity of this plastic-zone correction factor for the above case.

6.29 Is it possible to conduct a valid plane strain fracture toughness test for a CrMoV steel alloy under the following conditions: $K_{IC} = 53$ MPa$\sqrt{m}$, $\sigma_{ys} = 620$ MPa, $W = 6$ cm, and plate thickness $B = 2.5$ cm?

6.30 If the plate thickness in the previous problem were 1 cm, would the thickness be sufficient for a J_{IC} test? Assume these material properties: $E = 205$ GPa, $\nu = 0.25$.

6.31 A rod of soda-lime-silica glass is rigidly constrained at 400°K and then cooled rapidly to 300°K. Assume that $E = 70$ GPa, $\sigma_{ts} = 90$ MPa, $\alpha = 8 \times 10^{-6}$ K^{-1}, and $K_{IC} = 0.8$ MPa$\sqrt{m}$.

   a. With no visible surface damage, could you expect the rod to survive this quench?

   b. If the glass rod contained a 1-mm scratch that was oriented perpendicular to the axis of the rod, would your answer to part (a) be the same?

   c. Would failure occur if the temperature drop and the crack size were each half the values given above?

6.32 A set of double cantilever beam adhesion test specimens was fabricated with 6061-T6 aluminum alloy beams. Each beam had dimensions of $76.2 \times 12.7 \times 12.7$ mm. A bondline approximately 250 micrometers thick was created using either polyimide A or polyimide B, and a precrack of identical length was formed in each specimen.

   a. If the average $\mathcal{G}_{Ic}$ values for A and B were 19 and 63 J/m^2, respectively, what was the ratio of the critical loads?

   b. If the DCB specimen length was doubled, how would the critical load be expected to change? Explain.

Design

6.33 For the Ti-6A1-4V alloy test results given in Table 8.2, determine the sizes of the largest elliptical surface flaws ($a/2c = 0.2$) that would be stable when the design stress is 75% of σ_{ys}.

6.34 A plate of 4335 + V steel contains a semielliptical surface flaw, 0.8 mm deep and 2 mm long, that is oriented perpendicular to the design stress direction. Given that $K_{IC} = 72.5$ MPa$\sqrt{m}$, $\sigma_{ys} = 1340$ MPa, and the operating stress $= 0.4\sigma_{ys}$, determine whether the plate is safe for continued service, based on the requirement that the operative stress intensity level is below 0.5 K_{IC}.

6.35 A research group is beginning a new project to measure the fracture toughness of individual grain boundaries in Cu–Bi alloys. They plan to make tiny specimens that have one grain boundary across the center line, as shown (not to scale). The center section will be notched to ensure that the fracture occurs at the grain boundary. It will be tested in tension along its length. The initial design for the specimen calls for dimensions 100 μm long, 20 μm wide, and 0.5 μm thick.

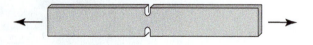

The notches will each be 3 μm long × 0.25 μm wide. The radius of curvature at the tip of each notch will be 0.125 μm. The elastic modulus of pure Cu is 110 GPa (assume isotropic behavior), and it has a yield strength of approximately 70 MPa. The fracture toughness previously measured for a Bi-embrittled grain boundary in Cu was around 2 MPa$\sqrt{\text{m}}$.

    a. Why are the researchers so confident that this sample *shape* will cause fracture at the grain boundary and not somewhere else? Please give a brief answer.

    b. Is this test likely to produce a K_{IC} value, or just a "generic" K_C value?

    c. The researchers need to choose a load cell with an appropriate range for this test. Would a load cell with a maximum force of 10 mN be sufficient to measure the load needed to fracture the doped boundary?

6.36 An unreinforced polymeric pressure vessel is constructed with a diameter $d = 0.44$ m and a length $L = 2$ m. The vessel is designed to be capable of withstanding an internal pressure of P = 7 MPa at a nominal hoop stress of 70 MPa. However, in service the vessel bursts at an internal pressure of only 3.5 MPa, and a failure investigation reveals that the fracture was initiated by a manufacturing-induced semicircular internal crack 2.5 mm in radius.

    a. Based on the original design criteria, what is the wall thickness of this pressure vessel? You may assume that it is a thin-walled vessel for this calculation, even though this may not be true.

    b. Calculate the fracture toughness (K_{Ic}) of the material used.

    c. Given the following materials to choose from, is it possible for this pressure vessel to meet a leak-before-break criterion at the original design stress without reinforcing the polymer or changing the vessel dimensions?

	PMMA	PS	PC	PET	PVC	PP
K_{IC} (MPa$\sqrt{\text{m}}$)	1.65	1.1	3.2	5.0	3.8	4.3

6.37 Table 1.12 lists thermal characteristics for several ceramic materials at room temperature, along with a thermal shock resistance parameter R'. Rank order the materials from highest shock resistance to lowest, assuming that each is in pristine condition, and is cooled very rapidly from 500 °C to 100 °C. If the materials are slightly damaged but are exposed only to thermal fluctuations of 100 °C, re-rank the likely order (you may assume that the surface energy is the same for all). For a certain application you may need to design with a single material to survive both scenarios. If you may select only one material for further investigation, and your decision must be made entirely on the basis of shock resistance, which material from the lists would you select?

Extend

6.38 Search for an image of the Pi adhesive joint and sketch it. Label it in such a way that it is clear why the design is superior to simpler T-joints.

6.39 Search for a published paper (other than one referenced in this chapter) that uses the 4-point bend adhesion test configuration. Summarize the motivation for the work, and the major finding of the paper. Provide a full reference and a copy of the paper abstract.

6.40 Find a published paper that describes a product failure analysis and that uses one of the fracture mechanics principles introduced in this chapter. Summarize the failure scenario and the major finding of the paper. Report on the fracture mechanics equation from this chapter that is used in the failure analysis. Provide a full reference and a copy of the paper abstract.

Chapter 7

Fracture Toughness

Lest the reader become too enamored with the continuum mechanics approach to fracture control, it should be noted that the profession of metallurgy predates to a considerable extent the mechanics discipline. To wit: "And Zillah she also bore Tubalcain, the forger of every cutting instrument of brass and iron" (Gen. 4:22).

7.1 SOME USEFUL GENERALITIES

As we have seen in Chapters 5 and 6, fracture toughness values can vary considerably both between the material classes and within a certain class of materials. In Fig. 5.14, the vast difference among the material classes is emphasized by the use of a log axis to depict fracture toughness. Even within a single class the variation is large (e.g., ~ 5 to $200\,\text{MPa}\sqrt{\text{m}}$ for metals). A closer look at the metals group shows that the fracture toughness of steel, alone, can change by a factor of four or more by altering the composition. However, what the figure does not depict is the substantial variation in fracture toughness that is possible even for a particular material composition. The missing component—the one that makes it possible to engineer improvements in toughness for a given type of material—is microstructure. Mother Nature has been aware of this secret for millennia, and the toughness of many natural load-bearing materials such as wood and bone is highly dependent on optimized microstructure. Engineers have also understood this connection, albeit for a much shorter period of time, and have discovered certain microstructural approaches to toughening that can be adapted to address multiple material classes. Increasingly, engineers are co-opting Mother Nature's tricks in an effort to optimize toughness of traditional structural materials in new and exciting ways. In the following sections, fundamental aspects of toughness will first be discussed, after which specific approaches to toughening of metals, ceramics/glasses, polymers, and their composites will be detailed. Also, conditions that lead to *embrittlement* (loss of toughness) will be described as a cautionary point that toughness levels can decline over time due to unintended microstructural changes.

7.1.1 Toughness and Strength

Certain trends appear when one considers the propensity for brittle fracture based on fundamental engineering properties such as yield strength, tensile strength, and tensile ductility. Recall from Chapter 1 that toughness, broadly speaking, is defined by the area under the stress–strain curve; consequently, toughness would be highest when an optimum combination of strength and ductility is developed. In Chapter 6 we saw that fracture toughness of a notched specimen depended on an optimum combination of yield strength and crack opening displacement [$V(c)$ (Eq. 6-65)]. Since $V(c)$ decreases sharply with increasing strength, a basic trend of *decreased toughness with increased strength* has been identified (see Table 6.3). It is apparent, then, that one is faced with a dilemma when selecting a material for a load-bearing application. Materials engineers can apply many strengthening procedures to enhance load-bearing capacity, but almost always to the detriment of the fracture toughness. It is also fairly easy to raise the fracture toughness level of a material; in the case of structural metals, for example, one only needs to alter the thermomechanical treatment to lower strength, and toughness naturally increases as a consequence (the solid line in Fig. 7.1). One might be satisfied with lower alloy strength in exchange for higher toughness if prevention of low-energy fracture were of paramount importance (for example, in the case of bridges or nuclear energy

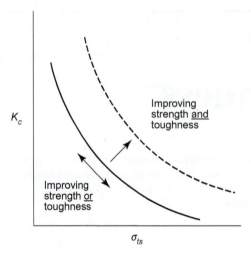

Figure 7.1 Typical inverse relation between fracture toughness and strength shown as the solid curve. Optimization of alloy properties would involve shifting the entire curve in the direction of the double arrow.

generating facilities). For other applications this approach may be impractical, or at least highly undesirable, since component size requirements would be increased as a result of the material's lower load-bearing capacity. As a result, engineers are constantly confronted with challenging decisions regarding materials selection and processing. To emphasize this point, one must recognize that factors such as energy conservation and cost containment (both of which can benefit from reduced materials use) often call for lighter or simpler structures that require materials with increased strength. And yet, safety considerations demand maintaining or enhancing toughness. What is the engineer to do?

The most desirable approach would involve shifting the curve in Fig. 7.1 up and toward the right (i.e., to the dashed line) so that the material might exhibit *both* higher strength and toughness for a given condition. It would appear that there are certain parameters that can be controlled to achieve this, since toughness levels can vary widely for the same material without any significant accompanying changes in strength (Fig. 7.2). These improvements may be realized in two general ways, by:

1. Developing optimum microstructures to maximize toughness.

2. Avoiding impurities, processing conditions, or environments that degrade toughness (i.e., that embrittle an otherwise tough material).

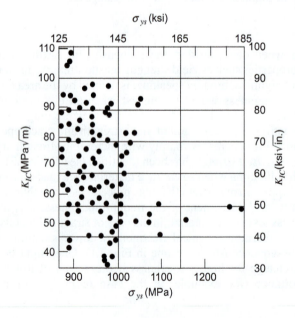

Figure 7.2 Fracture toughness data for Ti-6Al-4V alloy. A large variation in toughness but little difference in yield strength suggests different microstructures are present.[1] (From M. J. Harrigan, *Metals Engineering Quarterly,* May 1974, copyright American Society for Metals.)

7.1.2 Intrinsic Toughness

A number of mechanisms have been found acting in engineering solids that enhance toughness.[2] Before considering specific microstructural modifications that help in this regard, it is appropriate to take a closer look at certain aspects of a material's structure that have a fundamental influence on fracture resistance. These factors contribute to the *intrinsic toughness* of a material—that is, the resistance to damage mechanisms that are active in the region ahead of the crack tip at a given stress intensity. Materials with high intrinsic toughness are resistant to crack nucleation and crack propagation, and provide a consistent resistance to crack growth (i.e., show crack-length independent fracture toughness).[3] Furthermore, such materials possess excellent starting characteristics on which to build further microstructural toughening.

It has been pointed out that the baseline deformation and fracture characteristics of a given material will depend on the nature of the electron bond, the crystal structure, and the degree of order in the material.[4] The extent of brittle behavior based on these three factors is summarized in Table 7.1 for different types of materials. A key fact to keep in mind is that intrinsic brittleness is fundamentally associated with a lack of active energy dissipation mechanisms ahead of the crack tip—primarily those associated with plasticity. As such, it is worth recalling the trends connecting atomic structure and capacity for plastic deformation described in Chapters 1 and 2.

Atomic bonding provides the basis for the great differences in toughness among the different classes of materials. In general, it is seen that the more rigidly fixed a material's valence electrons are, the more brittle the material is likely to be. Since *covalent bonding* involves sharing of valence electrons between an atom and its nearest neighbors only, materials such as diamond, graphite, silicon, silicon carbide, and gallium arsenide tend to be very brittle. *Ionic bonding* is less restrictive to the location of valence electrons; the electrons are simply transferred from an electropositive anion to an electronegative cation. Furthermore, greater deformation capability is usually found in monovalent rather than multivalent ionic compounds. However, as mentioned in Chapter 2, plastic flow in ionic materials is also limited by the number of allowable slip systems that do not produce juxtaposition of like ions across the slip plane after a unit displacement. This tends to cause a strong tendency for brittle behavior at room temperature in ionic or partially ionic materials such as NaCl, metal oxides, and metal nitrides (i.e., compounds composed of elements with large differences in electronegativity). Materials with covalent, ionic, or mixed character therefore tend to have low intrinsic toughness, and little improvement is possible by altering chemical composition. *Metallic bonding*, on the other hand, provides the least restriction to valence electron movement; valence electrons are shared equally by all atoms in the solid. These materials generally have the greatest deformation capability and the greatest intrinsic toughness.

Within a material class, crystal structure can also play a significant role in differentiating tough materials from brittle materials. As seen in Table 7.1, brittle behavior is more prevalent in materials of low crystal symmetry for which slip is more difficult. In contrast, considerable plastic deformation is possible in close-packed metals of high crystal symmetry. As an example, the stabilization of the FCC form of iron rather than the BCC isomorph in a steel alloy contributes to greatly enhanced toughness. One way to accomplish this is by incorporating a certain amount of nickel to form an austenitic steel. The Peierls-Nabarro stress (a.k.a. lattice friction) level in the FCC lattice is much reduced compared to BCC (recall Chapter 2) and the low-energy cleavage fracture micromechanism is averted (recall Section 5.9.1).

Table 7.1   Relation Between Basic Structure of Solids and Their Effect on Brittle Behavior

Basic Characteristic	Increasing Tendency for Brittle Fracture		
Electron bond	Metallic	Ionic	Covalent
Crystal structure	Close-packed	Low-symmetry	Amorphous
Degree of order	Random solid solution	Short-range order	Long-range order

The ability of a given crystalline material to plastically deform (and therefore to exhibit high toughness) generally will decrease as the degree of order of atomic arrangement increases. Consequently, the addition of a solute to a metal lattice will cause greater suppression of plastic flow whenever the resulting solid solution changes from that of a random distribution to that of short-range order and finally to long-range order. A reduction in dispersoid and precipitate particle volume fraction, and an increase in particle spacing, enhances alloy ductility and represents another intrinsic toughening opportunity in crystalline metals.

At the extreme opposite end of the crystal structure trend, amorphous materials such as silicate glasses have no ability to slip via dislocation motion and therefore typically exhibit very low intrinsic toughness. Little can be done to alter this in a substantial way, although composition changes that modify the interconnectivity of a silicate glass network can have some influence (see Section 7.5.2). In an interesting departure from the simple assumption that glasses are always brittle, a mechanism called "shear banding" seen in *bulk metallic glasses* can provide an alternative plasticity mechanism to slip, so some significant intrinsic toughness can be achieved under certain loading conditions despite the lack of order. The fracture energy of amorphous metals[5] can range from $\mathscr{G} \approx 1\,\mathrm{J\,m^{-2}}$ to $> 10{,}000\,\mathrm{J\,m^{-2}}$. An analysis of many inorganic amorphous systems reveals that selection of elements with a low ratio of shear modulus to bulk modulus[i] (G/K) provides intrinsic toughening by inherently *minimizing the resistance to plastic deformation* and *maximizing the resistance to development of biaxial or triaxial tensile strain at the crack tip*. This ratio can be understood by recalling that $G/2\pi$ is one indicator of resistance to plastic yielding, as described in Chapter 2, and that K is a measure of resistance to volume expansion. This trend is depicted in Fig. 7.3 for a variety of amorphous

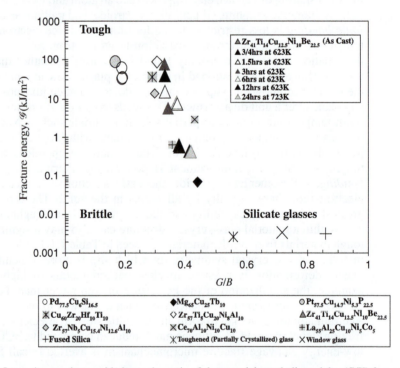

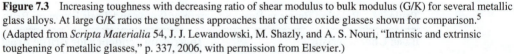

Figure 7.3 Increasing toughness with decreasing ratio of shear modulus to bulk modulus (G/K) for several metallic glass alloys. At large G/K ratios the toughness approaches that of three oxide glasses shown for comparison.[5] (Adapted from *Scripta Materialia* 54, J. J. Lewandowski, M. Shazly, and A. S. Nouri, "Intrinsic and extrinsic toughening of metallic glasses," p. 337, 2006, with permission from Elsevier.)

---

[i] Recall that in Chapter 1, the bulk modulus, K, was defined as the elastic resistance to uniform compression along three orthogonal axes. The symbol B is sometimes used instead of K to avoid confusion with the stress intensity factor or the Holloman strain hardening coefficient, both of which are also represented using K.

metal systems. Interestingly, a similar dependence on G/K was previous reported for the intrinsic fracture toughness of grain boundaries[6] and for the reduction in fracture toughness associated with the ductile-to-brittle transition in crystalline metal alloys,[7] so the fundamental significance of this metric is not limited to amorphous metals.

Finally, intrinsic toughness of polymers is somewhat more complicated to summarize using the preceding trends because multiple bond types are present (i.e., covalent and secondary) and there are no slip planes. Furthermore, temperature and strain rate can play major roles in energy-dissipating plasticity or a lack thereof. The influences that various conditions have on yielding processes and strain localization determine whether the fracture response is tough or brittle. These influences include molecular weight, crosslink density, crystallinity, plasticizers, and chain side group and chain branch chemistry. Nonetheless, the presence of secondary bonds and the network nature of polymer materials lead to low intrinsic toughness (when expressed as K_C) or moderate intrinsic toughness (when expressed as $\mathscr{G}_C$) on an absolute scale, motivating the wide use of extrinsic toughening mechanisms.

7.1.3   Extrinsic Toughening

Thus far we have seen that metals, as a class, tend to have a reasonable degree of intrinsic fracture toughness, and it has been implied that there are numerous approaches (to be discussed in the following sections) that can enhance this intrinsic toughness level. Ceramics and glasses, on the other hand, tend to have low intrinsic toughness that cannot be significantly improved by changes to chemistry or strengthening phases. The same can be said for many structural biological materials and certain polymers. Thankfully, a distinctly different set of toughening mechanisms, *extrinsic* in nature, has been identified. *Extrinsic toughening* occurs primarily by reducing the crack driving force (the stress intensity) rather than by dissipating energy. Perhaps surprisingly, many extrinsic mechanisms (summarized schematically in Fig. 7.4) actually act in the wake of the crack rather than at the crack tip. Each mechanism in some fashion "shields the crack tip" from the full impact of the macroscopic crack driving force (e.g., the applied stress intensity K_I, the strain energy release rate $\mathscr{G}$, or the J integral). This effect can be quantified by defining a local crack tip driving force K_{tip} that is reduced by a shielding stress intensity K_s such that $K_{\text{tip}} = K_I - K_s$ (or for fatigue, $\Delta K_{\text{tip}} = \Delta K - K_s$ as defined in Chapters 9 and 10).[8] These mechanisms have the greatest effect on crack extension rather than initiation, and lead to R-curve behavior (crack-length dependent toughness).[3] Although they may be most critical for improving the behavior of intrinsically brittle materials, these approaches can be applied to nearly any material, increasing fracture-toughness levels and lowering fatigue crack propagation (FCP) rates (see Chapter 10). A general description of these mechanisms follows here with more detailed discussions of some material-specific models featured in later sections of this chapter. Detailed discussion of those extrinsic mechanisms that are more effective at impeding crack growth under cyclic loading than under monotonic loading will be deferred to Chapters 9 and 10.

The first class of extrinsic toughening mechanisms enhances fracture toughness through *crack deflection and/or meandering* of the crack path. Recall from Fig. 6.18 that the Mode I stress intensity K_I is a function of crack angle with respect to the primary loading axis. It easily can be seen that as a crack is deflected from a path perpendicular to the maximum tensile stress, $K_{I,\text{tip}}$ is reduced. As long as the deflected portion is small compared to the total crack length, $K_{I,\text{tip}}$ will not go to zero even for a 90° deflection, but reductions on the order of 10–50% are possible.[9] Deflection can be accomplished either by interaction of the crack with particles or with weak planes in the material. The fracture of natural wood products is an especially noteworthy example of the latter case. When hard discrete particles in brittle matrices act to temporarily pin the advancing crack front and thereby attenuate its rate of advance, energy is also dissipated.[10,11] The crack is then forced to move around both sides of the particle before linking up behind the particle.

Several *zone-shielding* mechanisms have been identified, all of which involve inelastic deformation or dilatation (increase in volume) in a zone surrounding the crack tip and wake.

EXTRINSIC TOUGHENING MECHANISMS

1. CRACK DEFLECTION AND MEANDERING

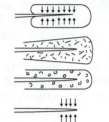

2. ZONE SHIELDING

— transformation toughening

— microcrack toughening

— crack wake plasticity

— crack field void formation

— residual stress fields

— crack tip dislocation shielding

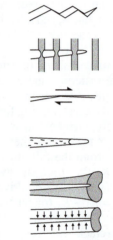

3. CONTACT SHIELDING

— wedging:

corrosion debris-induced crack closure

crack surface roughness-induced closure

— bridging:

ligament or fiber toughening

— sliding:

sliding crack surface interference

— wedging + bridging:

fluid pressure-induced crack closure

4. COMBINED ZONE AND CONTACT SHIELDING

— placticity-induced crack closure

— phase transformation-induced closure

Figure 7.4 Classes of extrinsic crack-tip shielding mechanisms in solids.[8] (Reprinted from *Materials Science and Engineering* A, 103, R. O. Ritchie, "Mechanisms of fatigue crack propagation in metals, ceramics and composites: Role of crack-tip shielding," p. 15, 1988, with permission from Elsevier.)

In one scenario—transformation toughening—this is accomplished when an unstable phase present in the microstructure transforms under stress to a more stable crystal form. This transformation causes an associated dissipation of energy, a volume expansion, and the development of a favorable compressive residual stress pattern. This generates a K_s component that partially counteracts the applied stress intensity K_I. Zone shielding also occurs through other means such as the development of a field of disconnected microcracks or microvoids that serve to relax crack-tip triaxiality and thereby reduce K_{tip}.

Another set of extrinsic toughening mechanisms involves crack surface *contact shielding* through direct or indirect physical contact of the crack faces in the crack wake. The crack wake behavior is therefore very different from that of a completely open crack. Contact shielding may, for instance, take the form of unbroken fibers in a composite material that bridge the crack, limiting the crack's ability to open (which introduces K_s), and then eventually pull out from the matrix with the expenditure of considerable amounts of energy (dissipation). As discussed in Section 10.4, wedging together of oxide debris and/or lateral sliding of adjacent regions on the fracture surface can bring about a significant reduction in fatigue crack propagation rates, well below those expected given the apparent crack driving force ΔK. In a related manner, combined zone and contact shielding can result from the wake of plasticity left behind the advancing crack tip, which also affects fatigue resistance.

7.2   INTRINSIC TOUGHNESS OF METALS AND ALLOYS

As previously discussed, intrinsic toughening mechanisms dominate the behavior of structural metals and alloys. Wide opportunities for improvements in toughness, or conversely for suffering catastrophic embrittlement, are therefore associated with changes in alloy crystal structure, chemistry, grain structure, and second phase characteristics. In the following sections, specific toughening processes associated with such changes will be discussed. Although the broad topic of environment-assisted embrittlement will be covered in detail in Chapter 8, here it is appropriate to examine two metallurgical embrittlement mechanisms that do not require exposure to an unfriendly environment at the time of loading: alterations in microstructure and/or solute redistribution as produced by improper heat treatment, and prolonged exposure to neutron irradiation. In a sense, these are also approaches to toughening insofar as they indicate processing and exposure conditions that have long-term effects on alloy behavior, and that are to be avoided if at all possible.

7.2.1   Improved Alloy Cleanliness

One of the simplest methods for improving intrinsic toughness, albeit not necessarily the easiest or cheapest to carry out, is to modify alloy chemistry to exclude certain common impurities. Although many elements are intentionally added to alloys to develop the best microstructures and associated properties, other "tramp" elements serve no such useful purpose and are, in fact, often very deleterious. For example, we see from Fig. 7.5 that small amounts of oxygen have a severe embrittling effect on the fracture toughness of diffusion-bonded Ti-6AL-4V alloy.[12] Also, hydrogen in solid solution is known to produce hydrogen embrittlement in a number of high-strength alloys and their weldments. The latter problem is examined in Chapter 8. For the moment we focus on those elements that contribute to the formation of undesirable second phases that serve as crack nucleation sites. Edelson and Baldwin[13] demonstrated convincingly that second-phase particles act to reduce alloy ductility (Fig. 7.6). The severe effect of sulfide inclusions on toughness in steel is shown in Fig. 7.7 , where the Charpy V-notch shelf energy drops appreciably as sulfur content increases. Since the yield strength of this material is greater than 965 MPa, it would be interesting to compute the fracture-toughness level K_{Ic} for these alloys with different sulfur content, using the Barsom-Rolfe relation illustrated in Fig. 6.43. This will be left to the reader as an exercise. By using K_{IC} as the

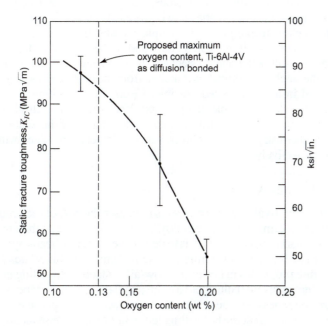

Figure 7.5   Effect of oxygen content on the fracture toughness of diffusion bonded Ti-6Al-4V.[12] (Copyright *Aviation Week & Space Technology*.)

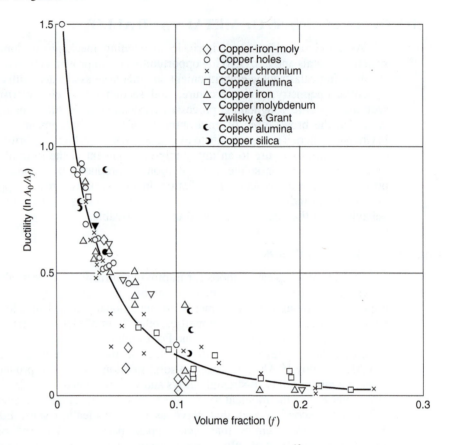

Figure 7.6   Effect of second-phase volume fraction on fracture ductility.[13] (Copyright American Society for Metals, 1962.)

measure of toughness, Birkle et al.[14] demonstrated the deleterious effect of sulfur content at all tempering temperatures in a Ni–Cr–Mo steel (Fig. 7.8).

The task, then, is to remove sulfur, phosphorous, and gaseous elements (such as hydrogen, nitrogen, and oxygen) from the melt before the alloy is processed further. This has been done with a number of sophisticated melting techniques. For example, melting in a vacuum rather than in air has contributed to a dramatic reduction in inclusion count and in the amount of trapped gases in the solidified ingot. To obtain a still better quality steel, steels are vacuum arc remelted (VAR). In this process, the electrode (the steel to be refined) is remelted from the heat generated by the arc and the molten metal is collected in a water-cooled crucible. Electroslag remelting (ESR) represents a variation of the consumable electrode remelting process: When the steel electrode is remelted, the molten metal droplets must first filter through a slag blanket floating above the molten metal pool. By carefully controlling the chemistry of the slag layer, various elements contained within the molten drops may be removed selectively.

7.2.1.1 Cleaning Up Ferrous Alloys

As one might expect, removal of tramp elements increases the cost of the product. Although these costs are justifiable in terms of improved alloy behavior, the price of the final product may not be competitive in the marketplace. The task is to devise inexpensive means by which the tramp elements are rendered more harmless. One truly excellent example of this relates to correction of inferior transverse fracture properties in hot-rolled, low-alloy steels. The rolling process severely deforms the grain structure, causing inclusions to align and flatten. This generates a considerable anisotropy in fracture toughness. In these alloys, the objectionable particles are manganese sulfide inclusions that become soft at the hot-rolling temperature and, consequently, smear out on

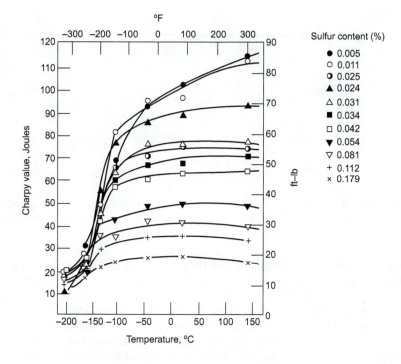

Figure 7.7   Effect of sulfur content on Charpy impact energy in steel plate (30R).[15] (Reprinted with permission from American Institute of Mining, Metallurgical, and Petroleum Engineers.)

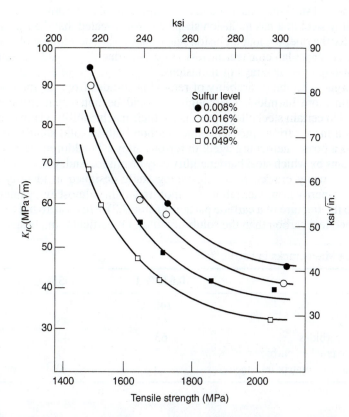

Figure 7.8   Influence of tensile strength and sulfur content on plane-strain fracture toughness of 0.45C–Ni–Cr–Mo steels. (Copyright American Society for Metals, 1966.)

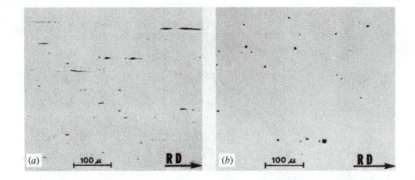

Figure 7.9   Longitudinal sections shown: (*a*) elongated manganese sulfide inclusions in quenched and tempered steel without inclusion shape control; (*b*) globular rare earth inclusions found in hot-rolled, low-alloy steel with inclusion shape control.

the rolling plane and in the rolling direction (Fig. 7.9*a*). The result: very poor transverse fracture properties. Since these alloys are used in automotive designs where components are bent in various directions for both functional and aesthetic reasons, poor transverse bending properties severely restrict the use of these materials. The objective, then, is to suppress the tendency for softening of the sulfide at the hot-rolling temperature and thereby preclude its smearing out on the rolling plane. This has been accomplished by very small additions of rare earth metals to the melt. (Certainly rare earth elements are not cheap, but the small amounts necessary result in limited additional cost per ton of steel produced.) Luyckx et al.[16] found that the manganese in the sulfide was replaced by rare earth elements (mostly cerium), which are thermodynamically favored to produce more stable and higher melting point sulfides. Since these did not deform during hot rolling, they maintained their globular shape (Fig. 7.9*b*), thus giving rise to greater isotropy in fracture properties. This is demonstrated in Fig. 7.10 by noting the rise in transverse Charpy shelf energy with increasing cerium/sulfur ratio. Plane-stress fracture toughness results from a quenched and tempered steel that has no inclusion shape control and from a rare-earth-modified, hot-rolled, low-alloy steel that has inclusion shape control revealed that the K_c anisotropy ratio was much higher for the steel without inclusion shape control.[17] The addition of rare earth metals to steel alloys also enhances lamellar tearing resistance since ductility would increase as a result of the elimination of large planar arrays of inclusions.[18]

Few would argue with the desirability of removing sulfides from the microstructure or at least making them more harmless. (Although it should be noted that sulfur is sometimes deliberately added to certain steel alloys to enhance their machinability.) However, removal of carbides presents a more serious problem, since carbon both in solid solution and in carbide particles serves as a potent hardening agent in ferrous alloys. In addition, carbon provides the most effective means by which steel hardenability may be raised. And, yet, carbides provide the nucleation sites for many cracks. In a painstaking study designed to identify the origin of microcracks in high-purity iron, McMahon[19] demonstrated that almost every microcrack found could be traced to the fracture of a carbide particle (Table 7.2). This was found true even for the alloy that contained less carbon than the solubility limit. Of particular importance in this study

Table 7.2   Initiation Sites of Surface Microcracks in Ferrite[19]

Material	0.035% C	0.035% C	0.005% C
Test temperature	−140°C	−180°C	−170°C
Total microcracks per 10^4 grains	66	43	17
Microcracks originating at cracked carbides	63	42	12
Microcracks probably originating at cracked carbides	3	1	4
Microcracks possibly originating at twin-matrix interface	0	0	1

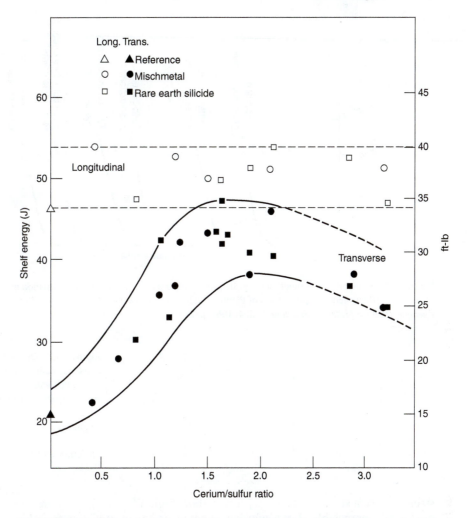

Figure 7.10 Relation between Charpy shelf energy and cerium/sulfur ratio in longitudinal and transverse oriented half-sized impact specimens of VAN-80 steel. (Copyright American Society for Metals, 1970.)

was a critical evaluation of the relative importance of various mechanisms of microcrack nucleation proposed by a number of investigators.

As we see from Fig. 7.11, McMahon found that cracks were much more likely to occur at embrittled grain boundaries and brittle second-phase particles than as a result of twin and/or slip band interactions. Consequently, although dislocation models proposed to account for microcrack formation, such as those shown in Fig. 7.12 , appear to be valid for certain ionic materials,[20] they are precluded by other crack nucleation events in metals. Indeed, McMahon stated that "there appears to be no direct evidence of initiation of cleavage by slip band blockage in metals."[19]

Because of the negative side effects of carbon solid solution and carbide strengthening in ferrous alloys, attempts have been made to develop alloys that derive their strength instead by precipitation-hardening processes involving various intermetallic compounds.[ii] Maraging

---

[ii] By contrast, a review by Lesuer et al. discussed the development of low-alloy plain carbon hypereutectoid steels containing 1–2.1% carbon; these high-carbon alloys exhibit high strength (800–1500 MPa) and generally good ductility (2.2–25%). This attractive combination of properties arises from the development of microstructures containing submicron-sized equiaxed ferrite grains and 15–32 v/o uniformly distributed spheroidized carbide particles. Of further interest, these fine-grained microstructures are amenable to net-shape forming via superplastic methods (recall Chapter 4). (See the review by D. R. Lesuer, C. K. Syn, A. Goldberg, J. Wadsworth, and O. D. Sherby, *J. Metals* 45 (8), 40 (1993)).

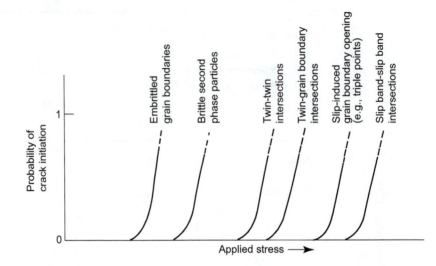

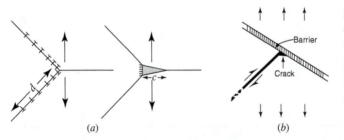

Figure 7.11 Probability as a function of applied stress that a particular microcrack formation mechanism will be operative.[19] (Reprinted with permission of Plenum Publishing Corporation, from C. J. McMahon, *Fundamental Phenomena in the Materials Sciences,* Vol. 4, 1967, p. 247.)

Figure 7.12 Dislocation models for crack nucleation: (*a*) Cottrell model.[21] (Reprinted with permission from American Institute of Mining, Metallurgical and Petroleum Engineers), (*b*) Zener model.[22] (Reprinted from *Fracturing of Metals,* © American Society for Metals, 1948.)

steels represent such a class of very low-carbon, high-alloy steels; they are soft upon quenching but harden appreciably after a subsequent aging treatment. It is seen from Fig. 7.13 that for all strength levels the toughness of maraging steels is superior to that of AISI 4340 steel, a conventional quenched and tempered steel. The chemistry of AISI 4340 steel and a typical maraging steel is given in Table 7.3 . It is felt that the lower carbon levels in maraging steels are partly responsible for their improved toughness and resistance to hydrogen, neutron, and temper embrittlement (see Section 7.8). Additional factors are discussed in Section 7.4.

7.2.1.2 Cleaning Up Aluminum Alloys

Striking improvements in the fracture toughness of aluminum alloys have also been achieved by eliminating undesirable second-phase particles. Since precipitation hardening is achieved by dislocation interaction with closely spaced submicron-sized particles, the very large, dark inclusions (e.g., Al_7Cu_2Fe, $(Fe, Mn) Al_6$, and Mg_2Si), or secondary micro-constituents (depending on your point of view), seen in Fig. 7.14 provide no strengthening increment. Instead, they provide sites for early crack nucleation.

Piper et al.[23] conducted an exhaustive study to determine how various elements affected the strength and toughness properties in 7178-T6 aluminum alloy.[iii] Their results are summarized in Table 7.4 . It is seen that some elements (copper and magnesium) provide a solid solution strengthening component to alloy strength, while zinc and magnesium contribute a

---

[iii] 7178 aluminum alloy: 1.6–2.4 Cu, 0.70 max. Fe, 0.50 max Si, 0.30 Mn, 2.4–3.1 Mg, 6.3–7.3 Zn, 0.18–0.40 Cr, 0.20 max Ti.

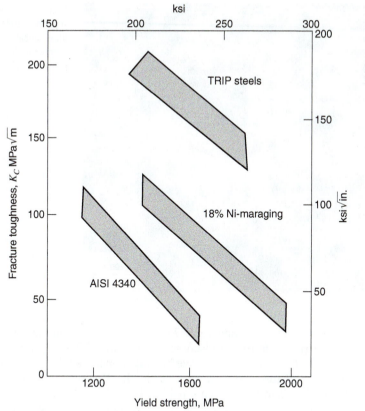

Figure 7.13 Fracture-toughness–tensile-strength behavior in AISI 4340, 18% Ni maraging, and TRIP steels.[15] (Reprinted with permission from V. F. Zackay and Elsevier Sequoia S. A.)

Table 7.3   Nominal Chemistry of Typical High-Strength Steels

| | Composition | | | | | | | | |
Material	C	Ni	Cr	Mo	Si	Mn	Co	Ti	Al
AISI 4340	0.40	1.65–2.00	0.70–0.90	0.20–0.30	0.20–0.35	0.60–0.80	—	—	—
Maraging steel	0.03 max.	18	—	5	0.20 max.	(0.20 max.)	8	0.4	0.1

precipitation-hardening increment. By comparing data from other investigators, Piper et al.[23] determined that strengthening of this alloy caused an expected reduction in fracture toughness. After examining strength and fracture-toughness data in 18 different alloys, all reasonably close to the composition of 7178, they were able to isolate the strength-toughness relation for the major alloying additions. Zinc was found to degrade K_C less for a given strength increment than that associated with the average response of the alloy. This would indicate that zinc is a desirable alloy-strengthening addition. Although yield-strength increments associated with copper and magnesium produced about average degradation in K_C, iron was found to degrade fracture toughness by the greatest amount ($3\frac{1}{2}$ times that associated with zinc additions). As expected, reduction in iron content brought about a significant improvement in alloy toughness and an associated reduction in the number of insoluble large particles. More recent results have confirmed the deleterious effect of iron *and* silicon content on fracture toughness (Fig. 7.15).[24–27] This has led to the development of alloys possessing the same general chemistry as previous ones with the exception that iron and silicon contents are kept to an absolute minimum. Examples of these newer materials include 2124 (the counterpart of 2024) and 7475 (the counterpart of 7075), which have the same strength as the older alloys but

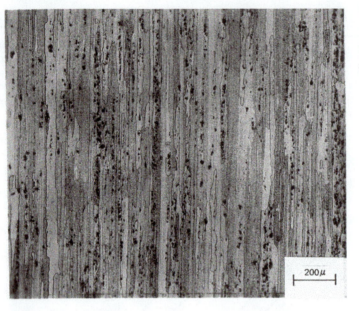

Figure 7.14 Metallographic section in 2024-T3 aluminum alloy revealing typically large number of Al_7Cu_2Fe second-phase particles.

enhanced toughness (see Fig. 7.15 and Table 7.8). Further significant improvements in fracture toughness, durability (i.e., safe-life, as discussed in Section 9.2.2)), and improved fatigue crack propagation resistance (see Chapter 10) have been demonstrated for alloy 2524-T3[28,29]; this alloy was chosen as the fuselage skin material in the Boeing 777 passenger airplane. At comparable strength levels to the conventional 2024-T3 alloy, the superior fracture toughness and fatigue resistance of alloy 2524 provide weight saving, longer service lifetime, and longer periods between inspections. These enhanced mechanical properties are derived in part by the very low Fe + Si content ($\leq$ 0.2%); in addition, precise control of alloy composition levels (especially for Cu and Mg) and carefully controlled solution treatment, and subsequent thermal mechanical processing, produce a microstructure that contains roughly half the volume fraction of Al_2CuMg and Al_2Cu particles than found in the conventional 2024 alloy.[28]

The role of inclusions in initiating microvoids in a wide variety of aluminum alloys has been examined by Broek,[30] who showed that microvoid dimple size was directly related to inclusion spacing (Fig. 7.16). Large particles that fractured at low stress levels allowed for considerable void growth prior to final failure, but smaller particles nucleated and grew spontaneously to failure. Using an analysis similar to that proposed by McClintock[31] and supported by the work of Edelson and Baldwin[13] (Fig. 7.6), Broek suggested that the fracture strain was related to some

Table 7.4 Function of Various Elements in 7178-T6 Aluminum Alloy[23]

Element	Function
Zinc	Found in Guinier-Preston zones and subsequently found in $MgZn_2$ precipitates. Element acts as precipitation-hardening agent.
Magnesium	Some Mg_2Si formation, but mostly found in $MgZn_2$ precipitates and in solid solution.
Copper	Exists in solid solution, in $CuAl_2$ and Cu–Al–Mg–type precipitates, and in Al_7Cu_2Fe intermetallic compounds.
Iron	Initially reacts to form Al–Fe–Si intermetallic compounds. Copper later replaces Si to form Al_7Cu_2Fe (the large black particles seen in Fig. 7.14).
Silicon	Initially reacts to form Al–Fe–Si compound prior to being replaced by Cu. Also forms Mg_2Si.
Manganese	Exact role not clear.
Chromium	Combines with Al and/or Mg to form fine precipitates, which serve to grain refine.

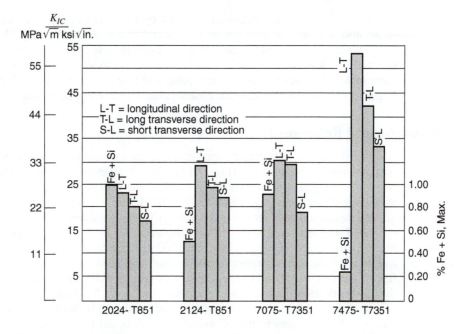

Figure 7.15   High-purity metal (low iron and silicon) and special processing techniques used to optimize toughness of 2xxx and 7xxx aluminum alloys.[26] (From R. Seng and E. Spuhler, *Metal Progress,* March 1975, copyright American Society for Metals.)

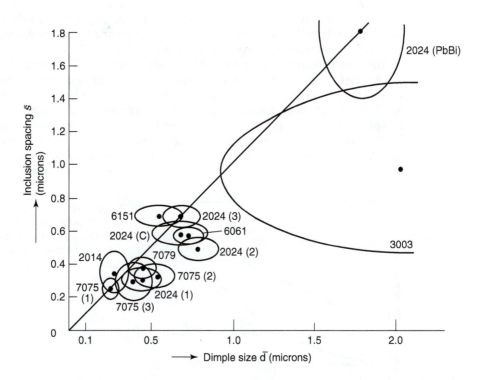

Figure 7.16   Observed relation between microvoid size and inclusion spacing. Numbers represent aluminum alloy designations; ellipses indicate scatter.[30] (Reprinted with permission from D. Broek, *Engineering Fracture Mechanics,* 1973, Pergamon Press.)

function of the volume fraction of particles or voids. Hence, $\varepsilon_f \propto f(1/V)$, where $\varepsilon_f =$ fracture strain and $V =$ volume fraction of the second phase. Consequently, the toughness of these alloys would be expected to rise with decreasing particle content. Therefore, one would predict that had Broek examined low iron and silicon alloys in his investigation, he would have found these materials to reveal larger microvoids and fracture strains. Indeed, Kaufman[25] confirmed larger microvoids in the tougher, cleaner aluminum alloys he examined.

7.2.2 Microstructural Refinement

Microstructural refinement represents a unique opportunity by which a metal may be *both* strengthened and toughened (Fig. 7.17). Boundary strengthening is therefore a particularly attractive strengthening mechanism in view of the generally observed inverse relation between strength and toughness (e.g., Figs. 7.1, 7.8, 7.13). The toughness and strength superiority of fine-grained metals has been recognized for many years, as evidenced by the well-accepted view that quenched and tempered steel alloy microstructures are superior to those associated with the normalizing process. (Quenched and tempered steels contain the finer transformation products, such as lower bainite and martensite, while normalizing produces coarser aggregates of proeutectoid ferrite and pearlite.) One beneficial effect of grain refinement is revealed by a reduction in the ductile–brittle transition temperature, as shown in Fig. 7.18. In addition to illustrating the beneficial effect of grain refinement on transition temperature, this figure reveals a shift to even lower transition temperatures in the "Controlled" as opposed to the "Standard" specimens. Kapadia et al.[32] demonstrated that the superior behavior of the controlled group of samples was attributable to enhanced delamination and associated stress relaxation in divider-type samples (see Section 7.3), which resulted from a thermomechanical treatment designed to accentuate mechanical fibering. More recent data reveal improvements also in K_{IC} levels with reduced grain size. Mravic and Smith[33] reported that a 0.3C–0.9Mn–3.2Ni–1.8Cr–0.8Mo steel produced by multiple-cycle rapid austenitizing with a prior austenite Ultra Fine Grain size (UFG) of ASTM No. 15 ($\sim$2-μm average grain "diameter") exhibited K_{IC} values in the range of 100 to 110 MPa$\sqrt{\text{m}}$ at a strength level of 1930 to 2000 MPa. By comparison, for the same strength level, 4340 steel with a conventional grain size of about ASTM No. 7 ($\sim$32 μm) exhibits a K_{IC} value of about 55 MPa$\sqrt{\text{m}}$. Further exploring the UFG regime of roughly 100–5000 nm, Song et al. have shown that a 0.2%C–Mn steel, processed by large strain warm deformation and subsequent annealing, shows a strong coupling between grain size, strength, and toughness.[34] A refinement from 6.8 μm to 1.3 μm average grain diameter resulted in an increase in yield strength, a decrease of the DBTT by approximately 40 °C, and an increase of the Charpy lower shelf energy from 0.5 J to 2 J. The Charpy upper shelf energy was also seen to decline, although to a lesser extent, due to the introduction of delaminations associated with planar arrays of cementite particles.

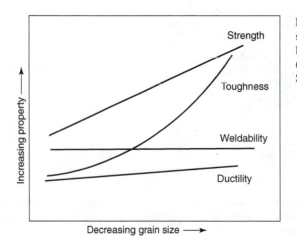

Figure 7.17 Simultaneous improvement in alloy strength and toughness with decreasing grain size. Ductility and weldability are not impaired. (Reprinted with permission from American Society of Agricultural Engineers.[35])

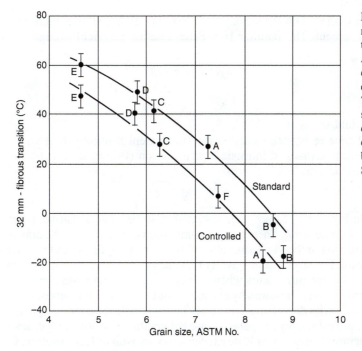

Figure 7.18 Effect of grain refinement in reducing transition temperature in hot-rolled steel. The ASTM grain size[iv] range of 4.5–9 corresponds approximately to an equivalent average diameter range of 75–16 μm. The "Controlled" specimens exhibited more delaminations in a crack "divider" configuration to account for superior behavior.[32] (Copyright American Society for Metals, 1962.)

To explain the beneficial role of structural refinement on toughness it may be argued that a microcrack will be stopped by an effective barrier (the grain boundary) more often the finer the grain size. As a result, the crack is forced to reinitiate repeatedly, and considerable energy is expended as it alters direction in search of the most likely propagation plane in the contiguous grain. Recall from Chapter 5 that this twisting of the crack front at the boundary gives rise to "river patterns" on cleavage fracture surfaces. One may argue, too, that fine-grained structures produce smaller potential flaws, thereby increasing the stress necessary for fracture (Eq. 6-6).

A number of investigators have attempted to describe the role of grain size in cleavage fracture for materials that undergo a temperature-sensitive fracture mechanism transition (see Chapter 8). Cottrell[21,36] and Petch[37] used dislocation theory to independently develop similar relations that could account for the effect temperature and various metallurgical factors have on the likelihood for cleavage failure. By using dislocation models and analyses analogous to those discussed previously with respect to Eqs. 4-8 to 4-11, they found that the fracture stress could be given by

$$\sigma_f \approx \frac{4G\gamma_m}{k_y} d^{-1/2} \tag{7-1}$$

where

$\sigma_f =$ fracture stress
$G =$ shear modulus
$\gamma_m =$ plastic work done around a crack as it moves through the crystal
$k_y =$ dislocation locking term from Hall-Petch relation (Eq. 7-2)
$d =$ grain size

[iv] The ASTM grain size number is a function of the number of grains per square inch measured using an image at a magnification of 100x. Thus a small grain size number corresponds to a population of large grains; increasing grain size number indicates decreasing average grain diameter. See ASTM Standard E 112 for details of the recommended measurement and calculation procedures.

The increase in σ_f with decreasing grain size parallels a similar increase in yield strength with grain refinement. The familiar Hall-Petch relation for yield strength is given by

$$\sigma_{ys} = \sigma_i + k_y d^{-1/2} \tag{7-2}$$

where

$\sigma_{ys} =$ yield strength
$\sigma_i =$ lattice resistance to dislocation movement resulting from various strengthening mechanisms and intrinsic lattice friction (Peierls stress)
$k_y =$ dislocation locking term
$d =$ grain size

As seen in Fig. 7.19 , Low[38] demonstrated σ_f to be more sensitive to grain size than the associated yield strength σ_{ys}. There are some important implications to be drawn from these data. First, the intersection of the yield-strength and fracture-strength curves represents a transition in material response. For large grains (greater than the critical size), failure must await the onset of plastic flow; hence, fracture occurs when $\sigma = \sigma_{ys} = \sigma_f$. For grains smaller than the critical size, yielding occurs first and is followed by eventual failure after a certain amount of plastic flow—the amount increasing with decreasing grain size. The latter situation reflects greater toughness with an increasing ratio σ_f/σ_{ys}. Since σ_f and σ_{ys} are temperature-sensitive properties, the critical grain size for the fracture transition would be expected to vary with test temperature. Consequently, the transition temperature is shown to decrease strongly with decreasing grain size (Fig. 7.20). As such, grain refinement serves to increase yield strength (Eq. 7-2) and fracture strength (Eq. 7-1), while lowering the ductile–brittle transition temperature (Fig. 7.20).

The significance of the terms in Eq. 7-1 has been treated at greater length by Tetelman and McEvily.[39] They argue that γ_m should increase with an increasing number of unpinned dislocation sources, temperature, and decreasing crack velocity. Obviously, the more dislocations that can be generated near the crack tip the more blunting can take place and the tougher the material will be. However, when these sources are pinned by solute interstitials, such as nitrogen and carbon in the case of steel alloys, or are highly immobile, as in ionic or

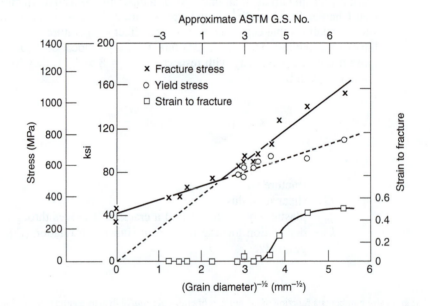

Figure 7.19   Yield and fracture strength and fracture strain dependence on grain size in low-carbon steel at $-196\,°C$.[38] (Reprinted from *Relation of Properties to Micro-structure*, copyright American Society for Metals, 1954.)

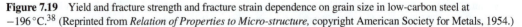

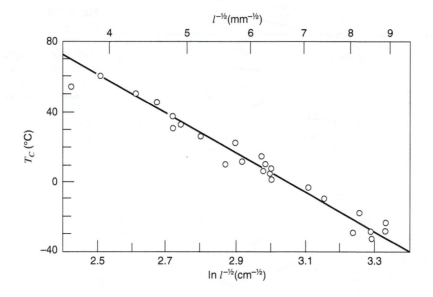

Figure 7.20 Dependence of transition temperature on grain size.[37] (Reprinted with permission from MIT Press.)

covalent materials, because of a high Peierls stress, y_m and σ_f are reduced. The beneficial effect of increased test temperature may be traced to a reduction in the Peierls stress and an increase in dislocation velocity. As we saw in Chapter 3, dislocation velocity was found to depend on the applied shear stress.

$$v = \left(\frac{\tau}{D}\right)^m \tag{3-19}$$

where

$$\begin{aligned}
v &= \text{dislocation velocity} \\
\tau &= \text{applied shear stress} \\
D, m &= \text{material properties}
\end{aligned}$$

With increasing temperature, D decreases so that for the same applied stress, dislocation velocity will increase, thereby enabling dislocations to move more rapidly to blunt the crack tip. In related fashion, a slower crack velocity will provide more time for dislocations to glide to the crack tip to produce blunting. In short, anything that enhances the number of mobile dislocations, their mobility and speed, and the time allowed for such movement will increase γ_m and σ_f and contribute to improved toughness. By comparing Eqs. 7-1 and 7-2, it should be noted that strengthening mechanisms (such as solid solution strengthening, precipitation hardening, dispersion hardening, and strain hardening) that restrict the number of free dislocations and their mobility contribute toward increasing σ_i while at the same time reducing the magnitude of γ_m. Therefore, attempts to increase yield strength by increasing σ_i are counterproductive, since γ_m and σ_f decrease. Likewise, k_y can be adjusted to improve σ_f or σ_{ys} but only at the expense of the other. Enhanced dislocation locking will increase σ_{ys} but will decrease σ_f directly (Eq. 7-1) and indirectly (since the number of mobile dislocations and γ_m decrease.). In this regard, the more brittle nature of nitrogen-bearing steel is attributed to its stronger dislocation locking character.

We may summarize this discussion by stating that the only way to improve σ_{ys}, σ_f, and toughness simultaneously is not by changing γ_m, σ_i, or k_y, but rather by decreasing grain size. By using the transition temperature as a measure of toughness (higher toughness corresponding to a lower transition temperature) the diagrams in Fig. 7.21 illustrate that only by grain refinement can you have both high yield strength and toughness. Equations 7-1 and 7-2 are instructive in identifying

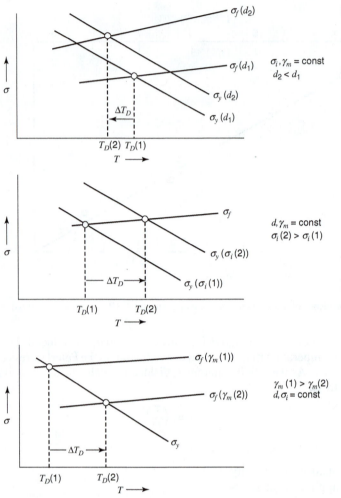

Figure 7.21 Diagrams showing effect of γ_m, σ_y, and grain size on σ_{ys}, $\sigma_{f'}$, and transition temperature. Only grain refinement produces simultaneous increase in σ_{ys}, σ_{f}, < and reduction in transition temperature. (Reprinted with permission from John Wiley & Sons, Inc.)

those parameters that affect the temperature of the fracture mechanism transition (from void coalescence to cleavage). However, the reader should recognize that these particular relations are not applicable for materials that do not cleave but which, however, may undergo a stress-state-controlled fracture energy transition.

7.3 TOUGHENING OF METALS AND ALLOYS THROUGH MICROSTRUCTURAL ANISOTROPY

As there are so many useful approaches to controlling the intrinsic toughness of metals, the only extrinsic approach to metal alloy toughness improvement that will be discussed in detail concerns the means by which cracks are deflected from their normal plane and direction of growth. Such crack deflections can occur at grain boundaries, flow lines, and inclusions that are aligned parallel to a particular processing direction. Regardless of the microstructural detail that causes the crack deflection, the behavior with regard to crack growth is similar.

7.3.1 Mechanical Fibering

To better understand the origin of microstructural alignment, and the associated mechanical anisotropy, we begin by presuming that we have taken a cube of equiaxed polycrystalline material and changed its shape by some mechanical process such as rolling, drawing, or

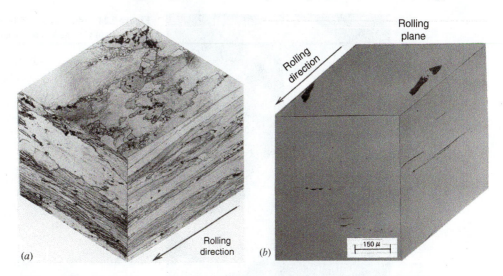

Figure 7.22   (*a*) Photomicrograph revealing mechanical fibering associated with rolling of 7075-T651 aluminum plate. (Courtesy J. Staley, Alcoa Aluminum Company.) (*b*) Alignment of manganese sulfide inclusions on rolling plane in hot-rolled steel plate. (After Heiser and Hertzberg[42]; reprinted with permission of the American Society of Mechanical Engineers.)

swaging (a combination of drawing and twisting). By the principle of similitude, the conversion of the cube into a thin plate or cylinder should be reflected by a change in the size and shape of the grains within the solid. In the case of a rolled plate, the originally-equiaxed grains should be flattened and spread out, as shown in Fig. 7.22a. A drawing operation will convert our reference cube into a long, thin wire or rod. In this case, the grains are found to be sausage-shaped and elongated in the drawing direction. In a transverse section normal to the rod axis the grains should appear equiaxed, but an orthogonal section would show highly elongated grains parallel to the rod axis. For very large draw ratios in BCC metals, however, the grains take on a ribbon-like appearance, because of the nature of the deformation process in the BCC lattice.[40] When a metal is swaged, the elongated grain structure along the rod axis is maintained (Fig. 7.23a), and the transverse section reveals a beautiful spiral nebula pattern, reflecting the twisting action of the rotating dies during the swaging process[41] (Fig. 7.23b).

The alignment of the grain structure in the direction of mechanical working is known as *mechanical fibering* and is exhibited most dramatically in forged products such as the one shown in Fig. 7.24 . Here the grains have been molded to parallel the contour of the forging dies. Engineers have found that the fracture resistance of a forged component can be enhanced

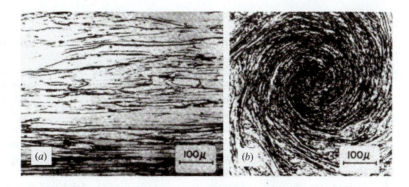

Figure 7.23   Longitudinal and transverse sections of swaged tungsten wire reduced by 87%. (After Peck and Thomas[41]; reprinted with permission of the Metallurgical Society of AIME.)

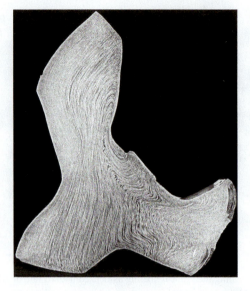

Figure 7.24 Flow lines readily visible in forged component. (Courtesy of George Vander Voort, Car Tech Corp.)

considerably when the forging *flow lines* are oriented parallel to the major tensile stress axis and normal to the path of a potential crack. As such, forged parts are considered to be superior to comparable castings because of the benefits derived from the deformation-induced microstructural anisotropy. Of course, when a forged product is used improperly, the flow lines act as readily available paths for easy crack propagation.

The reader should note that mechanical fibering involves not only alignment of grains but also alignment of inclusions. For example, in standard steel-making practice, the hot-rolling temperature for billet breakdown exceeds the softening point for manganese sulfide inclusions commonly found in most steels. Consequently, these inclusions are strung out in the rolling direction and flattened in the rolling plane, as shown in Fig. 7.22*b*. The deleterious nature of these aligned inclusions relative to the fracture properties of steel plate has already been discussed in Section 7.2.1.1, along with procedures aimed at inclusion shape control.

A similar layered microstructure can be created directly by bonding multiple layers of metal together to create a *laminated metal composite*. In a review of such materials, Wadsworth and Leseur[43] report that one of the earliest demonstrations of the impact toughness of a layered metal may be found in "The Iliad of Homer." In this epic tale, Achilles' shield stops the spear of Aeneas thanks to its special construction of multiple layers (two bronze, two tin, and one gold). Modern metallic laminates show a similar toughness and resistance to penetration by projectiles, albeit using much tougher constituent materials than were available in Achilles' day.[43]

Microstructural Toughening Case Study 7.1: The *Titanic*

Before concluding these remarks concerning fibered microstructures, the reader is reminded of mention made in Section 5.1 regarding the rapid sinking of the passenger ship *Titanic*. Here, Foecke[44,45] theorized that the impingement of the iceberg against the stern of the vessel caused rivet heads to pop off, thus opening a number of ship plate seams along the side of the vessel; water was then able to gush into the ship's interior, thereby leading to its ultimate and rapid sinking. What was wrong with these rivets? Before answering this question, we need a mini-review of rivet metallurgy, circa 1910. At that time, rivet materials were transitioning from being manufactured from wrought iron to steel. Indeed, the *Titanic*'s construction included both types of rivets, though wrought iron rivets were used in the ship's location where the iceberg hit the vessel. Accordingly, we will focus attention only on the wrought iron variety. This material is made in small batches by stirring a puddle of iron and iron silicate slag, the latter being retained from the melting process. The slag is designed to "pick up" and retain iron impurities such as C, Si, P, and Mn with carbon being burned off as CO and CO_2. Once solidified, optimal wrought iron would possess relatively pure iron and roughly 2–2.5 v/o slag inclusions. Since this rivet feedstock material was

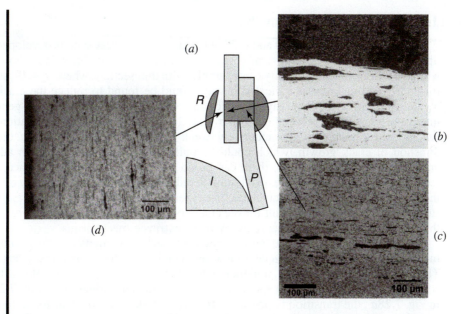

Figure 7.25   *Titanic* ship plate rivet details. (*a*) Schema shows that as the iceberg (*I*) impinges upon the ship plate (*P*), tensile stresses develop along the rivet shaft, leading to rivet failure at its hammered end (*R*); (*b*) micrograph revealing large v/o slag material (darker phase), especially at top, and slag inclusions aligned *parallel* to the rivet shaft axis; (*c*) slag particles aligned *parallel* to the rivet shaft axis; (*d*) the same rivet as in image (*c*), revealing slag particles oriented *perpendicular* to the rivet axis in the region adjacent to the on-site hammered rivet head. (Photos courtesy of T. Foecke, with permission.)

made in small batches, the quality of this product was variable at best. This was particularly true in the case of the rivets used to join the ship plates in the *Titanic* and her sister ships, *Olympic* and *Britannic*, where the construction of these three vessels, within the same time period, placed great pressure on both sound material availability and experienced construction and foundry labor.

The wrought stock material would then be hot formed into a mushroom-shaped rod, creating one of the rivet heads. At the construction site, such rods were reheated, the small end being placed in the holes of adjacent ship plates, and the rivet's tail end hand-driven to fill the hole and provide the opposite head so as to secure adjacent ship plates (see Fig. 7.25*a*). Now we have arrived at the proposed critical elements of the ship's demise. Forty-eight rivets[v] recovered from the wreckage site of the *Titanic* (some 3.9 km beneath the ocean surface) were analyzed, and nearly all of the wrought iron ones revealed their slag content to be coarsely distributed and in excess of 9 v/o, more than 3–4 times anticipated levels. Two rivets showed regions that were 40 v/o slag! (Fig. 7.25*b* shows a metallographic section from the shaft portion of a rivet with particularly high slag content.) As a result of the initial hot-forming process, these numerous slag inclusions were aligned parallel to the rivet shaft axis (Figs. 7.25*b* and 7.25*c*). Once the rod was reheated and hand-pounded to form the interior rivet head, these inclusions became reoriented parallel to the newly formed rivet head and *perpendicular to the rivet shaft axis* (see Fig. 7.25*d*). That means that when the iceberg pressed against one of the vessel's ship plates, large tensile stresses parallel to the rivet shaft would have been experienced by the rivets that were holding together this plate and its contiguous mate. Since these tensile stresses would now be perpendicular to the aligned and numerous slag inclusions in the vicinity of the interior rivet head, this theory explains why the failure of these rivet heads was inevitable. Once several rivet heads popped off, this ship plate seam would have opened enough to allow water to penetrate the vessel wall. Several of these events took place along the side of the ship as it passed by the massive iceberg. And so, after taking on water in several locations, down she went!

---

[v] T. Foecke, private communication.

7.3.2 Internal Interfaces and Crack Growth

As we saw from Eqs. 6-20 and 6-22, a triaxial tensile stress state is developed at the crack tip when plane-strain conditions are present. Since fracture toughness was shown to increase with decreasing tensile triaxiality (for example, with thin sections where $\sigma_z \approx 0$), some potential for improved toughness is indicated if ways could be found to reduce the crack-tip-induced σ_x and/or σ_z stresses. One way to reduce the σ_x stress would involve the generation of an internally free surface perpendicular to σ_x and the direction of crack propagation. This can be accomplished by providing moderately weak interfaces perpendicular to the anticipated direction of crack growth, which could be pulled apart by the σ_x stress in advance of the crack tip (Fig. 7.26a).[46] Since there can be no stress normal to a free surface, σ_x would be reduced to zero at this interface. In addition to reducing the crack-tip triaxiality by generation of the internally free surface, the crack becomes blunted when it reaches the interface (Fig. 7.26b). Both conditions make it difficult for the crack to reinitiate in the adjacent layer with the result that toughness is improved markedly by these extrinsic mechanisms.

Embury et al.[47] conducted laboratory experiments to demonstrate the dramatic improvement in toughness arising from delamination, which can effectively arrest crack propagation. These investigators soldered together a number of thin, mild steel plates to produce a standard-sized Charpy impact specimen with an "arrester" orientation (Fig. 7.27a). As seen in Fig. 7.28a, the transition temperature for the "arrester" sample was found to be more than 130 °C lower than that exhibited by homogeneous samples of the same steel. Additional confirmation of such favorable material response was reported by Leichter[48] who observed 163-J and 326-J Charpy impact energy absorption in "arrester" laminates of high-strength titanium and maraging steel alloys, respectively. Such energies are much higher than values expected from homogeneous samples of the same materials (Fig. 6.9).

The benefits of the crack-arrester geometry have been utilized for many years in a number of component designs. For example, one steel fabricator has developed a procedure for on-site construction of large pressure vessels using a number of tightly wrapped and welded concentric shells of relatively thin steel plate. This approach, though expensive to construct, boasts several advantages. First, only a *thin* layer of corrosion-resistant (more expensive) material would be needed (if at all) to contain an aggressive fluid within the vessel, as opposed to a full thickness vessel of the more expensive alloy. Second, the tightly wrapped layers are designed to create a favorable residual compressive stress on the inner layers, thus counteracting the hoop stresses of the pressure vessel. Third, the free surfaces between the layers act as crack arresters to a crack that might otherwise penetrate the vessel thickness. Finally, the metallurgical structure of thin plates (especially for low-hardenability steels) is generally superior to that of thicker sections.

In another example of crack-arrester design, large gun tubes often contain one or more sleeves shrunk fit into the outer jacket of the tube. Here, again, the procedure was originally developed to produce a favorable residual compressive stress in the inner sleeve(s), but serves also to introduce an internal surface for possible delamination and crack arrest. In one actual case

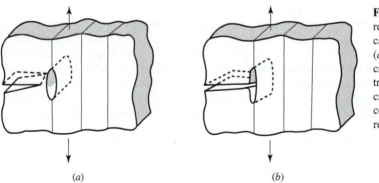

Figure 7.26 Delamination at relatively weak interface caused by σ_x stress. (a) Delamination ahead of the crack-tip reduces tensile triaxiality and (b) reduces crack-tip acuity. Both factors contribute to enhanced fracture resistance.

(a)   (b)

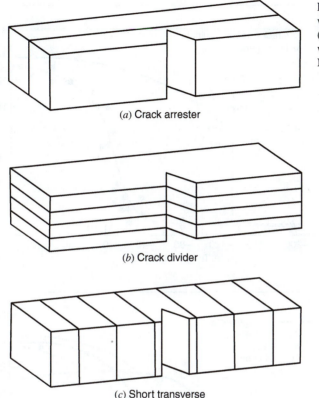

Figure 7.27   Specimens containing relatively weak interfaces: (*a*) arrester; (*b*) divider; and (*c*) short transverse configurations. (Reprinted with permission from American Society of Mechanical Engineers.)

(*a*) Crack arrester

(*b*) Crack divider

(*c*) Short transverse

history, a fatigue crack was found to have initiated at the inner bore of the sleeve, propagated to the sleeve–jacket interface, and proceeded around the interface, but *not* across the interface into the jacket itself. A similar crack-"arrester" response is found in conventional materials given thermo-mechanical treatments that produce layered microstructures. McEvily and Bush[49] showed that a Charpy specimen made from ausformed steel (warm rolled above the martensite transformation temperature) completely stopped a 325-J (240-ft-lb) impact hammer when the carbide-embrittled former austenite grain boundaries were oriented normal to the direction of crack propagation (Fig. 7.29).

Triaxiality can also be reduced by relaxing σ_z stresses brought about by delamination of interfaces positioned normal to the thickness direction. When delamination occurs, the effective thickness of the sample is reduced, since σ_z decreases to zero at each delamination interface. Consequently, the specimen acts like a series of thin-plane stress samples instead of one thick-plane strain sample. For this reason, the resulting shift in transition temperature will depend on the number of weak planes introduced in the specimen—the more planes introduced, the thinner the delaminated segments will be and the greater the tendency for plane-stress response. Embury et al.[47] conducted such tests with laminated samples in the "divider" orientation (Fig. 7.27*b*) and found the transition temperature to decrease with increasing number of weak interfaces (Fig. 7.28*b*). Note that at sufficiently low temperatures all the samples exhibited minimal toughness, indicating that even the thinnest layers were experiencing essentially plane-strain conditions at these low temperatures. Leichter[48] confirmed the beneficial character of the divider orientation in raising toughness. For example, the fracture toughness of a titanium alloy was improved six- to sevenfold for a laminated sample over that of the full-thickness sample made from the same material. It is interesting to note that the homogeneous samples exhibited an extensive amount of flat fracture, while each delaminated layer showed 100% full shear. As we saw in Chapter 5 this difference in fracture mode appearance also reflects the improvement in fracture toughness.

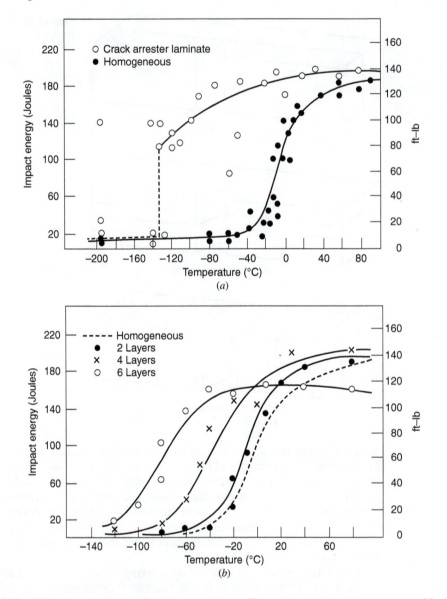

Figure 7.28 Effect of (a) arrester and (b) divider geometry on Charpy impact energy temperature transition.[47] (Reprinted with permission from American Institute of Mining, Metallurgical and Petroleum Engineers.)

The strength of the interface represents an important parameter in the delamination-induced toughening process. On one hand, the interface should not be so weak that the sample slides apart like a deck of playing cards. On the other hand, if the interface is too strong, delamination will not occur. In considering this point, Kaufman[50] demonstrated that the toughness of multilayered, adhesive-bonded panels of 7075-T6 aluminum alloy was significantly greater than that shown by homogeneous samples with the same total thickness. Conversely, when similar multilayered panels of the same material were metallurgically bonded (resulting in strong interfaces), no improvement in toughness was observed over that of the homogeneous sample.

We must also consider the third possible orientation of weakened interfaces relative to the stress direction. As one might expect, fracture-toughness properties are lowest in the short-transverse orientation. It is as though all the positive increments in toughness associated with crack "arrester" and "divider" orientations are derived at the expense of short-transverse

Figure 7.29 Extensive delaminations in ausformed steel with "arrester" orientation. Specimen absorbed 325-J energy.[49] (Copyright American Society for Metals, 1962.)

properties. The spalling fracture of solids, resulting from shock wave-material interactions, often represents a short-transverse fracture event. This arises from the fact that shock waves, produced by the impact of a high-velocity projectile, bounce off the back wall of an object (e.g., armor plate), reverse direction, and return as reflected tensile waves. If the associated tensile stresses of these reflected waves are great enough, nucleation, growth, and coalescence of voids and/or cracks will occur.[51] Often, a chunk of material (i.e., spall) breaks away from the surface opposite to the impacted surface. Studies have shown that spall formation is nucleated most readily at the interfaces between inclusions and the surrounding matrix. As expected, spall formation is significantly suppressed in steels that contain finer nonmetallic inclusions as a result of an electroslag remelt refining process than in a conventionally cast steel alloy[52] (see Section 7.3). Since inclusions tend to become aligned parallel to the plane of rolling, a reduction in inclusion content should enhance spall resistance in the short-transverse plane.

A potentially dangerous condition—lamellar tearing—can develop because of the poor short-transverse properties often found in rolled plate. Consider the consequences of a large T-joint weld such as the one shown in Fig. 7.30 . After the weld is deposited, large shrinkage stresses are developed that act in the thickness direction of the bottom plate. These stresses can be large enough to cause numerous microfissures at inclusion–matrix interfaces, which were aligned during the rolling operation. Clusters of these short-transverse cracks can seriously degrade the weld joint efficiency and should be minimized if at all possible.

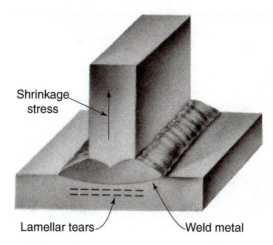

Figure 7.30 Lamellar tears generated along rolling planes as a result of weld shrinkage stresses.

Shrinkage stress

Lamellar tears   Weld metal

Table 7.5a   Plane-Strain Fracture-Toughness Anisotropy in Wrought, High-Strength Aluminum Alloys[53]

Alloy and Temper Designation	Product	$K_{IC}(MPa\sqrt{m})$		
		L-T	T-L	S-T
2014-T651	127-mm plate	22.9	22.7	20.4
7075-T651	45-mm plate	29.7	24.5	16.3
7079-T651	45-mm plate	29.7	26.3	17.8
7075-T6511	90 × 190-mm extruded bar	34.0	22.9	20.9
7178-T6511	90 × 190-mm extruded bar	25.0	17.2	15.4

Table 7.5b   Plane-Strain Fracture-Toughness Anisotropy in Wrought, High-Strength Aluminum Alloys[53]

Alloy and Temper Designation	Product	$K_{IC}(Ksi\sqrt{in.})$		
		L-T	T-L	S-T
2014-T651	5-in. plate	20.8	20.6	18.5
7075-T651	$1^3/_4$-in. plate	27.0	22.3	14.8
7079-T651	$1^3/_4$-in. plate	27.0	23.9	16.2
7075-T6511	$3^1/_2 \times 7^1/_2$-in. extruded bar	30.9	20.8	19.0
7178-T6511	$3^1/_2 \times 7^1/_2$-in. extruded bar	22.7	15.6	14.0

7.3.3 Fracture Toughness Anisotropy

Because of the anisotropy of wrought products, fracture-toughness values may be expected to vary with the type of specimen used to measure K_c or K_{IC}. This is not related to the specimen shape per se in terms of the K calibration but rather to the material anisotropy. For example, a rolled plate containing a surface flaw (arrester orientation) might be expected to exhibit somewhat higher toughness than the same material prepared in the form of an edge-notched plate, where the crack would propagate parallel to the rolling direction. To illustrate this behavior, the fracture-toughness anisotropy in a number of wrought aluminum alloys is shown in Table 7.5 with the fracture-toughness data given as a function of fracture plane orientation (first letter in code) and crack propagation (second letter in code) (Fig. 7.31). Additional data

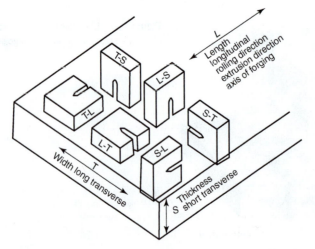

Figure 7.31   Code system for specimen orientation and crack propagation direction in plate.

EXAMPLE 7.1

Components were machined without consideration of direction from a plate of 7075-T651 aluminum alloy. If the design stress were set at 40% of the material's yield strength, would all components withstand fracture in the presence of internal 15-mm penny-shaped cracks, regardless of defect orientation?

The mechanical properties of the plate as a function of sample orientation are given in the following table.

	L-T	T-L	S-T
K_{IC} (MPa$\sqrt{m}$)	29.7	24.5	16.3
σ_{ys} (MPa)	515	510	460

To begin, we will assume that the internal circular crack in each component is oriented normal to the direction of the design stress. Since the stress intensity factor for this crack configuration is given by

$$K = \frac{2}{\pi}\sigma_d\sqrt{\pi a}$$

where σ_d = design stress = $0.4\sigma_{ys}$, and a = disc radius, therefore,

$$K = \frac{2}{\pi}(0.4)\sigma_{ys}\sqrt{\pi(.0075)}$$

By substituting yield strength values in the preceding equation, we see that the stress intensity factor level for the three orientations are 20.1, 19.9, and 18 MPa$\sqrt{m}$ for the *L-T*, *T-L*, and *S-T* orientations, respectively. Therefore, components machined in the *L-T* and *T-L* orientations could have sustained the design stress since $K < K_{IC}$, whereas the component prepared in the *S-T* orientation would have fractured abruptly because $K > K_{IC}$. From a design standpoint, it is important that the component have the highest fracture toughness values oriented parallel to the direction of greatest stress.

---

revealing fracture-toughness anisotropy in aluminum, steel, and titanium alloys are given in Table 7.8 at the end of this chapter.

7.4   OPTIMIZING TOUGHNESS OF SPECIFIC ALLOY SYSTEMS

7.4.1 Ferrous Alloys

Numerous studies have been conducted to determine which alloying elements and microstructures provide a given steel alloy with the best combination of strength and toughness. Since these structure–property correlations were established with many different properties (strength, ductility, impact energy, ductile–brittle transition temperature, and fracture toughness), it is difficult to make immediate data comparisons. There are, however, certain general statements that can be made with regard to the role of major alloying elements in optimizing mechanical properties. These are summarized in Table 7.6. In particular, the beneficial role of nickel in improving toughness and dramatically lowering the transition temperature (Fig. 7.32) has been recognized for many years and used in the development of new alloys with improved properties. For example, high-nickel steels (specifically 9% Ni steel) are generally the material of choice for cryogenic applications, such as in the construction of liquefied natural gas storage tanks. These materials have a certain amount of stable retained austenite even at temperatures as low as 77 K (–196 °C, –320 °F). Recall that austenite (FCC) has greater intrinsic toughness than

Table 7.6   Role of Major Alloying Elements in Steel Alloys

Element	Function
C	Extremely potent hardenability agent and solid solution strengthener; carbides also provide strengthening but serve to nucleate cracks.
Ni	Extremely potent toughening agent; lowers transition temperature; hardenability agent; austenite stabilizer.
Cr	Provides corrosion resistance in stainless steels; hardenability agent in quenched and tempered steels; solid solution strengthener; strong carbide former.
Mo	Hardenability agent in quenched and tempered steels; suppresses temper embrittlement; solid solution strengthener; strong carbide former.
Si	Deoxidizer; increases σ_{ys} and transition temperature when found in solid solution.
Mn	Deoxidizer; forms MnS, which precludes hot cracking caused by grain-boundary melting of FeS films; lowers transition temperature; hardenability agent.
Co	Used in maraging steels to enhance martensite formation and precipitation hardening kinetics.
Ti	Used in maraging steels for precipitation hardening; carbide and nitride former.
V	Strong carbide and nitride former.
Al	Strong deoxidizer; forms AlN, which pins grain boundaries and keeps ferrite grain size small. AlN formation also serves to remove N from solid solution, thereby lowering lattice resistance to dislocation motion and lowering transition temperature.

ferrite (BCC), and that the toughness of austenite is not strongly temperature dependent. In fact, several studies have shown an increase in toughness associated with an increase in fraction of retained austenite.[54] Not surprisingly, the cryogenic toughness of nickel steels is further improved by reductions of carbon, manganese, and the impurities phosphorus, and—in particular—sulfur. This effect can be profound: over a 15-year period in the 1970s and 1980s, improved control of composition in 9% nickel steel production heats resulted in a doubling of the Charpy notch toughness.[55]

Turning now from alloying elements to second phases, it is useful to examine the trends associated with steel transformation products. Low[57] examined the effect of typical alloy steel microconstituents on toughness and concluded that the finer ones—namely, lower bainite and martensite—provided greater fracture resistance than the coarser, high-temperature transformation products such as ferrite, pearlite, and upper bainite. (The question of structural refinement was already discussed in Section 7.2.2.)

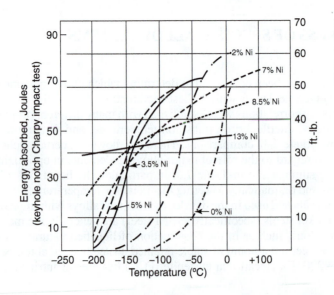

Figure 7.32   Effect of nickel content on transition temperature in steel.[56] (Reprinted with permission from the International Nickel Company, Inc., One New York Plaza, New York.)

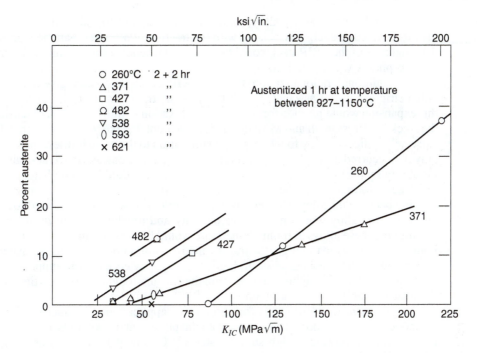

Figure 7.33   Relation between fracture toughness and retained austenite content in AFC 77 high-strength alloy. Retained austenite content varies with austenitizing and tempering temperatures.[58] (© American Society for Metals, 1968.)

Let us now also consider the fracture behavior of the parent austenite phase. As we will see, this is both a complex and intriguing task. For one thing, the stability of the austenite phase can be increased through judicious alloying so as to completely stabilize this high-temperature phase at very low cryogenic temperatures or partially stabilize it at room temperature. Low-temperature stability of austenite (γ) is highly beneficial in light of the general observation that austenitic steels are tougher than ferritic (α) or martensitic (α') steels because of the intrinsically tougher austenite FCC crystal structure. In a study of AFC 77, a high-strength steel alloy containing both martensite and austenite microconstituents, Webster[58] showed that the fracture-toughness level increased with an increasing amount of retained austenite in the microstructure (Fig. 7.33). It is believed that the retained austenite phase in this alloy serves as a crack arrester or crack blunter, since it is softer and tougher than the martensite phase. By sharp contrast, retained γ in high-carbon steels can damage overall material response when it undergoes an *ill-timed,* stress-induced transformation to untempered martensite, a much more brittle microconstituent. Several research groups and corporations have been experimenting with certain alloys that will undergo (in a carefully controlled manner) this mechanically induced phase transformation. The result has been the development of high-strength steels possessing remarkable ductility and toughness brought about by transformation-induced plasticity in the material. These TRIP[59] (an acronym for *transformation-induced plasticity*) steels compare very favorably with both quenched and tempered and maraging steels (Fig. 7.13).

How can this be? How can you transform a tough phase γ into a brittle phase α' and produce a tougher alloy? How does this crack-tip zone-shielding mechanism work (recall Fig. 7.4)? Antolovich[60] argued that a considerable amount of energy is absorbed by the system when the $\gamma \rightarrow \alpha'$ transformation takes place. Assuming for the moment that the fracture energy of γ and α' is the same, the total fracture energy of the system would be the elastic and plastic energies of fracture for each phase plus the energy required for the transformation itself. Since the fracture energy of α' is lower than that of the γ phase, the toughness of the unstable γ alloy would be greater than that of a stable γ alloy, so long as the energy of transformation more than made up the loss in fracture energy associated with the fracture of α' rather than the γ phase. Obviously, the toughness of the TRIP steel would be enhanced whenever the toughness difference between the two phases was minimized.

An additional rationalization for the TRIP effect has been given, based on the 3% volumetric expansion associated with the $\gamma \rightarrow \alpha'$ transformation. It has been argued[61] that this expansion would provide for some stress relaxation within the region of tensile triaxiality at the crack tip. Bressanelli and Moskowitz[62] pointed out that the *timing* of the transformation was extremely critical to alloy toughness. Transformation to the more brittle α' phase was beneficial only if it occurred during incipient necking. That is, if martensite formed at strains where plastic instability by necking was about to occur, the γ matrix could be strain hardened and, thereby, resist neck formation. If the $\gamma \rightarrow \alpha'$ transformation occurred at lower stress levels prior to necking because the alloy was very unstable, brittle α' would be introduced too soon, with the result that the alloy would have lower ductility and toughness. Note in this connection that prestraining these alloys at room temperature would be very detrimental. At the other extreme, if alloy stability were too high, the transformation would not occur and the material would not be provided with the enhanced strain-hardening capacity necessary to suppress plastic necking. Some success has been achieved in relating the fracture toughness of a particular TRIP alloy to the relative stability of the austenite phase.[63,64]

Employment of TRIP steels for automotive applications is attractive,[65] particularly for components such as door beams, windshield pillars, and roof rails that are already being fabricated from advanced high-strength steels.[vi] Excellent formability as compared to other advanced steels offers the automotive engineer new possibilities with regard to the design of components for reduced weight and improved structural performance. Although the TRIP process offers considerable promise, engineering usage of materials utilizing this mechanism is not widespread due to limited commercial availability and relatively high cost.

7.4.2 Nonferrous Alloys

Microstructural effects are also important when attempting to optimize the toughness of titanium-based alloys. For example, it has been found that toughness depends on the size, shape, and distribution of different phases that are present (Fig. 7.34). We see, for example, that metastable β (BCC phase) alloys possess the highest toughness with α (HCP phase) $+ \beta$ alloys generally being inferior. Furthermore, in these mixed-phase alloys, acicular α rather than equiaxed α within a β matrix is found to provide superior toughness. It is quite probable that the large amount of scatter in K_{IC} values shown in Fig. 7.2 for Ti-6Al-4V (an $\alpha + \beta$ alloy) was largely a result of variations in the microstructures just described. For further information concerning the effect of compositional and microstructural variables on the fracture behavior of titanium alloys, the reader is referred to the review by Margolin et al.[67]

Let us now reexamine the observation made in Chapter 6 (Table 6.3) that the 2024-T3 aluminum alloy possesses higher toughness than the 7075-T6 sister alloy. Although this is true when comparison is based on the *different* strength levels designated for these alloys, the 7075-T6 alloy actually is the tougher material when compared at the same strength level (Fig. 7.35).

Nock and Hunsicker[70] demonstrated that the superiority of 7000 series alloys was attributable to a relatively small amount of insoluble intermetallic phase and to a reduced precipitate size, which would be less likely to fracture (2.5 to 5.0 nm in 7075-T6 versus 50 to 100 nm in 2024-T86). Recent metallurgical studies have been concerned with optimizing the

---

[vi] An unintended consequence of the increased use of high-strength steels for automobile components is an increased difficulty in cutting open the passenger compartment by emergency personnel at a crash scene. Hydraulic shears intended to cut through ordinary steel pillars are unable to tackle high-strength steel, requiring either more powerful tools or alternative victim extraction methods. Thankfully, the use of advanced steel has also led to greater passenger protection.[66]

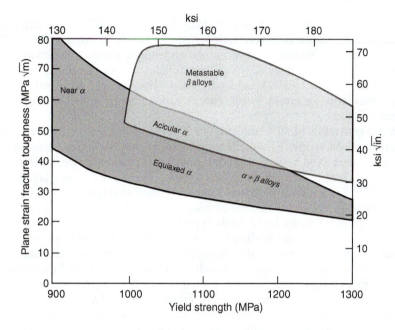

Figure 7.34 Schematic diagram showing effect of alloy strength and microconstituents on toughness in titanium alloys.[68] (Reprinted with permission from A. Rosenfield and A. J. McEvily, Jr., NATO AGARD Report 610, Dec. 1973, p. 23.)

fracture toughness, strength, and resistance to environmental attack of various aluminum alloys. For example, it has been shown[24,25,68] that while strength decreases, toughness is improved when the material is underaged, with a somewhat smaller improvement being associated with the overaged condition. However, the overaged alloy, with its greater resistance to stress corrosion cracking, is preferred, even though the toughness level is somewhat lower than that obtained in the underaged condition. Attempts are being made to combine mechanical deformation with variations in aging procedures to optimize material response.[25,71–73] Preliminary results have shown that toughness increases with decreasing size of Al_2CuMg particles and minimization of $Al_{12}Mg_2Cr$ dispersoids.[73] More studies of important aluminum alloys have been reported by Bucci.[74]

Aluminum-lithium alloys were developed, beginning in the 1950s, to take advantage of their low density and high alloy stiffness. For example, early alloys with 3.3% Li were approximately 10% lighter and 15% stiffer than conventional aluminum formulations.

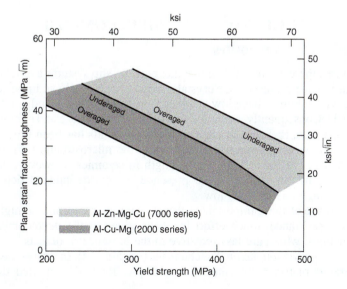

Figure 7.35 Toughness versus strength data for 2000 and 7000 series aluminum alloys revealing superior toughness in the latter at any given strength level.[68] (Reprinted with permission from A. Rosenfield and A. J. McEvily, Jr., NATO AGARD Report 610, Dec. 1973, p. 23, based on data from Develay.[69])

Unfortunately, these early versions such as alloy 2020 exhibited lower ductility, toughness, and stress-corrosion resistance than conventional Al-Cu-Mg-Zn aerospace alloys. These property deficiencies were attributable to severe, inhomogeneous planar slip deformation associated with inhomogeneous growth of Al$_3$Li (δ') precipitates[75,76] (recall Fig. 3.23a), which left soft precipitate-free zones (PFZs) around grain boundaries and subgrain boundaries. Inhomogeneous slip generates stress concentrations at grain boundaries and encourages intergranular fracture.

A second generation of Al-Li alloys (e.g., 2090) showed greatly improved performance, but sheet and plate products often exhibited a high degree of microstructural anisotropy and a strongly direction-dependent toughness. Not surprisingly, the *S-T* orientation was particularly poor (recall Section 7.3.2). Tramp element segregation (e.g., Na, K, H) also played a role in poor toughness.[77–80] For the case of several vacuum-refined experimental heats of the Al-Li 2090 alloy, fracture toughness levels increased dramatically when the Na + K content was reduced from then-commercial purity levels of 4–11 ppm Na + K to less than 1 ppm.[80]

Subsequent developments in alloy chemistry and thermomechanical treatment have largely eliminated the early problems with inhomogeneous deformation and cracking, and modern Al-Li alloys demonstrate strength and toughness properties that rival or exceed conventional Al alloys at room temperature.[81] As expected (recall Section 7.7.3), increases in toughness correlate with a shift in observed fracture micromechanisms from cleavage and intergranular separation to transgranular microvoid coalescence.[82] The increases in toughness are due largely to improvements in the homogeneity of the strengthening particles present in the microstructure (and the reduction in PFZs) that homogenize slip.[83,84] Third-generation alloys, such as 2099 and 2199, include Cu for generation of Al$_2$Cu (θ') and Al$_2$CuLi (T1), and Al$_6$CuLi$_3$ (T2) precipitates. Zr additions form coherent Al$_3$Zr (β') particles that contribute to toughening, as do Al$_{20}$Cu$_2$Mn$_3$ incoherent particles that come with Mn incorporation. Uniform precipitation of the δ' and T_1 strengthening phases is generally encouraged by the application of plastic deformation (cold work) prior to heat treatment.[83,84] In doing so, it discourages the grain boundary and subgrain boundary precipitation that leads to poor toughness.

Interestingly, very low temperature conditions can actually improve both strength and toughness properties of Al-Li alloys, making them particularly well suited for applications such as containment of cryogenic liquids (e.g., alloy 2195 in the Super Lightweight Tank version of the U.S. Space Shuttle main fuel tank).[82] This improvement in toughness has been attributed to extrinsic toughening (in the arrester and divider orientations) associated with through-thickness (*S-T*) delaminations,[3] and to the elimination of T_1 precipitation at subgrain boundaries.[82]

7.5   TOUGHNESS OF CERAMICS, GLASSES, AND THEIR COMPOSITES

7.5.1   Ceramics and Ceramic-Matrix Composites

In most instances, there are fewer than five independent slip systems available in ceramics to allow for arbitrary shape changes in the crystals (recall Chapter 2). In addition, such materials possess ionic or covalent bonds, have low symmetry crystal structures, and exhibit long-range order (see Table 7.1). Consequently, ceramics are typically very brittle and exhibit low fracture toughness. For this reason, their use in engineering components has been limited. Successful efforts to decrease the size and number of defects in the microstructure through improved fabrication procedures have led to improved strength in ceramics. However, these materials remain subject to premature failure in the presence of service-induced defects since the material's *intrinsic toughness* remains low.

To make greater use of these important materials with their exceptional high-temperature capability and wear resistance, much effort has been given to the development of tougher ceramic microstructures, which are less sensitive to the presence of defects.[85–89] This effort has concentrated on *extrinsic toughening* mechanisms (see Fig. 7.4). In many cases this requires fabrication of ceramic-matrix composites. For example, Becher[88] reported that when SiC

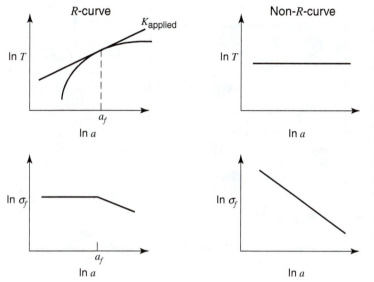

Figure 7.36 Toughness- and fracture strength-crack size dependence in R-curve and non-R-curve type ceramics. For R-curve materials, toughness increases with crack size up to a_f and fracture strength is insensitive to crack size when $a < a_f$. For non-R-curve type materials, toughness is constant and fracture strength decreases with crack size.[93] (Adapted from M. P. Harmer, H. M. Chan, and G. A. Miller, *J. Amer. Ceram. Soc.* 75 (7), 1715 (1992). Reprinted by permission.)

whiskers were added to an alumina matrix, the Weibull modulus of the material's flexural strength increased from 4.6 to 13.4. (Recall that higher values of the Weibull modulus correspond to decreased variability in the measured property.) As discussed previously, extrinsic mechanisms are responsible for the phenomenological development of R- or T-curve behavior[vii] (recall Section 6.12). For those ceramic materials that exhibit R-curve behavior as a result of crack-tip shielding phenomena, there is no single K_{IC} value to define the toughness of the material. Instead, toughness increases with increasing crack size and crack wake dimension until a steady-state fracture condition is achieved. Furthermore, for stable crack extension associated with crack lengths less than that associated with failure (i.e., a_f, where $\partial K/\partial a \leq \partial T/\partial a$), the material's fracture strength is insensitive to crack size (Fig. 7.36); accordingly, the material's structural reliability is improved. Conversely, non–R–curve materials possess a constant level of toughness and a fracture strength that decreases continuously with crack size.

Geometrical toughening involves *crack deflection* arising from crack-tip–grain boundary and/or second phase particle interactions (Figs. 7.4 and 7.26b).[87] For example, with decreasing grain boundary strength and/or increasing grain misorientation, one would expect a crack to become increasingly diverted along a grain boundary path and away from its current path along a cleavage plane within a particular grain. These perturbations in crack plane and directions reduce the local stress intensity factor and lead to a moderate improvement in toughness.[90,91] In addition, crack deflection can result from the interaction of an advancing crack front with a residual stress field, such as one generated by a thermal mismatch between the matrix and reinforcing particles. If the difference in coefficient of thermal expansion (CTE) between matrix and spherically-shaped second phase particles is positive (i.e., $\alpha_m - \alpha_p > 0$), a compressive radial stress (σ_r) is developed at the particle–matrix boundary along with a tensile tangential stress (σ_t) in the matrix; the crack is then "attracted" to the particle (Fig. 7.37a).[92] Conversely, when $\Delta\alpha < 0$, $\sigma_r > 0$ and $\sigma_t < 0$; in this instance, the crack is "rejected" and passes between the particles (Fig. 7.37b). This more tortuous crack path requires additional fracture energy and should result in greater composite toughness than that associated with crack passage through the particles.

[vii] As discussed in Section 6.12, R-curve behavior refers to the crack-dependent change in the material's resistance to fracture (R) (units of energy). For the case of metallic alloys, improved resistance to fracture is derived from the accumulation of plastic deformation. Crack instability occurs when $d\mathcal{G}/da = dR/da$. The T-curve provides an analogous display of the material's resistance to fracture with toughness (T) (units of stress intensity) plotted versus crack size. Accordingly, instability occurs when $dK/da = dT/da$.

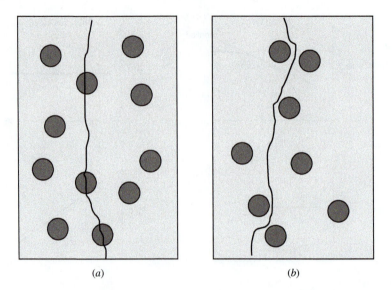

Figure 7.37   Schematic depiction of crack-path dependence on residual stress state in composites of glass reinforced with thoria spheres. (*a*) Crack is drawn toward thoria spheres when the coefficient of thermal expansion is larger in the matrix; (*b*) Crack is deflected away from thoria spheres, resulting in tortuous crack path and greater fracture energy, when the coefficient of thermal expansion is larger in the reinforcement phase. (Data from Springer Science+Business Media: *Journal of Materials Science* 3, p. 629, 1968, R. W. Davidge and T. J. Green, figure number 3.)

We now consider several *crack-tip shielding* mechanisms that contribute greatly to the toughening of ceramics and their composites. For simplicity, these are distinguished by those mechanisms that occur in the *frontal zone* (ahead of the advancing crack front) and those that act in the *crack wake* (bridging mechanisms) (Fig. 7.38).[87] Shielding zones in structural ceramics are approximately $1-1000\,\mu$m in length, whereas those in ceramic and cementitious composites are considerably larger.[87,94] Since dislocation mobility in ceramics is limited due to a large Peierls stress, dislocation cloud formation provides little toughening. Crack-tip stress fields and/or residual stresses can nucleate a cloud of microcracks at weakened microstructural sites such as grain boundaries. Such a cloud can serve to dilate the crack-tip region and reduce the effective stress level. This toughening mechanism is not very

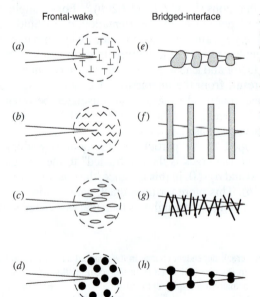

Figure 7.38   Crack-tip shielding mechanisms active in several classes of materials. Frontal zone: (*a*) dislocation cloud; (*b*) microcrack cloud; (c) phase transformation; (*d*) ductile second phase. Crack-wake bridging zone: (*e*) grain bridging; (*f*) continuous-fiber bridging; (*g*) short-whisker bridging; (*h*) ductile second phase bridging.[87] (B. Lawn, *Fracture of Brittle Solids*, 2nd ed., Cambridge University Press, 1993, p. 210. Reprinted with the permission of Cambridge University Press.)

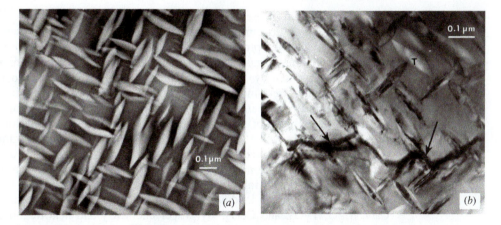

Figure 7.39 TEM images of partially stabilized zirconia alloy, containing 8.1 mole percent MgO. (*a*) Coherent tetragonal ZrO_2 particles embedded within a cubic MgO–ZrO_2 matrix; (*b*) ZrO_2 particles near crack plane are transformed from tetragonal to monoclinic form. Note that tetragonal particles away from crack plane are untrans-formed. (From Porter and Heuer[100], with permission from the American Ceramic Society, Inc.)

important in monophase ceramics; conversely, by increasing the volume fraction of second-phase particles, microcrack density in some two-phase ceramic composites is increased along with toughness.[89,95]

Transformation toughening is an extremely potent crack-tip zone-shielding mechanism that has attracted considerable attention because it is highly effective and is relatively easy to implement. It involves the presence of submicron-sized metastable particles of tetragonal zirconia (ZrO_2) that are embedded within a cubic fluorite ZrO_2 matrix or other ceramic matrix (e.g., alumina (Al_2O_3)).[96−105] The principal feature associated with toughening in these materials is analogous to the transformation-induced plasticity (TRIP) effects associated with certain high-strength steels (recall Section 7.4.1). In the case of partially stabilized zirconia-bearing ceramics, it is possible to bring about, near the crack tip, a stress-induced martensitic transformation of the tetragonal ZrO_2 particles to that of the monoclinic polymorph. Figure 7.39 shows a thin-film TEM image of coherent tetragonal particles before and after their stress-induced transformation to the monoclinic form. Note that tetragonal particles away from the crack plane had not transformed. It is generally believed that some of the toughening results from the fact that a portion of the energy available for fracture is dissipated during the stress-induced transformation process. In addition, the transformation process generates a favorable residual compressive stress environment as a result of the 3–5% volume expansion associated with the tetragonal to monoclinic phase change. McMeeking and Evans[106] have computed the contribution of transformation toughening to be

$$K_t = 0.3E\varepsilon^T V_f w^{1/2} \tag{7-3}$$

where

$K_t =$ toughness contribution due to phase transformation
$E =$ elastic modulus
$\varepsilon^T =$ unconstrained transformation strain of ZrO_2 particles
$V_f =$ volume fraction of ZrO_2
$w =$ width of the transformation zone on either side of the crack surface

Toughness is also enhanced by increasing the density of transformable ZrO_2 particles near the anticipated fracture plane, by choosing chemical systems that enhance the volume change of the transformation process, and by choosing a very rigid matrix (high E) so as to enhance the residual stress field in the transformed zone surrounding the crack.[107] Furthermore, studies have noted that the transformation-toughening contribution is greater than that predicted by Eq. 7-3 because of the presence of additional energy-absorbing processes.[107] For additional perspectives on transformation toughening in ceramics, see the recent reviews by Evans[89,102] and a five-paper set by Lange.[108] The ZrO_2 particle size controls the temperatures for spontaneous and stress-induced transformation; the optimum particle size has been found to be 0.1 to 1.0 μm.[97] Examples of the effects of aging time and amount of magnesia content on the fracture toughness of PSZ ceramics are shown in Fig. 7.40 . Note that in each case the fracture toughness value can be optimized with appropriate ceramic chemistry or heat treatment. Likewise, since toughening in zirconia-based alloys is achieved by the energy-consuming transformation of tetragonal-ZrO_2 to its monoclinic form, it is important that this phase transformation be inhibited during cooling from the processing temperature and, instead, delayed until the critical loading event; this is accomplished with processing additives such as CaO, MgO, Y_2O_3, and CeO_2. By proper chemistry and heat treatment, zirconia alloys can exhibit toughness levels as high as 20 MPa$\sqrt{m}$.

Crack bridging with either ductile (metal-toughened ceramics, called *cermets*) or brittle second-phase particles (e.g., fibers and whiskers) represents the second major group of toughening mechanisms in brittle ceramics (see Fig. 7.38). Energy is consumed when the interface separates ahead of the advancing crack and the triaxial stress state at the crack tip is relaxed (recall Fig. 7.26). As the crack extends, additional energy is consumed with progressive debonding of the ligaments. These unbroken ligaments produce tractions across the crack wake (i.e., closure

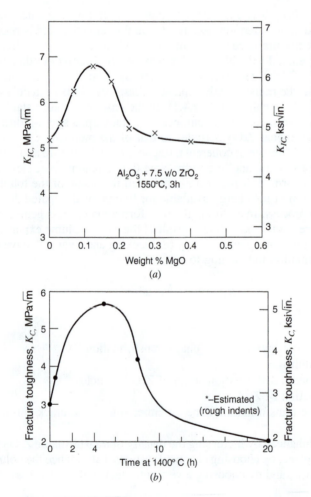

Figure 7.40   Fracture toughness of PSZ ceramic alloys, (*a*) K_{IC} values in Al_2O_3 +7.5 vol ZrO_2 as a function of MgO content. (From Claussen and Ruhle.[97]) (*b*) K_C values inferred from microhardness indentations in 8.1 mole percent MgO + ZrO_2 as a function of aging time at 1400 °C. (From Porter and Heuer[101]; reprinted with permission from the American Ceramic Society, Inc.)

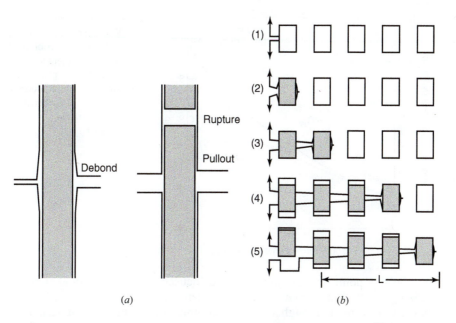

Figure 7.41 (*a*) Debonding and subsequent pull-out of broken fiber from matrix.[87] (B. Lawn, *Fracture of Brittle Solids*, 2nd ed., Cambridge University Press, 1993, p. 244. Reprinted with the permission of Cambridge University Press.) (*b*) Sequence of events involving (1) crack deflection, (2) particle debonding, (3) grain pull-out, (4) lengthening of bridging zone, (5) establishment of steady-state bridging zone of length *L*.[109] [N. P. Padture, Ph.D dissertation, Lehigh University, Bethlehem, PA, 1991. Reprinted by permission.]

forces), which diminish the local crack-tip stress level. Eventually, the ligaments fail and pull out of the matrix. (Fig. 7.41*a*).[87] As such, a steady-state bridging zone of length, *L*, is developed ahead of the advancing crack tip (Fig. 7.41*b*).

These phenomena are particularly important for fiber reinforcements. For optimal energy consumption, the fiber-matrix interfacial strength should not be too strong or too weak.[88,110,111] If the bond strength is high, fiber rupture will occur prior to debonding and no bridging will develop. Alternatively, if the interface is too weak, debonding will take place, but little energy will be consumed by the fiber-pull-out process. It has been reported that interfacial debonding and the absorption of high levels of fracture energy will generally occur when the ratio of interface/fiber toughness is < 0.25.[89] Additional efforts are being directed toward the development of optimal interfacial properties through the introduction of a separate interboundary phase.[88,89] Further analysis of the fiber pull-out process in fiber composites can be found in the context of polymer matrix composites, Section 7.6.

The extent of debonding, fiber pull-out, and associated energy consumption in ceramic-matrix composites is influenced by the presence of thermally induced residual stresses. For example, when CTE of the matrix is greater than that of the whisker or fiber, the matrix will "clamp" down on the whiskers and generate large frictional forces along the whisker–matrix interface. By reducing the CTE mismatch between the whisker and matrix, the resulting radial compressive stress is reduced and toughness increased by the enhanced ability of the whisker to pull out of the matrix.[88]

It has been shown that *grain bridging*–induced toughening can occur in noncubic monophase ceramics. Here, unbroken grains bridge across the crack plane; these ligaments are locked in place by thermally induced compressive forces, generated by the anisotropic thermal expansion mismatch between contiguous grains in these materials.[112–114] By contrast, CTE values in cubic ceramics are isotropic and no thermal-stress induced grain bridging is found.[114,115] Other studies show that toughness increases with the introduction of elongated grains[109,116] with the latter simulating the behavior of whiskers and/or short rods. In this instance, the materials are referred to as being "self-reinforced."

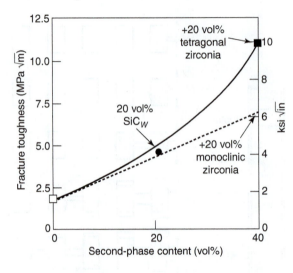

Figure 7.42   Fracture toughness in mullite at 800°K as a function of 20 v/o SiC whisker and 20 v/o zirconia. Superior toughness is associated with the presence of the tetragonal zirconia phase, which undergoes a phase transformation as compared with nontranforming monoclinic zirconia.[88] (P. F. Becher, *Journal of the American Ceramic Society* 74 (2), 255 (1991). Reprinted by permission.)

To this point, our discussions have focused on separate toughening mechanisms. It is striking to note that when multiple mechanisms are active (e.g., whisker bridging and zirconia-based transformation toughening), the combined level of toughening is multiplicative rather than additive. Such synergism arises from the fact that the crack surface tractions of the bridging ligaments serve to expand the size of the process zone where transformation toughening is occurring.[117] The limit for synergistic material response is found to occur when the ratio of the bridging zone length, L, to the process zone width, h, is $L/h \leq 10$. Figure 7.42 illustrates the combined influence of whisker-reinforcement and transformation toughening mullite composites.[88] The addition of 20 v/o of SiC whiskers doubles the toughness of mullite at 800 °C. When a third component is added to this composite (i.e., 20 v/o *monoclinic* zirconia), an additional contribution to overall toughness is achieved. However, the same volume addition of *tetragonal* zirconia leads to a significant increase in toughness, due to the synergistic interaction between whisker bridging and transformation toughening of the metastable tetragonal zirconia phase.

7.5.2 Glass

As discussed in Section 7.1.2, the intrinsic toughness of oxide glass is low because there are no plasticity mechanisms active under normal use conditions. Little can be done to alter this in a substantial way, but composition control does offer some possibility of modest control of intrinsic toughness by altering the coordination number and glass network coherency, and therefore the atomic-level response to the stress field at the tip of a crack. Specifically, Wiederhorn et al.[118] showed that in 3-point bending, borosilicate and aluminoborosilicate glasses are both tougher than pure amorphous SiO_2, while Eagan and Swearengen[119] showed that the toughness levels of binary Na_2O-SiO_2 glasses were the lowest. The K_{IC} of a binary Na_2O SiO_2 glass was only 0.58 MPa$\sqrt{m}$, as compared to 0.75 MPa$\sqrt{m}$ for pure fused silica. The aluminosilicate (Al_2O_3-Na_2O-SiO_2 with an Al:Na ratio of $\sim$1) was similar to the fused silica, while one of the borosilicate glasses (with a B:Na ratio of 0.67) came in at 0.94 MPa$\sqrt{m}$.[119] These trends can be understood by considering the effect of each addition to the network structure. The Na^+ ion acts as a network modifier in a silicate glass, creating non-bonding oxygen ions and therefore breaking up the structure. The addition of Al_2O_3 to a binary Na_2O-SiO_2 glass reduces the number of non-bridging oxygen ions until a 1:1 Al:Na ratio is achieved. The Al_2O_3 therefore "repairs" the coherency of the glass network. As shown by the fracture toughness results, even greater improvements in intrinsic toughness are available through the addition of B_2O_3, although the mechanism for this is less clear.[119]

Given the amorphous nature of glass, there is no microstructure to modify; typical extrinsic toughening mechanisms are therefore not possible to apply without significantly altering other desirable characteristics such as transparency. No matter what one does, glass does not have the forgiveness that metals offer for structural applications. As a result, a maximum-load design criterion, like that typically used for metals, is still relevant, but extra attention must be paid to design and material preparation details that could potentially cause stress concentrations.[120] Cumulative surface damage (e.g., scratches and chips) developed while in service must also be considered as part of a time-dependent probability of failure. Nevertheless, despite the challenges associated with using glass as a structural material, it is being used in this fashion in ever-increasing volume.[viii]

The ability to use glass as a structural material is directly due to relatively recent improvements in fracture resistance. Glass has been used as an architectural element since the days of the ancient Romans, but only in the past few decades has it been reliable enough to play a structural role. Modern fracture-resistant glasses are often called "strong glasses" because the only deviation from linear elastic loading is the fracture load (not the onset of plasticity), although the term "toughened glasses" is also used. If intrinsic toughness is poor and extrinsic toughening is not a viable option, what is left? In truth, the K_C or $\mathscr{G}_C$ toughness of these "toughened glasses" is not improved over that of conventional glass, but the effective fracture resistance and/or the damage tolerance is much greater, for reasons that will be explained below. Using strengthened glass, two different design approaches are possible: a no-break scenario in which the goal is complete prevention of fracture by raising strength and surface damage resistance as high as possible, and a safe-breakage scenario in which the main goal is post-breakage stability (and load transfer to unbroken portions of the glass).[120] The no-break scenario is best met by the use of *thermally* or *chemically tempered glass*, whereas a safe-breakage design is better achieved with *laminated glass*.

Thermally tempered soda lime silica (Na_2O-CaO-SiO_2) glass owes its fracture resistance to a permanent residual stress gradient that consists of a large compressive stress at the glass surface and tension stress in the interior. For fracture to occur, an applied tensile stress must completely counteract the residual compressive stress before it can begin to open a surface crack. A *crack opening stress* σ_{COS} can therefore be defined as the tensile component normal to the plane of a crack, which is equal to the sum of the applied normal stress, the built-in residual stress, and any additional stress imposed by constraints or prestressing: $\sigma_{COS} = \sigma_{appl} + \sigma_r + \sigma_{c/p}$. Because the fundamental fracture toughness of the glass as defined by Griffith (see Chapter 6) is not altered by the tempering process (~ 0.75 MPa$\sqrt{m}$ for soda-lime glass as noted in Section 7.1.2), our usual equations for determining the critical static fracture strength can be used for tempered glass with σ_{COS} inserted for the applied stress.[121] Alternatively, one can define an effective stress intensity factor K_{eff} that is equal to the sum of the applied stress intensity factor K_{appl} and an opposing stress intensity factor K_r associated with the residual stress field, $K_{eff} = K_{appl} + K_r$, so that the applied stress may be used without modification to determine the conditions under which unstable crack growth will occur.[122] Described in this way, the material has an apparent increase in toughness due to tempering.

The residual stress gradient in thermally tempered glass is achieved by heating the glass uniformly to a temperature approximately $100\,°C$ above the glass transition temperature, then cooling it rapidly using cool air or liquid jets. The surface solidifies first, forming a "shell." As the glass continues to cool and shrink, the constraint of the shell causes the still-hot interior to develop a tension stress that pulls the solid surface into compression. The classic stress profile created is parabolic in nature, as shown in Fig. 7.43. ASTM C 1048[123] calls for a surface compressive stress greater than 69 MPa for a "fully tempered" glass, or greater than 24 MPa for a "heat-strengthened" glass. The transition depth from compression to tension may be close to $600\,\mu m$ beneath the surface, providing significant protection against shallow surface flaws.[124] Because a strong temperature gradient and associated density gradient must be established during the cooling process, thermal tempering is difficult to achieve with borosilicate glass

---

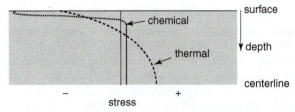

Figure 7.43 Comparison of residual stress profiles in a schematic cross section of thermally and chemically tempered glass plates. Compressive stress exists at the surface (−), tensile stress in the interior (+). Only the top half of the plate is shown; the stress profile in the bottom half would be symmetric. (Not to scale.)

(with its low thermal expansion coefficient) and the plate thickness cannot be less than 1/8 inch (∼3 mm). Dimensional changes during the thermal tempering process make it difficult to meet tight tolerances, and curved shapes are difficult to temper. Furthermore, one cannot grind, drill, or cut the glass after tempering because the surface compressive layer will be breached and the stored tension energy in the interior will be released explosively, fracturing the plate.[ix] Tempered glass is highly resistant to scratching and to overload fracture, but has no structural integrity once failure has occurred (recall Chapter 5). It is widely used for automotive side and rear windows, for architectural glazing, and for household goods such as transparent tables and shelves. Interestingly, an inhomogeneous strain pattern (also called a quench pattern or quench marks) associated with the locations of the cooling jets may be visible under certain light conditions; it is more likely to be seen when viewed with polarized light (e.g., through certain sunglasses or when illuminated by light reflected off of water).[123]

Sudden and unexpected failure of thermally tempered glass is known to occur on rare occasions if large NiS inclusions develop within the glass during processing. The rapid cooling required for tempering tends to encourage development of metastable α-phase NiS spheres or ellipsoids. Swain has shown that the 4% volume expansion associated with a change from the α-phase to the stable β-phase can be sufficient to create microcracks in the tensile interior that then lead to spontaneous (and undoubtedly shocking) failures long after the tempered glass is put into service.[125] The possibility of failure by this mechanism can be reduced by exposing the glass to a heat soak, a low-temperature treatment intended to speed the phase transformation and cause intentional failure of susceptible plates before they are placed into service.

Chemical tempering, also known as ion exchange strengthening, achieves high strength by the development of a compressive surface layer much like thermal tempering, but the residual stress profile is achieved by exchanging the Na^+ ions near the surface of the glass for K^+ ions. This occurs naturally by diffusion when the glass is immersed in molten potassium nitrate. The incoming K^+ ions are larger than the outgoing Na^+ ions by approximately 30%, so the exchange results in the development of a compressive residual stress in the exchange region. The compressive stress is at a maximum at or near the surface, reaching a level greater than 500 MPa (and stresses as high as 1000 MPa have been reported).[124,126,127] The depth of the K^+ exchange layer is dependent on the diffusion time and temperature, but also on the chemical composition of the glass. The maximum depth of the compressive layer may be only 10–15 μm for a standard soda-lime glass,[128] but can reach 50–300 μm (corresponding to ion exchanges ranging from 3 to 96 hours) in special sodium aluminosilicate glass such as Corning 0317 that has been modified to enhance the diffusion process.[129] A compensating tension stress is created in the interior of the glass, but it is very small and uniform as compared to thermally tempered glass (Fig. 7.43). Unlike thermally tempered glass, it is possible to chemically temper plates that are thinner than 1/8 inch, have high curvature, have varying thickness, or have stringent dimensional and optical requirements. Also, it can be safely ground, drilled, or cut to size because the internal tension is low. There is no spontaneous fracture mechanism like that associated with NiS in thermally tempered glass. These characteristics make chemical tempering well suited to

---

[ix] This can be demonstrated in a dramatic fashion by making Prince Rupert Drops—teardrop-shaped pieces of thermally tempered glass that are very hard to damage by impact on the main droplet region but easy to explode by breaking off the tail and thereby exposing the tensile interior. (M. M. Chaudhuri, "Explosive disintegration of thermally toughened soda–lime glass and Prince Rupert's drops," *Phys. Chem. Glasses: Eur. J. Glass Sci. Technol.* B, April 2006, 47 (2), 136–141.)

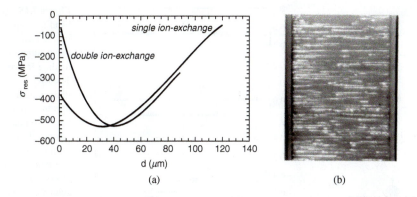

Figure 7.44   (*a*) Residual stress depth profiles produced by conventional single ion exchange and double ion exchange. (*b*) Stable surface cracks in double ion exchanged glass produced by bending loads.[130] (V. M. Sglavo and D. J. Green, *Journal of the American Ceramic Society* 84, 1832 (2001), with permission from John Wiley and Sons.)

applications for which the added value is worth the additional cost of the process, such as for optical components and handheld consumer electronics.

A clever variation on ion exchange strengthening has been introduced by Sglavo et al.[122,130] Instead of the usual single exchange step of K^+ for Na^+, a second (and less extensive) reverse-exchange step is carried out, out-diffusing some of the K^+ and replacing it once again with Na^+. This has the effect of moving the maximum compressive stress to a depth approximately 50 μm below the surface, and increasing the stress gradient between the surface and the maximum as shown in Fig. 7.44*a*. Together, these phenomena create an apparent fracture toughness that increases with distance from the surface, enabling the development of stable surface cracks (Fig. 7.44*b*).[130] Crack arrest and multiple cracking could offer an early warning of overstressing prior to catastrophic failure, as one can often do with metals.

Laminated glass is perhaps best known as a material for automotive front windshields because it offers a characteristic that is very different from tempered glass: retention of some structural integrity (and transparency) after fracture. For the same reason, it is used in architectural glazing when natural or human-generated conditions (e.g., hurricanes or explosions) endanger a building's inhabitants. A common automotive form of laminated glass is a pair of soda-lime panels bonded together by a layer of polyvinyl butyral (PVB) polymer,[131] while variations that use an interlayer of another material such as polycarbonate can be used for bullet and explosion resistance.[132,133] One key to the performance of laminated glass is that it is used in the crack arrest orientation (recall Section 7.3.1). A crack is initiated in the glass panel with the highest stress intensity factor. If the crack arrests at the glass/polymer interface, it leaves the second glass panel intact to continue in a load-bearing capacity. If fracture spreads to the second panel but the fragments are large, then they may lock together (held by the polymer layer) to provide some continued level of structural integrity. After both layers of glass have fragmented, polymer deformation also continues to provide some energy absorption.

Bennison et al.[131] have studied stress development and first-cracking under biaxial bending as shown in Fig. 7.45*a* (ring-loading on a 3-point support, a configuration that generates equibiaxial compression on the upper, ring-loaded side of a plate, and tension on the lower, point-supported side). They found two modes for the initiation of first cracking: at the internal glass/polymer interface of the ring-loaded glass, and at the outer glass surface of the point-supported side. The stress profile through a cross section of laminate is shown in Fig. 7.45*b* for two different loading rates. First, it is important to note that the glass plates act somewhat independently, each developing a compressive side and a tensile side despite being bonded together. This is because the stiffness of the polymer layer is much lower than that of the glass, so to some degree the two glass plates can slide past one another. Second, for the slower loading rate the stress distribution is similar for both plates but with a slightly greater degree of tension at the upper glass/polymer interface, whereas for the faster rate the tension stress at the outer

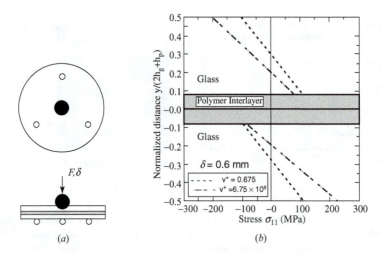

Figure 7.45   (a) Ring-loaded bending configuration. (b) Stress profile through a laminated glass panel subjected to equibiaxial loading at two different rates. The ring-loaded ("compressive") glass plate is on the top, and the 3-point supported ("tensile") plate is on the bottom.[131] (S. J. Bennison, A. Jagota, and C. A. Smith, *Journal of the American Ceramic Society* 82, 1761 (1999), with permission from John Wiley and Sons.)

surface of the point-supported side is the highest. The difference is due to the viscoelastic nature of the polymer layer, with its time-dependent elastic modulus. It was found experimentally that the probability of first-cracking at the upper ply glass/polymer interface was, indeed, higher for slow loading rates (or higher temperatures) as predicted by the stress profile, and fell to essentially zero at very high loading rates (or low temperatures), under which condition first-cracking always occurred at the bottom free surface. The actual probability distribution was affected by the Griffith flaw population present in the glass plates, and some failures at the free surface were found even for very slow loading rates and high temperatures. It was concluded that the initiation of first-fracture can be predicted with a satisfactory degree of accuracy using a Weibull statistical description of the glass behavior coupled with a finite-element-based stress analysis (recall the Weibull discussion in Section 5.3.1).

7.6 TOUGHNESS OF POLYMERS AND POLYMER-MATRIX COMPOSITES

7.6.1 Intrinsic Polymer Toughness

It is generally agreed that crystalline polymers tend to exhibit greater toughness than amorphous polymers owing to their folded chain conformation. There is, however, some controversy regarding the mechanism(s) responsible for whatever toughness amorphous materials possess. According to one proposal,[134,135] toughness should depend on the amount of free volume available for molecule segmental motion. With enhanced motions, toughness should be higher. Litt and Tobolsky[134] found that tough amorphous polymers generally contain a fractional free volume $\bar{f}$ (Eq. 6-6) greater than 0.09. In further support, Petrie[136] found that such polymers, when aged at $T < T_g$, suffered losses in impact energy absorption that were related to corresponding decreases in excess free volume after annealing. In a related finding, it has been shown that the $T < T_g(\beta)$ peak is also identified with impact toughness, the larger and broader the β peak the tougher a material generally tends to be.[135,137,138] For example, the β peak in polycarbonate is much greater than that shown for PMMA, consistent with the much greater toughness associated with the former material.

Finally, it is interesting to note that although a particular *polymer* possessing a low free volume and negligible β peak may be brittle, certain *plastics* based on this polymer may offer good toughness. This improvement in impact resistance is commonly achieved through the use of plasticizers. As discussed in Chapter 2, these high-boiling point, low-MW monomeric liquids serve to separate molecule chains from one another, thus decreasing their intermolecular attraction and providing chain segments with greater mobility. The ductility and toughness of plasticized polymers is found to increase, while their strength and T_g decrease. Care must be taken to add a sufficient amount of plasticizer to a particular polymer so that the material does not actually suffer a loss in toughness. This surprising reversal in material response is referred to as the *antiplasticizer* effect.[139] It has been argued that below a critical plasticizer content, the liquid serves mainly to fill some of the polymer's existing free volume with a concomitant loss in molecular chain segmental mobility.

7.6.2 Particle-Toughened Polymers

A brittle polymer may also be toughened with the addition of a finely dispersed rubbery phase in a diameter size range of 0.1 to 10 μm. For a comprehensive treatment of this subject area, the reader is referred to several monographs, conference proceedings, and key papers on the subject.[140–149] Rubber-toughened polymers can be prepared by using several different techniques. An amorphous plastic (typically polystyrene) and an elastomer (polybutadiene) can be mechanically blended using rollers or extruders to produce an amorphous matrix containing relatively large (5–10 μm), irregularly shaped, rubber particles. Such mixtures do not represent an efficient use of the rubbery phase from the standpoint of optimizing the toughness of the blend. Furthermore, the interface between the matrix and rubbery phase is weak and serves as a site for crack nucleation at relatively low stress levels. A far superior blend is achieved by using the solution-graft copolymer technique. In this method, the rubber phase is first dissolved in styrene monomer to a level of 5–10 w/o. When the polymerization process is about one-half completed, the agitated viscous mixture undergoes phase separation with the elastomeric phase separating in the form of discrete spherical particles from the continuous polystyrene matrix. The dissolved styrene monomer in the rubber phase then polymerizes in the form of discrete polystyrene droplets to form a cellular structure, as shown in Fig. 7.46. Some researchers have referred to this morphology as the "salami structure." The dark, skeletal-like structure is the polybutadiene phase that was revealed by staining with osmium tetroxide, to enhance contrast in the image from the transmission electron microscope;[150] the discrete occluded particles and the continuous matrix constitute the polystyrene phase. ABS (acrylonitrile–butadiene–styrene) polymers, a commercially important group of rubber-modified polymers, are produced by an emulsion polymerization method by which the plastic phase is polymerized onto seed-latex particles of the elastomeric phase. Cellular rubber particles are also found in these materials along with solid rubber particles, though the cellular particles in ABS are smaller (usually less than 1 μm in diameter) than those found in high-impact polystyrene (HiPS).[x] For the case of thermoset resins, successful composite manufacture has been achieved by dissolving the rubber

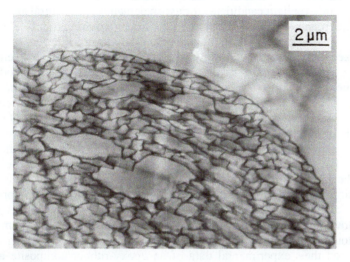

Figure 7.46 Photomicrograph revealing duplex structure of toughening phase in HiPS blend. Dark skeletal phase is polybutadiene elastomeric phase. Occluded particles and matrix are polystyrene. (Courtesy of Clare Rimnac.)

---

[x] See Case Study 9 in Chapter 11 for a discussion of a fatigue failure in HiPS.

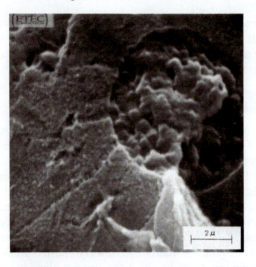

Figure 7.47 Matrix crazes emanating from the left side of a rubber particle as seen on fracture surface in high-impact polystyrene. (Reprinted with permission from John Wiley & Sons, Inc.)

phase in the liquid uncured resin. Upon commencement of the curing reaction, the rubber phase precipitates out from the homogeneous solution as a fine dispersion of rubber particles. More recently, preformed latex-core, hard shell particles have been developed that are added to the resin. Here, the shell material is chosen to facilitate formation of a strong bond with the matrix phase.

Many studies have demonstrated that amorphous, cross-linked, and semi-crystalline matrices can be toughened by various elastomeric additives. Increases of eight- to tenfold in fracture energy have been reported, for example, in polyester[151] and epoxy[143] resins. In such studies, toughness increases with volume fraction of the rubber phase up to a certain point, whereafter toughness remains constant or even decreases. Depending on the matrix phase, the mechanism responsible for toughening has been identified as massive crazing (as in HiPS),[140,141,152,153] or profuse shear banding as reported in ABS, PVC, and epoxy resins.[142,145,146,154–156] Overall, a synergistic interaction exists between the rubbery phase and the matrix. When loading HiPS, for example, a myriad of fine crazes develop at the interface between the glassy matrix and the micron-sized rubber particles (Fig. 7.47). Although craze formation itself is undesirable since it is the precursor of crack formation, the nucleation of many small crazes represents a large sink for strain energy release and serves to diffuse the stress singularity at the main crack front. In addition, the rubber particles act to arrest moving cracks.

Epoxy resins are excellent candidates for use in demanding structural applications, based on their good creep and solvent resistance. Their usage would certainly expand if it were possible to overcome the poor fracture toughness and fatigue crack propagation resistance of these cross-linked polymers. Accordingly, different approaches have been examined to improve the toughness of epoxy resins. The most widely studied method involves the addition of rubbery particles to the epoxy matrix, which facilitates three major toughening mechanisms:[142–148,157–161] (1) rubber particle cavitation and associated shear band formation; (2) matrix plastic void growth following rubber cavitation [analogous to microvoid growth in metal alloys in association with the microvoid coalescence fracture process (recall Section 5.9.1)]; and (3) rubber particle bridging (Fig. 7.48a). Several theoretical models[159,160,162,163] have been proposed to assess the relative contribution of these mechanisms to overall toughening. It is theorized that cavitation/shear banding and plastic void growth energy dissipation are proportional to the width of the damage zone and provide *additive* contributions to composite toughness. Since rubber particle bridging influences the size of this zone, it follows that the simultaneous occurrence of these three mechanisms may result in *multiplicative* toughening.[117,163] Huang and Kinloch's additive toughening model[159] successfully predicted their experimental data of an epoxy-rubber composite and noted that particle cavitation and shear band formation contributed a consistently major fraction to toughness over the temperature range from $-60°C$ to $40°C$ (Table 7.7). By contrast, the toughening contribution by particle bridging decreased with increasing temperature,

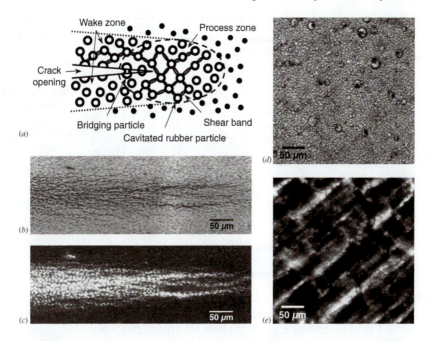

Figure 7.48 Toughening mechanisms in a rubber-modified epoxy. (*a*) Schematic representation of rubber particle cavitation and matrix shear banding in the deformation process zone along with crack bridging by rubber particles. (*b*) Optical micrograph of the process zone ahead of the crack tip showing cavitated particles in a CBTN rubber-modified epoxy. (*c*) Same view as (*b*) but imaged with cross-polarizers to show matrix plasticity (shear bands). (*d*) Optical micrograph taken near the necked region of a tensile specimen. Large cavitated rubber particles have a dimpled appearance. (*e*) Same area as (*d*) imaged with cross-polarizers showing the network of shear bands associated with the cavitated particles. (Parts (b)-(e) reproduced with permission from photographs provided by R. A. Pearson.)

providing 10% or less of overall composite toughness at temperatures of at least 0°C. Conversely, the toughening contribution associated with plastic void growth increased markedly with increasing temperature.

Figure 7.48*b* reveals cavitated rubber particles embedded within the epoxy resin; the extensive shear band network in the same system is made apparent by observing the material between cross-polarizers (Fig. 7.48*c*). Yee and Pearson have shown that rubber particle cavitation relaxes tensile triaxiality at the crack tip, which then facilitates the formation of a network of energy-absorbing shear bands.[142–147] This sequence of events—cavitation of rubber particles followed by shear banding of the matrix—has been confirmed by later experimental studies and modeling efforts as the major toughening mechanisms for many rubber-modified polymers.[165,166]

Rubber particle size has a significant effect on toughening. As explained by Azimi et al., in order to cavitate and enhance shear yielding the particles must be smaller than the plastic zone size.[167] Particles larger than the size of the plastic zone simply bridge the two crack surfaces, which is far less effective as a toughness enhancer. As a result, rubber particles larger than ~20 μm have been shown to be ineffective as epoxy toughening agents.[168] As size is reduced into the 1–5 μm range, toughness increases significantly. Although some initial explorations into even smaller particles indicated that cavitation might not be possible below a certain size threshold

Table 7.7 Contribution of Toughening Mechanisms in Rubber-Modified Epoxy[159]

Temperature (°C)	−60	−40	−20	0	23	40
$\mathscr{G}_{1c}$(kJm^{-2})	1.72	1.96	2.53	3.64	5.90	7.23
Rubber bridging (%)	0.36	0.26	0.14	0.11	0.10	0.05
Plastic shear banding (%)	0.64	0.74	0.68	0.60	0.61	0.47
Plastic void growth (%)	0.00	0.00	0.18	0.29	0.29	0.48

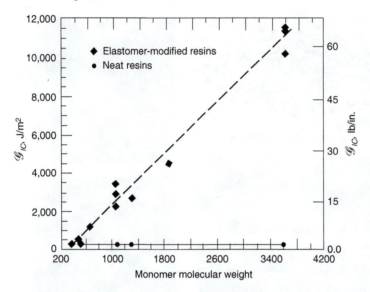

Figure 7.49 Fracture toughness in neat and rubber-modified epoxy resin as a function of molecular weight between cross-links.[143] Note dramatic improvement in composite toughness with increasing resin ductility as represented by increased M_w^c. (Reprinted with permission from A. F. Yee and R. A. Pearson, *Toughening in Plastics II*, 2/1 (1985), Plastics and Rubber Institute.)

(∼0.2 μm was claimed), it has since been demonstrated that block copolymers can be used to form self-assembled rubbery particles in epoxy that have a marked effect on toughness.[169] For example, Liu et al. mixed a poly(ethylene-alt-propylene)-b-poly-(ethylene oxide) (PEP-PEO) diblock copolymer into a standard epoxy resin to form nanoscale spherical micelles about 15 nm in diameter.[166] The fracture toughness of the epoxy increased from 0.96 to 2.73 MPa√m with the addition of the nanoparticles, a 180% increase over the neat resin. Nanocavitation was implied by the appearance of stress whitening in the composite (recall from Chapter 6 that the formation of voids is one cause of this phenomenon) and confirmed by TEM. At high resolution, nanocavities inside the micelles were observed, along with shape distortions that would accompany matrix shearing. An estimate of toughening based on the measured size of the shear zone led the authors to conclude that some crack-tip blunting associated with a reduction in the yield stress also contributed to the toughness increase. Nanoparticle toughening is attractive because it takes advantage of the natural desire of certain polymer blends to phase separate, coupled with careful control over the chemistry to encourage the separation to occur at the nanometer scale. It appears to have little effect on the glass transition temperature or the Young's modulus, unlike the addition of larger rubber particles.

It is important to note that the actual fracture toughness of a rubber-toughened epoxy can be strongly influenced by the properties of the matrix as well as the particles. For example, Yee and Pearson[143] showed that the cross-link density of the epoxy matrix plays a major role in determining overall fracture toughness. As expected, the toughness of the rubber-modified epoxy resin increased markedly with increasing molecular weight between cross-links, M_w^c, whereas the neat resin showed almost no influence of M_w^c (Fig. 7.49). (An increase in M_w^c corresponds to a decrease in cross-link density). Thus, epoxies with relatively high cross-link densities are poor candidates for rubber-toughening.

Another approach to epoxy matrix toughening involves the addition of rigid inorganic fillers such as glass spheres and short rods (typically 5–100 μm in size) that toughen without significantly changing the elastic modulus or T_g. The addition of such particles enhances the toughness of epoxy-matrix composites by enabling pinning and/or deflection of the crack front, facilitating particle debonding to enable the epoxy matrix to plastically deform, serving to generate a series of microcracks, and providing bridging elements across the wake of the crack.[170–173] These mechanisms serve to shield the crack tip from the applied stress, thereby reducing the effective crack-tip stress intensity factor (recall Fig. 7.4). For the case of glass spheres, the most important toughening mechanisms are believed to be crack pinning[148,171,172] and matrix plastic deformation in the vicinity of debonded glass spheres.[163] For rod-like particles, the most potent toughening mechanisms involve fiber debonding,[174] and pull-out,[175–178] crack deflection,[163] and fiber bridging.[175]

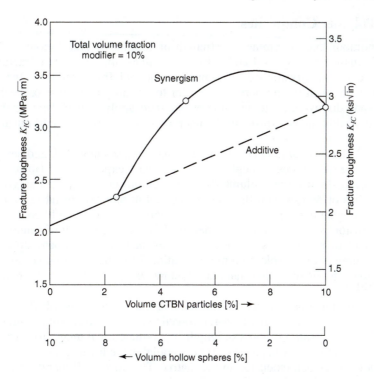

Figure 7.50   Synergistic toughening in CTBN rubber and hollow glass sphere hybrid-epoxy composite, corresponding to 3:1 volume ratio between CTBN rubber and hollow glass spheres.[186] (R. A. Pearson, A. K. Smith, and Y.W. Yee, Second International Conference on Deformation and Fracture of Composites, PRI, Manchester, UK, 9-1, 1993. Reprinted by permission.)

As in the case of the rubber-toughened epoxies, effects of rigid particle size on toughness have been explored, with the additional goal of reducing the effects rigid additions have on resin viscosity. Silica nanoparticles ∼20 nm in diameter have been shown to have a profound effect on epoxy toughness, increasing K_{IC} from 0.59 to 1.42 MPa$\sqrt{m}$ in one study (corresponding to $\mathcal{G}_{IC}$ fracture energies of 100 J/m^2 for the unmodified epoxy polymer and 460 J/m^2 for the composite) for which the filler made up 13.4 vol% of the material.[179] In this case, microscopic examination of the fracture surfaces pointed to nanoparticle debonding and plastic void growth as the critical mechanisms rather than crack pinning and crack deflection, as might be expected for much larger rigid particles. Other nanoparticle/epoxy combinations have shown somewhat different combinations of strengthening mechansisms, depending on particle size and composition.[180-182]

A third approach to the toughening of epoxy resins introduces both rubbery and hard inorganic particles to the epoxy matrix. As noted above, the combination of crack pinning and/or bridging, associated with hard inorganic particles, with energy absorbing process zone mechanisms (e.g., cavitation/shear band formation and plastic void growth), attributable to the rubber particles, provides a multiplicative interaction and synergistic toughening.[117,163] Reports of synergistic toughening in association with hybrid composites are cited in several references.[183-188] For example, Pearson et al.[186] noted synergistic toughening in an epoxy composite, containing both rubber particles and hollow glass spheres (Fig. 7.50). For this system, synergistic toughening was attributed to the multiplicative interaction between rubber particle cavitation/matrix shear banding and hollow glass sphere-induced microcracking. Hybrid toughened systems that include rigid nanoparticle phases and micrometer-scale rubber phases can also capitalize on the synergistic toughening effect.[189]

7.6.3 Fiber-Reinforced Polymer Composites

The conditions that determine the strength of continuous and discontinuous composites have been discussed in Chapters 1 and 3. Here we will assume at first that fracture is occurring in a common epoxy-matrix aligned-fiber laminate material. If fracture is likely to proceed in an *interlaminar* mode then the primary concern for toughening is the matrix itself. As such, the schemes already discussed for toughening of polymers apply. If *translaminar* fracture is more likely, then two additional extrinsic mechanisms come into play: fiber pull-out and fiber/matrix debonding.

The use of phase-separating epoxy matrices has been documented, and has been shown to have an influence on composite toughness, as one might expect.[190,191] However, increases in viscosity associated with large volume fraction additions are not welcome, given the need to infuse fiber plies or fabrics with the matrix material during certain fabrication processes.[192] Furthermore, toughening particles greater than a certain size may be filtered out by the fibers during the infusion process, eliminating their usefulness. The advent of nanoparticle toughening agents, and in particular synergistically-toughened epoxies, opens up possibilities for matrix toughening not available with more traditional toughening approaches. This approach has been shown to increase interlaminar fracture energies ($\mathcal{G}_C$) by as much as a factor of $\sim$3.[189,192,193]

The fiber pull-out process active in translaminar fracture is essentially the same for polymer matrix composites as for any other fiber-reinforced matrix (recall ceramic matrix composites in Section 7.5.1). Discontinuous fibers of length $l \leq l_c$ (the critical length to carry the maximum possible load, as defined in Chapter 2) that are parallel to the loading direction will not break at the load that causes crack extension in the matrix. The fibers will therefore bridge the crack, resisting its opening. In order for the crack to advance, the fibers must pull out of the matrix, which requires additional work. The energy (work) for pull-out of a single fiber is given by the shear force at the interface multiplied by the distance z over which the force acts. Recall from Chapter 2 that the total shear force is the integral of the product of the shear stress (τ) and the surface area ($\pi d_f dx$) on which it is imposed. Therefore,

$$\text{energy} = \int_0^z \tau \pi d_f z\, dz = \frac{\tau \pi d_f z^2}{2} \tag{7-4}$$

where

$\qquad d_f =$ fiber diameter
$\qquad \tau =$ shear stress at interface
$\qquad z =$ distance along fiber

The longest fiber that can be pulled out has length l_c. From Eq. 7-4, pull-out energy increases with fiber length and is therefore maximized when $z = l_c/2$.[194,195] Furthermore, we can assume that the shear stress is simply related to the matrix normal stress by $\tau = \sigma_m/2$. Thus we find

$$\text{Energy} = \frac{\pi d_f \sigma_m l_c^2}{16} \tag{7-5}$$

Ignoring for a moment any other influences, the total work required to advance the crack is simply scaled by the total number of fibers per unit area bridging the crack, N_f, that are being pulled out. Put in terms of the volume fraction of fibers perpendicular to the crack plane, V_f, such that $N_f = 4V_f/\pi d_f^2$, the work of fracture is therefore

$$\mathcal{G} = \frac{V_f \sigma_m l_c^2}{4 d_f} \tag{7-6}$$

and, because the critical length is related to the fiber fracture strength as $l_c = d_f\sigma_f/2\tau$ (Eq. 1-70) we can see that the maximum work associated with fiber pull-out is given by

$$\mathscr{G}_{\max} = \frac{V_f d_f \sigma_f^2}{4\sigma_m} \tag{7-7}$$

which favors high fiber strength, a high volume fraction of fibers oriented in the correct direction, and a sufficiently weak fiber/matrix interface to allow pull-out to occur.

In the case of continuous fibers, only those that fracture within $l_c/2$ of the crack plane will participate in pull-out. The fraction of fibers that do so is dependent on the fiber strength and the population of fiber defects. Those fibers that do not fracture in a way that contributes to pull-out can still play a role in toughening through crack deflection and the fiber/matrix debonding process. As the crack front reaches an unbroken fiber it will tend to run along the fiber/matrix interface (recall Fig. 7.41). This debonding process is aided by a difference in the Poisson contraction of the fiber and the surrounding matrix that creates a tensile stress component perpendicular to the fiber/matrix interface. Not only is there energy dissipated through the creation of exposed fiber and matrix surfaces, but the strain energy associated with subsequent fiber fracture (which is necessary for the crack to advance much beyond a bridging fiber) must be added on to the work of fracture. These contributions are important, but are generally much smaller than the pull-out energy.

A less expensive but more mechanically complex family of composites are discontinuous fiber reinforced thermoplastics used for injection molding. During fabrication the local fiber orientation is strongly influenced by the flow of the matrix melt, resulting in misaligned arrays of short fibers that are inhomogeneously arranged. Boundary layer flow of the melt near the mold walls may align the fibers in the mold fill direction, while fibers in the center section may be oriented perpendicular to those along the edges as shown in Fig. 7.51 (see also Fig. 3.41).[196] Thicker components may therefore have a higher fraction of fibers in a certain orientation than thinner components. This introduces a thickness dependence to K_{IC} on top of the usual trend associated with a plane stress to plane strain transition (see Fig. 7.52). Friederich[196] describes the microstructure contribution to toughness using a *mesostructural efficiency factor*, M, that linearly magnifies the toughness of the matrix: $K_{\text{composite}} = M K_{\text{matrix}}$. Factors that influence M include the volume fraction of fibers (V_f), the fiber orientation across the component thickness with respect to the crack growth direction, the fiber aspect ratio, the fiber/matrix bond strength, and fiber stiffness and strength. Although the interrelationships between these factors are complex, there is much opportunity for control of toughness (or lack of control!) in injection-molded fiber-reinforced thermoplastics.

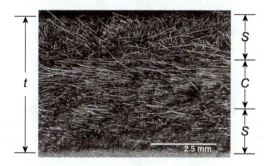

Figure 7.51 Variation in short-fiber alignment through the thickness (*t*) of an injection molded component, as seen on the fracture surface of a 30% glass short fiber reinforced polypropylene resin composite. Injection molding direction is normal to the plane of view. Mold walls are located along the top and bottom of the fracture surface shown in the micrograph. Note that in the skin areas (*S*) fibers are aligned parallel to the injection molding direction. Conversely, fibers are oriented perpendicular to the injection molding direction in the core region (*C*) of the component.

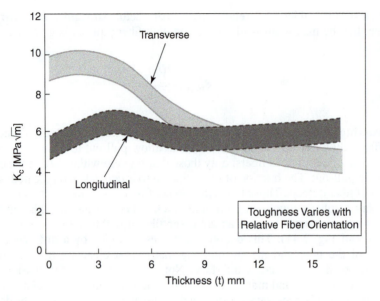

Figure 7.52 Variation in toughness with component thickness for an injection molded composite, and its dependence on crack growth orientation.[196] (Data from Springer Science+Business Media: *Journal of Materials Science* 33, p. 5535, (1998), K. Freidrich, figure number 12.)

7.7   NATURAL AND BIOMIMETIC MATERIALS

All of the microstructural toughening approaches discussed thus far in this chapter have been associated with "traditional" engineered materials. To varying degrees, intrinsic and extrinsic toughening mechanisms have been utilized to control and enhance the toughness of materials used in structural applications. Most of these mechanisms are active over a limited size scale determined by the small number of microstructural features being controlled. Nature has opted for a somewhat different approach, successfully employing a hierarchical scheme of multiple (mostly extrinsic) mechanisms that occur simultaneously over a wide size range (nm–μm), usually with a high degree of anisotropy. The individual mechanisms are similar in action to those that have been employed in engineered materials, but the ways in which they have been combined and optimized offer lessons to engineers. Furthermore, no natural materials that are used for load-bearing applications (i.e., those requiring high stiffness and toughness) are based on metals or metal alloys, unlike so many engineered structural materials. Instead, they often rely on fairly poor ceramic and polymer materials that are combined to form composites that can exhibit fracture energy dissipation three orders of magnitude greater than the ceramic phase alone, and elastic stiffness three orders of magnitude greater than the isolated polymer phase, as shown schematically in Fig. 7.53.[197,198] For these reasons, there is a strong motivation to understand the toughness of natural materials like mollusk shell, bone, and tooth enamel, and to attempt to use the lessons learned to create superior engineered materials. It is fair to say, however, that in comparison to more traditional engineered materials, the field of *biomimetic* (or *bio-inspired*) tough materials is still in its infancy.

7.7.1   Mollusk Shells

One of the most highly studied groups of tough natural materials is that of mollusk shells. These shells are built from brittle minerals including calcium carbonate ($CaCO_3$) in the form of either calcite (a trigonal polymoprh) or aragonite (an orthorhombic polymorph). The ceramic phase makes up 95–99% of the shell by volume, with the remaining fraction consisting of organic polymers that bind the mineral components together.[199] Toughness of the better mollusk shells is on the order of $K_C \approx 3$–10 MP$\sqrt{}$m (equivalent fracture energy ≈ 0.3–11.7 kJ/m^2).[200,201]

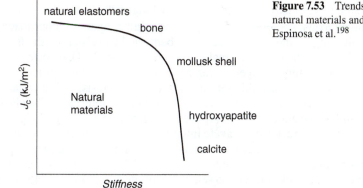

Figure 7.53 Trends between toughness and stiffness for natural materials and tough composites. Adapted from Espinosa et al.[198]

The toughness of these natural composite materials is no better than that of many engineered materials (e.g., $K_{IC} \approx 8.7$ MP$\sqrt{m}$ for Al_2O_3/SiC whisker composite[202]; $J_{IC} \approx 0.4$ kJ/m^2 for partially stabilized zirconia[87]), but their fracture behavior is remarkable given the relatively poor constituent materials and the very high volume fraction of brittle ceramic phase employed.[200] The optimization of shell toughness for a given material system stems from the complex arrangements of the ceramic and proteinaceous phases, and the details of the ceramic phase shapes. If the toughening mechanisms active in the mollusk shells can be duplicated with engineered materials, the resulting toughness levels should be impressive. Two particularly noteworthy mollusk shell types that can serve as models are those of the Pink (or Queen) Conch and the Red Abalone.

The Pink Conch (*Strombus gigas*) shell is made up of 99% aragonite crystals arranged in a *crossed-lamellar structure*. The crystal arrangement resembles a laminate or a "ceramic plywood," with three layers at alternating 90° angles to one another as shown in Fig. 7.54. It has the highest Mode I plane-strain fracture toughness of any mollusk structure.[203] The relatively high toughness of the conch shell in bending is primarily due to the action of two

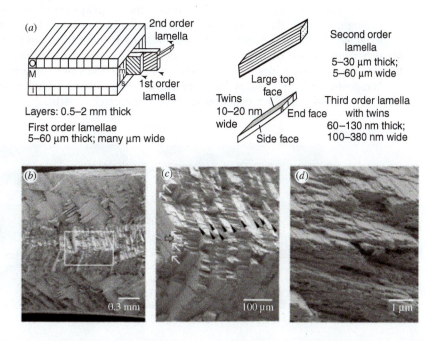

Figure 7.54 Schematic drawings (*a*) and SEM images (*b–d*) of the hierarchical micro-architecture of a Pink Conch shell, illustrating the alternating 90° arrangement of the Inner, Middle, and Outer macroscopic layers made of first-order lamellae, the alternating 45° arrangement of the second-order lamellae in the center layer, and the twinned third-order lamellae that are the basic building blocks of the system. The organic matrix that holds the lamellae together is not visible. [203] (Reprinted by permission from Macmillan Publishers Ltd: *Nature* 405, p. 1036, copyright 2000.)

energy-dispersion mechanisms: multiple microcracking in the tensile outer layer that develops at low mechanical loads, and crack bridging in the middle layer at higher loads.[203] In the first case, a crack initiating on the tensile side of the bent shell travels along the first-order lamellar interfaces (i.e., between the strips that make up layer O in Fig. 7.54), then arrests at the interface with the $\sim 4\times$ tougher middle layer (layer M in Fig 7.54; recall similar behavior of metal laminates in Section 7.3). Once this initial surface crack arrests, a somewhat higher stress is required to initiate a second surface crack along a parallel interface. Closely spaced parallel surface cracks shield one another, lowering the stress intensity at the crack tips and reducing the tendency of any surface crack to propagate into the next layer. Multiple surface cracks can then initiate, grow, and arrest, until a critical crack density is achieved, at which point it becomes possible to extend one of the cracks into the middle (M) layer. The second mechanism then comes into play, as the crack is forced to select one of the two possible $\pm 45°$ directions created by the alternating second-order lamellar interfaces. The result is that every other first-order lamellae in the middle layer can crack (e.g., $+45°$), but the crack is bridged by the intervening intact first-order lamellae (e.g., $-45°$). The contribution of bridging to the overall toughness is quite significant, and is analogous to the bridging of fibers in engineered composites.

Red Abalone (*Haliotis rufescens*) shell has an outer layer made of calcite mineral and an inner composite layer called nacre. Nacre appears as an iridescent layer on the inside of the shell, and is familiar to many as mother-of-pearl. It is this layer that is of great interest because of its complex structure and remarkable synergistic toughening mechanisms. Like the Pink Conch shell, nacre is made up of a combination of aragonite and organic polymer, although the ceramic phase of nacre makes up a slightly smaller fraction (~ 95 vol.%) of the material.[204] The columnar nacre micro-architecture is distinct from that of the crossed-lamellar Pink Conch shell, taking on the appearance of stacked tablets or platelets (Fig. 7.55a). The biopolymer phase fills the interfaces between the platelets in layers 20–30 nm thick.[204] Several toughening mechanisms are at work during fracture of nacre.[201,204] First, the weakest interfaces between the platelets can slide viscoplastically in shear at low loads, dissipating energy in the process zone immediately surrounding the crack tip. The biopolymer does not strongly resist shear strain, but it does adhere well to the aragonite, which maintains the integrity of the overall structure (Fig. 7.55b). As sliding continues, the platelet interfaces "harden" when rough features on their surfaces grind together.[204] As these interfaces lock up, sliding is induced in other nearby interfaces, thereby increasing the size of the process zone and reducing strain localization. As fracture proceeds, it is also impeded by tablet pullout as shown in Fig 7.55a (much like fiber pullout in engineered composites).[201,205] These phenomena lead to rising *R*-curve behavior (increasing toughness with crack extension), which can slow or even arrest crack growth.

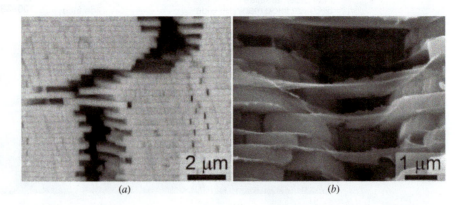

(a)   (b)

Figure 7.55 SEM images of partially fractured nacre showing (*a*) platelet aspect ratio and pullout, and (*b*) extensive inelastic deformation in the polymer phase.[205] (Reprinted from *Acta Materialia* 57, M. E. Launey, E. Munch, D. H. Alsem, H. B. Barth, E. Saiz, A. P. Tomsia, and R. O. Ritchie, "Designing highly toughened hybrid composites through nature inspired hierarchical complexity," p. 2919, 2009, with permission from Elsevier.)

7.7.2 Bone

As discussed in Chapter 1, from a structural point of view cortical (compact) bone is made up primarily of a mineral phase, hydroxyapatite ($Ca_{10}(PO_4)_6(OH)_2$), and an organic phase, type I collagen, as well as water.[206] In humans the ceramic phase makes up a much smaller volume fraction (~ 50 vol%) than in mollusk shells, leading to lower stiffness, but also to greater strength and toughness. Another significant difference is in the way in which the phases are arranged: in cortical bone the hydroxyapatite appears as nanoscale crystals coupled to collagen microfibrils. The collagen microfibrils are assembled into collagen fibers of micrometer diameter. These are grouped into osteons: cylindrical structures with concentric collagen/hydroxyapatite composite lamella surrounding open channels known as Haversian canals.[xi] The collagen fibers are aligned parallel to one another *within* each lamella, but change orientation from one lamellar layer to the next to achieve a cylindrical plywood effect (although not in quite as regular in fashion as the $\pm 45°$ secondary lamella in the Pink Conch shell).[206] In adult human bone, the osteons are tightly packed against one another, meeting at so-called *cement lines*, and are highly aligned as shown in Fig. 7.56 (and also in Fig. 1.14). In long bones (e.g., femur/thigh) this axis aligns with the long axis of the bone. The aligned lamellar arrangement leads to the anisotropic elastic behavior already discussed in Chapter 1. It should now come as no surprise that it also leads to significant toughness

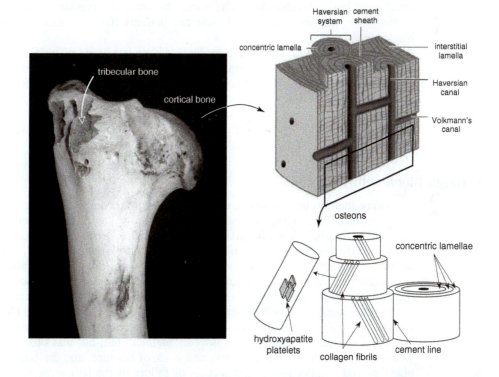

Figure 7.56 Photograph of a long bone from a caribou, with the porous tribecular interior structure visible where cortical surface bone has been removed. Accompanying schematic illustrations at different length scales demonstrate some of the hierarchical structures of human cortical bone. Cylindrical osteons are typically 200–300 μm in diameter. They are built from many concentric layers 3–7 μm thick, each of which contains aligned mineralized collagen fibrils that are ~100 nm in diameter.[209] Collagen fibrils consist of nm-scale collagen molecules and aligned platelets of the mineral hydroxyapatite.[210] (Caribou bone courtesy of Olga M. Stewart. Bone cross section reproduced with permission of Annual Reviews, Inc., from Annual Review of Materials Research, M. E. Launey, M. J. Buehler, and R. O. Ritchie, 40, 25 (2010); permission conveyed through Copyright Clearance Center, Inc.)

[xi] Bovine and equine bone arrange somewhat differently, with a more brick-like lamellar morphology known as plexiform bone.[206]

anisotropy, and promotes extrinsic toughening. It has been argued that the dominance of extrinsic toughening means that the single-value initiation toughness captured by K_{IC} values is not as useful as energy-based measures such as J_C, and that R-curves are needed to fully depict bone fracture.[207] Historically, however, both types of measurements have been used. Although the vast majority of real bone failures occur either as a result of high-speed loading or fatigue (also known as stress fractures), it has been shown that quasi-static loading (of the type usually used for fracture toughness testing) activates the same toughening mechanisms as impact loading, and that the measurements correlate reasonably well.[208]

Like other layered materials, bone is much tougher perpendicular to the osteon axes than along them, with $J_{C\parallel} \approx 0.4 \, \text{kJ/m}^2$ and $J_{C\perp} \approx 10 \, \text{kJ/m}^2$.[211] This matches the primary physiological need for fracture resistance, since most cracks would be initiated on the outer surface of a long bone and then propagate across the axis of the lamellae. The main toughening mechanisms identified are crack deflection, microcracking, and crack bridging, as shown in Fig. 7.57.[207,212,213] Crack deflection occurs when a propagating crack encounters the weak cement lines between osteons, and travels along the interface (similar to the process occurring in weakly-bonded fiber-reinforced composites). This dissipates energy and also reduces the stress intensity at the crack tip. Microcracks developing in the process zone also serve to shield the crack tip. Crack bridging can take two forms: uncracked-ligament bridging and collagen-fibril bridging. Uncracked bone ligaments can support significant loads, whereas bridging collagen fibrils alone provide only a small contribution to shielding. There may be other mechanisms at work at smaller length scales,[209,214] but those described here are perhaps the ones that are most accessible for biomimicry.

It should be appreciated by the reader that bone is a living material that evolves over time, repairing damage but also potentially undergoing detrimental degradation processes like those that underlie osteoporosis.[215] Furthermore, the collagen phase is a viscoelastic polymer, so there is a significant time-dependence associated with the elastic, plastic, and fracture processes that has been largely neglected here.[216] For much more complete descriptions of bone and its mechanical behavior, the reader is encouraged to explore review papers provided in the References.[206,209,217,218]

7.7.3 Tough Biomimetic Materials

Many attempts to reproduce the behavior of natural composites have been made, most with the intention of elucidating the critical features that lead to high toughness rather than creating practical materials.[xii] It is quite difficult to duplicate the structures created by Nature's "bottom-up" fabrication methods using typical engineering "top-down" methods, but many of the main architectural features can be explored and perhaps exploited. Several macroscale models of nacre have been developed using materials for the brittle phase such as glass slides, SiC ceramic tiles, and alumina plates, while the interlayer phase has varied from weakly sintered ceramic, to silicone elastomer, to adhesive tape.[219,220,221] Although none of the physical models truly matched the behavior of natural nacre (nor were they expected to), important lessons were learned about the architectures employed: "platelet" pull-out was observed, crack deflection was shown to improve the toughness and work of fracture, and the behavior of the adhesive phase played a major role in the success or failure of the structures.

A variety of techniques have also been used to produce materials that match nacre in scale as well as structure. In some cases, fabrication techniques have been employed that more closely resemble the natural growth process. Nacre-like materials have been made with feature sizes in the nm–μm range, using hard phases such as clay or alumina platelets. The soft phases attempted include poly(vinyl alcohol) and chitosan-based polymers.[222,223,224] In all cases, important lessons were learned regarding those aspects of the structures that promote toughness, but the fabrication methods were not conducive to large-scale fabrication of

---

xii A macroscale example of a practical system that somewhat resembles nacre is a damage-tolerant ceramic armor made from hexagonal tiles held together by rubber sheets.[231] However, it is not truly biomimicry in terms of the toughness mechanisms at play.

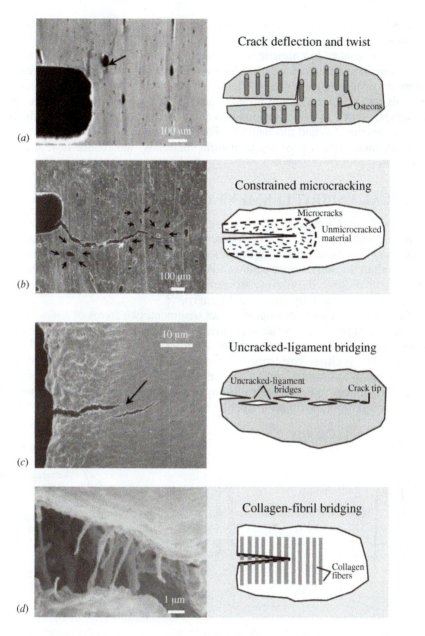

Figure 7.57 Toughening mechanisms in human cortical bone.[213] (Reprinted with permission from R. O. Ritchie, M. J. Buehler, and P. Hansma, *Physics Today*, 62 (6), 41 (2009). Copyright 2009, American Institute of Physics.)

structural components. Likewise, an effort to create a conch-like structure with silicon (the brittle phase) and photoresist (the ductile phase) demonstrated that it is possible to recreate some of the cross-lamellar toughening mechanisms using engineered materials.[225]

A particularly successful demonstration of relatively large components with microscale features resembling nacre was achieved by a novel "ice-templating" approach, in which alumina particles were suspected in water, and the mixture was directionally solidified.[226,205] This caused phase separation of ice and ceramic that resulted in a series of aligned plate-like ceramic regions partially connected by ceramic ligaments. When the water was removed and the ceramic sintered, a brick-like morphology remained behind, into which a polymer (PMMA) was introduced. The best fracture toughness of this material was reported as $K_{IC} \approx 5\,\mathrm{MP}\sqrt{m}$,

$K_{JC} \approx 32\,\mathrm{MP}\sqrt{\mathrm{m}}$, and $J_C \approx 8\,\mathrm{kJ/m^2}$, an enormous improvement over the individual constituent materials.

The true goal of biomimetic structural materials development is not to exactly duplicate Nature's materials, but to use her design principles along with engineered constituent materials to improve over the state of the art. Future progress in the development of biomimetic structural materials will require ongoing improvements in modeling[210] and innovations in fabrication techniques, as well as investigations into a wide variety of natural materials to gain additional insights on structures and mechanisms. For a broader view of attempts to measure and mimic the mechanical behavior of tough natural materials, the reader is referred to several key references.[197, 198, 218, 227–230]

7.8 METALLURGICAL EMBRITTLEMENT OF FERROUS ALLOYS

Throughout this chapter we have discussed ways in which materials may be intentionally toughened. Attention will now be focused on several undesirable circumstances that lead to serious loss of a metal's fracture toughness. As will be shown, these changes can be brought about by alterations in microstructure and/or solute redistribution as produced by improper heat treatment or prolonged exposure to neutron irradiation. For a comprehensive study of this subject, see the review by Briant and Banerji.[232]

Regarding the matter of improper heat treatment, two forms of *temper embrittlement* have been defined (Fig. 7.58). In the first case, high-strength martensitic steels may be embrittled following a short-time temper at low temperatures (in the range of 300 to 350°C). The "350°C embrittlement" is also referred to as "tempered martensite embrittlement" with a single tempering treatment being sufficient to induce embrittlement. The second type of embrittlement is found in lower strength steels and is brought about by a two-step heat treatment, as shown in Fig. 7.58, or by slowly cooling from the initial tempering temperature and through the embrittling temperature

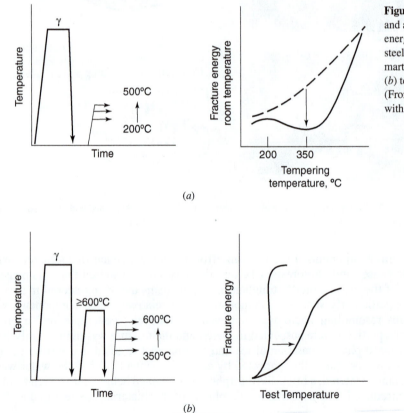

Figure 7.58 Heat treatments and associated fracture energies of temper embrittled steels. (*a*) Tempered martensite embrittlement, and (*b*) temper embrittlement. (From Briant and Banerji[232] with permission.)

range (around 500°C). Note that room temperature embrittlement in the one-step embrittled condition is found when the material is tempered at about 350°C. The fracture energies of samples embrittled by the two-step procedure are shifted to higher test temperatures relative to that of the unembrittled material (i.e., higher ductile–brittle transition temperature).

7.8.1 300 to 350°C or Tempered Martensite Embrittlement

Metallurgists have long recognized the potentially embrittling effects of tempering martensitic steels at about 300 to 350°C. Evidence for embrittlement has been found in this tempering temperature range by noting decreases in notched impact energy, ductility, and tensile strength[233] and a reduction in smooth bar tensile properties when unnotched samples are tested at subzero temperatures[234] (Fig. 7.59). Although precise models to account for all aspects of 300°C embrittlement have not been formulated as yet, certain facts are known. First, the embrittled condition coincides with the onset of cementite precipitation; second, segregation along grain boundaries of impurity elements such as phosphorus (P), sulfur (S), nitrogen (N), antimony (Sb), and tin (Sn) is essential for embrittlement to occur. For example, Capus and Mayer[235] observed no embrittlement trough in high-purity 1.5Ni–Cr–Mo steel when tempered at 300°C, whereas the commercial counterpart of this alloy was embrittled at the same tempering condition (Fig. 7.60). Similar results have been confirmed by Banerji and coworkers for the case of commercial and high-purity heats of 4340 steel.[236] Tempered martensite embrittlement (TME) represents a problem of intergranular embrittlement brought about as a result of precipitation of carbides along prior austenite grain boundaries that had been embrittled by the segregation of P and S during prior austenization. Indeed, Bandyopadhyay and McMahon[237] observed that the impact energy

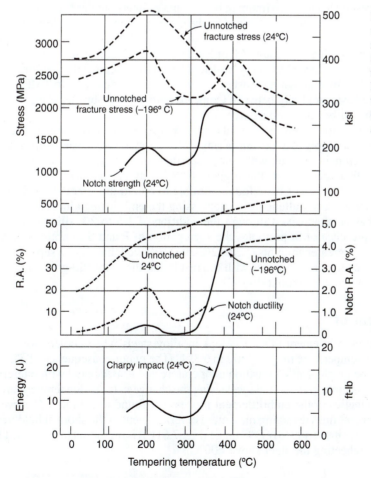

Figure 7.59 Notched and unnotched tensile properties at room and low temperatures for SAE 1340 steel, quenched and tempered at various temperatures. Poor properties associated with tempering in range of 300°C.[233] (Reprinted with permission of the American Society for Metals.)

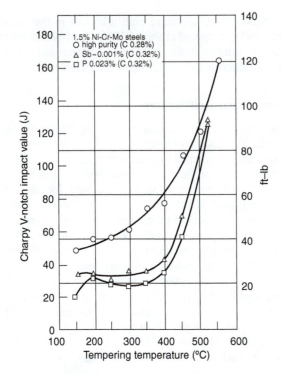

Figure 7.60 Influence of phosphorus and antimony on room temperature impact energy as a function of tempering temperature in 1.5% Ni–Cr–Mo steel.[235] (From J. M. Capus and G. Mayer, *Metallurgia* 62 (1960); with permission of Industrial Newspapers Ltd.)

minima corresponded to a maximum in the amount of intergranular fracture observed. The extent of such P and S segregation is enhanced by the presence of manganese (Mn) and silicon (Si) in the alloy; increased Mn or Si levels were found to increase the fraction of austenite grain boundaries that became embrittled as a result of P or S segregation.[238] Conversely, elimination of Mn and Si from a high-purity NiCrMo alloy restricted impurity segregation at prior austenite grain boundaries, thereby eliminating most of the material's susceptibility to TME.[237] In this context, there exists a basic similarity between tempered martensite embrittlement and temper embrittlement, which will be discussed shortly.

Short of preparing high-purity (but expensive) alloys, the most obvious way to avoid 300°C embrittlement is simply to avoid tempering at that temperature. Usually this involves tempering at a higher temperature but with some sacrifice in strength. However, there are material applications that arise where the higher strengths associated with tempering at 300°C are desired. Fortunately, it has been found possible to obtain the strength levels associated with a 300°C temper while simultaneously suppressing the embrittling kinetics. This optimization of properties has been achieved through the addition of 1.5 to 2% silicon to the alloy steel.[239] Surely, the presence of Si promotes the segregation of P and S to grain boundaries; however, when present in greater amounts (1.5 to 2%), it is believed that silicon suppresses the kinetics of the martensite tempering process with the result that the embrittling reaction shifts to a higher tempering temperature (about 400°C).

7.8.2 Temper Embrittlement

Temper embrittlement (TE) develops in alloy steels when cooled slowly or isothermally heated in the temperature range of 400 to 600°C. The major consequence of TE is found to be an increase in the tough–brittle transition temperature and is associated with intercrystalline failure along prior austenite grain boundaries. Using the change in transition temperature as the measure of TE, the kinetics of the embrittlement process are found to exhibit a *C*-curve response, with isoembrittlement lines depicting maximum embrittlement in the shortest hold time at intermediate temperatures in the 400 to 600°C range (Fig. 7.61). It is important to note that TE can be largely reversed by reheating the steel above 600°C.

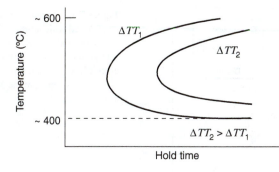

Figure 7.61   Isoembrittlement lines (fixed shift in tough–brittle transition temperature) as function of exposure temperature and hold time.

The catastrophic failure in 1969 of two forged alloy steel disks from the Hinkley Point nuclear power station steam turbine rotor offers dramatic proof that failure to understand the TE process can lead to major problems.[240] In this instance, failure was attributed to a combination of two factors: TE resulting from slowly cooling the disks during manufacture through the critical temperature range, and environment-assisted cracking resulting from the entrapment of condensate in the keyways of the disks. Although TE has been recognized for over 85 years (e.g., see the reviews by Holloman,[241] Woodfine,[242] Low,[243] McMahon,[244] and Briant and Banerji[232]), it is by no means under control.

Balajiva and co-workers[245,246] contributed much to our current understanding of temper embrittlement. They demonstrated that TE occurred only in alloy steels of commercial purity but not in comparable alloys of high purity (Fig. 7.62). The most potent embrittling elements were found to be antimony, phosphorus, tin, and arsenic. These results have been verified by others,[248] along with the additional finding that for a given impurity level, Ni–Cr–alloy steels are embrittled more than alloys containing nickel or chromium alone. It has generally been thought that embrittlement resulted from the segregation of impurity elements at prior austenite grain boundaries as a result of exposure to the 400 to 600°C temperature range. This has since been verified using Auger electron spectroscopy[249–251]—a technique by which the chemistry of the first few atomic layers of a material's surface is analyzed. Marcus and Palmberg[249,250] found, in a modified AISI 3340 steel alloy, antimony on the fracture surface (along prior austenite grain boundaries) in amounts exceeding 100 times that of the bulk concentration (0.03 a/o). Furthermore, the high antimony concentration layer was very shallow, extending only one to two atomic layers below the fracture surface.

The severity of embrittlement depends not only on the amount of poisonous elements present such as Sb, Sn, and P (Fig. 7.63), but also on the overall composition of the alloy. Regarding the latter, certain alloying elements may either enhance or suppress grain-boundary segregation of the embrittling species. The respective influence of alloying elements such as chromium (Cr), manganese (Mn), nickel (Ni), titanium (Ti), and molybdenum (Mo) on the

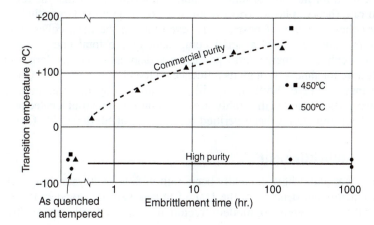

Figure 7.62   Effect of 450 and 500 °C exposure on temper embrittlement in commercial and high-purity nickel–chromium steels. (Data from Woodfine and Steven et al.[246, 247] Reprinted by permission of the American Society for Testing and Materials from copyrighted work.)

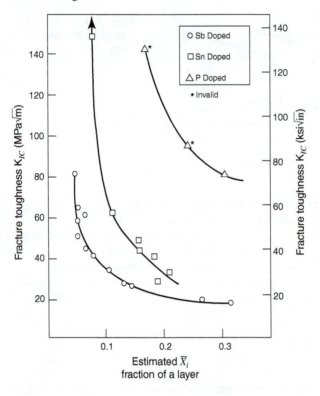

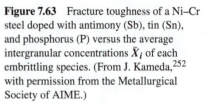

Figure 7.63 Fracture toughness of a Ni–Cr steel doped with antimony (Sb), tin (Sn), and phosphorus (P) versus the average intergranular concentrations $\bar{X}_i$ of each embrittling species. (From J. Kameda,[252] with permission from the Metallurgical Society of AIME.)

temper embrittlement process is reviewed by Briant and Banerji.[232] In another series of studies, the extent of TE in steels was reported to be reduced through the addition of lanthanide metal. In this instance, lanthanides served as scavengers by forming harmless compounds in the matrix with such embrittling impurities as P, As, Sn, and Sb;[253] as a result, the impurity content within the matrix was depleted along with the tendency for TE. [Also see Y. Jingsheng et al., *J. Metals,* 40 (5), 26 (1988).]

The actual mechanism for temper embrittlement remains unclear, though some potentially valuable models are being developed. For example, McMahon and co-workers[236,254–256] have argued that segregation of impurities such as Sb reduces the cohesive energy of the grain boundary, which, in turn, lowers the local stress necessary to generate an accelerating microcrack. Proceeding further, the lower stress necessary for fracture brings about a sharp drop in the plastic strain rate (i.e., dislocation activity) and associated plastic work term, since the plastic strain rate depends *exponentially* on the applied stress level (recall Eqs. 3-19 and 3-20). McMahon and Vitec[255] concluded for the case of temper embrittled ferritic steels that the relative decrease in plastic strain rate (hence, plastic work) is an order of magnitude larger than the reduction in intergranular cohesive energy. Consequently, even though the ideal cohesive energy term for intergranular fracture represents a small component of the total energy for fracture (recall Eq. 6-10a), the cohesive energy term possesses a disproportionately large influence on the material's fracture toughness through its influence on the plastic work term. On this basis, it is to be expected that temper embrittlement will be controlled by the maximum grain-boundary impurity concentration found in the highly stressed volume of material located near the notch root (Fig. 7.63). This postulate has been verified for the case of Sb-doped Ni–Cr steel.[257]

7.8.3 Neutron-Irradiation Embrittlement

Nuclear power plants have typically been built with a 30-40 year design lifetime in mind. The earliest reactors constructed in the United States were designed in the 1950s and early 1960s, before the full flowering of modern fracture mechanics. Furthermore, little was known at the time about the long-term effects of neutron bombardment on material behavior. Since then, our understanding of neutron-irradiation embrittlement has grown considerably, which has led directly to the closure of at least one facility (Yankee Rowe in 1992)[258] but, somewhat paradoxically, also to 20-year extensions of operating licenses for several others (e.g., Palisades, renewed to 2031).[259] Of particular concern has been the fracture resistance of the steel alloys and weld metals that have been used to fabricate the large reactor pressure vessels (RPVs) that contain the radioactive material and cooling water.[260] A pressurized water RPV might be 8 inches (0.2 m) thick, 35 feet (10 m) high, and 15 feet (4.5 m) in diameter, with an operating pressure of approximately 15 MPa (2200 psi); this creates hoop stresses in the cylindrical pressure vessel that can lead to disaster if a crack develops and the fracture toughness is inadequate to prevent its catastrophic growth. The so-called beltline region of the cylinder is exposed to the highest neutron fluence, so this is the critical region. Some early RPV designs used rolled plates with longitudinal and cylindrical welds as seen in Fig 7.64. More recent designs intentionally eliminate the longitudinal welds by using large ring forgings.[261] This is done to avoid welds that are oriented normal to the hoop stress direction (e.g., recall the stress distribution in a thin walled pressure vessel described in Chapter 1).

Early RPV steel alloy chemistry and heat treatment gave the material sufficient toughness at the outset of reactor service life and for many years thereafter. It was discovered, however, that these steel alloys—and in particular the weld alloys—had the potential for significant mechanical property degradation after exposure to neutron irradiation. For example, we see

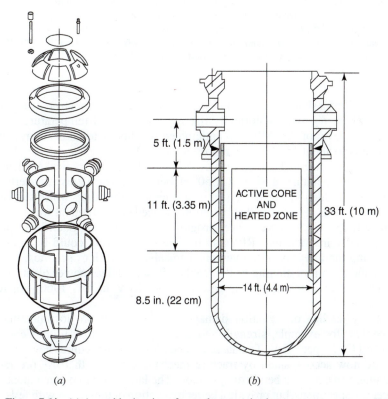

(a)   (b)

Figure 7.64   (a) Assembly drawing of an early pressurized water reactor pressure vessel. The plates are joined by welds. The beltline region of high neutron fluence is circled.[261] [Figure 8 on page 14 of International Atomic Energy Agency, *Integrity of Reactor Pressure Vessels in Nuclear Power Plants: Assessment of Irradiation Embrittlement Effects in Reactor Pressure Vessel Steels*, IAEA Nuclear Energy Series NP-T-3.11, IAEA, Vienna (2009).] (b) Schematic cross-section diagram of the Palisades Nuclear Power Plant RPV.[262] (Reprinted from *Nuclear Engineering Design*, 181, M. G. Vassilaros, M. E. Mayfield, and K. R. Wichman, "Annealing of nuclear reactor pressure vessels," p. 61, 1998, with permission from Elsevier.)

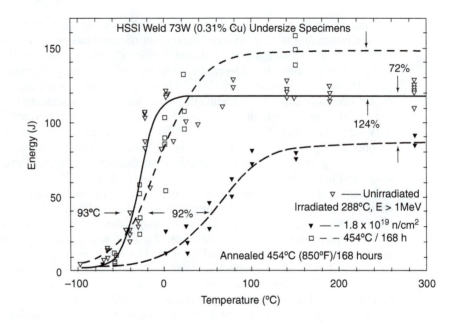

Figure 7.65 Unirradiated ($\triangleleft$), irradiated ($\blacktriangledown$), and irradiated+annealed ($\square$) Charpy V-notch specimens. Irradiation is seen to shift the curve to higher temperatures and to lower the upper shelf energy, while annealing after irradiation restores the transition temperature and also raises the upper shelf energy relative to the original unirradiated state.[264] (Reprinted, with permission, from S. K. Iskander, M. A. Sokolov, and R. K. Nanstad, *Deformation and Fracture Mechanics of Engineering Materials*, ASTM *STP 1270*, 277 (1996), copyright ASTM International, 100 Barr Harbor Drive, West Conshohocken, PA 19428.)

from Fig. 7.65 a sharp reduction in fracture toughness after neutron irradiation, especially in the region where K_{IC} values normally rise rapidly with test temperature.[263,264] As a result, the initial K_{IC} level of 200 MPa$\sqrt{\text{m}}$ anticipated for this material at room temperature is reduced drastically to about 45 MPa$\sqrt{\text{m}}$, representing a 20-fold *decrease* in the allowable flaw size for a given applied stress. This would not be a problem for a typical nuclear RPV at its normal operating temperature of $\sim$290°C(550°F), but it could be a grave concern if the reactor were to shut down suddenly, causing a significant temperature drop while still under pressure.[265]

Once it was discovered that radiation-induced embrittlement was occurring at a higher rate than anticipated, a surveillance program was initiated. Sacrificial pieces of steel were intentionally inserted into RPVs so that periodic testing could be conducted without compromising the integrity of the vessels themselves. A lack of such test coupons led to the shut down and decommissioning of the Yankee Rowe nuclear power plant in 1992; without them, the process of determining the safety of the RPV in this 1960s-era reactor was deemed infeasible.[258]

Early studies of radiation damage[266,267] focused on documenting changes in tensile properties (for example, strength and ductility) and Charpy impact energy absorption using standard CVN specimens. To a large extent, these test procedures have continued to the present, but are now accompanied by fracture mechanics studies that use precracked Charpy (PCC) specimens in a 3-point bend configuration. The latter are more convenient for surveillance than larger CT specimens, but provide a better evaluation of toughness with fewer measurements than the CVN specimens as long as a correction is made for loss of plane-strain constraint associate with their small size.[268] In the course of these studies, it has been found that neutron irradiation causes embrittlement primarily through hardening processes. The increasing yield strength is accompanied by an undesirable increase in the tough–brittle transition temperature. At the same time, there is a corresponding decrease in tensile ductility and Charpy impact upper shelf energy. A diagram showing irradiation-induced changes in Charpy impact response is given in Fig. 7.66,

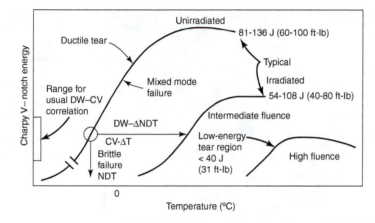

Figure 7.66 Diagram showing transition temperature shift resulting from neutron irradiation of typical reactor-vessel steels.[263] (Reprinted by permission of the American Society for Testing Materials from copyrighted work.)

where higher fluence[xiii] levels are seen to cause greater embrittlement. Studies have also shown that the extent of irradiation damage depends strongly on the irradiation temperature, with more damage accompanying low-temperature neutron exposures.[269]

Understanding of these phenomena has grown considerably since the 1970s.[270,271] For example, it has been shown that one mechanism of hardening is generation of point defects (interstitials and vacancies) and dislocation loops. Although diffusion quickly repairs most of these defects, over time enough accumulate to alter the yield strength. The second important hardening mechanism is the development of Cu-rich nanoclusters and precipitates. Their formation is accelerated by the neutron-induced high vacancy concentration. The amount of neutron embrittlement resulting from irradiation is found to depend strongly on the steel alloy content. For example, Hawthorne et al.[272,273] demonstrated that neutron embrittlement resulting from 288°C exposure could be eliminated completely by careful reduction in residual element content. Although the presence of phosphorus, sulfur, and vanadium in solid solution was identified as being objectionable, *copper* was singled out as the most harmful element. The combined effect of irradiation temperature and copper content on the increment in yield strength resulting from a given neutron fluence is shown in Fig. 7.67. Note that low copper levels (<0.03%) or high irradiation temperatures (288°C) cause the smallest increases in yield strength, while low-temperature (232°C) irradiation produces the greatest damage, regardless of the residual element level. This explains the particular sensitivity of the early weld metal, which was higher in copper than the base metal. In the 1980s, a project organized by the International Atomic Energy Agency resulted in a worldwide change to low-Cu alloys (below 0.1 wt%) for new reactors, significantly reducing the risk of reactor failure by embrittlement.[270] Since then, it has also been determined that the presence of Ni can exacerbate the formation of the Cu-rich nanoclusters, and that Mn also has a synergistic effect with Ni and Cu.[271]

Accurate prediction of the rate and extent of neutron embrittlement is critical to safe operation of nuclear reactors. Original assessments of damage used CVN results at a certain energy level (e.g., 41 J) to indicate the transition temperature and its change with irradiation. It was shown that

---

[xiii] The rate of neutron irradiation or neutron flux is defined as the number of neutrons crossing a one square centimeter area in one second. The neutron fluence is defined by the product of neutron flux and time. Hence, neutron fluence is given by the number of neutrons/cm^2. Studies have shown that these neutrons possess a spectrum of widely varying energies, with some capable of much greater damage than others. Since we do not know the amount of damage generated by a neutron with a particular energy level, it has become the interim accepted practice to define neutron fluence as the count of neutrons possessing more than a certain minimum energy.[194] This energy level is usually taken to be 1.6×10^{-13} J (1 MeV). This counting procedure, therefore, assumes that no damage results from neutrons with energies less than this threshold level (since they are not included in the neutron count) and that all neutron energies greater than this value produce the same damage. Certainly, the current neutron counting procedure is not very discriminating but does provide some means for quantitative analysis of damage.

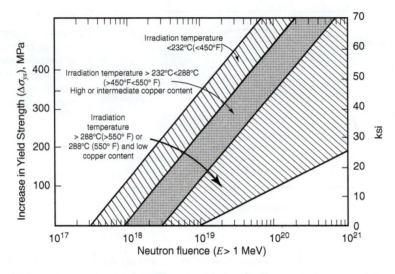

Figure 7.67 Effect of copper content and irradiation temperature on neutron-induced increase in room temperature yield strength of A302-B and A533-B steels.[263] (Reprinted by permission of the American Society for Testing and Materials from copyrighted work.)

this behavior correlates well with K_{IC} at the 100 MPa$\sqrt{m}$ level so a conversion from CVN results to fracture mechanics values is possible (recall Section 6.11.1).[270] In order to provide a conservative measure of safety, the U.S. Nuclear Regulatory Commission (NRC) requires the use of a lower-bound fracture toughness value, designated K_{IR}. In essence, this means that a large number of CVN specimens must be tested to convincingly indicate the lower bound behavior. Additional safety margins are added by using a conservative adjustment factor for the CVN to K_{IC} conversion, and for estimating the level of fluence.[274,275] The temperature dependence of the fracture toughness is then given by:

$$K_{IC} = A + B\exp\{C(T - DBTT)\} \tag{7-8}$$

where A, B, and C are material-specific constants and, T is the temperature of interest, and $DBTT$ is a reference ductile-to-brittle transition temperature.

These safety margins are so conservative, however, that they place potentially unnecessary limits on the lifetime and operating condition of a RPV. More recently, it has been shown that the transition temperature behavior of all RPV steels can be collapsed onto a Master Curve through the use of a material-specific reference temperature, T_0.[274,275] This allows evaluation of a median toughness rather than a lower bound, using as few as six PCC specimens. It assumes that brittle fracture is due to cleavage, and that the probability of cleavage can be described by a Weibull distribution (recall Section 5.31). The toughness as a function of temperature in the lower $DBTT$ region then can be predicted as[23]:

$$K_{jc} = 30 + 70\exp\{0.019(T - T_0)\} \tag{7-9}$$

where K_{jc} has units of MPa$\sqrt{m}$. T_0 is the reference temperature at which the fracture toughness for a fracture probability of 50% is 100 MPa$\sqrt{m}$. The technique for determining T_0 is described in ASTM E 1921, which has been repeatedly revised since 1997 as new data have become available.[276]

Once it has been shown that the fracture toughness of the most sensitive material in a RPV is approaching an unacceptable level, a power plant has several options for staying in operation. The first is to reduce neutron fluence to slow the rate of damage. This compromises the power output of the plant, but may be economically the best choice. However, the temperature dependence of irradiation damage suggests a means by which embrittlement may be reversed, thereby extending the lifetime of an RPV perhaps by decades. If RPV steel is exposed to

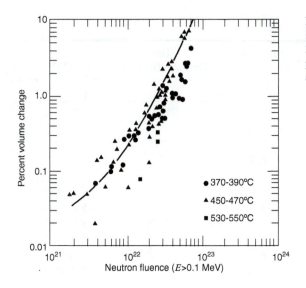

Figure 7.68   Effect of neutron fluence on swelling in annealed type 304 stainless steel in the 370 to 550 °C temperature range.[263] (Reprinted by permission of the American Society for Testing and Materials from copyrighted work.)

temperatures of ∼850°F (455°C) for one week, the transition temperature shifts back to its original, unirradiated value, and the Charpy upper shelf value also rises.[262, 264, 277−278] This annealing approach has been used in several Russian power plants and in U.S. naval reactors, and a successful demonstration project was carried out in the United States in 1998 at the never-completed Marble Hill power plant.[262,279] As of 2010, there were no working U.S. reactors that had undergone the annealing treatment, but it is an option that is acceptable to the NRC.[262]

Although the previous discussion focused on the embrittlement of reactor vessel low-alloy steels, it should be noted that neutron exposure in the temperature range $0.30 < T_h < 0.55$ also causes irradiation damage in stainless steels and other alloys used to contain the nuclear fuel. It is believed that hydrogen and helium—produced by nuclear transmutations—segregate to vacancy clusters and stabilize internal voids.[263] (As discussed, neutron irradiation produces both interstitials and vacancies, but there is preferential recombination of interstitials, leaving an excess of vacancies in the lattice.) Since the excess vacancy and gaseous element concentration increases with increasing neutron fluence, it is found that the relative volume change also increases (Fig. 7.68). The particular temperature regime $0.3 < T_h < 0.55$, where void-induced swelling is most prevalent, arises from the fact that the kinetics of vacancy condensation are too slow at temperatures below $T_h = 0.3$; above about $T_h = 0.55$, the supersaturation of vacancies is inadequate to sustain the voids.[263] Similarly, it has been found that swelling can be suppressed by addition of refractory elements to the alloy, which attenuate vacancy mobility. In addition to swelling of the fuel-cladding alloy, it is known that the uranium fuel itself undergoes volumetric expansion resulting from the precipitation of krypton and xenon gas bubbles, which form as a result of the fission process.[266] Both swelling of the fuel and cladding are highly undesirable and lead to reductions in the operating life of the fuel and metal cladding.

7.9   ADDITIONAL DATA

Additional K_{IC} data are provided in Table 7.8 to enable the reader to become more familiar with the fracture properties of a number of commercial alloys, ceramics, and polymers. The relatively brittle response of ceramic solids, which was anticipated in Table 7.1, is verified convincingly by the tabulated results.

Table 7.8a Strength and Fracture-Toughness Data for Selected Materials[280]*

Alloy	Material Supply	Specimen Orientation	Test Temperature (°C)	σ_{ys} (MPa)	K_{IC}(MPa$\sqrt{m}$)
Aluminum Alloys					
2014-T651	Plate	*L-T*	21–32	435–470	23–27
"	"	*T-L*	"	435–55	22–25
"	"	*S-L*	24	380	20
2014-T6	Forging	*L-T*	"	440	31
"	"	*T-L*	"	435	18–21
2020-T651	Plate	*L-T*	21–32	525–540	22–27
"	"	*T-L*	"	530–540	19
2024-T351	"	*L-T*	27–29	370–385	31–44
"	"	*T-L*	"	305–340	30–37
2024-T851	"	*L-T*	21–32	455	23–28
"	"	*T-L*	"	440–455	21–24
2124-T851	"	*L-T*	"	440–460	27–36
"	"	*T-L*	"	450–460	24–30
2219-T851	"	*L-T*	"	345–360	36–41
"	"	*T-L*	"	340–345	28–38
7049-T73	Forging	*L-T*	"	460–510	31–38
"	"	*T-L*	"	460–470	21–27
7050-T73651	Plate	*L-T*	"	460–510	33–41
"	"	*T-L*	"	450–510	29–38
"	"	*S-L*	"	430–440	25–28
7075-T651	Plate	*L-T*	21–32	515–560	27–31
"	"	*T-L*	"	510–530	25–28
"	"	*S-L*	"	460–485	16–21
7075-T7351	"	*L-T*	"	400–455	31–35
"	"	*T-L*	"	395–405	26–41
7475-T651	"	*T-L*	"	505–515	33–37
7475-T7351[a]	"	*T-L*	"	395–420	39–44
7079-T651	"	*L-T*	"	525–540	29–33
"	"	*T-L*	"	505–510	24–28
7178–T651	"	*L-T*	"	560	26–30
"	"	*T-L*	"	540–560	22–26
"	"	*S-L*	"	470	17
Ferrous Alloys					
4330V (275 °C temper)	Forging	*L-T*	21	1400	86–94
4330V (425 °C temper)	"	*L T*	"	1315	103–110
4340 (205 °C temper)	"	*L-T*	"	1580–1660	44–66
4340 (260 °C temper)	Plate	*L-T*	"	1495–1640	50–63
4340 (425 °C temper)	Forging	*L-T*	"	1360–1455	79–91
D6AC (540 °C temper)	Plate	*L-T*	"	1495	102
D6AC (540 °C temper)	"	*L-T*	−54	1570	62
9-4-20 (550 °C temper)	"	*L-T*	21	1280–1310	132–154
18 Ni(200)(480 °C 6 hr)	Plate	*L-T*	21	1450	110
18Ni(250)(480 °C 6 hr)	"	*L-T*	"	1785	88–97

18 Ni(300)(480 °C)	"	L-T	"	1905	50–64
18Ni(300)(480 °C 6 hr)	Forging	L-T	"	1930	83–105
AFC77 (425 °C temper)	"	L-T	24	1530	79
Titanium Alloys					
Ti-6Al 4V	(Mill anneal plate)	L-T	23	875	123
"	"	T-L	"	820	106
"	(Recryst anneal plate)	L-T	22	815–835	85–107
"	"	T-L	22	825	77–116
Ceramics[c]					
Mortar[281]	—	—	—	—	0.13–1.3
Concrete[282]	—	—	—	—	2–2.3
Al_2O_3[283–285]	—	—	—	—	3–5.3[b]
SiC[283]	—	—	—	—	3.4
SiN_4[284]	—	—	—	—	4.2–5.2
Soda-lime silicate glass[284]	—	—	—	—	0.7–0.8
Electrical porcelain ceramics[286]	—	—	—	—	1.03–1.25
WC(2.5–3 μm)—3 w/o Co[287]	—	—	—	—	10.6
WC(2.5–3 μm)—9 w/o Co[287]	—	—	—	—	12.8
WC(2.5–3.3 μm)—15 w/o Co[287,288]	—	—	—	—	16.5–18
Indiana limestone[289]	—	—	—	—	0.99
ZrO_2 (Ca stabilized)[290]	—	—	—	—	7.6
ZrO_2[290]	—	—	—	—	6.9
Al_2O_3/SiC whiskers[202]	—	—	—	—	8.7
SiC/SiC fibers[202]	—	—	—	—	25
Borosilicate glass/SiC fibers[202]	—	—	—	—	18.9
Polymer[d]					
PMMA[291]	—	—	—	—	0.8–1.75[b]
PS[292]	—	—	—	—	0.8–1.1[b]
Polycarbonate[293]	—	—	—	—	2.75–3.3[b]

[*] For additional fracture toughness data for metallic alloys, see C. M. Hudson and S. K. Seward, *Int. J. Fract.* **14**, R151 (1978); **20**, R59 (1982); **39**, R43 (1989). Also, C. M. Hudson and J. J. Ferrainolo, *Int. J. Fract.* **48**, R19(1991).
[a] Special processing.
[b] K_{Ic} is f(crack speed).
[c] For additional K_{IC} data, see Bradt et al.[294].
[d] For additional K_{IC} data, see Pearson et al.[186] and Azimi et al.[188] and Table 7.7.

Table 7.8b   Strength and Fracture-Toughness Data for Selected Materials[280]

Alloy	Material Supply	Specimen Orientation	Test Temperature (°F)	σ_{ys} (ksi)	K_{IC}(ksi$\sqrt{\text{in}}$)
Aluminum Alloys					
2014-T651	Plate	*L-T*	70–89	63–68	21–24
"	"	*T-L*	"	63–66	20–22
"	"	*S-L*	75	55	18
2014-T6	Forging	*L-T*	"	64	28
"	"	*T-L*	"	63	16–19
2020-T651	Plate	*L-T*	70–89	76–78	20–25
"	"	*T-L*	"	77–78	17–18
2024-T351	"	*L-T*	80–85	54–56	28–40
"	"	*T-L*	"	44–49	27–34
2024-T851	"	*L-T*	70–89	66	21–26
"	"	*T-L*	"	64–66	19–21
2124-T851	"	*L-T*	"	64–67	25–33
"	"	*T-L*	"	65–67	22–27
2219-T851	"	*L-T*	"	50–52	33–37
"	"	*T-L*	"	49–50	26–34
7049-T73	Forging	*L-T*	"	67–74	28–34
"	"	*T-L*	"	67–68	19–25
7050-T73651	Plate	*L-T*	"	67–74	30–37
"	"	*T-L*	"	65–74	26–35
"	"	*S-L*	"	62–64	22–26
7075-T651	Plate	*L-T*	70–89	75–81	25–28
"	"	*T-L*	"	74–77	23–26
"	"	*S-L*	"	67–70	15–19
7075-T7351	"	*L-T*	"	58–66	28–32
"	"	*T-L*	"	57–59	24–37
7475-T651	"	*T-L*	"	73–75	30–33
7475-T7351[a]	"	*T-L*	"	57–61	35–40
7079-T651	"	*L-T*	"	76–78	26–30
"	"	*T-L*	"	73–74	22–25
7178-T651	"	*L-T*	"	81	23–27
"	"	*T-L*	"	78–81	20–23
"	"	*S-L*	"	68	15
Ferrous Alloys					
4330V (525°F temper)	Forging	*L-T*	70	203	78–85
4330V (800°F temper)	"	*L-T*	"	191	94–100
4340 (400°F temper)	"	*L-T*	"	229–241	40–60
4340 (500°F temper)	Plate	*L-T*	"	217–238	45–57
4340 (800°F temper)	Forging	*L-T*	"	197–211	72–83
D6AC (1000°F temper)	Plate	*L-T*	"	217	93
D6AC (1000°F temper)	"	*L-T*	−65	228	56
9-4-20 (1025°F temper)	"	*L-T*	70	186–190	120–140
18Ni(200)(900°F 6 hr)	"	*L-T*	70	210	100
18Ni(250)(900°F 6 hr)	"	*L-T*	"	259	80–88
18Ni(300)(900°F)	"	*L-T*	"	276	45–58
18Ni(300)(900°F 6 hr)	Forging	*L-T*	"	280	75–95
AFC77 (800°F temper)	"	*L-T*	75	222	72

Titanium Alloys

Ti-6Al 4V	(Mill anneal plate)	L-T	74	127	112
"	"	T-L	"	119	96
"	(Recryst anneal plate)	L-T	72	118–121	77–97
"	"	T-L	72	120	70–105

Ceramics[c]

Mortar [281]	—	—	—	1.8–2.1
Concrete [282]	—	—	—	0.21–1.30
Al_2O_3 [283–285]	—	—	—	2.7–4.8
SiC [283]	—	—	—	3.1
Si_3N_4 [284]	—	—	—	3.8–4.7
Soda-lime silicate glass [284]	—	—	—	0.64–0.73
Electrical porcelain ceramics [286]	—	—	—	0.94–1.14
WC(2.5–3 μm)—3 w/o Co [287]	—	—	—	9.6
WC(2.5–3 μm)—9 w/o Co [287]	—	—	—	11.6
WC(2.5–3 μm)—15 w/o Co [287,288]	—	—	—	15–16.4
Indiana limestone [289]	—	—	—	0.9
ZrO_2 (Ca stabilized)[290]	—	—	—	6.9
ZrO_2 [290]	—	—	—	6.3
Al_2O_3/SiC whiskers[290]	—	—	—	7.9
SiC/SiC fibers[202]	—	—	—	22.7
Borosilicate glass/SiC fibers[202]	—	—	—	17.2

Polymer[d]

PMMA[291]	—	—	—	0.73–1.6[b]
PS[292]	—	—	—	0.73–1.0[b]
Polycarbonate[293]	—	—	—	2.5–3.0[b]

[a] Special processing.
[b] KIC is f (crack speed).
[c] For additional KIC data, see Bradt et al.[294]
[d] For additional K_{IC} data, see Pearson et al.[186] and Azimi et al.[188] and Table 7.7.

REFERENCES

1. M. J. Harrigan, *Met. Eng. Quart.*, (May 1974).

2. R. O. Ritchie, *Mechanical Behavior of Materials—5*, Proceedings, 5th International Conference, M. G. Yan, S. H. Zhang, and Z. M. Zheng, Eds., Pergamon, Oxford, 1988.

3. M. E. Launey and R. O. Ritchie, *Adv. Mater.* 21 (20), 2103 (2009).

4. R. A. Jaffe and G. T. Hahn, *Symposium on Design with Materials That Exhibit Brittle Behavior*, Vol. 1, MAB-175-M, National Materials Advisory Board, Washington, DC, Dec. 1960, p. 126.

5. J.J. Lewandowski, M. Shazly, and A. Shamimi Nouri, *Scripta Mat.* **54**, 337 (2006).

6. A. H. Cottrell, *Advances in Physical Metallurgy*, Eds. J. A. Charles and G. C. Smith, Institute of Metals, London, 1990, p. 181.

7. S. F. Pugh, *Philos. Mag.* **45**, 823 (1954).

8. R. O. Ritchie, *Mat. Sci. Eng.* **A103** (1), 15 (1988).

9. B. Cotterell and J. R. Rice, *Int. J. Fract.* **16**, 155 (1980).

10. F. F. Lange, *Philos. Mag.* **22**, 983 (1970).

11. A. G. Evans, *Philos. Mag.* **26**, 1327 (1972).

12. W. S. Hieronymus, *Aviat. Week Space Tech.*, **42** (July 26, 1971).

13. B. I. Edelson and W. M. Baldwin, Jr., *Trans. ASM* **55**, 230 (1962).

14. A. J. Birkle, R. P. Wei, and G. E. Pellissier, *Trans. ASM* **59**, 981 (1966).

15. J. M. Hodge, R. H. Frazier, and F. W. Boulger, *Trans. AIME* **215**, 745 (1959).

16. L. Luyckx, J. R. Bell, A. McClean, and M. Korchynsky, *Met. Trans.* **1**, 3341 (1970).

17. R. W. Hertzberg and R. Goodenow, *Microalloying 1975,* Oct. 1975, Washington, DC.

18. J. C. M. Farrar, J. A. Charles, and R. Dolby, *Effect of Second Phase Particles on the Mechanical Properties of Steel,* Proc. Conf. 1971, ISI, 171 (1972).

19. C. J. McMahon, Jr., *Fundamental Phenomena in the Material Sciences*, Vol. **4**, Plenum, New York, 1967, p. 247.

20. T. L. Johnston, R. J. Stokes, and C. H. Li, *Philos. Mag.* **7**, 23 (1962).

21. A. H. Cottrell, *Trans. Met. Soc. AIME* **212**, 192 (1958).

22. C. Zener, *Fracturing of Metals*, ASM, Cleveland, 1948, p. 3.

23. D. E. Piper, W. E. Quist, and W. E. Anderson, *Application of Fracture Toughness Parameters to Structural Metals*, Vol. **31**, Metallurgical Society Conference, 1966, p. 227.

24. R. E. Zinkham, H. Liebowitz, and D. Jones, *Mechanical Behavior of Materials*, Vol. **1**, Proceedings of the International Conference on Mechanical Behavior of Materials, The Society of Materials Science, Kyoto, Japan, 1972, p. 370.

25. J. G. Kaufman, Agard Meeting of the Structures and Materials Panel, Apr. 15, 1975.

26. R. R. Senz and E. H. Spuhler, *Met. Prog.* **107** (3), 64 (1975).

27. T. Ohira and T. Kishi, *Mater. Sci. Eng.* **78**, 9 (1986).

28. E. L. Colvin, J. I. Petit, R. W. Westerlund, and P. E. Magnusen, U.S. Patent No. 5,213,639, May 25, 1993.

29. W. Cassada, J. Liu, and J. Staley, *Adv. Mat. Proc.* **160** (12), 27 (2002).

30. D. Broek, *Eng. Fract. Mech.* **5**, 55 (1973).

31. F. A. McClintock, *Int. J. Fract. Mech.* **2**, 614 (1966).

32. B. M. Kapadia, A. T. English, and W. A. Backofen, *Trans. ASM* **55**, 389 (1982).

33. B. Mravic and J. H. Smith, "Development and Improved High-Strength Steels for Aircraft Structural Components," AFML-TR-71-213, Oct. 1971.

34. R. Song, D. Ponge, and D. Raabe, *Acta Mat.* **53**, 4881 (2005).

35. M. Korchynsky, American Society of Agricultural Engineers, Paper No. 70-682, Dec. 1970.

36. A. H. Cottrell, *Fracture*, Technology Press MIT and Wiley, New York, 1959, p. 20.

37. N. J. Petch, *Fracture*, Technology Press MIT and Wiley, New York, 1959, p. 54.

38. J. R. Low, Jr., *Relation of Properties to Microstructure*, ASM, Metals Park, OH, 1954, p. 163.

39. A. S. Tetelman and A. J. McEvily, Jr., *Fracture of Structural Materials*, Wiley, New York, 1967.

40. J. T. Michalak, *Metals Handbook*, Vol. **8**, ASM, Metals Park, OH, 1973, p. 220.

41. J. F. Peck and D. A. Thomas, *Trans Met. Soc. AIME*, 1240 (1961).

42. F. A. Heiser and R. W. Hertzberg, *J. Basic Eng.* **93**, 71 (1971).

43. J. Wadsworth and D. R. Lesuer, *Mater. Char.* **45**, 289 (2000).

44. T. Foecke, *Metallurgy of the RMS Titanic*, NIST, Gaithersburg, MD, 1998.

45. W. J. Broad, *New York Times,* Jan. 27, 1998, p. B9.

46. J. Cook and J. E. Gordon, *Proc. R. Soc. London*, **A282**, Ser. A 508 (1964).

47. J. D. Embury, N. J. Petch, A. E. Wraith, and E. S. Wright, *Trans. AIME* **239**, 114 (1967).

48. H. I. Leichter, *J. Spacecr. Rockets* **3** (7), 1113 (1966).

49. A. J. McEvily, Jr. and R. H. Bush, *Trans. ASM* **55**, 654 (1962).

50. J. G. Kaufman, *Trans. ASME, J. Basic Eng.* **89** (3), 503 (1967).

51. M. A. Meyers and C. T. Aimone, *Prog. Mater. Sci.* **28**, 1 (1983).

52. M. E. deMorton, R. L. Woodward, and J. M. Yellup, Proceedings, Fourth Tewksburg Symposium, Melbourne, Feb. 1979, p. 11.1.

53. J. G. Kaufman, P. E. Schilling, and F. G. Nelson, *Met. Eng. Quart.* **9** (3), 39 (1969).

54. N. Tsangarakis, *Mater. Sci. Engr.* **58**, 269 (1983).

55. R. D. Stout, S. J. Wiersma, *Adv. Cryo. Eng.* **32**, 389 (1986).

56. H. E. McGannon, *The Making, Shaping, and Treating of Steel*, 9th ed., United States Steel Corporation, Pittsburgh, PA, 1971.

57. J. R. Low, Jr., *Fracture*, Technology Press MIT and Wiley, New York, 1959, p. 68.

58. D. Webster, *Trans. ASM* **61**, 816 (1968).

59. V. F. Zackay, E. R. Parker, D. Fahr, and R. Busch, *Trans. ASM* **60**, 252 (1967).

60. S. Antolovich, *Trans. Met. Soc. AIME* **242**, 237 (1968).

61. E. R. Parker and V. F. Zackay, *Eng. Fract. Mech.* **5**, 147 (1973).

62. J. P. Bressanelli and A. Moskowitz, *Trans. ASM* **59**, 223 (1968).

63. W. W. Gerberich, G. Thomas, E. R. Parker, and V. F. Zackay, *Proceedings of the Second International Conference on the Strength of Metals and Alloys*, Asilomar, CA, 1970, p. 894.

64. D. Bhandarkar, V. F. Zackay, and E. R. Parker, *Met Trans.* **3**, 2619 (1972).

65. H. Fountain, "Many Faces, and Phases, of Steel in Cars," *New York Times*, Sept. 15, 2009.

66. H. Fountain, "Stronger Metal Poses Challenge to Rescuers", *New York Times*, Sept. 15, 2009.

67. H. Margolin, J. C. Williams, J. C. Chestnutt, and G. Luetjering, *Titanium 80*, H. Kimura and O. Izumi, Eds., AIME, Warrendale, PA, 1980, p. 169.

68. A.R. Rosenfield and A. J. McEvily, Jr., NATO AGARD Report No. 610, Dec. 1973, p. 23.

69. R. Develay, *Met Mater.*, **6**, 1972, p. 404.

70. J. A. Nock, Jr. and H. Y. Hunsicker, *J. Met.*, 216 (Mar. 1963).

71. D. S. Thompson, S. A. Levy, and D. K. Benson, Third International Conference on Strength of Metals and Alloys, Paper 24, 1973, p. 119.

72. N. E. Paton and A. W. Sommer, Third International Conference on Strength of Metals and Alloys, Paper 21, 1973, p. 101.

73. J. T. Staley, AIME Spring Meeting, Alcoa Report, May 23, 1974.

74. R. J. Bucci, *Eng. Fract Mech.* **12**, 407 (1979).

75. E. A. Starke, T. H. Sanders, Jr., and I. G. Palmer, *J. Metals* **33**, 24 (1981).

76. B. Nobel, S. J. Harris, and K. Dinsdale, *Metal Science* **16**, 425 (1982).

77. T. H. Sanders, Jr., NAVAIR Contract No. N62269-76-C-0271, Final Report, June 9, 1979.

78. A. K. Vasudevan, A. C. Miller, and M. M. Kersker, *Proceedings of the Second International Al-Li Conference*, E. A. Starke, Jr. and T. H. Sanders, Jr., Eds., AIME, 181 (1983).

79. D. Webster, *Met Trans.* **18A**, 2181 (1987).

80. D. Webster, *Advanced Materials and Processes* **145** (5), 18 (1994).

81. M. Romios, R. Tiraschi, C. Parrish, H.W. Babel, J.R. Ogren, and O.S. Es-Said, *J. Mater. Eng. Perf.* **14**, 641 (2005).

82. P. S. Chen, A. K. Kuruvilla, T. W. Malone, and W. P. Stanton, *J. Mater. Eng. Perf.* **7**, 682 (1998).

83. K. T. Venkateswara Rao and R. O. Ritchie, *Inter. Mater. Rev.* **37** (4), 153 (1992).

84. C. Giummarra, B. Thomas, and R. J. Rioja, Proc. Light Metals Tech. Conf. 2007 (2007).

85. A. G. Evans, *Ceramic Transactions* Vol I, *Ceram., Powder Sci.,* G. L. Messing, E. R. Fuller, Jr., and H. Hauser, Eds., American Ceramic Society, Westerville, OH, 989 (1989).

86. R. F. Cook, B. R. Lawn, and C. J. Fairbanks, *J. Amer. Ceram. Soc.* **68** (11), 604 (1985).

87. B. Lawn, *Fracture of Brittle Solids*, 2nd ed., Cambridge Solid State Science Series, Cambridge University Press, Cambridge, UK (1993).

88. P. F. Becher, *J Amer. Ceram. Soc.* **74** (2), 255 (1991).

89. A. G. Evans, *J Amer. Ceram. Soc.* **73** (2), 187 (1990).

90. S. Suresh and C. F. Shih, *Int. J. Fracture* **30**, 237 (1986).

91. K. T. Faber and A. G. Evans, *Acta Metall.* **31**, 577 (1983).

92. R. W. Davidge and T. J. Green, *J. Mater. Sci.* **3**, 629 (1968).

93. M. P. Harmer, H. M. Chan, and G. A. Miller, *J. Amer. Ceram. Soc.* **75** (7), 1715 (1992).

94. V. C. Li and T. Hashida, *J. Mater. Sci. Lett.*, **12**, 898 (1993).

95. M. Rühle, *Mater. Sci. Eng.* **A105/106**, 77 (1988).

96. A. H. Heuer and L. W. Hodds, Eds., *Advances in Ceramics*, Vol. 3, *Science and Technology of Zirconia*, ACS, Columbus, OH, 1981.

97. N. Claussen and M. Rühle, ibid., p. 137.

98. E. C. Subbarao, ibid., p. 1.

99. R. C. Garvie, R. H. Hanink, and R. T. Pascoe, Nature, *London* **258** (5537), 703 (1975).

100. D. L. Porter and A. H. Heuer, *J. Am. Ceram. Soc.* **60** (3–4), 183 (1977).

101. D. L. Porter and A. H. Heuer, *J. Am. Ceram. Soc.* **62** (5–6), 298 (1979).

102. A. G. Evans, *Mater. Sci. Eng.* **105/106**, 65 (1988).

103. A. G. Evans and R. M. Cannon, *Acta Metall.* **34** (5), 761 (1986).

104. D. Green, R. H. J. Hannink, and M. V. Swain, *Transformation Toughening of Ceramics*, CRC Press, Boca Raton, FL (1989).

105. A. G. Evans and A. H. Heuer, *J. Amer. Ceram. Soc.* **63** (5–6), 241 (1980).

106. R. M. McMeeking and A. G. Evans, *J. Am. Ceram. Soc.* **65** (5), 242 (1982).

107. S. M. Wiederhorn, *Annu. Rev. Mater. Sci.* **14**, 373 (1984).

108. F. F. Lange, *J. Mater. Sci.* **17**, 225–263 (1982).

109. N. Padture, Ph.D dissertation, Lehigh University, Bethlehem, PA (1991).

110. P. F. Becher, C. H. Hsueh, P. Angelini, and T. N. Tiegs, *J. Amer. Ceram. Soc.* **71** (12), 1050 (1988).

111. S. V. Nair, *J. Amer. Ceram. Soc.* **73** (10), 2839 (1990).

112. P. L. Swanson, C. J. Fairbanks, B. R. Lawn, and Y. W. Mai, *J. Amer. —Ceram. Soc*, **70** (4), 279 (1987).

113. Y. W. Mai and B. R. Lawn, *J. Amer. Ceram. Soc.* **70** (4), 289 (1987).

114. S. T. Bennison and B. R. Lawn, *Acta Metall*, **37** (10), 2659 (1989).

115. R. F. Cook, C. J. Fairbanks, B. R. Lawn, and Y. W. Mai, *J. Mater. Res.* **2**, 345 (1987).

116. Y. Tajima, K. Urashima, M. Watanabe, and Y. Matsuo, *Ceramic Transactions*, Vol. **1**, *Ceram. Powder Sci.-IIB*, G. L. Messing, E. R. Fuller, Jr., and H. Hauser, Eds., American Ceramic Society, Westerville, OH, 1034 (1988).

117. J. C. Amazigo and B. Budiansky, *J. Mech. Phys. Solids* **36**, 581 (1988).

118. S. M. Wiederhorn, A. G. Evans, and D. E. Roberts, *Fracture of Ceramics*, Vol. **2**. R. C. Bradt, D. P. H. Hasselman, and F. F. Lange, Eds., Plenum, New York, 1974, pp. 829–41.

119. R. J. Eagan and J. C. Swearekgen, *J. Am. Cer. Soc.* **61** (1–2), 27–30 (1978).

120. M. Haldimann, A. Luible, and M. Overend, *Structural Use of Glass*. International Association for Bridge and Structural Engineering IABSE, Zurich, 2008.

121. D. B. Marshall and B. R. Lawn, *J. Am. Cer. Soc.* **61** (1–2), 21 (1978).

122. V. M. Sglavo, L. Larentis and D. J. Green, *J. Am. Ceram. Soc.* **84** (8), 1827–31 (2001)

123. ASTM C1048-04, ASTM International, West Conshohocken, PA.

124. R. Tandon and R. F. Cook, *J. Am. Ceram. Soc.* **76** (4), 885–89 (1993).

125. M. V. Swain, *J. Mater. Sci.* **16**, 151–158 (1981).

126. ASTM C1422, ASTM International, West Conshohocken, PA.

127. A. L. Zijlstra and A. J. Burggraaf, *J. Non-Cryst. Solids* **1** (1), 49 (1968).

128. T. V. Permyakova et al., *Glass and Ceramics* **38** (5), 228–30 (1982).

129. J. E. Kooi, R. Tandon, S. J. Glass, and J. J. Mccholsky, *J. Mater. Res.* **23** (1), 214 (2008).

130. V. M. Sglavo and D. J. Green, *J. Am. Ceram. Soc.* **84** (8) 1832–38 (2001)

131. S. J. Bennison, A. Jagota, and C. A. Smith, *J. Am. Ceram. Soc.* **82** (7), 1761–70 (1999)

132. H. Chai and B. R. Lawn, *Acta Mat.* **50**, 2613–2625 (2002).

133. M. Richards, R. Clegg, and S. Howlett, 18th Int. Symp. Ballistics, San Antonio, TX, 1999.

134. M. H. Litt and A. V. Tobolsky, *J. Macromol. Sci. Phys. B* **1** (3), 433 (1967).

135. R. F. Boyer, *Rubber Chem. Tech.* **36**, 1303 (1963).

136. S. E. B. Petrie, *Polymeric Materials*, ASM, Metals Park, OH, 1975, p. 55.

137. J. Heijboer, *J. Polym. Sci.* **16**, 3755 (1968).

138. R. F. Boyer, *Polym. Eng. Sci.* **8** (3), 161 (1968).

139. W. J. Jackson, Jr., and J. R. Caldwell, *J. Appl. Polym. Sci.* **11**, 211 (1967).

140. J. A. Manson and L. H. Sperling, *Polymer Blends and Composites*, Plenum, New York, 1976.

141. C. B. Bucknall, *Toughened Plastics*, Applied Science, Barking, UK, 1977.

142. *Toughening of Plastics II*, Plast. Rub. Inst., London, 1985.

143. A. F. Yee and R. A. Pearson, *Toughening of Plastics II*, Rub. Inst., London, 1985, p. 2/1.

144. A. J. Kinloch and D. L. Hunston, *Toughening of Plastics II*, Plast. Rub. Inst., London, 1985, p. 4/1.

145. A. F. Yee and R. A. Pearson, *J. Mater. Sci.* **21**, 2462 (1986).

146. R. A. Pearson and A. F. Yee, *J. Mater. Sci.* **21**, 2475 (1986).

147. A.F. Yee, *ASTM STP* **937**, 1987, p. 383.

148. A. C. Garg and T. W. Mai, *Comp. Sci. and Tech.* **31**, 179 (1988).

149. *Rubber Toughened Engineering Plastics*, A. A. Collyer, Ed., Chapman and Hall, London, 1994.

150. C. E. Hall, *Introduction to Electron Microscopy*, 2nd ed., McGraw-Hill, New York, 1966.

151. G. A. Crosbie and M. G. Phillips, *J. Mater. Sci.* **20**, 182 (1985).

152. C. B. Bucknall and D. G. Street, *Soc. Chem. Ind. Monograph* No. 26, London, 1967, p. 272.

153. C. B. Bucknall, D. Clayton, and W. Keast, *J. Mater. Sci.* **7**, 1443 (1972).

154. H. Brewer, F. Haaf, and J. Stabenow, *J. Macromol. Sci. Phys. B* **14**, 387 (1977).

155. F. Haaf, H. Brewer, and J. Stabenow, *Angew Makromol. Chem.* **58/59**, 95 (1977).

156. A. M. Donald and E. J. Kramer, *J. Mater. Sci.* **17** (6), 1765 (1982).

157. J. N. Sultan, R. C. Liable, and F. J. McGarry, *Polym. Symp.* **16**, 127 (1971).

158. J. N. Sultan and F. J. McGarry, *Polym. Eng. Sci.* **13**, 29 (1973).

159. Y. Huang and A. J. Kinloch, *J. Mater. Sci. Lett.* **11**, 484 (1992).

160. A. J. Kinloch, *Rubber Toughened Plastics*, C. K. Riew, Ed., ACS **222**, 67 (1989).

161. Y. Huang and A. J. Kinloch, *Polymer* **33** (24), 5338 (1992).

162. S. Kunz-Douglass, P. W. R. Beaumont, and M. F. Ashby, *J. Mater. Sci.* **15**, 1109 (1980).

163. A.G. Evans, Z. B. Ahmed, D. G. Gilbert, and P. W. R. Beaumont, *Acta Metall.* **34**, 79 (1986).

164. H. Azimi, R. A. Pearson, and R. W. Hertzberg, *J. Mater. Sci. Lett.*, **13**, 1460 (1994).

165. F. J. Guild, A. J. Kinloch, and A. C. Taylor, *J. Mater. Sci.* **45**, 3882–3894 (2010).

166. J. Liu et al., *Macromolecules* **41**, 7616–7624 (2008).

167. H. Azimi, R. A. Pearson, and R. W. Hertzberg, *J. Mater. Sci.* **31**, 3777–3789 (1996).

168. R. A. Pearson and A. F. Yee, *J. Mater. Sci.* **26**, 3828 (1991).

169. A.-V. Ruzette and L. Leibler, *Nature Materials* **4**, 19 (2005).

170. A. G. Evans, S. Williams, and P. W. R. Beaumont, *J. Mater. Sci.* **20**, 3668 (1985).

171. F. F. Lange, *Phil. Mag.* **22**, 983 (1970).

172. A. C. Moloney, H. H. Kausch, T. Kaiser, and H. R. Beer, *J. Mater. Sci.* **22**, 381 (1987).

173. J. Spanoudakis and R. J. Young, *J. Mater. Sci.* **19**, 473 (1984).

174. J. K. Wells and P. W. R. Beaumont, *J. Mater. Sci.* **23**, 1274 (1988).

175. A.G. Evans and R. M. McMeeking, *Acta Met.* **34** (12), 2435 (1986).

176. A. H. Cottrell, *Proc. Roy. Soc. Lond.* **A282**, 2 (1964).

177. A. Kelly and W. R. Tyson, *J. Mech. Phys. Solids* **13**, 329 (1965).

178. M. D. Thouless, O. Sbaizero, L. S. Sigl, and A. G. Evans, *J. Amer. Cer. Soc.* **72** (4), 525 (1989).

179. B. B. Johnsen, A. J. Kinloch, R. D. Mohammed, A. C. Taylor, and S. Sprenger, *Polymer* **48**, 530–541 (2007).

180. B. Wetzel, P. Rosso, F. Haupert, and K. Friedrich, *Eng. Fract. Mech.* **73**, 2375–2398 (2006).

181. C. Chen, R. S. Justice, D. W. Schaefer, and J. W. Baur, *Polymer* **49**, 3805–3815 (2008).

182. C. Chen and A. B. Morgan, *Polymer* **50**, 6265–6273 (2009).

183. D. L. Maxwell, R. J. Young, and A. J. Kinloch, *J. Mater. Sci. Lett*, **3**, 9 (1984).

184. A. J. Kinlock, D. L. Maxwell, and R. J. Young, *J. Mater. Sci.*, **20**, 4169 (1985).

185. I. M. Low, S. Bandyopakhyay, and Y. W. Mai, *Polym. Inter.* **27**, 131 (1992).

186. R. A. Pearson, A. K. Smith, and Y. W. Yee, *Second International Conference on Deformation and Fracture of Comp.*, Manchester, 9-1 (1993).

187. T. Shimizu, M. Kamino, M. Miyagaula, N. Nishiwacki, and S. Kida, *Ninth International Conference on Deformation, Yield and Fracture of Polymers*, Churchill College, Cambridge, U.K., 76/1 April 1994.

188. H. Azimi, Ph.D. dissertation, Lehigh University, Bethlehem, PA (1994).

189. T. H. Hsieh et al., *J. Mater. Sci.* **45**, 1193–1210 (2010).

190. J. Kim, C. Baillie, J. Poh, and Y. W. Mai, *Comp. Sci. Tech.* **43** (3), 283–297 (1992).

191. M. D. Gilchrist and N. Svensson, *Comp. Sci. Tech.* **55**, 195–207 (1995).

192. A. J. Kinloch, R. D. Mohammed, A. C. Taylor, S. Sprenger, and D. Egan, *J. Mater. Sci.* **41**, 5043–5046 (2006).

193. A. J. Kinloch, K. Masania, A. C. Taylor, S. Sprenger, and D. Egan, *J. Mater. Sci.* **43**, 1151–1154 (2008).

194. A. Kelly, *Proc. R. Soc. London Ser. A* **319**, 95 (1970).

195. J. O. Outwater and M. C. Murphy, *Proc. 24th SPI/RP Conf.*, Paper 11-B, SPI, New York, 1969.

196. K. Friedrich, *J. Mater. Sci.* **33**, 5535–5556 (1998).

197. U. G. K. Wegst and M. F. Ashby, *Phil. Mag.* **84**, 2167–2186 (2004).

198. H. D. Espinosa, J. E. Rim, F. Barthelat, and M. J. Buehler, *Prog. Mater. Sci.* **54**, 1059–1100 (2009)

199. F. Barthelat, J. E. Rim, and H. D. Espinosa, *Applied Scanning Probe Methods XIII*, B. Bhushan and H. Fuchs,Eds., Springer, 2009, pp. 1059–1100.

200. A. P. Jackson, J. F. V. Vincent, and R. M. Turner, *J. Mater. Sci.* **25**, 3173 (1990).

201. F. Barthelat and H. D. Espinosa, *Experimental Mechanics* **47**, 311–324 (2007).

202. *Advanced Materials and Processes*, Vol. **2** (9), 1986, p. 32.

203. S. Kamat, X. Su, R. Ballarini, and A. H. Heuer, *Nature* **405**, 1036–1040 (2000)

204. F. Barthelat, H. Tang, P. D. Zavattieri, C. M. Li, and H. D. Espinosa, *J. Mech. Phys. Sol.* **55**, 306–337 (2007).

205. M. E. Launey, E. Munch, D. H. Alsem, H. B. Barth, E. Saiz, A. P. Tomsia, and R. O. Ritchie, *Acta Materialia* **57**, 2919–2932 (2009)

206. R. B. Martin, D. B. Burr, and N. A. Sharkey, *Skeletal Tissue Mechanics*, Springer, New York, 1998.

207. K. J. Koester, J. W. Ager, and R. O. Ritchie, *JOM* **60** (6), 33–38 (2008).

208. J. D. Currey, K. Brear, and P. Zioupos, *J. Biomechanics* **29** (2), 257–260 (1996).

209. M. E. Launey, M. J. Buehler, and R. O. Ritchie, *Annu. Rev. Mater. Res.* **40**, 9. 1–9. 29 (2010).

210. H. Gao, *Int. J. Fract.* **138**, 101–137 (2006).

211. H. Peterlik, P. Roschger, K. Klaushofer, and P. Fratzl, *Nature Materials* **5**, 52–55 (2006).

212. R. K. Nalla, J. S. Stölken, J. H. Kinney, and R. O. Ritchie, *J. Biomechanics* **38**, 1517–1525 (2005).

213. R. O. Ritchie, M. J. Buehler, and P. Hansma, *Physics Today*, June, 41–47 (2009).

214. K. Tai, F-J. Ulm, and C. Ortiz, *Nano Lett.* **6** (11), 2520–2525 (2006).

215. R. K. Nalla, J. J. Kruzic, J. H. Kinney, M. Balooch, J. W. Ager, and R. O. Ritchie, *Mat. Sci. Eng. C* **26**, 1251–1260 (2006).

216. R. R. Adharapurapu, F. Jiang, and K. S. Vecchio, *Mat. Sci. Eng. C* **26**, 1325–1332 (2006).

217. J.D. Currey, *Bones*, 2nd ed. Princeton University Press, Princeton, NJ, 2002.

218. M. A. Meyers, P-Y. Chen, A. Y-M. Lin, and Y. Seki, *Prog. Mater. Sci.* **53**, 1–206 (2008).

219. A. P. Jackson, J. F. V. Vincenta, and R. M. Turner, *Comp. Sci. Tech.* **36** (3), 255–266 (1989).

220. W. J. Clegg, K. Kendall, N. M. Alford, T. W. Button, and J. D. Birchall, *Nature* **347**, 455–457 (1990).

221. G. Mayer, *Mater. Sci. Eng. C* **26**, 1261–1268 (2006).

222. Z. Tang, N. A. Kotov, S. Magonov, and B. Ozturk, *Nature Materials* **2**, 413–418 (2003).

223. P. Podsiadlo, A. K. Kaushik, E. M. Arruda, A. M. Waas, B. S. Shim, J. Xu, H. Nandivada, B. G. Pumplin, J. Lahann, A. Ramamoorthy, and N. A. Kotov, *Science* **318**, 80–83 (2007).

224. L. J. Bonderer, A. R. Studart, and L. J. Gauckler, *Science* **319**, 1069–1073 (2008)

225. L. Chen, R. Ballarini, H. Kahn, and A. H. Heuer, *J. Mater. Res.* **22** (1), 124–131 (2007).

226. E. Munch, M. E. Launey, D. H. Alsem, E. Saiz, A. P. Tomsia, and R. O. Ritchie, *Science* **322**, 1516–1520 (2008).

227. Y. Seki, B. Kad, D. Benson, and M. A. Meyers, *Mater. Sci. Eng. C* **26**, 1412–1420 (2006).

228. P.-Y. Chen, A. Y.-M. Lin, A. G. Stokes, Y. Seki, S. G. Bodde, J. McKittrick, and M. A. Meyers, *JOM* **60** (6), 23–32 (2008).

229. K. Dubey and V. Tomar, *Annals of Biomedical Eng.* **38** (6), 2040–2055 (2010).

230. M.E. Launey, P.-Y. Chen, J. McKittrick, and R. O. Ritchie, *Acta Biomaterialia* **6**, 1505–1514 (2010).

231. Foster-Miller, Inc., *Cost Effective Advanced Ceramic Armor*, Waltham, MA, 1995.

232. C. L. Briant and S. K. Banerji, *Int. Met. Rev.,* Review 232, No. 4, 1978, p. 164.

233. W. F. Brown, Jr., *Trans. ASM* **42**, 452 (1950).

234. E. J. Ripling, *Trans. ASM* **42**, 439 (1950).

235. J. M. Capus and G. Mayer, *Metallurgia* **62**, 133 (1960).

236. S. K. Banerji, H. C. Feng, and C. J. McMahon, Jr., *Met. Trans.* **9A**, 237 (1978).

237. N. Bandyopadhyay and C. J. McMahon, Jr., *Met. Trans.* **14A**, 1313 (1983).

238. J. Yu and C. J. McMahon, Jr., *Met. Trans.* **11A**, 291 (1980).

239. C. H. Shih, B. L. Averbach, and M. Cohen, *Trans. ASM* **48**, 86 (1956).

240. D. Kalderon, *Proc. Inst. Mech. Eng.* **186**, 341 (1972).

241. J. H. Hollomon, *Trans. ASM* **36**, 473 (1946).

242. B. C. Woodfine, *JISI* **173**, 229 (1953).

243. J. R. Low, Jr., *Fracture of Engineering Materials*, ASM, Metals Park, OH, 1964, p. 127.

244. C. J. McMahon, Jr., ASTM *STP* **407**, 1968, p. 127.

245. K. Balajiva, R. M. Cook, and D. K. Worn, *Nature London* **178**, 433 (1956).

246. W. Steven and K. Balajiva, *JISI* **193**, 141 (1959).

247. J. M. Capus, ASTM *STP* **407**, 1968, p. 3.

248. J. R. Low, Jr., D. F. Stein, A. M. Turkalo, and R. P. LaForce, *Trans. Met. Soc. AIME* **242**, 14. (1968).

249. H. L. Marcus and P. W. Palmberg, *Trans. Met. Soc. AIME* **245**, 1665 (1969).

250. H. L. Marcus, L. H. Hackett, Jr., and P. W. Palmberg, ASTM *STP* **499**, 1972, p. 90.

251. D. F. Stein, A. Joshi, and R. P. LaForce, *Trans. ASM* **62**, 776 (1969).

252. J. Kameda, *Met. Trans.* **12A**, 2039 (1981).

253. C. I. Garcia, G. A. Ratz, M. G. Burke, and A. J. DeArdo, *J. Met.* **37** (9), 22 (1985).

254. C. J. McMahon, Jr., V. Vitek, and J. Kameda, *Developments in Fracture Mechanics*, Vol. **2**, G. G. Chell, Ed., Applied Science, New Jersey, 1981, p. 193.

255. C. J. McMahon, Jr., and V. Vitek, *Acta Met.* **27**, 507 (1979).

256. J. Kameda and C. J. McMahon, Jr., *Met. Trans.* **11A**, 91 (1980).

257. J. Kameda and C. J. McMahon, Jr., *Met. Trans.* **12A**, 31 (1981).

258. M. L. Wald, "U.S. Says 15 Reactors Need Testing," *New York Times*, Apr. 2, 1993.

259. NUREG-1871 Safety Evaluation Report, U.S. Nuclear Regulatory Commission (2007).

260. OTA-E-575, U.S. Congress, Office of Technology Assessment (1993).

261. NP-T-3.11, IAEA Nuclear Energy Series, International Atomic Energy Agency (2009).

262. M. G. Vassilaros, M. E. Mayfield, K. R. Wichman, *Nuc. Eng. Design* **181**, 61–69 (1998).

263. S. H. Bush, *J. Test. Eval.* **2** (6), 435 (1974).

264. S. K. Iskander, M. A. Sokolov, and R. K. Nanstad, ASTM STP **1270**, 277–293 (1996).

265. M. L. Wald, "Nuclear Accident Raises Doubt on Safety Margins," *New York Times*. Dec. 6 1981.

266. A. Tetelman and A. J. McEvily, *Fracture of Structural Materials*, Wiley, New York, 1967.

267. D. McClean, *Mechanical Properties of Metals*, Wiley, New York, 1962.

268. J. P. Petti and R. H. Dodds, *Eng. Fracture Mech.* **71**, 2677–2683 (2004).

269. L. E. Steele, *Nucl. Mater.* **16**, 270 (1970).

270. W. L. Server and R. K. Nanstad, *J. ASTM Int.* **6** (7), 1–17 (2009).

271. G. R. Odette and R. K. Nanstad, *JOM* **61** (7), 17–23 (2009).

272. U. Potapovs and J. R. Hawthorne, *Nucl. Appl.* **6** (1), 27 (1969).

273. J. R. Hawthorne, ASTM *STP* **484**, 1971, p. 96.

274. M. E. Mayfield, M. G. Vassilaros, E. M. Hackett, M. A. Mitchell, K. R. Wichman, J. R. Strosnider, and L. C. Shao, *Trans. 14th Int. Conf. Struct. Mech. Reactor Tech.*, 13–19 (1997).

275. H.-W. Viehrig, J. Boehmert and J. Dzugan, *Nuc. Eng. Design* **212**, 115–124 (2002).

276. ASTM E1921, ASTM International, West Conshohocken, PA.

277. J. A. Spitznagel, R. P. Shogan, and J. H. Phillips, ASTM *STP* **611**, 1976, p. 434.

278. J. R. Hawthorne, H. E. Watson, and F. J. Loss, ASTM *STP* **683**, 1979, p. 278.

279. E. Bish and B. Nugent, *Power Engineering* **103** (6), 30 (1999).

280. J. E. Campbell, W. E. Berry, and C. E. Feddersen, *Damage Tolerant Design Handbook, MCIC-HB-01*, Sept. 1973.

281. D. J. Nans, G. B. Batson, and J. L. Lott, *Fracture Mechanics of Ceramics*, Vol. **2**, R. C. Bradt, D. P. H. Hasselman, and F. F. Lange, Eds., Plenum, New York, 1974, p. 469.

282. R. Rossi, P. Acker, and D. Francois, *Advances in Fracture Research*, Vol. **4**, S. R. Valluri, D. M. R. Taplin, R. Ramo Rao, J. F. Knott, and R. Dubey, Eds., Pergamon, Oxford, England, 1984, p. 2833.

283. R. F. Pabst, ibid., p. 555.

284. S. M. Wiederhorn, ibid., p. 613.

285. S. W. Freiman, K. R. McKinney, and H. L. Smith, ibid., p. 659.

286. W. G. Clark, Jr., and W. A. Logsdon, ibid, p. 843.

287. R. C. Lueth, ibid., p. 791.

288. N. Ingelstrom and H. Nordberg, *Eng. Fract. Mech.* **6**, 597 (1974).

289. R. A. Schmidt, *Closed Loop* **5**, 3 (Nov. 1975).

290. *Guide to Selecting Engineered Materials*, Vol. **2** (1), ASM, Metals Park, OH, 1987, p. 83.

291. G. P. Marshall and J. G. Williams, *J. Mater. Sci.* **8**, 138 (1973).

292. G. P. Marshall, L. E. Culver, and J. G. Williams, *Int. J. Tract.* **9** (3), 295 (1973).

293. J. C. Radon, *J. Appl. Polym. Sci.* **17**, 3515 (1973).

294. R. C. Bradt, D. P. H. Hasselman, and F. F. Lange, Eds., *Fracture Mechanics of Ceramics*, Vols. **1–4**, Plenum, New York, 1978.

PROBLEMS

Review

7.1 Define *strength* and *toughness*. What is the typical trend between these two material properties when the microstructure of a given material is modified?

7.2 Under what circumstances might an engineer be willing to choose a low-strength material in exchange for high toughness? When might this not be the case?

7.3 Without consulting any source other than a periodic table, what relative level of toughness would you expect for Ti vs. TiN? Justify your answer based on the probable types of atomic bonding found in these two materials.

7.4 Summarize the differences between *intrinsic toughening processes* and *extrinsic toughening processes* with regard to (i) the basis for toughening, (ii) the zone of activity with respect to the crack tip, (iii) the effect on crack initiation and extension, and (iv) the influence on *R*-curve behavior.

7.5 What is *crack-tip shielding*, what role does it play in toughening, and what is the general means by which this role is accomplished?

7.6 What are the two general intrinsic toughening approaches that work for all metals?

7.7 Why would you expect a steel refined by the Bessemer process (air blown over or through the melt) to exhibit inferior fracture properties to a steel refined in a Basic Oxygen Furnace (in which pure oxygen is blown through the melt)?

7.8 Name five elements that should generally be removed from steel to improve toughness.

7.9 Explain why removing certain precipitate-forming elements from Al alloys is particularly beneficial for aircraft applications, but other precipitate-forming elements are critical and should not be removed. Please be specific about the elements in question, and be sure to explain why this is more relevant for aircraft than for many other possible applications.

7.10 Summarize the way in which fine-grain microstructures behave with regard to low temperature and room temperature strength and toughness. Why not make all metals as small-grained as possible?

7.11 Define *mechanical fibering* and list at least three metal forming processes that cause it. Sketch the type of fibering associated with each process. Why might this be beneficial?

7.12 Is the laminated glass that is used for automobile windshields best described in terms of the *arrester*, *divider*, or *short transverse* orientation?

7.13 Figure 7.31 depicts 6 possible Compact Tension specimens cut from the same rolled plate. Identify the pairs of CT specimen orientations that should exhibit nearly the same fracture toughness, and then identify which pair of orientations should be toughest and the least tough of the set.

7.14 What class of toughening mechanisms is usually at work in engineered ceramics, and why is this the case?

7.15 Identify one thing that *thermally tempered* and *chemically tempered* glass have in common, and one critical difference.

7.16 Identify which is typically tougher: a semi-crystalline or an amorphous polymer.

7.17 Describe two general approaches to toughening an amorphous polymer. Why do they work?

7.18 Which condition encourages the fiber pull-process in a fiber reinforced composite: high fiber/matrix interface strength or low fiber/matrix interface strength?

7.19 Certain mollusk shells can be much tougher than others, but are not extremely tough compared to many engineered materials. Why, then, are there attempts to mimic the toughening mechanisms at work in natural materials such as these?

7.20 Explain the structural origin behind the anisotropic toughness of cortical bone, and identify the toughest and least tough crack orientations for a long bone.

Practice

7.21 The severe effect of sulfide inclusions on toughness in steel is shown in Fig. 7.7, where the Charpy V-notch (CVN) upper shelf energy drops appreciably as sulfur content increases. Compute the fracture toughness level K_{lc} at approximately 25°C (77°F) for these 11 alloys using the Barsom-Rolfe relation described in Chapter 6. Plot the fracture toughness as a function of sulfur fraction.

7.22 For a stress level of 240 MPa compute the maximum stable radius of a semicircular surface flaw in 7075-T651 aluminum alloy plate when loaded in the *L-T*, *T-L*, and *S-L* orientations. Assume plane-strain conditions.

7.23 You are given a 10-cm-diameter cylindrically shaped extruded bar (length = 50 cm) of an aluminum alloy.

    a. What specimen configurations would you use to characterize the degree of anisotropy of the material's fracture toughness?

    b. If the lowest fracture toughness of the bar was found to be 20 MPa$\sqrt{m}$, what would be the load level needed to achieve this stress intensity level?

    c. Confirm the presence of plane-strain conditions, given that $\sigma_{ys} = 500$ MPa. (Assume specimen thickness = 1 cm, $a/W = 0.5$, and bar diameter = 1.2 W.)

7.24 A particular pressure vessel is fabricated by bending a rolled aluminum alloy plate into a cylinder then welding on end caps. The alloy used for the cylinder has a distinct layered structure from the rolling process. The rolling direction is around the circumference of the cylinder. The plate thickness is 3 mm. The measured fracture toughness values for the different orientations of this material are shown in Example 7.1 and Fig. 7.31. The internal pressure leads to a hoop stress of 300 MPa.

    a. If two identical semicircular cracks are initiated on the inner surface of the cylinder such that one is growing along the cylinder length and the other across the cylinder width, which would be more likely to lead to fast fracture?

    b. If a circular embedded penny crack was created internally during fabrication so that the crack lies in the S-T orientation, will this fail before either of the cracks in part (a)?

    c. If only the longitudinal crack was present, would this design meet a leak-before-break criterion?

7.25 For lack of a suitable material supply, a thin-walled cylinder is machined from a thick plate of 7178-T651 aluminum alloy such that the cylinder axis is oriented parallel to the rolling direction of the plate. If the cylinder's diameter and wall thickness are 5 cm and 0.5 cm, respectively, determine whether the cylinder could withstand a pressure of 50 MPa in the presence of a 0.2-cm-deep semicircular surface flaw.

7.26 Ordinary sheets of borosilicate glass are tested in bending, and are found to fracture at an average stress of 72 MPa. After thermal tempering, the stress at failure increases by 90%. What are the sign and magnitude of the stress induced in the glass by the tempering operation?

7.27 The quality of the bond between an epoxy and a metal is sometimes measured using a Double Cantilever Beam (DCB) test, as depicted below. In this case, two metal bars are epoxied together, a precrack is created, then the

DCB specimen is gradually pulled apart using pins inserted through holes drilled in the metal bars. Assume that the precrack is 10 mm long and the metal bars are each 150 mm long × 10 mm wide × 5 mm thick. Also assume that the fracture is cohesive. Imagine that in this case the epoxy has typical characteristics for this class of materials ($T_g > 20°C$, $E \sim 2.4$ GPa, $\nu \sim 0.3$, $\sigma_{TS} \sim 27$ MPa), and that the bars are hardened steel.

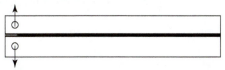

a. If the fracture toughness is determined to be 0.8 MPa$\sqrt{m}$, what is the strain energy release rate for this joint? Be sure that the units for your answer include an energy term.

b. If the temperature is raised high enough that some degree of plasticity becomes possible in the epoxy at the strain rate used in the test, would the fracture toughness probably *increase*, *decrease*, or *remain unchanged*? Explain, and provide a simple equation to support your answer.

c. Would you be better off under the circumstances adding an adhesion promoter to the epoxy/metal interface or adding rubber particles to the epoxy? Why?

d. If you were to add rubber particles, and the behavior of the reinforced epoxy is similar to that summarized in Table 7.7, what is the critical strain energy release rate expected for the DCB specimen tested at 30°C? For the scenario described above, what would be the failure load in this case?

e. A clever engineer has an idea to improve the fracture toughness of this joint by coating the metal with a layer of carbon nanotubes (CNTs) before flowing the epoxy. What would be the most advantageous orientation of the CNTs with respect to the joint if you assume they are rigid cylinders? Explain.

Design

7.28 A rectangular component is to be fabricated from the least expensive steel available (presumed to be the alloy with the least alloying additions). The final decision is to be made between 4330V (425°C temper) and 9-4-20 (550°C temper). Which alloy would you choose if the component is to experience a stress of half the yield strength in the presence of a quarter-circular corner crack with a radius of 10 mm? Would your answer be the same if the design stress were increased to 65% of the yield strength? The properties of these materials are given in Table 7.8.

7.29 A design for a pair of adjacent buildings calls for a ground-level, enclosed connecting walkway with large plates of glass for the walls and the ceiling. Considering the conditions associated with heavy use of the walkway, the potential for falling objects from above, and safety of the users, would you suggest using laminated or tempered glass for the walls? For the ceiling? If the architect calls for tempered glass for one of the locations, but wants a complicated saw-tooth prism cut deep into the glass before or after it is tempered, what is the right order for processing (tempering first vs. cutting first) and which is more likely to succeed: thermal tempering or chemical tempering?

Extend

7.30 Find an example of a product (or a proposed product) made from Partially Stabilized Zirconia. Explain why the properties associated with PSZ are important for your chosen application. Provide a clear reference to the source of your information.

7.31 Find an image of a quench pattern (a.k.a. quench marks) in a tempered glass plate (or, if ambitious, make such an image using a camera and an appropriate lens). Provide a full reference for the image.

7.32 Write a short report identifying the closest nuclear power plant to your hometown. Include information on the license, when it was last renewed, and its end-date.

Chapter 8

Environment-Assisted Cracking

The image of stress corrosion I see

Is that of a huge unwanted tree,

Against whose trunk we chop and chop,

But which outgrows the chips that drop;

And from each gash made in its bark

A new branch grows to make more dark

The shade of ignorance around its base,

Where scientists toil with puzzled face.

On Stress Corrosion, S. P. Rideout[i]

Much attention was given in preceding chapters to the importance of the plane-strain fracture-toughness parameter K_{IC} in material design considerations. It was argued that this value represents the lowest possible material toughness corresponding to the maximum allowable stress-intensity factor that could be applied short of fracture. And, yet, failures are known to occur when the *initial* stress-intensity-factor level is considerably below K_{IC}. How can this be? These failures arise because cracks are able to grow to critical dimensions with the initial stress-intensity level increasing to the point where $K = K_{IC}$ (Eq. 6-31). Such crack extension can occur by a number of processes. Subcritical flaw growth mechanisms involving a cooperative interaction between a static stress and the environment include stress corrosion cracking (SCC), hydrogen embrittlement (HE), and liquid-metal embrittlement (LME). The subject of fatigue and corrosion fatigue is examined in Chapter 10, while SCC, HE, and LME are considered in the present chapter. For those processes that have an environmental component, failure typically occurs after a period of time, rather than when exposure begins. The literature dealing with these topics is as staggering as is the history and significance of the problem. For example, according to a study released by the U.S. Federal Highway Administration in 2002, the combined direct and indirect costs of metallic corrosion in the United States are $552 billion *on an annual basis*. The study defined the total direct annual corrosion costs as those incurred by owners and operators of structures, manufacturers of products, and suppliers of services; indirect costs included factors such as lost productivity and litigation costs.[1] It is not within the scope of this book to cover this material in great depth, especially considering the complexity of the environmentally induced embrittlement phenomena itself. Indeed, as Staehle[2] has pointed out, "A general mechanism for stress corrosion cracking . . . seems to be an unreasonable and unattainable goal. Specific processes appear to operate under specific sets of metallurgical and environmental conditions." Latanision et al.[3] noted with appropriate sarcasm that "it is no surprise that evidence can be found to contradict virtually every point of view." A montage of some major SCC mechanisms is shown in Fig. 8.1, representing the cumulative results of many researchers. In addition to Ref. 1, the interested

---

[i] Reprinted with permission from *Fundamental Aspects of Stress Corrosion Cracking,* 1969, National Association of Corrosion Engineers.

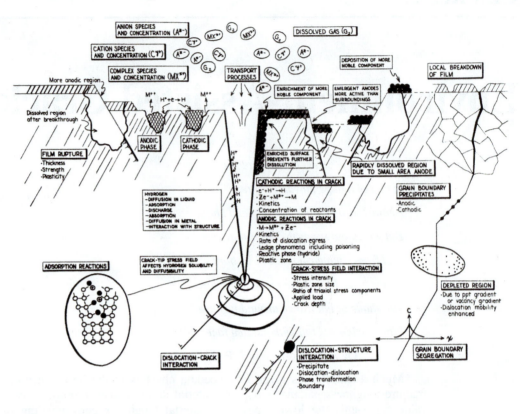

Figure 8.1 Montage of important stress corrosion cracking processes.[2] (Reprinted with permission from R. W. Staehle, *Fundamental Aspects of Stress Corrosion Cracking*, 1969, National Association of Corrosion Engineers.)

reader should find the several dozen other papers in this volume of particular interest regarding the specifics of SCC. In addition, the reader is referred to several comprehensive reviews.[3–12]

In dealing with the problems of SCC, HE, and LME, considerable discussion has surrounded both similarities and differences associated with these processes. For convenience of this discussion, these phenomena are referred to collectively as environment-assisted cracking (EAC). Indeed, Ford[7] has referred to such classifications of SCC, HE, LME, and corrosion fatigue as being "artificial and confining in terms of remedial actions." The objective of this chapter is to provide an overview of EAC that will enable the reader to better appreciate some of the major problems that befall many engineering materials.

Associated with the expanding EAC literature is a growing realization of the complex and interdependent nature of the various cracking processes. For example, Williams and co-workers[13] have advocated that a successful study of EAC requires an integrated inter-disciplinary approach involving the participation of fracture mechanics, chemistry, and materials science experts. To illustrate, the interrelated factors associated with hydrogen embrittlement (HE) are depicted in Fig. 8.2. Fracture mechanics tests can provide a characteri-zation of the phenomenology of EAC such as the rate of crack advance and the associated crack-velocity dependence on temperature, pressure, and concentration of aggressive species. Recent advances have been made regarding crack-tip damage mechanisms by applying cutting edge characterization tools to elucidate crack-tip damage at the nano-scale.[14] Surface chemis-try and electrochemistry studies are needed to identify the rate-limiting processes, whereas metallurgical investigations are important to identify what alloy compositions and micro-structures are susceptible to the cracking process and what fracture micromechanisms are operative (recall Section 5.9.1). A critical point to recognize with regard to Fig. 8.2—or for that matter with a comparable diagram that would describe the sequential processes associated with SCC—is that the slowest process represents the rate-controlling step in the embrittlement of the

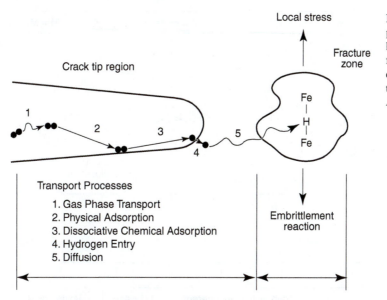

Figure 8.2   Various processes involved in the hydrogen embrittlement of ferrous alloys. (From Williams et al.,[13] with permission from the Metallurgical Society of AIME.)

Local stress

Fracture zone

Crack tip region

Fe | H | Fe

Embrittlement reaction

Transport Processes

1. Gas Phase Transport
2. Physical Adsorption
3. Dissociative Chemical Adsorption
4. Hydrogen Entry
5. Diffusion

material; this arises from the fact that these processes are mutually *dependent* on one another (recall Eq. 4-10). By contrast, final fracture can result from several mutually *independent* fracture mechanisms; in this instance, the fastest process will dominate the fracture mode (recall Eq. 4-12).

8.1   EMBRITTLEMENT MODELS

A growing number of models have been proposed to describe various SCC,[3,4,7] HE,[5,6,16,17] and LME[18,19] fracture processes. The need for so many models attests to the complexity of EAC phenomena. Yet, certain clear similarities and differences in proposed mechanisms are becoming apparent and have led some investigators to conclude that these embrittling processes are often interrelated.[3,5,6] For example, Thompson and Bernstein[5,6] have suggested that SCC may involve both HE and electrochemical processes that operate in *parallel*. Consequently, EAC may occur by either SCC or HE processes or by both.

8.1.1   Hydrogen Embrittlement Models

Hydrogen embrittlement involves a reduction in the bond strength of a metal or alloy. This is due to the presence of atomic hydrogen or hydrogen interaction with dislocation nucleation, mobility, and localization, which can manifest itself in a reduction in ductility and toughness, as well as subcritical crack growth. This damage mechanism affects many important alloy systems, most notably high-strength steels. Gangloff[12] suggests multiple models for crack propagation by HE, as distinguished by the source of hydrogen. In the hydrogen-environment-assisted-cracking (HEAC) model, hydrogen is introduced at or near the crack tip, potentially as a result of the corrosion reactions taking place at the crack tip. The difference between this cracking scenario and the SCC mechanism described in Section 8.1.2 is subtle and worthy of clarification. For the case of HEAC, the reactions that result in the formation of hydrogen atoms (i.e., cathodic reactions) and subsequent absorption of atomic hydrogen are believed to dominate the cracking process, rather than those that result in the corrosion of the metal at the crack tip (i.e., anodic reactions).

Based on the internal-hydrogen-assisted-cracking (IHAC) model, dissolved hydrogen is already present in the material, rather than being introduced during the corrosion process concurrent with the application of stress. The flow diagram shown in Fig. 8.3 provides a useful summary of hydrogen sources, transport paths, destinations, and induced fracture micro-mechanisms, very often exacerbated by the high hydrostatic stress field at the crack tip.

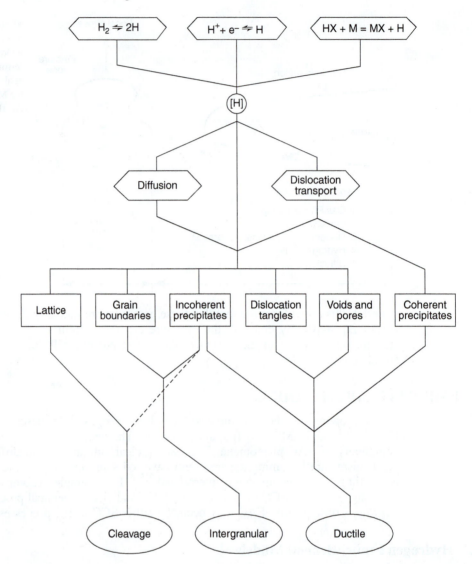

Figure 8.3   Flow diagram depicting hydrogen sources, transport paths, destinations, and induced fracture micromechanisms. (From A. W. Thompson and I. M. Bernstein,[6] *Advances in Corrosion Science and Technology*, Vol. 7, 1980, p. 145; with permission from Plenum Publishing Corporation.)

Hydrogen can enter the metal in a number of different ways. Before vacuum degassing techniques were developed, large steel castings were subject to a phenomenon called hydrogen flaking, wherein dissolved hydrogen in the molten metal would form entrapped gas pockets upon solidification. The large, localized pressures associated with these gas pockets generated many sharp cracks which, when located near the casting surface, caused chunks of steel to spall. A service failure of a large rotor forging caused by this type of defect was discussed previously in a case study in Chapter 6.

Hydrogen can also be picked up from the electrode cover material or from the breakdown of residual water or organic compounds during welding processes. After diffusing into the base plate while the weld is hot, embrittlement occurs after cooling by a process referred to as hydrogen-induced cracking or cold cracking.[20] Cold cracking is usually a delayed phenomenon, occurring possibly weeks or months after welding. This is in sharp contrast to solidification cracking or hot cracking, which occurs during or shortly after the actual welding process as a result of the inability of the weld region to accommodate thermal shrinkage strains during the last stages of cooling.

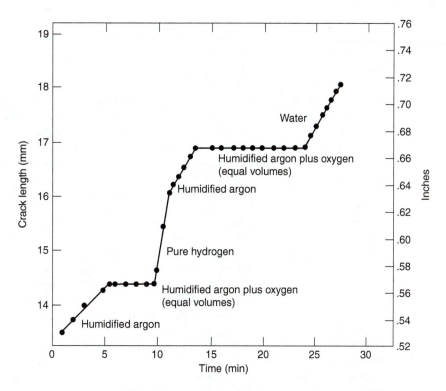

Figure 8.4   Subcritical crack growth of high-strength steel in water and hydrogen, but crack arrest in oxygen.[24] (Reprinted with permission of the American Institute of Mining, Metallurgical, and Petroleum Engineers.)

Hydrogen may also enter the material as a result of electroplating (i.e., cathodic charging), which contributes to early failure. It is ironic that the electroplating process, designed to protect a material against corrosion in aqueous environments as well as SCC, can actually undermine the fracture resistance of the component by simultaneously introducing another potential cracking process.[ii]

Atomic hydrogen can also be introduced into the metal whenever a sample is exposed to a hydrogen gas atmosphere; this occurs under stress or at elevated temperatures and when the component is cooled to temperatures where atomic hydrogen becomes trapped at microstructural features (e.g., dislocations or second phase particles). This is a recognized problem with petrochemical pressure vessels.[21] It should be noted that embrittlement does not occur as a result of prior exposure to hydrogen gas in the absence of stress at ambient temperature.[22,23] The dramatic difference in cracking behavior in H-11 steel stressed in the presence of oxygen, argon, and hydrogen gases and water is illustrated in Fig. 8.4.[24] Note the severe effect of dry hydrogen and moisture on the cracking rate, while oxygen causes total crack arrest. It is presently believed that molecular hydrogen is dissociated by chemical adsorption on iron,[25] allowing liberated atomic hydrogen to diffuse internally and embrittle the metal. Likewise, it has been shown that hydrogen is a product of the corrosion reaction between iron and water; this hydrogen then follows the same path as the chemically adsorbed hydrogen to the metal interior.[26] On the basis of this latter observation, it has been suggested that SCC and HE in steels are closely related.[5,6,27] Apparently, oxygen has a greater affinity for iron and forms a protective oxide barrier to block the chemisorption process.[22,24] It is believed that once the

---

[ii] It is possible to overcome many of the problems associated with cathodic charging by subjecting the electroplated material to a baking treatment. This involves heating the metal to a moderate temperature for a sufficient period of time to drive the hydrogen out of solid solution. Furthermore, weld-related cold cracking is suppressed by preheating the sample. This has the effect of lowering the postweld cooling rate, thereby allowing more time for the hydrogen to diffuse away from the weld zone. Post-weld heat treatments are also used to lower residual stresses associated with the welding process and can reduce the tendency for EAC.

oxygen is removed, hydrogen can reduce the oxide layer and thereby react again with a clean iron surface.

Referring again to Fig. 8.3, hydrogen can diffuse rapidly through the lattice because of its small size. In fact, fast EAC crack growth rates of various alloys scale with their associated hydrogen diffusivities over many orders of magnitude.[28] Complicating matters, hydrogen diffusivity can be greatly dependent upon the trapping ability of various microstructural features. Hydrogen tends to accumulate at grain boundaries, inclusions, voids, dislocation arrays, and solute atoms. To this extent, HE is controlled by those hydrogen accumulation sites that are most sensitive to fracture. Such microstructural features can be beneficial if homogeneously distributed, but detrimental if clustered or aligned to form an attractive crack path. From Fig. 8.3, we see that the cracking process can involve cleavage, intergranular, or ductile (microvoid coalescence) fracture micromechanisms. Beachem[17] reported all three mechanisms in the same steel alloy when tested at different stress levels. Consequently, there is no single fracture micromechanism that is characteristic of HE. For this reason, fractographic information is important to our understanding of HE, but it does not provide a unique characterization of the degree of embrittlement prevailing at a given time.

The hydrogen-embrittling process, therefore, depends on three major factors: (1) the original location and form of the hydrogen (internally charged versus atmospheric water, gaseous hydrogen, or aqueous electrolyte); (2) the transport reactions involved in moving the hydrogen from its source to the locations where it interacts with the metal to cause embrittlement; and (3) the embrittling mechanism itself. We may now ask what that embrittling mechanism is. Unfortunately, the answer is not a simple one, as evidenced by the number of theories that have been proposed. According to one model, called the "planar pressure mechanism," the high pressures developed within internal hydrogen gas pores of charged material cause cracking.[29,30] Although this mechanism appears valid for very high-fugacity hydrogen-charged steels, it cannot be operative for the embrittlement of steel by low-pressure hydrogen atmospheres. In the latter situation, there would be no thermodynamic reason for a low gas pressure external atmosphere to produce a high gas pressure within the solid. Troiano and co-workers[31,32] have argued that hydrogen diffuses under the influence of a stress gradient to regions of high tensile triaxiality just ahead of the crack tip in the damage process zone. The atomic hydrogen then interacts with the metal lattice to lower its cohesive strength, thereby facilitating local fracture and subsequent crack extension (see Figure 8.2). A third model to explain HE was proposed by Petch and Stables,[33] who suggested that hydrogen acts to reduce the surface energy of the metal at internally free surfaces. A significantly different HE model was proposed by Beachem[17] and discussed by Hirth,[16] among others. Beachem suggested that the presence of hydrogen in the metal lattice greatly enhances dislocation mobility at very low applied stress levels. Brittle behavior is then envisioned to occur as a result of extensive but highly localized plastic flow, which can occur at very low shear stress levels. Two variations of this concept have been advanced—namely, the adsorption induced dislocation emission (AIDE) and hydrogen enhanced localized plasticity (HELP) mechanisms. In the AIDE process,[34] enhanced dislocation emission takes place from crack-tip surfaces, whereas the HELP model relies upon enhanced dislocation mobility.[35] Finally, hydrogen embrittlement may result from the formation of metal hydrides in such materials as titanium, vanadium, and zirconium. Additional references pertaining to the subject of HE are found elsewhere.[12,36,37]

8.1.2 Stress Corrosion Cracking Models

Stress corrosion cracking (SCC) is a failure mechanism that requires three simultaneous conditions to be satisfied. These include a susceptible material, an aggressive corrosive environment, and a sustained (i.e., static) tensile stress. The susceptibility of a material is a function of alloy composition and microstructure. For example, the SCC susceptibility of copper alloys generally increases as a function of zinc content. In addition, the SCC resistance of certain austenitic stainless steels decreases as a result of grain boundary precipitation of chromium carbides. This is due to the formation of a chromium-depleted zone in the region immediately adjacent to the grain boundary that is susceptible to corrosion in many environments. A sustained tensile stress can be an applied stress or a residual stress generated as a result of assembly or manufacturing processes. For example, welding operations introduce residual tensile stresses in the weld metal as a result of differential cooling between the last-to-cool region and the remainder of a welded connection. In addition, riveting operations and tightened fasteners introduce both tensile and compressive residual stresses within a subassembly (e.g. see Chapter 9 pertaining to the prestressing of fasteners and thick-walled pressure vessels).

SCC commonly refers to a mechanism of crack propagation driven by an anodic reaction (i.e., corrosion) at a crack tip. This type of cracking can be intergranular or transgranular in nature, depending upon the material–environment combination, and usually occurs with alloys that form a protective surface film under certain environmental conditions. SCC typically occurs for these alloys under conditions whereby these films are, relatively speaking, "less stable." In other words, it is more challenging for the protective films to re-form in the event that they become compromised as a result of local rupture. This concept forms the basis of the film rupture model described below.

According to the film rupture model, a passive film forms on the walls of a crack as a result of exposure to a certain environment, thereby dramatically reducing the corrosion rate. As a result of high stresses at the crack tip, localized plastic flow leads to rupture of the protective layer in this location. Bare metal then is exposed to the aggressive environment at the crack tip, thus leading to a local increase in the corrosion rate; note that the walls of the crack and other surfaces remain protected by the passive film. As a result, the crack extends. The SCC process is maximized at intermediate passivation rates.[7] When the passivation rate is low, the crack tip becomes blunt because of excessive dissolution; when the passivation rate is high, the amount of crack-tip penetration per film-rupture event is minimized. Two versions of this model for crack propagation exist, as illustrated in Fig. 8.5. In one case, the crack propagates in a continuous manner, where the passive film at the crack tip does not re-form. In another case, it is postulated that the crack grows in a discontinuous manner in association with a series of film rupture and film formation steps at the crack tip. In aluminum-based alloys, for example, localized dissolution at the breached area of the passive film is invariably accompanied by cathodic hydrogen production. This demonstrates the difficulty in separating the coupled SCC and hydrogen damage processes.[38]

Bursle and Pugh[4] reviewed several different SCC models and concluded that the film-rupture model,[39,40] involving anodic dissolution at the crack tip, was capable of explaining most cases of intergranular SCC. Bursle and Pugh[4] further concluded that transgranular SCC could be

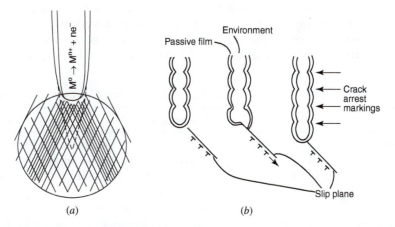

(a)   (b)

Figure 8.5   Schematic representation of vertical crack propagation by the film-rupture model. (a) Crack propagates in a continuous manner, where the passive film at the crack tip does not re-form. (b) Crack grows in a discontinuous manner, with a series of film rupture and film formation steps at the crack tip.[44] (Adapted from R. H. Jones, "Mechanisms of stress-corrosion cracking," in *Stress-Corrosion Cracking, Materials Performance and Evaluation*, 1992, p. 32. Reprinted with permission of ASM International, all rights reserved. www.asminternational.org)

Figure 8.6 (*a*) Transgranular stress corrosion cracking fracture surface in type 310S stainless steel. Fracture bands were produced in load pulse (10-s spacing) experiments and reflect a discontinuous cracking process. (From Hahn and Pugh[54]; copyright, American Society for Testing and Materials, 1916 Race Street, Philadelphia, PA 19103. Reprinted with permission.) (*b*) Optical image (4×) of the fracture surface of a high-strength steel appearing to have the characteristic beach-mark pattern of a fatigue fracture, discussed in Chapter 9. This fracture was a result of SCC, the pattern caused by differences in corrosion penetration rate.[55] (*Metals Handbook*, 8th ed., v. 9: *Fractography and Atlas of Fractography*, American Society for Metals, 1974, p.31. Reprinted with permission of ASM International, all rights reserved. www.asminternational.org)

described best by a discontinuous cleavage process based on a hydrogen–embrittlement-induced decohesion mechanism. The discontinuous cracking process was characterized by the presence of crack arrest lines on the fracture surface (Fig. 8.6*a*). While these crack arrest lines are visible only at high magnification, other fracture surface features are often visible with the unaided eye or at low magnification. Care must be exercised when interpreting these surface features, as they can sometimes suggest an inoperative failure mechanism. For example, Fig. 8.6*b* represents the fracture surface of a high-strength steel appearing to have the characteristic beach-mark pattern of a fatigue fracture, discussed in detail in Chapter 9. This fracture was, in fact, due to stress-corrosion cracking, with the pattern resulting from differences in the rate of corrosion penetration.

8.1.2.1 SCC of Specific Material–Environment Systems

There are many well-established material–environment combinations that are conducive to SCC, examples of which are shown in Table 8.1. Perhaps the most famous historical example of stress corrosion cracking from the 19th century relates to the season cracking of brass alloys. Brass cartridge cases used by the British army in India experienced this form of cracking as a

Table 8.1 Common Alloy/Environment Systems Exhibiting Stress Corrosion Cracking[41]

Alloy	Environment
Carbon steel	Hot nitrate, hydroxide, carbonate/bicarbonate solutions, seawater
High-strength steels	Aqueous electrolytes, particularly when containing hydrogen sulfide
Austenitic stainless steels	Hot, concentrated chloride solutions; chloride-contaminated steam
High-nickel alloys	High-purity steam
α-brass	Ammoniacal solutions
Aluminum alloys	Aqueous chlorides, bromides, and iodide solutions
Titanium alloys	Aqueous chlorides, bromides, and iodide solutions; organic liquids
Magnesium alloys	Aqueous chloride solutions
Zirconium alloys	Aqueous chloride solutions; organic liquids

Adapted from *Stress-Corrosion Cracking, Materials Performance and Evaluation*, edited by Russell H. Jones, ASM International, 1992, p. 2.

result of the combination of high residual tensile stresses and exposure to ammonia formed from the decomposition of organic matter during periods of heavy rainfall and warm weather.

In another example involving a brass–ammonia system, Fontana[42] reported a dramatic SCC problem that occurred in the southern region of the United States where farmers were using ammonia for fertilizer and liquid propane gas for heating and cooking. During the investigation, it was discovered that the gas distributors who sold both ammonia and liquid propane gas did not adequately clean their storage tanks and tank trucks before switching between these two products. As a result, residual ammonia was present in the liquid propane gas that then led to cracking of the gas tank's brass fittings. Fontana determined that merely *one part per million* of ammonia was required to cause SCC of the brass fittings, leading to their replacement at a cost of $80 million. This example clearly emphasizes the importance of both understanding and controlling intended and unintended service environments.

Another classic example of this insidious form of cracking can be found in the utilities industry. Dating back to the 1950s, Inconel 600 (Ni-16Cr-9Fe) served as the material of choice for pressurized water reactor steam generator tubing in the nuclear industry. A substantial amount of damage in these reactors was found to be due to SCC of the steam generator tubes that were exposed to primary cooling water, containing boric acid, lithium hydroxide, and hydrogen.[43] As a result, Portland General Electric Company determined it was necessary to close the 1.1 billion-watt Trojan plant in Oregon in the 1990s, after only half of its design life, rather than spend the necessary $200 million to replace the many cracked steam tubes.

While the current discussion of SCC is narrowly focused, the reader is reminded of other examples of SCC failures in this textbook. As discussed in Chapter 11, the catastrophic collapse of the Point Pleasant Bridge in the 1960s was determined to be a result of stress corrosion cracking. Likewise, a costly product recall involving ladders was a direct result of stress corrosion cracking of aluminum alloy rivets exposed to moisture and chlorides during service, as discussed in Chapter 12.

8.1.3 Liquid-Metal Embrittlement

When many ductile metals are coated with a micron-thin layer of certain liquid metals and then loaded in tension, the metal's fracture stress and strain are significantly reduced. Fracture times are extremely short, with crack velocities as high as 500 cm/s being reported for aluminum alloys and brass in the presence of liquid mercury (Hg). The reader is referred to several comprehensive review articles on the subject of liquid–metal embrittlement.[19,45−50] A large number of embrittlement systems have been identified (Table 8.2) that are highly specific in that a given liquid metal (e.g., Hg) may embrittle one metal (e.g., Al) but not another (e.g., Mg). Empirical guidelines for the existence of LME are discussed elsewhere.[45,46] It should be noted that embrittlement can also occur when the two metals are in contact and in the *solid* form. In this instance, the vapor phase of the embrittling metal migrates by surface

Table 8.2 Liquid-Metal Embrittlement Systems[50]

	Hg	Ga	Cd	Zn	Sn	Pb	Bi	Li	Na	Cs	In
Aluminum	×	×		×	×				×		×
Bismuth	×										
Cadmium		×			×					×	×
Copper	×	×		×	×	×	×	×	×		×
Iron	×	×	×	×		×		×			×
Magnesium				×					×		
Silver	×	×									
Tin	×	×									
Titanium	×		×								
Zinc	×	×			×	×					×

diffusion to the crack tip in the embrittled solid. Important practical examples of the solid-metal-induced embrittlement include cadmium–steel couples associated with cadmium-plated steel components, lead–steel couples as found in internally leaded steel alloys, and cadmium embrittlement of zirconium alloy reactor fuel cladding where cadmium is a product of UO_2 fission.[51–53]

Liquid-metal embrittlement is believed to result from liquid-metal chemisorption-induced reduction in the cohesive strength of atomic bonds in the region of a stress concentration.[49,56,57] The liquid-metal atom is believed to reduce the interatomic bond strength between solid atoms at the crack tip, thereby causing bond rupture to occur at reduced stress levels. Once the initial bond is broken, liquid-metal atoms then reduce the strength of the atomic bond between the next-nearest solid atoms with local fracture continuing at a rapid pace. Lynch[58] has proposed an alternative LME mechanism with premature fracture resulting from a reduction in shear strength rather than the cohesive strength of atomic bonds at the crack tip; in this manner, many dislocations can be nucleated at low stress levels that facilitate localized plastic deformation at much reduced stress levels (recall Refs. 16 and 17 as discussed in Section 8.1.1).

8.1.4 Dynamic Embrittlement

Dynamic embrittlement (DE) represents a form of time-dependent brittle, intergranular fracture that was first identified in the mid 1980s.[iii] This cracking mechanism is quasi-static in nature, whereby crack propagation is controlled by the diffusion of embrittling elements to grain boundaries. The source of the embrittling elements is either within the material (e.g., sulfur-induced cracking in alloy steels) or from the surrounding environment (e.g., oxygen-induced cracking of nickel-based superalloys). DE occurs by grain boundary decohesion due to the combined application of a tensile stress and the diffusion of embrittling elements to these regions. The nature of the mechanism is similar to that of diffusion controlled creep, aside from the presence of an embrittling element. In this manner, cavities form at the grain boundaries under constant tensile loading, even in alloys possessing a high creep resistance. It has been suggested that LME is merely a unique form of DE for the case of an extremely high concentration and mobility of an embrittling element species. Factors affecting DE include the magnitude of the tensile stress, the mobility of embrittling species, and the extent of grain boundary diffusivity.[iv] (Recall the potential effect of solute segregation on grain boundary diffusivity as discussed in Section 5.9.5.) While it is clear that atomic mobility will be affected by temperature, it is perhaps less appreciated that grain boundary diffusivity also is affected by its structure and orientation. For example, symmetrical grain boundaries exhibit relatively low diffusivity as compared with random, high-angle grain boundaries.

8.2   FRACTURE MECHANICS TEST METHODS

As already mentioned, the various manifestations of EAC have been long recognized. Consequently, different approaches to "solving" the problem have been developed, along with "standard" specimen types. For example, stress corrosion cracking studies of engineering materials had often made use of smooth test bars that were stressed in various aggressive environments. Here the *nucleation* kinetics of cracking, as well as its character (transgranular versus intergranular), were examined closely. Most often these studies focused on the nature of anodic dissolution in the vicinity of the crack tip. Recently, more attention has been given to the *propagation* stage, reflecting the more conservative and realistic philosophical viewpoint[59] that defects preexist in engineering components (recall the discussion in Chapter 6). These

---

[iii] C. T. Liu and C. L. White, "Dynamic embrittlement of boron-doped Ni₃Al alloys at 600C," *Acta Metall.* 35(3), 643–649 (1987).

[iv] U. Krupp and C. J. McMahon, Jr., "Dynamic embrittlement-time-dependent brittle fracture," *Journal of Alloys and Compounds* 378, 79–84 (2004).

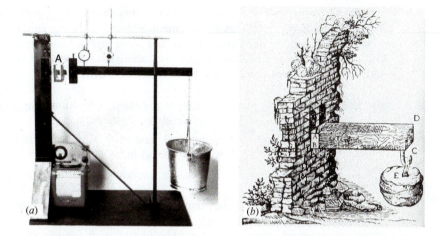

Figure 8.7   (*a*) Environment-assisted cracking test stand. Specimen is placed in environment chamber at A and loaded by weights placed in scrub bucket.[61] (Reprinted with permission from B. F. Brown and C. D. Beachem, *Corrosion Science* 5, 1965, Pergamon Press.) (*b*) Cantilever beam arrangement (adapted from Galileo).

propagation studies have been aided greatly by the discipline of fracture mechanics and are the focus of attention here.

In a dramatic series of experiments, researchers at the Naval Research Laboratories[60–62] showed that certain *precracked*, high-strength titanium alloys failed under load within a matter of *minutes* when exposed to both distilled water and saltwater environments. In all tests, the initial stress-intensity levels were below K_{IC}. Heretofore, it had been felt that these same alloys would represent a new generation of submarine hull materials, based on their resistance to general corrosion, which was vastly superior to steel alloys in these same environments. These initial experiments made use of the very simple loading apparatus shown in Fig. 8.7*a*. Precracked samples were placed in the environmental chamber and stressed in bending at different initial K levels by a loaded scrub bucket hung from the end of the cantilever beam. Note the strong similarity between the NRL test apparatus and the diagram attributed to Galileo some 400 years earlier (Fig. 8.7*b*). For each test condition associated with a different initial K value (always less than K_{IC}) the time to failure was recorded. A typical plot of such data is shown in Fig. 8.8 for the environment-sensitive Ti-8Al-1Mo-1V alloy. With an apparent fracture toughness level of about 100 MPa$\sqrt{m}$, test failures occurred at initial K levels of only 40 MPa$\sqrt{m}$ after a few minutes of exposure to a 3.5% NaCl solution. At slightly lower K levels, the time to failure increased rapidly, suggesting the existence of a threshold K level, originally designated K_{ISCC},[51] below which stress corrosion cracking would not occur. To be consistent

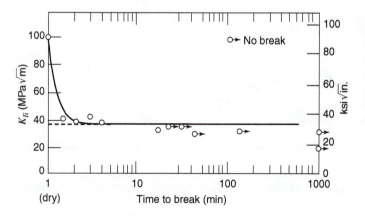

Figure 8.8   Initial stress-intensity level plotted versus time to break for Ti-8A1-1Mo-1V alloy in 3.5% NaCl solution. Note the threshold behavior.[60] (Reprinted by permission of the American Society for Testing and Materials from copyrighted work.)

with the philosophical viewpoint expressed in this chapter, K_{ISCC} will be redefined hereafter as K_{IEAC} where EAC represents environment-assisted cracking; furthermore, K_{IEAC} and K_{EAC} correspond to conditions of plane strain and plane stress, respectively. As a result, a new safe lower limit of the applied stress-intensity value (time-dependent in this case) was identified with fracture occurring according to the following criteria:

1. $K < K_{IEAC}$ No failure expected even after long exposures under stress to aggressive environments.

2. $K_{IEAC} < K < K_{IC}$ Subcritical flaw growth with fracture occurring after a certain loading period in an aggressive environment.

3. $K > K_{IC}$ Immediate fracture upon initial loading.

The reader must recognize that rigorous and scientific proof for the existence of an environmental threshold is lacking.[63,64] Therefore, the use of K_{IEAC} or K_{EAC} data in the design of structural components should be treated with caution. One should keep in mind that for the test conditions associated with Fig. 8.8 (criteria 2, above), stable crack extension causes the initial stress-intensity level to increase to the point where failure occurs when K approaches K_{IC}. Stated differently, the fracture toughness of the material is not affected by the environment; instead, small cracks grow under sustained loads to the point where the critical stress-intensity factor level is approached (Fig. 8.9).

Determination of K_{IEAC} values is not an easy matter. Since the environmental-threshold level depends on how long one chooses to conduct the test, K_{IEAC} values may vary from one laboratory to another, depending on the patience of the investigator. It may be that K_{IEAC} test times will have to be determined for each material-environment system on an individual basis. For example, an ASTM standard[65] for the determination of threshold stress intensity factor values for environment-assisted cracking suggests the following minimum test times for selected structural alloys:

steels ($\sigma_{ys} < 1200\,\text{MPa}$)	10,000 h
steels ($\sigma_{ys} > 1200\,\text{MPa}$)	5,000 h
aluminum alloys	10,000 h
titanium alloys	1,000 h

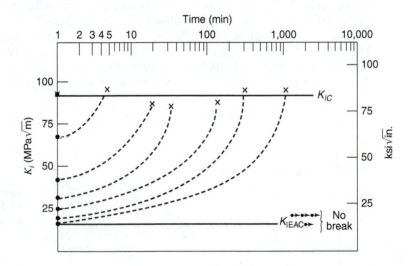

Figure 8.9   Change in K level with subcritical flaw growth. Regardless of $K_{initial}$ (K_i), failure in any sample occurs when K approaches K_{IC}.[61] (Reprinted with permission from B. F. Brown and C. D. Beachem, *Corrosion Science* 5, 1965, Pergamon Press.)

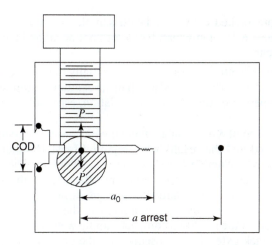

Figure 8.10 Modified compact tension sample with threaded bolt bearing on load pin. Initial crack opening displacement determined by extent to which bolt is engaged.

Some materials, such as high-strength steels and titanium alloys, exhibit a rather well-defined K_{IEAC} limit after a reasonable test time period, but in aluminum alloys this does not appear to be the case. Instead, K_{IEAC} values in high-strength aluminum alloys tend to decrease with increasing patience of the investigator. Consequently, an engineer must exercise extreme caution when utilizing K_{IEAC} data, especially in the design of components that will be stressed in an aggressive atmosphere for time periods longer than those associated with available K_{IEAC} data.

Recently, different specimen configurations and loading methods have been developed to determine K_{IEAC}. The aforementioned ASTM standard test method,[65] applicable to aqueous and other aggressive environments, utilizes fatigue precracked cantilever beam [SE(B)] or compact fracture [C(T)] specimens (see Appendix B) that are subjected to constant-load testing. For the determination of a material's K_{IEAC} value, it is necessary that SE(B) and/or C(T) specimen dimensions satisfy those corresponding to plane strain conditions (recall Eq. 6-54). It is recommended that tests be initiated at a minimum of four to six different K levels. K_{IEAC} or K_{EAC} is then determined by bracketing the two loading conditions that produce specimen failure after a relatively long time under load and where specimen failure does not occur prior to the specimen being loaded for the minimum test times, as previously noted.

EAC data have also been obtained with a modified compact specimen configuration[66] (Fig. 8.10). In this instance, a bolt, engaged in the top half of the sample, bears against the bottom crack surface. This produces a crack-opening displacement corresponding to some initial load. In this manner, the specimen is self-stressed and does not require a test machine for application of loads. As the crack extends by environment-assisted cracking, the load and, hence, the K level drop under the prevailing constant displacement condition. The crack finally stops when the K level drops below K_{IEAC}. Consequently, only one specimen is needed to determine K_{IEAC}. Such a test is very easy to conduct and very portable, since the self-stressed sample can be carried to any environment rather than vice versa. All one needs to do is to engage the bolt thread to produce a given crack-opening displacement and place the specimen in the environment. Samples are examined periodically to determine when the crack stops growing. The K_{IEAC} value is then defined by the residual applied load remaining after the crack has ceased growing and the final crack length as seen on the fracture surface. The major advantages of the modified compact tension sample relative to the precracked cantilever beam are:

1. The need for one sample versus many samples in determining K_{IEAC}.
2. The specimen is self-stressed and highly portable.
3. The method is less costly.
4. K_{IEAC} is determined directly by the arrest characteristics of the sample because of the continual decrease in K with increasing crack length. By comparison, the K_{IEAC} value

determined with the precracked cantilever beam samples represents an interpolated value between the highest K level at which EAC does not occur and the lowest K level where failures still occur.

5. The need for a sharp notch is not as great, since K is initially high, which results in early crack growth. By contrast, a poorly prepared notch in the cantilever beam specimen would involve a considerable period of time for crack initiation, especially at low K levels.

An interlaboratory comparison of K_{IEAC} data obtained with cantilever beam and modified compact samples was completed and the results were reported from 16 laboratories.[67] This study found that repeatable and reproducible K_{IEAC} values can be obtained with the use of the constant displacement-modified compact sample. On the other hand, the constant load–cantilever beam sample proved to be less useful because of difficulties with the time for the onset of cracking and the need for subjective judgment in the determination of K_{IEAC}. Problems with the stress analysis of the modified compact sample were also identified since the bolt unloads elastically with crack extension. Consequently, the crack-opening displacement (COD), which was assumed to remain constant, actually decreases slowly with test time. In addition, the volumetric expansion associated with corrosion product formation along the crack surface may occur so as to unload the bolt, increase COD, and, thereby, confound determination of the instantaneous stress-intensity level.

Another K_{IEAC} test procedure that employs a single sample has shown considerable promise. In this method, a conventional fracture-toughness sample is exposed to an aggressive environment and subjected to a rising load but at a loading rate lower than that associated with E399 procedures.[68–76] At the load level corresponding to the onset of environment-assisted cracking, the load-displacement trace deviates markedly from that associated with conventional loading rates in air (Fig. 8.11a). Hirano et al.[70] have noted excellent agreement between K_{IEAC} values in 4340 steel that were determined from conventional modified compact samples and from the inflection point in rising load tests so long as the loading rate $(\dot{K})$ in the rising load test was equal to or less than 0.25 MPa$\sqrt{m}$/min (Fig. 8.12). Using this test method, K_{IEAC} values can be determined readily after only a few hours of testing.[v] Finally, Raymond et al.[77,78] have demonstrated that deeply side-grooved Charpy samples can be used to assess the EAC susceptibility of structural materials; the associated test method involves the use of a series of rising step loads with crack initiation noted by a drop in load (Fig. 8.11b). It should be noted that

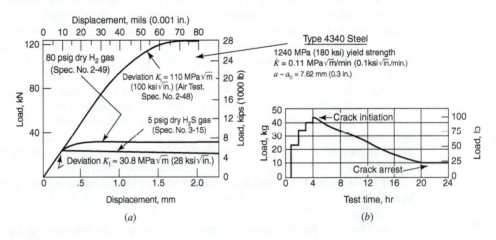

Figure 8.11 Rising load test method for the determination of K_{IEAC}. (a) Effect of environment on K_{IEAC} in 4340 steel.[69] (b) Onset of HE-induced cracking in steel based on rising step-load test.[78] (Reprinted with permission. Copyright ASTM.)

---

[v] For a recent summary of efforts to standardize slow strain rate test procedures, see Ref. 76.

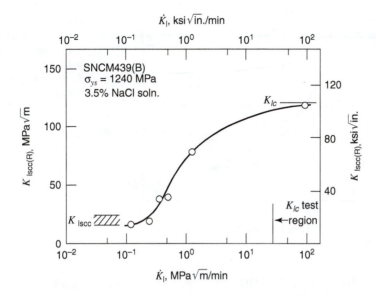

Figure 8.12 Influence of loading rate on K_{IEAC} in 4340 steel. Data bar corresponds to data from modified compact samples.[70] (Reprinted with permission. Copyright ASTM)

the influence of $\dot{K}$ on K_{IEAC} depends on the material–environment combination and loading method. For example, very high strength and susceptible materials have been shown to exhibit apparent thresholds that are independent of loading method, whereas low-strength alloys have been shown to exhibit greatly reduced thresholds under slow-rising CMOD.[12]

Although the modified compact tension sample and the rising load test procedure represent improvements in the method by which K_{IEAC} data are obtained, one must still contend with the fact that K_{IEAC} may not represent a true material property. As noted above, threshold values are often found to be a function of the length of the test (i.e., the patience of the technician) and the resolution of the crack growth detection instrument. For example, Lee and Gangloff[79] have performed accelerated K_{IEAC} testing by developing ultra-high resolution crack growth detection capabilities, with the goal of quantifying very low da/dt by resolving "low delta a" in a reasonable "delta t." By the time of this book's publication, the method was capable of resolving 5 μm of crack extension with stability over a period of days, resulting in a resolution limit of approximately 0.01 to 0.05 nm/s. This was accomplished by electrical potential resistance measurements, optimized with modern instrumentation and data analysis when applied to a short crack specimen. The resulting da/dt vs. K relationship can then be coupled with the standard K-similitude fracture mechanics life prediction. Using this approach, it was possible to utilize data generated in one week to predict the relationship between initial-applied K and time to failure in cantilever beam specimens; these constant load geometry tests were conducted over 20,000 hours for a steel–sodium chloride solution combination.

Studies have demonstrated that much lower K_{IEAC} values are obtained when a small pulsating load is superimposed on the static load applied during the EAC test[80] (Fig. 8.13). Furthermore, Fessler and Barlo[81] found that K_{IEAC} values decreased with decreasing frequency of the ripple loadings. From the load-time diagram given in Fig. 8.13, the superposition of ripple loading on the static load corresponds to a condition of corrosion fatigue under high mean stress conditions. Pao and Bayles[82] have taken both EAC and fatigue threshold data (see Chapter 10) into account to determine the material/environment system susceptibility to ripple loading. They determined that the percentage of degradation of a K_{IEAC} value is given by

$$\% \text{ degradation} = \left[1 - \frac{\Delta K_{th}}{K_{IEAC}(1-R)}\right]100 \tag{8-1}$$

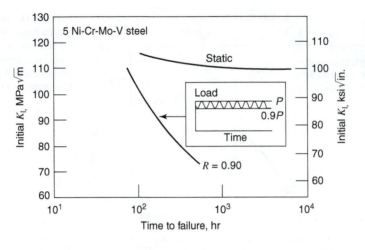

Figure 8.13 EAC response in 5Ni-Cr-Mo-V steel under static and ripple loading conditions in salt water at room temperature. $R = 0.9 = K_{min}/K_{max}$. (Adapted from Ref. 80 with permission.)

where ΔK_{th} = fatigue threshold stress intensity range ($K_{max} - K_{min}$) (see Section 10.4.2)

$$R = \frac{K_{min}}{K_{max}} = \frac{P_{min}}{P_{max}}$$

In the limit, no ripple loading effects are experienced when $\Delta K_{th}/(1-R) \geq K_{IEAC}$.

Some have suggested that such small amplitude cyclic loadings induce oxide-film rupture at the crack tip, which facilitates the cracking process. However, the potential effect of ripple loading on crack-tip surface reactions has not been isolated from the effect of such cyclic loading on crack-tip fatigue-plasticity damage.[83] Since some structures typically experience load fluctuations during their lifetime, it follows that K_{IEAC} values based on traditional static test methods may prove to be nonconservative in assessing the EAC susceptibility of a given material and component. Finally, it has been shown that environment-assisted cracking is influenced by both mechanical and electrochemical variables. Regarding the latter, Gangloff and Turnbull[84] reported that electrochemical driving forces are influenced by differences in crack size (especially short cracks), shape, and crack opening. At present, the true ripple load effects on crack-tip reactions including hydrogen uptake are poorly understood.

Engineers and scientists have sought other means of quantifying EAC processes. In this regard, considerable attention has been given to characterize the kinetics of the crack growth rate process by monitoring the rate of crack advance da/dt as a function of the instantaneous stress-intensity level. From the work of Wiederhorn[85–87] on the static crack growth of glass and sapphire in water, a log da/dt–K relation was determined, which took the form shown schematically in Fig. 8.14. Three distinct crack growth regimes are readily identified. In Region I, da/dt is found to depend strongly on the prevailing stress-intensity level, along with other factors including temperature, pressure, and the environment (e.g., pH and electrode potential, electrolyte composition). For some materials, the slope of this part of the curve is so steep as to allow for an alternative definition of K_{IEAC}; that is, the K level below which da/dt becomes vanishingly small. For aluminum alloys that do not appear to exhibit a true threshold level and that have a shallower Stage I slope, a "K_{IEAC}" value can be defined at a specific da/dt level much the same as the yield strength of a material exhibiting continuous work hardening behavior is defined by the 0.2% offset method. Environment-assisted crack growth is often relatively independent of the prevailing K level in Region II, but it is still affected strongly by temperature, pressure, and the environment. Finally, Region III reflects a second regime where da/dt varies strongly with K. In the limit, crack growth rates become unstable as K approaches K_{IC}.

In addition to the three *steady-state* crack growth regimes just described, a number of additional *transient* growth regions have been identified followed by a dormant or incubation period prior to steady-state growth.[88] Consequently, the total time to fracture is the summation

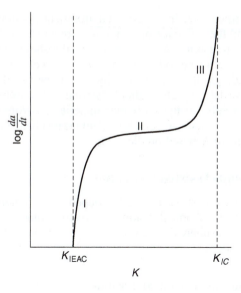

Figure 8.14 Diagram showing three stages of environment-assisted cracking under sustained loading in an aggressive atmosphere. Lower and upper K limits of plot determined by K_{IEAC} and K_{IC}, respectively.

of incubation, transient growth, and steady-state cracking periods. Hence

$$t_T = t_{inc} + t_{tr} + t_s \qquad (8\text{-}2)$$

where t_T = total time to failure

t_{inc} = incubation time

t_{tr} = total time during transient crack growth

t_s = time of Region I, II, and III steady-state crack growth

The transient time t_{tr} is usually small relative to t_{inc} and t_s and is often ignored in life computations. The relative importance of the other two regimes in affecting total life is shown schematically in Fig. 8.15. Note that the incubation time decreases rapidly with increasing initial K level.[88–91] Higher test temperatures decrease t_{inc}, as well.[90] Since the initial K level of the bolt-loaded, constant displacement type K_{IEAC} test sample is large, this configuration is preferred over

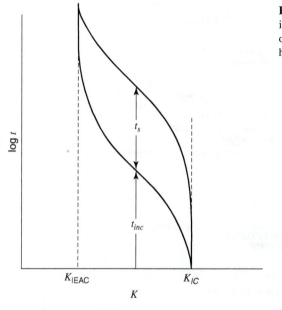

Figure 8.15 Diagram showing time for crack incubation and steady-state crack growth as a function of applied K level. Note the smaller incubation time at higher K values.

the cantilever beam geometry. Furthermore, in material–environment systems where da/dt is high, incubation times are short and the influence of $\dot{K}$ in rising load tests is limited. As noted previously, the very long times to failure at low K values suggest a threshold condition. It should be recognized, however, that the incubation period represents a large part of the time to failure. As a result, initial crack growth rate readings often are abnormally low, suggesting the existence of an erroneously high K_{IEAC} level. These data, therefore, should be used with extreme caution. These complicating effects depend sensitively on the particular material–environment system. An understanding of the crack environment chemistry, crack-tip damage mechanism, and kinetics are necessary foundational information for K-based predictions.

8.2.1 Major Variables Affecting Environment-Assisted Cracking

The degree to which materials are subject to EAC depends on a number of factors, including alloy chemistry and thermomechanical treatment, the environment, temperature, and pressure (for the case of gaseous atmospheres). The effect of these important variables on the cracking process will now be considered.

8.2.1.1 Alloy Chemistry and Thermomechanical Treatment

As one might expect, many studies have been conducted to examine the relative EAC propensity of different families of alloys and specific alloys thermomechanically treated to different specifications. For example, the log $da/dt–K$ plot for several high-strength aluminum alloys exposed to alternate immersion in a 3.5% NaCl solution reveals the 7079-T651 alloy to be markedly inferior relative to the response of the other alloys (Fig. 8.16).[92] For example,

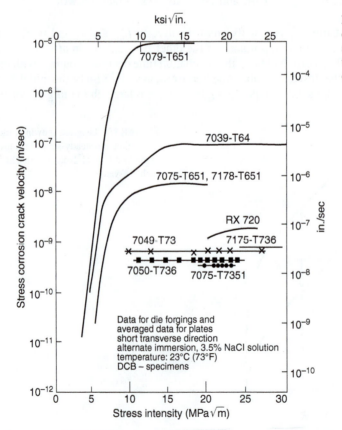

Figure 8.16 Environment-assisted cracking in 3.5% NaCl solution for several aluminum alloys and heat treatments.[92] (Reprinted with permission of the American Society for Metals.)

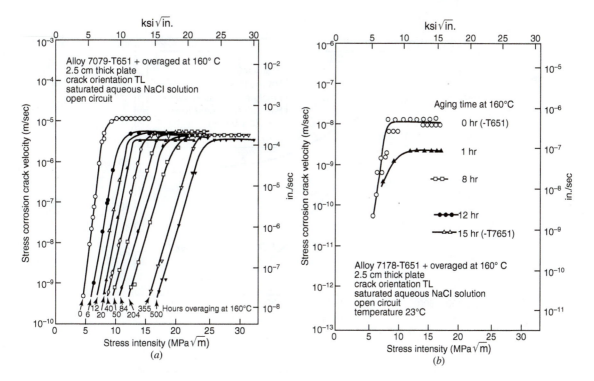

Figure 8.17 Effect of overaging on EAC (salt water) in 7xxx series aluminum alloys: (*a*) 7079 alloy shows pronounced shift of Stage I behavior to higher *K* levels while (*da/dt*)$_{\text{II}}$ remains relatively constant; (*b*) 7178 alloy shows sharp drop in (*da/dt*)$_{\text{II}}$. (With permission of Markus O. Speidel, Brown Boveri Co.)

Stage II cracking in the 7079 alloy occurs at a rate 1000 times greater than in the 7178-T651 alloy and corresponds to a Stage II crack growth rate of greater than 3 cm in 1 hr. No engineering component would be expected to resist final failure for long at that growth rate. Many investigations have been conducted to improve the EAC resistance of these materials. These studies indicate that overaging is the most effective way to accomplish this objective.[38,92–94] Coincidentally, toughness is often improved while strength decreases as a result of the overaging process. The effect of overaging (denoted by the –T7 temper designation) on 7079 and 7178 aluminum alloys is shown in Fig. 8.17. Although Stage I in the 7079 alloy is shifted markedly to higher *K* levels, reflecting a sharp increase in K_{IEAC}, the growth rates associated with Stage II cracking remain relatively unchanged. Consequently, the major problem of very high Stage II cracking rates in this material remains even after overaging. By contrast, data for the 7178 alloy show a marked decrease in Stage II crack growth rate with increasing aging time, while Stage I cracking is shifted to a much lesser extent.[38,92–94] Note the dramatic six order-of-magnitude difference in Stage II crack growth rates between these two alloys in the overaged condition. The difference in cracking behavior between these aluminum alloys has been attributed to the presence of copper, likely incorporated within grain boundary precipitates.[12] Upon reflection, it would be most desirable to have the overaging treatment effect a *simultaneous* lowering of the Stage II cracking rate and a displacement of the Stage I regime to higher *K* levels. This may prove to be the case in other alloy systems.

The reader might expect both K_{IEAC} and K_{IC} values to be greater in materials possessing lower yield strength, owing to correspondingly higher toughness values. For example, we see from Fig. 8.18(*a*) that K_{IEAC} and K_{IC} values decrease with increasing yield strength in a 4340 steel.[95] If the relative degree of susceptibility to environment-assisted cracking is defined by

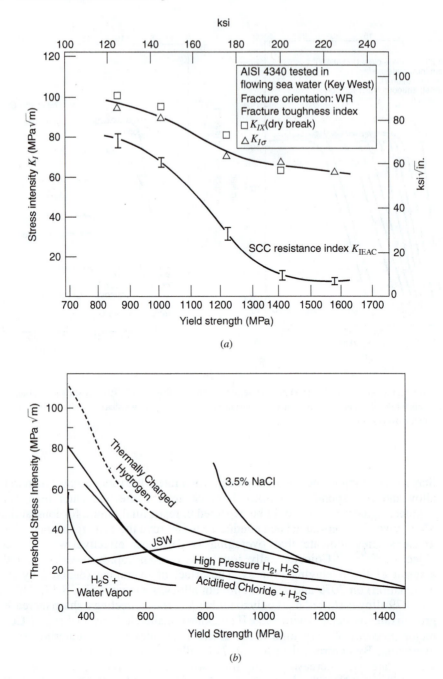

(a)

(b)

Figure 8.18 (*a*) Effect of yield strength on K_{IC} and K_{IEAC} (in water) in 4340 steel.[95] (Reprinted with permission from M. H. Peterson, B. F. Brown, R. L. Newbegin, and R. E. Grover, *Corrosion* 23, 1967, National Association of Corrosion Engineers.) (*b*) Lower-bound trend lines representing over 400 threshold stress-intensity factors for C-Mn and alloy steels exposed to various environments at 23°C. JSW (Japan Steel Works) refers to rising CMOD loading, whereas other thresholds were measured for quasi-static loading.[12] (Reprinted from *Comprehensive Structural Integrity*, J. Petit and P. Scott, Eds., R. P. Gangloff, *Hydrogen Assisted Cracking of High Strength Alloys*, 2003, p. 31, with permission from Elsevier.)

the ratio K_{IEAC}/K_{IC}, the generally observed trend is for K_{IEAC}/K_{IC} to decrease with increasing alloy strength. That is, K_{IEAC} values drop faster than K_{IC} values with increasing strength. However, Gangloff has found that the yield strength effect is very much dependent upon loading format. While a strong threshold dependence on yield strength is measured for quasi-static loading, such a threshold is uniformly low and relatively strength independent for the Japan Steel Works' (JSW) rising CMOD data shown in Fig. 8.18*b*.[12] The trend lines "JSW" and "Thermally Charged Hydrogen" both represent data for tempered bainitic Cr-Mo steels. Clearly, the rising CMOD load format promotes embrittlement in these alloys for yield strengths below 800 MPa. While the explanation for this behavior is unclear, it has been postulated that these differences are related to a combination of hydrogen loss during long-term laboratory testing, the damage role of active crack-tip plasticity in hydrogen embrittlement, and the effect of dK/dt on crack-tip driving force akin to an R-curve effect in stable-tearing ductile fracture.

Cracks associated with HE and SCC can follow an intergranular or transgranular path in the important high-strength steel, titanium, and aluminum alloys. Aluminum alloys often suffer stress corrosion cracking along grain-boundary paths, as discussed in Chapter 12 for the case of recalled ladders. Consequently, environment-assisted cracking in wrought alloys is usually of greater concern in the short transverse direction than in other orientations, owing to the elongated grain structure in the longitudinal direction. As such, EAC orientation sensitivity parallels K_{IC} orientation dependence as described in Section 7.3.

Takeda and McMahon[96] addressed the controversy regarding the fracture micromechanism associated with hydrogen embrittlement in steel. On the one hand, test conditions have been reported in which the presence of hydrogen allowed plastic flow (i.e., dislocation movement) to occur at abnormally low stress levels and resulted in transgranular fracture; the enhanced level of plastic flow in connection with HE represents a *strain*-controlled fracture mode. On the other hand, countless examples of hydrogen-induced intergranular failure have been reported in the literature. In this instance, the fracture mode is believed to be *stress*-controlled and influenced by the level of impurity elements segregated at grain boundaries (recall Section 7.8). Takeda and McMahon[96] observed *both* mechanisms in quenched and tempered steel samples possessing virtually the same microstructure and strength. However, the samples were distinguished by slight, but significant, differences in alloy composition. When samples were aged to segregate impurities to the prior austenite grain boundaries, these precracked samples exhibited intergranular failure when loaded in tension in the presence of gaseous hydrogen. In samples that did not contain embrittled prior austenite grain boundaries due to impurity segregation, failure was associated with transgranular fracture (cleavage and microvoid coalescence). Therefore, HE associated with an intergranular crack path was shown to be dependent on the presence of impurity elements at prior austenite grain boundaries, whereas transgranular failure was attributed to the intrinsic enhancement of dislocation motion by the presence of hydrogen in the metal lattice.

K_{IEAC} data for selected materials are listed in Table 8.3. Note the sharp disagreement in results for several aluminum alloys. Since Speidel's values are based on K levels associated with crack velocities less than about 10^{-10} m/s and were conducted over a long time period, they provide a more representative and conservative estimate of the material's environmental sensitivity. Additional K_{IEAC} information is provided in Ref. 97, which also contains numerous log $da/dt–K$ plots for aluminum, steel, and titanium alloys.

8.2.1.2 Environment

As one might expect, the kinetics of crack growth and the threshold K_{IEAC} level depend on the material–environment system. In fact, the reality of this situation is reflected by the characterization of various cracking processes, such as SCC (generally involving aqueous solution electrochemistry), HE, and LME. As mentioned at the beginning of this chapter, the complex aspects of the material–environment interaction can be greatly simplified by treating the problem from the phenomenological viewpoint in terms of a single mechanism, EAC. This concept is supported by Speidel's results shown in Fig. 8.19, which reveal parallel Stage I and II responses for the 7075 aluminum alloy in liquid mercury and aqueous potassium iodide environments.[95] Obviously, the liquid metal represents a more severe environment for this aluminum alloy (some five orders of magnitude difference in Stage II cracking rate), but

Table 8.3 Selected K_{IEAC} Data[97]

Metal	Environment	Test Orientation	Yield Strength MPa	ksi	K_{IC} or (K_{IX})[a] MPa$\sqrt{m}$	ksi$\sqrt{in.}$	K_{IEAC} MPa$\sqrt{m}$	ksi$\sqrt{in.}$	Test Time (hr)
Aluminum Alloys									
2014-T6	Synth. seawater	S-L	420	61	21	19	18	16	—
2014-T6	NaCl solution	S-L	—	—	—	—	≈8	≈7	≈10,000[b]
2024-T351	3.5% NaCl	S-L	325	47	(55)	(50)	11	10	—
2024-T351	NaCl solution	S-L	—	—	—	—	≈9	≈8	≈10,000[b]
2024-T852	Seawater	S-L	370	54	19	17.6	15	14	—
2024-T852	NaCl solution	S-L	—	—	—	—	≈17	≈15	≈10,000[b]
2024-T851	Dist. water	L-T	410	59	21	18.6	24	22	—
7075-T6	3.5% NaCl	S-L	505	73	25	23	21	19	—
7075-T6	NaCl solution	S-L	—	—	—	—	≈8	≈7	≈10,000[b]
7075-T7351	3.5% NaCl	S-L	360	52	26	24	23	21	—
7075-T7351	NaCl solution	S-L	—	—	—	—	≲22	≲20	≈10,000[b]
7075-T7351	3.5% NaCl	T-L	365	53	32	29	26	24	—
7175-T66	3.5% NaCl	—	525	76	32	29	≲6.6	≲6	—
7175-T66	NaCl solution	S-L	—	—	—	—	7	6	≈10,000[b]
7175-T736	NaCl solution	—	455	66	27	25	21	19	>1,029
Steel Alloys									
18 Ni(300)-maraging	"	T-L	1960	284	80	72	8	7.5	>150
4340	"	T-S	1335	194	79	72	9	8.5	>333
4340	"	L-T	1690	245	56	51	17	15	>58
4340	Seawater	T-L	1550	225	(69)	(63)	6	5	>20
"	"	"	1380	200	(65)	(59)	11	10	—
"	"	"	1205	175	(83)	(75)	30	27	—
"	"	"	1035	150	(94)	(85)	65	59	—
"	"	"	860	125	(98)	(89)	77	70	—
300M	3.5% NaCl	L-S	1735	252	70	64	22	20	—
"	"	T-L	1725	250	61	56	20	18	—
Titanium Alloys									
Ti-6Al-4V	3.5% NaCl	L-T	890	129	104	95	39 ± 10	35 ± 9	—
"	"	L-S	890	129	99	90	45 ± 8	41 ± 7	—
Ti-8Al-1Mo-1V	"	T-S	825	120	97	88	25	23	—
"	"	"	745	108	123	112	31	28	—
"	Water	T-L	855	124	(105)	(95)	29	26	—
"	Methanol	"	855	124	(105)	(95)	15	14	—
"	CCl$_4$	"	855	124	(105)	(95)	22	20	—
"	Water + 21000 ppm chloride	"	1035	150	(74)	(67)	15	14	—
"	Water + 100 ppm chloride	"	1035	150	(65)	(59)	23	21	—
"	Water + 0.1 ppm chloride	"	1035	150	(65)	(60)	27	24	—

[a] Numbers in parentheses are invalid K_{IC} values that do not satisfy Eq. 6-54.
[b] M. O. Speidel and M. W. Hyatt, *Advances in Corrosion Science and Technology*, Vol. 2, Plenum, New York, 1972, p. 115. (Ref. 98).

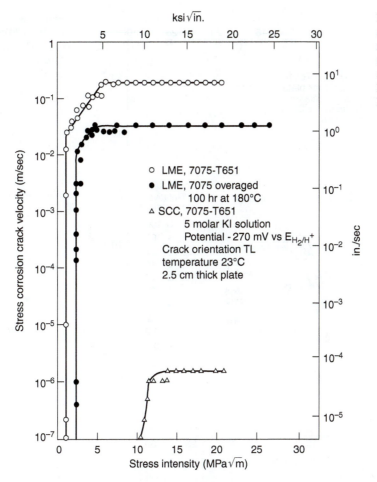

Figure 8.19 Environment-assisted cracking with liquid mercury and aqueous iodide solution in 7075 aluminum alloy.[93] (With permission of Markus O. Speidel, Brown Boveri Co.)

the phenomenology is the same. Furthermore, we see that the alloy in the overaged condition is more resistant to the liquid-metal EAC, as was the case for the salt solution results discussed above.

Environment-assisted cracking in dry gases does not appear to occur in aluminum alloys.[92] However, with increasing moisture content, cracking develops with increasing speed (Fig. 8.20). Consequently, EAC in aluminum alloys may take the form of stress corrosion cracking and liquid-metal embrittlement but not gaseous hydrogen embrittlement.[vi]

8.2.1.3 Temperature and Pressure

Since EAC processes involve electrochemical reactions, it is to be expected that temperature and pressure would be important variables. Test results, such as those shown in Fig. 8.21*a* for hydrogen cracking in a titanium alloy, show the strong effect of temperature on the Stage II cracking rate.[100] These data can be expressed mathematically in the form:

$$\left(\frac{da}{dt}\right)_{\text{II}} \propto e^{-\Delta H/RT} \tag{8-3}$$

where $\Delta H =$ activation energy for the rate-controlling process.

[vi] There is some debate, however, as to whether aqueous stress corrosion cracking in aluminum is related to hydrogen embrittlement. See N. J. H. Holyrod, in *Environment-Induced Cracking of Metals*, Eds. R. P. Gangloff and M. B. Ives, NACE, Houston, TX, 1990, pp. 311–345 (Ref. 99).

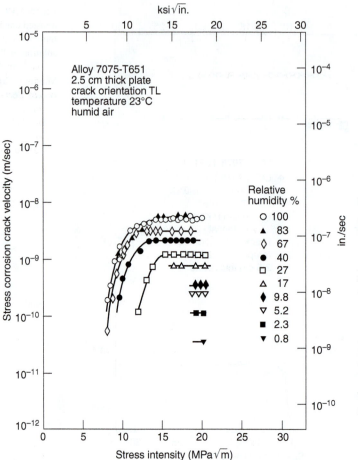

Figure 8.20 Effect of humidity on EAC in 7075-T651 aluminum alloy.[93] (With permission of Markus O. Speidel, Brown Boveri Co.)

The apparent activation energy may then be compared with other data to suggest the nature of the rate-controlling process. Recall from Chapter 4 that at $T_h > 0.5$, ΔH_{creep} was approximately equal to the activation energy for self-diffusion in many materials. In similar fashion, it has been found that the apparent activation energies for the cracking of high-strength steel in water and humidified gas are both about 38 kJ/mol,[4] which corresponds to the activation energy for hydrogen diffusion in the steel lattice.[101] The reader is referred to more detailed discussions on this topic.[102,103] On the other hand, recent studies have shown that the apparent activation energy for Stage II cracking in the presence of gaseous hydrogen is only 16 to 17 kJ/mol.[89,104] Since the embrittling mechanism appears to be the same for the two environments[104] (e.g., the fracture path is intergranular in both cases [see Section 5.9.1]), the change in ΔH probably reflects differences in the rate-controlling hydrogen-transport process. In this regard, note that the cracking rate in gaseous hydrogen is higher than that in water (Fig. 8.4).

The increase in Stage II crack growth rate with increasing pressure noted in Fig. 8.21b can be described mathematically in the form:

$$\left(\frac{da}{dt}\right)_{II} \propto P^n \tag{8-4}$$

Wei and Gangloff have explored the complex interactions of temperature and hydrogen gas pressure as they relate to the EAC of structural alloys.[15,105]

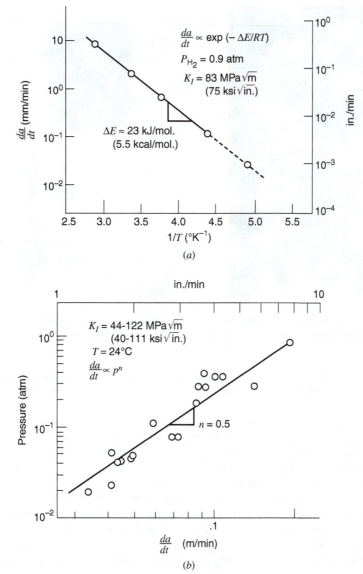

Figure 8.21 Effect of (*a*) temperature and (*b*) pressure on hydrogen-induced cracking in Ti-5A1-2.5Sn in Region II.[100] (From D. P. Williams, "A new criterion for failure of materials by environment-induced cracking." *International Journal of Fracture* 9, 63–74 (1973), published by Noordhoff, Leyden, The Netherlands.)

8.2.2 Environment-Assisted Cracking in Plastics

Engineering plastics are also susceptible to EAC with the extent of structural degradation dependent on the material–environment system, applied stress, and temperature. Although it can occur in many different polymers, including PVC, PE, PS, PC, and PMMA, it is a particular problem for amorphous thermoplastics. Examples of such damage include the development of an extensive network of crazes in a friend's plastic drinking glass and the generation of cracks in the showerhead used in one author's home (Fig. 8.22). The extensive array of crazes in the plastic drinking glass resulted from the presence of a residual tensile stress field (generated from thermal cycling in a dishwasher) and simultaneous exposure to alcohol (specifically, a few stiff gin and tonics). Cracking in the showerhead was caused by exposure to hot water in the presence of the constant stress produced by tightening the fitting to the water pipe.

Of far greater commercial importance is the premature failure of PVC and polyethylene water and gas transmission pipelines throughout the world. Numerous studies have shown that the rupture life of pipe resin increases with decreasing stress level and temperature (Fig. 8.23). (It is

Figure 8.22 Environment-assisted cracking in household plastic components. (*a*) Extensive craze formation in acrylic-based drinking glass (courtesy Elaine Vogel) and (*b*) crack formation in plastic showerhead (courtesy Linda Hertzberg).

also known that rupture life increases with increasing M_w.) At high applied stress levels (Region A), pressurized pipe samples fail by extensive plastic deformation associated with bulging or ballooning of the pipe's cross section. At lower stress levels (Region B), failure occurs in a brittle manner with little evidence of deformation (recall the definitions in Section 5.7).[106–109] It should be noted that such brittle fracture is typical of many service failures in gas pipeline systems.[107–108] The locus of failure times at different stress levels is found to be sensitive to the test environment; for example, the addition of a detergent to water exaggerates the level of EAC damage in the pipe resin as compared with that caused by exposure to water alone.

The applicability of fracture mechanics concepts to the study of the kinetics of environmental cracking has been reported.[110,111] As long as the applied K level does not result in extensive crack-tip deformation, *da/dt*–*K* plots are typically developed with a form similar to that shown in Fig. 8.14.[111] Furthermore, variation of the test temperature of the EAC test enables one to determine the activation energy of the rate-controlling EAC process (recall Fig. 8.21*a*). For example, Chan and

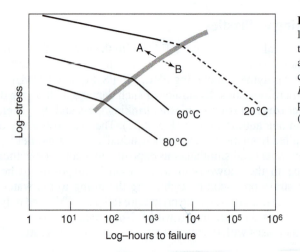

Figure 8.23 Environment-assisted cracking lifetime increases with decreasing stress and temperature. Region A corresponds to pipe bulging and Region B is associated with brittle cracking.[109] [Reprinted from R. W. Hertzberg, *Polymer Communications,* 26, 38 (1985), by permission of the publishers, Butterworth & Co. (Publishers) Ltd. Ⓒ.]

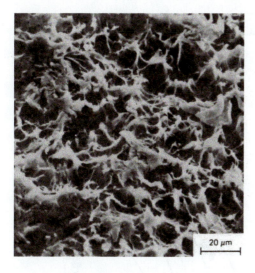

Figure 8.24   Environment-assisted cracking of polyacetal copolymer under stress in hot water. Tufted fracture surface appearance associated with elongated fibrils that eventually ruptured.

Williams[110] found the energy of cracking in high-density polyethylene to be approximately equal to the energy associated with the glass transition temperature (α-relaxation peak).

Environment-assisted cracking in polyethylene and polyacetal copolymer results in the development of a tufted fracture surface appearance (Fig. 8.24). These tufts are believed to represent fibril extension and subsequent rupture. For the case of polyethylene, tufting is observed over more than two orders of magnitude of crack growth rate, with the number and length of tufts decreasing at lower crack velocities in association with lower stress intensity levels.[110,112] The identification of this fracture feature is often a clue for the presence of environment-assisted cracking in water pipe systems.

From a predictive standpoint, the Bergen jig is a valuable screening method used in the polymer industry to evaluate environment-assisted cracking tendencies. Using this device, a test coupon is fixed at its ends and bent for a period of time to variable strain levels along its length. An engineer can then determine the relative susceptibilities of polymer-environment combinations from a single test and identify the critical strain level to cracking. For example, one of the authors used this approach during a root cause investigation of defective children's stuffed animal toys. A product recall was being considered due to the fracture of the painted plastic eye stems and the subsequent release of the plastic eye, which created a potential choking hazard for small children. Through the use of this test method, it was determined that the polycarbonate eye stems were susceptible to environment-assisted cracking as a result of their combined exposure to a specific paint thinner and the presence of residual tensile stresses introduced during the eye stem assembly process.

8.2.3 Environment-Assisted Cracking in Ceramics and Glasses

Ceramics, including glasses, are typically regarded as being chemically inert. For example, laboratory glass containers are often used to store corrosive liquids, with a notable exception of hydrofluoric acid. It may, therefore, be surprising to the reader that EAC of glass was first discovered in 1899 by Grenet[113] for the case of simultaneous exposure to both sustained tensile stress and water vapor. Furthermore, one can speculate as to whether EAC of glass had occurred in association with the famous fracture experiments of Griffith some 80 years ago! Six decades after Grenet observed that the strength of glass depended on the rate of loading as well as the length of time a load was applied, Wiederhorn examined the room temperature crack velocity in soda-lime glass as a function of applied force and relative humidity level (Fig. 8.25).[114] Crack growth rates as low as 10^{-8} m/s were measured, and three distinct regions of crack growth were identified, similar to those depicted in Fig 8.14. In Region I, crack growth increases

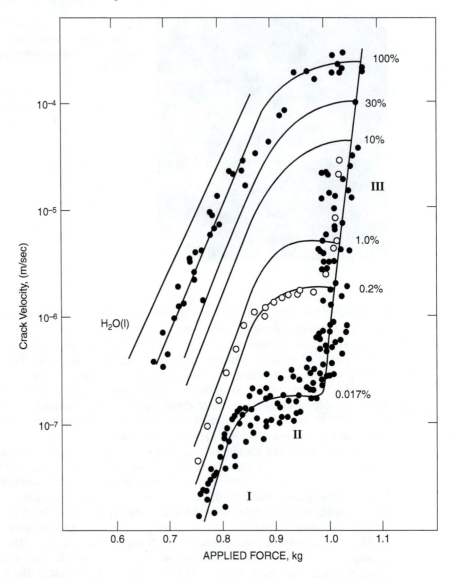

Figure 8.25 The dependence of crack velocity on applied force for soda-lime glass as a function of relative humidity.[114] (S. M. Wiederhorn, *Journal of the American Ceramic Society* 50, 407 (1967), with permission from John Wiley & Sons.)

exponentially with the level of applied force. This is believed to be due to corrosive attack of water vapor on the glass at the crack tip. Region II is relatively insensitive to load level and is believed to be limited by the rate of water vapor transport to the crack tip; for a given load level, note the increase in crack velocity with increasing humidity level. Region III reflects exponential behavior, with the slopes of the curves appearing similar for all levels of relative humidity. Thus, Region III seems to represent a propagation mechanism independent of water concentration. Wiederhorn later determined that crack velocity in room temperature water was a function of glass composition for a given stress intensity factor (see Fig. 8.26).[115] He also suggested the existence of a threshold stress intensity due to the dramatic drop in crack velocity for soda-lime silicate and borosilicate glasses at low stress intensity levels. Later, Kocer and Collins measured crack growth velocities as a function of K_I at room temperature for soda-lime glass, and measured rates as low as 10^{-14} m/s (Fig. 8.27).[116] The data provide strong experimental support for the existence of a subcritical crack growth limit in soda-lime glass.

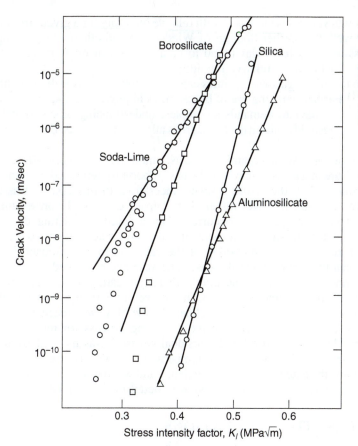

Figure 8.26   Fracture behavior of various glasses in 25°C water, revealing the influence of composition on crack velocity.[115] (S. M. Wiederhorn and L. H. Bolz, *Journal of the American Ceramic Society* 53, 543 (1970), with permission from John Wiley & Sons.)

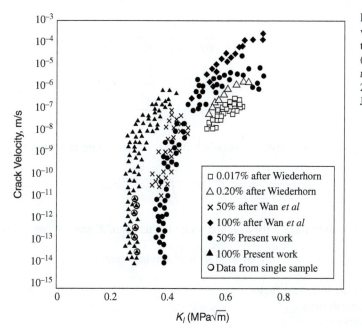

Figure 8.27   Crack growth velocities as a function of K_I at room temperature for soda-lime glass.[116] (C. Kocer and R. Collins, *Journal of the American Ceramic Society* 84, 2585 (2001), with permission from John Wiley & Sons.)

Here, crack growth velocities increased with relative humidity for a given stress-intensity factor and compared well with earlier findings of Wiederhorn and others. As one might expect, the threshold stress-intensity level was found to be a function of relative humidity, with higher values associated with lower relative humidity levels.

In addition to various glasses, ceramics can suffer from EAC as well, including alumina and zirconia.[117-121] Furthermore, the use of bioceramics in physiological environments associated with joint prosthetic devices demands a complete understanding of subcritical crack growth; determination of threshold limits in these materials allows for enhanced component reliability and patient safety.

As an example of EAC in ceramic products, one of the authors recalls a litigation matter involving the failure of a glass vase where the base reportedly separated from the sidewall while being lifted from a table; the result: a severe laceration. In this case, it was concluded that delayed failure was due to EAC rather than impact damage. Upon examination, it was determined that crack propagation occurred from a small preexisting defect in the glass. The high residual tensile stresses responsible for slow crack growth were created during fabrication as a result of uneven cooling of the vase due to a significant difference in wall thickness between the sidewall and the base. The vase had contained water (and flowers) for approximately one week prior to the injury, thereby creating a material/stress/environment combination wherein slow crack growth occurred over a relatively short period of time. Final separation resulted when the fracture toughness of the glass was reached. As a result of this failure analysis, it was determined that the annealing process for the vases (i.e., the oven temperature profile selected to relax the residual stresses) was most likely insufficient. The manufacturer subsequently reviewed the temperature profile histories for all of the factory ovens to determine the extent of product deficiency and whether a product recall was necessary. These important considerations are addressed in detail in Chapter 12.

8.3 LIFE AND CRACK-LENGTH CALCULATIONS

Kinetic crack growth data can be integrated to provide estimates of component life and crack length as a function of time.[88,100] For reactions occurring in parallel (Eq. 4-10), the effective steady-state cracking rate is controlled by the slowest process acting in Regions I, II, and III. If one ignores the contribution of $(da/dt)_{III}(= \dot{a}_{III})$, then the controlling crack growth rate is given by

$$\frac{1}{\dot{a}_T} = \frac{1}{\dot{a}_I} + \frac{1}{\dot{a}_{II}} \qquad (8\text{-}5)$$

or

$$\dot{a}_T = \frac{\dot{a}_I \dot{a}_{II}}{\dot{a}_I + \dot{a}_{II}} \qquad (8\text{-}6)$$

Upon rearrangement of terms, the time devoted to steady-state cracking is

$$t = \int_0^t dt = \int_{a_0}^{a_i} \frac{\dot{a}_I + \dot{a}_{II}}{\dot{a}_I \dot{a}_{II}} da \qquad (8\text{-}7)$$

To solve Eq. 8-7, expressions for $\dot{a}_{I,II}$ are needed in terms of K and the crack length a. In Stage I

$$\dot{a}_I = f(K, T, P, \text{environment}) \qquad (8\text{-}8)$$

where K = stress intensity factor
T = temperature
P = pressure

Since log da/dt–K plots are often linear, Williams[100] has suggested that

$$\dot{a}_I = C_1 e^{mK} \tag{8-9}$$

where C_1 and m are independent of K but may depend on T, P, and environment. For Region II

$$\dot{a}_{II} = f(T, P, \text{environment}) = C_2 \tag{8-10}$$

Note the lack of K dependence in $\dot{a}_{II}$ and the fact that C_2 depends on T, P, and environment. From the previous discussion in Section 8.2.1, C_2 can be evaluated by combining Eqs. 8-3 and 8-4, such that

$$\dot{a}_{II} = C_3 P^n e^{-\Delta H/RT} \tag{8-11}$$

By combining Eqs. 8-9 and 8-11 into Eq. 8-7, it is possible to calculate the length of a crack at any given time, once the various constants are determined from experimental data. The value of Eq. 8-7 lies in its potential to estimate failure times for conditions of T, P, and environments beyond those readily examined in a test program. This life-prediction procedure is now being used in the design of glass components.[122] One additional subtle point should be made regarding the life computation. It should be recognized that the life of a component or test specimen will depend on the rate of change of the stress-intensity factor with crack length dK/da. Consequently, for the same initial K level, the sample with the lowest dK/da characteristic will have the longest life. That is, changing specimen geometry would alter the time to failure. Much work is needed to determine the best method of loading and the best specimen geometry so that a standard EAC test procedure may be established. In this regard, modern computer controlled test systems that enable programmable dK/da and dK/dt testing conditions can serve an important role.

REFERENCES

1. "Corrosion Costs and Preventive Strategies in the United States," U.S. Federal Highway Administration Publication Number: FHWA-RD-01-156, released in 2002.

2. R. W. Staehle, Proceedings of Conference, *Fundamental Aspects of Stress Corrosion Cracking*, R. W. Staehle, A. J. Forty, and D. van Rooyen, Eds., NACE, Houston, TX, 1969.

3. R. M. Latanision, O. H. Gastine, and C. R. Conpeau, *Environment-Sensitive Fracture of Engineering Materials*, Z. A. Foroulis, Ed., AIME, Warrendale, PA, 1979, p. 48.

4. A. J. Bursle and E. N. Pugh, ibid., p. 18.

5. A. W. Thompson, ibid., p. 18.

6. A. W. Thompson and I. M. Bernstein, *Advances in Corrosion Science and Technology*, Vol. 7, R. W. Staehle and M. G. Fontana, Eds., Plenum, New York, 1980, p. 53.

7. F. P. Ford, General Electric Report No. 80CRD141, June 1980.

8. R. C. Newman and R. P. M. Proctor, *Brit. Corr. J.* **25**, 259 (1990).

9. W. W. Gerberich, P. Marsh, and H. Huang, *Parkins Symposium on Fundamental Aspects of Stress Corrosion Cracking*, S. M. Bruemmer et al., Eds., TMS, Warrendale, PA, **191** (1992).

10. H. K. Birnbaum, *Environment-Induced Fracture of Metals*, R. P. Gangloff and M. B. Ives, Eds., NACE, Houston, TX, **21** (1990).

11. R. N. Parkins, *J. Metals* **44** (12), 12 (1992).

12. R. P. Gangloff, "Hydrogen assisted cracking of high strength alloys," in *Comprehensive Structural Integrity*, I. Milne, R. O. Ritchie, and B. Karihaloo, Editors-in-Chief, J. Petit and P. Scott, Volume Editors, Vol. 6, Elsevier Science, New York, NY, 2003, pp. 31–101.

13. D. P. Williams, III, P. S. Pao, and R. P. Wei, *Environment-Sensitive Fracture of Engineering Materials*, Z. A. Foroulis, Ed., AIME, Warrendale, PA, 1979, p. 3.

14. *Gaseous Hydrogen Embrittlement of Metals in Energy Technologies*, R. P. Gangloff and B. P. Somerday, Eds., Woodhead Publishing Limited, Abington Hall, Cambridge, UK, in press (2011).

15. R. P. Gangloff and R. P. Wei, "Gaseous hydrogen embrittlement of high strength steels," *Metallurgical Transactions, A*, Vol. 8A (1977) pp. 1043–1053.

16. J. P. Hirth, *Met Trans.* **11A**, 861 (1980).

17. C. D. Beachem, *Met. Trans.* **3**, 437 (1972).

18. M. G. Nicholas and C. F. Old, *J. Mater. Sci.* **14**, 1 (1979).

19. N. S. Stoloff, *Environment-Sensitive Fracture of Engineering Materials*, Z. A. Foroulis, Ed., AIME, Warrendale, PA, 1979, p. 486.

20. G. E. Linnert, *Welding Metallurgy*, Vol. **2**, American Welding Society, New York, 1967.

21. Effect of Temper and Hydrogen Embrittlement on Fracture Mechanics and CVN Properties of 2.25Cr-1Mo Steel Grades—Application to Minimum Pressurizing Temperature (MPT) Issues, S. Pillot, P. Bourges, C. Chauvy, S. Corre, L. Coudreuse, and P. Toussaint, European Symposium on Pressure Equipment, 2010.

22. W. Hofmann and W. Rauls, *Weld. J.* **44**, 225s (1965).

23. J. B. Steinman, H. C. VanNess, and G. S. Ansell, *Weld. J.* **44**, 221s (1965).

24. G. G. Hancock and H. H. Johnson, *Trans. Met. Soc. AIME* **236**, 513 (1966).

25. D. O. Hayward and B. M. W. Trapnell, *Chemisorption*, 2d ed., Butterworths, Washington, DC, 1964.

26. F. J. Norton, *J. Appl. Phys.* **11**, 262 (1940).

27. G. L. Hanna, A. R. Troiano, and E. A. Steigerwald, *Trans. Quart. ASM* **57**, 658 (1964).

28. R. P. Gangloff, "Diffusion control of hydrogen environment embrittlement in high strength alloys," in *Hydrogen Effects on Material Behavior and Corrosion Deformation Interactions*, N. R. Moody, A. W. Thompson, R. E. Ricker, G. S. Was, and R. H. Jones, Eds., The Minerals, Metals & Materials Society, Warrendale, PA, 2003, pp. 477–497.

29. C. A. Zapffe, *JISI* **154**, 123 (1946).

30. A. S. Tetelman and W. D. Robertson, *Trans. AIME* **224**, 775 (1962).

31. A. R. Troiano, *Trans. ASM* **52**, 54 (1960).

32. J. G. Morlet, H. H. Johnson, and A. R. Troiano, *JISI* **189**, 37 (1958).

33. N. J. Petch and P. Stables, *Nature (London)* **169**, 842 (1952).

34. S. P. Lynch, "Mechanisms of Hydrogen Assisted Cracking—A Review" in *Hydrogen Effects in Materials*, N. R. Moody and A. W. Thompson, Eds., The Minerals, Metals & Materials Society, Warrendale, PA, 2003.

35. I. M. Robertson, "The effect of hydrogen on dislocation dynamics," *Engr. Frac. Mech.*, **68** (2001) pp. 671–692.

36. *Effect of Hydrogen on Behavior of Materials,* AIME, Sept. 7–11, 1975, Moran, Wyoming.

37. I. M. Bernstein and A. W. Thompson, Eds., *Hydrogen Effects in Metals*, AIME, Warrendale, PA, 1980.

38. *Microstructural Dependence of Aqueous-Environment Assisted Crack Growth and Hydrogen Uptake in AA7050*, Lisa M. Young, University of Virginia, Charlottesville, VA, 1999.

39. R. W. Staehle, *The Theory of Stress Corrosion Cracking in Alloys*, NATO, Brussels, 1971, p. 223.

40. R. W. Staehle, *Stress Corrosion Cracking and Hydrogen Embrittlement of Iron-Base Alloys*, NACE, Houston, 1977, p. 180.

41. *Stress-Corrosion Cracking, Materials Performance and Evaluation*, Russell H. Jones, Ed., ASM International, 1992, p. 2.

42. M. G. Fontana, *Corrosion Engineering*, 3rd ed., McGraw-Hill, Inc., New York, 1986, p. 372.

43. *Nuclear Engineering International*, Reed Business Publication, Sutton, England, January 1994, pp. 20–22.

44. *Stress-Corrosion Cracking, Materials Performance and Evaluation*, Russell H. Jones, Ed., ASM International, 1992, p. 32.

45. W. Rostoker, J. M. McCaughey, and M. Markus, *Embrittlement by Liquid Metals*, Van Nostrand-Reinhold, New York, 1960.

46. V. I. Likhtman, E. D. Shchukin, and P. A. Rebinder, *Physico-Chemical Mechanics of Metals*, Acad. Sci. USSR, Moscow, 1962.

47. M. H. Kamdar, *Prog. Mater. Sci.* **15**, 1 (1973).

48. N. J. Kelly and N. S. Stoloff, *Met. Trans.* **6A**, 159 (1975).

49. A. R. C. Westwood, C. M. Preece, and M. H. Kamdar, *Fracture*, Vol. **3**, H. Leibowitz, Ed., Academic, New York, 1971, p. 589.

50. M. H. Kamdar, *Treatise on Materials Science and Technology*, Vol. **25**, C. L. Briant and S. K. Banerji, Eds., Academic, New York, 1983, p. 362.

51. D. W. Fager and W. F. Spurr, *Corrosion-NACE* **27**, 72 (1971).

52. S. Mostovoy and N. N. Breyer, *Trans. Quart. ASM* **61**, 219 (1968).

53. R. P. Gangloff, "Solid cadmium embrittlement of textured zircaloy-2," in *Embrittlement by Liquid and Solid Metals*, M. H. Kamdar, Ed., TMS-AIME, Warrendale, PA, 1984, pp. 485–505.

54. M. T. Hahn and E. N. Pugh, ASTM, *STP 733*, 1981, p. 413.

55. Metals Handbook, 8th ed. Volume **9**, American Society for Metals, 1974, p. 31.

56. A. R. C. Westwood and M. H. Kamdar, *Philos. Mag.* **8**, 787 (1963).

57. N. S. Stoloff and T. L. Johnson, *Acta Met.* **11**, 251 (1963).

58. S. P. Lynch, *Acta Met.* **28**, 325 (1981).

59. H. H. Johnson and P. C. Paris, *Eng. Fract. Mech.* **1**, 3 (1967).

60. B. F. Brown, *Mater. Res. Stand.* **6**, 129 (1966).

61. B. F. Brown and C. D. Beachem, *Corr. Sci.* **5**, 745 (1965).

62. B. F. Brown, *Met. Rev.* **13**, 171 (1968).

63. B. F. Brown, *J. Materials,* JMLSA, **5** (4), 786 (1970).

64. R. A. Oriani and P. H. Josephic, *Acta Metall.* **22**, 1065 (1974).

65. *Standard Test Method for Determining Threshold Stress Intensity Factor for Environment-Assisted Cracking of Metallic Materials Under Constant Load,* ASTM, Philadelphia, December 1993.

66. S. R. Novak and S. T. Rolfe, *J. Mater.* **4**, 701 (1969).

67. R. P. Wei and S. R. Novak, *J. Test. Eval.* **15** (1), **38** (1987).

68. P. Mclntyre and A. H. Priest, Report MG/31/72, British Steel Corp., London, 1972.

69. W. G. Clark, Jr., and J. D. Landes, ASTM *STP 610*, 1976, p. 108.

70. K. Hirano, S. Ishizaki, H. Kobayashi, and H. Nakazawa, *J. Test. Eval* **13** (2), 162 (1985).

71. M. Khobaib, AFWAL-TR-4186, WPAFB, Ohio, 1982.

72. R. A. Mayville, T. J. Warren, and P. D. Hilton, *J. Engl Mater. Tech.* **109** (3), 188 (1987).

73. D. R. Mclntyre, R. D. Kane, and S. M. Wilhelm, *Corrosion* **44** (12), 920 (1988).

74. *Stress Corrosion Cracking, The Slow Strain-Rate Technique,* G. M. Ugiansky and J. H. Payer, Eds., ASTM *STP 665* (1979).

75. *Slow Strain Rate Testing for the Evaluation of Environmentally Induced Cracking: Research and Engineering Applications,* R. D. Kane, Ed., ASTM *STP 1210* (1993).

76. R. D. Kane, *ASTM Standardization News* **21** (5), 34 (1993).

77. L. Raymond, *Metals Handbook*, Vol. **13**, 9th ed. ASM, Metals Park, OH, 1987, p. 283.

78. D. L. Dull and L. Raymond, ASTM *STP 543*, 1974, p. 20.

79. Yongwon Lee and R. P. Gangloff, "Measurement and modeling of hydrogen environment assisted cracking of ultra-high strength steel," *Metallurgical and Materials Transactions, A,* vol. **38** (2007) pp. 2174-2190.

80. T. W. Crooker and J. A. Hauser II, NRL Memo Report 5763, April 3, 1986.

81. R. R. Fessler and T. J. Barlo, ASTM *STP 821*, 368 (1984).

82. P. S. Pao and R. A. Bayles, NRL Publication 190-6320 (1991).

83. M. Horstmann, J. K. Gregory, *Scripta Metallurgica et Materialia* (1991), vol. 25, (11), pp. 2503–2506.

84. R. P. Gangloff and A. Turnbull, *Modeling Environmental Effects on Crack Initiation and Propagation,* TMS-AIME, Warrendale, PA, 55 (1990).

85. S. M. Wiederhorn, *Environment-Sensitive Mechanical Behavior*, Vol. **35**, Metallurgical Society Conf., A. R. C. Westwood and N. S. Stoloff, Eds., Gordon & Breach, New York, 1966, p. 293.

86. S. M. Wiederhorn, *Int. J. Fract. Mech.* **42**, 171 (1968).

87. S. M. Wiederhorn, *Fracture Mechanics of Ceramics,* Vol. **4**, R. C. Bradt, D. P. H. Hasselman, and F. F. Lange, Eds., Plenum, New York, 1978, p. 549.

88. R. P. Wei, S. R. Novak and D. P. Williams, *Mater. Res. Stand.* **12** (9), 25 (1972).

89. S. J. Hudak, Jr., M. S. Thesis, Lehigh University, Bethlehem, PA, 1972.

90. J. D. Landes and R. P. Wei, *Int. J. Fract.* **9** (3), 277 (1973).

91. W. D. Benjamin and E. A. Steigerwald, *Trans. ASM,* **60**, 547 (1967).

92. M. O. Speidel, *Met. Trans.* **6A**, 631 (1975).

93. M. O. Speidel, *The Theory of Stress Corrosion Cracking in Alloys,* J. C. Scully, Ed., NATO, Brussels, Belgium, 1971, p. 289.

94. L. M. Young and R. P. Gangloff, "S-Phase effect on environmental cracking in AA7050," in *Advances in the Metallurgy of Aluminum Alloys,* M. Tiryakioglu, Ed., ASM International, Materials Park, OH, 2001, pp. 135-140.

95. M. H. Peterson, B. F. Brown, R. L. Newbegin, and R. E. Groover, *Corrosion* **23**, 142 (1967).

96. Y. Takeda and C. J. McMahon, Jr., *Met. Trans.* **12A**, 1255 (1981).

97. *Damage Tolerant Design Handbook,* MCIC-HB-01, Sept. 1973.

98. M. O. Speidel and M. W. Hyatt, *Advances in Corrosion Science and Technology*, Vol. 2, Plenum, New York, 1972, p. 115.

99. N. J. H. Holyrod, in *Environment-Induced Cracking of Metals,* R. P. Gangloff and M. B. Ives, Eds., NACE, Houston, TX, 1990, pp. 311–345.

100. D. P. Williams, *Int. J. Fract.* **9** (1), 63 (1973).

101. W. Beck, J. O'M. Bockris, J. McBreen, and L. Nanis, *Proc. R. Soc. London Ser. A.* **290**, 221 (1966).

102. R. P. Gangloff, in *Comprehensive Structural Integrity,* I. Milne, R. O. Ritchie and B. Karihaloo, Editors-in-Chief, J. Petit and P. Scott, Volume Editors, Vol. 6, 2003, Elsevier Science, New York, NY, pp. 31–101.

103. R. P. Gangloff, in *Hydrogen Effects on Material Behavior and Corrosion Deformation Interactions,* N. R. Moody, A.W. Thompson, R. E. Ricker,

G. S. Was, and R. H. Jones, Eds., The Minerals, Metals & Materials Society, Warrendale, PA, 2003, pp. 477–497.

104. D. P. Williams and F. G. Nelson, *Met. Trans.* **1**, 63 (1970).

105. R. P. Wei and R. P. Gangloff, in *Fracture Mechanics: Perspectives and Directions, ASTM STP 1020*, R. P. Wei and R. P. Gangloff, Eds., ASTM, Philadelphia, PA, 1989, pp. 233–264.

106. J. B. Price and A. Gray, Proceedings 4th Int. Conf. Plastic Pipes and Fittings, Plast. Rub. Inst., March 1979, p. 20.

107. E. Szpak and F. G. Rice, Proceedings, 6th Plastic Pipe Sym., April 4–6, 1978, Columbus, OH, p. 23.

108. F. Wolter and M. J. Cassady, ibid., p. 40.

109. R. W. Hertzberg, *Polym. Commun.* **26**, 38 (1985).

110. M. K. V. Chan and J. G. Williams, *Polymer* **24**, 234 (1983).

111. K. Tonyali and H. R. Brown, *J. Mater. Sci.* **21**, 3116 (1986).

112. C. S. Lee and M. M. Epstein, *Polym. Eng. Sci.* **22**, 549 (1982).

113. L. Grenet, "Mechanical strength of glass," *Bull. Soc. Enc. Industr. Nat. Paris*, (Ser. 5), 4 (1899), pp. 838–848.

114. S. M. Wiederhorn, *J. American Ceramic Society* **50**, 8 (1967), pp. 407–414.

115. S. M. Wiederhorn and L. H. Bolz, *J. American Ceramic Society* **53**, 10 (1970), pp. 543–548.

116. C. Kocer and R. Collins, *J. American Ceramic Society* **84**, 11 (2001), pp. 2585–2593.

117. A. H. De Aza, J. Chevalier, G. Fantozzi, M. Schehl, and R. Torrecillas, *Biomaterials* **23** (2002), pp. 937–945.

118. M. E. Ebrahimi, J. Chevalier, and G. Fantozzi, *J. Materials Research* **15**, 1 (2000).

119. M. K. Ferber and S. D. Brown, *J. American Ceramic Society* **63**, 7–8 (1980), pp. 424–429.

120. J. E. Ritter, Jr, and J. N. Humenik, *J. Materials Science* **14** (1979), pp. 626–632.

121. U. Seidelmann, H. Richter, and U. Soltesz, *J. Biomedical Materials Research* **16** (1982), pp. 705–713.

122. S. M. Wiederhorn, *Ceramics for High Performance Applications*, Brook Hill, Chestnut Hill, MA, 1974, p. 633.

PROBLEMS

Review

8.1 In the phrase "subcritical flaw growth mechanisms," what does *subcritical* mean?

8.2 What do the acronyms EAC, SCC, HE, and LME stand for?

8.3 What is the main difference between the HEAC model and the IHAC model?

8.4 What EAC phenomenon is sometimes associated with welding?

8.5 What is the objective of a post-weld heat treatment with regard to EAC?

8.6 What three major factors affect the hydrogen-embrittling process?

8.7 What three simultaneous conditions must be satisfied for SCC?

8.8 SCC is often described in terms of a film rupture model. What is the film the model refers to? Provide a specific example in support of your answer.

8.9 How is it that two solid metals in contact at a temperature below the melting point of either can lead to liquid-metal embrittlement?

8.10 What controls the rate of dynamic embrittlement, and what mathematical dependence on temperature does this suggest?

8.11 What are the similarities and differences between K_{ISCC}, K_{IEAC}, and K_{EAC}? How do they differ from K_{IC}?

8.12 When EAC processes are active, under what condition does final fracture occur?

8.13 Large transport aircraft make wide use of Al alloys with a −T6 temper (which indicates the peak strength condition). However, for certain locations in these aircraft it is preferred to use Al alloys with a −T7 temper (which indicates an overaged condition). What could possibly motivate this design decision even though it probably means an undesirable increase in vehicle weight?

8.14 EAC in metals causes cracks to form and grow. What other feature can form and grow in amorphous polymers suffering from EAC?

8.15 What two changes occur with regard to EAC behavior of PVC and polyethylene water and gas transmission pipelines as temperature is increased?

8.16 What fracture surface feature is often a clue for the presence of environment-assisted cracking in polyethylene water pipe systems?

8.17 Polycarbonate is sometimes used for protective helmets, such as motorcycle helmets. However, it is often not advisable to apply adhesive stickers to the helmet surface or to clean the surface with a solvent, as both can potentially degrade the helmet strength. Why might this be?

8.18 Ceramic and glass materials are routinely used to contain chemicals, including some that are quite aggressive. Are these materials therefore immune to EAC?

Practice

8.19 A wrought, high-strength steel known to be susceptible to hydrogen embrittlement is used to fabricate a component. The component is machined from thick plate with no regard as to the component orientation, relative to the plate's rolling plane and direction. It is determined that only a subset of components are failing in the field due to hydrogen embrittlement. How could this be?

8.20 The same alloy is used to fabricate two different components, both of which result in identical mechanical properties. Cold working yields a microstructure with a much higher incidence of symmetrical grain boundaries for Component 1 as compared with that of Component 2. Which item will most likely have a higher resistance to dynamic embrittlement?

8.21 An investigation was made of the rate of crack growth in a 7079-T651 aluminum plate exposed to an aggressive environment under a static stress σ. A large test sample was used with a single-edge notch placed in the T-L orientation. As indicated in the accompanying table, the rate of crack growth under sustained loading was found to vary with the magnitude of the applied stress and the existing crack length. The material exhibits Regions I and II EAC but not Region III. If the K_{IC} for the materials is 20 MPa$\sqrt{m}$, how long would it take to break a sample containing an edge crack 5 mm long under a load of 50 MPa? *Hint*: First establish the crack growth rate relations.

Cracking Rate (m/sec)	Applied Stress (MPa)	Crack Length (mm)
10^{-9}	35	5
32×10^{-9}	35	10
1×10^{-6}	70	5
1×10^{-6}	70	7.5

8.22 For the 18 Ni (300)-maraging steel listed in Table 8.3, calculate the stress level to cause failure in a center-notched sample containing a crack 5 mm long. What stress level limit would there have to be to ensure that EAC did not occur in a 3.5% NaCl solution?

8.23 How much faster than the room temperature value would a crack grow in a high-strength steel submerged in water if the temperature were raised 100 °C?

8.24 A metal plate is found to contain a single-edge notch and is exposed to a static stress in the presence of an aggressive environment. Representative data obtained from crack-growth measurements are given in the following table:

Measurement	Cracking Rate (m/s)	Applied Stress (MPa)	Crack Length (mm)
1	1×10^{-9}	30	5
2	4.1×10^{-9}	30	8
3	8×10^{-9}	30	10
4	6.4×10^{-8}	60	5
5	6.4×10^{-8}	60	6
6	6.4×10^{-8}	60	7

a. What is the growth rate relation among the cracking rate, stress, and crack size?

b. Does the relation change? If so, why?

c. What was controlling the cracking process in the regime associated with measurements 4, 5, and 6?

Design

8.25 As-welded, austenitic stainless steel connections are experiencing repeated failures in beachfront properties in Miami, Florida, due to chloride-induced stress corrosion cracking. No such failures have been reported to date in Chicago, Illinois, the other city where these connections were installed. Write a brief memo to your supervisor with an explanation of the regional dependence of this phenomenon, and propose two potential approaches to mitigate the risk for this type of cracking by changing both fabrication methods and material of construction.

8.26 You are a design engineer for a housewares company that is looking to produce glass flower vases. You are told that the factory that your company usually contracts to do this type of work has had intermittent problems in the past controlling their oven temperature. In addition, the manufacturer has asked you to choose between soda-lime and silica glass for your product. What are your concerns, and what decisions should you make in response?

Extend

8.27 Find a practical example of stress corrosion cracking failure of a component (or a class of components) made from a brass alloy. Summarize the failure circumstances and the failure analysis. Include copies of photographs, if they are available. Provide a full reference.

Chapter 9

Cyclic Stress and Strain Fatigue

Daydreamers have two options for supplementary entertainment: doodling or paper clip bending. The doodler is limited by the amount of paper available, while the paper clip bender's amusement is tragically short-lived—the clip breaks after only a few reversals! This simple example describes a most insidious fracture mechanism—failure does not occur when the component is loaded initially; instead, failure occurs after a certain number of similar load fluctuations have been experienced. The author of a book about metal fatigue[1] began his treatise by describing a photograph of his car's steel rear axle, which had failed: "the final fracture occurring at 6:00 AM just after setting out on holiday." Somewhat less expensive damage, but saddening nonetheless, was the failure of one of the authors' childhood vehicle (Fig. 9.1). From an examination of the fracture surfaces, it was concluded that this failure originated at several sites and traveled across the section, with occasional arrest periods prior to final separation. Another fatigue failure generated in an author household is shown in Fig. 9.2. The failure of this zinc die-cast doorstop nearly destroyed the lovely crystal chandelier in the front hall of the home. The light-colored and triangularly shaped region on the fracture surface shown in Fig. 9.2*b* reflects months/years of repeated loadings (i.e., door openings); the darker and larger region on the fracture surface was generated in a matter of milliseconds. Regardless of the material—steel paper clips and car axles, plastic tricycles, zinc doorsteps—fatigue failures will occur when the component experiences cyclic stresses or strains that produce permanent damage. Since the majority of engineering failures involve cyclic loading of one kind or another, it is appropriate to devote considerable attention to this subject in this chapter, and in Chapter 10. The reader is also referred to Chapter 11 for descriptions of case histories of fatigue failures.

9.1 MACROFRACTOGRAPHY OF FATIGUE FAILURES

A macroscopic examination of many service failures generated by cyclic loading reveals distinct fracture surface markings. For one thing, the fracture surface is generally *flat*, indicating the absence of an appreciable amount of gross plastic deformation during service life. In many cases, particularly failures occurring over a long period of time, the fracture surface contains lines referred to in the literature as "clam shell markings" arrest lines and/or "beach markings" (Fig. 9.3).[i] These markings have been attributed to different periods of crack extension, such as during one flight or one sequence of maneuvers of an aircraft or the operation of a machine during a factory work shift. It is to be emphasized that these bands reflect *periods* of growth and are not representative of *individual* load excursions. Unique markings associated with the latter are discussed in Chapter 10. It is believed that these alternate crack growth and dormant periods cause regions on the fracture surface to be oxidized and/or corroded by different amounts, resulting in the formation of a fracture surface containing concentric rings of nonuniform color. Similar bands resulting from variable amplitude block loading have been found on fracture

---

[i] The use of the expressions "beach markings" and "clamshell" markings has long been associated with the appearance of certain fatigue fracture surfaces. Clamshell markings on the outer surface of bivalve mollusks represent periods of shell growth and are analogous to macroscopic markings on fatigue fracture surfaces in metals and plastics that correspond to periods of crack growth. No such direct analogy exists for the case of ocean beach markings, which depend on the vagaries of wind and water flow patterns. Nevertheless, reference to "beach markings" provides a mental picture of a pattern of parallel lines that typifies the surface of many fatigue fractures.

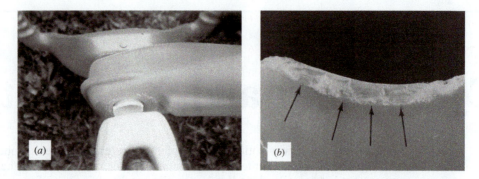

Figure 9.1   Fatigue fracture of plastic tricycle. (*a*) General location of failure; (*b*) several origins are identified by arrows. Note characteristic fatigue ring-like markings emanating from each origin, which represent periods of growth during life of component. (Courtesy Jason and Michelle Hertzberg.)

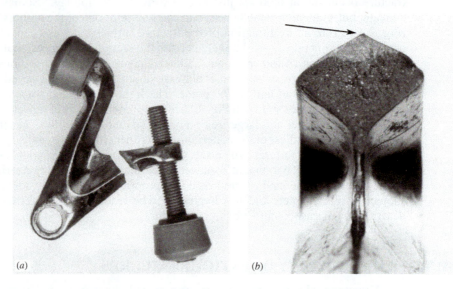

Figure 9.2   Zinc die-cast doorstop fatigue fracture. Fatigue crack grew from corner until reaching a critical size and causing failure. Arrow in (*b*) indicates crack origin.

surfaces (see Section 10.5.1). Since these "beach markings" often are curved, with the center of curvature at the origin, they serve as a useful guide to direct the investigator to the fracture initiation site.

A second set of fracture surface markings are seen in Fig. 9.3*b*. In addition to the set of horizontal clamshell markings (C), one also notes two vertically oriented curved black lines that separate parallel packets of clamshell markings. These lines are called *ratchet lines* (labeled R) and represent the junction surfaces between the three adjacent crack origins.[2] Since each microcrack is unlikely to form on the same plane, their eventual linkage creates a vertical step on the fracture surface. Once the initial cracks have linked together, the ratchet lines disappear. Hence, the ratchet lines connect contiguous regions where separate cracks had initiated. This point is best illustrated by examination of the fatigue fracture surface in Fig. 9.4. This specimen possessed a polished semicircular notch root from which three fatigue cracks initiated. Note the presence of two small horizontal lines at the edge of the

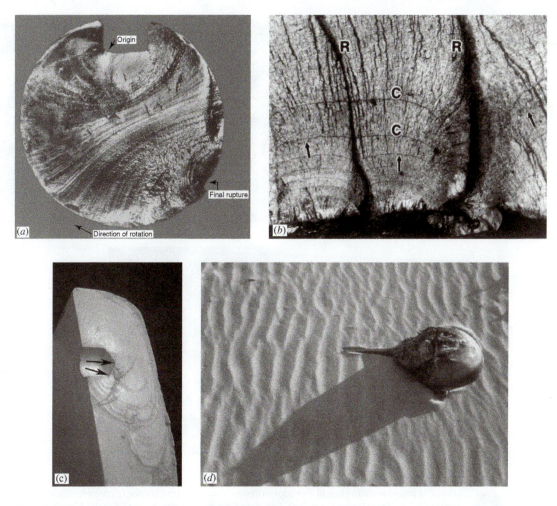

Figure 9.3 Fatigue fracture markings. (*a*) Rotating steel shaft. Center of curvature of earlier "beach markings" locate crack origin at corner of keyway.[2] (By permission from D. J. Wulpi, *How Components Fail*, copyright American Society for Metals, 1966.) (*b*) Clamshell markings (C) and ratchet lines (R) in aluminum. Arrows indicate the crack propagation direction (Photo courtesy of R. Jaccard). (*c*) Fatigue bands in high-impact polystyrene toilet seat. Two separate fatigue crack origin points are marked with black arrows. (*d*) Beach markings in South Carolina.

notch root (left side of Fig. 9.4*a*); these markings are ratchet lines that separate three different cracks that originated on three separate planes (Fig. 9.4*b*). (In this photo, the depth of field of the SEM was reduced to reveal the three different spatial elevation levels of the cracks.) Note the chevron-like pattern on the crack surface in the middle of the photograph. Such markings can be used to locate the crack origin, such as the inclusion that was found at the surface of the polished notch root (Fig. 9.4*c*).

As shown in Fig. 9.5, the fracture surface may exhibit any one of several patterns, depending on such factors as the applied stress and the number of potential crack nucleation sites. For example, we see that as the severity of a design-imposed stress concentration and/or the applied stress increases, the number of nucleation sites and associated ratchet lines increase. Either of these conditions should be avoided if at all possible. In fact, many service failures exhibit only one nucleation site, which eventually causes total failure. The size of this fatigue crack at the point of final failure is related to the applied stress level and the fracture toughness of the material (recall Eq. 6-31).

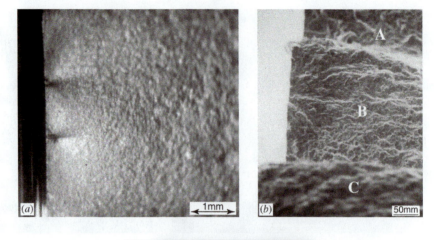

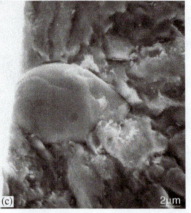

Figure 9.4 Fatigue fracture surface appearance in HSLA steel. (*a*) Two ratchet lines separate three crack origins; (*b*) SEM image revealing height elevation difference between three cracks; (*c*) cerium sulfide inclusion at origin of fatigue crack. (From Braglia et al.[3]; © American Society for Testing and Materials, 1916 Race Street, Philadelphia, PA 19103. Reprinted with permission.)

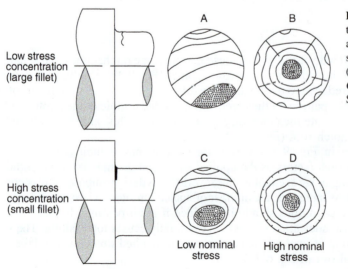

Figure 9.5 Diagrams showing typical fatigue fracture surface appearance for varying conditions of stress concentration and stress level.[2] (By permission from D. J. Wulpi, *How Components Fail*, © American Society for Metals, 1966.)

9.2 CYCLIC STRESS-CONTROLLED FATIGUE

Many engineering components must withstand numerous load or stress reversals during their service lives. Examples of this type of loading include alternating stresses associated with a rotating shaft, pressurizing and depressurizing cycles in an aircraft fuselage at takeoff and landing,[ii] and load fluctuations affecting the wings during flight. Depending on a number of factors, these load excursions may be introduced either between fixed strain or fixed stress limits; hence, the fatigue process in a given situation may be governed by a strain- or stress-controlled condition. Discussion in this section is restricted to stress-controlled fatigue; strain-controlled fatigue is considered in Section 9.3.

One of the earliest investigations of stress-controlled cyclic loading effects on fatigue life was conducted by Wöhler,[4] who studied railroad wheel axles that were plagued by an annoying series of failures. Several important facts emerged from this early investigation, as may be seen in the plot of stress versus the number of cycles to failure (a so-called *S-N* diagram) given in Fig. 9.6. First, the cyclic life of an axle increased with decreasing stress level, and below a certain stress level, it seemed to possess infinite life—fatigue failure did not occur (at least not before 10^6 cycles; as discussed below, no such fatigue limit appears when tests are conducted in the gigacycle range). Second, the fatigue life was reduced drastically by the presence of a notch. These observations have led many current investigators to view fatigue as a three-stage process involving initiation, propagation, and final failure stages (Fig. 9.7). When design defects or metallurgical flaws are preexistent, the initiation stage is shortened drastically or completely eliminated, resulting in a reduction in potential cyclic life.

Over the years, laboratory tests have been conducted in bending (rotating or reversed flexure), torsion, pulsating tension, or tension-compression axial loading. Such tests have been conducted under conditions of constant load or moment (to be discussed in this section), constant deflection or strain (Section 9.3), or a constant stress intensity factor (Chapter 7). Examples of different loading conditions are shown in Fig. 9.8. In rotating bending with a single load applied at the end of the cantilevered test bar (Fig. 9.8*a*), the bending moment increases with increasing distance

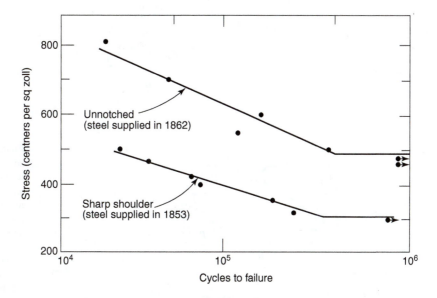

Figure 9.6 Wöhler's *S-N* curves for Krupp axle steel.[5] (Reprinted by permission of the American Society for Testing and Materials from copyrighted work.)

[ii] For example, recall the loss of much of the upper half of the fuselage section from one of Aloha Airlines' Boeing 737 fleet (see *New York Times*, May 1, 1988, p. 1.)

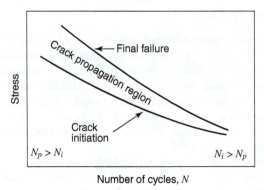

Figure 9.7   Fatigue life depends on the relative extent of initiation and propagation stages.

from the applied load point and precipitates failure at the base of the fillet at the end of the gage section. In effect, this represents a notched fatigue test, since the results will depend strongly on fillet geometry. The rotating beam-loaded case (Fig. 9.8b) produces a constant moment in the gage section of the test bar that can be used to generate either unnotched or notched test data. Notched test data are obtained by the addition of a circumferential notch in the gage section. Both specimen types generally represent zero mean load conditions or a load ratio of $R = -1$ ($R \equiv$ minimum load/maximum load). These test specimen configurations and modes of loading may be suitable for evaluating the fatigue characteristics of a component subjected to simple rotating loads. However, it is often more realistic to use the axially loaded specimen (Fig. 9.8c) to simulate service conditions that involve direct loading when mean stress is an important variable. Such is the case for aircraft wing loads where fluctuating stresses are superimposed on both a tensile (lower wing skin) and compressive (upper wing skin) mean stress.

Standard definitions regarding key load or stress variables are shown in Fig. 9.9 and defined by

$$\Delta\sigma = \sigma_{\max} - \sigma_{min} \tag{9-1}$$

$$\sigma_a = \frac{\sigma_{\max} - \sigma_{min}}{2} \tag{9-2}$$

$$\sigma_m = \frac{\sigma_{\max} + \sigma_{min}}{2} \tag{9-3}$$

$$R = \frac{\sigma_{min}}{\sigma_{\max}} \tag{9-4}$$

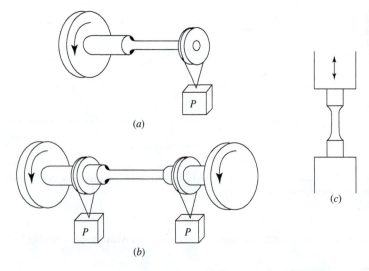

Figure 9.8   Various loading configurations used in fatigue testing. (*a*) Single-point loading, where bending moment increases toward the fixed end; (*b*) beam loading with constant moment applied in the gage section of sample; (*c*) pulsating tension or tension–compression axial loading. Test procedures to determine *S-N* diagrams are described in ASTM Standards E466-E468.[6]

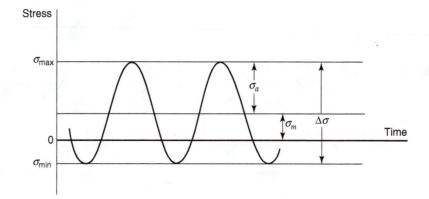

Figure 9.9   Nomenclature to describe test parameters involved in cyclic stress testing.

Most often, *S-N* diagrams (similar to that shown in Fig. 9.6) are plotted with the stress amplitude given as half the total stress range. Another example of constant load amplitude fatigue data for 7075-T6 aluminum alloy notched specimens is shown in Fig. 9.10. Note the considerable amount of scatter in fatigue life found among the 10 specimens tested at each stress level. The smaller scatter at high stress levels is believed to result from a much shorter initiation period prior to crack propagation. The existence of scatter in fatigue test results is common and deserving of considerable attention, since engineering design decisions must be based on recognition of the statistical character of the fatigue process. Consider what values of "fatigue limit" should be assigned at various lifetime levels for the data shown in Fig. 9.10. Traditionally, engineers have used various statistical methodologies to define a suitable fatigue limit; for example, the fatigue limit can be defined, assuming a Gaussian distribution of the data set, as the average alternating stress at a specified lifetime less three times the standard deviation of the data set. The latter characterizes a stress level where the probability of failure during that specified lifetime is close to zero. And yet, Bathias and co-workers experimentally determined that a more conservative estimate of the fatigue limit in the range of 10^6–10^9 cyclic life should be closer to one-half the average alternating stress level.[7] Clearly, more data are needed when fatigue property

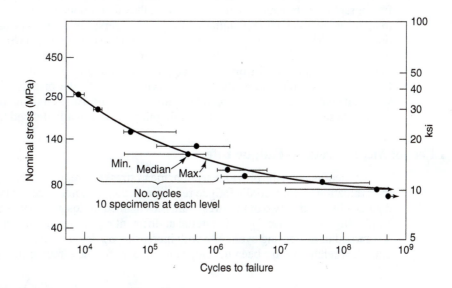

Figure 9.10   Constant load amplitude fatigue data for 7075-T6 aluminum alloy notched specimens (0.25-mm root radius).[10] (Reprinted from Hardrath et al., NASA TN D-210.)

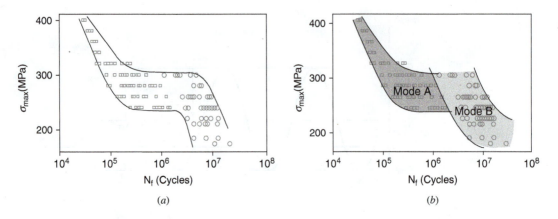

Figure 9.11 Fatigue lifetimes in 2024-T3 aluminum alloy. (*a*) All data points; (*b*) data segregated according to persistent slip band initiation mechanism (Mode A) and initiation from broken inclusions (Mode B).[7] (Data from C. Bathias and P. C. Paris, Gigacycle Fatigue in Mechanical Practice, Marcel Dekker, New York, 2004.)

determinations are based on statistical formulations, especially when seeking to extrapolate fatigue data from the 10^6 to 10^7 range into the 10^{10} to 10^{11} cyclic lifetime regime for very high cycle fatigue lifetime components. This proves to be easier said than done: with conventional testing machines, the acquisition time for one data point in the latter cyclic range might be a few years! In addition, there would be a heavy cost burden for such tests. On the other hand, Bathias and co-workers have used piezoelectric fatigue machines, operating at 20kHz, to complete 10^{10} load cycles in less than a week's time.[8,9] Hopefully, such expanded experimental data sets will improve the reliability of statistically defined fatigue property determinations.

The origins of test scatter are manifold. They include external variations in the testing environment, random defects present on the specimen surface, and alignment differences of the test machine. Regarding the latter, rotating bending machines generate the least amount of scatter as compared with that associated with axial loading machines. Internal material variables also contribute to the development of test scatter. In a recent study, Bathias et al.[12] examined the nature of the scatter associated with an *S-N* plot for Al alloy 2024-T3 (Fig. 9.11*a*) Following a fractographic examination of all test samples, they determined that test data could be segregated to identify specimen failures initiating from either persistent slip bands associated with dislocation motion (Mode A) or from fractured inclusions (Mode B) (Fig. 9.11*b*). By segregating the test data in this manner, scatter in test results for a given stress level and a particular cracking mechanism was reduced from two orders of magnitude to one. Mind you, though such post-fracture examination and analysis may serve to identify the origins of material-induced test scatter, it contributes little toward identifying *a priori* the overall amount of expected scatter for a given data set or toward improving the reliability of statistically defined fatigue property values.

9.2.1 Effect of Mean Stress on Fatigue Life

As mentioned in the previous section, mean stress can represent an important test variable in the evaluation of a material's fatigue response. It then becomes necessary to portray fatigue life data as a function of two of the stress variables defined in Eqs. 9-1 to 9-4. Sometimes this is done by plotting *S-N* data for a given material at different σ_m values, as shown in Fig. 9.12*a*. Here we see a trend of decreasing cyclic life with increasing σ_m level for a given σ_a level. Alternatively, empirical relations have been developed to account for the effect of mean stress on fatigue life.

$$\text{Goodman relation:} \quad \sigma_a = \sigma_{fat}\left(1 - \frac{\sigma_m}{\sigma_{ts}}\right) \tag{9-5}$$

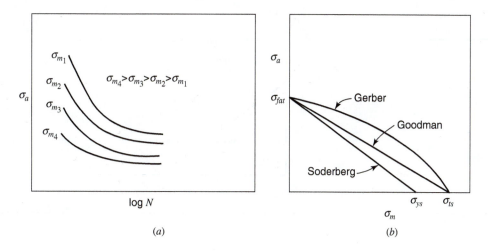

Figure 9.12   Representative plots of data showing effect of stress amplitude and mean stress on fatigue life. (a) Typical S-N diagrams with differing σ_m levels; (b) Gerber, Goodman, and Soderberg diagrams showing combined effect of alternating and mean stress on fatigue endurance.

$$\text{Gerber relation :} \quad \sigma_a = \sigma_{fat}\left[1 - \left(\frac{\sigma_m}{\sigma_{ts}}\right)^2\right] \tag{9-6}$$

$$\text{Soderberg relation :} \quad \sigma_a = \sigma_{fat}\left(1 - \frac{\sigma_m}{\sigma_{ys}}\right) \tag{9-7}$$

where $\sigma_a =$ fatigue strength in terms of stress amplitude, where $\sigma_m \neq 0$
$\sigma_m =$ mean stress
$\sigma_{fat} =$ fatigue strength in terms of stress amplitude, where $\sigma_m = 0$
$\sigma_{ts} =$ tensile strength
$\sigma_{ys} =$ yield strength

These relations are shown in Fig. 9.12b and illustrate the relative importance of σ_a and σ_m on fatigue endurance. Experience has shown that most data lie between the Gerber and Goodman diagrams; the latter, then, represents a more conservative design criteria for mean stress effects. (*Author Note: In a review of such constant life diagrams, Sendeckyj[13] pointed out that these two commonly labeled "Goodman and Gerber relations" were previously reported in the literature by others.*)

EXAMPLE 9.1

Suppose that Hertzalloy 100, a certain steel alloy, has an endurance limit (σ_{fat}) and a tensile strength of 700 and 1400 MPa, respectively. Would one expect fatigue failure if a component, manufactured from this alloy, were subjected to repeated loading from 0 to 600 MPa?

From Eq. 9-1, both alternating and mean stress levels are computed to be 300 MPa. The accompanying Goodman diagram shows that this cyclic loading condition (A) lies well within the safe region bounded by the line corresponding to the Goodman relation. Fatigue failure would not be expected. On the other hand, if the component were to possess a residual tensile stress of 700 MPa as a result of a prior welding procedure, the effective mean stress would be 1000 MPa (the sum of the internal residual stress [700 MPa] and the applied external mean stress [300 MPa]). Under these conditions (B), fatigue failure would be predicted.

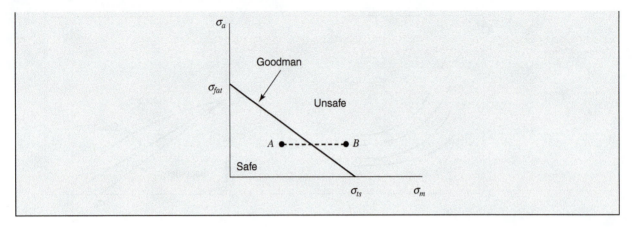

9.2.2 Stress Fluctuation, Cumulative Damage, and Safe-Life Design

Much of the fatigue data discussed thus far were generated from constant stress amplitude tests, but these results are not realistic in actual field service conditions. Many structures are subjected to a range of load fluctuations, mean levels, and frequencies. The task, then, is to predict, based on constant amplitude test data, the life of a component subjected to a variable load history. A number of cumulative damage theories, proposed during the past few decades, describe the relative importance of stress interactions and the amount of damage—plastic deformation, crack initiation, and propagation—introduced to a component. For example, if the same amount of damage is done to a component at any stress level as a result of a given fraction of the number of cycles required to cause failure, we see from Fig. 9.13 that $n_1/N_1 + n_2/N_2 + n_3/N_3 = 1$. This may be described in more general form by

$$\sum_{i=1}^{k} \frac{n_i}{N_i} = 1 \tag{9-8}$$

where $k =$ number of stress levels in the block loading spectrum
  $\sigma_i =$ ith stress level
  $n_i =$ number of cycles applied at σ_i
  $N_i =$ fatigue life at σ_i

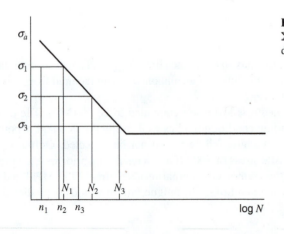

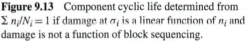

Figure 9.13   Component cyclic life determined from $\Sigma \, n_i/N_i = 1$ if damage at σ_i is a linear function of n_i and damage is not a function of block sequencing.

Equation 9-8 is the work of Palmgren[14] and Miner[15] and is often referred to as the Palmgren-Miner cumulative damage law. By combining Eq. 9-8 with standard *S-N* data, one can estimate the total or residual service lifetime of a structural component that experiences multiple load sequences.

EXAMPLE 9.2

A multipurpose traffic bridge has been in service for three years and each day carries a large number of trains, trucks, and automobiles. A subsequent highway analysis reveals a sharp difference between expected and actual traffic patterns that threatens to shorten the useful life span of the bridge. Fortunately, a nearby second bridge was recently completed that can assume all of the train traffic. Given the following fatigue information, estimate the remaining lifetime for the first bridge, assuming that it will carry only truck and automobile traffic.

Vehicle	Fatigue Lifetime	Vehicles/Day
Automobiles	10^8	5,000
Trucks	2×10^6	100
Trains	10^5	30

To determine the remaining service lifetime for the bridge, we first need to establish the amount of fatigue damage accumulated during the initial three-year service period.

For purposes of simplicity, we will assume that there are no load interaction effects corresponding to the three principal stress levels experienced by the bridge members. (A similar assumption is often made by civil engineers, regarding the fatigue life analysis of current bridge decks.) Accordingly, the amount of damage is estimated from the Palmgren-Miner relation where

$$\sum_{i=1}^{k} \frac{n_i}{N_i} = 1$$

For the initial three years of service (1095 days) involving train, truck, and automobile traffic, the amount of bridge damage that incurred, relative to its total design lifetime, is estimated to be

$$1095 \left[\frac{5000}{10^8} + \frac{100}{2 \times 10^6} + \frac{30}{10^5} \right] = 0.438$$

Therefore, the remaining fatigue lifetime corresponding to automobile and truck traffic alone will be 56.2% of the combined fatigue lifetime associated with these two stress levels. Therefore,

$$d \left(\frac{5000}{10^8} + \frac{100}{2 \times 10^6} \right) = 0.562$$

where d = remaining days of service.

It follows that an additional 5620 days or 15.4 years of useful fatigue lifetime remains before the bridge's fatigue design limit is reached. (By comparison, if train traffic were to be continued on this bridge, only 3.85 years of additional useful service life would remain.) At a later date, bridge lifetime could be extended still further by diverting truck traffic to a different route. The residual service life now available for automobile traffic alone could then be calculated in a manner similar to that previously shown, by first adding together the initial increment of life consumed for all traffic (i.e., 0.438) with the lifetime increment corresponding to the second phase of service (i.e., combined automobile and truck traffic). The remaining fraction, based on Eq. 9-8, would then be used to determine the remaining allowable lifetime for automobile traffic alone.

Such computations are used routinely to determine the "safe-life" or durability of numerous engineering components. By further illustration, let us assume that a design engineer assigns a "safe-life" of 1000 cycles to a particular component that experiences a uniform cyclic stress. Based on the *S-N* data for the component's material, failure after 1000 loading cycles would require an alternating stress of σ_1. The design stress for the component would then be σ_1/F, where *F* is the safety factor to account for such variables as batch-to-batch material property variations, installation-induced differences in component stress level, environmental changes, and existence of adventitious defects (also see Section 9.4). Alternatively, the safety factor could be used to define the allowable cyclic stress necessary for a service lifetime of $F \times 1000$ loading cycles. That is, an allowable stress is determined for a durability of perhaps "two or three" expected component lifetimes. When "safe-life"-designed components reach their lifetime limit, they are removed from service even though they may be defect-free and would surely survive many additional loading cycles. As a result of such conservative design practices, sound parts are often retired from service (or "trashed") with uneconomical consequences. By contrast, the fail-safe design criteria recognizes that cracks can develop in components and provides for a structure that will not fail prior to the time that the defect is discovered and repaired. More will be said about this fatigue design philosophy in Section 10.1.2.

As assumed in Example 9.2, we see that Eq. 9-8 shows no dependence on the order in which the block loads are applied and, as such, represents interaction-free behavior as well. In reality, the Palmgren-Miner law is unrealistic, since the amount of damage accumulated does depend on block sequencing and varies nonlinearly with n_i. For example, if a high-load block is followed by a low-load block, experimental data in *unnotched* specimens generally indicate $\Sigma n/N < 1$. (The reverse is true for the case of *notched* samples, as is described in Chapter 10; the opposite trend reflects different effects of load interactions on the initiation and propagation stages in the fatigue process.) Since crack propagation begins sooner at the higher stress levels, it is argued that the initial cycles at the second block of lower stress excursions would do more damage than normally anticipated, since the initiation process would have been truncated by the high-load block. The deleterious effect of overstressing in unnotched testing is shown in Fig. 9.14. Alternatively, when σ_1/σ_2 is less than unity, $\Sigma n/N > 1$ for some alloys. Such understressing is seen to "coax" the fatigue limit of certain steels that strain age to somewhat higher levels. The use of the Palmgren-Miner cumulative damage law in association with the prediction of fatigue lifetime under random loading conditions is discussed in Section 9.4.

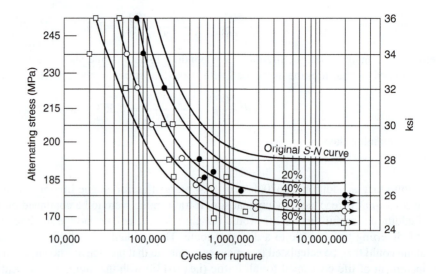

Figure 9.14 *S-N* diagrams showing decreased cyclic life after initial cycling at $\pm$ 250 MPa for 20, 40, 60, and 80% of anticipated life at that stress for SAE 1020 steel. (Reprinted by permission of the American Society for Testing and Materials from copyrighted work.)

9.2.3 Notch Effects and Fatigue Initiation

Fatigue failures often initiate at the surface of a component. What are some of the factors contributing to this behavior? First, many stress concentrations, such as surface scratches, dents, machining marks, and fillets, are unique to the surface, as is corrosion attack, which roughens the surface. In addition, cyclic slip causes the formation of surface discontinuities, such as intrusions and extrusions, that are precursors of actual fatigue crack formation. (The processes involved in intrusion and extrusion formation are discussed in Section 9.5.) The data shown in Fig. 9.15 clearly show the serious loss in fatigue limit associated with a deterioration in surface quality. Recall that a similar response was recognized by Wöhler more than 100 years ago (Fig. 9.6).

To quantitatively evaluate the severity of a particular stress concentration, many investigators adopted the stress concentration factor k_t as the comparative key parameter.[5] (From Chapter 6, it is not surprising to find the stress intensity factor also being used in this fashion; see Chapter 10.) Assuming elastic response, the fatigue strength at N cycles in a notched component would be expected to decrease by a factor equal to k_t. For example, if a material exhibits a smooth bar fatigue life of 210 MPa, the same material would have a fatigue life of 70 MPa if a theoretical stress concentration factor of 3 were present. In reality, the reduction in fatigue strength at N cycles is less than that predicted by the magnitude of k_t. Rather, the fatigue

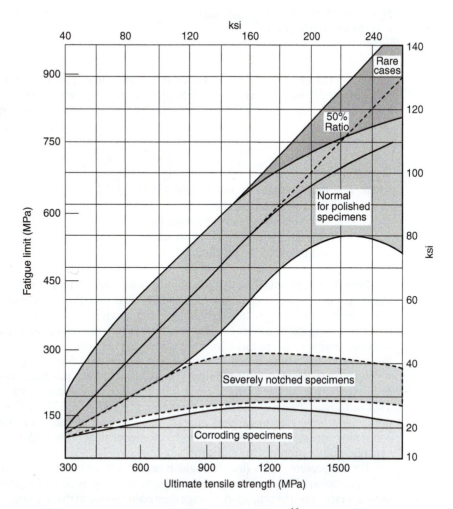

Figure 9.15   Effect of surface condition on fatigue limit in steel alloys.[16] (Reprinted with permission from John Wiley & Sons, Inc.)

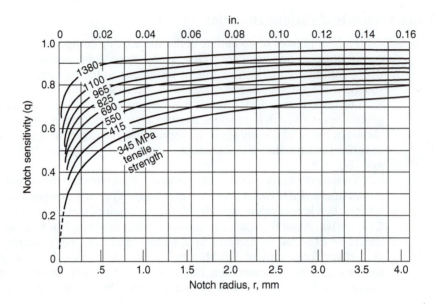

Figure 9.16   Effect of tensile strength and notch acuity on relative notch sensitivity.[18] (Reprinted with permission from McGraw-Hill Book Co.)

strength is reduced by a factor k_f, which represents the effective stress concentration factor as affected by plastic flow and by notch root surface area and volume considerations.[17] The relative notch sensitivity q for a given material and notch root detail may be given by[17]

$$q = \frac{1}{1 + p/r} = \frac{k_f - 1}{k_t - 1} \tag{9-9}$$

where q = notch sensitivity factor, wherein $0 \le q \le 1$
    k_t = theoretical stress concentration factor
    k_f = effective stress concentration factor
    p = characteristic material parameter
    r = radius of the notch root

    From Fig. 9.16, it is seen that the relative notch sensitivity increases with increasing tensile strength, since high-strength materials usually possess a limited capacity for deformation and crack-tip blunting. Of greater significance, the notch sensitivity factor q decreases markedly with decreasing notch root radii. This results from the fact that k_f increases more slowly than k_t with decreasing notch radius (Eq. 9-9); discrepancies between k_f and k_t as large as a factor of two or three have been noted in some cases. The reason for this apparent paradox (i.e., less severe fatigue damage susceptibility [k_f] as k_t increases) has been attributed to the lack of distinction made between fatigue crack initiation and fatigue crack propagation processes.[19,20] That is, fatigue cracks would be expected to initiate more readily with increasing k_t, but might not always propagate to failure; instead, one might find "nonpropagating cracks" in solids containing stress concentrations beyond some critical value. (See Section 10.4.3 for an expanded discussion of this subject.)

    For the present, let us first distinguish between these two stages in the fatigue process. We define initiation life by the number of loading or straining cycles N_i required to develop a crack of some specific size; the propagation stage then corresponds to that portion of the total cyclic life N_p which involves growth of that crack to some critical dimension at fracture. Hence, $N_T = N_i + N_p$ where N_T is the total fatigue life. Does the completion of the crack initiation process correspond to

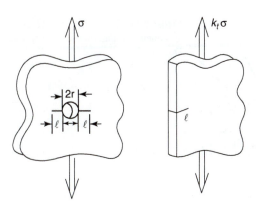

Figure 9.17 Growth of small cracks. (*a*) From hole with diameter $2r$ in infinitely wide plate under stress σ. (*b*) Edge crack in semi-infinite plate with remote stress equal to $k_t \sigma$.

the development of a crack with a length of 1 cm, 1 mm, or 1 μm? Alternatively, should initiation be defined when a newly formed crack attains a length equal to some multiple of the characteristic microstructural unit, such as the grain size in a metal or the spherulitic diameter in a semi-crystalline polymer? In general, when does a defect become a crack? These questions have provoked considerable discussion for many years and, yet, no precise definition for crack initiation has been or perhaps can be identified. Part of this difficulty arises from the fact that the fatigue life corresponding to the initiation and growth of a crack to some specified length often depends on the geometry of the test specimen. To illustrate, let us consider the development and growth of a crack from a circular hole in an infinitely wide plate that is subjected to an oscillating tensile stress (Fig. 9.17*a*). For this discussion, we should recognize that the range of the crack-tip stress intensity factor ΔK, which varies with the cyclic loading history, has a major impact on the rate of growth of the crack. (Chapter 10 deals at considerable length with this subject.) From Section 6.5, the stress intensity factor solution for this crack–hole configuration can be estimated in two distinctly different ways. At one extreme, when the crack length ℓ is small compared with the hole radius r, the stress intensity factor may be given by

$$K_s = 1.12 k_t \sigma \sqrt{\pi \ell} \tag{6-26}$$

where $K_s =$ stress intensity factor—short-crack solution where the crack tip is embedded within the stress field of the hole

$k_t =$ stress concentration factor for the hole in an infinite plate

$\sigma =$ remote stress

$\ell =$ crack length from the surface of the hole

$1.12 =$ surface flaw correction factor

At the other extreme, when $\ell \gg r$, the long-crack stress intensity factor solution is appropriate:

$$K_\ell = \sigma \sqrt{\pi a} \tag{6-27}$$

where $K_\ell =$ stress intensity factor—long-crack solution where the hole is considered to be part of a long crack

$a =$ crack length $= r + \ell$

$r =$ radius of hole

For the short-crack–hole condition, the stress intensity factor also can be estimated by a single-edge-notched solution where the remote stress is $k_t \sigma$ (i.e., the stress at the surface of the hole; Fig. 9.17*b*). Hence, if a very small crack or crack nucleus of length ℓ_1 is assumed to exist

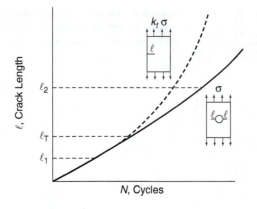

Figure 9.18 Difference in crack growth behavior due to specimen configuration when $l > l_T$. (Adapted from Dowling.[19])

and we wish to know how many cycles it would take for this defect to grow to a length ℓ_T, both crack configurations would provide the same estimate of cyclic life (Fig. 9.18). However, if one were to define crack initiation when a crack has grown to a length ℓ_2, then different fatigue initiation lives would be found for the two specimen configurations. To avoid such dependence of geometry on the fatigue initiation life, Dowling[19] has recommended that the crack length at initiation be on the order of ℓ_T or less, where ℓ_T corresponds to the transition between short-crack- and long-crack-controlled behavior.

A model for the determination of the transition crack length ℓ_T is suggested in Fig. 9.19. Shown here are the Y calibrations corresponding to Eqs. 6-26 and 6-27 versus the crack length to hole radius ratio ℓ/r. Also shown is the numerical solution for this crack–hole configuration. Note that the short-crack solution closely approximates the numerical solution where the crack length ℓ is small relative to the hole radius r; the long-crack solution provides good agreement when ℓ becomes large compared to ℓ_T. This means that, when $\ell < \ell_T$, fatigue behavior is controlled by the local stress field associated with the notch. Depending on the sharpness of the notch root radius and the magnitude of the applied load, the local stress field could be predominantly elastic or plastic in nature. As a result, the rate of crack growth in this region will be different (see Section 10.4.2 for further discussion). When $\ell > \ell_T$, fatigue behavior is controlled by the remote stress and concepts of linear elastic fracture mechanics. Both local and remote stress fields combine to influence fatigue behavior when $\ell \approx \ell_T$.

The transition point corresponding to the intersection of the curves in Fig. 9.19 depends on the specific stress intensity and stress concentration factors that, in turn, depend on the manner of loading and the crack geometry. For the case of a hole in a large plate, ℓ_T was found to be

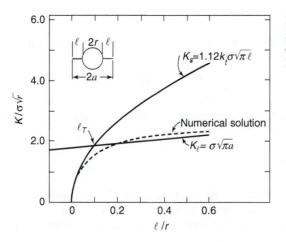

Figure 9.19 Y calibrations for short- and long-crack solutions for hole–crack configuration. The dashed curve is the numerical solution.[23] (Adapted from Dowling.[19])

approximately $r/10$ and was $r/5$ for a sharp notch configuration similar to that of a compact sample. Other values of ℓ_T/r for different crack configurations are given by Dowling.[19]

By estimating the cyclic life of a notched component from unnotched specimen test results, the choice of $\ell \leq \ell_T$ as a working definition of fatigue crack initiation avoids the complex problems associated with geometry dependence on initiation life.[iii] Unfortunately, when $\ell < \ell_T$, one may be faced with two additional problems: The size of the crack-tip plastic zone and the microstructural units (e.g., metal alloy grains) may no longer be small compared with ℓ. Both factors compromise the applicability of the stress intensity factor as the controlling parameter for short fatigue crack formation and early growth[21,22] (see Section 10.4.3). On the other hand, linear elastic fracture mechanics would be expected to characterize fatigue crack development when $\ell \gg \ell_T$. Total fatigue life can then be estimated by treating separately the initiation and propagation stages of crack formation. Initiation life can be estimated from notched specimen data characterized in terms of the local stress field $k_t\sigma$ or with the cyclic strain approach (see Section 9.3). In either case, initiation must be defined as the development of a crack whose size is on the order of ℓ_T. Linear elastic fracture mechanics can then be used to calculate the propagation portion of the total life by integrating the crack growth rate–stress intensity factor relation from a crack size of ℓ_T to the critical flaw size at fracture (see Section 10.1.1). The duration of the fatigue crack initiation stage relative to that of propagation depends on the notch root radius (Fig. 9.20). For a given stress level and for the case where the radius is small, total fatigue life is dominated by the propagation stage. On the other hand, when the notch root radius is large, the bulk of the total fatigue life involves initiation of the crack at the notch root.

Another factor that controls fatigue strength at N cycles is the size of the test bar. Although no size effect is observed in *axial* loading of *smooth* bars, a strong size effect is noted in smooth and notched samples subjected to *bending* and in *axially* loaded *notched* bars. In all cases, the section size effect is related to a stress gradient existing in the sample, which in turn controls the volume of material subjected to the highest stress levels (recall Section 5.3.1.1). For the case of

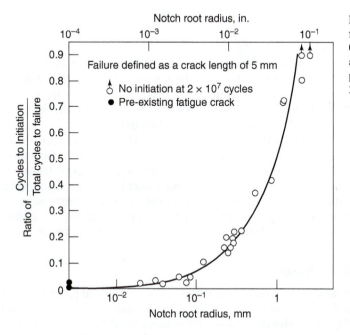

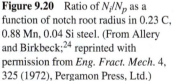

Figure 9.20   Ratio of N_i/N_p as a function of notch root radius in 0.23 C, 0.88 Mn, 0.04 Si steel. (From Allery and Birkbeck;[24] reprinted with permission from *Eng. Fract. Mech.* 4, 325 (1972), Pergamon Press, Ltd.)

[iii] It should be recognized that Dowling's working definition of the fatigue initiation stage combines both crack nucleation and early growth to a length ℓ_T. An alternative approach has been proposed wherein the "initiation" stage involves crack nucleation along with a more detailed analysis of the short-crack growth regime (see Section 10.4.3).

bending, the smaller the cross section of the test bar, the higher the stress gradient and the smaller the volume of material experiencing maximum stress. Comparing this situation to that of axially loaded smooth specimens where no stress gradient exists and the entire cross section is stressed equally, one finds bending fatigue strengths to be higher than values obtained from axially loaded samples. From a statistical viewpoint, the larger the volume of material experiencing maximum stress, the greater the probability of finding a weak area that would lead to more rapid failure. As a final note, it is helpful to consider a reappraisal by Findley[25] of the specimen size effect in fatigue testing. If one assumes that fatigue crack initiation will occur when cyclic slip develops over some critical region requiring a minimum driving force σ', then slip can occur only in the outer fibers, where the applied stress is greater than σ'. Since the stress at the outer fiber in bending is

$$\sigma = \frac{Mr}{I} \tag{9-10}$$

where $\sigma =$ flexural stress
$\quad M =$ bending moment
$\quad\quad I =$ moment of inertia
$\quad\quad r =$ radius of circular rod or distance from neutral axis to outer fiber

the flexural stress decreases to σ' when one moves a distance Δr from the outer surface, so that

$$\sigma' = \frac{M}{I}(r - \Delta r) \tag{9-11}$$

and

$$\sigma = \frac{\sigma'}{1 - \frac{\Delta r}{r}} \tag{9-12}$$

From Eq. 9-12, it is useful to re-examine Figs. 9.11a, b for additional insight as to the apparently high degree of test scatter with the reported data set. Recalling that Mode B test results represent fatigue crack initiation from subsurface fractured inclusions, it follows that these data should be shifted *downward* in the S-N plot to reflect lower flexural stress levels for these test results (i.e., see Eq. 9-12). Clearly, a portion of the test scatter can be attributed to the use of the nominal flexural stress level (Eq. 9-10) for all data, even though the Mode B subsurface initiation sites should be represented by stress levels associated with Eq. 9-12. Furthermore, we see that no size effect is predicted when uniaxial tension is applied to an unnotched specimen. Here $\sigma = \sigma'$. However, a size effect is anticipated in specimens possessing a large stress gradient (that is, in small specimens subjected to either bending or torsion, and in notched, axially loaded samples). Finally, it is apparent from Eq. 9-12 that the size effect disappears for large samples since $r \gg \Delta r$, whereupon σ' approaches σ.

9.2.4 Material Behavior: Metal Alloys

This section presents an overview of the effect of mechanical properties on material fatigue response. Since detailed discussions of the effect of microstructure and thermomechanical treatment on fatigue behavior in various alloy systems would be beyond the scope of this book, the reader is referred to numerous articles in the literature. Books by Bathias and Paris,[7] Forrest,[1] Sines and Waisman,[18] Fuchs and Stephens,[26] and Forsyth[27] should provide an excellent starting point for such an investigation.

Traditionally, materials have been described as exhibiting S-N plots of two basic shapes, based on data sets extending only to cyclic lifetimes of 10^6 to 10^7 cycles. These plots show

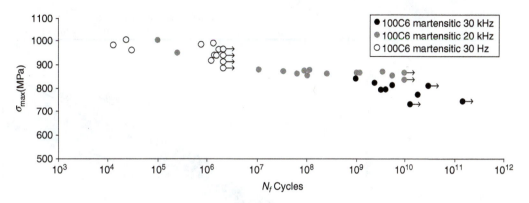

Figure 9.21   *S-N* plot of martensitic steel tested to $\sim 10^{11}$ cycles under $R = -1$ conditions. Note the absence of a well-defined endurance limit. (Adapted from *International Journal of Fatigue*, 32, C. Bathias, "Influence of the metallurgical instability on the gigacycle fatigue regime," p. 535, 2010, with permission from Elsevier.)[11]

either a well-defined fatigue limit (Fig. 9.6) below which the material would appear to be immune from cyclic damage, or a continually decreasing curve (Fig. 9.10) with no apparent lower stress limit below which the material could be considered completely "safe." (Note the strong resemblance to environment-assisted cracking behavior discussed in the previous chapter.) In materials that possess a "knee" in the *S-N* curve, the *fatigue limit* is readily determined as the stress associated with the horizontal portion of the *S-N* curve. It has been found that many steel alloys exhibit this type of behavior in the cyclic range up to 10^7 cycles. However, when test results are extended to the 10^9 to 10^{10} cyclic life range using piezoelectric fatigue machines, Bathias and co-workers[7,9,11,12] have noted that ferrous alloys do not exhibit a horizontal asymptote (i.e., a well-defined fatigue limit). Instead, such alloys demonstrate a finite lifetime even at low stress levels (Fig. 9.21). Note that conventional test data (i.e., 30 Hz test frequency) appear to reveal a fatigue limit of approximately 900–950 MPa for a cyclic lifetime of 10^6 cycles. By sharp contrast, when tested at 20–30 kHz, sample failures occur at stress levels below 800 MPa for cyclic lifetimes greater than 10^{10} cycles. (Recent studies by Bathias[11] have shown that certain alloys such as stainless steels may experience localized heating in the crack-tip zone, resulting in associated metallurgical instability, unless the test sample's temperature is maintained near ambient. Most other alloys do not demonstrate this phenomenon.) Apparently, fatigue property determinations may well be a function of both the investigator's equipment and patience!

Over the more limited and convenient testing range up to 10^6 to 10^7 cycles, where a well-defined *S-N* curve asymptote is observed for ferrous alloys, the fatigue limit often is estimated to be one-half the tensile strength of the material (Fig. 9.15). (Correspondingly, in the test region of roughly 10^9 cycles, investigators[7,28] have noted that the fatigue limit increases with increasing Vickers hardness level but decreases with increasing defect projected area as well as stress ratio.) However, it should be noted that the fatigue ratio σ_{fat}/σ_{ts} for such materials can vary between 0.35 and 0.60, as shown in Fig. 9.22a for the case of several carbon and alloy steels. Additional fatigue limit data for other alloys are given in Table 9.1. Initially, it would appear to be good design practice to use a material with as high a tensile strength as possible to maximize fatigue resistance. Unfortunately, this can cause a lot of trouble, since (as shown in Chapters 8 and 9) fracture toughness decreases and environmental sensitivity increases with increasing tensile strength. Since tensile strength and hardness are related, it is possible to estimate the fatigue limit in a number of steels simply by determining the hardness level—a very inexpensive test procedure, indeed. We see from Fig. 9.23 that a good correlation exists up to a hardness level of about $40R_C$. Above $40R_C$, test scatter becomes considerable and the fatigue limit–hardness relation becomes suspect.

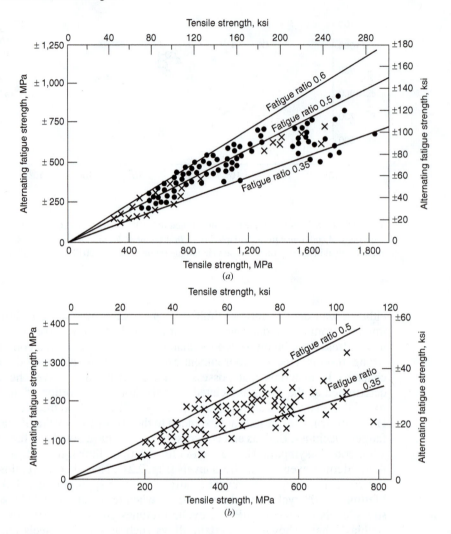

Figure 9.22   Relation between rotating, bending unnotched fatigue strength and tensile strength.[1] (*a*) Alloy (●) and carbon (×) steels; (*b*) wrought copper alloys. (P. J. Forrest, *Fatigue of Metals*, Addison-Wesley, Reading, MA, 1962, with permission.)

The fatigue behavior of nonferrous alloys, especially in the conventional data acquisition range, usually follows the second type of *S-N* plot, and no clear fatigue limit is defined. Consequently, the "fatigue limit" for any such alloy would have to be defined at some specific cyclic life—usually 10^7 cycles. Such an arbitrary definition does create uncertainty, especially when test data in the 10^6–10^7 cycle range are used to estimate fatigue limits for engineering components where cyclic lives exceed 10^8 cycles (e.g., automobile engines), 10^9 cycles (e.g., high-speed trains or ship's diesel engines), or even 10^{10} cycles (e.g., turbine engine components).[7] As noted earlier in Section 9.2, fatigue data generated with the use of piezoelectric machines can provide useful information in this regard.[7]

Examining various aluminum alloys reveals that "fatigue limit"/tensile strength ratios are lower than those found in steel alloys. The fatigue ratio for selected wrought copper alloys is shown in Fig. 9.22*b*, and fatigue limit data for nonferrous alloys are included in Table 9.1. Many studies have been conducted and theories proposed to account for the relatively poor fatigue response shown by this important group of engineering materials. It is presently felt[30] that extremely fine and atomically ordered precipitates, contained within Al–Cu alloys, are

Table 9.1   Fatigue Endurance Limit of Selected Engineering Alloys

Material	Condition	σ_{ts} MPa (ksi)	σ_{ys} MPa (ksi)	σ_f MPa	(ksi)
Steel Alloys[a] (Endurance limit based on 10^7 cycles)					
1015	Cold drawn—0%	455 (66)	275 (40)	240	(35)
1015	Cold drawn—60%	710 (102)	605 (88)	350	(51)
1040	Cold drawn—0%	670 (97)	405 (59)	345	(50)
1040	Cold drawn—50%	965 (140)	855 (124)	410	(60)
4340	Annealed	745 (108)	475 (69)	340	(49)
4340	Q & T (204°C)	1950 (283)	1640 (238)	480	(70)
4340	Q & T (427°C)	1530 (222)	1380 (200)	470	(68)
4340	Q & T (538°C)	1260 (183)	1170 (170)	670	(97)
HY140	Q & T (538°C)	1030 (149)	980 (142)	480	(70)
D6AC	Q & T (260°C)	2000 (290)	1720 (250)	690	(100)
9Ni–4Co–0.25C	Q & T (315°C)	1930 (280)	1760 (255)	620	(90)
300M	—	2000 (290)	1670 (242)	800	(116)
Aluminum Alloys[b] (Endurance limit based on 5×10^8 cycles)					
1100-0		90 (13)	34 (5)	34	(5)
2014-T6		483 (70)	414 (60)	124	(18)
2024-T3		483 (70)	345 (50)	138	(20)
6061-T6		310 (45)	276 (40)	97	(14)
7075-T6		572 (83)	503 (73)	159	(23)
Titanium Alloys[c] (Endurance limit based on 10^7 cycles)					
Ti–6Al–4V		1035 (150)	885 (128)	515	(75)
Ti–6Al–2Sn–4Zr–2Mo		895 (130)	825 (120)	485	(70)
Ti–5Al–2Sn–2Zr–4Mo–4Cr		1185 (172)	1130 (164)	675	(98)
Copper Alloys[c] (Endurance limit based on 10^8 cycles)					
70Cu–30Zn Brass	Hard	524 (76)	435 (63)	145	(21)
90Cu–10Zn	Hard	420 (61)	370 (54)	160	(23)
Magnesium Alloys[c] (Endurance limit based on 10^8 cycles)					
HK31A-T6	—	215 (31)	110 (16)	62–83	(9–12)
AZ91A	—	235 (34)	160 (23)	69–96	(10–14)

[a] *Structural Alloys Handbook*, Mechanical Properties Data Center, Traverse City, MI, 1977.
[b] *Aluminum Standards and Data 1976*, The Aluminum Association, New York, 1976. (See source for restrictions on use of data in design.)
[c] *Materials Engineering* **94** (6) (Dec. 1981), Penton/IPC Publication, Cleveland, OH.

penetrated by dislocations moving back and forth along active slip planes. This action produces an initial strain-hardening response followed by local softening, which serves to concentrate additional deformation in narrow bands and leads to crack initiation. Localized softening is believed to occur by a disordering process resulting from repeated precipitate cutting by the oscillating dislocations. To offset this, it has been suggested that additional platelike particles that are impenetrable by dislocations be added to the microstructure to arrest the mechanically induced disordering process. In this manner, the fine, ordered particles, penetrable by dislocations, would act as precipitation-hardening agents while relatively larger, platelike particles that are not cut by dislocations would enhance fatigue behavior.

Since the fatigue limit associated with long cyclic life is strongly dependent on tensile strength, it follows that fatigue behavior should be sensitive to alloy chemistry and thermo-mechanical treatment. Since large inclusions do not significantly alter tensile strength but do serve as potential crack nucleation sites, their presence in the microstructure is undesirable. By eliminating them through more careful melting practice and stricter alloy chemistry, one finds a reduction in early life failures and a concomitant reduction in the amount of scatter in test results. In this instance, a reduction in inclusion content has a favorable effect on both fatigue behavior and fracture toughness. The reader should recognize, however, that when a component's design stress level is low relative to the material's fatigue limit and when $K \ll K_{\text{IC}}$, consideration of inclusion content levels, as they relate to the fatigue life and ultimate failure of the component, is of little practical significance. (For example, examine the shotgun Failure Analysis Case History #1, discussed in Chapter 11.)

Although inclusions and certain other metallurgical microconstituents may have a deleterious effect on unnotched fatigue response, they have an interesting influence on notched fatigue behavior: The notch sensitivity associated with an external notch is lower in a material that already contains a large population of internal flaws. An example of this is seen from a comparison of relative notch sensitivity between flake-graphite gray cast iron and nodular cast iron.[1] The fatigue limit/tensile strength ratio is lowest in the flake-graphite cast iron (0.42 versus 0.48 for the nodular cast iron), which probably reflects the damaging effect of the sharp graphite flakes. Conversely, the notch sensitivity q to an external circumferential V-notch is lowest in the flake-graphite cast iron (0.06 versus 0.25). You might say that with all the graphite flakes present to create a multitude of stress concentrations, one more notch is not that harmful.

9.2.4.1 Surface Treatment

Although changes in overall material properties do influence fatigue behavior (for example, see Fig. 9.23), greater property changes are effected by localized modification of the specimen or component surface, since most fatigue cracks originate in this region. To this end, a number of surface treatments have been developed; they may be classified in three broad categories: mechanical treatments, including shot peening, cold rolling, grinding, and polishing; thermal

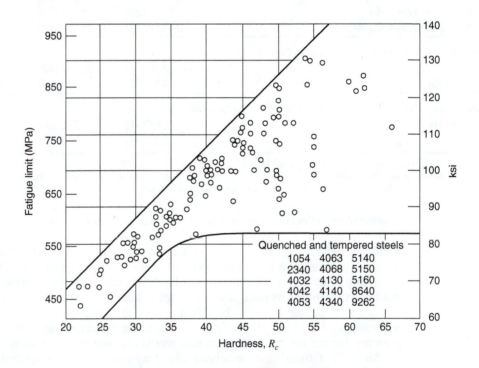

Figure 9.23 Fatigue limit of several quenched and tempered steels as a function of hardness level. Considerable uncertainty exists in determining fatigue limits at hardness levels in excess of $40R_c$ (about 1170 MPa).[29] (Reproduced with permission from *Metals Handbook*, Vol. 1, © American Society for Metals, 1961.)

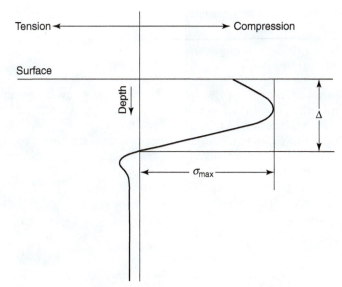

Figure 9.24 Diagram showing residual stress distribution after the shot peening process. Compressive residual stress extends from surface to a depth Δ.

treatments, such as flame and induction hardening; and surface coatings, such as case hardening, nitriding, and plating. One of the most widely used mechanical treatments involves the use of shot peening. In this process, small, hard particles (shot) about 0.08 to 0.8 mm diameter are blasted onto the surface that is to be treated. This action results in a number of changes to the condition of the material at and near the surface.[31] First, and most importantly, a thin layer of compressive residual stress is developed that penetrates to a depth of about one-quarter to one-half the shot diameter (Fig. 9.24). Since the peening process involves localized plastic deformation, it is believed that the surrounding elastic material forces the permanently strained peening region back toward its original dimensions, thereby inducing a residual compressive stress. Depending on the type of shot, shot diameter, pressure and velocity of shot stream, and duration of the peening process, the maximum compressive stress can reach about one-half the material yield strength. Consequently, the peening process benefits higher strength alloys more than the weaker ones. Since the peened region has a localized compressive mean stress, it acts to reduce the most damaging tensile portion of the applied alternating stress range (Fig. 9.9), resulting in a substantial improvement in fatigue life.

It should be emphasized that shot peening is most effective in specimens or components that contain a stress concentration or stress gradient; the peening process is also useful in unnotched components that experience a stress gradient (bending or torsion) such as leaf and coil springs, torsion bars, and axles. Shot peening is of limited use when high applied stresses are anticipated (that is, in the low cycle fatigue regime), since large stress excursions, particularly those in the plastic range, cause rapid "fading" of the residual stress pattern. On the other hand, shot peening is very useful in the high-cycle portion of fatigue life associated with lower stress levels.

Another beneficial effect of shot peening, though of secondary importance, is the work-hardening contribution in the peened material that results from plastic deformation. Particularly in cases involving low-strength alloys with high strain-hardening capacity, the material strain hardens, thereby contributing to a higher fatigue strength associated with the higher tensile strength. Finally, the shot peening process alters the surface by producing small "dimples" that, by themselves, would have a deleterious effect on fatigue life by acting as countless local stress concentrations. Fortunately, the negative aspect of this surface roughening is more than counterbalanced by the concurrent favorable residual compressive stress field. To be sure, the fatigue properties of a component may be improved still further if the part is polished after a shot peening treatment.

Surface rolling also produces a favorable residual stress that can penetrate deeper than that produced by shot peening and which does not roughen the component surface. Surface rolling finds extensive use in components possessing surfaces of rotation, such as in the practice of rolling machine threads.

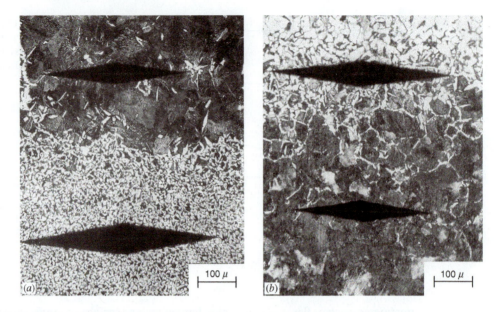

Figure 9.25 (*a*) Photomicrograph showing carburized layer (top) at surface of 1020 steel part. Microhardness impressions reveal considerable hardening in the case. (*b*) Photomicrograph showing decarburized layer (top) at surface of 1080 steel part. Microhardness impressions reveal softening in the decarburized zone.

Flame- and induction-hardening heat treatments in certain steel alloys are intended to make the component surface harder and more wear resistant. This is done by heating the surface layers into the austenite phase region and then quenching rapidly to form hard, untempered martensite. Since the tensile strength and hardness of this layer is markedly increased, the fatigue strength likewise is improved [though at the expense of reliability when the hardness exceeds $40R_c$ or about 1170 MPa (Fig. 9.23)]. In addition, since the austenite to martensite phase transformation involves a volume expansion that is resisted by the untransformed core, a favorable compressive residual stress is developed in this layer, which contributes an additional increment to the improved fatigue response of steel alloys heat treated in this manner.

Like flame and induction hardening, case hardening by either carburizing or nitriding is intended primarily to improve wear resistance in steels but simultaneously improves fatigue strength. Components to be carburized are treated in a high-temperature carbonaceous atmosphere to form a carbide-rich layer some 0.8 to 2.5 mm deep (Fig. 9.25*a*), while nitrided samples are placed in a high-temperature ammonia atmosphere, where nitrogen reacts with nitride-forming elements within the steel alloy to form a nitrided layer about 0.5 mm deep. In both instances, the improvement in fatigue strength results from the intrinsic strength increase within the carburized or nitrided case and also from the favorable residual compressive stress pattern that accompanies the process. The latter factor can be compared to similar residual stress patterns arising from the mechanical and thermal treatments described above, but it is in sharp contrast to the unfavorable residual tensile stresses from nickel and chromium plating procedures. In these two cases, fatigue resistance is definitely impaired. Such problems are not found with cadmium, zinc, lead, and tin platings, but one must be wary of any electrolytic procedure, since the component may become charged with hydrogen gas and be susceptible to hydrogen-embrittlement-induced premature failure (see Chapter 8).

The improvement in fatigue resistance afforded by case hardening is considerable enough to transfer the fatigue initiation site from the component surface to the case–core boundary region, where (1) the residual stress shifts abruptly to a tensile value, and (2) the intrinsic strength of the core is considerably less than that associated with the case material. As one

Table 9.2   Fatigue Strength in Threaded[a] Bolts[32]

Manufacturing Procedure	Fatigue Strength[b]	
	MPa	(ksi)
Thread rolling of unground stock + additional heat treatment	55–125	(8–18)
Machine cut threads	195–220	(28–32)
Thread rolling of ground stock with no subsequent heat treatment	275–305	(40–44)

[a] K_t 3.5–4.0
[b] Tensile strength of material = 760–895 MPa (110–130 ksi)

might expect, case hardening imparts a significant improvement in fatigue resistance to components experiencing a stress gradient, such as those in plane bending or in any notched sample. Here, the applied stress is much lower in the area of the weak link in a case hardened part—the case-core boundary. By contrast, less improvement is anticipated when an axially loaded unnotched part is case hardened, since failure can occur anywhere within the uniformly loaded cross section and will do so at the case–core boundary.

Although case hardening considerably improves fatigue resistance, inadvertent decarburizing in steel alloys during a heat treatment can seriously degrade hardness, strength, and fatigue resistance (Fig. 9.25b). Logically, decarburizing results in a loss of intrinsic alloy strength, since carbon is such a potent strengthening agent in most iron-based alloys (recall Section 3.5). In addition, the propensity for a volumetric contraction in the low carbon surface region, which is restrained by the higher carbon interior regions, may produce an unfavorable residual tensile stress pattern.

From the above, considerable improvement in fatigue properties may be achieved by introducing a favorable residual compressive stress field and avoiding any possibility for decarburization. In fact, Harris[32] showed that when decarburization was avoided and machine threads were rolled rather than cut, the fatigue endurance limit of threaded steel bolts increased by over 400% (Table 9.2).

A number of other conditions may degrade fatigue behavior. These include inadequate quenching, which produces local soft spots that have poorer fatigue resistance; excessive heating during grinding, resulting in reversion of the steel to austenite, which forms a brittle martensite upon quenching; and splatter from welding, which creates local hot spots and causes local metallurgical changes that adversely affect fatigue response.

9.2.5 Material Behavior: Polymers

The fatigue failures in polymers may be induced either by large-scale hysteretic heating, resulting in actual polymer melting, or by fatigue crack initiation and propagation to final failure[33] (Fig. 9.26). Over the years, the basic differences and importance of polymer fatigue failures induced by these two processes have become a source of controversy among researchers. Part of this difficulty is due to the nature of the ASTM recommended test procedure (ASTM Standard D671).[34]

The major cause of thermal failure is believed to involve the accumulation of hysteretic energy generated during each loading cycle. Since this energy is dissipated largely in the form of heat, an associated temperature rise will occur for every loading cycle when isothermal conditions are not met. As shown in Fig. 9.27, the temperature rise can be great enough to cause the sample to melt, thereby preventing it from carrying any load.[36] Failure is presumed, therefore, to occur by viscous flow, although the occurrence of some bond breakage cannot be excluded.

From the work of Ferry,[38] the energy dissipated in a given cycle may be described by

$$\dot{E} = \pi f J''(f, T)\sigma^2 \tag{9-13}$$

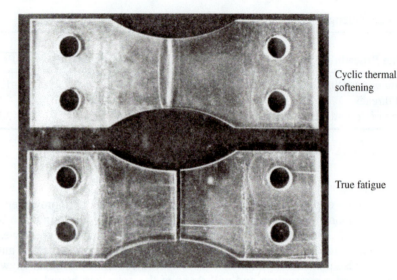

Cyclic thermal softening

True fatigue

Figure 9.26   Typical fatigue and cyclic thermal softening failures in poly(methyl methacrylate).[35] (Reproduced by courtesy of The Institution of Mechanical Engineers from an article by I. Constable, J. G. Williams, and D. J. Burns from *JMES* **12**, 20 (1970).)

where $\dot{E}$ = energy dissipated
$\quad f$ = frequency
$\quad J''$ = the loss compliance
$\quad \sigma$ = the peak stress

Neglecting heat losses to the surrounding environment, Eq. 9-13 may be reduced to show the temperature rise per unit time as

$$\Delta\dot{T} = \frac{\pi f J''(f,T)\sigma^2}{\rho c_p} \tag{9-14}$$

where $\Delta\dot{T}$ = temperature change/unit time
$\quad \rho$ = density
$\quad c_p$ = specific heat

Equation 9-14 is useful in identifying the major variables associated with hysteretic heating. For example, the temperature rise is seen to increase rapidly with increasing stress level. With increasing specimen temperature, the elastic modulus is found to decrease. For this reason, larger specimen deflections are required to allow for continuation of the test under constant stress conditions. These larger deflections then contribute to even greater hysteretic energy losses with continued cycling. A point is reached whereby the specimen is no longer capable of supporting the loads introduced by the test machine within the deflection limits of the test apparatus. For such test conditions, ASTM Standard D671 defines fatigue failure life (thermal fatigue, in this instance) as the number of loading cycles at a given applied stress range that brings about an "apparent modulus decay to 70% of the original modulus of the specimen determined at the start of the test."[34] Furthermore, the ASTM Standard calls for the investigator to "measure the temperature at failure unless it can be shown that the heat rise is insignificant for the specific material and test condition."[34] Figure 9.27a illustrates a typical curve of stress versus number of cycles to failure for poly(tetrafluoroethylene) (PTFE), along with the superposition of temperature rise curves corresponding to the various stress levels. Note

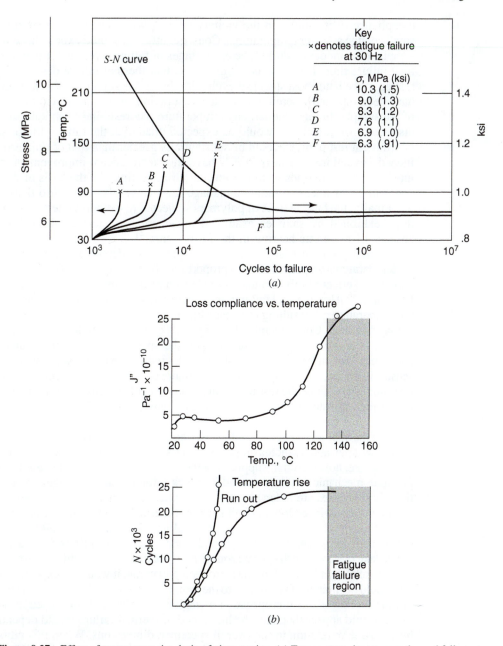

Figure 9.27 Effect of temperature rise during fatigue testing. (*a*) Temperature rise to cause thermal failure at different stress levels (no failure seen in *F* where sample temperature stabilized);[36] (*b*) loss compliance and temperature rise.[37] (Reprinted with permission of the Society of Plastics Engineers.)

that for all stress levels above the endurance limit (the stress level below which fatigue failure was not observed) the polymer heated to the point of melting (shown by the temperature rise curves *A*, *B*, *C*, *D*, and *E*). Evidently, heat was generated faster than it could be dissipated to the surrounding environment. When a stress level less than the endurance limit was applied, the temperature rise became stabilized at a maximum intermediate level below the point where thermal failure was observed (Curve *F*). These specimens did not fail after 10^7 cycles.

From Eq. 9-14, the rate of temperature rise depends on the magnitude of the loss compliance J'', which is itself a function of temperature. As we see in Fig. 9.27*b*, the loss compliance rises rapidly in the vicinity of the glass transition temperature after a relatively small change at lower temperatures. Consequently, one would expect the temperature rise in the sample to be moderate during the early stages of fatigue cycling but markedly greater near the final failure time. It is concluded, therefore, that thermal failure describes an event primarily related to the lattermost stages of cyclic life. In further support of this last statement, other tests have been reported wherein intermittent rest periods were interjected during the cyclic history of the sample. In this manner, any temperature rise resulting from adiabatic heating could be dissipated periodically. It would be expected, then, that the fatigue life of specimens allowed intermittent rest periods would be substantially greater than that of uninterrupted test samples. Indeed, several investigators[36,39,40] have shown significant improvement in fatigue life when intermittent rest periods were introduced during testing. On the basis of these test results, it is concluded that linear cumulative damage laws cannot be applied to thermal failures.

Finally, the fatigue lives of polymers subjected to isothermal test conditions are superior to those exhibited by samples examined under adiabatic test conditions, consistent with the absence of hysteretic heating in the former case.[41]

From Eq. 9-14, the fatigue life of a sample should decrease with increasing frequency, since the temperature rise per unit time is proportional to the frequency. Test results have confirmed the anticipated effect of this variable as evidenced by a decrease in endurance limit with increasing frequency.[36] Also, thermal failures are affected by specimen configuration. As mentioned above, the temperature rise resulting from each loading cycle depends on the amount of heat dissipated to the surroundings. Consequently, the fatigue life of a given sample should be dependent on the heat transfer characteristics of the sample and the specimen surface to volume ratio. Indeed, the endurance limit in PTFE has been shown to increase with decreasing specimen thickness.[36] This sensitivity to specimen shape constitutes a major drawback to unnotched specimen tests involving thermal failure, since the test results are a function of specimen geometry and, therefore, do not reflect the intrinsic response of the material being evaluated.

It is clear from the above that thermal fatigue may be suppressed by several factors, such as limiting the applied stress, decreasing test frequency, allowing for periodic rest periods, or cooling the test sample, and by increasing the sample's surface to volume ratio. It is extremely important to recognize, however, that suppressing thermal fatigue by any of the above procedures may not preclude mechanical failure caused by crack initiation and propagation. The corollary is true, though: If stresses are reduced to the point where mechanical failure does not occur, this stress level certainly will be low enough such that thermal failure will not occur either.

Although we choose to treat mechanical and thermal fatigue failures as distinctly different events, there are points of common ground. For example, it would be expected that hysteretic heating would take place within the plastic zone at the tip of a crack. Since this heat source is small compared to the much larger heat sink of the surrounding material, it would be expected that any temperature rise would be limited and restricted to the proximity of the crack tip. The influence of such localized heating on fatigue crack propagation in engineering plastics is discussed in Section 10.8.

It would appear then that the likelihood of thermal failure would depend on the size of the heated zone in relation to the overall specimen dimensions. When this ratio is large, as in the case of unnotched test bars, thermal failures are distinctly possible. When the ratio is very small—say, in the case of a notched bar—thermal failures would not be expected.

9.2.6 Material Behavior: Composites

9.2.6.1 Particulate Composites

The fatigue response of composite materials is dependent on the complex interaction between the mechanical properties and volume fraction of the matrix and reinforcing phases, fiber aspect ratio, and the strength of the bond between the two phases; in addition, the direction of loading (i.e., tensile and/or compressive), environment, temperature, and cyclic frequency also affect fatigue behavior. Within the context of this book, it will be possible to address these

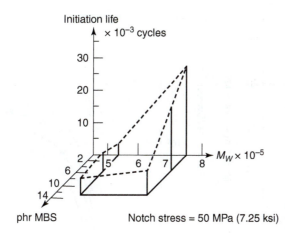

Figure 9.28   Three-dimensional plot revealing interactive effects of M_W and percent rubber content (MBS) on fatigue crack initiation lifetime in PVC.[45] (Reprinted with permission from *Deformation, Yield, and Fracture*, 1985, The Plastics and Rubber Institute.)

factors only in the briefest fashion. We begin this discussion by examining the fatigue response of impact-modified polymers. Consideration of overall fatigue life and the relative effects of rubber on crack initiation and propagation reveals interesting and perhaps unexpected behavior. For example, detailed examinations of several rubber-toughened polymers have shown that the high level of impact strength is not necessarily carried over into fatigue resistance; this is particularly true when compressive stresses are applied. At least with HIPS and ABS tested in tension–compression, the rubbery phase decreases both initiation and propagation lifetimes compared to typical unmodified polystyrenes (PS) or styrene–acrylonitrile copolymers (SAN).[42–44] Evidently, the lower yield strength in the toughened resins greatly facilitates initiation, and compressive loading severely damages craze fibrils during the crack closure part of the load cycle. In fact, if the stress levels in the *S-N* curves for HIPS and PS are normalized by their respective yield strengths, HIPS appears to be relatively superior to PS. The especially deleterious effect of compressive stresses is consistent with the observation that the fatigue life of HIPS and ABS is increased significantly by switching to tension-tension loading.[43,44]

Figure 9.28 is a three-dimensional plot showing the interaction of M_w and percentage rubber content on fatigue crack initiation lifetime (N_i) in impact-modified poly(vinyl chloride) (PVC).[45] The rubber phase in this blend is methacrylate–butadiene–styrene (MBS) polymer and N_i is defined as the number of loading cycles necessary to nucleate a crack 0.25 mm in length from the root of a polished round hole (radius, 1.59 mm) introduced at the end of the slot in a compact tension sample (recall Fig. 6.21*g*). Fatigue initiation life increases markedly with increasing molecular weight for all rubbery phase contents. Conversely, in high M_w PVC blends, rubber modification lowers fatigue crack initiation (FCI) resistance, although the addition of MBS in low M_w PVC slightly improves FCI resistance. The generally deleterious influence of rubbery phase on fatigue crack *initiation* resistance contrasts with the superior fatigue crack *propagation* resistance shown for this blend;[46] a 3- to 30-fold reduction in the crack propagation rate was observed when up to 14% MBS was added to the PVC matrix.

Thus, to avoid unpleasant surprises, caution should be used in subjecting rubber-modified plastics to fatigue loading, especially if compressive stresses are involved. Before application in such situations, careful tests should be run to simulate anticipated loading conditions. Furthermore, conclusions based on fatigue initiation studies may not necessarily be extrapolated to the realm of crack propagation.

9.2.6.2 Fiber Composites

Fatigue failure processes in fibrous composites are complex and include the following:[33,47,48] (1) Damage is progressive, physical integrity may be maintained for many decades of cycles, and criteria for failure may be based arbitrarily on the degradation of a property such as the elastic

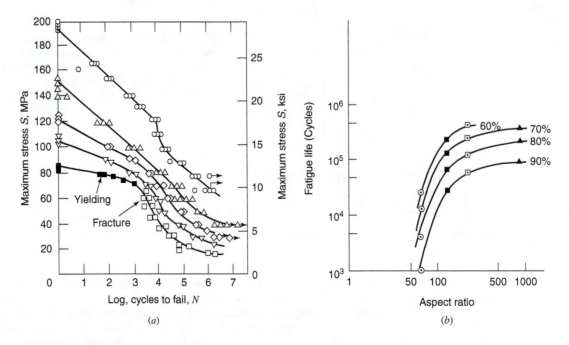

Figure 9.29 Fatigue response in polymer–matrix composites. (*a*) Injection-molded polysulfone matrix composites. $R = 0.1$, frequency = 5–10 Hz. ■ □, unreinforced; ▽, 10% glass; ◇, 20% glass; △, 40% glass; ○, 40% carbon.[50] (*Polymer Composites*, **4**, 32 (1983) (*b*) 50–50 vol% boron–fiber reinforced epoxy. Fatigue life to produce a 20% decrease in elastic modulus as a function of aspect ratio for applied stresses of 60, 70, 80, and 90% of ultimate failure stress.[51] (*Polymer Engineering and Science*, **9**, 365 (1969). Reprinted with permission from Society of Plastics Engineers.)

modulus. (2) Diverse micromechanisms of failure may occur and include fiber deformation and/or brittle fracture, fiber–matrix debonding, delamination of composite plies, and matrix cracking. (3) The balance of micromechanisms depends on such factors as hysteretic heating (modified by the presence of fibers), relative orientation of fiber and stress axes, mode of loading, and the presence and nature of preexisting flaws. (4) In principle, linear elastic fracture mechanics is not applicable to heterogeneous systems, and the concept of a crack requires redefinition in terms of a more diffuse zone of damage. (5) Under some circumstances, fatigue loading can result in effective crack blunting so that fracture toughness may actually increase, at least during part of the fatigue life.[48,49]

To illustrate the influence of fiber properties, geometry, and volume fraction on composite fatigue life, consider the results shown in Fig. 9.29. Typical *S-N* curves for tensile fatigue are shown in Fig. 9.29*a* for several injection-molded polysulfone composites.[50] The superiority of all composites to the unreinforced matrix is evident, as is the superiority of increased fiber fraction and carbon relative to glass fibers. Clearly, the greater stiffness and thermal conductivity of carbon constitute significant advantages in lowering the strains at a given stress and minimizing hysteretic heating. The importance of aspect ratio to fatigue performance is illustrated in Fig. 9.29*b* for a composite containing equal volume fractions of short boron fibers and an epoxy resin matrix.[51] The number of loading cycles required to produce a 20% decrease in elastic modulus at a frequency of 3 Hz (low enough to minimize hysteretic heating) increases sharply with aspect ratio up to a value of $l/d = 200$; little additional improvement in fatigue life is noted with further increases in aspect ratio. The latter is consistent with theoretical predictions as discussed in Chapter 1 (recall Eq. 1-72a).

The deleterious influence of compressive loading on continuous carbon fiber-reinforced epoxy resin is shown in Fig. 9.30.[52] Compression-induced effects, such as fiber buckling and delamination, and matrix shear reduce fatigue resistance;[53] cycling in flexure, torsion, and other shear modes is also especially deleterious.[33,47,52] In such cases, the dominance of the fibers in determining fatigue properties decreases, and the matrix and interface play more

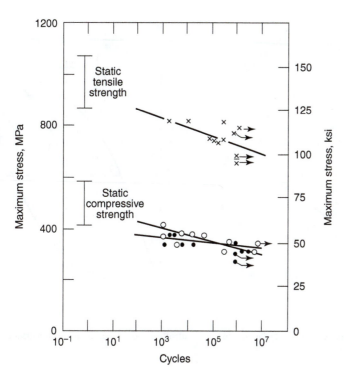

Figure 9.30 Axial-load fatigue results for unidirectional, surface-treated, carbon-fiber-reinforced epoxy resin.[52] Volume fraction of carbon fiber = 0.61. Test frequency = 117 Hz. ×, zero-tension; ○, zero-compression; ●, fully reversed. Compressive stress plotted positive (lower curve). (Reprinted from *Composite Materials*, **5**, 341 (1974).

important roles. Because such severe modes of loading are often more typical of actual service than is axial tension, more testing should be done under these more rigorous conditions.

9.3 CYCLIC STRAIN-CONTROLLED FATIGUE

Localized plastic strains can be generated by loading a component that contains a notch. Regardless of the external mode of loading (cyclic stress or strain controlled), the plasticity near the notch root experiences a strain-controlled condition dictated by the much larger surrounding mass of essentially elastic material. Scientists and engineers from the Society of Automotive Engineers (SAE) and the American Society for Testing and Materials (ASTM) have recognized this phenomenon and have developed strain-controlled test procedures to evaluate cumulative damage in engineering materials.[54–57] These procedures are particularly useful in evaluating component life where notches are present. Indeed, a reasonable assumption can be made that the same number of loading cycles is needed to develop a crack at the notch root of an engineering component and in an unnotched specimen, if the two cracked regions experience the same cyclic stress–strain history. Other examples of strain-controlled cyclic loading include thermal cycling, where component expansions and contractions are dictated by the operating temperature range, and reversed bending between fixed displacements, such as in the reciprocating motion shown in Fig. 9.31.

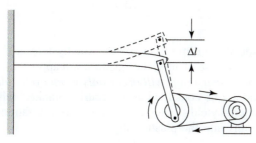

Figure 9.31 Reciprocating action produces fixed beam displacements. Compare this case to the stress-controlled condition shown in Fig. 9.8a.

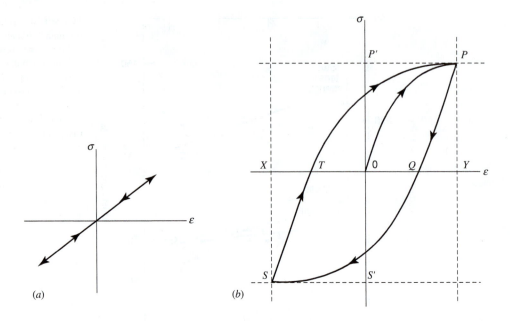

Figure 9.32   Hysteresis loops for cyclic loading in (*a*) ideally elastic material and (*b*) material undergoing elastic and plastic deformation.

By monitoring strain and stress during a cyclic loading experiment, the response of the material can be clearly identified. For example, for a material exhibiting stress–strain behavior involving only linear elastic deformation under the applied loads, a hysteresis curve like that shown in Fig. 9.32*a* is produced. Note that the material's stress–strain response is retraced completely; that is, the elastic strains are completely reversible. For behavior involving elastic–homogeneous plastic flow, the complete load excursion (positive and negative) produces a curve similar to Fig. 9.32*b* that reflects both elastic and plastic deformation. The area contained within the hysteresis loop represents a measure of plastic deformation work done on the material. Some of this work is stored in the material as cold work and/or associated with configurational changes (entropic changes), such as in polymer chain realignment; the remainder is emitted as heat. From Fig. 9.32*b*, the elastic strain range in the hysteresis loop is given by

$$\Delta\varepsilon_e = XT + QY = \frac{\Delta\sigma}{E} \tag{9-15}$$

The plastic strain range is equal to *TQ* or equal to the total strain range minus the elastic strain range. Hence

$$\Delta\varepsilon_p = \Delta\varepsilon_T - \frac{\Delta\sigma}{E} \tag{9-16}$$

Note that as the amount of plastic strain diminishes to zero, the hysteresis loop in Fig. 9.32*b* shrinks to that shown in Fig. 9.32*a*. Consequently, the elastic strain approaches the total strain. It is important to recognize that *fatigue damage will occur only when cyclic plastic strains are generated.* This basic rule should not be construed as a "security blanket" whenever nominal applied stresses are below the material yield strength, since stress concentrations readily elevate local stresses and associated strains into the plastic range.

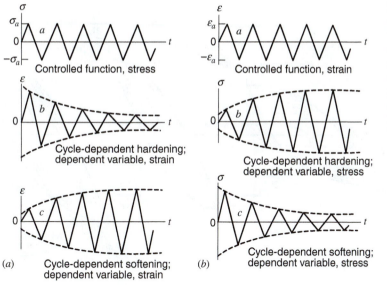

Figure 9.33   (a) Cycle-dependent material response under stress control.[57] (b) Cycle-dependent material response under strain control.[58] (Reprinted with permission of the University of Wisconsin Press and the Regents of the University of Wisconsin System.)

9.3.1   Cycle-Dependent Material Response

Cycle-dependent material responses under stress and strain control are shown in Figs. 9.33a and 9.33b, respectively, which reflect changes in the shape of the hysteresis loop. It is seen that, in both cases, the material response changes with continued cycling until cyclic stability is reached.[iv] That is, the material becomes either more or less resistant to the applied stresses or strains. Therefore, the material is said to cyclically strain harden or strain soften. Referring again to Fig. 9.32b for the case of stress control, where the fatigue test is conducted in a stress range between P' and S', the width of the hysteresis loop TQ (the plastic strain range) contracts when cyclic hardening occurs and expands during cyclic softening. Cyclic softening under stress control is a particularly severe condition because the constant stress range produces a continually increasing strain range response, leading to early fracture (Fig. 9.33a). Under cyclic strain conditions within limits of strains X and Y, the hysteresis loop expands above P and below S for cyclic hardening and shrinks below P and above S for cyclic softening (Fig. 9.33b). An example of cyclic strain hardening and softening under strain-controlled test conditions is shown in Fig. 9.34.

After cycling a material for a relatively short duration (often less than 100 cycles), the hysteresis loops generally stabilize and the material achieves an equilibrium condition for the imposed strain limits. The cyclically stabilized stress–strain response of the material may then be quite different from the initial monotonic response. Consequently, cyclically stabilized stress–strain curves are important characterizations of a material's cyclic response. These curves may be obtained in several ways. For example, a series of companion samples may be cycled within various strain limits until the respective hysteresis loops become stabilized. The cyclic stress–strain curve is then determined by fitting a curve through the tips of the various superimposed hysteresis loops, as shown in Fig. 9.35.[60] This procedure involves many samples and is expensive and time-consuming. A faster method for obtaining cyclic stress–strain curves is by multiple step testing, wherein the same sample is subjected to a series of alternating strains in blocks of increasing magnitude. In this manner, one specimen yields

---

[iv] Although most of our discussions will focus on symmetrical loading about zero, it is important to appreciate what happens to a sample when a nonzero mean stress is superimposed during a cyclic strain experiment. The specimen is found to accumulate strains as a result of each cycle. This accumulation has been termed "cyclic-strain-induced creep" and will contribute to either an extension or contraction of the sample, depending on the sense of the applied mean stress.

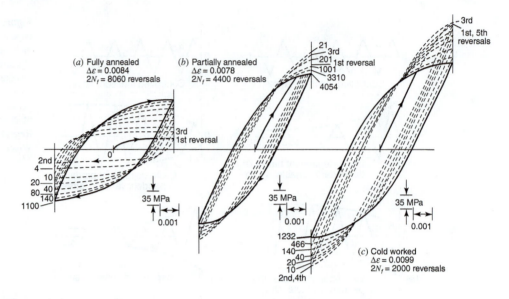

Figure 9.34 Strain-controlled fatigue response in OFHC copper. (*a*) Fully annealed sample exhibits cyclic strain hardening; (*b*) partially annealed sample exhibits relative cyclic stability; (*c*) severely cold-worked sample exhibits cyclic strain softening.[59] (Reprinted by permission of the American Society for Testing and Materials from copyrighted work.)

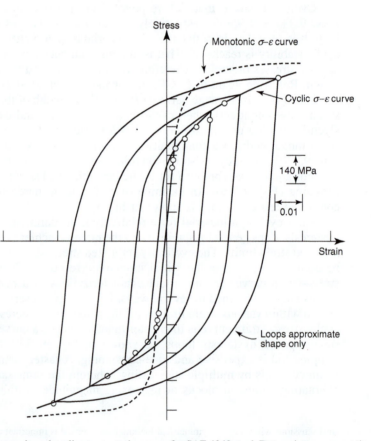

Figure 9.35 Monotonic and cyclic stress–strain curves for SAE 4340 steel. Data points represent tips of stable hysteresis loops from companion specimens.[62] (Reprinted by permission of the American Society for Testing and Materials from copyrighted work.)

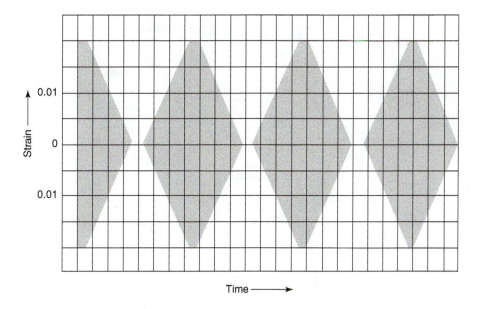

Figure 9.36 Incremental step test program showing strain–time plot.

several hysteresis loops, which may be used to construct the cyclic stress–strain curve.[61] An even quicker technique involving only one sample has been found to provide excellent results and is used extensively in current cyclic strain testing experiments. As seen in Fig. 9.36, the specimen is subjected to a series of blocks of gradually increasing and then decreasing strain excursions.[61] It has been found that after a relatively few such blocks (the greater the number of cycles within each block, the fewer the number of blocks needed for cyclic stabilization), the material reaches a stabilized condition. At this point, the investigator simply draws a line through the tips of each hysteresis loop, from the smallest strain range to the largest. As such, each loop contained within the hysteresis envelope represents the cyclically stabilized condition for the material at that particular strain range. By initiating the test with the maximum strain amplitude in the block, the monotonic stress–strain curve is automatically determined for subsequent comparison with the cyclically stabilized curve. In this manner, both the monotonic and cyclic stress–strain curves can be determined from the same sample. Obviously, this method results in savings in test time and money. It should be noted that if a specimen subjected to either multiple or incremental step testing were to be pulled to fracture after cyclic stabilization, the resulting stress–strain curve would be virtually coincident with the one generated by the locus of hysteresis loop tips.

By comparing monotonic and cyclically stabilized stress–strain curves, Landgraf et al.[61] demonstrated that certain engineering alloys will cyclically strain harden and others will soften (Fig. 9.37). From the Holloman relation given by $\sigma = K\varepsilon^n$, it is possible to mathematically describe the material's stress–strain response in either the monotonic or cyclically stabilized state. Consequently, one may define the strain-hardening exponent for both monotonic (n) and cyclic (n') conditions as well as the monotonic yield strength (σ_{ys}) and cyclic (σ'_{ys}) counterpart. Equation 9-17 describes the cyclically stabilized stress–strain curve where K' is the cyclic strength coefficient and $\Delta\sigma$ and $\Delta\varepsilon$ are true stress range and true strain range, respectively:

$$\frac{\Delta\varepsilon}{2} = \frac{\Delta\varepsilon_e}{2} + \frac{\Delta\varepsilon_p}{2} = \frac{\Delta\sigma}{2E} + \left(\frac{\Delta\sigma}{2K'}\right)^{1/n'} \tag{9-17}$$

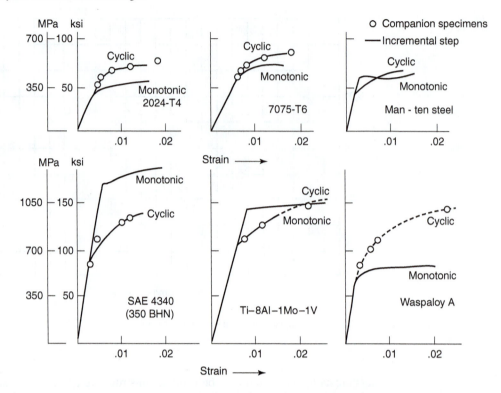

Figure 9.37   Monotonic and cyclic stress–strain curves for several engineering alloys.[61] (Reprinted by permission of the American Society for Testing and Materials from copyrighted work.)

Typical material property values for a number of metal alloys are given in Tables 9.3a,b. Although large changes occur in these properties as a result of cyclic hardening or softening, it is worth noting that, for most metals, n' ranges from 0.1 to 0.2. Beardmore and Rabinowitz[63–65] have conducted an extensive comparison of cyclic and monotonic stress–strain curves for several polymers. For all the materials examined, semicrystalline, amorphous, and two-phase, pronounced cyclic strain softening was observed; by comparison, no cyclic strain hardening took place.

Can one determine in advance which alloys will cyclically harden and which will soften? Manson et al.[60,66] observed that the propensity for cyclic hardening or softening depends on the ratio of monotonic ultimate strength to 0.2% offset yield strength. When $\sigma_{ult}/\sigma_{ys} > 1.4$, the material will harden, but when $\sigma_{ult}/\sigma_{ys} < 1.2$, softening will occur. For ratios between 1.2 and 1.4, forecasting becomes difficult, though a large change in properties is not expected. Also, if $n > 0.20$, the material is likely to strain harden, and softening will occur if $n < 0.10$. Therefore, *initially hard and strong materials will generally cyclically strain soften, and initially soft materials will harden*. Manson's rules make use of monotonic properties to determine *whether* cyclic hardening or softening will occur. However, the *magnitude* of the cyclically induced change can be determined only by comparison of the monotonic and cyclic stress–strain curves (see Tables 9.3a,b and Figs. 9.35 and 9.37).

But why do these materials cyclically harden or soften? The answer to this question appears to be related to the nature and stability of the dislocation substructure of the material. For an initially soft material, the dislocation density is low. As a result of plastic strain cycling, the dislocation density increases rapidly, contributing to significant strain hardening. At some point, the newly generated dislocations assume a stable configuration for that material and for the magnitude of cyclic strain imposed during the test. When a material is hard initially, subsequent strain cycling causes a rearrangement of dislocations into a new configuration that offers less resistance to deformation—that is, the material strain softens. The processes

Table 9.3a Monotonic and Cyclic Properties of Selected Engineering Alloys[a]

Material	Condition	σ_{ys}/σ'_{ys} (MPa)	n/n'	$\varepsilon_f/\varepsilon'_f$	σ_f/σ'_f (MPa)	b	c
Steel							
SAE 1015	Normalized, 80 BHN	225/240	0.26/0.22	1.14/0.95	725/825	−0.11	−0.64
SAE 950X	As received, 150 BHN	345/335	0.16/0.134	1.06/0.35	750/625	−0.075	−0.54
VAN-80	As received, 225 BHN	565/560	0.13/0.134	1.15/0.21	1220/1055	−0.08	−0.53
SAE 1045	Q + T (650°C), 225 BHN	635/415	0.13/0.18	1.04/1.0	1225/1225	−0.095	−0.66
SAE 1045	Q + T (370°C), 410 BHN	1365/825	0.076/0.146	0.72/0.60	1860/1860	−0.073	−0.70
SAE 1045	Q + T (180°C), 595 BHN	1860/1725	0.071/0.13	0.52/0.07	2725/2725	−0.081	−0.60
AISI 4340	Q + T (425°C), 409 BHN	1370/825	—/0.15	0.48/0.48	1560/2000	−0.091	−0.60
AISI 304 ELC	BHN 160	255/715	—/0.36	1.37/1.02	1570/2415	−0.15	−0.77
AISI 304 ELC	Cold drawn, BHN 327	745/875	—/0.17	1.16/0.89	1695/2275	−0.12	−0.69
AISI 305[b]	0% CW	250/405	—/0.05	—	—	—	—
AISI 305[b]	50% CW	850/710	—/0.11				
AM 350	Annealed	440/1350	—/0.13	0.74/0.33	2055/2800	−0.14	−0.84
AM 350	Cold drawn 30%, BHN 496	1860/1620	—/0.21	0.23/0.098	2180/2690	−0.102	−0.42
18 Ni maraging	ST(790°C)/1 h + 480°C (4 h), BHN 480	1965/1480	0.015—0.030 / 0.008	0.81/0.60	2240/2240	−0.07	−0.75
Aluminum							
2014-T6	BHN 155	460/415	—/0.16	0.29/0.42	600/850	−0.106	−0.65
2024-T4	—	305/440	0.20/0.08	0.43/0.21	635/1015	−0.11	−0.52
5456	H31, 95 BHN	235/360	—/0.16	0.42/0.46	525/725	−0.11	−0.67
7075-T6	—	470/525	0.113/0.146	0.41/0.19	745/1315	−0.126	−0.52
Copper							
OFHC[c]	Annealed	20/140	0.40/0.16	—	—	—	—
70/30 brass[b]	Annealed	140/240	—/0.08	—	—	—	—
70/30 brass[b]	82% CW	570/475	—/0.11	—	—	—	—
Nickel							
Waspalloy[c]	—	545/705	0.11/0.17	—	—	—	—
MP35N[b]	0% CW	350/625	—/0.06	—	—	—	—
MP35N[b]	20% CW	7910/745	—/0.10	—	—	—	—
MP35N[b]	40% CW	1180/1850	—/0.14	—	—	—	—

[a] L. E Tucker, R. W. Landgraf, and W. R. Brose, SAE Report 740279, Automotive Engineering Congress, Feb. 1974.
[b] Hickerson and Hertzberg[69]
[c] Landgraf[61]

Table 9.3b Monotonic and Cyclic Properties of Selected Engineering Alloys[a]

Material	Condition	σ_{ys}/σ'_{ys} (ksi)	n/n'	$\varepsilon_f/\varepsilon'_f$	σ_f/σ'_f (ksi)	b	c
Steel							
SAE 1015	Normalized, 80 BHN	33/35	0.26/0.22	1.14/0.95	105/120	-0.11	-0.64
SAE 950X	As received, 150 BHN	50/49	0.16/0.134	1.06/0.35	109/91	-0.075	-0.54
VAN-80	As received, 225 BHN	82/81	0.13/0.134	1.15/0.21	177/153	-0.08	-0.53
SAE 1045	Q + T (1200°F), 225 BHN	92/60	0.13/0.18	1.04/1.0	178/178	-0.095	-0.66
SAE 1045	Q + T (700°F), 410 BHN	198/120	0.076/0.146	0.72/0.60	270/270	-0.073	-0.70
SAE 1045	Q + T (360°F), 595 BHN	270/250	0.071/0.13	0.52/0.07	395/395	-0.081	-0.60
AISI 4340	Q + T (800°F), 409 BHN	199/120	—/0.15	0.48/0.48	226/290	-0.091	-0.60
AISI 304 ELC	BHN 160	37/104	—/0.36	1.37/1.02	228/350	-0.15	-0.77
AISI 304 ELC	Cold drawn, BHN 327	108/127	—/0.17	1.16/0.89	246/330	-0.12	-0.69
AISI 305[b]	0%CW	36/59	—/0.05	—	—	—	—
AISI 305[b]	50% CW	123/103	—/0.11	—	—	—	—
AM 350	Annealed	64/196	—/0.13	0.74/0.33	298/406	-0.14	-0.84
AM 350	Cold drawn 30%, BHN 496	270/235	—/0.21	0.23/0.098	316/390	-0.102	-0.42
18 Ni maraging	ST(1450°F/1h) + 900°F (4 h), BHN 480	285/215	$\frac{0.015—0.030}{0.008}$	0.81/0.60	325/325	-0.07	-0.75
Aluminum							
2014-T6	BHN 155	67/60	—/0.16	0.29/0.42	87/123	-0.106	-0.65
2024-T4	—	44/64	0.20/0.08	0.43/0.21	92/147	-0.11	-0.52
5456	H31, 95 BHN	34/52	—/0.16	0.42/0.46	76/105	-0.11	-0.67
7075-T6	—	68/76	0.113/0.146	0.41/0.19	108/191	-0.126	-0.52
Copper							
OFHC[c]	Annealed	3/20	0.40/0.16	—	—	—	—
70/30 brass[b]	Annealed	20/35	—/0.08	—	—	—	—
70/30 brass[b]	82% CW	83/69	—/0.11	—	—	—	—
Nickel							
Waspalloy[c]	—	79/102	0.11/0.17	—	—	—	—
MP35N[b]	0%CW	51/91	—10.06	—	—	—	—
MP35N[b]	20% CW	132/108	—/0.10	—	—	—	—
MP35N.[b]	40% CW	171/123	—/0.14	—	—	—	—

[a] L. E Tucker, R. W. Landgraf, and W. R. Brose, SAE Report 740279, Automotive Engineering Congress, Feb. 1974.
[b] Hickerson and Hertzberg[69]
[c] Landgraf[61]

associated with cyclic strain softening were referred to in earlier technical literature as the Bauschinger effect.[37] To characterize this effect, consider the yielding behavior of a metal alloy (typically strain hardened) that is subjected to a complete loading cycle. After exhibiting a particular strength level associated with initial yielding in tension, the yield-strength level under compressive loading is found to be reduced. Further cycling then leads to additional reductions in both tensile and yield-strength levels (recall Fig. 9.34c). The tendency for cold-worked metals to exhibit cyclic softening (i.e., the Bauschinger effect) is put to good use in certain metal-forming applications.[68] For example, if a strain-hardened metal sheet is passed through a series of roll pairs that are alternately slightly above and below the nominal plane of the workpiece, the alternating tensile and bending stresses associated with this roller-leveling operation induce cyclic strain softening and enhance alloy ductility.

As we saw in Chapters 2 and 3, dislocation mobility that strongly affects dislocation substructure stability depends on the material's stacking fault energy (SFE). Recall that when SFE is high, dislocation mobility is great because of enhanced cross-slip; conversely, cross-slip is restricted in low SFE materials. As a result, some materials cyclically harden or soften more completely than others. For example, in a relatively high SFE material like copper, initially hard samples cyclically strain soften, and initially soft samples cyclically harden; thus, the cyclically stabilized condition is the same *regardless* of the initial state of the material (Fig. 9.38a). In this case, the mechanical properties of the material in the stabilized state are *independent* of prior strain history. This is not true for a low stacking fault energy material, where restricted cross-slip will prevent the development of a common dislocation state from an initially hard and soft condition, respectively. In addition, the low SFE material will harden or soften more slowly than the high SFE alloy. We see from Fig. 9.38b that the material will cyclically soften and harden, but a final stabilized state is never completely achieved and is not equivalent for the two different starting conditions. For such materials, the "final" cyclically stabilized state is *dependent* on prior strain history.

One might then expect to find dislocation substructures in cyclically loaded samples similar to those found as a result of unidirectional loading. In fact, Feltner and Laird[71] observed that "those factors which give rise to certain kinds of dislocation structures in unidirectional deformation affect the cyclic structures in the same way." For example, we see from Fig. 9.39 that at high cyclic strains a cell structure is developed in high SFE alloys. If cyclic straining causes coarsening of a preexistent cell structure, then softening will occur. If the cell structure

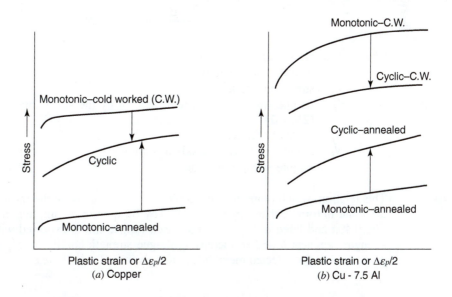

Figure 9.38   Cyclic response of (a) high stacking fault energy copper and (b) low stacking fault energy Cu-7.5% Al alloy. Cyclically stabilized state in high SFE alloy is path independent.[70] (Reprinted from C. E. Feltner and C. Laird, *Acta Metall.* 15 (1967), with permission of Pergamon Press.)

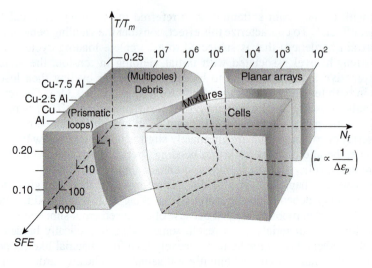

Figure 9.39 Schema showing dislocation substructures in FCC metals as a function of stacking fault energy, strain amplitude, and temperature.[71] (Reprinted with permission from the American Institute of Mining, Metallurgical and Petroleum Engineers.)

gets finer, then cyclic straining results in a hardening process. In low SFE alloys, dislocation planar arrays and stacking faults are present. These findings are similar to those discussed in Chapters 2 and 3 for monotonic loading. A parallel condition is found in monotonic and cyclically induced dislocation structures produced at low strains.

9.3.2 Strain Life Curves

Having identified the response of a solid to cyclic strains, it is now appropriate to consider how cyclically stabilized material properties affect the life[v] of a specimen or engineering component subjected to strain-controlled loading. To accomplish this, it is convenient to begin our analysis by considering the elastic and plastic strain components separately. The elastic component is often described in terms of a relation between the true stress amplitude and number of load reversals

$$\frac{\Delta \varepsilon_e E}{2} = \sigma_a = \sigma_f'(2N_f)^b \tag{9-18}$$

where $\frac{\Delta \varepsilon_e}{2}$ = elastic strain amplitude
   E = modulus of elasticity
   σ_a = stress amplitude
   σ_f' = fatigue strength coefficient, defined by the stress intercept at one load reversal ($2N_f = 1$)
   N_f = cycles to failure
   $2N_f$ = number of load reversals to failure
   b = fatigue strength exponent

This relation, which represents an empirical fit of data above the fatigue limit (see Fig. 9.19), is similar in form to that proposed in 1910 by Basquin.[72] A sampling of test results is shown in Fig. 9.40a and fitted to Eq. 9-18. Increased fatigue life is expected with a decreasing fatigue strength exponent b and an increasing fatigue strength coefficient σ_f'. Representative fatigue property data for selected metal alloys are given in Tables 9.3a,b.

---

[v] "Life" of a test bar can be defined in a number of ways dependent on the ultimate use of the fatigue data and the nature of the material being tested. Some criteria include total fracture, specific changes in the shape of the hysteresis loop, and the existence of microcracks of a certain size.

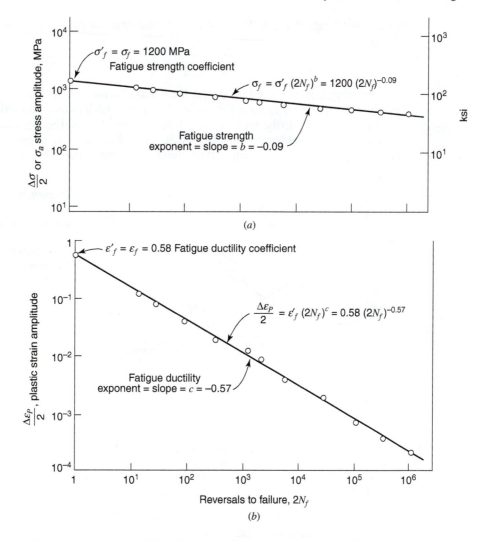

Figure 9.40   Fatigue properties of SAE 4340 steel. (*a*) Fatigue strength properties; (*b*) fatigue ductility properties.[59,60] (Reprinted by permission of the American Society for Testing and Materials from copyrighted work.)

The plastic component of strain is best described by the Manson-Coffin relation:[60,66,73,74]

$$\frac{\Delta \varepsilon_p}{2} = \varepsilon_f'(2N_f)^c \tag{9-19}$$

where $\frac{\Delta \varepsilon_p}{2}$ = plastic strain amplitude

ε_f' = fatigue ductility coefficient, defined by the strain intercept at one load reversal $(2N_f = 1)$

$2N_f$ = total strain reversals to failure

c = fatigue ductility exponent, a material property in the range -0.5 to -0.7

Data for SAE 4340 steel are plotted in Fig. 9.40*b* and are fitted to Eq. 9-19. In this instance, improved fatigue life is expected with a decreasing fatigue ductility exponent c and an increasing fatigue ductility coefficient ε_f'. Representative values for these qualities are given also in Tables 9.3*a,b*.

Manson et al.[66] argued that the fatigue resistance of a material subjected to a given strain range could be estimated by superposition of the elastic and plastic strain components. Therefore, by combining Eqs. 9-16, 9-18, and 9-19, the total strain amplitude may be given by

$$\frac{\Delta \varepsilon_T}{2} = \frac{\Delta \varepsilon_e}{2} + \frac{\Delta \varepsilon_p}{2} = \frac{\sigma'_f}{E}(2N_f)^b + \varepsilon'_f (2N_f)^c \tag{9-20}$$

By combining Eqs. 9-17 and 9-20, one finds that $n' = b/c$ and $K' = \sigma'_f / \varepsilon'^{n'}_f$. ASTM Standard E606 has been prepared to allow for the development of such fatigue properties.[57] The interested reader is referred also to the appropriate References.[54,55] It would be expected, then, that the total strain life curve would approach the plastic strain life curve at large strain amplitudes and approach the elastic strain life curve at low total strain amplitudes. This is shown in Fig. 9.41 for a

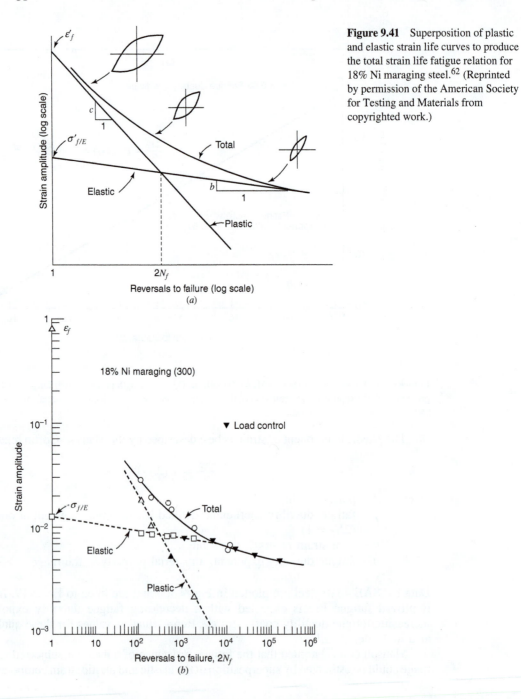

Figure 9.41 Superposition of plastic and elastic strain life curves to produce the total strain life fatigue relation for 18% Ni maraging steel.[62] (Reprinted by permission of the American Society for Testing and Materials from copyrighted work.)

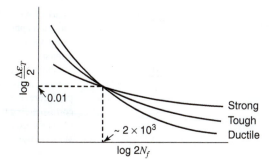

Figure 9.42 Schema showing optimum material for given total strain amplitude. Note lack of material property effect for cyclic life of about 10^3 (2×10^3 reversals) corresponding to a total strain range of about 0.01.[62] (Reprinted by permission of the American Society for Testing and Materials from copyrighted work.)

high-strength steel alloy. Loosely speaking, when the strain is largely elastic the lifetimes will be long, and so this portion of the total curve is often called the *high cycle fatigue* (HCF) regime. When the strain is largely plastic, lifetimes are relatively short and this portion of the total curve is called the *low cycle fatigue* (LCF) regime. (Note that there are no strict definitions of either term.) Finally, when the total strain life plots for strong, tough, and ductile alloys are compared, the aforementioned trends in material selection are verified. We see from Fig. 9.42 that ductile alloys are best for high cyclic strain applications (i.e., LCF), and strong alloys are superior in the region of low strains (i.e., HCF). Note that around 10^3 cycles (2×10^3 reversals), corresponding to a total strain of about 0.01, there appears to be no preferred material. That is to say, *it makes no sense to attempt to optimize material properties for applications if strains of about 0.01 are encountered—* just about any alloy will serve the purpose.

9.4   FATIGUE LIFE ESTIMATIONS FOR NOTCHED COMPONENTS

A number of different procedures dealing with the estimation of component fatigue life have been considered thus far in this chapter. For example, a typical *S-N* curve has been used to establish a "safe life" for an engineering component based on the endurance limit or the fatigue strength associated with a particular cyclic life. For more complex loading histories, Miner's law is applied also to account for the cumulative damage resulting from each block of loads. In this approach, no deliberate attempt is made to distinguish between fatigue crack initiation and propagation. Since most components contain some type of stress concentration, such as those shown in Fig. 5.9, "safe-life" values must be defined in terms of the local stress at the notch root rather than by the nominal applied stress. Indeed, Dowling and Wilson[75] have shown that specimens and components loaded with the same local stress field (defined by the product of the stress concentration and nominal stress field) exhibit similar crack initiation lives.

Special care is needed to equate the damage resulting from random loading cycles of varying amplitude with strain damage accumulated from constant amplitude loading events (e.g., see Figs. 9.40 and 9.41). To apply the linear damage accumulation concepts pertaining to Miner's law, it is necessary to identify the mean and amplitude values of each loading event so that the appropriate amount of damage can be properly "counted." The difficulty with this task lies in establishing a proper counting procedure that best accounts for the fatigue damage accumulated as a result of the random loading events.

Several counting procedures have been reviewed elsewhere, with the "rainflow method" accorded particular attention.[27,76] With reference to Fig. 9.43a, the random loading excursions result in the cyclic strains at the notch root as given in Fig. 9.43c.

To identify the effective strain cycles by the rainflow methodology, we begin with the highest peak strain (location 1) and follow the curve to the first strain reversal (at 2). Now proceed horizontally to the next downward segment of the strain–time history (at 2′) (as rainwater would flow off the edge of the roof) and proceed again to the next strain reversal site (at 4). Since a horizontal shift would not encounter another downward segment of the strain–time history sequence, the strain direction would reverse and proceed upward to the next strain

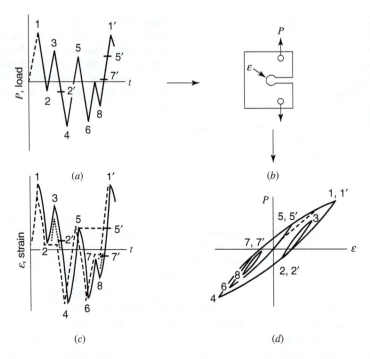

Figure 9.43 Random load spectrum applied to a notched component. (*a*) Load–time sequence; (*b*) notched component; (*c*) strain–time sequence at notch root and rain-flow analysis procedure; (*d*) hysteresis loops corresponding to rainfall analysis. (Adapted from N. E. Dowling, W. R. Brose, and W. K. Wilson, *Fatigue Under Complex Loading*, SAE, 1977, p. 55; reprinted with permission, © Society of Automotive Engineers, Inc.)

inflection point (at 5). Proceeding horizontally again to 5′ and continuing upward, we complete this strain segment at 1′. The initial load–strain hysteresis loop defined from this rainflow path would, therefore, correspond to strain and load limits defined by points 1, 4, and 1′, respectively (Fig. 9.43*d*).

Three additional hysteresis loops can be defined from rainflow analyses of the remaining positions of the original strain–time plot. The largest of these begins with the strain reversal at location 5. Proceeding downward to 6 and reversing direction to 7, we proceed horizontally to 7′ and continue the ascending strain segment until one reaches the location at 5′. The associated load–strain hysteresis loop defined from the second rainflow path corresponds to strain and load limits defined by points 5, 6, and 5′, respectively. Note the nonzero mean strain level corresponding to this hysteresis loop (Fig. 9.43*d*). The third and fourth hysteresis loops derived from the strain–time lot (Fig. 9.43*c*) correspond to the two remaining segments identified by points 2, 3, and 2′ and 7, 8, and 7′, respectively. Again note that these hysteresis loops (Fig. 9.43*d*) are not symmetrical about zero strain and involve various amounts of positive and negative cyclic-strain-induced creep. If the random loading sequence shown in Fig. 9.43*a* were to be repeated *n* times, the rainflow analysis procedure would characterize all of these random loading events in terms of the four hysteresis loops defined above, each repeated *n* times. As such, the random loading sequence can be reduced to a series of closed hysteresis loops of the type generated with laboratory samples.

It should also be noted that this simplification of the random loading pattern implies that there are no block sequence effects on component lifetime. This assumption of strain amplitude noninteraction parallels that implicit in Miner's law (recall Section 9.2.2); circumstances where fatigue life depends on load history are discussed in Section 10.5.

As discussed in the previous section, one can make a reasonable assumption that the same number of loading cycles is needed to develop a crack at various notch roots and in unnotched samples, if the cracked regions experience the same cyclic stress–strain history. Estimation of the finite life that involves such cumulative damage has been referred to as the "local stress–strain approach."[54,55] This procedure is not often easy to implement since it is not possible to monitor the elastoplastic stresses and strains near a notch root without resorting to advanced techniques such as elastic–plastic finite element analyses. Instead, estimation procedures have been developed such as the one proposed by Neuber.[77] Neuber's rule states that the *theoretical* stress concentration is equal to the geometric mean of the *actual* stress and strain concentrations (Eq. 9-21):

$$k_t = \sqrt{k_\sigma \cdot k_\varepsilon} \qquad (9\text{-}21)$$

where k_σ = actual stress concentration = $\Delta\sigma/\Delta S$
$\Delta\sigma$ = local stress range
ΔS = normal stress range
k_ε = actual strain concentration = $\Delta\varepsilon/\Delta e$
$\Delta\varepsilon$ = local strain range
Δe = normal strain range

After substitution of terms, Eq. 9-21 becomes

$$k_t = \sqrt{\frac{\Delta\sigma}{\Delta S} \cdot \frac{\Delta\varepsilon}{\Delta e}} \qquad (9\text{-}22)$$

If the nominal stress ΔS and nominal strain Δe are assumed to be elastic, then Eq. 9-22 can be rearranged to show

$$\Delta\sigma \cdot \Delta\varepsilon = (\Delta S \cdot k_t)^2 / E \qquad (9\text{-}23)$$

Topper et al.[78] modified this relation for cyclic loading applications by replacing k_t by the effective stress concentration factor k_f:

$$\Delta\sigma \cdot \Delta\varepsilon = (\Delta S \cdot k_f)^2 / E \qquad (9\text{-}24)$$

When Eqs. 9-17 and 9-24 are combined, the notch root stress and strain ranges can be determined as shown in Fig. 9.44. These values can then be used to obtain the cycle life of a notched component from Eq. 9-20 or from actual data (e.g., Fig. 9.10 or 9.41).

If the fatigue data corresponding to Fig. 9.44 reflect cyclic lives necessary to initiate a crack of approximate length l_T, then the fatigue crack initiation lifetime will have been defined (recall Section 9.2.3). Total life can then be given by this initiation time plus the number of cycles

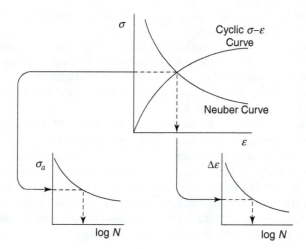

Figure 9.44 Determination of notch root stress and strain from simultaneous solution of Eq. 9-17 (cyclic stress–strain curve) and Eq. 9-24 (Neuber relation). Cyclic life associated with these notch root stress and strain values is determined from the experimental data of unnotched specimens.

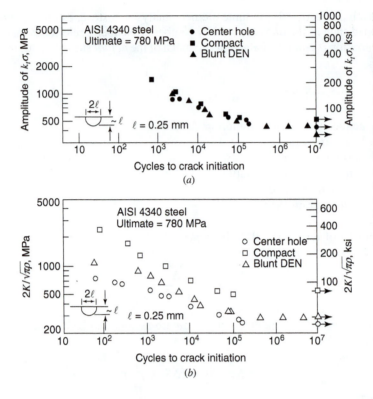

Figure 9.45 (*a*) Fatigue crack initiation life versus the amplitude of the local elastic stress of the notch root, $k_t\sigma$. Specimen shapes include center hole, compact, and double-edge-notched configurations. (After Dowling and Wilson;[75] reprinted with permission of Pergamon Press.) (*b*) Poor correlation in fatigue initiation lives is noted when the $\Delta K/\sqrt{\rho}$ parameter is used for different specimen-notch configurations. (After Dowling;[81] reprinted with permission, copyright 1982, Society of Automotive Engineers, Inc.)

necessary to propagate a crack from length l_T to the critical flaw size at final fracture (see Section 10.1.1). If one or more significant flaws are assumed to preexist in a structure, then the crack initiation stage is ignored and total fatigue life is represented only by crack propagation. Such circumstances form the basis of the "damage tolerant design philosophy," which is discussed at considerable length in Chapter 10.

As a final note, let us reexamine the effect of various notches on the fatigue life of a particular material. It is clear from Fig. 9.6 that fatigue life at a *nominal* stress σ is reduced because of the presence of the stress concentration at the notch root. In principle, one should be able to normalize these results by replotting cyclic life versus the *local* stress at the notch root, with the latter value being estimated by the quantity $k_t\sigma$ as long as general yielding does not occur. An example of such a normalized plot is shown in Fig. 9.45*a* for the case of samples of 4340 steel containing widely varying stress concentrations.[75] Some investigators have used an alternative parameter, $\Delta K/\sqrt{\rho}$, to normalize crack initiation lifetimes where ΔK is the stress intensity factor and ρ is the notch root radius.[3,79,80] In this instance, the computation of ΔK has been based on an idealized geometry in which the notch is equated with a narrow slit (see Eq. 6-27). The normalizing potential of the $\Delta K/\sqrt{\rho}$ term for the case of compact specimens containing a long, narrow notch and different notch root radii is shown in Fig. 9.46 with data for a high-strength low-alloy steel.[3] In this example, the fatigue initiation life was defined when a crack 0.25 mm in length had been developed.

It should be noted that the $\Delta K/\sqrt{\rho}$ parameter does not represent a new approach to our understanding of fatigue fracture.[19,75,81] Instead, it represents an alternative method for estimating the local elastic stress at the notch root. Indeed, this estimation procedure can prove to be inexact when widely varying specimen configurations are compared (Fig. 9.45*b*). This follows from the considerable errors that result when Eq. 6-27 is used to determine the stress intensity factor for the case of blunt notches (e.g., a circular hole). Consequently, the use of the $\Delta K/\sqrt{\rho}$ parameter in normalizing crack initiation data should be restricted to specimens that contain long, narrow notches (e.g., the compact specimen).

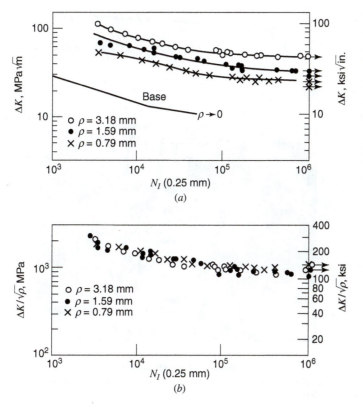

Figure 9.46 Fatigue initiation life in high-strength low-alloy steel associated with the development of a crack 0.25 mm in length. (a) Cyclic life versus ΔK as a function of notch root radius. (b) Cyclic life versus $\Delta K/\sqrt{\rho}$ (After Braglia, Hertzberg, and Roberts;[3] © American Society for Testing and Materials, 1916 Race Street, Philadelphia, PA 19103. Reprinted with permission.)

9.5   FATIGUE CRACK INITIATION MECHANISMS

Fatigue cracks are initiated at heterogeneous nucleation sites within the material whether they be preexistent (associated with inclusions, gas pores, or local soft spots in the microstructure) or are generated during the cyclic straining process itself (recall Fig. 9.4c). As one might expect, elimination of preexistent flaws can result in a substantial improvement in fatigue life. A good illustration of this is found when steels for roller bearings are vacuum melted as opposed to air melted. The much lower inclusion level in the vacuum-melted steel enables these bearings to withstand many more load excursions than the air-melted ones.

The most intriguing aspect of the fatigue crack initiation process relates to the generation of the nucleation sites. Although strains under monotonic loading produce surface offsets that resemble a staircase morphology, cyclic strains produce sharp peaks (extrusions) and troughs (intrusions)[82,83] resulting from nonreversible slip (Fig. 9.47). Many investigators have found that these surface notches serve as fatigue crack nucleation sites. It is probable that these extrusions and intrusions represent the initial stage in microcrack formation. (Recall the discussion of the phenomenology of microcrack formation in Section 9.2.3.) When the surface is periodically polished to remove these offsets, fatigue life is improved.[84] These surface upheavals represent the free surface terminations of dense bands of highly localized slip, the number of which increase with strain range. Careful studies have demonstrated these bands to be softer than the surrounding matrix material,[85] and it is believed that they cyclically soften relative to the matrix, resulting in a concentration of plastic strain. These bands are called *persistent slip bands* because of two main results. First, when a metallographic section is prepared from a damaged specimen, the deformation bands persist after etching, indicating the presence of local damage. Second, when the surface offsets are removed by polishing and the specimen is cycled again, new surface damage occurs at the same sites. Consequently, although cracking

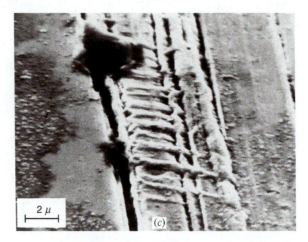

Figure 9.47   Plastic strain-induced surface offsets. (*a*) Monotonic loading giving rise to staircase morphology slip offsets; (*b*) cyclic loading, which produces sharp peaks (extrusions) and troughs (intrusions); (*c*) photomicrograph showing intrusions and extrusions on prepolished surface.[83] (Reprinted with permission from W. A. Wood and Academic Press.)

begins at the surface, it is important to recognize that the material within these persistent bands and below the surface is also damaged and will control the location of the surface nucleated cracks.[86] It should also be noted that in polycrystalline metals and alloys, persistent slip bands (PSB) can be arrested at grain boundaries, thereby contributing to crack nucleation at the PSB–grain-boundary junction.

Various dislocation models have been proposed to explain the fatigue crack nucleation process. It is now generally believed that the initial stage of cyclic damage is associated with homogeneous slip and rapid strain hardening. At the point where the cyclic stress–strain curve tends to level out (see Section 9.3.1), slip becomes concentrated along narrow bands (i.e., persistent slip bands) and the band zones become softer. Such localization of plastic strain has been found to be the precursor for fatigue crack initiation in the low strain cycling regime. For a detailed examination of this subject, see the extensive reviews by Laird[87] and Mughrabi.[88]

In a review, Kramer[89] has proposed an alternative mechanism for fatigue crack initiation. Based on a wide range of experimental findings, he concluded that the extent of work hardening near the free surface differs markedly from that found at the specimen interior. The development of such a surface layer is believed to influence the stress–strain response of a metal as well as characterize the extent of fatigue damage accumulation. In a number of instances, the dislocation density is considerably higher within a layer extending approximately 100 μm from the free surface than at the specimen interior. It is interesting to note that the improvement in fatigue life by periodic surface polishing, mentioned above, could also be rationalized by the surface layer model; the polishing action would remove the hardened layer, thereby arresting the formation of dislocation pileups believed responsible for the cracking process.

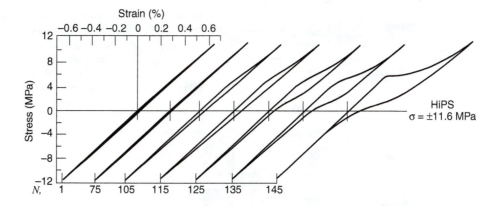

Figure 9.48 Hysteresis loops in high-impact polystyrene (HiPS). Note the progressive distortion of the tensile portion of the loop with increasing number of cyclic loads. (After Bucknall and Stevens,[90] from *J. Mater. Sci.* **15**, 295 (1980); with permission from Chapman & Hall.)

Only relatively recently has attention been given to an analysis of fatigue initiation mechanisms in engineering plastics. In some amorphous materials and rubber-toughened polymers such as high-impact polystyrene, crazing has been identified as the dominant fatigue damage mechanism.[90,91] Evidence of this effect has been obtained on the basis of volumetric measurements (recall that crazing involves a volume expansion; see Section 2.7.2) and the progressive change in shape of the cyclic stress–strain hysteresis loops. Regarding the latter, the *tensile* portion of the hysteresis loops tends to flatten and is characteristic of a crazed material (Fig. 9.48). By contrast, fatigue initiation in ABS was traced to a shear yielding process. In the latter material, the hysteresis loops generated during fatigue cycling remained symmetrical until near the end of the fatigue life of the sample. At this point, both shear yielding and crazing occurred.

9.6   AVOIDANCE OF FATIGUE DAMAGE

9.6.1 Favorable Residual Compressive Stresses

To this point, we have examined the influence of mean stress, load spectrum, and surface treatment on fatigue lifetime. Regarding the latter, shot peening and carburizing processes were found to introduce surface compressive stresses that extend the lifetime of service components (recall Section 9.2.4.1). Favorable residual compressive stresses may also be introduced by *mechanical* as well as metallurgical means. For example, cold-rolling fatigue-sensitive areas, such as thread roots in threaded fasteners, causes the material to plastically deform and spread laterally in the thread root area; however, such motion is constrained by the bulk elastic substrate, resulting in the development of compressive residual stresses at the notch root. The combination of this favorable residual stress pattern with the reduction in stress concentration brought about by the cold-rolling-induced enlargement of the thread root radius and the development of a favorable grain flow pattern (recall Fig. 7.24) leads to improved fatigue life in cold-rolled fasteners.

Compressive residual stresses can also be introduced around a fastener hole, a classical stress concentration site. In one such commercially successful procedure used extensively in the aircraft industry, an internally lubricated split-sleeve is placed over a mandrel and the assembly inserted into the plate hole (Fig. 9.49a, b).[92,93] When the mandrel is pulled through the split-sleeve, the hole is plastically deformed and expanded because the combined thickness of the split-sleeve and maximum dimension of the mandrel is greater than the hole diameter (Fig. 9.49c). The elastic springback of the plate then creates residual compressive stresses in the plastically deformed

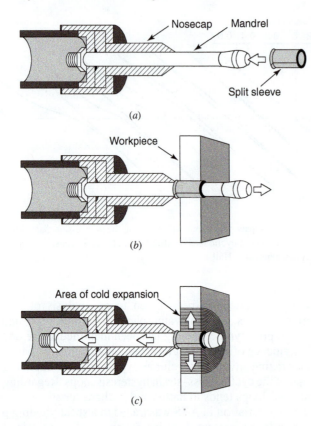

Nosecap Mandrel

Split sleeve

(a)

Workpiece

(b)

Area of cold expansion

(c)

Figure 9.49 Split-sleeve cold expansion process. (*a*) Internally lubricated split-sleeve is attached to the mandrel; (*b*) mandrel and split-sleeve are inserted into plate hole; (*c*) mandrel cold expands the hole to generate residual compressive stresses. (Courtesy of Fatigue Technology Inc.)

region surrounding the hole; the peak residual stress is approximately two-thirds of the tensile yield strength of the plate material. Minor distortions of the hole surface, resulting from the expansion process, can be eliminated by a final reaming operation. Since the compressive stress field extends one to two radii from the hole surface (see Fig. 9.50), the reaming operation does little to diminish the beneficial influence of this favorable stress field. Figure 9.51 confirms that fatigue lifetimes may be enhanced three- to ten-fold by the split-sleeve cold expansion process. Experience has shown that for best results, holes in aluminum alloy plates should be expanded

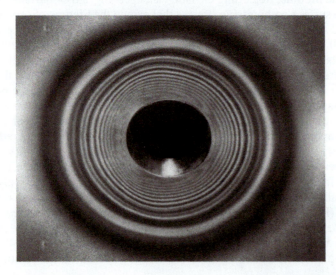

Figure 9.50 View of cold-expanded hole with the aid of polarized light through a birefringent plastic coating. Concentric rings correspond to compressive stress field surrounding the hole. (Courtesy of Fatigue Technology Inc.)

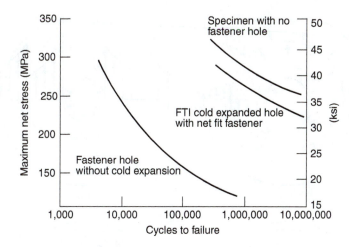

Figure 9.51 *S-N* plot of 7075-T6 aluminum specimens revealing the beneficial influence of cold expanded holes on total fatigue lifetime.[92] (From R. A. Feeler, *Aviat. Equip. Maint.*, 21 (March/April 1964), reprinted with permission.)

by at least 3%; in titanium and high strength steel alloys, holes should be expanded by at least $4\frac{1}{2}\%$.[93]

The fatigue resistance of thick-walled cylindrical pressure vessels is also improved by the introduction of a favorable residual compressive stress field at the inner bore surface where applied circumferential stresses are greatest. Three methods for the development of this stress field include: (1) shrinking an outer hollow cylinder (jacket) over the main cylinder (core); (2) constructing the thick-walled cylinder by nesting several prestressed thin sleeves over one another; and (3) a process called *autofrettage* or *self-hooping*. In the first method, an outer jacket—with inner radius slightly smaller than the outer radius of the core cylinder—is initially heated and placed over the core cylinder. As the jacket cools, it contracts, thereby generating a compressive circumferential stress field in the core cylinder[94] (Fig. 9.52*a*). Note that the outer jacket experiences a residual tensile stress field and that the magnitude of the circumferential stress in both the core and outer sleeve varies through the wall thickness. (Recall from Section 1.5.2.1 that the circumferential stress is essentially constant through the thickness of a *thin-walled* pressure vessel.) When the inner core is pressurized, a circumferential stress distribution is developed as shown in Fig. 9.52*b*. Note that the maximum applied stress at the inner bore surface (location A) is reduced from A' to B' by the superposition of the compressive residual stress at the inner bore. When a nest of thin-walled sleeves is constructed, a resulting residual stress pattern is developed as shown in Fig. 9.52*c*. Notice the development of a quasi-uniform stress field (*E–F*) when the internal-pressure-induced stress field (*C–D*) is superimposed on the latter residual stress pattern (Fig. 9.52*d*).

In the third pressure vessel prestressing procedure (autofrettage or self-hooping), the thick-walled cylinder is internally pressurized to cause yielding, beginning at the inner radius and spreading outward through the wall thickness; the extent of plastic deformation depends on the magnitude of the internal pressure. When the pressure is released, the spring-back action of the outer portion of the vessel generates a compressive stress field in the bore region of the vessel as the plastically deformed material is squeezed together. (Note the similarity of the autofrettage process with split-sleeve cold expansion of fastener holes.) A matching residual tensile stress field is created near the outer radius of the vessel. Since the maximum circumferential stress in thick-walled tube is located along the inner bore surface, there is a net positive influence of this residual stress field on the fatigue resistance of the cylinder. An example of the application of this technique to improve the fatigue resistance of a 175-mm gun tube is described in a failure analysis case history (see Chapter 11, Case 6). In this instance, the autofrettage-induced residual compressive stress at the bore surface was greater than the gas pressure-induced circumferential stress generated when the gun was fired!

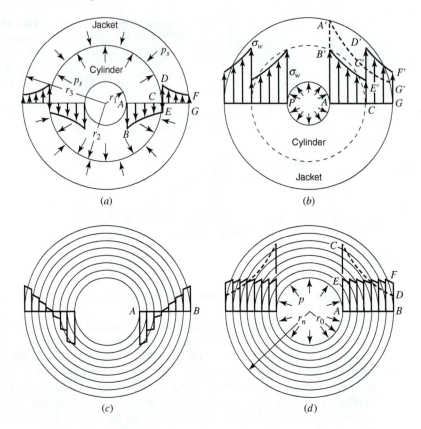

Figure 9.52   Introduction of residual stresses in thick-walled pressure vessels. (*a*) Residual stress pattern associated with shrink fitting a jacket onto the inner core hollow cylinder; (*b*) superposition of residual stress pattern from (*a*) and internally applied stress field. Maximum circumferential stress at bore surface is reduced from A' to B'; (*c*) residual stress pattern associated with nested cylinder construction; (*d*) superposition of residual stress pattern from (*c*) and internally applied stress field. Maximum circumferential stress at bore surface is reduced from C to E.[94] (From F. B. Seely and J. O. Smith, *Advanced Mechanics of Materials*, 2nd ed. (1952), with permission from John Wiley & Sons.)

9.6.2 Pretensioning of Load-Bearing Members

Residual *tensile* stresses also improve the fatigue lifetime of engineering components such as fasteners and associated fastener holes.[95,96] For example, when an interference fit is established between a bolt and the associated plate holes, the elastic stress pattern, shown in Fig. 9.53, is developed in the plate. For the case of an interference-fit bolt with a bolt-plate elastic modulus ratio of 3 (e.g., steel/aluminum), the repeated application of a remote stress, σ_a, will result in a local stress in the plate at $y = 0$ and $x = \pm R$ that fluctuates between 1.3 and 1.7 σ_a. (It is assumed here that the initial interference is sufficient such that contact between the bolt and plate is maintained at $x = 0$ and $y = \pm R$ when σ_a is applied.) For the case of a loose-fitting bolt, the maximum local stress where $y = 0$ and $x = \pm R$ is $3\sigma_a$, where 3 represents the stress concentration associated with a hole in a plate. It follows that if the external stress fluctuates between 0 and σ_a, the stress range at the hole in the plate is $3\sigma_a$ as compared with $0.4\sigma_a$ for the case of the interference bolt. Since the fatigue lifetime of a component depends strongly on the applied cyclic stress *range* (recall Section 9.2), the durability of the plate is significantly enhanced with the use of interference-fit fasteners. Note that the introduction of an interference-fit fastener of similar modulus (Fig. 9.53) generates local stresses at the plate hole that fluctuate between 1.0 and 2.0 σ_a. It follows that greater benefits from interference-fit fasteners result from an increased bolt/plate elastic modulii ratio, but that even interference fits with similar

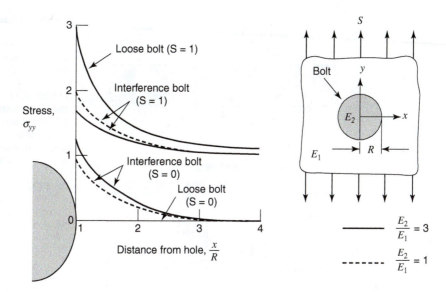

Figure 9.53   Elastic stress distribution surrounding plate hole with/without interference-fit bolt.[95] (From H. F. Hardrath, *J. Test. Eval.* **1** (1), 3 (1972). Copyright ASTM, reprinted with permission.)

materials provide improved fatigue resistance as compared to the case where no interference is present.

Tapered pins also provide the characteristics of interference-fit fasteners. As shown in Fig. 9.54a, tapered bolts (called *taperlock* fasteners) are inserted into tapered holes and then tightened into place.[97] This, again, results in an interference fit and the development of a tensile residual stress field in the plate; the latter effectively reduces the local stress range at the hole and increases service lifetime. For example, Fig. 9.54b demonstrates an approximate 20-fold improvement in fatigue lifetime associated with the use of taper pin fasteners.

In related manner, pretensioning of bolts results in their improved fatigue performance. Consider the case where two plates are attached with a bolt and nut (Fig. 9.55). If the nut is finger-tightened, then the entire load acting to separate the plates is carried by the bolt alone. If, on the other hand, the nut is tightly fastened onto the bolt, the bolt experiences a *tensile* preload P_i whereas the plates are squeezed together with a *compressive* preload of the same magnitude. This not only ensures a tight and probable leak-proof connection, but the fatigue resistance of the bolt is enhanced, as revealed by the following analysis. From Hooke's law we see that the bolt and plate deflections are

$$\delta_B = \frac{P_B}{k_B} \quad \text{and} \quad \delta_p = \frac{P_P}{k_P} \tag{9-25}$$

where $\delta_{B,P} =$ bolt and plate deflections, respectively
   $P_{B,P} =$ loads carried by bolt and plate, respectively
   $k_{B,P} =$ spring constants for bolt and plate, respectively

Since $\delta_B = \delta_P$, it follows that

$$\frac{P_B}{k_B} = \frac{P_P}{k_P} \tag{9-26}$$

If an external load, P_T, is applied so as to separate the plates, the applied load distribution is

$$P_T = P_B + P_P \tag{9-27}$$

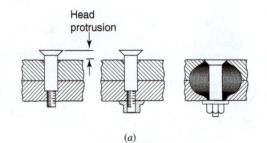

Head
protrusion

(a)

Installation sequence

1. Following hole preparation with tapered drill, tapered shank bolt is inserted in hole and firmly seated by hand pressure.

2. Full contact along entire shank of bolt and hole prevents rotation of bolt while tightening washer-nut. During tightening nut spins freely to the locking point; but washer remains stationary and provides a bearing surface against structure.

3. Torquing of washer-nut by conventional wrenching method produces a close tolerance interference fit, seats the bolt head, and creates an evenly balanced prestress condition within the bearing area of the structural joint.

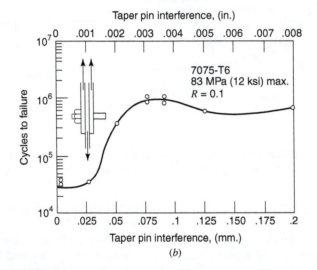

Taper pin interference, (in.)

7075-T6
83 MPa (12 ksi) max.
R = 0.1

Cycles to failure

Taper pin interference, (mm.)

(b)

Figure 9.54 (a) Sequence of installation and geometry of taper pin fasteners;[97] (Excerpted from *Assembly Engineering*, July, 1967. By permission of the Publisher © 1967. Hitchcock Publishing Co., all rights reserved.) (b) Influence of taper pin interference level on fatigue lifetime in 7075-T6 lugs. Applied stress = 83 MPa (12 ksi), R = 0.1.[98] (From C. R. Smith, *Exp. Mech.*, **5** (8), 19A (1965), with permission.)

By combining Eqs. 9-26 and 9-27, the total applied load is given by

$$P_T = P_B + \frac{P_B k_P}{k_B} = P_B \left[1 + \frac{k_P}{k_B} \right] \tag{9-28}$$

Alternatively,

$$P_B = \frac{P_T k_B}{k_B + k_P} \tag{9-29}$$

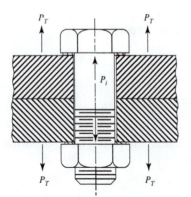

P_T P_T

P_i

P_T P_T

Figure 9.55 Preloading of bolt.

By combining the applied and pretensioning loads, the total load experienced by the bolt is

$$P_B = \frac{P_T k_B}{k_B + k_P} + P_i \tag{9-30}$$

Also, the total load experienced by the plate is

$$P_P = \frac{P_T k_P}{k_B + k_P} - P_i \tag{9-31}$$

EXAMPLE 9.3

From Fig. 9.55, assume that the stiffness of the plates and washers is four times greater than that of the bolt. If the applied cyclic load, P_T, and preload, P_i, are both 5000 N, find the cyclic loads carried by the bolt and the plate members.

From Eq. 9-30, the maximum load on the bolt is

$$P_{B_{max}} = \frac{5000(1)}{1+4} + 5000 = 6000\,\text{N}$$

Therefore, the cyclic load range experienced by the preloaded bolt is $6000 - 5000 = 1000$ N, rather than 5000 N for the case of a non-preloaded bolt. The maximum load experienced by the plates is

$$P_{P_{max}} = \frac{5000(4)}{1+4} - 5000 = -1000\,\text{N}$$

with the cyclic load being $-5000 - (-1000) = -4000$ N.

Note that the plates remain in compression even though the applied load is equal to the preload. Furthermore, most of the applied load and the major portion of the load fluctuation is carried by the stiffer plate members and *not* by the bolt. Hence, the fatigue resistance of the bolt should be greatly enhanced. Experimental data given in Table 9.4 confirm this scenario.[97] Therefore, to improve the fatigue life of a bolt, a high tensile preload should be applied (e.g., $P_i \approx 90\%\ \sigma_{ys}$) and the plates to be joined should be stiff relative to the stiffness of the bolt.[vi] (If a soft gasket or washer is used, the beneficial effect of the preload is significantly diminished.) It should be recognized that the level of bolt preloading can diminish with time due to the extrusion of dirt or paint particles from the contact surfaces and/or crushing of the joint members such as washers and gaskets. As expected, reduction in the preload level brings about a reduction in fatigue resistance of the bolt.

Table 9.4 Influence of Preload Level on Bolt Fatigue Lifetime[99]

Preload, N(lbs)	Mean Fatigue Lifetime, Cycles
6,315 (1,420)	6,000
26,330 (5,920)	36,000
32,115 (7,220)	215,000
37,450 (8,420)	5,000,000

[vi] It should be noted that tensile preloading of bolts can contribute to premature failure as a result of environmentally assisted crack extension, initiating at the stress concentration between the bolt head and shaft. Fastener failure occurs when the bolt head "pops off." (For example, see the Recall Case Study in Section 12.3.3 for the case of failed rivet heads.) Therefore, environmentally sensitive preloaded bolts should be shielded by the addition of a protective surface treatment, especially at the bolt shaft/head fillet.

REFERENCES

1. P. J. Forrest, *Fatigue of Metals*, Addison-Wesley, Reading, MA, 1962.

2. D. J. Wulpi, *How Components Fail*, American Society for Metals, Metals Park, OH, 1966.

3. B. L. Braglia, R. W. Hertzberg, and Richard Roberts, *Crack Initiation in a High-Strength Low-Alloy Steel*, ASTM *STP 677*, 1979, p. 290.

4. A. Wohler, *Zeitschrift fur Bauwesen* **10** (1860).

5. R. E. Peterson, Edgar Marbury Lecture, ASTM, 1962, p. 1.

6. ASTM E466-E468, ASTM International, West Conshohocken, PA.

7. C. Bathias and P. C. Paris, *Gigacycle Fatigue in Mechanical Practice*, Marcel Dekker, New York, 2004.

8. X. Kong, Ph.D. Dissertation, University of Technology, Compiegne, 1987.

9. J. Ni and C. Bathias, in *Recent Advances in Experimental Mechanics*, Silva Gomes et al., Eds., Balkerma, Rotterdam, 1121–1126 (1994).

10. H. F. Hardrath, E. C. Utley, and D. E. Guthrie, NASA *TN D-210*, 1959.

11. C. Bathias, *Int. J. Fat.* **32**, 535 (2010).

12. C. Bathias, C. Bin, I. Marines, *Int. J. Fract.* **25** (9), 1101–1107 (2003).

13. G. P. Sendeckyj, in *High Cycle Fatigue of Structural Materials*, W. O. Soboyejo and T. S. Srivatsan, Eds., TMS, Warrendale, PA, 1997, p. 95.

14. A. Palmgren, *Bertschrift des Vereines Ingenieure* **58**, 339 (1924).

15. M. A. Miner, *J. Appl. Mech.* **12**, A-159 (1954).

16. D. K. Bullens, *Steel and Its Heat Treatment* **1**, 37 (1938).

17. R. E. Peterson, *Stress Concentration Factors*, Wiley, New York, 1974.

18. G. Sines and J. L. Waisman, *Metal Fatigue*, McGraw-Hill, New York, 1959, p. 298.

19. N. E. Dowling, ASTM *STP 677*, 1979, p. 247.

20. S. J. Hudak, Jr., *J. Eng. Mater. Tech.* **103**, 26 (1981).

21. B. Leis, ASTM *STP 743*, 1981, p. 100.

22. D. Broek and B. Leis, *Materials Experimentation and Design in Fatigue*, Westbury House, Warwick, Surrey, England, 1981, p. 129.

23. J. C. Newman, Jr., NASA *TN D-6376*, Washington, DC, 1971.

24. M. B. P. Allery and G. Birkbeck, *Eng. Fract. Mech.* **4**, 325 (1972).

25. W. N. Findley, *J. Mech. Eng. Sci.* **14** (6), 424 (1972).

26. H. O. Fuchs and R. I. Stephens, *Metal Fatigue in Engineering*, Wiley-Interscience, New York, 1980.

27. P. J. E. Forsyth, *The Physical Basis of Metal Fatigue*, American Elsevier, New York, 1969.

28. Y. Murakami, *Metal Fatigue: Effects of Small Defects and Non-Metallic Inclusions*, Elsevier, 2002.

29. *Metals Handbook*, 8th ed., Vol. 1, *Properties and Selection of Metals*, ASM, Novelty, OH, 1961, p. 217.

30. C. Calabrese and C. Laird, *Mater. Sci. Eng.* **13** (2), 141 (1974).

31. F. Sherratt, "The Influence of Shot-Peening and Similar Surface Treatments on the Fatigue Properties of Metals," Part I, S&T Memo 1/66, Ministry of Aviation, U.S. Govt. Report 487487, Feb. 1966.

32. W. J. Harris, "The Influence of Decarburization on the Fatigue Behavior of Steel Bolts," S&T Memo 15/65, Ministry of Aviation, U.S. Govt. Report 473394, Aug. 1965.

33. R. W. Hertzberg and J. A. Manson, *Fatigue of Engineering Plastics*, Academic, New York, 1980.

34. ASTM D671, ASTM International, West Conshohocken, PA.

35. I. Constable, J. G. Williams, and D. J. Burns, *JMES* **12**, 20 (1970).

36. M. N. Riddell, G. P. Koo, and J. L. O'Toole, *Polym. Eng. Sci.* **6**, 363 (1966).

37. G. P. Koo, M. N. Riddell, and J. L. O'Toole, *Polym. Eng. Sci.* **7**, 182 (1967).

38. J. P. Ferry, *Viscoelastic Properties of Polymers*, Wiley, New York, 1961.

39. L. J. Broutman and S. K. Gaggar, Proceedings of the Twenty-Seventh Annual Technical Conference, 1972, Society of the Plastics Industry, Inc., Section 9-B, p. 1.

40. A. V. Stinskas and S. B. Ratner, *Plasticheskie Massey* **12**, 49 (1962).

41. L. C. Cessna, J. A. Levens, and J. B. Thomson, *Polym. Eng. Sci.* **9**, 399 (1969).

42. R. L. Thorkildsen, *Engineering Design for Plastics*, E. Baer, Ed., Van Nostrand-Reinhold, New York, 1964, p. 279.

43. J. A. Sauer and C. C. Chen, *Adv. Polym. Sci.* **52/53**, 169 (1983).

44. J. A. Sauer and C. C. Chen, *Toughening of Plastics*, Plastics and Rubber Institute, London, UK, 1985, paper 26.

45. J. T. Turkanis (Brennock), R. W. Hertzberg, and J. A. Manson, *Deformation, Yield and Fracture of Polymers*, Plastics and Rubber Institute, London, UK, 1985, paper 54.

46. M. D. Skibo, J. A. Manson, R. W. Hertzberg, S. M. Webler, and E. A. Collins, Jr., *Durability of Micromolecular Materials*, R. K. Eby, Ed., ACS Symposium

Series, No. 95, American Chemical Society, Washington, DC, 1979, p. 311.

47. R. W. Hertzberg and J. A. Manson, Fracture and Fatigue, *Encyclopedia of Polymer Science and Engineering*, Vol. 7, 2nd ed., Wiley, New York, 1986, p. 378.

48. K. L. Reifsnider, *Int. J. Fract.* **16**, 563 (1980).

49. E. M. Wu, *Composite Materials*, Vol. 5, L. J. Broutman and R. H. Krock, Eds., Academic, New York, 1974, Chapter 3

50. J. F. Mandell, F. J. McGarry, D. D. Huang, and C. G. Li, *Polym. Compos.* **4**, 32 (1983).

51. R. E. Lavengood and L. E. Gulbransen, *Polym. Eng. Sci.* **9**, 365 (1969).

52. S. Morris, Ph.D. Thesis, University of Nottingham, UK, 1970, quoted in *Composite Materials*, Vol. 5, L. J. Broutman and R. H. Krock, Eds., Academic, New York, 1974, p. 281.

53. C. C. Chamis, *Composite Materials*, Vol. 5, L. J. Broutman and R. H. Krock, Eds., Academic, New York, 1974, p. 94.

54. R. M. Wetzel (Ed.), *Advances in Engineering*, Vol. 6, *Fatigue under Complex Loading: Analysis and Experiments*, Society of Automotive Engineers, 1977.

55. *Manual on Low Cycle Fatigue Testing*, ASTM *STP 465*, 1969.

56. J. Morrow and D. F. Socie, *Materials, Experimentation and Design in Fatigue*, F. Sherratt and J. B. Sturgeon, Eds., Westbury House, Warwick, Surrey, England, 1981, p. 3.

57. ASTM E606, ASTM International, West Conshohocken, PA.

58. B. I. Sandor, *Fundamentals of Cyclic Stress and Strain*, University of Wisconsin Press, Madison, 1972.

59. J. D. Morrow, *Internal Friction, Damping and Cyclic Plasticity* ASTM *STP 378*, 1965, p. 45.

60. R. W. Smith, M. H. Hirschberg, and S. S. Manson, NASA *TN D-1574*, NASA, April 1963.

61. R. W. Landgraf, J. D. Morrow, and T. Endo, *J. Mater, JMLSA* **4** (1), 176 (1969).

62. R. W. Landgraf, *Achievement of High Fatigue Resistance in Metals and Alloys*, ASTM *STP 467*, 1970, p. 3.

63. P. Beardmore and S. Rabinowitz, *Treatise on Materials Science and Technology*, Vol. 6, R. J. Arsenault, Ed., Academic, New York, 1975, p. 267.

64. S. Rabinowitz and P. Beardmore, *J. Mater. Sci.* **9**, 81 (1974).

65. P. Beardmore and S. Rabinowitz, *Polymeric Materials*, ASM, Metals Park, OH, 1975, p. 551.

66. S. S. Manson and M. H. Hirschberg, *Fatigue: An Interdisciplinary Approach*, Syracuse University Press, Syracuse, NY, 1964, p. 133.

67. J. Bauschinger, *Zivilingur* **27**, 289 (1881).

68. S. T. Rolfe, R. P. Haak, and J. H. Gross, *Trans. ASME J. Bas. Eng.* **90**, 403 (1968).

69. J. P. Hickerson and R. W. Hertzberg, *Met. Trans.* **3**, 179 (1972).

70. C. E. Feltner and C. Laird, *Acta Met.* **15**, 1621 (1967).

71. C. E. Feltner and C. Laird, *Trans. AIME* **242**, 1253 (1968).

72. O. H. Basquin, *Proc. ASTM* **10**, Part II, 625 (1910).

73. L. F. Coffin, Jr., *Trans. ASME* **76**, 931 (1954).

74. J. F. Tavernelli and L. F. Coffin, Jr., *Trans. ASM* **51**, 438 (1959).

75. N. E. Dowling and W. K. Wilson, *Advances in Fracture Research*, D. Francois, Ed., Pergamon, Oxford, England, 1981, p. 518.

76. M. Matsuishi and T. Endo, *Fatigue of Metals Subjected to Varying Stress*, paper presented to Japan Society of Mechanical Engineers, Fukuoka, Japan, March 1968.

77. H. Neuber, *Trans. ASME, J. App. Mech.* **8**, 544 (1961).

78. T. H. Topper, R. M. Wetzel, and J. Morrow, *J. Mater, JMSLA* **4** (1), 200 (1969).

79. A. R. Jack and A. T. Price, *Int. J. Fracture Mech.* **6**, 401 (1970).

80. J. M. Barsom and R. C. McNicol, in *Fracture Toughness and Slow-Stable Cracking*, ASTM *STP 559*, 1974, p. 183.

81. N. E. Dowling, paper presented at SAE Fatigue Conference, April 14–16, 1982, Dearborn, MI.

82. W. A. Wood, *Fracture*, Technology Press of M.I.T. and Wiley, New York, 1959, p. 412.

83. W. A. Wood, *Treatise on Materials Science and Technology*, Vol. 5, H. Herman, Ed., Academic, New York, 1974, p. 129.

84. T. H. Alden and W. A. Backofen, *Acta Met.* **9**, 352 (1961).

85. O. Helgeland, *J. Inst. Met.* **93**, 570 (1965).

86. C. Roberts and A. P. Greenough, *Philos. Mag.* **12**, 81 (1965).

87. C. Laird, *Metallurgical Treatises*, J. K. Tien and J. F. Elliott, Eds., AIME, Warrendale, PA, 1981, p. 505.

88. H. Mughrabi, *Fifth International Conference on Strength of Metals and Alloys*, P. Haasen, V. Gerold, and G. Kostorz, Eds., Pergamon, Oxford, England, 1980, p. 1615.

89. I. R. Kramer, *Advances in the Mechanics and Physics of Surfaces*, Vol. 3, Gordon & Breach, New York, 1986, p. 109.

90. C. B. Bricknall and W. W. Stevens, *7. Mater. Sci.* **15**, 2950 (1980).

91. M. E. Mackay, T. G. Teng, and J. M. Schultz, *J. Mater. Sci.* **14**, 211 (1979).

92. R. A. Feeler, *Aviation Equipment Maintenance* **21**, March/April 1984.

93. *Fatigue Technology Inc.*, Seattle, Washington, corporate literature.

94. F. B. Seely and J. O. Smith, *Advanced Mechanics of Materials*, 2nd ed., John Wiley, New York, 1957.

95. H. F. Hardrath, *J. Test. Eval.* **1** (1), 3 (1972).

96. J. H. Crews, *NASA TND 6955*, NASA (August 1972).

97. C. R. Smith, *Assembly Engineering* **10** (7), 18 (1967).

98. C. R. Smith, *Experimental Mechanics* **5** (8), 19A (1965).

99. W. Orthwein, *Machine Component Design*, West Publishing Co., St. Paul, MN, 1990.

PROBLEMS

Review

9.1 What are clamshell/beach markings? Explain their significance in terms of the conditions experienced during the life of a failed component, and their value in failure analysis.

9.2 What are *ratchet lines*, and how are they arranged with respect to the fatigue crack front and any clamshell marks that may also be present?

9.3 Why is it safe to assume that the railroad wheel axles studied by Wöhler are better described by stress-controlled cyclic loading than by strain-controlled cyclic loading?

9.4 What role can a notch or a flaw play in determining the total fatigue life of a component?

9.5 What does it mean to conduct a fatigue test with $R = -0.5$?

9.6 What are the S and N axes on an S-N diagram, and what does the region to the left and below the plotted line indicate?

9.7 What is a *fatigue limit*, and what relevance does it have for lifetimes in the gigacycle (10^9) regime? Explain.

9.8 What is the typical effect of increasing mean stress on fatigue life?

9.9 Automobile manufacturers specify a certain number of miles at which the engine timing belt should be replaced. Is this an example of a "safe-life"-designed component or a "fail-safe"-designed component? What are the implications for the condition of the timing belt when it is removed?

9.10 What is the underlying assumption behind the Palmgren-Minor damage law that may not always be correct?

9.11 Given that the 10^6-cycle fatigue limit is often approximately half of the tensile strength of a metal, it would appear to be good design practice to use a material with as high a tensile strength as possible to maximize fatigue resistance. What is the potential shortcoming of this approach?

9.12 List three general categories of surface treatment that can increase fatigue life, and provide one example of a specific process for each category.

9.13 Polymer fatigue failures may include significant heating. Why is this more of a problem for polymers than for metals?

9.14 Explain why the fatigue life (N_f) of a polymer specimen should decrease with increasing frequency.

9.15 Why is fatigue generally less of a problem with ceramics and glasses than with metals and polymers?

9.16 Sketch a simplified version of Fig. 9.34 that shows only the first, last, and next to last numbered stress–strain loop for each case. Indicate the change in stress from one loop to the next so it is clear what the trend is with regard to cycle-dependent stress in each case.

9.17 If you have a material that is *initially hard and strong*, would you expect it to cyclically harden or soften? What would be a way of characterizing *how* strong it must be initially to make your answer a bit more quantitative?

9.18 What trend exists between the stacking fault energy of a metal and the rate at which is will cyclically strain harden or soften? Give an example of a common high SFE metal and a common low SFE metal.

9.19 Sketch a representative ε-N diagram, indicating the regions often called the Low Cycle and High Cycle Fatigue regimes.

9.20 What is the purpose of "rainflow counting"?

9.21 Sketch a persistent slip band, define what it is, and explain its role in fatigue.

Practice

9.22 Two investigators independently reported fatigue test results for Zeusalloy 300. Both reported their data in the form of σ–N curves for notched bars. The basic difference between the two results was that Investigator I reported inferior behavior of the material compared with the results of Investigator II (i.e., lower strength for a given fatigue life) but encountered much less scatter. Can you offer a possible explanation for this observation? Describe the macroscopic fracture surface appearance for the two sets of test bars.

9.23 Two different polymeric materials were evaluated to determine their respective fatigue endurance behavior. Both materials were tested separately in laboratory air and in flowing water. (Water was selected as a suitable liquid test environment since neither polymer was adversely affected by its presence.) Polymer A showed similar S-N plots in the two environments whereas Polymer B revealed markedly different results. Speculate as to which environment was associated with the superior fatigue response in Polymer B and characterize the structure and mechanical response of Polymers A and B.

9.24   For a steel alloy with a tensile strength of 1000 MPa, estimate the fatigue strength amplitude for this material when the mean stress is 200 MPa. Note that you must make an estimate of the fatigue strength using Fig. 9.15 in order to proceed.

9.25   Tensile and fully reversed loading fatigue tests were conducted for a certain steel alloy and revealed the tensile strength and endurance limit to be 1200 and 550 MPa, respectively. If a rod of this material supply were subjected to a static stress of 600 MPa and oscillating stresses whose total range was 700 MPa, would you expect the rod to fail by fatigue processes? *Hint:* You may want to plot a diagram to aid in presenting your answer.

9.26   The fatigue life for a certain alloy at stress levels of σ_1, σ_2, and σ_3 is 10,000, 50,000, and 500,000 cycles, respectively. If a component of this material is subjected to 2500 cycles of σ_1 and 10,000 cycles of σ_2, estimate the remaining lifetime in association with cyclic stresses at a level of σ_3.

9.27   A cylindrical component will be designed with a circumferential notch to accommodate a small snap ring. If the notch is 1 mm deep and has a radius of 0.5 mm, calculate the theoretical and effective stress concentration factors for this component (i) if it is made of annealed stainless steel alloy 405 with a tensile strength of 415 MPa, or (ii) if it is made with cold rolled stainless steel alloy 17-7PH with a tensile strength of 1380 MPa. What conclusion can you draw about plasticity, stress concentrations, and fatigue life?

9.28   In an effort to determine a material's resistance to fatigue crack initiation, two studies were undertaken. One investigator used a plate sample with a circular hole 1 cm in diameter, and the other investigator used a similar sample with a 4 cm circular hole. In both cases, the definition for fatigue life initiation was taken to be the number of loading cycles necessary to develop a crack 0.25 cm in length. The cyclic lives determined from these two investigations were not in agreement. Why? Based on the results of Dowling, what change in specimen geometry or initiation criteria would you recommend so that both investigators would report similar initiation lives?

9.29   The tensile strength of copper alloy C71500-H80 (copper-30%Ni) has been measured at 580 MPa. Estimate upper and lower bounds for the endurance limit at 10^8 cycles for this alloy.

9.30   Calculate the fatigue life of SAE 1015 and 4340 steel tempered at 425°C when the samples experience total strain ranges of 0.05, 0.01, and 0.001. Which alloy is best at each of these applied strain ranges?

Design

9.31   If the test depicted in Fig. 9.11 had only been carried out for 10^6 cycles, what approximate fatigue limit would have been claimed for 2024-T3 Al being tested? What assumption would this have led to with regard to designing for infinite fatigue life for a component made from this material? Is this a good assumption under some, all, or no conditions?

9.32   Cu-Ni alloy 71500 is specified for a certain seawater pump component because of its excellent corrosion resistance, particularly where chloride stress-corrosion may be a concern for steel alloys. With a safety factor already included, the component must be designed to last for 10^8 cycles. If the tensile strength of hot-rolled Cu-Ni alloy 71500 is measured at 380 MPa, estimate the maximum alternating stress that the component can withstand in order to achieve this lifetime.

Extend

9.33   What is a bone *stress fracture*, and how does it relate to the content of this chapter?

9.34   Find a photograph of a fracture surface associated with a fatigue failure; it must show beach marks and/or ratchet lines. If the image is not already labeled, add labels for the fatigue markings that are visible. Briefly summarize the circumstances behind the failure, and what can be learned from the appearance of the fracture surface.

Chapter 10

Fatigue Crack Propagation

As discussed in Chapter 5, a number of engineering system breakdowns can be attributed to preexistent flaws that caused failure when a certain critical stress was applied. In addition, these defects may have grown to critical dimensions prior to failure. Subcritical flaw growth is important to guard against for a number of reasons. First, if a structure or component contains a defect large enough to cause immediate failure upon loading, the defect quite likely could be detected by a number of nondestructive test (NDT) procedures and repaired before damaging loads are applied. If the defect is not detected, the procedure of proof testing (subjecting a structure, such as a pressure vessel, to a preservice simulation test at a stress level equal to or slightly higher than the design stress) would cause the structure to fail, but under controlled conditions with minimum risk to human lives and damage to other equipment in the engineering system. On the other hand, were the crack to be subcritical in size and undetected by NDT, a successful proof test would prove only that a flaw of critical dimensions did not exist at that time. *No guarantee could be given that the flaw would not grow during service to critical dimensions and later precipitate a catastrophic failure.* This chapter is concerned with factors that control the fatigue crack propagation (FCP) process in engineering materials.

10.1 STRESS AND CRACK LENGTH CORRELATIONS WITH FCP

Crack propagation data may be obtained from a number of specimens such as many of those shown in Fig. 6.21. Starting with a mechanically sharpened crack, cyclic loads are applied and the resulting change in crack length is monitored and recorded as a function of the number of load cycles. In most instances, crack length data acquisition is computer-controlled (Fig. 10.1) and employs the use of compliance measurements, eddy current techniques, electropotential measurements, and/or acoustic emission detectors. A typical plot of such data is shown in Fig. 10.2, where the crack length is seen to increase with increasing number of loading cycles. The fatigue crack growth rate is determined from such a curve either by graphical procedures or by computation. From these methods, the crack growth rates resulting from a given cyclic load are $(da/dN)a_i$ and $(da/dN)a_j$ when the crack is of lengths a_i and a_j, respectively.

It is important to note that the crack growth rate most often increases with increasing crack length. (This is generally the case, though not always, as will be discussed below.) It is most significant that the crack becomes longer at an increasingly rapid rate, thereby shortening component life at an alarming rate. An important corollary of this fact is that most of the loading cycles involved in the total life of an engineering component are consumed during the early stages of crack extension when the crack is small and, perhaps, undetected. The other variable that controls the rate of crack propagation is the magnitude of the stress level. It is evident from Fig. 10.2 that FCP rates increase with increasing stress level.

Since many researchers have probed the nature of the fatigue crack propagation process, it is not surprising to find in the literature a number of empirical and theoretical "laws," many of the form

$$\frac{da}{dN} \propto f(\sigma, a) \tag{10-1}$$

Figure 10.1 Monitoring an environmental fatigue test using computer-controlled instrumentation. (Photo courtesy of J. Keith Donald, Fracture Technology Associates.)

reflecting the importance of the stress level and crack length on FCP rates. Quite often, this function assumes the form of a simple power relation wherein

$$\frac{da}{dN} \propto \sigma^m a^n \tag{10-2}$$

where $m \approx 2 - 7$
$\quad\quad\ \ n \approx 1 - 2$

For example, Liu[1] theorized m and n to be 2 and 1, respectively, while Frost[2] found empirically for the materials he tested that $m \approx 3$ and $n \approx 1$. Weibull[3] accounted for the stress and crack length dependence of the crack growth rate by assuming the FCP rate to be dependent on the net section stress in the component. Paris[4] postulated that the stress intensity factor—itself a function of stress and crack length—was the overall controlling factor in the FCP process. This postulate appears reasonable, since one might expect the parameter K, which controlled static fracture (Chapter 6) and environment assisted cracking (Chapter 8) to control dynamic fatigue failures as well. From Chapter 6 the stress intensity factor levels corresponding to the crack growth rates identified in Fig. 10.2 would be $Y_i\sigma_1\sqrt{a_i}$ and $Y_j\sigma_1\sqrt{a_j}$ at σ_1 and $Y_i\sigma_2\sqrt{a_i}$ and $Y_j\sigma_2\sqrt{a_j}$ at σ_2, respectively. By plotting values of da/dN and ΔK at the associated values of $a_{i,j} \dots, n$, a strong correlation was observed (Fig. 10.3), which suggested a relation of the form

$$\frac{da}{dN} = A\,\Delta K^m \tag{10-3}$$

Figure 10.2 Crack propagation data showing the effect of the applied stress level. FCP rate increases with stress and crack length.

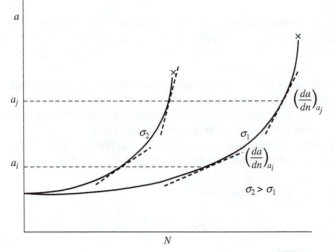

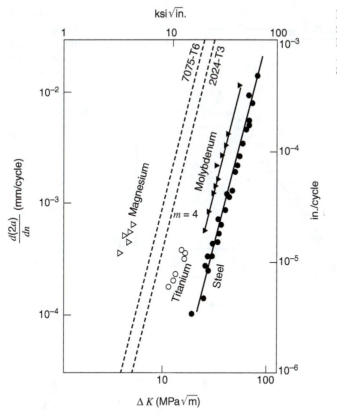

Figure 10.3 Fatigue crack propagation for various FCC, BCC, and HCP metals. Data verify the power relation between ΔK and da/dN.[4] (With permission of Syracuse University Press.)

where $\dfrac{da}{dN}$ = fatigue crack growth rate

ΔK = stress intensity factor range ($\Delta K = K_{max} - K_{min}$)

A, m = f(material variables, environment, frequency, temperature, stress ratio)

It is encouraging to note that the FCP response of many materials is correlated with the stress intensity factor range, even though FCC, BCC, and HCP metals all were included in the database.

Thus, for an interesting period during the early 1960s, the battle of the crack growth rate relations began. Although experimental evidence was mounting in favor of the stress intensity factor approach, it was not until two critical sets of experiments were reported and fully appreciated that the importance of the stress intensity factor in controlling the fatigue crack propagation rate was fully accepted. In one paper, Swanson et al.[5] reasoned that if any of the various relations between stress and crack length was truly the critical parameter controlling crack growth rate behavior, then keeping that parameter constant during a fatigue test would cause the crack to grow at a constant velocity. In other words, the crack length versus number of cycles curve would appear as a straight line. To achieve this condition, Swanson decreased the cyclic load level incrementally in varying amounts with increasing crack length to maintain constant magnitudes of $\sigma^3 a$, $\sigma^2 a$, σ_{net}, and ΔK, deemed to be the controlling parameters as discussed above. It was demonstrated clearly that a constant crack velocity throughout the entire fatigue test life was achieved *only* when ΔK was held constant. The second set of critical experiments was reported by Paris and Erdogan,[6] who analyzed test data from center cracked panels of a high-strength aluminum alloy. In one instance, the loads were applied uniformly and remote from the crack plane (Fig. 10.4). From Fig. 6.21c, the stress intensity factor for this configuration is found to increase with increasing crack length. Therefore, if a constant load range were applied during the life of the test, the stress

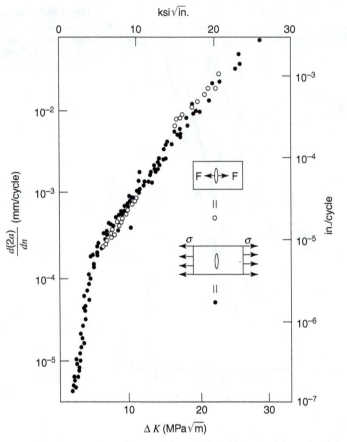

Figure 10.4 Fatigue crack propagation behavior in 7075-T6 for remote and crack line loading conditions.[4] (With permission of Syracuse University Press.)

intensity level would increase continually because of the increasing crack length. Correspondingly, crack growth rates would be low at first and increase gradually as the crack extended. In the other set of experiments, center notched panels were loaded by concentrated forces acting at the crack surfaces (Fig. 10.4). For this configuration, the stress intensity factor is found to be[7]

$$K = \frac{P}{\sqrt{\pi a}} \qquad (10\text{-}4)$$

Most interestingly, the stress intensity factor is observed to be large when the crack length is small but decreases with increasing crack length. Consequently, one would predict that if the stress intensity factor did control FCP rates, the crack growth rate would be large initially but would *decrease* with increasing crack length. Such a reversal in FCP behavior would not be predicted if crack growth were controlled by the net section stress, for example, since the latter term would increase with both specimen load configurations. In fact, the experimental results showed the crack growth rate reversal anticipated by the sense of the crack length-dependent change in stress intensity factor dK/da. The crack line-loaded sample exhibited the highest growth rates at the outset of the test when the crack was small, with progressively slower growth rates being monitored with increasing crack length. The opposite was true for the grip-loaded sample (Fig. 10.4). Saxena et al.[8] have demonstrated that both increasing and decreasing crack growth rates can be obtained from the same specimen if one part of the test is conducted with a positive dK/da gradient and the other part carried out with a negative dK/da gradient. The desired K and crack growth rate gradients are achieved by programming a computer to control cyclic loads to achieve a particular K gradient according to the relation

$$K_i = K_0 e^{C(a_i - a_0)} \tag{10-5}$$

where $K_i =$ instantaneous K level
$K_0 =$ initial K level
$C =$ normalized K gradient as $\dfrac{dK/da}{K}$
$a_i =$ instantaneous crack length
$a_0 =$ initial crack length

Therefore, if C is assigned a negative, zero, or positive value, the associated crack growth rates will decrease, remain the same, or increase, respectively, with increasing crack length. The important point to bear in mind is the fact that, regardless of the K gradient, the associated fatigue crack growth rates fall along the same da/dN–ΔK curve, with da/dN being controlled by ΔK. (The use of computers in the fatigue laboratory has gained increased importance since the early days of fatigue testing because unit labor costs per test can be reduced, and since the computer allows for more efficient utilization of the fatigue machines and allows for convenient implementation of experiments involving complex loading histories.)

Not only did these experiments verify the importance of the stress intensity factor in controlling the fatigue crack propagation process, they illustrated again the interchangeability of specimen geometry and load configuration in the determination of material properties such as crack growth rate or fracture toughness values. As a final note, over the growth range from about 2×10^{-6} to 2×10^{-3} mm/cyc, crack growth rate behavior in relatively inert environments as described by Eq. 10-3 is not strongly dependent on the mean stress intensity level. To a first approximation, crack growth rates are found to double when K_{mean} is doubled. By contrast, when the ΔK level is doubled, the crack would propagate 16 times faster (assuming $m = 4$). Consequently, mean stress effects in this growth rate range are considered to be of secondary importance. (Crack growth rate conditions where mean stress effects are more important are discussed in Section 10.4.) It is to be noted that the secondary importance of mean stress in controlling FCP response is similar to the smaller role played by mean stress in cyclic life tests as portrayed with Gerber and Goodman diagrams (Chapter 9).

Detailed procedures for conducting a fatigue crack propagation test have been established and are to be found in ASTM Standard E647.[9] The interested reader is referred to this document and to a collection of papers dealing with various laboratory experiences pertaining to this standard.[10] Furthermore, the reader *must* recognize that data generated in the low crack growth regime, according to conventional test methods described by this standard, may lead to nonconservative estimates of component life. This topic is considered at greater length in Section 10.4.

10.1.1 Fatigue Life Calculations

When conducting a failure analysis, it is often desirable to compute component life for comparison with the actually recorded service life. Alternatively, if one were designing a new part and wished to establish for it a safe operating service life, fatigue life calculations would be required. Such a computation can be performed by integrating Eq. 10-3 with the starting and final flaw size as limits of the integration.

$$\frac{da}{dN} = A \, \Delta K^m \tag{10-3}$$

$$\frac{da}{dN} = A \left(Y \Delta \sigma \sqrt{a} \right)^m$$

$$N_f = \int_{a_0}^{a_f} \left[\frac{1}{A (Y \Delta \sigma \sqrt{a})^m} \right] da = \frac{1}{A \Delta \sigma^m} \int_{a_0}^{a_f} \left(Y a^{1/2} \right)^{-m} da$$

When the geometrical correction factor Y does not change within the limits of integration (e.g., for the case of a circular flaw where $K = (2/\pi)\sigma\sqrt{\pi a}$, or when a_c/W is small), the cyclic life is given by

$$N_f = \frac{2}{(m-2)AY^m\Delta\sigma^m}\left[\frac{1}{a_0^{(m-2)/2}} - \frac{1}{a_f^{(m-2)/2}}\right] \quad \text{for } m \neq 2 \qquad (10\text{-}6)$$

where N_f = number of cycles to failure
    a_0 = initial crack size
    a_f = final crack size at failure
    $\Delta\sigma$ = stress range
   A, m = material constants
    Y = geometrical correction factor

Usually, however, this integration cannot be performed directly, since Y varies with the crack length. Consequently, cyclic life may be estimated by numerical integration procedures by using different values of Y held constant over a number of small crack length increments. It is seen from Eq. 10-6 that when $a_0 \ll a_f$ (the usual circumstance) the computed fatigue life is not very sensitive to the final crack length a_f but, instead, is strongly dependent on estimations of the starting crack size a_0.

EXAMPLE 10.1

Compare the differences in fatigue lifetimes for three components that experienced crack extension from 2 to 10 mm, versus where the initial crack length was four times smaller ($a_0 = 0.5$ mm), or where the final crack length was four times larger ($a_f = 40$ mm). Assume that crack growth rates follow a Paris relation where $m = 4$.
   We have three scenarios:

Case A: $a_0 = 2$ mm, $a_f = 10$ mm
Case B: $a_0 = 0.5$ mm, $a_f = 10$ mm
Case C: $a_0 = 2$ mm, $a_f = 40$ mm

   We are only interested in examining the relative influence of crack length on fatigue lifetimes. Hence, from Eq. 10-6 where $m = 4$,

$$N_f \propto \left[\frac{1}{a_0} - \frac{1}{a_f}\right]$$

Therefore, for Case A (2–10 mm):

$$\left[\frac{1}{0.002} - \frac{1}{0.010}\right] = 500 - 100 = 400$$

For Case B (0.5–10 mm):

$$\left[\frac{1}{0.0005} - \frac{1}{0.010}\right] = 2000 - 100 = 1900$$

For Case C (2–40 mm):

$$\left[\frac{1}{0.002} - \frac{1}{0.040}\right] = 500 - 25 = 475$$

For the case where the initial crack length is smaller, fatigue lifetime increases by 375%.

$$\frac{1900 - 400}{400} \times 100 = 375\%$$

By comparison, when the final crack length is increased by the same relative proportion, fatigue lifetime increases by only 18.75%.

$$\frac{475 - 400}{400} \times 100 = 18.75\%$$

Clearly, *fatigue lifetime is far more sensitive to initial than to final crack size.* In a practical sense, one finds that the major portion of fatigue lifetime in a service component is consumed prior to the point where the crack is discovered. Several additional examples[11] demonstrating this integration procedure are now considered.

In the first example, an extruded aluminum alloy is assumed to contain a semielliptical surface flaw with a and c dimensions of 0.15 and 10 mm, respectively (Fig. 10.5a). For a design stress of 128 MPa, a fatigue life computation reveals that the crack would penetrate the back surface after approximately 2,070,000 loading cycles. It should be noted that the numerical integration in this example requires different Y values to be computed for each intermediate crack front configuration and location (i.e., different ellipticity and back face corrections). (If the crack front shape were initially semicircular, the integration procedure would have been greatly simplified.) Several sets of crack growth rate material parameters (A and m values) were used in this computation, which corresponded to different regions of the da/dN versus ΔK curve. Note that most of the fatigue life of this component took place in extending the crack only a short distance from its original contour.

In the second example, the same material is assumed to contain a much deeper semi-elliptical surface flaw ($a = 1.5$ mm) and is subjected to the same stress level (128 MPa). From Fig. 10.5b we see that only 13,800 loading cycles were needed to cause the crack to penetrate the back face of the more deeply flawed component. As expected, a much shorter life was computed for the component with the larger initial flaw size.

A dramatically different conclusion is reached if life calculations are based on equal starting ΔK conditions (an alternative design assumption). For the first example, the 0.15-mm-deep elliptical flaw together with a stress of 128 MPa corresponds to a ΔK level of 3.1 MPa$\sqrt{\text{m}}$. If the component with the 1.5-mm-deep elliptical flaw were subjected to the same ΔK level, 22,000,000 loading cycles would be needed to grow the crack through the section. The reason for the 11-fold greater life in the component that contained the much larger flaw (10 times deeper!) is readily apparent when one recognizes the substantial difference in stress range experienced by the two components: The stress range in the part containing the shallow crack was 128 MPa, whereas the same ΔK level in the more deeply flawed component required a stress range of only 40 MPa. From Eq. 10-6, we see that the cyclic life varies inversely with the mth power of the cyclic stress range.

To further illustrate the use of Eq. 10-6 in the computation of fatigue life, let us reconsider the material selection problem described in Chapter 6 (Example 6.5). Let us suppose that the 0.45C-Ni-Cr-Mo steel is available in both the 2070 and 1520 MPa tensile strength levels, and a design stress level of one-half tensile strength is required. It is necessary to estimate the fatigue life of a component manufactured from the material in the two strength conditions. Using the design stress levels, a stress range of 1035 and 760 MPa would be experienced by the 2070 and 1520 MPa materials, respectively. It is immediately obvious from Eq. 10-5 that, all things being equal, the total fatigue life would be lower in the higher strength material because it would experience a higher stress range. Using a value of $m = 2.25$ as found by Barsom et al.[12] for

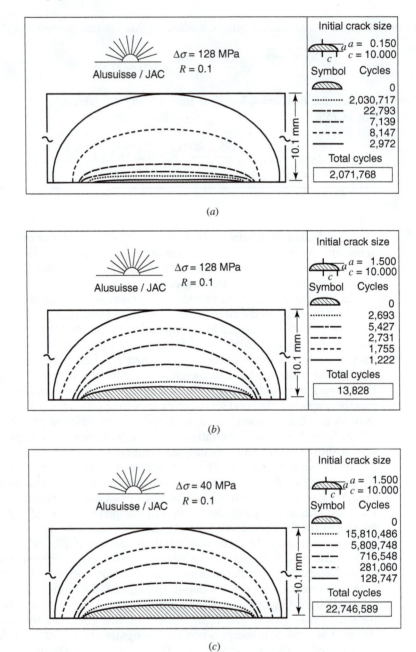

Figure 10.5   Crack contour changes associated with fatigue crack propagation in an extruded aluminum alloy. "Cycles" correspond to the number of loading cycles needed to grow the crack from one contour to the next: (a) $\Delta\sigma = 128$ MPa, $a = 0.150$ mm, $\Delta K = 3.1$ MPa $\sqrt{m}$; (b) $\Delta\sigma = 128$ MPa, $a = 1.50$ mm, $\Delta K = 10$ MPa $\sqrt{m}$; (c) $\Delta\sigma = 40$ MPa, $a = 1.50$ mm, $\Delta K = 3.1$ MPa $\sqrt{m}$. (Courtesy of Robert Jaccard, Alusuisse Ltd.)

19 steels, the fatigue life in the stronger material would be reduced by almost a factor of two. This should be considered as a minimum estimate of the reduction in fatigue life, since there is evidence to indicate that the exponent m increases with decreasing fracture toughness.[13] Furthermore, recalling that the critical flaw size in the 2070 MPa-level material was only one-fifth that found in the 1520 MPa alloy, the computed service life in the stronger alloy would be reduced further. This would be true especially if the initial crack was not much smaller than the

critical flaw size. Therefore, one concludes that the stronger material is inferior in terms of potential fatigue life, critical flaw size, associated fracture toughness, and environment-assisted cracking sensitivity (Fig. 8.14).

10.1.2 Fail-Safe Design and Retirement for Cause

As mentioned in the previous chapter, the "fail-safe" design philosophy recognizes that cracks can develop in components but ensures that the structure will not fail prior to the time that the defect is discovered and repaired. Several different design elements have been utilized to provide the structure with "fail-safe" characteristics. These include the presence of multiple load paths, which render the structure fail-safe as a result of load shedding from one component to its redundant loading path (e.g., see Section 11.2). Another fail-safe design feature involves the use of crack arresters such as stiffeners (e.g., "tear straps") in the fuselage section of aircraft; here, the fatigue crack growth rate of a fuselage crack is attenuated as the crack passes beneath each stiffener[14] (Fig. 10.6). A third fail-safe design feature, already discussed, pertains to the "leak-before-break" design of pressure vessels and pipes (recall Section 6.6). It is important to note that the "fail-safe" design approach requires periodic inspection of load-bearing components with sufficient flaw detection resolution so as to enable defective parts to be either repaired or replaced in a timely manner.

A specific application of the fail-safe design philosophy involves removal of a particular component only when there is clear evidence for the existence of a defect of critical dimensions. That is, based on a fracture mechanics analysis, the part would be replaced if failure were expected. As such, the component is subject to "retirement for cause." If no defect is found, the part is returned to service with the next inspection interval being determined from a fracture mechanics calculation, based on the existence of a crack whose length is just below the inspection resolution limit (Fig. 10.7). In this manner, components that would have been retired, based on attainment of their "safe-life" cyclic limit, would be allowed to continue in service, at considerable savings. To illustrate, the initial safe-life design of the F100 military jet engine involved the replacement of all engine disks after a service life when 1 in 1000 disks would be expected to have initiated a 0.75-mm-long crack. Harris[15] concluded that the replacement of safe-life design controls with retirement for cause procedures in 23 F100 military jet engine components would save the U.S. Air Force over $1 billion over a 20-year period of time, thereby representing the greatest savings to date for the Air Force Materials Laboratory on the basis of a technological development.

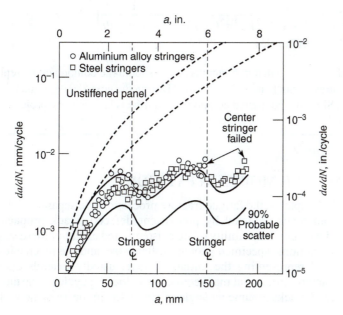

Figure 10.6 Attenuation of FCP rates in aluminum alloy sheet in the presence of stiffeners.[14]

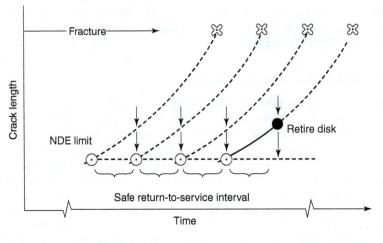

Figure 10.7
Characterization of "retirement for cause" procedures. Components are repeatedly inspected and returned to service until a quantifiable crack is detected, resulting in retirement of the component.[15]

EXAMPLE 10.2

An NDT examination of a steel component reveals the presence of a 5-mm-long edge crack. If plane-strain conditions prevail ($K_{IC} = 75\,\mathrm{MPa}\sqrt{m}$), what would be the residual service lifetime of the steel part if it were subjected to repeated stresses of 400 MPa?

Several pieces of information are needed before one can compute the fatigue lifetime of the component. These include the crack growth rate–ΔK relation and the size of the final crack length. Based on experimental test results, $da/dN = 4 \times 10^{-37}\,\Delta K^4$ where da/dN and ΔK have units of m/cyc and $\mathrm{Pa}\sqrt{m}$, respectively. The final crack length is calculated from the instability condition when $K = K_{IC}$. Accordingly,

$$K = K_{IC} = 1.1\sigma\sqrt{\pi a}$$

$$a_f = \left(\frac{75 \times 10^6}{1.1(400 \times 10^6)\sqrt{\pi}}\right)^2; \quad a = 9.2\,\mathrm{mm}$$

From Eq. 10-6

$$N_f = \frac{1}{4 \times 10^{-37}\,1.1^4\pi^2(400 \times 10^6)^4}\left[\frac{1}{a_0} - \frac{1}{a_f}\right]$$

$$N_f = 617\,\mathrm{cycles}$$

If the next inspection procedure is scheduled in 6 months' time, should the component be replaced now or retained in service, given that the stresses fluctuate every 4 hours? During the next 180 days, the part would have experienced 1080 loading cycles. Since the computed component lifetime is less than the projected stress cycle count, the part should be replaced now.

10.2 MACROSCOPIC FRACTURE MODES IN FATIGUE

As discussed in Chapter 9, the fatigue fracture process can be separated into three regimes: crack initiation (sometimes obviated by preexistent defects), crack propagation, and final fracture (associated with crack instability). The existence and extent of these stages depends on the applied stress conditions, specimen geometry, flaw size, and the mechanical properties of the material. Stage I, representing the initiation stage, usually extends over only a small percentage of the fracture surface but may require many loading cycles if the nucleation process is slow. Often, Stage I cracks assume an angle of about 45° in the xy plane with respect to the

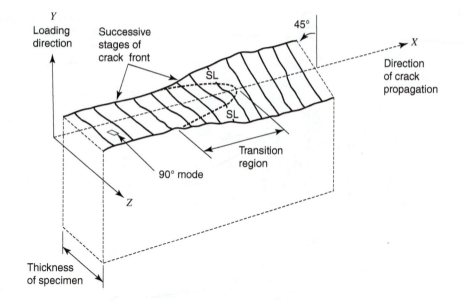

Figure 10.8   Diagram showing fracture mode transition from flat to slant fracture appearance. (Adapted from J. Schijve.)

loading direction.[16] After a relatively short distance, the orientation of a Stage I crack shifts to permit the crack to propagate in a direction normal to the loading direction. This transition has been associated with a changeover from single to multiple slip.[17,18] The plane on which the Stage II crack propagates depends on the relative stress state; that is, the extent of plane-strain or plane-stress conditions. When the stress intensity factor range is low (resulting from a low applied stress and/or small crack size), a small plastic zone is developed (Eq. 6-43). When the sheet thickness is large compared to this zone size, plane-strain conditions prevail and flat fracture usually results. With subsequent fatigue crack extension, the stress intensity factor and the plastic zone size increase. When the zone is large compared to specimen thickness, plane-stress conditions and slant fracture are dominant. Therefore, depending on the stress level and crack length, the fractured component will possess varying amounts of flat and slant fracture (see Chapter 5). Consequently, a fatigue crack may start out at 90° to the plate surface but complete its propagation at 45° to the surface (Fig. 10.8). Alternatively, the crack could propagate immediately at 45° if the plastic zone size to plate thickness ratio were high enough, reflecting plane-stress conditions.

It is important to recognize that both unstable, fast-moving cracks and stable, slow-moving fatigue cracks may assume flat, slant, or mixed macromorphologies. It should be noted, however, that a unique relation between the stress state and fracture mode was not observed by Vogelesang[19] in 7075-T6 and 2024-T3 aluminum alloys during fatigue crack propagation in aggressive environments. The influence of environment on the fracture mode transition (more flat fracture in corrosive atmosphere than in dry air at the same r_y/t ratio) was believed to be caused by a change in the fracture mechanism. As was just discussed, the plastic zone (defined by Eq. 6-43) can be used to estimate the relative amount of flat and slant fracture surface under both static and cyclic loading conditions. From Fig. 10.9, this plastic zone is developed by the application of a stress intensity factor of magnitude K_1. However, when the latter is reduced by h_k because the direction of loading is reversed, the local stress is reduced to a level corresponding to a stress intensity level of K_2. Since the elastic stress distribution associated with K_1 was truncated at σ_{ys} by local yielding, subtraction of an elastic stress distribution in going from K_1 to K_2 will cause the final crack-tip stress field to drop sharply near the crack tip and even go into compression. At K_2, a smaller plastic zone is formed in which the material undergoes compressive yielding. Paris[4] showed that the size of this smaller plastic zone, which experiences alternate tensile and

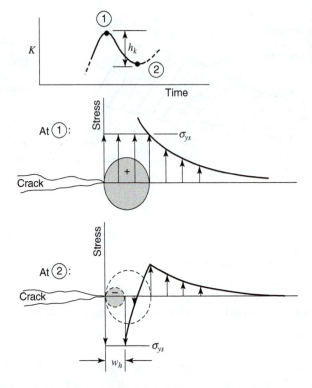

Figure 10.9 Monotonic and reversed plastic zone development at tip of advancing fatigue crack.[4] (With permission of Syracuse University Press.)

compressive yielding, may be estimated by substituting h_k for K and $2\sigma_{ys}$ for σ_{ys} in Eq. 6-43. As a result, the size of the reversed plastic zone may be given by

$$r_y = \frac{1}{8\pi}\left(\frac{h_k}{\sigma_{ys}}\right)^2 \tag{10-7}$$

or four times smaller than the comparable monotonic value. Hahn et al.[20] measured the reversed plastic zone dimension using an etch pitting technique and found it to be slightly different and given by

$$r_y^c = 0.033\left(\frac{\Delta K}{\sigma_{ys}}\right)^2 \tag{10-8}$$

where r_y^c is the reversed plastic zone when measured in the y direction from the crack plane to the elastic-plastic boundary.

Since the materials within this smaller plastic zone experiences reversed cyclic straining, it might be expected that cyclic strain hardening or softening would result, depending on the starting condition of the material. This has been borne out by microhardness measurements made by Bathias and Pelloux[21] near the tip of a fatigue crack in two different steels. The significance of material property changes on fatigue crack propagation in this zone is examined further in Section 10.7.

As previously discussed in Section 9.1 (e.g., recall Fig 9.3), fracture surface marker bands (i.e., clamshell or beach markings) are found on fatigue fracture surfaces when the failed component experienced different periods of crack extension and/or variable amplitude block loading histories. As long as there were no significant stress field interactions between adjacent blocks, the width of each marker band provides an estimate of crack advance associated with the number of loading cycles during each particular loading period.

Fatigue Failure Analysis Case Study 10.1: *Stress Intensity Factor Estimate Based on Fatigue Growth Bands*[22]

In Failure Analysis Case Study 6.2, two estimates of the stress intensity factor at failure were presented. These were based on an analysis of the crack configuration and an analysis of shear lip depth following fatigue failure of a 1.78-cm-thick steel plate with a drilled hole. The estimations of the prevailing critical stress intensity factor at fracture ranged from 101 to 136 MPa$\sqrt{m}$ in the first case, and ~106 MPa$\sqrt{m}$ in the second case (recall Eqs. 6-52 and 6-53). Now, let us identify two additional estimates of the critical stress intensity factor by using measurements of the fatigue growth bands. It was known that the last band was produced by 15 load fluctuations between stress levels of 137 and 895 MPa. This growth band measured 0.32 mm, and the average crack growth rate was found to be

$$\frac{da}{dn} \approx \frac{\Delta a}{\Delta n} \approx 3.2 \times 10^{-4}/15 \approx 2.1 \times 10^{-5}\,\text{m/cyc}$$

From the fatigue crack growth rate data of Carmen and Katlin[i] the corresponding ΔK level was determined to be about 77 MPa$\sqrt{m}$. The maximum K level was then given by

$$K_{max} = \Delta K \left(\frac{\sigma_{max}}{\Delta\sigma}\right)$$

$$K_{max} = 77\left(\frac{895}{758}\right)$$

$$K_{max} = 91\,\text{MPa}\sqrt{m}$$

A similar calculation was made for the next to last band, where

$$\Delta n = 2$$
$$\Delta a = 0.16\,\text{mm}$$
$$\sigma_{min} = 138\,\text{MPa}$$
$$\sigma_{max} = 992\,\text{MPa}$$
$$\frac{da}{dn} \approx 1.6 \times 10^{-4}/2 \approx 8 \times 10^{-5}\,\text{m/cyc}$$

From Carmen and Katlin's results, the ΔK level corresponding to this crack growth rate was found to be 82.5 MPa$\sqrt{m}$. Again using the expression for K_{max} from above

$$K_{max} = 82.5\left(\frac{992}{854}\right) = 95.8\,\text{MPa}\sqrt{m}$$

In both instances, the estimates of K_{max} from fatigue growth bands were in excellent agreement with values based on estimates of the prevailing stress intensity factor and shear lip measurements. As discussed in Chapter 7, the critical stress intensity factor of ~101 MPa$\sqrt{m}$ found by averaging all four estimates is very close to that of the known K_{IC} level for this material (see Table 7.8). As such, the multifaceted analyses of this laboratory failure demonstrates the simultaneous use of several different fracture mechanics approaches in solving a service failure. Ideally, a number of techniques should be utilized whenever possible to provide a cross-check for each estimate.

---

[i] C. M. Carmen and J. M. Katlin, ASME Paper No. 66-Met-3, *J. Basic Eng.*, 1966.

10.3 MICROSCOPIC FRACTURE MECHANISMS

A high-magnification examination of the clamshell markings found on service failure fracture surfaces (Figs. 9.1 and 9.3) reveals the presence of many smaller parallel lines, referred to as fatigue striations (Fig. 10.10). Several important facts are known about these markings. First, they appear on fatigue fracture surfaces in many materials, such as BCC, HCP, and FCC metals, and many engineering polymers, and are oriented parallel to the advancing crack front. In a quantitative sense, Forsyth and Ryder[24] provided critical evidence that each striation

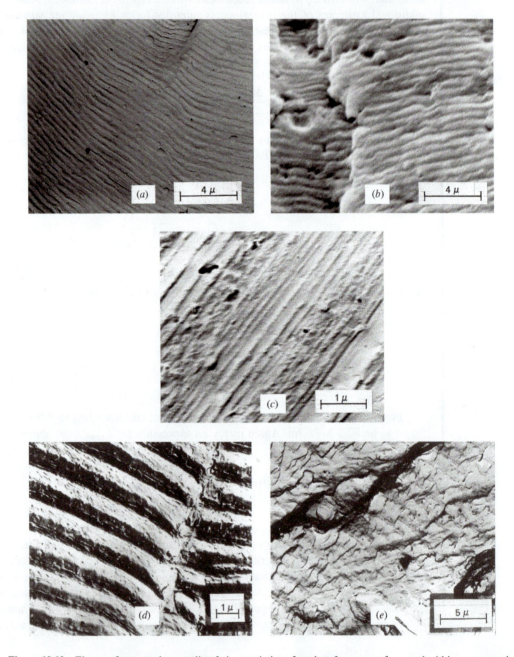

Figure 10.10 Electron fractographs revealing fatigue striations found on fracture surfaces and within macroscopic bands (Figs. 9.1, 9.3, 10.38). (*a*) TEM, constant load range; (*b*) SEM, constant load range; (*c*) TEM, random loading; (*d*) TEM, ductile striations.[23]; (*e*) TEM, brittle striations.[23] (Reprinted with permission of the American Society for Testing and Materials from copyrighted work.)

represents the incremental advance of the crack front as a result of one loading cycle and that the extent of this advance varies with the stress range. This is shown clearly in Fig. 10.10c, which reveals striations of differing width that results from a random loading pattern.

It is appropriate, then, to emphasize the clear distinction between macroscopically observed clamshell markings, which represent periods of growth during which thousands of loading cycles may have occurred, and microscopic striations, which represent the extension of the crack front during one load excursion. *There can be thousands or even tens of thousands of striations within a single clamshell marking.*

Although these striations provide evidence that fatigue damage was accumulated by the component during its service life, fatigue crack propagation can occur without their formation. Usually, microvoid coalescence occurs at high ΔK levels,[25] and a cleavage-like and/or rough faceted appearance dominates in many materials at very low ΔK levels[18] (Fig. 10.11). (For additional discussion of the rough fracture surface morphology associated with low ΔK test conditions, see further comments in Sections 10.5.1 and 10.7.) It is generally observed that the relative striation density found at intermediate ΔK values seems to vary with stress state and alloy content. Although striations are most clearly observed on flat surfaces associated with plane-strain conditions, elongated dimples and evidence of abrasion are the dominant fractographic features of plane-stress slant fracture surfaces. In terms of metallurgical factors, it is much easier to find striations on fatigue surfaces in aluminum alloys than in high-strength steels. In some cases, it is virtually impossible to identify clearly defined areas of striations in the latter material, thereby making fractographic examinations most difficult.

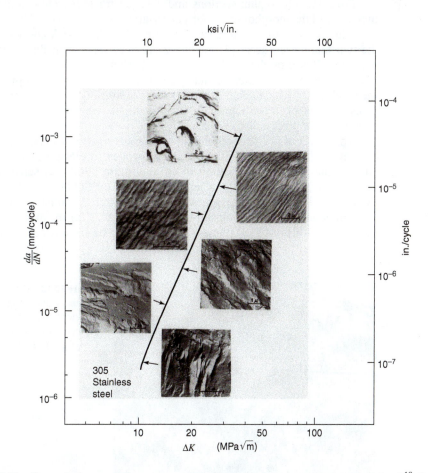

Figure 10.11 Change in fracture surface appearance in 305 stainless steel as a function of ΔK level.[18] (Reprinted with permission of the American Society for Testing and Materials from copyrighted work.)

Fatigue striations can assume many forms, such as the highly three-dimensional or flat ones seen in Fig. 10.10d, e. It is not absolutely clear why there are different morphologies, but they are often associated with the test environment during crack propagation. Fatigue striations are relatively flat and assume a cleavage-like appearance when formed in an aggressive environment, but tend to appear more ductile when formed in an inert environment. Although striation morphology may be affected by the service environment, definite and pronounced changes have been observed in the appearance of fatigue striations after exposure to oxidizing or corroding atmospheres. For example, fatigue striations can be completely obliterated as a result of exposure to a high-temperature, oxidizing environment. Even at room temperature, fatigue striations become increasingly more difficult to detect with time. As a result, the amount of fractographic information to be gleaned from a fracture surface decreases with time, particularly in the case of steel alloys; the fracture surface details in aluminum alloys are maintained for a longer period because of the protective nature of the aluminum oxide film that forms quickly on the fracture surface. Indeed, some of the fracture surface images of the aluminum alloys, described by Hertzberg and Mills,[18] were obtained fully twelve years after these fracture surfaces were originally generated.

The reader should recognize that even when striations are expected to form (Fig. 10.11), they are not always as clearly defined as those in Fig. 10.10. Whether due to environmental and/or metallurgical effects or related to service conditions, such as abrasion of the mating fracture surfaces, striations may appear either continuous or discontinuous, clearly or poorly defined, and straight or curved.

From metallographic sections and electron fractographic examination, three basic interpretations of the morphology of fatigue striations have evolved. The striations are considered to be undulations on the fracture surface with (1) peak-to-peak and valley-to-valley matching of the two mating surfaces, (2) matching crevices separating flat facets, or (3) peak-to-valley matching. Based on these interpretations of striation morphology, different mechanisms have been proposed for striation formation. One mechanism involves plastic blunting processes at the crack tip,[26] which occur regardless of material microscopic slip character; another model takes account of crystallographic considerations, wherein striations are thought to form by sliding off on preferred slip planes.[23,25,27]

The effect of crystallography on striation formation can be supported by both direct and circumstantial evidence. Pelloux[27] demonstrated in elegant fashion with the aid of etch pit studies that striation orientation in an aluminum alloy was sensitive to changes in crystal orientation and that striations tended to form on (111) slip planes and parallel to ⟨110⟩ directions. The latter is in agreement with theoretical considerations[25,27] (Fig. 10.12) and experimental findings, which show a strong tendency for the macroscopic fracture plane to lie parallel to {100} or {110} planes.[28] It might then be argued that when slip planes are oriented

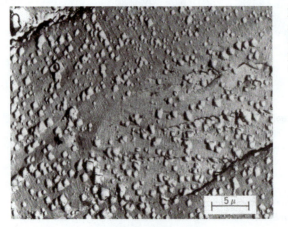

Figure 10.12 Fatigue striations in 2024-T3 aluminum alloy. Note concurrent change in striation and etch pit orientation.[27] (Reprinted with permission of the American Society of Metals. Copyright © 1969.)

5 μ

favorably with respect to the maximum resolved shear stresses at the advancing crack tip, a clearly defined striation could be formed. Alternatively, poorly defined striations or none at all might be found when slip planes are unfavorably oriented. It is quite probable that crystallographic considerations dominate striation formation at low ΔK levels where few slip systems are operative, while the plastic blunting model would provide a better picture of events at high ΔK levels.

10.3.1 Correlations with the Stress Intensity Factor

More quantitative information has been obtained from the measurement of fatigue striations than from any other fracture surface detail. Since the striation represents the position of the crack front after each loading cycle, its width can be used to measure the FCP rate at any given position on the fracture surface. It is not surprising, then, to find reasonable correlation at a given ΔK level between the macroscopically determined growth rate as measured with a traveling microscope and the microscopic growth rate as measured by the width of individual striae[29] (Fig. 10.13). Additional correlations[30] between striation spacing and ΔK have been found for a number of materials (Fig. 10.14). Here, ΔK has been normalized with respect to the elastic modulus of the respective materials examined (see Section 10.7).

The practical significance of the data correlations found in Figs. 10.13 and 10.14 cannot be overemphasized, since such data are very useful in failure analyses.

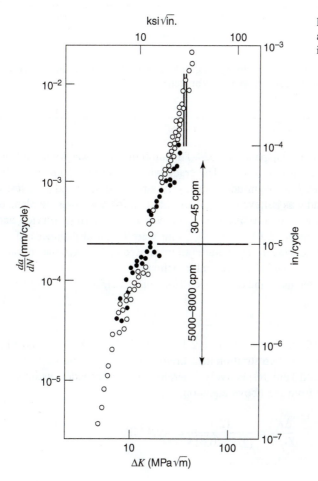

Figure 10.13 Correlation of macroscopic (○) and microscopic (●) crack growth rate with ΔK in 2024-T3 aluminum alloy.

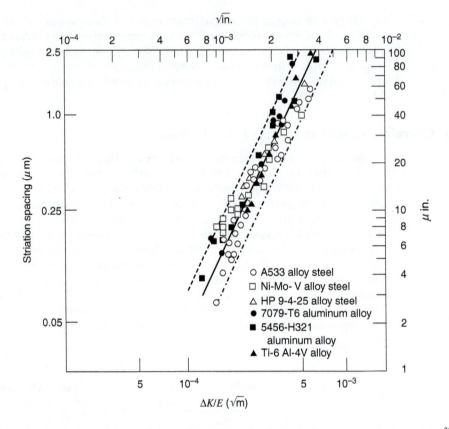

Figure 10.14 Correlation of fatigue striation spacing with ΔK normalized with respect to elastic modulus.[30] (Reprinted with permission of the American Society of Metals. Copyright 1969.)

EXAMPLE 10.3

After a certain period of service, a 15-cm-wide panel of 2024-T3 aluminum alloy was found to contain a 5-cm-long edge crack oriented normal to the stress direction. The crack was found to have nucleated from a small, preexistent flaw at the edge of the panel. The magnitude of the cyclic stress was analyzed to be less than 20% of the yield strength ($\sigma_{ys} \approx 345\,\text{MPa}$) and was believed to be distributed uniformly along the plane of the crack. Since the crack had reached dangerous proportions, the panel was removed from service and examined fractographically. Average striation widths of 10^{-4} and 10^{-3} mm were found at distances of 1.5 and 3 cm, respectively, from the origin of the crack. Was the premature failure caused by the existence of the surface flaw or related to a much higher cyclic stress level than originally estimated?

For this configuration, the stress intensity factor is given from Fig. 6.21f as

$$K = Y\sigma\sqrt{a}$$

with $Y(a/W)$ being defined at $Y(1.5/15)$ and $Y(3/15)$ to be 2.1 and 2.43, respectively. From Fig. 10.13, the apparent stress intensity range based on the two striation measurements is found to be 12.7 and 20.9 MPa$\sqrt{\text{m}}$, corresponding to crack lengths of 1.5 and 3 cm, respectively. Therefore, two independent estimates of the actual stress range can be obtained directly from the above equation:

$$\Delta\sigma = \frac{\Delta K}{Y\sqrt{a}} = \frac{12.7}{(2.1)\sqrt{0.015}} = 49.4\,\text{MPa}$$

and

$$\frac{20.9}{(2.43)\sqrt{0.03}} = 49.7\,\text{MPa}$$

Since both numbers are self-consistent and are in agreement with the original design estimates, the striation data appear valid. Therefore, it is concluded that premature failure was caused by early crack propagation from the small edge flaw.

Although this procedure is extremely useful, it should be implemented with deliberate caution. First, it is critical to accurately identify the crack length position where the striation spacing measurements were made. The stress level cannot be computed if the crack length is not known. In many service failure reports, striation photographs are presented without identification of the precise location of the region of the fracture surface. Without such information, the photograph serves only to identify the mechanism of failure, but does not enable the examiner to perform any meaningful calculations.

It has been shown repeatedly in laboratory experiments that for constant stress intensity conditions, striation spacings in a local region may vary by a factor of two to four. This results from the fact that striation formation is a highly localized event and is dependent on both the stress intensity factor and metallurgical conditions. In addition, the spacing between adjacent striations depends on the incident angle of the electron beam relative to the orientation of the replica if the transmission electron microscope is used for viewing (as opposed to the scanning electron microscope); that is, the investigator sees a *projected* image of the replica on the viewing screen. Stofanak et al.[31] concluded that minimal projection errors were found in striation width measurements when the support grid was oriented normal to the electron beam; this resulted from the fact that the carbon replicas had collapsed onto the support grid. The principal reason for the scatter in striation-spacing measurements within a given replica (at a given level of ΔK), therefore, was attributed primarily to metallurgical factors such as grain orientation variations and inclusion distribution. To arrive at a meaningful estimate of crack growth rate at a particular crack length, many measurements of striation spacing should be made. In addition, measurements should be made of different crack length positions, as done in Example 10.3, to serve as a comparative check on the computation.

The prevailing stress intensity factor range could also be estimated with the aid of an empirical correlation identified by Bates and Clark,[30] who showed that

$$\text{striation spacing} \approx 6\left(\frac{\Delta K}{E}\right)^2 \tag{10-9}$$

where ΔK = stress intensity factor range
E = modulus of elasticity

It is particularly intriguing that Eq. 10-9 can be used to estimate ΔK based on fractographic information for any metallic alloy (Fig. 10.14). Since the exponent in Eq. 10-9 is approximately two, while the exponent in the Paris-type Eq. 10-3 depends on material variables, environment, frequency, and temperature, and can vary from about 2 to 7, agreement between macroscopic and microscopic crack growth rates should not be expected in the majority of instances. Consequently, striation spacing measurements should be used in conjunction with Eq. 10-9 or compared with previously determined fractographic information, rather than macroscopic data, whenever estimations of the prevailing ΔK level are desired.

10.4 CRACK GROWTH BEHAVIOR AT ΔK EXTREMES

10.4.1 High ΔK Levels

Although Eq. 10-3 does provide a simple relation by which crack growth rates may be correlated with the stress intensity factor range, it does not account for crack growth characteristics at both low and high levels of ΔK. If enough data are obtained for a given material—say, four to five decades of crack growth rates—the da/dN versus ΔK curve assumes a sigmoidal shape, as shown in Fig. 10.15. That is, the ΔK dependence of crack growth rate increases markedly at both low and high ΔK values. At the high growth rate end of the spectrum, part of this deviation sometimes may be accounted for by means of a plasticity correction since the plastic zone becomes large at high ΔK levels. This has the effect of increasing ΔK_{eff} (Eq. 6-46) and thus tends to straighten out the curve. Another factor to be considered is that as $K_{\max}$ approaches K_c, local crack instabilities occur with increasing frequency, as evidenced by increasing amounts of microvoid coalescence and/or cleavage on the fracture surface. As might be expected, this effect is magnified with increasing mean stress. Characterizing the mean stress level by R, the ratio of minimum to maximum loads, it is seen from Fig. 10.16a that crack growth rates at high ΔK values increase with increasing mean stress, and diminished mean stress sensitivity is observed at lower ΔK levels. However, greater mean stress sensitivity on crack growth rates again is observed at very low ΔK levels (see Section 10.4.2). One relation expressing crack growth rates in terms of ΔK, K_c and a measure of K_{mean} was proposed by Forman et al.[33] in the form

$$\frac{da}{dN} = \frac{C\Delta K^n}{(1 - R)K_c - \Delta K} \tag{10-10}$$

where $C, n =$ material constants
$K_c =$ fracture toughness
$R =$ load ratio $\left(\dfrac{K_{\min}}{K_{\max}}\right)$

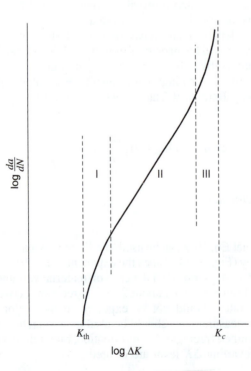

Figure 10.15 Diagram showing three regimes of fatigue crack growth response. Region I, crack growth rate decreases rapidly with decreasing ΔK and approaches lower limit at ΔK_{th}; Region II, midrange of crack growth rates where "power law" dependence prevails; Region III, acceleration of crack growth resulting from local fracture as $K_{\max}$ approaches K_c.

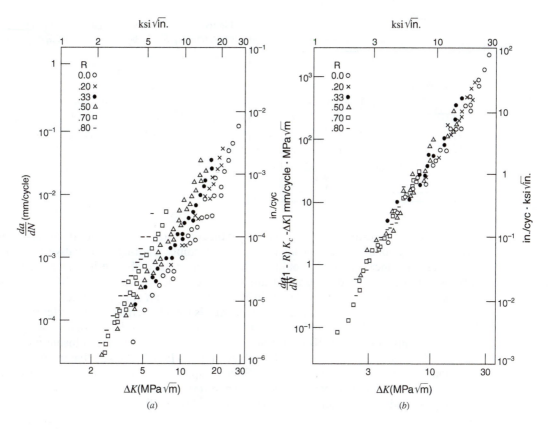

Figure 10.16 Fatigue crack propagation in 7075-T6 aluminum alloy showing effect of load ratio R and applicability of Forman, Kearney, and Engle relation: (*a*) ΔK vs. *da/dN*; (*b*) ΔK vs. $[(1 - R) K_c - \Delta K]$ *da/dN*. Note that there is less scatter in *b*. (Data from Hudson.)

From Eq. 10-10, we see that the simple power relation (Eq. 10-3) has been modified by the term $[(1 - R)K_c - \Delta K]$, which decreases with increasing load ratio R and decreasing fracture toughness K_c, both of which lead to higher crack growth rates at a given ΔK level. A typical plot of normalized data according to Eq. 10-10 is shown in Fig. 10.16*b*. Although Eq. 10-10 correctly identifies material FCP response under combinations of high ΔK and K_{mean} conditions, the relation is difficult to apply because of difficulties associated with the determination of the K_c value, which, as shown in Chapter 6, varies with planar and thickness dimensions of the test sample.

Other relations describing mean stress effects on FCP response have taken account of the plastic zones at the crack tip[34] and the plastic deformation process itself. With regard to the latter, Christensen[35] and Elber[36] proposed that the crack might be partially closed during part of the loading cycle, even when $R > 0$. Elber argued that residual tensile displacements, resulting from the plastic damage of fatigue crack extension, would interfere along the crack surface in the wake of the advancing crack front and cause the crack to close above the minimum applied load level. This hypothesis was verified with compliance measurements taken from fatigued test panels that showed that an *effective* change in crack length (i.e., change in compliance) occurred prior to any *actual* change in crack length. In other words, the crack was partially closed for a portion of the loading cycle and did not open fully until a certain opening K level, K_{op}, was applied. As a result, the damaging portion of the cyclic load excursion would be restricted to that part of the load cycle that acted on a fully opened crack. From Fig. 10.17, the effective stress intensity factor range ΔK_{eff} would be denoted by the opening level K_{op} to K_{max}, rather than by the applied ΔK level $K_{\mathrm{max}} - K_{\mathrm{min}}$. In this connection, it is interesting to note that crack growth rates are relatively insensitive to compressive loading excursions where $R < 0$. In

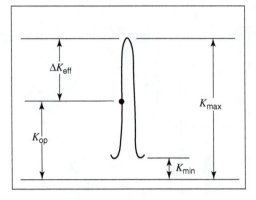

Figure 10.17   Crack surface interference results in crack opening K_{op} to be above zero. ΔK_{eff} defined as $K_{max} - K_{op}$.

fact, a number of investigators[32,37] have shown that the fatigue response of materials subjected to $R < 0$ loading conditions can be approximated by simply ignoring the negative portion of the load excursion, since the crack would be closed. The marked change in importance of mean stress when negative load excursions are encountered is shown in Fig. 10.18. We see in Fig. 10.18a, where $R \geq 0$, that FCP rates in 7075-T6 aluminum alloy are affected by changes in mean load level. By contrast, no significant change is found in crack growth rates when $R \leq 0$, (Fig. 10.18b). (Note that in the latter figure, ΔK was defined as K_{max}; that is, $K_{min} = 0$.)

The data given in Table 10.1 confirm the applicability of the crack closure model to the analysis of fracture surface markings. Hertzberg and von Euw[38] showed that the fracture mode transition (FMT) was dependent on ΔK_{eff}. From Chapter 6, this transition is related to a critical ratio of plastic zone size to panel thickness. Therefore, it was surprising to find the transition occurring instead at a constant crack growth rate[39,40] (Fig. 10.19). From Table 10.1, the FMT did not occur at a specified value of K_{max} or ΔK. However, it did occur at a constant ΔK_{eff} level, which would account for the FMT being observed at a constant growth rate.

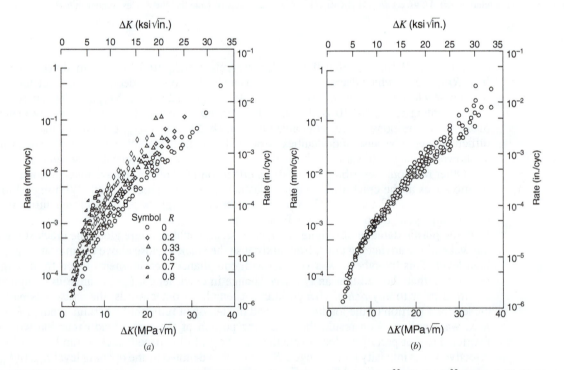

Figure 10.18   Variation of fatigue crack growth in 7075-T6 aluminum alloy, where (a) $R \geq 0$[32]; (b) $R < 0$.[32] (From C. M. Hudson, NASA TN D-5390, 1969.)

Table 10.1   Fracture Mode Transition in 2024-T3 Aluminum Alloy[38,39]

	K_{max}		ΔK_{app}		$\Delta K_{\mathrm{eff}}^{a}$	
R	MPa$\sqrt{\mathrm{m}}$	kg/mm$^{3/2}$	MPa$\sqrt{\mathrm{m}}$	kg/mm$^{3/2}$	MPa$\sqrt{\mathrm{m}}$	kg/mm$^{3/2}$
0.1	10.4	33.5	9.4	30.2	5.1	16.3
0.2	11.2	36	8.9	28.8	5.2	16.7
0.3	12.1	39	8.5	27.3	5.2	16.9
0.4	13	42	7.8	25.2	5.1	16.6
0.5	14.3	46	7.1	23	5.0	16.1

^{a} ΔK_{eff} in 2024-T3 aluminum alloy was calculated from the relation[36] $\Delta K_{\mathrm{eff}} = \Delta K_{\mathrm{app}}\,(0.5 + 0.4R)$

Though ΔK_{eff}, as defined from Fig. 10.17, contributes to our understanding of the fatigue process (e.g., see Fig. 10.18 and Table 10.1), the model falls short in more completely clarifying the roles of the many complex aspects of the closure phenomenon.[41,42] These issues include differences in plasticity between near surface (plane stress) and mid-thickness (plane strain) regimes, closure resulting from crack wake interference due to crack face asperity contact and oxide layer interference, and residual stresses both behind and in advance of the advancing crack front (e.g., recall Fig. 7.4). In this regard, recent investigators[43–48] have concluded that crack tip damage occurs at load levels below K_{op}. That is, they concluded that K_{op} represents a starting point where closure begins to occur but is not complete until a somewhat lower load is

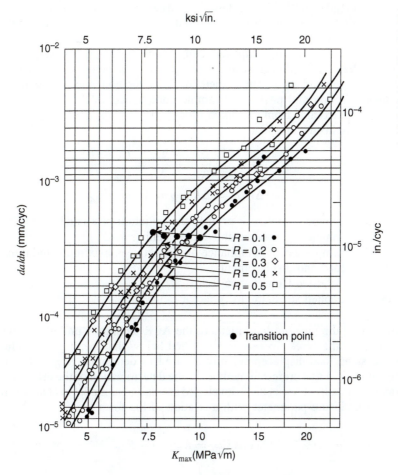

Figure 10.19 Crack propagation behavior in 2024-T3 Al clad sheet. Note fracture mode transition (●) at constant crack growth rate.[40] (Reprinted with permission of the American Society for Testing and Materials from copyrighted work.)

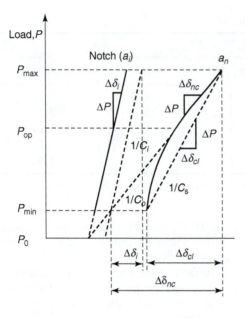

Figure 10.20   Determination of adjusted compliance ratio (ACR) to identify ΔK_{eff}, see Eq. 10.11a.[46] (J. K. Donald, private communication, with permission of J. K. Donald, Fracture Technology Associates.)

experienced. Hence, a more realistic measure of ΔK_{eff} would be somewhat larger than the value of ΔK_{eff}, based on Elber's formulation (i.e., $\Delta K_{eff} = K_{max} - K_{op}$).

Rather than characterizing an effective stress intensity factor range in terms of load levels, such as K_{op}, Donald et al.[43-45] have described an experimental method for estimating ΔK_{eff} in terms of the ratio between the actual crack tip displacement range, $\Delta\delta_{cl}$, and the displacement range that would have occurred had there been no closure, $\Delta\delta_{nc}$ (see Fig. 10.20). This parameter, referred to as the adjusted compliance ratio (ACR), was adopted based on observations that the nonlinear strain range provides a better estimate of cyclic damage than the effective load ratio range method. This method appears to be independent of measurement location and provides a remote compliance value due solely to the presence of the crack as it extends from the notch root.

From Fig. 10.20 we find that

$$ACR = \frac{\Delta\delta_{cl} - \Delta\delta_i}{\Delta\delta_{nc} - \Delta\delta_i} = \frac{C_s - C_i}{C_o - C_i}$$

$$\Delta K_{eff} = \Delta K_{ACR} = ACR \cdot \Delta K_{app}$$

(10-11)

where C_i = sample compliance prior to the initiation of a crack
C_o = sample compliance above the opening load
C_s = sample secant compliance drawn between maximum and minimum values of load and displacement
ACR = Adjusted Compliance Ratio
ΔK_{app} = applied stress intensity factor range

As distinct from the opening load method, the ACR technique provides a method by which strain activity at the crack tip can be quantified below the opening load. This methodology utilizes the same load-displacement plot as used to determine K_{op} and is presently being evaluated by ASTM as a preferred method for the determination of the effective stress intensity factor range.[47] Finally, Paris et al.[48] developed an analytical model to characterize ΔK_{eff}, using a partial crack closure parameter with a relation of the form

$$\Delta K_{eff} = K_{max} - \frac{2}{\pi} K_{op}$$

(10-12)

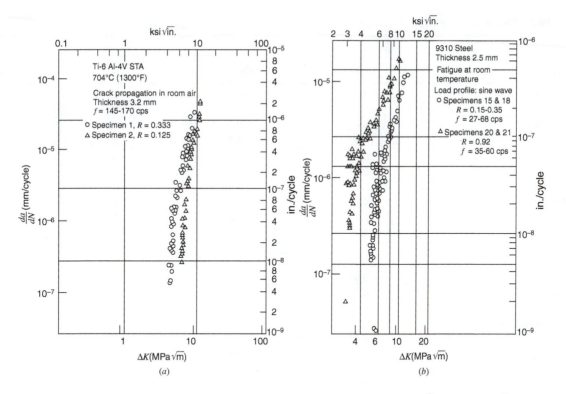

Figure 10.21   Threshold condition for crack growth as a function of stress ratio R in (a) Ti–6A1–4V[49]; (b) 9310 steel.[50] (Reprinted with permission of the American Society for Testing and Materials from copyrighted work.)

Early results, using this relation, have shown encouraging correlation with the ACR method, though uncertainties persist with this equation's use of the inexact measurement of K_{op}.

10.4.2 Low ΔK Levels

At the other end of the crack growth rate spectrum, the simple power relation (Eq. 10-3) is violated again for low ΔK conditions, where the FCP rate diminishes rapidly to a vanishingly small level (Fig. 10.15). From such data[49,50] as shown in Fig. 10.21, a limiting stress intensity factor range (the threshold level ΔK_{th}) is defined and represents a service operating limit below which fatigue damage is highly unlikely. In this sense, ΔK_{th} is much like K_{IEAC}, the threshold level for environment-assisted cracking (see Chapter 8). Designing a component such that $\Delta K \leq \Delta K_{th}$ would be a highly desirable objective, but it is sometimes not very realistic in the sense that ΔK_{th} for engineering materials often represents only 5 to 10% of anticipated fracture toughness values (Table 10.2). Therefore, to operate under $\Delta K \leq \Delta K_{th}$ conditions would require that virtually all defects be eliminated from a component and/or the design stress be extremely low. This is desirable in the design of nuclear power generation equipment where safety is of prime concern; however, designing an aircraft such that $\Delta K \leq \Delta K_{th}$ is highly impractical. Theoretically, you could design an airplane that would not fatigue, but the beefed-up structure necessary to reduce the stress intensity level to below the ΔK_{th} level would weigh so much that the plane would not be able to take off! Since many engineering structures do fulfill their intended service life without incident, it is apparent that some components do operate under $\Delta K \leq \Delta K_{th}$ conditions.

As seen in Table 10.2 and Fig. 10.21, the effect of K_{mean} (i.e., R ratio) on crack propagation becomes important once again at very low ΔK levels and has been the focus of considerable attention.[51–56] In this crack growth regime, different crack closure mechanisms than residual

Table 10.2 Threshold Data in Engineering Alloys

Material	R	ΔK_{th}		Ref.
		MPa$\sqrt{m}$	ksi$\sqrt{in.}$	
9310 Steel	0.25	~6.1	~5.5	a
	0.9	~3.3	~3	a
A533B Steel	0.1	8	7.3	b
	0.3	5.7	5.2	b
	0.5	4.8	4.4	b
	0.7	3.1	2.8	b
	0.8	3	2.75	b
A508	0.1	6.7	6.1	b
	0.5	5.6	5.1	b
	0.7	3.1	2.8	b
T-1	0.2	~5.5	~5	c
	0.4	~4.4	~4	c
	0.9	~3.3	~3	c
Ti–6A1–V	0.15	~6.6	~6	d
	0.33	~4.4	~4	d
18/8 Austenitic steel	0	6.1	5.5	e
	0.33	5.9	5.4	e
	0.62	4.6	4.2	e
	0.74	4.1	3.7	e
Copper	0	2.5	2.3	e
	0.33	1.8	1.6	e
	0.56	1.5	1.4	e
	0.80	1.3	1.2	e
60/40 Brass	0	3.5	3.2	e
	0.33	3.1	2.8	e
	0.51	2.6	2.4	e
	0.72	2.6	2.4	e
Nickel	0	7.9	7.2	e
	0.33	6.5	5.9	e
	0.57	5.2	4.7	e
	0.71	3.6	3.3	e
300-M Steel	0.05	8.5	7.6	f
(650°C temper-oil quench)	0.70	3.7	3.3	f
300-M Steel	0.05	6.2	5.6	f
(650°C temper-step cooled)	0.70	2.7	2.4	f
2024-T3 Aluminum	0.80	1.7	1.5	g
2219-T851 Aluminum	0.1	3.0	2.7	h
	0.5	1.7	1.5	h
A356 Cast aluminum	0.1	6.1	5.5	i
	0.8	2.4	2.1	i
AF42 Cast aluminum	0.5	3.4	3.1	j
	0.8	1.7	1.5	j

[a] P. C. Paris, *MTS Closed Loop Magazine*, **2**(5), 1970.
[b] P. C. Paris, et al., ASTM *STP 513*, 1972, p. 141.
[c] R. J. Bucci et al., op. cit., p. 177.
[d] R. J. Bucci et al., op. cit., p. 125.
[e] L. D. Pook, op. cit, p. 106.
[f] M. F. Carlson and R. O. Ritchie, *Metal Sci.* **11**, 368 (1977).
[g] R. A. Schmidt and P. C. Paris, ASTM *STP 536*, 1973, p. 79.
[h] R. J. Bucci, Alcoa Report No. 57-79-14 (1979).
[i] A. Saxena et al., *J. Test. Eval.* **6**, 167 (1978).
[j] R. J. Stofanak et al., *Eng. Fract. Mech.*, **17**(6), 527 (1983).

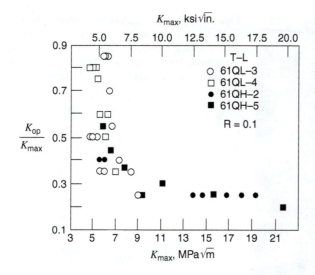

Figure 10.22   Increase in crack closure level in threshold regime for extruded aluminum alloy with crack propagation in the *T-L* orientation.[60] (Reprinted with permission, R. J. Stofanak, R. W. Hertzberg, G. Miller, R. Jaccard, and K. Donald, *Engineering Fracture Mechanics*, **17**(6) 527 (1983), Pergamon Journals Ltd.)

plasticity have been identified that strongly influence K_{op} and ACR levels, and associated ΔK_{eff} values. For example, crack-tip zone shielding (recall Fig. 7.4) occurs when an irregular crack path is generated, with the coarse facets on the mating fracture surfaces coming in contact during fatigue cycling.[57–59] With increasing surface roughness, K_{op} levels increase and ACR levels decrease, whereas ΔK_{eff} and the corresponding crack growth rates decrease (Fig. 10.22). It follows that the sensitivity of ΔK_{th} to R ratio for a given material depends on the observed level of crack closure. At one extreme where measured closure levels are minimal, no appreciable change in ΔK_{eff} would occur with increasing R ratio. Indeed, Minakawa et al.[61] reported no R ratio sensitivity on ΔK_{th} in closure-free IN9021-T4 P/M aluminum alloy (Fig. 10.23). Conversely, a significant decrease in ΔK_{th} with increasing R ratio was noted in the the 7090-T6 P/M aluminum alloy that exhibited pronounced roughness-induced closure in the threshold regime; this strong R ratio dependence of ΔK_{th} results from a sharp increase in ΔK_{eff} and associated FCP rates with increasing R ratio as K_{min} rises above K_{op}. Threshold conditions are then met only after the applied ΔK level is reduced.

An alternative crack closure mechanism has been proposed by Suresh et al.[62,63] to account for differences in ΔK_{th} for $2\frac{1}{4}$Cr–1Mo steel when tested in different gaseous atmospheres. The threshold fatigue value in this material decreased when the test atmosphere was changed from air to hydrogen. While it is tempting to rationalize this difference in terms of a hydrogen embrittlement-type argument, these authors pointed out that dry argon also accelerated near-threshold fatigue crack growth rates relative to air; in fact, dry argon behaved like dry hydrogen.

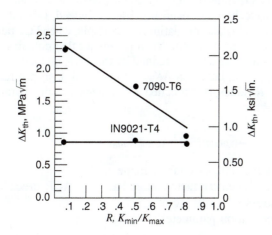

Figure 10.23   Influence of R ratio on ΔK_{th} in P/M IN9021-T4 and 7090-T6 aluminum alloys.[61] (Reprinted with permission from K. Minakawa, G. Levan, and A. J. McEvily, *Metallurgical Transactions*, 17A, 1787 (1986).)

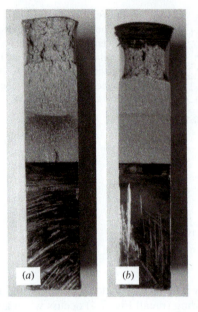

Figure 10.24   Dark bands of oxide debris formed on fracture surfaces of $2\frac{1}{2}$ Cr–1Mo steel in region associated with ΔK_{th}. (a) $\Delta K_{th} = 7.7 \, MPa\sqrt{m}$, $R = 0.05$; (b) $\Delta K = 3.1 \, MPa\sqrt{m}$, $R = 0.75$. Note that the oxide band almost disappears at high R ratios. (From Suresh et al.[62]; reprinted with permission from the Metallurgical Society of AIME.)

To explain the enhanced fatigue resistance of this material when tested in air as compared to dry argon and hydrogen, Suresh et al.[62,63] noted that fatigue testing in air creates an oxide layer on the fracture surface that thickens in the threshold regime as a result of closure-induced fretting (Fig. 10.24). Similar oxide layers in the threshold regime have been reported by others.[52,55,56] The thicker oxide layer, in turn, would be expected to increase K_{op} (and decrease ACR) and bring about a corresponding decrease in ΔK_{eff}, thereby leading to lower crack growth rates at a given applied ΔK. Tu and Seth[64] have also reported higher ΔK_{th} values in a seemingly more aggressive atmosphere (steam) than in air (Fig. 10.25). In this connection, they found more corrosion products on the steam-atmosphere fracture surfaces in the threshold regime (consistent with the Suresh et al. fretting oxide-induced closure model) than on other parts of the fracture surface. In addition, they observed more crack branching in the specimens tested in the steam atmosphere, which could have further reduced the effective crack-tip stress intensity factor.

It follows from Fig. 10.16 that the influence of such variables as R ratio and environment could be taken into account when FCP rates are compared on the basis of ΔK_{eff}. To accurately determine ΔK_{eff}, however, requires that precise measurements of closure be made throughout the test. Unfortunately, closure is often difficult to measure in a consistent manner and is subject to spurious interpretation, especially in the threshold regime.[65–69]

However, as discussed in the previous section, ΔK_{eff} may be more accurately defined by use of the adjusted compliance ratio technique (recall Eq. 10-11). Indeed, Donald et al.[43,70] have demonstrated marked improvement in data correlation, for example, when FCP rates in 2324-T39 aluminum alloy are described in terms of an empirical parameter that contains both K_{max} and ΔK_{eff} values (see Eq. 10-13), where the latter is defined by the ACR method (Fig. 10.26). Note how the considerable data dispersion, based on ΔK_{app} (Fig. 10.26a), shown for the 2324-T39 aluminum alloy over a range of R values, is reconciled with the normalizing parameter, ΔK_{norm}, given in Eq. 10-13 (Fig. 10.26b).

$$\Delta K_{norm} = \Delta K_{ACR}^{1-n} \cdot K_{max}^{n} \qquad (10\text{-}13)$$

where ΔK_{norm} = normalized stress intensity factor range
   ΔK_{ACR} = effective stress intensity factor range based on ACR procedure
   K_{max} = maximum stress intensity factor
   n = empirical correlation parameter (≈ 0.25)

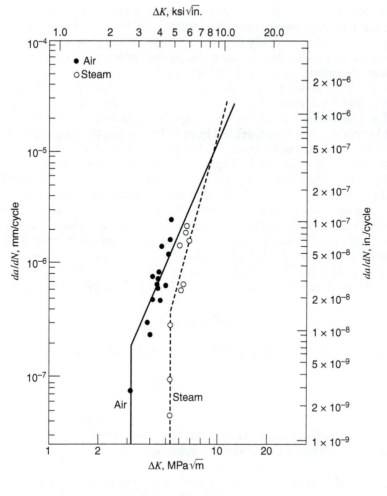

Figure 10.25 Near-threshold fatigue crack growth behavior in Ni–Cr–Mo–V A471 rotor steel-tested at $R = 0.35$, 100 Hz in air, and steam at 100 °C. Note higher ΔK_{th} when steel was tested in a steam environment. (After Tu and Seth[64]; © American Society for Testing and Materials, 1916 Race St., Philadelphia, PA 19103. Reprinted with permission.)

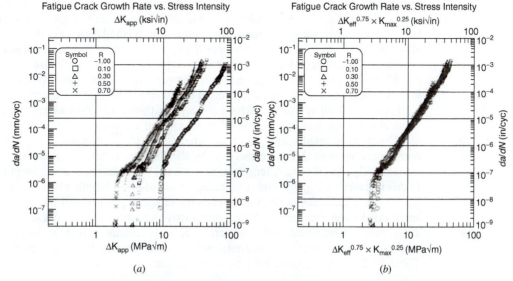

Figure 10.26 FCP rates in 2324-T39 aluminum alloy as a function of stress ratio, R.[70] (*a*) Correlated with ΔK_{app}; (*b*) correlated with $\Delta K_{eff}^{0.75} \times K_{max}^{0.25}$ (J. K. Donald, *Ten Questions About ACR*, Fracture Technology Associates, Sept. 7, 2009, with permisson of J. K. Donald, Fracture Technology Associates.)

Two final points regarding Eq. 10.13 are worth mentioning. First, as noted earlier in the chapter, the relatively small value of n reflects the fact that FCP rates are more strongly dependent on the stress intensity factor *range* than on K_{max}; second, where appropriate, the K_{max} term can be expanded to include the magnitude of any positive or negative residual stress component. Regarding the latter, FCP rates are found to increase for a given applied stress intensity factor range when there are residual tensile stresses present in the component; the opposite is true when residual compressive stresses are present.

Before concluding this section, it is appropriate to consider the similarity between the threshold stress intensity range, defined from FCP data, and the fatigue limit, determined from stress–cyclic life plots. Furthermore, it is intriguing to consider a possible correlation between these two fatigue parameters. To begin this analysis, consider the propagation of a small through-thickness crack in a large panel, for which $\Delta K = \Delta\sigma\sqrt{\pi a}$. At the threshold level where $\Delta K = \Delta K_{th}$, the corresponding stress range $\Delta\sigma_{th}$ is expected to vary with $1/\sqrt{a}$. A problem arises immediately when one attempts to define $\Delta\sigma_{th}$ where a approaches zero; the computed $\Delta\sigma_{th}$ is surely much larger than the experimentally determined value—the fatigue limit $\Delta\sigma_{fat}$—corresponding to test results from an unnotched sample (recall Section 9.2). In fact, several investigators have observed that when a is very small, $\Delta\sigma_{th}$ approaches an asymptotic limit corresponding to $\Delta\sigma_{fat}$.[71–76] To characterize this behavior, Haddad and his co-workers[73] proposed the existence of an "intrinsic crack length" for a given material, a_0, such that

$$\Delta K = \Delta\sigma\sqrt{\pi(a + a_0)} \tag{10-14}$$

We see from the above discussion that, when a approaches zero,

$$a_0 = \frac{\Delta K_{th}^2}{\pi\Delta\sigma_{fat}^2} \tag{10-15}$$

where $\Delta K_{th} =$ threshold ΔK from long-cracked panel experiments ($a \gg a_0$)

$\Delta\sigma_{fat} =$ endurance limit from unnotched samples

Haddad et al.[73] gave no physical significance to the intrinsic crack length a_0.[ii] However, Tanaka and co-workers[75,76] developed a similar model along theoretical grounds and concluded that a_0 was related to the combined influence of the materials' grain size and crack closure behavior. Of particular significance, they gathered published data for both ferrous and nonferrous alloys to show a general relation between the threshold stress $\Delta\sigma_{th}$, normalized by the endurance limit $\Delta\sigma_{fat}$, and the crack length a, normalized by the intrinsic crack size a_0 (Fig. 10.27a). Note that when long cracks are present in the test sample, $\Delta\sigma_{th}$ varies with $1/\sqrt{a}$. This is to be expected when crack growth behavior is controlled by linear elastic fracture mechanics (LEFM) considerations. At the other extreme where $a < a_0$, the threshold stress asymptotically approaches a value equal to the endurance limit of an unnotched sample.

Using their results and those of others,[71,73,77–81] Tanaka et al.[76] showed a similar correlation between $\Delta K'_{th}$ normalized by the threshold value corresponding to a long-crack sample $\Delta K'_{th}$, and the crack length, a, normalized by the intrinsic crack length a_0 (Fig. 10.27b). In this instance, $\Delta K'_{th}$ is found to be independent of crack length when $a > a_0$ (LEFM-control). On the other hand, $\Delta K'_{th}$ decreases when the crack length is small relative to a_0. That is, very small cracks can grow at ΔK levels previously thought safe (i.e., $\Delta K < \Delta K_{th}$). It follows that ΔK_{th} values associated with long-crack test specimens may lead to nonconservative life estimates of a component that contains very small cracks. For this reason, the data given in Table 10.2 should be used with extreme caution.

[ii] It is intriguing to note the analogous form of Eq. 6-43, in which the plastic zone size $r_y \propto K^2/\left(\pi\sigma_{ys}^2\right)$, as compared with Eq. 10-15 given above.

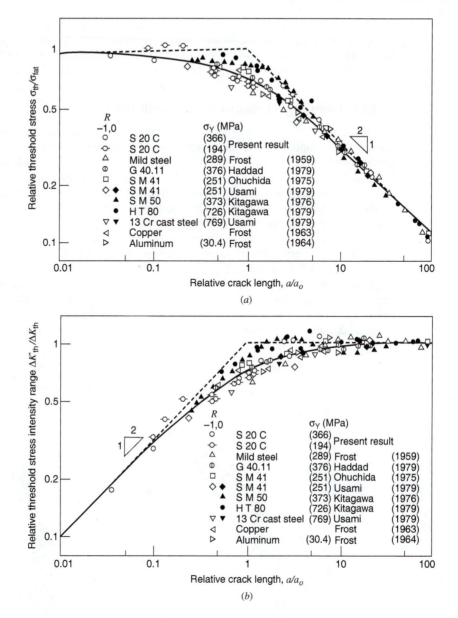

Figure 10.27　Normalized threshold behavior versus relative crack size in various materials. (*a*) Relative threshold stress; (*b*) relative stress intensity range. (After Tanaka et al.[76]; with permission from *Int. J. Fract.* **17**, 519 (1981).)

Considering the short-crack problem from a different perspective, we see that the concept of crack similitude is clearly violated. Similitude implies that different-sized cracks will possess the same plastic zone size, stress and strain distributions, and crack growth rates, if the stress intensity factor is the same.[82–84] The breakdown of similitude can be traced to several factors: (1) when continuum requirements are violated, that is, crack lengths are small compared with the scale of the microstructure; (2) when linear elastic fracture mechanics concepts are violated, that is, the length of the crack is small compared with the dimension of the crack-tip plastic zone; (3) when different crack propagation mechanisms are encountered; and (4) when different closure levels are found at the same applied ΔK level ($K_{max} - K_{min}$) for long and short cracks, respectively.

Regarding the latter point, crack-tip shielding is largely absent in small cracks since the crack wake has not yet developed. In the absence of associated closure, ΔK_{eff} in short cracks for a given ΔK_{app} is often significantly higher than that noted in long-crack samples and surely contributes markedly to the much higher growth rates observed in short-crack samples.

10.4.2.1 Estimation of Short-Crack Growth Behavior

Since real structures may initially contain short cracks without a wake zone, it could be argued that component lifetime is dominated by the behavior of such defects, rather than by predictions based on long-crack FCP information. In fact, the large closure levels encountered in the ΔK_{th} regime for long-crack samples may well be a consequence of the constant R ratio ($R^c = 0.1$) ΔK-decreasing test procedure (Standard E647) and not a fundamental characteristic of the material in question.[85,86] That is, long-crack data may provide an overly optimistic assessment of a material's FCP resistance.

Initial efforts to obtain more conservative FCP data for design life calculations have focused on the generation of large quantities of short-crack data.[87,88,iii] Unfortunately, the characterization of short-crack growth behavior is time-consuming, tedious, and subject to large amounts of experimental scatter.[87–89] For example, errors of up to 50% have been reported[90] in repetitive readings of crack length during the same short-crack fatigue experiment. Furthermore, several-fold differences in FCP rates have been observed at the same nominal ΔK level (solid line, ◇) (Fig. 10.28a). Also note that short-crack growth rates typically

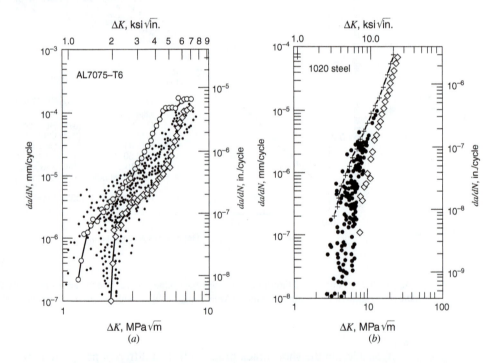

Figure 10.28 (a) Comparison of $R^c = 0.1$ K-decreasing data (solid line, ◇) with short-crack results (individual datum) for 7075-T6 aluminum alloy, along with $K^c_{\max} = 10$ MPa$\sqrt{\text{m}}$ test results (○). Note $K^c_{\max}$ curve is conservative relative to 85–90% of short crack data; (b) FCP data revealing the good agreement between $K^c_{\max}$ (+) and short-crack results for 1020 steel. R^c results (◇) are nonconservative. See Herman, Hertzberg, and Jaccard[96] for sources of short-crack data. (Reprinted with permission from W. A. Herman, R. W. Hertzberg, and R. Jaccard, *Fat. & Fract. Engng. Mater. & Struct.*, **11** (4), 303 (1988).)

[iii] A large collection of short-crack data is also included in the conference proceedings *Fatigue 87*, Vols. 1–3, R. O. Ritchie and E. A. Starke, Jr., Eds., EMAS Ltd., West Midlands, England, 1987.

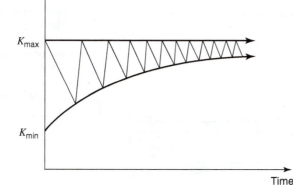

Figure 10.29 Schema showing constant K^c_{max} threshold test procedure.

exceed those associated with long-crack samples and that crack growth occurs at ΔK levels below ΔK_{th}. Attempts to correlate these short-crack results with corresponding long-crack data on the basis of ΔK_{eff} have met with partial success;[89,91,92] the major problem encountered is that such correlations are based on two experimental data bases—short-crack growth rates and crack closure measurements—that each possess large amounts of scatter.

Clearly, a different long-crack laboratory test method based on standard specimen geometries is needed to better simulate actual service loading conditions. Such a test method has been confirmed and is based on maintaining a constant maximum stress intensity (K^c_{max}) level during the ΔK-decreasing test procedure. By maintaining a constant K^c_{max} value (Fig. 10.29), mean stress and associated R ratios are found to increase markedly as ΔK decreases.[93–95] The development of such high mean stress and R ratio levels produces a long crack with no associated crack closure, which closely describes the behavior of short cracks.[96–98] The K^c_{max} data shown in Figs. 10.28a, b reveal compelling results for aluminum and iron alloys, respectively, that clearly demonstrate the utility of the K^c_{max} test procedure as a method by which a conservative estimate of short-crack growth rates may be obtained.[iv] In sharp contrast, the $R^c = 0.10$ curve anticipated almost none of the accelerated growth characteristics of short cracks.

Finally, the crack growth behavior of physically short cracks is complicated further when the crack is embedded within the stress field of a notch. Due to rapidly decreasing stresses, crack growth rates may *decrease* initially with *increasing* crack length. Depending on the local stress conditions, the crack could then either grow at an accelerating rate with increasing crack length or arrest[101–103] (Fig. 10.30). The existence of "nonpropagating cracks" has been confirmed by Frost,[104] as shown in

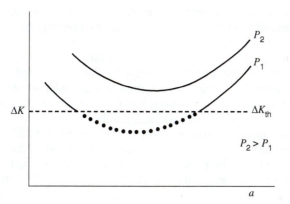

Figure 10.30 Effective ΔK level as a function of crack length at two load levels. At P_1, ΔK will decrease below ΔK_{th} as the crack moves through the notch root zone. Crack arrest will then occur.

[iv] It has also been shown that by precracking long-crack samples of 7475-T6 aluminum alloy in compression, a closure-free condition is established that generates FCP rates similar to those found under K^c_{max} test conditions. (H. Nowack and R. Marissen, *Fatigue 87*, Vol. 1, R. O. Ritchie and E. A. Starke, Jr., Eds., EMAS Ltd., West Midlands, England, 1987, p. 207.)

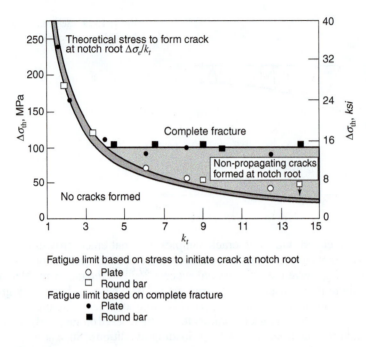

Figure 10.31 Effect of k_t on fatigue strength for crack initiation and complete fracture of mild steel. (After Frost[104]; reprinted by permission of the Council of the Institution of Mechanical Engineers from the *Journal of Mechanical Engineering Science.*)

Fatigue limit based on stress to initiate crack at notch root
    ○ Plate
    □ Round bar
Fatigue limit based on complete fracture
    ● Plate
    ■ Round bar

Fig. 10.31. These data reveal the change in long life fatigue strength of steel samples containing notches of the same length but different notch root radii. Note how the fatigue strength for crack initiation decreases continuously with increasing k_t and becomes *independent* of the stress concentration factor beyond some critical value of k_t. That is, beyond some critical k_t value a fatigue crack could initiate but then would not propagate to failure. The presence of these nonpropagating cracks is consistent with the fact that fatigue crack growth essentially ceases when the stress intensity factor decreases below the threshold level ΔK_{th}.

As a final note, the sharp decrease in the fatigue sensitivity factor q with increasing notch acuity (recall Fig. 9.16) can be interpreted as reflecting the development of nonpropagating cracks. That is, a sharp notch root radius would cause a fatigue crack to nucleate and grow a short distance from the notch. If this crack were to stop when $\Delta K < \Delta K_{th}$, then the influence of the notch on fatigue life would be diminished.

10.5 INFLUENCE OF LOAD INTERACTIONS

Much of the FCP data discussed thus far were gathered from specimens subjected to simple loading patterns without regard to load fluctuations. Although this may provide a reasonable simulation condition for components experiencing nonvarying load excursions, constant amplitude testing does not simulate variable load-interaction effects, which, in some cases, can either extend or shorten fatigue life measurably. In the most simple case, involving superposition of single-peak tensile overloads on a regular sinusoidal wave form, laboratory tests[21,86,105–114] have demonstrated significant FCP delay after each overload, with the amount of delay increasing with both magnitude and number of overload cycles.

The retarding effect of a peak overload is demonstrated clearly in Fig. 10.32 for a constant ΔK loading situation, where the crack growth rate associated with the invariant ΔK level is given by the constant slope b_2. Obviously, the FCP rate is depressed after the overload for a distance a^* from the point of the overload. Hertzberg and co-workers[107–110] have shown that this distance corresponds to the plastic zone dimension of the overload. Therefore, once the crack grows through the overload plastic zone, resumption of normal crack propagation is expected. Recent studies have shown that the extent of delay N_d depends on the effective overload ratio as defined by the ratio of the overload ΔK level and the *effective* ΔK base level corresponding to the prevailing closure value (Fig. 10.33). It follows that increased amounts of cyclic delay occur when

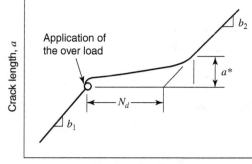

Figure 10.32 Crack growth rate plot illustrating effect of single-peak tensile overload. Note cyclic delay N_d and overload affected crack increment a^*.

a given overload ΔK_{OL} is applied in conjunction with conditions associated with large amounts of closure (i.e., low ΔK_{eff} levels). For example, increased amounts of delay were observed in 2024-T3 aluminum alloy with increasing base ΔK level and decreasing sheet thickness (Fig. 10.34a); both factors contribute to enhance plane-stress test conditions and greater levels of plasticity-induced crack closure. It follows, therefore, that the data in Fig. 10.34a can be normalized in terms of the overload plastic zone size/sheet thickness ratio (Fig. 10.34b). Note that the amount of delay increases dramatically as this ratio approaches unity.

Additional experimental findings lend further credence to the crack closure model. It has been shown that when the fracture surfaces in the wake of the fatigue crack are removed with a narrow grinding wheel, the crack growth rate upon subsequent load cycling is higher than before the machining operation.[108,110] Obviously, this effect may be rationalized in terms of the elimination of fracture surface material that was causing interference. It would be expected that if crack surface interference really does occur then some evidence of abrasion should be found on the fracture surface. As shown in Fig. 10.35 for both single-peak overload and high–low block loading sequences, extensive abrasion and obliteration of fracture surface detail is readily apparent. (Note the large striation or stretch band associated with the single-peak overload in Fig. 10.35a.)

The importance of large overload cycles in affecting fatigue life of engineering components is illustrated by the finding of Schijve et al.[115] They found that, under aircraft flight simulation conditions involving a random load spectrum, when the highest wind-related gust loads from the laboratory loading spectrum were eliminated, the specimens showed lower test sample fatigue life than did specimens that experienced some of the more severe load excursions. This fascinating load interaction phenomenon has led some investigators to conclude that an aircraft that logged some bad weather flight time could be expected to possess a longer service life than a plane having a less turbulent flight history.

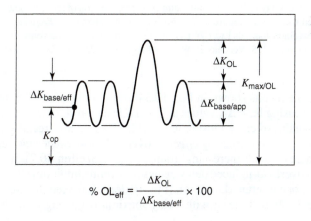

Figure 10.33 Definition of terms associated with single-peak tensile overload.

$$\% \, OL_{eff} = \frac{\Delta K_{OL}}{\Delta K_{base/eff}} \times 100$$

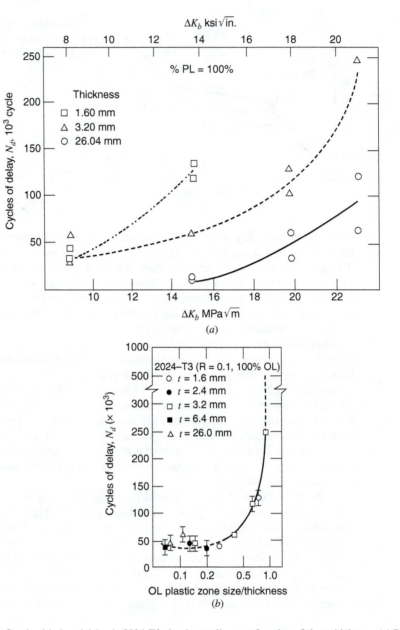

Figure 10.34 Overload-induced delay in 2024-T3 aluminum alloy as a function of sheet thickness. (*a*) Dependence on base *K* level.[109] (Reprinted with permission from W. J. Mills and R. W. Hertzberg, *Engineering Fracture Mechanics*, **7**, 705 (1975), Pergamon Journals Ltd.) (*b*) Dependence on overload plastic zone/sheet thickness ratio.[112] (Reprinted with permission, R. S. Vecchio, R. W. Hertzberg, and R. Jaccard, *Scripta Met.*, **17**, 343 (1983), Pergamon Journals Ltd.)

In sharp contrast to the results shown in Fig. 10.34, the amount of overload-induced cyclic delay *increases* with decreasing base ΔK values at low ΔK levels in conjunction with an increase in oxide- and roughness-induced crack closure levels (Fig. 10.36*a*; also recall Fig. 10.22); here, again, cyclic delay increases with increasing *effective* overload ratio. Since crack closure levels at low *K* values tend to increase with increasing grain size (see Section 10.7), it was possible to normalize the number of overload-induced delay cycles of several aluminum alloys in terms of the ratio of overload plastic zone to grain dimension (Fig. 10.36*b*). It is seen that trends in overload-induced delay follow a U-shaped curve with delay maxima occurring at both low and high

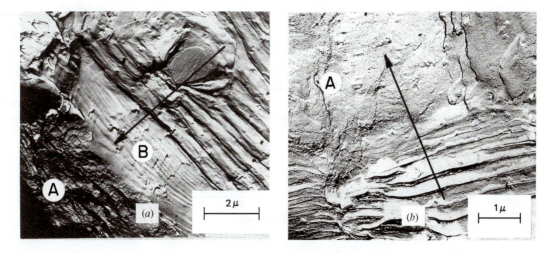

Figure 10.35 Abrasion in regions A resulting from overload cycling. (*a*) Single-peak overload. Note the stretch zone at B. (*b*) High–low block loading sequence. The arrow indicates crack direction.[107] (Reprinted with permission of the American Society for Testing and Materials from copyrighted work.)

ΔK levels in conjunction with enhanced closure levels. It should be noted, however, that the extensive amount of delay found at low ΔK levels may be illusory in that it is attributed to the excessive crack closure levels developed in the long-crack test samples used in the overload experiments. Since low ΔK levels in conjunction with short cracks exhibit less closure, the effective overload ratio and associated number of delay cycles in real structures are expected to be considerably smaller.

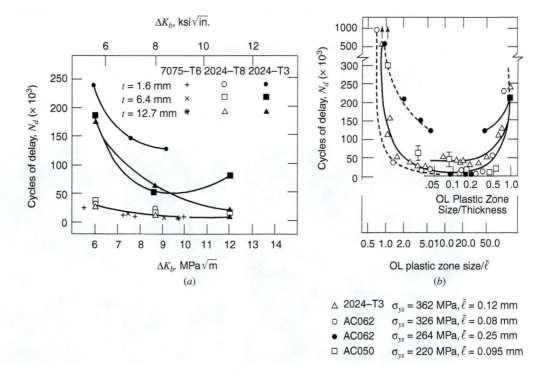

Figure 10.36 (*a*) Effect of sheet thickness and base line K on overload induced delay.[112] (*b*) U-shaped curve showing tensile overload-induced delay maxima at low and high K levels corresponding to small overload plastic zone size/grain size ratios and large overload plastic zone size/sheet thickness ratios, respectively.[69]

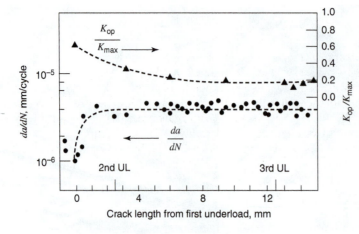

Figure 10.37 Influence of compressive underloads applied following a K-decreasing procedure in the threshold regime for an aluminum alloy. Note that FCP rates increase after the initial underload in conjunction with a decrease in crack closure level.[86]

When attempts are made to estimate the fatigue life of a component that is subjected to variable spectrum loading, it is necessary to consider the influence of *compressive* as well as tensile overloads. Whereas tensile overloads temporarily slow down the rate of crack growth or arrest altogether its advance, compressive overloads tend to accelerate crack growth[86,116−119] (Fig. 10.37). Furthermore, if a tensile overload is followed immediately by a compressive overload, the beneficial effect of the tensile overload may be significantly reduced.[117,120] To underscore the deleterious impact of compressive overloads on overall fatigue life, de Jonge and Nederveen[121] observed a 3.3-fold increase in fatigue life of 2024-T3 aluminum test samples when the ground–air–ground (GAG) cycles were removed from the loading spectrum. (The cycles correspond to the transition from compressive loading on the lower wing skin when the plane is on the ground to tensile loading when the aircraft is in flight.) Since aircraft do land from time to time, fatigue life predictions made without GAG cycles would be definitely unconservative!

10.5.1 Load Interaction Macroscopic Appearance

Similar to the macroscopic appearance of clamshell markings on many fatigue fracture surfaces (recall Fig. 9.3), macrobands are sometimes found that result from variable amplitude block loading[122] (Fig. 10.38a). The alternating dark and light bands reflect differences in the magnitude of the prevailing ΔK level associated with each loading block. Such marker bands

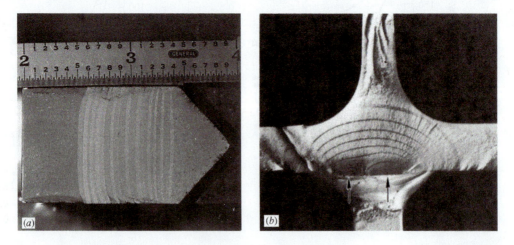

Figure 10.38 (a) Photograph showing macrobands on fatigue fracture surface in steel alloy resulting from variable amplitude block loading.[122] (Courtesy of H. I. McHenry.) (b) Photograph of marker band test on aluminum alloy. Arrows indicate crack origins. All block loads conducted with constant maximum stress. (Courtesy of R. Jaccard.)

can be used to characterize the size and shape of an advancing crack in a component where crack advance is predominantly internal (Fig. 10.38b). Although such marker bands can be formed by conventional constant low R ratio block loading test procedures, load interaction effects can interfere with the conduct of such a test; for example, crack growth delay can occur in cases where low level ΔK block loads follow high ΔK load excursions (recall the previous section). By contrast, a K_{max}-constant block loading procedure avoids load interaction effects and, as a result, provides a useful method for documenting the size and shape of a progressing internal flaw.[123] This information, along with crack growth rate data for the component's alloy, can be used to improve the accuracy of fatigue life predictions.

To illustrate, an Astroloy nickel-base alloy test bar was subjected to the constant K_{max} block load test profile shown in Fig. 10.39a.[123] The marker bands from each block load

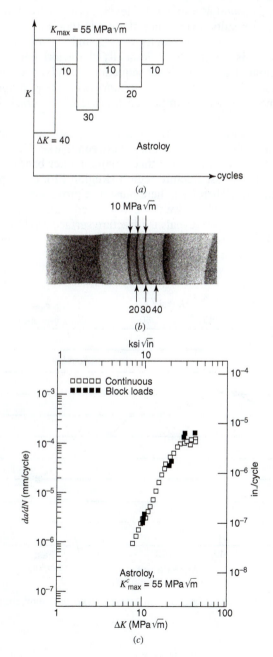

Figure 10.39 (*a*) Marker band test profile conducted on Astroloy. (*b*) Macroscopic appearance for $K^c_{max} = 55\ \text{MPa}\sqrt{\text{m}}$ load profile. Numbers correspond to ΔK level for each block, 2.7×. (*c*) FCP data for Astroloy nickel-base alloy. Open data points correspond to continuously decreasing ΔK test. Closed points represent block loading data. Note excellent agreement of results.[123] (Reprinted from C. Ragazzo, R. W. Hertzberg, and R. Jaccard, *J. Test Eval ASTM*, **23**(1), 19 (1995).)

segment were typically 0.25 mm to 0.75 mm wide in the low ΔK regions, and 1 mm or more in width in the higher ΔK regions. Each load block created a clearly defined marker band (crack growth from right to left), identified by its associated ΔK level (Fig. 10.39b). Additional tests[123] revealed that the development of contrast changes on fatigue fracture surfaces is dependent on the ΔK_{eff} level and is essentially independent of the K_{max} level. Therefore, changes in fracture surface brightness are the result of changes in the cyclic plastic zone size and not the monotonic plastic zone size. By contrast, crack growth delay is dependent on changes in the monotonic plastic zone size (recall Fig. 10.32). As such, when the monotonic plastic zone is held constant in a block loading test by maintaining K_{max} constant, no load interaction between load blocks would be expected. Indeed, the da/dN–ΔK data for a continuously decreasing ΔK test are shown in Fig. 10.39c along with crack growth information for the constant-ΔK block segments identified in Fig. 10.39a. Essentially no difference in crack growth rate is found at a given ΔK level, between continuously varying ΔK and block load sequence test results, even when the magnitude of the ΔK blocks was decreased by factors of two to four. (Similar results were found for the case of steel and aluminum alloys.)[123] Therefore, it is possible to periodically characterize the size and shape of growing internal cracks without affecting their subsequent crack advance behavior. Using this technique, it should be possible for an investigator to more accurately compute cyclic life intervals in structural components based on improved knowledge of crack front profiles during the period of stable crack extension.

We now conclude our consideration of the macroscopic appearance of fatigue fracture surfaces by determining the cause of contrast differences between marker bands generated by different ΔK levels. Recalling Fig. 10.39b, we see that fatigue marker bands become darker with decreasing ΔK level, though the fast fracture region (Region FF), corresponding to K_c conditions, is also dark in appearance. Therefore, there must be a progressive darkening in the contrast of the fracture surface at ΔK levels above those associated with the block loading test described in Fig. 10.39a. Indeed, Fig. 10.40 reveals the fracture surface brightness-ΔK relation in Astroloy for a $K_{max}^c = 85\,\mathrm{MPa}\sqrt{\mathrm{m}}$ test that shows the fracture surface to be dark at low and high ΔK levels, and relatively bright at intermediate ΔK levels.[122]

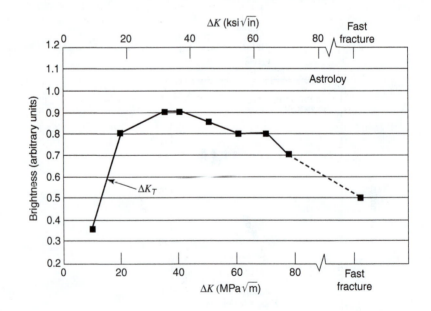

Figure 10.40 Effect of ΔK level on fracture surface brightness in Astroloy nickel-base alloy, corresponding to a test at $K_{max}^c = 85\,\mathrm{MPa}\sqrt{\mathrm{m}}$.[123]

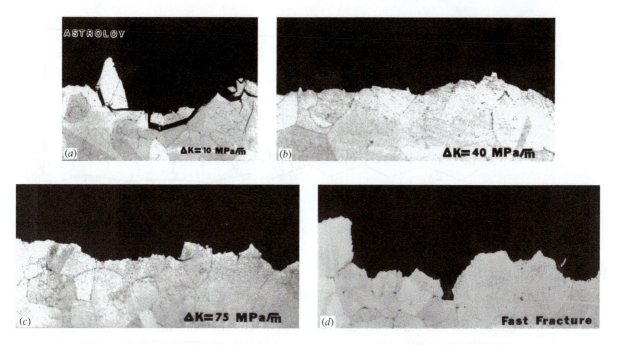

Figure 10.41   Crack profiles in Astroloy corresponding to ΔK levels of (a) 10, (b) 40, (c) 75 MPa$\sqrt{\text{m}}$, (d) fast fracture (400 ×).[123]

It is found that such contrast differences are related to ΔK-induced changes in fracture micromechanisms and their associated influence on fracture surface roughness. In the low ΔK regime, faceted growth dominates and a rough, dark surface is generated (Figs. 10.40 and 10.41a). Above the micromechanism transition point, a change occurs from the rough faceted mechanism to the relatively flat striation mode of fatigue crack advance (Fig. 10.41b). At progressively higher ΔK levels, the associated plastic zone size "sees" many second phase particles within the microstructure, which enables the crack to wander along a tortuous path (Path B) as compared with that corresponding to lower ΔK levels (Path A) where the plastic zone does not see the particles (Fig. 10.42). As such, the plastic zone acts as a "filter"; when the filter is large, many weak particles are encountered and the fracture surface is relatively rough (Fig. 10.41c, d) and dark in appearance (Fig. 10.40). It is the increased roughness at both low and high ΔK levels (associated with different micromechanisms) that causes the fracture surface to be relatively dark, whereas the flatter, striated region at intermediate ΔK levels generates a brighter fracture surface appearance (Fig. 10.43).

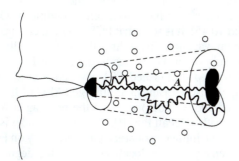

Figure 10.42   Dependence of crack-particle encounters on the size of the plastic zone. High ΔK conditions produce macroscopically rougher fracture path (B) than that generated at intermediate ΔK levels (A). (Rough appearance will again appear at low ΔK levels in association with faceted growth.)

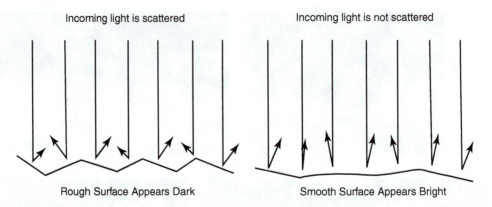

Figure 10.43 Schematic diagram illustrating the influence of fracture surface roughness on brightness level.[123]

10.6 ENVIRONMENTALLY ENHANCED FCP (CORROSION FATIGUE)

Recalling from Chapter 8 that cracks can grow in many materials as a result of sustained loading conditions in an aggressive environment, it is not surprising to find that fatigue crack propagation rates also are sensitive to environmental influences. The involvement of an aggressive environment in fatigue growth depends on a complex interaction between chemical, mechanical, and metallurgical factors. To this extent, corrosion fatigue studies benefit greatly from an interdisciplinary approach to test design and analysis of data.[124] Furthermore, the characteristics of a material–environment–stress system can be either simple or highly complex and may involve various combinations of failure mechanisms including stress corrosion cracking, hydrogen embrittlement, and corrosion fatigue.

Tests conducted on several aluminum alloys in various environments, such as wet and dry oxygen, wet and dry argon, and dry hydrogen, indicate that enhanced crack growth in aluminum alloys is due to the presence of moisture.[125–127,128–131] This is consistent with static test results reported in Section 8.1.1. This has led investigators to reexamine the fatigue behavior of many engineering alloys to determine the relative contribution of environmental effects. For example, earlier test results generated in uncontrolled laboratory environments (Fig. 10.3) indicated a marked superiority in fatigue performance of 2024-T3 versus 7075-T6 aluminum alloys. By conducting tests of these two alloys in both dry and wet argon atmospheres, Hartman[126] and Wei[127] determined that FCP differences in these alloys were minimized greatly by eliminating moisture from the test environment. Consequently, the superiority of 2024-T3 over 7075-T6 in uncontrolled test atmospheres is due mainly to a much greater environmental sensitivity in 7075-T6. This is consistent with the fact that 7075-T6 is more susceptible to environment-assisted cracking than is 2024-T3.

To further examine the influence of environment on fatigue crack growth behavior, Fig. 10.44 compares the FCP behavior in ultra high vacuum of two Al-Cu-X alloys (X = Li versus Mg) in both the naturally and peak aged conditions. Figure 10.45 shows the influence of high-pressure, high-purity water vapor on FCP behavior[128] for these four alloy/conditions. Overall (i.e., by comparing data in both Figs. 10.44 and 10.45), it is seen that FCP rates are consistently greater in the presence of moisture than in high vacuum. Regarding the latter condition, Fig. 10.44 demonstrates no influence of aging condition on FCP rates in the Al-Cu-Li alloy, whereas the peak aged Al-Cu-Mg alloy clearly revealed diminished FCP resistance at all ΔK levels. Not surprisingly, both naturally and peak aged Al-Cu-Li and naturally aged Al-Cu-Mg alloys exhibited similar fatigue crack paths (i.e., faceted and tortuous) whereas the peak aged Al-Cu-Mg alloy revealed a microscopically flatter crack path. (Recall Fig. 10.11; also, the reader is referred to the subsequent discussion in Section 10.7 of the influence of crack path tortuosity on FCP resistance.) Ro et al.[128] have argued that the propensity for slip localization and the degree of crack path tortuosity determine the degree of FCP resistance, rather than the aging condition, per se.

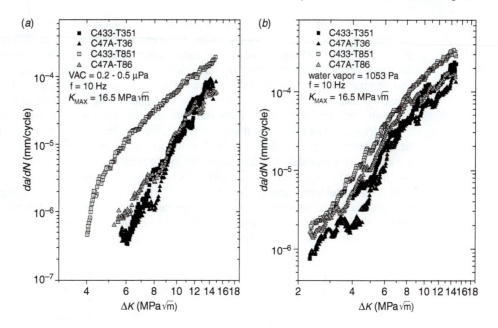

Figure 10.44   Fatigue crack propagation behavior for C47A (Al-Cu-Li) and C433 (Al-Cu-Mg) aluminum alloys in both underaged and peak aged treatments for decreasing ΔK test conditions. (*a*) Ultra high vacuum; (*b*) high-pressure, high-purity water vapor (1053 MPa).[128] (Reprinted from *Materials Science and Engineering*: A, 103, 468-470, Y. Ro, S. R. Agnew, and R. P. Gangloff, "Environment-exposure-dependent fatigue crack growth kinetics for Al–Cu–Mg/Li," p. 88, 2007, with permission from Elsevier.)

The strong influence of both moisture level and test frequency on fatigue response in these Al alloys is best revealed when FCP rates are compared as a function of the parameter P/f (units of Pa s) where P corresponds to the water partial pressure and f, the test frequency (Fig. 10.45). We see that crack growth rates vary by almost two orders of magnitude as a function of P/f and that there are four distinct growth rate regimes. At ultra low partial pressures and/or high test frequencies up to approximately 10^{-3} to 10^{-1} Pa s, cracking resistance in the Al-Cu-Li alloy for either aged condition is independent of moisture level. From 10^{-1} to approximately 2 Pa s, da/dN levels increase dramatically with increasing P/f; a more moderate increase in FCP levels with increasing P/f is revealed in regime three up to approximately 500 Pa s; a fourth regime

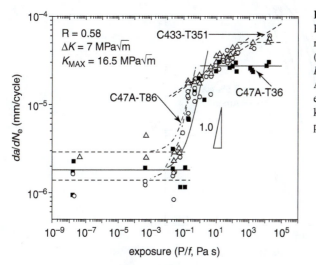

Figure 10.45   Dependence of FCP rate on P_{H_2O}/f for planar slip alloys. Growth rates reveal a four-stage cracking process.[128] (Reprinted from *Materials Science and Engineering*: A, 103, 468–470, Y. Ro, S. R. Agnew, and R. P. Gangloff, "Environment-exposure-dependent fatigue crack growth kinetics for Al–Cu–Mg/Li," p. 88, 2007, with permission from Elsevier.)

follows where FCP resistance again becomes independent of moisture level. These four different crack growth rate regimes have been explained in terms of different rate controlling processes.[128–134] These include: transport of environmental species between the bulk and the localized crack tip region; crack tip oxidation reaction leading to alumina on the crack surface and generation of atomic hydrogen; and diffusion of atomic hydrogen to damage sites in the crack-tip zone (recall Fig. 8.2). How these rate-controlling processes interact with crack path tortuosity and crack closure effects remains an issue for further clarification.

During the past few years, many more material–environment systems have been identified as being susceptible to corrosion fatigue. It is found that many aluminum, titanium, and steel alloys are adversely affected during fatigue testing by the presence of water, and titanium and steel alloys (but not aluminum alloys) are affected by dry hydrogen. In these studies, test frequency, load ratio, load profile, and temperature have been identified as major variables affecting FCP response of a material subjected to an aggressive environment. For example, the harmful effects of a 3.5% saline solution on fatigue performance in a titanium alloy are shown in Fig. 10.46 as a function of test frequency. As might be expected, FCP rates generally increase when more time is allowed for

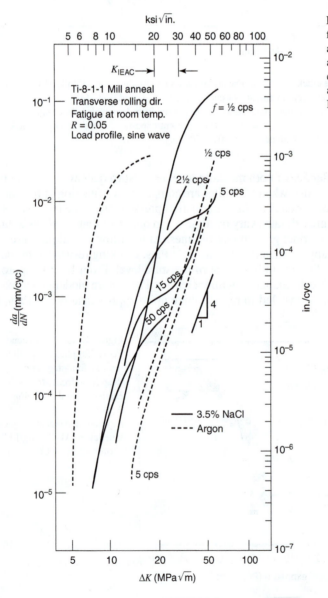

Figure 10.46   Effect of frequency on fatigue crack growth in Ti–8Al–1Mo–1V alloy in 3.5% NaCl and argon atmospheres.[135] The dashed curve at left corresponds to testing at 5 Hz in 3.5% NaCl and with $R = 0.75$. (Courtesy of R. J. Bucci, Alcoa Research Laboratories.)

environmental attack during the fatigue process (i.e., at lower frequencies), although this can depend upon the relationship between the rate of formation and stability of the passive film at the crack tip and the imposed loading rate (as shown by the lowest growth rates produced at the lowest f in the low ΔK regime of Figure 10.46). It should be pointed out that no important frequency effects are found in metals when tested in an inert atmosphere. Also note the negligible environmental effect on FCP at high crack growth rates where the mechanical process of fatigue damage is most likely taking place too quickly for electrochemical effects to be important.

Fatigue crack growth rate sensitivity to environment and test frequency has also been found in ferrous alloys (Fig. 10.47).[136–139] What is most intriguing about these data is the fact that they reveal a significant environmental sensitivity, even though all tests were conducted with

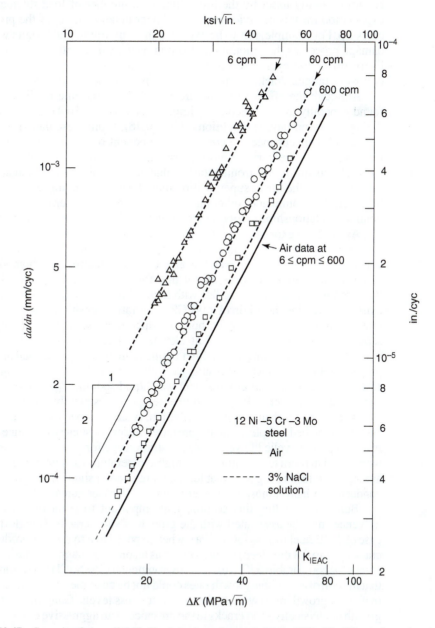

Figure 10.47   Corrosion fatigue (3% NaCl solution) crack growth data in 12Ni–5Cr–3Mo steel as a function of test frequency. All tests conducted with $\Delta K < K_{\text{IEAC}}$[137] (Reprinted with permission of the American Society for Testing and Materials from copyrighted work.)

K_{max} maintained *below* the K_{IEAC}[v] level for the material. Such behavior has been shown to be nearly universal for most alloy systems.

Why should there be any environmental effect during fatigue if the tests were conducted below K_{IEAC}? Perhaps a protective film, developed at the crack tip under sustained loading conditions, acts to protect the material from the environment but is ruptured by fatigue cycling, thereby permitting the corrodent to repeatedly attack the crack-tip region. Such a hypothesis is supported by observations made by Bucci,[135] who showed that environment-assisted cracking does occur below K_{IEAC} when the arrested crack is subjected to a period of load cycling and then reloaded below the previously established K_{IEAC} level. This would suggest that constant loading conditions used to obtain a K_{IEAC} value for a given material represent a *metastable* condition, easily upset by the imposition of a number of load fluctuations. A more accurate explanation for this behavior requires a better understanding of the processes occurring at the crack tip. For example, it is to be expected that environmental fatigue will progress by different damage processes than monotonic loading. Hence, dislocation structures at the crack tip most likely differ for cyclic versus monotonic loading.[140] As such, identification of operative corrosion fatigue cracking micromechanisms must await additional insight.

Although the effect of mean stress on FCP in the intermediate growth rate regime was found generally to be of secondary importance (Section 10.4), it does become a major variable during corrosion fatigue conditions (Fig. 10.46). From these data, it would appear that high R ratio conditions enhance the corrosion component of crack growth, while low R ratio testing reflects more of the intrinsic fatigue response of the material. The greater importance of mean stress effects during environmentally enhanced fatigue crack propagation may be rationalized with the aid of the linear superposition model described in the next section. It should be noted, however, that a number of other material–environment systems may require more complex and nonlinear relationships involving the influence of ΔK, R, and the material-environment couple.

As might be expected, the other major test variable relating to corrosion fatigue is that of test temperature. Many investigators have found FCP rates to increase with increasing temperature, which would seem logical given this effect of temperature on both anodic and cathodic reaction kinetics. (The latter, however, is not true for the case of H_2 gas environmental fatigue since hydrogen uptake and trapping fall with increasing temperature.) For many years, a controversy existed concerning the origin of this FCP temperature sensitivity. Is it due to a creep component or to an environmental component, both of which increase with increasing test temperature? In a series of experiments conducted at elevated temperatures in inert environments and in vacuum,[141–145] it was shown that neither temperature nor frequency had any effect on fatigue crack propagation rates. In fact, test results were comparable to room temperature results. This was the case even when the inert environment was liquid sodium.[144] On the basis of these results, it is concluded that higher FCP rates at higher temperatures mainly result from material–environment interactions, rather than a creep contribution. A similar conclusion was reached by Coffin for the case of unnotched samples, which were tested under constant strain range conditions. Figure 10.48 reveals a plot of cyclic life versus plastic strain range for many ferrous and nonferrous alloys that were tested in room temperature air or high-temperature vacuum or argon. The general agreement among these results suggests that temperature is not a significant variable when fatigue tests are conducted in inert atmospheres over a range of temperatures.

Before concluding this section, it is important to point out that unusual environmental influences may be associated with the growth of short cracks. Gangloff[146] reported that for the case of 4140 steel ($\sigma_{ys} = 1300$ MPa), when exposed to a hydrogen-producing environment, short cracks (0.1 to 0.8 mm deep) grew as much as an order of magnitude faster than 20- to 50-mm-long cracks for comparable ΔK levels and environmental (3% NaCl) conditions. Furthermore, he noted that the corrosion fatigue growth rates could not be described by a single-valued function of ΔK in that crack growth rates were *higher* at lower stress levels. Gangloff postulated that fatigue crack growth of physically short cracks in the presence of an aggressive environment could be described

---

[v] Recall from Chapter 8 that the threshold level for stress corrosion cracking, K_{ISCC}, has been redefined by the more general term for environment-assisted cracking, K_{IEAC}.

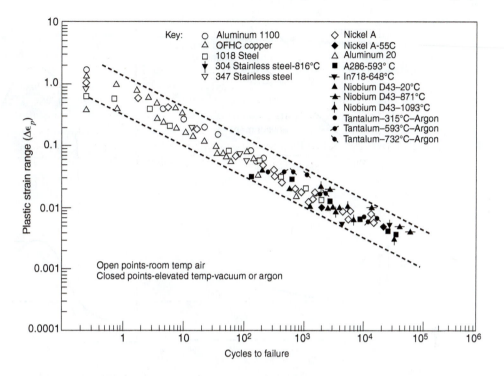

Figure 10.48   Plot of plastic strain range versus cyclic life in several alloys tested in room temperature air or in high-temperature vacuum or argon. (From Coffin[141]; reprinted with permission from the Metallurgical Society of AIME.)

in terms of a purely electrochemical mechanism based on processes responsible for the control of the hydrogen concentration levels at the crack tip. Surely, additional studies are needed to explore the generality of this shortcrack, cyclic load–environment interrelated phenomenon. For example, see additional studies by Gangloff and others.[147,148]

10.6.1 Corrosion Fatigue Superposition Model

Wei and Landes[149] and Bucci[135] developed a model to account for effects of environment, test frequency, waveform, and load ratio on corrosion fatigue crack propagation behavior. They approximated the total crack extension rate under corrosion fatigue conditions by a simple superposition of the intrinsic fatigue crack growth rate (determined in an inert atmosphere) and the crack extension rate due to a sustained load applied in an aggressive environment (determined as the environment-assisted crack growth rate). Therefore

$$\left(\frac{da}{dN}\right)_T = \left(\frac{da}{dN}\right)_{\text{fat}} + \int \frac{da}{dt}K(t)dt \tag{10-16}$$

where $\left(\frac{da}{dN}\right)_T$ = total corrosion fatigue crack growth rate

$\left(\frac{da}{dN}\right)_{\text{fat}}$ = fatigue crack growth rate defined in an inert atmosphere

$\frac{da}{dt}$ = crack growth rate under sustained loading

$K(t)$ = time-dependent change in stress intensity factor

Two important aspects of this model should be emphasized. First, its linear character implies that there is no interaction (or synergism) between the purely mechanical and environmental components. Second, the model also depends on the assumption that the

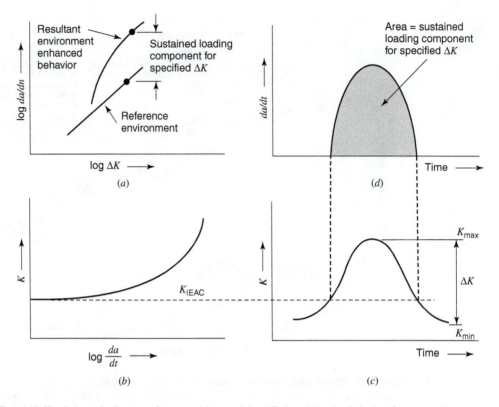

Figure 10.49   Schematic diagram of superposition model. (*a*) Fatigue behavior in both reference and aggressive environments; (*b*) sustained loading environment cracking behavior; (*c*) *K* versus time for one cycle of loading; (*d*) sustained loading crack velocity versus time. Note that the model cannot predict environment-enhanced fatigue crack propagation below K_{IEAC}.[135] (Courtesy of R. J. Bucci, Alcoa Research Laboratories.)

same mechanisms control the fracture process in both environment-assisted cracking and corrosion fatigue. For example, in the case of certain steels, both processes are considered to be controlled by hydrogen embrittlement. The application of the super-position model is demonstrated with the aid of Fig. 10.49. The FCP of the material in an inert environment $(da/dN)_{fat}$ is determined first and plotted in Fig. 10.49*a*. Next, the sustained loading crack growth rate component developed during one load cycle is obtained by integrating the product of the environment-assisted cracking rate *da/dt* and the time-dependent change in stress intensity level $K(t)$ over the time period for one load cycle (Fig. 10.49*b–d*). This increment then is added to $(da/dN)_{fat}$ to obtain $(da/dN)_T$, the corrosion fatigue crack growth rate.

Studies by Wei and co-workers[150,151] have attempted to account for environmentally enhanced fatigue crack growth behavior where $K_{max} < K_{IEAC}$ (see Fig. 10.47). To this end, a third component has been added to Eq. 10-16, which represents a cyclic-dependent contribution that involves a synergistic interaction between fatigue and environmental damage. A number of these more complex models have been reviewed by Gangloff.[152]

10.7 MICROSTRUCTURAL ASPECTS OF FCP IN METAL ALLOYS

A review of the literature reveals that there is a major influence of metallurgical variables on fatigue crack propagation at both low and high ΔK levels. Conversely, many studies conducted in the intermediate growth rate regime reveal that metallurgical variables such as yield strength, thermomechanical treatment, and preferred orientation do not have a pronounced effect on FCP rates in aluminum and steel alloys; that is, fatigue crack propagation at intermediate ΔK levels is relatively structure insensitive. In almost every case, the transition from structure-sensitive to

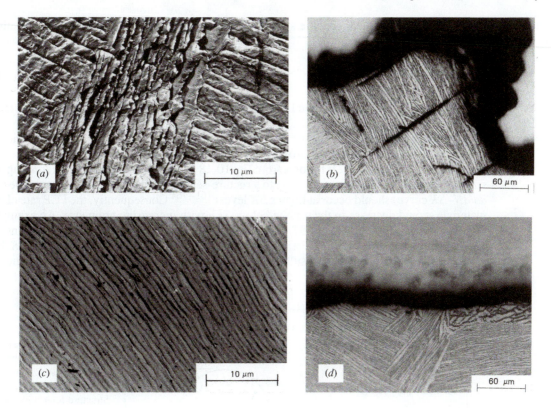

Figure 10.50   Fatigue fracture surface micromorphology in Ti-based alloy. (*a*) Faceted fracture surface; (*b*) multiple crack paths at $\Delta K < \Delta K_T$; (*c*) fatigue striations; (*d*) unperturbed crack profile at $\Delta K > \Delta K_T$. [(*a*) and (*c*) from Yoder et al.[161]; reprinted with permission from the Metallurgical Society of AIME. (*b*) and (*d*) from Yoder et al.[162]; reprinted with permission from *Eng. Fract. Mech.* **11**, 805 (1979), Pergamon Press, Ltd.]

structure-insensitive crack growth behavior is associated with a concomitant transition in fracture mechanisms (recall Fig. 10.11). These fractographic observations strongly suggest the role of microstructural influences in the determination of the operative micromechanisms of fracture. To confirm this hypothesis, let us consider some earlier results pertaining to the fatigue crack propagation response in numerous titanium alloys. Several investigators[153−158] reported that below some critical ΔK level, a highly faceted fracture surface was developed in conjunction with a highly branched crack front (Fig. 10.50*a, b*). Above this ΔK range, the fracture surface was much smoother overall and covered with fatigue striations. At the same time, the crack front was no longer bifurcated (Fig. 10.50*c, d*). In each instance, the fracture mechanism transition correlated with the development of a reversed plastic zone size (recall Section 10.2) equal to the grain size of the controlling phase in the alloy microstructure. For example, Yoder et al.[159−164] found that the transition from structure-sensitive to structure-insensitive FCP behavior occurred when the height of the cyclic plastic zone above the Mode I crack plane was equal to the average dimension of the Widmanstätten packet $\bar{l}_{wp}$. That is,

$$\bar{l}_{wp} = r_y^c = 0.033 \left(\frac{\Delta K_T}{\sigma_{ys}} \right)^2 \tag{10-17}$$

where $\bar{l}_{wp}$ = average Widmanstätten packet size
   r_y^c = cyclic plastic zone height above the Mode I plane
   ΔK_T = ΔK value at the fracture mechanism transition
   σ_{ys} = cyclic yield strength (recall Section 9.3.1), but often taken to be the monotonic σ_{ys}

When Eq. 10-17 is rearranged, the stress intensity factor range at the mechanism transition is given by[161]

$$\Delta K_T = 5.5\sigma_{ys}\sqrt{\bar{l}} \qquad (10\text{-}18)$$

where $\bar{l}$ is generalized to correspond to the controlling alloy phase associated with the fracture mechanism transition.

The transition from structure-sensitive to structure-insensitive behavior also influences the dependence of the macroscopic growth rate on the applied ΔK level as noted by the slope change drawn in Fig. 10.51. It follows from Eq. 10-18 and Fig. 10.51 that, with increasing grain size of the relevant phase, the transition to structure-sensitive behavior (i.e., steeper slope of the da/dN–ΔK curve) should occur at higher ΔK levels.[161–164] Consequently, the FCP rate of these materials should decrease in the regime below ΔK_T with *increasing* grain size. An example of this behavior is shown in Fig. 10.52. Note the increase in ΔK_T with increasing Widmanstätten packet size and the associated shift to lower fatigue crack growth rates at a given ΔK level. Yoder et al.[164] compared such results with those from other investigations and confirmed the increasing influence of grain size on fatigue crack growth rates with decreasing ΔK levels.

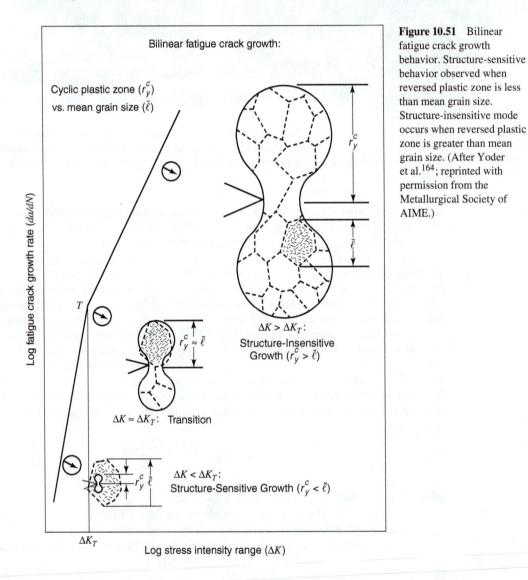

Figure 10.51 Bilinear fatigue crack growth behavior. Structure-sensitive behavior observed when reversed plastic zone is less than mean grain size. Structure-insensitive mode occurs when reversed plastic zone is greater than mean grain size. (After Yoder et al.[164]; reprinted with permission from the Metallurgical Society of AIME.)

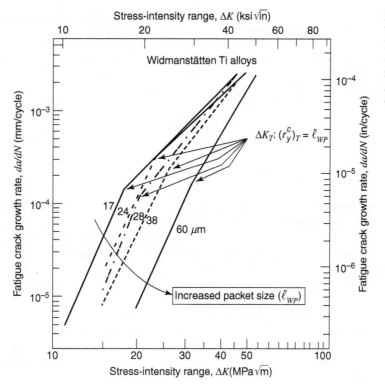

Figure 10.52 Influence of Widmanstätten packet size on FCP response in Ti alloy. (After Yoder et al.[162]; reprinted with permission from *Eng. Fract. Mech.* **11**, 805 (1979), Pergamon Press, Ltd.)

From the previous discussions, one would expect that metallurgical variables such as grain size also should influence the Region I fatigue threshold value. This follows from the fact that ΔK_{th} and ΔK_T do not differ greatly since the slope of the ΔK versus the da/dN curve in this regime is so steep.[165] Indeed, results reported for ferritic and pearlitic steels have shown that ΔK_{th} increases with the square root of grain size with a relation of the form

$$\Delta K_{\text{th}} = A + B\sqrt{d} \tag{10-19}$$

where A and B are material constants and d is the ferrite grain size,[74,75,166,167] Tanaka and co-workers[74,75] have proposed a model to explain this behavior in the materials they examined. They suggested that threshold conditions are established when the slip band at the crack tip is unable to traverse the nearby grain boundary. This should occur when the slip band size or the cyclic plastic zone dimension is approximately equal to the average grain diameter. Since the plastic zone size varies with the square of the stress intensity factor, it follows that the stress intensity level associated with the threshold condition should increase with increasing grain size.

It is important to recognize the similar dependence of ΔK_T and ΔK_{th} on structural size (Eqs. 10-18 and 10-19) and to recall that $\bar{l}$ corresponds to the controlling alloy phase or structural unit responsible for the fracture mechanism transition. Similarly, the term d in Eq. 10-19 should correspond to the dimension of the critical structural unit, which acts as the effective barrier to slip. Depending on the material, this barrier may correspond to the grain size, Widmanstätten packet dimension as in Ti alloys, subgrain size, or the dislocation cell dimension.

Since ΔK_{th} values in certain steel and titanium alloys increase with *increasing* grain size, one is faced with a design dilemma. Recall from earlier discussions in Chapters 3, 7, and 9 that the material's yield strength, fracture toughness, and smooth bar fatigue endurance limits, respectively, should increase with *decreasing* grain size. Consequently, some compromises become necessary when attempts are made to optimize simultaneously these four mechanical properties. This difficulty becomes readily apparent when one sets out to optimize the fatigue life of an

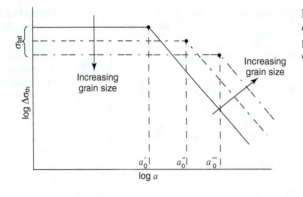

Figure 10.53 Influence of grain size on the determination of the characteristic crack length parameter a_0. Note that increasing grain size decreases σ_{fat} and increases σ_{th}.

engineering component. If one assumes that the component does not contain an initial flaw, then the fatigue life should depend strongly on the initiation stage of fatigue damage. As such, the fatigue life should increase with an increase in the endurance limit, σ_{fat}. Since σ_{fat} increases with increasing tensile strength (Fig. 9.15), then a reduction in grain size would be expected to enhance the material's fatigue resistance. On the other hand, if the component contains a preexistent flaw that can subsequently grow to failure, then the fatigue life can be improved by lowering crack growth rates at a given ΔK level (particularly at low ΔK levels) and/or by increasing ΔK_{th}. Consequently, if one changes the grain size to optimize σ_{fat}, the associated ΔK_{th} value will have been reduced, and vice versa. These conflicting trends can be shown in schematic form by superimposing the influence of grain size on σ_{fat} and $\Delta\sigma_{\text{th}}$, the latter being computed from the stress intensity factor formulation (Fig. 10.53). Note that the characteristic crack length parameter a_0, as described by Haddad et al.,[73] increases with increasing grain size.

Since fatigue crack initiation and propagation processes usually occur at component surfaces and in interior regions, respectively, the development of a duplex grain structure may lead to an optimization of fatigue performance. For example, if a component is thermomechanically treated so as to develop a fatigue-initiation resistant, fine-grained surface zone, coupled with a crack-propagation resistant, coarse-grained interior, then the component would be expected to exhibit superior overall fatigue resistance. Indeed, by shot peening and locally recrystallizing the surface zone of a Ti-8 Al alloy, a duplex grain structure was developed with fine grains located at the surface and coarse grains developed within the sample interior (Fig. 10.54a).[168] Preliminary S-N data attest to the superior fatigue response of this duplex-grain microstructure as compared with that of a completely coarse grained material (Fig. 10.54b).

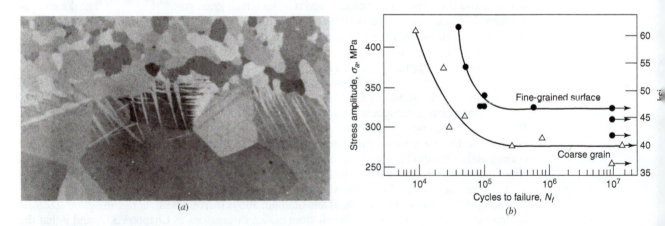

Figure 10.54 Shot peening-induced surface layer recrystallization in Ti-8A1. (a) Microstructure revealing fine grains in surface layer and coarse grains in substrate (250 x); (b) S-N plot revealing superior fatigue response of material containing duplex grain-size microstructure.[168] (Reprinted from L. Wagner and J. K. Gregory, *Advanced Materials and Processes*, **146** (1), 50 HH (1994), with permission from ASM International.)

The strong influence of grain size and slip character on ΔK_{th} may well reflect their impact on the development of roughness-induced crack closure; large grains and extended planar slip behavior would be expected to enhance crack surface interference and reduce ΔK_{eff}, thereby promoting higher ΔK_{th} levels as determined with conventional E647 test procedures. The emerging class of aluminum–lithium alloys are particularly noteworthy in this regard.[169,170] However, the beneficial influence of microstructure in the threshold regime is significantly reduced in the relatively closure-free environment associated with the growth of short cracks.[98,170] In fact, preliminary results suggest that the difference between short- and long-crack FCP rates is greatest in those alloys that exhibit the highest long-crack closure levels; hence, it is ironic that the most nonconservative estimates of fatigue life could be associated with the use of long-crack data from those alloys that "look the best," according to *low-R^c* threshold test procedures.

Recalling Fig. 10.11, the observed transition from faceted growth to striation formation in close-packed alloys could very well reflect the structure-sensitive to structure-insensitive transition discussed above. For the case of BCC steels, a similar set of fracture micromechanism transitions is observed. To wit, microvoids are found at the highest ΔK levels and gradually yield to fatigue striation formation with decreasing ΔK values. In most instances, the spacing between adjacent striations is found to vary with the second power of ΔK, consistent with the Bates and Clark relation (Eq. 10-9). At very low ΔK levels, the fracture surface micromorphology in certain steels has been shown to possess a highly faceted texture, similar to that found in FCC alloys (Fig. 10.55a). Overlapping the striation

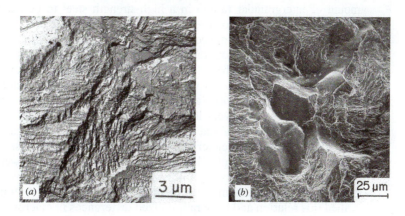

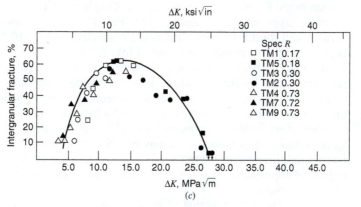

Figure 10.55   (*a*) Fractograph of AISI 9310 steel alloy (double-vacuum melted) revealing crisp faceted appearance in fatigue threshold region. $\Delta K \approx 8$ MPa $\sqrt{m}$. (*b*) A471 steel revealing localized evidence of intergranular failure as a result of fatigue loading. $\Delta K \approx 7$ MPa $\sqrt{m}$. Tested in water at 160 Hz. (*c*) Percentage of intergranular fracture in a medium carbon steel versus ΔK.[56] Tested in air. (Reprinted with permission from *Eng. Fract. Mech.* **7**, 69 (1975), Pergamon Press, Ltd.)

formation-faceted growth transition in BCC alloys is yet another fracture mechanism; in the ΔK range between 5 and 25 MPa $\sqrt{\mathrm{m}}$ many investigators have reported varying amounts of intergranular fracture (Fig. 10.55b).[55,56,171,180] The presence of intergranular fracture regions has been reported in low[173,179] and high-carbon steels,[55] high-strength steels,[171,175–178] and in microstructures composed of ferrite, bainite, martensite, and austenite. In some instances, investigators[55,56,153,172,173,179,181] have found that intergranular fracture takes place when the size of the reversed plastic zone is comparable to the relevant microstructural dimension, such as the prior austenite grain size. As such, there should be a maximum amount of intergranular fracture at some intermediate ΔK level with decreasing amounts of grain-boundary failure being observed at both higher and lower ΔK levels (Fig. 10.55c). Although this correlation is encouraging, other investigators[182] have questioned whether the maximum amount of intergranular failure should correspond instead to some critical K_{max} value. Furthermore, for the case of 4340 steel tested in air, Cheruvu[183] reported no correlation between the prior austenite grain size (20 to 200 μm) and the reverse plastic zone dimension at ΔK values corresponding to a maximum incidence of intergranular fracture. Surely more studies are needed to more fully identify the processes responsible for this fracture mechanism transition.

In a number of related studies, the amount of intergranular fracture observed tended to increase with the aggressiveness of the test environment.[174,177] Researchers also have found that the incidence of intergranular fracture can sometimes be traced to the presence of an embrittling grain-boundary film or solute segregation at certain grain boundaries.[174,175] Ritchie[171] found the amount of intergranular failure in a high-strength steel to be considerably greater when the material was heat treated to bring about a temper-embrittled condition. It appears, therefore, that intergranular facets are produced in various materials as a result of the *combined* influence of environmental and microstructural factors. In turn, the extent of the environmental sensitivity to intergranular fracture was found to be dependent on test frequency, with more intergranular facets being observed in 4340 steel (in 585 Pa water vapor) at 1 Hz than at 10 Hz.[184]

Metallurgical factors also affect fatigue crack propagation rates at high ΔK levels. This is because as ΔK becomes very large, K_{max} approaches K_{IC} or K_c where local fractures occur with increasing frequency and produce accelerated growth (recall Eq. 10-10). Since tougher materials are typically cleaner and will exhibit fewer local instabilities, their crack growth rates should be lower at high K levels. This is consistent with the well-established rule of thumb regarding low cycle fatigue (analogous to high FCP rates); low cycle fatigue resistance is enhanced by improvements in toughness and/or ductility. A number of investigators have verified this relation and have rationalized differences in macroscopic and microscopic crack growth rates. For example, in the high ΔK regime, FCP rates in a banded steel consisting of alternate layers of ferrite and pearlite (Fig. 10.56a) increased when tested in the arrester, divider, and short transverse directions, respectively (Fig. 10.56b).[185] Although little difference in fatigue response was found at low ΔK levels among the three orientations tested, a 40-fold difference in macroscopic crack growth rate was observed in going from the least to the most damaging loading direction. It is interesting to speculate whether the relative fatigue resistance associated with the three test orientations in the banded steel would be reversed at growth rates below 10^{-5} mm/cyc (the point where the FCP resistance was comparable for each test direction). Indeed, Mayes and Baker[186] reported that the highest ΔK_{th} value of a high-sulfur semifree machining steel was associated with the short transverse orientation. They interpreted these results in terms of increased fracture surface roughness associated with the rupture of nonmetallic inclusions; the enhanced roughness, in turn, resulted in a greater degree of crack closure.

It is important to note that FCP anisotropy was also found in homogenized samples of the banded steel when the layered microstructure was eliminated (but not the alignment of sulfide particles [see Fig. 7.22b]). Since the *microscopic* growth rate (i.e., fatigue striation spacings) was the same in the three crack plane orientations, it was concluded that the anisotropy in *macroscopic* FCP was related to different amounts of sulfide inclusion fracture in the three orientations. Consequently, macroscopic FCP is considered to be the summation of several

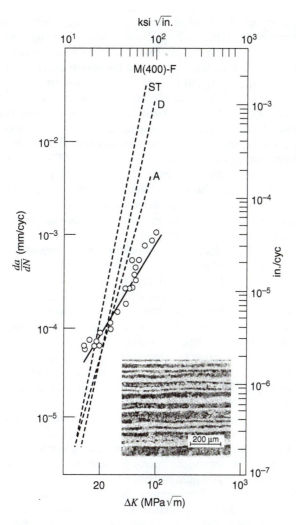

Figure 10.56 Fatigue crack propagation in banded steel (layers of ferrite and martensite) as a function of specimen orientation. (ST, short transverse; D, divider; A, arrester geometry.) Striation spacing (data points and solid line) is seen to be independent of specimen orientation. (*inset*) Microstructure in banded steel revealing alternating layers of ferrite and pearlite. (From F. Heiser and R. W. Hertzberg, *Transactions of the ASME*, 1971, with permission of the American Society of Mechanical Engineers.)

fracture mechanisms, the most important being striation formation and local fracture of brittle microconstituents. From this, macroscopic growth rates may be described by

$$\left(\frac{da}{dN}\right)_{\text{macro}} = \sum Af(K)_{\substack{\text{striation}\\\text{mechanism}}} + Bf'(K)_{\substack{\text{void}\\\text{coalescence}}} + Cf''(K)_{\text{cleavage}}$$
$$+ Df'''(K)_{\substack{\text{corrosion}\\\text{component}}} + \cdots \tag{10-20}$$

From Eq. 10-20, we know at least that $A \approx 6/E^2$ and $f(K) \approx \Delta K^2$.

In another attempt to correlate macroscopic and microscopic crack growth rates above the crossover point, Bates[187] adjusted Eq. 10-9 so that

$$\left(\frac{da}{dN}\right)_{\text{macro}} \approx \frac{6}{f_s}\left(\frac{\Delta K}{E}\right)^2 \tag{10-21}$$

where f_s is the percentage of striated area on the fracture surface. This relation is consistent with observations by Broek[188] and Pelloux et al.,[189] who also found increasing amounts of particle rupture and associated void coalescence on the fracture surface with increasing stress intensity levels. The latter investigators noted a marked increased in particle rupture when the plastic zone dimension grew to a size comparable to the particle spacing. Although Eq. 10-21 provides a rationale for differences in macroscopic and microscopic FCP rates, it may be too impractical to use because an extensive amount of fractographic information is required.

At intermediate ΔK levels, metallurgical factors do not appear to influence fatigue crack growth rates to a significant degree.[190] For example, only modest shifts in the slopes of log ΔK–log da/dN plots were found in studies of brass and stainless steel alloys in both cold-worked and annealed conditions where 4- to 10-fold differences in monotonic yield strength were reported.[28] Fairly strong crystallographic textures were developed in both the cold-worked and recrystallized conditions, but, again, little effect was noted on FCP response for both brass and steel specimens oriented so as to present maximum densities of {111}, {110}, and {100} crystallographic planes, respectively, on the anticipated crack plane. It was noted, however, that the actual fracture plane and crack direction were affected strongly by crystallographic texture in that the gross crack plane avoided a {111} orientation,[28,191] consistent with expectations of the striation formation model discussed in Section 10.3.

It has been suggested that FCP does not depend on typical tensile properties because monotonic properties are not the controlling parameter. Instead, cyclically stabilized properties may hold the key to fatigue crack propagation behavior. Starting or monotonic properties between two given alloys may differ widely, but their final or cyclically stabilized properties would not. For example, soft alloys would strain harden and hard ones would strain soften; as a result, the materials would be more similar in their final state than at the outset of testing. Consequently, if fatigue crack propagation response were dependent on cyclically stabilized properties, smaller differences in FCP behavior would be expected than that based on a comparison of monotonic values. A number of studies[192–196] have been conducted to establish correlations between cyclic strain and FCP data. For example, the slope m of the da/dN–ΔK plot is seen to decrease with increasing cyclic yield strength σ'_{ys} and cyclic strain-hardening exponent n'.[193] Furthermore, it has been shown that log A varies inversely with m (Eq. 10-3). Although it is encouraging to find such correlations, more work is needed before it will be possible to predict FCP rates from cyclic strain data. For one thing, it is not clear from Fig. 10.57 whether a high or low slope is desirable for optimum fatigue performance.[197] Obviously, the intercept A from Eq. 10-3 is equally important in this determination. For example, alloy A would be better than alloy C but alloy D, which has the same slope as A, would be worse than C. Furthermore, the choice between alloy B and C would depend on the anticipated crack growth rate regime for the engineering component. If many fatigue cycles were anticipated, the designer should opt for alloy B, since fatigue cracks would propagate more slowly over most of the component service life and allow for much greater fatigue life. In a low cycle fatigue situation, representative of conditions to the right of the crossover point, alloy C would be preferred.

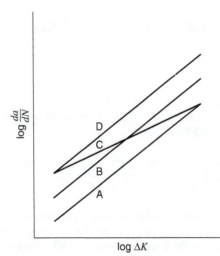

Figure 10.57 Diagram showing relative fatigue crack propagation behavior of several materials (A, B, C, D).[197] (Reprinted with permission of the American Society for Metals, copyright © 1974.)

In summary, fatigue crack propagation response in metal alloy systems is sensitive to structural variables at both low- and high-stress intensity value extremes. Both crack growth rates and fracture surface micromechanisms change with grain size, inclusion content, and mechanical properties such as yield strength. At low ΔK levels, structural sensitivity is observed when the crack-tip plastic zone is small as compared with the critical microstructural dimension. At high ΔK levels, fatigue crack growth rates are accelerated as K_{max} approaches K_c.

10.7.1 Normalization and Calculation of FCP Data

To this point, we have seen that microstructurally and/or crack length-induced differences in FCP data for a given alloy may be normalized by comparing crack growth rates as a function of ΔK_{eff}. It is interesting to note that differences in FCP rates between various metal alloy systems are dependent on the modulus of elasticity—a structure-insensitive property.[198] To illustrate, Ohta et al.[199] demonstrated excellent correlation of FCP data for aluminum, stainless steel, and low-carbon steel alloys when ΔK values were normalized by E (Fig. 10.58a); the minimal degree of scatter in Fig. 10.58b *is* attributed to their use of P_{max}^c-test conditions, which eliminated crack closure.

It is obvious, from earlier discussions, that the intrinsic fatigue crack propagation resistance of a metallic alloy may be characterized by its closure-free behavior with K_{max}^c data providing a convenient estimate of the latter. Regardless of the test method used (i.e., R^c or K_{max}^c), ASTM Standard 647 provides an operative definition of the ΔK threshold value (ΔK_{th}) at a crack growth rate of 10^{-10} m/cycle.[9] Based on the slip characteristics of a crystalline metal alloy, it may be reasonable to define a closely related ΔK value as that driving force corresponding to a

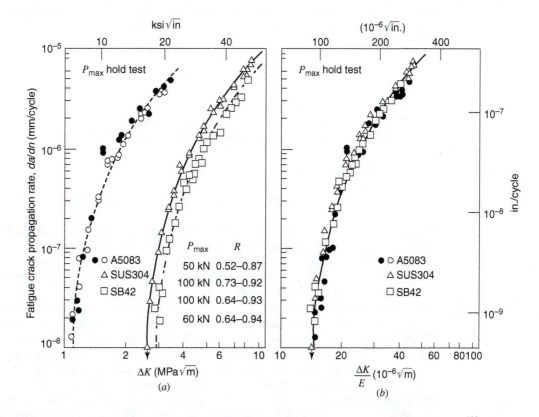

Figure 10.58 Fatigue crack propagation rates in 5083 aluminum, 304 stainless, and SB42 steel alloys[199], plotted as a function of: (a) ΔK; and (b) $\Delta K/E$. (Reprinted from *International Journal of Fatigue*, 14, A. Ohta, N. Suzuki, and T. Mawari, "Effect of Young's modulus on basic crack propagation properties near the fatigue threshold," p. 224, with permission from Elsevier.)

growth rate of a single Burgers vector. For purposes of identification, we may define this stress intensity factor driving force as ΔK_b. (For example, ΔK_b in steel and aluminum alloys would be identified at fatigue crack growth rates of 2.48 and 2.86×10^{-10} m, respectively.) One may then define ΔK_b as the limit of continuous damage accumulation with crack growth increments, $n\mathbf{b}$ ($n \geq 1$), occurring when $\Delta K \geq \Delta K_b$. Any growth increment less than the minimum unit of deformation (i.e., $\mathbf{b}$) would correspond to discontinuous crack extension.

If one examines the closure-free intrinsic fatigue crack propagation response for aluminum, nickel, titanium, and steel alloys, it is remarkable to note that $(\Delta K_b/E)^2$ values (units of m) correspond to the Burgers vector for each material[200,201] (Eq. 10-22).

$$\Delta K_b = E\sqrt{\mathbf{b}} \qquad (10\text{-}22)$$

where E = modulus of elasticity
$\quad\;\; \mathbf{b}$ = Burgers vector

Hence, a datum for the fatigue crack growth response of a crystalline metal in the near threshold regime would correspond to $da/dN = \mathbf{b}$/cyc and $\Delta K_b = E\sqrt{\mathbf{b}}$.[vi] This datum would identify the point above which crack growth would be continuous and occur in multiples of the Burgers vector and below which crack growth would be discontinuous in nature. Strong agreement was found between this computed datum (Point A) and K^c_{max}- or ΔK_{eff}-based experimental test results for several aluminum, steel, nickel, and titanium alloys[200,201] (e.g., see Fig. 10.59a, b).

Proceeding further, it was found for a wide range of metallic alloys that K^c_{max}- generated da/dN–ΔK plots tended to assume a log–log relation[200,201] with crack growth rates being approximately dependent on ΔK^3. As such, additional data points could be calculated directly at FCP rates above $\mathbf{b}$/cyc wherein

$$da/dN = \mathbf{b}(\Delta K/\Delta K_b)^3 \qquad (10\text{-}23)$$

where $\quad \mathbf{b}$ = Burgers vector
$\quad\quad \Delta K$ = closure-free stress intensity factor range
$\quad\quad \Delta K_b$ = closure-free ΔK level associated with $da/dN = \mathbf{b}$/cyc

By combining Eqs. 10-22 and 10-23, it follows that

$$da/dN = \mathbf{b}(\Delta K/E\sqrt{\mathbf{b}})^3 = (\Delta K/E)^3(1/\sqrt{\mathbf{b}}) \qquad (10\text{-}24)$$

The dashed lines shown in Figs. 10.59b–d represent data points calculated from Eq. 10-24 within an arbitrarily defined ΔK range from ΔK_b to $10\Delta K_b$ (Point B). The agreement between experimental and computed data points is most encouraging. Furthermore, one of the steel alloys shown in Figure 10.59 is replotted along with associated short-crack test data (Fig. 10.59c). One may readily conclude that both experimental and calculated closure-free curves provide an upper bound estimate of short crack behavior (recall Section 10.4.2.1).

Finally, an attempt was made to predict ΔK_b and associated ΔK data, corresponding to Eq. 10-24, for the case of a material (Cu-Be alloy 25) for which no closure-free data were known to the author. Using values of E and $\mathbf{b}$ of 125 GPa and 2.55×10^{-10} m, respectively, and assuming that closure-free crack growth rates are dependent on ΔK^3, Points A and B were computed (Fig. 10.59d). Two K^c_{max} fatigue tests were then performed at 35 and 45 MPa $\sqrt{m}$, respectively. Excellent agreement is seen between the experimental and computed data. (Note that the slopes of the $K^c_{max} = 35$ and 45 MPa $\sqrt{m}$ data plots were 2.97 and 3.03, respectively, in excellent

---

[vi] Note the agreement between the form of the $\Delta K_b = E\sqrt{\mathbf{b}}$ relation and the theoretical models by Sadananda and Shahinian,[202] and Yu and Yan[203] for ΔK_{th} where the latter were found to be proportional to $G\sqrt{\mathbf{b}}$ and $Ee_f\sqrt{\mathbf{b}}$, respectively.

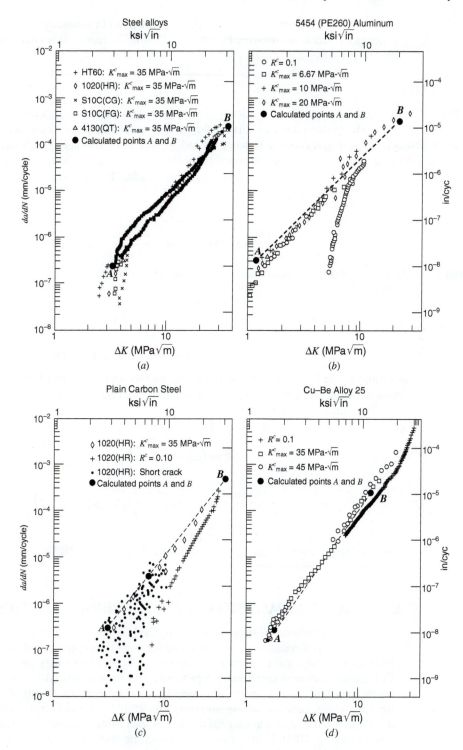

Figure 10.59 Experimental and calculated (line *A-B*) FCP data for several metallic alloys.[200,201] (*a*) Several steel alloys (R. W. Hertzberg, *Int. Journal of Fracture*, 64 (3), 135 (1993). Reprinted by permission of Kluwer Academic Publishers.); (*b*) 5454 aluminum; (*c*) long- and short-crack data in hot-rolled 1020 steel; (*d*) Cu-Be Alloy 25. (R. W. Hertzberg, *Matls. Sci. Eng.*, **A190**, 25 (1995). Reprinted by permission.)

agreement with the assumed $da/dN/\Delta K^3$ dependence.) In summary, this simple model provides a means by which closure-free FCP data in monolithic metal alloys may be predicted in the ΔK range near ΔK_{th} and above.

EXAMPLE 10.4

A 30-cm-diameter and 1.5-cm-thick, cylindrically shaped steel pressure vessel contains a small surface crack, oriented normal to the hoop stress direction. (a) Confirm that a leak-before-break condition exists, given that the vessel experiences a cyclic gas pressure between 40 and 50 MPa and the material's fracture toughness is 180 MPa $\sqrt{m}$. (b) At the point where the crack breached the wall thickness, estimate the rate of fatigue crack advance.

The maximum and minimum hoop stresses are given from Eq. 1-43 where

$$\sigma = \frac{P \cdot r}{t}$$

$$\sigma = \frac{50 \cdot 15}{1.5} = 500\,\text{MPa} \quad \text{and} \quad \frac{40 \cdot 15}{1.5} = 400\,\text{MPa}$$

From Eq. 6-32 the leak-before-break condition will occur when

$$K_c > K = \sigma\sqrt{\pi a} = \sigma\sqrt{\pi t}$$

In this instance,

$$K = 500 \cdot 10^6 \cdot \sqrt{\pi \cdot 0.015} = 108.5\,\text{MPa}\sqrt{m} < 180\,\text{MPa}\sqrt{m}$$

This vessel satisfies the leak-before-break condition. At this stage of damage development, the vessel contains a through-thickness crack with a total length of 3 mm. Since the cyclic hoop stress varies from 400 to 500 MPa (i.e., $R = 0.8$), little or no crack closure would be expected. Accordingly, crack growth rates can be estimated from Eq. 10-24. Since $E = 205$ GPa and $\mathbf{b} = 2.48 \times 10^{-10}$ m, the crack growth rate is given by

$$da/dN = (\Delta K/E)^3 (1/\sqrt{\mathbf{b}}) \tag{10-24}$$

$$da/dN = \left(\frac{(500-400)\sqrt{\pi 0.015}}{E} \cdot 10^6 \right)^3 \cdot \left(1/\sqrt{2.48 \times 10^{-10}} \right)$$
$$da/dN = 7.54 \times 10^{-5}\,\text{mm/cycle}$$

which is in very good agreement with the experimental data shown in Fig. 10.59a.

10.8 FATIGUE CRACK PROPAGATION IN ENGINEERING PLASTICS

A growing number of studies concerning the FCP behavior of engineering plastics have been conducted and information is now available for more than two dozen different materials.[204–211] With such a body of data, certain conclusions and generalities may be drawn. As in metals, the FCP rates of polymers are strongly dependent on the magnitude of the stress intensity factor range, regardless of polymer chemistry or molecular arrangement (Fig. 10.60). Note the data correlation for amorphous, semicrystalline, and rubber-modified polymers on the same plot of ΔK versus da/dN. In a sense, this is analogous to plots of data from metal alloys possessing various crystal structures (Fig. 10.3). Figure 10.60 also shows the relative ranking of the fatigue resistance of metals and plastics when compared as a function of ΔK and $\Delta K/E$. It is clear that plastics will exhibit superior or inferior FCP resistance as compared with metal alloys, depending on whether cycling is conducted under strain-controlled or stress-controlled conditions, respectively.

On the basis of these results, it is concluded that superior FCP resistance is exhibited by semi-crystalline polymers such as nylon 66, ST801 (rubber modified nylon 66), nylon 6, polyvinylidiene fluoride (PVDF), polyacetal (PA), and polyethylene terephthalate (PET). In all

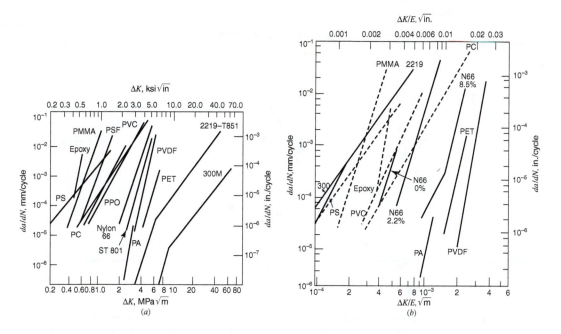

Figure 10.60   FCP behavior in various metal and polymeric materials, (a) Data plotted versus ΔK; (b) data plotted versus $\Delta K/E$. (Adapted from J. A. Manson et al., *Advances in Fracture Research*, D. Francois et al., Eds., Pergamon Press, Oxford, 1980.)

likelihood, the superior fatigue resistance of crystalline polymers relative to that associated with amorphous structures is not fortuitous. Koo et al.[212,213] and Meinel and Peterlin[214] have pointed out that crystalline polymers not only can dissipate energy when crystallites are deformed, but they can also apparently re-form a crystalline structure that is extremely strong. To further illustrate this point, the remarkably superior FCP resistance of amorphous PET was traced to strain-induced crystallization that took place within the plastic zone ahead of the crack tip.[215] Furthermore, the percent crystallinity increased with increasing ΔK level (i.e., increasing plastic zone size); for a given level of crystallinity, FCP resistance was greater in thermal treatments that lead to enhanced resistance to cyclically induced polymer chain disentanglement through the development of higher tie molecule densities.[207,208]

The adverse influence of cross-linking and the beneficial role of rubbery additions on FCP resistance have been examined and are described elsewhere.[205,206] For the present discussion, it is relevant to note that the fatigue crack propagation resistance of engineering plastics increases directly with the material's fracture toughness. Indeed, a striking correlation is evident in a large number of polymeric systems between values of ΔK^* (the value of ΔK required to drive a crack at a constant value of da/dN) and K_{cf} (the maximum value of K during the fatigue test preceding unstable crack extension) (Fig. 10.61). Since K_{cf} represents some measure of the material's fracture toughness, it is seen that the greater the toughness of a polymer, the greater the driving force required to advance the crack at a constant speed. A similar correlation is found for the case of rubber and hollow glass sphere-toughened hybrid composites.[210] Evidence for synergistic toughening at intermediate combinations of CTBN rubber particles and hollow glass spheres was shown in Fig. 7.50; such behavior was attributed to the multiplicative interaction between rubber particle cavitation/matrix shear banding and hollow glass sphere-induced microcracking. Figure 10.62a reveals that the fatigue crack propagation response of these blends possesses a similar synergistic response at the same rubber/glass ratio. Azimi[210] concluded that this behavior was attributable to the development of an enlarged plastic zone consisting of cavitated rubber particles with associated shear banding and distinct secondary deformation branches generated by the interaction between the elastic stress fields of the advancing crack front and the hollow glass spheres (Fig. 10.62b). Finally, it is worth noting that the fatigue crack propagation resistance of rubber-modified epoxy blends is

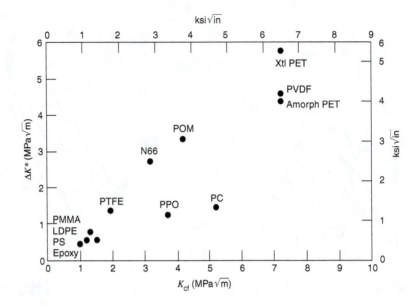

Figure 10.61   Relation between ΔK^* (the ΔK driving force to generate a crack velocity of 5×10^{-4} mm/cycle) and K_{cf} (the maximum value of K during the fatigue test preceding unstable crack extension) for selected amorphous and semi-crystalline polymers. (Reprinted from Pecorini).[207]

relatively insensitive to modifier content at low ΔK levels (Fig. 10.63).[210,211] Above the transition level (referred to as ΔK_T, where the plastic zone size is on the order of the rubber particle diameter), rubber-modified blends reveal superior FCP resistance.[vii] For example, Azimi determined that when the plastic zone dimension is smaller than the rubber particle size (i.e., where $\Delta K < \Delta K_T$), little influence of rubber particles on FCP response is expected. Conversely, when the size of the plastic zone is large enough to engulf many rubber particles, cavitation/shear banding mechanisms are activated, thereby leading to improved FCP resistance (Fig. 10.64). Note the lack of rubber cavitation and matrix dilation when $\Delta K = 0.5\,\text{MPa}\sqrt{\text{m}} < \Delta K_T$ and the correspondingly pronounced activity of these two deformation mechanisms when $\Delta K = 2.5\,\text{MPa}\sqrt{\text{m}} > \Delta K_T$.

Perhaps the greatest change in FCP resistance occurs when the molecular weight is modified. For example, Rimnac et al.[216] found a thousand-fold decrease in FCP rates when the molecular weight (MW) of polyvinyl chloride (PVC) was increased by little more than a factor of three (Fig. 10.65). Similarly, major improvements in FCP resistance with increasing MW have been found in polyacetal,[217] polycarbonate,[218] polyethylene,[219] nylon 66,[217] and poly (methyl methacrylate).[220] It is suggested that cyclic loading tends to disentangle whatever molecular network exists, and that this disentanglement is easier at lower MW. In addition, there may be positive contributions from enhanced orientation hardening with the higher MW species. Other investigations have shown a strong influence of molecular weight distribution (MWD) on FCP resistance.[221] Michel et al.[222–224] developed a theoretical model to show that the strong sensitivity of FCP rate to MW and MWD is related to the fraction of molecules that can form effective entanglement networks. It follows that longer chains lead to the development of more fracture-resistant entanglement networks.

10.8.1 Polymer FCP Frequency Sensitivity

One is faced with an interesting challenge when trying to explain the effect of test frequency on polymer fatigue performance. Although hysteretic heating arguments appear sufficient to

---

[vii] Recall an earlier discussion of ΔK-dependent transitional behavior (see Fig. 10.51).

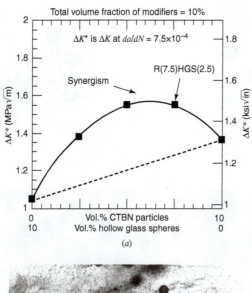

Total volume fraction of modifiers = 10%

ΔK^* is ΔK at $da/dN = 7.5 \times 10^{-4}$

Synergism

R(7.5)HGS(2.5)

Vol.% CTBN particles

Vol.% hollow glass spheres

(a)

Figure 10.62 Synergistic fatigue response of CTBN rubber particle-hollow glass sphere hybrid epoxy composites. (*a*) Maximum FCP resistance associated with epoxy blend containing 7.5 v/o CTBN rubbery particles and 2.5 v/o hollow glass spheres. Maximum fracture toughness also associated with same blend composition (recall Fig. 7.50); (*b*) thin-section of same hybrid epoxy resin composite, viewed under transmitted and bright field conditions. The plastic zone contains rubber cavitation/matrix shear banding and secondary branching due to interaction of elastic stress fields between crack tip and hollow glass spheres. (Reprinted from H. Azimi.[210])

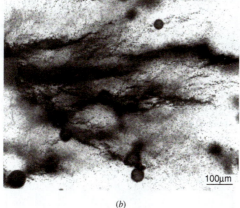

100μm

(b)

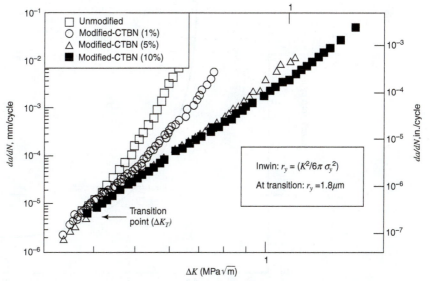

ΔK (ksi$\sqrt{\text{in}}$)

Unmodified
Modified-CTBN (1%)
Modified-CTBN (5%)
Modified-CTBN (10%)

Transition point (ΔK_T)

Inwin: $r_y = (K^2/6\pi\,\sigma_y^2)$

At transition: $r_y = 1.8\mu m$

ΔK (MPa$\sqrt{\text{m}}$)

Figure 10.63 Influence of rubber particle additions on FCP resistance in epoxy resins. Above ΔK_T, the plastic zone size is large compared with rubber particle size, thereby resulting in improved fatigue performance. (Reprinted from H. Azimi et al[211].)

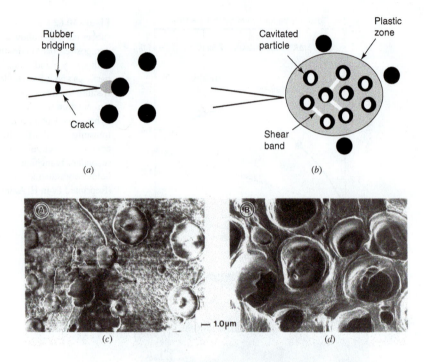

Figure 10.64 Mechanism associated with improved FCP resistance in epoxy-CTBN rubber particle blends. Little cavitation and matrix dilation occur when $\Delta K < \Delta K_T$ and the plastic zone is small compared with rubber particle diameter (see a and c). When $\Delta K > \Delta K_T$, the plastic zone is large compared with particle dimension and much cavitation and matrix dilation occur (see b and d). (Reprinted from Azimi.[210])

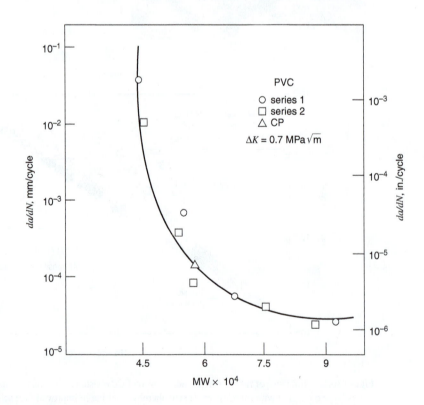

Figure 10.65 Fatigue crack propagation rates in PVC at $\Delta K = 0.7\,\mathrm{MPa}\sqrt{\mathrm{m}}$ as a function of MW (From Rimnac et al.[216]; reprinted from *J. Macromol. Sci. Phys. B* **19**, 351, (1981), by courtesy of Marcel-Dekker, Inc.)

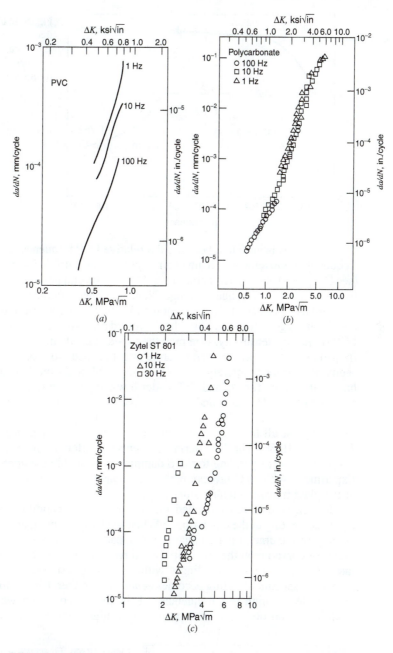

Figure 10.66 Effect of test frequency on fatigue crack growth rate. Crack growth can decrease, increase, or remain unchanged with increasing test frequency, (a) PVC (Skibo[225]); (b) polycarbonate (Hertzberg et al.[226]; with permission from the Society of Plastics Engineers, Inc.); (c) impact-modified nylon 66 (Hertzberg et al.[227]; © American Society for Testing and Materials, 1916 Race Street, Philadelphia, PA 19103. Reprinted with permission.)

explain a *diminution* of fatigue resistance with increasing cyclic frequency in *unnotched* polymer test samples (recall Section 9.2.5), the fatigue resistance of several polymers in the *notched* condition is *enhanced* with increasing cyclic frequency. Note the pronounced decrease in FCP rate with increasing test frequency for a given ΔK level in PVC (Fig. 10.66a). Similar attenuation of FCP rates with increasing test frequency has been reported in several other polymeric solids.[205,226,228−230] Other polymers, such as polycarbonate (PC) and polysulfone (PSF), showed no apparent sensitivity of FCP rate to test frequency (Fig. 10.66b). An intriguing

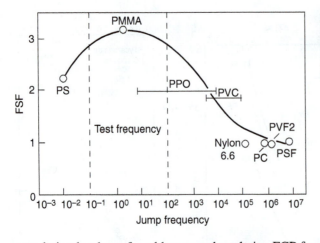

Figure 10.67 Relation between FCP frequency sensitivity and the room temperature jump frequency for several polymers.[226] (Reprinted with permission of the Society of Plastics Engineers.)

correlation has been found between the relative FCP frequency sensitivity in polymers and the frequency of movement of main chain segments responsible for generating the β transition peak (see Chapter 4) at room temperature.[226] Data for several polymers are shown in Fig. 10.67, along with the fatigue test frequency range. Note the greatest frequency sensitivity in the material that revealed its β peak at a frequency comparable to the fatigue test frequency range. This resonance condition suggests the possibility that localized crack-tip heating may be responsible for polymer FCP frequency sensitivity. One may then speculate whether other materials that were not FCP frequency sensitive at room temperature might be made so at other temperatures, if the necessary segmental motion jump frequency were comparable to the mechanical test frequency. Indeed, this has been verified for PC and PSF under low-temperature test conditions; conversely, the FCP response of PMMA was found to be *less* frequency sensitive at –50°C than at room temperature, which is consistent with expectations.[228,229] Of great significance, the overall frequency sensitivity for all the engineering plastics tested thus far has been shown to be dependent on $T - T_\beta$ (Fig. 10.68). This latter term represents the difference between the test temperature and the temperature corresponding to the β damping peak within the appropriate test frequency range. Experiments with PVC have confirmed a similar relation for fatigue tests conducted in the vicinity of the glass transition temperature (T_α).[231]

Recent studies have verified that the resonance condition, noted above, contributes to localized heating at the crack tip.[232] When the temperature increases, yielding processes in the vicinity of the crack tip are enhanced and lead to an increase in the crack-tip radius. A larger radius of curvature at the crack tip should result in a lower effective ΔK; fatigue crack growth rates should decrease accordingly with increasing test frequency. On the other hand, if the amount of specimen heating becomes *generalized* rather than *localized*, higher FCP rates would be expected at high test frequencies. This special condition was found to exist in impact-modified nylon 66, a material possessing a high degree of internal damping (Fig. 10.66*c*).

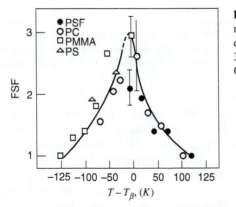

Figure 10.68 Frequency sensitivity factor (FSF) relative to normalized β-transition temperature $T - T_\beta$. (From Hertzberg et al.[229]; reproduced from R. W. Hertzberg et al., *Polymer* **19**, 359 (1978), by permission of the publishers, Butterworth & Co. Ltd.)

Temperature measurements obtained with an infrared microscope and thermocouples revealed crack-tip temperatures in the range of 130°C and substantial heating throughout the specimen's unbroken ligament.[227,232] Such major temperature rises in the specimen decrease the specimen stiffness and enhance damage accumulation (recall Section 9.2.5). It is seen then that the antipodal behavior of rubber-toughened nylon 66 with that of PMMA, PVC, or polystyrene reflects a different balance between gross hysteretic heating (which lowers the elastic modulus overall) and localized crack-tip heating (which involves crack-tip blunting). Materials like PC and PSF, which do not reveal FCP rate frequency sensitivity over a large ΔK range at room temperature, exhibited no significant localized crack-tip heating.[233]

10.8.2 Fracture Surface Micromorphology

At least two distinctly different sets of parallel markings have been found on the fatigue fracture surfaces of amorphous plastics, such as PMMA, PS, and PC. At relatively large ΔK levels, striations are found that correspond to the incremental advance of the crack as a result of one load cycle[234] (Fig. 10.69a). Similar markings have been reported for rubber.[235] The dependence of fatigue striation spacing on the stress intensity factor range and the excellent correlation with associated macroscopic crack growth rates in epoxy, PS, and PC may be seen in Fig. 10.70a.[234] The essentially exact correlation between macroscopic and microscopic growth rates reflects the fact that 100% of the fracture surface in this ΔK regime is striated; that is, only one micromechanism is operative. Contrast this with the results for metals, where several micromechanisms operate simultaneously (Eq. 10-20). In the latter case, the two measurements of crack growth rate do not always agree (see Fig. 10.56).

The other sets of parallel fatigue markings have been found at low ΔK levels and at high test frequencies in PS, PC, PSF, PMMA, and at all stress intensity levels in PVC.[205,206,234−240] These bands (Fig. 10.69b) are too large to be caused by the incremental extension of the crack during one loading cycle. Instead, they correspond to discontinuous crack advance following several hundred loading cycles during which the crack tip remains stationary. The fatigue fracture sequence that produces these markings is shown in Fig. 10.71a. The plastic zone—actually a long, thin craze—is seen to grow continuously, although it is characterized by a decreasing rate with increasing craze length. When the craze reaches a critical length, the crack advances abruptly across the entire craze and arrests. With further cycling, a new craze is developed and the process is repeated. The sequence involving continuous craze growth and discontinuous crack growth is modeled in Fig. 10.71b. Close examination of the fracture surface reveals that the growth bands consist of equiaxed dimples, which decrease in diameter from the beginning to the end of each band (Fig. 10.72). The variable dimple size is believed to reflect the void size distribution within the craze prior to crack instability; it also parallels the extent of crack opening displacement with increasing distance from the crack tip.[237]

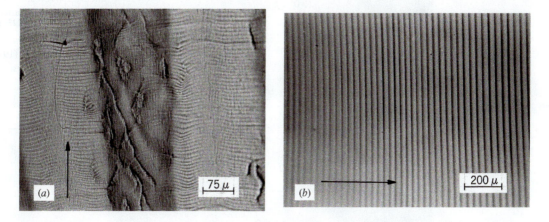

Figure 10.69 Fatigue fracture surface markings in amorphous plastics. (*a*) Striations associated with crack advance during one load cycle; (*b*) discontinuous growth bands equal in size to crack-tip plastic zone. The arrows indicates crack direction.

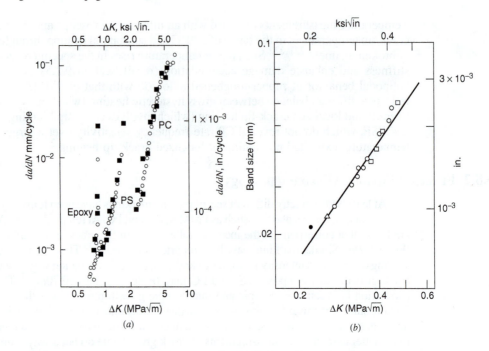

Figure 10.70 (*a*) Macroscopic (○) and microscopic (■) fatigue crack growth in epoxy, polystyrene, and polycarbonate.[234] (*b*) Size of discontinuous growth bands as a function of ΔK in polystyrene.[236]

The size of these bands increases with ΔK (Fig. 10.70*b*) and corresponds to the dimension of the crack-tip plastic zone[236,238,239] as computed by the Dugdale plastic strip model

$$R \approx \frac{\pi}{8}\frac{K^2}{\sigma_{ys}^2} \tag{6-51}$$

This calculation can be reversed to compute the yield strength that controls the crack-tip deformation process. By setting R equal to the band width for a given ΔK level, an inferred yield

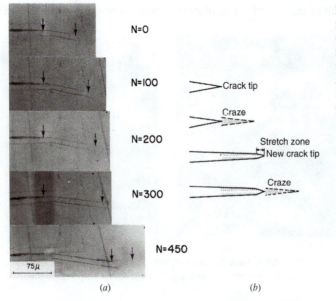

Figure 10.71 (a) Composite micrograph revealing position of craze (⇣) and crack (↓) tip after fixed cyclic increments in PVC.[227] (Reproduced with permission from *Journal of Materials Science* 8 (1973).) (*b*) Schema showing model of discontinuous cracking process.

Figure 10.72   Transmission electron fractograph showing variable equiaxed dimple size in PVC discontinuous growth bands. Largest dimples are found near beginning of band. Region A is the arrest line between bands. Arrow indicates crack propagation direction.[227] (Reproduced with permission from *Journal of Materials Science* **8** (1973).)

strength was determined in PS and other materials and found equal to the craze stress for that material. The number of loading cycles required for craze development and sudden breakdown is estimated by dividing the bandwidth dimension by the corresponding macroscopic crack growth rate; depending on the material and ΔK level, band stability can extend from 10^2 to 10^5 loading cycles (Fig. 10.73).

In summary, two different sets of fatigue markings may be found on the fracture surfaces of the same plastic, each band corresponding to either one or as many as 100,000 load cycles.

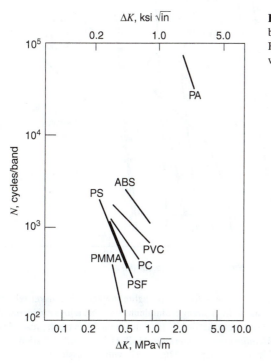

Figure 10.73   Cyclic stability of discontinuous growth bands in several polymers as a function of ΔK. (From Hertzberg et al.,[241] from *J. Mater. Sci.* **13**, 1038 (1978); with permission from Chapman and Hall.)

Without accurate macroscopic growth rate data or fractographic analysis, it is difficult to distinguish between them. To do so is highly desirable, since the markings would provide valuable information concerning the number of cycles associated with the fatigue crack propagation process.

10.9 FATIGUE CRACK PROPAGATION IN CERAMICS

Scientific thought, including that expressed in earlier editions of this text, once held that without an aggressive environment, ceramic materials were not subject to pure cyclic damage. This view was based mainly on the negligible amount of plastic deformation capacity believed to be available at the crack tip within simple ceramic microstructures. More recent studies,[242–254] however, have demonstrated clearly that fatigue damage does occur under both cyclic tensile and compressive loading conditions, and that stable FCP takes place. Figure 10.74 shows representative crack growth data for transformation-toughened, whisker-reinforced, and single-phase ceramics and composites; typical data for high-strength aluminum and steel alloys are added for comparative purposes. For these and other reported results, brittle solids reveal a trend toward higher ΔK_{th} values with increasing K_c (i.e., $\Delta K_{th}/K_c \approx 0.6$). Furthermore, a growing literature in ceramic and ceramic composites reveals a Paris-type relation (Eq. 10-3) with a crack growth rate-ΔK dependence between 15 and 42 as compared with 2–4 for the case of monolithic metal alloys. If one were to compute the change in fatigue lifetime for a metal versus ceramic component in association with a twofold increase in nominal stress level (recall Eq. 10-6), one would find lifetime decreasing by

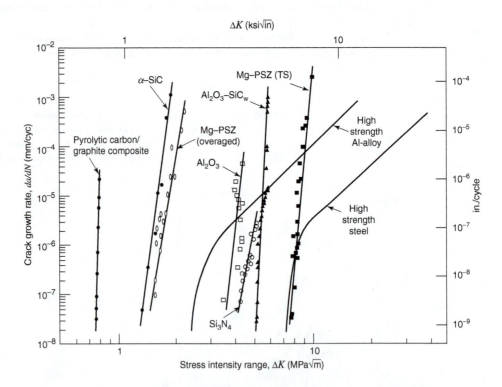

Figure 10.74 Fatigue crack propagation response for transformation-toughened, whisker-reinforced, and single-phase ceramics and composites.[243] Data for high-strength aluminum and steel alloys are included for comparative purposes. Apart from the threshold region, note that da/dN dependence on ΔK is much greater in ceramics compared to metallic alloys. (Reprinted from *Metallurgica et Materialia*, 41, R. H. Dauskardt, "A frictional-wear mechanism for fatigue-crack growth in grain bridging ceramics," p. 2765, 1993, with permission from Elsevier.)

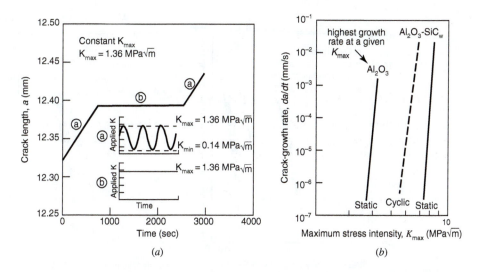

Figure 10.75 Comparison of crack propagation rates under sustained and cyclic loading conditions. (*a*) Sustained vs. cyclic-loading ($R = 0.1$ (50 Hz)) response in pyrolytic-carbon coated graphite tested in Ringer's solution at 37°C. Note much higher rates under cyclic loading conditions (region a) than under sustained loading conditions (region b).[242] (R. O. Ritchie and R. H. Dauskardt, *J. Ceram. Soc. Japan*, **99**(10), 1047 (1991). Reprinted with permission of Fuji Technology Press, Ltd.); (*b*) time-corrected comparison of static vs. cyclic loading rates in Al_2O_3 and Al_2O_3-SiC composites. Composite with cyclic loading exhibits lower crack growth rates than with unreinforced Al_2O_3, but higher crack growth rates than with statically loaded conditions.[247] (Data from R. H. Dauskardt, B. J. Dalgleish, D. Yao, R. O. Ritchie, and P. F. Becker, *Journal of Materials Science*, 28, p. 3258, 1993.)

factors of 16 and 4×10^{12}, respectively![viii] Alternatively, a 16-fold decrease in fatigue lifetime in a ceramic (the same as that for the metal alloy but assuming $m = 42$) would correspond to less than a 7% increase in stress level, hardly discernible within the range of residual stresses present in ceramic components. As a result, one must anticipate highly unreliable fatigue life predictions in ceramic components.

When assessing component lifetime, it is important to recognize that cyclic-loading induced damage accumulation in brittle solids can lead to much higher crack growth rates (i.e., shorter lifetimes) than those associated with environmental cracking under sustained loading conditions (i.e., static fatigue); for example, when pyrolitic-carbon coated graphite is tested in Ringer's solution, crack growth rates are much greater when cyclic rather than static loads are applied (Fig. 10.75*a*). Figure 10.75*b* shows a time-corrected comparison of cyclic versus static fatigue crack growth in Al_2O_3-28v/oSiC$_w$ in ambient air. Note that cyclic fatigue rates far exceed those of the statically loaded composite but are slower than static fatigue rates in monolithic aluminum oxide.

The crack growth resistance in ceramics and their composites depends strongly on crack-tip shielding mechanisms (e.g., transformation toughening in partially stabilized zirconia systems and whisker/fiber bridging in reinforced composites). To illustrate, Steffen et al.[245] accounted for the transformation-toughened shielding contribution on ΔK_{tip} ($K - K_s$), where K_s is found from Eq. 7-3; as a result, they were able to significantly normalize FCP data in a Mg-PSZ alloy subjected to different heat treatments[245] (Fig. 10.76).

Given the presence of assorted shielding mechanisms in toughened and/or reinforced ceramics, one may expect that short crack behavior will occur.[245,246] As discussed in Section 10.4.2.1, this arises from the limited wake in small cracks, which minimizes the crack closure level. An example of differences in long versus short crack behavior is shown in Fig. 10.77 for the case of the Al_2O_3-SiC$_w$ composite. Note that small cracks, introduced from

---

[viii] Values of $m = 50$–100 have also been reported. [See R. H. Dauskardt, R. O. Ritchie, and B. N. Cox, *Advanced Materials & Processes*, **144** (2), 30 (1993).] Accordingly, a two-fold increase in stress would decrease projected lifetime by as much as 30 orders of magnitude!

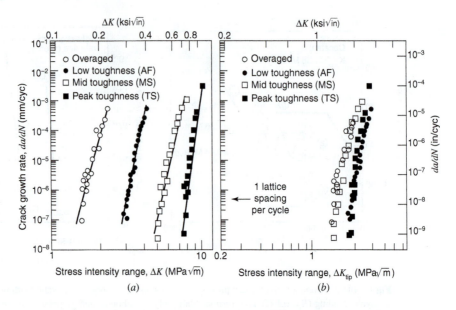

Figure 10.76   FCP response in transformation-toughened Mg-PSZ.[245] (*a*) Data plotted vs. ΔK; (*b*) data plotted vs. ΔK_{tip} ($K_{max} - K_s$). Note that FCP resistance increases with magnitude of transformation toughening. (A. A. Steffen, R. H. Dauskardt, and R. O. Ritchie, *J. Amer. Ceram. Soc.*, **76** (6), 1259 (1991). Reprinted by permission.)

microindentations, grew at ΔK levels about 2–3 times lower than ΔK_{th} values associated with long crack test results. In addition, short crack growth rates tended to *decrease* with increasing stress intensity level. These differences in growth rate behavior between long and short cracks in this ceramic composite were largely attributed to residual stresses introduced into the short crack sample during the indent process. When the crack-tip K level was corrected for residual stresses, the data were normalized; the latter suggests that a stabilized level of shielding was achieved after a limited advance of the indent crack.

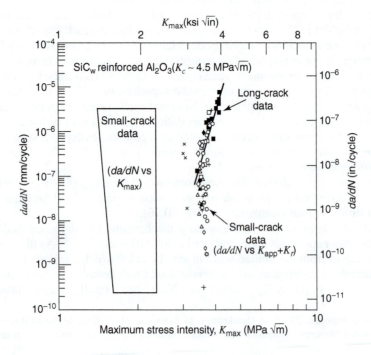

Figure 10.77   Short- and long-crack growth rate data in Al₂O₃-SiC composite.[246] Short crack growth data in box corresponds to applied ΔK test conditions whereas individual short crack datum refer to ΔK values corrected for residual stresses generated during the indentation process. (R. H. Dauskardt, M. R. James, J. R. Porter, and R. O. Ritchie, *J. Amer. Ceram. Soc.* **75** (4), 759 (1992). Reprinted by permission.)

Recent studies[242,244] have identified a large number of crack growth mechanisms in ceramics and composites that relate to both intrinsic crack-tip-microstructural interactions and extrinsic crack-tip shielding mechanisms (Fig. 10.78) (also recall Section 7.1.3). A growing number of studies[243,244] have shown cyclic-induced frictional wear of bridging zones and

Mechanisms of Cyclic Fatigue Crack Growth in Ceramics

a) Extrinsic mechanisms

1. Degradation of transformation
   toughening

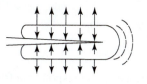

   - degree of reversability of transformation
   - cyclic accommodation of transformation
     strain
   - cyclic modification of zone morphology

2. Damage to bridging zone
   - fraction and wear degradation of:

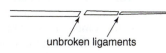

   unbroken ligaments

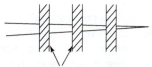

   whisker/fiber reinforcements

   - crushing of asperities and
     interlocking zones

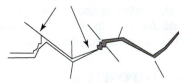

3. Fatigue of ductile reinforcing phase

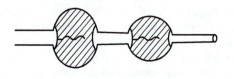

b) Intrinsic mechanisms

1. Accumulated (damage) localized
   microplasticity/microcracking

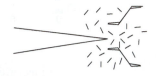

2. Mode II and III crack propagation
   on unloading

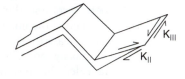

3. Crack tip blunting/resharpening

   a) Continuum

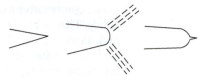

   b) Alternating shear

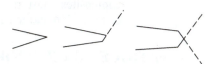

4. Relaxation of residual stresses

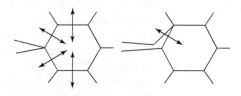

Figure 10.78   Examples of proposed fatigue crack advance mechanisms in polycrystalline ceramics and composites. (R. O. Ritchie, R. H. Dauskardt, W. Yu, and A. M. Brendzel, *Journal of Biomedical Materials Research* **24**, 189 (1990), with permission from John Wiley & Sons.)

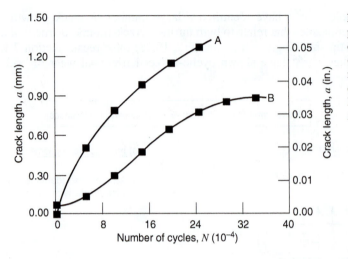

Figure 10.79 Fatigue crack growth in polycrystalline alumina under cyclic compression. Crack wake retained (Curve B) and removed (Curve A) to reveal the influence of crack closure on crack extension.[249] (Data adapted from L. Ewart and S. Suresh, *J. Mater. Sci. Lett.*, **5**, 774 (1986).)

crushing of asperities on interlocking interfaces to be responsible for the major portion of the degradation process (see Fig. 10.78a). With progressive wear, crack-tip shielding is attenuated, whereas the effective stress intensity factor increases along with the crack growth rate.

Suresh and co-workers[249,250,254] have also confirmed the existence of crack growth in polycrystalline alumina under cyclic compression. It is believed that residual tensile stresses, induced at the notch tip by the applied compressive loads, contributes to the accumulation of damage associated with grain boundary failure. Crack growth rates were found to decrease with increasing crack length before arresting (Fig. 10.79). An examination of the crack wake revealed the development of debris that enhanced closure and contact between the mating fracture surfaces. When the debris was periodically removed by ultrasonic cleaning, crack growth rates increased (Curve *A*) relative to that associated with the retention of crack wake debris (Curve *B*).

The existence of fatigue damage accumulation in ceramic and ceramic composites under cyclic compressive loading raises considerable concern regarding life prediction for such brittle solids. That is, brittle solids are up to ten times stronger in uniaxial compression than in tension and, therefore, are believed to be much safer when loaded in compression than in tension. However, it has been shown[249-254] that subcritical crack growth occurs in the presence of stress concentrations with cyclic compressive stress levels far below the material's compressive strength; such results have been reported for the case of monolithic, transformable, and reinforced ceramics. Therefore, a reliable design methodology for ceramics and ceramic composites must recognize the potential for subcritical crack growth under *both* cyclic compression and tensile loading as well as static loading conditions.

10.10 FATIGUE CRACK PROPAGATION IN COMPOSITES

As might be expected, the addition of reinforcing fibers and whiskers leads to a reduction in FCP rates for a given composite material. This results from the reduction in cyclic strain within the matrix (characterized by $\Delta K/E$ (recall Fig. 10.58b)) and the transfer of cyclic loads to the fibers. Furthermore, as cracking proceeds within the matrix, unbroken fibers remain behind the advancing crack front and restrict crack opening. This crack-tip shielding mechanism, involving fiber bridging (recall Fig. 7.4), leads to vastly reduced FCP rates as demonstrated in Fig. 10.80 for the case of the ARALL hybrid composite[255-257] (recall Fig. 3.33a). The conventional FCP behavior of 7075-T6 is contrasted with the behavior of a 3/2-ARALL hybrid composite containing three layers of 7075-T6 aluminum alloy separated by two layers of an adhesive/aramid fiber composite. The crack growth resistance of prestrained ARALL is of special interest in that a 1000-fold reduction in crack growth rates is achieved from that of the unreinforced aluminum alloy matrix. In this instance, when the composite is prestrained into the

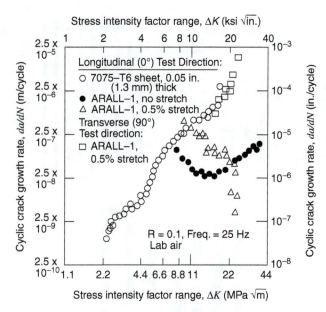

Figure 10.80 FCP response in prestrained and unprestrained ARALL hybrid composite as compared with that of 7075-T6 base metal. Note attenuated crack growth due to presence of aramid fibers. (With permission from R. J. Bucci, Alcoa Technology Center.)

elastic-plastic regime (elastic aramid fiber response versus plastic deformation in the aluminum alloy) a significant residual compressive stress is developed in the aluminum layers upon removal of the prestrain load. As a result, crack initiation and propagation resistance in the composite increases dramatically. In fact, crack growth rates are found to *decrease* with increasing crack length in the aluminum layers as more aramid fibers bridge the crack wake. The overall efficiency of this crack bridging mechanism depends on the rate of damage accumulation within the adhesive/aramid fiber layers, resulting from adhesive shear deformation and delamination; thinner layers and a greater number of laminates per unit thickness will reduce damage accumulation by these processes and suppress FCP rates. Studies have shown that the length of the crack-tip bridging zone is on the order of 3–5 mm.[258] Since ARALL composites and their cousins, the GLARE composites, are lighter than aluminum sheets of the same thickness and possess vastly improved FCP resistance while maintaining the bending, milling, drilling, riveting, and bolting characteristics of conventional monolithic aluminum alloys, hybrid composites are being put to use in fatigue critical airplane components. (See Section 3.9.2 for more information about this class of materials.)

The action of these aramid fibers as numerous *microscopic* stiffeners within the laminate may be compared with the action of conventional *macroscopic* stiffeners used in large panel structures.[14] These stiffeners serve to take up the load from the main panel, as the crack passes, thereby reducing the ΔK level at the crack tip. The associated reduction in crack growth rate is readily apparent each time the crack in the main panel passes beneath a panel stiffener (recall Fig. 10.6).

Chan and Davidson[259] recently completed a review of the fatigue response of continuous fiber-reinforced metal matrix composites and showed that the prevailing fatigue fracture micromechanisms were strongly dependent on the respective properties of the matrix, interface, and fibers (Fig. 10.81). For the case of strong interfaces and weak fibers, failure is typically dominated by fiber and matrix fracture events. Conversely, when strong fibers are weakly bonded to the matrix, crack branching occurs as the crack moves through the matrix. Furthermore, crack bridging occurs when the latter scenario is present along with residual compressive stresses at fiber–matrix interfaces. As discussed previously, crack branching and crack bridging lead to enhanced FCP resistance.

Studies of fatigue damage accumulation processes in short-fiber reinforced plastics have identified the need to redefine the crack length and the prevailing ΔK level at the crack tip.[260–263] First, the crack is not a simple entity that can be directly characterized in terms of its

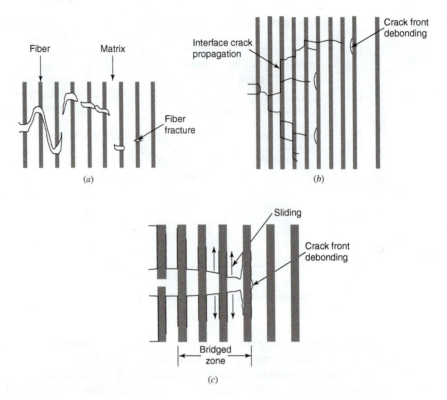

Figure 10.81   Fatigue fracture micromechanisms in metal–matrix composites associated with conditions of: (*a*) strong interfaces and weak fibers—fiber fracture dominated; (*b*) strong fibers and weak interfaces—interfaces decohesion dominated; and (c) strong fibers and weak interfaces, coupled with residual compressive stresses at fiber–matrix interfaces—fiber bridging dominated.[259] (K. S. Chan and D. L. Davidson, *Proc. of Engineering Foundation, International Conference on Fatigue of Advanced Materials*. R. O. Ritchie, R. H. Dauskardt, and B. N. Cox, Eds., 1991. Reprinted by permission.)

length. Instead, the crack should be viewed as a main crack surrounded by many secondary cracks that lie away from the plane of the main crack; the crack-tip region, therefore, contains many microcracks with overall crack extension characterized by the development and growth of this diffuse damage zone. In addition to microcrack-induced crack-tip zone shielding, the crack tip is also influenced by the bridging of fibers across the crack surfaces. Both micro-cracking and fiber bridging in the crack-tip region shield the crack tip from the full influence of the applied loads (recall Fig. 7.4), thereby reducing the effective ΔK level. The prevailing ΔK level is also profoundly affected by out-of-plane growth of the crack (e.g., crack deflection parallel to the fiber axis and stress direction).

The improvement in FCP resistance of injection-molded nylon 66 with the addition of short glass fibers is shown clearly in Fig. 10.82[261] and may be traced to the following: (1) load transfer from the matrix to the much stronger fibers along with the overall increase in specimen stiffness; (2) additional energy dissipation mechanisms associated with fiber debonding and pullout, and local plastic deformation in the matrix around the fibers; and (3) reduction in the effective ΔK level in associaton with microcracking and fiber bridging-crack-tip shielding.

The influence of fiber orientation on FCP rates deserves additional comment. One would typically expect crack growth resistance to be greater in samples containing fibers oriented parallel to the stress axis (i.e., perpendicular to the anticipated crack plane) than in the plane of the notch. The situation is complex for the case of injection-molded parts since fibers are

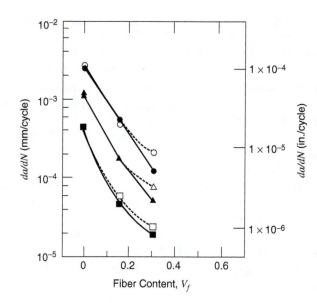

Figure 10.82 Effect of fiber volume fraction v_f on fatigue crack growth rates in sgfr Nylon 66 at various levels of ΔK for two specimen orientations (frequency = 10 Hz, minimum/maximum load ratio = 0.1). $\Delta K =$ 4.5 MPa $\sqrt{m}$: ●, L direction; ○, T direction. $\Delta K = 4.0$ MPa $\sqrt{m}$: ▲, L direction; △, T direction. $\Delta K = 3.5$ MPa $\sqrt{m}$: ■, L direction; □, T direction. (Reprinted with permission from R. W. Lang, J. A. Manson, and R. W. Hertzberg in *Polymer Blends and Composites in Multiphase Systems*, C. D. Hans, Ed., *Advances in Chemistry Series, No. 206*, 1984, p. 261.)

oriented parallel to the injection-molding direction along the mold surfaces and transverse to the molding direction in the core region because of the divergent flow pattern in the middle of the component (recall Fig. 3.41). Depending on the thickness of the two skin layers of the injection-molded plaque relative to that of the core region, FCP resistance of the material may be superior when loaded either parallel or perpendicular to the injection molding direction.[260–263] For example, the same FCP rates in longitudinal- and transverse-oriented samples containing 18 v/o glass fibers implies a microstructural balance, whereas 31 v/o glass fiber FCP results imply domination of the skin layers in the determination of FCP resistance of this composite (Fig. 10.82). Furthermore, the influence of orientation on FCP behavior depends on the thickness of the injection-molded plaque in that the relative influence of the surface layers will vary[264]; the FCP resistance in the *L* and *T* directions should decrease and increase, respectively, with increasing plaque thickness.

REFERENCES

1. H. W. Liu, *Appl Mater. Res.* **3** (4), 229 (1964).

2. N. E. Frost, *J. Mech. Eng. Sci.* **1** (2), 151 (1959).

3. W. Weibull, *Acta Met.* **11** (7), 725 (1963).

4. P. C. Paris, *Fatigue—An Interdisciplinary Approach*, Proceedings, 10th Sagamore Conference, Syracuse University Press, Syracuse, NY, 1964, p. 107.

5. S. R. Swanson, F. Cicci, and W. Hoppe, ASTM *STP 415*, 1961, p. 312.

6. P. C. Paris and F. Erdogan, *J. Basic Eng. Trans. ASME, Series D* **85** (4), 528 (1963).

7. P. C. Paris and G. C. M. Sih, ASTM *STP 381*, 1965, p. 30.

8. L. Saxena, S. J. Hudak, Jr., J. K. Donald, and D. W. Schmidt, *J. Test Eval.* **6**, 167 (1978).

9. ASTM *E647* ASTM International, West Conshohocken, PA.

10. *Fatigue Crack Growth Measurement and Data Analysis*, ASTM *STP 738*, S. J. Hudak, Jr., and R. J. Bucci, Eds., ASTM, Philadelphia, PA, 1981.

11. R. Jaccard, *Fatigue Life Prediction of Aluminum Structures Based on SN-Curve Simulation*, 2nd Int. Conf. on Aluminum Weldments, Aluminum-Verlag, Dusseldorf, Germany, 1982.

12. J. M. Barsom, E. J. Imhof, and S. T. Rolfe, *Engl. Fract. Mech.* **2** (4), 301 (1971).

13. G. Miller, *Trans. Quart. ASM* **61**, 442 (1968).

14. C. C. Poe, AFFDL-TR-70-144, 207 (1970).

15. J. A. Harris, Jr., *Engine Component Retirement for Cause, Vol. I. Executive Summary*, AFWAL-TR-87-4069, Wright-Patterson Air Force Base, Ohio, 1987.

16. P. J. E. Forsyth, *Acta Met.* **11** (7), 703 (1963).

17. D. O. Swenson, *J. Appl. Phys.* **40**, 3467 (1969).

18. R. W. Hertzberg and W. J. Mills, ASTM *STP 600*, 1976, p. 220.

19. L. B. Vogelesang, *Report LR-286*, Delft, The Netherlands, 1979.

20. G. T. Hahn, R. G. Hoagland, and A. R. Rosenfield, *Met. Trans.* **3**, 1189 (1972).

21. C. Bathias and R. M. Pelloux, *Met. Trans.* **4** (5), 1265, (1973).

22. R. J. Gran, F. D. Orazio, Jr., P. C. Paris, G. R. Irwin, and R. W. Hertzberg, AFFDL-TR-70-149, Mar. 1971.

23. J. C. McMillan and R. W. Hertzberg, ASTM *STP 436*, 1968, p. 89.

24. P. J. E. Forsyth and D. A. Ryder, *Metallurgia* **63**, 117 (1961).

25. R. W. Hertzberg, ASTM *STP 415*, 1967, p. 205.

26. C. Laird and G. C. Smith, *Philos. Mag.* **7**, 847 (1962).

27. R. M. Pelloux, *Trans. Quart. ASM* **62** (1), 281 (1969).

28. J. H. Weber and R. W. Hertzberg, *Met. Trans.* **4**, 595 (1973).

29. R. W. Hertzberg and P. C. Paris, Proceedings, International Fracture Conference, Sendai, Japan, Vol. 1, 1965, p. 459.

30. R. C. Bates and W. G. Clark, Jr., *Trans. Quart. ASM* **62** (2), 380 (1969).

31. R. J. Stofanak, R. W. Hertzberg, and R. Jaccard, *Int. J. Fract.* **20**, R145 (1982).

32. C. M. Hudson, NASA Tech. Note *D-5390*, 1969.

33. R. G. Forman, V. E. Kearney, and R. M. Engle, *J. Basic Eng. Trans. ASME* **89**, 459 (1967).

34. R. Roberts and F. Erdogan, *J. Basic Eng. Trans. ASME* **89**, 885 (1967).

35. R. H. Christensen, *Appl. Mater. Res.*, **X** (X), 207 (1963).

36. W. Elber, ASTM *STP 486*, 1971, p. 230.

37. W. Illg and A. J. McEvily, Jr., NASA *TND-52*, 1959.

38. R. W. Hertzberg and E. F. J.von Euw, *Int. J. Fract. Mech.* **7**, 349 (1971).

39. D. Broek and J. Schijve, National Aeronautical and Astronautical Research Institute, Amsterdam, Holland, NLR-TR-M 2.111, 1963.

40. J. Schijve, C ASTM *STP 415*, 1967, p. 415.

41. D. L. Chen, B. Weiss, and R. Stickler, *Int. J. Fat.* **13**, 327 (1991).

42. D. L. Chen, B. Weiss, and R. Stickler, *Mater. Sci. and Eng.* **A20S**, ISI (1996).

43. J. K. Donald, G. H. Bray, and R.W. Bush, *High Cycle Fatigue of Structural Materials*, W. O. Soboyejo and T. S. Srivatsan, Eds., TMS, Warrendale, PA, 1997, p. 123.

44. J. K. Donald, G. H. Bray, and R.W. Bush, ASTM *STP 1332*, 1998, p. X.

45. J. K. Donald, G. M. Connelly, P. C. Paris, and H. Tada, ASTM *STP 1360*, 2000, p. 185.

46. J. K. Donald, private communication.

47. Draft, Recommended Practice for Determination of ACR-Based Effective Stress-Intensity Factor Range, ASTM *Standard E647*, 11/24/2008.

48. P. C. Paris, H. Tada, and J. K. Donald, *Int. J. Fat.* **21**, S35 (1999).

49. R. J. Bucci, P. C. Paris, R. W. Hertzberg, R. A. Schmidt, and A. F. Anderson, ASTM *STP 513*, 1972, p. 125.

50. R. J. Bucci, W. G. Clark, Jr., and F P. C. Paris, ASTM *STP 513*, 1972, p. 177.

51. M. Klesnil and P. Lukas, *Mater. Sci. Eng.* **9**, 231 (1972).

52. R. J. Stofanak, R. W. Hertzberg, G. Miller, R. Jacard, and J. K. Donald, *Eng. Fract. Mech.* **17** (6), 527 (1983).

53. R. O. Ritchie, *J. Eng. Mater. Tec.* **99**, 195 (1977).

54. J. L. Robinson and C. J. Beevers, *Met. Sci.* **7**, 153, (1973).

55. R. J. Cooke and C. J. Beevers, *Mater. Sci. Eng.* **13**, 201 (1974).

56. R. J. Cooke, P. E. Irving, C. S. Booth, and C. J. Beevers, *Eng. Fract Mech.* **7**, 69 (1975).

57. N. Walker and C. J. Beevers, *Fat. Eng. Mater. Struct.* **1**, 135 (1979).

58. K. Minakawa and A. J. McEvily, *Scripta Met.* **15**, 663 (1981).

59. S. Suresh and R. O. Ritchie, *Met. Trans.* **13A**, 1627 (1982).

60. R. J. Stofanak, R. W. Hertzberg, G. Miller, R. Jaccard, and K. Donald, *Eng. Fract. Mech.* **17** (6), 527 (1983).

61. K. Minakawa, G. Levan, and A. J. McEvily, *Met. Trans.* **17A**, 1787 (1986).

62. S. Suresh, G. F. Zamiski, and R. O. Ritchie, *Met. Trans.* **12A**, 1435 (1981).

63. S. Suresh, D. M. Parks, and R. O. Ritchie, *Fatigue Thresholds*, J. Backhand, A. F. Blom, and C. J. Beevers, Eds., EMAS Publ. Ltd., Warley, England, 1982.

64. L. K. L. Tu and B. B. Seth, *J. Test. Eval.* **6** (1), 66 (1978).

65. ASTM Committee E-24 Round-Robin Study on Closure Measurements, Bal Harbour, Florida, November 10, 1987.

66. R. S. Vecchio, J. Crompton, and R. W. Hertzberg, *Int. J. Fract.* **31** (2), R29 (1986).

67. C. H. Newton and R. W. Hertzberg, ASTM *STP 982*, 139 (1988).

68. J. K. Donald, op cit., p. 222.

69. R. S. Vecchio, Ph.D. dissertation, Lehigh University, Bethlehem, PA, 1985.

70. J. K. Donald, *Ten Questions about ACR*, Fracture Technology Associates, Sept. 7, 2009.

71. H. Kitagawa and S. Takahashi, *Trans. Japan Soc. Mech. Eng.* **45**, 1289 (1979).

72. H. Ohuchida, S. Usami, and A. Nishioka, *Trans. Japan Soc. Mech. Eng.* **41**, 703 (1975).

73. M. H. El Haddad, K. N. Smith, and T. H. Topper, *Trans. ASME* **101H**, 42 (1979).

74. S. Taira, K. Tanaka, and M. Hoshina, ASTM *STP 675*, 1979, p. 135.

75. Y. Nakai and K. Tanaka, Proc. 23rd Japan Cong. Mater. Res., 1980, p. 106.

76. K. Tanaka, Y. Nakai, and M. Yamashita, *Int. J. Fract.* **17**, 519 (1981).

77. H. Kitagawa and S. Takahashi, Proc. 2nd. Int. Conf. on Mech. Beh. Maters., ASM, Metals Park, Ohio, 1976, p. 627.

78. M. H. El Haddad, T. H. Topper, and K. N. Smith, *Eng. Fract. Mech.* **11**, 573 (1979).

79. N. E. Dowling, ASTM *STP 637*, 1977, p. 637.

80. N. E. Frost, *J. Mech. Eng. Sci.* **5**, 15 (1963).

81. N. E. Frost, *J. Mech. Eng. Sci.* **6**, 203 (1964).

82. D. Broek and B. N. Leis, *Materials, Experimentation and Design in Fatigue*, F. Sherratt and J. B. Sturgeon, Eds., Westbury House, Guildford, 1981, p. 129.

83. J. Schijve, *Fatigue Thresholds*, Vol. 2, EMAS Ltd., West Midlands, England, 1982, p. 881.

84. R. O. Ritchie and S. Suresh, *Mater. Sci. Eng.* **57**, 1983, L27.

85. J. C. Newman, AGARD Conf. Proc. No. 328, paper 6, 1983.

86. C. H. Newton, Ph. D. Dissertation, Lehigh University, Bethlehem, PA, 1984.

87. R. O. Ritchie and J. Lankford, Eds., *Small Fatigue Cracks*, AIME, Warrendale, PA, 1986.

88. K. J. Miller and E. R. de los Rios, Eds., *The Behavior of Short Fatigue Cracks*, Mechanical Engineering Publications Ltd., London, 1986.

89. S. Suresh and R. O. Ritchie, *Int. Metals Rev.* **29** (6), 445 (1984).

90. B. N. Leis and T. P. Forte, ASTM *STP 743*, 1981, p. 100.

91. E. R. de los Rios, E. Z. Tang, and K. J. Miller, *Fat Eng. Mater. Struct.* **7** (2), 97 (1984).

92. P. K. Liaw and W. A. Logsdon, *Eng. Fract. Mech.* **22** (1), 115 (1984).

93. D. E. Castro, G. Marci, and D. Munz, *Fat. Fract. Eng. Mater.*, in press

94. H. Doker and M. Peters, *Fatigue 84*, C. Beevers, Ed., EMAS Ltd., West Midlands, England, 1984, p. 275.

95. H. Doker and G. Marci, *Int. J. Fat.* **5** (4), 187 (1983).

96. W. A. Herman, R. W. Hertzberg, and R. Jaccard, *J. Fat. Fract. Eng. Mater. Struct.* **11** (4), 303 (1988).

97. W. A. Herman, R. W. Hertzberg, C. H. Newton, and R. Jaccard, *Fatigue 87*, R. O. Ritchie and E. A. Starke, Jr., Eds., 2, EMAS Ltd., West Midlands, England, 1987, p. 819.

98. R. W. Hertzberg, W. A. Herman, T. R. Clark, and R. Jaccard, *ASTM 1149*, J. M. Larsen and J. E. Allison, Eds., ASTM Philadelphia (1992) 197.

99. R. Jaccard, 3rd Int. Conf. Alum. Weld., Munich, FRG, 1985.

100. R. Jaccard, 2nd Int. Conf. Alum. Weld., Dusseldorf, FRG, 1982.

101. M. H. El Haddad, K. N. Smith, and T. H. Topper, ASTM *STP 677*, 1979, p. 274.

102. R. A. Smith and K. J. Miller, *Int. J. Mech. Sci.* **20**, 201 (1978).

103. M. M. Hammouda, R. A. Smith, and K. J. Miller, *Fat. Eng. Mater. Struct.* **2**, 139 (1979).

104. N. E. Frost, *J. Mech. Eng. Sci.* **2** (2), 109 (1960).

105. J. Schijve and D. Broek, *Aircr. Eng.* **34**, 314 (1962).

106. J. Schijve, D. Broek, and P. deRijk, NLR Report *M2094*, Jan. 1962.

107. E. F. J.von Euw, R. W. Hertzberg, and R. Roberts, ASTM *STP 513*, 1972, p. 230.

108. V. W. Trebules, Jr., R. Roberts, and R. W. Hertzberg, ASTM *STP 536*, 1973, p. 115.

109. W. J. Mills and R. W. Hertzberg, *Eng. Fract. Mech.* **7**, 705 (1975).

110. W. J. Mills and R. W. Hertzberg, *Eng. Fract. Mech.* **8**, 657 (1976).

111. W. J. Mills, R. W. Hertzberg, and R. Roberts, ASTM *STP 637*, 1977, p. 192.

112. R. S. Vecchio, R. W. Hertzberg, and R. Jaccard, *Scripta Met.* **17**, 343 (1983).

113. R. S. Vecchio, R. W. Hertzberg, and R. Jaccard, *Fat. Eng. Mater. Struct.* **7** (3), 181 (1984).

114. C. H. Newton, R. S. Vecchio, R. W. Hertzberg, and R. Jaccard, *Fatigue Crack Growth Threshold Concepts*, AIME, Warrendale, PA, 1984, p. 379.

115. J. Schijve, F. A. Jacobs, and P. J. Tromp, NLR *TR 69050 U*, June 1969.

116. R. I. Stephens, D. K. Chen, and B. W. Horn, ASTM *STP 595*, 1976, p. 27.

117. B. M. Hillberry, W. X. Alzos, and A. C. Skat, AFFDL-TR-75-96, Aug. 1975.

118. M. K. Himmelein and B. M. Hillberry, ASTM *STP 590*, 1975, p. 321.

119. E. Zaiken and R. O. Ritchie, *Eng. Fract. Mech.* **22** (1), 35 (1985).

120. W. S. Johnson, ASTM *STP 748*, 1981, p. 85.

121. J. B. deJonge and A. Nederveen, ASTM *STP 714*, 1980, p. 170.

122. H. I. McHenry, Ph.D. Dissertation, Lehigh University, Bethlehem, PA (1970).

123. C. Ragazzo, R. W. Hertzberg, and R. Jaccard, ASTM *J. Test. Eval* **23** (1), 19 (1995).

124. D. P. Williams, III, P. S. Pao, and R. P. Wei, *Environment-Sensitive Fracture of Engineering Materials*, Z. A. Foroulis, Ed., AIME, Warrendale, PA, 1979, p. 3.

125. A. Hartman, F. J. Jacobs, A. Nederveen, and P. deRijk, NLR *TN/M 2182*, 1967.

126. A. Hartman, *Int. J. Fract. Mech.* **1** (3), 167 (1965).

127. R. P. Wei, *Eng. Fract. Mech.* **1**, 633 (1970).

128. Y. Ro, S. R. Agnew and R. P. Gangloff, *Mater. Sci. and Eng.*, Vol. A468-470, 88 (2007).

129. R. P. Piascik and R. P. Gangloff, *Met. Trans A.* **22 A**, (1991) p. 2415.

130. R. P. Piascik and R. P. Gangloff, *Met. Trans A.* **24 A**, (1993) p. 2751.

131. G. H. Bray, M. Glazov, R. J. Rioja, D. Li, and R. P. Gangloff, *Int. J. Fatigue*, **23**, (2002) p. S265.

132. R. P. Wei, P. S. Pao, R. G. Hart, T. W. Weir, and G. W. Simmons, *Metall. Trans. A* **11** (1980) p. 151.

133. P. S. Pao, M. Gao, and R. P. Wei, *Basic Questions in Fatigue*, R. P. Wei and R. P. Gangloff, Eds., ASTM *STP 924*, 1988, p. 182.

134. T. H. Shih and R. P. Wei, *Eng. Fract. Mech.*, **18**, (1983) p. 827.

135. R. J. Bucci, Ph.D. Dissertation, Lehigh University, Bethlehem, PA, 1970.

136. J. M. Barsom, *Eng. Fract. Mech.* **3** (1), 15 (1971).

137. E. J. Imhof and J. M. Barsom, ASTM *STP 536*, 1973, p. 182.

138. R. P. Gangloff, *Hydrogen Effects on Materials*, B. P. Somerday, P. Sofronis, and R. H. Jones, Eds., ASM International, OH, 2009, p. 1.

139. R. Krishnamurthy, C. N. Marzinsky, and R. P. Gangloff, *Hydrogen Effects on Material Behavior*, N. R. Moody and A. W. Thompson, Eds., TMS-AIME, Warrendale, PA, 1990, p. 891.

140. Y. J. Ro, S. R. Agnew, and R. P. Gangloff, 4th International Conference on Very High Cycle Fatigue, J. E. Allison, J. W. Jones, J. M. Larsen, and R. O. Ritchie, Eds., TMS-AIME, Warrendale, PA, 2007, pp. 409–420.

141. L. F. Coffin, Jr., *Met. Trans.* **3**, 1777 (1972).

142. H. D. Solomon and L. F. Coffin, ASTM *STP 520*, 1973, p. 112.

143. M. W. Mahoney and N. E. Paton, *Nucl. Tech.* **23**, 290 (1974).

144. L. A. James and R. L. Knecht, *Met. Trans.* **6A**, 109 (1975).

145. H. W. Liu, and J. J. McGowan, AFWAL-TR-81-4036, June 1981.

146. R. P. Gangloff, *Met Trans.*, **16A**, (1985) p. 953.

147. R. P. Gangloff and R. P. Wei, *Small Fatigue Cracks*, R. O. Ritchie and J. Langford, Eds., TMS-AIME, Warrendale, OH, 1986, p. 239.

148. R. P. Gangloff and A. Turnbull, *Modeling Environmental Effects on Crack Initiation and Propagation*, R. H. Jones and W. W. Gerberich, Eds., TMS-AIME, Warrendale, OH, 1986, p. 55.

149. R. P. Wei and J. D. Landes, *Mater. Res. Stand.* **9**, 25 (1969).

150. T. W. Weir, G. W. Simmons, R. G. Hart, and R. P. Wei, *Scripta Met.* **14**, 357 (1980).

151. R. P. Wei and G. W. Simmons, *Int. J. Fract.* **17**, 235 (1981).

152. R. P. Gangloff, *Environment Induced Cracking of Metals*, R. P. Gangloff and M. B. Ives, Eds., NACE, Houston, TX, 1990, p. 55.

153. P. E. Irving and C. J. Beevers, *Mater. Sci. Eng.* **14**, 229 (1974).

154. J. L. Robinson and C. J. Beevers, 2nd Inter. Conf. Titanium, Cambridge, England, 1972.

155. A. Yuen, S. W. Hopkins, G. R. Leverant, and C. A. Rau, *Met. Trans.* **5**, 1833 (1974).

156. M. F. Amateau, W. D. Hanna, and E. G. Kendall, *Mechanical Behavior, Proceedings of the International Conference on Mechanical Behavior of Materials*, Vol. 2, Soc. Mater. Sci., Japan, 1972, p. 77.

157. M. J. Harrigan, M. P. Kaplan, and A. W. Sommer, *Fracture Prevention and Control*, D. W. Hoeppner, Ed., Vol. 3, ASM, 1974, p. 225.

158. J. C. Chestnutt, C. G. Rhodes, and J. C. Williams, ASTM *STP 600*, 1976, p. 99.

159. G. R. Yoder, L. A. Cooley, and T. W. Crooker, *Proc. 2nd Int. Conf. Mech. Beh. of Mater.*, ASM, Metals Park, OH, 1976, p. 1010.

160. G. R. Yoder, L. A. Cooley, and T. W. Crooker, *J. Eng. Mater. Tech.* **99**, 313 (1977).

161. G. R. Yoder, L. A. Cooley, and T. W. Crooker, *Met. Trans.* **8A**, 1737 (1977).

162. G. R. Yoder, L. A. Cooley, and T. W. Crooker, *Eng. Fract. Mech.* **11**, 805 (1979).

163. G. R. Yoder, L. A. Cooley, and T. W. Crooker, *Trans. ASME* **101**, 86 (1979).

164. G. R. Yoder, L. A. Cooley, and T. W. Crooker, *Titanium 80*, H. Kimura and O. Izumi, Eds., Vol. 3, AIME, Warrendale, PA, 1980, p. 1865.

165. G. R. Yoder, L. A. Cooley, and T. W. Crooker, ASTM *STP 791*, 1983, p. I-348.

166. J. Masounave and J. P. Bailon, *Proc. 2nd Int. Conf. Mech. Beh. Master.*, ASM, Boston, 636 (1976).

167. J. Masounave and J. P. Bailon, *Scripta Met.* **10**, 165 (1976).

168. L. Wagner and J. K. Gregory, *Advanced Materials and Processes*, **146** (1), 50HH (1994).

169. K. T. Venkateswara Rao, W. Yu, and R. O. Ritchie, *Met. Trans.* **19A** (3), 549–561 (1988).

170. K. T. Venkateswara Rao, W. Yu, and R. O. Ritchie, *Met. Trans.* **19A** (3), 563–569 (1988).

171. R. O. Ritchie, *Metal Sci.* **11**, 368 (1977).

172. E. J. Prittle, *Fracture* 1977, **2** ICF 4, p. 1249.

173. G. Birkbeck, A. E. Inckle, and G. W. J. Waldron, *J. Mater. Sci.* **6**, 319 (1971).

174. T. C. Lindley, C. E. Richards, and R. O. Ritchie, *Metallurgia and Metal Forming*, 268 (Sept. 1976).

175. M. W. Lui and I. LeMay, *IMS Proceedings*, 1971, p. 227.

176. R. O. Ritchie, *Met. Trans.* **8A**, 1131 (July 1977).

177. R. D. Zipp and G. H. Walter, *Metallography*, **7**, 77 (1974).

178. I. LeMay and M. W. Lui, *Metallography*, **8**, 249 (1975).

179. G. W. J. Waldron, A. E. Inckle, and P. Fox, 3rd Scanning Electron Microscope Symposium, 1970, p. 299.

180. P. R. V. Evans, N. B. Owen, and B. E. Hopkins, *Eng. Frac. Mech.* **3**, 463 (1971).

181. P. E. Irving and C. J. Beevers, *Met. Trans.* **5**, 391 (1974).

182. G. Clark, A. C. Pickard, and J. F. Knott, *Engineering Fracture Mechanics*, Vol. 8, Pergamon, Oxford, England, 1976, pp. 449–451.

183. N. S. Cheruvu, *ASTM Symposium on Fractography in Failure Analysis of Metals and Ceramics*, ASTM, April 1982, Philadelphia.

184. P. S. Pao, W. Wei, and R. P. Wei, *Environment-Sensitive Fracture of Engineering Material*, Z. A. Fouroulis, Ed., AIME, Warrendale, PA, 1979, p. 3.

185. F. A. Heiser and R. W. Hertzberg, *J. Basic Eng. Trans. ASME* **93**, 71 (1971).

186. I. C. Mayes and T. J. Baker, *Fat. Eng. Mater. Struct.* **4** (1), 79 (1981).

187. R. C. Bates, Westinghouse Scientific Paper *69-1D9-RDAFC-P2*, 1969.

188. D. Broek, Paper 66, Second International Fracture Conference, Brighton, 1969, p. 754.

189. S. M. El-Soudani and R. M. Pelloux, *Met. Trans.* **4**, 519 (1973).

190. J. M. Barsom, *J. Eng. Ind., ASME, Series B* **92** (4), 1190 (1971).

191. J. H. Weber and R. W. Hertzberg, *Met. Trans.*, **2**, 3498 (1971).

192. B. Tomkins, *Philos. Mag.* **18**, 1041 (1968).

193. J. P. Hickerson, Jr., and R. W. Hertzberg, *Met. Trans.* **3**, 179 (1972).

194. S. Majumdar and J. D. Morrow, ASTM *STP 559*, 1974, p. 159.

195. S. D. Antolovich, A. Saxena, and G. R. Chanani, *Eng. Fract. Mech.* **7**, 649 (1975).

196. A. Saxena and S. D. Antolovich, *Met. Trans.* **6A**, 1809 (1975).

197. R. W. Hertzberg, *Met. Trans.* **5**, 306 (1974).

198. S. Pearson, *Nature (London)* **211**, 1077 (1966).

199. A. Ohta, N. Suzuki, and T. Mawari, *Int. J. Fat.* **14** (4), 224 (1992).

200. R. W. Hertzberg, *Int. J. Frac.* **64** (3), 135 (1993).

201. R. W. Hertzberg, *J. Mater. Sci. Eng.* **A190**, 25 (1995).

202. K. Sadananda and P. Shahinian, *Int. J. Fract.* **13**, 585 (1977).

203. B. Yu and M. Yan, *Fat. Engng. Mater. Struct.* **3**, 189 (1980).

204. J. A. Manson and R. W. Hertzberg, *CRC Crit. Rev. Macromol Sci.* **1** (4), 433 (1973).

205. R. W. Hertzberg and J. A. Manson, *Fatigue of Engineering Plastics*, Academic, New York, 1980.

206. R. W. Hertzberg and J. A. Manson, *Encyl. Polym. Sci. Eng.*, Vol. 7, 2nd ed., Wiley, New York, 1986, p. 378.

207. T. Pecorini, Ph. D. Dissertation, Lehigh University, Bethlehem, PA 1992.

208. T. Pecorini and R. W. Hertzberg, *Polymer*, **34** (24), 5053 (1993).

209. T. Clark, Ph. D. Dissertation, Lehigh University, Bethlehem, PA 1993.

210. H. R. Azimi, Ph. D. Dissertation, Lehigh University, Bethlehem, PA 1994.

211. H. R. Azimi, R. A. Pearson, and R. W. Hertzberg, *J. Mater. Sci. Lett.* **13**, 1460 (1994).

212. G. P. Koo, *Fluoropolymers, High Polymers*, Vol. 25, L. A. Wall, Ed., Wiley-Interscience, New York, 1972, p. 507.

213. G. P. Koo and L. G. Roldan, *J. Polym. Sci. Polym. Lett. Ed.* **10**, 1145 (1972).

214. G. Meinel and A. Peterlin, *J. Polym. Sci. Polym. Lett. Ed.* **9**, 67 (1971).

215. A. Ramirez, Ph.D. Dissertation, Lehigh University, Bethlehem, PA, 1982.

216. C. M. Rimnac, J. A. Manson, R. W. Hertzberg, S. M. Webler, and M. D. Skibo, *J. Macromol. Sci., Phys.* **B19** (3), 351 (1981).

217. P. E. Bretz, R. W. Hertzberg, and J. A. Manson, *J. Appl. Polym. Sci.* **27**, 1707 (1982).

218. G. Pitman and I. M. Ward, *J. Mater. Sci.* **15**, 635 (1980).

219. F. X. de Charentenay, F. Laghouati, and J. Dewas, *Deformation Yield and Fracture of Polymers*, Plastics and Rubber Institute, Cambridge, England, 1979, p. 6.1.

220. S. L. Kim, M. D. Skibo, J. A. Manson, and R. W. Hertzberg, *Polym. Eng. Sci.* **17** (3), 194 (1977).

221. S. L. Kim, J. Janiszewski, M. D. Skibo, J. A. Manson, and R. W. Hertzberg, *ACS Org. Coatings Plast. Chem.* **38** (1), 317 (1978).

222. J. C. Michel, J. A. Manson, and R. W. Hertzberg, *Org. Coatings Plast. Chem.* **45**, 622 (1981).

223. J. C. Michel, J. A. Manson, and R. W. Hertzberg, *Polym. Prepr. Am. Chem. Soc. Div. Polym. Chem.* **26** (2), 141 (1985).

224. J. C. Michel, J. A. Manson, and R. W. Hertzberg, *Polymer* **25**, 1657 (1984).

225. M. D. Skibo, Ph.D. Dissertation, Lehigh University, Bethlehem, PA, 1977.

226. R. W. Hertzberg, J. A. Manson, and M. D. Skibo, *Polym. Eng. Sci.* **15**, 252 (1975).

227. R. W. Hertzberg, M. D. Skibo, and J. A. Manson, ASTM *STP 700*, 1980, p. 49.

228. M. D. Skibo, R. W. Hertzberg, and J. A. Manson, *Fracture 1977*, **3**, 1127 (1977).

229. R. W. Hertzberg, J. A. Manson, and M. D. Skibo, *Polymer* **19**, 359 (1978).

230. R. W. Hertzberg, M. D. Skibo, J. A. Manson, and J. K. Donald, *J. Mater. Sci.* **14**, 1754 (1979).

231. J. D. Phillips, R. W. Hertzberg, and J. A. Manson, unpublished research.

232. M. T. Hahn, R. W. Hertzberg, R. W. Lang, J. A. Manson, J. C. Michel, A. Ramirez, C. M. Rimnac, and S. M. Webler, *Deformation, Yield and Fracture of Polymers*, Plastics and Rubber Institute, Cambridge, England, 1982, p. 19.1.

233. R. W. Lang, unpublished research.

234. R. W. Hertzberg, M. D. Skibo, and J. A. Manson, ASTM *STP 675*, 1979, p. 471.

235. E. H. Andrews, *J. Appl. Phys.* **32** (3), 542 (1961).

236. M. D. Skibo, R. W. Hertzberg, and J. A. Manson, *J. Mater. Sci.* **11**, 479 (1976).

237. R. W. Hertzberg and J. A. Manson, *J. Mater. Sci.* **8**, 1554 (1973).

238. J. P. Elinck, J. C. Bauwens, and G. Homes, *Int. J. Fract. Meek* **7** (3), 227 (1971).

239. C. M. Rimnac, R. W. Hertzberg, and J. A. Manson, ASTM *STP 733*, 1981, p. 291.

240. J. Janiszewski, R. W. Hertzberg, and J. A. Manson, ASTM *STP 743*, 1981, p. 125.

241. R. W. Hertzberg, M. D. Skibo, and J. A. Manson, *J. Mater. Sci.* **13**, 1038 (1978).

242. R. O. Ritchie and R. H. Dauskardt, *J. Ceram. Soc. Japan* **99** (10), 1047 (1991).

243. R. H. Dauskardt, *Acta Metall. Mater.* **41** (9), 2765 (1993).

244. S. Lathabai, J. Rodel, and B. R. Lawn, *J. Amer. Ceram. Soc.* **74**, 1340 (1991).

245. A. A. Steffen, R. H. Dauskardt, and R. O. Ritchie, *J. Amer. Ceram. Soc.* **76** (6), 1259 (1991).

246. R. H. Dauskardt, M. R. James, J. R. Porter, and R. O. Ritchie, *J. Amer. Ceram. Soc.* **75** (4), 759 (1992).

247. R. H. Dauskardt, B. J. Dalgleish, D. Yao, R. O. Ritchie, and P. F. Becher, *J. Mater. Sci.* **28**, 3258 (1993).

248. T. Hoshide, T. Ohara, and T. Yamada, *Int. J. Fract.* **37**, 47 (1988).

249. M. Ewart and S. Suresh, *J. Mater, Sci, Lett.* **5**, 774 (1986).

250. L. Ewart and S. Suresh, *J. Mater, Sci,* **22**, 1173 (1987).

251. S. Suresh and J. R. Brockenbrough, *Act Met.* **36**, 1455 (1988).

252. S. Suresh, *Int. J. Fract.* **42**, 41 (1990).

253. S. Suresh, *Fatigue of Materials*, Cambridge Solid State Science Series, Cambridge University Press, Cambridge, England, Chapter 13 (1991).

254. L. A. Sylva and S. Suresh, *J. Mater. Sci.* **24**, 1729 (1989).

255. R. Marissen, *Eng. Fract. Mech.* **19** (2), 261 (1984).

256. R. Marissen, *Int. Conf Aero. Set.*, Vol. 2.6.2, 1986, p. 801.

257. R. J. Bucci, L. N. Mueller, R. W. Schultz and J. L. Prohaska, 32nd Int. SAMPE Symp., Anaheim, CA, April 6, 1987.

258. R. O. Ritchie, W. Yu, and R. J. Bucci, *Engineering Fracture Mechanics* **32** (3), 361 (1989).

259. K. S. Chan and D. L. Davidson, *Proc. of Engineering Foundation, International Conference, Fatigue of Advanced Materials*, R. O. Ritchie, R. H. Dauskardt, and B. N. Cox, Eds., Santa Barbara, CA, 325 (1991).

260. R. W. Lang, J. A. Manson, and R. W. Hertzberg, *Polym. Eng. Sci.* **22** (15), 982 (1982).

261. R. W. Lang, J. A. Manson, and R. W. Hertzberg, *Polymer Blends and Composites in Multiphase Systems*, C. D. Han, Ed., ACS Adv. Chem. Ser. No. 206, ACS, New York, 1984, p. 261.

262. K. Friedrich, *Colloid. Polym. Sci.* **259**, 808 (1981).

263. K. Friedrich, *Deformation, Yield and Fracture of Polymers*, Plast. Rub. Inst., London, 1982, p. 26.1.

264. K. Friedrich, private communication, 1982.

FURTHER READING

S. Suresh, *Fatigue of Materials*, 2nd ed., Cambridge Solid State Science Series, Cambridge University Press, Cambridge, U.K., 1998.

J. A. Bannantine, J. J. Comer, and J. L. Handrock, *Fundamentals of Metal Fatigue Analysis*, Prentice Hall, Englewood Cliffs, NJ, 1990.

R. I. Stephens, A. Fatemi, R. R. Stephens, and H. O. Fuchs, *Metal Fatigue in Engineering*, 2nd ed., Wiley-Interscience, John Wiley & Sons, New York, 2000.

C. Bathias and P.C. Paris, *Gigacycle Fatigue in Mechanical Practice*, Marcel Dekker, New York, 2004.

PROBLEMS

Review

10.1 Dye penetrant inspection, eddy-current testing, and radiographic testing are examples of NDT techniques. What is NDT?

10.2 Assuming that a given component received a thorough proof test and was found to be satisfactory, would you still be concerned about failure of this component after it was placed into service?

10.3 As crack length gradually increases due to fatigue, what typically happens to the fatigue crack growth rate?

10.4 How does fatigue crack growth rate vary with applied stress level?

10.5 How is it that different styles of fracture mechanics specimen, like those described in Appendix B, can be used interchangeably when determining the fatigue properties of a material?

10.6 As a fatigue crack grows from its initial length to its final length, does it spend most of its life in the shorter end or the longer end of the crack size range?

10.7 For the case of a center notched panel that is loaded by concentrated forces acting at the crack surfaces, what would happen to the crack growth rate as the crack size increased?

10.8 The elliptical cracks depicted in Fig. 10.5 change aspect ratio as they grow. Why do they become deeper at a greater rate than they become wider?

10.9 What is the main difference between the safe-life and fail-safe design philosophies?

10.10 What are the macroscopic shape, orientation, and direction of propagation of the fracture surface of a typical fatigue crack growing in a metal component? What can be learned from the shape? If the shape depends on certain factors, be clear about what these are.

10.11 What are two differences between *clamshell/beach markings* and *fatigue striations*?

10.12 Are *fatigue striations* always visible on a fatigue fracture surface? If not, under what conditions are they likely to be present and under what conditions are they likely to be absent?

10.13 What important quantity (or quantities) can be determined by measuring striation spacing?

10.14 Is the Paris power law (Eq. 10-3) relevant for all conditions of fatigue crack growth?

10.15 When does the mean stress level affect fatigue crack propagation rates in metal alloys?

10.16 What does the *crack closure model* say about fatigue damage developed during the compressive portion of an $R < 0$ loading cycle? Also, what is ΔK_{eff} in the context of this model?

10.17 What discovery regarding K_{op} motivated the development of the adjusted compliance ratio (ACR) parameter?

10.18 What is the meaning of ΔK_{th}, and what implications, if any, does this definition have for design?

10.19 Under what circumstance could a fatigue crack become nonpropagating, even though the component experiences the same stress cyclic levels?

10.20 If a constant-stress load cycle is periodically interrupted by an anomalously large stress cycle, does the overall fatigue lifetime of the component remain the same, diminish, or increase?

10.21 What is the difference between EAC and corrosion fatigue? What role does K_{IEAC} play in both processes?

10.22 Does metal corrosion fatigue FCP rate change with increasing temperature? If so, does it typically increase or decrease?

10.23 Is the nominal fatigue crack propagation process similar for all types of metals and engineering plastics? If so, what is the relationship that best describes the process?

10.24 When choosing a polymer for fatigue crack propagation resistance, would you typically select an amorphous thermoplastic, a semicrystalline polymer, or a heavily cross-linked material? Other than crystallinity and cross-linking, what characteristic of a polymer can have a profound effect on FCP rate?

10.25 Do fatigue striations have the same interpretation in polymers as in metals?

10.26 Which fatigue marker bands are larger in polymeric solids—*fatigue striations* or *discontinuous growth bands* (DGB)? If one were to examine discontinuous growth band markings at a high magnification, what would you expect to see?

10.27 How does the crack growth rate–ΔK dependence of ceramic material compare with that of metals? What effect does this have on our ability to predict the fatigue life of ceramic components?

10.28 If crack-tip plasticity is not to blame for ceramic fatigue, what mechanism is responsible?

10.29 What is the general effect on FCP rate of adding reinforcing fibers and whiskers to form a composite?

Practice

10.30 A component was manufactured in 1950, according to best design practices associated with cyclic loading conditions. By 1960, that part was removed from service. What was the probable cause for removal of the part from service?

10.31 Imagine that you have two cracked components that are identical to one another except that Component A has a preexisting crack that is twice as long as that found in Component B. Does that mean that the fatigue lifetime of Component A will be 50% that of Component B?

10.32 A 10-cm-square, 20-cm-long extruded bar of 7075-T6511 is hollowed out to form a thin-walled cylinder (closed at one end), 20 cm long with an outer diameter of 9 cm. The cylinder is fitted with a 7-cm-diameter piston designed to increase pressure within the cylinder to 55 MPa.

a. On one occasion, a malfunction in the system caused an unanticipated pressure surge of unknown magnitude, and the cylinder burst. Examination of the fracture surface revealed a metallurgical defect in the form of an elliptical flaw 0.45 cm long at the inner diameter wall and 0.15 cm deep. This flaw was oriented normal to the hoop stress of the cylinder. Compute the magnitude of the pressure surge responsible for failure. (For mechanical property data see Chapter 7.)

b. Assume that another cylinder had a similarly oriented surface flaw but with a semicircular (a = 0.15 cm). How many pressure cycles could the cylinder withstand before failure? Assume normal operating conditions for this cylinder and that the material obeys a fatigue crack propagation relation

$$\frac{da}{dn} = 5 \times 10^{-39}(\Delta K)^4$$

where da/dn and ΔK have the units of m/cyc and Pa$\sqrt{\mathrm{m}}$, respectively.

10.33 A large steel plate is used in an engineering structure. A radical metallurgy graduate student intent on destroying this component decides to cut a very sharp notch in the edge of the plate (perpendicular to the applied loading direction). If he walks away from the scene of his dastardly deed at a rate of 5 km/h, how far away will he get by the time his plan succeeds? Here are hallowed hints for the hunter:

a. The plate is cyclically loaded uniformly from zero to 80 kN at a frequency of 25 Hz.

b. The steel plate is 20 cm wide and 0.3 cm thick.

c. The yield strength is 1400 MPa and the plane-strain fracture toughness is 48 MPa$\sqrt{\mathrm{m}}$.

d. The misled metallurgist's mutilating mark was measured to be 1 cm long (through thickness).

e. A janitor noted, in subsequent eyewitness testimony, that the crack was propagating at a velocity proportional to the square of the crack-tip plastic zone size. (The janitor had just completed a correspondence course entitled "Relevant Observations on the Facts of Life" and was alerted to the need for such critical observations.)

f. Post-failure fractographic examination revealed the presence of fatigue striations 2.5×10^{-4} mm in width where the crack was 2.5 cm long.

10.34 If the plate in the previous problem had been 0.15 or 0.6 cm thick, respectively, would the villain have been able to get farther away before his plan succeeded? (Assume that the load on the plate was also adjusted so as to maintain a constant stress.)

10.35 Estimate the stress intensity factor range corresponding to an observed striation spacing of 10^{-4} mm/cyc in the steel alloy shown in Fig. 10.56. Compare the results you would get when ΔK is determined from the striation data and the *macroscopic* data in the same figure. Also, compute ΔK from Eq. 10-9.

10.36

a. A material with a plane-strain fracture toughness of $K_{IC} = 55$ MPa$\sqrt{\mathrm{m}}$ has a central crack in a very wide panel. If $\sigma_{ys} = 1380$ MPa and the design stress is limited to 50% of that value, compute the maximum allowable fatigue flaw size that can grow during cyclic loading. (Assume that plane-strain conditions prevail.)

b. If the initial crack had a total crack length of 2.5 mm, how many loading cycles (from zero to the design stress) could the panel endure? Assume that fatigue crack growth rates varied with the stress intensity factor range raised to the fourth power. The proportionality constant may be taken to be 1.1×10^{39}

10.37 A thin-walled cylinder of a high-strength aluminum alloy ($K_{IC} = 24$ MPa$\sqrt{m}$) has the following dimensions: length = 20 cm; outer diameter = 9 cm; inner diameter 7 cm. A semicircular crack of depth $a = 0.25$ cm is discovered on the inner diameter and oriented along a line parallel to the cylinder axis. If the cylinder is repeatedly pressurized, how many pressure cycles could the cylinder withstand before failure? The pressure within the cylinder reaches 75 MPa, and the material obeys a fatigue crack propagation relation of the form

$$\frac{da}{dN} = 5 \times 10^{-39}(\Delta K)^4$$

where da/dN and ΔK have the units of m/cycle and Pa$\sqrt{m}$, respectively.

10.38 A 2-cm-long through thickness crack is discovered in a steel plate. If the plate experiences a stress of 50 MPa that is repeated at a frequency of 30 cpm, how long would it take to grow a crack, corresponding to the design limit where $K_{limit} = K_{IC}/3$? Assume that $K_{IC} = 90$ MPa $\sqrt{m}$ and the material possesses a growth rate relation where $da/dN = 4 \times 10^{-37} \Delta K^4$, with da/dN and ΔK being given in units of m/cycle and Pa$\sqrt{m}$, respectively.

10.39 An 8-cm-square bar of steel is found to contain a 1-mm corner crack, oriented perpendicular to the length of the bar. If an axial stress is applied from 0 to 420 MPa at a frequency of once every 10 minutes, how long will it take for the rod to fracture? The properties of the bar are: $K_{IC} = 90$ MPa$\sqrt{m}$, $\sigma_{ys} = 1500$ MPa, and the crack growth relation is $da/dN = 2 \times 10^{-37} \Delta K^4$, with da/dN and ΔK being given in units of m/cycle and Pa$\sqrt{m}$, respectively.

10.40 The presence of striations on the fatigue fracture surface of an aluminum alloy is used to determine the magnitude of an overload cycle. Striations immediately before the overload have a width of 2×10^{-4} mm, corresponding to 50% crack closure loading conditions; the overload cycle produced a striation width of 10^{-3} mm. What was the magnitude of the overload cycle?

Design

10.41 A certain steel alloy has been chosen for use in a fatigue limiting service application. Experimental test results provide the following mechanical properties for this material:

K_{IC} = 50 MPa$\sqrt{m}$

ΔK_{th} = 4 MPa$\sqrt{m}$

σ_{ts} = 1000 MPa

$da/dN = 4 \times 10^{-37} \Delta K^4$

A plate of this material 2 m wide is expected to experience a cyclic stress range of 200 MPa ($R = 0.1$) during component operation. You must develop a set of guidelines for the inspection and analysis of this component. If no defect is discovered as a result of NDT inspection capable of detecting cracks as small as 1 mm:

a. Is it safe to say that the user need not worry about either sudden or progressive failure of this plate?

b. If fatigue failure is a possibility, estimate the minimum service lifetime for this plate to determine the next inspection interval.

10.42 Your company designed a structure that includes an aluminum plate for which you specified routine inspection after every 50,000 loading cycles. The NDT procedure that is employed by the operator of the structure possesses a resolution limit of 1 mm. Through a mix-up, the inspection team calibrated their instrument to yield a crack resolution limit of 1 cm. No crack was found on this occasion, but unstable fracture took place following 34,945 additional loading cycles in association with the development of an edge crack, oriented normal to the major stress direction. Injuries to clients of the operator resulted from the failure. The operator is being sued, but is trying to pass the blame onto your design team. Are there grounds for a lawsuit against you for improper design and inspection specifications, or against the operator based on improper inspection procedures? The key stress level fluctuates between 50 and 100 MPa. The properties of the alloy are: $K_{IC} = 30$ MPa$\sqrt{m}$, $\sigma_{ys} = 550$ Mpa, $E = 70$ GPa, and the crack growth rate relation is given by $da/dN = 5 \times 10^{-35} \Delta K^4$, with da/dN and ΔK being given in units of m/cycle and Pa$\sqrt{m}$, respectively.

10.43 A disgruntled former employee has attempted to sabotage a local shipping firm's fleet of trucks. He was caught exiting the company grounds with a hacksaw in hand. He had already cut a long groove across one leaf spring of every truck (as shown below), intending to induce failures once the trucks were out on the road.

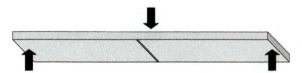

A leaf spring is loaded in 3-point bending. The maximum stress experienced when a truck is fully loaded and bouncing along is 250 MPa; under the worst-case scenario, this bouncing occurs at 1 Hz. The material has an elastic modulus of 205 GPa, a yield strength of 593 MPa, and a plane-strain fracture toughness value of 54 MPa$\sqrt{m}$. The deepest cut is 3 mm in a 6-cm-thick spring. The support span is 48 cm.

a. The company owner wants to know if it would be safe to drive the trucks temporarily at the maximum load so he can stay in business without risk of immediate failure. Please show clearly if this is the case. A safety factor of 50% is required (i.e., it must survive 1.5× the expected maximum stress).

b. Assume that while driving the stress usually alternates between ($\sigma_{MAX}-50$ MPa) and σ_{MAX} due to vibration. If this material has fatigue crack propagation parameters of $m = 3.0$ and $A = 1.0 \times 10^{-12}$ (where stress is in MPa and dimensions are in meters), how many load cycles and how many 8-hour days could you expect the leaf spring to survive when the truck is full? (Assume a 5× safety factor here.)

Extend

10.44 Find a drawing or a photograph that illustrates the appearance and location of a tear strap used in the fuselage section of an aircraft. Reproduce the image and supply a complete reference for the source.

10.45 The FAA Federal Airworthiness Requirements (FAR) 25.571 describes the requirements regarding failure resistance of commercial aircraft. In 1964, Amendment 25-0 required that aircraft possess "fail-safe" features. In 1978, Amendment 25-45 added a "damage tolerance" demonstration requirement. What is the difference between the fail-safe design philosophy and the damage tolerance design philosophy?

10.46 Find a recent journal paper that addresses the topic of "ultra high cycle fatigue" and copy its abstract. Also get a copy of the paper so you can extract one figure that has a connection to the contents of this chapter. Write one short paragraph summarizing the point of the figure, making a clear connection to the chapter topics. Give a citation for the paper in proper reference format.

10.47 Find a recent journal paper that addresses the topic of "fatigue in dental ceramics" or "fatigue in ceramic implants." Write a short review summarizing the point of the paper, making a clear connection to this chapter (and perhaps earlier chapters as well). Provide a copy of the abstract and give a citation for the paper in proper reference format.

Chapter 11

Analyses of Engineering Failures

We have come to the moment of truth—we must now use our knowledge and understanding of fracture mechanics and the relation between mechanical properties and microstructure to analyze actual service failures. However, before discussing recommended procedures for failure analyses and the details of several case histories, it is best to stand back for a moment and view component failures in a broader sense. To begin, we must ask who bears responsibility for these failures? Is it the company or individual that manufactured the component or engineering system, or the company or person that operated it when it failed? Such is the basis for debate in many product liability lawsuits. The chapter following this one examines the historical development of product liability litigation along with product recall experiences aimed at protecting and enhancing public safety. Regarding contested legal matters, opposing lawyers might ask of manufacturer and user the following questions:

1. Were engineering factors such as stress, potential flaw size, material, and environment considered in the design of the part?
2. Was the part underdesigned?
3. Was a proper material selection made for the manufacture of the part?
4. Was the part manufactured properly?
5. What limits were placed upon the use of the part and what, if any, service life was guaranteed?
6. Were operating instructions properly identified and component parts clearly labeled?
7. Were these limits conservative or unconservative?
8. Were these limits respected during the operation of the part?

A product liability case often becomes entangled in a number of ambiguities arising from incomplete or unsatisfactory answers to these questions. As such, it is important for the practicing engineer called in to analyze a failure and, perhaps, testify in court, to identify the major variables pertaining to the design and service life of the component. Because an individual from one field may be reluctant to challenge the conclusions drawn by an expert in another discipline, it becomes difficult to reconcile the two points of view without an overview of the facts involved. In many cases, these differences contrast the importance of the continuum versus the microstructural approach to the understanding of the component response (or failure). The most valuable expert witness is one who can appreciate and evaluate the input from different disciplines and educate the court as to their respective significance in the case under study. On the basis of such expert testimony, the courts are able to draw conclusions and render judgments. A delightful statement, made by the auditor for the 1919 Boston molasses tank lawsuit, relates to this decision-making process:

Weeks and months were devoted to evidence of stress and strain, of the strength of materials, of the force of high explosives, of the bursting power of gas and of similar technical problems. . . . I have listened to a demonstration that piece "A" could have been carried into the playground only by the force of a high explosive. I have thereafter heard an equally forcible demonstration that the same result could be and in this case was produced by the pressure caused by the weight of the molasses alone. I have heard that the presence of Neumann bands[i] in the steel herein considered along the line of fracture proved an explosion. I have heard that Neumann bands proved nothing. I have listened to men upon the faith of whose judgment any capitalist might well rely in the expenditure of millions in structural steel, swear that the secondary stresses in a structure of this kind were negligible and I have heard from equally authoritative sources that these same secondary stresses were undoubtedly the cause of the accident. Amid this swirl of polemical scientific waters it is not strange that the auditor has at times felt that the only rock to which he could safely cling was the obvious fact that at least one-half the scientists must be wrong. By degrees, however, what seem to be the material points in the case have emerged.[1]

A more recent service failure has had an even greater impact on our understanding of fracture and has led to the development of design procedures to guard against such future accidents. In this instance, a key structural member in the wing assembly of an F-111 fighter-bomber fractured, thereby leading to the loss of the plane and the death of the two pilots.[2] A post-fracture examination of the broken wing section revealed that a large crack, suspected of having been introduced during the heat-treatment procedure, had gone undetected during the various stages of fabrication and assembly of the wing component. The darkened appearance of the elliptically shaped defect on the fracture surface was believed to represent oxidation, which occurred during the normal heat treatment cycle of this forged part (Fig. 11.1). This preexistent flaw was surrounded by a narrow, shiny band that represented the extent of fatigue crack extension during the 109-h service life of the component. Beyond this point, fracture proceeded in an unstable fashion. Based on this aircraft accident and its associated failure analysis, the United States Air Force changed procedures regarding the safe design of aircraft; this led to the development of military specification MIL-A-83444.[3] Embodied within this document are requirements for the *damage tolerance* of a given component. That is, damage is assumed to exist in each component (e.g., a crack located at a rivet hole) and the structure is designed to ensure that the crack will not grow to a critical size within a specified period of time (recall Section 10.1.2). To perform the computations needed for this design procedure, it is necessary to know how the crack growth behavior of the specified material depends on such variables as the stress intensity factor range, load ratio, test temperature and frequency, environment, and complex load interactions. To this end, data described in Chapter 10 assume great importance in the implementation of this military specification.

In the following sections (as well as discussed earlier in Chapters 6, 9, and 10), attention is given to the identification of typical defects, consideration of fracture surface examination techniques, and identification of data needed for a successful failure analysis. A discussion of numerous service failures then follows.

Figure 11.1 Fracture surface of F-111 wingbox area. Dark, semielliptical surface flaw preexisted the flight service. Smooth bright band at boundary of dark flaw represents fatigue crack propagation zone prior to unstable fracture. (After Wood[2]; reprinted with permission from *Eng. Fract. Mech.* **7**, 557 (1975), Pergamon Press, Ltd.)

---

[i] Deformation twin bands in BCC iron.

11.1   TYPICAL DEFECTS

A wide variety of defects can be found in a given engineering component.[4] These flaws may result from such sources as material imperfections, defects generated during service, and defects introduced as a result of faulty design practice. Regarding the first source mentioned, defects can be found within the original material supply or can be introduced during the manufacturing process. Typical material defects include porosity, shrinkage cavities, and quench cracks. Other microstructural features can trigger crack formation if the applied stresses exceed some critical level. These include nonmetallic inclusions, unfavorably oriented forging flow lines (recall Section 7.3), brittle second phases, grain-boundary films, and microstructural features resulting from 300°C and temper embrittlement. The list of manufacturing defects includes machining, grinding and stamping marks (such as gouges, burns, tears, scratches, and cracks), laps, seams, delaminations, decarburization, improper case hardening, and defects due to welding (e.g., porosity, hot cracking, cold cracking, lack of penetration, and poor weld bead profile).

Defects can be introduced into the component during service conditions as a result of excessive fretting and wear. Environmental attack can also cause material degradation as a result of general corrosion damage, liquid metal and hydrogen embrittlement, stress corrosion cracking, and corrosion fatigue. Surely, cyclic loading can initiate fatigue damage without an aggressive environment and may lead to serious cracking of a component.

Finally, defects can be introduced into a component through faulty design. These human errors include the presence of severe stress concentrations, improper selection of material properties and surface treatments, failure to take remedial actions (such as baking a steel part after it has been cadmium plated to remove charged hydrogen gas), inadequate or inaccurate stress analysis to identify stress fields in the component, and improper attention to important load and environmental service conditions as they relate to material performance.

11.2   MACROSCOPIC FRACTURE SURFACE EXAMINATION

The functions of a macroscopic fracture surface examination are to locate the crack origin, determine its size and shape, characterize the texture of the fracture surface, and note any gross markings suggestive of a particular fracture mechanism. To begin, one should attempt to identify whether there are one or more crack origins, since this may provide an indication of the magnitude of stress in the critical region. In general, the number of crack nuclei increases with increasing applied stress and magnitude of an existing stress concentration factor (Fig. 9.5). Even when one crack grows to critical dimensions, secondary cracks can develop before final failure because of load adjustments that may accommodate the presence of the primary defect.

Whether there are one or more fracture nucleation sites, it is of utmost importance to locate them and identify precisely the reason for their existence. When the fracture mechanism(s) responsible for growth of the initial defect to critical proportions is known, the engineer can recommend "fixes" or changes in component design.

The task at this point is to find the origin. This was not difficult in the case of the tricycle and doorstop failures mentioned in Chapter 9, but one can well imagine the difficulty of sifting through the wreckage of a molasses tank, ship, or bridge failure (see Chapter 5) to find their respective fracture origins. For these situations, there could be literally thousands of linear meters of fracture surface to examine. Where does one begin? Once begun, how does one know the direction in which to proceed to locate the origin? As discussed in Section 5.8.1, the fracture surface often reveals contour lines that point back to the crack origin. These features, referred to as chevron markings, are found on the fracture surfaces in many engineering solids and aid the investigator in locating the region where the crack had formed or preexisted. The microscopist is then able to focus attention on the micromorphological features of the origin and gain insight into the cause of failure. Sometimes, however, these markings may be obscured by other fracture markings, such as by secondary fractures in anisotropic materials. A crack "divider" orientation Charpy specimen of banded steel (Fig. 10.56) reveals many fracture surface

Figure 11.2 Crack "divider" fracture surface. Delaminations obscure anticipated chevron pattern.[5] (Reprinted with permission from American Society for Metals, *JISI* **209**, 975 (1971).)

delaminations caused by σ_z stresses acting parallel to the crack front, but they cloud the expected "chevron" pattern (Fig. 11.2). As discussed earlier in Section 6.9, macroscopic shear lip development on the component's fracture surface sometimes can identify the extent of plane stress vs. plane strain conditions and the local stress intensity level.

It is possible that a crack may initiate by one mechanism and propagate by one or more different ones. For example, a crack may begin at a metallurgical defect, propagate for a certain distance by a fatigue process, and then continue growing by a combination of fatigue crack propagation and environment-assisted cracking when the stress intensity factor exceeds K_{IEAC}. Such mechanism changes may be identified by changes in texture of the fracture surface. For example, the fracture surface shown in Fig. 11.3a reveals the different textures associated with fatigue and stress corrosion subcritical flaw growth. In the broken wing strut (Fig. 11.3b), we find the regions of FCP (shiny areas) interrupted by two separate localized crack instabilities (dull areas), which probably were caused by two high-load excursions during the random loading life history of the strut. Another example of a plane-strain "pop-in" is found on the fracture surface of a fracture-toughness test sample that exhibited Type II (Fig. 6.36) load-deflection response (Fig. 11.3c).

"Pop-in" can also result from the presence of local residual stresses. If a crack is embedded within a localized stress concentration region, the application of a moderate load could develop a stress intensity level (magnified by the local stress concentration) equal to K_{IC} or K_C (depending on the prevailing stress state), which would cause the crack to run unstably through the component. However, the crack would soon run out of the region of high stress concentration and arrest (and produce a fracture surface marking), since the moderate load without the stress concentration does not possess the necessary driving force to sustain crack growth.

Other macroscopic arrest lines, such as the fatigue crack propagation "beach mark" and load block band, were discussed in Chapter 9 (Fig. 9.3a) and Chapter 10 (Figs. 10.38a, b and 10.39b). A more critical examination of Fig. 9.3a reveals a striking feature. As expected, the crack initiation site in the steel shaft is located in the vicinity of a stress concentration at the base of the shaft key way. The exceptional aspect of this fracture is the extent to which the fatigue crack was able to grow prior to the onset of unstable fracture. Indeed, the crack is seen to have grown more than 90% across the component width. The fatigue fracture in a magnesium housing provides another example of this unusual pattern (Fig. 11.4a).[7] In this instance, multiple cracks had initiated (note the ratchet lines) at the center hole and propagated across more than 95% of the section width. For cracks to grow to the extent noted in Figs. 9.3a and 11.4a, one of two scenarios must exist: either the fracture toughness of the material must be extremely high or the stress level must be extremely low in association with a normal value of fracture toughness. The first scenario can be dismissed since the steel and magnesium alloys

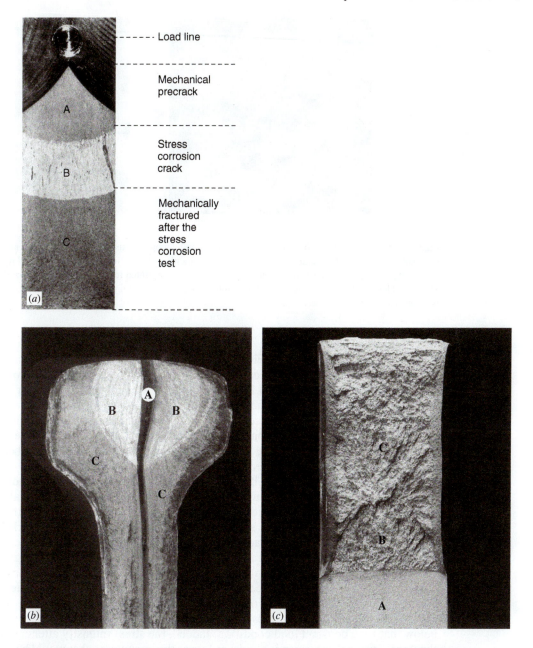

------ Load line

Mechanical precrack

A

Stress corrosion crack

B

Mechanically fractured after the stress corrosion test

C

(a)

(b)

(c)

Figure 11.3 Macrofractographs revealing fracture mechanism transitions. (*a*) Transition from fatigue A to stress corrosion cracking B to fast fracture C;[6] (*b*) wing strut with metallurgical delamination A, fatigue B, and static fracture C; (*c*) fracture-toughness sample revealing fatigue precracking zone A, pop-in instability B, and fast fracture C. (Figure 11.3*a* reprinted with permission of Markus O. Speidel, Brown Boveri Co.)

shown do not possess unusual fracture properties. Therefore, fracture in both instances must have taken place in association with a very low stress level. A dichotomy is immediately apparent. If the stress level was so low, then the resultant cyclic stress intensity levels at nascent cracks would be below the threshold stress intensity for the material, and any preexisting crack would not grow: hence, failure would never have occurred.

The rationalization for this dilemma is that the load level was, indeed, high enough to cause the crack to initiate but dropped progressively as the crack lengthened. There are two ways in

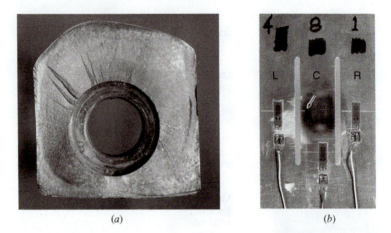

(a)   (b)

Figure 11.4   (*a*) Fatigue fracture surface in a magnesium helicopter housing. Note that fatigue markings are observed over 90% of the fracture surface. (*b*) three-ligament specimen [left (L), center (C), and right (R)] containing small crack (see arrow) in central (C) load path.[7] (Reprinted from *International Journal of Fatigue*, **15**, R. W. Hertzberg and T. J. Pecorini, "An examination of load shedding during fatigue fracture," p. 509, 1993, with permission from Elsevier.)

which the load level can drop. First, loads can shed if the component is loaded under fixed displacement conditions (e.g., recall the bolt-loaded compact sample used to generate stress corrosion cracking data in Fig. 8.10). Alternatively, load shedding can occur when there is a redundant load path such that the load in the cracked segment would be transferred to the unbroken ligament(s), much as load transfer occurs in a composite material from a low-stiffness to a high-stiffness component. That is, as the crack grows in one load path, the stiffness of that segment decreases relative to the unbroken load path(s) and the load will shed to the stiffer members. As the number of redundant paths increases, the response of the component approaches that of a component being loaded under fixed displacement conditions.

To confirm a load shedding scenario, a three-ligament 2024-T3 aluminum alloy specimen (Fig. 11.4*b*) was tested in fatigue with a crack introduced into the 23-mm-wide center load path.[7] Figure 11.5*a* shows the relation between load and crack length for the actual data measured from the cracked central path of the three-ligament specimen, as well as the theoretical curves based on constant load and the three-ligament loading configurations. Clearly, the measured decrease in the actual load in the cracked ligament is in reasonable agreement with theoretical expectations. Figure 11.5*b* depicts the theoretical *K* levels for conditions of constant load, constant displacement, and multiple-path load shedding, respectively. Load shedding greatly reduces the stress intensity in this specimen geometry below that for a constant load condition. Indeed, the stress intensity attenuation produced in the three-ligament specimen is almost as low as the attenuation that would be produced under constant displacement. Also note that the *K* level, corresponding to an *a/W* level of 0.95, remains beneath the fracture toughness value for the 2024-T3 aluminum alloy three-legged specimen.

The fracture surface of the cracked central ligament progressed entirely across its width in a flat fracture mode, corresponding to plane-strain conditions associated with minimal escalation in the *K* level with increasing crack length (Fig. 11.6). This appearance is strikingly similar to load-shedding service failures shown in Figs. 9.3*a* and 11.4*a*. (No clamshell markings are observed in the present sample since periodic load fluctuations were not introduced.) By comparison, a crack in a single-ligament sample (with no load shedding) developed shear lips at *a/W* ≈ 0.25 and subsequently ruptured abruptly (see the arrow); the latter fracture surface mode reflects a significant increase in stress intensity with increasing crack length (Fig. 11.5*b*).

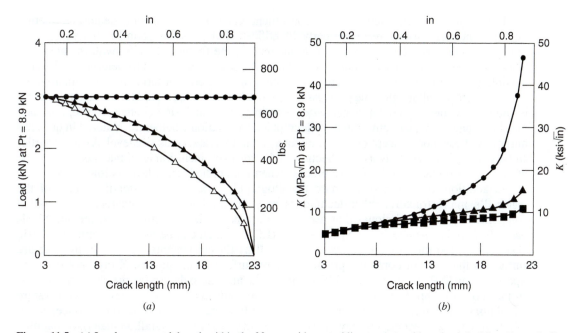

Figure 11.5   (*a*) Load versus crack length within the 23-mm-wide central ligament: ●, without load shedding (theoretical); ▲, with load shedding (theoretical); △, measured during test (*b*) theoretical relationship between stress intensity and crack length within a 23-mm-wide central ligament: ●, without load shedding; ▲, with load shedding; ■, constant displacement.[7] (Reprinted from *International Journal of Fatigue*, 15, R. W. Hertzberg and T. J. Pecorini, "An examination of load shedding during fatigue fracture," p. 509, 1993, with permission from Elsevier.)

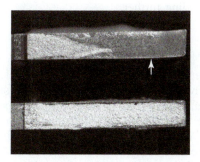

Figure 11.6   Macroscopic appearance of fatigue failures in single-ligament (top) and three-ligament (bottom) samples. Arrow indicates onset of fast fracture in the single-load path sample. Crack growth direction is from left to right. Note lack of shear-lip development in the three-ligament sample (bottom) in association with a modest increase in ΔK level with increasing crack length.[7] (Reprinted from *International Journal of Fatigue*, **15**, R. W. Hertzberg and T. J. Pecorini, "An examination of load shedding during fatigue fracture," p. 509, 1993, with permission from Elsevier.)

11.3   METALLOGRAPHIC AND FRACTOGRAPHIC EXAMINATION

Having located the crack origin, a typical failure analysis would proceed normally with two main interim objectives: (1) identification of the micromechanism(s) of subcritical flaw growth, and (2) estimation of the stress intensity conditions prevailing at the crack tip when failure occurred. As was discussed in Chapter 5, metallographic techniques have often been used to determine the crack path relative to the component microstructure. In addition to identifying the microscopic fracture path (transgranular versus intergranular), metallographic sections are useful in establishing the metallurgical condition of the material. Grain size and shape offer important clues to the thermomechanical history of the component (see Chapter 7). For example, a coarse-grained structure is indicative of a very high-temperature annealing process, while an elongated grain structure indicates not only the application of a deformation process in the history of the material, such as rolling, forging, and drawing, but also the deformation direction. Such information allows one to anticipate the presence of anisotropic mechanical properties that must be identified in relation to the predominant stress direction. Examples of mechanical property anisotropy were given in Tables 7.5 and 7.8.

Determination of the microstructural constituents enables the examiner to determine whether a component has been heat treated properly. Identification of a possible grain-boundary phase, for example, can explain the occurrence of an intercrystalline fracture. Finally, with the aid of an inclusion count, the relative cleanliness of the metallurgical structure can be determined. It is not possible to reliably express the fracture toughness of a material in terms of some measure of inclusion content [although some progress has been achieved in this regard (see Chapter 7)], but it is known that fracture toughness decreases with increasing inclusion content. Hence, a trained metallographer may ascertain from metallographic examination whether the material in question might have been more susceptible to failure for a given stress and crack size level. As noted earlier and discussed in Case Study #1 (see Section 11.5), however, inclusion level may not prove to be of primary importance in the failure process when the applied stress level is low.

In addition to the use of standard metallographic procedures to identify details of the material's microstructure, other techniques may be used to further characterize a failed component. Several techniques that use X-rays are particularly useful. These include SEM-based Energy Dispersive X-ray Spectroscopy (EDS), which can be used to identify the basic chemical composition of the component as well as those elements present on the fracture surface in the form of corrosion products and/or surface contaminants. X-ray Photoelectron Spectroscopy (XPS or sometimes ESCA) allows the analyst to determine the chemical state of surface corrosion products; in some cases, XPS can confirm exposure to elevated temperature based upon the nature of a high temperature oxide film present on the fracture surface. X-ray Diffraction (XRD) techniques may be used to identify unknown crystalline materials that may be present. Finally, radiographic analysis of a component/assembly can often detect internal damage, potentially a result of a manufacturing defect.

A fractographic study with either a scanning or transmission electron microscope reveals the microscopic character of the fracture surface. Since this topic has been discussed at considerable length in earlier chapters, the reader is referred specifically to Chapter 5, Chapter 10, and Appendix A before proceeding further.

11.4 COMPONENT FAILURE ANALYSIS DATA

Having been introduced to the fundamentals of fracture mechanics analysis, stress intensity analysis of cracks, macroscopic and microscopic features of the fracture surface, and the pertinent mechanical property data necessary to adequately characterize the performance of a given material, the reader should be able to synthesize this information and solve a given service failure problem. A checklist will often assist the investigator in this task. Such a checklist suitable for failure analysis of metallic components is provided in Appendix D in a format that could be copied and taken to a failure investigation. (It could also be modified to include the key features of failures expected for other material systems.) The checklist, which takes into account the component geometry, stress state, flaw characterization, fractographic observations, metallurgical information (including component manufacture and other service information), summarizes the raw data desirable for a complete failure analysis of a fractured component. Experience has shown, however, that in the majority of instances, certain facts are never determined, and educated guesses or estimates (such as the K calibration estimations described in Section 6.5) must be introduced to complete the analysis. The objective of a checklist is to minimize the amount of guesswork while maximizing the opportunity for firm quantitative analysis. The reader may wish to "follow along" with the Appendix D checklist as the following case histories are analyzed.

11.5 CASE HISTORIES

Failure analysis case studies often serve as fascinating (and enlightening) sources of practical information. The curious reader is referred to compilations of more than 100 failure analyses that have been reported elsewhere.[8-11] In addition, several journals and handbooks that focus on failure analyses and fracture are cited at the conclusion of this chapter. The collection of papers[11] from ASTM is of particular interest because fracture mechanics analyses are used extensively. In one material selection case history from this collection, Reid and Baikie[12] analyzed the static and cyclic loading service conditions for high-pressure water pipes and concluded that one particular steel alloy was preferred over another, even though the less favorable one was approximately 50% cheaper than the alloy chosen. It was found that the considerably greater potential service life and greater margin of safety associated with the steel alloy selected more than compensated for its relatively higher unit cost. The reader should recognize that the total price tag for a given component or engineering system depends on both initial material and fabrication costs, along with expenses associated with maintenance and repair. Often, initial material costs are of secondary importance and are subordinate to other financial considerations. The paper by Pearson and Dooman[13] in the same collection is of interest because the fracture mechanics analysis demonstrated that failure of a truck-mounted propane tank was more complicated than first assumed. A fractographic analysis revealed that a girth weld crack had existed for several years prior to failure and was initially thought to be responsible for the explosion of the tank. However, a fracture mechanics analysis revealed that failure could not have occurred unless the preexisting crack had experienced internal gas pressures far in excess of the allowable level. Further investigation confirmed that such high gas pressures were present: The relief valve of the tank was badly corroded and unable to open, therefore producing excessive gas pressures within the tank. Also contributing to higher gas pressures were solar heating and the proximity of the tank to the truck's hot tailpipe. In the remainder of this section, nine additional case histories are discussed in detail; all involve the application of fracture mechanics principles to failure analysis. Although not all component failures require a fracture mechanics analysis, the latter approach is emphasized here to demonstrate the applicability of the fundamental concepts introduced in Chapters 5 through 10.

Case 1: *Shotgun Barrel Failures*

This case history is unique among the nine in this set insofar as it describes several product liability lawsuits in addition to presenting the technical details of multiple, related shotgun failures. For a more in-depth discussion of the history of product liability law and the emerging societal focus on product recall, the reader is referred to the next chapter. It is often beneficial for technical experts to become well informed regarding the legal and ethical aspects of these matters.

Overview of Failure Events and Background Information   During recreational trap/skeet shooting outings at different firing ranges, several shotgun barrel failures occurred over a period of time that were all characterized by a pronounced "petal-like" or splayed open appearance (see Fig. 11.7). Failures had initiated in the barrel's breech area in the form of multiple, longitudinal through-wall cracks, with barrel segments having peeled back and away from their original positions. Of note, these barrels never fragmented; instead, fracture paths were predominantly restricted to those planes oriented parallel to the barrel axis. What happened? Was there

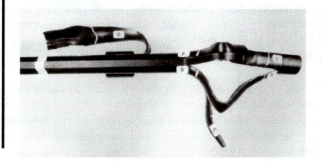

Figure 11.7   Fractured shotgun barrel revealing splayed appearance with fractures occurring essentially along planes parallel to the barrel axis.

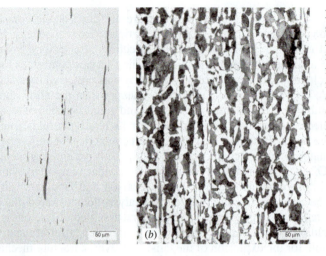

Figure 11.8 Longitudinal section of modified 1140 steel revealing elongated MnS inclusions embedded within a ferrite and pearlite microstructure. (*a*) Unetched; (*b*) etched 2% Nital. (Photographs courtesy of MariAnne Sullivan, Lehigh University).

something wrong with the gun barrel material? Was there a defect in the gun? Finally, was there some other cause, such as an obstruction in the barrel or a problem with the ammunition used by the gun owner?

Referring to the Failure Analysis Checklist in Appendix D, critical information to be gathered includes the following: the material (including its method of manufacture, microstructure and mechanical properties), the possible presence of a preexisting defect in the gun barrel (based on metallographic and fractographic examination), and finally, the likely stress level at the time of the fractures.

The gun barrels were produced from AISI modified 1140 steel; additional amounts of manganese (Mn) (1.05–1.2%) and sulfur (S) (0.16–0.23%) were introduced to enhance alloy machinability.[14] Raw material stock initially was hot worked and then warm rotary forged to produce hollow tubes with an outer diameter of 2.5 cm and a wall thickness of 3 mm. The microstructure of this tube product contained a mechanically fibered and fine-grained mixture of ferrite and pearlite along with manganese sulfide (MnS) stringers (recall Section 7.2.1); these stringers were encapsulated primarily within ferrite regions and were oriented parallel to the forging direction (i.e., the axis of the tube blank [e.g., see Fig. 11.8]).

The stress histories for these guns were reported as follows. As part of an internal quality assurance program, each barrel was initially fired at least once, using a factory "proof round" before the gun was approved for sale to the public. The burning gunpowder from the "proof" cartridge generates a pneumatic pressure of 117–152 MPa that corresponds to a hoop stress range of 370–480 MPa in the nominally thin-walled cylindrical barrel (recall Eq. 1-43). In ordinary use, gas pressures of 55–83 MPa are generated when a factory-made or properly prepared home-reloaded cartridge is fired; the latter gas pressure creates a hoop stress of 175–260 MPa. (Gun owners sometimes reuse spent cartridges that are then reloaded at home using a reloading device.) No specific information was available as to the total number of factory and/or reloaded cartridges that were fired other than to note that final failure had occurred in conjunction with the use of a reloaded spent cartridge. (Even that information was disputed in some cases in which the user of the gun denied using reloaded shells and claimed that a factory-made shell was in use at the time of the failure.) Information as to the possible preexistence of a barrel defect will be discussed below.

Proposed Causation Theories Firearm failures are typically caused by either: (1) use of improper ammunition; (2) existence of a defect in the firearm; or (3) misuse of the firearm. Opposing parties in several different lawsuits associated with the shotgun failures advanced two fundamentally different fracture theories. Several different litigants (those parties attempting to collect damages) speculated that MnS *inclusion clusters* were present along the length of these gun barrels, along with randomly dispersed MnS inclusions (the latter being commonly observed in certain steels). These litigants argued that fatigue cracks would initiate somewhere within these presumed MnS clusters and subsequently grow each time the shotgun was fired. It was speculated that total barrel failure would occur once one of these fatigue cracks reached a critical size. In addition, they argued that since these MnS inclusions were aligned perpendicular to the major stress axis (i.e., the hoop stress direction within the gun barrel), the material's toughness and maximum ductility levels would be reduced to unacceptable levels by the presence of the aligned MnS inclusions. Alternatively, the opposing litigant theorized that these barrel failures were caused by individual overloading events associated with the use of faulty reloaded cartridges that contained an excessive gunpowder/shot charge. In this scenario, it was argued that errors were made when the gun owner reloaded spent cartridges at home. Furthermore, it was argued that the somewhat reduced toughness and ductility levels exhibited by the barrel material were not relevant to normal design parameters, since *design operating stresses*, associated with the use of normal rounds of ammunition, are many times lower than those required for a catastrophic burst scenario.

Fractographic Evidence of Failed Gun Barrels   The fracture surfaces of the litigants' gun barrels revealed a mixture of microvoid coalescence (MVC) and cleavage (Fig. 11.9). The existence of cleavage markings is consistent with an overload-induced high strain rate event in this BCC steel alloy. In an attempt to simulate this type of appearance in the laboratory, an exemplar gun was fired with a deliberately overloaded shell. (Courtroom testimony was presented to confirm that an overloaded cartridge could be produced with the reloading apparatus used by the various plaintiffs.) The macroscopic appearance of this gun tube exhibited a splayed open appearance, strongly resembling that found in the litigants' guns (recall Fig. 11.7); the microscopic fracture surface appearance of the deliberately overloaded barrel revealed the same mixture of MVC and cleavage.

Also present on the fracture surfaces of both the actual failures and the deliberately overloaded gun were highly aligned channels, corresponding to the location of individual MnS stringers in the microstructure (Fig. 11.9). Since these stringers separated interfacially from the matrix, some channels revealed the presence of MnS particles while others presented an imprint of where the stringer had been stripped away from that fracture surface. *Importantly, no MnS inclusion clusters were found either on the actual fracture surfaces of any of these failed gun barrels or on the deliberately overloaded barrel surfaces.* The presence of cleavage and the absence of MnS clusters strongly suggest that the litigants' barrel failures were associated with a high strain rate fracture event, consistent with the use of faulty overloaded ammunition.

Estimation of the Material's Fatigue Endurance Limit   Efforts were made to estimate the fatigue properties of gun barrel material and to document the related fractographic evidence for the

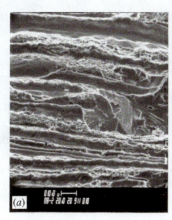 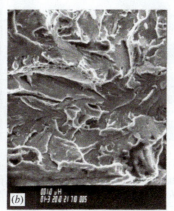

Figure 11.9   Representative micrographs from a failed shotgun and a deliberately overloaded gun: (*a*) microvoid coalescence and cleavage adjacent to MnS inclusions; (*b*) cleavage.

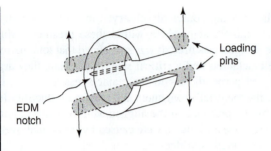

Figure 11.10 Fatigue C-shaped test sample and loading configuration. (Reprinted with permission from R. W. Hertzberg, K. S. Vecchio, and F. E. Schmidt, Jr., *Journal of Testing and Evaluation*, **17**, 261 (1989), copyright ASTM International, 100 Barr Harbor Drive, West Conshohocken, PA 19428.)

fatigue cracking process. To that end, fatigue tests were conducted using full-scale gun barrels as well as notched and unnotched C-shaped ring samples removed from barrel material. Regarding the latter, 1.9-cm-wide rings were removed from actual gun barrels and then slotted in the longitudinal direction so as to create a C-shaped test specimen (Fig. 11.10).[15] This specimen configuration, consistent with that described in ASTM E 399, has been recommended for the evaluation of the fatigue and fracture toughness response of cylindrical pressure vessels. [Also, see Appendix B and the arc-shaped specimen A(T)]. Details of the various test procedures and material prior loading history are described by Hertzberg et al.[15] An elastic-plastic finite element analysis was conducted to determine each stress level corresponding to the various loads used in the C-ring fatigue experiments that were performed. From these fatigue tests, it was determined that, after at least five million loading cycles in any sample (including the notched samples), no fatigue failures occurred when the C-ring was subjected to maximum stress levels up to 450 MPa.[15] Since that stress level corresponds to the gun barrel factory proof round, it follows that no fatigue fractures would be expected in the gun barrel when either normal factory or properly prepared reloaded cartridges are used.

Microfractography of Fatigue Fracture in Gun Barrel Material Fractographic examination of the fatigued C-ring samples (stresses in excess of the proof stress) revealed clear evidence for the presence of fatigue striations. Depending on the local direction of the advancing fatigue crack front, fatigue striations were observed both parallel to and perpendicular to the aligned MnS inclusions (Fig 11.11). No cleavage and microvoid coalescence markings were found on the fracture surfaces of these fatigued samples. Furthermore, no evidence was found for MnS inclusion *clustering* on any of the fatigue samples' fracture surfaces. Instead, individual MnS stringers were found randomly dispersed across the fracture surface. Since no fatigue failure

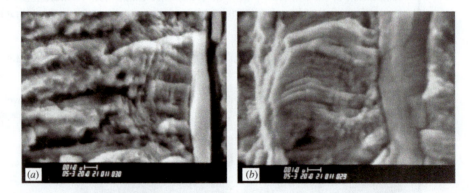

Figure 11.11 SEM micrographs from fatigue test samples, revealing fatigue striations oriented both (*a*) parallel and (*b*) perpendicular to MnS inclusions. (Reprinted with permission from K. S. Vecchio, R. W. Hertzberg, and F. E. Schmidt, Jr., *Journal of Testing and Evaluation*, **17**, 267 (1989), copyright ASTM International, 100 Barr Harbor Drive, West Conshohocken, PA 19428.)

would be expected with the use of normal rounds and no evidence found for the fatigue process (i.e., fatigue striations) on the plaintiff gun barrels' fracture surfaces, one could logically conclude that a fatigue fracture process did not occur in the actual failures. Instead, as noted above, the presence of cleavage and microvoid coalescence on the fracture surface would suggest an overloaded cartridge-induced fast fracture event.

Based on observed fatigue striation width measurements (e.g., 5×10^{-7} m/cyc) and using the Bates and Clark relation (Eq. 10-9), the gun barrel material is capable of sustaining a stress intensity level of at least 60 MPa$\sqrt{m}$. As such, one may use a leak-before-break computation (recall Eq. 6-32) to determine a stable crack length prior to final failure. Therefore,

$$K = \sigma\sqrt{\pi a}$$

where $K = \geq 60$ MPa$\sqrt{m}$
$\sigma =$ normal round hoop stress (207 MPa)
$a =$ half-crack length

It follows that a total crack length ($2a$) of approximately 5 cm, aligned along the length of the gun barrel, would have been tolerated before final failure. Since the barrel wall thickness was only 3 mm, it is obvious that a leak-before-break condition would have existed prior to final failure. Furthermore, were there to have been such a stable through-thickness crack, then as this hypothetical crack were to have grown, one would expect that smoke would have likely exited from the crack each time the gun was fired. No such evidence was reported. Furthermore, abundant evidence for fatigue striations would have been found on the barrels' fracture surfaces. No such electron fractographic evidence was found on a plaintiff's barrel fracture surface.

To further explore a leak-before-break scenario, a 22-mm-long slot (oriented parallel to the gun barrel axis and along the inner diameter) was machined into two different new gun barrels; each slot extended approximately 75% through the barrel wall thickness. These slots were introduced to simulate the existence of a "pseudo-cluster" of MnS inclusions in a hypothetically defective gun barrel. When these gun barrels were examined after being subjected to more than 2800 and 3200 normal factory rounds, respectively, fatigue cracks were observed along the longitudinal axis of each barrel (Fig. 11.12). Clearly, these loading cycles had caused a fatigue crack to grow from the machined slot's root and through the barrel's wall thickness to the outer surface. Unlike the actual failures and the barrel that had experienced a deliberately over-charged round, these barrels did not splay open in a catastrophic manner (recall Fig. 11.7);

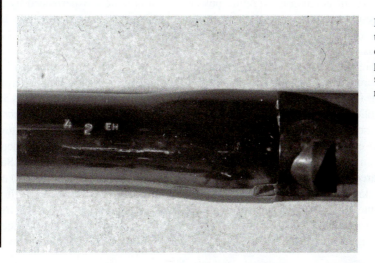

Figure 11.12   Profile of through-thickness ∼25-mm-long fatigue crack seen along outer diameter of pre-cracked barrel. Gun barrel subjected to almost 2900 factory rounds of ammunition.

instead, the barrel's overall appearance was unaltered, except for the development of the fatigue crack. Furthermore, when one of these pre-cracked and subsequently fatigued barrel's fracture surface was examined, extensive evidence for fatigue striations was found.

Clearly, the postulate that inclusion clusters had caused fatigue failure of the accident barrels was unsupported by post-failure experimentation and analysis. Instead, the overloaded barrel experiments and associated fractographic observations demonstrated that the failed gun barrel appearance was consistent with an overloading event such as one might find in connection with the use of an overcharged round of ammunition.

As highlighted in the fatigue test results described above, one of the design aspects of the gun barrels included a leak-before-break design feature. Furthermore, the absence of barrel fragmentation in association with the fatigue fracture events provided an additional measure of safety for those nearby individuals who might otherwise have been injured by flying shrapnel from a burst barrel; the latter is attributable to the directional fracture characteristics of the fibrous barrel material, including the randomly distributed MnS inclusions.

The Verdicts Even though the basic technical arguments presented by the gun users and the shotgun manufacturer were essentially the same in each case, in some cases the juries returned verdicts for the gun users and in other cases for the shotgun manufacturer. Why? The authors believe that these different jury verdicts may well have resulted not from differences in the technical evidence, but rather from differences arising within the courtroom. These include different jury pools with potentially different capacities to determine key facts from speculation; differences in individual juror's backgrounds and personal biases; different plaintiffs with different abilities to invoke sympathy from a jury; different lawyers with different procedural approaches; and different expert witnesses with different technical strengths (and weaknesses) and different communication skills. Regarding the latter, the reader, having hopefully digested the technical material presented thus far in this book, is clearly more informed about the technical nuances of a given case than is a typical lay juror. We cannot, however, overemphasize the importance of a technical expert's ability to both understand the technical issues at hand but, also, to communicate them in a simple and direct manner to the jury. Objectively correct technical determinations are of little influence on jurors who cannot understand what the expert is trying to communicate. You see, not just math and science skills are required in the courtroom; one also must possess the communication skills to educate the jurors and enable them to reach a sound and unprejudiced verdict.

Case 2: *Analysis of Aileron Power Control Cylinder Service Failure*[16]

Several failures of an aileron hydraulic power control unit were experienced by a certain fighter aircraft. These units consisted of four parallel chambers, pressurized by two separate pumps. Failures occurred by cracking through either the inner or the outer chamber walls. In either case, the resulting loss of pressure contributed to an aircraft malfunction. Test results indicated the normal mean pressure in these chambers to be about 10.3 MPa, with fluctuations between 5.2 and 15.5 MPa caused by aerodynamic loading fluctuations. Furthermore, during an in-flight aileron maneuver, the pressure was found to rise sometimes to 20.7 MPa, with transient pulses as high as 31 MPa resulting from hydraulic surge conditions associated with rapid commands for aileron repositioning. In one particular case, an elliptical surface flaw grew from the inner bore of one cylinder toward the bore of the adjacent cylinder. A series of concentric markings suggested the initial fracture mode to be fatigue. At this point, the crack had grown to be 0.64 cm deep and 1.42 cm long. Subsequently, the crack appeared to propagate by a different mechanism (macroscopic observation) until it became a through-thickness flaw 2.7 cm long, at which time unstable fracture occurred. It was considered likely that the latter stage of subcritical flaw growth was controlled by an environment-assisted cracking process that would account for the change in fracture surface appearance, similar to that shown in Fig. 11.3a. The component was made from 2014-T6 aluminum alloy and was manufactured in such a way that the hoop stress within each chamber acted perpendicular to the short transverse direction of the original forging. From the *Damage Tolerant Design Handbook*,[17] the yield strength and fracture toughness of the material in this direction are given as 385 MPa and 19.8 MPa$\sqrt{m}$ respectively. Additional data concerning the geometry of the power control unit are:

chamber wall thickness $(t) = 0.84$ cm
elliptical crack depth $(a) = 0.64$ cm
elliptical crack length $(2c) = 1.42$ cm
$a/2c = 0.445$
elliptical flaw correction factor $(Q) \cong 2.2$
bore diameter $(D) = 5.56$ cm
through-thickness crack length $(2a_1) \cong 2.7$ cm

To use the plane-strain fracture-toughness value in subsequent fracture calculations, it is necessary to verify that t and $a \geq 2.5(K_{IC}/\sigma_{ys})^2$. This condition is met for this case history and supported by the observation that the fracture surface was completely flat. The stress necessary to fracture the unit may be computed by the formula for a through-thickness flaw where

$$K_{IC} = \sigma\sqrt{\pi a}$$

Setting $K_{IC} = 19.8$ MPa $\sqrt{m}$ and $a = 1.35$ cm

$$19.8 = \sigma\sqrt{\pi(1.35 \times 10^{-2})}$$
$$\sigma = 96.1 \text{ MPa}$$

The chambers have a large diameter-to-thickness ratio so that pressurization could be analyzed in terms of a thin-walled cylinder formulation. Since both cylinders are pressurized, the hoop stress between cylinder bores is estimated to be

$$\sigma_{\text{hoop}} = 2\left(\frac{PD}{2t}\right)$$

where $P = $ internal fluid pressure.

Using the component dimensions and the calculated stress level at fracture (i.e., 96.1 MPa), the pressure level at fracture P is calculated to be

$$96.1 = \frac{2P(5.56 \times 10^{-2})}{2(8.4 \times 10^{-3})}$$
$$P = 14.5 \text{ MPa}$$

Since the normal mean pressure in the cylinder bores is about 10.3 MPa and reaches a maximum of about 15.5 MPa, unstable fracture could have occurred during either normal pressurization or during pressure buildups associated with an aileron repositioning maneuver.

As mentioned above, the change in fracture mechanism when the elliptical crack reached a depth and length of 0.64 and 1.42 cm, respectively, could have been due to the onset of static environment-assisted cracking at a stress intensity where the cracking rate became independent of the K level (i.e., Stage II behavior). For such an elliptical flaw

$$K^2 = \left[1 + 0.12\left(1 - \frac{a}{c}\right)\right]^2 \sigma^2 \frac{\pi a}{Q}\left(\sec\frac{\pi a}{2t}\right) \tag{11-1}$$

with the result that

$$K^2 = \left[1 + 0.12\left(1 - \frac{0.64}{0.71}\right)\right]^2 \sigma^2 \frac{\pi(6.4 \times 10^{-3})}{2.2}\left[\sec\frac{\pi(0.64)}{2(0.84)}\right]$$

$$K = 0.14\sigma$$

Assuming that the major stresses associated with static environment-assisted cracking were those associated with the mean pressure level of about 10.3 MPa, the associated hoop stress is calculated to be

$$\sigma_{hoop} = \frac{2(10.3)(5.56 \times 10^{-2})}{2(8.4 \times 10^{-3})}$$

$$= 68.2\,\text{MPa}$$

Using this stress level, the stress intensity level for the onset of static environment-assisted cracking is estimated to be

$$K = 0.14\sigma$$

$$= 0.14(68.2)$$

$$= 9.5\,\text{MPa}\sqrt{\text{m}}$$

Unfortunately, no environment-assisted cracking (EAC) data for this material–environment system are available to check whether the number computed above is reasonable. It is known, however, that EAC rates in this alloy become appreciable in a saltwater environment when the stress intensity level approaches 11 MPa$\sqrt{\text{m}}$. Further material evaluations would be needed to determine whether hydraulic fluid has a similar effect on the cracking response of this alloy at stress intensity levels of about 11 MPa$\sqrt{\text{m}}$.

Case 3: *Failure of Pittsburgh Station Generator Rotor Forging*[18,19]

A steam turbine-generator rotor at the Pittsburgh Station power plant failed on March 18, 1956, during an overspeed check. (Overspeed checks were conducted routinely after a shutdown period and before the rotor was returned to service.) The rotor was designed for 3600 rpm service and failed when being checked at 3920 rpm. The cylindrical rotor body split down its central axis, then broke into smaller fragments. It is important to note that on 10 previous occasions during its two-year life the rotor satisfactorily endured similar overspeed checks above 3920 rpm. Surely, failure during the 11th check must have come as a rude shock to the plant engineers. One may conclude, therefore, that some subcritical flaw growth must have taken place during the two-year service life to cause the rotor to fail during the eleventh overspeed test but not during any of the other 10 tests, even though these tests were conducted at higher stress levels.

The Pittsburgh Station rotor was similar in design and material selection to the Arizona rotor described in a Chapter 6 case history, except that it did not contain a borehole. Consequently, the centrifugal tangential stresses at the innermost part of this forged steel rotor were roughly half those found in the Arizona rotor (see Case Study 6.3). On the other hand, the lack of a borehole increased the likelihood of potentially damaging microconstituents located along the rotor centerline. As we will see, this potential condition was realized and did contribute to the fracture.

Macrofractographic examination revealed the probable initiation site to be an array of nonmetallic inclusions in the shape of an ellipse 5×12.5 cm and located nearly on the rotor centerline (Fig. 11.13).[18] The maximum bore tangential stress at burst speed was found to be 165 MPa and the temperature at burst equal to 29°C. The tensile properties of the rotor material were

Figure 11.13   Cluster of inclusions contributing to fracture of a Pittsburgh turbine rotor.[18] (Reprinted with permission from Academic Press.)

given as 510 and 690 MPa for the yield and tensile strength, respectively, with the room temperature Charpy impact energy equal to 9.5 J.

If we take the critical flaw to be equivalent to a 5×12.5-cm elliptical crack—assuming that all the inclusions had linked up prior to catastrophic failure (possibly as a result of subcritical flaw growth)—the stress intensity factor at fracture could be given by

$$K = \sigma\sqrt{\pi a/Q}$$

(see Fig. 6.21). The elliptical flaw shape factor Q for the condition where $a/2c = 2.5/12.5 = 0.2$ and $\sigma/\sigma_{ys} = 165/510 = 0.32$ is found from Fig. 6.21h to be 1.28. The fracture toughness of the material is then calculated to be

$$K_{IC} = 165\sqrt{\frac{\pi(2.5 \times 10^{-2})}{1.28}}$$
$$K_{IC} \approx 41\,\text{MPa}\sqrt{\text{m}}$$

This result compares very favorably with K_{IC} estimates based on the Barsom-Rolfe[20] and Sailors-Corten[21] K_{IC}–CVN correlations (Eqs. 6-60 and 6-61), where values of 37 and 45 MPa$\sqrt{\text{m}}$ may be computed, respectively.

Although the estimated K_{IC} value derived from the crack configuration and stress information was remarkably close to the values determined from the empirical K_{IC}–CVN correlations, it must be kept in mind that the latter values represent only a crude approximation of K_{IC}. Such derived values can vary widely because of the considerable scatter associated with Charpy energy measurements. Nevertheless, the basic merits of using the fracture mechanics approach to analyze this failure have been clearly demonstrated.

Case 4: *Stress Corrosion Cracking Failure of the Point Pleasant Bridge*[22]

The failure of the Point Pleasant, West Virginia, bridge in December 1967 occurred without warning, resulting in the loss of 46 lives. Several studies were conducted immediately afterward to determine the cause(s) of failure, since the collapse caused considerable anxiety about the safety of an almost identical bridge built around the same time and possessing a similar design and structural steel. Failure was attributed to brittle fracture of an eyebar (Fig. 11.14) that was about 17 m long, 5.1 cm thick, and 30.5 cm wide in the shank section. The ends of the bar were 70 cm in diameter and contained 29.2-cm-diameter holes. It was determined that a crack had traversed one of the ligaments

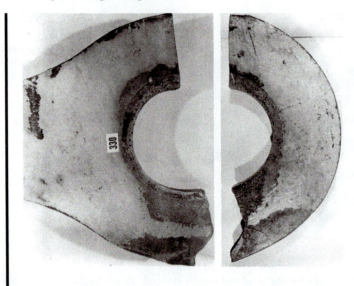

Figure 11.14   Fractured eyebar responsible for failure of Point Pleasant Bridge.[22] (Reprinted from *Journal of Testing and Evaluation* with permission from American Society for Testing and Materials.)

(the one on the top in Fig. 11.14) of the eye (along the transverse center line) with little apparent energy absorption (the fracture surface was very flat with little shear lip formation). The ligament on the opposite side of the hole suffered extensive plastic deformation before it failed, probably as a result of a bending overload.

After removing the rust from the fracture surface, investigators found two discolored regions covered with an adherent oxide layer.[22] These regions were contiguous and in the shape of two elliptical surface flaws (Fig. 11.15). The size of the large flaw was

$$a = 0.3 \, \text{cm}$$
$$2c = 0.71 \, \text{cm}$$
$$a/2c = 0.43$$

The smaller flaw had the dimensions

$$a \approx 0.1 \, \text{cm}$$
$$2c = 0.51 \, \text{cm}$$
$$a/2c \approx 0.2$$

Portions of the hole surface were heavily corroded, and some secondary cracks were parallel to the main fracture surface but initiated only in those regions where corrosion damage was extensive.

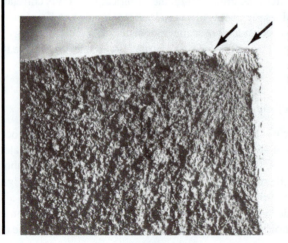

Figure 11.15   Fracture surface of broken eyebar from Point Pleasant Bridge showing two elliptical surface flaws.[22] (Reprinted from *Journal of Testing and Evaluation* with permission from American Society for Testing and Materials.)

These findings suggested the strong possibility that stress corrosion and/or corrosion fatigue mechanism(s) were involved in the fracture process. The hypothesis was further substantiated by metallographic sections which showed that the secondary cracks contained corrosion products and propagated in an irregular pattern from corrosion pits at the hole surface. Furthermore, some of these secondary cracks were opened in the laboratory, examined in the SEM and electron microprobe, and found to contain high concentrations of sulfur near the crack origin.[23] The presence of sulfur on the fracture surface was believed to be from H_2S in the air near the bridge rather than associated with manganese sulfide inclusions (commonly found in this material). The sensitivity of the bridge steel to H_2S stress corrosion cracking was verified by several tests performed on notched specimens. Fatigue crack propagation data were also obtained and used to examine the possibility that the two surface flaws had propagated instead by corrosion fatigue. Taking the maximum alternating stress on the bridge to be ± 100 MPa, Bennett and Mindlin[22] estimated that it would require over half a million load cycles to propagate a crack from a depth of 0.05 cm to one 0.25 cm deep. Since this was considered to be an unrealistically large number, it was concluded that the actual fracture mechanism was stress corrosion cracking.

Attention was then given to an evaluation of the steel's fracture-toughness capacity. Using both Charpy V-notch and fracture-toughness test procedures, the SAE 1060 steel (0.61 C, 0.65 Mn, 0.03 S), which had been austenitized, water quenched, and tempered for 2 h at 640°C, was shown to be brittle. For example, the material was found to exhibit an average plane-strain fracture toughness level of 51 MPa$\sqrt{m}$ at 0°C, the temperature of fracture. This low value is consistent with the fact that the material displayed a strong stress corrosion cracking tendency—something usually found only in more brittle engineering alloys (see Chapter 8). Based on a measured yield strength of 550 MPa, these results were found to reflect valid plane-strain test conditions for the specimen dimensions chosen.

Estimating the stress intensity level in the usual way, Bennett and Mindlin computed the stress level at fracture by considering only the larger surface flaw:

$$K = 1.1\sigma\sqrt{\pi a/Q}$$
$$= 1.1\sigma\sqrt{\frac{\pi(3 \times 10^{-3})}{1.92}}$$
$$= 7.7 \times 10^{-2}\sigma$$

or

$$\sigma = 13\,K$$

Using the range of experimentally determined K_{IC} values (47.3 to 56.1 MPa$\sqrt{m}$, the stress level at fracture was found to be

$$\sigma = 615 - 730\,\text{MPa}$$

This represents an upper bound range of the fracture stress, since allowance was not made for the presence of the smaller contiguous elliptical flaw. If one assumes the crack to be elliptical with a maximum depth of 0.3 cm but with $2c = 1.6$ cm, then $a/2c \approx 0.19$ and $Q = 1.05$. This assumption should lead to a slight underestimate of the stress level:

$$K = 1.1\sigma\sqrt{\frac{\pi(3 \times 10^{-3})}{1.05}}$$
$$\sigma = 9.6K$$

Again using the K_{IC} range of 47.3 to 56.1 MPa$\sqrt{m}$, a lower stress range is found to be

$$\sigma = 455 - 540\,\text{MPa}$$

It is concluded that the actual stress range for failure was

$$455 - 540 < \sigma_{\text{actual}} \ll 615 - 730 \, \text{MPa}$$

It is seen that the failure stress is approximately equal to the material yield strength. Since the shank section of the eyebar was recommended for a design stress of 345 MPa, Bennett and Mindlin concluded that stresses on the order of the yield strength could exist at the considerable stress concentration associated with this region.

On the basis of this detailed examination, it was concluded that the critical flaw was developed within a region of high stress concentration and progressed by a stress corrosion cracking mechanism to a depth of only 0.3 cm before fracture occurred. Consequently, the hostile environment, the inability to adequately paint the eyebar and thus protect it from atmospheric attack, the low fracture toughness of the material, and the high design stress all were seen to contribute to failure of the bridge. It should come as no surprise that the combination of low toughness and high stress would result in a small critical flaw size (see Eq. 6-31).

Case 5: *Weld Cold Crack-Induced Failure of Kings Bridge, Melbourne, Australia*[24]

On a cold winter morning in July 1962, while a loaded truck with a total weight of 445 kN was crossing the bridge, a section of the 700-m-long elevated four-lane Kings Bridge freeway fractured, causing a portion of the bridge to drop 46 cm. Examination of the four main support girders that broke revealed that all had suffered some cracking *prior* to installation (Fig. 11.16). Indeed, subsequent welding tests established that a combination of poor detail design of the girder flange cover plate, poor weldability of the steel, poor welding procedure, and failure to properly dry low-hydrogen electrodes before use contributed to the formation of weld cold cracks located at the toe of transverse welds at the ends of the cover plates. In three of these girders, 10-cm-long through-thickness cracks had developed before erection but none were ever discovered during inspection. In addition, it was determined that girder W14-2 was almost completely broken before the span failed. (The crack in this girder extended across the bottom flange and 1.12 m up the web.)

The collapse of the span was traced to failure of girder W14-3, which contained a T-shaped crack extending 12.5 cm across the bottom flange and 10 cm up the web (Fig. 11.16). Madison[24] postulated that the stress intensity condition at instability could be approximated by the superposition of two major components. One major K component was attributed to uniform bending loads acting along the flange and perpendicular to the 12.5-cm-long flange crack.

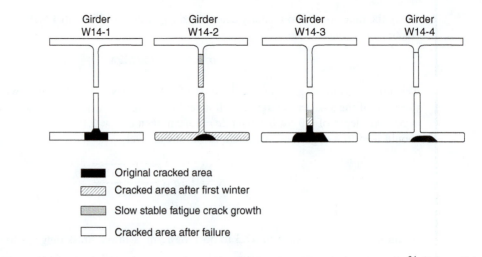

Figure 11.16   Diagram showing extent of cracking of girders from Kings Bridge, Australia.[24] (Courtesy of Dr. Ronald Madison.)

Accordingly

$$K = \sigma\sqrt{\pi a}\sqrt{\sec \pi a / W} \qquad (6\text{-}24)$$

where $\sqrt{\sec \pi a / W}$ = finite width correction

σ = bending stress, 83 MPa

a = 6.25 cm

W = 41 cm

$$= 83\sqrt{\pi(6.25 \times 10^{-2})\sec \pi(6.25/41)}$$

K = 39 MPa$\sqrt{\text{m}}$

The second K component was related to load transfer from the web, which produced wedge force loads extending 10 cm along both sides of the flange crack. These loads reflect residual stresses generated by the flange to web welds. For this configuration the K calibration is[23,26]

$$K = \frac{\sigma\sqrt{a}}{\sqrt{\pi}}\left[\sin^{-1}\frac{c}{a} - \left(1 - \frac{c^2}{a^2}\right)^{1/2} + 1\right] \qquad (11\text{-}2)$$

where $2a$ = crack length, 12.5 cm

σ = wedge force, 262 MPa

$2c$ = length of wedge force, 10 cm

$$K = \frac{262\sqrt{6.25 \times 10^{-2}}}{\sqrt{\pi}}\left\{\sin^{-1}\frac{5}{6.25} - \left[1 - \left(\frac{5}{6.25}\right)^2\right]^{1/2} + 1\right\}$$

K = 49 MPa$\sqrt{\text{m}}$

Therefore, $K_T = 39 + 49 = 88$ MPa$\sqrt{\text{m}}$. Note the significant contribution of the residual stresses. This value was found to be in reasonably good agreement with the dynamic fracture toughness of samples prepared from the bridge steel.

Case 6: *Failure Analysis of 175-mm Gun Tube*[27]

In April 1966, U.S. Army gun tube No. 733 failed catastrophically after a crack located near the breech end of the tube reached critical proportions. Brittle fracture was suspected since little evidence could be found for plastic deformation. The gun barrel, manufactured from a high-strength steel alloy, broke into 29 pieces that were hurled over distances up to 1.25 km from the firing site (Fig. 11.17). Davidson and co-workers reported this to be the first such brittle fracture of a 175-mm gun tube.[27] Previously, large-caliber gun tubes manufactured from medium-strength, high-toughness steel had been reported typically to fail by excessive wear and erosion of the barrel bore, with such wear resulting in a loss of projectile accuracy.[28-30] Since gun tube No. 733 had been manufactured to a higher strength but lower toughness specification, these latter properties were immediately called into question as being responsible for the catastrophic failure.

To analyze the cause of this fracture, we follow the outline of the Appendix D Failure Analysis Checklist and define the component configuration, the prevailing stresses prior to and at the time of the fracture, the details of the critical flaw, and the material properties. For stress analysis purposes, the gun barrel can be thought of as being a thick-walled tube, 10.5 m in length, with outer and inner

Figure 11.17 Fragments from exploded 175-mm gun tube. (After Davidson et al.[27])

diameters of 37.3 and 17.8 cm, respectively. At the time of failure, the gun was being fired at two-minute intervals, with the final round generating a nominal pressure of 345 MPa. Altogether, the gun tube experienced 373 rounds at a nominal peak pressure of 345 MPa and 227 rounds at a pressure of 152 MPa. The fracture surfaces of the many broken segments revealed a predominantly flat-fracture appearance, indicative of plane-strain fracture conditions.

The critical flaw was found to be semielliptical in shape, as denoted by its darkened appearance (presumably a result of the deposition of combustion products during firing), with half-minor axis and major axis dimensions of 0.94 and 2.79 cm, respectively (Fig. 11.18a). The material was forged AISI 4335 steel, modified with respect to the overall Cr and Mo content and by the addition of 0.14% V.[27] Selected tensile and fracture properties of this material are shown in Table 11.1.

Figure 11.18 Fracture surfaces of broken 175-mm gun tubes. (a) Fracture surface of gun tube No. 733. Note the small semielliptical surface flaw representing critical crack size. (b) Fracture surface of autofrettaged gun tube revealing leak-before-break condition. (After Davidson et al.[27])

Table 11.1   Mechanical Properties of Gun Tube No. 733

Property	Undefined	Near Failure	Toward Muzzle	Toward Breech
Yield strength, MPa (ksi)	1180(171)			
Tensile strength, MPa (ksi)	1385 (201)			
Elongation, %	10			
21°C reduction area, %		9–28	17–22	18–34
−40°C Charpy energy, J (ft-lb)		6.1–8.8 (4.5–6.5)	10.2–11.5 (7.5–8.5)	5.4–11.5 (4.0–8.5)
21°C fracture toughness, MPa$\sqrt{m}$(ksi$\sqrt{in.}$)		89–91 (81–83)	74–99 (67–90)	—

Davidson and co-workers initially considered the possibility that failure had occurred as a result of higher than expected pressure during firing; it was thought that this condition would account for the early development and growth of the critical flaw and its small final dimensions.[27] Subsequent examination of the gun tube fragments, however, revealed no evidence of overpressure. Furthermore, nothing abnormal was found when tests were conducted of the ammunition being fired at the time of the failure. The possibility of the environment-assisted cracking under sustained loading conditions was also ruled out since the time under service load (during actual firing) was too short (about 20×10^{-3} s) and the magnitude of residual tensile stresses in the tube was too low. Finally, loading rate effects on the material fracture toughness were not considered to be of any consequence for this high-strength steel.

After further analysis of the fracture surface markings, the character of the steel's microstructure, and the prevailing stress intensity levels, the following fracture scenario was identified. Crack initiation was believed to have occurred on the inner bore of the gun tube from a thermally induced cracking process known as "heat checking." This results in the development of a random network of cracks that typically penetrate up to 0.13 cm below the inner bore surface that is in contact with the hot combustion gases. For the firing conditions associated with this gun tube, the heat checking pattern was found to be fully developed after only ten rounds of ammunition were fired. The total life of the gun tube was then assumed to reflect only fatigue crack propagation (one round = one loading cycle) during which time the crack grew from a presumed depth of 0.13 cm to the 0.94 × 2.79-cm semielliptical configuration at fracture.

Judging from the low fracture-toughness properties of the steel near the failure site (Table 11.2) and evidence for intergranular and cleavage micromechanisms on the fatigue fracture surfaces,

Table 11.2   Fracture Data for 175-mm Gun Tubes with 170–190 ksi Yield Strength

Tube No.	Total Cycles to Failure	σ_{ys}		Charpy		K_{IC}		Critical Flaw	
		MPa	(ksi)	J	(ft-lb)	MPa$\sqrt{m}$	(ksi$\sqrt{in.}$)	cm	(in.)
733	373	1180	(171)	8.1	(6)	88	(80)	0.94	(0.37)
863	1011	1270	(184)	12.2	(9)	103	(94)	4.3	(1.7)
1131	9322	1255	(182)	19	(14)	142	(129)	4.3	(1.7)
1382	1411	1275	(185)	14.9	(11)	108	(98)	3.8	(1.5)
1386	4697	1250	(181)	19	(14)	116	(106)	4.6	(1.8)
Typical values for 35 tubes	4000	1240	(180)	16.3	(12)	121	(110)	3.8	(1.5)

Davidson et al. concluded that a condition of temper embrittlement had contributed to both accelerated fatigue crack growth and premature final fracture of tube No. 733.[27] A study of other gun tubes confirmed the relation between gun tube life and material fracture properties. Note in Table 11.2 that the total cycles to failure (at 345 MPa) and the final flaw depth increased with increasing Charpy energy and fracture toughness.

The stress intensity factor in an internally pressurized thick-walled tube containing a long, straight surface flaw located in the inner bore is given by Bowie and Freese[31] in the form

$$K = f(a/W,\ r_2/r_1)P\sqrt{\pi a} \tag{11-3}$$

where a/W = crack depth to tube thickness ratio
r_2/r_1 = outer-to-inner radius ratio
P = internal pressure
a = crack depth with crack plane being normal to hoop stress direction

At final failure, where $a = 0.94\,\text{cm}$,

$$K = 2.7P\sqrt{\pi a} \tag{11-4}$$

Since the crack shape at fracture was semielliptical, Eq. 11-4 was modified[32] for the appropriate a/W and $a/2c$ values such that

$$K = 1.7P\sqrt{\pi a} \tag{11-5}$$

The stress intensity factor at fracture in association with $P = 345$ MPa and $a = 0.94$ cm is therefore computed to be 99 MPa$\sqrt{\text{m}}$. This value is in fairly good agreement with the reported toughness for the tube material (Table 11.1). To estimate the service life of gun tube No. 733, the crack growth rate expression (Eq. 11-6) was integrated

$$\frac{da}{dN} = 6.49 \times 10^{-12} \Delta K^3 \tag{11-6}$$

where da/dN is m/cycle. (This relation was derived from laboratory tests conducted on a material with 50% higher toughness.) Since the calibration factor Y for the changing crack front configuration in the tube varied with the crack length, the integration should most properly be carried out numerically or in parts where Y is held constant over the various intervals of integration. As a first approximation, the integration was performed assuming that Y possesses a constant value of 2.2, corresponding to a simple average between the values of 2.7 and 1.7 in Eqs. 11-4 and 11-5, respectively. The computed life, assuming only stress fluctuations with a range of 345 MPa, was found to be 2070 cycles, between 5 and 6 times greater than the number of 345 MPa stress fluctuations experienced by the gun tube prior to failure.

Several reasons can be given to show that the actual and computed gun tube lives are actually in much closer agreement. A more realistic determination of Eq. 11-6 should reflect the temper-embrittled nature of the material. For example, Ritchie[33] reported FCP rates 2.5 times greater in a temper-embrittled 43XX type steel than in properly treated samples of the same material. Also, the low fracture toughness of the material in gun tube No. 733 would be expected to result in higher crack growth rates at a given ΔK level (recall Eq. 10-10). Finally, Eq. 11-6 was based on test results from laboratory air-test conditions and not from experiments conducted in the presence of more aggressive hot combustion gas products. Taken together, these factors would all be expected to lower the estimated fatigue life below the 2070-cycle value initially computed. Furthermore, the effective service life is most likely greater than 373 cycles at a nominal pressure of 345 MPa since no damage was attributed to the 227 rounds fired at a pressure of 152 MPa. (It is estimated that the life of gun tube No. 733 would have been about 10% greater in the absence of the 227 lower stress rounds.)

The failure analysis report contained additional information pertaining to the avoidance of future gun tube fractures. As a short-range interim procedure, all gun tubes possessing a Charpy impact energy less than 13.5 J were immediately withdrawn from the field. Other tubes were assigned a reduced service life of 300 rounds at 345 MPa instead of the original 800 rounds. Following these changes, no additional field failures occurred. Gun tubes currently in the manufacturing process were heat treated to a lower strength level so that both impact and fracture properties could be increased. Indeed, the cyclic life of these gun tubes increased to about 10,000 rounds, while the final crack depth at fracture was twice that shown in Table 11.2. To further minimize the risk of brittle fracture, gun tubes were subsequently heat treated to a lower yield strength in the range of 965 to 1100 MPa and given an autofrettage treatment. In the autofrettage treatment used in this case (recall Section 9.6.1), the gun tube is subjected to a hydrostatic internal pressure sufficiently high to produce plastic deformation about halfway across the tube thickness. When this pressure is removed, the yielded zone experiences a compressive residual stress gradient with the highest compressive stress located at the inner surface of the gun tube (Fig. 11.19). Note that the compressive residual stress is numerically greater than the hoop stress at the inner bore. As a result, the fatigue life should increase appreciably. To wit, autofrettaged tubes withstood more than 20,000 firing cycles at 345 MPa, representing a 50-fold improvement in fatigue life over that experienced by gun tube No. 733! Associated with this vast improvement in the fatigue life of the gun tube was a trend toward stable fatigue crack propagation completely through the tube wall (Fig. 11.18b); hence, the

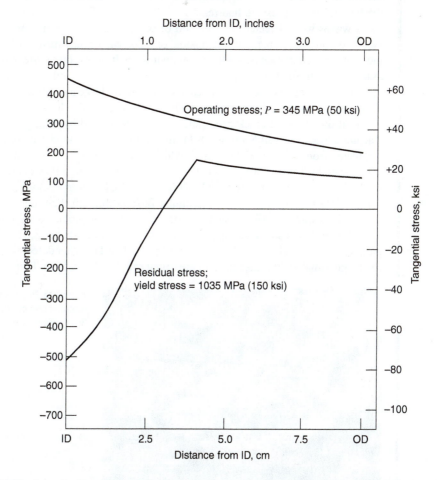

Figure 11.19   Operating hoop stress gradient in gun tube versus residual stress profile resulting from autofrettage treatment. Note overall compressive stress at the inner wall of the tube. (After Davidson et al.[27])

combination of an increase in fracture toughness, because of a reduction in yield strength, and the development of a favorable residual compressive stress, a result of the autofrettage treatment, created a leak-before-break failure condition (recall Section 6.6).

Case 7: *Hydrotest Failure of a 660-cm-Diameter Rocket Motor Casing*[34]

This failure analysis describes the catastrophic rupture of a 660-cm rocket motor casing that fractured prematurely during a hydrotest at an internal pressure of only 56% that of the planned value. Experiments of this type were being performed to demonstrate the feasibility of designing solid-propellant rocket casings with a thrust capacity of 27×10^6 N. This particular case had been fabricated by welding together many sections of a 250-grade air-melted maraging steel. Nominal yield- and tensile-strength values for the base plate (1.85 cm thick at the fracture origin) were 1585 and 1725 MPa, respectively, and the weld efficiency was assumed to be 90%. Initially, approximately 300 m of longitudinal and circumferential welds were prepared by a submerged arc process. Subsequent nondestructive inspections revealed the presence of numerous weld defects that were removed by grinding and repaired using a manual TIG welding process. In turn, some of these weld repairs were found to be defective and in need of repair. Altogether, approximately 100 m of weld repairs and re-repairs were required.

It was planned that the hydrotest be extended to a water pressure of 6.6 MPa, 10% above the maximum expected operating pressure of the rocket motor casing. Instead, failure occurred when the internal water pressure had reached only 3.7 MPa. During the course of the test, a number of stress waves were detected with the aid of several accelerometers and strain gages that had been mounted onto the casing. Although some of these waves may have reflected the relative motion of motor casing components, such as bolts within bolt holes, other stress waves, including the ones associated with final fracture, most likely were associated with subcritical crack growth.

On fracturing, the rocket motor casing broke into a large number of pieces that were subsequently collected and reassembled as shown in Fig. 11.20. The relative locations of the fracture segments and the local directions of crack propagation were determined by noting the chevron markings on the fracture surfaces (recall Section 5.8.1). In addition, the chevron pattern revealed that fracture had originated from two preexistent flaws, which were fairly close to one another. Both of these cracks were located within the heat-affected zone (HAZ) beneath the TIG weld repair and within the coarse-grained heat-affected zone of the submerged arc weld. Electron fractographic studies revealed that the fracture had progressed in an intergranular fashion through these coarse-grained regions. For this reason, it is quite possible that these defects had been produced by cold cracking in the HAZ. The primary flaw had a clean appearance, was elliptical in shape (3.6×0.25 cm), and was oriented parallel to a longitudinal weld centerline (Fig. 11.21*a*). The other defect was oriented perpendicular to the longitudinal weld centerline and measured roughly 4×0.5 cm. Furthermore, part of the fracture

Figure 11.20 Reassembled fragments from ruptured 660-cm rocket motor casing. (From Srawley and Esgar[34]).]

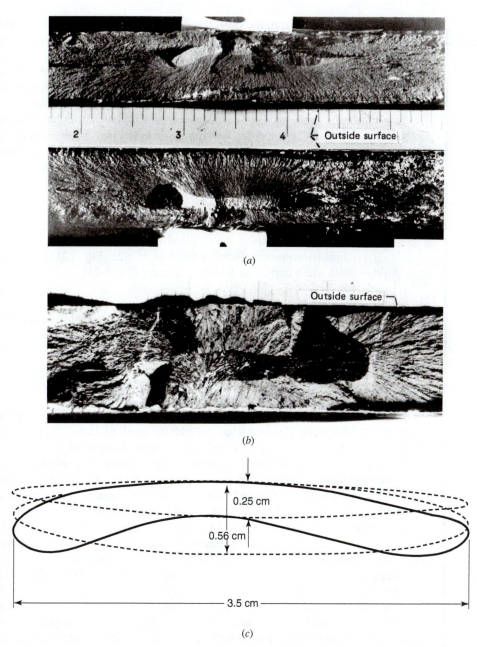

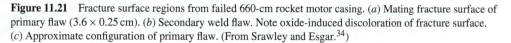

Figure 11.21 Fracture surface regions from failed 660-cm rocket motor casing. (*a*) Mating fracture surface of primary flaw (3.6×0.25 cm). (*b*) Secondary weld flaw. Note oxide-induced discoloration of fracture surface. (*c*) Approximate configuration of primary flaw. (From Srawley and Esgar.[34])

surface of this defect was black and extended to the surface of the casing (Fig. 11.21*b*). X-ray diffraction analysis determined that the black debris was a combination of Fe_2O_3 and Fe_3O_4; presumably, this preexistent defect, which extended to the free surface, became oxidized during the 450°C aging treatment.

Investigators concluded from an examination of the chevron pattern on the fracture surface that final failure had initiated at the "clean" defect. Furthermore, by triangulating the stress wave signals from the accelerometer devices, the stress wave origin was located much closer to the "clean" flaw than to the "black" flaw. It was reasoned that, after the "clean" crack began to propagate unstably, the changing stress distribution resulted in independent growth from the "black" origin. It is interesting to note that had the "clean" defect been detected and removed prior to the hydrotest, premature failure still would have occurred; in this instance, the "black" defect would have provided the initiation site.

To estimate the stress intensity factor associated with the catastrophic fracture of this rocket motor casing, the "clean" crack origin was approximated by two different elliptical configurations; lower and upper bound values of K_{max} at failure were estimated by assuming the minor axis of the flaw to be 0.25 and 0.56 cm, respectively (Fig. 11.21c). In both instances,

$$K = \sigma\sqrt{\pi a/Q} \cdot f(a/t) \tag{11-7}$$

where
$\sigma =$ design stress, approximated by the hoop stress
$a =$ one-half minor axis of the elliptical flaw
$Q =$ ellipticity correction factor, a function of $a/2c$ (Fig. 6.21h)
$f(a/t) =$ finite width correction factor, $\sqrt{\sec \pi a/t}$ (Eq. 6-24)

The applied stress σ, neglecting any additional stress component from manufacturing-induced residual stresses, was computed to be approximately 690 MPa. (Since two separate investigative teams reached different conclusions regarding the possible existence and sense of a residual stress pattern, no attempt was made to include such a stress component in the stress estimate.) From Eq. 11-7, the lower- and upper-bound estimates of K_{max} are 44 and 65 MPa$\sqrt{m}$, respectively. These estimates are then compared with measured values of the material's fracture toughness, corresponding to the microstructure surrounding the "clean" crack. The most accurate estimate of K_{IC} in this region would have required testing a sample with the crack tip embedded in the overlapping heat-affected zones of the submerged arc and the TIG manual repair welds. Unfortunately, no fracture-toughness specimens were prepared from such a location. Instead, K_{IC} was based on values obtained from specimens located in the HAZ of a submerged arc weld that had not been TIG-weld repaired. For this location, K_{IC} was measured to be 85 MPa$\sqrt{m}$, which is higher than the computed estimates of the maximum stress intensity factor at fracture. It should be noted, however, that investigators estimated the fracture toughness in the overlapping heat-affected zones to be considerably less than 85 MPa$\sqrt{m}$, based on microstructural and fractographic evidence. As such, one would have expected much better agreement between the computed stress intensity level at fracture and the material's fracture toughness, had the latter been determined in the relevant region of the microstructure.

One of the major conclusions drawn from the analysis of this fracture was the fact that the NDT techniques used in the manufacture of this rocket motor casing (dye penetrant, ultrasonic, radiographic, and visual) were much less sensitive and reliable than had been expected. (Recall the similar circumstances surrounding the F-lll aircraft accident that were discussed in Case Study 6.2.) In addition to the two overlooked defects already discussed, 11 other defects were discovered during a postfailure reinspection of the welds. To be sure, it is uncertain how many of these defects initiated and/or grew to detectable dimensions during the hydrotest. At any rate, it is instructive to compare the loading conditions necessary to fracture a test specimen containing one of these defects with that value based on a fracture mechanics computation. To this end, a section of the casing containing a weld defect was removed and tested in tension to fracture ($\sigma_{max} = 793$ MPa). The fracture surface revealed a defect, 1.5 cm long × 0.3 cm wide, with the long dimension oriented parallel to the plate surface. This defect was located in the heat-affected zone of a TIG repair weld, which was embedded, in turn, in the center of submerged arc weld metal. For a fracture-toughness specimen containing a crack in the submerged arc weld metal zone, the best estimate of K_{IC} was 54.1 MPa$\sqrt{m}$. Using Eq. 11-7, the maximum stress intensity factor associated with fracture of the section containing the 1.5 cm × 0.3 cm flaw is 53.2 MPa$\sqrt{m}$, in excellent agreement with the material's intrinsic resistance to fracture.

Case 8: *Premature Fracture of Powder-Pressing Die*[35]

Sintered metal powder rods were to be compacted in the die shown schematically in Fig. 11.22. After the powder was placed in the die, the charge was compressed from both ends with two movable plungers. During the die's initial compaction cycle, sudden failure occurred at an applied stress level roughly 20% *below* the rated value. Why?

On close examination, the fracture origin was traced to a small semicircular flaw located at the surface of the inner bore near one end of the cylinder (Fig. 11.23). This defect experienced the full effect of the hoop stress σ_t in the thick-walled cylinder as given by

$$\sigma_t = P\frac{\left(\dfrac{1}{r^2} + \dfrac{1}{r_o^2}\right)}{\left(\dfrac{1}{r_i^2} - \dfrac{1}{r_o^2}\right)} \tag{11-8}$$

where r_i = inner radius of thick-walled cylinder
$\quad\ r_o$ = outer radius of thick-walled cylinder
$\quad\ r$ = radius
$\quad\ P$ = internal pressure
$\quad\ \sigma_t$ = hoop stress (tangential)

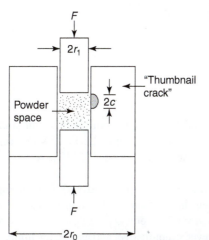

Figure 11.22 Schematic drawing of powder-pressing die. (From Ashby and Jones[35]; with permission from Pergamon Press.)

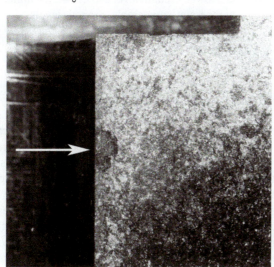

Figure 11.23 Photograph showing semicircular surface flaw at inner bore surface of die. (From Ashby and Jones[35]; with permission from Pergamon Press.)

For this die, the relevant dimensions of the cylinder and initial flaw size are

$$r_i = 6.4 \, \text{mm}$$
$$r_o = 38 \, \text{mm}$$
$$a = 1.2 \, \text{mm}$$
$$2c = 2.4 \, \text{mm}$$

Neglecting the presence of the flaw, the tangential stress acting at the surface of the inner bore is found to be 1.06 P. The maximum allowable pressure for this die, based on a failure criterion of incipient plastic deformation in conjunction with a safety factor of three, is then given by

$$P = \frac{\sigma_{ys}}{1.06 \times 3} = 630 \, \text{MPa}$$

where $\sigma_{ys} = 2000 \, \text{MPa}$.

Instead of experiencing plastic deformation, the die failed in a brittle fashion associated with a flat fracture appearance. For this specimen-crack configuration, the prevailing stress intensity factor is given by

$$K = 1.1 \frac{2}{\pi} \sigma \sqrt{\pi a} \tag{11-9}$$

where σ = tangential stress = $1.06 \, P$

1.1 = surface flaw correction

$\dfrac{2}{\pi}$ = a semicircular flaw correction factor

(Note the similarity between this relation and Eq. 6-29.) From independent studies, the fracture toughness of this medium carbon chromium steel was determined to be $22 \, \text{MPa}\sqrt{\text{m}}$. Therefore, from Eq. 11-9, the internal pressure at fracture was found to be 512 MPa, roughly 20% below the allowable pressure level.

Surely, the combined effects of low fracture toughness and very high strength level for die material contributed to the premature failure. In addition, evidence of hydrogen cracking was also reported, which could account for the presence of the preexistent flaw. An improved design would involve either the use of this material at a lower strength level and associated higher toughness or the use of another alloy with superior toughness at comparable strength levels.

Case 9: *A Laboratory Analysis of a Lavatory Failure*[36]

The reader will certainly agree that the sudden collapse of a toilet seat is enough to distract one's concentration! To think, the thoughts of the day on one's mind and the pieces of a broken polymer literally under one's behind! A segment of such a broken seat was shown previously (Figs. 5.2 and 9.3). Note the clear evidence for clamshell markings (Fig. 11.24a) indicative of repeated loadings and unloadings as well as variable hold-time periods associated with irregular biological functioning and/or sustained periods of thought or literature review. An analysis of the fractured toilet seat reveals the origin at a surface flaw and the development of a semielliptical fatigue corner crack, which extended to a depth of 0.95 cm and 1.9 cm along the bottom edge of the seat.

The last person to use the seat weighed approximately 980 N (220 lbs.) with an estimated 20–25% of that amount being supported by the individual's legs. From a careful reassessment of the load on the toilet seat at the time of fracture, it was ascertained that the total buttock and thigh loads were applied along only one of the two leaves of the toilet seat. The latter body force distribution was associated with

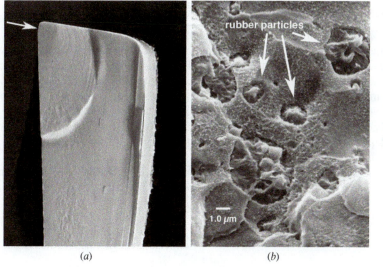

Figure 11.24   Failed toilet seat. (*a*) Fracture surface showing clamshell markings associated with loading and unloading (also recall Fig, 9.3c). The crack origin is marked with a white arrow. (*b*) Typical fracture surface micromorphology, consistent with appearance of high-impact polystyrene (HiPS).

repositioning of the skeletal frame in conjunction with the use of sanitary tissue. Anatomical considerations provide an approximation of the force distribution on this leaf as a linearly decreasing load between the back support and the front tab. The flexural stress at the bottom (tensile) surface of the seat was estimated to be 11 MPa.[36] The stress intensity factor at final failure was estimated from the relationship given in Eq. 11-10:

$$K = 1.1 M_B \sigma \sqrt{\pi a / Q} \cdot \sqrt{\sec \pi a / 2t} \qquad (11\text{-}10)$$

where $M_B =$ correction factor for the flexural stress at the crack front away from the bottom surface of the seat[37] (0.9)

$\sigma =$ flexural stress at bottom surface of seat (11 MPa)

$a =$ crack depth (0.95 cm)

$Q =$ ellipticity correction factor (1.3)

$\sqrt{\sec \frac{\pi a}{2t}} =$ finite width correction factor[38] (1.19)

Substitution of the appropriate values into Eq. 11-10 reveals $K_{\max} \approx 2.2 \, \mathrm{MPa}\sqrt{\mathrm{m}}$.

Laboratory studies were then conducted to identify the material used in the manufacture of the toilet seat. First, a fragment of the seat was burned; the resultant trail of smoke was consistent with the presence of a styrene-based polymer. The fractured surface was then suitably coated and examined in the SEM. Figure 11.24*b* reveals a representative region typical of the appearance of high-impact polystyrene (HiPS).[39] This rubber-modified polymer is a blend of spherical polybutadiene-polystyrene particles embedded within a polystyrene matrix.

From an earlier study,[40] the maximum K level associated with the fatigue fracture of HiPS was found to be approximately 2.4 MPa$\sqrt{\mathrm{m}}$ in good agreement with the estimated $K_{\max}$ of 2.2 MPa$\sqrt{\mathrm{m}}$ for the toilet seat. Finally, the last few fatigue bands seen on the fracture surface were approximately 10 to 20 μm in width and consistent with the macroscopic growth rate in HiPS near final failure. As such, we conclude that these last fracture bands are fatigue striations resulting from individual load excursions associated with attempts at the fulfillment of certain biological functioning. Whether subcritical flaw growth was enhanced by the prevailing aqueous and gaseous environments remains a topic for future study. In conclusion, it is suggested that the reader be mindful of the potential for subcritical flaw growth and premature fracture of toilet seat leaves. A simple nondestructive visual examination is recommended, time permitting.

11.6   ADDITIONAL COMMENTS REGARDING WELDED BRIDGES

Before concluding this section, it is important to comment further on the general problem of fatigue and fracture in welded bridges. Since these structures usually are very complicated because many cover plates, stiffeners, attachments, and splices are added to the basic beam, it is important to recognize the potential danger associated with a particular weld detail. In addition to the King's Bridge failure described in Case 5, prominent structural failures of welded bridges in relatively recent years include the Seongsu Bridge in Seoul, South Korea, in 1994, and the Hoan Bridge in Milwaukee, Wisconsin, in 2000. While the Seongsu Bridge collapse led to the tragic deaths of 32 people,[41] the cracks in the Hoan Bridge were fortuitously found in time to prevent a similar disaster.[42] In light of the seriousness of weld details in bridges, the construction of the new San Francisco-Oakland Bay Bridge was halted in 2005 (at great expense) when some of the construction crew claimed that their employer had forced them to

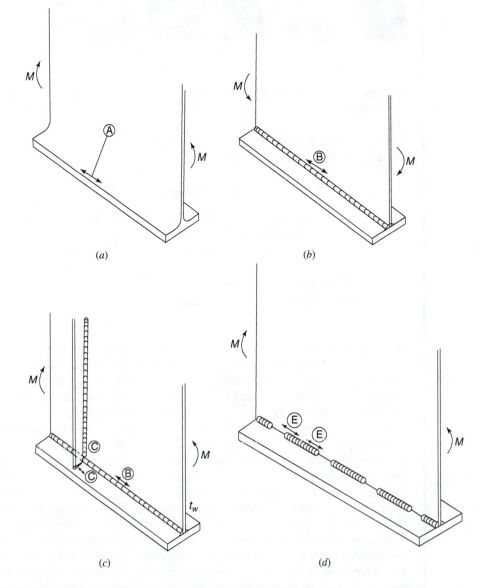

Figure 11.25   Drawings showing various welded beam details and their relative stress concentration severity (increasing from category A to E).[44] (Reprinted from *Research Results Digest 59*, with permission of the National Cooperative Highway Research Program.)

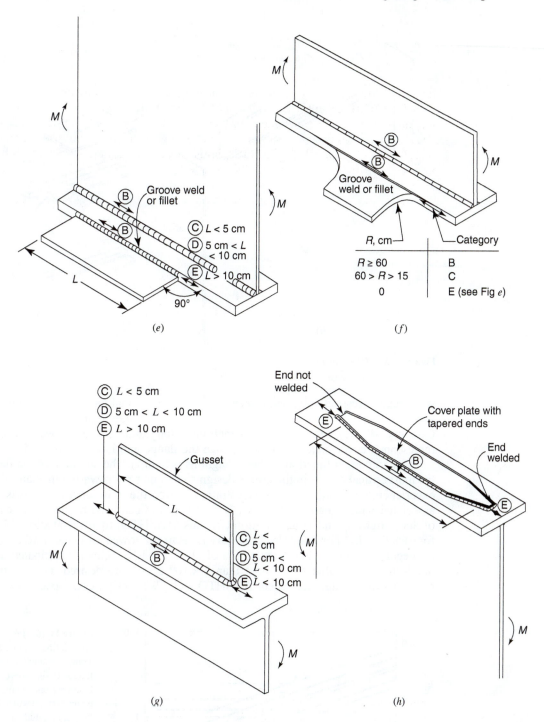

Figure 11.25   (*Continued*)

make poor welds for financial reasons. After careful inspection and analysis, it was determined that the welds were of high quality and the claims were dismissed.[43]

Fisher et al.[44-48] have conducted an extensive study of welded bridge connections, and have proposed several categories of relative attachment detail severity as shown schematically in Fig. 11.25. Categories E and A represent the potentially most damaging and least damaging

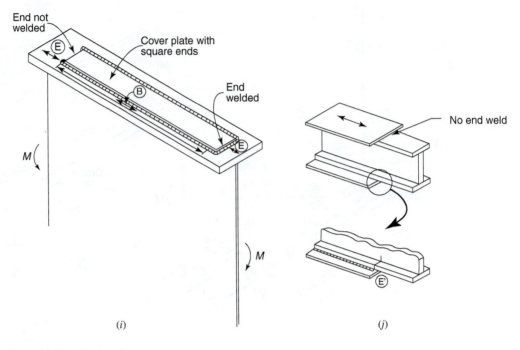

(i)

(j)

Figure 11.25   (*Continued*)

weld details, respectively. More recently,[48] two additional categories, designated B′ and E′, were identified that correspond to more severe conditions than those shown in Fig. 11.25. Category B′ represents a partial-penetration longitudinal groove weld, and category E′ represents a cover plate that is wider than the flange and has no end weld.

One important function of these diagrams is to direct the attention of the field engineer to the most critical details in the bridge design so that no time is wasted in examining those areas experiencing a lower stress concentration. Also, the differences in stress concentration associated with categories A, B, C, D, and E have been used by the American Association of State Highway and Transportation Officials (AASHTO) to arrive at allowable stress ranges for each detail.[44] For example, a category E detail is allowed only one-third the stress range of a category A region when a cyclic life of up to 2×10^6 cycles is anticipated and only about one-fifth of that value when more than 2×10^6 cycles are desired. For additional specifics regarding this matter, see recent AASHTO and AWS design code specifications.[49,50]

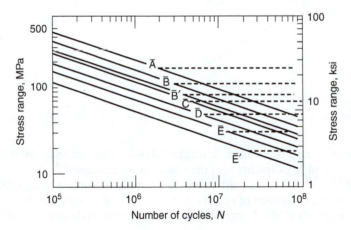

Figure 11.26   Proposed American Association of State Highway and Transportation Officials (AASHTO) fatigue design curves.[49] Note: Detail E′ corresponds to partial-length cover plates that are wider than the beam flange, longitudinally welded along the flange edge but not across the cover plate ends. Cracking occurs in the beam flange at the end of the cover plate. (With permission.)

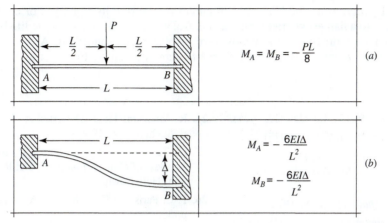

Figure 11.27 Beams fixed at both ends. (*a*) Load controlled; (*b*) displacement controlled.

Based on more than 2000 test results, fatigue design curves were defined for each of the welded steel bridge details described above and are shown in Fig. 11.26.[48] Each design curve corresponds to 95% confidence limits for 95% survival based on regression analysis of each respective database. Furthermore, it was concluded that straight-line extensions at a slope of -3.0 to low stress levels of each *S-N* plot represented a more conservative lower bound estimate of fatigue life for these welded details than one based on discrete endurance limits (dashed horizontal lines) for each detail. Note that Fig. 11.26 essentially shows *S-N* plots for welded details containing different stress concentration factors.

As a final note, the engineer should remember the subtle but important difference between load- and displacement-controlled conditions governing the behavior of a given structure. For example, for the case of the two fixed-ended beams shown in Fig. 11.27, the maximum flexural stress for load-induced and displacement-induced conditions are

$$\text{load-induced: } \sigma_{\max} = \frac{PLc}{8I} \tag{11-11}$$

where $\sigma_{\max}$ = maximum flexural stress
P = point load
L = beam length
c = distance to outermost fiber
I = moment of inertia

and

$$\text{displacement-induced: } \sigma_{\max} = \frac{6Ec\Delta}{L^2} \tag{11-12}$$

where E = modulus of elasticity
Δ = displacement

The passage of a fully loaded truck across a bridge might represent a situation that could be analyzed by Eq. 11-11, but if one of the bridge foundations had settled by an amount Δ then Eq. 11-12 would prove more correct. Identification of the importance of load or displacement in controlling the bridge response is critical, since the design changes one would make to improve either load or displacement resistance of a given beam are mutually incompatible. For instance, we see from Eq. 11-11 that for a given load *P*, the load-bearing capacity of a beam is enhanced by *decreasing* its unsupported length and/or *increasing* its moment of inertia or rigidity. By sharp contrast, we see from Eq. 11-12 that for a given displacement Δ, the flexural stress may be reduced by *increasing* the beam length and/or *decreasing* its rigidity. For example, adding a cover plate to a flange experiencing load control would enhance its fatigue life but would prove to be counterproductive if the beam were displacement controlled. A discussion of such a dilemma with regard to the Lehigh Canal Bridge is given by Fisher et al.[45]

REFERENCES

1. *Engineering News-Record* **94** (5), 188 (Jan. 29, 1925).

2. H. A. Wood, *Eng. Fract. Mech.* **7**, 557 (1975).

3. Military Specification, Mil-A-83444, *Damage Tolerance Design Requirements for Aircraft Structures,* USAF, July 1974.

4. T. J. Dolan, *Met. Eng. Quart.* **32** (Nov. 1972).

5. F. A. Heiser and R. W. Hertzberg, *JISI* **209**, 975 (1971).

6. M. O. Speidel, *The Theory of Stress Corrosion Cracking in Alloys,* J. C. Scully,Ed., NATO, Brussels, 1971.

7. R. W. Hertzberg and T. J. Pecorini, *Int. J. Fatigue* **15** (6), 509 (1993).

8. *Metals Handbook*, Vol. 10, ASM, Metals Park, OH, 1975.

9. *Source Book in Failure Analysis,* American Society for Metals, Metals Park, OH, Oct. 1974.

10. T. P. Rich and D. J. Cartwright, Eds., *Case Histories in Fracture Mechanics,* AMMRC MS 77-5, 1977, p. 3.9.1.

11. C. M. Hudson and T. P. Rich, Eds., *Case Histories Involving Fatigue and Fracture Mechanics,* ASTM, *STP 918,* Philadelphia, PA, 1986.

12. C. N. Reid and B. L. Baikie, op. cit., p. 102.

13. H. S. Pearson and R. G. Dooman, op. cit., p. 65.

14. K. S. Vecchio, R. W. Hertzberg and F. E. Schmidt, Jr., *J. Test and Eval.* **17** (5), 267 (1989).

15. R. W. Hertzberg, K. S. Vecchio and F. E. Schmidt, Jr., *J. Test and Eval.* **17** (5), 261 (1989).

16. R. J. Gran, F. D. Orazio, Jr., P. C. Paris, G. R. Irwin, and R. W. Hertzberg, AFFDL-TR-70-149, Mar. 1971.

17. J. E. Campbell, W. E. Berry, C. E. Feddersen, *Damage Tolerant Design Handbook,* MCIC-HB-01, Sept. 1973.

18. S. Yukawa, D. P. Timio, and A. Rubio, *Fracture,* Vol. 5, H. Liebowitz, Ed., Academic, New York, 1969, p. 65.

19. R. J. Bucci and P. C. Paris, Del Research Corporation, Hellertown, PA, Oct. 23, 1973.

20. J. M. Barsom and S. T. Rolfe, *Eng. Fract. Mech.* **2** (4), 341 (1971).

21. R. H. Sailors and H. T. Corten, ASTM *STP 514,* Part II, 1972, p. 164.

22. J. A. Bennett and H. Mindlin, *J. Test. Eval.* **1** (2), 152 (1973).

23. D. B. Ballard and H. Yakowitz, *Scanning Electron Microscope 1970,* IITRI, Chicago, April 1970, p. 321.

24. R. Madison, Ph.D. Dissertation, Lehigh University, Bethlehem, PA, 1969.

25. H. Tada, P. C. Paris, and G. R. Irwin, *The Stress Analysis of Cracks Handbook*, Del Research Corporation, Hellertown, PA, 1973.

26. P. C. Paris and G. C. M. Sih, ASTM *STP 381* (1965), p. 30.

27. T. E. Davidson, J. F. Throop, and J. H. Underwood, *Case Histories in Fracture Mechanics,* T. P. Rich and D. J. Cartwright, Eds., AMMRC MS 77-5, 1977, p. 3.9.1.

28. J. C. Ritter and M. E. deMorton, *J. Austral. Inst. Met.* **22** (1), 51 (1977).

29. R. S. Montgomery, *Wear* **33** (2), 359 (1975).

30. R. B. Griffin et al., *Metallography* **8**, 453 (1975).

31. O. L. Bowie and C. E. Freese, *Eng. Fract. Mech.* **4** (2), 315 (1972).

32. I. S. Raju and J. C. Newman, Jr., *Eng. Fract. Mech.* **11** (4), 817 (1979).

33. R. O. Ritchie, *Int. Metals Rev. Nos. 5 and 6*, Review 245, 1979, p. 205.

34. J. E. Srawley and J. B. Esgar, NASA *RM-X-1194,*1967.

35. M. F. Ashby and D. R. H. Jones, *Engineering Materials*, Pergamon, Oxford, England, 1980.

36. R. W. Hertzberg, M. T. Hahn, C. M. Rimnac, J. A. Manson, and P. C. Paris, *Int. J. Fract.* **22**, R57 (1983).

37. B. J. Gross and J. E. Srawley, NASA *TND-2603,* 1965.

38. C. E. Feddersen, ASTM *STP 410* (1967) p. 77.

39. J. A. Manson and R. W. Hertzberg, *J. Polym. Sci. (Phys.)* **11**, 2483 (1973).

40. R. W. Hertzberg, J. A. Manson, and W. C. Wu, ASTM *STP 536* (1973) p. 391.

41. C. B. Yun, J. J. Lee, S. K. Kim, J. W. Kim, "Recent R&D Activities on Structural Health Monitoring for Civil Intrastructures in Korea," *KSCE Journal of Civil Engineering,* vol. 7, no. 6 (2003) pp. 637–645.

42. *Hoan Bridge Forensic Investigation Failure Analysis Final Report,* Lichtenstein Consulting Engineers, June 2001.

43. Oakland Bay Bridge Pile Connection Plate Welding Investigation Report (PDF): Federal Aid Project ACIM- 080-1 (085) 8N, MTE File No. S5021 (Mayes Testing Engineers, Inc., May 3, 2005).

44. J. W. Fisher, *NCHRP Research Results Digest* **59** (Mar. 1974).

45. J. W. Fisher, B. T. Yen, and N. V. Marchica, Fritz Engineering Laboratory Report No. 386.1, Lehigh University, Bethlehem, PA, Nov. 1974.

46. J. W. Fisher, K. H. Frank, M. A. Hirt, and B. M. McNamee, NCHRP *Report 102,* 1970.

47. J. W. Fisher, P. A. Albrecht, B. T. Yen, D. J. Klingerman, and B. M. McNamee, NCHRP *Report 147,*1974.

48. P. B. Keating and J. W. Fisher, NCHRP *Report 286,* 1986.

49. *AASHTO LRFD Bridge Design Specification*, 2nd ed., Am. Assoc. of State Highway and Transportation Officials, Wash., DC, 1998; see also 2009 interim update.

50. *Structural Welding Code: Steel*, 19th ed. AWS D1.1/D1.1M, American Welding Society, Miami, FL, 2004.

FURTHER READING

ASM Handbook, Volume 11: Failure Analysis and Prevention, R. J. Shipley and W. T. Becker, Eds., ASM International, Metals Park, OH, 2002.

ASM Handbook, Volume 19: Fatigue and Fracture, A. C. Ruffin, Ed., ASM International, Metals Park, OH, 1996.

ASM Handbook, Volume 12: Fractography, ASM International, Metals Park, OH, 1987.

Engineering Failure Analysis, Elsevier Science.

Handbook of Case Histories in Failure Analysis, K. A. Esaklul, ASM International, Metals Park, OH, 1993.

J. Failure Analysis and Prevention, Springer, Boston.

P. R. Lewis and C. Gagg, *Forensic Polymer Engineering*, Woodhead Publishing, Cambridge, 2010.

Chapter 12

Consequences of Product Failure

12.1 INTRODUCTION TO PRODUCT LIABILITY

To this point, we have discussed the deformation and fracture response of materials as influenced by various stress, time–temperature, and environmentally dependent conditions. Throughout the text, we examined case histories of actual failures associated with various material and design inadequacies. The purpose of this chapter is to address the consequences of these failures. Did such failures cause bodily harm? What was the extent of the resulting property damage? Who should make financial restitution for both personal and property damage arising from the failure, and what is an appropriate financial payment? These are specific legal questions addressed by trained lawyers in both state and federal courtrooms, as technical experts do *not* engage in such legal arguments. Engineers and scientists, however, do play a critical role in litigation and potential product recalls. Often, they are called upon to examine the physical evidence, perform necessary testing and analysis, draw conclusions based on their technical expertise, report their findings, and occasionally render their expert opinions in the courtroom (e.g., the shotgun failure litigation discussed in the previous chapter.) To better define an engineer's potential role in such matters, the developing landscape of product liability[i] law is introduced here along with overviews of product recall and an engineering approach to both avoiding and addressing recalls.

The branch of the law that typically deals with matters of product liability is referred to as the "civil law," as distinguished from the "criminal law." The civil law in this country is initially derived from the "common law" of England, which has been built upon by our "case law." Case law is straightforward in nature: a judge hears a case, reviews opinions of prior cases, and then renders a ruling. As such, judges decide what the law should be for a given situation. Judges by and large adhere to the principle of *stare decisis,* meaning that they will stand by decisions made in earlier cases and thereby provide predictability; hence, later decisions typically reflect the rules of law enunciated in rulings from earlier cases. As will be discussed, judges don't always follow *stare decisis,* rendering a different ruling from that articulated in prior cases. In such instances, the judge might decide that the original reason for the earlier ruling did not apply to the current case before the judge. More profoundly, the judge might conclude that the earlier ruling was erroneous and, instead, create new law. Hence, case law has been elastic, dynamic, and constantly changing. Such conflicting court opinions create confusion that can sometimes be resolved when a higher court decides which opinion is "correct" and, therefore, should become the new "law"; also, "new law" arises when the various state courts adopt new guidelines, periodically set forth by groups of legal experts (see Section 12.2.7).

A typical product liability case involves a claim brought by an allegedly injured plaintiff for damages against a retailer, distributor, or manufacturer of a product used by the plaintiff.[1] The plaintiff seeks to demonstrate to the court that the product was defective and caused his/her injuries. In addition, the plaintiff may also claim that the product was used properly. Damages sought will include those associated with medical expenses, temporary/permanent injuries, pain and suffering, lost wages, and property damage where indicated. By contrast, the defendant may attempt to demonstrate to the court that the product was not defective and that it was manufactured properly and marketed in a reasonable manner. In addition, the defendant may seek to

---

[i] Note that the term *products liability* is sometimes used as an alternative to *product liability.*

demonstrate that the plaintiff used the product in an unsafe and unreasonably foreseeable manner. Often, engineers are called in by both parties to assist in the characterization of the product and to render opinion(s) as to the soundness of the design and manufacture of the product, and to identify whether or not a critical defect was present in the product at the time of the sale or was generated as a result of the product's use. The financial stakes for both parties in such cases are often high.

12.2 HISTORY OF PRODUCT LIABILITY

One of the earliest recorded documents involving matters of "failed" products is found in the Code of Hammurabi, established by King Hammurabi, the first King of the Babylonian Empire almost 3800 years ago.[2] Aside from the Hebrew Torah, the set of 282 laws contained in the Code of Hammurabi represents one of the first written codes of law in human history.[ii] Each code contains mention of a specific crime and its associated punishment. (The stele tablet shown in Fig. 12.1 is inscribed with these pronouncements.) It is believed that these rulings were established to please the gods that the King worshipped. Note that there was little or no opportunity for the accused to defend their action. Five codes (specifically #229–#233) are noteworthy in the present discussion:

> *#229: If a builder build a house for someone, and does not construct it properly, and the house which he built fall in and kill its owner, then that builder shall be put to death.*
>
> *#230: If it kill the son of the owner the son of the builder shall be put to death.*
>
> *#231: If it kill a slave of the owner, then he shall pay slave for slave to the owner of the house.*
>
> *#232: If it ruin goods, he shall make compensation for all that has been ruined, and in as much as he did not construct properly this house which built and it fell, he shall re-erect the house from his own means.*
>
> *#233: If a builder build a house for someone, even though he has not completed it, if then the walls seem toppling, the builder must make the walls solid from his own means.*

Clearly, these stern rulings were established during a different era in human history and required modification for the modern period. Accordingly, we shall initiate our discussion with a historical

Figure 12.1   The stele of Hammurabi. The text of the Code appears in the center section of the stele, beginning just below the figures. (The sculpted figure is Hammurabi, standing before Shamash, the sun god and Lord of Judgment; photo copyright Ivy Close Images/Alamy Ltd.)

[ii] The Code of Lipit-Ishtar predates the Code of Hammurabi by approximately 200 years. (See F. R. Steele, *The Code of Lipit-Ishtar,* Reprinted from *American J. Archaeology* Vol. **LII,** No. 3, July–Sept. 1948.)

review of the development of product liability law. This law has changed over time to reflect the needs of society, giving greater protection to the perceived weaker party. For example, during the early stages of the industrial revolution, the courts tended to protect fledgling manufacturers. More recently, the courts have determined that the consumer needs greater protection since the consumer is generally unable to identify product defects that would likely do them harm.

Product liability law typically constitutes a mixture of *contract* and *tort* law. Contract law typically pertains to claims for breach of warranty for a product being not as originally represented. Tort law represents a body of law that focuses on legal theories of negligence and strict liability, terms that will be described below. In each instance, product liability litigation provides an opportunity for a plaintiff who believes he has been wronged to seek redress by passing along injury-related expenses to others.

12.2.1   Caveat Emptor and Express Warranty

Over 400 years ago, aside from agricultural products, there were few items entering the stream of commerce; those included farm utensils, kitchen items, and clothing. Also, vendors were often itinerant peddlers who came to town to sell their wares of sometimes questionable quality and who were gone by day's end. A classic example of such a peddler was the horse-trader. Seldom would such an individual openly proclaim the various deficiencies of his stallion or mare. Many buyers recognized the dubious nature of dealing with such individuals, but many did not. However, most buyers knew horses or were expected to be knowledgeable about them. Also, buyers recognized that they were on their own since the courts at that time tended to side with the sellers. Indeed, Roman law established the concept of *caveat emptor* (i.e., buyer beware). Essentially, the buyer had an opportunity and was obliged to use his knowledge in selecting an object for purchase and had to accept the consequences for his inattention to the purchase of a defective item.[3] In other words, unless the seller provided an express warranty, specifically stating that the product was both sound and as represented, then the buyer was on his own. The following case history is cited often as a primary example of product litigation based on the concept of *caveat emptor*.

> *Chandler v. Lopus*[4]: Around 1600, a jeweler, Chandler, sold for 100 pounds a precious stone to Lopus. The buyer claimed that the jeweler affirmed that the stone was a bezoar-stone. The latter is found in the stomach of certain animals and was believed to have certain medicinal properties. Subsequently, the buyer learned that the jewel was not a bezoar-stone and sued the jeweler for damages. The trial court ruled in favor of the buyer. The appellate court reversed, finding that the jeweler had not warranted that the stone was a bezoar-stone, explaining that " . . . the bare affirmation that it was a bezoar-stone, without warranting it to be so [in writing], is no cause of action: and although he knew it to be no bezoar-stone, it is not material; for everyone in selling his wares will affirm that his wares are good, or the horse which he sells is sound; yet if he does not warranty them to be so, it is no cause of action . . . "

An *express warranty* is an oral or written guarantee by the seller as to the specific functionality of the product and in absence thereof, a statement of conditions under which the product may be repaired, returned, and/or exchanged. When advertising claims are made about the product, such claims may represent express warranties; the product must substantially conform to such advertising statements. For example, if the seller states that an automobile can accelerate from 0 to 100 kph in under 4 seconds, this constitutes an express warranty provided the car has been used properly and for its intended purpose. On the other hand, companies or individuals may state opinions of value that, for example, their automobile is the fastest vehicle in the world. Such a *vague* statement represents "puffery" upon which the buyer cannot rely as part of the basis for a purchasing decision.[5,6]

12.2.2   Implied Warranty

Over the next several hundred years following the Chandler-Lopus decision, new industries developed, making available a much broader assortment of products to consumers who often resided a distance from the product's point of origin. Though the courts generally continued to accept the concept of *caveat emptor* in product litigation, they slowly recognized that buyers might purchase products that did not meet with their expectations. In response to the need to facilitate trade between merchants who did so at great distances, courts began to identify an *implied* warranty in sales contracts. As a remedy for consumers who did not possess an *express* warranty, the courts began to hold that sales contracts had an *implied* warranty or promise that the goods sold met some minimum standard of manufacture and performance. As such, the existence of an implied warranty became more desirable and acceptable in the courts. The 1815 English case of *Gardiner v. Gray*[7] illustrates the imposition of an implied warranty in a sales contract.

> *Gardiner v. Gray*[7]: The buyer purchased 12 bags of what was represented to be "waste silk." When the plaintiff took possession he discovered the bags were not waste silk. The buyer sought to rescind the contract but the seller objected on the grounds that there was no express warranty that the bags were waste silk. The court agreed that the contract had no express warranty but found that the "intention of both parties must be taken to be, that the goods shall be saleable in the market under the denomination mentioned in the contract between them." The court further held that inherent in the sales contract was a promise by the seller that the sale merchandise did, in fact, possess the characteristics of the material mentioned in the sales contract. If the material in the sales contract was "waste silk," then the consumer had the right to expect that the delivered material conformed to the marketplace's definition of "waste silk." Accordingly, the court found for the buyer.

12.2.3 Privity of Contract

As time passed and industry expanded, the stream of commerce took on new characteristics. New industries sprouted up across the lands that were increasingly remote from buyers. As a consequence, consumers were less likely to purchase their goods directly from a craftsman at a local marketplace; instead, consumers bought their goods from town and city shopkeepers, who acted as intermediaries between the companies that made the products and the consumers who purchased them (e.g., the distributors and retailers of today's marketplace). Accordingly, the consumer entered into a sales contract with the storekeeper who, in turn, had a sales agreement with the manufacturer. As such, it became increasingly rare for consumers to contract directly with manufacturers.

The courts closely examined the relationship between the consumer, who allegedly suffered personal injury and/or property damage, and the defendant who manufactured the allegedly defective product but who had not entered into a sales agreement with the plaintiff. Buyers tried to hold the manufacturer accountable under an implied warranty theory; however, implied warranty requires a contractual relationship. In the 19th century, the courts found that if no direct contractual agreement existed between the plaintiff (usually the consumer) and the defendant (e.g., the manufacturer) (i.e., the plaintiff was not in "privity of contract" with the defendant), then the plaintiff could not bring charges against the manufacturer, even though the product in question might be found to be defective. The courts did so to protect these new and often financially fragile manufacturing companies from costly litigation.

For example, in *Winterbottom v. Wright*,[8] the court endorsed the privity obstacle that barred a plaintiff from seeking damages from a defendant where no contract existed between the two.

> *Winterbottom v. Wright*[8]: In this case, Wright entered into a sales agreement with the Postmaster General to provide a mail-coach and maintain same in good working order. Subsequent to delivery of the coach, Winterbottom was hired by the Postmaster General to drive the coach to deliver the mail. On August 8, 1840, while Winterbottom was driving the coach, it " . . . gave way and broke down, whereby the plaintiff was thrown from his seat, and

in consequence of injuries then received, had become lamed for life."[8] Winterbottom brought suit against Wright for recovery of his damages, claiming that " . . . said mail-coach being then in a frail, weak and infirm, and dangerous state and condition . . . " was the cause of his injuries. At the conclusion of the trial, the court found for the *defendant*. Lord Abinger stated that "we ought not to permit a doubt to rest upon this matter, for our doing so might be the means of letting in upon us an infinity of actions. There is no privity of contract between these parties and if the plaintiff can sue, every passenger, or even any person passing along the road, who was injured by the upsetting of the coach, might bring a similar action. Unless we confine the operation of such contracts as this to the parties who entered into them, the most absurd and outrageous consequences, to which I can see no limit, would ensue . . . " With a hint of conscience, Rolfe, in a concurring opinion, added " . . . this is one of those unfortunate cases . . . it is, no doubt, a hardship upon the plaintiff to be without a remedy, but by that consideration we ought not be influenced. Hard cases, it has been frequently observed, are apt to introduce bad law."

It also should be noted that Winterbottom sued Wright, the provider or manufacturer of the coach, because he was barred by law of sovereign immunity from suing the Postmaster General, who was part of the English government.

12.2.4 Assault on Privity Protection

While the court's focus of attention in the 19th century would appear to be protective of the manufacturer, by the turn of the century some exceptions to the privity defense argument were beginning to be advanced. In *Huset v. J. I. Case Threshing Machine Co.*,[9] Judge Sanborn noted three exceptions: one involved products that could be classified as "imminently" or "inherently" dangerous; another exception involved a situation where someone was injured, having been invited to use the owner's defective product on the owner's premises; the third exception permitted a plaintiff to recover from a seller when the seller sold or delivered a product that was known to be defective without notifying the buyer and which caused an injury that could have been reasonably anticipated. Such exceptions were limited to dangerous products such as drugs, poisons, explosives, guns, and certain foodstuffs.

What about the full range of new products that were entering the stream of commerce? What if someone had purchased an electric table lamp from a local store, used the product with proper care for several weeks, but then received a terrible electric shock and serious burns when next attempting to light the room. What if it was later determined that the lamp contained an electrical wiring defect that directly contributed to the accident? Given that the consumer had suffered severe burns but had not entered into a sales agreement with the company that manufactured the lamp, what legal action could he have pursued to seek redress for his wrong? At the turn of the 20th century, the injured consumer couldn't have done very much. Furthermore, government agencies like the Consumer Product Safety Commission (CPSC), Food and Drug Administration (FDA), and National Highway Transportation and Safety Administration (NHTSA) were not in place at the time to protect the public (see Section 12.3).

Consider another scenario. What if someone purchased an automobile from a local car dealership, drove the car for a few months and then while driving one day, experienced a complete fracture of the front wheel axle? As a result, one wheel fell off, causing the car to veer sharply off the road and into a large tree. Both passenger and driver were thrown against the windshield and received severe head injuries. Examination of the axle revealed that the fracture was directly attributable to a preexisting but previously undetected forging defect that had grown by a fatigue cracking mechanism to a critical size during those few months of normal operation. What could those car passengers do to seek compensation for their injuries when they had no privity with the automobile manufacturer?

Similar circumstances occurred when MacPherson purchased a Buick automobile from a local dealer. Following an accident involving an allegedly defective wheel, MacPherson brought suit against Buick Motor Company.[10] MacPherson had no direct sales contract with Buick. The outcome of this case profoundly changed how courts viewed damage claims by plaintiffs and finally pierced the manufacture's protective shield of privity. In addition, it provides a good example of the roles of various types of engineers as expert witnesses in the legal process, and shows that different opinions can be reached by different experts when presented with the same physical evidence and set of facts surrounding the incident.

MacPherson v. Buick Motor Co.[10]: MacPherson purchased a Buick automobile from Close Brothers in 1910 in Saratoga, New York. (Figure 12.2 reveals a comparable model, circa 1910, of this vehicle.) One year later, while he was driving at 15 mph (the car possessed a 22-horsepower engine), MacPherson suddenly felt the rear end skid. That was soon followed by the sound of breaking wood, whereupon the left rear of the car dropped and swerved. Unable to control the direction of the car's progress, MacPherson's car smashed into a telephone pole and overturned. When the accident scene was examined, it was found that the left rear wheel had separated from the car and the wooden wheel spokes were shattered and strewn about.

After recovering from his injuries, MacPherson retained the leading lawyer in the county, State Senator Edgar Brackett, and sued Buick Motor Company. Brackett's task was to convince the jury and the court that Buick Motor should be held liable for MacPherson's losses *even though MacPherson had no direct dealings with Buick Motor.* For their part, Buick Motor viewed this case with great alarm in that it provided the potential for a major shift in how the current and future courts would view the validity of the privity doctrine. To that end, Buick retained the top automotive lawyer in Detroit, William Van Dyke, to defend them in this matter. His task was to convince the court that Buick had done all it could reasonably do to ensure the production safety of the car and to suggest that the car company would experience dire financial consequences were they to be held responsible for plaintiff's damages. Indeed, the entire automobile industry (and many other industries for that matter) risked potentially catastrophic financial damages and possible ruin.

The battle was then joined over a five-year period of time and involved three separate trials. Brackett argued that the wheel was defective along with its wooden spokes and that Buick failed to perform tests to ensure the integrity of said wheels; consequently, Buick should bear responsibility for MacPherson's losses even though the plaintiff was not in privity with the defendant, Buick Motor. Brackett engaged several expert witnesses who were

Figure 12.2 Buick Runabout Vehicle, circa 1910. Note the wooden wheel spokes. (M. Hyman and S. Dougan, Hyman Ltd. Classic Cars, St. Louis, MO; photo reproduced with permission from Mark Hyman.)

experienced in the manufacture of wheels (a wheelwright), the nature of hickory wood as used in wheel spokes (a materials engineer), and in the testing of wheels to establish their integrity (a test engineer). It was initially determined that the wheel had been manufactured by Imperial Wheel Co. and subsequently purchased by Buick for use in their cars; Imperial, a subcontractor to Buick, was recognized in the automobile industry as an experienced wheel manufacturer.

The first expert witness was George Palmer, who claimed 30 years of carriage-making experience along with recent experience with both the making and repairing of automobile wheels. Palmer testified that the spokes had broken off square, indicating the wood to be of inferior quality. He testified that " . . . apparently, the wood was brittle, coarse grained such as you find in old trees. I never saw a spoke broke off as square as they were. Good quality of sound hickory, when it breaks, brooms up, 'slivers up'. The fact that they [the broken spokes in the MacPherson case] are broken square across indicates that they were brittle, of poor quality, trash."

Brackett then presented to the court Adelbert Payne, a local expert with 20-years experience in the manufacture of carriages and wheels. Payne concurred with Palmer's assessment that the MacPherson wheel was of poor quality and " . . . light and poor, no heft, broken off too squarely for good hickory." Payne further testified " . . . the integrity of hickory can be determined from an examination of the surface [of the sides of a hickory spoke] by an experienced eye. A sound, good quality of hickory has a very fine grain that shows distinctly through the outer surface." He further offered that if the spokes were already painted, one could scrape off sufficient paint to see the grain and fiber of the wood. That would be " . . . a perfectly feasible and easy test to determine the character of a spoke that way, and make sure."

Brackett then went to the Thomas Motor Co. to engage a wheel tester. This individual testified that his company bought wheels in the natural state, had them visually inspected in the receiving department and subsequently subjected to a hydraulic pressure applied on the wheel hub to ensure their soundness. It was then noted in court that Buick had not conducted such a test. In summary, Brackett had made the case that MacPherson's car wheel was defective and that Buick had not conducted a proper examination to ensure the safety of their automobile (i.e., an argument based on inadequate quality control that Buick had not met the reasonable standard of care for the industry).

Now it was time for Van Dyke to present his case to the jury. Van Dyke sought to counter the testimony of MacPherson's experts. He retained the director of the Purdue University materials testing laboratory for his expertise and opinions regarding the testing and analysis of wood products; this expert had extensive testing experience for the United States Forestry Service. He argued that it was unfounded prejudice [i.e., the plaintiff's expert opinion] that wood from young trees was superior to that from older trees, irrespective of whether the wood grain was coarse or fine. Instead, he argued that fast-growing hickory trees had the greatest "shock resisting capacity" *[Author note: Was this an early allusion to the concept of a material's "fracture toughness"?]* It was his opinion that good hickory had a tree ring count ranging from 5 to 25 rings per inch. With that and having been shown the broken wheel spokes, he proclaimed that they were " . . . first class, running about fifteen rings per inch." He further explained to the jury the virtual impossibility of examining the side surfaces of the spokes to identify the quality of the wood; the latter would require an examination of the tree rings—an impossible task with a finished wheel.

[Author note: There you have it. Two sets of experts with completely different conclusions concerning a given set of observations. Whose testimony was to be accepted by the court? The auditor for the Boston molasses tank law suit (see Chapter 5), when faced with conflicting expert opinions from both plaintiff and defendant, reached the following conclusion: "Amid this swirl of polemical scientific waters it is not strange that the auditor has at times felt that the only rock to which he could safely cling was the obvious fact that at least one-half the scientists must be wrong."[11]]

Proceeding further, Van Dyke presented representatives from several major automobile manufacturers and wheel makers who testified that no test was conducted or could be conducted to evaluate the wooden spokes in wheels purchased from wheel suppliers. Further, they opined that the hydraulic test performed by the Thomas Motor Co. was worthless and that automobile manufacturers were not in a position to evaluate the integrity of the spokes since they did not employ wood experts who could address this issue. With this opinion Buick disputed the position that Thomas Motors' methods should be determined as the industry standard of care by which Buick should be judged. *[Author note: Can you imagine in today's technical and legal environments, an automobile manufacturer not employing teams of material scientists?]* Finally, Van Dyke presented witnesses from Buick Motor and Imperial Wheel who testified that Buick had produced 125,000 cars with 500,000 wheels having been supplied by Imperial Wheel. Neither firm had previously received a single complaint of a broken-down wheel.

The case was then sent to the jury with the following instruction: the jury should return a verdict for the plaintiff if: (1) the accident was caused by the crumbling of the wheel spokes and (2) Buick had failed to conduct such tests " . . . that a person skilled in the manufacture of cars . . . ought to have used to discover a defect." After due deliberation, the jury found for the plaintiff and awarded him $5,000.

This initial legal battle had focused on facts in the case: to wit, the wheel was defective and Buick Motor was at fault. There were two more court battles to be fought that focused on the legal consequences of the lower court's decision. The initial appeal by Buick Motors was defeated.[12] Van Dyke then appealed to the New York Court of Appeals, the highest court in New York. At issue was whether the doctrine of privity, established in *Winterbottom v. Wright*,[8] would be upheld or whether the law would change so as to place a higher burden of responsibility on the manufacturer for an improved level of protection for the ultimate user.

In his argument before the New York Court of Appeals,[10] Van Dyke laid out his interpretation of the law as follows: "We believe that the limitation of the liability of the manufacturer of an innocuous article is based upon sound reasoning. The maker of an innocuous article, such as a chair, table or buggy, cannot follow it through every hand for all time. He cannot know as to whether it goes into prudent or imprudent hands, or whether it has had minor accident, or undue exposure, which have weakened it. He cannot trace the history of his article through every subsequent and remote user for all time and in all places. Therefore, the law recognizes this inability and holds him liable only where it is definitely shown that he created or had knowledge of the defect and was willfully careless or practiced a fraud."

Senator Brackett then presented his position concerning the injuries sustained by his client but also focused on a broader issue: "the decision of this question . . . is of the utmost interest, not simply for this little tombstone dealer [MacPherson] in the remote hamlet of Galway in Saratoga County, but to every person, everywhere, who shall hereafter use this modern means of locomotion." He then added the following statement of great import: "What substantial difference, what difference in morals, should be made in testing this defendant's liability for this maiming of the plaintiff, whether, having first manufactured, it sells this machine directly to MacPherson or sells it to Close Brothers, with the expectation and intention that Close Brothers will sell it to some MacPherson? The defendant was equally culpable in either case. The plaintiff has suffered equal damage; and it is close to, if not quite, immoral and wicked to allow the recovery in one case and deny it in the other . . . "

Justice Cardozo, speaking for the Appeals Court, summarized the positions of both parties and concluded that the overall charge " . . . is one not of fraud, but of negligence. The question to be determined is whether the defendant owed a duty of care and vigilance to anyone but the immediate purchaser."[10] Cardozo referred to earlier cases (e.g., *Thomas v. Winchester*[13]) that involved exclusion of the defendant's use of privity (this case involved a mislabeled bottle of poison). (Recall Judge Sanborn's first stated exception for a valid privity defense by a manufacturer.) Cardozo then pointed out that " . . . the defendant argues that things imminently dangerous to life are poisons, explosives, deadly weapons—things whose normal function it is to injure or destroy . . . But whatever the rule in *Thomas v. Winchester*[13] may once have been, it has no longer that restricted meaning." He continued that other products that are not obviously dangerous become so if imperfectly constructed and/or negligently made. As such, the definition of a "dangerous" product had been expanded greatly to include any product if the latter contained a defect.

As to the matter of the privity defense, notwithstanding the definition of a "dangerous" product, Cardozo concluded " . . . we have put aside the notion that the duty to safeguard life and limb, when the consequences of negligence may be foreseen, grows out of contract and nothing else." Hence, the defendant's privity wall of protection was forever breached when the New York Court of Appeals affirmed the judgment of the lower courts.

12.2.5 Negligence

Over a period of time, the courts increasingly accepted negligence as a valid cause of action for product liability. Negligence, defined as conduct involving an unreasonably large risk of causing either personal injury or property damage, with respect to products would be identified if the following four conditions are satisfied.[14]

1. Does the product contain a defect?
2. Could reasonable means have been used to eliminate the defect?
3. Did the defect cause injury?
4. Did the plaintiff use the product in a reasonable manner?

It should be noted that negligence could also be identified in association with faulty or inadequate wording as related to instruction guidelines, warnings of potential danger, labels, and other literature that pertains to product use and associated potential harm, as will be discussed in the Product Recall section of this chapter.

It is clear that the pivotal term, pertaining to the concept of negligence, is "reasonable." From the plaintiff's perspective, how "reasonable" was their use of the product that may have led to the injury? On the other hand, did the defendant exercise "reasonable" care in the design, manufacturing, and marketing of their product? As will be discussed in the Product Recall section of this chapter, consideration of what is "unreasonable" versus what is "reasonable" (e. g., unreasonable risk of serious injury or death, reasonably foreseeable use and misuse of a product) is of great importance in determining whether a product should be recalled.

In the 1930s and 1940s, Judge Learned Hand developed a formula that formed the framework for assessment of the "reasonableness" of a defendant's actions.[15]

$$B < PL \tag{12-1}$$

where $B =$ the cost **B**urden to prevent or minimize the damage
$P =$ the **P**robability that harm will occur
$L =$ the gravity and financial extent of the **L**oss

According to this equation (essentially the metric for a cost-benefit analysis), when the magnitude of the loss, L, times the probability, P, of the event is greater than the cost burden to prevent such damage, B, the accident should have been prevented. (It should be noted that the magnitude of the losses might include potential litigation costs.) As such, the formula defines an economic meaning of negligent conduct. In addition, a judgment of negligence implies society's moral disapproval of the defendant's behavior, implying there was a cheaper alternative to the accident. On the other hand, it might be argued that if $B > PL$, the condition leading to the accident need not be avoided.[16] Here it is argued that if either or both the probability of damage and the magnitude of the potential damage are relatively small compared to the cost burden to prevent such an event, then that burden is unjustified and there is no moral indignation. Instead, such costs should be redirected so as to improve the overall net wealth or welfare of society. That is, the net worth or welfare of society is maximized by preventing only those accidents when $B < PL$. The Hand formula thus attempts to describe the balance between safety and economic efficiency.

Equation 12-1 provides a possible framework for juries to apportion blame. Posner,[16] however, pointed out that juries typically are not provided the information they need to properly examine the consequences of the Hand formula; often, such information is unavailable. To assist in this matter, Grossman et al.[17] suggest that the size of insurance premiums may estimate the probability of a particular event occurring and costs for the associated damages. Regarding potential *property* damage, one would need to consider the burden (i.e., costs) associated with the replacement of a presumably defective component versus the benefit (i.e., avoidance) of such expenditures such as those associated with avoidance of defective component-induced property damages and potential component recall.

For the case of *bodily* harm, such expenses are strikingly more difficult to define on a dispassionate basis. For example, consider the following scenario: An engineer is asked to design the safest passenger vehicle on the road. Accordingly, she must strike a balance between various cost/benefit decisions, dealing with a multitude of material selection and design element considerations, leading to the development of a safe but commercially viable automobile. An initial design concept might incorporate many of the features of the latest military armored vehicle. To be sure, such a vehicle would generate an impressive road safety record. Consider, however, a few of the associated costs for such a product: fuel consumption would be excessive; additional road and bridge damage would result from the far greater vehicle weight, thereby placing a considerable burden on municipal budgets; and the far greater material utilization and manufacturing costs would price such vehicles beyond the means of the typical buyer. Clearly, a second design iteration, based on more commercially viable considerations, would shift the cost/benefit balance point in the direction of lower total costs. Implicit in this decision, one would anticipate a potentially greater number of crash-related injuries/deaths associated with the use of this commercial vehicle. Society must, therefore, decide on the location of the appropriate cost/benefit balance point. More specifically, a jury panel often will be charged with that responsibility. And yet, it might prove difficult for a juror to assess the "societal cost," involving human injury or loss of life, without viewing the plaintiff's loss through their own emotionally tinged lens: as a proxy for the plaintiff's loss, what monetary value might the juror attribute to a loved one's life or for the juror's own limb if they or she were, instead, the injured party? How can a juror's judgment not be prejudiced by such thoughts? There is another issue to consider: Is it morally and/or ethically appropriate or even possible to assign a monetary value for a broken bone, a lost limb, loss of an eye, or ultimately the loss of life? Furthermore, should the jury assign for such damages a greater value for a corporate executive, community leader or single parent than for a vagrant or a person with a criminal record? Then again, what alternative metric should be used to "quantify" the loss of life and/or limb? To be sure, a cost-benefit analysis, involving human injury and death, is imperfect at best and, some might argue, ethically unsound at worst. Given this difficulty, how else besides a cost-benefit analysis that includes a quantification of personal injury can a corporation or design engineer identify that elusive balance point between risk and reward?

It should be noted that virtually all human beings engage almost daily in some form of cost-benefit analysis. Bungee jumpers accept the risk of severe injury for the reward (i.e., "adrenalin rush") of an exciting plunge into the river gorge; others would not. The same is true for certain skiers who race down black diamond slopes for that special thrill while others are content to engage in an après skiing rendezvous. Then there is the driver who ignores the warning bells at the railroad crossing and who tries to race across the tracks before the gate descends so as to arrive home to play a few minutes longer with their children. Where is the risk/reward balance point drawn in these instances or, for that matter, in the manufacture of a commercial vehicle? That very issue was addressed with events associated with rear-end collision-induced fires associated with the Ford Pinto subcompact automobile.

Cases Involving the Ford Motor Co. Pinto: In response to fierce marketplace conditions, the Ford Motor Co. set out in 1967 to design and build a new subcompact automobile, called the Pinto. This model was brought to market with an accelerated design and production schedule. Initially, the model met with commercial success; however, it was soon determined that when the car was involved in a rear-end collision, the fuel tank could be damaged, leading to excessive gas leakage and fatal car fires.[18-22]

Was the Ford Motor Co. aware of this deficient crash worthiness? Apparently, yes. Subsequent court proceedings[23] and separate news investigations[19,21] identified that Ford's crash tests, performed between 1969 and 1972 (a number of which were performed prior to the launching of the Pinto in 1970), identified several problem areas: when the car was rear-ended at moderate speeds up to 48 kph (30 mph), the gas tank (located 7.5 cm (3 in) *behind* the rear axle and 15 cm (6 in) in front of the rear bumper) was pushed against the rear axle where it could be punctured by bolt heads protruding from the differential housing, thereby leading to fuel leakage; the fuel filler pipe, which was inserted into the gas tank through a gasket-lined hole, could separate from the tank during the collision and lead to additional fuel leakage; location of the gas tank *above rather than behind* the rear axle was determined to be a safer option; insertion of a plastic shield prevented differential bolt heads from piercing the tank; and installation of either a rubber bladder within the tank or longitudinal side rails reduced the potential for rear-end collision-induced fuel leakage. None of these remedies were implemented prior to the sales launch of this vehicle, even though Ford engineers had estimated that many of the design remedies would have cost the company only $2 to $6 per vehicle![24]

By 1977, a number of reported rear-end collision-induced fires led to the publication of an exposé article entitled "Pinto Madness." Dowie[19] alleged design deficiencies and Ford's reckless decision to reject their engineer's recommendation to implement necessary changes to the Pinto's fuel system. Furthermore, Dowie[19] alleged that Ford had used a cost-benefit analysis that assigned monetary values for human life and burn injuries to justify *not* making design changes. [*In fact, this reported cost-benefit analysis did not specifically address the issue of rear end collision-induced gas tank fires, but rather addressed a different accident scenario.*] This emotionally charged article, which gained national attention, was written during a time of heightened consumer advocacy and examination of alleged corporate malfeasance; the Ford Pinto model and the associated crash-induced fires became the focus of such attention. Within a month of its publication, the Dowie report prompted the National Highway Traffic Safety Administration (NHTSA) to launch an investigation into the crash worthiness of the Pinto and other cars. NHTSA's own crash test results (conducted in the spring of 1978) confirmed earlier Ford findings of rear-end crash-induced excessive fuel leakage and resulting fires. In a preliminary report[25] in May 1978, NHTSA confirmed for the Pinto " . . . an initial determination of the existence of a safety-related defect." The following month, Ford voluntarily recalled approximately 1.5 million Pintos that had been manufactured between 1971 and 1976 and made necessary changes to these cars. Ford engineers had proposed similar design changes during the pre-launching date in 1970 and thereafter but none were implemented until the voluntary recall in 1978.

Why were such changes not incorporated earlier? Dowie[19] maintained that Ford avoided making necessary design changes because such modifications were not cost effective. Though Ford denied using a cost-benefit analysis to justify their inaction, Birsch[24] argued otherwise. First, he pointed out that government regulators and auto companies in the 1960s had agreed that cost-benefit analyses (i.e., application of the Hand formula) would form the basis for making decisions concerning auto safety. Indeed, a 1972 NHTSA report,[26] completed with the encouragement of the auto industry, contained a table that identified the societal cost components for fatalities and assigned a monetary value of approximately $200,000 for an auto-related human fatality. Second, Ford separately had submitted to the NHTSA a cost-benefit analysis concerning burn-induced deaths resulting from rollover accidents. In this report,[18] Ford used NHTSA figures to estimate that the avoidance (i.e., cost benefit) of 180 burn deaths (at $200,000), 180 serious burn injuries (at $67,000), and 2100 burned vehicles (at $700) would amount to a beneficial saving of $49.5 million. By contrast, were suggested design changes to be made for 11 million cars and 1.5 million light trucks to mitigate rollover propensity at a cost of $11 per vehicle, the overall cost would be $137 million. On that basis, Ford argued that making such changes was not economically justified. Third, two separate internal Ford documents were discovered that stipulated the delay of gas tank modifications until 1976 would realize cost savings of $10M and $20M, respectively.[18] Presumably, the changes to be made in 1976 would have been mandated after the anticipated approval of NHTSA Standard 301,[27] which outlined maximum permissible gas tank leakage rates, resulting from a rear-end crash. The final piece of evidence was the testimony of one of Ford's own engineers, Harley Copp, who testified that Ford's decision not to upgrade the Pinto's fuel system was based on cost savings that would result from delaying the improvements.[28]

One such accident and its associated court case then received widespread national attention. In May 1972, a 1972 model Pinto hatchback automobile was struck in the rear at a speed of roughly 48 kph (30 mph). At impact, the Pinto was engulfed in flames, resulting in the death of the driver (Mrs. Gray) and seriously burning and injuring a 13-year-old boy (Richard Grimshaw). Based on the findings and assertions noted above, a jury verdict in February 1978 found for the plaintiffs in the amounts of roughly $3 million for compensatory damages and *$125 million in punitive damages!*[23,29] That punitive damage judgment was subsequently reduced to several million dollars with the judge asserting that "Ford's institutional mentality was shown to be one of callous indifference to public safety."[22] Despite this verdict, Ford delayed making safety modifications until the 1971–1976 model recalls in June 1978 that followed release of the aforementioned 1978 NHTSA preliminary report.[25]

In another accident and trial of notoriety,[30] three women from the Ulrich family were burned to death in August 1978 as a result of a rear-end collision of their 1973 Pinto. Based on the previously discussed Dowie report,[19] findings reported in 1979 by the *Chicago Tribune*[21], NHTSA findings,[25] and prior trial findings,[23] the Indiana Elkhart County State Attorney, Michael Cosentino, concluded that Ford had failed in its responsibility to provide a reasonably safe product to the marketplace but had, instead, intentionally endangered its victims. Accordingly, he concluded that Ford was liable for acts of *commission* of building a defective car and *omission* by ignoring its responsibility to remedy a known design defect. Cosentino convened a special grand jury that agreed that *criminal charges of "reckless homicide"* be brought against Ford. This astounding criminal charge generated front-page headlines in newspapers across the nation. In its defense, Ford immediately countered that the Ford *Corporation* could not be criminally charged, based on this statute since: (a) the company was not a "person"; (b) the 1973 Pinto was delivered to the marketplace several years *before* the Indiana criminal statute was enacted; thus, ex post facto application of the law constitutionally barred prosecution; and (c) Ford could not be charged in Indiana in this matter since Congress had created the National Traffic and Motor Vehicle Safety Act to oversee such allegations. The 1980 trial did proceed, though with a more limited scope.[20] Ultimately, the jury returned a verdict of not guilty; however, as a result of the damaging publicity to the brand, Pinto sales plunged and the model was subsequently removed from the company's product line. Following the trial, Wheeler[31] examined the issues of monetization of human injury "costs" as they relate to a design imposed cost-benefit analysis and the application of criminal law in product liability cases. As noted earlier in this section, the need to quantify the extent of a loss, even the loss of life or limb, is necessary, albeit emotionally charged. Furthermore, he opined that " . . . the use of general criminal laws to regulate product design will ill serve the public's interest."[31] Clearly, there are differences of opinion regarding these issues.

12.2.6 Strict Liability

A significant new concept in product liability law, which would later be referred to as "strict liability," was enunciated in a 1944 case involving a waitress and a broken bottle of soda.

Escola v. Coca Cola Bottling Co. of Fresno[32]: The plaintiff, a restaurant waitress, claimed that a soda bottle exploded in her hand when she was transferring it from its case to the refrigerator. However, expert testimony was unable to identify any specific glass defect or excessive gas pressure in the bottle. To that end, plaintiff relied on the legal argument of *res ipsa loquitur* (i.e., the facts speak for themselves)—namely, that when a person is injured in a manner that but for negligence on the part of the manufacturer an injury like this would not occur, then the manufacturer is deemed to be liable. Since the defendant was responsible for the manufacture, charging, and inspection of the bottle and since the bottle did explode, there was an *inference* of defendant's negligent conduct. Indeed, the court found for the plaintiff. Judge Traynor, in a concurring opinion, concluded that the manufacturer's negligence should not be singled out as the basis for plaintiff's efforts to recover their damages. Instead, Traynor argued " . . . even if there is no negligence, however, public policy demands that responsibility be fixed wherever it will most effectively reduce the hazards to life and health inherent in defective products that reach the market. It is evident that the manufacturer can anticipate some hazards and guard against the recurrence of others, as the public cannot. . . . " Traynor went on to state that the jury need not establish the manufacturer's negligence. Specifically, " . . . If such products nevertheless find their way into the market it is to the public interest to place the responsibility for whatever injury they may cause upon the manufacturer, who, even if he is not negligent in the manufacture of the product, is responsible for reaching the market. . . . Against such a risk there should be general and constant protection and the manufacturer is best suited to afford such protection."[32]

Judge Traynor's concurring opinion of strict liability in *Escola v. Coca Cola* became the majority opinion 19 years later in *Greenman v. Yuba Power Products, Inc.*[33]

Greenman v. Yuba Power Products, Inc.: In this case, the plaintiff brought suit for damages against both the retailer and manufacturer of a Shopsmith combination power tool. The latter is a home shop device that can be used as a saw, drill, or wood lathe. The plaintiff purchased the Shopsmith in 1955 and subsequently purchased in 1957 a lathe attachment so that he might turn a large piece of wood into a chalice. After working on this wood piece on a few occasions without incident, it suddenly flew off the lathe, striking him in the forehead. He sustained serious injuries from this accident and subsequently sued on negligence and breach of warranty theories. The plaintiff's experts provided strong evidence that the Shopsmith was designed and constructed in a defective manner. For example, the plaintiff contended that the attachment screws were inadequate and that there existed better ways to attach the various components of the Shopsmith device. After a mixed finding, both plaintiff and defendant appealed the verdict. The California Supreme Court ruled that the plaintiff was entitled to damages based on the doctrine of strict liability, regardless of whether negligence or breach of warranty had been proven. To wit, " . . . A manufacturer is strictly liable in tort when an article he places on the market, knowing that it is to be used without inspection for defects, proves to have a defect that causes injury to a human being. Recognized first in the case of unwholesome food products, such liability has now been extended to a variety of other products that create as great or greater hazards if defective. . . . The purpose of such liability is to ensure that the cost of injuries resulting from defective products are borne by the manufacturers that put such products on the market rather than by the injured persons who are powerless to protect themselves."[33]

There is a fundamental difference between negligence and strict liability theories. Negligence focuses on the *conduct* of the manufacturer as related to what the manufacturer did or didn't do with regard to the product's design, manufacture, inspection, and manner by which the product's usage was described as in its operating instructions. By contrast, strict liability focuses on the *product* and its potential for doing harm. Here the court must distinguish between the product's reasonable safety and its unreasonable danger. That is, the risk of injury caused by the product must be balanced against the product's usefulness to society.[34]

A judgment of strict liability does not require proof of negligence. Instead, the plaintiff need only establish that:

1. The product contained an unreasonably dangerous defect.
2. The defect was under the control of the defendant or that the defect was present when the product left the possession of the defendant.
3. The plaintiff suffered injuries.
4. The defect caused the injuries.

Therefore, even though the defendant may have exercised due care in the manufacture of the product and not demonstrated any negligence, the defendant would be held liable, nonetheless, if these four conditions were satisfied.[35]

12.2.7 Attempts to Codify Product Liability Case Law

To this point, we have briefly examined several landmark cases that framed the ever-changing landscape of product liability "law." Hundreds of other reported cases have supported and contradicted those major rulings; these subsequent different court positions from many different jurisdictions were nominally coherent but contributed to a mixed fabric of rulings. To bring greater order, coherence, reason, and consistency to the plethora of such legal findings, the American Law Institute (ALI) was incorporated in 1923 for the purpose of "restating" areas of common law, such as that pertaining to product liability.[36] Over the past 80 years, the ALI, consisting of the United States' most prestigious judges, lawyers, and academicians, has worked to reconcile the diverse views of different state courts; these efforts resulted in a series of three restatements of tort law. Of significance, the Restatement (Second) of Torts embraced the concept of strict liability, as defined in *Greenman v. Yuba Power Products Inc.*[33] For example, Section 402A stated:

1. One who sells any product in a defective condition unreasonably dangerous to the user or consumer or to his property is subject to liability for physical harm thereby caused to the ultimate user or consumer, or to his property, if
   a. The seller is engaged in the business of selling such a product, and
   b. It is expected to and does reach the user or consumer without substantial change in the condition in which it is sold.
2. The rule stated in Subsection (1) applies although
   a. The seller has exercised all possible care in the preparation and sale of his product, [*Author note: There is no negligence*] and
   b. The user or consumer has not bought the product from or entered into any contractual relationship with the seller. [*Author note: No privity between consumer and seller is required.*]

Plaintiffs in numerous product liability cases made extensive use of such strict liability arguments. With time, however, Section 402A became the source of great confusion, particularly with regard to the meaning of the phrases "defective condition" and "unreasonably dangerous." In addition, some courts were now identifying three separate and distinct classes of defects: those associated with manufacturing defects, design defects, and those associated with inadequate instructions or warnings. Accordingly, the Restatement (Third) of Torts[37] did subsequently formalize three distinct types of defects:

"A product is defective when, at the time of sale or distribution, it contains a manufacturing defect, is defective in design, or is defective because of inadequate instructions or warnings. A product:

a. *Contains a manufacturing defect when the product departs from its intended design even though all possible care was exercised in the preparation and marketing of the product;*

b. *Is defective in design when the foreseeable risks of harm posed by the product could have been reduced or avoided by the adoption of a reasonable alternative design by the seller or other distributor, or a predecessor in the commercial chain of distribution, and the omission of the alternative design renders the product not reasonably safe;*

c. *Is defective because of inadequate instructions or warnings when the foreseeable risks of harm posed by the product could have been reduced or avoided by the provision of reasonable instructions or warnings by the seller or other distributor, or a predecessor in the commercial chain of distribution, and the omission of the instructions or warnings renders the product not reasonably safe."[37]*

Restatement (Third) followed the goal and purpose of Section 402A from Restatement (Second) by imposing strict liability on manufacturers for "manufacturing" defects, even if all possible care was exercised. By sharp contrast, however, Restatement (Third) followed a body of more recent case law in utilizing fault-based liability for "design" and "warning" cases. In the matter of alleged design defects, for example, Restatement (Third) requires a plaintiff to demonstrate a reasonable alternative design that could have resulted in superior product safety in order for the plaintiff to recover damages.[36,37] Proceeding further, " . . . design defects and defects based on inadequate instructions or warnings are predicated on a different concept of responsibility . . . such defects cannot be determined by reference to the manufacturer's own design or marketing standards because those standards are the very ones that plaintiffs attack as unreasonable. Some sort of independent assessment of advantages and disadvantages to which some attach the label "risk-utility balancing" is necessary. Products are not generically defective merely because they are dangerous."[37] This concept will be discussed further in the Product Recall section of this chapter. [*Author note: Clearly, Restatement (Third) broadens the scope of the technical expert's responsibilities in product liability litigation.*]

The reader should recognize that various political and social forces along with special interest groups are brought to bear on the articulation of these Restatements.[38] For example, some would argue " . . . tort law's expansion in helping make injured people whole is needed to hold greedy corporations accountable for their deeds . . . On the other side are those who believe tort law today encourages plaintiff's lawyers to sue indiscriminately, playing the odds and turning courthouses into casinos."[39] Such opposing points of view are reflected by the fact that individual state jurisdictions may vary in their respective conformity to these tort restatements and quite possibly to those that will probably follow.[40]

12.3   PRODUCT RECALL

As we have just discussed in matters pertaining to product liability litigation, the courts have witnessed a general shift in *judicial* findings, from those protecting fledgling manufacturers (e.g., by establishing privity barriers) to those protecting consumers (e.g., by introducing negligence and strict liability arguments). In addition to product liability law that becomes relevant in the event of a lawsuit, engineers must also be familiar with mandatory federal regulations, regardless of whether a lawsuit has been filed. In the United States, these are included in the Code of Federal Regulations, which is a codification of all federal government regulations published in the Federal Register[iii] pertaining to a wide range of issues. Now we will consider *regulatory* actions stemming from the activities of various governmental agencies created to protect the general public. Examples of these governing bodies include the National Highway Traffic Safety Administration (NHTSA) (e.g., automobiles, as discussed in Section 12.2.5 regarding the Ford Pinto recall), Food and Drug Administration (FDA) (e.g., foods, drugs and medical devices), Federal Aviation Administration (FAA) (e.g., airplanes), and the Consumer Product Safety Commission (CPSC) (e.g., consumer products).

Product recalls can occur for a wide variety of engineered products that are regulated by these governmental agencies. For ease of discussion, our attention will focus on the many thousands of product types sold to the general public that fall under the jurisdiction of the CPSC as well as similar governmental agencies around the globe. It is important to note that many of the considerations discussed below for consumer products also pertain to recall investigations regarding products that fall under the jurisdiction of other governmental agencies.

The distinction between product liability and product recall generally is that tort law comes into play after an injury has occurred, whereas federal safety standards and product recalls are intended to prevent injuries from occurring. As stated by the United States Government Accountability Office with respect to CPSC:

> "CPSC was designed as a complement to tort law, under which one may seek compensation for harm caused by another's wrongdoing. The threat of legal action under tort law plays an important role in assuring that companies produce safe products. However, tort law is primarily a post injury mechanism; and foreign manufacturers are usually outside of the U.S.

[iii] The Federal Register includes rules, proposed rules, and notices of Federal agencies and organizations.

tort law system. Therefore, CPSC has certain authorities intended to prevent unsafe consumer products from entering the market in the first place."[41]

In recent years, extraordinary attention has been focused on the occurrence of consumer product recalls. For example, 465 consumer product recalls, involving 230 million product units, were conducted in the United States in 2009.[42] According to the CPSC, deaths, injuries, and property damage associated with consumer products place a societal burden within the United States of more than $700 billion annually. Estimates for these costs include medical expenses, work losses, pain and suffering, legal costs, property damage estimates and other related factors.[43] In an effort to address the growing number of recalled products, Congress increased the maximum civil penalties for failing to report a potential substantial product hazard from $1.825 million to $15 million in 2009.[44] This dramatic escalation in monetary fine, coupled with the added potential for criminal penalties of up to five years imprisonment and asset forfeiture, signaled a new chapter in the enforcement of consumer product safety in the United States.

Similarly, a raised awareness exists in the European Union. European manufacturers and distributors are obligated to notify regulatory authorities and to take necessary action (e.g., sales bans and recalls) if they become aware that they have placed a product in the market that might present unacceptable risks to consumers (see Section 12.3.1.2). From 2003 to 2008, the total number of such notifications increased from 139 to 1866, reflecting more than an order of magnitude increase.[45]

Given these recent developments, it is necessary for engineers who design and manufacture products to be familiar with regulatory requirements and product recall considerations. The following discussion addresses these matters.

12.3.1 Regulatory Requirements and Considerations

12.3.1.1 Consumer Product Safety Commission

As discussed earlier, the CPSC is an independent agency of the United States federal government with jurisdiction over consumer products. The basic charter of the CPSC as described in the Consumer Product Safety Act is to:

1. Protect the public against unreasonable risks of injury associated with consumer products;

2. Help consumers to evaluate the comparative safety of consumer products;

3. Develop uniform safety standards for consumer products and minimize conflicting local and state regulations;

4. Encourage research into the causes and associated prevention of product-related injuries, illnesses, and deaths.

The Consumer Product Safety Act also contains reporting requirements under Sections 15 and 37.[iv] For example, Section 37 requires manufacturers of consumer products to report information about settled or adjudicated lawsuits. The following discussion will focus on the more generally applicable reporting requirements set forth in Section 15.

According to Section 15(b), manufacturers, importers, distributors and retailers must notify the CPSC (i.e., report) immediately if they obtain information that reasonably supports the conclusion that a product exhibits one or more of the following conditions:

1. Fails to comply with a *mandatory* consumer product safety standard or banning regulation (i.e., a rule that bans a product from having certain characteristics or meeting specified criteria),

---

[iv] Title 15 of the United States Code (U.S.C.) Sections 2064(b) and 2084. Section 102 of Public Law 103-267 requires that companies report certain choking incidents.

2. Fails to comply with any other rule, regulation standard or ban under any Act enforced by the Commission,

3. Contains a *defect* that could create a *substantial product hazard* to consumers,

4. Creates an *unreasonable risk* of serious injury or death, or

5. Fails to comply with a *voluntary* consumer product safety standard upon which the CPSC has relied on under the Consumer Product Safety Act.

In 2006, the CPSC provided the following additional factors for consideration in an effort to further clarify these four reporting conditions: an evaluation of the obviousness of the risk, the adequacy of warnings and instructions in mitigating risk, the role and "foreseeability" of consumer misuse, and the potential for reduced risk as the number of products in use declines over time.

In considering these overall reporting requirements, one might conclude that the failure to meet a mandatory or voluntary standard or comply with a banning regulation is straightforward. However, "defect," "substantial product hazard," and "unreasonable risk" are vague terms and warrant further discussion.

12.3.1.1.1 *Defect*

According to the CPSC's regulation interpreting the reporting requirements, at a minimum:

> " . . . *a defect is a fault, flaw, or irregularity that causes weakness, failure, or inadequacy in form or function.*"[46]

A product defect could exist in a product's materials, design, construction, finish, packaging, warnings, and/or instructions. To assist companies in understanding the concept of a defect as it pertains to the Consumer Product Safety Act, the Code of Federal Regulations provides specific examples of product defects, including the following:

> "*A kite made of electrically conductive material presents a risk of electrocution if it is long enough to become entangled in power lines and be within reach from the ground. The electrically conductive material contributes both to the beauty of the kite and the hazard it presents. The kite contains a design defect.*"[47]

It is important to note that not all products that present a risk of injury are defective. For example, the cutting edge of a kitchen knife is not a product defect for two reasons: (1) the risk associated with a sharp blade is considered reasonable and *obvious,* and (2) a dull blade inherently lacks utility. Likewise, as Judge Cardozo suggested (recall Section 12.2.4), some products are not obviously dangerous, or may even appear to be innocuous, but can become dangerous if a defect exists that can result in a failure with safety implications. The latter concept is explored in Section 12.3.1.1.3 for the case of drawstrings on children's clothing.

In determining whether a risk of injury makes a product defective, the CPSC considers the following questions in their analysis:[v]

1. What is the utility of the product (i.e., what is it supposed to do)?

2. What is the nature (potential) of the risk of the injury that might occur?

3. What is the need for the product?

4. What is the exposed population and demographic(s)?

5. What is the obviousness of the risk?

6. Are the warnings and instructions adequate to mitigate the risk?

7. What is the role of consumer misuse of the product and the foreseeability of such misuse?

---

[v] Title 16 of the Code of Federal Regulations, Section 1115.4(e).

8. What is the CPSC's experience with the product?

9. What other information and factors are relevant to the determination of defect including product liability case law and cases interpreting public health and safety statutes?

12.3.1.1.2 Substantial Product Hazard

In determining whether a substantial product hazard exists, the CPSC considers:

1. The pattern of defect,

2. The number of defective products distributed in commerce,

3. The severity of the risk of injury, and

4. The vulnerability of the population at risk, e.g., children, the elderly.[vi]

It is important to understand that product failures and associated injuries are not a prerequisite for a product recall. Consider the delayed fracture of tempered glass containing nickel sulfide inclusions.[48] In annealed glass, these undesirable and unintended inclusions transform to a low temperature form as the glass is cooled slowly during the manufacturing process. However, during the tempering process, glass is rapidly cooled and the tendency for this transformation to occur will be a function of the time–temperature history of the product during its service life; potentially, this can occur after long periods of time at room temperature. When the transformation does occur, the associated volumetric expansion can result in glass breakage without warning, depending on the size and location of these inclusions. (See Section 7.5.2 for more information about tempered glass and about NiS inclusions.) If a glass panel manufacturer determines that nickel sulfide inclusions are most likely present in a specific product population, a recall may be warranted prior to any reported failures. In this case, a substantial product hazard may exist due to the high likelihood of delayed spontaneous fracture and the associated risk of injury, even if the product satisfies the mandatory safety standard relating to tempered glass.[49]

12.3.1.1.3 Unreasonable Risk

Whether a product creates an unreasonable risk of injury or death may often be ascertained by considering available resources including CPSC's National Electronic Injury Surveillance System (NEISS). NEISS is a publicly available and searchable CPSC database that contains statistically valid national estimates and specific information about product-related injuries that are treated in hospital emergency rooms. According to the CPSC, NEISS annually supplies over 360,000 product-related cases from a sampling of approximately 100 hospitals.[50]

Drawstrings on children's upper outerwear represent one example of a product character-istic that creates a substantial and unreasonable risk of injury to children. Drawstrings on children's clothing are recognized as a *hidden* (not *obvious*) hazard that can lead to injuries and deaths when they catch on items such as playground slides, bus doors, or cribs. In the case of playground slides, strangulations have resulted from a knot or toggle on a child's sweatshirt drawstring becoming entangled in a gap at the top of the slide as the child descends. As a result, the drawstring pulls the garment taut around the neck, thereby strangling the child. From January 1985 through January 1999, CPSC received reports of 22 deaths and 48 non-fatal incidents involving the entanglement of children's clothing drawstrings.[51] In 1996, the CPSC issued guidelines to help prevent this hazard and ASTM adopted a voluntary standard that incorporated CPSC's guidelines in 1997.[vii] The CPSC has announced numerous recalls of children's upper outerwear due to this hazard, independent of whether injuries had been

---

[vi] Title 15 of the United States Code, Section 2064(b); Title 16 of the Code of Federal Regulations, Section 1115.12(g).

[vii] ASTM F1816-97, Standard Safety Specification for Drawstrings on Children's Upper Outerwear.

reported for the specific product (recall the example of glass panels containing nickel sulfide inclusions).

On the other hand, an estimated 495 adult eye injuries associated with pens and pencils occurred annually between 2001 and 2004, according to the NEISS database.[52] While these writing implement–induced injuries are unfortunate, pens and pencils have not been removed from the marketplace and will likely continue to be the writing implements of choice in the future. As is the case for a kitchen knife, just because injuries are associated with the use of a product does not necessarily make that product unreasonably dangerous, nor does it necessarily imply that the product requires a design change.

12.3.1.2 International Governmental Landscape

Organizational bodies and associated regulations pertaining to consumer product safety vary widely around the globe. In Japan, for example, the Ministry of Economy, Trade and Industry is the entity responsible for consumer product safety policy that includes the Japanese Consumer Product Safety Law. In the European Union, the Directorate General for Health and Consumers is the primary agency responsible for consumer product safety.

Let us briefly consider the European approach to consumer product safety. The key regulatory piece of legislation is the General Product Safety Directive (GPSD),[53] originally introduced in 1992 and most recently revised in 2010.[54] The overall aim of the GPSD is to ensure that only "safe" consumer products enter the marketplace. A product is considered "safe" if it does not present any risk to users or only minimal risks that are compatible with the product's use, when used under normal or reasonably foreseeable conditions. During such a determination, factors similar to those discussed in Section 12.3.1.1 are taken into consideration.[55]

In addition to complying with relevant European standards, producers are expected to perform a risk assessment of their products before they bring them to market. A methodological framework exists to establish consistency in how risk assessments are performed. Hazards are characterized by type (e.g., mechanical, electrical, thermal), severity of injury (e.g., slight, serious, and very serious), probability (e.g., $>50\%$, $<10^{-6}$), and vulnerability of the exposed population (e.g., severely disabled, very old, and/or very young persons). Producers must also provide consumers with necessary product information to warn of risks associated with normal and reasonably foreseeable use, especially where such risks are not immediately obvious.[56]

In the event of a product investigation, the risk assessment framework serves as a valuable tool for a company determining whether notification to the European regulatory authorities is necessary. The decision to notify the national authorities can result in warning consumers of a potential problem or, where appropriate, recalling products from the marketplace. To facilitate compliance, a rapid alert system called RAPEX exists for reporting dangerous consumer products and allows for the RAPid EXchange of information among European Union member countries. Even if notice is provided through RAPEX, companies still must comply with the requirements of each country.

12.3.2 Technical Considerations Regarding Potential Recalls

Thus far, we have discussed various regulatory requirements and considerations when evaluating the safety of engineered products and the potential need for a product recall. While basic requirements exist, the decision to report and potentially recall a consumer product is made on a case-by-case basis.[57] We will now outline a basic engineering approach that can help to determine, from a *technical* perspective, when and if to report a "condition" to the CPSC or other similar agency, and whether it is necessary to initiate a product recall. It is important for the reader to understand that "reporting" a potential issue to the CPSC signals the beginning of an investigative period and does not necessarily mean that a product recall is imminent. CPSC strongly encourages reporting, even if it is unclear whether a real danger or hazard exists. It is clear that a company must consider technical as well as legal and regulatory perspectives when faced with a potential product recall.

In an engineering analysis of this kind, one can readily envision the need to address the following factors:

1. Determination of the failure process
2. Identification of the affected product population
3. Assessment of risk associated with product failure
4. Generation of an appropriate corrective action plan

12.3.2.1   Determination of the Failure Process

To determine why a product has failed, a multidisciplinary approach typically is required. The range of expertise may include materials scientists, corrosion experts, mechanical, chemical and electrical engineers, and thermal scientists (i.e., fire cause and origin specialists). In addition, human factors (i.e., study of human–machine interactions) must also be considered, including an evaluation of warnings, labels, and instructions for the proper use of a given product.

To determine why a product has failed, it is necessary to identify the steps involved in the failure process and, if possible, the root cause that triggered that process. It is important to first distinguish between the terms *root cause* and *failure process*, as they are sometimes used interchangeably. The *root cause* is the fundamental, underlying reason for the failure. A *failure process* is the process by which final failure takes place. As such, if the *root cause* is removed or avoided, the *failure process* does not occur.

As an example of this distinction, consider the following fictitious scenario: A leading parts manufacturer, Metals-r-Us, sells thousands of seemingly identical rods to a medical device company, Body by Metal. The rod is part of an implanted medical device that provides mechanical support in the body for an extended period of time. The rod design was used successfully for many years, with the exception of a single recent failure. Upon microscopic examination of the explanted rod, it was determined that the rod failed by a fatigue process, as evidenced by the presence of both clamshell markings and fatigue striations on the fracture surface. Furthermore, it was determined that a single fatigue crack (i.e., no ratchet lines) initiated at a metal forming-induced surface discontinuity. In this hypothetical case, the *root cause* of the failure was determined to be a manufacturing defect; in turn, the *failure process* consisted of fatigue crack initiation at this forming-induced defect, fatigue crack propagation, and subsequent overload fracture of the rod's remaining unbroken ligament. It is important to note that an appropriate corrective action plan sometimes can be constructed even if the root cause is unknown.

There are many aspects to an engineering failure analysis investigation. The more information that is available for review, the more likely a complete understanding of the failure can be achieved. Certainly, informative responses to the detailed checklist provided in Appendix C can facilitate a comprehensive failure analysis of a specific component; however, when faced with the failure of a product containing many components, "system level" data are often required. Examples of such additional information include:

1. A summary of events leading up to the incident, eyewitness accounts, incident reports/photographs, and documentation of product handling after the incident.
2. Application specific information including how and where the product was used.
3. The service/maintenance history of incident product(s) and similar models, including reports of previous issues.
4. The "fingerprint" of the product including: the product manufacturing date, serial number, specific model and batch numbers, and the manufacturing facility where the product was assembled.
5. Any available product documentation including both design and manufacturing assembly drawings, operation manuals, warnings, and instructions.

6. The timeline of product evolution, including changes in product design, materials, components, construction, packaging, warnings, and instructions.

7. The quality control procedures incorporated by raw material vendors through the various manufacturing and assembly operations, including material certification sheets.

After reviewing these available file materials, the first step in determining the root cause and associated failure process involves a non-destructive analysis of the incident unit(s). This typically consists of a visual examination and photographic documentation of the product and, if possible, the surrounding environment at the time of the incident. All available markings and labels on the product should be recorded, as this information can sometimes be useful in limiting the extent of affected product, as discussed in Section 12.3.2.2. It is also prudent to perform a thorough examination of damage to the product.

Mechanical damage (e.g., impact, wear) can provide insight into the use or misuse of the product and the loads experienced during its lifetime. For example, during an investigation of water damage from failed pipe joints, one of the authors observed deep wrench tooth gouges on numerous pipe fittings, suggesting the occurrence of environmentally assisted cracking. In this case, the failure was caused by a combination of residual tensile stresses created as a result of overtightening, the presence of an aqueous environment, and a susceptible plastic piping material.

Patterns of thermal damage can provide insight into the origin of heat, smoke, or flame. In some cases, it is possible to pinpoint the region or even the component responsible for a fire, depending upon the degree of damage to the product. It is important to note that thermal damage patterns can be marred or inadvertently destroyed during handling or product removal from the incident site. Therefore, careful attention should be given to preserving both the product's thermal damage patterns as well as the surrounding environment for subsequent analysis.

X-ray imaging represents a powerful non-destructive method to examine the product's interior without the need for disassembly. Depending upon the densities of the materials of construction, it is sometimes possible to determine the extent of damage to inner components without disturbing their condition or relative position. Other potentially insightful non-destructive techniques include liquid dye penetrant and magnetic particle inspection which can be used to detect the presence of surface cracks.[58]

In most cases, non-destructive evaluations provide an incomplete assessment of the failure event; accordingly, a destructive examination is typically in order. However, once a product is altered from its original condition, valuable information can be permanently lost if proper procedures are not followed. Therefore, it is very important to carefully document product disassembly and the individual steps associated with the destructive examination protocol. In addition, if legal action has been taken or is pending as a consequence of the product failure, it is necessary to perform any destructive analysis in the presence of interested parties in order to avoid problems related to evidence *spoliation* (i.e., modification that destroys its evidentiary value during ongoing or anticipated investigation or litigation).

After the product has been disassembled, it is often useful to examine the components at higher magnifications using optical and/or scanning electron microscopy. As discussed in Chapters 5, 9, and 10, it is often possible to identify macroscopic and microscopic features that indicate the nature of a failure (e.g., clamshell markings and striations for the case of a fatigue failure) or the existence of a material or manufacturing defect (e.g., porosity). In addition, it is also possible to evaluate the chemical composition of a component using various methods, including energy dispersive X-ray spectroscopy. This can be helpful in determining the materials of construction, as well as the existence of contamination or corrosion products that may have contributed to the product failure. If chemical state information is required in order to determine thermal history or environmental effects, X-ray diffraction methods can be used to identify unknown crystalline phases. In some cases, it is possible to estimate the temperatures achieved during an overheating event by the color of a metal oxide.

Important insight into the foreseeable use and/or misuse of a product can also be gained by reviewing the associated warnings, labels, and instruction manuals. Some of these deficiencies include:

1. Unclear or inaccurate description of required steps and/or precautions involved in assembly, operation, or maintenance of the product.
2. Ineffective placement of warning labels on the product.
3. Ineffective or inappropriate use of textual and pictorial components of the warning labels. For example, "Danger," "Warning," and "Caution" are terms used to convey decreasing levels of hazard.[59]
4. Ineffective usage of background color or foreground text font and/or color for warning labels.

Exemplar products, specifically new and identical products with respect to construction and function, serve as useful tools during these types of investigations. Laboratory testing of exemplar products makes it possible to support or refute proposed failure processes by evaluating various potential incident scenarios. (Recall the use of exemplar gun tubes from Case History 1 in Chapter 11.) The engineer can then determine whether the test results are consistent with reported failures and the condition of available incident products.

12.3.2.2 Identification of the Affected Product Population

Once the failure process has been identified, it is important to determine whether the product's failure is an isolated event or if a "pattern of defect" is identified. In this regard, it is prudent to carefully review available databases including product returns, warranty claims, customer complaints, and reported incidents. This exercise should be conducted on an ongoing basis to help determine recent trends in the nature of product failures and to compare this information with historical data. For example, is this the first product to be returned or is this the 20th product to be returned in the past three weeks? If there is a spike in the number of returns, are these failures of a similar nature? To answer these questions, a detailed examination of available failed products is often required. If none are available to examine, it is beneficial to review all available photographs and reports associated with these incidents. If a pattern of defect is established, it is then necessary to determine the population that is potentially affected. Some of the important questions for an engineer to consider in this determination include:

1. Is there an inherent design issue? For example, are the material properties or performance characteristics inadequate? Is the problem associated with a defective component purchased from a supplier? Is this component single-sourced or multiple-sourced? If the problem is with only one component supplier, is it possible to distinguish the affected population from the total product universe?
2. Within a given production facility, is there a batch problem tied to a manufacturing process deviation?
3. Is this problem associated only with certain models of the product? For example, are there differences in product construction that influence product susceptibility to a specific failure mode or unsafe condition? Can these performance differences be verified empirically?
4. Are there geographical considerations that make it possible to rule out certain populations of product? For example, are electronic products more likely to overheat and cause fires in certain countries based on standard power outlet voltages (e.g., a 110-volt supply is standard for wall outlets in the United States, whereas 220-volt service is standard in many parts of Europe)?

In many cases, the answer to the question, "How many products are affected?" can have a profound impact on a company's financial stability, especially if a recall is ultimately required. As discussed in the Recall Case Study below, the answer to this question can sometimes determine whether a company can emerge relatively unscathed from a product crisis or, instead, be forced into bankruptcy.

12.3.2.3 Assessment of Risk Associated with Product Failure

The risk associated with a product failure is generally expressed as a function of both the frequency of occurrence within the product population and the severity of its associated consequences (recall Eq. 12-1). Accordingly, it is possible to assign a higher risk level to 1,000 products in the field that have a slight chance of failure resulting in serious harm as compared with 1 million products in the field that will likely fail but would cause little or no harm.

Analytical tools are helpful in evaluating the risk associated with various failure scenarios. To illustrate, Failure Modes and Effects Analysis (FMEA) is a *qualitative* risk assessment framework typically used to prioritize risk before a product is introduced to the marketplace. A traditional FMEA quantifies risk in terms of three categories: severity, probability of occurrence, and detection (i.e., the likelihood that existing process controls will detect this failure). Each category is rated on a relative scale (e.g., 1 to 5), with a lower rating corresponding to lower risk. The ratings for the three categories are multiplied together to calculate the risk priority number (RPN). The higher the RPN value, the higher the overall risk of a potential failure. By assessing risk using this framework, it is possible to logically and consistently compare relative risks that can be difficult to otherwise evaluate, similar to the European Union approach to risk assessment discussed in Section 12.3.1.2.

During an actual recall investigation, it is useful to review prior FMEA findings to challenge the assumptions made during the initial product development stage. When conducting such an investigation, there is usually little time to perform a complete FMEA to document all possible failure modes for all components; besides, analyses of many components may well be irrelevant to the problem at hand. In cases where a very specific failure mode is being investigated (e.g., an electrical fire), a targeted FMEA could be considered that is focused on all possible ways in which relevant components can cause that specific failure mode to occur. This represents a worst-case analysis of consequences resulting from individual component failures, but does not address system level effects. Clearly, though a focused FMEA does not provide perspective on the overall risk profile for the product, it may represent an effective way to quickly identify the issues relevant to the problem that precipitated the current product crisis.

A Fault Tree Analysis (FTA) represents another valuable tool that can be used during a product recall investigation. Rather than being limited to individual component failures, FTA is a *quantitative* risk assessment methodology whereby the engineer identifies combinations of component or subassembly failures that can result in an incident or "end point." Using this approach, one begins with the final result (e.g., a fire) and works backwards from that event. The engineer assigns probabilities of failure to the individual steps required to create the specific "end point"; this makes it possible to assess the overall risk of a failure. To enable the engineer to assign (i.e., estimate) the probabilities of events in the failure process, references can be made to existing databases that contain field failure data as well as component failure rates[60] for various parts (e.g., fans, power supplies, and pumps).[61] Since FTA is a quantitative method, the probability of a specific end point then can be calculated by multiplying together the probabilities of each step. For example, assume there is a 1% (0.01) probability that each of three separate, sequential steps will occur. If all three steps are required to produce an end point, the probability of this resulting hazard is calculated to be 10^{-6}.

In order to fully understand the potential consequences of a product failure, targeted "worst-case scenario" laboratory tests should be performed, if possible, to simulate a "perfect storm" with regard to foreseeable conditions. These tests can be more aggressive than those addressing proposed failure process(es), as discussed in Section 12.3.2.1. The purpose of these targeted tests is to determine the *most* extreme consequences in the event of a catastrophic failure, rather than just those observed in the reported failures. This approach can be used to determine whether the malfunction of a product will lead to a potential safety hazard or merely a nuisance to the user. This type of information can be useful in deciding whether or not to recall a product.

As such, FMEA and FTA analytical tools enable an engineer to determine relative and actual risk values associated with a specific failure event; laboratory testing can certainly provide additional valuable insight. Nevertheless, the ultimate question remains: What level of risk is *reasonable*? Edwards (op. cit., p. 4) has provided a valuable distinction between *reasonable* and *unreasonable* risk. To wit,

> *"Risks of bodily harm to users are not unreasonable when consumers understand that risks exist, can appraise their probability and severity, know how to cope with them, and voluntarily accept them to get benefits that could not be obtained in less risky ways . . . But preventable risk is not reasonable (a) when consumers do not know that it exists; or (b) when, though aware of it, consumers are unable to estimate its frequency and severity; or (c) when consumers do not know how to cope with it, and hence are likely to incur harm unnecessarily; or (d) when risk is unnecessary in . . . that it could be reduced or eliminated at a cost in money or in the performance of the product that consumers would willingly incur if they knew the facts and were given the choice."*

To illustrate this point, recall the annual number of eye injuries reported in the NEISS database for pens and pencils, as discussed in Section 12.3.1. Pens and pencils have very sharp points that can find their way into one's eye if care is not exercised. Consumers willingly accept the inherent risk associated with these indispensable consumer products given the benefits received in the form of their low cost and functionality; therefore, this particular risk is deemed reasonable by society.

An insightful view of reasonable risk was published earlier by Tetelman and Starr.[62] These investigators sought to understand acceptable societal risk in terms of the fatality rate in the United States from all causes; primarily using old age and disease factors, they calculated death events to occur at a rate of about 1% per year or approximately 10^{-6} per exposure hour. Based on this estimation, the probability of a "natural" death was determined to be an approximately "one in a million" per exposure hour event. According to Tetelman et al., risk data suggested that a 10^{-6} level might be considered as a baseline for assessing the risk posed by activities (and associated products) such as driving an automobile and/or by using commercial aviation.[viii]

Tetelman et al. pointed out that risks associated with certain activities and/or products during their initial introduction into the marketplace are relatively high and that, over time, these risk levels tend to decrease to this 10^{-6} baseline level. For example, during the early 20th century, the risk of fatality as a result of an automobile accident was far greater than the risk level due to natural causes. Over a period of time, the number of drivers increased dramatically while significant improvements were made with regard to vehicle safety and driver awareness. As a result, the risk level associated with driving approached an equilibrium value of 10^{-6}, the risk level due to natural causes. By contrast, the probability of death from "pure chance" events like lightning strikes was determined to be 10^{-9} per exposure hour, significantly lower than that associated with natural causes, as would be expected. While the reader may assume that there is universal agreement that the range of 10^{-6} to 10^{-9} per exposure hour would define the spectrum of acceptable risk (defined by the fatality rates associated with all causes and pure chance, respectively), this is not necessarily the case. For example, some individuals choose to lower their pure chance risk levels by never venturing outdoors when thunderstorms are predicted. The distinction between *reasonable* and *unreasonable* risk is clearly difficult to pinpoint or define; however, it is clear from both a judicial and regulatory perspective that the threshold for acceptable risk has been reduced considerably over the course of time (e.g., comparing societies from before the industrial revolution to those corresponding to the 21st century).

[viii] Ironically, Tetelman was killed in a commercial airline mid-air crash in 1978 while en route to an air safety conference.

12.3.2.4 Generation of an Appropriate Corrective Action Plan

If the decision is made to recall a product, it is necessary to determine the most appropriate remedial action, commonly referred to as a corrective action plan. This plan, including notification of the customer, can take one of several different forms including: a customer- or service technician-implemented fix using a repair kit sent from the manufacturer; complete removal of all products from the marketplace; or an exchange for a new model. Though the original hazard(s) will be sufficiently mitigated or eliminated by implementing an appropriate fix, it is critical to avoid the introduction of new hazards as a result of a proposed corrective action. Therefore, it is prudent to carefully review potential correction actions, especially if a customer-implemented fix is the preferred solution. This evaluation should assess the ease of implementation and determine what can go wrong during the process. Detailed instructions along with visual aids must be provided to the individuals performing the repair. Testing these procedures with individuals unfamiliar with the product can also be an insightful exercise during the development of remedial procedures. The bottom line with any corrective action plan is to consider the KISS principle of basic design: Keep-It-Simple-Stupid.

12.3.3 Proactive Considerations

Thus far, we have discussed both the regulatory and technical aspects that must be considered when addressing a potential product recall. Having reviewed these aspects relating to the "back end" (i.e., product recall investigations), the reader can now understand and appreciate the importance of focusing considerable effort on the "front end" (i.e., product design and development stage) to avoid or, at the least, minimize the impact of a product recall. The elements of a product recall prevention and management initiative may vary, but an effective approach should consider the following components:

12.3.3.1 Think Like a Consumer

During the product design process and when generating instructions and warnings, it is critical for engineers to consider how consumers might interpret written instructions and warnings and how they might conceivably use their products. Thus, questions such as "Are the supplied warnings and instructions clear?" or "How might they be interpreted or misinterpreted?" are relevant to those charged with designing or evaluating products. As discussed, faulty product instructions, warnings or labels alone can constitute a product defect worthy of a recall, even if the product itself is not defective. Thinking like a consumer is a fundamentally important concept; incorporating this view into the product design process can raise awareness of important human factors before the product ever reaches the consumer.

12.3.3.2 Test Products Thoroughly

Risk analysis can take place in a variety of forms, including FMEA and FTA, as discussed in Section 12.3.2.3. These are powerful tools in determining how a product can fail, especially when they reflect the experience of seasoned engineers and subject matter experts. After performing an FMEA, for example, a company can rank the risks associated with a product and then test accordingly to validate design assumptions and determine if additional action is required.

Problems can result when companies do not perform comprehensive testing of their prototype products before they arrive in their customers' hands. In some cases, products can fail in a manner that was not anticipated, thereby creating safety hazards. Various types of design validation tests can shed light on possible failure scenarios and resulting consequences. These include normal-use tests, in which the product is used in accordance with the instructions; misuse tests, in which the product is tested in ways that are reasonably foreseeable, (i.e., not precisely according to the instructions but in a manner that might be reasonably extrapolated from conventional use); and abuse tests, whereby the product is purposely abused with an aggressive testing approach to see what might happen. In addition, forced-failure tests can be performed to simulate the failure of specific components to determine the associated consequences, similar in concept to the "perfect storm" approach discussed in Section 12.3.2.3. If accelerated life testing is contemplated, care should be exercised to ensure that only operative failure modes are explored. After completing this type of analysis, instructions can be revised and the product design can be modified accordingly.

In many cases, consumer products are evaluated by a certified testing laboratory (e.g., Underwriters Laboratory) to verify compliance with voluntary or mandatory standards; however, these tests should be considered as minimum requirements to be satisfied, as they do not necessarily prove that the product is entirely safe (e.g., recall the discussion of tempered glass in Section 12.3.1.1.2). In fact, testing to standards developed for classes of products is no substitute for product-specific testing. Testing should be considered for those scenarios not covered by voluntary or mandatory standards. It is easier and far less costly to understand how your product might fail during prototype or qualification testing in a laboratory than when in use.

12.3.3.3 Ensure Adequate Traceability

Good recordkeeping will increase the likelihood of traceability if a problem occurs. Creating a product "fingerprint" can help narrow or define the extent of a problem and help clarify which products might be affected (recall Section 12.3.2.2). In some cases, an engineering-based argument can be made for why a product recall need only be conducted for a specific population. Comprehensive documentation and product traceability then can allow a company to put a fence around the problem and limit costs associated with a total recall. *You cannot limit the scope of a problem if you cannot confirm the problem's boundaries.*

One example of a high level of traceability can be found in the watch industry. Schwab-Feller AG is one of the few companies in the business of manufacturing mainsprings that power mechanical watches. Their entire process involves a large number of individual steps, including numerous rolling operations, cutting, finishing, cleaning, polishing, heat treatment, spot welding, coating, and baking. Every spring lot is marked with a bar code identifier that enables the company to track the spring back to the raw material stage. In addition, it is possible to identify when the mainspring was fabricated and the individual who performed each step of the process.[63]

12.3.3.4 Manage Change Carefully

In some cases, a safe product is produced for a period of time and then something is changed. This modification may be a result of consumer feedback, a new component supplier, a manufacturing process change to increase efficiency, or perhaps new environmental regulations. Unfortunately, even minor changes can have enormous safety implications. In these circumstances, it is important to conduct those tests that will evaluate the effect of the proposed change(s) on product performance. For products without an historical risk analysis, this may be the time to update files and revisit technical assumptions.

Recall Case Study: *The "Unstable" Ladder*

Even when a company makes a concerted effort to perform qualification testing and exercise due diligence, product recalls are sometimes necessary. In one matter involving an investigation by one of the authors, a ladder manufacturer made a business decision to change rivet suppliers from supplier A to supplier B. As a result, ladders assembled with supplier B rivets were tested to confirm that they met existing performance specifications. The test matrix included torque, tensile, and shear tests, all performed to failure in order to determine the weakest link in the ladder's assembly. In all cases, the vertical rails of the fiberglass ladder were damaged as a result of the different test conditions; by contrast, the supplier B rivets that attached the rungs to these members remained intact. Based on these test results, the ladder manufacturer began to purchase supplier B rivets for future production. During the transition period from supplier A to supplier B, rivets were used interchangeably. It is critical to note that supplier B rivets were easily distinguishable from supplier A rivets by the presence of a dimple on the supplier B rivet heads. After a period of time, customers began reporting that large percentages of rivet heads were "popping off." As a consequence of these failures, ladder rungs were separating from their rails and workers were being injured. The manufacturer performed inspections of the incident ladders and found that many of the affected ladders contained both supplier A and supplier B rivets (the latter being distinguished by the dimple on the rivet head); of particular note, *all* of the fractured rivets were from supplier B.

As part of the failure analysis investigation, the chemical compositions of the two rivets were compared. The original rivets, from supplier A, were found to have been fabricated from a 2xxx series aluminum alloy, whereas supplier B used a 5xxx series aluminum alloy, containing approximately 5% magnesium. A scanning electron microscope investigation revealed that the fracture surfaces of the supplier B rivets were entirely intergranular in nature. Accordingly, the rivet fracture mode was consistent with stress corrosion cracking (SCC) in 5xxx series aluminum alloys. (See Fig. 12.3.)

As discussed in Chapter 8, the SCC mechanism requires three simultaneous conditions to be present: sufficient tensile stresses, an aggressive environment, and a susceptible material. The manufacturing and assembly steps involved in the fabrication of the ladders resulted in the generation of residual tensile stresses in the rivets from the riveting process, thereby satisfying the requirement of a sustained tensile stress. Furthermore, the ladders were typically exposed to humid and wet conditions, sometimes in the presence of chloride ions (e.g., road salt), thereby, satisfying the aggressive environment requirement for SCC in aluminum alloys.[64]

Finally, the insidious nature of these rivet failures was attributed to a unique characteristic of this susceptible material. Unlike the 2xxx series alloy, the replacement 5xxx series material is metallurgically unstable in that an SCC-susceptible modified microstructure can develop over time.[65] Indeed, 5xxx series aluminum alloys with high levels of magnesium are highly supersaturated solid solutions. As a result, excess levels of magnesium in the alloy can lead to the precipitation of Mg_2Al_3 particles that are anodic (more reactive) with respect to the matrix. The presence of such precipitates at the grain boundaries was confirmed during a metallographic examination of these failed rivets. For 5xxx aluminum alloys containing greater than approximately 3% magnesium, SCC can occur in the cold-worked condition upon exposure to elevated temperatures; furthermore, as the level of magnesium is increased to approximately 4% and above, SCC can occur after long periods of time at room temperature.[66] Given that all the qualification tests were short term in nature (i.e., completed in a

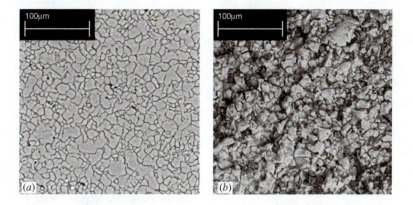

Figure 12.3 (*a*) Etched metallographic specimen of a supplier B rivet revealing the grain structure, and (*b*) the fracture surface of a failed supplier B rivet displaying the features of intergranular fracture.

matter of minutes), the time-dependent change in microstructure of supplier B rivets and the material's resulting environmental susceptibility were never identified.

Had the company performed a literature search on the characteristics of 5xxx series aluminum alloys, they likely would have found that Alcoa had identified in the late 1920s the stress corrosion cracking tendencies of aluminum–magnesium alloys.

> "... saucepans which had been made of sheet containing 10% Mg were subjected to a corrosion test by filling them with New Kensington, Pennsylvania, tap water. After a period of a few months, these pans began to open up much like the unfolding of rose petals. This experience started a rather intensive study of susceptibility to stress-corrosion cracking of these alloys and the mechanism involved."[67]

Another reference specifically warned against the use of 5xxx series aluminum alloys for rivets, stating:

> "Aluminum-magnesium alloy rivets should contain less than 3.0% magnesium, otherwise the heavily cold worked heads suffer SCC."[68]

In summary, the ladder manufacturer made a very unfortunate mistake. To make matters worse, given that an unknown number of their ladders had both types of rivets, they could not put a fence around the problem. If they had a system in place to ensure traceability, the company could have potentially limited the size of the affected population. Their idea of performing qualification testing of a new rivet material was prudent; however, their incomplete characterization and testing along with a lack of a literature review resulted in the need for a total recall of the ladder population. The result: The company went bankrupt. Until that time, they were just plain lucky.

REFERENCES

1. W. P. Keeton, D. G. Owen, J. E. Montgomery, and M. D. Green, *Products Liability and Safety*, 2nd ed. The Foundation Press, Westbury, NY, 1989, p. 15.

2. *Mesopotamia: The Code of Hammurabi*, R. Hooker, Ed., translated by L. W. King (1910) 1996, Washington State University.

3. W. H. Hamilton, *Yale Law J*. **40**, 1133, 1133–53.

4. *Chandler v. Lopus,* 79 Eng. Rep. 3 (1603).

5. D. A. Hoffman, *Ioma Law Rev.*, July 1, 2006, p. 102.

6. J. Russell Jackson, *National Law Rev.*, Apr. 9, 2007.

7. *Gardiner v. Gray*, 4 Camp. 144, 171 Eng. Rep. 46, Court of King's Bench, 1815.

8. *Winterbottom v. Wright,* 10 M & W. 109, 152 Eng. Rep. 402 (Ex. 1842).

9. *Huset v. J. I. Case Threshing Machine Co.*, 120 F. 865 (8th Cir. 1903).

10. *MacPherson v. Buick Motor Co.*, Court of Appeals of New York, 1916. 217 N.Y. 382, 111 N.E. 1050.

11. *Engineering News-Record*, **94** (5), 188 (Jan. 29, 1925).

12. MacPherson, 160 App. Div. 55, 145 N.Y.S. 462 (1914).

13. *Thomas v. Winchester*, 6 N.Y. 397, 1852.

14. W. L. Prosser, *Handbook of the Law of Torts*, 4th ed. West Publishing Co., St. Paul, MN, 1971, p. 646.

15. *United States v. Carroll Towing Co.*, 159 F. 2d 169 (2nd Cir. 1947).

16. Richard A. Posner, *Tort Law: Cases and Economic Analysis*, 1982.

17. P. Z. Grossman, R. W. Cearley, and D. H. Cole, *Law, Probability, and Risk*, **5** (1), March 2006, pp. 1–18.

18. *The Ford Pinto Case*, D. Birsch and J. H. Fielder, Eds., SUNY Press, New York, 1994.

19. Mark Dowie, "Pinto Madness," *Mother Jones*, 18, Sept./Oct. 1977.

20. L. Stroebel, *Reckless Homicide? Ford's Pinto Trial*, South Bend, IN, 1980.

21. *Chicago Tribune*, Oct 13, 14, 1979.

22. F. T. Cullen, W. J. Maakestad, and G. Cavender, *Corporate Crime under Attack*, Anderson Publishing Co., Chicago, 1987.

23. *Grimshaw v. Ford Motor Co.* (1981) 119 Cal. App. 3d 757.

24. *The Ford Pinto Case*, D. Birsch and J. H. Fielder, Eds., SUNY Press, New York, 1994, Part I, Chap. 4.

25. Investigation Report, Phase I: Alleged Fuel Tank and Filler Neck Damage in Rear-End Collision of Subcompact Passenger Cars, Office of Defects Investigation Enforcement, NHTSA, May 1978.

26. Societal Costs of Motor Vehicle Accidents, NHTSA, Preliminary Report 1972.

27. Motor Vehicle Safety Standard, Part 571; S301, NHTSA, 1976.

28. West's *California Reporter*, 176, p. 361.

29. *The Ford Pinto Case*, D. Birsch and J. H. Fielder, Eds., SUNY Press, New York, 1994, Part IV, Chap. 17.

30. F. T. Cullen, W. J. Maakestad, and G. Cavender,"The Ford Pinto Case and Beyond: Corporate Crime, Moral Boundaries, and the Criminal Sanction" from *Corporations as Criminals*, Ellen Hochstedler, Ed., Sage Publications, Beverly Hills, CA, 1984, p. 107.

31. M. E. Wheeler, *The Forum*, Section of Tort and Insurance Practice, American Bar Association, Vol. XVII, Number 2, Fall 1981.

32. *Escola v. Coca Cola Bottling Co. of Fresno,* Supreme Court of California, 1944, 24 Cal. 2d 453, 150.

33. *Greenman v. Yuba Power Products, Inc.*, Supreme Court of California, 1963, 59 Cal. 2d. 57, 27 Cl. Rptr. 697, 377 p. 2d, 897.

34. *Forensic Engineering*, S. Brown, Ed., Part I, ISI Publications, Humble, TX, 1995, Chap 1.

35. R. A. Moll, *J. Engineering Education*, **66** (4), January 1976, p. 326.

36. V. E. Schwartz, *Hofstra Law Rev.* **26** (1998) 743.

37. Restatement of the Law, Third, Torts: Products Liability, 1998, The American Law Institute.

38. P. Lavelle, *Duquesne Law Rev.* **38** (2000) p. 1059

39. T. Carter, American Bar Association, **92** (2006), p. 42.

40. S. D. Sugarman, *UCLA Law Rev.* **50** (2002) 585.

41. GAO Report to Congressional Committees, p. 6, August 2009 report Consumer Safety-GAO-09-803.

42. United States Consumer Product Safety Commission's 2009 Performance and Accountability Report, www.cpsc.gov, November 2009, p. 6.

43. 15 U.S.C. § 2064(b).

44. The Consumer Product Safety Improvement Act of 2008.

45. Keeping European Consumers Safe, 2008 Annual Report on the Operation of the Rapid Alert System for Non-food Consumer Products, RAPEX, The Directorate-General for Health and Consumers of the European Commission, Office for Official Publications of the European Communities, 2009.

46. 16 C.F.R. § 1115.4.

47. 16 C.F.R. § 1115.4(c).

48. R. Huet, J. Wolf, and P. Moncarz."Delayed fracture of tempered glass panels due to nickel sulfide inclusions." In *Handbook of Case Histories in Failure Analysis*. ASM International, 1992.

49. 16 CFR Part 1201—Safety Standard for Architectural Glazing Materials.

50. CPSC Performance and Accountability Report, November 2008.

51. Guidelines for Drawstrings on Children's Upper Outerwear, Consumer Product Safety Commission, CPSC 208, September 1999.

52. http://www.cpsc.gov/library/neiss.html

53. Directive 2001/95/EC of the European Parliament and the Council of the European Union.

54. *Official Journal of the European Union*, L22, Decision 2010/15/EU, ISSN 1725-2555, Volume 53, January 26, 2010.

55. Directive 2001/95/EC of the European Parliament and of the Council of 3 December 2001 on General Product Safety, Official Journal of the European Parliament and of the Council 15.1.2002.

56. Commission Decision of 14 December 2004, Official Journal of the European Union, 28.12.2004.

57. 16 Part 1115.

58. *ASM Handbook*, Non-destructive Evaluation and Quality Control Section, Volume 17, 1989.

59. ANSI Z535.4-2002, American National Standard For Product Safety Signs and Labels, American National Standards Institute, Inc.

60. Military Handbook, Reliability Prediction of Electronic Equipment, MIL-HDBK-217F, Department of Defense, 1991.

61. Nonelectronic Parts Reliability Data (NPRD), Reliability Information Analysis Center, Department of Defense, 1995.

62. Social Consequences of Engineering, "Chapter Ten—Public risk and engineering safety: How safe is safe enough?" Hayrettin Kardestuncer, Ed., Boyd & Fraser Publishing Company, San Francisco, 1979.

63. *Watch Time Magazine*, December 2008, p. 237.

64. *Stress Corrosion Cracking*, R. Jones, Ed., ASM International, 1992, pp. 243–246.

65. Ibid.

66. Alcoa Research Laboratories, Technical Paper No. 14, Development of Wrought Aluminum-Magnesium Alloys, E. H. Dix, W. A. Anderson and M. Byron Shumaker, Aluminum Company of America, 1958, p. 58.

67. Alcoa Research Laboratories, Technical Paper No. 14, Development of Wrought Aluminum-Magnesium Alloys, E. H. Dix, W. A. Anderson and M. Byron Shumaker, Aluminum Company of America, 1958, p. 2.

68. *Aluminum: Properties and Physical Metallurgy*, John E. Hatch, Ed., American Society for Metals, 1984, p. 281.

PROBLEMS

Review

12.1 How recently have the concepts of product liability been introduced into our social awareness?

12.2 What is the meaning of *caveat emptor*?

12.3 How is the concept of *stare decisis* applied as it relates to the court's rulings in product liability matters?

12.4 Can a judge render a judgment that is different from earlier rulings of similar cases?

12.5 Do engineers and scientists ever get involved in product liability and product recall matters? If so, in what way?

12.6 What is the difference between an express warranty and "puffery" pertaining to a product's characteristics?

12.7 In the matter of *Gardiner v. Gray*, what type of warranty was established?

12.8 Assume that you purchased a nationally known lawn mower from your local hardware store. After adding oil and gasoline to the engine and pulling the starter cord, the mower made a loud grinding noise, caught fire, and was then totally destroyed in the flames. In addition, you sustained severe hand burns when attempting to douse the flames. What tort case ruling enables you to sue the manufacturer? In that regard, what concept was struck down when the "law" changed?

12.9 Define the conditions that reflect negligent behavior.

12.10 Define those conditions associated with a judgment, based on a strict liability interpretation of the law.

12.11 What is the purpose of the Restatement of Torts as it has evolved over the years?

12.12 List four basic questions of an engineering approach that can help to determine, from a *technical* perspective, when and if to report a "condition" to the CPSC or other similar agency, and whether it is necessary to initiate a product recall.

12.13 True or False:

    a. All products that present a risk of injury are defective.

    b. Hidden hazards and obvious hazards are both hazards, and are, therefore, considered to be equivalent.

    c. Root cause and the failure process are not one and the same.

12.14 List two analytical tools that can assist in evaluating risk associated with a product and explain the differences between these methods.

12.15 List proactive considerations that can be incorporated into a product recall prevention and management initiative.

12.16 A returned product that has been involved in a fire has become available for analysis as part of a product recall investigation. Name some of the non-destructive and destructive techniques that could be used in an effort to determine the most likely root cause.

Extend

12.17 Assume that several members of your family were passengers on an ill-fated commercial airliner that crashed upon takeoff. Several deaths were reported and many survivors, including your relatives, sustained serious injuries. A subsequent NTSB examination revealed that the airline's maintenance crew had improperly reinstalled engine components that would have enabled the plane to achieve proper lift characteristics so as to obviate the crash. Based solely on the legal opinion set forth by Lord Abinger in the *Winterbottom v. Wright* case, how successful do you believe your relatives and the rest of the passengers would be in receiving compensation for their damages?

12.18 Today's consumer is provided with information concerning many food products' contents including their caloric content, sugar, sodium, and fat levels. Society has determined that such information is desirable so that the consumer can make a more intelligent decision regarding the health-related appropriateness of such food products. Recalling Eq. 12.1, should non-food product labeling include information pertaining to product failure rates and risk factors to enable the consumer, prior to product purchase, to assess the risk/reward balance for these items?

12.19 Using resources provided by the Consumer Product Safety Commission, examine examples of recent product recalls that involve mechanical and/or environmental degradation or damage resulting in product failure. Useful search keywords include *cracking*, *failure*, *fracture*, and *corrosion*. Select one recall example, provide the recall title and a copy of the official hazard description, and briefly explain the cause of the problem in your own words. Finally, propose a solution to the problem that involves a change of materials, manufacturing method, or design.

12.20 State your opinion as to the appropriateness of a cost-benefit analysis when a case involves personal injury and/or death. What is your reasoning?

12.21 Do you feel that it was appropriate for criminal charges to have been raised in the Pinto crashworthiness case? What is your reasoning?

Appendix A

Fracture Surface Preservation, Cleaning and Replication Techniques, and Image Interpretation

A.1 FRACTURE SURFACE PRESERVATION

The first step to be taken prior to a fractographic examination—whether by SEM or TEM imaging—involves preservation of the fracture surface; if the fine details on the fracture surface are mechanically or chemically attacked, a meaningful fractographic analysis becomes suspect, if not impossible. Therefore, unless a fresh fracture surface is examined immediately, some successful method of surface retention must be employed. For example, if specimens are sufficiently small, they may be stored in a desiccator or vacuum storage jar. Alternatively, the fresh fracture surfaces can be protected with a lacquer spray that dries to form a transparent protective layer. In this manner, one is able to perform a macroscopic examination of the various fracture features even though the surface is protected. Field fractures can be protected by applying a coating of fresh oil or axle grease as long as these oil-based products do not react chemically with the metal surface. Boardman et al.[1] have also reported success in protecting fracture surfaces with a commercially available petroleum-base compound (Tectyl 506). To be sure, these coatings would have to be removed prior to both macroscopic and microscopic examination of the fracture surfaces. Finally, protection can be achieved by applying a strip of cellulose acetate replicating tape to the fracture surface. The tape is initially softened with acetone and then pressed onto the surface and permitted to dry. Unfortunately, the dried tape does not always adhere well to the fracture and has a tendency to pop off or peel away with time and with any significant degree of handling.

A.2 FRACTURE SURFACE CLEANING

In many instances, broken components are received in the laboratory in a condition that precludes their immediate examination in the electron microscope. Fracture surface details may have been obliterated by repeated rubbing together of the mating pieces or with other hard objects. A fire resulting indirectly from the component fracture can also seriously damage fracture surfaces and even lead to localized melting of surface detail.

In most other situations, fracture surfaces are found to be covered with loose foreign particles and grease or a layer of corrosion or oxidation product. Depending on the tenacity of the debris and the electrochemical layer, various cleaning methods are suggested as outlined in Table A.l. It is suggested that cleaning methods 1 to 3 be attempted initially since they are easy to perform and do not degrade the fracture surface. On the other hand, methods 4 to 6, when carelessly employed, can result in permanent damage to the fracture surface. When using

713

Table A.1   Cleaning Procedures for Typical Surface Debris[2]

Cleaning Method	Matter Removed
1. Dry air blast; soft natural bristle brush	Loosely adhering particles

2. Organic solvent cleaning[a]
    Toluene, xylene
    Ketones (acetone, methylethylketone)
    Alcohols (methyl, ethyl, isopropyl)
	Hard and viscous debris
	Oils, greases
	Preformed lacquers, varnishes, gums
	Dyes, fatty acids

3. Repeated stripping of successive cellulose acetate replicas	Debris, oxidation and corrosion products
4. Water-based detergent cleaning in conjunction with ultrasonic agitation[b]	Oxidation and corrosion products
5. Cathodic cleaning[c] (surface to be cleaned is cathode with inert anode of carbon or platinum to prevent electroplating on fracture surface)	Oxidation and corrosion products
6. Chemical etch cleaning	Very adherent oxidation and corrosion products

[a] A general purpose organic solvent would contain 40% toluene, 40% acetone, and 20% denatured alcohol.
[b] Suggested cleaning solution[2]: 15 g Alconox powder + 350 mL water. Solution is heated to 95°C and agitated ultrasonically for 15 to 30 min, but not to the point where fracture surface becomes etched (R. S. Vecchio and R. W. Hertzberg, ASTM *STP 827,* 1984, p. 267).
[c] For suggested electrolytes, see Refs. 3, 4, and 5.

method 3—the repeated replica stripping technique—it is suggested that the first and, perhaps, the second replica be kept for possible further study. For example, the fracture surface debris retained on the plastic strip can be analyzed chemically using an energy dispersive X-ray device. (For this purpose, the tape must first be coated with a conductive layer prior to insertion in the SEM.) An example showing the results of method 4 is shown in Fig. A.1. The fatigue fracture surface of a steel alloy was initially sprayed with a saltwater solution that resulted in the accumulation of corrosion debris on the fracture surface (Fig. A.1*a*). Individual fatigue striations and occasional secondary fissures (the widely separated black lines) were made visible in the SEM after the specimen was ultrasonically cleaned in a heated Alconox solution for 30 minutes (Fig. A.1*b*).

Since the formation of an oxide or corrosion layer involves atoms from the base metal, subsequent removal of this layer during any cleaning procedure simultaneously removes some detail from the metal substrate. The extent of this damage depends on the thickness of the electrochemical layer relative to the depth of the fracture surface marking. For example, when the thickness of the corrosive layer is small compared to the depth of the fracture surface contours, removal of the corrosive layer should not significantly impair the fractographic

Figure A.1   SEM images of fatigue fracture surface in steel alloy. (*a*) Before cleaning. (*b*) After ultrasonic cleaning in heated Alconox solution for 30 minutes.[2] (R. D. Zipp, *Scanning Electron Microscopy I*, 1979, p. 355, with permission.)

analysis. On the other hand, when fractographic features are themselves shallow, such as fatigue striations in a high-strength steel alloy, removal of the electrochemical layer will seriously reduce the overall depth of those fracture markings to the point where they become extremely difficult to discern.

A.3   REPLICA PREPARATION AND IMAGE INTERPRETATION

Several replication procedures have been developed in the metallurgical laboratory. In the one-step process, a carbon film is vacuum deposited directly onto the fracture surface and subsequently floated free by placing the sample in an acid bath to dissolve the sample. Although this technique produces a high-resolution replica, it is not employed often since the specimen is destroyed in the replication process. The most commonly used technique is a nondestructive, two-stage process that produces a replica with reasonably good resolution. A presoftened strip of cellulose acetate is pressed onto the fracture surface and allowed to dry. (To produce replicas from the fracture surfaces of polymeric materials that would dissolve in acetone, one author has

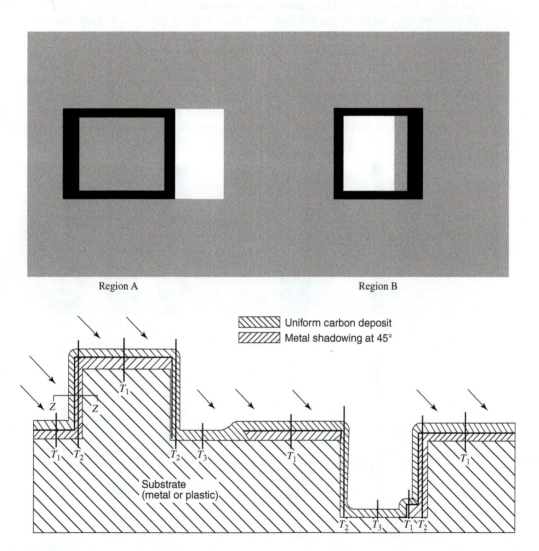

Figure A.2   Schema revealing stages in preparation of a fracture surface replica. Replica appears opaque to electron beam when effective thickness is T_2. Brightest image occurs in "shadow" where replica has thickness T_3.[1] (Courtesy of Cedric D. Beachem, Naval Research Laboratory.)

found a water slurry of polyacrylic acid to be an effective replicating media.) When it is stripped from the specimen, the tape carries an impression of the fracture surface topography. Since this tape is opaque to the electron beam, further steps in the replication procedure are necessary. A layer of heavy metal is deposited at an acute angle on the side of the tape bearing the fracture impression. This is done to improve the eventual contrast of the replica. Finally, a thin layer of carbon is vapor deposited onto the tape. The plastic tape, heavy metal, carbon composite is then placed in a bath of acetone where the plastic is dissolved. (The polyacrylic acid replica would be dissolved in a water bath.) In the final step, the heavy metal, carbon replica is removed from the acetone bath and placed on mesh screens for viewing in the electron microscope. Since the viewing screens are only 3 mm in diameter, selection of the critical region(s) for examination is most important. This factor emphasizes the need for a carefully conducted macroscopic examination that should direct the examiner to the primary fracture site. By adhering to the recommended procedures for the preparation of replicas, little difficulty should be encountered.

An important part of electron fractography is the interpretation of electron images in terms of actual fracture mechanisms. In addition to the references just cited, the reader is referred to the work of Beachem.[6,7] These results are summarized briefly. The fluorescent screen of the microscope (and the photographic film) will react to those electrons that are able to penetrate the replica; that is, the more electrons that penetrate the replica, the brighter the image and vice versa. We see from Fig. A.2 that the vertical walls of the raised and depressed regions at A and B, respectively, show up as black lines, since the electrons cannot penetrate the replica segment with thickness T_2. On the other hand, "shadows" are produced in the absence of the heavy metal (more resistant to electron penetration), which had been deposited at an angle to the fracture surface. These regions appear as white areas *behind* a raised particle and *within* a depressed region associated with a replica thickness T_3. The *location* of the shadow helps to determine whether the

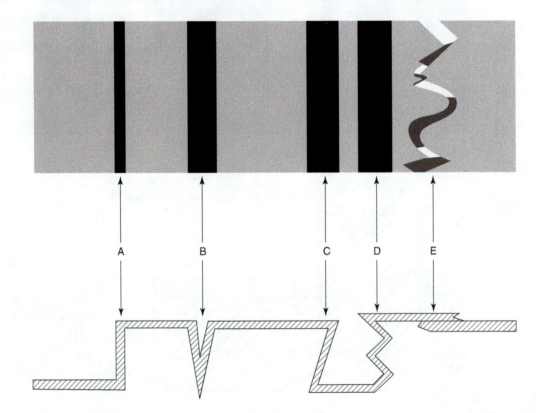

Figure A.3 Schema showing different interpretations of opaque black bands.[1] (Courtesy of Cedric D. Beachem, Naval Research Laboratory.)

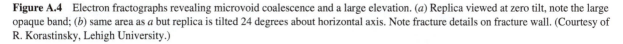

Figure A.4   Electron fractographs revealing microvoid coalescence and a large elevation. (*a*) Replica viewed at zero tilt, note the large opaque band; (*b*) same area as *a* but replica is tilted 24 degrees about horizontal axis. Note fracture details on fracture wall. (Courtesy of R. Korastinsky, Lehigh University.)

region being examined is above or below the general fracture plane. At this point, it is crucial to recognize the image reversal between one- and two-stage replicas. If the substrate in Fig. A.2 were the actual fracture surface, then region A would be correctly identified as lying above the fracture plane while region B would lie below this surface. If the substrate were the plastic, however, the previous conclusion would be completely incorrect. That is, an elevated region on the replica really represents a depressed region on the actual fracture surface. Therefore, when two-stage replicas are prepared, the images produced should be interpreted in reverse fashion—*everything that looks up is really down and vice versa.*

Finally, care should be exercised when attempting to determine the meaning of black bands on the photographs. For the example shown in Fig. A.2, these bands represent simple vertical elevations associated with regions A and B. As seen in Fig. A.3, other more complex explanations are possible. In any case, these opaque bands reflect an elevation on the fracture surface that should reveal fracture mechanism details if reoriented with respect to the electron beam (Fig. A.4).

REFERENCES

1. B. E. Boardman, R. Zipp, and W. A. Goering, *SAE Automotive Engineering Meeting,* Detroit, Michigan Oct. 1975, Paper No. 750967.

2. R. D. Zipp, *Scanning Electron Microscopy*, Scanning Electron Microscopy, Inc., Chicago, 1979, p. 355.

3. H. DeLeiris, E. Mencarelli, and P. A. Jacquet, *Mem. Scient. Rev. Met.* **63** (5) 463 (1966).

4. P. M. Yuzawich and C. W. Hughes, *Prac. Metall.* **15**, 184 (1978).

5. R. Löhberg, A. Gräder, and J. Hickling, *Microstructural Sci.* **9**, 421 (1981).

6. C. D. Beachem, *NRL Report* **6360**, U. S. Naval Research Laboratory, Washington, DC, Jan. 21, 1966.

7. C. D. Beachem, *Fracture*, Vol. 1, H. Liebowitz, Ed., Academic, New York, 1968, p. 243.

Appendix **B**

K Calibrations for Typical Fracture Toughness and Fatigue Crack Propagation Test Specimens

Type	Stress Intensity Formulation	Configuration
1. Compact Tension specimen[a] C(T)	$K = \dfrac{P}{BW^{1/2}}f(a/W)$ where $f(a/W) = \dfrac{(2 + a/W)}{(1 - a/W)^{3/2}}$ $\times \left[0.886 + 4.64a/W - 13.32(a/W)^2 + 14.72(a/W)^3 - 5.6(a/W)^4\right]$	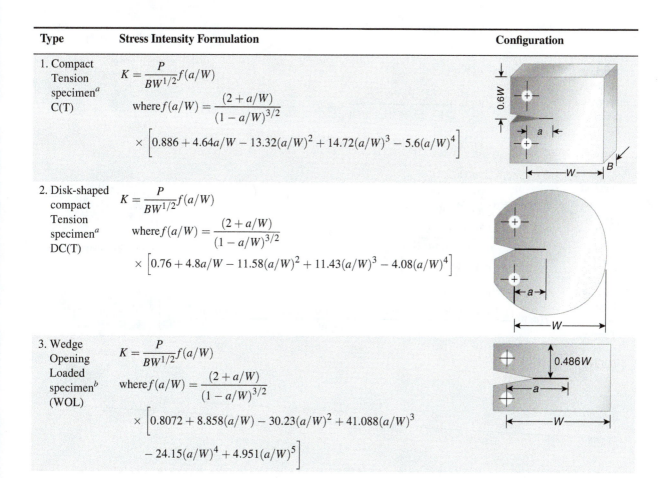
2. Disk-shaped compact Tension specimen[a] DC(T)	$K = \dfrac{P}{BW^{1/2}}f(a/W)$ where $f(a/W) = \dfrac{(2 + a/W)}{(1 - a/W)^{3/2}}$ $\times \left[0.76 + 4.8a/W - 11.58(a/W)^2 + 11.43(a/W)^3 - 4.08(a/W)^4\right]$	
3. Wedge Opening Loaded specimen[b] (WOL)	$K = \dfrac{P}{BW^{1/2}}f(a/W)$ where $f(a/W) = \dfrac{(2 + a/W)}{(1 - a/W)^{3/2}}$ $\times \left[0.8072 + 8.858(a/W) - 30.23(a/W)^2 + 41.088(a/W)^3\right.$ $\left. - 24.15(a/W)^4 + 4.951(a/W)^5\right]$	

4. Center-Cracked Tension specimen[a] (CCT)

$$K = \frac{P\sqrt{\pi a}}{BW} f(a/W)$$

where $f(a/W) = \sqrt{\sec\frac{\pi a}{W}}$

0.3% for any $a/W \leq 0.7$

1.0% for $a/W = 0.8$

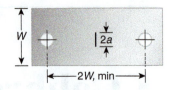

5. Arc-shaped Tension specimen A(T)

$$K = \frac{P}{BW^{1/2}}\left(3\frac{X}{W} + 1.9 + 1.1 a/W\right)$$

$$\times \left[1 + 0.25(1 - a/W)^2(1 - r_1/r_2)\right] f(a/W)$$

where $f(a/W) = \left[(a/W)^{1/2}/(1 - a/W)^{3/2}\right]$

$$\times \left[3.74 - 6.3(a/W) + 6.32(a/W)^2 - 2.43(a/W)^3\right]$$

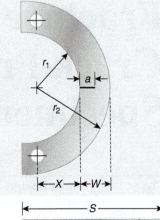

6. Single Edge-notched Bend specimen[a] SE(B)

$$K = \frac{PS}{BW^{3/2}} f(a/W)$$

where $f(a/W) = \dfrac{3(a/W)^{1/2}}{2(1 + 2a/W)(1 - a/W)^{3/2}}$

$$\times \left[1.99 - (a/W)(1 - a/W)(2.15 - 3.93 a/W + 2.7 a^2/W^2)\right]$$

7. Single Edge-notched Tension specimen[d] SE(T)

$$K_I = f(a/W)\sigma\sqrt{\pi a}$$

$$f(a/W) = \sqrt{\frac{2W}{\pi a}\tan\frac{\pi a}{2W}} \times$$

$$\left(\frac{0.752 + 2.02(a/W) + 0.37(1 - \sin(\pi a/2W))^3}{\cos(\pi a/2W)}\right)$$

$\leq 0.5\%$ for any a/W

8. Double Edge-notched Tension specimen[d] (DENT)

$$K_I = f(a/W)\sigma\sqrt{\pi a}$$

$$f(a/W) = \frac{1.122 - 0.561(a/W) - 0.205(a/W)^2 + 0.471(a/W)^3 - 0.190(a/W)^4}{\sqrt{1 - (a/W)}}$$

0.5% for any a/W

9. Notched cylinder in tension[d]

$$K_I = f(a/b) \frac{P}{\pi b^2} \sqrt{\pi a}$$

$$f(a/b) = \frac{1}{(1 - a/b)^{3/2}} \left[1.122 - 1.302(a/b) + 0.988(a/b)^2 - 0.308(a/b)^3 \right]$$

10. Notched cylinder in bending[d]

$$K_{I,\max} = f(a/b) \frac{4M}{b^3} \sqrt{\pi a}$$

$$f(a/b) = \frac{1}{(1 - a/b)^{5/2}} \times$$

$$\left\{ 0.563 - 0.188(a/b) + (1 - a/b)^2 \left[0.559 - 1.47(a/b) + 2.72(a/b)^2 - 2.40(a/b)^3 \right] \right\}$$

[a] ASTM Standard E 399-81, *Annual Book of ASTM Standards*, Part 10, 1981.

[b] A. Saxena and S. Hudak, *Int. J. Fract.* **14**, 453 (1978).

[c] C. E. Feddersen, ASTM *STP 410*, 1976, p. 77.

[d] H. Tada, P. C. Paris, and G. R. Irwin, *The Stress Analysis of Cracks Handbook*, 3rd ed., ASME, New York, 2000.

Appendix C

Y Calibration Factors For Elliptical and Semicircular Surface Flaws

Elliptical and semicircular surface flaws are very common, particularly for ceramics and glasses. Analytical expressions for geometric Y factors similar to those already presented for through-cracks in Appendix B are therefore quite useful. As discussed in Section 6.5, K varies along the crack front of an elliptical surface crack from the surface intersection point to the point deepest in the interior of the material. As surface cracks are usually shallow, and are generally wider than they are deep (i.e., $c > a$ in the notation of Fig. 6.21*h*), this means that often $Y_{surface} < Y_{interior}$. The opposite is true when c $\approx$ a. A widely-used solution for elliptical and semicircular surface cracks was proposed by Newman and Raju,[i] which for bending takes the form

$$Y_{surface} = SH_1 \frac{M}{\sqrt{Q}} \sqrt{\pi}$$

$$Y_{interior} = H_2 \frac{M}{\sqrt{Q}} \sqrt{\pi}$$

where Q is the elliptical flaw shape parameter[ii]

$$Q \approx 1 + 1.464 \left(\frac{a}{c}\right)^{1.65}$$

and S, H_1, H_2, and M are additional geometric terms. Following Quinn,[iii] the form of the equations given here assumes that $2c \ll W$, where W is the width of the component along the $2c$ face, so no finite specimen size term need be included. The geometric terms are defined using the same a and c dimensions shown in Fig. 6.21*h*, with the addition of a plate thickness term, h, measured in the same direction as a. (Note that this is different from the notation in Fig. 6.21*e*, for which W is used to represent the thickness in the crack depth direction.) These equations are also suitable for modeling uniform tensile loading if the component thickness h is set to a value much larger than crack depth a, so that the bending stress gradient along a is inconsequential. A value of $h = 1000a$ is sufficient for this purpose.[iii]

[i] J. C. Newman and I. S. Raju, *Eng. Fract. Mech.* **15** (1-2), 185–192 (1981).

[ii] J. C. Newman and I. S. Raju, in *Computational Methods in the Mechanics of Fracture*, S. N. Atluri, Ed., Elsevier Science Publishers, 1986.

[iii] G. D. Quinn, *Fractography of Ceramics and Glasses*, NIST Special Pub 960-16, 2007.

$$M = \left[1.13 - 0.09\left(\frac{a}{c}\right)\right] + \left[-0.54 + \frac{0.89}{0.2 + \left(\frac{a}{c}\right)}\right]\left(\frac{a}{h}\right)^2 + \left[0.5 - \frac{1}{0.65 + \left(\frac{a}{c}\right)} + (14)\left(1 - \left(\frac{a}{c}\right)\right)^{24}\right]\left(\frac{a}{h}\right)^4$$

$$H_1 = 1 - \left[0.34 + 0.11\left(\frac{a}{c}\right)\right]\left(\frac{a}{h}\right)$$

$$H_2 = 1 - \left[1.22 + 0.12\left(\frac{a}{c}\right)\right]\left(\frac{a}{h}\right) + \left[0.55 - 1.05\left(\frac{a}{c}\right)^{0.75} + 0.47\left(\frac{a}{c}\right)^{1.5}\right]\left(\frac{a}{h}\right)^2$$

$$S = \left[1.1 + 0.35\left(\frac{a}{h}\right)^2\right]\sqrt{\frac{a}{c}} \approx 1.1\sqrt{\frac{a}{c}} \text{ for } a \ll h$$

Appendix D

Suggested Checklist of Data Desirable for Complete Failure Analysis

I. Description of Component Size, Shape, and Use _____

    A. Specify areas of design stress concentrations _____

        1. Magnitude of stress concentration at failure site _____

II. Stress State for Component_____

    A. Type of stresses_____

        1. Magnitude of design stress levels _____

            a. Mean stress _____

            b. Stress range _____

        2. Type of stress (e.g., Mode I, II, III, or combinations)_____

        3. Presence of stress gradient _____

    B. State of stress: plane strain vs. plane stress_____

        1. Fracture surface appearance: percent shear lip_____

        2. Estimation from calculated plastic zone size to thickness ratio ___

        _____

    C. Nature of load variations _____

        1. Hours of component operation _____

        2. Load cycle frequency_____

        3. Type of loading pattern _____

            a. Random loading_____

            b. Existence of overloads resulting from abnormal service life events

III. Details of Critical Flaw_____

    A. Date of previous inspection _____

        1. Findings of previous inspection _____

    B. Nature of critical flaw leading to fracture (make use of clearly labeled sketches and/or macrophotographs with accurate magnifications) ___

    _____

1. Location of critical flaw by macroscopic examination _____
2. Critical flaw size, shape, and orientation at instability _____
3. Surface or imbedded flaw _____
4. Direction of crack propagation as determined by
   Chevron markings   _____
   Pop-in   _____
   Beach markings   _____

C. Manufacturing flaws related to crack initiation
   Scratches _____   Misfits _____
   Undercuts _____   Others _____
   Weld defects _____

D. Metallurgical flaws related to crack initiation
   Inclusions _____   Voids _____
   Second-phase particles _____   Weak interfaces _____
   Entrapped slag _____   Others _____

E. Fractographic observations
   1. Qualitative observations
      Dimpled rupture _____   Fatigue striations _____
      Cleavage _____   Corrosion _____
      Intercrystalline _____   Fretting _____

   2. Quantitative observations
      a. Striation spacings at known crack length positions _____
      b. Striation spacing evidence of uniform or random loading _____
         _____

      c. Stretch zone width at onset of unstable crack extension _____

IV. Component Material Specifications
   A. Alloy designation _____
   B. Mechanical properties

	σ_{ys}	σ_{ts}	% Elong.	% R.A.	K_{IC}	K_{IEAC}	Fatigue Characterization
Specified							
Actual							

C. Alloy chemistry

Elements

	A	B	C	D	E	F	G
Specified							
Actual							

D. Melting practice
Air melted_____
Vacuum melted_____
Other _____

E. Ingot breakdown
Hot rolled _____
Cold rolled _____
Cross rolled_____

F. Thermomechanical treatment

1. Annealing or solution treatment condition _____

2. Tempering or aging treatment_____

3. Intermediate mechanical working _____

G. Component manufacture

1. Forged _____   Spun _____

   Cast _____   Extruded _____

   Machined _____   Other _____

2. Joint detail
   Welded _____   Bolted _____

   Brazed_____   Other _____

   Adhesive bonded_____

H. Surface treatment
Shot peened_____   Flame or induction
                                      hardened _____

Cold rolled _____   Pickled _____
Carburized_____   Plated _____
Nitrided_____
Other _____

I. Component microstructure

   1. Presence of mechanical fibering and/or banding from chemical segregation

       _____

   2. Grain size and shape _____

      a. Elongated with respect to stress axis _____

      b. Grain run-out in forgings _____

   3. Inclusion count and classification _____

Author Index

Abdel-Latif, A. I. A.   294
Acker, P.   459
Adams, R. D.   378
Adharapurapu, R. R.   458
Ager, J. W.   457–458
Agnew, S.R.   638
Agricola, G.   56
Ahmed, Z. B.   457
Aimone, C. T.   454
Albrecht, P. A.   681
Alden, T. H.   244, 555
Alexander, G. B.   185
Alford, N. M.   458
Alfredsson, K. S.   378
Alfrey, T.   122, 138, 245
Allery, M. B. P.   554
Alsem, D. H.   457–458
Alzos, W. X.   638
Amateau, M. F.   638
Amazigo, J. C.   456
Anderson, A. F.   636
Anderson, W. A.   711
Anderson, W. E.   454
Andrade, E. N.   137, 243
Andrews, E. H.   138, 640
Andrews, R. D.   245
Andrews, W. R.   377
Angelini, P.   455
Angeliu, T. M.   295
Ansell, G. S.   494
Anstis, G. R.   376
Antolovich, S.   454
Antolovich, S. D.   639
Ardell, A. J.   243
Argon, A. S.   138
Armstrong, R. W.   184
Arruda, E. M.   458
Arsenault, R. J.   185
Asaro, R. J.   377
Ashby, M. F.   185, 243–244,
   294–295, 457, 680
Ashman, R. B.   56
Athey, R. L.   245
Atkinson, M.   110, 138

Averbach, B. L.   458
Avery, D. H.   244
Azimi, H.   457, 639

Babel, H. W.   455
Backensto Jr., A. B.   185
Backofen, W. A.   109, 138, 244,
   454, 555
Baikie, B. L.   680
Baillie, C.   457
Bailon, J. P.   639
Baker, T. J.   639
Balajiva, K.   458
Baldwin Jr., W. M.   453
Ballard, D. B.   680
Ballarini, R.   457, 458
Balooch, M.   458
Bandyopadhyay, N.   457–458
Banerji, S. K.   458
Bannantine, J. A.   641
Baratta, F. I.   376
Barker, L. M.   376
Barlo, T. J.   495
Barrett, C. R.   243
Barsom, J. M.   353, 376, 555,
   635, 638–639, 680
Barth, H. B.   457
Barthelat, F.   457
Basquin, O. H.   555
Bassim, M. N.   377
Bassin, M. N.   184
Batchelder, D. N.   45
Bates, R. C.   377, 636, 639
Bathias, C.   554, 636, 641
Batson, G. B.   459
Baur, J. W.   457
Bauschinger, J.   555
Bauwens, J. C.   640
Bayles, R. A.   495
Bayoumi, M. R.   377
Beachem, C. D.   294, 493, 495,
   717
Beardmore, P.   134, 138, 295,
   555

Beaumont, P. W. R.   457
Becher, P. F.   455, 640
Beck, W.   495
Becker, W. T.   294
Beer, H. R.   457
Beevers, C. J.   636, 638
Begley, J. A.   376–377
Bei, H.   55, 68
Bell, J. R.   454
Benjamin, J. S.   185
Benjamin, W. D.   495
Benn, R. C.   185, 245
Bennett, J. A.   680
Bennett, P. E.   137
Bennison, S. J.   456
Benson, D.   45
Benson, D. J.   184
Benson, D. K.   455
Bernstein, I. M.   493–494
Berry, J. P.   295
Berry, W. E.   459, 680
Bhandarkar, D.   455
Bilby, B. A.   138
Bin, X.   554
Birchall, J. D.   458
Bird, J. E.   243
Birkbeck, G.   554, 639
Birkle, A. J.   295, 454
Birnbaum, H. K.   493
Bish, E.   459
Bishop, J. F. W.   137
Boardman, B. E.   717
Boas, W.   137
Bockris, J. O'M.   495
Bodde, S. G.   458
Boehmert, J.   459
Boening, H. V.   138
Bolz, L. H.   496
Bonderer, L. J.   458
Booth, C. S.   636
Boulger, F. W.   454
Bourges, P.   494
Bowden, P. B.   129, 138
Bowen, A. W.   137

Bowie, O. L.   376, 680
Boyer, R. F.   138, 245, 456
Bradley, E. F.   244
Bradt, R. C.   244, 294, 375–376, 459
Braglia, B. L.   554
Bray, G. H.   636, 638
Brear, K.   457
Brenner, S. S.   55
Breslaur, M.   295
Bressanelli, J. P.   454
Bretz, P. E.   640
Brewer, H.   456
Breyer, N. N.   494
Briant, C. L.   458
Bricknall, C. B.   555
Bricknell, R. H.   244
Bridgman, P. W.   98, 137
Brittain, J. O.   243
Broad, W. J.   294, 454
Brockenbrough, J. R.   640
Broek, D.   376, 454, 554, 636–637, 639
Brose, W. R.   535
Broutman, L. J.   554
Brown Jr., W. F.   376, 458
Brown, B. F.   495
Brown, H. R.   496
Brown, L. M.   184
Brown, N.   137
Brown, S. D.   496
Brown, W. L.   191
Bucci, R. J.   185, 376, 455, 584, 635–636, 638, 640, 680
Buckley, C. P.   42, 56
Bucklin, A. G.   243
Bucknall, C. B.   42, 56, 456
Budiansky, B.   456
Buehler, M. J.   376, 457–458
Bülfinger, G. B.   137
Bullens, D. K.   554
Burggraaf, A. J.   456
Burke, M. G.   458
Burke, P. M.   243
Burns, D. J.   554
Burr, D. B.   294, 457
Bursle, A. J.   493
Busch, R.   454
Bush, R. H.   454, 459
Bush, R.W.   636
Button, T. W.   458

Caddell, R. M.   109, 138
Cadek, J.   295

Cahn, R. W.   137–138
Calabrese, C.   554
Caldwell, J. R.   456
Calnan, E. A.   137
Campbell, J. E.   459, 680
Cannon, R. M.   455
Capus, J. M.   458
Carlson, M. F.   584
Carmen, C. M.   571
Carr, F. L.   137
Carter, T.   711
Cartwright, D. J.   680
Cassada, W.   454
Cassady, M. J.   496
Castro, D. E.   637
Catsiff, E.   245
Cavender, G.   710–711
Cearley, R. W.   710
Cessna, L. C.   554
Chai, H.   377–378, 456
Chamis, C. C.   555
Chan, H. M.   455
Chan, K. S.   641
Chan, M. K. V.   496
Chanani, G. R.   639
Chang, V. S.   238
Chantikul, P.   376
Charalambides, P. G.   378
Charles, J. A.   453, 454
Chaturvedi, M. C.   244
Chaudhuri, A. R.   184
Chaudhuri, M. M.   424
Chauvy, C.   494
Chen, C.   458
Chen, C. C.   138, 554
Chen, C. Q.   377
Chen, D. K.   637
Chen, D. L.   363
Chen, L.   458
Chen, P. S.   455
Chen, P.-Y.   458
Cheruvu, N. S.   639
Chestnutt, J. C.   455, 638
Chevalier, J.   496
Chin, G. Y.   107, 137
Chou, Y. T.   184
Christensen, R. H.   636
Christian, J. W.   138
Christman, T.   185
Cicci, F.   635
Clarebrough, L. M.   184
Clark Jr., W. G.   377, 459, 495, 636
Clark, E. S.   138

Clark, G.   639
Clark, R.   138
Clark, T.   639
Clark, T. R.   637
Clarke, G. A.   377
Claussen, N.   455
Clayton, D.   456
Clegg, R.   456
Clegg, W. J.   458
Clifton, K. B.   294, 376
Cline, H. E.   245
Cline, R. S.   137
Coble, R. L.   243
Coffin, L. F.   555, 638
Cohen, M.   458
Cohen, R. E.   138
Cole, D. H.   710
Collins Jr., E. A.   554
Collins, R.   496
Collyer, A. A.   456
Colvin, E. L.   454
Comer, J. J.   641
Connelly, G. M.   636
Conpeau, C. R.   493
Conrad, H.   184
Constable, I.   554
Conway, J. B.   243
Cook, J.   454
Cook, R. F.   455, 456
Cook, R. M.   458
Cooke, R. J.   636
Cooley, L. A.   638
Cooper, A. H.   245
Coppola, J. A.   375
Cordill, M. J.   377
Corre, S.   494
Corten, H. T.   353, 376, 680
Cotterell, B.   453
Cottrell, A. H.   136–137, 184, 243, 453–454, 457
Coudreuse, L.   494
Cowap, J. W.   378
Cowin, S. C.   56
Cox, B. N.   629
Cox, H. L.   45
Cox, T. B.   377
Craig, G. B.   138
Crews, J. H.   378, 556
Crocker, A. G.   138
Crocker, L.   138
Crompton, J.   636
Crompton, J. S.   245
Crooker, T. W.   495, 638–639

Crosbie, G. A. 456
Crossman, F. W. 244
Cullen, F. T. 710–711
Cullity, B. D. 103
Culver, L. E. 459
Currey, J. D. 457, 458
Curwick, L. R. 185
Cutler, C. P. 244

Dahoun, A. 133, 138
Dalgleish, B. J. 640
Dash, W. C. 184
Dauskardt, R. H. 377–378, 640, 641
Davidge, R. W. 56, 294, 455
Davidson, D. L. 641
Davidson, T. E. 680
Davies, G. J. 137
Davies, R. G. 245
De Aza, A. H. 496
de Beer, G. 56
de Charentenay, F. X. 640
de los Rios, E. R. 637
Dean, G. 138
Deanin, R. D. 138, 245
DeArdo, A. J. 458
Decker, R. F. 245
deJonge, J. B. 638
DeLeiris, H. 717
deMorton, M. E. 454, 680
deRijk, P. 637–638
Develay, R. 455
Dewas, J. 640
Dieter, G. 110, 138
Dillamore, I. L. 137
Dinsdale, K. 455
Dix, E. H. 711
Dodds, R. H. 459
Doker, H. 637
Dolan, M. T. 137
Dolan, T. J. 680
Dolby, R. 454
Donald, A. M. 456
Donald, J.K. 635–637, 640
Donald, K. 636
Donoso, J. R. 184
Dooman, R. G. 680
Dorn, J. E. 243
Dowie, M. 710
Dowling, N. E. 377, 554–555, 637
Drucker, D. C. 377
Dubey, K. 458
Dugdale, D. S. 376

Duhl, D. N. 245
Dull, D. L. 495
Dunlop, G. L. 244
Dupre, 377

Eagan, R. J. 459
Eastman, J. A. 184
Ebrahimi, M. E. 496
Eckelmeyer, K. E. 138
Edelson, B. I. 456
Edington, J. W. 244
Egan, D. 453, 457
Ehrenstein, G. W. 294
El Haddad, M. H. 637
Elam, C. F. 137
Elber, W. 636
Elinck, J. P. 640
Elliott, J. F. 244
El-Soudani, S. M. 639
Endo, T. 555
Engelstad, S. P. 378
Engle, R. M. 636
Englel, L. 294
English, A. T. 137, 454
Ensign, C. R. 244
Epstein, M. M. 496
Erdogan, F. 635–636
Ernst, H. 377
Esgar, J. B. 680
Eshelby, J. D. 184
Espinosa, H. D. 454, 457
Es-Said, O. S. 455
Estevez, R. 138
Evans, A. G. 378, 453, 455–457
Evans, K. R. 184
Evans, P. R. V. 639
Ewart, L. 640
Ewart, M. 640

Faber, K. T. 455
Fager, D. W. 494
Fahr, D. 454
Fairbanks, C. J. 455–456
Fantozzi, G. 496
Farquharson, G. 378
Farrar, J. C. M. 454
Fatemi, A. 641
Feddersen, C. 376
Feddersen, C. E. 459, 680, 721
Feeler, R. A. 556
Feltner, C. E. 555
Feng, H. C. 458
Ferber, M. K. 496

Ferrainolo, J. J. 451
Ferry, J. D. 245
Ferry, J. P. 554
Fessler, R. R. 495
Field, F. A. 377
Findley, W. N. 554
Fisher, J. C. 185
Fisher, J. W. 681
Fisher, R. A. 184
Fisher, R. M. 185
Flansburg, B. D. 378
Fleischer, R. L. 184
Flom, D. G. 137
Flory, P. J. 185
Foecke, T. 294, 454
Fogelman, E. L. 376
Fontana, M. G. 493
Ford, F. P. 493
Forman, A. J. 137
Forman, R. G. 636
Forrest, P. J. 554
Forsyth, P. J. E. 554, 635–636
Forte, T. P. 637
Fountain, H. 455
Fox, P. 639
Francois, D. 459
Frank, F. C. 184, 185
Frank, K. H. 681
Fratzl, P. 458
Frazier, R. H. 454
Frechette, V. D. 295
Freeman, J. W. 245
Freese, C. E. 680
Freiman, S. W. 294, 377, 459
Frenkel, J. 136
Friedel, J. 138
Friedrich, K. 295, 457, 641
Froes, F. H. 185
Frost, H. J. 243
Frost, N. E. 635, 637
Fuchs, H. O. 554, 641

G'Sell, C. 138
Gagg, C. 681
Gaggar, S. K. 554
Galiotis, C. 45
Gandhi, C. 244
Gangloff, R. P. 493–496, 638
Gao, H. 377, 458
Gao, M. 638
Garcia, C. I. 458
Garg, A. C. 456
Garofalo, F. 243

Garstone, J.   184
Garvie, R. C.   455
Garzke, W.   294
Gastine, O. H.   493
Gauckler, L. J.   458
Gebizlioglu, O. S.   138
Geil, P. H.   138
Gell, M.   244–245
Gensamer, M.   138
George, E. P.   55
Gerald, V.   184
Gerberich, W. W.   377, 455, 493
German, M. D.   377
Gertsch, W. J.   185
Gilbert, D. G.   457
Gilchrist, M. D.   457
Gilman, J. J.   137, 184, 295
Giummarra, C.   455
Glass, S. J.   456
Glazov, M.   638
Gleiter, H.   184–185
Goering, W. A.   717
Goldberg, A.   393
Golden, J. H.   185
Goldhoff, R. M.   244
Goldstein, J. I.   185
Goodenow, R.   454
Goodenow, R. H.   376
Goodier, J. N.   377
Goodman, S. R.   137
Gordon, J. E.   454
Gosline, J. M.   185
Gräder, A.   717
Gran, R. J.   376, 636, 680
Grant, N. J.   185, 243
Gray, A.   496
Green, D.   137, 455
Green, D. J.   456
Green, M. D.   710
Green, T. J.   455
Greenough, A. P.   555
Gregory, E.   185
Gregory, J. K.   495, 639
Grenet, L.   496
Griffin, R. B.   680
Griffith, A.   375
Groover, R. E.   495
Gross, B. J.   680
Gross, J.   375
Gross, J. H.   555
Grossman, P. Z.   710
Grove, R. A.   295
Groves, G. W.   137

Guard, R. W.   184
Guerette, P. A.   185
Guild, F. J.   457
Gulbransen, L. E.   555
Guo, J. S.   238
Gupta, D. K.   245
Gurnee, E. F.   138, 245
Guthrie, D. E.   554
Guy, A. G.   136

Haaf, F.   456
Haak, R. P.   555
Haasen, P.   184
Haberkorn, H.   184
Hack, G. A. J.   185
Hackett Jr., L. H.   458
Hackett, E. M.   459
Hadley, D. W.   55
Hagel, W. C.   244
Hahn, G. T.   184, 376–377, 453, 636
Hahn, M. T.   494, 640, 680
Haldimann, M.   456
Hall, C. E.   456
Hall, E. O.   138, 184
Ham, R. K.   184–185
Hamilton, W. H.   710
Hammant, B. L.   185
Hammouda, M. M.   637
Hanaki, Y.   376
Hancock, G. G.   494
Handrock, J. L.   641
Hanink, R. H.   455
Hanna, G. L.   494
Hanna, W. D.   638
Hannink, R. H. J.   455
Hannoosh, J. G.   138
Hansma, P.   376, 458
Hardrath, H. F.   554, 556
Hargreaves, M. E.   184
Harmer, M. P.   455
Harrigan, M. J.   453, 638
Harris Jr., J. A.   635
Harris, S.   294
Harris, S. J.   455
Harris, W. J.   554
Hart, E. W.   185, 191
Hart, R. G.   638
Hartbower, C. E.   376
Hartman, A.   638
Hart-Smith, L. J.   377
Hashida, T.   455
Hasselman, D. P. H.   56, 375, 459

Haupert, F.   457
Hauser II, J. A.   495
Hawthorne, J. R.   459
Hayward, D. O.   494
Hazell, E. A.   185
Heijboer, J.   245, 456
Heiser, F. A.   454, 639, 680
Helgeland, O.   555
Herman, W. A.   637
Herring, C.   243
Hertzberg, J. L.   294
Hertzberg, R. W.   138, 245, 295, 376, 454, 457, 496, 554–555, 636–641, 680, 714
Hetnarski, R. B.   375
Heuer, A. H.   455, 457–458
Heuler, P.   262
Hickerson, J. P.   555, 639
Hickling, J.   717
Hieronymus, W. S.   453
Hill, D. C.   295
Hill, R.   137
Hillberry, B. M.   638
Hilton, P. D.   495
Himmelein, M. K.   638
Hirano, K.   495
Hirsch, P. B.   137
Hirschberg, M. H.   555
Hirt, M. A.   681
Hirth, J. P.   184, 493
Hirthe, W. M.   243
Hiver, J. M.   138
Hoagland, R. G.   184, 636
Hodds, L. W.   455
Hodge, J. M.   454
Hoffman, D. A.   710
Hofmann, W.   494
Holick, A. S.   295
Hollomon, J. H.   137, 458
Holyrod, N. J. H.   485, 495
Homes, G.   640
Hondros, E. D.   295
Honeycombe, R. W. K.   184
Hooghan, T. K.   191
Hooke, R.   55
Hopkins, B. E.   639
Hopkins, S. W.   638
Hoppe, W.   635
Horn, B. W.   637
Hornbogen, E.   184–185
Horstmann, M.   495
Hosford, W. F.   138
Hoshide, T.   640

Hoshina, M. 637
Howie, A. 137
Howlett, S. 456
Howson, T. E. 185, 245
Hren, J. J. 184
Hsiao, C. C. 295
Hsieh, T. H. 457
Hsueh, C. H. 455
Hu, H. 137
Huang, D. D. 555
Huang, H. 493
Huang, Y. 456, 457
Hubert, J. F. 244
Hudak Jr., S. J. 495, 554, 635
Hudak, S. 721
Hudson, C. M. 636, 680
Huet, R. 711
Hughes, C. W. 717
Hull, D. 137–138, 184–185, 295
Humenik, J. N. 496
Hunsicker, H. Y. 455
Hunston, D. L. 456
Hutchinson, J. W. 377
Hyatt, M. W. 484
Hyun, S. 191

Illg, W. 636
Imhof, E. J. 635, 638
Inagaki, H. 137
Inckle, A. E. 639
Ingelstrom, N. 459
Inglis, C. E. 294
Inglis, E. 375
Irmann, I. 185
Irving, P. E. 636, 638–639
Irwin, G. R. 375–376, 378, 636, 680, 721
Ishizaki, S. 495
Iskander, S. K. 459
Izumi, O. 244

Jaccard, R. 636–638
Jack, A. R. 555
Jackson Jr., W. J. 456
Jackson, A. P. 457–458
Jackson, J. R. 710
Jacobs, F. A. 637
Jacobs, F. J. 638
Jacobson, L. A. 376
Jacquet, P. A. 717
Jaeger, J. C. 56
Jaffe, R. A. 453
Jagota, A. 456
James, L. A. 638

James, M. R. 640
Janiszewski, J. 640
Jaswon, M. A. 137
Jena, A. K. 244
Jenkins, A. D. 245
Jensen, R. R. 244
Jiang, F. 458
Jillson, D. C. 137
Jin, Z. H. 184
Johnsen, B. B. 457
Johnson, H. H. 494
Johnson, M. A. 377
Johnson, R. H. 244
Johnson, T. L. 494
Johnson, W. S. 638
Johnston, T. L. 245, 454
Johnston, W. G. 137, 184
Jonas, J. J. 138
Jones, D. 454
Jones, D. R. H. 185, 294–295, 680
Jordan, W. 294
Josephic, P. H. 495
Joshi, A. 458
Jukes, D. A. 138
Jung, S. H. 185
Justice, R. S. 457
Juvinall, R. C. 137

Kad, B. 458
Kahn, H. 458
Kaiser, T. 457
Kalderon, D. 458
Kallend, J. S. 137
Kamat, S. 457
Kambour, R. P. 138, 245, 295
Kamdar, M. H. 494
Kameda, J. 458
Kamino, M. 457
Kane, R. D. 495
Kang, S. K. 245
Kanninen, M. F. 375
Kapadia, B. M. 454
Kaplan, M. P. 638
Karlak, R. F. 185
Katlin, J. M. 571
Katti, S. S. 185
Kaufman, J. G. 454
Kaufman, M. 138
Kausch, H. H. 138, 294, 457
Kaushik, A. K. 458
Kay, R. C. 244
Kê, T.-S. 191
Kear, B. H. 244–245

Kearney, V. E. 636
Keast, W. 456
Keating, P. B. 681
Keeton, W. P. 710
Kelly, A. 137, 244, 457
Kelly, N. J. 494
Kelly, P. 137
Kendall, E. G. 638
Kendall, K. 458
Kent, W. B. 244
Kerlins, V. 294
Kersker, M. M. 455
Khobaib, K. 184
Khobaib, M. 495
Kida, S. 457
Kies, J. A. 376
Kim, D. K. 376
Kim, J. 457
Kim, J. W. 680
Kim, S. C. 185
Kim, S. K. 680
Kim, S. L. 640
Kimura, H. 244, 455
Kinloch, A. J. 375, 456–457
Kinney, J. H. 458
Kishi, T. 454
Kitagawa, H. 637
Klassen, R. J. 184, 377
Klaushofer, K. 458
Klesnil, M. 636
Klingele, H. 294
Klingerman, D. J. 681
Knecht, R. L. 638
Knight, D. P. 185
Knott, J. F. 377, 639
Kobayashi, H. 377, 495
Kocer, C. 496
Kocks, U. F. 137
Koehler, J. S. 137, 184
Koester, K. J. 376, 457
Komatsubara, T. 137
Kong, X. 554
Koo, G. P. 554, 640
Kooi, J. E. 456
Korchynsky, M. 454
Kortovich, C. S. 244
Kotov, N. A. 458
Kramer, E. J. 138, 456
Kramer, I. R. 555
Krishna, N. 378
Krishna, Y. 244
Krishnamurthy, R. 638
Krupp, U. 472
Kruzic, J. J. 45, 376

Kübler, J. J.   376
Kuhlmann-Wilsdorf, D.   184
Kukta, R. V.   184
Kumar, V.   377
Kunz-Douglass, S.   457
Kuruvilla, A. K.   455
Kusy, R. P.   295
Kutumba Rao, V.   244

LaForce, R. P.   458
Laghouati, F.   640
Lahann, J.   458
Laird, C.   554–555, 636
Lakes, R. S.   55
Landel, R. F.   245
Landes, J. D.   377, 495, 638
Landgraf, R. W.   535, 555
Lane, M.   377, 378
Lang, R. W.   295, 640–641
Langdon, T. G.   244
Lange, F. F.   453, 455, 457
Langford, G.   185
Larentis, L.   456
Larson, F. R.   137, 244
Larson, R.   375
Latanision, R. M.   493
Lathabai, S.   640
Launey, M. E.   453, 457–458
Lauterwasser, B. D.   138
Lavelle, P.   711
Lavengood, R. E.   555
Lawn, B. R.   376, 455–456, 640
Lee, C. S.   496
Lee, J. J.   680
Lee, S.   377
Lee, Y.   495
Lee, Y. P.   238
Leibler, L.   457
Leichter, H. I.   454
Leis, B.   554
Leis, B. N.   637
LeMay, I.   639
Lemkey, F. D.   245
Lenel, F. V.   185
Lesuer, D. R.   393, 454
Levan, G.   636
Levens, J. A.   554
Leverant, G. R.   638
Levy, S. A.   455
Lewandowski, J. J.   453
Lewis, P. R.   681
Li, C. G.   555
Li, C. H.   454
Li, C.-M.   457

Li, C.-Y.   191
Li, D.   638
Li, V. C.   455
Liable, R. C.   456
Liang, Y.-M.   377
Liaw, P. K.   637
Liebowitz, H.   454
Liechti, K. M.   377
Lifshitz, L. M.   244
Likhtman, V. I.   494
Lin, A. Y.-M.   458
Lindley, T. C.   639
Linnert, G. E.   494
Litt, M. H.   245, 456
Liu, C. T.   472
Liu, H. W.   635, 638
Liu, J.   454, 457
Livy   56
Logsdon, W. A.   459, 637
Löhberg, R.   717
Loss, F. J.   459
Lott, J. L.   459
Lovat, N.   184
Low, I. M.   457
Low, J. R.   184, 377, 454, 458
Lowell, C. E.   56
Lu, K.   184
Lu, L.   184
Lu, Q. H.   184
Lua, J.   378
Lucas, F.   185
Lueth, R. C.   459
Luetjering, G.   455
Lui, M. W.   639
Luible, A.   456
Lukas, P.   636
Lund, J.   378
Luyckx, L.   454
Lynch, S. P.   494
Lytton, J. L.   243

Ma, Q.   378
Maakestad, W. J.   710, 711
Mackay, M. E.   556
Madison, R.   680
Magnusen, P. E.   454
Magonov, S.   458
Mahajan, S.   138
Mahoney, M. W.   638
Mai, T. W.   456
Mai, Y. W.   456, 457
Majno, L.   377
Majumdar, S.   639
Mall, S.   378

Malone, T. W.   455
Mammel, W. L.   137
Mandell, J. F.   555
Manson, J. A.   295, 456, 554, 555, 639–641, 680
Manson, S. S.   244, 555
Marchica, N. V.   681
Marci, G.   637
Marcus, H. L.   458
Marder, A. R.   185
Margary, G. M.   378
Margolin, H.   455
Marin, J.   295
Marines, I.   554
Marissen, R.   185, 591, 640
Markus, M.   494
Marsh, P.   493
Marshall, D. B.   376, 456
Marshall, G. P.   459
Martin, R. B.   294, 457
Marzinsky, C. N.   638
Masania, K.   457
Masounave, J.   639
Matlock, D. K.   244
Matsuishi, M.   555
Matsuo, Y.   456
Matthews, W. T.   375
Mauer, G. E.   244
Mawari, T.   639
Maxwell, B.   295
Maxwell, D. L.   457
Mayer, G.   458
Mayes, I. C.   639
Mayfield, M. E.   459
Mayville, R. A.   495
McBreen, J.   495
McCarty, J. H.   294
McCaughey, J. M.   494
McClean, A.   454
McClean, D.   459
McClintock, F. A.   376–377, 454
McCrum, N. G.   42, 56, 245
McDanels, D. L.   245
McEvily, A. J.   454–455, 459, 636
McGannon, H. E.   454
McGarry, F. J.   456, 555
McGowan, J. J.   638
McHenry, H. I.   638
McIntyre, D. R.   495
McIntyre, P.   495
McKinney, K. R.   459
McKittrick, J.   458

McMahon, C. J.   137, 454, 458, 472, 495
McMeeking, R. M.   378, 455, 457
McMillan, J. C.   636
McNamee, B. M.   681
McNicol, R. C.   375–376, 555
Mecholsky, J. J.   294, 376–377, 456
Meinel, G.   640
Melton, K. N.   244
Mencarelli, E.   717
Merkle, J. G.   377
Meshii, M.   184
Meyers, M. A.   184, 454, 458
Michalak, J. T.   137, 454
Michel, J. C.   640
Miller, A. C.   455
Miller, G.   635–636
Miller, G. A.   455
Miller, J.   244
Miller, K. J.   637
Miller, R. A.   56
Mills, N. J.   185
Mills, W. J.   636–637
Minakawa, K.   636
Mindlin, H.   680
Miner, M. A.   554
Mishra, A.   184
Misra, A.   184
Mitchell, M. A.   459
Mitchell, T. E.   137
Miyagaula, M.   457
Moffatt, W.   294
Mohamed, F. A.   244
Mohammed, R. D.   457
Moll, R. A.   711
Moloney, A. C.   457
Moncarz, P.   711
Monkman, F. C.   243
Montgomery, J. E.   710
Montgomery, R. S.   680
Moore, J. B.   245
Morgan, A. B.   457
Morlet, J. G.   494
Morrell, R.   376
Morris, S.   555
Morrogh, H.   55
Morrow III, H.   244
Morrow, J.   555
Morrow, J. D.   555, 639
Morscher, G. N.   227
Moskowitz, A.   454
Mostovoy, S.   494

Mravic, B.   454
Mueller, L. N.   185, 640
Mughrabi, H.   555
Mukherjee, A. K.   243
Munch, E.   457, 458
Munz, D.   637
Murakami, Y.   376, 554
Muratoglu, O. K.   56
Murphy, M. C.   457
Murray, J.   295
Muzyka, D. R.   244

Nabarro, F. R. N.   184, 243
Nair, S. V.   455
Nakai, Y.   637
Nakamura, H.   377
Nakazawa, H.   377, 495
Nalla, R. K.   458
Nandivada, H.   458
Nanis, L.   495
Nans, D. J.   459
Nardone, V. C.   245
Nederveen, A.   638
Neite, G.   185
Nelson, F. G.   454, 496
Nembach, E.   185
Neuber, H.   294, 555
Newbegin, R. L.   495
Newkirk, J. B.   137
Newman, J. C.   376, 554, 637, 680, 723
Newman, R. C.   493
Newton, C.   376
Newton, C. H.   636, 637
Ni, J.   554
Nicholas, M. G.   494
Nicholson, R. B.   244
Nieh, T. G.   185
Nielsen, L. E.   138, 245
Nir, N.   191
Nishida, T.   376
Nishioka, A.   637
Nishiwacki, N.   457
Nix, W. D.   243, 244
Nobel, B.   455
Nock, J. A.   455
Nordberg, H.   459
Norstrom, L. A.   185
Norton, F. J.   494
Notis, M. R.   244
Novak, S. R.   353, 495
Nowack, H.   591
Nugent, B.   459
Nutting, J.   137

O'Toole, J. L.   554
Odette, G. R.   459
Ogorkiewicz, R. M.   138, 245
Ogren, J. R.   455
Ohara, T.   640
Ohira, T.   454
Ohta, A.   639
Ohuchida, H.   637
Oikawa, H.   244
Old, C. F.   494
Oliver, W. C.   55
Orazio, F. D.   376, 636, 680
Oriani, R. A.   495
Orner, G. M.   376
Orowan, E.   185, 376
Orr, L.   294
Orr, R. L.   243
Orthwein, W.   556
Ortiz, C.   458
Ortlepp, C. S.   185
Osgood, W. R.   91
Otte, H. M.   184
Outwater, J. O.   457
Overend, M.   456
Owen, D. G.   710
Owen, N. B.   639
Ozturk, B.   458

Pabst, R. F.   459
Padgett, R. A.   295
Padture, N.   455
Palmberg, P. W.   458
Palmer, I. G.   455
Palmgren, A.   554
Pao, P. S.   493, 495, 638–639
Paraventi, D. J.   295
Paris, P. C.   376–378, 494, 554, 584, 635–636, 641, 680, 721
Parker, E. R.   294, 454–455
Parkins, R. N.   493
Parks, D. M.   636
Parrish, C.   455
Parsons, W. B.   294
Partridge, P. G.   137–138
Pascoe, R. T.   455
Passoja, D. R.   295
Patel, J. R.   184
Paton, N. E.   455, 638
Pavan, A.   138
Pearson, D. D.   245
Pearson, H. S.   680
Pearson, R. A.   456–457, 639
Pearson, S.   639
Peck, J. F.   454

Pecorini, T. 639
Pecorini, T. J. 680
Pellegrino, J. V. 376
Pellissier, G. E. 295, 454
Pelloux, R. M. 636, 639
Perez, J. 138
Permyakova, T. V. 456
Petch, N. J. 184, 454, 494
Peterlik, H. 458
Peterlin, A. 138, 640
Peters, M. 637
Peterson, M. H. 495
Peterson, R. E. 294, 554
Petit, J. I. 454
Petrasek, D. W. 245
Petrie, S. E. B. 245, 456
Petrovic, J. J. 376
Petti, J. P. 459
Pezzotti, G. 376
Pharr, G. M. 55
Phillips, A. 294
Phillips, J. D. 640
Phillips, J. H. 459
Phillips, M. G. 456
Phillips, R. 184
Piascik, R.P. 638
Pickard, A. C. 639
Pickering, F. B. 185
Piearcey, B. J. 244–245
Piehler, H. R. 137
Piercy, G. R. 137
Piggott, M. R. 46
Pillot, S. 494
Piper, D. E. 454
Pitman, G. 640
Pliny 56
Pocius, A. V. 377
Podsiadlo, P. 458
Poe, C. C. 635
Poh, J. 457
Ponge, D. 454
Pook, L. D. 584
Porter, D. L. 455
Porter, J. R. 640
Porter, R. S. 185
Posner, R. A. 710
Potapovs, U. 459
Preece, C. M. 494
Preston, J. 185
Price, A. T. 555
Price, J. B. 496
Priest, A. H. 495
Prittle, E. J. 639
Proctor, R. P. M. 493

Prohaska, J. L. 185, 640
Prosser, W. L. 710
Pry, R. H. 185
Psioda, J. A. 377
Pugh, E. N. 493, 494
Pugh, S. F. 453
Pumplin, B. G. 458

Quesnel, D. J. 184
Quinn, G. D. 294, 376, 723
Quinson, R. 138
Quist, W. E. 454

Raabe, D. 454
Rabinowitz, S. 138, 295, 555
Radavich, J. F. 244
Radon, J. C. 459
Raffo, P. L. 309
Ragazzo, C. 638
Rahm, L. F. 295
Raj, R. 244
Raju, I. S. 376, 680, 723
Rama Rao, P. 244
Ramamoorthy, A. 458
Ramberg, W. 91
Ramirez, A. 640
Ramo Rao, R. 459
Rankin, A. W. 376
Rao, M. 244
Ratner, S. B. 554
Ratz, G. A. 458
Rau, C. A. 638
Rauls, W. 494
Raymond, L. 495
Read, B. 138
Read, B. E. 245
Read, W. T. 136, 184
Rebinder, P. A. 494
Reeder, J. R. 378
Reed-Hill, R. E. 138, 184
Reep, R. L. 294, 376
Reid, C. N. 680
Reifsnider, K. L. 555
Rhodes, C. G. 638
Rice, F. G. 496
Rice, J. C. 56
Rice, J. R. 377, 453
Rice, R. W. 294
Rich, T. P. 680
Richards, C. E. 639
Richards, M. 456
Richter, H. 496
Riddell, M. N. 554
Riew, C. K. 457

Rim, J. E. 457
Rimnac, C. M. 640, 680
Rink, M. 138
Rioja, R. J. 455, 638
Ripling, E. J. 458
Ritchie, R. O. 376, 377, 453,
    455, 457, 458, 584, 590–591,
    629, 636–641, 680
Ritter Jr., J. E. 496
Ritter, J. C. 680
Ro, Y. 638
Roberts, C. 555
Roberts, D. E. 456
Roberts, R. 554, 636–637
Roberts, W. T. 137
Robertson, I. M. 494
Robertson, R. E. 245, 295
Robertson, W. D. 494
Robinson, I. M. 45
Robinson, J. L. 636, 638
Robinson, P. M. 137
Robinson, S. L. 243
Rodel, J. 640
Rodriguez, F. 245
Roldan, L. G. 640
Rolfe, S. T. 353, 376, 495, 555,
    635, 680
Romios, M. 455
Roschger, P. 458
Roscoe, R. 137
Rose, M. V. 185
Rosen, B. W. 137
Rosen, S. L. 138
Rosenberg, J. M. 137
Rosenfield, A. R. 376–377, 455,
    636
Rossi, R. 459
Rossi, R. C. 375
Rosso, P. 457
Rossow, M. 377
Rostoker, W. 494
Rubio, A. 376, 680
Rühle, M. 455
Russell, J. D. 378
Ruzette, A.-V. 457
Ryder, D. A. 636

Sadananda, K. 639
Sailors, R. H. 353, 376,
    680
Saiz, E. 457, 458
Sanders Jr., T. H. 455
Sanders, P. G. 184
Sandor, B. I. 555

Sato, A.   184
Sauer, J. A.   138, 295, 554
Savage, K. N.   185
Saxena, A.   584, 639, 721
Saxena, L.   635
Sbaizero, O.   457
Schabtach, C.   376
Schaefer, D. W.   457
Schaper, H.   294
Schehl, M.   496
Schijve, J.   636, 637
Schilling, P. E.   454
Schmid, E.   137
Schmidt, D. W.   378, 635
Schmidt, F. E.   680
Schmidt, R. A.   459, 584,
    636
Schneibel, J. H.   295
Schultz, J. M.   185, 556
Schultz, R. W.   185, 640
Schwartz, V. E.   711
Schwier, C. E.   138
Scott, H. G.   137
Seah, M. P.   295
Seeger, A.   184
Seeger, T.   262
Seely, F. B.   556
Seidelmann, U.   496
Seki, Y.   458
Sendeckyj, G. P.   554
Senz, R. R.   454
Server, W. L.   377, 459
Seth, B. B.   636
Seward, S. K.   451
Sglavo, V. M.   456
Shahinian, P.   639
Shamimi Nouri, A.   453
Shand, E. B.   294
Shank, M. E.   245
Shao, L. C.   459
Sharkey, N. A.   294, 457
Shazly, M.   453
Shchukin, E. D.   494
Sheffler, K. D.   245
Shen, Y. F.   184
Shenoy, V. B.   184
Sherby, O. D.   243, 393
Sherratt, F.   554
Shih, C. F.   377, 455
Shih, C. H.   458
Shih, T. H.   638
Shim, B. S.   458
Shim, S.   55
Shimizu, T.   457

Shipley, R. J.   294
Shogan, R. P.   459
Shumaker, M. B.   711
Sigl, L. S.   457
Sih, G. C. M.   376, 635, 680
Simmons, G. W.   638
Simnad, M. T.   243
Simonen, E. P.   295
Sims, C. T.   244
Sinclair, G. M.   137
Sines, G.   554
Singh, J. P.   56, 375
Skat, A. C.   638
Skibo, M. D.   554, 640
Slagter, W. J.   185
Smallman, R. E.   137, 184
Smith, A. K.   457
Smith, B. W.   295
Smith, C. A.   456
Smith, C. R.   556
Smith, G. C.   453, 636
Smith, H. L.   459
Smith, J. H.   454
Smith, J. O.   556
Smith, K. N.   637
Smith, R. A.   637
Smith, R. W.   555
Socie, D. F.   555
Sokolov, M. A.   459
Solomon, H. D.   638
Soltesz, U.   496
Somerday, B. P.   493
Sommer, A. W.   455, 638
Song, R.   454
Sorenson, E. P.   377
Souahi, A.   138
Southern, J. H.   185
Spanoudakis, J.   457
Speidel, M. O.   495, 680
Sperling, L. H.   238, 456
Spitznagel, J. A.   459
Sprenger, S.   457
Spuhler, E. H.   454
Spurr, W. F.   494
Srawley, J. E.   376, 680
Stabenow, J.   456
Stables, P.   494
Staehle, R. W.   493–494
Staley, J.   454
Staley, J. T.   455
Stanton, W. P.   455
Starke, E. A.   455
Steele, F. R.   684
Steele, L. E.   459

Steffen, A. A.   640
Steigerwald, E. A.   494–495
Stein, D. F.   184, 458
Steinman, J. B.   494
Stephans, J. R.   245
Stephens, R. I.   554, 637, 641
Stephens, R. R.   641
Sternstein, S. S.   295
Steven, W.   458
Stevens, W. W.   555
Stickler, R.   636
Stinskas, A. V.   554
Stock, S. R.   103
Stofanak, R. J.   584, 636
Stokes, A. G.   458
Stokes, R. J.   454
Stölken, J. S.   458
Stoloff, N. S.   138, 244, 494
Stout, R. D.   454
Street, D. G.   456
Stroebel, L.   710
Strosnider, J. R.   459
Studart, A. R.   458
Stulga, J. E.   185, 245
Su, X.   457
Subbarao, E. C.   455
Sugarman, S. D.   711
Sullivan, C. P.   245
Sultan, J. N.   456
Sunderesan, R.   185
Suresh, S.   185, 455, 636–637,
    640, 641
Sutton, W. H.   137
Suzuki, N.   639
Svensson, N.   457
Swain, M. V.   455–456
Swanson, P. L.   456
Swanson, S. R.   635
Swearekgen, J. C.   456
Swenson, D. O.   635
Sylva, L. A.   640
Syn, C. K.   393
Szpak, E.   496

Tada, H.   376–378, 636, 680,
    721
Tadmor, Z.   185
Tai, K.   458
Taira, S.   637
Tajima, Y.   456
Takahashi, S.   637
Takeda, Y.   495
Tanaka, K.   637
Tandon, R.   456

Tang, E. Z.   637
Tang, H.   457
Tang, Z.   458
Taplin, D. M. R.   244, 377
Tavernelli, J. F.   555
Taylor, A. C.   457
Taylor, G. I.   137, 184
Tegart, W. J. McG.   136–137, 184, 294
Teng, T. G.   556
Terao, N.   294
Tetelman, A.   459
Tetelman, A. S.   454, 494
Thaveeprungsriporn, V.   295
Thomas, B.   455
Thomas, D. A.   454
Thomas, G.   455
Thompson, A. W.   377, 493–494
Thompson, D. S.   455
Thompson, E. R.   245
Thomson, J. B.   554
Thorkildsen, R. L.   554
Thornton, P. H.   245
Thornton, P. R.   137
Thouless, M. D.   457
Throop, J. F.   680
Tiegs, T. N.   455
Tien, J. K.   185, 244–245
Tijssens, M. G. A.   138
Timio, D. P.   376, 680
Tiraschi, R.   455
Tobolsky, A. V.   245, 456
Tomar, V.   458
Tomkins, B.   639
Tomsia, A. P.   457–458
Tonyali, K.   496
Topper, T. H.   555, 637
Torrecillas, R.   496
Toussaint, P.   494
Trapnell, B. M. W.   494
Trebules Jr., V. W.   637
Treloar, L. R. G.   55
Tressler, R. E.   244, 294
Trinkhaus, H.   295
Troiano, A. R.   494
Tromp, P. J.   637
Trozera, T. A.   243
Tsangarakis, N.   454
Tu, L. K. L.   636
Tucker, L. E.   535
Turkalo, A. M.   184, 458
Turkanis, J. T.   554
Turnbull, A.   495, 638

Turner, D. T.   295
Turner, I. R.   244
Turner, R. M.   457–458
Tvergaard V, V.   378
Tyson, W. R.   457

Ulm, F-J.   458
Underwood, J. H.   680
Urashima, K.   456
Usami, S.   637
Utley, E. C.   554

Vainchtein, A.   377
Valluri, S. R.   377, 459
Van Buskirk, W. C.   56
Van der Giessen, E.   138
Van Stone, R. H.   377
Van Uitert, L. G.   51, 56
VanLandingham, M. R.   55
VanNess, H. C.   494
Vassilaros, M. G.   459
Vasudevan, A. K.   455
Vaughn, D.   378
Vecchio, K. S.   458, 680
Vecchio, R. S.   637, 714
Venkateswara Rao K. T.   455, 639
Verall, R. A.   244
Verheulpen-Heymans, N.   138
Verolme, J. L.   185
Versnyder, F. L.   244–245
Viehrig, H.-W.   459
Vincent, J. F. V.   457
Vincent, P. I.   375
Vincenta, J. F. V.   458
Vinci, R.P.   191
Vitek, V.   458
Vliegar, H.   376
Vogelesang, L. B.   185, 636
Vogelsang, M.   185
Vollrath, F.   185
von Mises, R.   137, 138
von Euw, E. F. J.   636, 637

Waas, A. M.   458
Wadsworth, J.   393, 454
Wagner, L.   639
Waisman, J. L.   554
Wald, M. L.   458, 459
Waldron, G. W. J.   639
Walker, N.   636
Walter, G. H.   639
Ward, I. M.   55, 640

Warren, T. J.   495
Was, G. S.   295
Wassermann, G.   137
Watanabe, M.   456
Watanabe, T.   295
Watson, H. E.   459
Weber, J. H.   636, 639
Webler, S. M.   554, 640
Webster, D.   454–455
Weeks, N.   185
Weertman, J.   184, 243–244
Weertman, J. R.   184
Wegst, U. G. K.   457
Wei, R. P.   295, 454, 493, 495–496, 638–639
Wei, W.   639
Weibull, W.   294, 635
Weir, T. W.   638
Weiss, B.   636
Wells, A. A.   377
Wells, J. K.   457
Wernick, J. H.   137
Westergaard, H. M.   376
Westerlund, R. W.   454
Westfall, L. J.   245
Westwood, A. R. C.   494
Wetzel, B.   457
Wetzel, R. M.   555
Wheeler, M. E.   711
White, C. L.   295, 472
Whiteson, B. V.   294
Wichman, K. R.   459
Wiederhorn, S. M.   455–456, 459, 495–496
Wiersma, S. J.   454
Wilhelm, S. M.   495
Williams, D. F.   138
Williams, D. P.   493, 495–496, 638
Williams, G.   245
Williams, J. C.   455, 638
Williams, J. G.   459, 496, 554
Williams, M. L.   245
Williams, S.   457
Williamson, G. K.   184
Wilsdorf, H. G. F.   377
Wilshire, B.   244
Wilson, W. K.   555
Winnie, D. H.   376
Wlodek, S. T.   244
Wohler, A.   554
Wolf, J.   711
Wolter, F.   496

Wood, H. A.   680
Wood, J. K.   137
Wood, W. A.   555
Woodfine, B. C.   458
Woodford, D. A.   244
Woodward, R. L.   454
Worn, D. K.   458
Wraith, A. E.   454
Wright, E. S.   454
Wright, L.   138
Wu, E. M.   555
Wu, G.   185
Wu, H. F.   185
Wu, L. L.   185
Wu, W. C.   680
Wulff, J.   294
Wullaert, R. A.   377
Wulpi, D. J.   554

Xu, J.   458

Yakowitz, H.   680
Yamada, T.   640
Yamashita, M.   637
Yan, J.   377
Yan, M.   639
Yan, M. G.   453
Yang, J.-M.   185
Yao, D.   640
Yee, A. F.   456, 457
Yee, Y. W.   457
Yeh, G. S. Y.   138
Yellup, J. M.   454
Yen, B. T.   680–681
Yoder, G. R.   638–639
Yoerger, D.   294
Yoo, M. H.   295
Young, L. M.   494–495
Young, R. J.   45, 375, 457
Youngdahl, J.   184

Yu, B.   639
Yu, J.   458
Yu, W.   639–640
Yuen, A.   638
Yukawa, S.   376, 680
Yun, C. B.   680
Yuzawich, P. M.   717

Zackay, V. F.   454–455
Zahoor, A.   377
Zaiken, E.   638
Zamiski, G. F.   636
Zapffe, C. A.   494
Zavattieri, P. D.   457
Zener, C.   454
Zijlstra, A. L.   456
Zinkham, R. E.   454
Zioupos, P.   457
Zipp, R.   717
Zipp, R. D.   639

Materials Index

Acetal:
  applications, 125
  characteristics, 125
  composite strength, 182
  environment assisted cracking, 489
  fatigue crack propagation, 619, 627
  repeat unit, 125
  tensile strength of, 20–21
Acrylonitrile-butadiene-styrene (ABS):
  fatigue crack propagation, 627
  tensile strength, 20–21
  toughness, 428
Alumina silicate:
  compressive strength, 22
  elastic modulus, 22
  flexural strength, 22
  tensile strength, 22
Aluminum:
  activation energy for creep, 198–199
  composite properties, 175–177
  creep data, 198–199, 219
  creep rate of, 198
  dislocation junctions, 151
  dislocation-solute interaction, 161
  elastic anisotropy, 38
  elastic compliances, 38
  elastic constants, 38
  elastic modulus, 10–12, 38, 65
  failure analyses, 658–660
  rolling texture of, 108
  stacking fault energy of, 80, 93
  strain hardening coefficient, 92, 93
  texture, thin film, 107
  theoretical and experimental yield strength, 65, 67
  thermal stress, thin film, 51–52
  thin section of, 153, 169, 176
  twin elements, 115
  wire texture, 106–107

Aluminum alloys:
  aging curves, 166
  alloy element effect on, 396
  Charpy impact energy, 309, 315, 348
  creep data, 220
  environment assisted cracking, 480–486, 708–710
  failure analyses, 658–660
  fatigue crack propagation, 561–562, 566, 567, 575, 576, 579–581, 584–587, 589–591, 594–596, 601, 605, 615, 617, 619, 633
  fatigue data, 505–506, 519, 535–536
  fatigue life, 505
  fracture mode appearance, 270
  fracture toughness, 268, 345, 394–398, 410–411, 414–416, 449–453
  iron and silicon effect on, 394–398
  liquid metal embrittlement, 485
  mechanical fibering of, 396, 402–403
  rivet failure, 708–710
  strain controlled-cyclic life response, 534–536
  stress corrosion cracking, 708–710
  tensile and yield strength, 20–21
  Ultra High Cycle Fatigue, 506
Aluminum-lithium:
  precipitate morphology, 166, 169
  toughness, 415–416
Aluminum oxide:
  activation energy for creep, 199
  compressive strength, 22
  elastic modulus, 10–11, 22
  environment assisted cracking, 492
  fatigue crack propagation, 628–632

flexural strength, 22
fracture toughness, 268, 371, 449–453
slip system, 83
tensile strength, 22
thermal shock resistance, 55, 303
whisker properties, 41, 67, 183, 254
Anthracene, yield behavior, 85–86
Antimony, temper embrittlement effect of, 441–444
ARALL, 176–177, 632–633
Aramid (Kevlar 49):
  fiber properties, 41, 183
  general description, 14
Ausformed steel:
  fiber strength, 254
  fracture anisotropy, 407–409

Bakelite, 14
Beryllium:
  interplanar angles in, 118
  rolling texture of, 108
  slip plane, 83
  strain ellipsoid of, 118
  theoretical and experimental yield strength, 65
  twin elements, 115
Beryllium oxide, thermal shock resistance, 55
Bezoar-stone, 685
Bone:
  elastic constants, 35–36, 181
  degradation, age-related, 329
  fracture mechanisms, 439
  fracture, spiral, 275
  manatee, 327
  toughness, 181, 434–435, 437–438
  strength, 181
  structure, 34, 437
Boron:
  fiber properties, 41
  surface stabilizer, 220–223

Boron carbide (B$_4$C), whisker
    strength, 67
Brass:
    deformation markings in, 81,
        112
    fatigue crack propagation, 584,
        589
    fatigue properties, 518–519
    rolling texture of, 108
    stacking fault energy of, 80, 93
    strain controlled-cycle life
        response, 535–537
    strain hardening coefficient, 92

Cadmium:
    elastic compliances, 38
    elastic constants, 38
    elastic modulus, 10, 11
    rolling texture of, 108
    slip plane, 83
    theoretical and experimental
        yield strength, 65, 67
    twin elements, 115
Calcium fluoride (CaF$_2$), slip
    system, 83
Carbides, role in fracture of steel,
    392
Carbon:
    fiber properties, 41, 183
    solid solution strengthening,
        158–161, 170
    whisker strength, 67
Cast iron, fatigue, 519–520
Ceramic Matrix Composite
    (CMC), 227, 416
Cesium chloride, slip system, 83
Chalk, torsional fracture, 275
Cheese, Velveeta®, creep, 238
Chromium:
    elastic modulus, 10–11
    surface stabilizer, 220–223
Cobalt:
    activation energy for creep, 199
    elastic compliances, 38
    elastic constants, 38
    high temperature strength,
        222–224
    rolling texture of, 108
    slip plane, 83
    wire texture, 106–107
Collagen, 181
Concrete, fracture toughness,
    450–453
Copolymer, 125

dynamic response, 242
fracture surface, 287, 427–428
Copper:
    activation energy for creep, 199
    creep data, 199
    cyclic strain hardening,
        softening, 532
    dislocation-solute interaction,
        160, 161
    elastic anisotropy, 38
    elastic compliances, 38
    elastic constants, 38
    elastic modulus, 10–11, 38–39
    fatigue crack propagation, 584,
        617
    fatigue data, 518–519,
        534–538
    fatigue life, 518
    nanocrystalline, 157
    nanotwinned, 157
    neutron irradiation,
        embrittlement effect of,
        447–448
    multilayers, Cu-Nb, 158
    rolling texture of, 108
    solid solution strengthening in,
        161
    stacking fault energy of, 80, 93
    strain controlled-cyclic life
        response, 532, 534–538
    strain hardening coefficient,
        92–93
    theoretical and experimental
        yield strength, 65, 67
    thin section of, 153
    twin elements, 115
    whisker strength, 67
    wire texture, 107

Dispersion strengthened alloys,
    170–172
Diamond:
    elastic modulus, 10–11
    slip system, 83

Elastin, 181
Epoxy:
    composite properties, 41, 183,
        528–529
    fatigue crack propagation,
        619–620, 626
    fracture surface, 290
Ethane, conformation potential
    energy, 121–122

Gallium, 471
Gallium Arsenide, elastic
    constants, 38
Germanium:
    deformation mechanism map,
        208, 209
    slip system, 83
GLARE, 176–177, 633
Glass:
    elastic modulus, 10–11
    environment assisted cracking,
        489, 492
    fiber properties, 41, 183
    fracture appearance,
        273–275
    fracture, bottle, 694
    fracture, thermal, 274
    fracture surface of, 275–277
    fracture toughness, 263, 268,
        301, 344, 370–371, 386–387,
        416, 418, 422–426, 451, 453
    ion-exchanged (chemically
        strengthened),424–425
    metallic, 386
    spheres, 430, 431
    tempered, 273–274, 423–425
Gold:
    activation energy for creep, 199
    creep data, 199
    elastic anisotropy, 38
    elastic compliances, 38
    elastic constants, 38
    elastic modulus, 10, 11, 38–39
    stacking fault energy of, 80
    twin elements, 115
    wire texture, 107

Hafnium, surface stabilizer,
    220–223
High impact polystyrene, 427–428,
    547
Hydrogen, embrittlement, 389,
    449, 463–468, 469, 470,
    482–483, 485, 486, 487

Iron:
    activation energy for creep, 199
    creep data, 202, 216
    deformation twins in, 117
    dislocation-solute interaction,
        160–161
    elastic anisotropy, 38
    elastic compliances, 38
    elastic constants, 38

elastic modulus, 10–11, 38–39
liquid metal embrittlement, 471
Lüder strain of, 99–100, 161
rolling texture of, 108
solid solution strengthening in, 160–161
steady state creep rate of, 194, 201, 220
strain hardening coefficient, 92
stress rupture, 193
theoretical and experimental yield strength, 65, 67, 254
twin elements, 115
whisker strength, 67, 254
wrought, rivets, 251, 404
yield point of, 99–100

Kevlar, 14, 41, 181, 183

Laminated metal composite, 404
Lead:
    activation energy for creep, 199
    creep data, 199
    hot work, 154
    liquid metal embrittlement, 471
    superplastic behavior, 206
Liquid crystal polymer, 14, 180, 181–182
    composite strength, 182
Lithium fluoride:
    dislocation etch pits in, 67–68
    dislocation-solute interaction, 160–161
    elastic compliances, 38
    elastic constants, 38
    slip system, 83
    stress-strain curves of, 163

Magnesium:
    deformation twins in, 117–118
    elastic compliances, 38
    elastic constants, 38
    elastic modulus, 10–11, 38
    rolling texture of, 108
    slip plane, 83
    theoretical and experimental yield strength, 65
    twin elements, 115
Magnesium alloys:
    fatigue crack propagation, 561
    fatigue data, 519
    fracture surface, 650
    tensile and yield strengths, 20–21

Magnesium oxide:
    elastic anisotropy, 39
    elastic compliances, 38
    elastic constants, 38
    elastic modulus, 12, 38–39
    fracture toughness, 268
    slip system, 83
Manatee bone, 327
Melamine resin, 14
Mercury, liquid metal embrittlement, 471, 483
Metallic glass, 386
Molybdenum:
    creep data, 202
    elastic compliances, 38
    elastic constants, 38
    fatigue crack propagation, 561
    micropillars, 68
    rolling texture of, 108
    strain rate sensitivity, 100
    superalloys, addition to, 220
    theoretical and experimental yield strength, 65, 67–68

Nacre, 436, 438–439
Nickel:
    activation energy for creep, 199
    creep data, 199
    deformation mechanism map, 208–213
    dislocation-solute interaction, 161
    elastic compliances, 38
    elastic constants, 38
    elastic modulus, 10–11
    rolling texture of, 108
    stacking fault energy of, 80
    theoretical and experimental yield strength, 65, 67
    twin elements, 115
    whisker strength, 67
    wire texture, 107
Nickel alloys:
    corrosion fatigue, 605
    creep data, 220, 222
    deformation mechanism map, 208–213
    fatigue crack propagation, 584, 597–599
    fatigue data, 534–536
    high temperature strength, 222–227
    strain controlled-cycle life response, 535–536

Nickel aluminide:
    crystal form, 167–168
    slip, 168
    superlattice dislocations in, 167–168
Nickel sulfide, in tempered glass, 424, 700
Niobium:
    activation energy for creep, 199
    dislocation-solute interaction, 161
    elastic modulus, 10–11
    fatigue data, 605
    multilayers, Cu-Nb, 158
    Ni-based superalloy, in, 164
    rolling texture of, 108
    solid solution strengthening in, 160–161
    theoretical and experimental yield strength, 65
Nitrogen:
    dislocation locking by, 160–163, 173–174
    yield point and strain aging, role in, 161–162
Nylon 66:
    applications, 125
    characteristics, 125
    composite properties, 41, 182–183, 289, 634
    crystallinity, 124
    drawn, 179
    elastic modulus, 10–11, 181
    extensibility, 181
    fatigue crack propagation, 618–620, 623–625, 634–635
    fracture surface, 289
    plasticized, 128
    repeat unit, 125
    tensile strength of, 20–21, 81
    toughness, 181, 268, 273

Oxide-dispersion hardened alloys, 170–172, 201, 224–226
Oxygen, embrittlement of titanium alloys, 389–390

Partially stabilized zirconia, 419–420, 422, 435, 629
Phenol-formaldehyde resin (PF), 15
Phosphorus:
    effect on fracture toughness in steel, 390
    temper embrittlement, effect on, 441–444

Piano wire, strength, 254
Plasticizers, 14, 128, 387, 426
   antiplasticizer, 427
   toughness, effect on, 426
Platinum creep data, 202
Polybutylene terephthalate (PBT),
      182
   composite strength, 182
Polycarbonate (PC):
   composite strength, 182
   elastic modulus, 10–11
   fatigue crack propagation,
      619–620, 623–625,627
   fracture toughness, 268, 426,
      451, 453
   fracture surface, 625
   general description, 13
   glass laminate, 425
   load drop, 133
   properties, effect of MW, 178
   tensile behavior, 132–133
   tensile strength of, 20–21, 181
   yield, 135
Polychlorotrifluoroethylene
      (PCTFE), spherulites, 127
Polydimethylsiloxane (PDMS),
      melt viscosity, 179
Polyester:
   composite properties, 41
   shear modulus, 15
Polyetheretherketone (PEEK), 132
   composite strength, 182
   load drop, 133
Polyethylene (PE):
   applications, 125
   characteristics, 125
   elastic modulus, 10–11
   environment assisted cracking,
      487
   extended chain conformation, 121
   general description, 14, 123, 125
   load drop, 132–133
   repeat unit, 123
   single crystal, 126
   tensile strength of, 20–21, 124
Polyethylene terephthalate (PET):
   composite strength, 182
   fatigue crack propagation, 619
   general description, 13–14
Polyisobutylene (PIB), time-
      temperature superposition, 232
Polymer additions:
   blowing agents, 128
   crosslinking agents, 128

fillers, 128
pigments and dyestuffs, 127
plasticizers, 128
stabilizers, 128
Poly(methyl methacrylate)
      (PMMA):
   applications, 125
   characteristics, 125
   deformation map, 284
   discontinuous growth bands, 627
   elastic modulus, 10–11
   fatigue crack propagation, 619,
      620, 624–625, 627
   fatigue failure, 524
   fracture surface, 625
   fracture toughness, 272, 426,
      451, 453
   general description, 13–14, 125
   glass transition temperature, 231
   repeat unit, 125
   tensile behavior, 132
   shear yielding, 136
   yield, 133–136
Polyphenylene oxide (PPO):
   crazes, 283
   fatigue crack propagation, 619
Polyphenylene sulfide (PPS),
      composite strength, 182
Polypropylene (PP):
   applications, 125
   characteristics, 125
   cold drawing, 129
   elastic modulus, 10–11
   fracture surface, 287–288
   load drop, 133
   repeat unit, 123, 125
   tensile behavior, 133
   toughness, 268, 273
   stereoregularity, 124
Polystyrene (PS):
   elastic modulus, 10–11
   fatigue crack propagation,
      619–620, 624, 626
   fracture surface, 286–287
   fracture toughness, 451, 453
   general description, 13, 123
   glass transition temperature, 231
   repeat unit, 123
   strength-orientation dependence,
      180
   tensile strength of, 20–21
   toughness, 268, 272
Polysulfone:
   composite fatigue life, 528

fatigue crack propagation,
      619–620, 623–625, 627
   fracture surface, 625
   yield strength, 20, 21
Poly(tetrafluoroethylene) (Teflon)
      (PTFE):
   applications, 125
   characteristics, 125
   fatigue crack propagation, 620
   general description, 14, 125
   load drop, 133
   repeat unit, 125
   tensile strength, 20–21
   thermal fatigue, 524–526
Polyurethane (PU), 14
Poly(vinyl chloride) (PVC):
   applications, 125
   characteristics, 125
   crazing, 626
   environment assisted cracking,
      487
   fatigue crack
      propagation, 619–620,
      622–625, 627
   fatigue life, 527
   fracture surface, 625, 627
   general description, 13–14, 123,
      125
   load drop, 132–133
   pipe resin, 234
   repeat unit, 123, 125
   shear banding, 129
   toughness, 272–273
Poly(vinyl fluoride) (PVF), repeat
      unit, 123
Poly(vinylidene fluoride) (PVDF):
   fatigue crack propagation, 619,
      624
   tensile strength of, 20–21
Pyrolitic graphite, fatigue crack
      propagation, 628

Quartz, elastic modulus, 10–11

Rare earth, additions to steel, 392,
      502
Resilin, 181
Rubber, cavitation, 428–431
Rubber, synthetic, 181

Shell:
   mollusk, 434
   pink conch, 435
   red abalone, 436

Silica, whisker strength, 254
Silicon:
  elastic constants, 38
  Frank-Read source in, 145
  slip system, 83
  stress-strain curves of, 163
  thin section, 145
  whisker strength, 254
Silicon carbide:
  elastic constants, 38
  elastic modulus, 10–11
  fatigue crack propagation, 628
  fracture toughness, 268, 385,
    451, 453
  tensile strength, 258
  thermal expansion, coefficient
    of, 52
  thermal shock resistance, 52, 55
  whisker properties, 41, 67, 183
  whisker reinforcement,
    175–176, 227
Silicon nitride:
  compressive strength, 55
  elastic modulus, 55
  fatigue crack propagation, 628
  flexural strength, 55
  fracture toughness, 268, 451, 453
  tensile strength, 55
  thermal expansion, coefficient
    of, 55
  thermal shock, 55
  whisker properties, 41
Silk:
  cocoon, 181
  spider, 14, 180–182
  waste, bags of, 686
Silver:
  creep data, 199
  deformation mechanism map,
    209
  elastic modulus, 10–11, 65
  rolling texture of, 107–108
  stacking fault energy of, 80
  theoretical and experimental
    yield strength, 65
  thermal expansion, coefficient
    of, 52
  twin elements, 115
  wire texture of, 106
Silver bromide, activation energy
  for creep, 199
Sodium chloride:
  dislocation-solute interaction,
    161

elastic compliances, 38
elastic constants, 38
slip system, 83
Spider silk, 14, 180–182
Spinel ($MgAl_2O_4$):
  elastic anisotropy, 38–39
  elastic compliances, 38
  elastic constants, 38
  elastic modulus, 38
  slip system, 83
Stainless steel:
  crack tip heating during fatigue,
    517
  creep data, 220
  fatigue crack propagation, 573
  fracture toughness, 345, 413,
    450–453
  Larson-Miller parameter
    constants, 217, 220
  neutron irradiation damage, 449
  rolling texture of, 108
  stacking fault energy of, 80, 93
  strain controlled-cyclic life
    response, 535–536
  strain hardening coefficient,
    92–93
  stress corrosion cracking, 468,
    470
  tensile and yield strengths of,
    20–21
  thin section of, 69, 79
Steels:
  alloying elements, effect on, 412
  bainite strengthening, 172–175
  brazed joint, 265
  carburizing, 522–523
  Charpy impact energy, 309–311,
    313–314, 323, 334, 338, 348,
    350, 355, 391–393, 398, 406,
    408, 412, 441–442, 446–447
  chevron markings, 270–271
  cleavage, 281–282
  creep data, 220
  environment assisted cracking,
    465–470, 472, 474–478,
    482–484, 486
  failure analyses, 339, 354, 653,
    660, 661, 664, 665, 670
  fatigue crack propagation, 561,
    571, 573, 576, 583–584, 587,
    589, 590, 603, 611–613, 615,
    617, 619, 628
  fatigue data, 518–520, 534–536,
    592, 605

fatigue fracture surface, 501,
    502, 586, 596
  fatigue life, 503, 510–512, 515,
    518–520, 544–547, 678
  ferrite strengthening, 172–175
  fracture appearance, 97–98, 310,
    339, 354, 611, 622, 646, 648,
    649, 651, 655, 666, 671
  fracture mode transition, 339
  fracture toughness, 345, 346,
    348, 366, 350, 391, 395,
    411–414, 444, 446, 450–453
  high temperature strength, 220
  intergranular fracture, 281
  Larson-Miller parameter
    constants, 217, 220
  Lüder bands, 99–100, 161–164
  martensite strengthening,
    172–175
  MnS inclusions in, 390, 392,
    403–404, 412
  neutron irradiation
    embrittlement, 444–449
  notch strengthening, 264–265,
    307
  strain controlled-cyclic life
    response, 534–536
  stress-rupture life of, 193, 216
  temper embrittlement, 412,
    440–444
  tempered martensite
    embrittlement, 440–442
  tensile and yield strengths of,
    20–21, 101, 400
  transition temperature data, 309,
    311, 313–315, 349, 351
  TRIP, 395, 413–414
  Ultra High Cycle Fatigue, 517
Sulfur, effect on fracture toughness
  in steel, 389–394, 403
Superalloys:
  creep data, 211, 220–227
  creep rate, 201, 208
  oxide dispersion strengthened,
    171
  precipitation hardening, 164,
    166, 167–168, 170

Tantalum:
  activation energy for creep, 199
  creep data, 199
  elastic modulus, 10–11
  fatigue data, 605
  rolling texture of, 108

Thorium oxide (ThO_2), slip
system, 83
Tin:
    activation energy for creep, 199
    temper embrittlement, effect on,
        441
Titanium:
    Charpy impact energy, 309
    elastic compliances, 38
    elastic constants, 38
    elastic modulus, 10–11
    rolling texture of, 108
    slip plane, 83
    theoretical and experimental
        yield strength, 65
    twin elements, 115
Titanium alloys:
    environment assisted cracking,
        468, 470, 473–475,
        483–484
    fatigue crack propagation, 561,
        576, 583–584, 602, 607
    fatigue data, 519, 534
    fracture toughness, 268, 345,
        384, 389, 414–415, 451, 453
    Larson-Miller parameter
        constants, 217
    liquid metal embrittlement, 471
    strain controlled-cyclic life
        response, 534
    tensile and yield strength, 20–21
Titanium carbide:
    elastic anisotropy, 38
    elastic constants, 38

elastic modulus, 38
Titanium oxide (TiO_2):
    activation energy for creep, 197,
        199
    slip system, 83
    steady state creep rate, 197
Tungsten:
    activation energy for creep,
        199
    cold work, 154
    creep data, 202
    elastic compliances, 38
    elastic constants, 38
    elastic isotropy, 38
    elastic modulus, 10–11
    rolling texture of, 108
    swaged wire microstructure of,
        403
Tungsten carbide:
    elastic modulus, 10–11
    fracture toughness, 451, 453
    thermal shock resistance,
        51, 55

Ultra High Molecular Weight
    Polyethylene (UHMWPE), 14
Uranium oxide (UO_2):
    activation energy for creep,
        199
    slip system, 83

Vanadium, elastic modulus, 10–11
Vinyl polymers, 122–123,
    125

Wood:
    toughness, 268
    wheel spokes, 688–691
Wool, 181
Wrought iron, rivet, 251, 404

Zinc:
    deformation markings in,
        112
    elastic compliances, 38
    elastic constants, 38
    fatigue fracture, 500
    interplanar angles in, 118
    rolling texture of, 108
    slip plane, 83
    strain ellipsoid of, 119
    twin elements, 115
Zinc sulfide:
    elastic constants, 38
Zirconia:
    compressive strength, 22
    elastic modulus, 22
    environment assisted cracking,
        492
    flexural strength, 22
    fracture toughness, 451, 453
    tensile strength, 22
Zirconium:
    fracture surface, 288
    rolling texture of, 108
    slip plane, 82–83
    surface stabilizer, 221, 223
    twin elements, 115
    twinning modes in, 119

Subject Index

Achilles' shield, 404
Activation energy:
  creep, 195–200, 215–217
  hydrogen embrittlement, 485, 488
  self-diffusion, 198
  temperature, effect of, 197
Adhesion:
  adherend, 371
  adhesive, 371
  adhesive failure, , 289, 371
  cohesive failure, 289, 371
  joint failure, 372
  mixed mode of fracture, 372
  mode mixity, 373
  Pi joint, 373
  test specimens, 373
  thermodynamic work of, 372
Adjusted compliance ratio, 582
Aircraft failure, 503, 646, 649–650, 658
Anelasticity, 12, 191
Anisotropy:
  biomimetic materials, 434
  elastic, 34
  bone, 34, 437
  fracture toughness, 390, 392, 402–411, 416, 434
  mechanical fibering, 402
  metallic laminates, 404
  mollusk shell, 434
  plastic, 108
  Titanic, rivet failure, 404
Antiphase domains, 167
Antiplasticizer effect, 427
Apparent fracture toughness, 358
Arc-shaped Tension specimen A(T), 719
Arrester orientation, laminate, 406
Auger electron spectroscopy, 443
Autofrettage, 549, 665
Auxetic materials, 28

Ballistic strain rate, 101
Beach markings, 499, 570, 648, 726

Beer barrel fracture, 252
Bend testing, 23
  adhesion, 37
  cell phone, 4
  elastic modulus, 23–24
  bending proof strength, 25
  modulus of rupture, 25, 55
  probability of survival, 260
  stress distribution, 24
Bergen jig, 489
Beta (β) transition, 242, 624
Biaxial loading, 29
Biomimetic materials, 434
Bone:
  cancellous, 34
  cortical, 34
  elastic constants, 35–36, 181
  degradation, age-related, 329
  fracture mechanisms, 439
  fracture, spiral, 275
  manatee, 327
  toughness, 181, 434–435, 437–438
  tribecular, 34
  strength, 181
  structure, 34, 437
Brass-Ammonia EAC, 470
Brazed joints, plastic constraint, 265
Bridge failure, 251, 255, 265, 269, 661, 664, 676
Bridgman relation, 96
Buckling, Euler, 17
Bulk modulus, 28
Burgers circuit, 66, 72
Burgers vector, 66, 72, 77, 80, 82, 84
  fatigue crack propagation, effect on, 616

Carbon bond rotation, 121
Carburizing, fatigue resistance, 522, 523, 547

Case Studies/Case Histories, See Failure Analyses
Caveat emptor, 685
Cavitation:
  metal irradiation, voiding, 449
  polymers, 428
  rubber particle, 428
Center-Cracked Tension specimen (CCT), 719
Ceramic Matrix Composite (CMC), see Composites
Chain branching, 124
Charpy impact specimen, 307
  fracture toughness correlations, 347
  precracked, 347, 351, 446
Checklist for failure analysis, 725
Chemically strengthened glass (ion-exchanged), 424–425
Chevron:
  fracture surface markings, 270–271, 275, 647, 648, 670, 726
  notch fracture specimen, 343
Clam shell markings, 499, 570, 572–573, 596, 674
Cleaning, fracture surface, 713
Cleavage, 281–282, 289–290, 416, 448
  fracture theory, 368
  grain size effect on, 399
Coble creep, 203, 208, 214,
Coffin-Manson relation, 539
Cohesive:
  failure, 371
  strength, theoretical, 253
Cold drawing, polymers, 129, 131, 132, 178, 180
  draw ratio, 180
Cold working, 113, 133, 144, 151, 154
Compact Tension Specimen C(T), 320, 341, 358, 719
Compatibility constraint, 87

Compliance:
  cracked plate, 305, 356, 365
  creep, 229
Composites:
  adhesion testing, 371
  anisotropy, 34, 36,
  ARALL, 176
  biomimetic, 434, 438
  bone, 327, 437
  ceramic matrix, 227, 416
  creep behavior, 224, 227
  elastic modulus, 41
  eutectic, 227
  fiber fracture, 45, 276
  fiber length influence of, 45
  fiber orientation influence of, 41,
    43, 48, 49
  fiber pullout, 278, 421, 432
  fracture appearance, 277,
    289–290
  GLARE, 176
  isostrain analysis, 41
  isostress analysis, 43
  laminated, 176, 404
  metal matrix, 175
  metal multilayer, 158
  polymer, 182, 426
  short fibers, 44
  strength, 41, 47, 177, 182–183
  stress-strain curves, 177
  thermal fracture, 303
  toughening mechanisms, 388,
    416, 426–434, 439
  toughness, 268
  whisker reinforced, 175
Compression testing, 17
Consumer Product Safety
    Commission (CPSC), 687
Contact shielding mechanisms,
    387, 388
Copolymer, 125, 242, 287, 427
Cottrell-Lomer locks, 147, 150,
    151
Crack
  branching, glass and ceramic,
    273
  closure, 580, 582–583, 585, 588,
    591, 593–594, 611, 629, 632
  deflection mechanisms, 387,
    388, 417
  line-wedge-loaded C(W)
    specimen, 358
  opening displacement, 320, 331,
    358, 365, 367, 369

Crack tip:
  plastic zone size, 332, 336, 341,
    358, 359, 372, 569–570, 578,
    580, 588, 592, 594, 598, 607,
    614, 619, 621, 626
  shielding, 387–388, 413,
    417–418, 438–439
  stress field, 316–317
Craze:
  definition, 129, 282
  embrittlement, 130
  fatigue induced, 625
  fracture of, 180, 272, 282
  in polyphenylene oxide, 283
  strength of, 283
  structure, 130
Creep:
  cavity formation, 293
  Coble, 203
  compliance, 229
  crosslinking density, effect of,
    232–233
  deformation maps, 208
  deformation mechanisms,
    202
  diffusional, 202
  dislocation, 203
  fracture micromechanisms, 287,
    291
  grain boundary sliding, 204
  graphical optimization
    procedure, 219
  Larson-Miller parameter, 215,
    217, 220
  logarithmic, 195
  metallurgical instabilities,
    192
  minimum commitment method,
    219
  modulus, 229
  molecular weight, effect of,
    232–233
  Nabarro-Herring, 203
  parabolic, 195
  parametric relations, 218
  primary, 192
  power law, 201
  rate, see Creep rate
  rupture life, 192–194, 216, 221,
    223, 226
  Sherby-Dorn parameter, 218
  tertiary, 192
  test conditions, 191
  transient, 192

Creep rate:
  crystal structure, effect of, 196,
    200
  diffusivity, effect of, 200, 208, 220
  elastic modulus, effect of, 196,
    200, 220
  grain size, effect of, 196, 203,
    205–206, 208
  stack fault energy, effect of, 196,
    201
  steady state, 191–192
  stress, effect of, 195, 225
  temperature effect of, 195
Critical resolved shear stress, 85
Cross:
  linking, 13–15, 120, 128, 177
  slip, see Dislocation
Cryogenic compatibility, 416
Crystallographic texture (preferred
    orientation):
  brass type, 107
  copper type, 107
  development of, 105
  fatigue, effect on, 614
  fiber texture, 107
  plastic strain anisotropy, effect
    on, 108
  pole figures, 106
  rolling texture, 108
  sheet texture, 107
  silver type, 107
  stacking fault influence, 107
  wire texture, 106
Cup, drawing, 109
Cup-cone fracture morphology
  hydrostatic pressure effect on, 98
  origin, 96
  microvoid coalescence in, 279
Cyclic strain
  fatigue, 529
  hardening, 531
  softening, 532, 534, 537
Cyclic stress fatigue, 503

Damping:
  anelastic, 191
  loss modulus, 241
  polymer, 239
  spectrum, 242
  storage modulus, 241
  tan δ, 241
Defect population, 254, 647
Deformation mechanism maps,
    208, 284

Degree of polymerization, 121
Delamination, 278, 339, 371, 398, 406, 416, 528, 633
Delayed retardation, 592
Design philosophy, 304, 307, 309, 312, 315, 328, 474, 493, 505, 507–508, 517, 544, 563, 565, 567, 583, 609, 632
Die failure, 673
Diffusivity, 200, 208, 220
  crystal structure, effect of, 196
  melting point, effect of, 196
  valence, effect of, 196
Disk-shaped Compact Tension specimen DC(T), 719
Dislocation:
  bowing stress, 78, 144
  Burgers vector of, 66
  cell structure, 152
  climb, 73
  combination, 75
  conservative motion of, 72
  crack nucleation models, 394, 368
  creep, 203
  cross slip, 74, 80
  density, 143, 152, 154, 156, 162, 175
  dissociation, 75, 77–78
  dipoles, 149
  double cross-slip mechanism, 145
  edge, 72
  elastic stress field of, 75, 158
  etch pits, 67
  fatigue induced, 534
  force between, 78
  forest, 146
  Frank-Read source, 144
  interactions of, 146
  jogs, 147
  junctions, 146
  kinks, 71
  line, 65
  line tension of, 77
  locks, 147, 150
  loop, 74, 144, 160, 167
  mixed, 74
  mobility, 73, 147
  multiplication of, 144
  node, 75
  nonconservative motion of, 73
  partial, 78

Peierls stress, *see* Peierls-Nabarro stress
  pencil glide, 87
  pileup, 144, 146, 155, 157, 173
  planar arrays of, 69, 152
  planar glide, 72, 81, 152
  screw, 73
  sessile, 146
  Shockley partial, 78
  solute atmospheres, 99, 161
  stacking faults, 79, 112–113, 120, 151
  stair rod, 151
  strain energy of, 76
  superlattice, 167
  tilt boundary, 78
  velocity stress sensitivity, 162
  wavy glide, 81
Dispersion strengthening, 170, 172, 175, 221
  oxide-dispersion strengthened (ODS), 171, 201, 224
Divider orientation, laminate, 407
Double
  Cantilever Beam (DCB) test specimen, 373
  Edge notched Tension specimen (DENT), 320, 720
Draw ratio, 180
Drop weight:
  specimen, 311
  tear test (DWTT), 311
Ductility, 8
Dugdale plastic strip model, 335, 359, 626
Dynamic:
  embrittlement, 472
  mechanical testing, 240
  recovery (parabolic hardening), 152

Easy glide, 152
Elastic anisotropy, *See also* Anisotropy:
  bone, 34, 36
  compliance, 34, 38
  stiffness, 34, 38
Elastic
  constants, 10–11, 38
  deformation, 3
  energy release rate, 302, 305
  limit, 8
Elastic modulus (*see also* Back Cover Lookup Table):

fatigue crack propagation, effect on, 577, 613, 616
  interatomic force dependence, 9
  temperature dependence of, 12–13, 15, 229
Electroslag remelting, 390
Embrittlement models, 465
End-Notched Flexure (ENF) test specimen, 374
Energy, damping spectra, 239
  loss modulus, 241
  storage modulus, 241
Environment assisted cracking:
  aluminum alloys, 469, 471, 474, 478, 480, 483–485, 708
  ceramics and glasses, 489
  plastics, 487
  steel alloys, 465, 468, 470, 472, 474–475, 482, 484
  test specimen, 475
  threshold level (K_{IEAC}), 474
  titanium alloys, 468, 470, 473, 483–484
Etch pits, dislocations, 67
Eutectic composites, 224
Express warranty, 685
Extension ratio (λ), 17
Extrinsic toughening, 387, 402, 406, 416, 422, 432, 434, 438

Failsafe design, 510, 567
Failure
  analyses, 327, 339, 354, 404, 571, 653, 658, 660–661, 664–665, 670, 673, 674
  analysis checklist, 725
  theories, 50
Fatigue:
  composites, 526, 619, 628, 631, 632
  cumulative damage, 508, 526, 529, 541
  cyclic hardening, 531
  cyclic softening, 531
  ductility coefficient, 539
  durability, 510, 550
  extrusions, 511, 545
  Goodman diagram, 506
  initiation, 500, 503, 506, 508, 511, 523, 527, 543, 545
  intrusions, 511, 545
  limit, 503
  macroscopic markings, 499, 596–597, 646

Fatigue (*continued*)
  mean stress, effect on, 504, 506
  Miner's Law, 508, 541
  Neuber analysis, 543
  notch effects, 503, 510–511,
    520, 529, 541
  polymers, 523, 527, 534, 547,
    572, 618
  pretensioning, effect of, 550
  rainflow analysis, 541
  residual stress effect on,
    521–522, 547, 551
  safe-life, 508, 510, 541, 567
  strain-life curves, 538
  strength coefficient, 533, 538
  striations, *see* Fatigue crack
    propagation
  surface treatment, effect on, 520
  test scatter, 506, 516, 527
  test specimens, 504, 514, 524,
    526, 529, 544
  thermal, 524
  Wöhler diagram, 503
Fatigue crack propagation:
  ceramics, 628
  composites, 628, 632
  crack closure, 580, 582–583,
    585, 588, 591, 593–594, 611,
    629, 632
  crack length, effect of, 559
  discontinuous growth bands,
    625
  elastic modulus, effect on, 577,
    615, 616, 625
  environment, effect of, 561, 569,
    574, 586, 600
  fracture mode transition, 569,
    580–581, 607
  frequency sensitivity, 561, 578,
    601–605, 612, 620
  grain size, effect of, 588, 594,
    608–612, 615
  life calculations, 563
  load interactions, effect of, 592
  mean stress, effect of, 563,
    578–580, 591, 604
  microstructure, effects of, 606
  plastic zone estimation, 332,
    568, 570, 607, 626
  polymers, 572, 618
  short crack, effect of, 583
  specimens, 319–320, 719
  striations, 572–575, 577, 607,
    611, 625

threshold, 583–592, 607–610,
    615–618
Fibrous fracture:
  Charpy specimen, 310
  tensile test, 96, 279
Flow lines, forging, 402, 404
Fracture:
  brittle, 268
  cohesive strength, theoretical, 253
  creep micromechanisms, 287,
    289, 291
  crystal structure, effect of, 385
  degree of order, effect of, 385
  ductile, 268
  electron bond, effect of, 385
  energy, 309
  energy release rate, 302, 305
  Leonardo da Vinci studies, 255
  mechanisms, 278
  mode, 315, 372, 375
  probability, 257
  size effect, 254, 313, 336, 366
  thermal, 273
  tough, 269
  toughness, *see* Fracture
    toughness
  volume dependence, 259
Fracture mode transition, 336, 338,
    569, 580–581, 607
Fracture stress, true, 22
Fracture surface:
  bone, 327, 439
  ceramics, macroscopic, 273
  ceramics, microscopic, 287
  cleaning, 713
  composites, macroscopic
    appearance, 277
  composites, microscopic
    appearance, 289
  glass, macroscopic appearance,
    273
  glass, microscopic appearance,
    275
  hackle bands, 284, 286
  hackle zone, 272, 275
  metals, macroscopic appearance,
    97, 270–271, 310, 338–339,
    354, 409, 470, 501–502, 586,
    596–597, 599, 646,
    648–651, 661–662, 666,
    671, 673
  metals, microscopic appearance,
    279, 369, 470, 573–574, 607,
    611, 655–656, 709

mirror zone, 272, 275
  mist zone, 272, 275
  polymers, macroscopic
    appearance, 271, 500, 675
  polymers, microscopic
    appearance, 282, 428, 489,
    500, 621–622, 625, 627, 675
  preservation, 713
  quantitative analysis of ceramics
    and glasses, 276
  replica, 715
  shell, 435–436
  wake hackle, 290
Fracture toughness:
  aluminum alloys, 345, 394, 410,
    414, 449–452
  anisotropy, 397, 402
  ASTM Standard, 8, 311,
    342–344, 358, 363–365, 367,
    373, 375
  biomimetic materials, 438
  bone, 437, 439
  ceramics, concrete, stone, 416,
    451, 453
  cleanliness, effect of, 389
  comparisons among material
    classes, 268
  composites, 416, 421, 426
  contact shielding mechanisms,
    387–388, 438
  crack deflection mechanisms,
    387–388, 438
  critical stress intensity level, 326
  dynamic (K_{ID}), 349
  extrinsic, 387 (*see also* Extrinsic
    toughening)
  grain refinement, effect of, 398
  impact energy correlations, 347,
    661
  instrinsic, 385 (*see also* Intrinsic
    toughening)
  J-integral, 360, 369, 435
  microhardness, determination
    of, 344
  plane stress, 300, 317, 331, 336
  polymers, 387, 426, 432, 451,
    543
  shell, 435
  specimen size requirement, 341,
    366
  steel alloys, 345, 391, 395, 411,
    444, 446, 450–453
  stretch zone correlation, 369,
    595

tensile property correlations, 368
titanium alloys, 345, 384, 389, 415, 451, 453
zone shielding mechanisms, 387–388, 438
Frank-Read source, *see* Dislocation
Free volume:
  excess, 228, 231
  toughness dependence on, 426

Generator rotor failures, 354, 660
Geometrical
  hardening, 103
  softening, 103
Gigacycle fatigue testing, 506, 517
Glass:
  cantilever curl, 274
  elastic modulus, 10, 11
  environment assisted cracking, 489, 492
  fiber properties, 41, 183
  fracture appearance, 273– 275
  fracture, bottle, 694
  fracture, thermal, 274
  fracture surface of, 275–277
  fracture toughness, 263, 268, 301, 344, 370–371, 386–387, 416, 418, 422–426, 451, 453
  ion-exchanged,424–425
  metallic, 386
  spheres, 430, 431
  stable cracking, 425
  tempered, 273–274, 423–425
  thermal fracture, 273–274
Glass transition temperature:
  chain mobility, 128
  definition, 13, 123
  free volume, 228
  molecular weight, 231
  plasticizer effect on, 128
  PMMA, 231
  polystyrene, 231
  reduced, 15
  rubber elasticity, 15
  viscoelasticity and creep, 190
Goodman diagram, 507
Grain boundary:
  sliding, 157, 204, 205, 208, 212, 214, 221, 223
  strengthening, 155, 398
Griffith crack theory, 299
Gun tube failure, 665
Gusset plate failure, 269

Hall-Petch relationship, 155, 399, 400
Hammurabi, Code of, 682
Hannibal, 50
High temperature alloys, 167, 171, 201, 211, 220
Hollomon strain hardening relation, 91
Hooke's law, 9, 12–13, 26, 29, 35
Hot working, 154
Hydrogen:
  embrittlement (*See also* Environment assisted cracking), 389, 464–465, 470, 483, 485
  flakes, 354, 466

Impact strength, 311–312
Inclusions:
  alignment, 65, 403
  shape control, 392
  spacing-microvoid size relation, 396
  sulfide, 389, 654–655
Injection molding, 44, 182, 433
Instrumented indentation, 33
Intergranular fracture, 281, 287–288, 291, 416, 441, 466, 468–469, 472, 483, 486, 611, 667, 670, 709
I-35 bridge failure, Minneapolis, Minnesota, 269
Implied warranty, 686
Indentation fracture (IF) test, 344
Intrinsic toughening, 385
  metals, 389, 411,
  ceramics and ceramic-matrix composites, 416
  glass, 422
  polymers, 426
Ion exchange strengthening, glass, 424
Isochronous stress-strain curves, 233
Isometric stress-strain curves, 234–235
Isostrain analysis, 41
Isostress analysis, 43
Izod impact specimen, 308, 311

J-integral, 360, 369, 435

Lamellar structures:
  biomimetic, 439

osteon (bone), 437
pink conch shell, 435
Lamellar tearing, 392, 409
Laminated-beam (sandwich) test specimens, 373
Laminated glass, 423, 425
  bending, 425
  design with, 423
Laminated structures, 36, 43, 175–176, 278, 373, 404, 406, 408, 423, 425, 432, 435, 633
Larson-Miller parameter, 215, 217, 220
Latent hardening, 105
Lattice friction, *see* Peierls-Nabarro stress
Layered pressure vessels, 406, 547
Law:
  contract, 685
  tort, 685
Leak before break criteria, 329, 567, 618, 657, 666, 670
Leonardo da Vinci, fracture studies, 255
Liberty Bell, xvii, 261
Life calculations, 477, 492, 563, 590
Limiting drawing ratio (LDR), 100
Linear hardening, 152
Liquid metal embrittlement (*See also* Environment assisted cracking), 281, 328, 471, 485
Logarithmic creep, 195
Lüders band, 99, 161

Mackerel patterns (bands), 284
Maraging steels, 164, 175, 412, 394, 395, 406
Martensite, 156, 160, 170, 172, 175, 264, 398, 412, 413, 440
Maxwell model, 235
Mechanical alloying, 171
Mechanical analogs, *see* Viscoelasticity
Mechanical fibering, 398, 402
  inclusion alignment in, 404, 409
Mer, 120
Metal forming:
  limiting drawing ratio, 110
  plastic strain anisotropy, 109
Microvoid coalescence, 96, 279, 573, 578, 611, 655
Middle-Cracked Tension M(T) specimen, 358

Miner's law, 509
Mixed Mode Bending (MMB)
    adhesion test, 375
Mode of loading, 316
Mode mixity, 373
Modulus:
    biaxial, 33
    bulk, 28
    creep, 229
    loss, 241
    plane strain, 33
    relaxation, 230
    rupture, 25
    tangent, 11
    secant, 11
    storage, 241
    Weibull, 257
    Young's, 9
Molasses tank failures, 251, 645
Molecular weight, 121
    fatigue, effect on, 527, 620
    relaxation modulus, effect on,
        232
    stiffness, effect on, 178
    strength, effect on, 177
    T_g, effect on, 231
    viscosity, effect on, 178–179
Molecular weight distribution, 121
Multilayer metals, 156

Nabarro-Herring creep (diffusional
    creep), 202, 205, 207–208
Nanocrystalline (nc) metals, 156
Nanoindentation, see Instrumented
    indentation
Necking, tensile, 6, 8, 17, 91, 93,
    95, 98, 110, 129, 131
Negligence, 685, 691
Neuber analysis, 543
Neumann bands, 120, 645
Neutron irradiation:
    copper, effect of, 446–447
    damage reversal, 448
    embrittlement, 444
    fuel cladding alloy, 449
    hardening, 446
    neutron flux effect of, 445
    precracked Charpy (PCC), 446
    reactor pressure vessel (RPV),
        445
Notch:
    sensitivity, 264, 266, 307, 312,
        511, 523, 543
    strengthening, 264, 307

Notched cylinder:
    bending, 721
    tension, 721

Orowan looping, 144, 169
Orthotropic, 36
Ostwald ripening, 165
Overloads, see Fatigue crack
    propagation
Overshoot, 104
Oxide-dispersion strengthened
    (ODS), 171, 201, 222,
    224–226

Parabolic:
    creep (β flow), 195
    hardening (dynamic recovery),
        152
Parametric relations, creep, 215
Partial dislocation, see Dislocation
Pearlite, 156, 172
Peierls-Nabarro stress:
    atomic bonding, effect of, 70
    crystal structure dependence, 71,
        82
    definition, 69
    dislocation orientation,
        dependence of, 71
    dislocation width, effect of, 70
    fracture, effect on, 266, 307
    intrinsic strength, 156, 173
    slip plane, dependence on,
        72, 82
    temperature, dependence on, 70,
        71
Perfect crystal, strength of, 63, 67
Petch-Hall relationship, 155, 399,
    400
Phase angle of loading, 373
Planar glide, 81
Plane strain:
    criterion, 336, 341, 367
    fracture appearance, 337
    fracture toughness, 336
    fracture toughness testing , 341
    fracture toughness data , 345,
        450–453
    plastic zone size, effect on, 333
    stress state, 300, 317, 318
Plane stress:
    crack tip plastic zone, 333, 335
    criterion, 336
    fracture appearance, 337–339
    fracture toughness, 336

fracture toughness testing , 355
    stress state, 300, 317–318, 336
Plastic constraint:
    notched specimens, 264, 310,
        317
    tensile test, 4, 8–9, 13, 90,
        93–96, 99–100, 113, 161, 255,
        368
Plastic deformation:
    constant-volume process of, 6
    energy, 304
    heterogeneous, 99, 113, 161
    homogeneous, 90
    strain hardening models, 90
Plastic instability, 6, 17, 22, 93
Plastic zone size, 332, 568, 570,
    607, 626
Poisson's ratio, 28–29, 34
    data, 10–11, 55
Pole figures, 105
Polymer additions:
    blowing agents, 128
    cross-linking agents, 128
    fillers, 128
    pigments, 127
    plasticizers, 128
    stabilizers, 128
Polymer deformation mechanisms:
    crazing, 129, 180, 271–272,
        282–283, 428
    shear yielding, 128, 136, 180,
        271, 283, 429
Polymer structure, 120
    chain mobility, influence on, 121
    crystallinity, 123, 126
    degree of polymerization (DP),
        121
    repeat unit, 123, 125
    side group configuration, 123
Polymer strengthening, 143, 177
    molecular weight, influence of,
        178
    orientation, effect of, 178
Portevin-Le Chatelier effect, 164
Power law creep (dislocation
    creep), 203, 205–206, 208,
    211
Precipitation hardening, 164
    aging, 164
    antiphase domain boundary
        mechanism, 167
    coherent precipitate, 166
    cutting mechanism, 167
    incoherent precipitate, 167

looping mechanism, 169
misfit strain, 166–167, 169
overaging, 170
solution treatment, 164
Preferred orientation, *see*
    Crystallographic texture
Privity of contract, 686
Product:
    design and development, 707
    failure process, determination of,
        702
    liability, 683
    recall, 697
    risk assessment methods, 705
    safety, regulatory requirements
        regarding, 698
    Safety Commission, Consumer,
        687
    testing, 3, 707
Proof test, 330, 559, 670
Proportional limit, 8, 18, 19

R-curve analysis, 328, 356, 364,
    367, 387, 417, 436
Radial fracture, tensile test, 96
Rainflow analysis, 541
Ratchet lines, 500–502, 648
Relaxation time, 236
Replica techniques and
    interpretation, 278, 280, 713
Residual stresses:
    fatigue, effects on, 520, 547,
        648, 665, 669, 672
    thermal, 50
Resilience, 19
Resolved shear stress, 85
Retirement for cause, 567
River patterns, in cleavage,
    281–282, 289–290
Rocket motor casing failure, 670
Root cause, product failure 702
Rubber elasticity:
    elastomers, 9, 11, 15
    kinetic theory model, 17
    Mooney-Rivlin constants, 17
    temperature dependence of, 16

Safety factor, 257
Schmid's law, 85
Secant modulus, 11
Shear lip:
    fracture mode transition, 338–339,
        568, 571, 648, 650–651, 662
    tensile test, 96

Sherby-Dorn parameter, 196, 218
Ship fractures, 251
    Titanic, 252, 404
Shockley partial dislocation, *see*
    Dislocation
Short-transverse orientation,
    laminate, 407, 410
Shotgun barrel failure, 653
Shot peening, 520
Single Edge notched
    Bend specimen SE(B), 720
    Tension specimen SE(T), 320,
        720
Size effect, *see* Fracture toughness
Slip:
    basal, 65, 84–85
    ceramics, 83, 88
    conjugate, 104–105
    directions and planes, 81
    easy glide, 152
    independent slip systems, 83
    induced crystal rotation, 102
    metals, 81
    multiple, 87
    planar, 81, 93
    plane, 63
    primary, 104
    prism, 65, 82, 84
    steps, 82
    systems, 82–83
    wavy, 81, 93
Solid solution strengthening, 158
Solvus line, solvus temperature,
    155
Spherulites, 126, 130, 287–288
Stacking fault:
    annealing twins, origin of, 112,
        120
    dislocation cell, influence on, 152
    energy, 80, 93
    fatigue, effect on, 537
    hardening, influence on, 93, 152,
        154
    Lomer-Cottrell dislocation, 151
    partial dislocation separation
        distance, 80
    planar defect, 79–80
    probability, 103
    slip character, influence on, 81
    texture, influence on, 106
Stare decisis, 683
Steady state creep, *see* Creep rate
Steel, strengthening of, 156, 160,
    163–164, 172

Stereographic projection, 103
Stereoregularity, 124
Steric hindrance, 121
Strain:
    distribution in tensile specimen,
        94
    energy density, 19
    energy release rate, 305
    engineering, 5
    shear, 26
    true, 5
    uniform, 95
Strain controlled-cyclic life
    response, *see* Fatigue
Strain hardening:
    coefficient, cyclic, 533, 535–536
    coefficient, monotonic, 91–93
    coefficient, polymer, 132
    creep, 192
    cyclic, 531, 535, 536
    dislocation loops, role in, 169
    exponent, Ramberg-Osgood, 92
    Holloman relation, 91
    metals, 90, 151
    models of, 91, 92
    necking, effect on, 93
    polycrystal vs. single crystal, 155
    polymers, 132
    Ramberg-Osgood relation, 91
    strength coefficient, 91, 92
    shot peening, 521
    stacking fault energy,
        dependence on, 93
    uniform strain, effect on, 95
Strain rate:
    creep, 195, 208, 234
    sensitivity, 100, 132, 190
    superplasticity, 206
    toughness, effect on, 307, 311,
        347, 351, 387
    viscoelastic, 235
Strength, *see* Yield strength
Strength coefficient, strain
    hardening, 91–92
Strengthening mechanisms:
    boundary strengthening, 155
    dispersion strengthening, 171
    extrinsic, 143, 175
    intrinsic, 143
    molecular alignment
        (orientation), 178
    multiple, 172
    polymer architecture, 177
    polymer chemistry, 177

Strengthening mechanisms
(*continued*)
precipitation hardening, 164
solid solution strengthening, 158
strain hardening, 151
Stress:
critical octahedral shear, 89
critical resolved shear, 85
engineering, 5
hoop, 31
normal, 5, 85
octahedral shear, 89
resolved shear, 85
shear, 26
torsion, 26
true, 5
true, corrected for necking, 95
Stress concentration factor, 260,
324, 326, 511, 543, 592
Stress corrosion cracking, *See also*
Environment assisted
cracking, 281, 287, 463–464,
468, 470, 472, 483, 648, 661,
709
Stress intensity factor, 318, 661,
668, 672, 674–675
effective, 333, 342, 357, 578, 582,
586, 590–591, 598, 611, 616
striations, correlation with, 575,
625
test specimens, calculations for,
719
Stress relaxation:
linear viscoelastic, 229
mechanical analogs, 235
temperature dependence, 228
time shift factor (WLF relation),
232
viscoelastic, 227
Stress rupture, 192
Stress-strain curves:
composites, 177
elastic-heterogeneous plastic,
113, 162
elastic-heterogeneous plastic-
homogeneous plastic, 99, 162
elastic-homogeneous plastic, 90,
162
isochronous, 234
isometric, 234
metals, 8
micropillars, 68
nanocrystalline and
nanotwinned Cu, 157

necking, 94
polymers, 8, 132, 133–134
rubber, 8
serrated, 113, 162
single crystal, 86, 152
stress concentration (notch),
effect of, 266, 307
temperature and strain rate,
dependence on, 100, 266
toughness, 23
twinning, 113
yield point, 99, 162
Stress, thermal
rock fracture, 50
thin film, 51
Stretch:
ratio (λ), 17
zone width, 369, 595
Strict liability, 685, 694
Superplasticity, 205
Surface:
crack in flexure (SCF) specimen,
344
energy, 299
flaws, elliptical and semicircular,
723

Tangent modulus, 11
Taylor factor, 87, 155
Tearing modulus, 367
Temper embrittlement, 440, 442
Temperature-compensated time
parameter, 196
Tempered martensite
embrittlement, 440–441
Tenacity, 181
Tensile impact test, 311
Tensile strength:
ceramics, 22, 25, 67, 254,
450–453
composites, 10–11, 41
defined, 8
fibers, 41, 181
metals, 20–21, 65, 67, 254,
450–453
polymers, 20, 21, 450–453,
124
temperature and strain rate,
dependence on, 100
Tertiary creep, 192
Texture, crystallographic, *See*
Crystallographic Texture
Theoretical crystal strength, 64, 65,
67, 253

Thermal:
barrier coating, 53, 227
expansion coefficients, 52, 55
fatigue, 524
fracture, propagation-controlled,
301
Thermal shock, 50, 301
material selection for resistance,
304
Thermal stress, 50
cooling rate dependence, 54
damage resistance parameter,
301
superposition on mechanical
stress, 51
thin film, 51
Thermally tempered (strengthened)
glass, 423
Thermodynamic work of adhesion,
372
Thermoplastics, 13, 120
Thermoplastics, oriented, 14
Thermosets, 13, 120
Thick-walled pressure vessels, 307,
329, 549,
Thin-walled pressure vessels,
30–32, 307, 329
300°C embrittlement, 441
Tilt boundary, 78
Time-temperature superposition, 232
Titanic ship fracture, 251, 404
Toilet seat fracture, 252, 674
Toughening mechanisms,
see Fracture toughening
Toughness, tensile specimen, *See
also* Fracture toughness, 22
Transformation:
induced plasticity (TRIP), 395,
413
toughening, 419
Transient creep, 191, 195
Transition temperature
crack length, effect of, 313,
data, steels, 311
definition, 309
energy criterion, 309
fracture appearance criterion,
310
lateral expansion or contraction
criterion, 309
limitations of, 312
polymers, 311
size, effect of, 313
Tresca yield criterion, 88, 133

Triaxial loading, 30
Twin:
    annealing, 112
    appearance, 112
    composition plane, 114, 116
    criterion for formation, 113
    cubic crystals, 115, 120
    deformation, 111
    dislocation array, 116
    hexagonal crystals, 115, 117
    shape, 116
    shear strain, 114, 116, 119
    strain ellipsoid, 118
    stress-strain curve, 113
    thickness, 116

Ultra-fine grained materials, 398
Ultra high cycle fatigue, *see*
    Gigacycle fatigue

Vacancy diffusion:
    dislocation climb, 73
    formation and creep, 203
    self-diffusion and creep, 198
Vacuum arc remelting, 390
Viscoelasticity:
    creep modulus, 229
    linear, 229
    loss modulus, 241
    mechanical analogs, 235
    relaxation modulus, 230
    storage modulus, 241
    time-temperature superposition,
    232
Viscoplasticity, 191
Viscosity, 191, 233, 235, 238

Voigt model, 235
Von Mises' yield criterion, 88,
    135

Wake hackle, 290
Wallner lines, 272
Wavy glide, 81, 93
Wedge Opening Load (WOL) test
    specimen, 374, 719
Weld cracking, 445, 466–467, 469,
    653, 664, 670, 676
Weibull analysis, 255
Whisker
    dislocation free, 67
    reinforcement, 175
Wöhler diagram, 503

Y, stress intensity parameter:
    analytical expressions, 719
    definition, 318
    elliptical and spherical
        correction, 321, 723
    finite width correction, 324
    multiplicity, 323
    plots, 319–321
Yield criterion:
    distortional energy, 88–89
    Drucker-Prager, 135
    Mohr-Coulomb, 133
    modified Tresca, 133
    normal stress, 88
    pressure dependent, 133
    pressure-modified von Mises,
    135
    Tresca, 88
    von Mises, 88

Yield point:
    aluminum-based alloys, 100,
    163
    carbon and nitrogen, effect of,
    99, 161
    ceramic and ionic crystals, 100,
    162
    dynamic strain aging, 164
    elongation, 100
    iron-based alloys, 99
    Lüders bands, 99
    Portevin-Le Chatelier effect, 164
    strain aging, effect of, 100
    stress-strain curve, 99, 162
    stretcher strains, 100
    titanium-based alloys, 162
    upper and lower, 99
Yield strength:
    critical resolved shear, relation
        to, 85
    cyclic, 531, 607
    data, 20–21, 65
    dislocation-free micropillars, 68
    flexural, 25
    grain size, effect of, 155
    offset, 19
    polycrystal, 87
    polymers, 128, 131
    strain rate, effect of, 101
    temperature, effect of, 70–71,
    101
    theoretical, 63, 65, 67
    twinning, 113

Zone shielding mechanisms,
    387–388